Matrices

Graphs and trees

Counting and probability

Relations

Algebraic Systems

Discrete Mathematics

FIFTH EDITION

Discrete Mathematics

Kenneth A. Ross
Charles R. B. Wright
University of Oregon

Pearson Education, Inc.
Upper Saddle River, New Jersey 07458

Editor-in-Chief: *Sally Yagan*
Acquisitions Editor: *George Lobell*
Vice President/Director of Production and Manufacturing: *David W. Riccardi*
Executive Managing Editor: *Kathleen Schiaparelli*
Senior Managing Editor: *Linda Mihatov Behrens*
Production Editor: *Bob Walters*
Manufacturing Buyer: *Michael Bell*
Manufacturing Manager: *Trudy Pisciotti*
Marketing Assistant: *Rachel Beckman*
Assistant Managing Editor, Math Media Production: *John Mathews*
Editorial Assistant/Supplements Editor: *Jennifer Brady*
Art Director: *Wanda Espana*
Interior / Cover Designer: *Maureen Eide*
Creative Director: *Carole Anson*
Art Editor: *Thomas Benfatti*
Director of Creative Services: *Paul Belfanti*
Cover Photo: *Rolph Scarlett* (1889-1984) "Untitled" Circa 1937. Oil on Canvas, 36 × 48 inches/Courtesy David Findlay Jr. Fine Art, New York.
Art Studio: *Laserwords*

Pearson Education, Inc.
Upper Saddle River, New Jersey 07458

Printed in the United States of America

10 9 8 7 6 5 4 3 2 1

ISBN 0-13-065247-4

Pearson Education LTD., *London*
Pearson Education Australia PTY, Limited, *Sydney*
Pearson Education Singapore, Pte. Ltd
Pearson Education North Asia Ltd., *Hong Kong*
Pearson Education Canada, Ltd., *Toronto*
Pearson Educación de Mexico, S.A. de C.V.
Pearson Education–Japan, *Tokyo*
Pearson Education Malaysia, Pte. Ltd.

To our Grandchildren
Matthew, Mark,
Maggie, Jeremy, . . .

Contents

4 Induction and Recursion 128

5 Counting 181

6 Introduction to Graphs and Trees 225

7 Recursion, Trees, and Algorithms 269

8 Digraphs 318

Preface to the Fifth Edition

In writing this book we have had in mind both computer science students and mathematics majors. We have aimed to make our account simple enough that these students can learn it and complete enough that they won't have to learn it again.

The most visible changes in this edition are the 274 new supplementary exercises and the new chapters on probability and on algebraic structures. The supplementary exercises, which have complete answers in the back of the book, ask more than 700 separate questions. Together with the many end-of-section exercises and the examples throughout the text, these exercises let students practice using the material they are studying.

One of our main goals is the development of mathematical maturity. Our presentation starts with an intuitive approach that becomes more and more rigorous as the students' appreciation for proofs and their skill at building them increase.

Our account is careful but informal. As we go along, we illustrate the way mathematicians attack problems, and we show the power of an abstract approach. We and our colleagues at Oregon have used this material successfully for many years to teach students who have a standard precalculus background, and we have found that by the end of two quarters they are ready for upperclass work in both computer science and mathematics. The math majors have been introduced to the mathematics culture, and the computer science students have been equipped to look at their subject from both mathematical and operational perspectives.

Every effort has been made to avoid duplicating the content of mainstream computer science courses, but we are aware that most of our readers will be coming in contact with some of the same material in their other classes, and we have tried to provide them with a clear, *mathematical* view of it. An example of our approach can be seen first in Chapter 4, where we give a careful account of **`while`** loops. We base our discussion of mathematical induction on these loops, and also, in Chapter 4 and subsequently, show how to use them to design and verify a number of algorithms. We have deliberately stopped short of looking at implementation details for our algorithms, but we have provided most of them with time complexity analyses. We hope in this way to develop in the reader the habit of automatically considering the running time of any algorithm. In addition, our analyses illustrate the use of some of the basic tools we have been developing for estimating efficiency.

The overall outline of the book is essentially that of the fourth edition, with the addition of two new chapters and a large number of supplementary exercises. The first four chapters contain what we regard as the core material of any serious discrete mathematics course. These topics can readily be covered in a quarter. A semester course can add combinatorics and some probability or can pick up graphs, trees, and recursive algorithms.

We have retained some of the special features of previous editions, such as the development of mathematical induction from a study of **`while`** loop invariants, but we have also looked for opportunities to improve the presentation, sometimes by changing notation. We have gone through the book section by section looking for ways to provide more motivation, with the result that many sections now begin where they used to end, in the sense that the punch lines now appear first as questions or goals that get resolved by the end of the section.

We have added another "Office Hours" section at the end of Chapter 1, this one emphasizing the importance of learning definitions and notation. These sections, which we introduced in the fourth edition, allow us to step back a bit from our role as text authors to address the kinds of questions that our own students have asked. They give us a chance to suggest how to study the material and focus on what's important. You may want to reinforce our words, or you may want to take issue with them when you talk with your own students. In any case, the Office Hours provide an alternative channel for us to talk with our readers without being formal, and perhaps they will help your students open up with their own questions in class or in the office.

We have always believed that students at this level learn best from examples, so we have added examples to the large number already present and have revised others, all to encourage students to read the book. Our examples are designed to accompany and illustrate the mathematical ideas as we develop them. They let the instructor spend time on selected topics in class and assign reading to fill out the presentation. Operating in this way, we have found that we can normally cover a section a day in class. The instructor's manual, available from Prentice Hall, indicates which sections might take longer and contains a number of suggestions for emphasis and pedagogy, as well as complete answers to all end-of-section exercises.

The end-of-chapter supplementary questions, which are a new feature of this edition, are designed to give students practice at thinking about the material. We see these exercises as representative of the sorts of questions students should be able to answer after studying a chapter. We have deliberately not arranged them in order of difficulty, and we have deliberately also not keyed them to sections—indeed, many of the exercises bring together material from several sections. To see what we mean, look at the supplementary exercises for Chapter 5, on combinatorics, where we have included an especially large number of problems, many of which have a variety of essentially different parts. A few of the supplementary questions, such as the ones in Chapter 12 on algorithms to solve the Chinese Remainder and Polynomial Interpolation problems, also extend the text account in directions that would have interrupted the flow of ideas if included in the text itself. Some of the questions are very easy and some are harder, but none of them are meant to be unusually difficult. In any case, we have provided complete answers to all of them, not just the odd-numbered ones, in the back of the book, where students can use them to check their understanding and to review for exams.

The new chapters on probability and algebraic structures respond to requests from current and past users who were disappointed that we had dropped these topics in going from the third edition to the fourth. Since those were two of our favorite chapters, we were happy to reinstate them and we have taken this opportunity to completely revise each of them. In Chapter 9 we now work in the setting of discrete probability, with only tantalizing, brief allusions to continuous probability, most notably in the transition to normal distributions from binomial distributions. The material on semigroups, rings, and fields in Chapter 12 is not changed much from the account in the third edition, but the discussion of groups is dramatically different. The emphasis is still on how groups act on sets, but in the context of solving some intriguing combinatoric problems we can develop basic abstract ideas of permutation group theory without getting bogged down in the details of cycle notation. As another response to reader feedback, we have moved the section on matrix multiplication from Chapter 3 to Chapter 11, which is the first place we need it.

Naturally, we think this edition is a substantial improvement and worth all of the effort it has taken. We hope you will agree. We welcome any comments and of course especially welcome reports of errors or misprints that we can correct in subsequent printings.

Supplements

The Instructor's Resource Manual, which course instructors may obtain gratis from Prentice Hall, contains complete answers to all exercises in the text. In addition, Prentice Hall publishes inexpensive student workbooks of practice problems on discrete mathematics, with full solutions to all exercises. The Prentice Hall Companion Web site for this text contains information about such materials.

Acknowledgments

This is a better book because of the many useful suggestions from our colleagues and correspondents Patrick Brewer, William Kantor, Richard M. Koch, Eugene Luks, George Matthews, Christopher Phillips, and Brad Shelton. Dick Koch's gentle questions and suggestions, based on his incisive ideas and point of view, were especially helpful. We also benefitted from suggestions provided by the following reviewers and other anonymous reviewers: Dr. Johannes Hattingh, Georgia State University; Bharti Temkin, Texas Tech; Marjorie Hicks, Georgia State University; and Timothy Ford, Florida Atlantic University.

Thanks are also due to our wonderful production editor, Bob Walters, and to the superb compositors at Laserwords. Our editor for this edition was George Lobell, whose suggestions for improvements and overall support have been as helpful as his guidance through the production process.

KENNETH A. ROSS
ross@math.uoregon.edu

CHARLES R. B. WRIGHT
wright@math.uoregon.edu

To the Student Especially

You may find this course the hardest mathematics class you have ever taken, at least in part because you won't be able to look through the book to find examples just like your homework problems. That's the way it is with this subject, and you'll need to use the book in a way that you may not be used to. You'll have to read it. [In computer science there's an old saying, usually abbreviated as RTFM, which stands for "read the friendly manual."] When you do read the book, you'll find that some places seem harder than others. We've done our best to write clearly, but sometimes what we think is clear may not be so obvious to you. In many cases, if you don't find a passage obvious or clear, you are probably making the situation too complicated or reading something unintended into the text. Take a break; then back up and read the material again. Similarly, the examples are meant to be helpful. In fact, in this edition we have made a special effort to put even more examples early in each section to help you see where the discussion is leading. If you are pretty sure you know the ideas involved, but an example seems much too hard, skip over it on first reading and then come back later. If you aren't very sure of the ideas, though, take a more careful reading.

Exercises are an important part of the book. They give you a chance to check your understanding and to practice thinking and writing clearly and mathematically. As the book goes on, more and more exercises ask you for proofs. We use the word "show" most commonly when a calculation is enough of an answer and "prove" to indicate that some reasoning is called for. "Prove" means "give a convincing argument or discussion to show why the assertion is true." What you write should be convincing to an instructor, to a fellow student, and to yourself the next day. Proofs should include words and sentences, not just computations, so that the reader can follow your thought processes. Use the proofs in the book as models, especially at first. The discussion of logical proofs in Chapter 2 will also help. Perfecting the ability to write a "good" proof is like perfecting the ability to write a "good" essay or give a "good" oral presentation. Writing a good proof is a lot like writing a good computer program. Using words either too loosely or extensively (when in doubt, just write) leads to a very bad computer program and a wrong or poor proof. All this takes practice and plenty of patience. Don't be discouraged when one of your proofs fails to convince an expert (say a teacher or a grader). Instead, try to see what failed to be convincing.

Now here's some practical advice, useful in general, but particularly for this course. The point of the homework is to help you learn by giving you practice thinking correctly. To get the most out of it, keep a homework notebook. It'll help you review and will also help you to organize your work. When you get a problem wrong, rework it in your notebook. When you get a problem right, ask yourself what your method was and why it worked. Constantly retracing your own successes will help to embed correct connections in the brain so that when you need them the right responses will pop out.

Read ahead. Look over the material before class to get some idea of what's coming and to locate the hard spots. Then, when the discussion gets to the tough points, you can ask for clarification, confident that you're not wasting class time on things that would be obvious after you read the book. If you're prepared, you can take advantage of your instructor's help and save yourself a lot of struggling. After class, rewrite your class notes while they're still fresh in your mind.

At strategic places in the book, we have inserted very short "Office Hours" sections with the kinds of questions our own students have asked us. These informal sections address questions about how to study the material and what's important to get out of it. We hope they will lead to your own questions that you may want to raise in class or with your instructor.

Study for tests, even if you never did in high school. Prepare review sheets and go over them with classmates. Try to guess what will be on the exams. That's one of the best ways to think about what the most important points have been in the course. Each chapter ends with a list of the main points it covers and with some suggestions for how to use the list for review. One of the best ways to learn material that you plan to use again is to tie each new idea to as many familiar concepts and situations as you can and to visualize settings in which the new fact would be helpful to you. We have included lots of examples in the text to make this process easier. The review lists can be used to go over the material in the same way by yourself or with fellow students.

The supplementary exercises at the ends of the chapters are also a good way to check your command of the chapter material. You don't need to work them all, though it's a good idea to look them all over just to see what kinds of questions one can be expected to answer. Our best advice on doing exercises is that doing a few thoughtfully is better than trying to do a lot in a hurry, and it will probably take you less time overall.

Answers or hints to most odd-numbered exercises in the sections, as well as complete answers to all supplementary exercises, are given in the back of the book. Wise students will look at the answers only after trying seriously to do the problems. When a proof is called for, we usually give a hint or an outline of a proof, which you should first understand and then expand upon. A symbols index appears on the inside front cover of the book. At the back of the book there is an index of topics. After Chapter 13 there is a brief dictionary of some terms that we use in the text without explanation, but which readers may have forgotten or never encountered. Look at these items right now to see where they are and what they contain and then join us for Chapter 1.

KENNETH A. ROSS
ross@math.uoregon.edu

CHARLES R. B. WRIGHT
wright@math.uoregon.edu

Discrete Mathematics

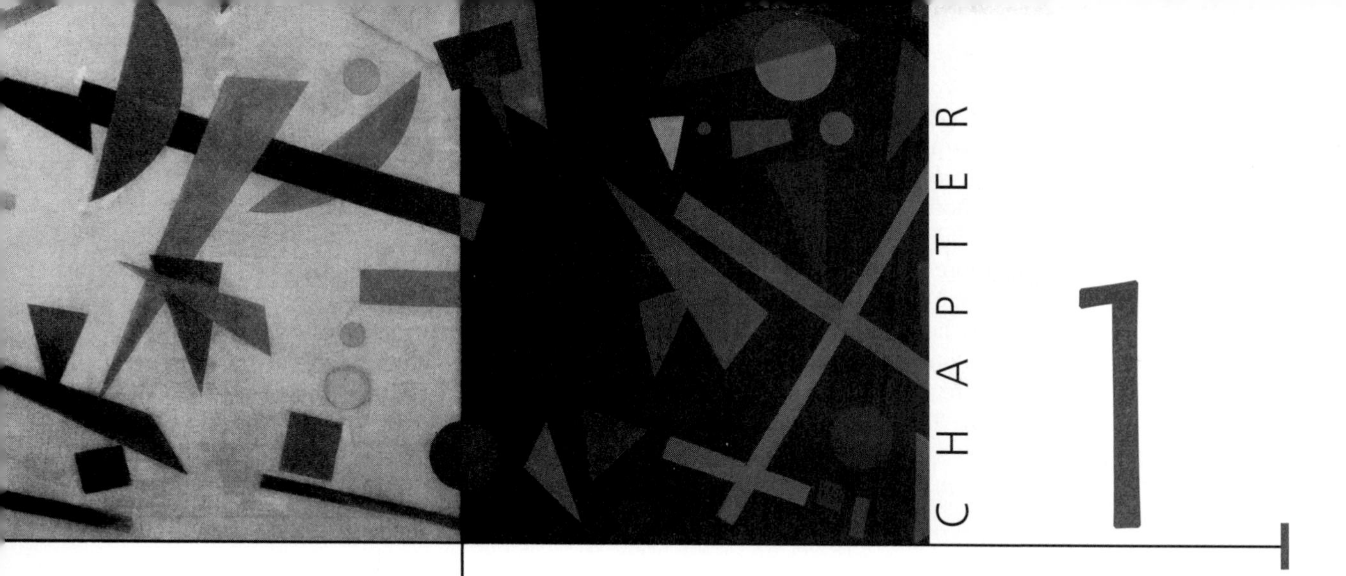

Sets, Sequences, and Functions

This chapter is introductory and contains a relatively large number of fundamental definitions and notations. Much of the material may be familiar, though perhaps with different notation or at a different level of mathematical precision. Besides introducing concepts and methods, this chapter establishes the style of exposition that we will use for the remainder of the book.

1.1 Some Warm-up Questions

In this section we point out the importance of precision, abstraction and logical thinking. We also introduce some standard terms and notation. The mathematical content is less important than the ways of thinking about things.

How many numbers are there between 1 and 10? Between 1 and 1000? Between 1000 and 1,000,000? Between 6357 and 924,310? These questions are all just special cases of the more general question, "How many numbers are there between m and n?" For example, the question about how many numbers there are between 1000 and 1,000,000 is the case where $m = 1000$ and $n = 1{,}000{,}000$. If we learn how to answer the general question, then we'll be able to apply our method to solve the four specific [and fairly artificial] problems above and also any others like them that we come across in practice. By the end of this section we'll even be able to answer some substantially more challenging questions by using what we've learned about these simpler ones.

The process of going from specific cases to general problems is called **abstraction**. One of our goals in this book is to convince you that *abstraction is your friend*. This assertion may seem hard to believe right now, but you will see lots of examples to prove our point. For now, we claim that abstraction is valuable for at least three reasons.

First, the process of abstraction strips away inessential features of a specific problem to focus attention on the core issues. In this way, abstraction often makes a problem easier to analyze and hence easier to solve. Indeed, one can view abstraction as just one standard step in analyzing any problem.

Second, the solution to the abstracted problem applies not just to the problem we started with but to others like it as well, i.e., to all problems that have the same abstract type. By solving the abstract problem, we solve a whole class of specific problems at no extra cost.

Moreover, using abstraction means that you don't have to do a lot of essentially similar homework problems. A few well-chosen exercises, together with abstraction, can provide all the practice necessary to master—really master—new ideas. We will have more to say later about how you can use abstraction to get the most out of your homework.

So how many numbers *are* there between the two numbers m and n? That depends on what we mean by "number." If we allow rational numbers like $311/157$, then there are already infinitely many of them between any two different numbers m and n. The answer is always the same and not very interesting in this case.

Let's make the question interesting by asking how many *integers* there are between m and n. [Recall that an **integer** is one of the numbers in the list

$$\ldots, -5, -4, -3, -2, -1, 0, 1, 2, 3, 4, 5, \ldots$$

which extends without bound in both directions.] How many integers are there between 1 and 10? If we mean "strictly between," then the numbers are 2, 3, 4, 5, 6, 7, 8, 9, and there are 8 of them, but if we mean "between, but allowing 1 and 10, too," then the answer is $8+2 = 10$. We can't answer the question until we have a clear, precise statement of it. Usually, when people say "think of a number between 1 and 10," they mean to allow both 1 and 10, so for the rest of this section let's agree that "between" includes both ends of the range. Then our general problem is

How many integers i are there with $m \le i \le n$?

Lots of people will guess that the answer is $n - m$. To show that this guess is wrong in general, it's enough to exhibit one pair of numbers m and n for which it's wrong. How about $m = 1$ and $n = 10$, i.e.,

How many integers i are there with $1 \le i \le 10$?

The right answer is 10, as we observed above, not $10 - 1$, so the guess $n - m$ is wrong.

In fact, our example with $m = 1$ and $n = 10$ shows how to get the correct answer in lots of cases. If $m = 1$ and n is any positive integer, then the integers i with $m = 1 \le i \le n$ are just $1, 2, \ldots, n$, and there are exactly n of them. We record this modest piece of information.

Fact 1 If n is a positive integer, then there are n integers i such that $1 \le i \le n$.

Now to find out how many integers i there are with $1000 \le i \le 1{,}000{,}000$ we could list them all and then count them, but there's an easier way that leads to a general method: count the i's with $1 \le i \le 1{,}000{,}000$ and then throw away the ones with $i < 1000$, i.e., with $i \le 999$. Fact 1 gives us our answer: $1{,}000{,}000 - 999 = 999{,}001$. This method leads us to another general fact.

Fact 2 If m and n are positive integers with $m \le n$, then the number of integers i such that $m \le i \le n$ is $n - (m - 1) = n - m + 1$.

Before you read on, stop and make sure that you see where the $m - 1$ came from here and why this result really does follow from Fact 1.

Fact 2 is a statement in the form "if ... , then .... " The part between "if" and "then" contains its **hypotheses**, that m and n are positive integers and that $m \le n$, and the part after "then" is its **conclusion**, in this case a statement about a certain set of integers. If ... then ... statements of this sort come up in all kinds of situations, not just in mathematics, and the ability to handle them correctly can be crucial. We will work with such statements throughout this book, and we will pay special attention to them in Chapter 2 on logic and arguments.

Given an implication such as Fact 2, it's natural to ask whether the hypotheses are really necessary. Would the conclusion still be true if we dropped a hypothesis? If not, how would it change? Can we say anything at all useful under weaker hypotheses?

If we leave out the second hypothesis of Fact 2 and allow $m > n$, then we get an easy answer, since there are no i's at all with *both* $m \leq i$ *and* $i \leq n$. If a computer program with inputs m and n prints something for each integer between m and n, and if we want to know how many things it prints, then we may need to consider the case $m > n$. For our discussion now, though, let's keep $m \leq n$ as a hypothesis.

What would happen if we allowed either m or n to be negative or 0? For example, how many integers i are there with $-10 \leq i \leq 90$? If we simply add 11 to everything, then the question becomes how many integers $i + 11$ are there with $-10 + 11 = 1 \leq i + 11 \leq 90 + 11 = 101$? Then Fact 2 says that there are $101 - 1 + 1$ possible $i + 11$'s, so that's how many i's there are, too. This answer is $(90 + 11) - (-10 + 11) + 1 = 90 - (-10) + 1 = n - m + 1$ still. It didn't really matter that we chose to add 11; any integer bigger than 10 would have let us use Fact 2 and would have canceled out at the end. The method also didn't depend on our choice of $m = -10$. The hypothesis that m and n are positive is unnecessary, and we have the following more general result.

Fact 3 If m and n are integers with $m \leq n$, then there are $n - m + 1$ integers i with $m \leq i \leq n$.

Let's try a harder question now. How many integers between 1 and 100 are even? How many between 10 and 100? Between 11 and 101? Between 17 and 72? Recall that an integer is **even** if it's twice some integer. The even integers, then, are the ones in the list

$$\ldots, -8, -6, -4, -2, 0, 2, 4, 6, 8, 10, \ldots$$

and the **odd** integers are the ones that aren't even, i.e., the ones in the list

$$\ldots, -7, -5, -3, -1, 1, 3, 5, 7, 9, \ldots.$$

Our new general question is, "How many even integers i are there that satisfy $m \leq i \leq n$?" Let's assume first that m and n are positive and worry later about negative possibilities. Since Fact 2 says that there are $n - m + 1$ integers between m and n, and since about half of them are even, we might guess that the answer is $(n - m + 1)/2$. We see that this guess can't be right in general, though, because it's not a whole number unless $n - m + 1$ is even. Still, we can figure that $(n - m + 1)/2$ is probably *about* right.

The strategy that worked before to get the correct answer still seems reasonable: count the even integers from 1 to n and subtract the number of even integers from 1 to $m - 1$. Without actually listing the integers from 1 to n, we can describe the even ones that we'd find. They are $2, 4, \ldots, 2s$ for some integer s, where $2s \leq n$ but $n < 2(s + 1)$, i.e., where $s \leq n/2 < s + 1$. For example, if $n = 100$, then $s = 50$, and if $n = 101$, then $n/2 = 50.5$ and again $s = 50$. Our answer is s, which we can think of as "the integer part" of $n/2$.

There's a standard way to write such integer parts. In general, if x is any real number, the **floor** of x, written $\lfloor x \rfloor$, is the largest integer less than or equal to x. For example $\lfloor \pi \rfloor = \lfloor 3.14159265 \cdots \rfloor = 3$, $\lfloor 50.5 \rfloor = 50$, $\lfloor 17 \rfloor = 17$, and $\lfloor -2.5 \rfloor = -3$. The **ceiling** of x, written $\lceil x \rceil$, is similarly defined to be the smallest integer that is at least as big as x, so $\lceil \pi \rceil = 4$, $\lceil 50.5 \rceil = 51$, $\lceil 72 \rceil = 72$, and $\lceil -2.5 \rceil = -2$. In terms of the number line, if x is not an integer, then $\lfloor x \rfloor$ is the nearest integer to the left of x and $\lceil x \rceil$ is the nearest integer to the right of x.

We've seen that the number of even integers in $1, 2, \ldots, n$ is $\lfloor n/2 \rfloor$, and essentially the same argument shows the following.

Fact 4 Let k and n be positive integers. Then the number of multiples of k between 1 and n is $\lfloor n/k \rfloor$.

Again, stop and make sure that you see how the reasoning we used for the even numbers with $k = 2$ would work just as well for any positive integer k. What does Fact 4 say for $k = 1$?

It follows from Fact 4 that if m is a positive integer with $m \le n$, then the number of multiples of k in the list $m, m+1, \ldots, n$ is $\lfloor n/k \rfloor - \lfloor (m-1)/k \rfloor$. It's natural to wonder whether there isn't a nicer formula that doesn't involve the floor function. For instance, with $k = 2$ the number of even integers between 10 and 100 is $\lfloor 100/2 \rfloor - \lfloor 9/2 \rfloor = \lfloor 50 \rfloor - \lfloor 4.5 \rfloor = 50 - 4 = 46$, and the number between 11 and 101 is $\lfloor 101/2 \rfloor - \lfloor 10/2 \rfloor = \lfloor 50.5 \rfloor - \lfloor 5 \rfloor = 50 - 5 = 45$. In both cases $(n - m + 1)/2 = 91/2 = 45.5$, which is close to the right answer.

In fact, as we will now show, $(n - m + 1)/2$ is *always* close to our answer $\lfloor n/k \rfloor - \lfloor (m-1)/k \rfloor$, which is the actual number of multiples of k between m and n. Notice that we will need to use a little algebra here to manipulate inequalities. As we go along in the book, we'll need more algebraic tools, though probably nothing harder than this. If your algebra skills are a little rusty, now would be a good time to review them. Here's the argument.

In general, $\lfloor x \rfloor \le x < \lfloor x \rfloor + 1$, so $x - 1 < \lfloor x \rfloor \le x$. Thus, with n/k for x, we get

$$\frac{n}{k} - 1 < \left\lfloor \frac{n}{k} \right\rfloor \le \frac{n}{k} \tag{1}$$

and similarly, with $(m-1)/k$ for x, we get

$$\frac{m-1}{k} - 1 < \left\lfloor \frac{m-1}{k} \right\rfloor \le \frac{m-1}{k}. \tag{2}$$

Multiplying (2) through by -1 reverses the inequalities to give

$$-\frac{m-1}{k} \le -\left\lfloor \frac{m-1}{k} \right\rfloor < -\frac{m-1}{k} + 1. \tag{3}$$

Finally, adding inequalities (1) and (3) term by term gives

$$\frac{n}{k} - 1 - \frac{m-1}{k} < \left\lfloor \frac{n}{k} \right\rfloor - \left\lfloor \frac{m-1}{k} \right\rfloor < \frac{n}{k} - \frac{m-1}{k} + 1,$$

which just says that

$$\frac{n-m+1}{k} - 1 < \text{our answer} < \frac{n-m+1}{k} + 1.$$

We've shown that the number of multiples of k between m and n is always an integer that differs from $(n - m + 1)/k$ by less than 1, so $(n - m + 1)/k$ was a pretty good guess at the answer after all.

To finish off our general question, let's allow m or n to be negative. How does the answer change? For example, how many even integers are there between -11 and 72? The simple trick is to add a big enough even integer to everything, for instance to add 20. Then, because i is even with $-11 \le i \le 72$ if and only if $i + 20$ is even with $-11 + 20 \le i + 20 \le 72 + 20$, and because the number of even integers between $-11 + 20 = 9$ and $72 + 20 = 92$ is $\lfloor 92/2 \rfloor - \lfloor 8/2 \rfloor = 46 - 4 = 42$, this is the answer to the question for -11 and 72, too.

The same trick will work for multiples of any positive integer k. Moreover, if we add a multiple of k to everything, say tk for some integer t, then we have

$$\begin{aligned}\left\lfloor \frac{n+tk}{k} \right\rfloor - \left\lfloor \frac{m-1+tk}{k} \right\rfloor &= \left\lfloor \frac{n}{k} + t \right\rfloor - \left\lfloor \frac{m-1}{k} + t \right\rfloor \\ &= \left\lfloor \frac{n}{k} \right\rfloor + t - \left(\left\lfloor \frac{m-1}{k} \right\rfloor + t \right) \qquad \text{[why?]} \\ &= \left\lfloor \frac{n}{k} \right\rfloor - \left\lfloor \frac{m-1}{k} \right\rfloor,\end{aligned}$$

so the answer to our question is still the same, even if m is negative. Altogether, then, we have the following.

Theorem Let m and n be integers with $m \leq n$, and let k be a positive integer. Then the number of multiples of k between m and n is

$$\left\lfloor \frac{n}{k} \right\rfloor - \left\lfloor \frac{m-1}{k} \right\rfloor,$$

which differs from $(n - m + 1)/k$ by at most 1.

When we dropped the little question "[why?]" into the argument just before the theorem, we hoped that you would pause for a moment and not just rush through the symbols to get to the end. Reading mathematics slowly enough takes lots of practice, especially in view of our natural tendency to read as quickly as possible. In this instance, something changes with each "=" sign, and to be convinced that the whole argument has actually proved something, we need to examine each step for correctness. To answer the "[why?]" we need some sort of argument to show the fact that

$$\text{if } t \text{ is an integer, then } \lfloor x + t \rfloor = \lfloor x \rfloor + t \quad \text{for every number } x.$$

One way to visualize this fact on the number line is to observe that adding t to numbers simply slides points t units to the right on the line. The numbers $\lfloor x \rfloor$ and x get slid to $\lfloor x \rfloor + t$ and $x + t$, respectively, and so $\lfloor x \rfloor + t$ is the closest integer to the left of $x + t$.

Convinced? Here's another argument based on the definition of floor. If t is an integer, then so is $\lfloor x \rfloor + t$. Since $\lfloor x \rfloor \leq x < \lfloor x \rfloor + 1$, we have $\lfloor x \rfloor + t \leq x + t < \lfloor x \rfloor + t + 1$. But $\lfloor x + t \rfloor$ is the *only* integer k with $k \leq x + t < k + 1$, so $\lfloor x \rfloor + t$ must be $\lfloor x + t \rfloor$.

The theorem gives us a formula into which we can plug numbers to get answers. It tells us what sorts of input numbers we are allowed to use, and it tells us what the answers mean. It's a fine theorem, but the theorem itself and its formula are not as important as the analysis we went through to get them. Nobody is going to remember that formula for very long, but the ideas that went into deriving it are easy to understand and can be applied in other situations as well. At the end of this section we list some generally useful methods, most of which came up on the way to the theorem. It's worthwhile right now to look back through the argument and to review the ideas that went into it. As for the formula itself, it will be helpful for working some of the easiest homework exercises.

Now how about *really* changing the problem? How many prime integers are there between m and n? [Recall that an integer greater than 1 is **prime** if it is not the product of two smaller positive integers.] Suddenly the problem gets a lot harder. We'd still be done if we could tell how many primes there are in $1, 2, \ldots, n$, but unfortunately there is no general way known of counting these primes other than actually going through $1, \ldots, n$ and listing all of them. A list of the primes less than 1000 is given in Figure 1. The Prime Number Theorem, a nontrivial result from analytic number theory, says that for large enough values of n the fraction of numbers between 1 and n that are prime is approximately $1/\ln n$, where ln is the natural logarithm function; so we can estimate that there are something like $n/\ln n$ primes less than n. If $100 \leq m \leq n$, then the answer to our question is roughly

$$\frac{n}{\ln n} - \frac{m-1}{\ln(m-1)}.$$

How big is "large enough" here? In particular, is 100 anywhere near large enough for these estimates to be valid? And even if the difference between the fraction we want and $1/\ln n$ is very small, multiplying it by n can give a big difference between the actual number of primes and $n/\ln n$. The percentage error stays the

Figure 1 ▶
The Primes Less Than 1000

2, 3, 5, 7, 11, 13, 17, 19, 23, 29, 31, 37, 41, 43, 47, 53, 59, 61, 67, 71, 73, 79, 83, 89, 97, 101, 103, 107, 109, 113, 127, 131, 137, 139, 149, 151, 157, 163, 167, 173, 179, 181, 191, 193, 197, 199, 211, 223, 227, 229, 233, 239, 241, 251, 257, 263, 269, 271, 277, 281, 283, 293, 307, 311, 313, 317, 331, 337, 347, 349, 353, 359, 367, 373, 379, 383, 389, 397, 401, 409, 419, 421, 431, 433, 439, 443, 449, 457, 461, 463, 467, 479, 487, 491, 499, 503, 509, 521, 523, 541, 547, 557, 563, 569, 571, 577, 587, 593, 599, 601, 607, 613, 617, 619, 631, 641, 643, 647, 653, 659, 661, 673, 677, 683, 691, 701, 709, 719, 727, 733, 739, 743, 751, 757, 761, 769, 773, 787, 797, 809, 811, 821, 823, 827, 829, 839, 853, 857, 859, 863, 877, 881, 883, 887, 907, 911, 919, 929, 937, 941, 947, 953, 967, 971, 977, 983, 991, 997

same, though, and despite these valid concerns the Prime Number Theorem is still a handy and commonly used tool for making rough estimates.

Questions such as the one we have just considered come up in the design of encryption codes and fast algorithms that use random choices. If we're looking for a fairly large prime, say one with 30 decimal digits, we can calculate $\ln 10^{30} = 30 \ln 10 \approx 70$ and conclude that roughly $1/70$ of the integers less than 10^{30} are prime. Most of the primes in this range are bigger than 10^{29} [see Exercise 15]. It turns out that on average we should expect to try about 70 random numbers between 10^{29} and 10^{30} before we hit a prime. When you think about it, 70 trials is not very many, considering the sizes of the numbers involved. There are *lots* of primes out there.

In this section we've taken some fairly elementary questions about integers as examples to illustrate various problem-solving methods that will appear again and again throughout the book. Some of these methods are:

Removing ambiguity by determining precisely what the problem is about.
Abstracting general problems from specific cases.
Using specific examples as a guide to general methods.
Solving special cases first and then using them as parts of the general solution.
Changing hypotheses to see how the conclusions change.
Using suitable notation to communicate economically and precisely.
Counting a set without actually listing its members.
Counting a set by counting a larger set and then throwing out some members.

It's worth looking back through the section to see where these methods came up. Some were used several times; can you spot them all? One reason for hunting for such usage is to get practice in identifying and describing standard steps so that you will, in time, automatically think of using them yourself. The techniques listed above are just a few examples of ones that will help you to analyze and solve problems quickly; using them should help give you confidence in your answers.

Exercises 1.1

1. How many integers are there between the following pairs of numbers?
(a) 1 and 20 (b) 2 and 21 (c) 2 and 20
(d) 17 and 72 (e) -6 and 4 (f) 0 and 40
(g) -10 and -1 (h) -10^{30} and 10^{30} (i) 10^{29} and 10^{30}

2. (a) How many 4-digit numbers are there, i.e., numbers from 1000 to 9999?
(b) How many 5-digit numbers are there that end in 1?
(c) How many 5-digit numbers end in 0?
(d) How many 5-digit numbers are multiples of 10?

3. Give the value of each of the following.
(a) $\lfloor 17/73 \rfloor$ (b) $\lfloor 1265 \rfloor$ (c) $\lfloor -4.1 \rfloor$ (d) $\lfloor -4 \rfloor$

4. Give the value of each of the following.
(a) $\lceil 0.763 \rceil$ (b) $2\lceil 0.6 \rceil - \lceil 1.2 \rceil$
(c) $\lceil 1.1 \rceil + \lceil 3.3 \rceil$ (d) $\lceil \sqrt{3} \rceil - \lfloor \sqrt{3} \rfloor$
(e) $\lceil -73 \rceil - \lfloor -73 \rfloor$

5. How many even integers are there between the following pairs of numbers?
(a) 1 and 20 (b) 21 and 100 (c) 21 and 101
(d) 0 and 1000 (e) -6 and 100 (f) -1000 and -72

6. How many odd integers are there between the following pairs of numbers?

(a) 1 and 20 (b) 21 and 100

(c) 21 and 101 (d) 0 and 1000

7. How many multiples of 6 are there between the following pairs of numbers?

(a) 0 and 100 (b) 9 and 2967 (c) -6 and 34

(d) -600 and 3400

8. How many multiples of 10 are there between the following pairs of numbers?

(a) 1 and 80 (b) 0 and 100

(c) 9 and 2967 (d) -6 and 34

(e) 10^4 and 10^5 (f) -600 and 3400

9. Explain *in words* how you could use a simple pocket calculator to find $\lfloor n/k \rfloor$ for integers n and k. What if n/k is negative?

10. (a) Describe a situation in which you would prefer to use $\lfloor x \rfloor$ notation rather than the phrase "the greatest integer at most equal to x."

(b) Describe a situation in which you would prefer to use words rather than the $\lfloor \quad \rfloor$ notation.

11. In the argument for the theorem on page 5 we used the fact about $\lfloor \quad \rfloor$ that if t is an integer, then $\lfloor x + t \rfloor = \lfloor x \rfloor + t$ for every number x.

(a) State a similar fact about $\lceil \quad \rceil$.

(b) Explain how the fact in part (a) follows from the definition of ceiling.

12. How many 4-digit numbers [see Exercise 2] are multiples of your age?

13. (a) How many numbers between 1 and 33 are prime?

(b) How does this number compare with $33/\ln 33$?

(c) Would you like to try this problem with 33 replaced by 3333? Maybe with a computer?

14. (a) Write a generalized problem that is an abstraction of the following specific problem. How many integers between 1 and 5000 are multiples of 5 but not multiples of 10?

(b) Write another.

(c) What is the answer to the specific problem in part (a)?

15. Use the estimate in the text based on the Prime Number Theorem to give approximate values of the following.

(a) The number of primes between 1 and 10^{30}.

(b) The number of primes between 1 and 10^{29}.

(c) The number of 30-digit primes.

(d) The percentage of 30-digit numbers that are primes.

16. Give a convincing argument that Fact 2 is true in general, and not just when m = 1000 and n = 1,000,000, as was shown in the text.

17. Give a convincing argument that Fact 3 is true, given that Fact 2 is true. Your argument should work for all integers m and n, and not just for one or two examples.

18. (a) What does Fact 4 say for $k = 1$? Is this statement obvious?

(b) What does Fact 4 say for $k > n$? Is this statement obvious?

19. (a) Give a specific example of numbers x and y for which $\lfloor x \rfloor + \lfloor y \rfloor < \lfloor x + y \rfloor$.

(b) Give a specific example of numbers x and y for which $\lfloor x \rfloor + \lfloor y \rfloor = \lfloor x + y \rfloor$.

(c) Give a convincing argument that $\lfloor x \rfloor + \lfloor y \rfloor \le \lfloor x + y \rfloor$ for every pair of numbers x and y. *Suggestion:* Use the fact that $\lfloor x + y \rfloor$ is the *largest* integer less than or equal to $x + y$.

20. Let x and y be any numbers at all with $x \le y$.

(a) Show that the number of integers between x and y is $\lfloor y \rfloor - \lceil x \rceil + 1$.

(b) Show that the number of integer multiples of the positive integer k between x and y is $\lfloor y/k \rfloor - \lceil x/k \rceil + 1$.

1.2 Factors and Multiples

When we divide one integer by another, the answer is generally not an integer. If it *is* an integer, then this fact is usually worth noting. In this section we look at questions in which getting an integer answer is the main concern. To keep the discussion simple, we will restrict our attention here to the **natural numbers**, i.e., the nonnegative integers, but much of what we say will also be true for integers in general.

If m and n are integers, then n is a **multiple of** m if $n = km$ for some integer k. One could also say that n is an **integer multiple** of m to stress the important fact that k is an integer. Other ways to say that n is a multiple of m are that n is **divisible by** m, that m **divides** n, that m is a **divisor of** n, or that m is a **factor of** n. We write $\boldsymbol{m|n}$, which we read as "m divides n," to signify any of these equivalent statements, and we write $\boldsymbol{m \nmid n}$ in case $m|n$ is false.

If n is a multiple of m, then so is every multiple of n, because if $n = km$ for some integer k, then $ln = (lk) \cdot m$ for every integer l. Similarly, if $d|m$, then $d|lm$ for every integer l.

EXAMPLE 1

(a) We have $3|6$, $4|20$, $15|15$, and $91|2002$, since $6 = k \cdot 3$ for $k = 2$, and similarly $20 = 5 \cdot 4$, $15 = 1 \cdot 15$, and $2002 = 22 \cdot 91$. Also $11|2002$, since $11|22$ and $22|2002$.

(b) For every nonzero integer n, we have $1|n$ and $n|n$, since $n = n \cdot 1 = 1 \cdot n$. ■[1]

If $n = km$ and $m \neq 0$, then k must be $\frac{n}{m}$, because $\frac{n}{m}$ is *defined* to be the number x for which $xm = n$. There is only one such number, since if $xm = ym$ then $x = y$. [Either think of canceling the m's or rewrite the equation as $(x - y) \cdot m = 0$ to see that $x - y$ must be 0 because $m \neq 0$.] Thus, if $m \neq 0$, then $m|n$ if and only if $\frac{n}{m}$ is an integer.

EXAMPLE 2

(a) We could have approached Example 1(a) by observing that $\frac{6}{3} = 2$, $\frac{20}{4} = 5$, $\frac{15}{15} = 1$, and $\frac{2002}{91} = 22$.

(b) Similarly $1|n$ and $n|n$ because $\frac{n}{1}$ and $\frac{n}{n}$ are integers.

(c) Since $\frac{7}{4}$, $\frac{11}{12}$, and $\frac{2002}{17}$ are not integers, we have $4 \nmid 7$, $12 \nmid 11$, and $17 \nmid 2002$. ■

EXAMPLE 3

The integer 0 needs special treatment. Since $0 = 0 \cdot n$ whenever n is an integer, 0 is always a multiple of n, i.e., every integer is a divisor of 0. In particular, 0 divides 0. On the other hand, if 0 divides n, i.e., if n is a multiple of 0, then $n = k \cdot 0 = 0$, so the only multiple of 0 is 0 itself. We can't define $\frac{n}{0}$ if $n \neq 0$, because we would need to have $\frac{n}{0} \cdot 0 = n$, which is impossible. We can't define $\frac{0}{0}$ either without violating some of the laws of algebra. [For example, the nonsense equation $\frac{0}{0} + \frac{1}{1} = \frac{0 \cdot 1 + 1 \cdot 0}{0 \cdot 1} = \frac{0}{0}$ would imply $1 = 0$.] ■

There is a well-known link between divisibility and size.

Proposition If m and n are positive integers such that $m|n$, then $m \leq n$ and $\frac{n}{m} \leq n$.

Proof Let $k = \frac{n}{m}$. Then k is an integer and $n = km$. Since $n \neq 0$ and since n and m are both positive, k can't be 0 or negative. The smallest positive integer is 1, so $1 \leq k$. Hence $m = 1 \cdot m \leq k \cdot m = n$. Since $k|n$, the same argument shows that $k \leq n$. ■

Note that if we were working with real numbers, instead of restricting ourselves to integers, then the conclusion of the proposition might well be false. For example, in the real number setting we have $5 = \frac{5}{7} \cdot 7$, so we could view 7 as a divisor of 5, but of course $7 \not\leq 5$. What goes wrong with the proof in this case? Think about it.

As we said in Section 1.1, an integer n is **even** if it is a multiple of 2. Thus n is even if and only if $n = 2k$ for some integer k. In particular, 0 is an even integer. An integer is **odd** if it is not even. Every odd integer can be written as $2k + 1$ for some integer k. One way to see this is to imagine listing the integers in increasing order. Every second integer is a multiple of 2 and the odd integers lie between the even ones, so adding 1 to the even integers must give the odd ones. It is common to refer to even integers as **even numbers** and to odd integers as **odd numbers**. This terminology should cause no confusion, since the terms "even" and "odd" are only applied to numbers that are integers.

As we noted in Examples 1 and 2, both 1 and n are always divisors of n. The next theorem characterizes the primes as the positive integers n for which these two are the only divisors.

Theorem 1 An integer n greater than 1 is a prime if and only if its only positive divisors are 1 and n.

[1] We will use ■ to signify the end of an example or proof.

Proof We will show the logically equivalent statement that n is *not* a prime if and only if it has at least one positive divisor other than 1 and n.

Suppose that n is not a prime. Since the primes are the integers greater than 1 that are not products of two smaller integers, there must be positive integers s and t with $s < n$, $t < n$, and $st = n$. Then s is a positive divisor of n that's not n. It is also not 1, since $st = n \neq t = 1 \cdot t$.

In the other direction, suppose that n has some positive divisor m different from 1 and n. Then $\frac{n}{m}$ is an integer, and $n = \frac{n}{m} \cdot m$. By the proposition above, both m and $\frac{n}{m}$ are smaller than n, so n is not a prime. ■

Note that s and t in the proof of the theorem might be equal. For example, if $n = 9$, then they would have to be equal. We used two different symbols, s and t, because we had no reason to believe that the two factors would be the same in general.

Theorem 1 may seem too familiar and obvious to be worth mentioning. One reason we have included it is to illustrate its pattern of proof, called an *indirect* proof. Later on, in Chapter 2, we will look closely at how to construct logical arguments. When we get there, you may want to recall this proof as an elementary example.

The first few positive integers are

$$1,\ 2,\ 3,\ 4 = 2 \cdot 2,\ 5,\ 6 = 2 \cdot 3,\ 7,\ 8 = 2 \cdot 2 \cdot 2,$$
$$9 = 3 \cdot 3,\ \ldots,\ 29,\ 30 = 2 \cdot 3 \cdot 5.$$

The primes in this list, 2, 3, 5, 7, ..., 29, are the ones greater than 1 that have no interesting factors. In that sense, they are the leftovers after we factor everybody else. See page 6 for a list of the primes less than 1000.

So how do we factor the nonprimes? Suppose that we are given n to factor. We can try dividing n by all smaller positive integers. If none of them divide n, then n is prime. Otherwise, when we find a factor, say m, then we can attack m and $\frac{n}{m}$ and try to find their factors. We can keep this up until nothing we look at has smaller factors, i.e., until we have found a collection of primes whose product is n. For example, $60 = 6 \cdot 10 = (2 \cdot 3) \cdot (2 \cdot 5)$ gives primes 2, 2, 3, 5 with $60 = 2 \cdot 2 \cdot 3 \cdot 5$.

Actually, as shown in Exercise 17, we don't need to try all numbers smaller than n as factors, but only the primes p with $p \leq \sqrt{n}$. Lists of thousands of primes are readily available to make the computation easier. [Note that such a list only has to be computed once and then stored.] The trouble with this method is that it is only effective for small values of n. If n is large—integers with 100 or more decimal digits are routinely used in encryption programs—then, even if we somehow had a list of all smaller primes, there would just be too many to try.

A great deal of recent progress has been made on alternative methods for factoring integers. Symbolic computation programs make short work of factoring quite large numbers, but factorization for really giant integers is still considered to be exceedingly difficult in practice. Nevertheless, the approach that we have just described has theoretical value and is the essence of the proof of the following theorem.

Theorem 2 Every positive integer n can be written as a product of primes. Moreover, there is only one way to write n in this form [except for rearranging the order of the terms].

This fact is probably familiar; it may even be so familiar that you think it is obvious. Its proof is not trivial, though. In this book we prove the theorem in two stages; see Example 1 on page 167 and Theorem 5 on page 177. Note that the statement "there is only one way to write n in this form" would not be true if we allowed the integer 1 to be a prime; for example, we could write $6 = 1 \cdot 1 \cdot 2 \cdot 3$ or simply $6 = 2 \cdot 3$. This is why 1 is not considered a prime.

EXAMPLE 4

We can, at least conceptually, display factorization information graphically. For example, Figure 1 shows the complete set of divisors of 60 arranged in a suggestive way. The arrows display the divisor relation, since $m|n$ if and only if we can go from m to n along a path made by following arrows. Each arrow corresponds to some prime, and the way we've drawn the figure the edges that are parallel all go with the same prime. The divisors of 30, or of any other divisor of 60, form a piece of this figure in the same style as the whole figure. Later, when we have more graph-theoretic tools at our disposal, we will come back to pictures like this. ■

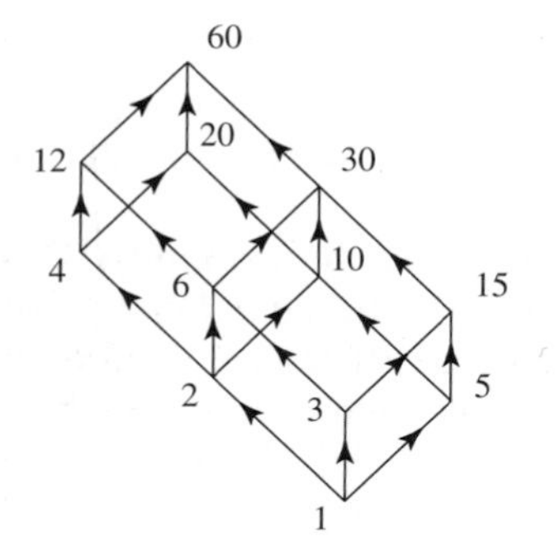

Figure 1 ▲

We saw at the beginning of this section that

$$\text{if } m|n \text{ and } n|q, \text{ then } m|q.$$

Then we saw in the Proposition on page 8 that if m and n are positive and if $m|n$, then $m \le n$. It's natural to think of another familiar fact:

$$\text{if } m \le n \text{ and } n \le q, \text{ then } m \le q.$$

There is an important similarity here that we will come back to in §§3.1 and 11.1. As Figure 1 reminds us, though, | and $\le$ are not the same. Indeed, $4|60$, $30|60$, and $4 \le 30$, but $4 \nmid 30$.

If we need to compute the prime factors of a large integer n, then we may be in for a long wait. Often, though, one really just wants to know the factors that two integers m and n have in common. This problem may seem more complicated than factorization of a single integer, but it turns out to be much easier. Later on, in §4.7, we will give a complete account of Euclid's algorithm for finding largest common factors. To discuss the problem and its applications more completely, we need to define some more terminology.

Consider two positive integers m and n. Any integer that divides both of them is called a **common divisor** or a **common factor** of m and n. Of course, 1 is always a common divisor, and often the hope is that it will be the only common divisor. In any case, among the common divisors of m and n, there is a largest one, called their **greatest common divisor** and denoted by $\mathbf{gcd}(\boldsymbol{m}, \boldsymbol{n})$. [Some people use the term "highest common factor" and write $\text{hcf}(m, n)$.]

EXAMPLE 5

(a) The divisors of 18 are 1, 2, 3, 6, 9, and 18, and the divisors of 25 are 1, 5, and 25. The only *common* divisor of 18 and 25 is 1, so $\gcd(18, 25) = 1$.

(b) The divisors of 24 are 1, 2, 3, 4, 6, 8, 12, and 24, so the common divisors of 18 and 24 are 1, 2, 3, and 6, and thus $\gcd(18, 24) = 6$.

(c) The common divisors of 168 and 192 are 1, 2, 3, 4, 6, 8, 12, and 24, so $\gcd(168, 192) = 24$.

(d) The only common divisors of 168 and 175 are 1 and 7, so $\gcd(168, 175) = 7$.

(e) The only common divisor of the integers 192 and 175 is 1, and hence we have $\gcd(192, 175) = 1$. ■

In Example 5 we made some claims that would be easy, but tedious, to verify. Although we didn't say so, our approach relied upon finding all factors of m and all factors of n, and then finding the largest number common to both lists. Theorem 2 assures us that this approach will succeed, but our comments above indicate that finding all the factors may take a long time.

EXAMPLE 6

We revisit Example 5 and write each of 168, 192, and 175 as a product of primes:

$$168 = 2 \cdot 2 \cdot 2 \cdot 3 \cdot 7 = 2^3 \cdot 3 \cdot 7;$$

$$192 = 2 \cdot 2 \cdot 2 \cdot 2 \cdot 2 \cdot 2 \cdot 3 = 2^6 \cdot 3;$$

$$175 = 5 \cdot 5 \cdot 7 = 5^2 \cdot 7.$$

A common divisor of 168 and 192 will also be a product of primes that divide both 168 and 192, and it can have at most three occurrences of 2, at most one occurrence of 3, and no factors 7. So the greatest common divisor must be $2 \cdot 2 \cdot 2 \cdot 3 = 2^3 \cdot 3 = 24$, i.e., $\gcd(168, 192) = 24$.

Similar observations confirm that $\gcd(168, 175) = 2^0 \cdot 3^0 \cdot 5^0 \cdot 7^1 = 7$ and $\gcd(192, 175) = 2^0 \cdot 3^0 \cdot 5^0 \cdot 7^0 = 1$. ■

We can describe our thinking in Example 6 with the help of some new notation. If a and b are numbers, then their **minimum**, written **min{a, b}**, is defined to be a if $a \le b$ and to be b if $b < a$. Using this notation, we could write the calculation of $\gcd(168, 192)$ as

$$\gcd(2^3 \cdot 3^1 \cdot 7^1, 2^6 \cdot 3^1 \cdot 7^0) = 2^{\min\{3,6\}} \cdot 3^{\min\{1,1\}} \cdot 7^{\min\{1,0\}} = 2^3 \cdot 3^1 \cdot 7^0.$$

In Example 5 we observed that $\gcd(192, 175) = 1$ because 192 and 175 have no common factors—except 1 of course—a situation sufficiently important to warrant a name. Two positive integers m and n are **relatively prime** [to each other] if $\gcd(m, n) = 1$. We emphasize that "relatively prime" is a property of *pairs* of integers, neither of which needs to be a prime. For example, 192 and 175 are relatively prime, but neither of them is a prime. In particular, it makes no sense to ask whether a single integer, such as 192, is relatively prime.

Verifying that two numbers are relatively prime turns out to have practical importance, not only in mathematics itself but also in applications of mathematics to computer science, in particular to encryption methods. It is quite remarkable that the Euclidean algorithm is able to compute greatest common divisors and hence can check whether two integers are relatively prime, using a number of arithmetic steps that only increases in proportion to the number of digits in the numbers involved. That is, the number of steps for the worst 100-digit numbers is only about twice the number of steps for 50-digit numbers or, indeed, only 50 times the number of steps for 2-digit numbers.

Not only is this 2300-year-old algorithm fast, it can also be made to produce integers s and t so that

$$\gcd(m, n) = sm + tn. \qquad (*)$$

One consequence of $(*)$, which we will justify in §4.7, is the useful and familiar fact that if a prime divides a product of two positive integers, then it divides at least one of the factors. See the proof of Lemma 4.7 on page 177 for the details.

Here's another consequence of $(*)$. You probably noticed in Example 5 that the greatest common divisor was always not only the *largest* of the common divisors, but it was also a *multiple* of all of the others. This phenomenon is true in general, because if d is any divisor of both m and n, say $m = ad$ and $n = bd$, then

$$\gcd(m, n) = sm + tn = s \cdot ad + t \cdot bd = (sa + tb) \cdot d,$$

so $\gcd(m, n)$ is a multiple of d.

Sometimes one can rule out potential factors. Suppose that we are trying to find the factors of 100. We can see that the factors of 99 are *not* going to divide 100, because if d is a divisor of both 100 and 99, say $100 = ad$ and $99 = bd$, then $1 = 100 - 99 = (a - b) \cdot d$, so d must be 1. We have deduced that 100 and 99 are relatively prime without finding the factors of 100. Similarly, we can see that $\gcd(8750, 8751) = 1$ without factoring either number. The same kind of reasoning shows that $\gcd(8750, 8752) = 2$, and that $\gcd(8751, 8753) = 1$ [not 2, because both numbers are odd]. Euclid's ancient algorithm organizes this kind of argument into a constructive scheme.

A notion closely related to greatest common divisor is "least common multiple." You are already familiar with this idea, possibly under the different name "lowest common denominator." Suppose that we wish to simplify

$$\frac{3}{10} + \frac{4}{15}.$$

The idea is to rewrite both fractions with the same denominator. The smallest common multiple of both 10 and 15 is 30, so the lowest common denominator is 30 and we calculate

$$\frac{3}{10}+\frac{4}{15}=\frac{3\cdot 3}{10\cdot 3}+\frac{4\cdot 2}{15\cdot 2}=\frac{9}{30}+\frac{8}{30}=\frac{17}{30}.$$

The idea of lowest common denominator only makes sense in the context of fractions, but the new term that we will define makes sense in a wider context.

Given two positive integers m and n, any integer that is a multiple of both integers is called a **common multiple**. One such common multiple is the product mn. The smallest positive common multiple of m and n is called their **least common multiple** and is written $\mathbf{lcm}(\boldsymbol{m}, \boldsymbol{n})$. Note that

$$1 \le \operatorname{lcm}(m, n) \le mn.$$

EXAMPLE 7

Recall from Example 6 that

$$168 = 2\cdot 2\cdot 2\cdot 3\cdot 7 = 2^3\cdot 3\cdot 7,$$
$$192 = 2\cdot 2\cdot 2\cdot 2\cdot 2\cdot 2\cdot 3 = 2^6\cdot 3, \quad \text{and}$$
$$175 = 5\cdot 5\cdot 7 = 5^2\cdot 7.$$

A positive integer is a multiple of 168 if and only if its prime factorization has at least three 2's, one 3, and one 7. It is a multiple of 192 if and only if its factorization has at least six 2's and one 3. To be a multiple of both 168 and 192, a number must have at least six 2's, one 3, and one 7 as factors. Therefore,

$$\operatorname{lcm}(168, 192) = 2^6\cdot 3\cdot 7 = 1344.$$

In symbols,

$$\operatorname{lcm}(2^3\cdot 3^1\cdot 7^1, 2^6\cdot 3^1\cdot 7^0) = 2^{\max\{3,6\}}\cdot 3^{\max\{1,1\}}\cdot 7^{\max\{1,0\}},$$

where $\mathbf{max}\{\boldsymbol{a}, \boldsymbol{b}\}$ is the larger of a and b. ■

In the last example we had

$$\gcd(168, 192)\cdot \operatorname{lcm}(168, 192) = 24\cdot 1344 = 168\cdot 192.$$

Equality here is not an accident, as we now show.

Theorem 3 For positive integers m and n, we always have

$$\gcd(m, n)\cdot \operatorname{lcm}(m, n) = mn.$$

Proof We could give a proof based on prime factorization and the fact that $\min\{a, b\} + \max\{a, b\} = a + b$ [Exercise 21]. However, there's an easier way. For short, let's write g for $\gcd(m, n)$ and l for $\operatorname{lcm}(m, n)$. Since l is a multiple of both m and n, there are positive integers a and b with $l = am = bn$. As we said above, the Euclidean algorithm produces integers s and t with $g = sm + tn$. This is not an obvious fact now, but will be easy when we study it later. Thus

$$\begin{aligned} g\cdot l &= sm\cdot l + tn\cdot l \\ &= smbn + tnam \\ &= (sb + ta)\cdot mn \\ &\ge mn. \end{aligned}$$

The last inequality holds because $sb + ta$ is a positive integer, so it's at least 1.

Now also $\frac{m}{g}$ and $\frac{n}{g}$ are integers, and

$$\frac{mn}{g} = m\cdot\frac{n}{g} = \frac{m}{g}\cdot n,$$

so $\frac{mn}{g}$ is a multiple of both m and n. Since l is the *smallest* common multiple of m and n, we must have $\frac{mn}{g} \geq l$, and hence

$$mn \geq g \cdot l.$$

We've shown that $g \cdot l \geq mn$ and $mn \geq g \cdot l$, so $g \cdot l = mn$, as desired. ■

This theorem has a practical consequence. Since

$$\text{lcm}(m, n) = \frac{mn}{\gcd(m, n)},$$

in order to compute least common multiples we just need to compute greatest common divisors (using the fast Euclidean algorithm) and then multiply and divide.

As a side note, one of our students once told us that she saw grade-school children in Guatemala being taught the Euclidean algorithm so that they could use the theorem to calculate lowest common denominators.

We remarked above that every common divisor of m and n is a divisor of $\gcd(m, n)$, and it's natural to suspect that every common multiple of m and n must be a multiple of $\text{lcm}(m, n)$, too. One might even hope to use Theorem 3 somehow to show this, but that doesn't seem to be so easy. Luckily, we can get a simple proof by using another fundamental idea.

So far, this section has emphasized cases in which $\frac{m}{n}$ is an integer, but of course, in general, when we divide m by n there will be something left over. For example, if we divide any odd integer by 2, there's a remainder of 1. Here is the precise result that we need, which we will return to in §3.5.

The Division Algorithm Let m be a positive integer. For each integer n there are unique integers q and r satisfying

$$n = m \cdot q + r \text{ and } 0 \leq r < m.$$ ■

The numbers q and r are called the **quotient** and **remainder**, respectively, **when** n is **divided by** m. For example, if $m = 7$ and $n = 31$, then $31 = 7 \cdot 4 + 3$, so $q = 4$ and $r = 3$. If $m = 73$ and $n = 1999$, then $1999 = 73 \cdot 27 + 28$, so $q = 27$ and $r = 28$.

It may seem odd to call this theorem an algorithm, since the statement doesn't explain a procedure for finding either q or r. The name for the theorem is traditional, however, and in most applications the actual method of computation is unimportant. A constructive algorithm to divide integers will be given in §4.1; see the theorem on page 132. For now, let's just use the theorem, since we all believe it.

If we divide the conditions $n = mq + r$ and $0 \leq r < m$ through by m and rewrite them as

$$\frac{n}{m} = q + \frac{r}{m} \quad \text{with} \quad 0 \leq \frac{r}{m} < 1,$$

we see that q is just $\lfloor n/m \rfloor$. Then $r = (n/m - \lfloor n/m \rfloor) \cdot m$, which is easy to compute with a pocket calculator if n and m are reasonably small. For example,

$$\frac{8{,}324{,}576}{51{,}309} \approx 162.24397 \quad \text{and} \quad 0.24397 \times 51{,}309 \approx 12{,}518,$$

yielding $8{,}324{,}576 = 162 \cdot 51{,}309 + 12{,}518$.

The Division Algorithm lets us complete the proof of the following.

Theorem 4 Let m and n be positive integers.

(a) Every common divisor of m and n is a divisor of $\gcd(m, n)$.
(b) Every common multiple of m and n is a multiple of $\text{lcm}(m, n)$.

Proof We showed (a) earlier, on page 11. To show (b), consider a common multiple s of m and n. By the Division Algorithm, $s = q \cdot \text{lcm}(m, n) + r$ for some integers q and r with $0 \le r < \text{lcm}(m, n)$. Now s and $q \cdot \text{lcm}(m, n)$ are both common multiples of m and n, so r must be too, because it's the difference between two common multiples. But $r < \text{lcm}(m, n)$, and $\text{lcm}(m, n)$ is the *least* common multiple. The only way out is for r to be 0, which means that $s = q \cdot \text{lcm}(m, n)$, a multiple of $\text{lcm}(m, n)$ as desired. ■

Exercises 1.2

1. True or False. Explain briefly.

(a) $8|4$ (b) $4|15$ (c) $22|374$

(d) $11|1001$ (e) $3|10^{400}$

2. True or False. Explain briefly.

(a) $n|1$ for all positive integers n.

(b) $n|n$ for all positive integers n.

(c) $n|n^2$ for all positive integers n.

3. (a) Find the following:

$\gcd(27, 28)$ $\gcd(6, 20)$ $\gcd(15, 30)$
$\gcd(16, 27)$ $\gcd(13, 91)$

(b) Find the following:

$\text{lcm}(27, 28)$ $\text{lcm}(6, 20)$ $\text{lcm}(15, 30)$
$\text{lcm}(16, 27)$ $\text{lcm}(13, 91)$

4. Check your answers to Exercise 3 using Theorem 3.

5. (a) Find the following:

$\gcd(8, 12)$ $\gcd(52, 96)$ $\gcd(22, 374)$
$\gcd(56, 126)$ $\gcd(37, 37)$

(b) Find the following:

$\text{lcm}(8, 12)$ $\text{lcm}(52, 96)$ $\text{lcm}(22, 374)$
$\text{lcm}(56, 126)$ $\text{lcm}(37, 37)$

6. True or False. Explain briefly.

(a) The integer 33,412,363 is a prime.

(b) $33{,}412{,}363 = 4649 \cdot 7187$.

7. Since every integer is a divisor of 0, it makes sense to talk about the common divisors of 0 and n for n a positive integer.

(a) Find $\gcd(0, 10)$, $\gcd(1, 10)$, and $\gcd(10, 10)$.

(b) What is $\gcd(0, n)$ in general?

(c) Discuss the problem of defining $\text{lcm}(0, n)$.

8. To emphasize the meaning of $m|n$, some authors say that "m divides n evenly." We don't. Do you see why?

9. Consider positive integers m and n.

(a) Suppose that $\text{lcm}(m, n) = mn$. What can you say about m and n? Explain.

(b) What if $\text{lcm}(m, n) = n$?

(c) How about if $\gcd(m, n) = m$?

10. What are $\gcd(p, q)$ and $\text{lcm}(p, q)$ if p and q are different primes?

11. For each of the following pairs of positive integers, state whether they are relatively prime and explain your answer.

(a) 64 and 729 (b) 27 and 31

(c) 45 and 56 (d) 69 and 87

12. Can two even integers be relatively prime? Explain.

13. (a) Describe the positive integers that are relatively prime to 2.

(b) Describe the positive integers that are relatively prime to 3.

(c) Describe the positive integers that are relatively prime to 4.

(d) Describe the positive integers that are relatively prime to some fixed prime p.

(e) Describe the positive integers that are relatively prime to a fixed positive power p^k of a prime p.

14. Suppose that m and n are integers that are multiples of d, say $m = ad$ and $n = bd$.

(a) Explain why $d|lm$ for every integer l.

(b) Show that $m + n$ and $m - n$ are multiples of d.

(c) Must d divide $17m - 72n$? Explain.

15. (a) Show that $\gcd(m, n)$ is a divisor of $n - m$.

(b) Show that the only possible common divisors of m and $m + 2$ are 1 and 2.

(c) Find $\gcd(1000, 1002)$ and $\gcd(1001, 1003)$.

(d) Find $\gcd(1000, 1003)$ and $\gcd(1000, 1005)$.

16. (a) List the positive integers less than 30 that are relatively prime to 30.

(b) List the positive integers less than 36 that are relatively prime to 36.

17. Let n be a positive integer.

(a) Show that if $n = kl$, with $1 \le k \le l < n$, then $k \le \sqrt{n}$.

(b) Give an example for which $n = kl$ and $k = \sqrt{n}$.

(c) Show that if $n \ge 2$ and n is not a prime, then there is a prime p such that $p \le \sqrt{n}$ and $p|n$.

(d) Conclude that if $n \ge 2$ and n has no prime divisors p with $p \le \sqrt{n}$, then n is a prime.

18. Draw diagrams like the one in Figure 1 that show the following:

(a) The divisors of 90.

(b) The divisors of 8.

(c) The divisors of 24.

19. Suppose that m and n are positive integers and that s and t are integers such that $\gcd(m, n) = sm + tn$. Show that s and t cannot both be positive or both negative.

20. Least common multiples let us find least common denominators, but some cancellation may still be required to reduce the answer to lowest terms.

(a) Illustrate this fact by adding $\frac{1}{10}$ and $\frac{1}{15}$.

(b) Find two other positive integers m and n such that $\frac{1}{m} + \frac{1}{n} = \frac{1}{s}$ with $1 < s < \text{lcm}(m, n)$.

21. Show that $\min\{x, y\} + \max\{x, y\} = x + y$ for all real numbers x and y. *Suggestion:* Split the argument into two cases.

22. For $m \neq 0$, explain why $\frac{0}{m}$ is meaningful and equal to 0.

Office Hours 1.2

Section 1.1 seemed pretty easy, and I thought I knew the stuff in Section 1.2, but this book looks different from my other math books, so maybe I don't know what's expected.

Lots of the homework exercises say "Explain." What do they want? I can't find examples in the book.

Some of the exercises say "Show." Does that mean "Prove"? I thought that's what we were going to learn, not something I should already know.

Those are good questions. First off, the book probably *is* different from your other math books, because the course itself is different. As you can see by flipping through the book, there's a lot of subject matter to cover, but even more important than the content itself is developing logical thinking. That's one of the main skills you'll get out of this course, but it won't come without lots of practice. We'll emphasize the kinds of logic that go into mathematical proofs or into program verification, but the training will also be useful in whatever you do, to help you think through complicated situations.

I certainly don't expect you to be a logic expert already, but I would guess that you've had quite a bit of experience explaining ideas. That's where we'll start out, and then we'll work our way up to fancier proofs. So when an exercise says "Explain," the authors really just want you to write out, in your own words, the reasons for your answer. Write as if you were trying to explain to a fellow student in our class. One of the skills we can always practice more is writing ideas clearly. That and problem-solving ability are the two things employers always mention right off the bat when they're asked what they look for. An idea is no good if you can't explain it. In fact, one reason for making a habit of writing clearly is so you can decipher your own notes later on. What seems obvious once may be impossible to recall later.

So that's what the book means when it says "Explain." "Show" probably calls for a little more logical argument. Usually you're given some piece of information or an assumption, and the point of the exercise is to show that, as a consequence of what you're given, something else must be true. When we get into working on methods of proof in Sections 2.3 and 2.4, we'll call what we're given the hypothesis and what we're trying to show the conclusion, and we'll look at how to build a logical derivation of the conclusion from the hypothesis. But that's not until Chapter 2. Meanwhile, just use your own words to give an argument that you think is logical.

Now how about examples? There actually *are* a fair number of examples in the book, labeled as such. It's also full of unlabeled examples of arguments, because after all that's a main part of what the book is about. In addition, the answers in the back of the book give a rough idea of what's expected. Of course the authors are sometimes pretty brief, and you shouldn't be surprised if you find your own answers longer than theirs, particularly on the "Explain" or "Show" problems.

What you'll notice most, though, is that the exercises aren't just like the examples in the book. That's the nature of a discrete math class, and it's probably a lot different from what you're used to. When you do the homework, except for the very simplest problems, you won't be able to look back through the section and find an example to imitate. That's actually good, but it takes a different style of studying. The idea is to use the examples as *abstract* models, and in order to do that you'll need to read them in a new way.

Here's what I'd recommend.

- Read through the section fairly quickly before class, to get some idea of what's going to be covered. Usually you should read math v e r y s l o w l y, which is sometimes hard to remember to do, but in this case a quick overview is appropriate.
- Listen up in class. If we come to a point that you found confusing in your reading, and if I don't explain it, stop me. If you've thought about it and my explanation hasn't straightened you out, then chances are very good that a lot of other students will also need help. I need your questions in order to know how best to use our class time.
- After class reread the section slowly before you start on the homework. Here's where the new way of reading kicks in. Read each example carefully, of course, and then ask yourself how the answer or the method in the example would change if you changed some of the conditions a little. If the 73 were a 17, what difference would that make? If the set had only one element, would the answer be the same? What if one number were a lot bigger than the other? What if there were three numbers instead of two? Ask yourself questions like this, and answer them in your own words. Try to describe to yourself in a general way what the method was in the example. For instance, "We found that every number was a divisor of 0 because that's the same as saying that 0 is a multiple of every number, and that's obvious because every number times 0 is 0." The point of talking to yourself in this way as you read is to abstract the essence of the example—to see what's really important in it and what's just inessential detail. Reading in this way makes one example do the work of many, and means that when you get to the homework you'll have a better understanding of the basic ideas.
- When you do the homework, do the same thing as you did with the examples. Just after you finish a problem ask yourself, while the exercise is as fresh in your mind as it will ever be, "What did I just do?" and "How would the answer have changed if . . . ?" You'll be able to abstract the exercise and get as much out of it as if you'd done dozens of similar ones.

That's probably enough for today. Give these ideas a try, and let me know how it goes. And speak up in class if you see that we need to go over something.

1.3 Some Special Sets

In the first two sections we discussed some properties of real numbers, with a focus on integers and fractions. It is convenient to have names, i.e., symbols, for some of the sets that we will be working with. We will reserve the symbol $\mathbb{N}$ for the set of **natural numbers**:

$$\mathbb{N} = \{0, 1, 2, 3, 4, 5, 6, \dots\}.$$

Note that we include 0 among the natural numbers. We write $\mathbb{P}$ for the set of **positive integers**:

$$\mathbb{P} = \{1, 2, 3, 4, 5, 6, 7, \dots\}.$$

Many mathematicians call this set $\mathbb{N}$, but in discrete mathematics it makes sense to include 0 among the natural numbers. The set of all **integers**, positive, zero, or negative, will be denoted by $\mathbb{Z}$ [for the German word *Zahl*].

The symbol $\in$ will be used frequently to signify that an object belongs to a set. For example, $73 \in \mathbb{Z}$ simply asserts that the object [number] 73 belongs to the set $\mathbb{Z}$. In fact, we also have $73 \in \mathbb{N}$. On the other hand, $-73 \notin \mathbb{N}$, though of course $-73 \in \mathbb{Z}$. The symbol $\in$ can be read as a verb phrase "is an element of," "belongs to," or "is in." Depending on context, it can also be read as the preposition "in." For example, we often want statements or definitions to be true "for all $n \in \mathbb{N}$," and this would be read simply as "for all n in $\mathbb{N}$." The symbol $\notin$ can be read as "is not an element of," "does not belong to," or "is not in."

Numbers of the form m/n, where $m \in \mathbb{Z}$, $n \in \mathbb{Z}$, and $n \neq 0$, are called **rational numbers** [since they are ratios of integers]. The set of all rational numbers is denoted by $\mathbb{Q}$. The set of all **real numbers**, rational or not, is denoted by $\mathbb{R}$. Thus $\mathbb{R}$ contains all the numbers in $\mathbb{Q}$, and $\mathbb{R}$ also contains $\sqrt{2}$, $\sqrt{3}$, $\sqrt[3]{2}$, $-\pi$, e and many, many other numbers. Small finite sets can be listed using braces $\{\ \}$ and commas. For example, $\{2, 4, 6, 8, 10\}$ is the set consisting of the five positive even integers less than 12, and $\{2, 3, 5, 7, 11, 13, 17, 19\}$ consists of the eight primes less than 20. Two sets are **equal** if they contain the same elements. Thus

$$\{2, 4, 6, 8, 10\} = \{10, 8, 6, 4, 2\} = \{2, 8, 2, 6, 2, 10, 4, 2\};$$

the order of the listing is irrelevant and there is no advantage [or harm] in listing elements more than once. We will consistently use braces $\{\ \}$, not brackets [ ] or parentheses ( ), to describe sets.

Large finite sets and even infinite sets can be listed with the aid of the mathematician's *et cetera*, that is, three dots $\dots$, provided the meaning of the three dots is clear. Thus $\{1, 2, 3, \dots, 1000\}$ represents the set of positive integers less than or equal to 1000 and $\{3, 6, 9, 12, \dots\}$ presumably represents the infinite set of positive integers that are divisible by 3. On the other hand, the meaning of $\{1, 2, 3, 5, 8, \dots\}$ may be less than perfectly clear. The somewhat vague use of three dots is not always satisfactory, especially in computer science, and we will develop techniques for unambiguously describing such sets without using dots.

Sets are often described by properties of their elements using the notation

$$\{\ :\qquad \}.$$

A variable [n or x, for instance] is indicated before the colon, and the properties are given after the colon. For example,

$$\{n : n \in \mathbb{N} \text{ and } n \text{ is even}\}$$

represents the set of nonnegative even integers, i.e., the set $\{0, 2, 4, 6, 8, 10, \dots\}$. The colon is always read "such that," and so the above is read "the set of all n such that n is in $\mathbb{N}$ and n is even." Similarly,

$$\{x : x \in \mathbb{R} \text{ and } 1 \le x < 3\}$$

represents the set of all real numbers that are greater than or equal to 1 and less than 3. The number 1 belongs to the set, but 3 does not. Just to streamline notation, the last two sets can be written as

$$\{n \in \mathbb{N} : n \text{ is even}\} \quad \text{and} \quad \{x \in \mathbb{R} : 1 \le x < 3\}.$$

The first set is then read "the set of all n in $\mathbb{N}$ such that n is even."

Another way to list a set is to specify a rule for obtaining its elements using some other set of elements. For example, $\{n^2 : n \in \mathbb{N}\}$ represents the set of all integers that are the squares of integers in $\mathbb{N}$, i.e.,

$$\{n^2 : n \in \mathbb{N}\} = \{m \in \mathbb{N} : m = n^2 \text{ for some } n \in \mathbb{N}\} = \{0, 1, 4, 9, 16, 25, 36, \dots\}.$$

Note that this set also equals $\{n^2 : n \in \mathbb{Z}\}$. Similarly, $\{(-1)^n : n \in \mathbb{N}\}$ represents the set obtained by evaluating $(-1)^n$ for all $n \in \mathbb{N}$, so

$$\{(-1)^n : n \in \mathbb{N}\} = \{-1, 1\}.$$

This set has only two elements, even though $\mathbb{N}$ is infinite.

As with numbers and functions, we need to be able to work with general sets, not just specific examples. We will usually denote generic sets by capital letters such as A, B, S, or X. Generic members of sets are usually denoted by lowercase letters such as a, b, s, or x.

Now consider two sets S and T. We say that S is a **subset** of T provided every element of S belongs to T, i.e., if $s \in S$ implies $s \in T$. If S is a subset of T, we write $S \subseteq T$. The symbol $\subseteq$ can be read as "is a subset of." We also will occasionally say "S is contained in T" in case $S \subseteq T$, but note the potential for confusion if we also say "x is contained in T" when we mean "$x \in T$." Containment for a subset and containment for elements mean quite different things.

Two sets S and T are **equal** if they have exactly the same elements. Thus $S = T$ if and only if $S \subseteq T$ and $T \subseteq S$.

EXAMPLE 1

(a) We have $\mathbb{P} \subseteq \mathbb{N}$, $\mathbb{N} \subseteq \mathbb{Z}$, $\mathbb{Z} \subseteq \mathbb{Q}$, and $\mathbb{Q} \subseteq \mathbb{R}$. As with the familiar inequality $\leq$, we can run these assertions together:

$$\mathbb{P} \subseteq \mathbb{N} \subseteq \mathbb{Z} \subseteq \mathbb{Q} \subseteq \mathbb{R}.$$

This notation is unambiguous, since $\mathbb{P} \subseteq \mathbb{Z}$, $\mathbb{N} \subseteq \mathbb{R}$, etc.

(b) Since 2 is the only even prime, we have

$$\{n \in \mathbb{P} : n \text{ is prime and } n \geq 3\} \subseteq \{n \in \mathbb{P} : n \text{ is odd}\}.$$

(c) Consider again any set S. Obviously, $x \in S$ implies $x \in S$ and so $S \subseteq S$. That is, we regard a set as a subset of itself. This is why we use the notation $\subseteq$ rather than $\subset$. This usage is analogous to our usage of $\leq$ for real numbers. The inequality $x \leq 5$ is valid for many numbers, such as 3, 1, and -73. It is also valid for $x = 5$, i.e., $5 \leq 5$. This last inequality looks a bit peculiar because we actually know more, namely $5 = 5$. But $5 \leq 5$ says that "5 is less than 5 or else 5 is equal to 5," and this is a true statement. Similarly, $S \subseteq S$ is true even though we know more, namely $S = S$. Statements like "$5 = 5$," "$5 \leq 5$," "$S = S$," or "$S \subseteq S$" do no harm and are often useful to call attention to the fact that a particular case of a more general statement is valid. ■

We will occasionally write $T \subset S$ to mean that $T \subseteq S$ and $T \neq S$, i.e., T is a subset of S different from S. This usage of $\subset$ is analogous to our usage of $<$ for real numbers. If $T \subset S$, then we say that T is a **proper subset** of S.

We next introduce notation for some special subsets of $\mathbb{R}$, called **intervals**. For $a, b \in \mathbb{R}$ with $a < b$, we define

$$[a, b] = \{x \in \mathbb{R} : a \leq x \leq b\}; \qquad (a, b) = \{x \in \mathbb{R} : a < x < b\};$$

$$[a, b) = \{x \in \mathbb{R} : a \leq x < b\}; \qquad (a, b] = \{x \in \mathbb{R} : a < x \leq b\}.$$

The general rule is that brackets [,] signify that the endpoints are to be included and parentheses (,) signify that they are to be excluded. Intervals of the form $[a, b]$ are called **closed**; ones of the form (a, b) are **open**. It is also convenient to use the term "interval" for some unbounded sets that we describe using the symbols ∞ and $-\infty$, which do *not* represent real numbers but are simply part of the notation for the sets. Thus

$$[a, \infty) = \{x \in \mathbb{R} : a \leq x\}; \qquad (a, \infty) = \{x \in \mathbb{R} : a < x\};$$

$$(-\infty, b] = \{x \in \mathbb{R} : x \leq b\}; \qquad (-\infty, b) = \{x \in \mathbb{R} : x < b\}.$$

Set and interval notation must be dealt with carefully. For example, [0, 1], (0, 1) and {0, 1} all denote different sets. In fact, the intervals [0, 1] and (0, 1) are infinite sets, while {0, 1} has only two elements.

Think about the following sets:

$$\{n \in \mathbb{N} : 2 < n < 3\}, \qquad \{x \in \mathbb{R} : x^2 < 0\},$$

$$\{r \in \mathbb{Q} : r^2 = 2\}, \qquad \{x \in \mathbb{R} : x^2 + 1 = 0\}.$$

What do you notice? These sets have one property in common: They contain no elements at all. From a strictly logical point of view, they all contain the same elements, so they are equal in spite of the different descriptions. This unique set having no elements at all is called the **empty set**. We will use two notations for it, the standard $\varnothing$ and very occasionally the suggestive $\{\ \ \}$. The symbol $\varnothing$ is not a Greek phi ϕ; it is borrowed from the Norwegian alphabet and non-Norwegians should read it as "empty set." We regard $\varnothing$ as a subset of every set S, because we regard the statement "$x \in \varnothing$ implies $x \in S$" as logically true in a vacuous sense. You can take this explanation on faith until you study §2.4.

The set of all subsets of a set S is called the **power set** of S and will be denoted $\mathcal{P}(S)$. Since $\varnothing \subseteq S$ and $S \subseteq S$, both the empty set $\varnothing$ and the set S itself are elements of $\mathcal{P}(S)$, i.e., $\varnothing \in \mathcal{P}(S)$ and $S \in \mathcal{P}(S)$.

EXAMPLE 2

(a) We have $\mathcal{P}(\varnothing) = \{\varnothing\}$, since $\varnothing$ is the only subset of $\varnothing$. Note that this one-element set $\{\varnothing\}$ is different from the empty set $\varnothing$.

(b) Consider a typical one-element set, say $S = \{a\}$. Then $\mathcal{P}(S) = \{\varnothing, \{a\}\}$ has two elements.

(c) If $S = \{a, b\}$ and $a \neq b$, then $\mathcal{P}(S) = \{\varnothing, \{a\}, \{b\}, \{a, b\}\}$ has four elements.

(d) If $S = \{a, b, c\}$ has three elements, then

$$\mathcal{P}(S) = \{\varnothing, \{a\}, \{b\}, \{c\}, \{a, b\}, \{a, c\}, \{b, c\}, \{a, b, c\}\}$$

has eight members.

(e) Let S be a finite set. If S has n elements and if $n \leq 3$, then $\mathcal{P}(S)$ has 2^n elements, as shown in parts (a)–(d) above. This fact holds in general, as we show in Example 4(b) on page 141.

(f) If S is infinite, then $\mathcal{P}(S)$ is also infinite, of course. ■

We introduce one more special kind of set, denoted by Σ^*, that will recur throughout this book. Our goal is to allow a rather general, but precise, mathematical treatment of languages. First we define an **alphabet** to be a finite nonempty set Σ [Greek capital sigma] whose members are symbols, often called **letters** of Σ, and which is subject to some minor restrictions that we will discuss at the end of this section.

Given an alphabet Σ, a **word** is any finite string of letters from Σ. Finally, we denote the set of all words using letters from Σ by $\boldsymbol{\Sigma^*}$ [sigma-star]. Any subset of Σ^* is called a **language** over Σ. This is a very general definition, so it includes many languages that a layman would not recognize as languages. However, this is a place where generalization and abstraction allow us to focus on the basic features that different languages, including computer languages, have in common.

EXAMPLE 3

(a) Let $\Sigma = \{a, b, c, d, \ldots, z\}$ consist of the twenty-six letters of the English alphabet. *Any* string of letters from Σ belongs to Σ^*. Thus Σ^* contains *math, is, fun, aint, lieblich, amour, zzyzzoomph, etcetera*, etc. Since Σ^* contains *a, aa, aaa, aaaa, aaaaa,* etc., Σ^* is clearly an infinite set. To provide a familiar example of a language, we define the **American language** L to be the subset of Σ^* consisting of the words in the latest edition of *Webster's New World Dictionary of the American Language.* Thus

$$L = \{a, aachen, aardvark, aardwolf, \ldots, zymurgy\},$$

a very large but finite set.

(b) Here is another alphabet:

$\Sigma = \{$А, Б, В, Г, Д, Е, Ж, З, И, Й, К, Л, М, Н, О, П, Р, С, Т, У, Ф, Х, Ц, Ч, Ш, Щ, Ъ, Ы, Ь, Э, Ю, Я$\}$.

This is the Russian alphabet and could be used to define the Russian language or any other language that uses Cyrillic letters. Two examples of Russian words are МАТЕМАТИКА and ХРУЩЕВ. Note that Ы looks like two adjacent symbols; however, it is regarded as a single letter in the alphabet Σ and it is used that way in the Russian language. We will discuss other instances where letters involve more than one symbol. In particular, see the discussion of the ALGOL language in Example 4(d) below.

(c) To get simple examples and yet illustrate the ideas, we will frequently take Σ to be a 2-element set $\{a, b\}$. In this case Σ^* contains *a, b, ab, ba, bab, babbabb,* etc.; again Σ^* is infinite.

(d) If $\Sigma = \{0, 1\}$, then the set B of words in Σ^* that begin with 1 is exactly the set of binary notations for positive integers. That is,

$$B = \{1, 10, 11, 100, 101, 110, 111, 1000, 1001, \ldots\}.$$ ■

There is a special word in Σ^* somewhat analogous to the empty set, called the **empty word**, **null word**, or **null string**; it is the string with no letters at all and is denoted by λ [Greek lowercase lambda].

EXAMPLE 4

(a) If $\Sigma = \{a, b\}$, then

$$\Sigma^* = \{\lambda, a, b, aa, ab, ba, bb, aaa, aab, aba, abb, baa, bab, bba, \ldots\}.$$

(b) If $\Sigma = \{a\}$, then

$$\Sigma^* = \{\lambda, a, aa, aaa, aaaa, aaaaa, aaaaaa, \ldots\}.$$

This example doesn't contain any very useful languages, but like Σ^*, where $\Sigma = \{a, b\}$, it will serve to illustrate concepts later on.

(c) If $\Sigma = \{0, 1, 2\}$, then

$$\Sigma^* = \{\lambda, 0, 1, 2, 00, 01, 02, 10, 11, 12, 20, 21, 22, 000, 001, 002, \ldots\}.$$

(d) Various computer languages fit our definition of language. For example, the alphabet Σ for one version of ALGOL has 113 elements; Σ includes letters, the digits $0, 1, 2, \ldots, 9$, and a variety of operators, including sequential operators such as "go to" and "if." As usual, Σ^* contains all possible finite strings of letters from Σ, without regard to meaning. The subset of Σ^* consisting of those strings accepted for execution by an ALGOL compiler on a given computer is a well-defined and useful subset of Σ^*; we could call it the ALGOL language determined by the compiler. ■

As promised, we now discuss the restrictions needed for an alphabet Σ. A problem can arise if the letters in Σ are themselves built out of other letters, either from Σ or from some other alphabet. For example, if Σ contains as letters the symbols *a, b,* and *ab,* then the string *aab* could be taken to be a string of three letters *a, a,* and *b* from Σ or as a string of two letters *a* and *ab*. There is no way to tell which it should be, and a machine reading in the letters *a, a, b* one at a time would find it impossible to assign unambiguous meaning to the input. To take another example, if Σ contained *ab, aba,* and *bab,* the input string *ababab* could be interpreted as either (*ab*)(*ab*)(*ab*) or (*aba*)(*bab*). To avoid these and related problems, we will not allow Σ to contain any letters that are themselves strings beginning with one or more letters from Σ. Thus $\Sigma = \{a, b, c\}$, $\Sigma = \{a, b, ca\}$, and $\Sigma = \{a, b, Ab\}$ are allowed, but $\Sigma = \{a, b, c, ac\}$ and even $\Sigma = \{a, b, ac\}$ are not.

With this agreement, we can unambiguously define **length**(w) for a word w in Σ^* to be the number of letters from Σ in w, counting each appearance of a letter. For example, if $\Sigma = \{a, b\}$, then length(*aab*) = length(*bab*) = 3. If $\Sigma = \{a, b, Ab\}$, then length(*ab*) = length(*AbAb*) = 2 and length(*aabAb*) = 4. We also define length(λ) = 0. A more precise definition is given in §7.1.

One final note: We will use w, w_1, etc., as generic or variable names for words. This practice should cause no confusion, even though w also happens to be a letter of the English alphabet.

EXAMPLE 5

If $\Sigma = \{a, b\}$ and $A = \{w \in \Sigma^* : \text{length}(w) = 2\}$, then $A = \{aa, ab, ba, bb\}$. If

$$B = \{w \in \Sigma^* : \text{length}(w) \text{ is even}\},$$

then B is the infinite set $\{\lambda, aa, ab, ba, bb, aaaa, aaab, aaba, aabb, \ldots\}$. Observe that A is a subset of B. ■

Exercises 1.3

1. List five elements in each of the following sets.
(a) $\{n \in \mathbb{N} : n \text{ is divisible by } 5\}$
(b) $\{2n + 1 : n \in \mathbb{P}\}$
(c) $\mathcal{P}(\{1, 2, 3, 4, 5\})$
(d) $\{2^n : n \in \mathbb{N}\}$
(e) $\{1/n : n \in \mathbb{P}\}$
(f) $\{r \in \mathbb{Q} : 0 < r < 1\}$
(g) $\{n \in \mathbb{N} : n + 1 \text{ is prime}\}$

2. List the elements in the following sets.
(a) $\{1/n : n = 1, 2, 3, 4\}$
(b) $\{n^2 - n : n = 0, 1, 2, 3, 4\}$
(c) $\{1/n^2 : n \in \mathbb{P},\ n \text{ is even and } n < 11\}$
(d) $\{2 + (-1)^n : n \in \mathbb{N}\}$

3. List five elements in each of the following sets.
(a) Σ^* where $\Sigma = \{a, b, c\}$
(b) $\{w \in \Sigma^* : \text{length}(w) \le 2\}$ where $\Sigma = \{a, b\}$
(c) $\{w \in \Sigma^* : \text{length}(w) = 4\}$ where $\Sigma = \{a, b\}$
Which sets above contain the empty word λ?

4. Determine the following sets, i.e., list their elements if they are nonempty, and write $\emptyset$ if they are empty.
(a) $\{n \in \mathbb{N} : n^2 = 9\}$ (b) $\{n \in \mathbb{Z} : n^2 = 9\}$
(c) $\{x \in \mathbb{R} : x^2 = 9\}$ (d) $\{n \in \mathbb{N} : 3 < n < 7\}$
(e) $\{n \in \mathbb{Z} : 3 < |n| < 7\}$ (f) $\{x \in \mathbb{R} : x^2 < 0\}$

5. Repeat Exercise 4 for the following sets.
(a) $\{n \in \mathbb{N} : n^2 = 3\}$
(b) $\{x \in \mathbb{Q} : x^2 = 3\}$
(c) $\{x \in \mathbb{R} : x < 1 \text{ and } x \ge 2\}$
(d) $\{3n + 1 : n \in \mathbb{N} \text{ and } n \le 6\}$
(e) $\{n \in \mathbb{P} : n \text{ is prime and } n \le 15\}$ [Remember, 1 isn't prime.]

6. Repeat Exercise 4 for the following sets.
(a) $\{n \in \mathbb{N} : n | 12\}$ (b) $\{n \in \mathbb{N} : n^2 + 1 = 0\}$
(c) $\{n \in \mathbb{N} : \lfloor \frac{n}{3} \rfloor = 8\}$ (d) $\{n \in \mathbb{N} : \lceil \frac{n}{2} \rceil = 8\}$

7. Let $A = \{n \in \mathbb{N} : n \le 20\}$. Determine the following sets, i.e., list their elements if they are nonempty, and write $\emptyset$ if they are empty.
(a) $\{n \in A : 4 | n\}$
(b) $\{n \in A : n | 4\}$
(c) $\{n \in A : \max\{n, 4\} = 4\}$
(d) $\{n \in A : \max\{n, 14\} = n\}$

8. How many elements are there in the following sets? Write ∞ if the set is infinite.
(a) $\{n \in \mathbb{N} : n^2 = 2\}$
(b) $\{n \in \mathbb{Z} : 0 \le n \le 73\}$
(c) $\{n \in \mathbb{Z} : 5 \le |n| \le 73\}$
(d) $\{n \in \mathbb{Z} : 5 < n < 73\}$
(e) $\{n \in \mathbb{Z} : n \text{ is even and } |n| \le 73\}$
(f) $\{x \in \mathbb{Q} : 0 \le x \le 73\}$
(g) $\{x \in \mathbb{Q} : x^2 = 2\}$
(h) $\{x \in \mathbb{R} : x^2 = 2\}$

9. Repeat Exercise 8 for the following sets.
(a) $\{x \in \mathbb{R} : 0.99 < x < 1.00\}$
(b) $\mathcal{P}(\{0, 1, 2, 3\})$
(c) $\mathcal{P}(\mathbb{N})$
(d) $\{n \in \mathbb{N} : n \text{ is even}\}$
(e) $\{n \in \mathbb{N} : n \text{ is prime}\}$
(f) $\{n \in \mathbb{N} : n \text{ is even and prime}\}$
(g) $\{n \in \mathbb{N} : n \text{ is even or prime}\}$

10. How many elements are there in the following sets? Write ∞ if the set is infinite.
(a) $\{-1, 1\}$ (b) $[-1, 1]$
(c) $(-1, 1)$ (d) $\{n \in \mathbb{Z} : -1 \le n \le 1\}$
(e) Σ^* where $\Sigma = \{a, b, c\}$
(f) $\{w \in \Sigma^* : \text{length}(w) \le 4\}$ where $\Sigma = \{a, b, c\}$

11. Consider the sets
$A = \{n \in \mathbb{P} : n \text{ is odd}\}$
$B = \{n \in \mathbb{P} : n \text{ is prime}\}$
$C = \{4n + 3 : n \in \mathbb{P}\}$
$D = \{x \in \mathbb{R} : x^2 - 8x + 15 = 0\}$
Which of these sets are subsets of which? Consider all 16 possibilities.

12. Consider the sets $\{0, 1\}$, $(0, 1)$ and $[0, 1]$. True or False.
(a) $\{0, 1\} \subseteq (0, 1)$ (b) $\{0, 1\} \subseteq [0, 1]$
(c) $(0, 1) \subseteq [0, 1]$ (d) $\{0, 1\} \subseteq \mathbb{Z}$
(e) $[0, 1] \subseteq \mathbb{Z}$ (f) $[0, 1] \subseteq \mathbb{Q}$
(g) $1/2$ and $\pi/4$ are in $\{0, 1\}$
(h) $1/2$ and $\pi/4$ are in $(0, 1)$
(i) $1/2$ and $\pi/4$ are in $[0, 1]$

13. Consider the following three alphabets: $\Sigma_1 = \{a, b, c\}$, $\Sigma_2 = \{a, b, ca\}$, and $\Sigma_3 = \{a, b, Ab\}$. Determine to which of Σ_1^*, Σ_2^*, and Σ_3^* each word below belongs, and give its length as a member of each set to which it belongs.
(a) aba (b) bAb (c) cba
(d) cab (e) $caab$ (f) $baAb$

14. Here is a question to think about. Let $\Sigma = \{a, b\}$ and imagine, if you can, a dictionary for all the nonempty words of Σ^* with the words arranged in the usual alphabetical order. All the words *a, aa, aaa, aaaa*, etc., must appear before the word *ba*. How far into the dictionary will you have to dig to find the word *ba*? How would the answer change if the dictionary contained only those words in Σ^* of length 5 or less?

15. Suppose that w is a nonempty word in Σ^*.

(a) If the first [i.e., leftmost] letter of w is deleted, is the resulting string in Σ^*?

(b) How about deleting letters from both ends of w? Are the resulting strings still in Σ^*?

(c) If you had a device that could recognize letters in Σ and could delete letters from strings, how could you use it to determine if an arbitrary string of symbols is in Σ^*?

1.4 Set Operations

In this section we introduce operations that allow us to create new sets from old sets. We define the **union** $A \cup B$ and **intersection** $A \cap B$ of sets A and B as follows:

$$\boldsymbol{A \cup B} = \{x : x \in A \text{ or } x \in B \text{ or both}\};$$

$$\boldsymbol{A \cap B} = \{x : x \in A \text{ and } x \in B\}.$$

We added "or both" to the definition of $A \cup B$ for emphasis and clarity. In ordinary English, the word "or" has two interpretations. Sometimes it is the **inclusive or** and means one or the other or both. This is the interpretation when a college catalog asserts: A student's program must include 2 years of science or 2 years of mathematics. At other times, "or" is the **exclusive or** and means one or the other but not both. This is the interpretation when a menu offers soup or salad. In mathematics we always interpret **or** as the "inclusive or" unless explicitly specified to the contrary. Sets A and B are said to be **disjoint** if they have no elements in common, i.e., if $A \cap B = \varnothing$.

EXAMPLE 1

(a) Let $A = \{n \in \mathbb{N} : n \le 11\}$, $B = \{n \in \mathbb{N} : n \text{ is even and } n \le 20\}$, and $E = \{n \in \mathbb{N} : n \text{ is even}\}$. Then we have

$$A \cup B = \{0, 1, 2, 3, 4, 5, 6, 7, 8, 9, 10, 11, 12, 14, 16, 18, 20\},$$

$$A \cup E = \{n \in \mathbb{N} : n \text{ is even or } n \le 11\}, \text{ an infinite set,}$$

$$A \cap B = \{0, 2, 4, 6, 8, 10\} \quad \text{and} \quad A \cap E = \{0, 2, 4, 6, 8, 10\}.$$

Since $B \subseteq E$, we have $B \cup E = E$ and $B \cap E = B$.

(b) Consider the intervals $[0, 2]$ and $(0, 1]$. Then $(0, 1] \subseteq [0, 2]$ and so

$$(0, 1] \cup [0, 2] = [0, 2] \quad \text{and} \quad (0, 1] \cap [0, 2] = (0, 1].$$

(c) Let $\Sigma = \{a, b\}$, $A = \{\lambda, a, aa, aaa\}$, $B = \{\lambda, b, bb, bbb\}$, and $C = \{w \in \Sigma^* : \text{length}(w) \le 2\}$. Then we have

$$A \cup B = \{\lambda, a, b, aa, bb, aaa, bbb\} \quad \text{and} \quad A \cap B = \{\lambda\},$$

$$A \cup C = \{\lambda, a, b, aa, ab, ba, bb, aaa\} \quad \text{and} \quad A \cap C = \{\lambda, a, aa\},$$

$$B \cup C = \{\lambda, a, b, aa, ab, ba, bb, bbb\} \quad \text{and} \quad B \cap C = \{\lambda, b, bb\}.$$

■

Given sets A and B, the **relative complement** $A \setminus B$ is the set of objects that are in A and not in B:

$$\boldsymbol{A \setminus B} = \{x : x \in A \text{ and } x \notin B\} = \{x \in A : x \notin B\}.$$

It is the set obtained by removing from A all the elements of B that happen to be in A. Note that the set $(A \setminus B) \cup B$ may be bigger than the set A.

The **symmetric difference** $A \oplus B$ of the sets A and B is the set

$$\boldsymbol{A \oplus B} = \{x : x \in A \text{ or } x \in B \text{ but not both}\}.$$

Note the use of the "exclusive or" here. It follows from the definition that

$$A \oplus B = (A \cup B) \setminus (A \cap B) = (A \setminus B) \cup (B \setminus A).$$

EXAMPLE 2

(a) Let $A = \{n \in \mathbb{N} : n \leq 11\}$, $B = \{n \in \mathbb{N} : n \text{ is even and } n \leq 20\}$, and $E = \{n \in \mathbb{N} : n \text{ is even}\}$ as in Example 1(a). Then we have

$$A \setminus B = \{1, 3, 5, 7, 9, 11\}, \quad B \setminus A = \{12, 14, 16, 18, 20\}, \quad \text{and}$$
$$A \oplus B = \{1, 3, 5, 7, 9, 11, 12, 14, 16, 18, 20\}.$$

Since $B \subseteq E$, we also have $B \setminus E = \emptyset$. Some other equalities are

$$E \setminus B = \{n \in \mathbb{N} : n \text{ is even and } n \geq 22\} = \{22, 24, 26, 28, \dots\},$$
$$\mathbb{N} \setminus E = \{n \in \mathbb{N} : n \text{ is odd}\} = \{1, 3, 5, 7, 9, 11, \dots\},$$
$$A \oplus E = \{1, 3, 5, 7, 9, 11\} \cup \{n \in \mathbb{N} : n \text{ is even and } n \geq 12\}$$
$$= \{1, 3, 5, 7, 9, 11, 12, 14, 16, 18, 20, 22, \dots\}.$$

(b) for the intervals $[0, 2]$ and $(0, 1]$, where $(0, 1] \subseteq [0, 2]$, we have

$$(0, 1] \setminus [0, 2] = \emptyset,$$
$$[0, 2] \setminus (0, 1] = \{0\} \cup (1, 2] \quad \text{and} \quad [0, 2] \setminus (0, 2) = \{0, 2\}.$$

(c) Again let $\Sigma = \{a, b\}$, $A = \{\lambda, a, aa, aaa\}$, $B = \{\lambda, b, bb, bbb\}$, and $C = \{w \in \Sigma^* : \text{length}(w) \leq 2\}$. Then we have

$$A \setminus B = \{a, aa, aaa\}, \qquad B \setminus A = \{b, bb, bbb\},$$
$$A \cap C = \{\lambda, a, aa\}, \qquad B \setminus C = \{bbb\},$$
$$C \setminus A = \{b, ab, ba, bb\}, \qquad A \setminus \Sigma = \{\lambda, aa, aaa\}.$$

■

It is sometimes convenient to illustrate relations between sets with pictures called **Venn diagrams**, in which sets correspond to subsets of the plane. See Figure 1, where the indicated sets have been shaded in.

Figure 1 ▶

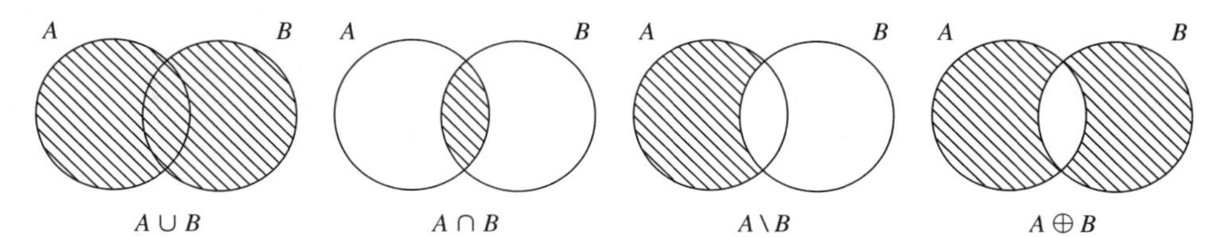

It is useful to have notation for the set of elements that are *not* in some set. We wouldn't really want *all* elements not in the set, for that would include everything imaginable: matrices, graphs, and nonmathematical objects such as stars, ideas, snails, and puppy dogs' tails. We really want all elements that are not in the set but are in some larger set U that is of current interest, which in practice might be $\mathbb{N}$, $\mathbb{R}$, or Σ^*, for example. In this context, we will call the set of interest the **universe** or **universal set** and focus on its elements and subsets. For example, depending upon the choice of the universe, the interesting objects that are not in the set $\{2, 4, 6, 8, \dots\}$ of positive even integers might be the other nonnegative integers, might be the other integers [including all the negative integers], might be all the other real numbers, etc. For $A \subseteq U$ the relative complement $U \setminus A$ is called in this setting the **absolute complement** or simply the **complement** of A and is denoted by $\boldsymbol{A^c}$. Thus we have

$$A^c = \{x \in U : x \notin A\}.$$

Note that the relative complement $A \setminus B$ can be written in terms of the absolute complement: $A \setminus B = A \cap B^c$. In the Venn diagrams in Figure 2, we have drawn the universe U as a rectangle and shaded in the indicated sets.

Figure 2 ▶

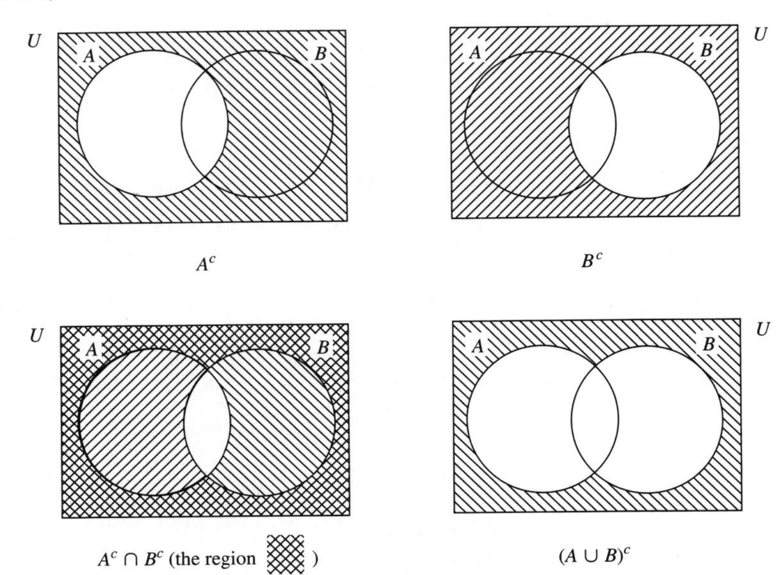

EXAMPLE 3

(a) If the universe is $\mathbb{N}$, and if A and E are as in Example 1(a) and Example 2(a), then

$$A^c = \{n \in \mathbb{N} : n \geq 12\} \quad \text{and} \quad E^c = \{n \in \mathbb{N} : n \text{ is odd}\}.$$

(b) If the universe is $\mathbb{R}$, then $[0, 1]^c = (-\infty, 0) \cup (1, \infty)$, while

$$(0, 1)^c = (-\infty, 0] \cup [1, \infty) \quad \text{and} \quad \{0, 1\}^c = (-\infty, 0) \cup (0, 1) \cup (1, \infty).$$

For any $a \in \mathbb{R}$, we have $[a, \infty)^c = (-\infty, a)$ and $(a, \infty)^c = (-\infty, a]$. ■

The last two Venn diagrams in Figure 2 illustrate that $A^c \cap B^c = (A \cup B)^c$. This set identity and many others are true in general. Table 1 lists some basic identities for sets and set operations. Don't be overwhelmed by them; look at them one at a time. Except perhaps for laws 3 and 9, they should be apparent with a moment's thought. If not, drawing Venn diagrams might help.

As some of the names of the laws suggest, many of them are analogues of laws from algebra, though there is only one distributive law for numbers. Compare the identity laws with the ones in algebra:

$$x + 0 = x, \qquad x \cdot 1 = x \qquad \text{and} \qquad x \cdot 0 = 0 \qquad \text{for } x \in \mathbb{R}.$$

The idempotent laws are new [certainly $a + a = a$ fails for most numbers], and the laws involving complementation are new. All sets in Table 1 are presumed to be subsets of some universal set U. Because of the associative laws, we can write the sets $A \cup B \cup C$ and $A \cap B \cap C$ without any parentheses and cause no confusion.

The identities in Table 1 can be verified in at least two ways. One can shade in the corresponding parts of Venn diagrams for the sets on the two sides of the equations and observe that they are equal. One can also show that two sets S and T are equal by showing that $S \subseteq T$ and $T \subseteq S$, verifying these inclusions by showing that $x \in S$ implies $x \in T$ and by showing that $x \in T$ implies $x \in S$. We illustrate both sorts of arguments by verifying two of the laws that are not so apparent. However, proving set-theoretic identities is not a primary focus in discrete mathematics, so the proofs are just here for completeness.

EXAMPLE 4

The De Morgan law 9a is illustrated by Venn diagrams in Figure 2.

Here is a proof in which we first show $(A \cup B)^c \subseteq A^c \cap B^c$, and then we show $A^c \cap B^c \subseteq (A \cup B)^c$. To show $(A \cup B)^c \subseteq A^c \cap B^c$, we consider an element x in $(A \cup B)^c$. Then $x \notin A \cup B$. In particular, $x \notin A$ and so we must have $x \in A^c$. Similarly, $x \notin B$ and so $x \in B^c$. Therefore, $x \in A^c \cap B^c$. We have shown that $x \in (A \cup B)^c$ implies $x \in A^c \cap B^c$; hence $(A \cup B)^c \subseteq A^c \cap B^c$.

TABLE 1 Laws of Algebra of Sets

1a.	$A \cup B = B \cup A$	commutative laws
b.	$A \cap B = B \cap A$	
2a.	$(A \cup B) \cup C = A \cup (B \cup C)$	associative laws
b.	$(A \cap B) \cap C = A \cap (B \cap C)$	
3a.	$A \cup (B \cap C) = (A \cup B) \cap (A \cup C)$	distributive laws
b.	$A \cap (B \cup C) = (A \cap B) \cup (A \cap C)$	
4a.	$A \cup A = A$	idempotent laws
b.	$A \cap A = A$	
5a.	$A \cup \emptyset = A$	identity laws
b.	$A \cup U = U$	
c.	$A \cap \emptyset = \emptyset$	
d.	$A \cap U = A$	
6.	$(A^c)^c = A$	double complementation
7a.	$A \cup A^c = U$	
b.	$A \cap A^c = \emptyset$	
8a.	$U^c = \emptyset$	
b.	$\emptyset^c = U$	
9a.	$(A \cup B)^c = A^c \cap B^c$	De Morgan laws
b.	$(A \cap B)^c = A^c \cup B^c$	

To show the reverse inclusion $A^c \cap B^c \subseteq (A \cup B)^c$, we consider y in $A^c \cap B^c$. Then $y \in A^c$ and so $y \notin A$. Also $y \in B^c$ and so $y \notin B$. Since $y \notin A$ and $y \notin B$, we conclude that $y \notin A \cup B$, i.e., $y \in (A \cup B)^c$. Hence $A^c \cap B^c \subseteq (A \cup B)^c$. ■

EXAMPLE 5 The Venn diagrams in Figure 3 show the distributive law 3b. The picture of the set $A \cap (B \cup C)$ is double-hatched.

Figure 3 ▶

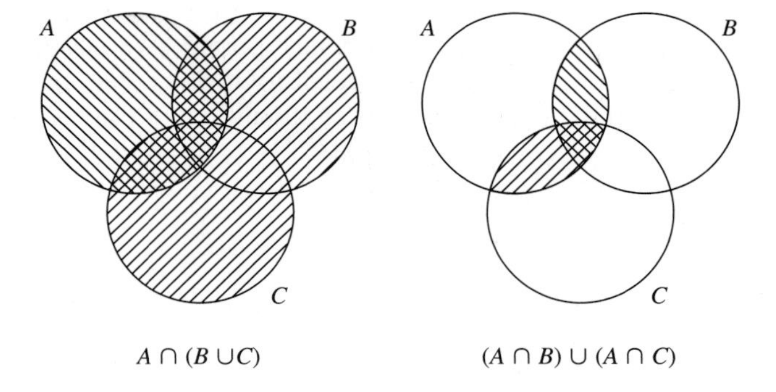

Here is a proof where we show that the sets are subsets of each other. First consider $x \in A \cap (B \cup C)$. Then x is in A for sure. Also x is in $B \cup C$. So either $x \in B$, in which case $x \in A \cap B$, or else $x \in C$, in which case $x \in A \cap C$. In either case, we have $x \in (A \cap B) \cup (A \cap C)$. This shows that $A \cap (B \cup C) \subseteq (A \cap B) \cup (A \cap C)$.

Now consider $y \in (A \cap B) \cup (A \cap C)$. Either $y \in A \cap B$ or $y \in A \cap C$; we consider the two cases separately. If $y \in A \cap B$, then $y \in A$ and $y \in B$, so $y \in B \cup C$ and hence $y \in A \cap (B \cup C)$. Similarly, if $y \in A \cap C$, then $y \in A$ and $y \in C$, so $y \in B \cup C$ and thus $y \in A \cap (B \cup C)$. Since $y \in A \cap (B \cup C)$ in both cases, we've shown that $(A \cap B) \cup (A \cap C) \subseteq A \cap (B \cup C)$. We already proved the opposite inclusion and so the two sets are equal. ■

The symmetric difference operation $\oplus$ is at least as important in computer science as the union $\cup$ and intersection $\cap$ operations. It will reappear in various places in the book. Like $\cup$ and $\cap$, $\oplus$ is also an associative operation:

$$(A \oplus B) \oplus C = A \oplus (B \oplus C),$$

but this fact is not so obvious. We can see it by looking at the Venn diagrams in Figure 4. On the left we have hatched $A \oplus B$ one way and C the other. Then $(A \oplus B) \oplus C$ is the set hatched one way or the other *but not both.* Doing the same sort of thing with A and $B \oplus C$ gives us the same set, so the sets $(A \oplus B) \oplus C$ and $A \oplus (B \oplus C)$ are equal.

Figure 4 ▶

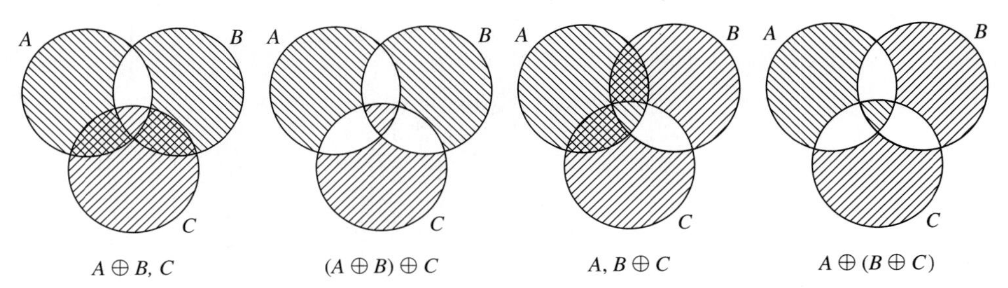

Of course it is also possible to prove this fact without appealing to the pictures. You may want to construct such a proof yourself. Be warned, though, that a detailed argument will be fairly complicated [see Exercise 18 on page 49].

Since $\oplus$ is associative, the expression $A \oplus B \oplus C$ is unambiguous. Note that an element belongs to this set provided it belongs to exactly one or to all three of the sets A, B, and C; see again Figure 4.

Consider two sets S and T. For each element s in S and each element t in T, we form an **ordered pair (s, t)**. Here s is the first element of the ordered pair, t is the second element, and the order is important. Thus $(s_1, t_1) = (s_2, t_2)$ if and only if $s_1 = s_2$ and $t_1 = t_2$. The set of all ordered pairs (s, t) is called the **product** of S and T and written $S \times T$:

$$\boldsymbol{S \times T} = \{(s, t) : s \in S \text{ and } t \in T\}.$$

If $S = T$, we sometimes write $\boldsymbol{S^2}$ for $S \times S$.

EXAMPLE 6

(a) Let $S = \{1, 2, 3, 4\}$ and $T = \{a, b, c\}$. Then $S \times T$ consists of the 12 ordered pairs listed on the left in Figure 5. We could also depict these pairs as corresponding points in labeled rows and columns, in the manner shown on the right in the figure. The reader should list or draw $T \times S$ and note that $T \times S \neq S \times T$.

Figure 5 ▶

(1,c)	(2,c)	(3,c)	(4,c)
(1,b)	(2,b)	(3,b)	(4,b)
(1,a)	(2,a)	(3,a)	(4,a)

List of {1,2,3,4} × {a,b,c}

c	*	*	*	*
b	*	*	*	*
a	*	*	*	*
	1	2	3	4

Picture of {1,2,3,4} × {a,b,c}

(b) If $S = \{1, 2, 3, 4\}$, then $S^2 = S \times S$ has 16 ordered pairs; see Figure 6. Note that $(2, 4) \neq (4, 2)$; these ordered pairs involve the same two numbers, but in different orders. In contrast, the *sets* $\{2, 4\}$ and $\{4, 2\}$ are the same. Also note that $(2, 2)$ is a perfectly good ordered pair in which the first element happens to equal the second element. On the other hand, the set $\{2, 2\}$ is just the one-element set $\{2\}$ in which 2 happens to be written twice.

Figure 6 ▶

(1,4)	(2,4)	(3,4)	(4,4)
(1,3)	(2,3)	(3,3)	(4,3)
(1,2)	(2,2)	(3,2)	(4,2)
(1,1)	(2,1)	(3,1)	(4,1)

List of $\{1,2,3,4\}^2$

4	*	*	*	*
3	*	*	*	*
2	*	*	*	*
1	*	*	*	*
	1	2	3	4

Picture of $\{1,2,3,4\}^2$

Our notation for ordered pairs, for instance $(2, 4)$, is in apparent conflict with our earlier notation for intervals in §1.3, where the symbol $(2, 4)$ represented the set $\{x \in \mathbb{R} : 2 < x < 4\}$. Both uses of this notation are standard, however. Fortunately, the intended meaning is almost always clear from the context. ■

For any finite set S, we write $|S|$ for the number of elements in the set. Thus $|S| = |T|$ precisely when the finite sets S and T are of the same size. Observe that

$$|\emptyset| = 0 \text{ and } |\{1, 2, \ldots, n\}| = n \quad \text{for } n \in \mathbb{P}.$$

Moreover, $|S \times T| = |S| \cdot |T|$. You can see where the notation $\times$ for the product of two sets came from. It turns out that $|\mathcal{P}(S)| = 2^{|S|}$, as noted in Example 2 on page 19, so some people also use the notation 2^S for the set $\mathcal{P}(S)$.

We can define the product of any finite collection $S_1, S_2, \ldots, S_n$ of sets. The **product set** $S_1 \times S_2 \times \cdots \times S_n$ consists of all **ordered n-tuples** $(s_1, s_2, \ldots, s_n)$, where $s_1 \in S_1$, $s_2 \in S_2$, etc. That is,

$$S_1 \times S_2 \times \cdots \times S_n = \{(s_1, s_2, \ldots, s_n) : s_k \in S_k \text{ for } k = 1, 2, \ldots, n\}.$$

Just as with ordered pairs, two ordered n-tuples $(s_1, s_2, \ldots, s_n)$ and $(t_1, t_2, \ldots, t_n)$ are regarded as equal if all the corresponding entries are equal: $s_k = t_k$ for $k = 1, 2, \ldots, n$. If the sets $S_1, S_2, \ldots, S_n$ are all equal, say all equal to S, then we may write S^n for the product $S_1 \times S_2 \times \cdots \times S_n$. For example, if $S = \{0, 1\}$, then $S^n = \{0, 1\}^n$ consists of all n-tuples of 0's and 1's, i.e., all $(s_1, s_2, \ldots, s_n)$, where each s_k is 0 or 1.

Exercises 1.4

1. Let $U = \{1, 2, 3, 4, 5, \ldots, 12\}$, $A = \{1, 3, 5, 7, 9, 11\}$, $B = \{2, 3, 5, 7, 11\}$, $C = \{2, 3, 6, 12\}$, and $D = \{2, 4, 8\}$. Determine the sets

(a) $A \cup B$ (b) $A \cap C$ (c) $(A \cup B) \cap C^c$
(d) $A \setminus B$ (e) $C \setminus D$ (f) $B \oplus D$
(g) How many subsets of C are there?

2. Let $A = \{1, 2, 3\}$, $B = \{n \in \mathbb{P} : n \text{ is even}\}$, and $C = \{n \in \mathbb{P} : n \text{ is odd}\}$.

(a) Determine $A \cap B$, $B \cap C$, $B \cup C$, and $B \oplus C$.
(b) List all subsets of A.
(c) Which of the following sets are infinite? $A \oplus B$, $A \oplus C$, $A \setminus C$, $C \setminus A$.

3. In this exercise the universe is $\mathbb{R}$. Determine the following sets.

(a) $[0, 3] \cap [2, 6]$ (b) $[0, 3] \cup [2, 6]$ (c) $[0, 3] \setminus [2, 6]$
(d) $[0, 3] \oplus [2, 6]$ (e) $[0, 3]^c$ (f) $[0, 3] \cap \emptyset$
(g) $[0, \infty) \cap \mathbb{Z}$ (h) $[0, \infty) \cap (-\infty, 2]$
(i) $([0, \infty) \cup (-\infty, 2])^c$

4. Let $\Sigma = \{a, b\}$, $A = \{a, b, aa, bb, aaa, bbb\}$, $B = \{w \in \Sigma^* : \text{length}(w) \geq 2\}$, and $C = \{w \in \Sigma^* : \text{length}(w) \leq 2\}$.

(a) Determine $A \cap C$, $A \setminus C$, $C \setminus A$, and $A \oplus C$.
(b) Determine $A \cap B$, $B \cap C$, $B \cup C$, and $B \setminus A$.
(c) Determine $\Sigma^* \setminus B$, $\Sigma \setminus B$, and $\Sigma \setminus C$.
(d) List all subsets of Σ.
(e) How many sets are there in $\mathcal{P}(\Sigma)$?

5. In this exercise the universe is Σ^*, where $\Sigma = \{a, b\}$. Let A, B, and C be as in Exercise 4. Determine the following sets.

(a) $B^c \cap C^c$ (b) $(B \cap C)^c$ (c) $(B \cup C)^c$
(d) $B^c \cup C^c$ (e) $A^c \cap C$ (f) $A^c \cap B^c$
(g) Which of these sets are equal? Why?

6. The following statements about sets are false. For each statement, give an example, i.e., a choice of sets, for which the statement is false. Such examples are called **counterexamples**. They are examples that are counter to, i.e., contrary to, the assertion.

(a) $A \cup B \subseteq A \cap B$ for all A, B.
(b) $A \cap \emptyset = A$ for all A.
(c) $A \cap (B \cup C) = (A \cap B) \cup C$ for all A, B, C.

7. For any set A, what is $A \oplus A$? $A \oplus \emptyset$?

8. For the sets $A = \{1, 3, 5, 7, 9, 11\}$ and $B = \{2, 3, 5, 7, 11\}$, determine the following numbers.

(a) $|A|$ (b) $|B|$ (c) $|A \cup B|$
(d) $|A| + |B| - |A \cap B|$
(e) Do you see a general reason why the answers to (c) and (d) have to be the same?

9. The following statements about sets are false. Give a counterexample [see Exercise 6] to each statement.

(a) $A \cap B = A \cap C$ implies $B = C$.
(b) $A \cup B = A \cup C$ implies $B = C$.
(c) $A \subseteq B \cup C$ implies $A \subseteq B$ or $A \subseteq C$.

10. (a) Show that relative complementation is not commutative; that is, the equality $A \setminus B = B \setminus A$ can fail.

(b) Show that relative complementation is not associative: $(A \setminus B) \setminus C = A \setminus (B \setminus C)$ can fail.

11. Let $A = \{a, b, c\}$ and $B = \{a, b, d\}$.

(a) List or draw the ordered pairs in $A \times A$.
(b) List or draw the ordered pairs in $A \times B$.
(c) List or draw the set $\{(x, y) \in A \times B : x = y\}$.

12. Let $S = \{0, 1, 2, 3, 4\}$ and $T = \{0, 2, 4\}$.

(a) How many ordered pairs are in $S \times T$? $T \times S$?
(b) List or draw the elements in the set $\{(m, n) \in S \times T : m < n\}$.

(c) List or draw the elements in the set $\{(m, n) \in T \times S : m < n\}$.
(d) List or draw the elements in the set $\{(m, n) \in S \times T : m + n \geq 3\}$.
(e) List or draw the elements in the set $\{(m, n) \in T \times S : mn \geq 4\}$.
(f) List or draw the elements in the set $\{(m, n) \in S \times S : m + n = 10\}$.

13. For each of the following sets, list all elements if the set has fewer than seven elements. Otherwise, list exactly seven elements of the set.

(a) $\{(m, n) \in \mathbb{N}^2 : m = n\}$
(b) $\{(m, n) \in \mathbb{N}^2 : m + n \text{ is prime}\}$
(c) $\{(m, n) \in \mathbb{P}^2 : m = 6\}$
(d) $\{(m, n) \in \mathbb{P}^2 : \min\{m, n\} = 3\}$
(e) $\{(m, n) \in \mathbb{P}^2 : \max\{m, n\} = 3\}$
(f) $\{(m, n) \in \mathbb{N}^2 : m^2 = n\}$

14. Draw a Venn diagram for four sets A, B, C, and D. Be sure to have a region for each of the 16 possible sets such as $A \cap B^c \cap C^c \cap D$. *Note:* This problem cannot be done using just circles, but it can be done using rectangles.

1.5 Functions

You are undoubtedly already familiar with the term "function" and its usage. Here are some functions that you have probably encountered before:

$$f_1(x) = x^2; \quad f_2(x) = \sqrt{x}; \quad f_3(x) = \log(x); \quad f_4(x) = \sin(x).$$

In fact, each of these functions appears on all scientific calculators, where log is shorthand for $\log_{10}$. Here we have named the functions f_1, f_2, f_3 and f_4. The letter x, called a variable, represents a typical input, and the equations above tell us what the corresponding output is. The most crucial feature of a function is that this output is uniquely determined; that is, for each x there is exactly one output. For example, the function f_1 assigns the value [or output] 81 to the input $x = 9$, while the function f_2 assigns the value 3 to the input $x = 9$. More briefly, we have $f_1(9) = 81$ and $f_2(9) = 3$. Incidentally, note that the output $\sqrt{9} = 3$ solves the equation $x^2 = 9$, but that it is not the only solution. Nevertheless, the notation $\sqrt{x}$ is reserved for the *nonnegative* square root; the reason for this is that it is very useful to have $\sqrt{\ }$ represent a function. Try to imagine what your calculator would do if this symbol represented two values rather than one!

The second feature of a function is that one needs to know what values of x are allowed or intended. The functions above are so familiar that you undoubtedly expected the allowed values of x to be real numbers. We want to be more careful than this. In the first place, it turns out that it makes sense to square other objects than real numbers, like complex numbers and certain matrices [whatever they are], so how can we be sure that the inputs for f_1 are supposed to be real numbers? The answer is that we should have specified the possible values *when we defined* f_1, though we sort of gave this away when we mentioned calculators. We also want our functions to be meaningful, which in the case of the four functions above means that we want real numbers for outputs. So, in fact, the input values for the function f_2 should be restricted to $x \geq 0$, and the input values for f_3 should be restricted to $x > 0$. In this book we will encounter many functions whose input values are *not* real numbers—sometimes not numbers at all—so this issue of specifying allowable values will be important.

The restrictions for input values of f_2 and f_3 above simply allowed them to be meaningful functions. But we might want to restrict the input values of a function for any number of reasons. For example, the area of a circle of radius r is given by a function

$$A(r) = \pi r^2.$$

Viewed as a function, A makes sense for all real values of r, but in this context we would surely restrict r to be positive. This simple function illustrates a couple of other points. We usually use f, g, etc., for the names of generic functions, but we will often use a suggestive name if the function has some particular significance. In this case, A is a sensible name, since the outputs represent areas. Likewise, x, y, and z, and also lowercase letters from the beginning of the alphabet, are used for generic variables [inputs] unless they have some particular significance. Here, of course, r represents radius.

Here, then, is a working definition of "function." A **function** f assigns to each element x in some set S a unique element in a set T. We say such an f is **defined on** S with **values in** T. The set S is called the **domain** of f and is sometimes written **Dom(f)**. The element assigned to x is usually written $\boldsymbol{f(x)}$. Care should be taken to avoid confusing a function f with its functional values $f(x)$, especially when people write, as we will later, "the function $f(x)$." A function f is completely specified by:

(a) the set on which f is defined, namely $\text{Dom}(f)$;
(b) the assignment, rule, or formula giving the value $f(x)$ for each x in $\text{Dom}(f)$.

We will usually specify the domain of a function, but it is standard practice to omit mentioning the domain in case it's either unimportant or it's the "natural" domain of all input values for which the rule makes sense. For x in $\text{Dom}(f)$, $f(x)$ is called the **image** of x **under** f. The set of all images $f(x)$ is a subset of T called the **image** of f and written $\text{Im}(f)$. Thus we have

$$\mathbf{Im}(\boldsymbol{f}) = \{f(x) : x \in \text{Dom}(f)\}.$$

EXAMPLE 1

(a) Here are the functions f_1, f_2, f_3, and f_4 again, defined properly. Let $f_1(x) = x^2$, where $\text{Dom}(f_1) = \mathbb{R}$. Let $f_2(x) = \sqrt{x}$, where $\text{Dom}(f_2) = [0, \infty) = \{x \in \mathbb{R} : x \geq 0\}$. Let $f_3(x) = \log(x)$, where $\text{Dom}(f_3) = (0, \infty) = \{x \in \mathbb{R} : x > 0\}$. Let $f_4(x) = \sin(x)$, where $\text{Dom}(f_4) = \mathbb{R}$.

While it is often interesting to know the image sets of functions, they are not needed for the definition of the functions. For our functions we have $\text{Im}(f_1) = [0, \infty)$, $\text{Im}(f_2) = [0, \infty)$, $\text{Im}(f_3) = \mathbb{R}$, and $\text{Im}(f_4) = [-1, 1]$. You won't be needing the function f_4 in this book, but if the other image sets are not evident to you, you should look at a sketch of the functions.

(b) Consider the function g with $\text{Dom}(g) = \mathbb{N}$ and $g(n) = n^3 - 73n + 5$ for $n \in \mathbb{N}$. It is not clear what the image set is for this function, though of course we can write $\text{Im}(g) = \{n^3 - 73n + 5 : n \in \mathbb{N}\}$. This equation is true, but doesn't really tell us which numbers are in $\text{Im}(g)$. Is 4958 in this set? We don't know. What's clear is that $\text{Im}(g) \subseteq \mathbb{Z}$. For many functions, such as this one, there is no need to determine the image set exactly. ■

It is often convenient to specify a set of allowable images for a function f, i.e., a set T containing $\text{Im}(f)$. Such a set is called a **codomain** of f. While a function f has exactly one domain $\text{Dom}(f)$ and exactly one image $\text{Im}(f)$, any set containing $\text{Im}(f)$ can serve as a codomain. Of course, when we specify a codomain we will try to choose one that is useful or informative in context. The notation $f : S \to T$ is shorthand for "f is a function with domain S and codomain T." We sometimes refer to a function as a **map** or **mapping** and say that f **maps** S into T. When we feel the need of a picture, we sometimes draw sketches such as those in Figure 1.

You will often encounter sentences such as "Consider a function $f : \mathbb{N} \to \mathbb{Z}$." What does it mean? It means that $\text{Dom}(f) = \mathbb{N}$ and, for each $n \in \mathbb{N}$, $f(n)$ represents a unique number in $\mathbb{Z}$. Thus $\mathbb{Z}$ is a codomain for f, but the image $\text{Im}(f)$ may be a much smaller set.

Figure 1 ▶

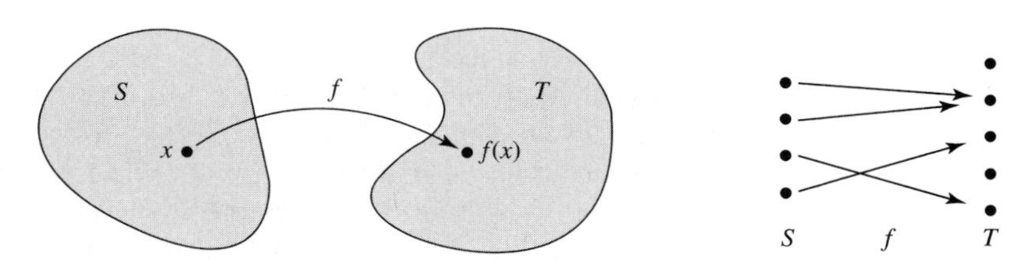

EXAMPLE 2

(a) Consider again the function g from Example 1(b) defined by $g(n) = n^3 - 73n + 5$ for $n \in \mathbb{N}$. Note that $g\colon \mathbb{N} \to \mathbb{Z}$. This tells us all that we need to know. The domain is $\mathbb{N}$, the values $g(n)$ are all in $\mathbb{Z}$, and we have the assignment rule.

(b) Recall that the **absolute value** $|x|$ of x in $\mathbb{R}$ is defined by the rule

$$|x| = \begin{cases} x & \text{if } x \geq 0, \\ -x & \text{if } x < 0. \end{cases}$$

This rule defines the function f given by $f(x) = |x|$ with domain $\mathbb{R}$ and image set $[0, \infty)$; note that $|x| \geq 0$ for all $x \in \mathbb{R}$. Absolute value has two important properties that we will return to in §2.4: $|x \cdot y| = |x| \cdot |y|$ and $|x + y| \leq |x| + |y|$ for all $x, y \in \mathbb{R}$.

(c) We already encountered the **floor function** $\lfloor x \rfloor$ and the **ceiling function** $\lceil x \rceil$ back in §1.1, though we avoided the word "function" in that section. Recall that, for each $x \in \mathbb{R}$, $\lfloor x \rfloor$ is the largest integer less than or equal to x, while $\lceil x \rceil$ is the smallest integer greater than or equal to x. These functions have natural domain $\mathbb{R}$ and their image sets are $\mathbb{Z}$.

Here we have used the notations $\lfloor x \rfloor$ and $\lceil x \rceil$ for functions, when their real names should have been $\lfloor\ \rfloor$ and $\lceil\ \rceil$. As we mentioned before, it's common to write $f(x)$ instead of f, but one must be careful not to mistake $f(x)$ for a specific *value* of f.

(d) Consider a new function f_2 defined by

$$f_2(x) = \begin{cases} 1 & \text{if } x \geq 0, \\ 0 & \text{if } x < 0. \end{cases}$$

Then $\text{Im}(f_2) = \{0, 1\}$, and we could write $f_2\colon \mathbb{R} \to [0, \infty)$ or $f_2\colon \mathbb{R} \to \mathbb{N}$ or even $f_2\colon \mathbb{R} \to \{0, 1\}$, among other choices.

(e) Consider again the function $f_1\colon \mathbb{R} \to \mathbb{R}$, where $f_1(x) = x^2$. In this case, we have $\text{Im}(f_1) = [0, \infty)$ and we could write $f_1\colon \mathbb{R} \to [0, \infty)$. ■

EXAMPLE 3

Functions need not be defined on $\mathbb{R}$ or subsets of $\mathbb{R}$. The rule $f((m, n)) = \lfloor \frac{n}{2} \rfloor - \lfloor \frac{m-1}{2} \rfloor$ defines a function from the product set $\mathbb{Z} \times \mathbb{Z}$ into $\mathbb{Z}$. When functions are defined on ordered pairs, it is customary to omit one set of parentheses. Thus we will write

$$f(m, n) = \left\lfloor \frac{n}{2} \right\rfloor - \left\lfloor \frac{m-1}{2} \right\rfloor.$$

Here are some sample calculations: $f(3, 7) = \lfloor 3.5 \rfloor - \lfloor 1 \rfloor = 2$; $f(7, 3) = \lfloor 1.5 \rfloor - \lfloor 3 \rfloor = -2$; $f(3, 12) = \lfloor 6 \rfloor - \lfloor 1 \rfloor = 5$; $f(12, 3) = \lfloor 1.5 \rfloor - \lfloor 5.5 \rfloor = -4$; $f(4, 16) = \lfloor 8 \rfloor - \lfloor 1.5 \rfloor = 7$; and $f(16, 4) = \lfloor 2 \rfloor - \lfloor 7.5 \rfloor = 2 - 7 = -5$.

This function appears in the theorem on page 5, with $k = 2$, but it's only meaningful there for $m \leq n$. So if our focus is that theorem, we should redefine the function by restricting its domain to $D = \{(m, n) \in \mathbb{Z} \times \mathbb{Z} : m \leq n\}$. For $(m, n) \in D$, the theorem tells us that there are $f(m, n)$ even integers between m and n. ■

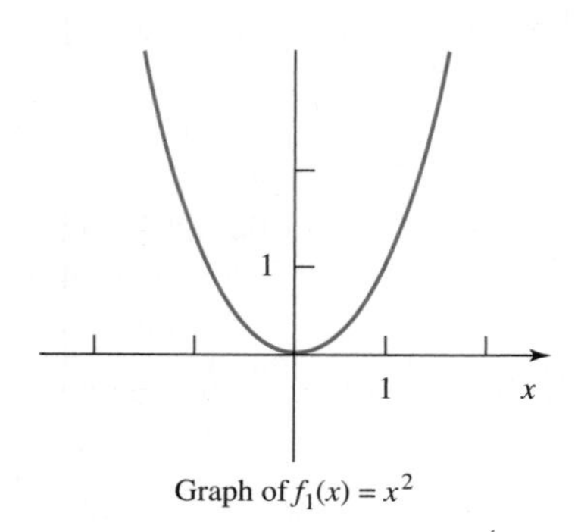

Graph of $f_1(x) = x^2$

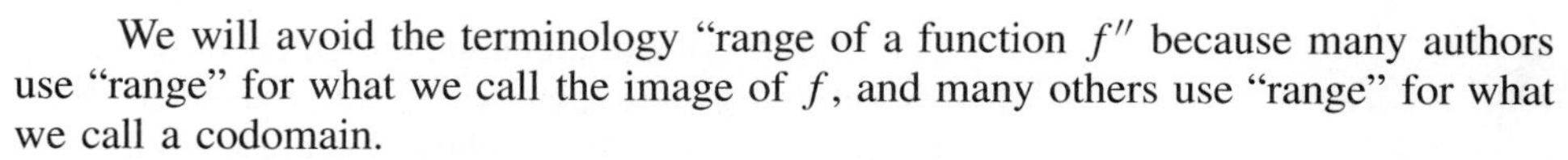
We will avoid the terminology "range of a function f" because many authors use "range" for what we call the image of f, and many others use "range" for what we call a codomain.

Consider a function $f\colon S \to T$. The **graph** of f is the following subset of $S \times T$:

$$\text{Graph}(f) = \{(x, y) \in S \times T : y = f(x)\}.$$

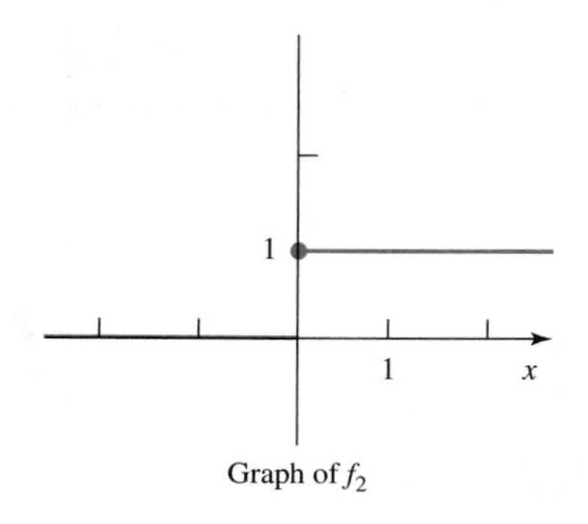

Graph of f_2

Figure 2 ▲

This definition is compatible with the use of the term in algebra and calculus. The graphs of the functions in Examples 2(d) and 2(e) are sketched in Figure 2.

Our working definition of "function" is incomplete; in particular, the term "assigns" is undefined. A precise set-theoretical definition can be given. The key observation is this: Not only does a function determine its graph, but a function can be recovered from its graph. In fact, the graph of a function $f: S \to T$ is a subset G of $S \times T$ with the following property:

$$\text{for each } x \in S \text{ there is exactly one } y \in T \text{ such that } (x, y) \in G.$$

Given G, we can see that $\text{Dom}(f) = S$ and, for each $x \in S$, $f(x)$ is the unique element in T such that $(x, f(x)) \in G$. The point to observe is that nothing is lost if we regard functions and their graphs as the same, and we gain some precision in the process. From this point of view, we can say that a **function** with domain S and codomain T is a subset G of $S \times T$ satisfying

$$\text{for each } x \in S \text{ there is exactly one } y \in T \text{ such that } (x, y) \in G.$$

If S and T are subsets of $\mathbb{R}$ and if $S \times T$ is drawn so that S is part of the horizontal axis and T is part of the vertical axis, then a subset G of $S \times T$ is a function [or the graph of a function] if every vertical line through a point in S intersects G in exactly one point.

We next introduce a useful notation for functions that take on only the values 0 and 1. Consider a set S and a subset A of S. The function on S that takes the value 1 at members of A and the value 0 at the other members of S is called the **characteristic function** of A and is denoted χ_A [Greek lowercase chi, sub A]. Thus

$$\chi_A(x) = \begin{cases} 1 & \text{for} \quad x \in A, \\ 0 & \text{for} \quad x \in S \setminus A. \end{cases}$$

We can think of the characteristic function χ_A as a way to describe the set A by tagging its members. The elements of S that belong to A get tagged with 1's, and the remaining elements of S get 0's.

EXAMPLE 4

(a) Recall the function f_2 defined in Example 2(d) by

$$f_2(x) = \begin{cases} 1 & \text{if } x \geq 0, \\ 0 & \text{if } x < 0. \end{cases}$$

The function f_2 is the characteristic function $\chi_{[0,\infty)}$, defined on $\mathbb{R}$.

(b) For each $n \in \mathbb{N}$, let $g(n)$ be the remainder when n is divided by 2, as in the Division Algorithm on page 13. For example, $g(52) = 0$ since $52 = 2 \cdot 26 + 0$, and $g(73) = 1$ since $73 = 2 \cdot 36 + 1$. Here $g: \mathbb{N} \to \{0, 1\}$, and the first few values are

$$0, 1, 0, 1, 0, 1, 0, 1, 0, 1, 0, 1, 0, 1, 0, 1, 0, 1, \ldots.$$

In fact, g is the characteristic function of the set $A = \{n \in \mathbb{N} : n \text{ is odd}\}$; i.e., we have $g = \chi_A$. ■

Now consider functions $f: S \to T$ and $g: T \to U$; see Figure 3. We define the **composition** $g \circ f: S \to U$ by the rule

$$(g \circ f)(x) = g(f(x)) \quad \text{for all} \quad x \in S.$$

One might read the left side "g circle f of x" or "g of f of x." Complicated operations that are performed in calculus or on a calculator can be viewed as the composition of simpler functions.

Figure 3 ▶

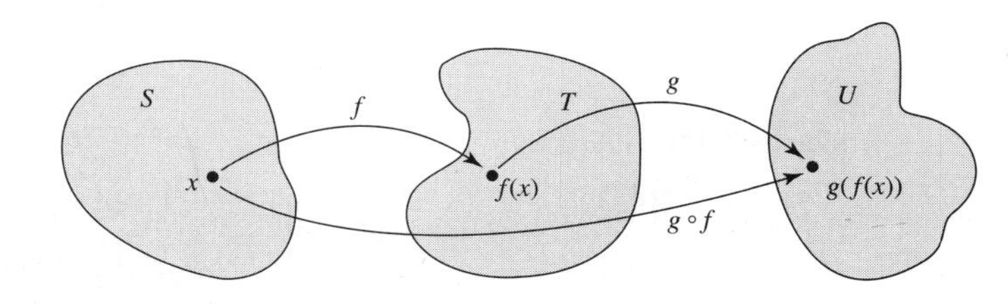

EXAMPLE 5

(a) Consider the function $h: \mathbb{R} \to \mathbb{R}$ given by

$$h(x) = (x^3 + 2x)^7.$$

The value $h(x)$ is obtained by first calculating $x^3 + 2x$ and then taking its seventh power. We write f for the first or inside function: $f(x) = x^3 + 2x$. We write g for the second or outside function: $g(x) = x^7$. The name of the variable x is irrelevant; we could just as well have written $g(y) = y^7$ for $y \in \mathbb{R}$. Either way, we see that

$$(g \circ f)(x) = g(f(x)) = g(x^3 + 2x) = (x^3 + 2x)^7 = h(x) \quad \text{for} \quad x \in \mathbb{R},$$

[or, equivalently, $(g \circ f)(x) = (f(x))^7 = (x^3 + 2x)^7$]. Thus $h = g \circ f$. The ability to view complicated functions as the composition of simpler functions is a critical skill in calculus, but won't be needed in this book. Note that the order of f and g is important. In fact,

$$(f \circ g)(x) = f(x^7) = (x^7)^3 + 2(x^7) = x^{21} + 2x^7 \quad \text{for} \quad x \in \mathbb{R}.$$

(b) Suppose that one wishes to find $\sqrt{\log a}$ on a hand-held calculator for some positive number a. The calculator has keys labeled $\sqrt{\ }$ and **log**, which stands for $\log_{10}$. One works from the inside out, first keying in the input a, say $a = 73$, then pressing the **log** key to get $\log_{10} a$, approximately 1.8633 in this case, then pressing the $\sqrt{\ }$ key to get 1.3650. This sequence of steps calculates $h(a) = \sqrt{\log a}$, where $h = g \circ f$ with $f(x) = \log x$ and $g(x) = \sqrt{x}$. As in part (a), order is important: $h \neq f \circ g$, i.e., $\sqrt{\log x}$ is not generally equal to $\log \sqrt{x}$. For example, if $x = 73$, then $\sqrt{x}$ is approximately 8.5440 and $\log \sqrt{x}$ is approximately 0.9317.

(c) Of course, some functions f and g do **commute** under composition, i.e., satisfy $f \circ g = g \circ f$. For example, if $f(x) = \sqrt{x}$ and $g(x) = 1/x$ for $x \in (0, \infty)$, then $f \circ g = g \circ f$ because

$$\sqrt{\frac{1}{x}} = \frac{1}{\sqrt{x}} \quad \text{for} \quad x \in (0, \infty).$$

For example, for $x = 9$ we have $\sqrt{1/9} = 1/3 = 1/\sqrt{9}$. ■

We can compose more than two functions if we wish.

EXAMPLE 6

Define the functions f, g, and h that map $\mathbb{R}$ into $\mathbb{R}$ by

$$f(x) = x^4, \quad g(y) = \sqrt{y^2 + 1}, \quad h(z) = z^2 + 72.$$

We've used the different variable names x, y, and z to help clarify our computations below. Let's calculate $h \circ (g \circ f)$ and $(h \circ g) \circ f$ and compare the answers. First, for $x \in \mathbb{R}$ we have

$$\begin{aligned}
(h \circ (g \circ f))(x) &= h(g \circ f(x)) && \text{by definition of } h \circ (g \circ f)\\
&= h(g(f(x))) && \text{by definition of } g \circ f\\
&= h(g(x^4)) && \text{since } f(x) = x^4\\
&= h(\sqrt{x^8 + 1}) && y = x^4 \text{ in definition of } g\\
&= (\sqrt{x^8 + 1})^2 + 72 && z = \sqrt{x^8 + 1} \text{ in definition of } h\\
&= x^8 + 73 && \text{algebra.}
\end{aligned}$$

On the other hand,

$$\begin{aligned}
((h \circ g) \circ f)(x) &= (h \circ g)(f(x)) && \text{by definition of } (h \circ g) \circ f\\
&= h(g(f(x))) && \text{by definition of } h \circ g\\
&= x^8 + 73 && \text{exactly as above.}
\end{aligned}$$

We conclude that

$$(h \circ (g \circ f))(x) = ((h \circ g) \circ f)(x) = x^8 + 73 \quad \text{for all} \quad x \in \mathbb{R},$$

so the functions $h \circ (g \circ f)$ and $(h \circ g) \circ f$ are exactly the same function. This is no accident, as we observe in the next general theorem. ■

Associativity of Composition Consider functions $f: S \to T$, $g: T \to U$, and $h: U \to V$. Then $h \circ (g \circ f) = (h \circ g) \circ f$.

The proof of this basic result amounts to checking that the functions $h \circ (g \circ f)$ and $(h \circ g) \circ f$ both map S into V and that, just as in Example 6, for each $x \in S$ the values $(h \circ (g \circ f))(x)$ and $((h \circ g) \circ f)(x)$ are both equal to $h(g(f(x)))$.

Since composition is associative, we can write $h \circ g \circ f$ unambiguously without any parentheses. We can also compose any finite number of functions without using parentheses.

Exercises 1.5

1. Let $f(n) = n^2 + 3$ and $g(n) = 5n - 11$ for $n \in \mathbb{N}$. Thus $f: \mathbb{N} \to \mathbb{N}$ and $g: \mathbb{N} \to \mathbb{Z}$. Calculate

(a) $f(1)$ and $g(1)$ (b) $f(2)$ and $g(2)$

(c) $f(3)$ and $g(3)$ (d) $f(4)$ and $g(4)$

(e) $f(5)$ and $g(5)$

(f) To think about: Is $f(n) + g(n)$ always an even number?

2. Consider the function $h: \mathbb{P} \to \mathbb{P}$ defined by $h(n) = |\{k \in \mathbb{N} : k|n\}|$ for $n \in \mathbb{P}$. In words, $h(n)$ is the number of divisors of n. Calculate $h(n)$ for $1 \le n \le 10$ and for $n = 73$.

3. Let Σ^* be the language using letters from $\Sigma = \{a, b\}$. We've already seen a useful function from Σ^* to $\mathbb{N}$. It is the length function, which already has a name: length. Calculate

(a) length(bab) (b) length($aaaaaaaa$)

(c) length(λ)

(d) What is the image set Im(length) for this function? Explain.

4. The codomain of a function doesn't have to consist of numbers either. Let Σ^* be as in Exercise 3, and define

$$g(n) = \{w \in \Sigma^* : \text{length}(w) \le n\} \quad \text{for} \quad n \in \mathbb{N}.$$

Thus $g: \mathbb{N} \to \mathcal{P}(\Sigma^*)$. Determine

(a) $g(0)$ (b) $g(1)$ (c) $g(2)$

(d) Are all the sets $g(n)$ finite?

(e) Give an example of a set in $\mathcal{P}(\Sigma^*)$ that is not in the image set Im(g).

5. Let f be the function in Example 3.

(a) Calculate $f(0, 0)$, $f(8, 8)$, $f(-8, -8)$, $f(73, 73)$, and $f(-73, -73)$.

(b) Find $f(n, n)$ for all (n, n) in $\mathbb{Z} \times \mathbb{Z}$. *Hint:* Consider the cases when n is even and when it is odd.

6. The greatest common divisor gcd defines a function on the product set $\mathbb{P} \times \mathbb{P}$. It already has a fine name: gcd.

(a) Calculate gcd(7, 14), gcd(14, 28), and gcd(1001, 2002).

(b) What is gcd(n, $2n$) for all $n \in \mathbb{P}$?

(c) What is the image set Im(gcd)?

7. We define $f: \mathbb{R} \to \mathbb{R}$ as follows:

$$f(x) = \begin{cases} x^3 & \text{if } x \ge 1, \\ x & \text{if } 0 \le x < 1, \\ -x^3 & \text{if } x < 0. \end{cases}$$

(a) Calculate $f(3)$, $f(1/3)$, $f(-1/3)$, and $f(-3)$.

(b) Sketch a graph of f.

(c) Find Im(f).

8. Let $S = \{1, 2, 3, 4, 5\}$ and consider the functions 1_S, f, g, and h from S into S defined by $1_S(n) = n$, $f(n) = 6 - n$, $g(n) = \max\{3, n\}$, and $h(n) = \max\{1, n - 1\}$.

(a) Write each of these functions as a set of ordered pairs, i.e., list the elements in their graphs.

(b) Sketch a graph of each of these functions.

9. For $n \in \mathbb{Z}$, let $f(n) = \frac{1}{2}[(-1)^n + 1]$. The function f is the characteristic function for some subset of $\mathbb{Z}$. Which subset?

10. Consider subsets A and B of a set S.

(a) The function $\chi_A \cdot \chi_B$ is the characteristic function of some subset of S. Which subset? Explain.

(b) Repeat part (a) for the function $\chi_A + \chi_B - \chi_{A \cap B}$.

(c) Repeat part (a) for the function $\chi_A + \chi_B - 2 \cdot \chi_{A \cap B}$.

11. Here we consider two functions that are defined in terms of the floor and ceiling functions.

(a) Let $f(n) = \lfloor \frac{n}{2} \rfloor + \lfloor \frac{n}{3} \rfloor$ for $n \in \mathbb{N}$. Calculate $f(n)$ for $0 \le n \le 10$ and for $n = 73$.

(b) Let $g(n) = \lceil \frac{n}{2} \rceil - \lfloor \frac{n}{2} \rfloor$ for $n \in \mathbb{Z}$. g is the characteristic function of some subset of $\mathbb{Z}$. What is the subset?

12. In Example 5(b), we compared the functions $\sqrt{\log x}$ and $\log \sqrt{x}$. Show that these functions take the same value for $x = 10{,}000$.

13. We define functions mapping $\mathbb{R}$ into $\mathbb{R}$ as follows: $f(x) = x^3 - 4x$, $g(x) = 1/(x^2+1)$, $h(x) = x^4$. Find

(a) $f \circ f$ (b) $g \circ g$ (c) $h \circ g$

(d) $g \circ h$ (e) $f \circ g \circ h$ (f) $f \circ h \circ g$

(g) $h \circ g \circ f$

14. Repeat Exercise 13 for the functions $f(x) = x^2$, $g(x) = \sqrt{x^2+1}$, and $h(x) = 3x - 1$.

15. Consider the functions f and g mapping $\mathbb{Z}$ into $\mathbb{Z}$, where $f(n) = n - 1$ for $n \in \mathbb{Z}$ and g is the characteristic function χ_E of $E = \{n \in \mathbb{Z} : n \text{ is even}\}$.

(a) Calculate $(g \circ f)(5)$, $(g \circ f)(4)$, $(f \circ g)(7)$, and $(f \circ g)(8)$.

(b) Calculate $(f \circ f)(11)$, $(f \circ f)(12)$, $(g \circ g)(11)$, and $(g \circ g)(12)$.

(c) Determine the functions $g \circ f$ and $f \circ f$.

(d) Show that $g \circ g = g \circ f$ and that $f \circ g$ is the negative of $g \circ f$.

16. Several important functions can be found on hand-held calculators. Why isn't the identity function, i.e., the function $1_{\mathbb{R}}$, where $1_{\mathbb{R}}(x) = x$ for all $x \in \mathbb{R}$, among them?

1.6 Sequences

Many of the important functions in discrete mathematics are defined on "discrete" sets, such as $\mathbb{N}$ and $\mathbb{P}$. We have already seen some examples, where we used the normal function notation. However, functions on these sets have a different character than functions on $\mathbb{R}$, for example. They are often viewed as lists of things. Accordingly, they are called "sequences" and are handled somewhat differently. In particular, their input variables are often written as subscripts. Before pursuing sequences further, we will make some general comments about subscripts.

Subscript notation comes in handy when we are dealing with a large collection of objects; here "large" often means "more than 3 or 4." For example, letters x, y, and z are adequate when dealing with equations involving three or fewer unknowns. But if there are ten unknowns or if we wish to discuss the general situation of n unknowns for some unspecified integer n in $\mathbb{P}$, then $x_1, x_2, \ldots, x_n$ would be a good choice for the names of the unknowns. Here we distinguish the unknowns by the little numbers $1, 2, \ldots, n$ written below the x's, which are called **subscripts**. As another example, a general nonzero polynomial has the form

$$a_n x^n + a_{n-1}x^{n-1} + \cdots + a_2 x^2 + a_1 x + a_0,$$

where $a_n \neq 0$. Here n is the degree of the polynomial and the $n+1$ possible coefficients are labeled $a_0, a_1, \ldots, a_n$ using subscripts. For example, the polynomial $x^3 + 4x^2 - 73$ fits this general scheme, with $n = 3$, $a_3 = 1$, $a_2 = 4$, $a_1 = 0$, and $a_0 = -73$.

We have used the symbol Σ as a name for an alphabet. In mathematics the big Greek sigma $\sum$ has a standard use as a general summation sign. The terms following it are to be summed according to how $\sum$ is decorated. For example, the decorations "$k = 1$" and "10" in the expression

$$\sum_{k=1}^{10} k^2$$

tell us to add up the numbers k^2 obtained by successively setting $k = 1$, then $k = 2$, then $k = 3$, etc., on up to $k = 10$. That is,

$$\sum_{k=1}^{10} k^2 = 1 + 4 + 9 + 16 + 25 + 36 + 49 + 64 + 81 + 100 = 385.$$

The letter k is a variable [it varies from 1 to 10] that could be replaced here by any other variable. Thus

$$\sum_{k=1}^{10} k^2 = \sum_{j=1}^{10} j^2 = \sum_{r=1}^{10} r^2.$$

We can also consider more general sums like

$$\sum_{k=1}^{n} k^2$$

in which the stopping point n can take on different values. Each value of n gives a particular value of the sum; for each choice of n the variable k takes on the values from 1 to n. Here are some of the sums represented by $\sum_{k=1}^{n} k^2$.

Value of *n*	**The sum**
$n = 1$	$1^2 = 1$
$n = 2$	$1^2 + 2^2 = 1 + 4 = 5$
$n = 3$	$1^2 + 2^2 + 3^2 = 14$
$n = 4$	$1^2 + 2^2 + 3^2 + 4^2 = 30$
$n = 10$	$1^2 + 2^2 + 3^2 + 4^2 + 5^2 + 6^2 + 7^2 + 8^2 + 9^2 + 10^2 = 385$
$n = 73$	$1^2 + 2^2 + 3^2 + 4^2 + \cdots + 73^2 = 132{,}349$

We can also discuss even more general sums such as

$$\sum_{k=1}^{n} x_k \quad \text{and} \quad \sum_{j=m}^{n} a_j.$$

Here it is understood that $x_1, x_2, \ldots, x_n$ and $a_m, a_{m+1}, \ldots, a_n$ represent numbers. Presumably, $m \le n$, since otherwise there would be nothing to sum.

In analogy with $\sum$, the big Greek pi $\prod$ is a general product sign. For $n \in \mathbb{P}$ the product of the first n integers is called ***n* factorial** and is written $n!$. Thus

$$\boldsymbol{n!} = 1 \cdot 2 \cdot 3 \cdots n = \prod_{k=1}^{n} k.$$

The expression $1 \cdot 2 \cdot 3 \cdots n$ is somewhat confusing for small values of n like 1 and 2; it really means "multiply consecutive integers until you reach n." The expression $\prod_{k=1}^{n} k$ is less ambiguous. Here are the first few values of $n!$:

$$1! = 1, \quad 2! = 1 \cdot 2 = 2, \quad 3! = 1 \cdot 2 \cdot 3 = 6, \quad 4! = 1 \cdot 2 \cdot 3 \cdot 4 = 24.$$

More values of $n!$ appear in Figures 1 and 2. For technical reasons $n!$ is also defined for $n = 0$; $0!$ is defined to be 1. The definition of $n!$ will be reexamined in §4.1.

An infinite string of objects can be listed by using subscripts from the set $\mathbb{N} = \{0, 1, 2, \ldots\}$ of natural numbers [or from $\{m, m+1, m+2, \ldots\}$ for some integer m]. Such strings are called **sequences**. Thus a sequence on $\mathbb{N}$ is a list $s_0, s_1, \ldots, s_n, \ldots$ that has a specified value s_n for each integer $n \in \mathbb{N}$. We frequently call s_n the nth **term** of the sequence. It is often convenient to denote the sequence itself by (s_n) or $(s_n)_{n \in \mathbb{N}}$ or $(s_0, s_1, s_2, \ldots)$. Sometimes we will write $s(n)$ instead of s_n. Computer scientists commonly use the notation $s[n]$, in part because it is easy to type on a keyboard.

The notation $s(n)$ looks like our notation for functions. In fact, a sequence *is* a function whose domain is the set $\mathbb{N} = \{0, 1, 2, \ldots\}$ of natural numbers or is $\{m, m+1, m+2, \ldots\}$ for some integer m. Each integer n in the domain of the sequence determines the value $s(n)$ of the nth term.

You've already seen a lot of examples of sequences. Every function with domain $\mathbb{N}$ or $\mathbb{P}$ in the last section is a sequence. Only the notation is a bit different. Our next examples will illustrate sequences of real numbers.

EXAMPLE 1

(a) The sequence $(s_n)_{n \in \mathbb{N}}$, where $s_n = n!$, is just the sequence $(1, 1, 2, 6, 24, \dots)$ of factorials. The *set* of its values is $\{1, 2, 6, 24, \dots\}$, i.e., the set $\{n! : n \in \mathbb{P}\}$.

(b) The sequence $(a_n)_{n \in \mathbb{N}}$ given by $a_n = (-1)^n$ for $n \in \mathbb{N}$ is the sequence $(1, -1, 1, -1, 1, -1, 1, \dots)$ whose *set* of values is $\{-1, 1\}$.

(c) Consider the sequence $(b_n)_{n \in \mathbb{P}}$ defined by $b_n = \lfloor \frac{n}{3} \rfloor$. This is the sequence $(0, 0, 1, 1, 1, 2, 2, 2, \dots)$. The *set* of values is $\mathbb{N}$. ■

As the last example suggests, it is important to distinguish between a sequence and its set of values. We always use braces { } to list or describe a set and never use them to describe a sequence. The sequence $(a_n)_{n \in \mathbb{N}}$ given by $a_n = (-1)^n$ in Example 1(b) has an infinite number of terms, even though their values are repeated over and over. On the other hand, the set of values $\{(-1)^n : n \in \mathbb{N}\}$ is exactly the set $\{-1, 1\}$ consisting of two numbers.

Sequences are frequently given suggestive abbreviated names, such as SEQ, FACT, SUM, and the like.

EXAMPLE 2

(a) Let $\text{FACT}(n) = n!$ for $n \in \mathbb{N}$. This is exactly the same sequence as in Example 1(a); only its name [FACT, instead of s] has been changed. Note that $\text{FACT}(n+1) = (n+1) * \text{FACT}(n)$ for $n \in \mathbb{N}$, where $*$ denotes multiplication of integers.

(b) For $n \in \mathbb{N}$, let $\text{TWO}(n) = 2^n$. Then TWO is a sequence. Observe that we have $\text{TWO}(n+1) = 2 * \text{TWO}(n)$ for $n \in \mathbb{N}$. ■

Our definition of sequence allows the domain to be any set that has the form $\{m, m+1, m+2, \dots\}$, where m is an integer.

EXAMPLE 3

(a) The sequence (b_n) given by $b_n = 1/n^2$ for $n \geq 1$ clearly needs to have its domain avoid the value $n = 0$. The first few values of the sequence are 1, $\frac{1}{4}$, $\frac{1}{9}$, $\frac{1}{16}$, $\frac{1}{25}$.

(b) Consider the sequence whose nth term is $\log_2 n$. Note that $\log_2 0$ makes no sense, so this sequence must begin with $n = 1$ or some other positive integer. We then have $\log_2 1 = 0$ since $2^0 = 1$; $\log_2 2 = 1$ since $2^1 = 2$; $\log_2 4 = 2$ since $2^2 = 4$; $\log_2 8 = 3$ since $2^3 = 8$; etc. The intermediate values of $\log_2 n$ can only be approximated. See Figure 1. For example $\log_2 5 \approx 2.3219$ is only an approximation since $2^{2.3219} \approx 4.9999026$. ■

Figure 1 ▶

$\log_2 n$	$\sqrt{n}$	n	n^2	2^n	$n!$	n^n
0	1.0000	**1**	1	2	1	1
1.0000	1.4142	**2**	4	4	2	4
1.5850	1.7321	**3**	9	8	6	27
2.0000	2.0000	**4**	16	16	24	256
2.3219	2.2361	**5**	25	32	120	3125
2.5850	2.4495	**6**	36	64	720	46,656
2.8074	2.6458	**7**	49	128	5040	823,543
3.0000	2.8284	**8**	64	256	40,320	$1.67 \cdot 10^7$
3.1699	3.0000	**9**	81	512	362,880	$3.87 \cdot 10^8$
3.3219	3.1623	**10**	100	1024	3,628,800	10^{10}

EXAMPLE 4

(a) We will be interested in comparing the growth rates of familiar sequences such as $\log_2 n$, $\sqrt{n}$, n^2, 2^n, $n!$, and n^n. Even for relatively small values of n it seems clear from Figure 1 that n^n grows a lot faster than $n!$, which grows a lot faster than 2^n, etc., although $\log_2 n$ and $\sqrt{n}$ seem to be running close to each other. In §4.3 we will make these ideas more precise and give arguments that don't rely on appearances based on a few calculations.

(b) We are primarily interested in comparing the sequences in part (a) for large values of n. See Figure 2.[2] It now appears that $\log_2 n$ does grow [a lot] slower than $\sqrt{n}$, a fact that we will verify in §4.3. The growth is slower because 2^n grows [a lot] faster than n^2, and $\log_2 x$ and $\sqrt{x}$ are the inverse functions of 2^x and x^2, respectively. Inverses of functions are discussed in the next section. ■

Figure 2 ▶

$\log_2 n$	$\sqrt{n}$	n	n^2	2^n	$n!$	n^n
3.32	3.16	**10**	100	1024	$3.63 \cdot 10^{6}$	10^{10}
6.64	10	**100**	10,000	$1.27 \cdot 10^{30}$	$9.33 \cdot 10^{157}$	10^{200}
9.97	31.62	**1000**	10^{6}	$1.07 \cdot 10^{301}$	$4.02 \cdot 10^{2567}$	10^{3000}
13.29	100	**10,000**	10^{8}	$2.00 \cdot 10^{3010}$	$2.85 \cdot 10^{35659}$	10^{40000}
16.61	316.2	**100,000**	10^{10}	$1.00 \cdot 10^{30103}$	$2.82 \cdot 10^{456573}$	10^{500000}
19.93	1000	$\mathbf{10^{6}}$	10^{12}	$9.90 \cdot 10^{301029}$	$8.26 \cdot 10^{5565708}$	$10^{6000000}$
39.86	10^{6}	$\mathbf{10^{12}}$	10^{24}	big	bigger	biggest

So far, all of our sequences have had real numbers as values. However, there is no such restriction in the definition and, in fact, we will be interested in sequences with values of other sorts.

EXAMPLE 5

The following sequences have values that are sets.

(a) A sequence $(D_n)_{n\in\mathbb{N}}$ of subsets of $\mathbb{Z}$ is defined by

$$D_n = \{m \in \mathbb{Z} : m \text{ is a multiple of } n\} = \{0, \pm n, \pm 2n, \pm 3n, \ldots\}.$$

(b) Let Σ be an alphabet. For each $k \in \mathbb{N}$, Σ^k is defined to be the set of all words in Σ^* having length k. In symbols,

$$\Sigma^k = \{w \in \Sigma^* : \text{length}(w) = k\}.$$

The sequence $(\Sigma^k)_{k\in\mathbb{N}}$ is a sequence of subsets of Σ^* whose union, $\bigcup_{k\in\mathbb{N}} \Sigma^k$, is Σ^*. Note that the sets Σ^k are disjoint, that $\Sigma^0 = \{\lambda\}$, and that $\Sigma^1 = \Sigma$. In case $\Sigma = \{a, b\}$, we have $\Sigma^0 = \{\lambda\}$, $\Sigma^1 = \Sigma = \{a, b\}$, $\Sigma^2 = \{aa, ab, ba, bb\}$, etc. ■

In the last example we wrote $\bigcup_{k\in\mathbb{N}} \Sigma^k$ for the union of an infinite sequence of sets. As you probably guessed, a word is a member of this set if it belongs to one of the sets Σ^k. The sets Σ^k are disjoint, but in general we consider unions of sets that may overlap. To be specific, if $(A_k)_{k\in\mathbb{N}}$ is a sequence of sets, then we define

$$\bigcup_{k\in\mathbb{N}} A_k = \{x : x \in A_k \text{ for at least one } k \text{ in } \mathbb{N}\}.$$

This kind of definition makes sense, of course, if the sets are defined for k in $\mathbb{P}$ or in some other set. Similarly, we define

$$\bigcap_{k\in\mathbb{N}} A_k = \{x : x \in A_k \text{ for all } k \in \mathbb{N}\}.$$

The notation $\bigcup_{k=0}^{\infty} A_k$ has a similar interpretation except that ∞ plays a special role. The notation $\bigcup_{k=0}^{\infty}$ signifies that k takes the values $0, 1, 2, \ldots$ without stopping; but k does *not* take the value ∞. Thus

$$\bigcup_{k=0}^{\infty} A_k = \{x : x \in A_k \text{ for at least one integer } k \geq 0\} = \bigcup_{k\in\mathbb{N}} A_k,$$

---

[2] We thank our colleague Richard M. Koch for supplying the larger values in this table. He used *Mathematica*.

whereas

$$\bigcap_{k=1}^{\infty} A_k = \{x : x \in A_k \text{ for all integers } k \geq 1\} = \bigcap_{k \in \mathbb{P}} A_k.$$

In Example 5 we could just as well have written $\Sigma^* = \bigcup_{k=0}^{\infty} \Sigma^k$.

Some lists are not infinite. A **finite sequence** is a string of objects that are listed using subscripts from a finite subset of $\mathbb{Z}$ of the form $\{m, m+1, \ldots, n\}$. Frequently, m will be 0 or 1. Such a sequence $(a_m, a_{m+1}, \ldots, a_n)$ is a function with domain $\{m, m+1, \ldots, n\}$, just as an infinite sequence $(a_m, a_{m+1}, \ldots)$ has domain $\{m, m+1, \ldots\}$.

EXAMPLE 6

(a) At the beginning of this section we mentioned general sums such as $\sum_{j=m}^{n} a_j$. The values to be summed are from the finite sequence $(a_m, a_{m+1}, \ldots, a_n)$.

(b) The digits in the base 10 representation of an integer form a finite sequence. The digit sequence of 8832 is (8, 8, 3, 2) if we take the most significant digits first, but is (2, 3, 8, 8) if we start at the least significant end. ■

Exercises 1.6

1. Calculate

(a) $\frac{7!}{5!}$ (b) $\frac{10!}{6!4!}$ (c) $\frac{9!}{0!9!}$

(d) $\frac{8!}{4!}$ (e) $\sum_{k=0}^{5} k!$ (f) $\prod_{j=3}^{6} j$

2. Simplify

(a) $\frac{n!}{(n-1)!}$ (b) $\frac{(n!)^2}{(n+1)!(n-1)!}$

3. Calculate

(a) $\sum_{k=1}^{n} 3^k$ for $n = 1, 2, 3,$ and 4

(b) $\sum_{k=3}^{n} k^3$ for $n = 3, 4,$ and 5

(c) $\sum_{j=n}^{2n} j$ for $n = 1, 2,$ and 5

4. Calculate

(a) $\sum_{i=1}^{10} (-1)^i$ (b) $\sum_{k=0}^{3} (k^2+1)$ (c) $\left(\sum_{k=0}^{3} k^2\right) + 1$

(d) $\prod_{n=1}^{5} (2n+1)$ (e) $\prod_{j=4}^{8} (j-1)$

5. (a) Calculate $\prod_{r=1}^{n} (r-3)$ for $n = 1, 2, 3, 4,$ and 73.

(b) Calculate $\prod_{k=1}^{m} \frac{k+1}{k}$ for $m = 1, 2,$ and 3. Give a formula for this product for all $m \in \mathbb{P}$.

6. (a) Calculate $\sum_{k=0}^{n} 2^k$ for $n = 1, 2, 3, 4,$ and 5.

(b) Use your answers to part (a) to guess a general formula for this sum.

7. Consider the sequence given by $a_n = \frac{n-1}{n+1}$ for $n \in \mathbb{P}$.

(a) List the first six terms of this sequence.

(b) Calculate $a_{n+1} - a_n$ for $n = 1, 2, 3$.

(c) Show that $a_{n+1} - a_n = \frac{2}{(n+1)(n+2)}$ for $n \in \mathbb{P}$.

8. Consider the sequence given by $b_n = \frac{1}{2}[1 + (-1)^n]$ for $n \in \mathbb{N}$.

(a) List the first seven terms of this sequence.

(b) What is its set of values?

9. For $n \in \mathbb{N}$, let $\text{SEQ}(n) = n^2 - n$.

(a) Calculate $\text{SEQ}(n)$ for $n \leq 6$.

(b) Show that $\text{SEQ}(n+1) = \text{SEQ}(n) + 2n$ for all $n \in \mathbb{N}$.

(c) Show that $\text{SEQ}(n+1) = \frac{n+1}{n-1} * \text{SEQ}(n)$ for $n \geq 2$.

10. For $n = 1, 2, 3, \ldots,$ let $\text{SSQ}(n) = \sum_{i=1}^{n} i^2$.

(a) Calculate $\text{SSQ}(n)$ for $n = 1, 2, 3,$ and 5.

(b) Observe that $\text{SSQ}(n+1) = \text{SSQ}(n) + (n+1)^2$ for $n \geq 1$.

(c) It turns out that $\text{SSQ}(73) = 132{,}349$. Use this to calculate $\text{SSQ}(74)$ and $\text{SSQ}(72)$.

11. For the following sequences, write the first several terms until the behavior of the sequence is clear.

(a) $a_n = [2n - 1 + (-1)^n]/4$ for $n \in \mathbb{N}$.

(b) (b_n), where $b_n = a_{n+1}$ for $n \in \mathbb{N}$ and a_n is as in part (a).

(c) $v_n = (a_n, b_n)$ for $n \in \mathbb{N}$.

12. Find the values of the sequences $\log_2 n$ and $\sqrt{n}$ for $n =$ 16, 64, 256, and 4096, and compare.

13. (a) Using a calculator or other device, complete the table in Figure 3. [Write E if the calculation is beyond the capability of your computing device.]

n	n^4	4^n	n^{20}	20^n	$n!$
5			$9.54 \cdot 10^{13}$	$3.2 \cdot 10^6$	
10				$1.02 \cdot 10^{13}$	$3.63 \cdot 10^6$
25	$3.91 \cdot 10^5$				
50		$1.27 \cdot 10^{30}$			

Figure 3 ▲

(b) Discuss the apparent relative growth behaviors of n^4, 4^n, n^{20}, 20^n, and $n!$.

14. Repeat Exercise 13 for the table in Figure 4.

n	$\log_{10} n$	$\sqrt{n}$	$20 \cdot \sqrt[4]{n}$	$\sqrt[4]{n} \cdot \log_{10} n$
50	1.70	7.07	53.18	4.52
100				
10^4				
10^6				

Figure 4 ▲

1.7 Properties of Functions

Some of the most important functions we will study are ones that match up the elements of two sets. Even if a function fails to match up elements, we will see that it still decomposes its domain into disjoint sets in a useful way. This section introduces the ideas and terminology that we will use to describe matching and decomposition of domains. We will come back to these topics in more detail in Chapter 3.

Before we turn to mathematical examples, we illustrate the ideas in a nonmathematical setting.

EXAMPLE 1

Suppose that each student in a class S is assigned a seat number from the set $N = \{1, 2, \ldots, 75\}$. This assignment determines a function $f: S \to N$ for which $f(s)$ is the seat number assigned to student s. If we view the function f as a set of ordered pairs, then it will consist of pairs in $S \times T$ like (Pat Hand, 73).

If no two students are assigned to the same seat—equivalently, if each seat gets at most one student—then we say that the assignment is **one-to-one**. If f is one-to-one, then S can't have more members than N, so the class must have at most 75 students in this case.

In general, there will be unassigned seat numbers. In case every number is assigned, we say that f maps S **onto** N. For this case to happen, the class must have at least 75 students. The function f would be neither one-to-one nor onto N if it assigned more than one student to some seat and left at least one seat number unassigned. The function f could be one-to-one but not onto N if the class had fewer than 75 students. If the class had more than 75 students, then f could not be one-to-one; it might or might not be onto N, depending on whether every seat were assigned. If f is both one-to-one and onto N, which can only occur if there are exactly 75 students, then we call f a **one-to-one correspondence** between S and N. In that case the students s match up with the seats $f(s)$, with one student per seat and with every seat taken. ■

Here are the general definitions. A function $f: S \to T$ is called **one-to-one** in case distinct elements in S have distinct images in T under f:

$$\text{if } x_1, x_2 \in S \quad \text{and} \quad x_1 \neq x_2, \quad \text{then } f(x_1) \neq f(x_2).$$

This condition is logically equivalent to

$$\text{if } x_1, x_2 \in S \quad \text{and} \quad f(x_1) = f(x_2), \quad \text{then } x_1 = x_2,$$

a form that is often useful. In terms of the graph $G = \{(x, y) \in S \times T : y = f(x)\}$ of f, f is one-to-one if and only if

$$\text{for each } y \in T \text{ there is at most one } x \in S \text{ such that } (x, y) \in G.$$

If S and T are subsets of $\mathbb{R}$ and f is graphed so that S is part of the horizontal axis and T is part of the vertical axis, this condition states that horizontal lines intersect G at most once.

Given $f: S \to T$, we say that f maps **onto** a subset B of T provided $B = \text{Im}(f)$. In particular, we say f maps **onto** T provided $\text{Im}(f) = T$. In terms of the graph G of f, f maps S onto T if and only if

$$\text{for each } y \in T \text{ there is at least one } x \in S \text{ such that} (x, y) \in G.$$

A function $f: S \to T$ that is one-to-one and maps onto T is called a **one-to-one correspondence** between S and T. Thus f is a one-to-one correspondence if and only if

$$\text{for each } y \in T \text{ there is exactly one } x \in S \text{ such that } (x, y) \in G.$$

These three kinds of special functions are illustrated in Figure 1.

Figure 1 ▶

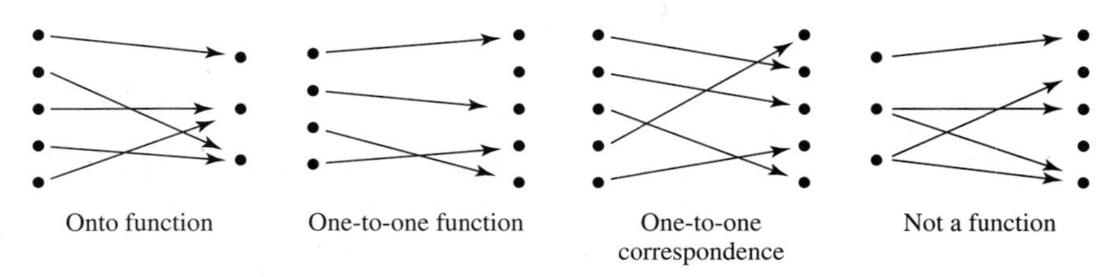

EXAMPLE 2

(a) We define $f: \mathbb{N} \to \mathbb{N}$ by the rule $f(n) = 2n$. Then f is one-to-one, since

$$f(n_1) = f(n_2) \quad \text{implies} \quad 2n_1 = 2n_2 \quad \text{implies} \quad n_1 = n_2.$$

However, f does not map $\mathbb{N}$ onto $\mathbb{N}$ since $\text{Im}(f)$ consists only of the even natural numbers.

(b) Let Σ be an alphabet. Then $\text{length}(w) \in \mathbb{N}$ for each word w in Σ^*; see §1.3. Thus "length" is a function from Σ^* to $\mathbb{N}$. [Recall that functions can have fancier names than "f."] We claim that the function length maps Σ^* *onto* $\mathbb{N}$. Recall that Σ is nonempty, so it contains some letter, say a. Then $0 = \text{length}(\lambda)$, $1 = \text{length}(a)$, $2 = \text{length}(aa)$, etc., so $\text{Im}(\text{length}) = \mathbb{N}$. The function length is not one-to-one unless Σ has only one element.

(c) Recall that the characteristic function χ_A of a subset A of a set S is defined by

$$\chi_A(x) = \begin{cases} 1 & \text{for} \quad x \in A, \\ 0 & \text{for} \quad x \in S \setminus A. \end{cases}$$

The function $\chi_A: S \to \{0, 1\}$ is rarely one-to-one and is usually an onto map. In fact, χ_A maps S onto $\{0, 1\}$ unless $A = S$ or $A = \emptyset$. Moreover, χ_A is not one-to-one if either A or $S \setminus A$ has at least two elements. ■

The next example is an important one for understanding the remainder of this section.

EXAMPLE 3

We indicate why the function $f: \mathbb{R} \to \mathbb{R}$ defined by $f(x) = 3x - 5$ is a one-to-one correspondence of $\mathbb{R}$ onto $\mathbb{R}$. To check that f is one-to-one, we need to show that

$$\text{if} \quad f(x) = f(x') \quad \text{then} \quad x = x',$$

i.e.,

$$\text{if} \quad 3x - 5 = 3x' - 5 \quad \text{then} \quad x = x'.$$

But if $3x - 5 = 3x' - 5$, then $3x = 3x'$ [add 5 to both sides], and this implies that $x = x'$ [divide both sides by 3].

To show that f maps $\mathbb{R}$ onto $\mathbb{R}$, we consider an element y in $\mathbb{R}$. We need to find an x in $\mathbb{R}$ such that $f(x) = y$, i.e., $3x - 5 = y$. So we solve for x and obtain

$$x = \frac{y+5}{3}.$$

Thus, given y in $\mathbb{R}$, the number $\frac{y+5}{3}$ belongs to $\mathbb{R}$ and

$$f\left(\frac{y+5}{3}\right) = 3\left(\frac{y+5}{3}\right) - 5 = y.$$

This shows that every y in $\mathbb{R}$ belongs to $\mathrm{Im}(f)$, so f maps $\mathbb{R}$ onto $\mathbb{R}$. ■

When we argued in the last example that f maps $\mathbb{R}$ *onto* $\mathbb{R}$, we began with a number y in $\mathbb{R}$ and found the unique x in $\mathbb{R}$ that f mapped to y. We did this by setting $y = f(x)$ and then solving for x. In fact, we found that

$$x = \frac{y+5}{3}.$$

This provides us with a new function g, defined by $g(y) = \frac{y+5}{3}$, that undoes what f does. When f takes a number x to y, then g takes that number y back to x. In other symbols, $g(f(x)) = x$ for all $x \in \mathbb{R}$. You can readily check that $f(g(y)) = y$ for all $y \in \mathbb{R}$, so f also undoes what g does. The functions f and g are "inverses" of each other.

In general, an inverse for the function f is a function, written f^{-1}, that undoes the action of f. Applying f first and then the inverse restores every member of the domain of f to where it started. And vice versa. See Figure 2 for a schematic representation of this.

Figure 2 ▶

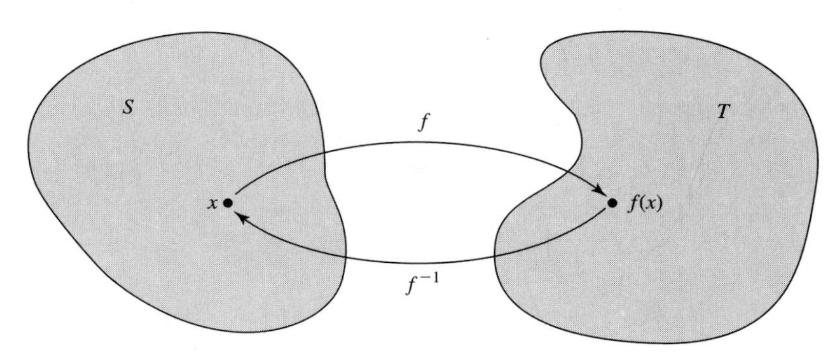

EXAMPLE 4

(a) The functions x^2 and $\sqrt{x}$ with domains $[0, \infty)$ are inverses to each other. If you apply these operations in either order to some value, the original value is obtained. Try it on a calculator! In symbols,

$$\sqrt{x^2} = x \quad \text{and} \quad (\sqrt{x})^2 = x \quad \text{for } x \in [0, \infty).$$

(b) The function h defined on $\mathbb{R} \setminus \{0\}$ by $h(x) = 1/x$ is its own inverse. If you apply h twice to some number, you get the original number. That is,

$$\frac{1}{1/x} = x \quad \text{for all nonzero } x \text{ in } \mathbb{R}.$$

In this case, we can write $h^{-1} = h$. Incidentally, the -1 in the notation h^{-1} is *not* an exponent. We could write $1/x$ as x^{-1}, but not $1/h$ for h^{-1}.

(c) As we discussed after Example 3, the functions defined on $\mathbb{R}$ by $f(x) = 3x - 5$ and $f^{-1}(y) = \frac{y+5}{3}$ are inverses to each other. ■

Here is the precise definition. An **inverse** of a function $f: S \to T$ is a function $f^{-1}: T \to S$ such that

$$f^{-1}(f(x)) = x \quad \text{for all } x \in S$$

and

$$f(f^{-1}(y)) = y \quad \text{for all } y \in T.$$

Not all functions have inverses; those that do are called **invertible** functions. The next theorem tells us which functions are invertible.

Theorem The function $f: S \to T$ is invertible if and only if f is one-to-one and maps S onto T.

Proof Suppose that f maps S onto T. Then for each y in T there is at least one x in S with $f(x) = y$. If f is also one-to-one, then there is exactly one such x, and so we can define a function f^{-1} by the rule

$$f^{-1}(y) = \text{that unique } x \in S \text{ such that } f(x) = y. \qquad (*)$$

The function defined by $(*)$ actually is the inverse of f, since $f(f^{-1}(y)) = y$ by $(*)$ and $f^{-1}(f(x))$ is the unique member of S that f maps onto $f(x)$, namely x.

To argue in the other direction, suppose that f is known to have an inverse. Then for each y in T we have $y = f(f^{-1}(y)) \in f(S)$, so $T = f(S)$, i.e., f maps S onto T. Moreover, f must be one-to-one because, if x_1 and x_2 are members of S such that $f(x_1) = f(x_2)$, then $x_1 = f^{-1}(f(x_1)) = f^{-1}(f(x_2)) = x_2$. ■

The key fact here is that when f^{-1} exists it is given by $(*)$. Thus to find $f^{-1}(y)$ one simply needs to solve for x in terms of y in the equation $f(x) = y$, just as we did in Example 3.

EXAMPLE 5

(a) If we were handed the functions $f(x) = 3x - 5$ and $g(y) = \frac{y+5}{3}$, we could verify that they are inverses by simply noting that

$$g \circ f(x) = g(3x - 5) = \frac{(3x-5)+5}{3} = x \quad \text{for } x \in \mathbb{R}$$

and

$$f \circ g(y) = f\left(\frac{y+5}{3}\right) = 3 \cdot \left(\frac{y+5}{3}\right) - 5 = y \quad \text{for } y \in \mathbb{R}.$$

(b) A function $f: S \to S$ is its own inverse if and only if

$$f \circ f(x) = f(f(x)) = x \quad \text{for all } x \in S. \qquad ■$$

EXAMPLE 6

Consider a positive real number b with $b \neq 1$. The function f_b given by $f_b(x) = b^x$ for $x \in \mathbb{R}$ is a one-to-one function of $\mathbb{R}$ onto $(0, \infty)$. It has an inverse f_b^{-1} with domain $(0, \infty)$, which is called a **logarithm function**. We write $f_b^{-1}(y) = \mathbf{log}_b\, y$; by the definition of an inverse, we have

$$\log_b b^x = x \quad \text{for every } x \in \mathbb{R}$$

and

$$b^{\log_b y} = y \quad \text{for every } y \in (0, \infty).$$

The important examples of b are 2, 10, and the number e that appears in calculus and is approximately 2.718. In particular, e^x and $\log_e x$ are inverse functions. The function $\log_e x$ is called the **natural logarithm** and is often denoted $\mathbf{ln}\, x$. The functions 10^x and $\log_{10} x$ are inverses, and so are 2^x and $\log_2 x$. The functions $\log_{10} x$ and $\log_e x$ appear on many calculators, usually with the names `log` and `ln`, respectively. Such calculators also allow one to compute the inverses 10^x and e^x of these functions. To compute $\log_2 x$ on a calculator, use one of the formulas

$$\log_2 x = \frac{\log_{10} x}{\log_{10} 2} \approx 3.321928 \cdot \log x$$

or

$$\log_2 x = \frac{\log_e x}{\log_e 2} = \frac{\ln x}{\ln 2} \approx 1.442695 \cdot \ln x.$$

■

EXAMPLE 7

Consider the function $g\colon \mathbb{Z} \times \mathbb{Z} \to \mathbb{Z} \times \mathbb{Z}$ given by $g(m,n) = (-n,-m)$. We will check that g is one-to-one and onto, and then we'll find its inverse. To see that g is one-to-one, we need to show that

$$g(m,n) = g(m',n') \quad \text{implies} \quad (m,n) = (m',n').$$

If $g(m,n) = g(m',n')$, then $(-n,-m) = (-n',-m')$. Since these ordered pairs are equal, we must have $-n = -n'$ and $-m = -m'$. Hence $n = n'$ and $m = m'$, so $(m,n) = (m',n')$, as desired.

To see that g maps onto $\mathbb{Z} \times \mathbb{Z}$, consider (p,q) in $\mathbb{Z} \times \mathbb{Z}$. We need to find (m,n) in $\mathbb{Z} \times \mathbb{Z}$ so that $g(m,n) = (p,q)$. Thus we need $(-n,-m) = (p,q)$, and this tells us that n should be $-p$ and m should be $-q$. In other words, given (p,q) in $\mathbb{Z}\times\mathbb{Z}$ we see that $(-q,-p)$ is an element in $\mathbb{Z}\times\mathbb{Z}$ such that $g(-q,-p) = (p,q)$. Thus g maps $\mathbb{Z} \times \mathbb{Z}$ onto $\mathbb{Z} \times \mathbb{Z}$.

To find the inverse of g, we need to take (p,q) in $\mathbb{Z} \times \mathbb{Z}$ and find $g^{-1}(p,q)$. We just did this in the last paragraph; g maps $(-q,-p)$ onto (p,q), and hence $g^{-1}(p,q) = (-q,-p)$ for all (p,q) in $\mathbb{Z} \times \mathbb{Z}$.

It is interesting to note that $g = g^{-1}$ in this case. ■

Consider a general function $f\colon S \to T$. If A is a subset of S, we define

$$\boldsymbol{f(A)} = \{f(x) : x \in A\}.$$

Thus $f(A)$ is the set of images $f(x)$ as x varies over A. We call $f(A)$ the **image of the set** A **under** f. We are also interested in the **inverse image of a set** B in T:

$$\boldsymbol{f^{\leftarrow}(B)} = \{x \in S : f(x) \in B\}.$$

The set $f^{\leftarrow}(B)$ is also called the **pre-image of the set** B **under** f.

If f is invertible, then the pre-image of the subset B of T under f equals the image of B under f^{-1}, i.e., in this case

$$f^{\leftarrow}(B) = \{f^{-1}(y) : y \in B\} = f^{-1}(B).$$

If f is not invertible, it makes no sense to write $f^{-1}(y)$ or $f^{-1}(B)$, of course. Because $f^{-1}(B)$ can't have any meaning *unless* it means $f^{\leftarrow}(B)$, many people extend the notation and write $f^{-1}(B)$ for what we denote by $f^{\leftarrow}(B)$, even if f is not invertible. Beware!

For $y \in T$ we write $f^{\leftarrow}(y)$ for the set $f^{\leftarrow}(\{y\})$. That is,

$$\boldsymbol{f^{\leftarrow}(y)} = \{x \in S : f(x) = y\}.$$

This *set* is the **pre-image of the element** y **under** f. Note that solving the equation $f(x) = y$ for x is equivalent to finding the set $f^{\leftarrow}(y)$. That is, $f^{\leftarrow}(y)$ is the **solution set** for the equation $f(x) = y$. As with equations in algebra, the set $f^{\leftarrow}(y)$ might have one element, more than one element, or no elements at all.

EXAMPLE 8

(a) Consider $f: \mathbb{R} \to \mathbb{R}$ given by $f(x) = x^2$. Then

$$f^{\leftarrow}(4) = \{x \in \mathbb{R} : x^2 = 4\} = \{-2, 2\},$$

which is the solution set of the equation $x^2 = 4$. The pre-image of the set $[1, 9]$ is

$$f^{\leftarrow}([1, 9]) = \{x \in \mathbb{R} : x^2 \in [1, 9]\} = \{x \in \mathbb{R} : 1 \le x^2 \le 9\} = [-3, -1] \cup [1, 3].$$

Also we have $f^{\leftarrow}([-1, 0]) = \{0\}$ and $f^{\leftarrow}([-1, 1]) = [-1, 1]$.

(b) Consider the function $g: \mathbb{N} \times \mathbb{N} \to \mathbb{N}$ defined by $g(m, n) = m^2 + n^2$. Then $g^{\leftarrow}(0) = \{(0, 0)\}$, $g^{\leftarrow}(1) = \{(0, 1), (1, 0)\}$, $g^{\leftarrow}(2) = \{(1, 1)\}$, $g^{\leftarrow}(3) = \emptyset$, $g^{\leftarrow}(4) = \{(0, 2), (2, 0)\}$, $g^{\leftarrow}(25) = \{(0, 5), (3, 4), (4, 3), (5, 0)\}$, etc. ■

EXAMPLE 9

(a) Let Σ be an alphabet and for this example let L be the length function on Σ^*; $L(w) = \text{length}(w)$ for $w \in \Sigma^*$. As we noted in Example 2(b), L maps Σ^* onto $\mathbb{N}$. For each $k \in \mathbb{N}$,

$$L^{\leftarrow}(k) = \{w \in \Sigma^* : L(w) = k\} = \{w \in \Sigma^* : \text{length}(w) = k\}.$$

Note that the various sets $L^{\leftarrow}(k)$ are disjoint and that their union is Σ^*:

$$\bigcup_{k \in \mathbb{N}} L^{\leftarrow}(k) = L^{\leftarrow}(0) \cup L^{\leftarrow}(1) \cup L^{\leftarrow}(2) \cup \cdots = \Sigma^*.$$

These sets $L^{\leftarrow}(k)$ are exactly the sets Σ^k defined in Example 5(b) on page 37. Henceforth we will use the notation Σ^k for these sets.

(b) Consider $h: \mathbb{Z} \to \{-1, 1\}$, where $h(n) = (-1)^n$. Then

$$h^{\leftarrow}(1) = \{n \in \mathbb{Z} : n \text{ is even}\} \quad \text{and} \quad h^{\leftarrow}(-1) = \{n \in \mathbb{Z} : n \text{ is odd}\}.$$

These two sets are disjoint and their union is all of $\mathbb{Z}$:

$$h^{\leftarrow}(1) \cup h^{\leftarrow}(-1) = \mathbb{Z}.$$ ■

It is not a fluke that the pre-images of elements cut the domains into slices in these last two examples. We will see in §3.4 that something like this always happens.

Exercises 1.7

1. Let $S = \{1, 2, 3, 4, 5\}$ and $T = \{a, b, c, d\}$. For each question below: if the answer is YES, give an example; if the answer if NO, explain briefly.
   (a) Are there any one-to-one functions from S into T?
   (b) Are there any one-to-one functions from T into S?
   (c) Are there any functions mapping S onto T?
   (d) Are there any functions mapping T onto S?
   (e) Are there any one-to-one correspondences between S and T?
2. The functions sketched in Figure 3 have domain and codomain both equal to [0,1].
   (a) Which of these functions are one-to-one?
   (b) Which of these functions map [0, 1] onto [0, 1]?
   (c) Which of these functions are one-to-one correspondences?
3. The function $f(m, n) = 2^m 3^n$ is a one-to-one function from $\mathbb{N} \times \mathbb{N}$ into $\mathbb{N}$.
   (a) Calculate $f(m, n)$ for five different elements (m, n) in $\mathbb{N} \times \mathbb{N}$.
   (b) Explain why f is one-to-one.
   (c) Does f map $\mathbb{N} \times \mathbb{N}$ onto $\mathbb{N}$? Explain.
   (d) Show that $g(m, n) = 2^m 4^n$ defines a function on $\mathbb{N} \times \mathbb{N}$ that is not one-to-one.
4. Consider the following functions from $\mathbb{N}$ into $\mathbb{N}$: $1_{\mathbb{N}}(n) = n$, $f(n) = 3n$, $g(n) = n + (-1)^n$, $h(n) = \min\{n, 100\}$, $k(n) = \max\{0, n - 5\}$.
   (a) Which of these functions are one-to-one?
   (b) Which of these functions map $\mathbb{N}$ onto $\mathbb{N}$?
5. Here are two "shift functions" mapping $\mathbb{N}$ into $\mathbb{N}$: $f(n) = n + 1$ and $g(n) = \max\{0, n - 1\}$ for $n \in \mathbb{N}$.
   (a) Calculate $f(n)$ for $n = 0, 1, 2, 3, 4, 73$.
   (b) Calculate $g(n)$ for $n = 0, 1, 2, 3, 4, 73$.
   (c) Show that f is one-to-one but does not map $\mathbb{N}$ onto $\mathbb{N}$.
   (d) Show that g maps $\mathbb{N}$ onto $\mathbb{N}$ but is not one-to-one.
   (e) Show that $g \circ f(n) = n$ for all n, but that $f \circ g(n) = n$ does not hold for all n.

Figure 3 ▶

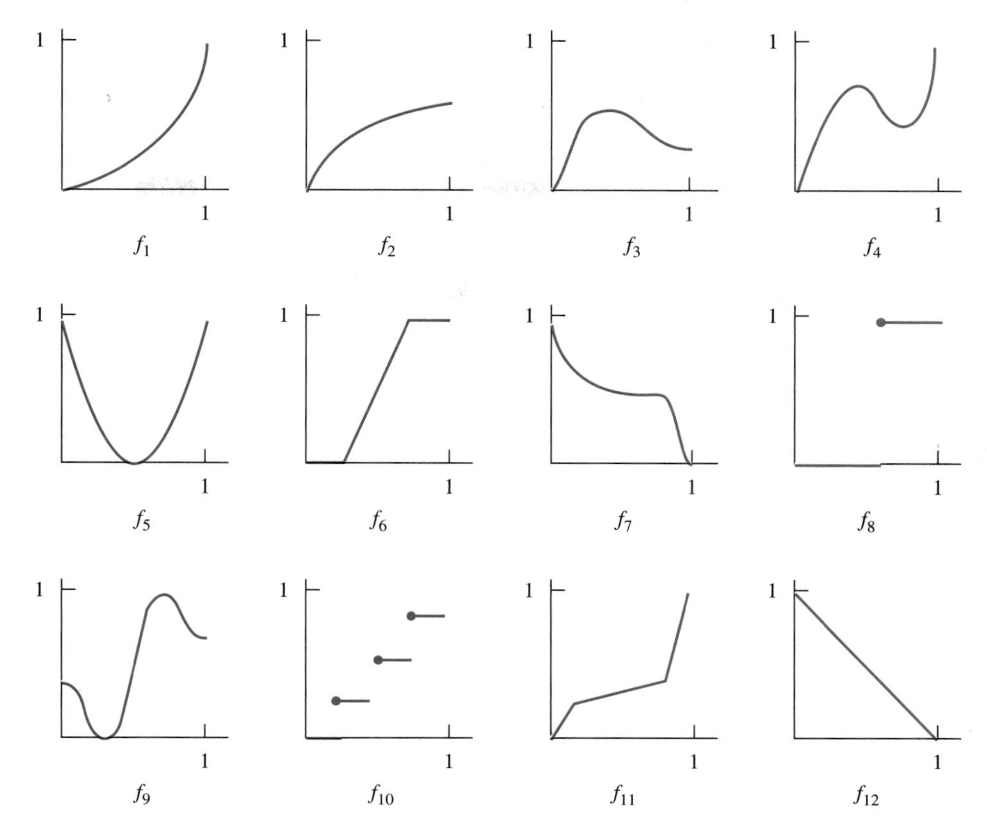

6. Let $\Sigma = \{a, b, c\}$ and let Σ^* be the set of all words w using letters from Σ; see Example 2(b). Define $L(w) = \text{length}(w)$ for all $w \in \Sigma^*$.
   (a) Calculate $L(w)$ for the words $w_1 = cab$, $w_2 = ababac$, and $w_3 = \lambda$.
   (b) Is L a one-to-one function? Explain.
   (c) The function L maps Σ^* into $\mathbb{N}$. Does L map Σ^* onto $\mathbb{N}$? Explain.
   (d) Find all words w such that $L(w) = 2$.

7. Find the inverses of the following functions mapping $\mathbb{R}$ into $\mathbb{R}$.
   (a) $f(x) = 2x + 3$
   (b) $g(x) = x^3 - 2$
   (c) $h(x) = (x - 2)^3$
   (d) $k(x) = \sqrt[3]{x} + 7$

8. Many hand-held calculators have the functions $\log x$, x^2, $\sqrt{x}$, and $1/x$.
   (a) Specify the domains of these functions.
   (b) Which of these functions are inverses of each other?
   (c) Which pairs of these functions commute with respect to composition?
   (d) Some hand-held calculators also have the functions $\sin x$, $\cos x$, and $\tan x$. If you know a little trigonometry, repeat parts (a), (b), and (c) for these functions.

9. Show that the following functions are their own inverses.
   (a) The function $f\colon (0, \infty) \to (0, \infty)$ given by $f(x) = 1/x$.
   (b) The function $\phi\colon \mathcal{P}(S) \to \mathcal{P}(S)$ defined by $\phi(A) = A^c$ ∗ .
   (c) The function $g\colon \mathbb{R} \to \mathbb{R}$ given by $g(x) = 1 - x$.

10. Let A be a subset of some set S and consider the characteristic function χ_A of A. Find $\chi_A^{\leftarrow}(1)$ and $\chi_A^{\leftarrow}(0)$.

11. Here are some functions from $\mathbb{N} \times \mathbb{N}$ to $\mathbb{N}$: $\text{SUM}(m, n) = m + n$, $\text{PROD}(m, n) = m * n$, $\text{MAX}(m, n) = \max\{m, n\}$, $\text{MIN}(m, n) = \min\{m, n\}$; here $*$ denotes multiplication of integers.
    (a) Which of these functions map $\mathbb{N} \times \mathbb{N}$ onto $\mathbb{N}$?
    (b) Show that none of these functions are one-to-one.
    (c) For each of these functions F, how big is the set $F^{\leftarrow}(4)$?

12. Consider the function $f\colon \mathbb{R} \times \mathbb{R} \to \mathbb{R} \times \mathbb{R}$ defined by
$$f(x, y) = (x + y, x - y).$$
This function is invertible. Show that the inverse function is given by
$$f^{-1}(a, b) = \left(\frac{a+b}{2}, \frac{a-b}{2}\right)$$
for all (a, b) in $\mathbb{R} \times \mathbb{R}$.

13. Let $f\colon\ S \to T$ and $g\colon\ T \to U$ be one-to-one functions. Show that the function $g \circ f\colon S \to U$ is one-to-one.

14. Let $f\colon\ S \to T$ be an invertible function. Show that f^{-1} is invertible and that $(f^{-1})^{-1} = f$.

15. Let $f\colon\ S \to T$ and $g\colon\ T \to U$ be invertible functions. Show that $g \circ f$ is invertible and that $(g \circ f)^{-1} = f^{-1} \circ g^{-1}$.

Office Hours 1.7

Chapter 1 was just lots of definitions and basic stuff. Is the rest of the course going to be mostly more definitions, or are we going to get to things we can use?

Now wait a minute! Who says definitions aren't useful? But I understand your question. There *were* a fair number of definitions at first, and we'll see lots more, but that's not the point of the course. The definitions are there to make life easier for us and to bring out the main ideas. I've always thought of definitions as handy ways to describe the basic things we want to think about. A friend of mine tells his students they're the rules of the game—if you don't know the rules, then you can't play the game. Whichever way you want to think of definitions is fine, if it helps you learn and use them.

Let's take a simple example. In Section 1.7 we said that a function was "one-to-one" in case distinct elements had distinct images. That description defines the term "one-to-one." We could get along without this word. Every time we wanted to point out that a function had the property that distinct elements have distinct images we could say that whole, long phrase. Instead, we've agreed to use the word "one-to-one." It's convenient.

We could also make a new definition and say that a function is "at-most-five-to-one" in case no more than five distinct elements have the same image. Why don't we do that? Mainly because experience and common sense tell us that this special property isn't likely to come up ever again, much less in an important way. Having distinct images for distinct elements, on the other hand, has turned out to be a useful property everywhere in math, so everybody in the game uses the term "one-to-one." As you run into new definitions in the book, you can be pretty sure they're there because they describe important ideas. Always ask yourself why a new term is introduced. Is it just temporary, or does it look as if it will get used a lot? If it's really a puzzle, feel free to ask me about the new term.

Also, to help remember a new definition, it's a good idea to think of some examples of things that fit the conditions and also of some examples that almost do, but not quite. How about a function that's almost one-to-one?

OK. Or maybe something like a one-to-one function that's not quite a one-to-one correspondence?

Exactly.

You didn't mention it, but symbolic notation is something else we have to learn if we expect to play the game. For instance, you'd be in big trouble if you didn't know the difference between the symbols for union and intersection. What sort of memory trick you use to keep them straight is up to you, but it has to be absolutely reliable. Incidentally, one indication that a notation is extremely important is if there's a special printed symbol for it. Certainly, anything in the symbols index inside the front and back book covers is going to get a fair amount of use.

I don't know a magic way to learn definitions or notation. The best way I've found to embed them in the brain is to practice using them myself. Just reading them in the book doesn't work for me. The good news is that learning definitions is not actually all that hard. I'm sure you can think of all kinds of strange stuff that you know really well, like words to songs or names of book characters or how to win at computer games. You've learned these things because you had a reason to, and you can learn math terms just as easily.

Now how's the homework going?

Chapter Highlights

To check your understanding of the material in this chapter, we recommend that you consider each item listed below and:

(a) Satisfy yourself that you can define each concept and notation and can describe each method.
(b) Give at least one reason why the item was included in this chapter.
(c) Think of at least one example of each concept and at least one situation in which each fact or method would be useful.

This chapter is introductory and contains a relatively large number of fundamental definitions and notations. It is important to be comfortable with this material as soon as possible, since the rest of the book is built on it.

CONCEPTS

hypotheses
  conclusion
counterexample
prime numbers
factors, divisors, greatest common divisors
multiples, least common multiples
even and odd numbers
relatively prime [pairs of numbers]
set
  member = element,
  subset of a set, equal sets, disjoint sets
  set operations
  universe [universal set], complement [in the universe], relative complement
  Venn diagram
ordered pair, product of sets
alphabet, language, word, length of a word
function = map = mapping
  domain, codomain
  image of x, $\text{Im}(f)$ = image of a function f
  graph of a function
  one-to-one, onto, one-to-one correspondence
  composition of functions
  inverse of a function
  image $f(A)$ of a set A under a function f
  pre-image $f^{\leftarrow}(B)$ of a set B under $f^{\leftarrow}$
sequence, finite sequence

EXAMPLES AND NOTATION

floor function $\lfloor\ \rfloor$, ceiling function $\lceil\ \rceil$
$m|n$ [m divides n]
$\gcd(m,n)$, $\text{lcm}(m,n)$, $\min\{a,b\}$, $\max\{a,b\}$
$\mathbb{N}$, $\mathbb{P}$, $\mathbb{Z}$, $\mathbb{Q}$, $\mathbb{R}$
$\in$, $\notin$, $\{\ :\ \}$, $\subseteq$, $\subset$
$\emptyset$ = empty set
notation for intervals: $[a,b]$, (a,b), $[a,b)$, $(a,b]$
$\mathcal{P}(S)$, $\cup$, $\cap$, $A \setminus B$, $A \oplus B$, A^c
(s,t) notation for ordered pairs, $S \times T$, S^n
$|S|$ = number of elements in the set S
$\text{Dom}(f)$, $\text{Im}(f)$, $f: S \to T$, $f \circ g$

$\sum$ notation for sums, $\prod$ notation for products
$n!$ for n factorial
Σ^*, λ = empty word
special functions: $\log_b$, characteristic function χ_A of a set A

FACTS

Division Algorithm [$n = m \cdot q + r$ where $0 \le r < m$].
Basic laws of set algebra [Table 1 on page 25].
Composition of functions is associative.
A function is invertible if and only if it is a one-to-one correspondence.

METHODS

Abstracting from the specific.
Use of Venn diagrams.
Reasoning from definitions and previously established facts.

Supplementary Exercises

1. (a) Find gcd(555552, 555557).

(b) How about gcd(55555555553, 55555555558)?

2. (a) Show that $(A \setminus B) \setminus C$ and $A \setminus (B \setminus C)$ are not in general equal for sets A, B, and C by giving an explicit counterexample. A Venn diagram is not a counterexample.

(b) Give an example of sets A, B, and C for which $(A \setminus B) \setminus C = A \setminus (B \setminus C)$.

3. For $n \in \mathbb{P}$, let $D_n = \{k \in \mathbb{Z} : k$ is divisible by $n\}$. List three members of each of the following sets:

(a) $D_3 \cap D_5$ (b) D_3^c (c) $D_3 \oplus D_5$

4. (a) Calculate $\lfloor \frac{k}{5} \rfloor$ for $k = 1, 2, 3, 4, 5, 9, 10, 13$, and 73.

(b) For which integers k do we have $\lfloor \frac{k}{5} \rfloor = \lceil \frac{k}{5} \rceil$?

5. The summary of methods at the end of §1.1 lists "counting a set without actually listing its members" as one method. What sets got counted that way in §1.1? Can you think of other examples in which you have counted sets without listing them?

6. The function $f: [0, 2] \to [-4, 0]$ defined by $f(x) = x^2 - 4x$ is invertible. Give the domain, codomain, and a formula for the inverse function f^{-1}. [Be careful.]

7. Let $f: \mathbb{Z} \to \mathbb{Z}$ be defined by $f(x) = x^2 - 3$. Find $f^{\leftarrow}(\{0, 1\})$.

8. Define $f: \mathbb{R} \to \mathbb{R}$ by $f(x) = x^2 - 1$. Determine the following sets:

(a) $f([0, 1])$ (b) $f^{\leftarrow}([0, 1])$ (c) $f^{\leftarrow}(f([0, 1]))$

9. (a) Convince yourself that if a positive integer n ends with digit 7, then n^2 ends with digit 9. More generally, if n ends with digit 1, 2, 3, 4, 5, 6, 7, 8, 9, or 0, then n^2 ends with digit 1, 4, 9, 6, 5, 6, 9, 4, 1, or 0, respectively.

(b) How many squares n^2 (of positive integers n) that are less than or equal to 1,000,000 end with digit 9?

(c) Repeat part (b) for the other digits 0, 1, 2, 3, 4, 5, 6, 7, and 8.

10. One approach to finding the greatest common divisor gcd(m, n) of m and n is to find all divisors of m, find which of those are divisors of n, and then take the largest of them.

(a) What looks like the hardest part of this outline?

(b) Why can't we just replace "divisors" by "multiples" and "largest" by "smallest" in this approach to get lcm(m, n)?

11. For positive integers l, m, and n, define gcd(l, m, n), just as for two positive integers, to be the largest integer that divides all three of l, m, and n.

(a) Find:

gcd(8, 12, 21) gcd(9, 21, 56)
gcd(35, 45, 63) gcd(77, 91, 119)

(b) Generalize Theorems 3 and 4 on pages 12 and 13, if possible, or show that the natural generalizations fail. [If you get stuck, think about prime factorizations.]

12. The discussion on page 6 leads to the estimate that roughly $\frac{1}{70}$ of the integers with 30 decimal digits are primes. Using the same methods, roughly what proportion of 40-digit integers are primes? [*Hint:* Exercise 15 on page 7 may help.]

13. The theorem on page 5 says that the number of multiples of k between m and n is about $\frac{n}{k} - \frac{m-1}{k}$, which is itself about $\frac{n-m}{k}$. When would this sort of rough answer be good enough?

14. Bob says that if he just knows gcd(m, n) and lcm(m, n) with $m \le n$, then he can figure out what m and n are. Ann says he's wrong and that she has two different examples with gcd$(m, n) = 2$ and lcm$(m, n) = 12$. Is either Ann or Bob correct?

15. The definitions of gcd(m, n) and lcm(m, n) make sense even if m or n is a negative integer.

(a) Find $\gcd(-4, 6)$ and $\text{lcm}(10, -6)$.

(b) Is it still true that $\gcd(m, n) \cdot \text{lcm}(m, n) = m \cdot n$ if m and n are allowed to be negative? That is, does Theorem 3 on page 12 still hold? Explain.

16. Let Σ be the alphabet $\{a, b\}$.

(a) Is there a function mapping Σ onto Σ^*? If yes, give an example. If no, explain.

(b) Is there a function mapping Σ^* onto Σ? If yes, give an example. If no, explain.

17. Let $S = \{x \in \mathbb{R} : x \geq -1\}$ and $T = \{x \in \mathbb{R} : x \geq 0\}$, and define $f(x) = \sqrt{x+1}$ for $x \in S$. Then $f: S \to T$.

(a) Show that f is one-to-one.

(b) Show that f maps S onto T.

(c) Does f have an inverse? If so, find it.

(d) Find a formula for the function $f \circ f$, and give its domain and codomain.

(e) Is $f \circ f$ one-to-one? Explain.

18. (a) Prove the associative law $(A \oplus B) \oplus C = A \oplus (B \oplus C)$ without using Venn diagrams.

(b) By part (a), the notation $A \oplus B \oplus C$ is unambiguous. Give a description of the elements of $A \oplus B \oplus C$ that doesn't involve either of the sets $A \oplus B$ or $B \oplus C$.

(c) Use the ideas in part (b) to describe the elements in $A \oplus B \oplus C \oplus D$.

(d) Generalize to n sets $A_1, \ldots, A_n$.

19. Note the exponents that appear in the 2^n column of Figure 2 on page 37. Note also that $\log_{10} 2 \approx 0.30103$. What's going on here?

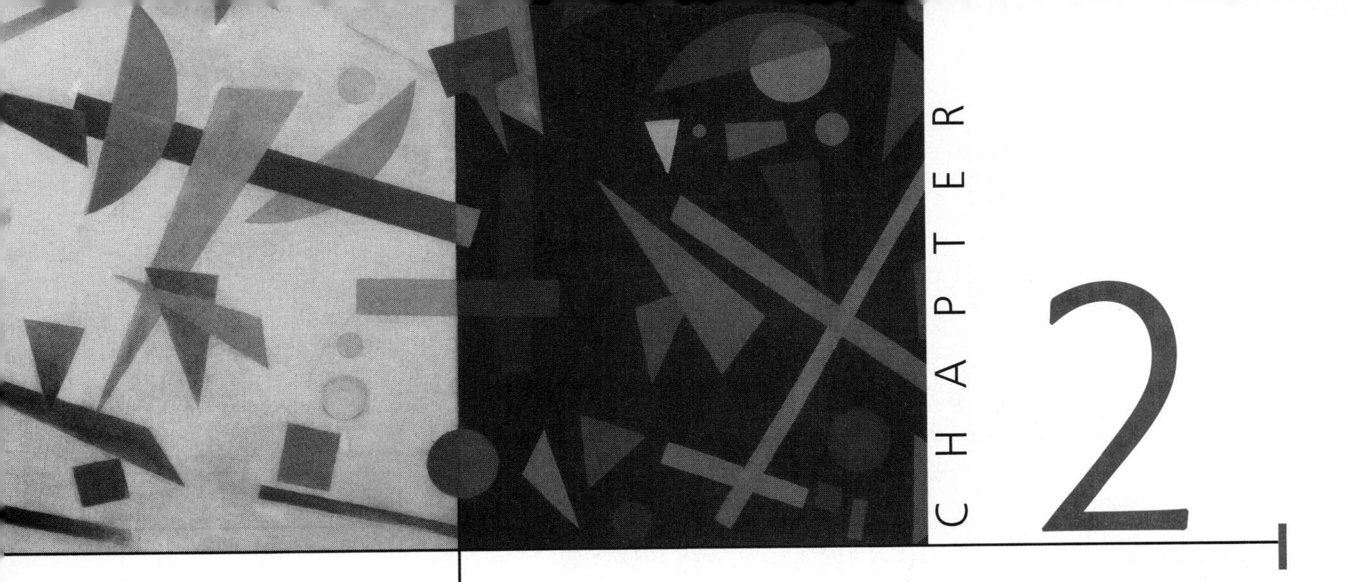

Elementary Logic

This chapter contains an informal introduction to logic, both as a set of tools for understanding and building arguments (i.e., *proofs*) and as an object of study in its own right. Everyone who depends on making inferences needs to be able to recognize and distinguish between valid and invalid arguments. Mathematicians place a great deal of emphasis on creating proofs that are logically watertight. Computer scientists need to be able to reason logically, of course, but in addition they need to know the formal rules of logic that their machines follow. Our emphasis will be on logic as a working tool. We will try to be informal, but we will also hint at what would be involved in a careful formal development of the subject. We will also develop some of the symbolic techniques required for computer logic. The connection with hardware and logical circuits will be made more explicit in Chapter 10.

Section 2.1 introduces some terminology and common notation, including the useful quantifiers $\forall$ and $\exists$. In §2.2 we give the basic framework of the propositional calculus. The general concepts are important and should be mastered. Some rather intimidating tables, which need not be memorized, are provided for easy reference. Ultimately, the purpose of proofs is to communicate by providing convincing arguments. In §§2.3 and 2.4 we start to discuss proofs as encountered in practice. In particular, §2.3 contains some suggestions and advice to assist in learning how to write proofs. Finally, we formalize the ideas in §2.5 and then return to analyze informal arguments in §2.6.

2.1 Informal Introduction

We plan to study the logical relationships among propositions, which are usually interpretable as meaningful assertions in practical contexts. For us, a **proposition** will be any sentence that is either true or false, but not both. That is, it is a sentence that can be assigned the truth value **true** or the truth value **false**, and not both. We do not need to know what its truth value is in order to consider a proposition.

EXAMPLE 1

The following are propositions:

(a) Julius Caesar was president of the United States.

(b) $2 + 2 = 4$.

(c) $2 + 3 = 7$.
(d) The number 4 is positive and the number 3 is negative.
(e) If a set has n elements, then it has 2^n subsets.
(f) $2^n + n$ is a prime number for infinitely many n.
(g) Every even integer greater than 2 is the sum of two prime numbers.

Note that propositions (b) and (c) are mathematical sentences, where "=" serves as the verb "equals" or "is equal to." Proposition (d) is false, since 3 is not negative. If this is not clear now, it will become clear soon, since (d) is the compound proposition "4 is positive *and* 3 is negative." We [the authors] have no idea whether proposition (f) is true or false, though some mathematicians may know the answer. On the other hand, as of the writing of this book *no one knows* whether proposition (g) is true; its truth is known as "Goldbach's conjecture." ■

EXAMPLE 2

Here are some more propositions:

(a) $x + y = y + x$ for all $x, y \in \mathbb{R}$.
(b) $2^n = n^2$ for some $n \in \mathbb{N}$.
(c) It is not true that 3 is an even integer or 7 is a prime.
(d) If the world is flat, then $2 + 2 = 4$.

Proposition (a) is really an infinite set of propositions covered by the phrase "every" or "for all." Proposition (b) is a special sort of proposition because of the phrase "for some." Propositions of these types will be discussed later in this section and studied systematically in Chapter 13. Proposition (c) is a somewhat confusing compound proposition whose truth value will be easy to analyze after the study of this chapter. Our propositional calculus will allow us to construct propositions like the one in (d) and to decide whether they are true or false, even when they may appear silly or even paradoxical. ■

EXAMPLE 3

The following sentences are not propositions:

(a) Your place or mine?
(b) Why is induction important?
(c) Go directly to jail.
(d) Knock before entering!
(e) $x - y = y - x$.

The reason that sentence (e) is not a proposition is that the symbols are not specified. If the intention is

(e′) $x - y = y - x$ for all $x, y \in \mathbb{R}$,

then this is a false proposition. If the intention is

(e″) $x - y = y - x$ for some $x, y \in \mathbb{R}$,

or

(e‴) $x - y = y - x$ for all x, y in the set $\{0\}$,

then this is a true proposition. The problem of unspecified symbols will be discussed in Example 12. ■

EXAMPLE 4

Of course, in the real world there are ambiguous statements:

(a) Teachers are overpaid.
(b) Math is fun.
(c) $A^2 = 0$ implies $A = 0$ for all A.

The difficulty with sentence (c) is that the set of allowable A's is not specified. The statement in (c) is true for all $A \in \mathbb{R}$. It turns out that (c) is meaningful, but false, for the set of all 2×2 matrices A. Ambiguous statements should either be made unambiguous or abandoned. We will not concern ourselves with this process, but assume that our propositions are unambiguous. [See Exercises 17 and 18, though, for more potentially ambiguous examples.] ■

Although much of the logic we will discuss has perfectly good applications to everyday life, our main goals are quite narrow. We want to develop a logical framework and structures that are adequate to handle the questions that come up in mathematics and computer science.

We begin by developing some symbolic notation for manipulating the logic of propositions. In this context we will generally use lowercase letters such as p, q, $r, \ldots$ to stand for propositions, and we will combine propositions to obtain compound propositions using standard connective symbols:

$\neg$ for "not" or negation;

$\wedge$ for "and";

$\vee$ for "or" [inclusive];

$\rightarrow$ for "implies" or the conditional implication;

$\leftrightarrow$ for "if and only if" or the biconditional.

The inclusive and exclusive "or"s are discussed at the beginning of §1.4. Other connectives, such as $\oplus$, appear in the exercises of §§2.2 and 2.5. Sometimes we will be thinking of building compound propositions from simpler ones, and sometimes we will be trying to analyze a complicated proposition in terms of its constituent parts.

In the next section we will carefully discuss each of the connective symbols and explain how each affects the truth values of compound propositions. At this point, though, we will just treat the symbols informally, as a kind of abbreviation for English words or phrases, and we will try to get a little experience using them to modify propositions or to link propositions together. Our first illustration shows how some of the propositions in Examples 1 and 2 can be viewed as compound propositions.

EXAMPLE 5

(a) Recall proposition (d) of Example 1: "The number 4 is positive and the number 3 is negative." This can be viewed as the compound proposition $p \wedge q$ [read "p and q"], where $p =$ "4 is positive" and $q =$ "3 is negative."

(b) Proposition (d) of Example 2, "If the world is flat, then $2 + 2 = 4$," can be viewed as the compound proposition $r \rightarrow s$ [read "r implies s"], where $r =$ "the world is flat" and $s =$ "$2 + 2 = 4$."

(c) Proposition (c) of Example 2 says "It is not true that 3 is an even integer or 7 is a prime." This is $\neg(p \vee q)$, where $p =$ "3 is even" and $q =$ "7 is a prime." Actually, the nonsymbolic version of this proposition is poorly written and can also be interpreted to mean $(\neg p) \vee q$. When we read (c) aloud as "not p or q," we need to make it clear somehow where the parentheses go. Is it not p, or is it not both?

(d) The English word "unless" can be interpreted symbolically in several ways. For instance, "We eat at six unless the train is late" means "We eat at six or the train is late," or "If the train is not late, then we eat at six," or even "If we do not eat at six, then the train is late." ■

The compound proposition $p \rightarrow q$ is read "p implies q," but it has several other English-language equivalents, such as "if p, then q." In fact, in Example 5(b) the compound proposition $r \rightarrow s$ was a translation of "if r, then s." That proposition could have been written: "The world is flat implies that $2 + 2 = 4$." Other English-language equivalents for $p \rightarrow q$ are "p only if q" and "q if p." We will usually avoid these; but see Exercises 15 to 18.

Compound propositions of the form $p \to q$, $q \to p$, $\neg p \to \neg q$, etc., appear to be related and are sometimes confused with each other. It is important to keep them straight. The proposition $q \to p$ is called the **converse** of the proposition $p \to q$. As we will see, it has a different meaning from $p \to q$. It turns out that $p \to q$ *is* equivalent to $\neg q \to \neg p$, which is called the **contrapositive** of $p \to q$.

EXAMPLE 6

Consider the sentence "If it is raining, then there are clouds in the sky." This is the compound proposition $p \to q$, where $p =$ "it is raining" and $q =$ "there are clouds in the sky." This is a true proposition. Its converse $q \to p$ reads "If there are clouds in the sky, then it is raining." Fortunately, this is a false proposition. The contrapositive $\neg q \to \neg p$ says, "If there are no clouds in the sky, then it is not raining." Not only is this a true proposition, but most people would agree that this follows "logically" from $p \to q$, without having to think again about the physical connection between rain and clouds. It does follow, and this logical connection will be made more precise in Table 1 on page 62, item 9. ■

In logic we are concerned with determining the truth values of propositions from related propositions. Example 6 illustrates that the truth of $p \to q$ does not imply that $q \to p$ is true, but it suggests that the truth of the contrapositive $\neg q \to \neg p$ follows from that of $p \to q$. Here is another illustration of why one must be careful in manipulating logical expressions.

EXAMPLE 7

Consider the argument "If the pig is not outdoors, then the house is not quiet. The pig is outdoors, so the house is quiet." We are not concerned with whether the pig makes a lot of noise, but just with whether the house's being quiet follows *logically* from the previous two assertions, "If the pig is not outdoors, then the house is not quiet" and "The pig is outdoors." It turns out that this reasoning is not valid. The first sentence only tells us that things inside will be loud if the pig *is* there; it tells us nothing otherwise. Perhaps the inhabitants like to party, whether or not there's a pig around. If the reasoning above were valid, the following would also be: "If Carol doesn't buy a lottery ticket, then she will not win $1,000,000. Carol does buy a lottery ticket, so she will win $1,000,000."

Symbolically, these invalid arguments both take this form: If $\neg p \to \neg q$ and p are true, then q is true. The propositional calculus we develop in §§2.2 and 2.5 will provide a formal framework with which to analyze the validity of arguments such as these. ■

The truth of $p \to q$ is sometimes described by saying that p is a **sufficient condition** for q or that q is a **necessary condition** for p. Saying that p is a necessary and sufficient condition for q is another way of saying $q \to p$ and $p \to q$ are true, i.e., that $p \leftrightarrow q$ is true.

EXAMPLE 8

(a) For a person to have grandchildren it is necessary to have or to have had children. That is, "grandchildren $\to$ children" is true. Having children is not sufficient for having grandchildren, though, since the children might not have children themselves.

(b) Hitting a fly with a cannonball is sufficient to kill it, but not necessary. Thus "cannonball $\to$ splat" is true, but "splat $\to$ cannonball" is not.

(c) A necessary and sufficient condition for a prime number p to be even is that $p = 2$. ■

We have been considering "if ... then ... " statements in the context of implication and logical proof. In everyday usage, such statements commonly also carry along a meaning of causality, that q is true *because* p is true, or of time dependence, so that q only follows from p if it happens later. In the English language, the "if ... then ... " construction is often used to analyze situations that may be contrary to fact or that may not yet be decided, such as "If my grandmother had wheels, then she would be a bicycle," or "If I lifted that box, then I might hurt myself." In some other languages it is actually not possible to make statements of these sorts without rather elaborate constructions. So far as we are concerned, though, none of these natural-language considerations matter. As we said earlier, we just want to be able to handle the questions that come up in mathematics and computer science. If our tools apply more generally, then that's fine too.

We should also mention another usage of "if" in computer science, as a flow control instruction. In some programming languages, both **if** and **then** are key words, as in program segments

```
if A then ( do B ).
```

In some languages the "then" is omitted, leading to segments

```
if A ( do B ).
```

The interpretation in either case is "Check to see if A is true and, if it is, then do B. If it's not, then move on to the next instruction." Here A is some condition that's either true or false, but B is just some program segment, so it's neither true nor false, but simply a list of instructions. Thus the flow control usage is quite different from our logical "if ... then ... " interpretation.

Many statements in mathematics are general statements about members of a large, possibly infinite, set of objects. They may be statements about natural numbers, such as those in parts (e)–(g) of Example 1 and part (b) of Example 2. They may be statements about real numbers, as in part (a) of Example 2, or about sets or functions or matrices or some other classes of objects. Statements like this often include the phrase "for all" or "for every" and can be viewed as compound propositions of simpler statements.

EXAMPLE 9

Consider again Goldbach's conjecture from Example 1: "Every even integer greater than 2 is the sum of two prime numbers." This proposition turns out to be decomposable as

$$\text{“}p(4) \wedge p(6) \wedge p(8) \wedge p(10) \wedge p(12) \wedge \cdots\text{”}$$

or

$$\text{“}p(n) \text{ for every even } n \text{ in } \mathbb{N} \text{ agreater than 4,”}$$

where $p(n)$ is the simple proposition "n is the sum of two prime numbers." However, the rules for the connectives $\wedge$ and $\vee$ in the propositional calculus, which we will develop in §2.2, do not allow constructions, such as these, that involve more than a finite number of propositions. ■

There are two useful connectives, $\forall$ and $\exists$, from what is called the "predicate calculus" that will allow us to symbolize propositions like those in Example 9. We will discuss the logic of these connectives, called **quantifiers**, in more detail in Chapter 13; for now we just treat them as informal abbreviations.

In Example 9, the variable n in $p(n)$ takes integer values, but in general the variable values could be real numbers or other sorts of objects. For the general setting, we will use U to denote a generic set of possible values; U is sometimes called a "universal set." Thus we suppose that $\{p(x) : x \in U\}$ is a family of propositions, where U is a set that may be infinite. In other words, p is a proposition-valued function defined on the set U. The **universal quantifier** $\boldsymbol{\forall}$, an upside-down A as in "for **A**ll," is used to build compound propositions of the form

$$\forall x\, p(x),$$

which we read as "for all x, $p(x)$." Other translations of $\forall$ are "for each," "for every," and "for any." Beware of "for any," though, as it might be misinterpreted as meaning "for some." The phrase "if you have income for any month" is not the same as "if you have income for every month."

The **existential quantifier** $\exists$, a backward E as in "there **E**xists," is used to form propositions such as

$$\exists x\ p(x),$$

which we read as "there exists an x such that $p(x)$," "there is an x such that $p(x)$," or "for some x, $p(x)$."

EXAMPLE 10

(a) For each n in $\mathbb{N}$, let $p(n)$ be the proposition "$n^2 = n$." Then $\forall n\ p(n)$ is shorthand for the statement "for all $n \in \mathbb{N}$, $n^2 = n$," which we'd probably write as "$n^2 = n$ for all $n \in \mathbb{N}$." Similarly, $\exists n\ p(n)$ is shorthand for "$n^2 = n$ for some $n \in \mathbb{N}$." The universal set U in this case is $\mathbb{N}$.

(b) Let $p(x)$ be "$x \leq 2x$" and $q(x)$ be "$x^2 \geq 0$" for $x \in \mathbb{R}$. Since $\mathbb{R}$ cannot be listed as a finite sequence, it would really be impossible to symbolize $\exists x\ p(x)$ or $\forall x\ q(x)$ in the propositional calculus. The expression $\exists x\ p(x)$ is shorthand for "$x \leq 2x$ for some $x \in \mathbb{R}$," and $\forall x\ q(x)$ is shorthand for "$x^2 \geq 0$ for all $x \in \mathbb{R}$." Here the universal set U is $\mathbb{R}$.

(c) Quantifiers are also useful when one is dealing with a finite, but large, set of propositions. Suppose, for example, that we have propositions $p(n)$ for n in the set $\{n \in \mathbb{N} : 0 \leq n \leq 65{,}535\}$. The notation $\forall n\ p(n)$ is clearly preferable to

$$p(0) \wedge p(1) \wedge p(2) \wedge p(3) \wedge \cdots \wedge p(65{,}535)$$

though we might invent the acceptable $\bigwedge_{n=0}^{65{,}535} p(n)$. ■

In Section 13.1 we will formalize the intended meaning that the compound proposition $\forall x\ p(x)$ is true provided every proposition $p(x)$ is true. Otherwise, we regard this compound proposition as false. Thus

$\forall x\ p(x)$ is true if $p(x)$ is true for every x in the universal set U;
otherwise, $\forall x\ p(x)$ is false.

Similarly, the compound proposition $\exists x\ p(x)$ is true provided $p(x)$ is true for at least one value of x in the universal set. Thus

$\exists x\ p(x)$ is true if $p(x)$ is true for at least one x in U;
$\exists x\ p(x)$ is false if $p(x)$ is false for every x in U.

EXAMPLE 11

(a) As in Example 10(a), $p(n)$ represents the proposition "$n^2 = n$" for each $n \in \mathbb{N}$. Then $\forall n\ p(n)$ is false because, for example, $p(3)$, i.e., "$3^2 = 3$," is false. On the other hand, $\exists n\ p(n)$ is true because at least one proposition $p(n)$ is true; in fact, exactly two of them are true, namely $p(0)$ and $p(1)$.

(b) As in Example 10(b), let $p(x)$ be "$x \leq 2x$" and $q(x)$ be "$x^2 \geq 0$" for $x \in \mathbb{R}$. Clearly, $\exists x\ p(x)$ is true, but $\forall x\ p(x)$ is false because $p(x)$ is false for negative x. Both $\forall x\ q(x)$ and $\exists x\ q(x)$ are true.

(c) The Goldbach conjecture can now be written as $\forall n\ p(n)$ is true, where $p(n) =$ "if n in $\mathbb{N}$ is even and greater than 2, then n is the sum of two prime numbers." It is known that $p(n)$ is true for a huge number of values of n, but it isn't known whether $p(n)$ is true for *all* n in the universal set $\mathbb{N}$. ■

EXAMPLE 12

(a) The quantifiers $\forall$ and $\exists$ are often used informally as abbreviations. The first two propositions in Example 2 might be written as "$x + y = y + x\ \forall x, y \in \mathbb{R}$" and "$\exists n \in \mathbb{N}$ so that $2^n = n^2$."

(b) In practice, it is common to omit understood quantifiers. The associative and cancellation laws for $\mathbb{R}$ are often written

(A) $(x + y) + z = x + (y + z)$,
(C) $xz = yz$ and $z \neq 0$ imply $x = y$.

The intended meanings are

$$\text{(A)} \quad \forall x\, \forall y\, \forall z[(x + y) + z = x + (y + z)],$$
$$\text{(C)} \quad \forall x\, \forall y\, \forall z[(xz = yz \wedge z \neq 0) \rightarrow x = y],$$

where x, y, and z are in $\mathbb{R}$. In everyday usage (A) might also be written as

$$(x + y) + z = x + (y + z) \quad \forall x\, \forall y\, \forall z,$$

or

$$(x + y) + z = x + (y + z) \quad \forall x, y, z \in \mathbb{R},$$

or

$$(x + y) + z = x + (y + z) \quad \text{for all } x, y, z \in \mathbb{R}.$$ ■

We will often be able to prove propositions $\forall n\, p(n)$, where $n \in \mathbb{N}$, by using an important technique, mathematical induction, which we describe in Chapter 4.

A compound proposition of the form $\forall x\, p(x)$ will be false if any one [or more] of its propositions $p(x)$ is false. So, to **disprove** such a compound proposition, it is enough to show that one of its propositions is false. In other words, it is enough to supply an example that is counter to, or contrary to, the general proposition, i.e., a **counterexample**.

Goldbach's conjecture is still unsettled because no one has been able to prove that *every* even integer greater than 2 is the sum of two primes, and no one has found a counterexample. The conjecture has been verified for a great many even integers.

EXAMPLE 13

(a) The number 2 provides a counterexample to the assertion "All prime numbers are odd numbers."

(b) The number 7 provides a counterexample to the statement "Every positive integer is the sum of three squares of integers." Incidentally, it can be proved that every positive integer is the sum of four squares of integers, e.g., $1 = 1^2 + 0^2 + 0^2 + 0^2$, $7 = 2^2 + 1^2 + 1^2 + 1^2$, and $73 = 8^2 + 3^2 + 0^2 + 0^2$, but just checking a large number of cases is not a proof.

(c) The value $n = 3$ provides a counterexample to the statement "$n^2 \leq 2^n$ for all $n \in \mathbb{N}$," which we may write as "$n^2 \leq 2^n\ \forall n \in \mathbb{N}$." There are no other counterexamples, as we will see in Example 1(c) on page 146.

(d) Gerald Ford is a counterexample to the assertion: "All American presidents have been right-handed." Do you know of other counterexamples? ■

Given a general assertion whose truth value is unknown, often the only strategy is to make a guess and go with it. If you guess that the assertion is true, then analyze the situation to see why it always seems to be true. This analysis might lead you to a proof. If you fail to find a proof and you can see why you have failed, then you might discover a counterexample. Then again, if you can't find a counterexample, you might begin to suspect once more that the result is true and formulate reasons why it must be. It is very common, especially on difficult problems, to spend considerable efforts trying to establish *each* of the two possibilities, until one wins out. One of the authors spent a good deal of energy searching for a counterexample to a result that he felt was false, only to have a young mathematician later provide a proof that it was true.

Exercises 2.1

1. Let p, q, and r be the following propositions:

$p =$ "it is raining,"
$q =$ "the sun is shining,"
$r =$ "there are clouds in the sky."

Translate the following into logical notation, using p, q, r, and logical connectives.

(a) It is raining and the sun is shining.

(b) If it is raining, then there are clouds in the sky.

(c) If it is not raining, then the sun is not shining and there are clouds in the sky.

(d) The sun is shining if and only if it is not raining.

(e) If there are no clouds in the sky, then the sun is shining.

2. Let p, q, and r be as in Exercise 1. Translate the following into English sentences.

(a) $(p \wedge q) \rightarrow r$ (b) $(p \rightarrow r) \rightarrow q$

(c) $\neg p \leftrightarrow (q \vee r)$ (d) $\neg(p \leftrightarrow (q \vee r))$

(e) $\neg(p \vee q) \wedge r$

3. (a) Give the truth values of the propositions in parts (a) to (e) of Example 1.

(b) Do the same for parts (a) and (b) of Example 2.

4. Which of the following are propositions? Give the truth values of the propositions.

(a) $x^2 = x \ \forall x \in \mathbb{R}$.

(b) $x^2 = x$ for some $x \in \mathbb{R}$.

(c) $x^2 = x$.

(d) $x^2 = x$ for exactly one $x \in \mathbb{R}$.

(e) $xy = xz$ implies $y = z$.

(f) $xy = xz$ implies $y = z \ \forall x, y, z \in \mathbb{R}$.

(g) $w_1 w_2 = w_1 w_3$ implies $w_2 = w_3$ for all words $w_1, w_2, w_3 \in \Sigma^*$.

5. Consider the ambiguous sentence "$x^2 = y^2$ implies $x = y \ \forall x, y$."

(a) Make the sentence into an (unambiguous) proposition whose truth value is true.

(b) Make the sentence into an (unambiguous) proposition whose truth value is false.

6. Give the converses of the following propositions.

(a) $q \rightarrow r$.

(b) If I am smart, then I am rich.

(c) If $x^2 = x$, then $x = 0$ or $x = 1$.

(d) If $2 + 2 = 4$, then $2 + 4 = 8$.

7. Give the contrapositives of the propositions in Exercise 6.

8. (a) In connection with Goldbach's conjecture, verify that some small even integers, such as 4, 6, 8, and 10, are sums of two primes.

(b) Do the same for 98.

9. (a) Show that $n = 3$ provides one possible counterexample to the assertion "$n^3 < 3^n \ \forall n \in \mathbb{N}$."

(b) Can you find any other counterexamples?

10. (a) Show that $(m, n) = (4, -4)$ gives a counterexample to the assertion "If m, n are nonzero integers that divide each other, then $m = n$."

(b) Give another counterexample.

11. (a) Show that $x = -1$ is a counterexample to "$(x + 1)^2 \geq x^2 \ \forall x \in \mathbb{R}$."

(b) Find another counterexample.

(c) Can a nonnegative number serve as a counterexample? Explain.

12. Find counterexamples to the following assertions.

(a) $2^n - 1$ is prime for every $n \geq 2$.

(b) $2^n + 3^n$ is prime $\forall n \in \mathbb{N}$.

(c) $2^n + n$ is prime for every positive odd integer n.

13. (a) Give a counterexample to "$x > y$ implies $x^2 > y^2$ $\forall x, y \in \mathbb{R}$." Your answer should be an ordered pair (x, y).

(b) How might you restrict x and y so that the proposition in part (a) is true?

14. Let S be a nonempty set. Determine which of the following assertions are true. For the true ones, give a reason. For the false ones, provide a counterexample.

(a) $A \cup B = B \cup A \quad \forall A, B \in \mathcal{P}(S)$.

(b) $(A \setminus B) \cup B = A \quad \forall A, B \in \mathcal{P}(S)$.

(c) $(A \cup B) \setminus A = B \quad \forall A, B \in \mathcal{P}(S)$.

(d) $(A \cap B) \cap C = A \cap (B \cap C) \quad \forall A, B, C \in \mathcal{P}(S)$.

15. Even though we will normally use "implies" and "if . . . , then . . . " to describe implication, other word orders and phrases often arise in practice, as in the examples below. Let p, q, and r be the propositions

p = "the flag is set,"

q = "$I = 0$,"

r = "subroutine S is completed."

Translate each of the following propositions into symbols, using the letters p, q, and r and the logical connectives.

(a) If the flag is set, then $I = 0$.

(b) Subroutine S is completed if the flag is set.

(c) The flag is set if subroutine S is not completed.

(d) Whenever $I = 0$, the flag is set.

(e) Subroutine S is completed only if $I = 0$.

(f) Subroutine S is completed only if $I = 0$ or the flag is set.

Note the ambiguity in part (f); there are two different answers, each with its own claim to validity. Would punctuation help?

16. Consider the following propositions: r = "ODD$(N) = T$," m = "the output goes to the monitor," and p = "the output goes to the printer." Translate the following, as in Exercise 15.

(a) The output goes to the monitor if ODD$(N) = T$.

(b) The output goes to the printer whenever ODD$(N) = T$ is not true.

(c) ODD$(N) = T$ only if the output goes to the monitor.

(d) The output goes to the monitor if the output goes to the printer.

(e) ODD$(N) = T$ or the output goes to the monitor if the output goes to the printer.

17. Each of the following sentences expresses an implication. Rewrite each in the form "If p, then q."

(a) Touch those cookies if you want a spanking.

(b) Touch those cookies and you'll be sorry.

(c) You leave or I'll set the dog on you.

(d) I will if you will.

(e) I will go unless you stop that.

18. Express the contrapositive of each sentence in Exercise 17 in the form "If p, then q."

2.2 Propositional Calculus

We have two goals in this section. We want to develop a set of formal rules for analyzing and manipulating propositions, a sort of algebra of propositions similar in some ways to the algebra of numbers and the algebra of sets. We also want a mechanical way to calculate truth values of complicated propositions. It is the calculation aspect that has given the name "calculus" to the subject.

If a proposition is constructed from other propositions by using logical connectives, then its truth or falsity is completely determined by the truth values of the simpler propositions, together with the way the compound proposition is built up from them. Given propositions p and q, the truth values of the compound propositions $\neg p$, $p \wedge q$, $p \vee q$, $p \rightarrow q$, and $p \leftrightarrow q$ will be determined by the truth values of p and q. Since there are only four different combinations of truth values for p and q, we can simply give tables to describe the truth values of these compound propositions for all of the possible combinations.

One way to indicate truth values in a table would be to use the letters T and F. We have chosen instead to be consistent with the usage for Boolean variables in most computer languages and to use 1 for true and 0 for false.

The proposition $\neg p$ should be false when p is true and true when p is false. Thus our table for the connective $\neg$ is

p	$\neg p$
0	**1**
1	**0**

The column to the left of the vertical line lists the possible truth values of p. To the right of the line are the corresponding truth values for $\neg p$.

The truth table for $\wedge$ is

p	q	$p \wedge q$
0	0	**0**
0	1	**0**
1	0	**0**
1	1	**1**

Here the four possible combinations of truth values for p and q are listed to the left of the line, and the corresponding truth values of $p \wedge q$ are shown to the right. Note that $p \wedge q$ has truth value true exactly when both p *and* q are true. No surprise here!

As we explained at the beginning of §1.4, the use of "or" in the English language is somewhat ambiguous, but our use of $\vee$ will not be ambiguous. We define $\vee$ as follows:

p	q	$p \vee q$
0	0	**0**
0	1	**1**
1	0	**1**
1	1	**1**

Most people would agree with the truth-value assignments for the first three lines. The fourth line states that we regard $p \vee q$ to be true if both p and q are true. This is the "inclusive or," sometimes written "and/or." Thus $p \vee q$ is true if p is true or q is true *or both.* The "exclusive or," symbolized $\oplus$, means that one or the other is true but not both; see Exercise 13.

The **conditional implication** $p \rightarrow q$ means that the truth of p implies the truth of q. In other words, if p is true, then q must be true. The only way that this implication can fail is if p is true while q is false.

p	q	$p \to q$
0	0	**1**
0	1	**1**
1	0	**0**
1	1	**1**

The first two lines of the truth table for $p \to q$ may bother some people because it looks as if false propositions imply anything. In fact, we are simply defining the *compound proposition* $p \to q$ to be true if p is false. This usage of implication appears in ordinary English. Suppose that a politician promises "If I am elected, then taxes will be lower next year." If the politician is not elected, we would surely not regard him or her as a liar, no matter how the tax rates changed.

We will discuss the biconditional $p \leftrightarrow q$ after we introduce general truth tables.

A **truth table** for a compound proposition built up from propositions p, q, $r, \ldots$ is a table giving the truth values of the compound proposition in terms of the truth values of $p, q, r, \ldots$. We call $p, q, r, \ldots$ the **variables** of the table and of the compound proposition. One can determine the truth values of the compound proposition by determining the truth values of subpropositions working from the inside out, just as one calculates algebraic expressions like $(7 + 8 \cdot (9 - 4))^3$ from the inside out.

EXAMPLE 1

Now let's consider the truth table for the compound proposition $(p \wedge q) \vee \neg(p \to q)$. Note that there are still only four rows, because there are still only four distinct combinations of truth values for p and q.

column	1	2	3	4	5	6
	p	q	$p \wedge q$	$p \to q$	$\neg(p \to q)$	$(p \wedge q) \vee \neg(p \to q)$
	0	0	0	1	0	**0**
	0	1	0	1	0	**0**
	1	0	0	0	1	**1**
	1	1	1	1	0	**1**

The values in columns 3 and 4 are determined by the values in columns 1 and 2. The values in column 5 are determined by the values in column 4. The values in column 6 are determined by the values in columns 3 and 5. The sixth column gives the truth values of the complete compound proposition.

One can use a simpler truth table, with the same thought processes, by writing the truth values under the connectives, as follows:

	p	q	$(p \wedge q)$	$\vee$	$\neg$	$(p \to q)$
	0	0	0	**0**	0	1
	0	1	0	**0**	0	1
	1	0	0	**1**	1	0
	1	1	1	**1**	0	1
step	1	1	2	4	3	2

The truth values at each step are determined by the values at earlier steps. Moreover, they are listed directly under the last connectives used to determine the truth values. Thus the values at the second step were determined for $(p \wedge q)$ and $(p \to q)$ and placed under the connectives $\wedge$ and $\to$. The values at the third step for $\neg(p \to q)$ were determined by the values in the last column on the right and placed under the connective $\neg$. The values at the fourth step for the compound proposition $(p \wedge q) \vee \neg(p \to q)$ were determined by the values in the third and fifth columns and placed under the connective $\vee$. The column created in the fourth and last step gives the truth values of the compound proposition. ■

The simpler truth tables become more advantageous as the compound propositions get more complicated.

EXAMPLE 2

Here is the truth table for

$$(p \to q) \wedge [(q \wedge \neg r) \to (p \vee r)].$$

The rows of a truth table can, of course, be given in any order. We've chosen a systematic order for the truth combinations of p, q, and r partly to be sure we have listed them all.

	p	q	r	$(p \to q)$	$\wedge$	$[(q$	$\wedge \neg r)$	$\to$	$(p \vee r)]$
	0	0	0	1	**1**	0	1	1	0
	0	0	1	1	**1**	0	0	1	1
	0	1	0	1	**0**	1	1	0	0
	0	1	1	1	**1**	0	0	1	1
	1	0	0	0	**0**	0	1	1	1
	1	0	1	0	**0**	0	0	1	1
	1	1	0	1	**1**	1	1	1	1
	1	1	1	1	**1**	0	0	1	1
step	1	1	1	2	5	3	2	4	2

■

The **biconditional** $p \leftrightarrow q$ is defined by the truth table for $(p \to q) \wedge (q \to p)$:

	p	q	$(p \to q)$	$\wedge$	$(q \to p)$
	0	0	1	**1**	1
	0	1	1	**0**	0
	1	0	0	**0**	1
	1	1	1	**1**	1
step	1	1	2	3	2

That is,

p	q	$p \leftrightarrow q$
0	0	**1**
0	1	**0**
1	0	**0**
1	1	**1**

Thus $p \leftrightarrow q$ is true if both p and q are true or if both p and q are false. The following are English-language equivalents to $p \leftrightarrow q$: "p if and only if q," "p is a necessary and sufficient condition for q," and "p precisely if q."

It is worth emphasizing that the compound proposition $p \to q$ and its converse $q \to p$ are quite different; they have different truth tables.

An important class of compound propositions consists of those that are always true no matter what the truth values of the variables p, q, etc., are. Such a compound proposition is called a **tautology**. Why would we ever be interested in a proposition that is always true and hence is pretty boring? One answer is that we are going to be dealing with some rather complicated-looking propositions that we hope to show are true. We will show their truth by using other propositions that are known to be true always. We'll also use tautologies to analyze the validity of arguments. Just wait. We begin with a very simple tautology.

EXAMPLE 3

(a) The classical tautology is the compound proposition $p \to p$:

p	$p \to p$
0	**1**
1	**1**

If p is true, then p is true. Obvious and boring!

(b) The compound proposition $[p \wedge (p \to q)] \to q$ is a tautology:

	p	q	$[p \wedge$	$(p \to q)]$	$\to$	q
	0	0	0	1	**1**	0
	0	1	0	1	**1**	1
	1	0	0	0	**1**	0
	1	1	1	1	**1**	1
step	1	1	3	2	4	1

If p is true and $p \to q$ is true, then q is true. The logic here is pretty obvious, too, when stated so explicitly, but careless attempts at similar reasoning in practice can lead to trouble. We'll return to this topic in §2.3, where we discuss other potential misuses of logic.

(c) $\neg(p \vee q) \leftrightarrow (\neg p \wedge \neg q)$ is a tautology:

	p	q	$\neg$	$(p \vee q)$	$\leftrightarrow$	$(\neg p$	$\wedge$	$\neg q)$
	0	0	1	0	**1**	1	1	1
	0	1	0	1	**1**	1	0	0
	1	0	0	1	**1**	0	0	1
	1	1	0	1	**1**	0	0	0
step	1	1	3	2	4	2	3	2

The proposition $\neg(p \vee q)$ is true if and only if $\neg p \vee \neg q$ is true. Think about this, or see Example 5(a). ■

A compound proposition that is always false is called a **contradiction**. Clearly, a compound proposition P is a contradiction if and only if $\neg P$ is a tautology.

EXAMPLE 4

The classical contradiction is the compound proposition $p \wedge \neg p$:

p	p	$\wedge$	$\neg p$
0	0	**0**	1
1	1	**0**	0

In other words, one cannot have p and $\neg p$ both true. ■

Two compound propositions P and Q are regarded as **logically equivalent** if they have the same truth values for all choices of truth values of the variables p, q, etc. In other words, the final columns of their truth tables are the same. When this occurs, we write $P \Longleftrightarrow Q$. Since the table for $P \leftrightarrow Q$ has truth values true precisely where the truth values of P and Q agree, we see that

$$P \Longleftrightarrow Q \textit{ if and only if } P \leftrightarrow Q \textit{ is a tautology}.$$

Observing that $P \Longleftrightarrow Q$ will be especially useful in cases where P and Q look quite different from each other. See, for instance, the formulas in Table 1.

TABLE 1 Logical Equivalences

Law	Name
1. $\neg\neg p \Longleftrightarrow p$	double negation
2a. $(p \vee q) \Longleftrightarrow (q \vee p)$	commutative laws
b. $(p \wedge q) \Longleftrightarrow (q \wedge p)$	
c. $(p \leftrightarrow q) \Longleftrightarrow (q \leftrightarrow p)$	
3a. $[(p \vee q) \vee r] \Longleftrightarrow [p \vee (q \vee r)]$	associative laws
b. $[(p \wedge q) \wedge r] \Longleftrightarrow [p \wedge (q \wedge r)]$	
4a. $[p \vee (q \wedge r)] \Longleftrightarrow [(p \vee q) \wedge (p \vee r)]$	distributive laws
b. $[p \wedge (q \vee r)] \Longleftrightarrow [(p \wedge q) \vee (p \wedge r)]$	
5a. $(p \vee p) \Longleftrightarrow p$	idempotent laws
b. $(p \wedge p) \Longleftrightarrow p$	
6a. $(p \vee \mathbf{0}) \Longleftrightarrow p$	identity laws
b. $(p \vee \mathbf{1}) \Longleftrightarrow \mathbf{1}$	
c. $(p \wedge \mathbf{0}) \Longleftrightarrow \mathbf{0}$	
d. $(p \wedge \mathbf{1}) \Longleftrightarrow p$	
7a. $(p \vee \neg p) \Longleftrightarrow \mathbf{1}$	
b. $(p \wedge \neg p) \Longleftrightarrow \mathbf{0}$	
8a. $\neg(p \vee q) \Longleftrightarrow (\neg p \wedge \neg q)$	De Morgan laws
b. $\neg(p \wedge q) \Longleftrightarrow (\neg p \vee \neg q)$	
c. $(p \vee q) \Longleftrightarrow \neg(\neg p \wedge \neg q)$	
d. $(p \wedge q) \Longleftrightarrow \neg(\neg p \vee \neg q)$	
9. $(p \rightarrow q) \Longleftrightarrow (\neg q \rightarrow \neg p)$	contrapositive
10a. $(p \rightarrow q) \Longleftrightarrow (\neg p \vee q)$	implication
b. $(p \rightarrow q) \Longleftrightarrow \neg(p \wedge \neg q)$	
11a. $(p \vee q) \Longleftrightarrow (\neg p \rightarrow q)$	
b. $(p \wedge q) \Longleftrightarrow \neg(p \rightarrow \neg q)$	
12a. $[(p \rightarrow r) \wedge (q \rightarrow r)] \Longleftrightarrow [(p \vee q) \rightarrow r]$	
b. $[(p \rightarrow q) \wedge (p \rightarrow r)] \Longleftrightarrow [p \rightarrow (q \wedge r)]$	
13. $(p \leftrightarrow q) \Longleftrightarrow [(p \rightarrow q) \wedge (q \rightarrow p)]$	equivalence
14. $[(p \wedge q) \rightarrow r] \Longleftrightarrow [p \rightarrow (q \rightarrow r)]$	exportation law
15. $(p \rightarrow q) \Longleftrightarrow [(p \wedge \neg q) \rightarrow \mathbf{0}]$	reductio ad absurdum

In this table, **1** represents any tautology and **0** represents any contradiction.

EXAMPLE 5

(a) In view of Example 3(c), the compound propositions $\neg(p \vee q)$ and $\neg p \wedge \neg q$ are logically equivalent. That is, $\neg(p \vee q) \Longleftrightarrow (\neg p \wedge \neg q)$; see 8a in Table 1. To say that I will not go for a walk *or* watch television is the same as saying that I will not go for a walk *and* I will not watch television.

(b) The very nature of the connectives $\vee$ and $\wedge$ suggests that $p \vee q \Longleftrightarrow q \vee p$ and $p \wedge q \Longleftrightarrow q \wedge p$. Of course, one can verify these assertions by showing that $(p \vee q) \leftrightarrow (q \vee p)$ and $(p \wedge q) \leftrightarrow (q \wedge p)$ are tautologies. ■

It is worth stressing the difference between $\leftrightarrow$ and $\Longleftrightarrow$. The expression "$P \leftrightarrow Q$" simply represents some compound proposition that might or might not be a tautology. The expression "$P \Longleftrightarrow Q$" is an *assertion about propositions*, namely that P and Q are logically equivalent, i.e., that $P \leftrightarrow Q$ *is* a tautology.

Table 1 lists a number of logical equivalences selected for their usefulness. To obtain a table of tautologies, simply replace each $\Longleftrightarrow$ by $\leftrightarrow$. These tautologies can all be verified by truth tables. However, most of them should be intuitively reasonable.

Many of the entries in the table have names, which we have given, but there is no need to memorize most of the names. The logical equivalences 2, 3, and 4 have familiar names. Equivalences 8, the De Morgan laws, and 9, the **contrapositive** rule, come up often enough to make their names worth remembering.

Notice that in the De Morgan laws the proposition on one side of the $\Longleftrightarrow$ has an $\wedge$, while the one on the other side has an $\vee$. We will see in §2.5 that we can use the De Morgan laws to replace a given proposition by a logically equivalent one in which some or all of the $\wedge$'s have been converted into $\vee$'s, or vice versa.

Given compound propositions P and Q, we say that P **logically implies** Q in case Q has truth value true whenever P has truth value true. We write $P \Longrightarrow Q$ when this occurs. Thus

$P \Longrightarrow Q$ *if and only if the compound proposition* $P \rightarrow Q$ *is a tautology.*

Equivalently, $P \Longrightarrow Q$ means that P and Q never simultaneously have the truth values 1 and 0, respectively; when P is true, Q is true, and when Q is false, P is false.

EXAMPLE 6

(a) We have $[p \wedge (p \rightarrow q)] \Longrightarrow q$, since $[p \wedge (p \rightarrow q)] \rightarrow q$ is a tautology by Example 3(b).

(b) The statement $(A \wedge B) \Longrightarrow C$ means that $(A \wedge B) \rightarrow C$ is a tautology. Since $(A \wedge B) \rightarrow C \Longleftrightarrow A \rightarrow (B \rightarrow C)$ by Rule 14 [exportation], $(A \wedge B) \rightarrow C$ is a tautology if and only if $A \rightarrow (B \rightarrow C)$ is a tautology, i.e., if and only if $A \Longrightarrow (B \rightarrow C)$. Thus the statements $(A \wedge B) \Longrightarrow C$ and $A \Longrightarrow (B \rightarrow C)$ mean the same thing. ■

In Table 2 we list some useful logical implications. Each entry becomes a tautology if $\Longrightarrow$ is replaced by $\rightarrow$. As with Table 1, many of the implications have names that need not be memorized.

To check the logical implication $P \Longrightarrow Q$, it is only necessary to look for rows of the truth table where P is true and Q is false. If there are any, then $P \Longrightarrow Q$ is not true. Otherwise, $P \Longrightarrow Q$ is true. Thus we can ignore the rows in which P is false, as well as the rows in which Q is true.

TABLE 2 Logical Implications

16.	$p \Longrightarrow (p \vee q)$	addition
17.	$(p \wedge q) \Longrightarrow p$	simplification
18.	$(p \rightarrow \mathbf{0}) \Longrightarrow \neg p$	absurdity
19.	$[p \wedge (p \rightarrow q)] \Longrightarrow q$	modus ponens
20.	$[(p \rightarrow q) \wedge \neg q] \Longrightarrow \neg p$	modus tollens
21.	$[(p \vee q) \wedge \neg p] \Longrightarrow q$	disjunctive syllogism
22.	$p \Longrightarrow [q \rightarrow (p \wedge q)]$	
23.	$[(p \leftrightarrow q) \wedge (q \leftrightarrow r)] \Longrightarrow (p \leftrightarrow r)$	transitivity of $\leftrightarrow$
24.	$[(p \rightarrow q) \wedge (q \rightarrow r)] \Longrightarrow (p \rightarrow r)$	transitivity of $\rightarrow$ *or* hypothetical syllogism
25a.	$(p \rightarrow q) \Longrightarrow [(p \vee r) \rightarrow (q \vee r)]$	
b.	$(p \rightarrow q) \Longrightarrow [(p \wedge r) \rightarrow (q \wedge r)]$	
c.	$(p \rightarrow q) \Longrightarrow [(q \rightarrow r) \rightarrow (p \rightarrow r)]$	
26a.	$[(p \rightarrow q) \wedge (r \rightarrow s)] \Longrightarrow [(p \vee r) \rightarrow (q \vee s)]$	constructive dilemmas
b.	$[(p \rightarrow q) \wedge (r \rightarrow s)] \Longrightarrow [(p \wedge r) \rightarrow (q \wedge s)]$	
27a.	$[(p \rightarrow q) \wedge (r \rightarrow s)] \Longrightarrow [(\neg q \vee \neg s) \rightarrow (\neg p \vee \neg r)]$	destructive dilemmas
b.	$[(p \rightarrow q) \wedge (r \rightarrow s)] \Longrightarrow [(\neg q \wedge \neg s) \rightarrow (\neg p \wedge \neg r)]$	

EXAMPLE 7

(a) We verify the implication $(p \wedge q) \Longrightarrow p$. We need only consider the case where $p \wedge q$ is true, i.e., both p and q are true. Thus we consider the cut-down table

p	q	$(p \wedge q)$	$\rightarrow$	p
1	1	1	1	1

(b) To verify $\neg p \Longrightarrow (p \rightarrow q)$, we need only look at the cases in which $\neg p$ is true, i.e., in which p is false. The reduced table is

p	q	$\neg p$	$\rightarrow$	$(p \rightarrow q)$
0	0	1	**1**	1
0	1	1	**1**	1

Even quicker, we could just consider the case in which $p \rightarrow q$ is false, i.e., the case p true and q false. The only row is

p	q	$\neg p$	$\rightarrow$	$(p \rightarrow q)$
1	0	0	**1**	0

(c) We verify the implication 26a. The full truth table would require 16 rows. However, we need only consider the cases where the implication $(p \vee r) \rightarrow (q \vee s)$ might be false. Thus it is enough to look at the cases for which $q \vee s$ is false, that is, with both q and s false.

	p	q	r	s	$[(p \rightarrow q)$	$\wedge$	$(r \rightarrow s)]$	$\rightarrow$	$[(p \vee r)$	$\rightarrow$	$(q \vee s)]$
	0	0	0	0	1	1	1	**1**	0	1	0
	0	0	1	0	1	0	0	**1**	1	0	0
	1	0	0	0	0	0	1	**1**	1	0	0
	1	0	1	0	0	0	0	**1**	1	0	0
step	1	1	1	1	2	3	2	4	2	3	2

■

Just from the definition of $\Longleftrightarrow$ we can see that if $P \Longleftrightarrow Q$ and $Q \Longleftrightarrow R$, then $P \Longleftrightarrow R$ and the three propositions P, Q, and R are logically equivalent to each other. If we have a chain $P_1 \Longleftrightarrow P_2 \Longleftrightarrow \cdots \Longleftrightarrow P_n$, then all the propositions P_i are equivalent. Similarly [Exercise 25(a)], if $P \Longrightarrow Q$ and $Q \Longrightarrow R$, then $P \Longrightarrow R$. The symbols $\Longleftrightarrow$ and $\Longrightarrow$ behave somewhat analogously to the symbols $=$ and $\geq$ in algebra.

EXAMPLE 8

You may have noticed the similarity between the laws for sets on page 25 and the laws for logic on page 62. In fact, we presented the laws of algebra of sets in Chapter 1 in the hope that they would make the laws of logic seem less strange. In any case, there is a close connection between the subjects. In fact, the laws of sets are consequences of the corresponding laws of logic. First, the correspondence between the operations $\cup$, $\cap$ and set complementation and the logical connectives $\vee$, $\wedge$, and $\neg$ can be read from the following definitions:

$$A \cup B = \{x : x \text{ is in } A \quad \vee \quad x \text{ is in } B\},$$

$$A \cap B = \{x : x \text{ is in } A \quad \wedge \quad x \text{ is in } B\},$$

$$A^c = \{x : \ \neg \ (x \text{ is in } A)\}.$$

The laws of sets 1–5, 6, 7, and 9 on page 25 correspond to the laws of logic 2–6, 1, 7, and 8, respectively, on page 62.

We illustrate the connection by showing why the De Morgan law $(A \cup B)^c = A^c \cap B^c$ in set theory follows from the De Morgan law $\neg(p \vee q) \Longleftrightarrow \neg p \wedge \neg q$ in logic. Fix an element x, and let $p =$ "x is in A" and $q =$ "x is in B." Then

$$x \text{ is in } (A \cup B)^c \Longleftrightarrow \neg\, (x \text{ is in } A \cup B) \Longleftrightarrow \neg\, (x \text{ is in } A \text{ or } x \text{ is in } B)$$
$$\Longleftrightarrow \neg\, (p \vee q) \Longleftrightarrow \neg p \wedge \neg q \Longleftrightarrow \neg\, (x \text{ is in } A) \text{ and } \neg\, (x \text{ is in } B)$$
$$\Longleftrightarrow (x \text{ is in } A^c) \text{ and } (x \text{ is in } B^c) \Longleftrightarrow x \text{ is in } A^c \cap B^c.$$

This chain of double implications is true for all x in the universe, so the sets $(A \cup B)^c$ and $A^c \cap B^c$ are equal. ■

Exercises 2.2

1. Give the converse and contrapositive for each of the following propositions.
   (a) $p \to (q \wedge r)$.
   (b) If $x + y = 1$, then $x^2 + y^2 \geq 1$.
   (c) If 2 + 2 = 4, then 3 + 3 = 8.
2. Consider the proposition "if $x > 0$, then $x^2 > 0$." Here $x \in \mathbb{R}$.
   (a) Give the converse and contrapositive of the proposition.
   (b) Which of the following are true propositions: the original proposition, its converse, its contrapositive?
3. Consider the following propositions:

   $p \to q$, $\quad \neg p \to \neg q$, $\quad q \to p$, $\quad \neg q \to \neg p$,
   $q \wedge \neg p$, $\quad \neg p \vee q$, $\quad \neg q \vee p$, $\quad p \wedge \neg q$.

   (a) Which proposition is the converse of $p \to q$?
   (b) Which proposition is the contrapositive of $p \to q$?
   (c) Which propositions are logically equivalent to $p \to q$?
4. Determine the truth values of the following compound propositions.
   (a) If 2 + 2 = 4, then 2 + 4 = 8.
   (b) If 2 + 2 = 5, then 2 + 4 = 8.
   (c) If 2 + 2 = 4, then 2 + 4 = 6.
   (d) If 2 + 2 = 5, then 2 + 4 = 6.
   (e) If the earth is flat, then Julius Caesar was the first president of the United States.
   (f) If the earth is flat, then George Washington was the first president of the United States.
   (g) If George Washington was the first president of the United States, then the earth is flat.
   (h) If George Washington was the first president of the United States, then 2 + 2 = 4.
5. Suppose that $p \to q$ is known to be false. Give the truth values for
   (a) $p \wedge q$ (b) $p \vee q$ (c) $q \to p$
6. Construct truth tables for
   (a) $p \wedge \neg p$ (b) $p \vee \neg p$
   (c) $p \leftrightarrow \neg p$ (d) $\neg\neg p$
7. Construct truth tables for
   (a) $\neg(p \wedge q)$ (b) $\neg(p \vee q)$
   (c) $\neg p \wedge \neg q$ (d) $\neg p \vee \neg q$
8. Construct the truth table for $(p \to q) \to [(p \vee \neg q) \to (p \vee q)]$.
9. Construct the truth table for $[(p \vee q) \wedge r] \to (p \wedge \neg q)$.
10. Construct the truth table for $[(p \leftrightarrow q) \vee (p \to r)] \to (\neg q \wedge p)$.
11. Construct truth tables for
    (a) $\neg(p \vee q) \to r$ (b) $\neg((p \vee q) \to r)$
    This exercise shows that one must be careful with parentheses. We will discuss this issue further in §7.2, particularly in Exercise 15 of that section.
12. In which of the following statements is the "or" an "inclusive or"?
    (a) Choice of soup or salad.
    (b) To enter the university, a student must have taken a year of chemistry or physics in high school.
    (c) Publish or perish.
    (d) Experience with C++ or Java is desirable.
    (e) The task will be completed on Thursday or Friday.
    (f) Discounts are available to persons under 20 or over 60.
    (g) No fishing or hunting allowed.
    (h) The school will not be open in July or August.
13. The **exclusive or** connective $\oplus$ [the computer scientists' **XOR**] is defined by the truth table

p	q	$p \oplus q$
0	0	**0**
0	1	**1**
1	0	**1**
1	1	**0**

    (a) Show that $p \oplus q$ has the same truth table as $\neg(p \leftrightarrow q)$.
    (b) Construct a truth table for $p \oplus p$.
    (c) Construct a truth table for $(p \oplus q) \oplus r$.
    (d) Construct a truth table for $(p \oplus p) \oplus p$.
14. Sometimes it is natural to write truth tables with more rows than absolutely necessary.
    (a) Show that $(p \wedge q) \vee (p \wedge \neg q)$ is logically equivalent to p. [Here the table for $(p \wedge q) \vee (p \wedge \neg q)$ seems to require at least 4 rows, although p can be described by a 2-row table.]
    (b) Write an 8-row (p, q, r)-table for $p \vee r$.
    (c) Write a 4-row (p, q)-table for $\neg q$.

15. We could define a logical connective **unless** with symbol $\sim$ by saying that "p unless q" is false if both p and q are false and is true otherwise. [Recall Example 5(d) on page 52.]

(a) Give the truth table for $\sim$.

(b) Identify $\sim$ as a logical connective that we have already defined.

(c) Show that "p unless q" is logically equivalent to "if not p, then q" and also to "q unless p."

16. (a) Write a compound proposition that is true when exactly one of the three propositions p, q, and r is true.

(b) Write a compound proposition that is true when exactly two of the three propositions p, q, and r are true.

17. (a) Rewrite parts (g) and (h) of Exercise 12 using De Morgan's law.

(b) Does De Morgan's law hold using the exclusive or $\oplus$ instead of $\vee$? Discuss.

18. Prove or disprove.

(a) $[p \to (q \to r)] \Longleftrightarrow [(p \to q) \to (p \to r)]$

(b) $[p \oplus (q \to r)] \Longleftrightarrow [(p \oplus q) \to (p \oplus r)]$

(c) $[(p \to q) \to r] \Longleftrightarrow [p \to (q \to r)]$

(d) $[(p \leftrightarrow q) \leftrightarrow r] \Longleftrightarrow [p \leftrightarrow (q \leftrightarrow r)]$

(e) $(p \oplus q) \oplus r \Longleftrightarrow p \oplus (q \oplus r)$

19. Verify the following logical equivalences using truth tables.

(a) Rule 12a (b) The exportation law, rule 14

(c) Rule 15

20. Verify the following logical implications using truth tables.

(a) Modus tollens, rule 20

(b) Disjunctive syllogism, rule 21

21. Verify the following logical implications using truth tables and shortcuts as in Example 7.

(a) Rule 25b (b) Rule 25c (c) Rule 26b

22. Prove or disprove the following. Don't forget that only *one* line of the truth table is needed to show that a proposition is *not* a tautology.

(a) $(q \to p) \Longleftrightarrow (p \wedge q)$ (b) $(p \wedge \neg q) \Longrightarrow (p \to q)$

(c) $(p \wedge q) \Longrightarrow (p \vee q)$

23. A logician told her son "If you don't finish your dinner, you will not get to stay up and watch TV." He finished his dinner and then was sent straight to bed. Discuss.

24. Consider the statement "Concrete does not grow if you do not water it."

(a) Give the contrapositive.

(b) Give the converse.

(c) Give the converse of the contrapositive.

(d) Which among the original statement and the ones in parts (a), (b), and (c) are true?

25. (a) Show that if $A \Longrightarrow B$ and $B \Longrightarrow C$, then $A \Longrightarrow C$.

(b) Show that if $P \Longleftrightarrow Q$, $Q \Longrightarrow R$, and $R \Longleftrightarrow S$, then $P \Longrightarrow S$.

(c) Show that if $P \Longrightarrow Q$, $Q \Longrightarrow R$, and $R \Longrightarrow P$, then $P \Longleftrightarrow Q$.

2.3 Getting Started with Proofs

The constant emphasis on logic and proofs is what sets mathematics apart from other pursuits. The proofs used in everyday working mathematics are based on the logical framework that we have introduced using the propositional calculus. In §2.5 we will develop a concept of formal proof, using some of the symbolism of the propositional calculus, and in §2.6 we will apply the formal development to analyze typical practical arguments. In this section, we address some common concerns that students have when they are first learning how to construct their own proofs, and we look at two proof methods. Section 2.4 will continue the discussion of methods of proof.

You may have asked yourself why we need proofs anyway. If something's true, can't we just use it without seeing a proof? We could, if there were a big book that contained all the facts we'd ever need, and if we were willing to take their truth on faith. Although it's not a bad idea sometimes to accept a statement without proof, it's generally preferable to have at least some idea of why the statement is true—of how it could be proved—even if you're just willing to believe it. The key reason that we look so hard at proofs in this book, though, is that proofs are the only way to justify statements that are not obvious. Whether it's a matter of convincing ourselves that we've considered all possibilities or of convincing somebody else that our conclusions are inescapable, proofs provide the justification, the logical framework that reasoning requires.

You are already familiar with proofs, of course. For example, suppose that you are asked to solve the quadratic equation $x^2 + x - 12 = 0$ and to check your answer.

By some means, probably by factoring or using the quadratic formula, you find that the solutions are 3 and -4. When you check your answers, you are verifying, i.e., *proving* a statement, that $x = 3$ and $x = -4$ are solutions of the equation. In discrete mathematics and most advanced areas of mathematics, more abstract assertions need to be dealt with. This is where more abstract proofs and methods of proof come in.

Before embarking on a proof, we need a statement to prove. How do we know what is true? In mathematics and computer science it is often not clear what is true. This fact of life is why many exercises in this book ask you to "Prove or disprove." Such exercises are harder and require extra work, since one has to first decide whether the statements are true, but they give good training in how proofs are actually created and used in practice.

In general, one begins with a more or less well-formulated question and wonders whether the result is true. At the beginning, the key is to look at examples. If any example contradicts the general assertion, then you're done: The general assertion is not true and your counterexample is sufficient to prove this fact. If the examination of many examples seems to confirm the general result, then it is time to try to figure out why it holds. What are the essential features of each example? If the question is somewhat difficult, this can turn into a back-and-forth process. If you can't find a counterexample, try to construct a proof. If this doesn't work, look at more examples. And so on.

This section contains exercises along the way, as well as at the end. To get the most benefit from the section, it's essential that you work the exercises in the text as you come to them. Look at each answer in the back of the book only after you have thought carefully about the exercise. Here, right now, are some exercises to practice on.

Two Exercises

1. For each statement, make a serious effort to decide whether it is true or false. If it is false, provide a counterexample. Otherwise, state that it *seems* to be true and give some supporting examples.

(a) The sum of three consecutive integers is always divisible by 3.

(b) The sum of four consecutive integers is always divisible by 4.

(c) The sum of any five consecutive integers is divisible by 5.

2. Repeat Exercise 1 for the following.

(a) If the product of positive integers m and n is 1,000,000, then 10 divides either m or n.

(b) If p is prime, then so is $p^2 - 2$. [The primes less than 1000 are listed on page 6.]

Now how does one start to write a proof? How do you translate a problem into mathematics? What is appropriate notation, and how do you choose it? The translation problem is part of figuring out what is to be proved. Translation often consists just of writing out in symbols what various words in the problem mean, and this is where you get to choose your notation. If the proof is of a general statement, then the notation needs to be general, though sometimes we can give a descriptive proof without using technical notation at all, as in many of the graph theory proofs in Chapter 6. In general, though, mathematical notation is unavoidable. The choice of good suggestive notation is essential. Avoid overly complicated notation. For example, subscripts are often necessary, but they are clutter that should be avoided when possible.

Experience shows that if you get started right on a proof, then the rest will often go easily. By the time you have formulated the question properly and introduced useful notation, you understand the problem better and may well see how to proceed.

Two More Exercises

3. We are asked to prove a statement about two positive odd integers. Which of the following are reasonable beginnings of a proof?

(a) Let m and n be positive odd integers. Then m can be written as $2k + 1$ and n can be written as $2j + 1$ for nonnegative integers k and j.

(b) The odd integers can be written as $2k + 1$ and $2j + 1$ for nonnegative integers k and j.

(c) Let m and n be positive odd integers. Then m can be written as $2k + 1$ for a nonnegative integer k. For the same reason, n can be written as $2k + 1$.

(d) Let m and n be positive odd integers. Then $m = 2j + 1$ and $n = 2k + 1$ for nonnegative integers j and k.

(e) The odd integers can be written as $2x + 1$ and $2y + 1$ for nonnegative integers x and y.

(f) Some positive odd integers are $7, 17, 25, 39$, and 73. We'll check the result for all pairs of these numbers [there are ten of them]. That ought to be convincing enough.

4. We are asked to prove a statement involving a rational number and an irrational number. Which of the following are reasonable beginnings of a proof?

(a) Let x be a rational number, and let y be an irrational number.

(b) Let x be a rational number, and let y be an irrational number. The number x can be written as $\frac{p}{q}$, where $p, q \in \mathbb{Z}$ and $q \geq 1$.

(c) Let p be a rational number, and let q be an irrational number.

(d) Consider a rational number $\frac{p}{q}$ and an irrational number x.

(e) Let $x = \frac{p}{q}$ be a rational number, and let $y \neq \frac{p}{q}$ be an irrational number.

EXAMPLE 1

We prove that the product of two odd numbers is always an odd number. [We'll start out as in Exercise 3(b).] The odd integers can be written as $2k + 1$ and $2j + 1$ for integers k and j. Their product is

$$(2k + 1)(2j + 1) = 4kj + 2k + 2j + 1 = 2(2kj + k + j) + 1.$$

This number has the form $2m + 1$, where m is an integer, so the product is an odd number. ■

Now let's start out wrong.

Silly Conjecture For positive odd integers m and n, we have $\gcd(m, n) = 1$.

We can test this conjecture using the positive odd integers $7, 17, 25, 39$, and 73 in Exercise 3(f). In fact, we find that $\gcd(m, n) = 1$ for all ten pairs of distinct integers from this set. Does this information confirm the conjecture? Indeed not. Moreover, in this case it is easy to find counterexamples. For instance, we have $\gcd(9, 15) = 3$. The fact that the counterexample was easy to find is a hint as to why this conjecture was silly. The real reason it was silly is that we never had any reason to expect the conjecture to be true. There is nothing in the structure of odd numbers [which simply tells us that they aren't divisible by 2] to suggest this conjecture.

The preceding paragraph illustrates a general fact:

To prove a result, it is *not* sufficient to give some examples.

Sometimes the selected examples are misleading. Moreover, even if the general result is true, it is often not clear from a few examples why it is true. A general proof will clarify what it is about the hypotheses that cause the conclusion to hold.

EXAMPLE 2

Let's return to the question in Exercise 2(a): If the product mn of positive integers is 1,000,000, must 10 divide either m or n? Testing examples could have led you to believe that this is true. In fact, it is false, but there is only one pair of integers m and n that can serve as a counterexample. One of them needs to be $2^6 = 64$ and the other needs to be $5^6 = 15{,}625$.

There's more than trial and error to this problem. Here's how we solved it. As noted in Theorem 2 on page 9, every integer can be factored into a product of primes, and in essentially only one way. Since $1{,}000{,}000 = 10^6 = (2 \cdot 5)^6 = 2^6 \cdot 5^6$, the factors of 1,000,000 must have the form $2^a \cdot 5^b$, where a and b are nonnegative integers. The factor will be a multiple of 10 if both a and b are positive, but otherwise it will not be. To get a counterexample to the original question, we took one of the numbers m and n to be just a power of 2 and the other to be just a power of 5. ■

To close this section, we illustrate two methods of proof that we will discuss again in the next section. Sometimes the easiest way to handle hypotheses is to break them into separate possibilities. A proof organized in this way is called a **proof by cases**.

EXAMPLE 3

We prove that $n^2 - 2$ is not divisible by 5 for any positive integer n.

We will write out the proof the way we did it at first, before we had a chance to reflect on it and polish it; see Example 4 below for a refined version. Divisibility by 5 reminds us of the Division Algorithm on page 13 and suggests that we may have five pertinent cases, depending on what the remainder is after n is divided by 5. In other words, the argument may depend on whether n has the form $5k$, $5k + 1$, $5k + 2$, $5k + 3$, or $5k + 4$. We will write out all five cases, since this is the only way to be sure we have a proof.

Case (i). $n = 5k$. Then $n^2 - 2 = 25k^2 - 2$. Since 5 divides $25k^2$, but 5 does not divide -2, 5 does not divide $n^2 - 2$. [We are using here the little fact that if $m = pl + j$, then p divides m if and only if it divides j. See Exercise 12.]

Case (ii). $n = 5k+1$. Then $n^2-2 = 25k^2+10k-1$. Since 5 divides $25k^2+10k$ but does not divide -1, 5 does not divide $n^2 - 2$.

Case (iii). $n = 5k+2$. Then $n^2-2 = 25k^2+20k+2$. Since 5 divides $25k^2+20k$ but not 2, 5 does not divide $n^2 - 2$.

Case (iv). $n = 5k + 3$. Then $n^2 - 2 = 25k^2 + 30k + 7$ and, as before, 5 does not divide $n^2 - 2$.

Case (v). $n = 5k + 4$. Then $n^2 - 2 = 25k^2 + 40k + 14$ is not divisible by 5. ■

In the next example, we will give a more polished proof of the result above. Proofs in books are generally just presented in finished form, after the authors have spent some time perfecting them. Such presentations have the advantage that they get to the heart of the matter, without distractions, and focus on what makes the proof work. They are also efficient, since the polished proofs are often shorter than what the authors thought of first. They have the disadvantage, though, that it can be harder for students to use polished proofs for models to see how proofs are constructed. Some students get discouraged too easily, because their first efforts don't look at all like the proofs in the books. Be assured that our wastebaskets are full of scratchwork in which we struggled to fully understand the proofs. Serious students will follow our example and think through and revise their proofs before submitting the final product. Don't turn in your scratchwork!

It is worth emphasizing at this point that we expect our readers to follow and understand far more complex proofs than we expect them to create or construct. In fact, we're professional mathematicians, yet we've read and understood much more complex proofs than we've created ourselves.

EXAMPLE 4

We return to the assertion that $n^2 - 2$ is not divisible by 5 for all positive integers n.

(a) We could have streamlined our proof in Example 3 by doing all the cases at once [at least at the beginning], since each case depends on the remainder obtained when n is divided by 5. By the Division Algorithm, we can write $n = 5k + r$, where $r = 0, 1, 2, 3$, or 4. Then $n^2 - 2 = 25k^2 + 10rk + r^2 - 2$. Since $25k^2 + 10rk$ is divisible by 5, it suffices to show that $r^2 - 2$ is *not* divisible by 5. Now we are reduced to five cases involving r. [They are so simple that it would just be clutter for us to spell out the cases formally as we did in Example 3.] Substituting $r = 0, 1, 2, 3$, and 4, we see that $r^2 - 2$ is either $-2, -1, 2, 7$, or 14. In each case, $r^2 - 2$ is not divisible by 5, so the proof is complete.

The paragraph above, without the first sentence and the bracketed sentence, is an excellent short proof. It gets right to the point. If this were turned in to us, it would get an A and the remark "Nice proof!"

(b) The assertion that $n^2 - 2$ is not divisible by 5 is actually a statement in modular arithmetic. In the language of §3.5, it says that n^2 is never equal to 2 modulo 5, i.e., $n^2 \not\equiv 2 \pmod 5$ for all n. See Exercise 22 on page 126. ■

Here are some exercises that are similar to the last two examples, but easier.

Three More Exercises

5. Prove that 3 does not divide $n^2 - 2$ for integers $n \geq 1$.
6. Prove that 2 does not divide $n^2 - 2$ if the integer n is odd.
7. (a) Use the method of the last two examples and Exercise 5 to try to prove that 7 does not divide $n^2 - 2$ for all integers $n \geq 1$. What goes wrong?
   (b) Use your work in part (a) to find a counterexample to the claim.
   (c) Use your work in part (a) to find two counterexamples that are odd numbers.

Example 4 and Exercises 5 and 6 together tell us the following: For n odd, $n^2 - 2$ is *not* divisible by 2, 3, or 5. This illustrates why it may have taken you a little time searching to answer the question, in Exercise 2(b), of whether $p^2 - 2$ is prime for all primes p. The smallest counterexample to that assertion is $11^2 - 2 = 119 = 17 \cdot 7$.

Sometimes the only [or best] way to show why some conclusion holds is to assume that it doesn't. If this leads to nonsense "contrary" to reality or a result "contrary" to the hypotheses, then we have reached a contradiction. Such a proof is called a **proof by contradiction**.

EXAMPLE 5

We prove by contradiction that there are infinitely many primes. Thus assume that there are finitely many primes, say k of them. We write them as $p_1, p_2, \ldots, p_k$ so that $p_1 = 2$, $p_2 = 3$, etc. Over 2000 years ago, Euclid had the clever idea to consider $n = 1 + p_1 p_2 \cdots p_k$. Since $n > p_j$ for all $j = 1, 2, \ldots, k$, the integer n is bigger than all primes, so it is itself not prime. However, n can be written as a product of primes, as noted in Theorem 2 on page 9. Therefore, at least one of the p_j's must divide n. Since each p_j divides $n - 1$, at least one p_j divides both $n - 1$ and n, which is impossible. Indeed, if p_j divides both $n - 1$ and n, then it divides their difference, 1, which is absurd. ■

Proofs by contradiction can be especially natural when the desired conclusion is that some statement be *not* true. The proof then starts by assuming that the statement *is* true and aims for a contradiction. Negations of some fairly straightforward statements are often quite complicated to write out—indeed §13.2 is largely concerned with how to do so—and there can be a definite advantage to avoiding negations. The next exercise illustrates what we mean.

A number is rational if it's of the form $\frac{p}{q}$ with p and q integers and $q > 0$, and a real number is irrational if it's not rational. What form does an irrational number have? All we can say is what it *doesn't* look like, and that's not much help. To prove that some number is irrational, it seems promising to suppose that it is rational—a condition that we can describe—and then to try to get a contradiction.

Yet Another Exercise

8. Prove that the sum of a rational number and an irrational number is always irrational.

This section started out with a discussion of general proof-building strategy and has included two particular plans of attack: proof by cases and proof by contradiction. The next section looks some more at these two approaches and introduces others as well. You will see more examples of all the material in this section, but to fully appreciate them you should try to write some more proofs yourself now.

Exercises 2.3

9. Prove that the product of two even integers is a multiple of 4.

10. Prove that the product of an even integer and an odd integer is even.

11. Prove that 5 does not divide $n^2 - 3$ for integers $n \geq 1$.

12. Suppose that integers m, l, j, and p satisfy $m = pl + j$. Show that p divides m if and only if p divides j.

13. (a) Prove that $n^4 - n^2$ is divisible by 3 for all $n \in \mathbb{N}$.

(b) Prove that $n^4 - n^2$ is even for all $n \in \mathbb{N}$.

(c) Prove that $n^4 - n^2$ is divisible by 6 for all $n \in \mathbb{N}$.

14. (a) Prove that the product of a nonzero rational number and an irrational number is an irrational number.

(b) Why is the word "nonzero" in the hypothesis of part (a)?

15. Prove or disprove.

(a) If 3 divides the product mn of positive integers m and n, then $3|m$ or $3|n$.

(b) If 4 divides the product mn of positive integers m and n, then $4|m$ or $4|n$.

(c) Why are things different in parts (a) and (b)? Can you conjecture a general fact? That is, for which positive integers d is it true that if d divides the product mn, then $d|m$ or $d|n$?

2.4 Methods of Proof

This section continues the discussion of common methods of proof that we began in §2.3 and introduces some standard terminology. It also ties these ideas in with some of the notation we developed in §2.2 to handle logical arguments. Try to keep your eye on the broad outlines and not get bogged down in the details, even though the details are essential to the proofs.

We are typically faced with a set of hypotheses $H_1, \ldots, H_n$ from which we want to infer a conclusion C. One of the most natural sorts of proof is the **direct proof** in which we show

$$H_1 \wedge H_2 \wedge \cdots \wedge H_n \Longrightarrow C. \tag{1}$$

Many of the proofs that we gave in Chapter 1—for example, proofs about greatest common divisors and common multiples—were of this sort.

Proofs that are not direct are called **indirect**. The two main types of indirect proof both use the negation of the conclusion, so they are often suitable when that negation is easy to state. The first type we consider is **proof of the contrapositive**

$$\neg C \Longrightarrow \neg(H_1 \wedge H_2 \wedge \cdots \wedge H_n). \tag{2}$$

According to rule 9 in Table 1 on page 62, the implication (2) is true if and only if the implication (1) is true, so a proof of (2) will let us deduce C from $H_1, \ldots, H_n$.

EXAMPLE 1

Let $m, n \in \mathbb{N}$. We wish to prove that if $m + n \geq 73$, then $m \geq 37$ or $n \geq 37$. There seems to be no natural way to prove this fact directly; we certainly couldn't consider all possible pairs m, n in $\mathbb{N}$ with $m+n \geq 73$. Instead, we prove the contrapositive: not "$m \geq 37$ or $n \geq 37$" implies not "$m+n \geq 73$." By De Morgan's law, the negation of "$m \geq 37$ or $n \geq 37$" is "not $m \geq 37$ and not $n \geq 37$," i.e., "$m \leq 36$ and $n \leq 36$." So the contrapositive proposition is this: If $m \leq 36$ and $n \leq 36$, then $m + n \leq 72$. This proposition follows immediately from a general property about inequalities: $a \leq c$ and $b \leq d$ imply that $a + b \leq c + d$ for all real numbers a, b, c, and d. ■

EXAMPLE 2

One way to show that a function $f : S \to T$ is one-to-one is to consider elements $s_1, s_2 \in S$, to suppose that $s_1 \neq s_2$, and to show directly, somehow, that $f(s_1) \neq f(s_2)$. The contrapositive approach would be to suppose that $f(s_1) = f(s_2)$ and then to show directly that $s_1 = s_2$. Depending on the way f is defined, one of these two ways may seem easier than the other, though of course the two are logically equivalent. ■

Another type of indirect proof is a **proof by contradiction**:

$$H_1 \wedge H_2 \wedge \cdots \wedge H_n \wedge \neg C \Longrightarrow \text{a contradiction.} \qquad \textbf{(3)}$$

Rule 15 in Table 1 on page 62 tells us that (3) is true if and only if the implication (1) is true. We already encountered proofs of this type in §2.3 and saw an example of one in Exercise 8 on page 71. Here are some more.

EXAMPLE 3

We wish to prove that $\sqrt{2}$ is irrational. That is, if x is in $\mathbb{R}$ and $x^2 = 2$, then x is not a rational number. Since the property of *not* being irrational, i.e., of being rational, is easy to describe, we assume, for purposes of obtaining a contradiction, that $x \in \mathbb{R}$, that $x^2 = 2$, and that x is rational. Our goal is to deduce a contradiction. By the definition of rational number, we have $x = \frac{p}{q}$ with $p, q \in \mathbb{Z}$ and $q \neq 0$. By reducing this fraction if necessary, we may assume that p and q have no common factors. In particular, p and q are not both even. Since $2 = x^2 = \frac{p^2}{q^2}$, we have $p^2 = 2q^2$ and so p^2 is even. This implies that p is even, since otherwise p^2 would be odd, by Example 1 on page 68. Hence $p = 2k$ for some $k \in \mathbb{Z}$. Then $(2k)^2 = 2q^2$ and therefore $2k^2 = q^2$. Thus q^2 and q are also even. But then p and q are both even, contradicting our earlier statement. Hence $\sqrt{2}$ is irrational.

This well-known proof is moderately complicated. Whoever originally created this proof used a bit of ingenuity. Given the basic idea, though, it isn't hard to modify it to prove similar results. Exercises 2 and 4 ask for more such proofs. ■

EXAMPLE 4

We wish to prove now that $\sqrt{4}$ is irrational. This is nonsense, of course, since $\sqrt{4} = 2$, but we're interested in what goes wrong if we try to imitate the proof in Example 3.

Assume that $x \in \mathbb{R}$, that $x^2 = 4$, and that x is rational. Then, by the definition of rational number, we have $x = \frac{p}{q}$ with $p, q \in \mathbb{Z}$, and $q \neq 0$. By reducing the fraction if necessary, we may assume that p and q have no common factors. In particular, p and q are not both even. Since $4 = x^2 = \frac{p^2}{q^2}$, we have $p^2 = 4q^2 = 2(2q^2)$, and so p^2 is even, which implies that p is even. Hence $p = 2k$ for some $k \in \mathbb{Z}$. Then $(2k)^2 = 4q^2$ and therefore $2k^2 = 2q^2$. So what? This was the place where we could smell the contradiction coming in Example 3, but now there's no contradiction. We could have $k = q = 1$, for instance, if we'd started with $x = \frac{2}{1}$.

We have done nothing wrong here. We have simply observed the place where showing that $\sqrt{n}$ is irrational in this way depends on what n is. One could imagine that some other method of proof would get around the difficulty, but of course that can't be true, since $\sqrt{n}$ is not always irrational. ■

One should avoid artificial proofs by contradiction such as in the next example.

EXAMPLE 5

We prove by contradiction that the sum of two odd integers is an even integer. Assume that $m, n \in \mathbb{Z}$ are odd integers, but that $m + n$ is not even, i.e., that $m + n$ is odd. There exist k and l in $\mathbb{Z}$ so that $m = 2k + 1$ and $n = 2l + 1$. Then

$$m + n = 2k + 1 + 2l + 1 = 2(k + l + 1),$$

an even number, contradicting the assumption that $m + n$ is odd.

This proof by contradiction is artificial because we did not use the assumption "$m + n$ is odd" until after we had established directly that $m + n$ was even. The following direct proof is far preferable.

Consider odd integers m, n in $\mathbb{Z}$. There exist $k, l \in \mathbb{Z}$ so that $m = 2k + 1$ and $n = 2l + 1$. Then

$$m + n = 2k + 1 + 2l + 1 = 2(k + l + 1),$$

which is even. ■

Similarly, a natural proof of the contrapositive does not need the excess baggage of a contradiction.

EXAMPLE 6

We prove that if x^2 is irrational, then x is irrational. [The converse is false, by the way; do you see an example?] Suppose that x is not irrational, i.e., that it's rational. Then $x = \frac{m}{n}$ for some $m, n \in \mathbb{Z}$ with $n \neq 0$. Then $x^2 = (\frac{m}{n})^2 = \frac{m^2}{n^2}$ with $m^2, n^2 \in \mathbb{Z}$ and $n^2 \neq 0$ [since $n \neq 0$]. Thus x^2 is rational, i.e., not irrational.

We could have turned this contrapositive proof into a proof by contradiction by adding "Suppose that x^2 is irrational" at the beginning and "which is a contradiction" at the end, but these phrases would just have added clutter without changing at all the basic direct proof of the contrapositive. ■

An implication of the form

$$H_1 \vee H_2 \vee \cdots \vee H_n \Longrightarrow C$$

is equivalent to

$$(H_1 \Longrightarrow C) \quad \text{and} \quad (H_2 \Longrightarrow C) \quad \text{and} \quad \cdots \quad \text{and} \quad (H_n \Longrightarrow C)$$

[for $n = 2$, compare rule 12a, Table 1 on page 62] and hence can be proved by **cases**, i.e., by proving each of $H_1 \Longrightarrow C, \ldots, H_n \Longrightarrow C$ separately. We saw proof by cases in Example 3 on page 69. The next example illustrates again how boring and repetitive such a proof can be, yet sometimes there is no better way to proceed.

EXAMPLE 7

Recall that the **absolute value** $|x|$ of x in $\mathbb{R}$ is defined by the rule

$$|x| = \begin{cases} x & \text{if } x \geq 0 \\ -x & \text{if } x < 0. \end{cases}$$

Assuming the familiar order properties of $\leq$ on $\mathbb{R}$, we prove the "triangle inequality"

$$|x + y| \leq |x| + |y| \quad \text{for} \quad x, y \in \mathbb{R}.$$

We consider four cases: (i) $x \geq 0$ and $y \geq 0$; (ii) $x \geq 0$ and $y < 0$; (iii) $x < 0$ and $y \geq 0$; (iv) $x < 0$ and $y < 0$.

Case (i). If $x \geq 0$ and $y \geq 0$, then $x + y \geq 0$, so $|x + y| = x + y = |x| + |y|$.

Case (ii). If $x \geq 0$ and $y < 0$, then

$$x + y < x + 0 = |x| \leq |x| + |y|$$

and

$$-(x + y) = -x + (-y) \leq 0 + (-y) = |y| \leq |x| + |y|.$$

Either $|x + y| = x + y$ or $|x + y| = -(x + y)$; either way we conclude that $|x + y| \leq |x| + |y|$ by the inequalities above.

Case (iii). The case $x < 0$ and $y \geq 0$ is similar to Case (ii).

Case (iv). If $x < 0$ and $y < 0$, then $x + y < 0$ and $|x + y| = -(x + y) = -x + (-y) = |x| + |y|$.

So in all four cases, $|x + y| \leq |x| + |y|$. ■

EXAMPLE 8

For every $n \in \mathbb{N}$, $n^3 + n$ is even. We can prove this fact by cases.

Case (i). Suppose n is even. Then $n = 2k$ for some $k \in \mathbb{N}$, so

$$n^3 + n = 8k^3 + 2k = 2(4k^3 + k),$$

which is even.

Case (ii). Suppose n is odd; then $n = 2k + 1$ for some $k \in \mathbb{N}$, so

$$n^3 + n = (8k^3 + 12k^2 + 6k + 1) + (2k + 1) = 2(4k^3 + 6k^2 + 4k + 1),$$

which is even.

Here is a more elegant proof by cases. Given n in $\mathbb{N}$, we have $n^3+n = n(n^2+1)$. If n is even, so is $n(n^2 + 1)$. If n is odd, then n^2 is odd by Example 1 on page 68. Hence $n^2 + 1$ is even, and so $n(n^2 + 1)$ is even.

Here is a third argument. If n is even, so are n^3 and the sum $n^3 + n$. If n is odd, so is $n^3 = n \cdot n^2$. Hence the sum $n^3 + n$ is even [Example 5] in this case too. ■

An implication $P \Longrightarrow Q$ is sometimes said to be **vacuously true** or **true by default** if P is false. Example 7(b) on page 64 showed that $\neg p \Longrightarrow (p \rightarrow q)$, so if $\neg P$ is true, then $P \rightarrow Q$ is also true. A **vacuous proof** is a proof of an implication $P \Longrightarrow Q$ by showing that P is false. Such implications rarely have intrinsic interest. They typically arise in proofs that consider cases. A case handled by a vacuous proof is usually one in which P has been ruled out; in a sense, there is no hypothesis to check. Although $P \Longrightarrow Q$ is true in such a case, we learn nothing about Q.

EXAMPLE 9

(a) Consider finite sets A and B and the assertion

If A has fewer elements than B, then there is a one-to-one mapping of A onto a proper subset of B.

The assertion is vacuously true if B is the empty set, because the hypothesis must be false. A vacuous proof here consists of simply observing that the hypothesis is impossible.

(b) We will see in Example 2(b) on page 146 the assertion that $n \geq 4m^2$ implies $n^m < 2^n$. This implication is vacuously true for $n = 0, 1, 2, \ldots, 4m^2 - 1$. For these values of n, its truth does not depend on—or tell us—whether $n^m < 2^n$ is true. For instance, $n^m < 2^n$ is true if $n = 10$ and $m = 2$, and it is false if $n = 10$ and $m = 4$, while $n \geq 4m^2$ is false in both cases. ■

An implication $P \Longrightarrow Q$ is sometimes said to be **trivially true** if Q is true. In this case, the truth value of P is irrelevant. A **trivial proof** of $P \Longrightarrow Q$ is one in which Q is shown to be true without any reference to P.

EXAMPLE 10

If x and y are real numbers such that $xy = 0$, then $(x + y)^n = x^n + y^n$ for $n \in \mathbb{P}$. This proposition is trivially true for $n = 1$; $(x + y)^1 = x^1 + y^1$ is obviously true, and this fact does not depend on the hypothesis $xy = 0$. For $n \geq 2$, the hypothesis is needed. ■

One sometimes encounters references to **constructive proofs** and **nonconstructive proofs** for the existence of mathematical objects satisfying certain properties. A constructive proof either specifies the object [a number or a matrix, say] or indicates how it can be determined by some explicit procedure or algorithm. A nonconstructive proof establishes the existence of objects by some indirect means, such as a proof by contradiction, without giving directions for how to find them.

EXAMPLE 11

In Example 5 on page 70 we proved by contradiction that there are infinitely many primes. We did not construct an infinite list of primes. The proof can be revised to give a constructive procedure for building an arbitrarily long list of distinct primes, provided we have some way of factoring integers. [This is Exercise 10.] ■

EXAMPLE 12

If the infinite sequence $a_1, a_2, a_3, \ldots$ has all its values in the finite set S, then there must be some s in S such that $a_i = s$ for infinitely many values of i, since otherwise each s would only equal a finite set of a_i's and altogether the members of S could only account for a finite number of a_i's. We can only say that at least one such s exists; we can't pick out such an s, nor specify which a_i's it is equal to, without knowing more about the sequence. Still, the existence information alone may be useful. For example, we can assert that some two members of the sequence must have the same value.

For instance, we claim that some two members of the sequence $1, 2, 2^2, 2^3, 2^4, \ldots$ must differ by a multiple of the prime 7. That's pretty easy: $2^3 - 1 = 7$ by inspection, and in fact $2^4 - 2 = 16 - 2 = 2 \cdot 7$, $2^5 - 2^2 = 4 \cdot 7$, and $2^6 - 1 = 9 \cdot 7$ as well. What if we take a larger prime p, though, say 8191? Must some two members of the sequence differ by a multiple of p? Yes. Think of taking each power 2^k, dividing it by p, and putting it in a box according to what the remainder is. The ones with remainder 1 go in the 1's box, the ones with remainder 2 in the 2's box, and so on. There are only p boxes, corresponding to the possible remainders $0, 1, 2, \ldots, p-1$. [There aren't any members of the sequence in the 0 box, but that doesn't matter.] At least one box has at least two members in it. [This is our existential statement.] Say 2^k and 2^l are both in the m's box, so $2^k = sp + m$ and $2^l = tp + m$ for some integers s and t. Then $2^k - 2^l = sp + m - tp - m = (s - t)p$, a multiple of p.

This nonconstructive argument doesn't tell us how to find k and l, but it does convince us that they must exist. Given the existence information, we can say somewhat more. If $2^l < 2^k$, say, then $2^k - 2^l = 2^l \cdot (2^{k-l} - 1)$. Since p is an odd prime that divides this difference, $2^{k-l} - 1$ must be a multiple of p, so there must be at least two numbers in the 1's box, namely 1 and 2^{k-l}. ■

EXAMPLE 13

Every positive integer n has the form $2^k m$, where $k \in \mathbb{N}$ and m is odd. This fact can be proved in several ways that suggest the following constructive procedure. If n is odd, let $k = 0$ and $m = n$. Otherwise, divide n by 2 and apply the procedure to $n/2$. Continue until an odd number is reached. Then k will equal the number of times division by 2 was necessary. Exercise 11 asks you to check out this procedure, which will be given as an algorithm in Exercise 24 on page 137. ■

We began this chapter with a discussion of a very limited system of logic, the propositional calculus. In the last section and this one, we relaxed the formality in order to discuss several methods of proof encountered in this book and elsewhere. We hope that you now have a better idea of what a mathematical proof is. In particular, we hope that you've seen that, although there are various natural outlines for proofs, *there is no standard method* or formula for constructing proofs. Of course this variability is part of what allows proofs to be so widely useful.

In §§2.5 and 2.6 we return to the propositional calculus and its applications. Chapter 13 deals with some more sophisticated aspects of logic.

Outside the realm of logic, and in particular in most of this book, proofs are communications intended to convince the reader of the truths of assertions. Logic will serve as the foundation of the process, but will recede into the background. That is, it should usually not be necessary to think consciously of the logic presented in this chapter; but if a particular proof in this book or elsewhere is puzzling, you may wish to analyze it more closely. What are the exact hypotheses? Is the author using hidden assumptions? Is the author giving an indirect proof?

Finally, there is always the possibility that the author has made an error or has not stated what was intended. Maybe you can show that the assertion is false—or at least show that the reasoning is fallacious. Even some good mathematicians have made the mistake of trying to prove $P \Longleftrightarrow Q$ by showing both $P \Longrightarrow Q$ and $\neg Q \Longrightarrow \neg P$, probably in some disguise. We will come back to these themes after the next section.

Exercises 2.4

In all exercises with proofs, indicate the methods of proof used.

1. Prove or disprove:

(a) The sum of two even integers is an even integer.

(b) The sum of three odd integers is an odd integer.

(c) The sum of two primes is never a prime.

2. Prove that $\sqrt{3}$ is irrational. *Hint:* Imitate the proof in Example 3, using the fact that if p is not divisible by 3, then p^2 is not divisible by 3.

3. Prove or disprove [yes, these should look familiar from §2.3]:

(a) The sum of three consecutive integers is divisible by 3.

(b) The sum of four consecutive integers is divisible by 4.

(c) The sum of five consecutive integers is divisible by 5.

4. Prove that $\sqrt[3]{2}$ is irrational.

5. Prove that $|xy| = |x| \cdot |y|$ for $x, y \in \mathbb{R}$.

6. Prove the result in Example 10. Use the fact that if $xy = 0$, then $x = 0$ or $y = 0$.

7. Prove that there are two different primes p and q whose last six decimal digits are the same. Is your proof constructive? If so, produce the primes.

8. (a) It is not known whether there are infinitely many **prime pairs**, i.e., odd primes whose difference is 2. Examples of prime pairs are (3, 5), (5, 7), (11, 13), and (71, 73). Give three more examples of prime pairs.

(b) Prove that (3, 5, 7) is the only "prime triple." *Hint:* Given $2k + 1$, $2k + 3$, and $2k + 5$, where $k \in \mathbb{N}$, show that one of these must be divisible by 3.

9. (a) Prove that, given n in $\mathbb{N}$, there exist n consecutive positive integers that are not prime; i.e., the set of prime integers has arbitrarily large gaps. *Hint:* Start with $(n + 1)! + 2$.

(b) Is the proof constructive? If so, use it to give six consecutive nonprimes.

(c) Give seven consecutive nonprimes.

10. Suppose $p_1, p_2, \ldots, p_k$ is a given list of distinct primes. Explain how one could use an algorithm that factors integers into prime factors to construct a prime that is not in the list. *Suggestion:* Factor $1 + p_1 p_2 \cdots p_k$.

11. Use the procedure in Example 13 to write the following positive integers in the form $2^k m$, where $k \in \mathbb{N}$ and m is odd.

(a) 14 (b) 73 (c) 96 (d) 1168

12. (a) The argument in Example 12 applies to $p = 5$. Find two powers of 2 that differ by a multiple of 5.

(b) Do the same for $p = 11$.

13. Floor and ceiling notation were defined in Section 1.1.

(a) Show that $\lfloor \frac{n}{3} \rfloor + \lceil \frac{2n}{3} \rceil = n$ for all positive integers $n \geq 1$.

(b) Show that $\lfloor \frac{m}{2} \rfloor + \lceil \frac{m}{2} \rceil = m$ for all integers m.

14. Trivial and vacuous proofs are, by their very nature, easy and uninteresting. Try your hand by proving the following assertions for a real number x and $n = 1$. Don't forget to indicate your method of proof.

(a) If $x \geq 0$, then $(1 + x)^n \geq 1 + nx$.

(b) If $x^n = 0$, then $x = 0$.

(c) If n is even, then $x^n \geq 0$.

Office Hours 2.4

We did proofs in high school, but not this way. Can I do them the way I'm used to?

Well, I don't know how you were taught to write proofs, and I certainly don't want to claim that there's only one correct way to do it. Let me tell you, though, why I think you should try to write your proofs the way the book suggests.

What's most important is that the reader can understand what you're trying to say and see how it's all connected logically. A string of statements or equations with no connecting words is essentially useless. In a proof, it needs to be clear at all times what you've assumed or deduced and what is still only a goal.

Some students I've talked with say that there are high schools where they teach you to write what you want to prove and then keep changing it until you get a true statement. Even if you had some sort of logical rules for making the changes, that kind of proof pattern would make it very easy to get confused and end up assuming what you're trying to prove. If you were taught to write proofs that way, I'd say that you should switch over now to go from hypothesis to conclusion, especially since some of the arguments will get pretty complicated as we go along. Of course, you might work from conclusions to hypotheses in your scratchwork, but your final proof should *logically* deduce conclusions *from* hypotheses.

I don't mean that it's a bad idea to start a proof by stating what you want to show. Looking ahead to the conclusion is a good way to discover a proof, and it's also a way to help the reader see where you're headed. You just have to make it clear what you're doing—especially, what you're assuming and what you're showing.

Another proof pattern that used to be fairly commonly taught, and maybe still is, consists of writing two columns, with a chain of statements in the left column and the reasons for them on the right. At the top of the left column you put what you're assuming, and at the bottom of it you put the conclusion. Then you put in between enough steps, with reasons, to fill in the argument. The book uses a format like this quite a bit in Chapter 2, and there's nothing wrong with it if the reasons for the steps aren't too complicated to write out. The format does look a little artificial, though, and probably isn't the way you'd want to write up an important argument to convince somebody. Professional mathematicians don't write that way in public.

The main point, as I said before, is to write up your proofs so the steps in the arguments come through clearly. Use plenty of words. If you think there's a step that won't be obvious to your reader, fill it in. In the process, you may discover that the alleged step is, in fact, false. That happens, and when it does you can try to fix the argument with another approach.

Maybe I haven't answered your question, though. How *did* you do proofs in high school?

Actually, this has been helpful. We did lots of two-column proofs, and I wanted to know if you'd accept them. It sounds as if you will, but probably on most proofs I should just write out my argument in sentences. OK?

Exactly. Good luck, and let me know how it goes.

2.5 Logic in Proofs

Logical thinking is important for everybody, but it is especially important in mathematics and in fields such as computer science, where an error in understanding a complicated logical situation can have serious consequences. In this section we look at what makes a logical argument "valid" and describe some kinds of arguments that truth tables tell us should be acceptable.

The proofs in Chapter 1 and the arguments that we just saw in §2.4 were based on logical thinking, but without explicitly defining what that meant. As those proofs illustrate, it is perfectly possible to use deductive logic and the idea of a "logical argument" without ever mentioning truth tables. Here is a very simple example.

EXAMPLE 1

Suppose that we are told "If this entry is the last, then the table is full" and "This entry is the last." We reasonably conclude that the table is full.

The form of this little argument can be given schematically as

$$\begin{array}{rl} & L \to F \\ & L \\ \hline \therefore & F \end{array}$$

where the hypotheses are listed above the line, and the symbol ∴ is read as "therefore" or "hence." ■

The fact that we all regard the short argument in this example as "logical" or "valid" has nothing to do with entries or tables, of course, but simply with the *form* of the argument. No matter what the propositions P and Q are, everybody would agree that if P is true and if we know that P implies Q, then we can conclude that Q is true.

This argument is an example of the sort of reasoning that we built into our definition of the truth table for $p \to q$, in the sense that we arranged for q to have truth value 1 whenever p and $p \to q$ did. We made sure that our definition of $\to$ gave us the modus ponens logical implication $p \wedge (p \to q) \Longrightarrow q$ and the tautology $(p \wedge (p \to q)) \to q$. The way truth tables work, we can even be sure that $(P \wedge (P \to Q)) \to Q$ is a tautology for all compound propositions P and Q, no matter how complicated, because we only need to consider those rows in the table for which P and $P \to Q$ have the truth value 1, and in those rows Q *must* have truth value 1.

It doesn't matter at all which hypotheses we listed first in Example 1. The arguments symbolized by

$$\begin{array}{rl} & P \to Q \\ & P \\ \hline \therefore & Q \end{array} \qquad \text{and} \qquad \begin{array}{rl} & P \\ & P \to Q \\ \hline \therefore & Q \end{array}$$

are both equally valid. What does matter is that the conclusion comes after the hypotheses.

EXAMPLE 2

Here's another short argument.

If the table is full, then it contains a marker.
The table does not contain a marker.
Therefore, the table is not full.

The form of the argument this time is

$$\begin{array}{rl} & F \to M \\ & \neg M \\ \hline \therefore & \neg F \end{array}$$

which corresponds to the modus tollens logical implication

$$[(p \to q) \wedge \neg q] \Longrightarrow \neg p$$

[number 20 in Table 2 on page 63]. Again, we would all agree that this short argument—and *every* argument of this form—is "logical," whatever that term means. Its legitimacy depends only on its form and not on what the propositions F and M are actually saying. ■

Table 3 lists a few short arguments with names that match their implication counterparts in Table 2 on page 63 [except rule 34, which corresponds to the trivial implication $p \wedge q \Longrightarrow p \wedge q$]. Look at each of these arguments now and convince yourself that you would accept as "logical" any argument that had that form. Indeed, these little arguments capture most of the logical reasoning—the drawing of inferences—that people do.

TABLE 3 Rules of Inference

	Premises	Conclusion	Name		Premises	Conclusion	Name
28.	P	$\therefore P \vee Q$	addition	29.	$P \wedge Q$	$\therefore P$	simplification
30.	P; $P \to Q$	$\therefore Q$	modus ponens	31.	$P \to Q$; $\neg Q$	$\therefore \neg P$	modus tollens
32.	$P \vee Q$; $\neg P$	$\therefore Q$	disjunctive syllogism	33.	$P \to Q$; $Q \to R$	$\therefore P \to R$	hypothetical syllogism
34.	P; Q	$\therefore P \wedge Q$	conjunction				

EXAMPLE 3

It's possible to combine several simple arguments into longer ones. For instance, suppose that we have the following three hypotheses.

If this entry is the last, then the table is full.
If the table is full, then it contains a marker.
The table contains no marker.

We can quickly deduce that this entry is not the last. Symbolically, it seems that the argument would look like

$$\begin{array}{rl} & L \to F \\ & F \to M \\ & \neg M \\ \hline \therefore & \neg L \end{array}$$

but there is nothing quite like this in Table 3. We see that the argument can be broken down into small steps, which we can show by inserting more terms in the chain. We give reasons, too, as we go along; the numbers 1, 2, 3, and 4 in the reasons refer to previous lines in the argument.

		Proof	**Reasons**
1.		$L \to F$	hypothesis
2.		$F \to M$	hypothesis
3.	$\therefore$	$L \to M$	1, 2; hypothetical syllogism rule 33
4.		$\neg M$	hypothesis
5.	$\therefore$	$\neg L$	3, 4; modus tollens rule 31

Notice that each proposition in the chain is either a hypothesis or a consequence of propositions above it, using one of the rules of inference in Table 3.

We could have moved the hypothesis $\neg M$ to any place earlier in the chain without harm. Indeed, we could have rearranged the argument even more radically, for instance to the following.

		Proof	**Reasons**
1.		$\neg M$	hypothesis
2.		$F \to M$	hypothesis
3.	$\therefore$	$\neg F$	1, 2; modus tollens, as in Example 2
4.		$L \to F$	hypothesis
5.	$\therefore$	$\neg L$	3, 4; modus tollens

■

When we reason, we're certainly allowed to bring in facts that everybody agrees are always true, even if they're not explicitly listed as hypotheses. For instance, everyone agrees that $P \to \neg(\neg P)$ is always true, regardless of what P is, so if it seemed helpful we could insert a line of the form

$$P \to \neg(\neg P) \qquad \text{tautology}$$

in an argument without harm to its validity. Of course, we would not want to do this sort of thing using complicated tautologies, since the whole point is to see that the argument can be broken down into short steps that everyone agrees are legal.

EXAMPLE 4

The arguments in Example 3 could have been replaced by

		Proof	Reasons
1.		$L \to F$	hypothesis
2.		$F \to M$	hypothesis
3.		$\neg M$	hypothesis
4.	$\therefore$	$(L \to F) \wedge (F \to M)$	1, 2; rule 34
5.	$\therefore$	$((L \to F) \wedge (F \to M)) \wedge \neg M$	3, 4; rule 34
6.		$[((L \to F) \wedge (F \to M)) \wedge \neg M] \to \neg L$	tautology [!!]
7.	$\therefore$	$\neg L$	5, 6; rule 30

but verifying by truth table that the expression in line 6 is a tautology looks like more work than just following the reasoning in Example 3. ■

The examples that we have seen so far provide a general idea of what we should mean when we talk about "logical" reasoning. The key to the deductive approach is to have a careful definition of the concept of "proof." Suppose that we are given some set of propositions, our **hypotheses**, and some **conclusion** C. A **formal proof** of C from the hypotheses consists of a chain $P_1, P_2, \ldots, P_n, C$ of propositions, ending with C, in which each P_i is either

(i) a hypothesis, or
(ii) a tautology, or
(iii) a consequence of previous members of the chain by using an allowable rule of inference.

The word *formal* here doesn't mean that the proof is stiff or all dressed up. The term simply comes from the fact that we can tell that something is a proof because of its *form*.

A **theorem** is a statement of the form "If H, then C," where H is a set of hypotheses and C is a conclusion. A formal proof of C from H is called a formal proof of the theorem.

The **rules of inference**, or logical rules, that we allow in formal proofs are rules based on logical implications of the form $H_1 \wedge H_2 \wedge \cdots \wedge H_m \Longrightarrow Q$. The rules in Table 3 are of this kind, so we can allow their use in formal proofs. If $H_1, H_2, \ldots, H_m$ have already appeared in the chain of a proof, and if the implication $H_1 \wedge H_2 \wedge \cdots \wedge H_m \Longrightarrow Q$ is true, then we are allowed to add Q to the chain.

One way to list the members of the chain for a formal proof is just to write them in a sequence of sentences, like ordinary prose. Another way is to list them vertically, as we did in Examples 2 to 4, so that we have room to keep a record of the reasons for allowing members into the chain.

EXAMPLE 5

Here is a very short formal proof written out vertically. The hypotheses are $B \wedge S$ and $B \vee S \to P$, and the conclusion is P. The chain gives a formal proof of P from $B \wedge S$ and $B \vee S \to P$. To give some meaning to the symbolism, you might think of B as "I am wearing a belt," S as "I am wearing suspenders," and P as "my pants stay up."

		Proof	Reasons
1.		$B \wedge S$	hypothesis
2.	$\therefore$	B	1; simplification rule 29
3.	$\therefore$	$B \vee S$	2; addition rule 28
4.		$B \vee S \rightarrow P$	hypothesis
5.	$\therefore$	P	3, 4; modus ponens rule 30

■

Here are two additional fairly complicated formal proofs to illustrate how steps can be justified. Read each of them slowly.

EXAMPLE 6

(a) We conclude $s \rightarrow r$ from the hypotheses $p \rightarrow (q \rightarrow r)$, $p \vee \neg s$ and q.

	Proof	Reasons
1.	$p \rightarrow (q \rightarrow r)$	hypothesis
2.	$p \vee \neg s$	hypothesis
3.	q	hypothesis
4.	$\neg s \vee p$	2; commutative law 2a [see below]
5.	$s \rightarrow p$	4; implication rule 10a
6.	$s \rightarrow (q \rightarrow r)$	1, 5; hypothetical syllogism rule 33
7.	$(s \wedge q) \rightarrow r$	6; exportation rule 14
8.	$q \rightarrow [s \rightarrow (q \wedge s)]$	tautology; rule 22 [see below]
9.	$s \rightarrow (q \wedge s)$	3, 8; modus ponens rule 30
10.	$s \rightarrow (s \wedge q)$	9; commutative law 2b
11.	$s \rightarrow r$	7, 10; hypothetical syllogism rule 33

Some of the lines here require explanation. The reason for line 4 is the logical implication $(p \vee \neg s) \Longrightarrow (\neg s \vee p)$ that follows from the commutative law 2a in Table 1 on page 62. The corresponding little rule of inference would be

$$\begin{array}{rc} & P \vee Q \\ \hline \therefore & Q \vee P \end{array}$$

which seems unnecessary to list separately. The proposition in line 8 is the tautology corresponding to the logical implication 22 in Table 2 on page 63. Lines 5, 7, and 10 are similarly based on known logical implications. We have also stopped putting in the $\therefore$ signs, which would go in every line except the ones that contain hypotheses.

(b) Here we conclude $\neg p$ by contradiction from the hypotheses $p \rightarrow (q \wedge r)$, $r \rightarrow s$, and $\neg(q \wedge s)$.

	Proof	Reasons
1.	$p \rightarrow (q \wedge r)$	hypothesis
2.	$r \rightarrow s$	hypothesis
3.	$\neg(q \wedge s)$	hypothesis
4.	$\neg(\neg p)$	negation of the conclusion
5.	p	4; double negation rule 1
6.	$q \wedge r$	1, 5; modus ponens rule 30
7.	q	6; simplification rule 29
8.	$r \wedge q$	6; commutative law 2b
9.	r	8; simplification rule 29
10.	s	2, 9; modus ponens rule 30
11.	$q \wedge s$	7, 10; conjunction rule 34
12.	$(q \wedge s) \wedge \neg(q \wedge s)$	3, 11; conjunction rule 34
13.	contradiction	12; rule 7b

■

Recall Example 3, in which we deduced $\neg L$ from $L \rightarrow F$, $F \rightarrow M$, and $\neg M$. Regardless of which steps we use to fill in the argument in that example, the overall inference corresponds to the single logical implication

$$(p \rightarrow q) \wedge (q \rightarrow r) \wedge \neg r \Longrightarrow \neg p,$$

i.e., to the tautology

$$[(p \rightarrow q) \wedge (q \rightarrow r) \wedge \neg r] \rightarrow \neg p,$$

which showed up in line 6 of Example 4. This tautology is quite complicated. Since our lists of tautologies and logical implications are deliberately short and contain only very simple propositions, it's not surprising that we have not seen this implication and tautology before, even written in terms of p, q, and r, rather than L, F, and M. We may never see them again, either.

The names $p, q, r, \ldots$ that we used for the variables in our lists of tautologies and implications don't really matter, of course. We could replace them with others, so long as we don't try to replace the same name by two different things. We have the following general principle, which we have already observed in special cases.

Substitution Rule (a) If all occurrences of some variable in a tautology are replaced by the same, possibly compound, proposition, then the result is still a tautology.

EXAMPLE 7

(a) The argument for this example illustrates the idea behind Substitution Rule (a). According to the modus ponens rule 19 of Table 2 on page 63, the proposition $[p \wedge (p \rightarrow q)] \rightarrow q$ is a tautology; its truth-table entries are all 1 regardless of the truth values for p and q. [See the truth table in Example 3(b) on page 61.] If we replace each occurrence of p by the proposition $q \rightarrow r$ and each q by $s \vee t$, we obtain the new proposition

$$[(q \rightarrow r) \wedge ((q \rightarrow r) \rightarrow (s \vee t))] \rightarrow (s \vee t).$$

This proposition must itself be a tautology, because its truth value will be 1 no matter what truth values q, r, s, and t give to $q \rightarrow r$ and $s \vee t$. We know this without writing out any rows at all in the 16-row q, r, s, t truth table.

(b) We can get into trouble if we try to replace a variable by two different things. For example, $p \wedge q \rightarrow q$ is a tautology, but $p \wedge r \rightarrow s$ is certainly not. ■

Substitution Rule (a) can be used to produce new logical equivalences from old.

EXAMPLE 8

Many of the logical equivalences in Table 1 on page 62 can be derived from others in the table using substitution. We illustrate this by showing how rule 8a leads to rule 8d. Rule 8a yields the tautology

$$(\neg p \wedge \neg q) \leftrightarrow \neg(p \vee q).$$

We replace each occurrence of p by $\neg p$ and each occurrence of q by $\neg q$ to obtain another tautology:

$$(\neg\neg p \wedge \neg\neg q) \leftrightarrow \neg(\neg p \vee \neg q).$$

This has the same truth table as

$$(p \wedge q) \leftrightarrow \neg(\neg p \vee \neg q),$$

so this must also be a tautology. Thus the logical equivalence 8d in Table 1 on page 62 follows from 8a. ■

The replacement of $\neg\neg p \wedge \neg\neg q$ by $p \wedge q$ that we just made in Example 8 is an illustration of a second kind of substitution, somewhat like replacing quantities by equal quantities in algebra. In the case of logic, the replacement proposition only needs to be equivalent to the one it substitutes for, and the result will just be equivalent to what we started with. Here is a precise statement.

Substitution Rule (b) If a compound proposition P contains a proposition Q, and if Q is replaced by a logically equivalent proposition, then the resulting compound proposition is logically equivalent to P.

Our next example illustrates again why Substitution Rule (b) is valid.

EXAMPLE 9

(a) Consider the proposition

$$P = \neg[(p \to q) \wedge (p \to r)] \to [q \to (p \to r)]$$

which is not a tautology. By Substitution Rule (b), we obtain a proposition logically equivalent to P if we replace $Q = (p \to q)$ by the logically equivalent $(\neg p \vee q)$. Similarly, we could replace one or both occurrences of $(p \to r)$ by $(\neg p \vee r)$; see rule 10a. We could also replace $[(p \to q) \wedge (p \to r)]$ by $[p \to (q \wedge r)]$ thanks to rule 12b. Thus P is logically equivalent to each of the following propositions, among many others:

$$\neg[(\neg p \vee q) \wedge (p \to r)] \to [q \to (p \to r)],$$

$$\neg[(p \to q) \wedge (\neg p \vee r)] \to [q \to (p \to r)],$$

$$\neg[p \to (q \wedge r)] \to [q \to (\neg p \vee r)].$$

(b) Let's see why $P = \neg[(p \to q) \wedge (\boldsymbol{p \to r})] \to [q \to (p \to r)]$ is equivalent to $P^* = \neg[(p \to q) \wedge (\boldsymbol{\neg p \vee r})] \to [q \to (p \to r)]$. We chose to analyze this substitution so that we could show that the appearance of *another* $(p \to r)$, which was not changed, does not matter. Think about the columns of the truth tables for P and P^* that correspond to $\boldsymbol{\to}$ and $\boldsymbol{\vee}$ in the boldfaced appearances of $\boldsymbol{p \to r}$ and $\boldsymbol{\neg p \vee r}$. Since $p \to r$ and $\neg p \vee r$ are equivalent, the truth values in these two columns are the same. The rest of the truth values in the tables for the propositions P and P^* will be identical, so the final values will agree.

This sort of reasoning generalizes to explain why Substitution Rule (b) is valid. ■

It is worth emphasizing that, unlike Substitution Rule (a), Substitution Rule (b) does not require that *all* occurrences of some proposition be replaced by the same equivalent proposition.

EXAMPLE 10

(a) We use the De Morgan law 8d and substitution to find a proposition that is logically equivalent to $(p \wedge q) \to (\neg p \wedge q)$ but that does not use the connective $\wedge$. Since $p \wedge q$ is equivalent to $\neg(\neg p \vee \neg q)$, and $\neg p \wedge q$ is equivalent to $\neg(\neg\neg p \vee \neg q)$, the given proposition is equivalent by Substitution Rule (b) to

$$\neg(\neg p \vee \neg q) \to \neg(\neg\neg p \vee \neg q)$$

and so, again by Substitution Rule (b), to

$$\neg(\neg p \vee \neg q) \to \neg(p \vee \neg q).$$

Thus we have a logically equivalent proposition that does not involve $\wedge$.

(b) If desired, we could apply rule 10a to obtain the equivalent

$$\neg(p \to \neg q) \to \neg(q \to p),$$

which uses neither $\wedge$ nor $\vee$. On the other hand, we could avoid the use of the connective $\to$ by applying rule 10a.

The sort of rewriting we have done here to eliminate one or more symbols has important applications in logical circuit design, as we will see in Chapter 10.

(c) Notice that in parts (a) and (b) we are *not* claiming that any of the three propositions

$$(p \wedge q) \to (\neg p \wedge q) \quad \text{or} \quad \neg(\neg p \vee \neg q) \to \neg(p \vee \neg q)$$
$$\text{or} \quad \neg(p \to \neg q) \to \neg(q \to p)$$

are tautologies. We have simply asserted that these three propositions are logically equivalent to each other. ■

One could use the substitution rules to deal with incredibly complicated propositions, but we won't. Our point here is simply to show that there are techniques, analogous to familiar ones from algebra, that let us manipulate and rewrite logical expressions to get them into more convenient forms. Just as in algebra, what is "convenient" depends, of course, on what we intend to use them for.

This section began by studying arguments that would be valid no matter what meanings were attached to the propositions in them. The Substitution Rules bring us back to tautologies and logical equivalence, notions that we defined in terms of truth tables. It's natural to ask what the connection is between truth and validity, i.e., between semantics and syntax, to use technical terms. We have already seen that the rules of inference in Table 3 correspond to logical implications in Table 2 on page 63, and we have explicitly allowed tautologies as elements of valid proofs, so we would expect to find validity and truth tables closely linked. The next theorem describes the connection.

Theorem Let H and C be propositions. There is a valid proof of C from H if, and only if, $H \to C$ is a tautology, i.e., if, and only if, $H \Longrightarrow C$ is true.

Proof Suppose first that there is a valid proof of C from H, but that $H \to C$ is *not* a tautology. There is some assignment of truth values to the variables in H and C that makes H true and C false. Someplace along the chain of statements that form a proof of C from H there must be a first statement, say P, that is false for this assignment. Perhaps P is C; it's certainly not H. Now P isn't a hypothesis, and it's not a tautology either [why not?]. By the definition of a formal proof, P must be a consequence of previous statements by using an allowable rule of inference. But all of our allowable rules are based on implications of the form $H_1 \wedge \cdots \wedge H_m \Longrightarrow Q$, which force Q to be true if all of $H_1, \ldots, H_m$ are true. Since all statements before P are true for our chosen truth assignment, P must be also true for that assignment, contrary to its choice. Thus P cannot exist, so $H \to C$ must be a tautology after all.

In the other direction, if $H \to C$ *is* a tautology, then

	H	hypothesis
	$H \to C$	tautology
$\therefore$	C	modus ponens

is a valid proof of C from H. ■

Corollary There is a formal proof of C from $A \wedge B$ if, and only if, there is a formal proof of $B \to C$ from A.

Proof By the Theorem, there is a formal proof of C from $A \wedge B$ if, and only if, $(A \wedge B) \to C$ is a tautology, and there is a formal proof of $B \to C$ from A if, and only if, $A \to (B \to C)$ is a tautology. Rule 14 [exportation] says that $(A \wedge B) \to C$ and $A \to (B \to C)$ are logically equivalent, so if either one is a tautology, then so is the other. ■

Exercises 2.5

1. Give reasons for each of the following. Try not to use truth tables.

(a) $(p \vee q) \wedge s \Longleftrightarrow (q \vee p) \wedge s$

(b) $s \to (\neg(p \vee q)) \Longleftrightarrow s \to [(\neg p) \wedge (\neg q)]$

(c) $(p \to q) \wedge (p \vee q) \Longleftrightarrow (\neg p \vee q) \wedge (\neg\neg p \vee q)$

(d) $t \wedge (s \vee p) \Longleftrightarrow t \wedge (p \vee s)$

2. Repeat Exercise 1 for the following.

(a) $s \wedge p \Longleftrightarrow s \wedge (p \wedge p)$

(b) $[(a \vee b) \leftrightarrow (p \to q)] \Longleftrightarrow [(a \vee b) \leftrightarrow (\neg p \vee q)]$

(c) $[(a \vee b) \leftrightarrow \neg(p \wedge q)] \Longleftrightarrow [(b \vee a) \leftrightarrow (\neg p \vee \neg q)]$

3. Give a reason for each equivalence in the following chain.

$$(p \to s) \vee (\neg s \to t)$$

(a) $\Longleftrightarrow (\neg p \vee s) \vee (s \vee t)$

(b) $\Longleftrightarrow [(\neg p \vee s) \vee s] \vee t$

(c) $\Longleftrightarrow [\neg p \vee (s \vee s)] \vee t$

(d) $\Longleftrightarrow (\neg p \vee s) \vee t$

(e) $\Longleftrightarrow \neg p \vee (s \vee t)$

(f) $\Longleftrightarrow p \to (s \vee t)$

4. Repeat Exercise 3 with the following.

$$[(a \wedge p) \vee p] \to p$$

(a) $\Longleftrightarrow \neg[(a \wedge p) \vee p] \vee p$

(b) $\Longleftrightarrow [\neg(a \wedge p) \wedge \neg p] \vee p$

(c) $\Longleftrightarrow [(\neg a \vee \neg p) \wedge \neg p] \vee p$

(d) $\Longleftrightarrow p \vee [(\neg a \vee \neg p) \wedge \neg p]$

(e) $\Longleftrightarrow [p \vee (\neg a \vee \neg p)] \wedge (p \vee \neg p)$

(f) $\Longleftrightarrow [(\neg a \vee \neg p) \vee p] \wedge \mathbf{1}$ [$\mathbf{1}$ for tautology here]

(g) $\Longleftrightarrow (\neg a \vee \neg p) \vee p$

(h) $\Longleftrightarrow \neg a \vee (\neg p \vee p)$

(i) $\Longleftrightarrow \neg a \vee t$

(j) $\Longleftrightarrow t$

5. Give an explanation for each step in the following formal proof of $\neg s$ from $(s \vee g) \to p$, $\neg a$, and $p \to a$.

1. $(s \vee g) \to p$ 2. $\neg a$

3. $p \to a$ 4. $s \to (s \vee g)$

5. $s \to p$ 6. $s \to a$

7. $\neg s$

6. Rearrange the steps in Exercise 5 in three different ways, all of which prove $\neg s$ from $(s \vee g) \to p$, $\neg a$, and $p \to a$.

7. Use Substitution Rule (a) with $p \to q$ replacing q to make a tautology out of each of the following.

(a) $\neg q \to (q \to p)$ (b) $[p \wedge (p \to q)] \to q$

(c) $p \vee \neg p$ (d) $(p \vee q) \leftrightarrow [(\neg q) \to p]$

8. Give the rules of inference corresponding to the logical implications 23, 26a, and 27b.

9. Find a logical equivalence in Table 1 on page 62 from which the use of Substitution Rule (a) yields the indicated equivalence.

(a) $[(p \wedge (q \wedge r)) \to r] \Longleftrightarrow [p \to ((q \wedge r) \to r)]$

(b) $[p \vee (q \wedge (r \wedge s))] \Longleftrightarrow [(p \vee q) \wedge (p \vee (r \wedge s))]$

(c) $\neg[(\neg p \wedge r) \vee (q \to r)] \Longleftrightarrow [\neg(\neg p \wedge r) \wedge \neg (q \to r)]$

10. Find a logical implication in Table 2 on page 63 from which Substitution Rule (a) yields the indicated implication.

(a) $[\neg p \vee q] \Longrightarrow [q \to ((\neg p \vee q) \wedge q)]$

(b) $[(p \to q) \wedge (r \to q)] \Longrightarrow [(\neg q \vee \neg q) \to (\neg p \vee \neg r)]$

(c) $[((p \to s) \to (q \wedge s)) \wedge \neg(q \wedge s)] \Longrightarrow \neg(p \to s)$

11. Show by the methods of Example 8 that rule 8c in Table 1 on page 62 follows from rule 8b. Give reasons for your statements.

12. As in Exercise 11, show that rule 10b in Table 1 on page 62 follows from rules 8c and 10a.

13. Let P be the proposition $[p \wedge (q \vee r)] \vee \neg[p \vee (q \vee r)]$. Replacing all occurrences of $q \vee r$ by $q \wedge r$ yields

$$P^* = [p \wedge (q \wedge r)] \vee \neg[p \vee (q \wedge r)].$$

Since $q \wedge r \Longrightarrow q \vee r$, one might suppose that $P \Longrightarrow P^*$ or that $P^* \Longrightarrow P$. Show that neither of these is the case.

14. Show that, if the first p in the tautology $p \to [q \to (p \wedge q)]$ is replaced by the proposition $p \vee q$, then the new proposition is not a tautology. This shows that Substitution Rule (a) must be applied with care.

15. (a) Show that, if A is a tautology and if there is a formal proof of C from A, then there is a proof of C with no hypotheses at all.

(b) Show that, if there is a formal proof of C from B, then there is a proof of $B \to C$ with no hypotheses at all.

16. Every compound proposition is equivalent to one that uses only the connectives $\neg$ and $\vee$. This fact follows from the equivalences $(p \to q) \Longleftrightarrow (\neg p \vee q)$, $(p \wedge q) \Longleftrightarrow \neg(\neg p \vee \neg q)$, and $(p \leftrightarrow q) \Longleftrightarrow [(p \to q) \wedge (q \to p)]$. Find propositions logically equivalent to the following, but using only the connectives $\neg$ and $\vee$.

(a) $p \leftrightarrow q$ (b) $(p \wedge q) \to (\neg q \wedge r)$

(c) $(p \to q) \wedge (q \vee r)$ (d) $p \oplus q$

17. (a) Show that $p \vee q$ and $p \wedge q$ are logically equivalent to propositions using only the connectives $\neg$ and $\to$.

(b) Show that $p \vee q$ and $p \to q$ are logically equivalent to propositions using only the connectives $\neg$ and $\wedge$.

(c) Is $p \to q$ logically equivalent to a proposition using only the connectives $\wedge$ and $\vee$? Explain.

18. The **Sheffer stroke** is the connective $|$ defined by the truth table

p	p	$p \vert q$
0	0	**1**
0	1	**1**
1	0	**1**
1	1	**0**

Thus $p|q \Longleftrightarrow \neg(p \wedge q)$ [hence the computer scientists' name **NAND** for this connective]. All compound propositions are equivalent to ones that use only $|$, a useful fact that follows from the remarks in Exercise 16 and parts (a) and (b) below.

(a) Show that $\neg p \Longleftrightarrow p|p$.

(b) Show that $p \vee q \Longleftrightarrow (p|p)|(q|q)$.

(c) Find a proposition equivalent to $p \wedge q$ using only the Sheffer stroke.

(d) Do the same for $p \to q$.

(e) Do the same for $p \oplus q$.

19. Verify the following **absorption laws**.

(a) $[p \vee (p \wedge q)] \Longleftrightarrow p$ [Compare with Exercise 4.]

(b) $[p \wedge (p \vee q)] \Longleftrightarrow p$

2.6 Analysis of Arguments

Section 2.5 described a formalization of the notion of "proof." Our aim in this section is to use that abstract version as a guide to understanding more about the proofs and fallacies that we encounter in everyday situations. We will look at examples of how to construct arguments, and we will see some errors to watch out for. The same kind of analysis that lets us judge validity of proofs will also help us to untangle complicated logical descriptions.

Our formal rules of inference are sometimes called **valid inferences**, and formal proofs are called **valid proofs**. We will extend the use of the word "valid" to describe informal arguments and inferences that correspond to formal proofs and their rules. A sequence of propositions that fails to meet the requirements for being a formal proof is called a **fallacy**. We will use this term also to mean an argument whose formal counterpart is not a valid proof.

We will see a variety of examples of fallacies. First, though, we construct some valid arguments. Rules 1 to 34 can be found in the tables on pages 62, 63, and 79.

EXAMPLE 1

We analyze the following. "If the entry is small or giant, then the output is predictable. If the output is predictable, then it is negative. Therefore, if the output is not negative, then the entry is not giant." Let

s = "The entry is small,"

g = "The entry is giant,"

p = "The output is predictable,"

n = "The output is negative."

We are given the hypotheses $s \vee g \to p$ and $p \to n$ and want a proof of $\neg n \to \neg g$. Here is our scratchwork. Since $\neg n \to \neg g$ is the contrapositive of $g \to n$, it will be enough to get a proof of $g \to n$ and then quote the contrapositive law, rule 9. From $s \vee g \to p$ we can surely infer $g \to p$ [details in the next sentence] and then combine this with $p \to n$ to get $g \to n$ using rule 33. To get $g \to p$ from $s \vee g \to p$, rule 33 shows that it is enough to observe that $g \to s \vee g$ by the addition rule. Well, almost: The addition rule gives $g \to g \vee s$, so we need to invoke commutativity, too. Here is the corresponding formal proof.

	Proof	Reasons
1.	$s \vee g \to p$	hypothesis
2.	$p \to n$	hypothesis
3.	$g \to g \vee s$	addition (rule 16)
4.	$g \to s \vee g$	3; commutative law 2a
5.	$g \to p$	4, 1; hypothetical syllogism (rule 33)
6.	$g \to n$	5, 2; hypothetical syllogism (rule 33)
7.	$\neg n \to \neg g$	6; contrapositive (rule 9)

So there is a valid proof of the conclusion from the hypotheses. Isn't that reassuring? Of course, we are accustomed to handling this sort of simple reasoning in our heads every day, and the baggage of a formal argument here just seems to complicate an easy problem. ■

Still, we continue.

EXAMPLE 2

"If the entry is small or giant, then the output is predictable. The output is not negative. If the output is predictable, then it must be negative. Therefore, the entry is not small." With the notation of Example 1, we want a proof of this: If $s \vee g \to p$, $\neg n$ and $p \to n$, then $\neg s$. The only rule of inference in Table 3 on page 79 that might help is rule 31 [modus tollens]. Given the hypothesis $\neg n$, this rule would allow us to infer $\neg s$ if we could just get $s \to n$, which we can deduce from the first and third hypotheses, as in Example 1. Here is a formal version.

	Proof	**Reasons**
1.	$s \vee g \to p$	hypothesis
2.	$s \to s \vee g$	addition (rule 16)
3.	$s \to p$	2, 1; hypothetical syllogism (rule 33)
4.	$p \to n$	hypothesis
5.	$s \to n$	3, 4; hypothetical syllogism (rule 33)
6.	$\neg n$	hypothesis
7.	$\neg s$	5, 6; modus tollens (rule 31)

In words, the argument would go:

> If the entry is small, then, since either a small or giant entry means that the output is predictable, the output must be predictable and hence negative. So if the entry is small, then the output must be negative. Since the output is not negative, the entry must not be small.

In this argument, rule 16 is only used implicitly, and the three inferences made are clearly "logical." ■

Our scratchwork in these examples illustrates a general strategy that can be helpful in constructing proofs. Look at the conclusion C. What inference would give it? For example, do we know of a proof of something like $B \to C$? How hard would it be to prove B? And so on, working backward from C to B to A to . . . . Alternatively, look at the hypotheses. What can we quickly deduce from them? Do any of the new deductions seem related to C? If you have ever had to prove trigonometric identities, you will recognize the strategy: You worked with the right-hand side, then the left-hand side, then the right-hand, etc., trying to bring the two sides of the identity together.

EXAMPLE 3

We can also obtain the conclusion in Example 2 by contradiction, i.e., by assuming the negation of the conclusion, that the entry *is* small, and reaching an absurdity. The formal proof looks a little more complicated. Given hypotheses $s \vee g \to p$, $\neg n$, $p \to n$, and now $\neg(\neg s)$, we want to get a contradiction like $s \wedge (\neg s)$, $n \wedge (\neg n)$, $g \wedge (\neg g)$, or $p \wedge (\neg p)$. Since we already have $\neg n$ and s [from $\neg(\neg s)$], the first two contradictions look easiest to reach. We aim for $n \wedge (\neg n)$, because we already have $p \to n$. We can get $s \to p$ from $s \vee g \to p$, and we have $p \to n$, so we can get $s \to n$. Since we have s, we then can infer n. We have just outlined the following formal proof. See Exercise 11 for a proof with modus ponens instead of hypothetical syllogism; Exercise 10 asks for proofs reaching different contradictions.

	Proof	**Reasons**
1.	$s \to s \vee g$	addition (rule 16)
2.	$s \vee g \to p$	hypothesis

3.	$s \to p$	1, 2; hypothetical syllogism (rule 33)
4.	$p \to n$	hypothesis
5.	$s \to n$	3, 4; hypothetical syllogism (rule 33)
6.	$\neg(\neg s)$	negation of the conclusion
7.	s	6; double negation (rule 1)
8.	n	7, 5; modus ponens (rule 30)
9.	$\neg n$	hypothesis
10.	$n \wedge (\neg n)$	8, 9; conjunction (rule 34)
11.	contradiction	10; rule 7b

The English version goes like this:

As in Example 2, if the entry is small, then the output is negative. Suppose that the entry is small. Then the output is negative. But it's not, by hypothesis. This contradiction means that our supposition must be false; the entry *is* small.

In general, a proof by contradiction begins with a supposition and runs for a while until it reaches a contradiction. At that point we look back to the last "suppose" and say, "Well, that supposition must have been wrong." Since the hypotheses and $\neg C$ are mutually inconsistent, the conclusion C must hold when the hypotheses hold; i.e., the hypotheses imply C.

The formal proof by contradiction in this example is longer than the proof in Example 2, but the English version is about the same length, and it may be conceptually more straightforward. In practice, one resorts to proofs by contradiction when it is easier to use $\neg C$ *together with* the hypotheses than it is to derive C *from* the hypotheses. We illustrated this point in §2.4. ■

In general, it is not easy to construct proofs. Still, with practice and by looking closely at the proofs you read, you can get quite good at building your own proofs and at spotting holes in other people's.

This book is full of examples of correct proofs; for example, look back at the arguments in Chapter 1. Let's now examine a few fallacies.

EXAMPLE 4

(a) Suppose that we want to show that life is tough. A standard sort of argument goes like this.

Assume that life is tough.
Blah, blah, blah.
Therefore, life is tough.

Assuming the conclusion in this way seems too obviously wrong to be worth mentioning, but experience teaches us that this fallacy, perhaps disguised, is a common one.

(b) Suppose that we want to show that "If I am an adult, then life is tough." Try this, with $L =$ "Life is tough," and $A =$ "I am an adult."

Assume L. Since $L \to (A \to L)$ is a tautology, then $A \to L$ must be true.

Here again we start out assuming an additional hypothesis, conveniently strong enough to get the conclusion.

EXAMPLE 5

(a) We have already mentioned in §2.4 the fallacy of arguing $A \Longleftrightarrow B$ by showing $A \Longrightarrow B$ and $\neg B \Longrightarrow \neg A$.

(b) A related fallacy is proving the converse of what's called for, i.e., trying to show $A \Longrightarrow B$ by showing $B \Longrightarrow A$.

(c) The false inference $A, B \to A, \therefore B$ is a similar fallacy. It is amazing how many people fall into these traps. For example, it has been observed that after criminals commit certain crimes they profess their innocence. Therefore, defendants who profess their innocence must have committed the crimes.

Sometimes confusing wording trips people up. Thus, suppose that I am rich and that people are rich whenever they live in big houses. It does not follow that I live in a big house. I might live on board my yacht. Here we have R and $H \to R$ and want to deduce H. To do so, though, we would need $R \to H$; that is, all rich people live in big houses.

(d) Exercise 3 examines the popular false inference $A \to Z, B \to Z, \dots, Y \to Z, \therefore Z$, which is related to the idea that if a result is true in a few cases, then it must always be true. Many mathematics students will try to *prove* a result by giving a few illustrations; sometimes this is because authors and teachers use the word *show* instead of *prove* and the students interpret this to mean *illustrate*.

Younger children also try this sort of reasoning. For example, to prove that $Z =$ "(reasonable) parents let their children watch TV until midnight," they will observe that $A \to Z$, $B \to Z$, etc., where $A =$ "Tommy has parents," $B =$ "Susie has parents," etc. ∎

As we suggested in the preface to the student, one should be careful in using and reading the key words "clearly" and "obviously." They should indicate that the result is an immediate consequence of preceding material or can be verified by a short straightforward argument. They should not be used to hide a large gap in the argument! Obviously, there are no mistakes in *this* book, but previous editions did have one or two logical errors.

If you recognize a standard fallacy, such as assuming the conclusion, proving the converse, or misusing modus ponens, then you can see immediately that an alleged argument is not valid. Mistakes in reasoning are often subtle, however, and require close examination to detect. If you just want to show that an argument *must* be wrong, without finding out precisely where it goes astray, one way to do so is to exhibit a situation in which the hypotheses are true but the conclusion is false. We did this in Example 5(c) when we pointed out the possibility of rich people living on yachts. Here are some more illustrations.

EXAMPLE 6

We are given that if a program does not fail, then it begins and terminates, and we know that our program began and failed. We conclude that it did not terminate. Is this reasoning valid? Let

$B =$ "the program begins,"

$T =$ "the program terminates,"

$F =$ "the program fails."

Our hypotheses are $\neg F \to (B \wedge T)$ and $B \wedge F$, and the conclusion is $\neg T$. Here is an attempt at a formal proof.

	Proof	**Reasons**
1.	$\neg F \to (B \wedge T)$	hypothesis
2.	$B \wedge F$	hypothesis
3.	$(B \wedge T) \to T$	simplification (rule 17)
4.	$\neg F \to T$	1, 3; hypothetical syllogism (rule 33)
5.	F	2; simplification (rule 29)
6.	$\neg T$	4, 5; **??**

How can we infer proposition 6, $\neg T$, from propositions 4 and 5? It appears that the best hope is modus tollens, but a closer look shows that modus tollens does not apply. What is needed is $P \to Q, \neg P, \therefore \neg Q$, and this is *not* a valid rule of inference. The alleged proof above is not valid; it is a fallacy. It could be that we simply went about the proof in the wrong way. However, this is not the situation in this case. In fact, according to the Theorem on page 84, no correct proof exists because the conclusion does not follow from the hypotheses. That is, the proposition

$$\{[\neg F \to (B \wedge T)] \wedge (B \wedge F)\} \to \neg T$$

is not a tautology. To see this, consider the row in its truth table where propositions B, F, and T are all true. In terms of the original hypotheses, the program might begin and terminate and yet fail for some other reason. ■

EXAMPLE 7

(a) I am a famous basketball player. Famous basketball players make lots of money. If I make lots of money, then you should do what I say. I say you should buy Pearly Maid shoes. Therefore, you should buy Pearly Maid shoes.

We set

B = "basketball player,"

M = "makes lots of money,"

D = "do what I say," and

S = "buy those shoes!"

The hypotheses are B, $B \to M$, $M \to D$, and $D \to S$, and the conclusion S surely follows by three applications of modus ponens.

(b) Suppose we leave out $M \to D$ in part (a). After all, why should I do what you say just because you make lots of money? Then there is no valid way to conclude S from B, $B \to M$, and $D \to S$. [Look at the case S and D false, B and M true, for example.] The argument would require filling in the hidden assumption that we should do what people with lots of money tell us to.

(c) Suppose that instead of $M \to D$ in the original version we have the appeal to sympathy: "I will make money only if you buy our shoes." The hypotheses now are B, $B \to M$, $M \to S$, and $D \to S$. Again, S is a valid conclusion. The hypothesis $M \to S$ seems suspect, but, given the truth of this hypothesis, the argument is flawless.

(d) We could replace $M \to D$ in the original by the more believable statement $B \vee S \to M$: "I will make money if I am a famous player *or* if you buy our shoes." We already have the stronger hypothesis $B \to M$, though, so the analysis in part (b) shows that we can't deduce S in this case either. ■

Sometimes a situation is so complicated that it is not at all clear at first whether one should be trying to show that an alleged conclusion does follow or that it doesn't. In such a case it is often a good plan to see what you can deduce from the hypotheses, in the hope that you'll either learn a great deal about the possibilities or arrive at a contradiction. Either way, you win. If nothing else, this approach forces you to look at the problem slowly and methodically, which by itself is often enough to unravel it.

EXAMPLE 8

(a) Let

A = "I am an adult,"

B = "I am big and brave,"

Y = "I am young,"

L = "life is tough," and

N = "nobody loves me."

Here is part of an argument. If I am an adult, then I am big and brave. If I am big and brave or I am young, then life is tough and nobody loves me. If I am big and brave and nobody loves me, then I am an adult. If I am big and brave or I am not young, then I am an adult. If I am big and brave or I am young, then nobody loves me. Either I am young, or I'm not. If I am big and brave and somebody loves me, then I am an adult. Therefore, I am an adult [though this is a sad example of adult reasoning].

What a mess! Does the conclusion A follow? Let's see. The hypotheses are $A \to B$, $(B \vee Y) \to (L \wedge N)$, $(B \wedge N) \to A$, $(B \vee \neg Y) \to A$, $(B \vee Y) \to N$ [a consequence of $(B \vee Y) \to (L \wedge N)$], $Y \vee \neg Y$ [a tautology], and $(B \wedge \neg N) \to A$. For all of these to be true and yet A to be false, we must have p false for every hypothesis $p \to A$. Thus, if A is false, then $B \wedge N$, $B \vee \neg Y$, and $B \wedge \neg N$ must be false. If $B \vee \neg Y$ is false, then B must be false and Y true. Then, since $(B \vee Y) \to (L \wedge N)$ is true, so is $Y \to (L \wedge N)$, and thus L and N are also true. At this point we know that if A is false, then so is B, but Y, L, and N must be true. We check that these truth values satisfy all the hypotheses, so A does not follow from the hypotheses by any valid argument.

We did not need to go through all the analysis of the last paragraph. All we *really* needed was a set of truth values that satisfied the hypotheses with A false. But how were we to find such values? The way we went about the analysis meant that we might be led to such truth values or we might be led to a proof that no such values are possible. In either case we would get a useful answer.

The original argument here was really just a lot of hypotheses and a conclusion, with no steps shown in between. It is possible for an argument to have lots of steps filled in correctly, but still to be a fallacy if there is just one gap somewhere. ■

The thought patterns that help us verify proofs can also be useful for untangling complicated logical formulations.

EXAMPLE 9

We look at a fairly typical excerpt from the manual of a powerful computer operating system. Let

$A =$ "a signal condition arises for a process,"

$P =$ "the signal is added to a set of signals pending for the process,"

$B =$ "the signal is currently blocked by the process,"

$D =$ "the signal is delivered to the process,"

$S =$ "the current state of the process is saved,"

$M =$ "a new signal mask is calculated,"

$H =$ "the signal handler is invoked,"

$N =$ "the handling routine is invoked normally,"

$R =$ "the process will resume execution in the previous context,"

$I =$ "the process must arrange to restore the previous context itself."

The manual says "$A \to P$, $(P \wedge \neg B) \to D$, $D \to (S \wedge M \wedge H)$, $(H \wedge N) \to R$, $(H \wedge \neg R) \to I$." It really does. We have just translated the manual from English into letters and symbols and left out a few words.

What can we conclude from these hypotheses? In particular, what will happen if $A \wedge \neg B \wedge \neg R$ is true, i.e., if a signal condition arises, the signal is not blocked by a process, but the process does not resume execution in its previous context?

We can deduce $P \wedge \neg B$ from $A \to P$ and $A \wedge \neg B$ [Exercise 13(a)]. Hence, using $(P \wedge \neg B) \to D$ and $D \to (S \wedge M \wedge H)$, we have $S \wedge M \wedge H$. In particular, H is true. Another short proof [Exercise 13(b)] deduces $\neg N$ from $H \wedge \neg R$ and $(H \wedge N) \to R$. Thus, since $(H \wedge \neg R) \to I$, $H \wedge \neg R$ yields $I \wedge \neg N$. We have been able to show that if A is true and B and R are false, then I is true and N is false;

i.e., signal handling is not invoked normally, and the process must restore its own previous context. Along the way we also showed that P, D, S, M, and H are true.

Of course, one would not usually write this analysis out as a formal proof. It is helpful, though, to dissect the verbal description into separate statements, to give the statements labels, and to write out symbolically what the hypotheses are, as well as any conclusions that follow from them. ■

When you are constructing proofs to answer exercises in this book or elsewhere, you should check for unused hypotheses. Although in practice problems often have more than enough information, textbook exercises *usually* have none to spare. If you haven't used some of what was given, perhaps your argument has a flaw, for instance an overlooked case.

Exercises 2.6

1. We observe C and observe that A implies C. We reason that this means that if A were false, then C would be false too, which it isn't. So A must be true. Is this argument valid? Explain.

2. Give two examples of fallacies drawn from everyday life. Explain why the arguments are fallacies. *Suggestion:* Advertising and letters to the editor are good sources.

3. (a) Show that there is no valid proof of C from the hypothesis $A \to C$.

(b) Is there a valid proof of C from $A \to C$ and $B \to C$? Explain.

(c) How strong is the argument for C if $A_1 \to C$, $A_2 \to C, \ldots, A_{1,000,000} \to C$ are true? Explain.

4. (a) If we leave out the hypothesis $B \to M$ in Example 7(a), do the remaining hypotheses imply S? Explain.

(b) Would the argument for S be stronger in part (a) with the added hypothesis $M \to S$, but still without $B \to M$? Explain.

5. For the following sets of hypotheses, state a conclusion that can be inferred and specify the rules of inference used.

(a) If the TV set is not broken, then I will not study. If I study, then I will pass the course. I will not pass the course.

(b) If I passed the midterm and the final, then I passed the course. If I passed the course, then I passed the final. I failed the course.

(c) If I pass the midterm or the final, then I will pass this course. I will take the next course only if I pass this course. I will not take the next course.

6. Consider the following hypotheses. If I take the bus or subway, then I will be late for my appointment. If I take a cab, then I will not be late for my appointment and I will be broke. I will be on time for my appointment. Which of the following conclusions *must* follow, i.e., can be inferred from the hypotheses? Justify your answers.

(a) I will take a cab.

(b) I will be broke.

(c) I will not take the subway.

(d) If I become broke, then I took a cab.

(e) If I take the bus, then I won't be broke.

7. Assume the hypotheses of Example 8(a). Which of the following are valid conclusions? Justify your answer in each case.

(a) I am an adult if and only if I am big and brave.

(b) If I am big and brave, then somebody loves me.

(c) Either I am young or I am an adult.

(d) Life is tough and nobody loves me.

8. Either Pat did it or Quincy did. Quincy could not have been reading and done it. Quincy was reading. Who did it? Explain, using an appropriate formal proof with the variables P, Q, and R.

9. Convert each of the following arguments into logical notation using the suggested variables. Then provide a formal proof.

(a) "If my computations are correct and I pay the electric bill, then I will run out of money. If I don't pay the electric bill, the power will be turned off. Therefore, if I don't run out of money and the power is still on, then my computations are incorrect." (c, b, r, p)

(b) "If the weather bureau predicts dry weather, then I will take a hike or go swimming. I will go swimming if and only if the weather bureau predicts warm weather. Therefore, if I don't go on a hike, then the weather bureau predicts wet or warm weather." (d, h, s, w)

(c) "If I get the job and work hard, then I will get promoted. If I get promoted, then I will be happy. I will not be happy. Therefore, either I will not get the job or I will not work hard." (j, w, p, h)

(d) "If I study law, then I will make a lot of money. If I study archaeology, then I will travel a lot. If I make a lot of money or travel a lot, then I will not be disappointed. Therefore, if I am disappointed, then I did not study law and I did not study archaeology." (l, m, a, t, d)

10. (a) Revise the proof in Example 3 to reach the contradiction $s \wedge (\neg s)$. *Hint:* Use lines 5 and 9 and rearrange steps.

(b) Revise the proof in Example 3 to reach the contradiction $p \wedge (\neg p)$. *Hint:* Use lines 3 and 7, then lines 4 and 9, and rearrange.

11. Give an alternative formal proof for Example 3 that does not use rule 33.

12. For each of the following, give a formal proof of the theorem or show that it is false by exhibiting a suitable row of a truth table.

(a) If $(q \wedge r) \to p$ and $q \to \neg r$, then p.

(b) If $q \vee \neg r$ and $\neg(r \to q) \to \neg p$, then p.

(c) If $p \to (q \vee r)$, $q \to s$ and $r \to \neg p$, then $p \to s$.

13. Give formal proofs of the following.

(a) $P \wedge \neg B$ from $A \to P$ and $A \wedge \neg B$.

(b) $\neg N$ from $H \wedge \neg R$ and $(H \wedge N) \to R$. This little theorem is mentioned in Example 9. *Suggestion:* Proof by contradiction.

14. When one of the authors was young, he was told that "a job worth doing is worth doing well." He agreed and still did the yardwork poorly. Discuss.

Chapter Highlights

To check your understanding of the material in this chapter, we recommend that you consider the items listed below and:

(a) Satisfy yourself that you can define each concept and notation and can describe each method.
(b) Give at least one reason why the item was included in this chapter.
(c) Think of at least one example of each concept and at least one situation in which each fact or method would be useful.

CONCEPTS AND NOTATION

propositional calculus
- proposition
- logical connectives $\neg, \vee, \wedge, \to, \leftrightarrow$ $\forall, \exists$ notation
- converse, contrapositive, counterexample
- necessary condition, sufficient condition
- compound proposition
  - truth table
  - tautology, contradiction
  - logical equivalence $\Longleftrightarrow$, logical implication $\Longrightarrow$

methods of proof
- direct, indirect, by contradiction, by cases
- vacuous, trivial
- constructive, nonconstructive

formal proof
- hypothesis, conclusion, theorem
- rule of inference
- vertical listing of proof

analysis of arguments
- valid inference, valid proof, fallacy

FACTS

Basic logical equivalences [Table 1 on page 62].
Basic logical implications [Table 2 on page 63].
Substitution Rules (a) and (b) on pages 82 and 83.
Basic rules of inference [Table 3 on page 79].
There is a formal proof of C from H if and only if $H \Longrightarrow C$ is true.

METHODS

Use of:
- Truth tables, especially to verify logical equivalences and implications.
- De Morgan laws to eliminate $\vee$ or $\wedge$.
- Rules of inference to construct formal proofs.
- Formal symbolism to analyze informal arguments.

Supplementary Exercises

1. Find a proposition that is logically equivalent to $p \oplus q$ and that uses only the connectives $\wedge$ and $\neg$.

2. Let n be an integer. Show that $n^2 - 3$ is *not* a multiple of 4. *Hint:* Consider cases $n = 2k$ and $n = 2k + 1$.

3. Show that if a, b, and c are real numbers such that $a + b + c > 300$, then $a > 100$ or $b > 100$ or $c > 100$.

4. Give the negation of the proposition "I did not hear anything that could have been written by nobody but Stravinsky."

5. Let m and n be positive integers. Show that if $mn > 56$, then $m \geq 8$ or $n \geq 8$.

6. Show each of the following by using cases.

(a) $\max\{x, y\} \cdot \min\{x, y\} = xy$.

(b) $(\max\{x, y\} - \min\{x, y\})^2 = (x - y)^2$.

7. Give reasons for the logical equivalences in the following chain.

$$p \to (q \vee r)$$

(a) $\Longleftrightarrow \neg p \vee (q \vee r)$

(b) $\Longleftrightarrow (\neg p \vee q) \vee r$

(c) $\Longleftrightarrow \neg(\neg\neg p \wedge \neg q) \vee r$

(d) $\Longleftrightarrow \neg(p \wedge \neg q) \vee r$

(e) $\Longleftrightarrow (p \wedge \neg q) \to r$

8. Provide a formal proof of "I will take the bus" from the hypotheses "if it is raining, then I will drive or I will take the bus," "I will not drive," and "it is raining." Use b, d, and r for notation. [Compare with Exercise 7.]

9. What is wrong with the following proof of the proposition "$x \geq 5$ if and only if $10x \geq 44 - x$"?

Proof(??) If $x \geq 5$, then $-x \leq -5$, so $44 - x \leq 44 - 5 = 39 < 50 \leq 10x$, as desired. In the other direction, if $10x \geq 44 - x$ is false, i.e., if $10x < 44 - x$, then $11x < 44$, so $x < 4 < 5$, and thus $x \geq 5$ is false. ■

10. What is wrong with the following proof of the false proposition "$x \geq 6$ if $x^2 \geq 25$"?

Proof(??) If $x \geq 6$, then $x^2 = x \cdot x \geq 6 \cdot 6 = 36 \geq 25$. ■

11. Prove that if m and n are positive integers and $n + 2m \geq 30$, then $n \geq 11$ or $m \geq n$.

12. (a) Give the complete truth table for the proposition $(\neg p \wedge q) \to (p \vee r)$.

(b) Is this proposition a tautology? Explain.

13. Prove or disprove each statement about sets A, B, and C.

(a) $A \cup B = A \cup C$ implies $B = C$.

(b) $A \oplus B = A \oplus C$ implies $B = C$.

14. Which of the following are tautologies? Prove your assertions. It is fair to create (partial) truth tables or to cite rules in the book.

(a) $[p \wedge (p \to q)] \to q$ (b) $[(p \vee q) \wedge \neg p] \to \neg q$

(c) $[(p \to q) \wedge \neg p] \to \neg q$

(d) $[(p \to q) \wedge \neg q] \to \neg p$

(e) $[(p \to q) \vee (q \to r)] \to (p \to r)$

(f) $(p \wedge q) \to q$

15. Here is a proof by contradiction for the following theorem: If $p \vee (q \to r)$, $q \vee r$, and $r \to p$, then p. For each step, supply an explanation.

1. $p \vee (q \to r)$
2. $q \vee r$
3. $r \to p$
4. $\neg p$
5. $\neg r$
6. $r \vee q$
7. q
8. $q \to r$
9. r
10. $r \wedge \neg r$
11. contradiction

16. Suppose that n is a positive integer that is not divisible by 2 or by 3. Show that 12 divides $n^2 - 1$.

17. Let n be an integer.

(a) Show that n is not divisible by 3 if and only if $n^2 - 3n + 2$ is a multiple of 3.

(b) Show that n is a multiple of 3 if and only if $n^2 - 3n + 2$ is not divisible by 3.

18. Give a formal proof of $q \to r$ from the hypotheses $p \to (q \to r)$ and $q \to p$.

19. The logical equivalence $p \to (q \vee r) \Longleftrightarrow (p \wedge \neg q) \to r$ of Exercise 7 is often used to prove $p \to (q \vee r)$ by proving $(p \wedge \neg q) \to r$ instead. Use this idea to prove the following propositions about positive integers m and n.

(a) If $m \cdot n \geq 8$, then $m \geq 3$ or $n \geq 4$.

(b) If $m^2 + n^2 \geq 25$, then $m \geq 3$ or $n \geq 5$.

(c) If $m^2 + n^2 \geq 25$ and n is a multiple of 3, then $m \geq 4$ or $n \geq 6$.

20. What is wrong with the following proof of the false proposition "n is a multiple of 3 whenever $n^2 - n$ is even"?

Proof(??) If n is a multiple of 3, then $n = 3k$ for some integer k, so $n^2 - n = 9k^2 - 3k = 3k(3k - 1)$. If k is even, then so are $3k$ and $3k(3k - 1)$. If k is odd, then $3k$ is odd, so $3k - 1$ is even, and again $n^2 - n$ is even. ■

21. Prove or disprove the following statements for a function $f: S \to T$.

(a) $f(A_1) = f(A_2)$ implies $A_1 = A_2$. Here A_1 and A_2 represent subsets of S.

(b) $f(f^{\leftarrow}(B)) \subseteq B$ for all subsets B of T.

Relations

One frequently wants to link elements of two different sets or to compare or contrast various members of the same set, e.g., to arrange them in some appropriate order or to group together those with similar properties. The mathematical framework to describe this kind of organization of sets is the theory of relations. This chapter introduces relations and connects them with digraphs and matrices.

3.1 Relations

Given sets S and T, a **(binary) relation** from S to T is just a subset R of $S \times T$, i.e., a set of ordered pairs (s, t). This is a general, all-purpose definition. As we go through the book we will see a variety of interesting special kinds of relations.

EXAMPLE 1 A mail-order music company has a list L of customers. Each customer indicates interest in certain categories of music: classical, easy-listening, Latin, religious, folk, popular, rock, etc. Let C be the set of possible categories. The set of all ordered pairs (name, selected category) is a relation R from L to C. This relation might contain such pairs as (K. A. Ross, classical), (C. R. B. Wright, classical), and (K. A. Ross, folk). ■

EXAMPLE 2 A university would be interested in the database relation R_1 consisting of all ordered pairs whose first entries are students and whose second entries are the courses the students are currently enrolled in. This relation is from the set S of university students to the set C of courses offered. For a given student s in S, the set of courses taken by s is $\{c \in C : (s, c) \in R_1\}$. On the other hand, if course c in C is fixed, then $\{s \in S : (s, c) \in R_1\}$ is the class list for the course.

Another relation R_2 consists of all ordered pairs whose first entries are courses and whose second entries are the departments for which the course is a major requirement. Thus R_2 is a relation from C to the set D of departments in the university. For fixed $c \in C$, the set of departments for which c is a major requirement is $\{d \in D : (c, d) \in R_2\}$. For fixed d in D, the set $\{c \in C : (c, d) \in R_2\}$ is the list of courses required for that department's majors. A computerized degree-checking program would need to use a data structure that contained enough information to determine the relations R_1 and R_2. ■

EXAMPLE 3

(a) Consider a set P of programs written to be carried out on a computer and a catalog C of canned programs available for use. We get a relation from C to P if we say that a canned program c is related to a program p in P provided p calls c as a subroutine. A frequently used c might be related to a number of p's, while a c that is never called is related to no p.

(b) A translator from decimal representations to binary representations can be viewed as the relation consisting of all ordered pairs whose first entries are allowable decimal representations and whose second entries are the corresponding binary representations. Actually, this relation is a function. ■

We sometimes say that s is ***R*-related** to t or that s is related to t by R in case $(s, t) \in R$. It seems almost as natural, though, to say that t is R-related to s in this case and, to avoid getting things backward, it's probably best not to use such terminology at all unless both (s, t) and (t, s) are in R. To say that "s and t satisfy R" is a little less risky. There is no ambiguity, however, in the notation $\boldsymbol{s\,R\,t}$, which one can verbalize as "s is R-related to t."

By its very nature, "relation" is a very general concept. We next discuss some special sorts of relations, some of which will already be familiar. Recall that in §1.5 we indicated how functions can be identified with their graphs and hence with sets of ordered pairs. In fact, if $f : S \to T$, we identified the function f with the set

$$R_f = \{(x, y) \in S \times T : y = f(x)\},$$

which is a relation from S to T. From this point of view, a **function from S to T** is a special kind of relation R from S to T, one such that

$$\text{for each } x \in S \text{ there is exactly one } y \in T \text{ with } (x, y) \in R.$$

This characterization is simply a restatement of the definition in §1.5. Thus functions are the relations for which functional notation f makes sense: $f(x)$ is that unique element in T such that $(x, f(x))$ belongs to R_f.

Relations of the sort we have just been considering have obvious applications to databases that link objects of different types. There is a natural generalization of the idea to organize data of more than two types. An ***n*-ary relation** for $n \geq 2$ is just a set of n-tuples in $S_1 \times \cdots \times S_n$. We will stick to $n = 2$, though. The basic ideas are already present in this simpler case.

Relations that link objects of the same type—two integers, two classes, two people, etc.—have quite a different flavor from the relations we have just considered and often have nothing to do with handling data. We will say that a subset R of $S \times S$ is a **relation on S**.

EXAMPLE 4

(a) For the set $\mathbb{R}$ the familiar inequality relation $\leq$ can be viewed as a subset R of $\mathbb{R} \times \mathbb{R}$, namely the set $R = \{(x, y) : x \leq y\}$. Since $(x, y) \in R$ if and only if $x \leq y$, we normally write the relation as $\leq$. Note that the familiar properties

(R) $x \leq x$ for all $x \in \mathbb{R}$,
(AS) $x \leq y$ and $y \leq x$ imply that $x = y$,
(T) $x \leq y$ and $y \leq z$ imply that $x \leq z$,

can be written as

(R) $(x, x) \in R$ for all $x \in \mathbb{R}$,
(AS) $(x, y) \in R$ and $(y, x) \in R$ imply that $x = y$,
(T) $(x, y) \in R$ and $(y, z) \in R$ imply that $(x, z) \in R$.

Here the labels (R), (AS), and (T) refer to "reflexive," "antisymmetric," and "transitive," terms that we will be using in connection with arbitrary relations on a set S.

(b) The strict inequality relation $<$ on $\mathbb{R}$ is also a relation and corresponds to the set $R = \{(x, y) : x < y\}$. This relation satisfies the following properties:

(AR) $x < x$ *never* holds,
(T) $x < y$ and $y < z$ imply that $x < z$.

These can be rewritten as

(AR) $(x, x) \notin R$ for all $x \in \mathbb{R}$,
(T) $(x, y) \in R$ and $(y, z) \in R$ imply that $(x, z) \in R$.

Here (AR) refers to "antireflexive" and (T) again refers to "transitive." ■

Let p be a fixed integer greater than 1. Consider integers m and n. We say that m **is congruent to** n **modulo** p, and we write $m \equiv n \pmod p$ provided $m - n$ is a multiple of p. This condition defines what is called a **congruence relation** on the set $\mathbb{Z}$ of integers. We will return to this relation just prior to Theorem 1 on page 121, where we will see that

(R) $m \equiv m \pmod p$ for all $m \in \mathbb{Z}$,
(S) $m \equiv n \pmod p$ implies $n \equiv m \pmod p$,
(T) $m \equiv n \pmod p$ and $n \equiv r \pmod p$ imply $m \equiv r \pmod p$.

For the corresponding formal relation

$$R = \{(m, n) \in \mathbb{Z} \times \mathbb{Z} : m \equiv n \pmod p\},$$

these properties become

(R) $(m, m) \in R$ for all $m \in \mathbb{Z}$,
(S) $(m, n) \in R$ implies $(n, m) \in R$,
(T) $(m, n) \in R$ and $(n, r) \in R$ imply that $(m, r) \in R$.

Here the labels (R), (S), and (T) refer to "reflexive," "symmetric" and "transitive." Note that this usage of reflexive and transitive is consistent with that in Example 4.

In general, we define a relation R on the set S to be **reflexive**, **antireflexive**, **symmetric**, **antisymmetric**, or **transitive** if it satisfies the corresponding condition:

(R) $(x, x) \in R$ for all $x \in S$,
(AR) $(x, x) \notin R$ for all $x \in S$,
(S) $(x, y) \in R$ implies $(y, x) \in R$ for all $x, y \in S$,
(AS) $(x, y) \in R$ and $(y, x) \in R$ imply that $x = y$,
(T) $(x, y) \in R$ and $(y, z) \in R$ imply that $(x, z) \in R$.

These conditions may also be written as:

(R) $x\,R\,x$ for all $x \in S$,
(AR) $x\,R\,x$ fails for all $x \in S$,
(S) $x\,R\,y$ implies $y\,R\,x$ for all $x, y \in S$,
(AS) $x\,R\,y$ and $y\,R\,x$ imply that $x = y$,
(T) $x\,R\,y$ and $y\,R\,z$ imply that $x\,R\,z$.

Observe that the antireflexive condition (AR) is different from nonreflexivity, since a relation R is nonreflexive provided $x\,R\,x$ fails for *some* $x \in S$. Similarly, the antisymmetric condition (AS) is different from nonsymmetry.

EXAMPLE 5

Every set S has the very basic **equality relation**

$$E = \{(x, x) : x \in S\}.$$

Two elements in S are E-related if and only if they are identical. We normally write $=$ for this relation. Thus $(x, y) \in E$ if and only if $x = y$. It would be correct, but very strange, to write $(2, 2) \in =$.

The relation E is reflexive, symmetric, and transitive, assertions that you should verify mentally. For example, transitivity asserts in this case that if $x = y$ and if

$y = z$, then $x = z$. In §3.4, we will study relations that are reflexive, symmetric, and transitive, which are called "equivalence relations." The relation E is the most basic equivalence relation, and it is the prototype of all such relations where objects are related if we are viewing them as essentially the same, even though they may not be equal.

The relation E is antisymmetric, as well as symmetric, so antisymmetry is certainly different from nonsymmetry. Is E antireflexive? ■

Consider again an arbitrary relation R from a set S to a set T. That is, $R \subseteq S \times T$. The **converse relation $R^{\leftarrow}$** is the relation from T to S defined by

$$(t, s) \in R^{\leftarrow} \quad \text{if and only if} \quad (s, t) \in R.$$

Since every function $f\colon S \to T$ is a relation, its converse $f^{\leftarrow}$ always exists:

$$\textit{As a relation} \quad f^{\leftarrow} = \{(y, x) \in T \times S : y = f(x)\}.$$

This relation is a function precisely when f is an invertible function as defined in §1.7, in which case we have $f^{\leftarrow} = f^{-1}$.

EXAMPLE 6

(a) Recall that if $f\colon S \to T$ is a function and $A \subseteq S$, then the image of A under f is

$$f(A) = \{f(s) : s \in A\} = \{t \in T : t = f(s) \text{ for some } s \in A\}.$$

If we view f as the relation R_f, then the set $f(A)$ is equal to

$$\{t \in T : (s, t) \in R_f \text{ for some } s \in A\}.$$

Similarly, for any relation R from S to T we can define

$$\boldsymbol{R(A)} = \{t \in T : (s, t) \in R \text{ for some } s \in A\}.$$

Since $R^{\leftarrow}$ is a relation from T to S, for $B \subseteq T$ we also have

$$\begin{aligned} R^{\leftarrow}(B) &= \{s \in S : (t, s) \in R^{\leftarrow} \text{ for some } t \in B\} \\ &= \{s \in S : (s, t) \in R \text{ for some } t \in B\}. \end{aligned}$$

If R is actually R_f for a function f from S to T, then we have

$$R_f^{\leftarrow}(B) = \{s \in S : t = f(s) \text{ for some } t \in B\} = \{s \in S : f(s) \in B\},$$

which is exactly the definition we gave for $f^{\leftarrow}(B)$ in §1.7.

(b) For a concrete example of part (a), let S be a set of suppliers and T a set of products, and define $(x, y) \in R$ if supplier x sells product y. For a given set A of suppliers, the set $R(A)$ is the set of products sold by at least one member of A. For a given set B of products, $R^{\leftarrow}(B)$ is the set of suppliers who sell at least one product in B. The relation R is R_f for a function f from S to T if and only if each supplier sells exactly one product. ■

It is sometimes handy to draw pictures of relations on small sets.

EXAMPLE 7

(a) Consider the relation R_1 on the set $\{0, 1, 2, 3\}$ defined by $\leq$; thus $(m, n) \in R_1$ if and only if $m \leq n$. A picture of R_1 is given in Figure 1(a). Observe that we have drawn an arrow from m to n whenever $(m, n) \in R_1$, though we left off the arrowheads on the "loops" $0 \to 0$, $1 \to 1$, etc.

(b) Let R_2 be the relation on $\{1, 2, 3, 4, 5\}$ defined by $m\, R_2\, n$ if and only if $m - n$ is even. A picture is given in Figure 1(b).

(c) The picture of the converse relation $R_1^{\leftarrow}$ is obtained by reversing all the arrows in Figure 1(a). The loops are unchanged.

(d) The picture of the converse $R_2^{\leftarrow}$ is also obtained by reversing all the arrows [in Figure 1(b)], but this time we obtain the same picture. This is because R_2 is symmetric and so $R_2^{\leftarrow} = R_2$; see Exercise 15(a). ■

Figure 1 ▶

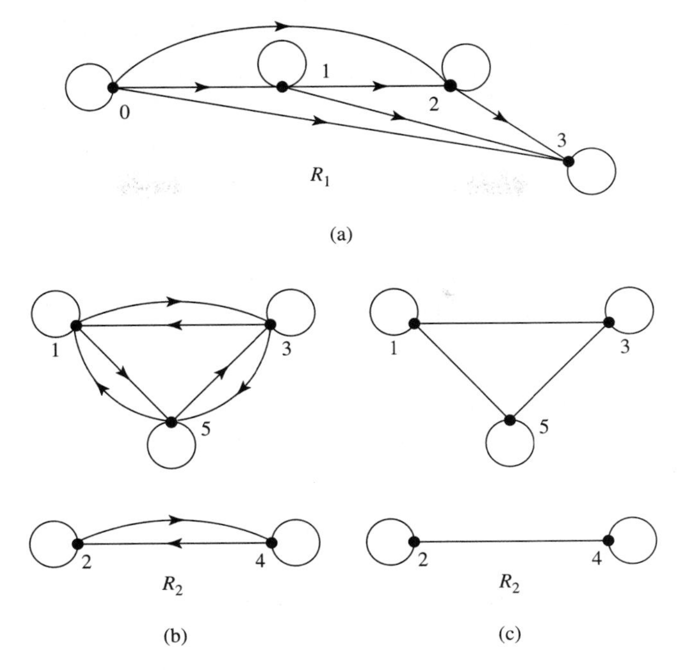

When a relation is symmetric, such as R_2 in Figure 1(b), for every arrow drawn there is a reverse arrow, so it is redundant to draw each such pair of arrows. An equally informative picture is drawn in Figure 1(c). Just as with city maps, arrows signify one-way streets and plain lines signify two-way streets.

Exercises 3.1

1. For the following relations on $S = \{0, 1, 2, 3\}$, specify which of the properties (R), (AR), (S), (AS), and (T) the relations satisfy.
   (a) $(m, n) \in R_1$ if $m + n = 3$
   (b) $(m, n) \in R_2$ if $m - n$ is even
   (c) $(m, n) \in R_3$ if $m \leq n$
   (d) $(m, n) \in R_4$ if $m + n \leq 4$
   (e) $(m, n) \in R_5$ if $\max\{m, n\} = 3$
2. Let $A = \{0, 1, 2\}$. Each of the statements below defines a relation R on A by $(m, n) \in R$ if the statement is true for m and n. Write each of the relations as a set of ordered pairs.
   (a) $m \leq n$ (b) $m < n$
   (c) $m = n$ (d) $mn = 0$
   (e) $mn = m$ (f) $m + n \in A$
   (g) $m^2 + n^2 = 2$ (h) $m^2 + n^2 = 3$
   (i) $m = \max\{n, 1\}$
3. Which of the relations in Exercise 2 are reflexive? symmetric?
4. The following relations are defined on $\mathbb{N}$.
   (a) Write the relation R_1 defined by $(m, n) \in R_1$ if $m + n = 5$ as a set of ordered pairs.
   (b) Do the same for R_2 defined by $\max\{m, n\} = 2$.
   (c) The relation R_3 defined by $\min\{m, n\} = 2$ consists of infinitely many ordered pairs. List five of them.
5. For each of the relations in Exercise 4, specify which of the properties (R), (AR), (S), (AS), and (T) the relation satisfies.
6. Consider the relation R on $\mathbb{Z}$ defined by $(m, n) \in R$ if and only if $m^3 - n^3 \equiv 0 \pmod 5$. Which of the properties (R), (AR), (S), (AS), and (T) are satisfied by R?
7. Define the "divides" relation R on $\mathbb{N}$ by
   $$(m, n) \in R \quad \text{if} \quad m|n.$$
   [Recall from §1.2 that $m|n$ means that n is a multiple of m.]
   (a) Which of the properties (R), (AR), (S), (AS), and (T) does R satisfy?
   (b) Describe the converse relation $R^{\leftarrow}$.
   (c) Which of the properties (R), (AR), (S), (AS), and (T) does the converse relation $R^{\leftarrow}$ satisfy?
8. What is the connection between a relation R and the relation $(R^{\leftarrow})^{\leftarrow}$?
9. (a) If S is a nonempty set, then the empty set $\emptyset$ is a subset of $S \times S$, so it is a relation on S, called the the **empty relation**. Which of the properties (R), (AR), (S), (AS), and (T) does $\emptyset$ possess?
   (b) Repeat part (a) for the **universal relation** $U = S \times S$ on S.
10. Give an example of a relation that is:
    (a) antisymmetric and transitive but not reflexive,
    (b) symmetric but not reflexive or transitive.

11. Do a relation and its converse always satisfy the same conditions (R), (AR), (S), and (AS)? Explain.

12. Show that a relation R is transitive if and only if its converse relation $R^{\leftarrow}$ is transitive.

13. Let R_1 and R_2 be relations on a set S.

(a) Show that $R_1 \cap R_2$ is reflexive if R_1 and R_2 are.

(b) Show that $R_1 \cap R_2$ is symmetric if R_1 and R_2 are.

(c) Show that $R_1 \cap R_2$ is transitive if R_1 and R_2 are.

14. Let R_1 and R_2 be relations on a set S.

(a) Must $R_1 \cup R_2$ be reflexive if R_1 and R_2 are?

(b) Must $R_1 \cup R_2$ be symmetric if R_1 and R_2 are?

(c) Must $R_1 \cup R_2$ be transitive if R_1 and R_2 are?

15. Let R be a relation on a set S.

(a) Prove that R is symmetric if and only if $R = R^{\leftarrow}$.

(b) Prove that R is antisymmetric if and only if $R \cap R^{\leftarrow} \subseteq E$, where $E = \{(x, x) : x \in S\}$.

16. Let R_1 and R_2 be relations from a set S to a set T.

(a) Show that $(R_1 \cup R_2)^{\leftarrow} = R_1^{\leftarrow} \cup R_2^{\leftarrow}$.

(b) Show that $(R_1 \cap R_2)^{\leftarrow} = R_1^{\leftarrow} \cap R_2^{\leftarrow}$.

(c) Show that if $R_1 \subseteq R_2$ then $R_1^{\leftarrow} \subseteq R_2^{\leftarrow}$.

17. Draw pictures of each of the relations in Exercise 1. Don't use arrows if the relation is symmetric.

18. Draw pictures of each of the relations in Exercise 2. Don't use arrows if the relation is symmetric.

3.2 Digraphs and Graphs

You are already familiar with the idea of a graph as a picture of a function. The word "graph" is also used to describe a different kind of structure that arises in a variety of natural settings. In a loose sense these new graphs are diagrams that, properly interpreted, contain information. The graphs we are concerned with are like road maps, circuit diagrams, or flow charts in the sense that they depict connections or relationships between various parts of a diagram. They also give us ways to visualize relations.

Figure 1 ▶

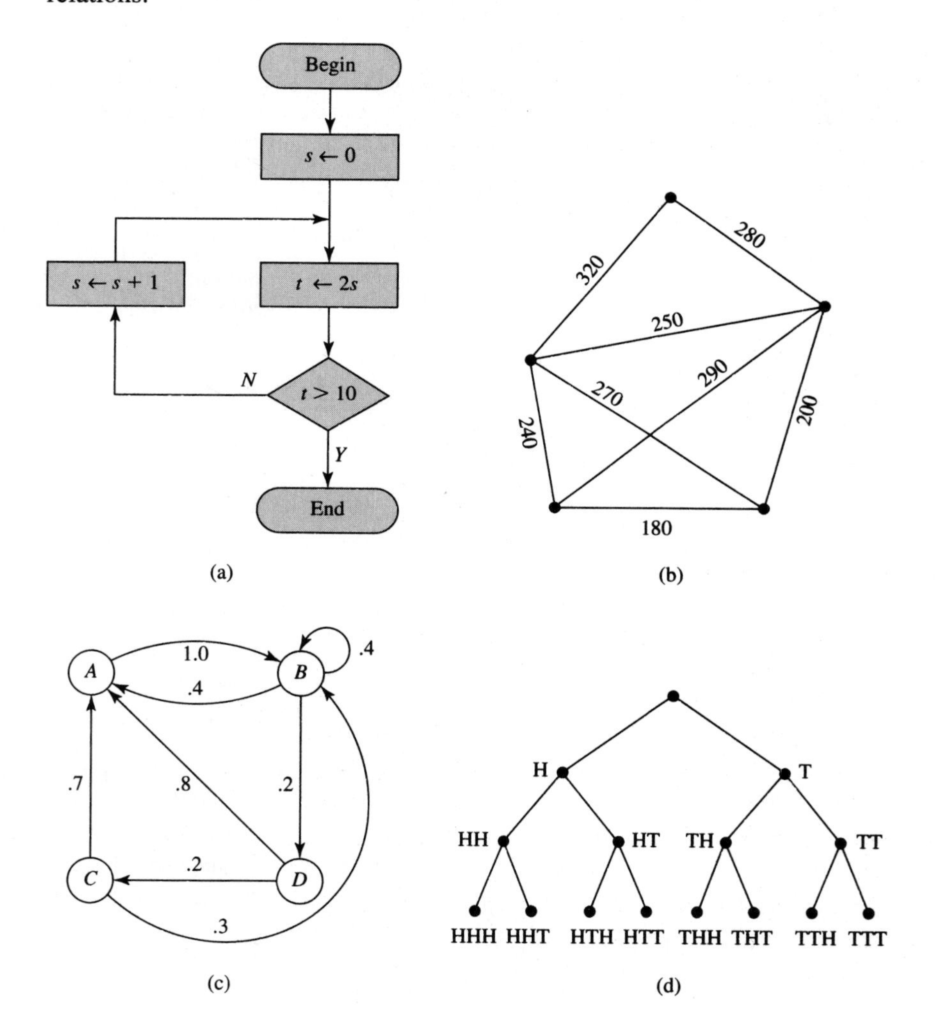

The diagrams in Figure 1 are from a variety of settings. Figure 1(a) shows a simple flow chart. Figure 1(b) might represent five warehouses of a trucking firm and truck routes between them, labeled with their distances. Figure 1(c) could be telling us about the probability that a rat located in one of four cages will move to one of the other three or stay in its own cage. Figure 1(d) might depict possible outcomes of a repeated experiment such as coin tossing [Heads or Tails]. The diagrams in Figure 1 on page 99 tell us which pairs of vertices belong to the relations R_1 and R_2. What do all of these diagrams have in common? Each consists of a collection of objects—boxes, circles, or dots—and some lines, possibly curved, that connect objects. Sometimes the lines are directed; that is, they are arrows.

The essential features of a directed graph [digraph for short] are its objects and directed lines. Specifically, a **digraph** G consists of two sets, the nonempty set $\boldsymbol{V(G)}$ of **vertices of** G and the set $\boldsymbol{E(G)}$ of **edges** of G, together with a function $\boldsymbol{\gamma}$ [Greek lowercase gamma] from $E(G)$ to $V(G) \times V(G)$ that tells where the edges go. If e is an edge of G and $\gamma(e) = (p, q)$, then we say e **goes from** p **to** q, and we call p the **initial vertex** of e and q the **terminal vertex** of e. This definition makes sense even if $V(G)$ or $E(G)$ is infinite, but we will assume in this section that $V(G)$ and $E(G)$ are finite because our applications are to finite sets.

A **picture** of the digraph G is a diagram consisting of points, corresponding to the members of $V(G)$, and arrows, corresponding to the members of $E(G)$, such that if $\gamma(e) = (p, q)$ then the arrow corresponding to e goes from the point labeled p to the point labeled q.

EXAMPLE 1

Consider the digraph G with vertex set $V(G) = \{w, x, y, z\}$, edge set $E(G) = \{a, b, c, d, e, f, g, h\}$, and γ given by the table in Figure 2(a). The diagrams in Figure 2(b) and 2(c) are both pictures of G. In Figure 2(b) we labeled the arrows to make the correspondence to $E(G)$ plain. Figure 2(c) is a picture of the same digraph even though the positions of the vertices are different, which at first glance makes the graph look different. Also, we simply labeled the points and let the arrows take care of themselves, which causes no confusion because, in this case, there are no **parallel edges**; i.e., there is at most one edge with a given initial vertex and terminal vertex. In other words, the function γ is one-to-one. Note also that we omitted the arrowhead on edge d, since z is clearly both the initial and terminal vertex. ■

Figure 2 ▸

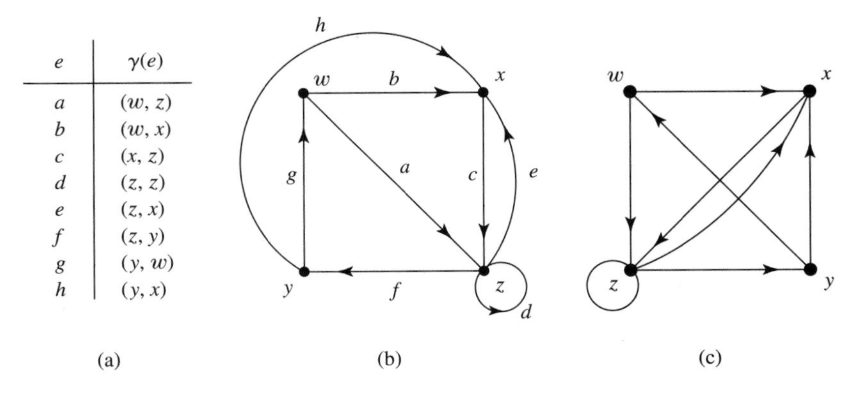

e	$\gamma(e)$
a	(w, z)
b	(w, x)
c	(x, z)
d	(z, z)
e	(z, x)
f	(z, y)
g	(y, w)
h	(y, x)

(a) (b) (c)

If $\gamma\colon E(G) \to V(G) \times V(G)$ is one-to-one, as it will be in many cases that we encounter, then we can identify the edges e with their images $\gamma(e)$ in $V(G) \times V(G)$ and consider $E(G)$ to *be* a subset of $V(G) \times V(G)$. In such a case, there's no need to introduce the function γ or give names to the edges. The list of vertices $\{w, x, y, z\}$ and the list of edges in the second column of Figure 2(a) contain all the information about the digraph, as does the corresponding picture in Figure 2(c). By the way, some people define digraphs to have $E(G) \subseteq V(G) \times V(G)$ and call the more general digraphs we are considering "directed multigraphs."

Given any picture of G, we can reconstruct G itself, since the arrows tell us all about γ. We will commonly describe digraphs by giving pictures of them, rather than tables of γ, but the pictorial description is chosen just for human convenience. A computer stores a digraph by storing the function γ in one way or another.

Many of the important questions connected with digraphs can be stated in terms of sequences of edges leading from one vertex to another. A **path** in a digraph G is a sequence of edges such that the terminal vertex of one edge is the initial vertex of the next. Thus, if $e_1, \ldots, e_n$ are in $E(G)$, then $e_1\, e_2 \cdots e_n$ is a path, provided there are vertices $x_1, x_2, \ldots, x_n, x_{n+1}$ so that $\gamma(e_1) = (x_1, x_2)$, $\gamma(e_2) = (x_2, x_3)$, and, in general, $\gamma(e_i) = (x_i, x_{i+1})$ for $i = 1, 2, \ldots, n$. We say that $e_1\, e_2 \cdots e_n$ is a path of **length** n **from** x_1 **to** x_{n+1}. The path is **closed** if it starts and ends at the same vertex, i.e., if $x_1 = x_{n+1}$.

EXAMPLE 2

In the digraph G in Figure 2, the sequence $f\, g\, a\, e$ is a path of length 4 from z to x. The sequences $c\, e\, c\, e\, c$ and $f\, g\, a\, f\, h\, c$ are also paths, but $f\, a$ is not a path, since $\gamma(f) = (z, y)$, $\gamma(a) = (w, z)$, and $y \neq w$. The paths $f\, g\, a\, f\, h\, c$, $c\, e\, c\, e$, and d are closed; $f\, h\, c\, e$ and $d\, f$ are not. ■

A path $e_1 \cdots e_n$ with $\gamma(e_i) = (x_i, x_{i+1})$ has an associated sequence of vertices $x_1\, x_2 \cdots x_n\, x_{n+1}$. If each e_i is the only edge from x_i to x_{i+1}, then this sequence of vertices uniquely determines the path, and we can describe the path simply by listing the vertices in order. This will be the case whenever we are able to consider $E(G)$ as a subset of $V(G) \times V(G)$.

EXAMPLE 3

(a) In Figure 2 the path $f\, g\, a\, e$ has vertex sequence $z\, y\, w\, z\, x$. Observe that this vertex sequence alone determines the path. The path can be recovered from $z\, y\, w\, z\, x$ by looking at Figure 2(b) or (c) or by using the table of γ in Figure 2(a). Since the digraph has no parallel edges, we may regard $E(G)$ as a subset of $V(G) \times V(G)$ so that all its paths are determined by their vertex sequences.

(b) The digraph pictured in Figure 3 has parallel edges a and b, so γ is not one-to-one and we cannot view $E(G)$ as a subset of $V(G) \times V(G)$. While the the vertex sequence $y\, z\, z\, z$ corresponds only to the path $f\, g\, g$, the sequence $y\, v\, w\, z$ belongs to both $c\, a\, e$ and $c\, b\, e$. ■

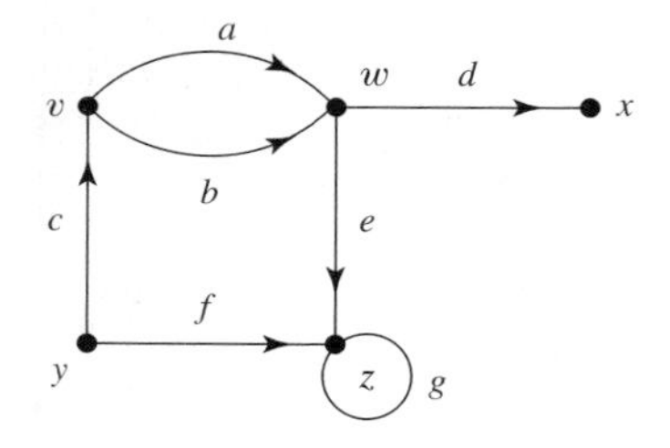

Figure 3 ▲

A closed path of length at least 1 with vertex sequence $x_1\, x_2 \cdots x_n\, x_1$ is called a **cycle** if $x_1, \ldots, x_n$ are all different. The language of graph theory has not been standardized: various authors use "circuit" and "loop" for what we call a cycle, and "cycle" is sometimes used as a name for a closed path. Sometimes it's important that a digraph contain a cycle, and sometimes it's important that it have none. A digraph with no cycles is called **acyclic**, a definition whose meaning turns out to be the same whichever definition of "cycle" is used [by Corollary 1 on page 319]. Some people call an acyclic digraph a **DAG**, short for "directed acyclic graph." A path is acyclic if the digraph consisting of the vertices and edges of the path is acyclic.

EXAMPLE 4

In Figure 2 the path $a\, f\, g$ is a cycle, since its vertex sequence is $w\, z\, y\, w$. Likewise, the paths $c\, f\, h$ and $c\, f\, g\, b$, with vertex sequences $x\, z\, y\, x$ and $x\, z\, y\, w\, x$, are cycles. The short path $c\, e$ and the loop d are also cycles, since their vertex sequences are $x\, z\, x$ and $z\, z$, respectively. The path $c\, f\, g\, a\, e$ is not a cycle, since its vertex sequence is $x\, z\, y\, w\, z\, x$ and the vertex z is repeated. ■

As we saw in Example 7 on page 98, a relation R on a set S determines a digraph G in a natural way: Let $V(G) = S$ and put an edge from v to w whenever $(v, w) \in R$. We can reverse this procedure. Given a digraph G and vertices v and w in $V(G)$, call v **adjacent** to w if there is an edge in $E(G)$ from v to w. The **adjacency relation** A on the set $V(G)$ of vertices is defined by $(v, w) \in A$ if and only if v is adjacent to w. The relation A can be perfectly general and does not need to have any special properties. It will be reflexive only if G has a loop at each vertex and symmetric only if there is an edge from v to w whenever there is one from w to v.

EXAMPLE 5

(a) Consider the digraph in Figure 2. The adjacency relation consists of the ordered pairs (w, z), (w, x), (x, z), (z, z), (z, x), (z, y), (y, w), and (y, x). In other words, A consists of the images of γ listed in Figure 2(a). In general, $A = \gamma(E(G)) \subseteq V(G) \times V(G)$.

(b) The adjacency relation A for the digraph in Figure 3 consists of the ordered pairs (v, w), (w, x), (w, z), (z, z), (y, z) and (y, v). We cannot recover the digraph from A because A does not convey information about multiple edges. Since (v, w) belongs to A, we know that the digraph has *at least one* edge from v to w, but we cannot tell that it has exactly two edges.

(c) The digraph obtained from Figure 3 by removing the edge a has the same adjacency relation as the original digraph. ■

The previous example shows that different digraphs may have the same adjacency relation. However, if we restrict our attention to digraphs without multiple edges, then there is a one-to-one correspondence between such digraphs and relations.

If we ignore the arrows on our edges, i.e., their directions, we obtain what are called "graphs." Figure 1(b) and (d) of this section and Figure 1(c) on page 99 are all pictures of graphs. Graphs can have multiple edges; to see an example, just remove the arrowheads from Figure 3. The ideas and terminology for studying graphs are similar to those for digraphs.

Instead of being associated with an ordered pair of vertices, as edges in digraphs are, an undirected edge has an unordered set of vertices. Following the pattern we used for digraphs, we define an [undirected] **graph** to consist of two sets, the set $\boldsymbol{V(G)}$ of vertices of G and the set $\boldsymbol{E(G)}$ of **edges** of G, together with a function γ from $E(G)$ to the set $\big\{\{u, v\} : u, v \in V(G)\big\}$ of all subsets of $V(G)$ with one or two members. For an edge e in $E(G)$, the members of $\gamma(e)$ are called the **vertices** of e or the **endpoints** of e; we say that e **joins** its endpoints. A **loop** is an edge with only one endpoint. Distinct edges e and f with $\gamma(e) = \gamma(f)$ are called **parallel** or **multiple** edges.

These precise definitions make it clear that a computer might view a graph as two sets together with a function γ that specifies the endpoints of the edges.

What we have just described is called a **multigraph** by some authors, who reserve the term "graph" for those graphs with no loops or parallel edges. If there are no parallel edges, then γ is one-to-one and the sets $\gamma(e)$ uniquely determine the edges e. That is, there is only one edge for each set $\gamma(e)$. In this case, we often dispense with the set $E(G)$ and the function γ and simply write the edges as sets, like $\{u, v\}$ or $\{u\}$, or as vertex sequences, like $u\,v$, $v\,u$, or $u\,u$. Thus we will commonly write $e = \{u, v\}$; we will also sometimes write $e = \{u, u\}$ instead of $e = \{u\}$ if e is a loop with vertex u.

A **picture** of a graph G is a diagram consisting of points corresponding to the vertices of G and arcs or lines corresponding to edges, such that if $\gamma(e) = \{u, v\}$, then the arc for the edge e joins the points labeled u and v.

EXAMPLE 6

(a) The graphs in Figures 4(a) and 4(b) have 5 vertices and 7 edges. The crossing of the two lines in Figure 4(b) is irrelevant and is just a peculiarity of our drawing. The graph in Figure 4(c) has 4 vertices and 6 edges. It has multiple edges joining v and w and has a loop at w.

(b) If we take the arrowheads off the edges in Figure 2, we get the picture in Figure 5(a). A table of γ for this graph is given in Figure 5(b). This graph has parallel edges c and e joining x and z and has a loop d with vertex z. The graph pictured in Figure 5(c) has no parallel edges, so for that graph a description such as "the edge $\{x, z\}$" is unambiguous. The same phrase could not be applied unambiguously to the graph in Figure 5(a). ■

Figure 4 ▶

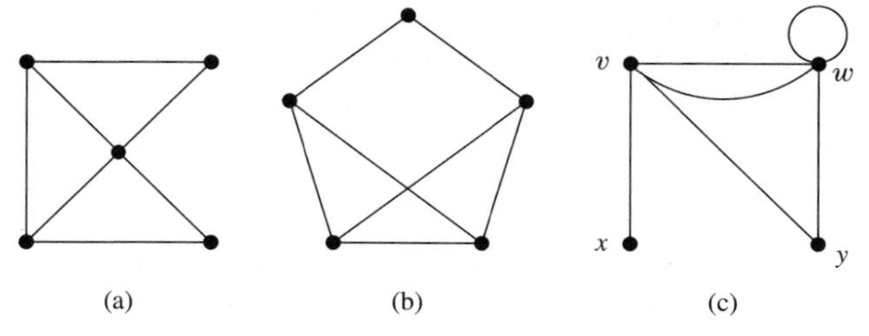

Figure 5 ▶

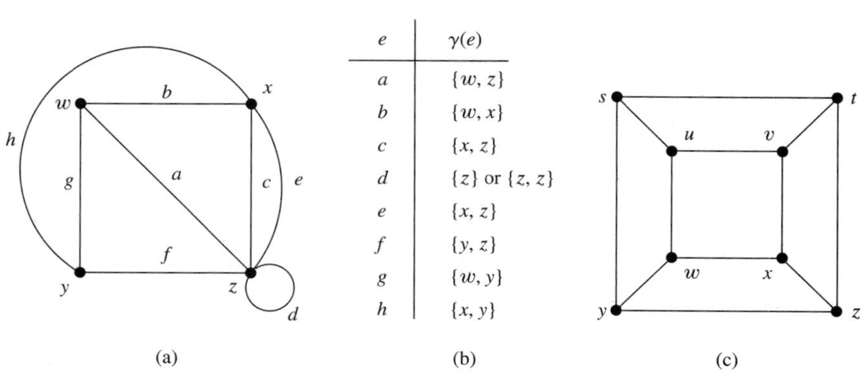

e	$\gamma(e)$
a	$\{w, z\}$
b	$\{w, x\}$
c	$\{x, z\}$
d	$\{z\}$ or $\{z, z\}$
e	$\{x, z\}$
f	$\{y, z\}$
g	$\{w, y\}$
h	$\{x, y\}$

A sequence of edges that link up with each other is called a **path**. Examples of paths in Figure 5(a) are $g\,a\,d\,e\,b$, $c\,f\,h\,e\,a\,g$, and $g\,b\,c\,d\,d\,a\,g$. The **length** of a path is the number of edges in the path. Thus $g\,b\,c\,d\,d\,a\,g$ has length 7. Consecutive edges in a path must have a vertex in common, so a path determines a sequence of vertices. The vertex sequences for the paths just discussed are $y\,w\,z\,z\,x\,w$, $x\,z\,y\,x\,z\,w\,y$, and $y\,w\,x\,z\,z\,z\,w\,y$. Note that we cannot always recover a path from its vertex sequence: Which edge corresponds to the segments $x\,z$ and $z\,x$?

Note that the number of vertices in a vertex sequence is 1 larger than the number of edges in the path. When a loop appears in a path, its vertex is repeated in the vertex sequence. Vertex sequences treat parallel edges the same, and so different paths, such as $b\,c\,a$ and $b\,e\,a$, can have the same vertex sequence. If a graph has no parallel edges or multiple loops, then vertex sequences do uniquely determine paths. In this case the edges can be described by just listing the two vertices that they connect, and we may describe a path by its vertex sequence.

In general, a **path** of **length** n from the vertex u to the vertex v is a sequence $e_1 \cdots e_n$ of edges together with a sequence $x_1 \cdots x_{n+1}$ of vertices with $\gamma(e_i) = \{x_i, x_{i+1}\}$ for $i = 1, \ldots, n$ and $x_1 = u$, $x_{n+1} = v$. One has to decide whether to consider paths of length 0. We will not have occasion to use them, so we will only permit paths to have length at least 1. If $e_1\,e_2 \cdots e_n$ is a path from u to v with vertex sequence $x_1\,x_2 \cdots x_{n+1}$, then $e_n \cdots e_2\,e_1$ with vertex sequence $x_{n+1}\,x_n \cdots x_1$ is a path from v to u. We may speak of either of these paths as a **path between** u and v. If $u = v$, then the path is said to be **closed**.

EXAMPLE 7

In Figure 5(c), we didn't bother to name the edges of the graph, because there are no parallel edges. All the edges and paths can be described by sequences of vertices. A nice path is $y\,w\,u\,v\,x\,z\,t\,s\,y$. Its length is 8 (not 9) and it's a closed path because the first and last vertices are the same. Moreover, it visits every vertex and only repeats the vertex y where it begins and ends. Such paths will be studied in §6.5 and are called "Hamilton circuits." Two paths between w and v are $w\,x\,v$ and $v\,t\,z\,y\,w$. Of course, paths can repeat edges or vertices; two examples are $w\,x\,z\,t\,s\,u\,w\,x\,v$ and $w\,y\,s\,u\,w\,x\,v$. ■

The edge sequence of a path usually determines the vertex sequence, and we will sometimes use phrases such as "the path $e_1\,e_2 \cdots e_n$" without mentioning vertices. There is a slight fuzziness here, since $e\,e$, $e\,e\,e$, etc., don't specify which end of e to start at. A similar problem arises with $e\,f$ in case e and f are parallel. If a graph has no parallel edges, then the vertex sequence completely determines the edge sequence. In that setting, or if the actual choice of edges is unimportant, we will commonly use vertex sequences as descriptions for paths.

Just as with digraphs, we can define an **adjacency relation** A for a graph: $(u, v) \in A$ provided $\{u, v\}$ is an edge of the graph. For the remainder of this section, A will denote an adjacency relation either for a digraph or for a graph. To get a transitive relation from A, we must consider chains of edges, from u_1 to u_2, u_2 to u_3, etc. The appropriate notion is reachability. Define the **reachable relation** R on $V(G)$ by

$$(v, w) \in R \text{ if there is a path in } G \text{ from } v \text{ to } w.$$

Then R is a transitive relation. Since we require all paths to have length at least 1, R might not be reflexive.

EXAMPLE 8

(a) All the vertices in the digraph of Figure 2 are reachable from all other vertices. Hence the reachability relation R consists of all possible ordered pairs; R is called the **universal relation**.

(b) The reachability relation for the digraph in Figure 3 consists of (v, w), (v, x), (v, z), (w, x), (w, z), (y, v), (y, w), (y, x), (y, z), and (z, z). Note that every vertex can be reached from y except y itself. Also, z is the only vertex that can be reached from itself.

(c) All the graphs in Figures 4 and 5 are connected, in the sense that every vertex can be reached from every other vertex. So in each case the reachability relation is the universal relation.

(d) For the graph labeled R_2 in Figure 1 on page 99, every odd-numbered vertex can be reached from every other odd-numbered vertex, and the even-numbered ones can be reached from each other. ■

Exercises 3.2

1. Give a table of the function γ for each of the digraphs pictured in Figure 6.

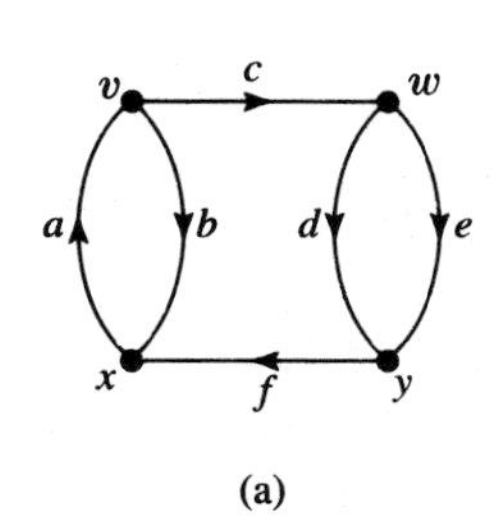

(a)

(b)

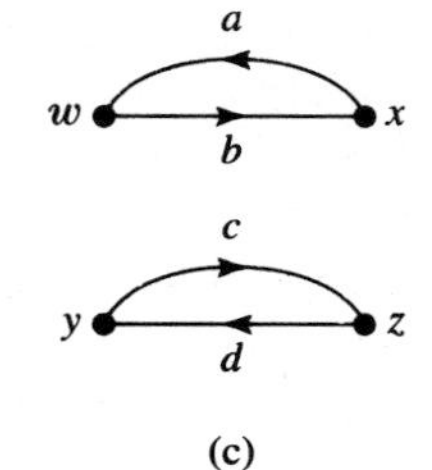

(c)

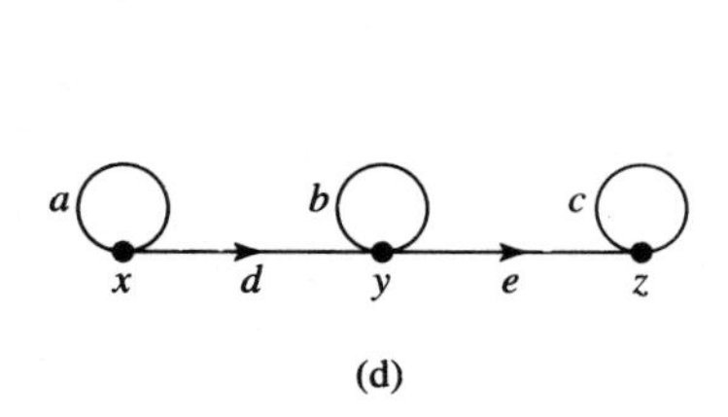

(d)

Figure 6 ▲

2. Draw a picture of the digraph G with vertex set $V(G) = \{w, x, y, z\}$, edge set $E(G) = \{a, b, c, d, e, f, g\}$, and γ given by the following table.

e	a	b	c	d	e	f	g
$\gamma(e)$	(x, w)	(w, x)	(x, x)	(w, z)	(w, y)	(w, z)	(z, y)

3. Which of the following vertex sequences describe paths in the digraph pictured in Figure 7(a)?

(a) $z\ y\ v\ w\ t$

(b) $x\ z\ w\ t$

(c) $v\ s\ t\ x$

(d) $z\ y\ s\ u$

(e) $x\ z\ y\ v\ s$

(f) $s\ u\ x\ t$

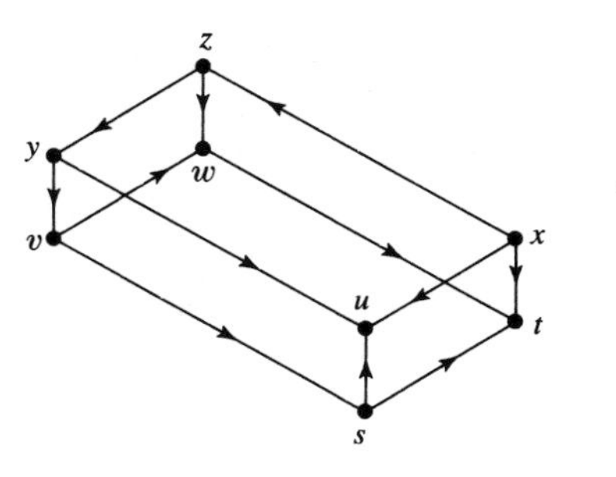

(a)

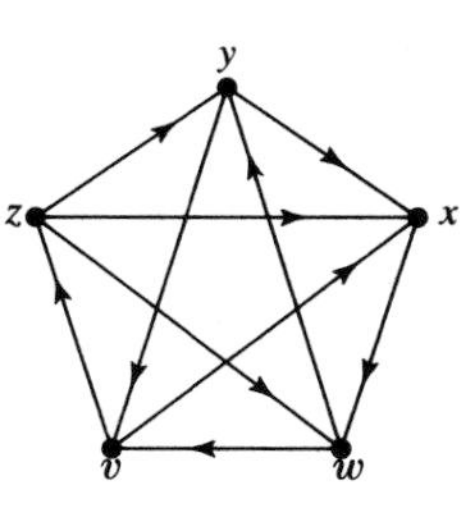

(b)

Figure 7 ▲

4. Find the length of a shortest path from x to w in the digraph shown in Figure 7(a).
5. Consider the digraph pictured in Figure 7(b). Describe an acyclic path
   (a) from x to y (b) from y to z (c) from v to w
   (d) from x to z (e) from z to v
6. There are four basic blood types: A, B, AB, and O. Type O can donate to any of the four types, A and B can donate to AB as well as to their own types, but type AB can only donate to AB. Draw a digraph that presents this information. Is the digraph acyclic?
7. Give an example of a digraph with vertices x, y, and z in which there is a cycle with x and y as vertices and another cycle with y and z, but there is no cycle with x and z as vertices.
8. Determine the reachability relation for the digraphs in Figures 6(a), (c), and (d).
9. (a) Which of the following belong to the reachability relation for the digraph in Figure 6(b)? (v, u), (v, v), (v, w), (v, x), (v, y), (v, z)
   (b) Which of the following belong to the reachability relation for the digraph in Figure 7(a)? (v, s), (v, t), (v, u), (v, v), (v, w), (v, x), (v, y), (v, z)
   (c) Which of the following belong to the reachability relation for the digraph in Figure 7(b)? (v, v), (v, w), (v, x), (v, y), (v, z)
10. Which of the following vertex sequences correspond to paths in the graph of Figure 8(a)?
    (a) $z\ x\ w$ (b) $w\ x\ z\ x\ w\ y\ w\ w$
    (c) $w\ w\ x\ z$ (d) $w\ x\ z\ z$
    (e) $z\ x\ w\ y\ y\ w\ z$ (f) $w\ x\ w$
    (g) $y\ y\ w\ w$
11. Give the length of each path in Exercise 10.
12. Which paths in Exercise 10 are closed paths?
13. List the parallel edges in the graphs of Figure 8.

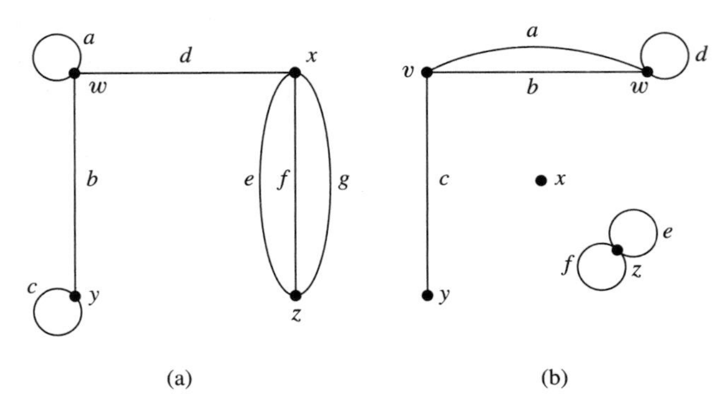

Figure 8 ▲

14. How many loops are there in each of the graphs of Figure 8?
15. Give the adjacency relation A and the reachable relation R for each of the graphs of Figure 8.
16. For the graph in Figure 8(a), give an example of each of the following.
    Be sure to specify the edge sequence and the vertex sequence.
    (a) a path of length 2 from w to z.
    (b) a path of length 4 from z to itself.
    (c) a path of length 5 from z to itself.
    (d) a path of length 3 from w to x.
17. For the graph in Figure 8(b), give an example of each of the following.
    Be sure to specify the edge sequence and the vertex sequence.
    (a) a path of length 3 from y to w.
    (b) a path of length 3 from v to y.
    (c) a path of length 4 from v to y.
    (d) a path of length 3 from z to itself.

3.3 Matrices

We saw the close connection between digraphs and relations in §3.2. Matrices, the subject of this section, are an important tool for describing both digraphs and relations, as well as for many other purposes. We will see that the topics of the first three sections of this chapter give us three different ways of viewing equivalent structures: relations, digraphs without parallel edges, and matrices whose entries are 0's and 1's.

In general a **matrix** is a rectangular array. It is traditional to use capital letters, such as **A**, for matrices. If we denote the entry in the ith row and jth column by a_{ij}, then we can write

$$\mathbf{A} = \begin{bmatrix} a_{11} & a_{12} & \cdots & a_{1n} \\ a_{21} & a_{22} & \cdots & a_{2n} \\ a_{31} & a_{32} & \cdots & a_{3n} \\ \vdots & \vdots & & \vdots \\ a_{m1} & a_{m2} & \cdots & a_{mn} \end{bmatrix},$$

or simply $\mathbf{A} = [a_{ij}]$. This matrix has m horizontal rows and n vertical columns and is called an $\boldsymbol{m \times n}$ **matrix**. Whenever double indexes are used in matrix theory, rows precede columns! Sometimes we denote the entry in the ith row and jth column by $\mathbf{A}[i, j]$; this notation is preferable in computer science since it avoids subscripts. In this book the entries of a **matrix** are real numbers unless otherwise specified.

EXAMPLE 1

(a) The matrix

$$\mathbf{A} = \begin{bmatrix} 2 & -1 & 0 & 3 & 2 \\ 1 & -2 & 1 & -1 & 3 \\ 3 & 0 & 1 & 2 & -3 \end{bmatrix}$$

is a 3×5 matrix. If we write $\mathbf{A} = [a_{ij}]$, then $a_{11} = 2$, $a_{31} = 3$, $a_{13} = 0$, $a_{35} = -3$, etc. If we use the notation $\mathbf{A}[i, j]$, then $\mathbf{A}[1, 2] = -1$, $\mathbf{A}[2, 1] = 1$, $\mathbf{A}[2, 2] = -2$, etc.

(b) If $\mathbf{B}$ is the 3×4 matrix defined by $\mathbf{B}[i, j] = i - j$, then $\mathbf{B}[1, 1] = 1 - 1 = 0$, $\mathbf{B}[1, 2] = 1 - 2 = -1$, etc., so

$$\mathbf{B} = \begin{bmatrix} 0 & -1 & -2 & -3 \\ 1 & 0 & -1 & -2 \\ 2 & 1 & 0 & -1 \end{bmatrix}.$$ ■

Matrices are used in all the mathematical sciences. Because they provide a convenient way to store data, they also have important uses in business, economics, and computer science. Matrices arise in solving systems of linear equations. Many physical phenomena in nature can be described, at least approximately, by means of matrices. In other words, matrices are important objects. In addition, the set of all $n \times n$ matrices has a very rich algebraic structure, which is of interest in itself and is also a source of inspiration in the study of more abstract algebraic structures. We will introduce various algebraic operations on matrices in this section.

For positive integers m and n, we write $\mathfrak{M}_{m,n}$ for the set of all $m \times n$ matrices. Two matrices $\mathbf{A}$ and $\mathbf{B}$ in $\mathfrak{M}_{m,n}$ are **equal** provided all their corresponding entries are equal; i.e., $\mathbf{A} = \mathbf{B}$ provided that $a_{ij} = b_{ij}$ for all i and j with $1 \le i \le m$ and $1 \le j \le n$. Matrices that have the same number of rows as columns are called **square matrices**. Thus $\mathbf{A}$ is a square matrix if $\mathbf{A}$ belongs to $\mathfrak{M}_{n,n}$ for some $n \in \mathbb{P}$. The **transpose** $\mathbf{A}^T$ of a matrix $\mathbf{A} = [a_{ij}]$ in $\mathfrak{M}_{m,n}$ is the matrix in $\mathfrak{M}_{n,m}$ whose entry in the ith row and jth column is a_{ji}. That is, $\mathbf{A}^T[i, j] = \mathbf{A}[j, i]$. For example, if

$$\mathbf{A} = \begin{bmatrix} 2 & -1 & 0 & 4 \\ 3 & 2 & -1 & 2 \\ 4 & 0 & 1 & 3 \end{bmatrix}, \quad \text{then} \quad \mathbf{A}^T = \begin{bmatrix} 2 & 3 & 4 \\ -1 & 2 & 0 \\ 0 & -1 & 1 \\ 4 & 2 & 3 \end{bmatrix}.$$

The first row in $\mathbf{A}$ becomes the first column in $\mathbf{A}^T$, etc.

Two matrices $\mathbf{A}$ and $\mathbf{B}$ can be added if they are the same size, that is, if they belong to the same set $\mathfrak{M}_{m,n}$. In this case, the **matrix sum** is obtained by adding corresponding entries. More explicitly, if both $\mathbf{A} = [a_{ij}]$ and $\mathbf{B} = [b_{ij}]$ are in $\mathfrak{M}_{m,n}$, then $\mathbf{A} + \mathbf{B}$ is the matrix $\mathbf{C} = [c_{ij}]$ in $\mathfrak{M}_{m,n}$ defined by

$$c_{ij} = a_{ij} + b_{ij} \quad \text{for} \quad 1 \le i \le m \quad \text{and} \quad 1 \le j \le n.$$

Equivalently, we define

$$(\mathbf{A} + \mathbf{B})[i, j] = \mathbf{A}[i, j] + \mathbf{B}[i, j] \quad \text{for} \quad 1 \le i \le m \quad \text{and} \quad 1 \le j \le n.$$

EXAMPLE 2

(a) Consider

$$\mathbf{A} = \begin{bmatrix} 2 & 4 & 0 \\ -1 & 3 & 2 \\ -3 & 1 & 2 \end{bmatrix}, \quad \mathbf{B} = \begin{bmatrix} 1 & 0 & 5 & 3 \\ 2 & 3 & -2 & 1 \\ 4 & -2 & 0 & 2 \end{bmatrix}, \quad \mathbf{C} = \begin{bmatrix} 3 & 1 & -2 \\ -5 & 0 & 2 \\ -2 & 4 & 1 \end{bmatrix}.$$

Then we have

$$\mathbf{A} + \mathbf{C} = \begin{bmatrix} 5 & 5 & -2 \\ -6 & 3 & 4 \\ -5 & 5 & 3 \end{bmatrix},$$

but $\mathbf{A}+\mathbf{B}$ and $\mathbf{B}+\mathbf{C}$ are not defined. Of course, the sums $\mathbf{A}+\mathbf{A}$, $\mathbf{B}+\mathbf{B}$ and $\mathbf{C}+\mathbf{C}$ are also defined; for example,

$$\mathbf{B}+\mathbf{B}=\begin{bmatrix} 2 & 0 & 10 & 6 \\ 4 & 6 & -4 & 2 \\ 8 & -4 & 0 & 4 \end{bmatrix}.$$

(b) Consider the 1-row matrices

$$\mathbf{v}_1=\begin{bmatrix} -2 & 1 & 2 & 3 \end{bmatrix},\quad \mathbf{v}_2=\begin{bmatrix} 4 & 0 & 3 & -2 \end{bmatrix},\quad \mathbf{v}_3=\begin{bmatrix} 1 & 3 & 5 \end{bmatrix}$$

and the 1-column matrices

$$\mathbf{v}_4=\begin{bmatrix} 1 \\ 2 \\ -3 \\ 2 \end{bmatrix},\quad \mathbf{v}_5=\begin{bmatrix} 0 \\ 3 \\ -2 \end{bmatrix},\quad \mathbf{v}_6=\begin{bmatrix} 4 \\ 1 \\ 5 \end{bmatrix}.$$

The only sums of distinct matrices here that are defined are

$$\mathbf{v}_1+\mathbf{v}_2=\begin{bmatrix} 2 & 1 & 5 & 1 \end{bmatrix} \quad\text{and}\quad \mathbf{v}_5+\mathbf{v}_6=\begin{bmatrix} 4 \\ 4 \\ 3 \end{bmatrix}.$$ ■

Before listing properties of addition, we give a little more notation. Let $\mathbf{0}$ represent the $m \times n$ matrix all entries of which are 0. [Context will always make plain what size this matrix is.] For $\mathbf{A}$ in $\mathfrak{M}_{m,n}$ the matrix $-\mathbf{A}$, called the **negative of A**, is obtained by negating each entry in $\mathbf{A}$. Thus, if $\mathbf{A}=[a_{ij}]$, then $-\mathbf{A}=[-a_{ij}]$; equivalently, $(-\mathbf{A})[i,j]=-\mathbf{A}[i,j]$ for all i and j.

Theorem For all $\mathbf{A}$, $\mathbf{B}$, and $\mathbf{C}$ in $\mathfrak{M}_{m,n}$

(a) $\mathbf{A}+(\mathbf{B}+\mathbf{C})=(\mathbf{A}+\mathbf{B})+\mathbf{C}$ [associative law]
(b) $\mathbf{A}+\mathbf{B}=\mathbf{B}+\mathbf{A}$ [commutative law]
(c) $\mathbf{A}+\mathbf{0}=\mathbf{0}+\mathbf{A}=\mathbf{A}$ [additive identity]
(d) $\mathbf{A}+(-\mathbf{A})=(-\mathbf{A})+\mathbf{A}=\mathbf{0}$ [additive inverses]

Proof These properties of matrix addition are reflections of corresponding properties of the addition of real numbers and are easy to check. We check (a) and leave the rest to Exercise 14.

Say $\mathbf{A}=[a_{ij}]$, $\mathbf{B}=[b_{ij}]$, and $\mathbf{C}=[c_{ij}]$. The (i,j) entry of $\mathbf{B}+\mathbf{C}$ is $b_{ij}+c_{ij}$, so the (i,j) entry of $\mathbf{A}+(\mathbf{B}+\mathbf{C})$ is $a_{ij}+(b_{ij}+c_{ij})$. Similarly, the (i,j) entry of $(\mathbf{A}+\mathbf{B})+\mathbf{C}$ is $(a_{ij}+b_{ij})+c_{ij}$. Since addition of real numbers is associative, the corresponding entries of $\mathbf{A}+(\mathbf{B}+\mathbf{C})$ and $(\mathbf{A}+\mathbf{B})+\mathbf{C}$ are equal, so the matrices are equal. ■

Since addition of matrices is associative, we can write $\mathbf{A}+\mathbf{B}+\mathbf{C}$ without causing ambiguity.

Matrices can be multiplied by real numbers, which in this context are often called **scalars**. Given $\mathbf{A}$ in $\mathfrak{M}_{m,n}$ and c in $\mathbb{R}$, the matrix $c\mathbf{A}$ is the $m \times n$ matrix whose (i,j) entry is $c \cdot a_{ij}$; thus $(c\mathbf{A})[i,j]=c\cdot\mathbf{A}[i,j]$. This multiplication is called **scalar multiplication** and $c\mathbf{A}$ is called the **scalar product**.

EXAMPLE 3

(a) If

$$\mathbf{A}=\begin{bmatrix} 2 & 1 & -3 \\ -1 & 0 & 4 \end{bmatrix},$$

then

$$2\mathbf{A}=\begin{bmatrix} 4 & 2 & -6 \\ -2 & 0 & 8 \end{bmatrix} \quad\text{and}\quad -7\mathbf{A}=\begin{bmatrix} -14 & -7 & 21 \\ 7 & 0 & -28 \end{bmatrix}.$$

(b) In general, the scalar product $(-1)\mathbf{A}$ is the negative $-\mathbf{A}$ of $\mathbf{A}$. ■

It is sometimes possible to multiply two matrices together, using a product definition that appears complicated, but that is completely natural for many applications. Here is the recipe. If **A** is an $m \times n$ matrix and **B** is an $n \times p$ matrix, then the **product AB** is the $m \times p$ matrix **C** defined by

$$c_{ik} = \sum_{j=1}^{n} a_{ij} b_{jk} \quad \text{for} \quad 1 \le i \le m \quad \text{and} \quad 1 \le k \le p.$$

In subscript-free notation,

$$(\mathbf{AB})[i, k] = \sum_{j=1}^{n} \mathbf{A}[i, j] \cdot \mathbf{B}[j, k].$$

For the sum to make sense here, the rows of **A** must have the same number of entries as the columns of **B**.

In the simple case $n = m = p = 2$, we obtain the equation

$$\begin{bmatrix} a_{11} & a_{12} \\ a_{21} & a_{22} \end{bmatrix} \begin{bmatrix} b_{11} & b_{12} \\ b_{21} & b_{22} \end{bmatrix} = \begin{bmatrix} a_{11}b_{11} + a_{12}b_{21} & a_{11}b_{12} + a_{12}b_{22} \\ a_{21}b_{11} + a_{22}b_{21} & a_{21}b_{12} + a_{22}b_{22} \end{bmatrix},$$

so, for example,

$$\begin{bmatrix} 2 & -1 \\ -6 & 3 \end{bmatrix} \begin{bmatrix} 1 & 2 \\ 2 & 4 \end{bmatrix} = \begin{bmatrix} 0 & 0 \\ 0 & 0 \end{bmatrix} \ne \begin{bmatrix} -10 & 5 \\ -20 & 10 \end{bmatrix} = \begin{bmatrix} 1 & 2 \\ 2 & 4 \end{bmatrix} \begin{bmatrix} 2 & -1 \\ -6 & 3 \end{bmatrix}.$$

As these calculations illustrate, the product of two nonzero matrices can be the zero matrix, and the order of the factors matters. We will have more to say about matrix multiplication in §11.3.

We end this section by describing a matrix that is useful for studying finite relations, digraphs, and graphs. First consider a finite graph or digraph G with vertex set $V(G)$. Let $v_1, v_2, \ldots, v_n$ be a list of the vertices in $V(G)$. The **adjacency matrix** is the $n \times n$ matrix **M** such that each entry $\mathbf{M}[i, j]$ is the number of edges from v_i to v_j. Thus $\mathbf{M}[i, j] = 0$ if there is no edge from v_i to v_j, and $\mathbf{M}[i, j]$ is a positive integer otherwise.

EXAMPLE 4

(a) The digraph in Figure 1 has adjacency matrix

$$\mathbf{M} = \begin{bmatrix} 0 & 1 & 0 & 0 \\ 1 & 0 & 0 & 0 \\ 1 & 2 & 0 & 0 \\ 3 & 0 & 0 & 1 \end{bmatrix}.$$

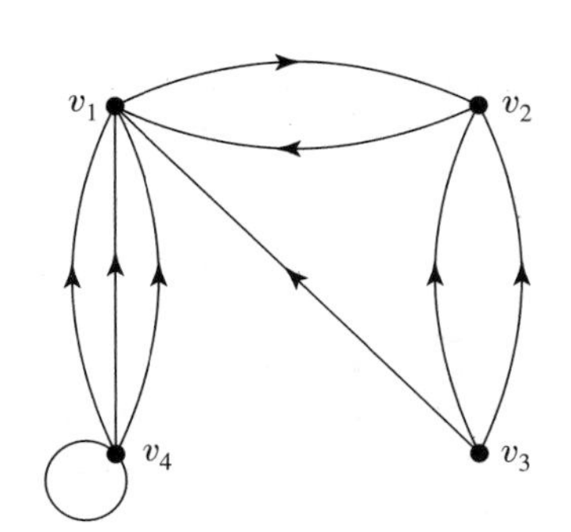

Figure 1 ▲

Note that the matrix **M** contains all the information about the digraph. It tells us that there are four vertices, and it tells us how many edges connect each pair of vertices.

(b) Don't look for it now, but here is the adjacency matrix for a digraph in §3.2:

$$\mathbf{M} = \begin{bmatrix} \mathbf{0} & 1 & 0 & 1 \\ 0 & \mathbf{0} & 0 & 1 \\ 1 & 1 & \mathbf{0} & 0 \\ 0 & 1 & 1 & \mathbf{1} \end{bmatrix}.$$

Let's see how much information we can gather from the matrix alone. There are four vertices because the matrix is 4×4. There are eight edges because the sum of the entries is 8. Since all the entries are 0's and 1's, there are no multiple edges. There is one loop since there is one 1 on the **main diagonal** [in boldface]. Now look at Figure 2 in §3.2 for the digraph, where we have mentally relabeled the vertices w, x, y, z as v_1, v_2, v_3, v_4. For example, the loop is at $z = v_4$. ■

Recall that every relation R on a finite set S corresponds to a finite digraph G with no multiple edges. Hence it also corresponds to a matrix $\mathbf{M}_R$ of 0's and 1's. Since the vertex set of G is the set S, if $|S| = n$, then the matrix is $n \times n$.

EXAMPLE 5

We return to the relations of Example 7 on page 98.

(a) The relation R_1 on $\{0, 1, 2, 3\}$ is defined by $m \le n$. The matrix for R_1 is

$$\mathbf{M}_{R_1} = \begin{bmatrix} 1 & 1 & 1 & 1 \\ 0 & 1 & 1 & 1 \\ 0 & 0 & 1 & 1 \\ 0 & 0 & 0 & 1 \end{bmatrix}.$$

We have again mentally relabeled the set of vertices; this time 0, 1, 2, 3 correspond to v_1, v_2, v_3, v_4.

(b) The relation R_2 on $S = \{1, 2, 3, 4, 5\}$ is defined by requiring $m - n$ to be even. If we keep 1, 2, 3, 4, 5 in their usual order, the matrix for R_2 is

$$\mathbf{M}_{R_2} = \begin{bmatrix} 1 & 0 & 1 & 0 & 1 \\ 0 & 1 & 0 & 1 & 0 \\ 1 & 0 & 1 & 0 & 1 \\ 0 & 1 & 0 & 1 & 0 \\ 1 & 0 & 1 & 0 & 1 \end{bmatrix}.$$

If we reorder S as 1, 3, 5, 2, 4, we obtain

$$\begin{bmatrix} 1 & 1 & 1 & 0 & 0 \\ 1 & 1 & 1 & 0 & 0 \\ 1 & 1 & 1 & 0 & 0 \\ 0 & 0 & 0 & 1 & 1 \\ 0 & 0 & 0 & 1 & 1 \end{bmatrix}.$$

From this matrix it is clear that the first three elements of S [1, 3, and 5] are all related to each other and the last two are related to each other. ■

The matrix for the converse relation $R^{\leftarrow}$ of a relation R is the transpose of the matrix for R. In symbols, $\mathbf{M}_{R^{\leftarrow}} = \mathbf{M}_R^T$.

EXAMPLE 6

(a) The matrix for $R_1^{\leftarrow}$, where R_1 is in Example 5, is

$$\begin{bmatrix} 1 & 0 & 0 & 0 \\ 1 & 1 & 0 & 0 \\ 1 & 1 & 1 & 0 \\ 1 & 1 & 1 & 1 \end{bmatrix}.$$

In other words, this is the matrix for the relation defined by $m \ge n$.

(b) Consider the relation R_2 in Example 5. Once the order of the set S is fixed, the matrix for $R_2^{\leftarrow}$ is the same as for R_2, because $R_2 = R_2^{\leftarrow}$ [since R_2 is symmetric]. ■

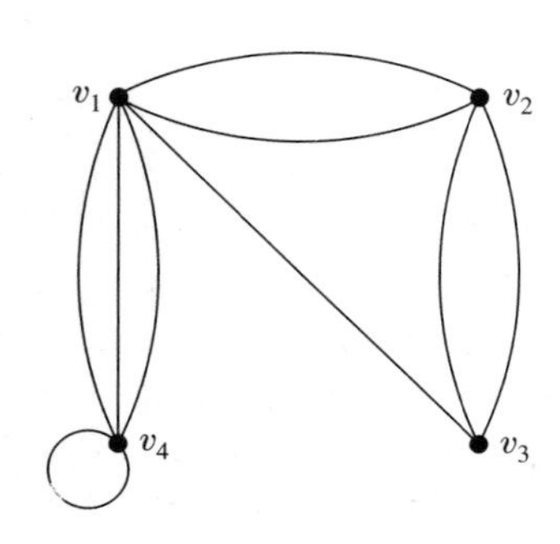

Figure 2 ▲

In general, a relation R is symmetric if and only if its matrix $\mathbf{M}_R$ equals its transpose. Matrices $\mathbf{M}$ such that $\mathbf{M}^T = \mathbf{M}$ are called **symmetric**, so a relation is symmetric if and only if its matrix is symmetric. The definition of adjacency matrix that we have given for a digraph makes sense for an undirected graph as well and yields a symmetric matrix whose entries are nonnegative integers.

EXAMPLE 7

Figure 2 is a picture of the graph obtained from Figure 1 by ignoring the arrowheads. Its adjacency matrix is

$$\mathbf{M} = \begin{bmatrix} \mathbf{0} & 2 & 1 & 3 \\ 2 & \mathbf{0} & 2 & 0 \\ 1 & 2 & \mathbf{0} & 0 \\ 3 & 0 & 0 & \mathbf{1} \end{bmatrix}.$$

Note that $\mathbf{M}$ is symmetric, i.e., $\mathbf{M}^T = \mathbf{M}$, as expected. As before, the graph is completely determined by the matrix, and a lot of information about the graph can be read off from the matrix. For example, we can see the one loop from the main diagonal. Counting edges is a bit trickier now, however, because each nonloop appears twice in the matrix. The number of edges is now the sum of the entries on and below the diagonal. ■

Exercises 3.3

1. Consider the matrix

$$\mathbf{A} = \begin{bmatrix} 1 & -2 & 5 \\ 3 & -2 & 3 \\ 2 & 0 & 1 \end{bmatrix}.$$

Evaluate

(a) a_{11} (b) a_{13} (c) a_{31} (d) $\sum_{i=1}^{3} a_{ii}$

2. Consider the matrix

$$\mathbf{B} = \begin{bmatrix} 1 & 2 & -2 & 1 \\ 3 & 0 & 1 & 2 \\ 2 & -1 & 4 & 1 \\ 0 & -3 & 1 & 3 \end{bmatrix}.$$

Evaluate

(a) b_{12} (b) b_{21} (c) b_{23} (d) $\sum_{i=1}^{4} b_{ii}$

3. Consider the matrices

$$\mathbf{A} = \begin{bmatrix} -1 & 0 & 2 \\ 1 & 3 & -2 \\ 4 & 2 & 3 \end{bmatrix}, \quad \mathbf{B} = \begin{bmatrix} 6 & 8 & 5 \\ 4 & -2 & 7 \\ 3 & 1 & 2 \end{bmatrix},$$

$$\mathbf{C} = \begin{bmatrix} 1 & 3 \\ 2 & -4 \\ 5 & -2 \end{bmatrix}.$$

Calculate the following when they exist.

(a) $\mathbf{A}^T$ (b) $\mathbf{C}^T$ (c) $\mathbf{A}+\mathbf{B}$
(d) $\mathbf{A}+\mathbf{C}$ (e) $(\mathbf{A}+\mathbf{B})^T$ (f) $\mathbf{A}^T+\mathbf{B}^T$
(g) $\mathbf{B}+\mathbf{B}^T$ (h) $\mathbf{C}+\mathbf{C}^T$

4. For the matrices in Exercise 3, calculate the following when they exist.

(a) $\mathbf{A}+\mathbf{A}$ (b) $2\mathbf{A}$
(c) $\mathbf{A}+\mathbf{A}+\mathbf{A}$ (d) $4\mathbf{A}+\mathbf{B}$

5. Let $\mathbf{A} = [a_{ij}]$ and $\mathbf{B} = [b_{ij}]$ be matrices in $\mathfrak{M}_{4,3}$ defined by $a_{ij} = (-1)^{i+j}$ and $b_{ij} = i + j$. Find the following matrices when they exist.

(a) $\mathbf{A}^T$ (b) $\mathbf{A}+\mathbf{B}$ (c) $\mathbf{A}^T+\mathbf{B}$
(d) $\mathbf{A}^T+\mathbf{B}^T$ (e) $(\mathbf{A}+\mathbf{B})^T$ (f) $\mathbf{A}+\mathbf{A}$

6. Let $\mathbf{A}$ and $\mathbf{B}$ be matrices in $\mathfrak{M}_{3,3}$ defined by $\mathbf{A}[i,j] = ij$ and $\mathbf{B}[i,j] = i + j^2$.

(a) Find $\mathbf{A}+\mathbf{B}$.

(b) Calculate $\sum_{i=1}^{3} \mathbf{A}[i,i]$.

(c) Does $\mathbf{A}$ equal its transpose $\mathbf{A}^T$?

(d) Does $\mathbf{B}$ equal its transpose $\mathbf{B}^T$?

7. Consider the matrices

$$\mathbf{A} = \begin{bmatrix} 3 & 9 \\ 1 & 3 \end{bmatrix} \quad \text{and} \quad \mathbf{B} = \begin{bmatrix} 1 & -3 \\ -2 & 6 \end{bmatrix}.$$

Calculate the following

(a) $\mathbf{AB}$ (b) $\mathbf{BA}$
(c) $\mathbf{A}^2 = \mathbf{AA}$ (d) $\mathbf{B}^2$

8. (a) For the matrices in Exercise 7, calculate

$$(\mathbf{A}+\mathbf{B})^2 \quad \text{and} \quad \mathbf{A}^2 + 2\mathbf{AB} + \mathbf{B}^2.$$

(b) Are the answers to part (a) the same? Discuss.

9. (a) List all the 3×3 matrices whose rows are

$$[\ 1\ \ 0\ \ 0\ ],\ [\ 0\ \ 1\ \ 0\ ],\ \text{and}\ [\ 0\ \ 0\ \ 1\ ].$$

(b) Which matrices obtained in part (a) are equal to their transposes?

10. In this exercise, $\mathbf{A}$ and $\mathbf{B}$ represent matrices. True or false?

(a) $(\mathbf{A}^T)^T = \mathbf{A}$ for all $\mathbf{A}$.

(b) If $\mathbf{A}^T = \mathbf{B}^T$, then $\mathbf{A} = \mathbf{B}$.

(c) If $\mathbf{A} = \mathbf{A}^T$, then $\mathbf{A}$ is a square matrix.

(d) If $\mathbf{A}$ and $\mathbf{B}$ are the same size, then $(\mathbf{A}+\mathbf{B})^T = \mathbf{A}^T + \mathbf{B}^T$.

11. For each $n \in \mathbb{N}$, let

$$\mathbf{A}_n = \begin{bmatrix} 1 & n \\ 0 & 1 \end{bmatrix} \quad \text{and} \quad \mathbf{B}_n = \begin{bmatrix} 1 & (-1)^n \\ -1 & 1 \end{bmatrix}.$$

(a) Give $\mathbf{A}_n^T$ for all $n \in \mathbb{N}$.

(b) Find $\{n \in \mathbb{N} : \mathbf{A}_n^T = \mathbf{A}_n\}$.

(c) Find $\{n \in \mathbb{N} : \mathbf{B}_n^T = \mathbf{B}_n\}$.

(d) Find $\{n \in \mathbb{N} : \mathbf{B}_n = \mathbf{B}_0\}$.

12. For $\mathbf{A}$ and $\mathbf{B}$ in $\mathfrak{M}_{m,n}$, let $\mathbf{A} - \mathbf{B} = \mathbf{A} + (-\mathbf{B})$. Show that

(a) $(\mathbf{A}-\mathbf{B}) + \mathbf{B} = \mathbf{A}$

(b) $-(\mathbf{A}-\mathbf{B}) = \mathbf{B} - \mathbf{A}$

(c) $(\mathbf{A}-\mathbf{B}) - \mathbf{C} \neq \mathbf{A} - (\mathbf{B}-\mathbf{C})$ in general

13. Consider $\mathbf{A}$, $\mathbf{B}$ in $\mathfrak{M}_{m,n}$ and a, b, c in $\mathbb{R}$. Show that

(a) $c(a\mathbf{A} + b\mathbf{B}) = (ca)\mathbf{A} + (cb)\mathbf{B}$

(b) $-a\mathbf{A} = (-a)\mathbf{A} = a(-\mathbf{A})$

(c) $(a\mathbf{A})^T = a\mathbf{A}^T$

14. Prove parts (b), (c), and (d) of the theorem on page 108.

15. Give the matrices for the digraphs in Figure 3.

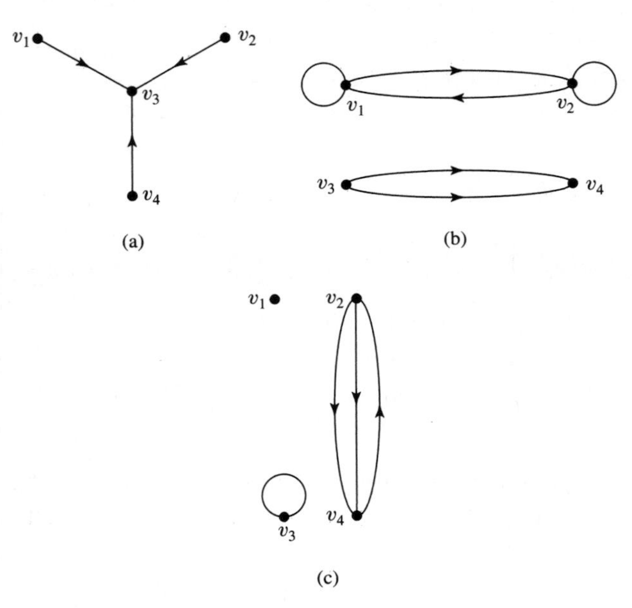

Figure 3 ▲

16. Write matrices for the graphs in Figure 4.

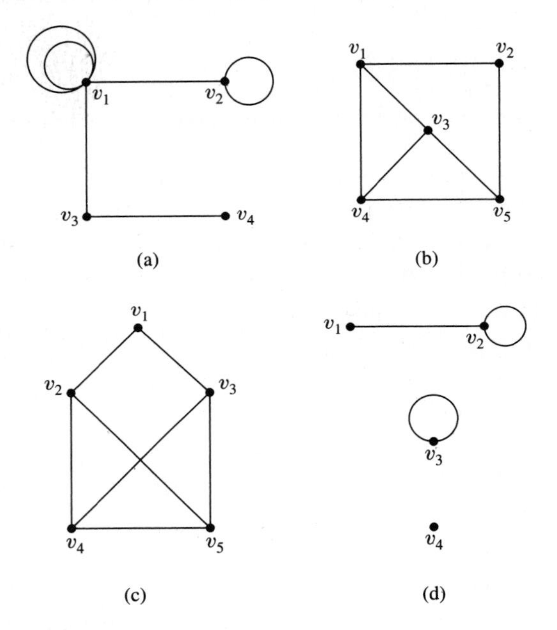

Figure 4 ▲

17. For each matrix in Figure 5, draw a digraph having the matrix.

$$\begin{bmatrix} 0 & 0 & 2 & 1 \\ 3 & 0 & 1 & 0 \\ 0 & 1 & 0 & 0 \\ 0 & 0 & 0 & 1 \end{bmatrix} \qquad \begin{bmatrix} 1 & 1 & 0 & 1 \\ 0 & 3 & 0 & 2 \\ 0 & 0 & 0 & 0 \\ 2 & 0 & 0 & 1 \end{bmatrix} \qquad \begin{bmatrix} 0 & 1 & 0 & 0 & 0 \\ 0 & 0 & 1 & 0 & 0 \\ 0 & 0 & 0 & 1 & 0 \\ 0 & 0 & 0 & 0 & 1 \\ 1 & 0 & 0 & 0 & 0 \end{bmatrix}$$

(a) (b) (c)

Figure 5 ▲

18. For each matrix in Figure 6, draw a graph having the matrix.

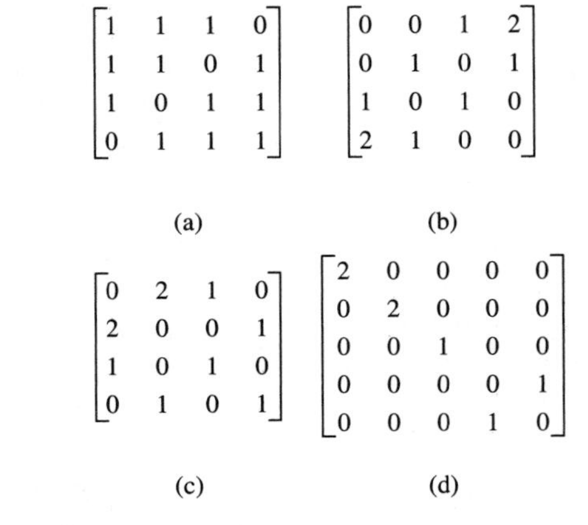

$$\begin{bmatrix} 1 & 1 & 1 & 0 \\ 1 & 1 & 0 & 1 \\ 1 & 0 & 1 & 1 \\ 0 & 1 & 1 & 1 \end{bmatrix} \qquad \begin{bmatrix} 0 & 0 & 1 & 2 \\ 0 & 1 & 0 & 1 \\ 1 & 0 & 1 & 0 \\ 2 & 1 & 0 & 0 \end{bmatrix}$$

(a) (b)

$$\begin{bmatrix} 0 & 2 & 1 & 0 \\ 2 & 0 & 0 & 1 \\ 1 & 0 & 1 & 0 \\ 0 & 1 & 0 & 1 \end{bmatrix} \qquad \begin{bmatrix} 2 & 0 & 0 & 0 & 0 \\ 0 & 2 & 0 & 0 & 0 \\ 0 & 0 & 1 & 0 & 0 \\ 0 & 0 & 0 & 0 & 1 \\ 0 & 0 & 0 & 1 & 0 \end{bmatrix}$$

(c) (d)

Figure 6 ▲

19. Give a matrix for each of the relations in Exercise 1 on page 99.

20. Draw a digraph having the matrix in Figure 6(b).

21. Give a matrix for each of the relations in Exercise 2 on page 99.

3.4 Equivalence Relations and Partitions

In this section we study special relations, called equivalence relations, that group together elements that have similar characteristics or share some property. Equivalence relations occur throughout mathematics and other fields, often without being formally acknowledged, when we want to regard two different objects as essentially the same in some way. They provide a mathematically careful way of focusing on common properties and ignoring inessential differences.

EXAMPLE 1

(a) Let S be a set of marbles, little, round, colored-glass objects. We might regard marbles s and t as equivalent if they have the same color, in which case we might write $s \sim t$. Note that the relation $\sim$ satisfies three properties:

(R) $s \sim s$ for all marbles s,
(S) if $s \sim t$, then $t \sim s$,
(T) if $s \sim t$ and $t \sim u$, then $s \sim u$.

For example, (T) asserts that if marbles s and t have the same color and t and u have the same color, then s and u have the same color. We can break S up into disjoint subsets of various colors—put the marbles in different bags—so that elements belong to the same subset if and only if they are equivalent, i.e., if and only if they have the same color.

(b) For the same set S of marbles, we might regard marbles s and t as equivalent if they are of the same size and write $s \approx t$ in this case. The statements in part (a) apply just as well to $\approx$, with obvious changes. ■

More generally, let S be any set and let $\sim$ be a relation on S. Then $\sim$ is called an **equivalence relation** provided it satisfies the reflexive, symmetric, and transitive laws:

(R) $s \sim s$ for every $s \in S$;
(S) if $s \sim t$, then $t \sim s$;
(T) if $s \sim t$ and $t \sim u$, then $s \sim u$.

If $s \sim t$, we say that s and t are **equivalent**. Depending on the circumstances, we might also say that s and t are **similar** or **congruent** or some fancy word like **isomorphic**. Other notations sometimes used for equivalence relations are $s \approx t$, $s \cong t$, $s \equiv t$, and $s \leftrightarrow t$. All these notations are intended to convey the idea that s and t have equal [or equivalent] status, a reasonable view because of the symmetry law (S).

EXAMPLE 2

Triangles T_1 and T_2 in the plane are said to be **similar**, and we write $T_1 \approx T_2$, if their angles can be matched up so that corresponding angles are equal. If the corresponding sides are also equal, we say that the triangles are **congruent**, and we write $T_1 \cong T_2$. In Figure 1 we have $T_1 \cong T_2$, $T_1 \approx T_3$, and $T_2 \approx T_3$, but T_3 is not congruent to T_1 or to T_2. Both $\approx$ and $\cong$ are equivalence relations on the set of all triangles in the plane. All the laws (R), (S), and (T) are evident for these relations. ■

Figure 1 ▶

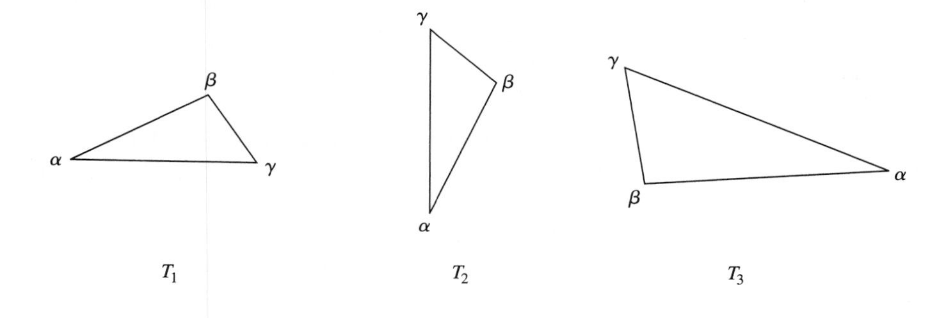

EXAMPLE 3

(a) Consider a machine that accepts input strings in Σ^* for some alphabet Σ and that generates output strings. We can define an equivalence relation $\sim$ on Σ^* by letting $w_1 \sim w_2$ if the machine generates the same output string for either w_1 or w_2 as input. To see whether two words are equivalent, we can ask the machine.

(b) We can also talk about equivalent machines. Given a set S of machines, we define the relations $\approx_1, \approx_2, \approx_3, \ldots$ on S by writing $B \approx_k C$ for machines

B and C if B and C produce the same output for every choice of input word of length k. Define $\approx$ by letting $B \approx C$ if $B \approx_k C$ for all $k \in \mathbb{P}$. Then all the relations $\approx_k$ and $\approx$ are equivalence relations on the set of machines, and two machines are equivalent under $\approx$ if and only if they produce the same response to all input words with letters from Σ. ■

EXAMPLE 4

Let $\mathcal{S}$ be a family of sets, and for $S, T \in \mathcal{S}$ define $S \sim T$ if there is a one-to-one function mapping S onto T. The relation $\sim$ is an equivalence relation on $\mathcal{S}$. Indeed:

(R) $S \sim S$ because the identity function $1_S : S \to S$, defined by $1_S(x) = x$ for every x in S, is a one-to-one mapping of S onto S.

(S) If $S \sim T$, then there is a one-to-one mapping f of S onto T. Its inverse f^{-1} is a one-to-one mapping of T onto S, so $T \sim S$.

(T) If $S \sim T$ and $T \sim U$, then there are one-to-one correspondences $f : S \to T$ and $g : T \to U$. It is easy to check that the composition $g \circ f$ is a one-to-one correspondence of S onto U. This also follows from Exercise 13 on page 45. In any event, we conclude that $S \sim U$.

Observe that if S is finite, then $S \sim T$ if and only if S and T are the same size. If S is infinite, then $S \sim T$ for some of the infinite sets T in $\mathcal{S}$, but probably not all of them, because not all infinite sets are equivalent to each other. This last assertion is not obvious. A brief glimpse at this fascinating story is available in the last section of the book, §13.3. ■

We normally use a notation like $\sim$ or $\equiv$ only when we have an equivalence relation. Of course, the use of such notation does not automatically guarantee that we have an equivalence relation.

EXAMPLE 5

(a) Let G be a graph. Define $\sim$ on $V(G)$ by $v \sim w$ if v and w are connected by an edge. Thus $\sim$ is the adjacency relation. It is reflexive only if there is a loop at each vertex. We can build a new relation $\simeq$ on $V(G)$ that is reflexive by defining $v \simeq w$ if and only if either $v = w$ or $v \sim w$. This relation $\simeq$ is reflexive and symmetric, but it still need not be transitive. Think of an example of vertices u, v, w with $u \simeq v$ and $v \simeq w$, but with $u \neq w$ and with no edge from u to w. In general, neither $\sim$ nor $\simeq$ are equivalence relations on $V(G)$.

(b) In §3.2 we defined the reachable relation R on $V(G)$ by $(v, w) \in R$ if there is a path from v to w. This R is what we get if we try to make a transitive relation out of the adjacency relation $\sim$. The new relation R is transitive and symmetric; but if G has isolated vertices without loops, then it is not reflexive. We can make an equivalence relation $\cong$ from R with the trick from part (a). Define $v \cong w$ in case either $v = w$ or $(v, w) \in R$. ■

In Example 1 we explicitly observed that our set S of marbles can be viewed as a disjoint union of subsets, where two marbles belong to the same subset if and only if they are equivalent. The original collection is cut up into disjoint subsets, each consisting of objects that are equivalent to each other. In fact, a similar phenomenon occurs in each of the examples of this section, though it may not be so obvious in some cases. There is a technical term for such a family of sets. A **partition** of a nonempty set S is a collection of nonempty subsets that are disjoint and whose union is S.

EXAMPLE 6

(a) Let f be a function from a set S onto a set T. Then the set $\{f^{\leftarrow}(y) : y \in T\}$ of all inverse images $f^{\leftarrow}(y)$ partitions S. In the first place, each $f^{\leftarrow}(y)$ is nonempty, since f maps *onto* T. Every x in S is in exactly one subset of the form $f^{\leftarrow}(y)$, namely the set $f^{\leftarrow}(f(x))$, which consists of all s in S with $f(s) = f(x)$. If $y \neq z$, then we have $f^{\leftarrow}(y) \cap f^{\leftarrow}(z) = \emptyset$. Also, the union

$\bigcup_{y \in T} f^{\leftarrow}(y)$ of all the sets $f^{\leftarrow}(y)$ is S, so $\{f^{\leftarrow}(y) : y \in T\}$ partitions S. We saw this sort of partition in Example 9 on page 44.

(b) Let's return to the set S of marbles in Example 1(a). If we let C be the set of colors of the marbles and define the function $f \colon S \to C$ by $f(s) =$ "the color of s" for each s in S, then the partition $\{f^{\leftarrow}(c) : c \in C\}$ is exactly the partition of S mentioned in Example 1. That is, two marbles have the same image under f and hence belong to the same subset if and only if they have the same color. The function $g \colon S \to \mathbb{R}$ given by $g(s) =$ "the diameter of s" puts two marbles in the same set $g^{\leftarrow}(d)$ if and only if they have the same diameter. The connection between equivalence relations and partitions given by inverse images is a general phenomenon, as we will see shortly in Theorem 2. ■

Consider again an equivalence relation $\sim$ on a set S. For each s in S, we define

$$[s] = \{t \in S : s \sim t\}.$$

The set $[s]$ is called the **equivalence class** or **$\sim$-class** containing s. For us, "class" and "set" are synonymous; so $[s]$ could have been called an "equivalence set," but it never is. The set of all equivalence classes of S is denoted by $\mathbf{[S]}$, i.e., $[S] = \{[s] : s \in S\}$. We will sometimes attach subscripts to $[s]$ or $[S]$ to clarify exactly which of several possible equivalence relations is being used.

EXAMPLE 7

(a) In the marble setting of Example 1(a), the equivalence class $[s]$ of a given marble s is the set of all marbles that are the same color as s; this includes s itself. The equivalence classes are the sets of blue marbles, of red marbles, of green marbles, etc.

(b) Consider the equivalence relation $\cong$ on the set V of vertices of a graph, built from the reachable relation in Example 5(b). Two vertices are equivalent precisely if they belong to the same connected part of the graph. [In fact, this is how we will define "connected" in §6.2.] For example, the equivalence classes for the graph in Figure 2 are $\{v_1, v_6, v_8\}$, $\{v_2, v_4, v_{10}\}$, and $\{v_3, v_5, v_7, v_9, v_{11}, v_{12}\}$. If a graph is connected, then the only $\cong$-class is the set V itself. ■

Figure 2 ►

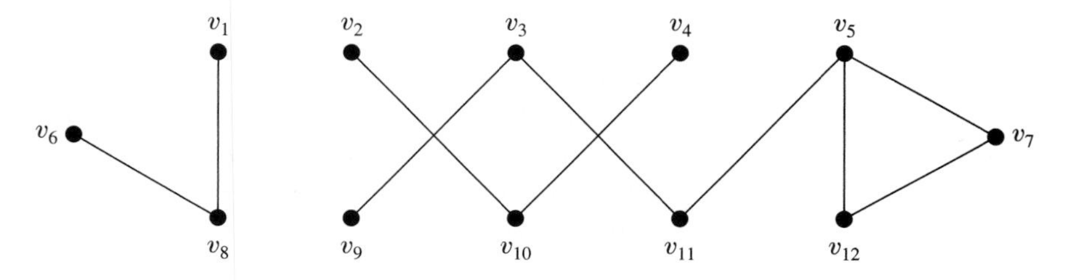

The equivalence classes in Example 7 also form a partition of the underlying set. Before we give Theorem 1, which states that something like this always happens, we prove a lemma that gives the key facts.

Lemma Let $\sim$ be an equivalence relation on a set S. For s and t in S, the following assertions are logically equivalent:

(i) $s \sim t$;

(ii) $[s] = [t]$;

(iii) $[s] \cap [t] \neq \emptyset$.

Proof "Logically equivalent" means, just as in §2.2, that for every particular choice of s and t all three assertions are true or else all are false; if any one is true, then they all are. We prove (i) $\Longrightarrow$ (ii), (ii) $\Longrightarrow$ (iii), and (iii) $\Longrightarrow$ (i).

$(s \sim t) \Longrightarrow ([s] = [t])$. Suppose that $s \sim t$. By symmetry $t \sim s$. Consider an element s' in the set $[s]$. Then $s \sim s'$. Since $t \sim s$ and $s \sim s'$, transitivity implies that $t \sim s'$. Thus s' is in $[t]$. We've shown that every s' in $[s]$ belongs to $[t]$, and hence $[s] \subseteq [t]$. Similarly, $[t] \subseteq [s]$.

$([s] = [t]) \Longrightarrow ([s] \cap [t] \neq \emptyset)$ is obvious, since each set $[s]$ is nonempty [why?].

$([s] \cap [t] \neq \emptyset) \Longrightarrow (s \sim t)$. Select an element u in the set $[s] \cap [t]$. Then $s \sim u$ and $t \sim u$. By symmetry, $u \sim t$. Since $s \sim u$ and $u \sim t$, we have $s \sim t$ by transitivity. ■

Here is the link between equivalence relations and partitions.

Theorem 1

(a) If $\sim$ is an equivalence relation on a nonempty set S, then $[S]$ is a partition of S.

(b) Conversely, if $\{A_i : i \in I\}$ is a partition of the set S, then the subsets A_i are the equivalence classes corresponding to some equivalence relation on S. ■

Proof

(a) To show that $[S]$ partitions S, we need to show that

$$\bigcup_{s \in S} [s] = S \tag{1}$$

and that

$$\text{for} \quad s, t \in S \quad \text{either} \quad [s] = [t] \quad \text{or} \quad [s] \cap [t] = \emptyset. \tag{2}$$

Clearly, $[s] \subseteq S$ for each s in S, so $\bigcup_{s \in S} [s] \subseteq S$. Given any s in S, we have $s \sim s$, so $s \in [s]$; hence $S \subseteq \bigcup_{s \in S} [s]$. Therefore, (1) holds.

Assertion (2) is logically equivalent to

$$\text{if } [s] \cap [t] \neq \emptyset \quad \text{then} \quad [s] = [t], \tag{3}$$

which follows from the lemma.

(b) Given a partition $\{A_i : i \in I\}$ of S, define the relation $\sim$ on S by $s \sim t$ if s and t belong to the same set A_i. Properties (R), (S), and (T) are obvious, so $\sim$ is an equivalence relation on S. Given a nonempty set A_i, we have $A_i = [s]$ for all $s \in A_i$, so the partition consists precisely of all equivalence classes $[s]$. ■

Sometimes an equivalence relation is defined in terms of a function on the underlying set. In a sense, which we will make precise in Theorem 2, every equivalence relation arises in this way.

EXAMPLE 8

Define $\sim$ on $\mathbb{N} \times \mathbb{N}$ by $(m, n) \sim (j, k)$ provided $m^2 + n^2 = j^2 + k^2$. It is easy to show directly that $\sim$ is an equivalence relation. However, here we take a slightly different approach. Define $f : \mathbb{N} \times \mathbb{N} \to \mathbb{N}$ by the rule

$$f(m, n) = m^2 + n^2.$$

Then ordered pairs are $\sim$-equivalent exactly when they have equal images under f. The equivalence classes are simply the nonempty sets $f^{\leftarrow}(r)$, where $r \in \mathbb{N}$. Some of the sets $f^{\leftarrow}(r)$, such as $f^{\leftarrow}(3)$, are empty, but this does no harm. ■

Here is the link between equivalence relations and functions.

Theorem 2

(a) Let S be a nonempty set. Let f be a function with domain S, and define $s_1 \sim s_2$ if $f(s_1) = f(s_2)$. Then $\sim$ is an equivalence relation on S, and the equivalence classes are the nonempty sets $f^{\leftarrow}(t)$, where t is in the codomain T of S.

(b) Every equivalence relation $\sim$ on a set S is determined by a suitable function with domain S, as in part (a). ■

Proof We check that $\sim$ is an equivalence relation.

(R) For all $s \in S$, we have $f(s) = f(s)$, so $s \sim s$.

(S) If $f(s_1) = f(s_2)$, then $f(s_2) = f(s_1)$, so $s_1 \sim s_2$ implies $s_2 \sim s_1$.

(T) If $f(s_1) = f(s_2)$ and $f(s_2) = f(s_3)$, then $f(s_1) = f(s_3)$, and so $\sim$ is transitive.

The statement about equivalence classes is just the definition of $f^{\leftarrow}(t)$.

To prove (b), we define the function $\nu\colon S \to [S]$, called the **natural mapping** of S onto $[S]$, by

$$\nu(s) = [s] \quad \text{for} \quad s \in S.$$

[That's not a v; it's a lowercase Greek nu, for natural.] By the lemma before Theorem 1, $s \sim t$ if and only if $[s] = [t]$, so $s \sim t$ if and only if $\nu(s) = \nu(t)$. That is, $\sim$ is the equivalence relation determined by ν. Note that ν maps S onto $[S]$ and that $\nu^{\leftarrow}([s]) = [s]$ for every s in S. ■

Theorems 1 and 2 tell us that equivalence classes, partitions, and sets $f^{\leftarrow}(t)$ are three ways to look at the same basic idea.

EXAMPLE 9

(a) If S is our familiar set of marbles and $f(s)$ is the color of s, then $\nu(s)$ is the class $[s]$ of all marbles the same color as s. We could think of $\nu(s)$ as a bag of marbles, with ν the function that puts each marble in its proper bag. With this fanciful interpretation, $[S]$ is a collection of bags, each with at least one marble in it. The number of bags is the number of colors used.

Switching to the function g for which $g(s)$ is the diameter of s would give a new ν and a partition $[S]$ consisting of new bags, one for each possible diameter of marble.

(b) Define the function $f\colon \mathbb{R} \times \mathbb{R} \to \mathbb{R}$ by $f(x, y) = x^2 + y^2$. Then f gives an equivalence relation $\sim$ defined by $(x, y) \sim (z, w)$ in case $x^2 + y^2 = z^2 + w^2$. The equivalence classes are the circles in the plane $\mathbb{R} \times \mathbb{R}$ centered at $(0, 0)$, because $x^2 + y^2 = z^2 + w^2$ implies $\sqrt{x^2 + y^2} = \sqrt{z^2 + w^2}$; i.e., (x, y) and (z, w) are the same distance from $(0, 0)$. Thus $[\mathbb{R} \times \mathbb{R}]$ consists of these circles, including the set $\{(0, 0)\}$ [the circle of radius 0]. The function ν maps each point (x, y) to the circle it lives on. There is a one-to-one correspondence between the set of circles and the set of values of f. ■

EXAMPLE 10

The set $\mathbb{Q}$ of rational numbers consists of numbers of the form $\frac{m}{n}$ with $m, n \in \mathbb{Z}$ and $n \neq 0$. Each rational number can be written in lots of ways; for instance,

$$\frac{2}{3} = \frac{4}{6} = \frac{8}{12}, \quad -5 = \frac{-5}{1} = \frac{10}{-2} \quad \text{and} \quad 0 = \frac{0}{1} = \frac{0}{73}.$$

We can view the members of $\mathbb{Q}$ as equivalence classes of pairs of integers so that $\frac{m}{n}$ corresponds to the class of (m, n). Here's how.

We will want $(m, n) \sim (p, q)$ in case $\frac{m}{n} = \frac{p}{q}$, so for n and q not 0 we set

$$(m, n) \sim (p, q) \quad \text{in case} \quad m \cdot q = n \cdot p,$$

a definition that just involves multiplying integers. No fractions here. It is easy to check [Exercise 13] that $\sim$ is an equivalence relation, and we see that (m, n)

and (p, q) are equivalent in case the numerical ratio of m to n is the ratio of p to q. [Hence the term "*ratio*nal."] The class $[(2, 3)] = [(4, 6)] = [(8, 12)]$ corresponds to the number $\frac{2}{3}$, and we could even think of $\frac{2}{3}$, $\frac{4}{6}$, etc., as other names for this class. ■

A very subtle problem can occur when one attempts to define a function on equivalence classes. You probably wouldn't make the mistake in part (b) of the next example, but we give this simple example to illustrate what could go wrong if one were not paying attention.

EXAMPLE 11

(a) We can define the function $f : \mathbb{Q} \to \mathbb{Q}$ by $f\left(\frac{m}{n}\right) = \frac{m^2}{n^2}$, because if $\frac{m}{n} = \frac{p}{q}$, then $\frac{m^2}{n^2} = \frac{p^2}{q^2}$. In the notation of Example 10, $\frac{m}{n}$ corresponds to the equivalence class $[(m, n)]$, and if $(m, n) \sim (p, q)$, then $m \cdot q = n \cdot p$, $m^2 \cdot q^2 = n^2 \cdot p^2$, and thus $(m^2, n^2) \sim (p^2, q^2)$. The definition of f is unambiguous (hence well defined) because the output value $\frac{m^2}{n^2}$ does not depend on the choice of representative in the equivalence class $[(m, n)]$ consisting of all pairs corresponding to the value $\frac{m}{n}$.

(b) We can't define the function $g : \mathbb{Q} \to \mathbb{Q}$ by $g\left(\frac{m}{n}\right) = m + n$. If we could, we would want $g\left(\frac{1}{2}\right) = 1 + 2 = 3$, but also $g\left(\frac{2}{4}\right) = 2 + 4 = 6$. The trouble comes because we have two different names, $\frac{1}{2}$ and $\frac{2}{4}$, for the same object, and our rule for g is based on the name, not on the object itself. When we look at the problem in terms of equivalence classes, we see that $[(1, 2)] = [(2, 4)]$. We would have no trouble defining $g(m, n) = m + n$, but this g does not respect the equivalence relation, since $(m, n) \sim (p, q)$ does not imply that $m + n = p + q$. Our original g was not well defined. ■

Example 11 shows that defining functions on sets of equivalence classes can be tricky. One always has to check that the function definition does not really depend on which representative of the class is used. In the case of f in part (a), there is no problem, because if $[\frac{m}{n}] = [\frac{p}{q}]$, then $f(\frac{m}{n}) = f(\frac{p}{q})$, so we get the same value of $f([\frac{m}{n}])$ whether we think of $[\frac{m}{n}]$ as $[\frac{m}{n}]$ or as $[\frac{p}{q}]$. The example in part (b) shows how one can go wrong, though. To ask whether a function f is **well-defined** is to ask whether each value $f(x)$ depends only on what x is and not just on what name we've given it. Does the rule defining f give the same value to $f(x)$ and to $f(y)$ if $x = y$?

The more abstract the setting, the more care that's needed.

EXAMPLE 12

Consider again the equivalence relation $\sim$ determined by a function f on a set S, as in Theorem 2. The mapping θ defined by $\theta([s]) = f(s)$ is well-defined because $[s] = [t]$ implies that $f(s) = f(t)$. In fact, θ is a one-to-one correspondence between the set $[S]$ of equivalence classes and the set $f(S)$ of values of f. ■

Exercises 3.4

1. Which of the following describe equivalence relations? For those that are not equivalence relations, specify which of (R), (S), and (T) fail, and illustrate the failures with examples.

(a) $L_1 \| L_2$ for straight lines in the plane if L_1 and L_2 are the same or are parallel.

(b) $L_1 \perp L_2$ for straight lines in the plane if L_1 and L_2 are perpendicular.

(c) $p_1 \sim p_2$ for Americans if p_1 and p_2 live in the same state.

(d) $p_1 \approx p_2$ for Americans if p_1 and p_2 live in the same state or in neighboring states.

(e) $p_1 \approx p_2$ for people if p_1 and p_2 have a parent in common.

(f) $p_1 \cong p_2$ for people if p_1 and p_2 have the same mother.

2. For each example of an equivalence relation in Exercise 1, describe the members of some equivalence class.

3. Let S be a set. Is equality, i.e., "=", an equivalence relation?

4. Define the relation $\equiv$ on $\mathbb{Z}$ by $m \equiv n$ in case $m - n$ is even. Is $\equiv$ an equivalence relation? Explain.

5. If G and H are both graphs with vertex set $\{1, 2, \ldots, n\}$, we say that G is **isomorphic** to H, and write $G \simeq H$, in case there is a way to label the vertices of G so that it becomes H. For example, the graphs in Figure 3, with vertex set $\{1, 2, 3\}$, are isomorphic by relabeling $f(1) = 2$, $f(2) = 3$, and $f(3) = 1$.

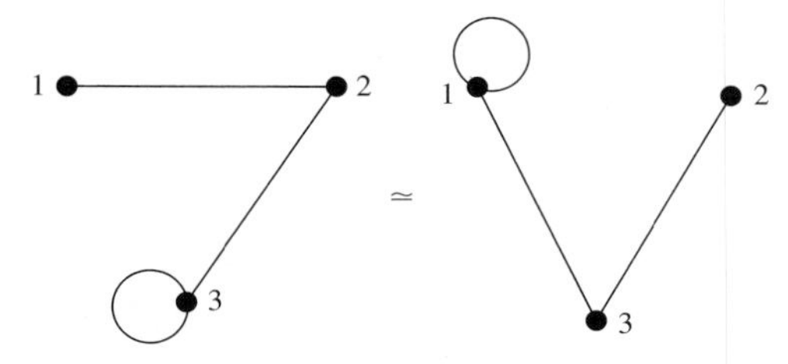

Figure 3 ▲

(a) Give a picture of another graph isomorphic to these two.

(b) Find a graph with vertex set $\{1, 2, 3\}$ that is not isomorphic to the graphs in Figure 3, yet has three edges, exactly one of which is a loop.

(c) Find another example as in part (b) that isn't isomorphic to the one you found in part (b) [or the ones in Figure 3].

(d) Show that $\simeq$ is an equivalence relation on the set of all graphs with vertex set $\{1, 2, \ldots, n\}$.

6. Can you think of situations in life where you'd use the term "equivalent" and where a natural equivalence relation is involved?

7. Define the relation $\approx$ on $\mathbb{Z}$ by $m \approx n$ in case $m^2 = n^2$.

(a) Show that $\approx$ is an equivalence relation on $\mathbb{Z}$.

(b) Describe the equivalence classes for $\approx$. How many are there?

8. (a) For $m, n \in \mathbb{Z}$, define $m \sim n$ in case $m - n$ is odd. Is the relation $\sim$ reflexive? symmetric? transitive? Is $\sim$ an equivalence relation?

(b) For a and b in $\mathbb{R}$, define $a \sim b$ in case $|a - b| \le 1$. One could say that $a \sim b$ in case a and b are "close enough" or "approximately equal." Answer the questions in part (a).

9. Consider the functions g and h mapping $\mathbb{Z}$ into $\mathbb{N}$ defined by $g(n) = |n|$ and $h(n) = 1 + (-1)^n$.

(a) Describe the sets in the partition $\{g^{\leftarrow}(k) : k \text{ is in the codomain of } g\}$ of $\mathbb{Z}$. How many sets are there?

(b) Describe the sets in the partition $\{h^{\leftarrow}(k) : k \text{ is in the codomain of } h\}$ of $\mathbb{Z}$. How many sets are there?

10. On the set $\mathbb{N} \times \mathbb{N}$ define $(m, n) \sim (k, l)$ if $m + l = n + k$.

(a) Show that $\sim$ is an equivalence relation on $\mathbb{N} \times \mathbb{N}$.

(b) Draw a sketch of $\mathbb{N} \times \mathbb{N}$ that shows several equivalence classes.

11. Let Σ be an alphabet, and for w_1 and w_2 in Σ^* define $w_1 \sim w_2$ if $\text{length}(w_1) = \text{length}(w_2)$. Explain why $\sim$ is an equivalence relation, and describe the equivalence classes.

12. Let P be a set of computer programs, and regard programs p_1 and p_2 as equivalent if they always produce the same outputs for given inputs. Is this an equivalence relation on P? Explain.

13. Consider $\mathbb{Z} \times \mathbb{P}$ and define $(m, n) \sim (p, q)$ if $mq = np$.

(a) Show that $\sim$ is an equivalence relation on $\mathbb{Z} \times \mathbb{P}$.

(b) Show that $\sim$ is the equivalence relation corresponding to the function $\mathbb{Z} \times \mathbb{P} \to \mathbb{Q}$ given by $f(m, n) = \frac{m}{n}$; see Theorem 2(a).

14. In the proof of Theorem 2(b), we obtained the equality $\nu^{\leftarrow}([s]) = [s]$. Does this mean that the function ν has an inverse and that the inverse of ν is the identity function on $[S]$? Discuss.

15. As in Exercise 7 define $\approx$ on $\mathbb{Z}$ by $m \approx n$ in case $m^2 = n^2$.

(a) What is wrong with the following "definition" of $\le$ on $[\mathbb{Z}]$? Let $[m] \le [n]$ if and only if $m \le n$.

(b) What, if anything, is wrong with the following "definition" of a function $f: [\mathbb{Z}] \to \mathbb{Z}$? Let $f([m]) = m^2 + m + 1$.

(c) Repeat part (b) with $g([m]) = m^4 + m^2 + 1$.

(d) What, if anything, is wrong with the following "definition" of the operation $\oplus$ on $[\mathbb{Z}]$? Let $[m] \oplus [n] = [m + n]$.

16. Let $\mathbb{Q}^+ = \{\frac{m}{n} : m, n \in \mathbb{P}\}$. Which of the following are well-defined definitions of functions on $\mathbb{Q}^+$?

(a) $f\left(\frac{m}{n}\right) = \frac{n}{m}$ (b) $g\left(\frac{m}{n}\right) = m^2 + n^2$

(c) $h\left(\frac{m}{n}\right) = \frac{m^2 + n^2}{mn}$

17. (a) Verify that the relation $\cong$ defined in Example 5(b) is an equivalence relation on $V(G)$.

(b) Given a vertex v in $V(G)$, describe in words the equivalence class containing v.

18. Let S be the set of all sequences (s_n) of real numbers, and define $(s_n) \approx (t_n)$ if $\{n \in \mathbb{N} : s_n \ne t_n\}$ is finite. Show that $\approx$ is an equivalence relation on S.

19. Show that the function θ in Example 12 is a one-to-one correspondence between the set $[S]$ of equivalence classes and the set $f(S)$ of values of f.

3.5 The Division Algorithm and Integers Mod p

This section is devoted to those equivalence relations on $\mathbb{Z}$ that are tied to the algebraic operations addition and multiplication. The key to defining the relations is integer division, so we start by examining what division in $\mathbb{Z}$ actually means.

When we divide 42 by 7 to get 6, there's no mystery; $42 \div 7 = 6$ is just another way of saying $42 = 6 \cdot 7$. When we divide 61 by 7, though, it doesn't "come out even." There is no integer q with $61 = q \cdot 7$. The best we can do is to get eight 7's out of 61, with a remainder of 5; $61 = 8 \cdot 7 + 5$. In general, when we try to divide an integer n by a nonzero integer p, this sort of outcome is all we can expect. The following theorem, which we already used in §1.2, says that we can always get a quotient and remainder, and there's only one possible answer.

The Division Algorithm Let $p \in \mathbb{P}$. For each integer n there are unique integers q and r satisfying

$$n = p \cdot q + r \qquad \text{and} \qquad 0 \le r < p. \qquad \blacksquare$$

The numbers q and r are called the **quotient** and **remainder**, respectively, **when** n **is divided by** p. For example, if $p = 7$ and $n = 31$, then $q = 4$ and $r = 3$ because $31 = 7 \cdot 4 + 3$. If $p = 1$ and $n = 31$, then $q = 31$ and $r = 0$.

We will not stop now to prove the Division Algorithm. It is not too hard to show that q and r are unique [Exercise 19], and their existence can be proved fairly quickly using a nonconstructive argument. In §4.1 we will develop an algorithm that produces q and r from n and p, which of course yields a constructive proof of the Division Algorithm; see the theorem on page 132. In any case, we all believe it.

So how do we find the values of q and r? Given a calculator, we can add or multiply integers to get integers, but when we divide, we typically get some decimal expression, rather than q and r. No problem. We can easily get q and r from the decimal output, as we observed in §1.2. Rewrite the conditions $n = p \cdot q + r$ and $0 \le r < p$ as

$$\frac{n}{p} = q + \frac{r}{p} \qquad \text{with} \qquad 0 \le \frac{r}{p} < 1.$$

Then $q = \lfloor n/p \rfloor$ and r/p is the fractional part of n/p, i.e., the number to the right of the decimal point in n/p. To get q, we calculate $\lfloor n/p \rfloor$, and then we compute $r = (n/p - \lfloor n/p \rfloor) \cdot p$.

EXAMPLE 1

(a) Take $n = 31$ and $p = 7$. A pocket calculator gives $31 \div 7 \approx 4.429$. Thus $q = \lfloor 4.429 \rfloor = 4$ and $r \approx (4.429 - 4) \cdot 7 = 0.429 \cdot 7 = 3.003$, so $r = 3$. We check that $31 = 4 \cdot 7 + 3$.

(b) Now consider $m = 7$ and $n = -31$. Then $31 = 7 \cdot 4 + 3$, so we have $-31 = 7 \cdot (-4) + (-3)$. Does this mean that $q = -4$ and $r = -3$? No, because r must be nonnegative. Recall that q is always the largest integer less than or equal to n/p. In our present case, $q = -5$, since $-5 < -4.429 < -4$; so $r = -31 - (-5) \cdot 7 = 4$. ■

Some calculators and most computer languages will do these manipulations for us, so we can ask for q and r directly, using the two built-in integer functions DIV and MOD. The definitions are

$$n \text{ DIV } p = \left\lfloor \frac{n}{p} \right\rfloor \qquad \text{and} \qquad n \text{ MOD } p = \left(\frac{n}{p} - n \text{ DIV } p\right) \cdot p,$$

so we have

$$n = (n \text{ DIV } p) \cdot p + n \text{ MOD } p \qquad \text{and} \qquad 0 \le n \text{ MOD } p < p.$$

We will suppose that DIV and MOD satisfy these conditions even if n is negative, but one should double-check that assumption in practice, as definitions for the negative case vary from computer language to computer language. Exercise 18 shows how to use functions DIV and MOD that work for $n \in \mathbb{N}$ to handle $n \le 0$ as well.

The integers $n \text{ DIV } p$ and $n \text{ MOD } p$ are the unique q and r guaranteed by the Division Algorithm. Given a positive integer p, the number $n \text{ MOD } p$ is the remainder obtained when n is divided by p and is called the **remainder mod p**. Both $\text{DIV}\ p$ and $\text{MOD}\ p$ are functions of n, even though we write $n\ \text{DIV}\ p$ and $n\ \text{MOD}\ p$ instead of $(\text{DIV}\ p)(n)$ and $(\text{MOD}\ p)(n)$; the variable n has appeared in unusual places before, as with $|n|$ and $n!$. The values of $\text{MOD}\ p$ are in the set $\{0, 1, \ldots, p-1\}$, which we will call $\mathbb{Z}(p)$, and so $\text{MOD}\ p\colon \mathbb{Z} \to \mathbb{Z}(p)$. In fact, $\text{MOD}\ p$ maps $\mathbb{Z}$ onto $\mathbb{Z}(p)$, because $n\ \text{MOD}\ p = n$ for every n in $\mathbb{Z}(p)$.

EXAMPLE 2

(a) We have $31\ \text{DIV}\ 7 = 4$ and $31\ \text{MOD}\ 7 = 3$. Also, $(-31)\ \text{DIV}\ 7 = -5$ and $(-31)\ \text{MOD}\ 7 = 4$. Note that $(-31)\ \text{DIV}\ 7 \neq -(31\ \text{DIV}\ 7)$ and $(-31)\ \text{MOD}\ 7 \neq -(31\ \text{MOD}\ 7)$, so we must be careful about writing $-n\ \text{DIV}\ p$ or $-n\ \text{MOD}\ p$.

(b) $n\ \text{MOD}\ 2$ is 0 if n is even and 1 if n is odd.

(c) $n\ \text{MOD}\ 10$ is the last decimal digit of n. ■

The sets $\mathbb{Z}(p)$ and the mappings $\text{MOD}\ p$ for p in $\mathbb{P}$ play important roles in a host of applications, ranging from signal transmission to hashing to random number generators to fast computer graphics, and they are associated with fundamental equivalence relations on $\mathbb{Z}$.

Consider p in $\mathbb{P}$ with $p \geq 2$. As we remarked on page 97, the congruence relation "$\equiv \pmod{p}$" satisfies

(R) $m \equiv m \pmod{p}$ for all $m \in \mathbb{Z}$,

(S) $m \equiv n \pmod{p}$ implies $n \equiv m \pmod{p}$,

(T) $m \equiv n \pmod{p}$ and $n \equiv r \pmod{p}$ imply $m \equiv r \pmod{p}$.

That is, the relation "$\equiv \pmod{p}$" is an equivalence relation. This fact can be proved directly, but the verification is a bit tedious and not very illuminating. Instead, we give a proof in the next theorem based on Theorem 2 on page 117.

Theorem 1 Let p be in $\mathbb{P}$.

(a) For $m, n \in \mathbb{Z}$, we have $m\ \text{MOD}\ p = n\ \text{MOD}\ p$ if and only if $m \equiv n \pmod{p}$.

(b) The relation "$\equiv \pmod{p}$" is an equivalence relation.

Proof

(a) We use the facts that

$$m = (m\ \text{DIV}\ p) \cdot p + m\ \text{MOD}\ p \quad \text{and} \quad n = (n\ \text{DIV}\ p) \cdot p + n\ \text{MOD}\ p.$$

If $m\ \text{MOD}\ p = n\ \text{MOD}\ p$, then $m - n = [m\ \text{DIV}\ p - n\ \text{DIV}\ p] \cdot p$, a multiple of p, so $m \equiv n \pmod{p}$. Conversely, suppose that $m - n$ is a multiple of p. Then

$$m\ \text{MOD}\ p - n\ \text{MOD}\ p = (m - n) + [n\ \text{DIV}\ p - m\ \text{DIV}\ p] \cdot p,$$

which is also a multiple of p. Since $m\ \text{MOD}\ p$ and $n\ \text{MOD}\ p$ are both in the set $\{0, 1, \ldots, p-1\}$, their difference is at most $p - 1$. Thus they must be equal.

(b) Consider the function $f\colon \mathbb{Z} \to \mathbb{Z}(p)$ defined by $f(n) = n\ \text{MOD}\ p$. By Theorem 2 on page 117, f defines an equivalence relation on $\mathbb{Z}$, where m and n in $\mathbb{Z}$ are equivalent if $f(m) = f(n)$. By part (a), $f(m) = f(n)$ if and only if $m \equiv n \pmod{p}$, so the equivalence relation defined by f is exactly the relation "$\equiv \pmod{p}$." ■

The equivalence relation "$\equiv \pmod{p}$" is called **congruence mod p**. The expression $m \equiv n \pmod{p}$ is read as "m [is] congruent to n mod p [or modulo p]." To avoid confusing "mod p" with "$\text{MOD}\ p$," remember that $\text{MOD}\ p$ is the name for a *function* [like FACT], whereas mod p is *part* of the notation for a *relation*. Since $m - (m\ \text{MOD}\ p) = (m\ \text{DIV}\ p) \cdot p$, we have $m\ \text{MOD}\ p \equiv m \pmod{p}$. This is where the MOD notation came from historically. Because distinct elements of $\mathbb{Z}(p)$ are too close together to be congruent mod p, we see that $r = m \text{ MOD } p$ is the unique number in $\mathbb{Z}(p)$ for which $r \equiv m \pmod{p}$.

The equivalence class of n with respect to the equivalence relation $\equiv \pmod{p}$ is called its **congruence class mod** $\boldsymbol{p}$ and denoted by $[\boldsymbol{n}]_p$, or sometimes just by $[\boldsymbol{n}]$ if p is understood. Thus

$$[\boldsymbol{n}]_p = \{m \in \mathbb{Z} : m \equiv n \pmod{p}\}.$$

The case $p = 1$ is quite special and somewhat boring [Exercise 13], so unless we explicitly say otherwise we will be thinking of $p \geq 2$ in what follows. Many of the arguments will still work for $p = 1$, though.

EXAMPLE 3

(a) Two integers are congruent mod 2 if they are both even or if they are both odd, i.e., if both have the same remainder when divided by 2. The congruence classes are $[0]_2 = [2]_2 = \cdots = \{n \in \mathbb{Z} : n \text{ is even}\}$ and $[1]_2 = [-3]_2 = [73]_2 = \{n \in \mathbb{Z} : n \text{ is odd}\}$.

(b) The integers that are multiples of 5,

$$\ldots, -25, -20, -15, -10, -5, 0, 5, 10, 15, 20, 25, \ldots,$$

are all congruent to each other mod 5, since the difference between any two numbers on this list is a multiple of 5. These numbers all have remainders 0 when divided by 5.

If we add 1 to each member of this list, we get a new list

$$\ldots, -24, -19, -14, -9, -4, 1, 6, 11, 16, 21, 26, \ldots.$$

The *differences* between numbers haven't changed, so the differences are all still multiples of 5. Thus the numbers on the new list are also congruent to each other mod 5. They all have remainder 1 when divided by 5. The integers

$$\ldots, -23, -18, -13, -8, -3, 2, 7, 12, 17, 22, 27, \ldots$$

form another congruence class. So do the integers

$$\ldots, -22, -17, -12, -7, -2, 3, 8, 13, 18, 23, 28, \ldots,$$

and the integers

$$\ldots, -21, -16, -11, -6, -1, 4, 9, 14, 19, 24, 29, \ldots.$$

Each integer belongs to exactly one of these five classes; i.e., the classes form a partition of $\mathbb{Z}$, and each of the classes contains exactly one of the numbers 0, 1, 2, 3, 4. We could list the classes as $[0]_5$, $[1]_5$, $[2]_5$, $[3]_5$, and $[4]_5$. ■

Our next theorem shows the link between congruence mod p and the arithmetic in $\mathbb{Z}$.

Theorem 2 Let $m, m', n, n' \in \mathbb{Z}$ and let $p \in \mathbb{P}$. If $m' \equiv m \pmod{p}$ and $n' \equiv n \pmod{p}$, then

$$m' + n' \equiv m + n \pmod{p} \quad \text{and} \quad m' \cdot n' \equiv m \cdot n \pmod{p}.$$

Proof By hypothesis, $m' = m + kp$ and $n' = n + lp$ for some $k, l \in \mathbb{Z}$. Thus

$$m' + n' = m + n + (k + l) \cdot p \equiv m + n \pmod{p}$$

and

$$m' \cdot n' = m \cdot n + (kn + ml + kpl) \cdot p \equiv m \cdot n \pmod{p}. \quad ■$$

Taking $m' = m \text{ MOD } p$ and $n' = n \text{ MOD } p$ in Theorem 2 gives the following useful consequence.

Corollary Let $m, n \in \mathbb{Z}$ and $p \in \mathbb{P}$. Then

(a) $m \text{ MOD } p + n \text{ MOD } p \equiv m + n \pmod{p}$.
(b) $(m \text{ MOD } p) \cdot (n \text{ MOD } p) \equiv m \cdot n \pmod{p}$.

$+_2$	0	1
0	0	1
1	1	0

$*_2$	0	1
0	0	0
1	0	1

$\mathbb{Z}(2)$

Figure 1 ▲

We can use this corollary to help us transfer the arithmetic of $\mathbb{Z}$ over to $\mathbb{Z}(p)$. First, we define two operations, $+_p$ and $*_p$, on $\mathbb{Z}(p)$ by

$$a +_p b = (a + b) \text{ MOD } p \quad \text{and} \quad a *_p b = (a \cdot b) \text{ MOD } p$$

for $a, b \in \mathbb{Z}(p)$. Since $m \text{ MOD } p \in \mathbb{Z}(p)$ for all $m \in \mathbb{Z}$, $a +_p b$ and $a *_p b$ are in $\mathbb{Z}(p)$.

EXAMPLE 4

(a) A very simple but important case is $\mathbb{Z}(2)$. The addition and multiplication tables for $\mathbb{Z}(2)$ are given in Figure 1.

(b) For $\mathbb{Z}(6) = \{0, 1, 2, 3, 4, 5\}$, we have $4 +_6 5 = (4 + 5) \text{ MOD } 6 = 9 \text{ MOD } 6 = 3$. Similarly, $4 *_6 4 = (4 \cdot 4) \text{ MOD } 6 = 16 \text{ MOD } 6 = 4$, since $16 \equiv 4 \pmod{6}$. The complete addition and multiplication tables for $\mathbb{Z}(6)$ are given in Figure 2. Notice that the product of two nonzero elements under $*_6$ can be 0.

Figure 2 ▶

$+_6$	0	1	2	3	4	5
0	0	1	2	3	4	5
1	1	2	3	4	5	0
2	1	3	4	5	0	1
3	3	4	5	0	1	2
4	4	5	0	1	2	3
5	5	0	1	2	3	4

$*_6$	0	1	2	3	4	5
0	0	0	0	0	0	0
1	0	1	2	3	4	5
2	0	2	4	0	2	4
3	0	3	0	3	0	3
4	0	4	2	0	4	2
5	0	5	4	3	2	1

$\mathbb{Z}(6)$

Figure 3 ▶

$+_5$	0	1	2	3	4
0	0	1	2	3	4
1	1	2	3	4	0
2	2	3	4	0	1
3	3	4	0	1	2
4	4	0	1	2	3

$*_5$	0	1	2	3	4
0	0	0	0	0	0
1	0	1	2	3	4
2	0	2	4	1	3
3	0	3	1	4	2
4	0	4	3	2	1

$\mathbb{Z}(5)$

(c) Figure 3 gives the tables for $\mathbb{Z}(5)$. ■

The new operations $+_p$ and $*_p$ are consistent with the old + and · on $\mathbb{Z}$ in the following sense.

Theorem 3 Let $m, n \in \mathbb{Z}$ and $p \in \mathbb{P}$. Then

(a) $(m + n) \text{ MOD } p = (m \text{ MOD } p) +_p (n \text{ MOD } p)$.
(b) $(m \cdot n) \text{ MOD } p = (m \text{ MOD } p) *_p (n \text{ MOD } p)$.

Proof

(a) By Theorem 2 or its corollary, $m + n \equiv m \text{ MOD } p + n \text{ MOD } p \pmod{p}$. This means that

$$(m + n) \text{ MOD } p = (m \text{ MOD } p + n \text{ MOD } p) \text{ MOD } p,$$

which is $(m \text{ MOD } p) +_p (n \text{ MOD } p)$ by definition.

The proof of (b) is similar. ■

Thus the function $\text{MOD } p$ carries sums in $\mathbb{Z}$ to sums [under $+_p$] in $\mathbb{Z}(p)$ and products in $\mathbb{Z}$ to products [under $*_p$] in $\mathbb{Z}(p)$.

EXAMPLE 5

(a) $(6 + 3) \text{ MOD } 2 = 9 \text{ MOD } 2 = 1$, but also $6 \text{ MOD } 2 +_2 3 \text{ MOD } 2 = 0 +_2 1 = 1$. In fact, in general

$$(\text{even} + \text{odd}) \text{ MOD } 2 = \text{odd MOD } 2 = 1 = 0 +_2 1 = \text{even MOD } 2 +_2 \text{odd MOD } 2.$$

(b) $(8 \cdot 3) \text{ MOD } 6 = 24 \text{ MOD } 6 = 0$, and $8 \text{ MOD } 6 *_6 3 \text{ MOD } 6 = 2 *_6 3 = 0$, too. ■

Theorem 3 also lets us show that $+_p$ and $*_p$ satisfy some familiar algebraic laws.

Theorem 4 Let $p \in \mathbb{P}$ and let $m, n, r \in \mathbb{Z}(p)$. Then:

(a) $m +_p n = n +_p m$ and $m *_p n = n *_p m$;
(b) $(m +_p n) +_p r = m +_p (n +_p r)$ and $(m *_p n) *_p r = m *_p (n *_p r)$;
(c) $(m +_p n) *_p r = (m *_p r) +_p (n *_p r)$.

Proof

We show the distributive law (c). The other proofs are similar. [See Exercise 21.]

Since $(m+n) \cdot r = mr + nr$ in $\mathbb{Z}$, we have $((m+n) \cdot r) \text{ MOD } p = (mr+nr) \text{ MOD } p$. By Theorem 3,

$$\begin{aligned}((m + n) \cdot r) \text{ MOD } p &= (m + n) \text{ MOD } p *_p (r \text{ MOD } p) \\ &= (m \text{ MOD } p +_p n \text{ MOD } p) *_p (r \text{ MOD } p).\end{aligned}$$

Since m, n and r are already in $\mathbb{Z}(p)$, it follows that $m \text{ MOD } p = m$, $n \text{ MOD } p = n$, and $r \text{ MOD } p = r$, so this equation just says that

$$((m + n) \cdot r) \text{ MOD } p = (m +_p n) *_p r.$$

Similarly,

$$\begin{aligned}(mr + nr) \text{ MOD } p &= (mr \text{ MOD } p) +_p (nr \text{ MOD } p) \\ &= ((m \text{ MOD } p) *_p (r \text{ MOD } p)) +_p ((n \text{ MOD } p) *_p (r \text{ MOD } p)) \\ &= (m *_p r) +_p (n *_p r).\end{aligned}$$ ■

The set $\mathbb{Z}(p)$ with its operations $+_p$ and $*_p$ acts, in a way, like a finite model of $\mathbb{Z}$. We need to be a little careful, however, because, although Theorem 4 shows that many of the laws of arithmetic hold in $\mathbb{Z}(p)$, cancellation may not work as expected. We saw that $3 *_6 5 = 3 *_6 3 = 3 *_6 1 = 3$, but $5 \neq 3 \neq 1$ in $\mathbb{Z}(6)$. Moreover, $3 *_6 2 = 0$ with $3 \neq 0$ and $2 \neq 0$.

EXAMPLE 6

(a) We can try to define operations $+$ and $\cdot$ on the collection $[\mathbb{Z}]_p$ of congruence classes $[m]_p$. The natural candidates are

$$[m]_p + [n]_p = [m + n]_p \quad \text{and} \quad [m]_p \cdot [n]_p = [m \cdot n]_p.$$

We know from §3.4 that we need to be careful about definitions on sets of equivalence classes. To be sure that $+$ and $\cdot$ are well-defined on $[\mathbb{Z}]_p$, we need to check that if $[m]_p = [m']_p$ and $[n]_p = [n']_p$, then

(1) $[m + n]_p = [m' + n']_p$ and
(2) $[m \cdot n]_p = [m' \cdot n']_p$.

Now $[m]_p = [m']_p$ if and only if $m' \equiv m \pmod p$ and, similarly, $[n]_p = [n']_p$ means $n' \equiv n \pmod p$. Condition (1) translates to $(m'+n') \equiv (m+n) \pmod p$, which follows from Theorem 2. The proof of (2) is similar. Thus our new operations $+$ and $\cdot$ are well-defined on $[\mathbb{Z}]_p$.

(b) For instance, $[3]_6 + [5]_6 = [8]_6 = [2]_6$ and $[3]_5 \cdot [2]_5 = [6]_5 = [1]_5$. Notice for comparison that $3 +_6 5 = 2$ and $3 *_5 2 = 1$. Our operations on $[\mathbb{Z}]_6$ and $[\mathbb{Z}]_5$ look a lot like our operations on $\mathbb{Z}(6)$ and $\mathbb{Z}(5)$. [See Exercise 20 for the full story.]

If we write EVEN for $[0]_2$ and ODD for $[1]_2$ in $[\mathbb{Z}]_2$, then EVEN + ODD = ODD, ODD · ODD = ODD, etc.

(c) Let's try to define $f\colon [\mathbb{Z}]_6 \to \mathbb{Z}$ by the rule $f([m]_6) = m^2$. For example, $f([2]_6) = 2^2 = 4$, $f([3]_6) = 3^2 = 9$, $f([8]_6) = 64 =$ OOPS! The trouble is that $[8]_6 = [2]_6$, but $4 \neq 64$. Too bad. Our f is not well-defined. ■

Exercises 3.5

1. Use any method to find q and r as in the Division Algorithm for the following values of n and m.

(a) $n = 20, \quad m = 3$

(b) $n = 20, \quad m = 4$

(c) $n = -20, \quad m = 3$

(d) $n = -20, \quad m = 4$

(e) $n = 371{,}246, \quad m = 65$

(f) $n = -371{,}246, \quad m = 65$

2. Find n DIV m and n MOD m for the following values of n and m.

(a) $n = 20, \quad m = 3$

(b) $n = 20, \quad m = 4$

(c) $n = -20, \quad m = 3$

(d) $n = -20, \quad m = 4$

(e) $n = 371{,}246, \quad m = 65$

(f) $n = -371{,}246, \quad m = 65$

3. List three integers that are congruent mod 4 to each of the following.

(a) 0 (b) 1 (c) 2

(d) 3 (e) 4

4. (a) List all equivalence classes of $\mathbb{Z}$ for the equivalence relation congruence mod 4.

(b) How many different equivalence classes of $\mathbb{Z}$ are there with respect to congruence mod 73?

5. For each of the following integers m, find the unique integer r in $\{0, 1, 2, 3\}$ such that $m \equiv r \pmod 4$.

(a) 17 (b) 7 (c) -7

(d) 2 (e) -88

6. Calculate

(a) $4 +_7 4$

(b) $5 +_7 6$

(c) $4 *_7 4$

(d) $0 +_7 k$ for any $k \in \mathbb{Z}(7)$

(e) $1 *_7 k$ for any $k \in \mathbb{Z}(7)$

7. (a) Calculate $6 +_{10} 7$ and $6 *_{10} 7$.

(b) Describe in words $m +_{10} k$ for any $m, k \in \mathbb{Z}(10)$.

(c) Do the same for $m *_{10} k$.

8. (a) List the elements in the sets A_0, A_1, and A_2 defined by

$$A_k = \{m \in \mathbb{Z} : -10 \le m \le 10 \text{ and } m \equiv k \pmod 3\}.$$

(b) What is A_3? A_4? A_{73}?

9. Give the complete addition and multiplication tables for $\mathbb{Z}(4)$.

10. Use Figure 2 to solve the following equations for x in $\mathbb{Z}(6)$.

(a) $1 +_6 x = 0$ (b) $2 +_6 x = 0$

(c) $3 +_6 x = 0$ (d) $4 +_6 x = 0$

(e) $5 +_6 x = 0$

11. Use Figure 3 to solve the following equations for x in $\mathbb{Z}(5)$.

(a) $1 *_5 x = 1$ (b) $2 *_5 x = 1$

(c) $3 *_5 x = 1$ (d) $4 *_5 x = 1$

12. For m, n in $\mathbb{N}$, define $m \sim n$ if $m^2 - n^2$ is a multiple of 3.

(a) Show that $\sim$ is an equivalence relation on $\mathbb{N}$.

(b) List four elements in the equivalence class $[0]$.

(c) List four elements in the equivalence class $[1]$.

(d) Do you think there are any more equivalence classes?

13. The definition of $m \equiv n \pmod p$ makes sense even if $p = 1$.

(a) Describe this equivalence relation for $p = 1$ and the corresponding equivalence classes in $\mathbb{Z}$.

(b) What meaning can you attach to m DIV 1 and m MOD 1?

(c) What does Theorem 3 say if $p = 1$?

14. (a) Prove that if $m, n \in \mathbb{Z}$ and $m \equiv n \pmod p$, then $m^2 \equiv n^2 \pmod p$.

(b) Is the function $f\colon [\mathbb{Z}]_p \to [\mathbb{Z}]_p$ given by $f([n]_p) = [n^2]_p$ well defined? Explain.

(c) Repeat part (b) for the function $g\colon [\mathbb{Z}]_6 \to [\mathbb{Z}]_{12}$ given by $g([n]_6) = [n^2]_{12}$.

(d) Repeat part (b) for the function $h\colon [\mathbb{Z}]_6 \to [\mathbb{Z}]_{12}$ given by $h([n]_6) = [n^3]_{12}$.

15. (a) Show that the four-digit number $n =$ **abcd** is divisible by 9 if and only if the sum of the digits **a** + **b** + **c** + **d** is divisible by 9.

(b) Is the statement in part (a) valid for every n in $\mathbb{P}$ regardless of the number of digits? Explain.

16. (a) Show that the four-digit number $n =$ **abcd** is divisible by 2 if and only if the last digit **d** is divisible by 2.

(b) Show that $n = \mathtt{abcd}$ is divisible by 5 if and only if $\mathtt{d}$ is divisible by 5.

17. Show that the four-digit number $n = \mathtt{abcd}$ is divisible by 11 if and only if $\mathtt{a} - \mathtt{b} + \mathtt{c} - \mathtt{d}$ is divisible by 11.

18. (a) Show that if $n \text{ MOD } p = 0$, then $(-n) \text{ MOD } p = 0$ and $(-n) \text{ DIV } p = -(n \text{ DIV } p)$.

(b) Show that if $0 < n \text{ MOD } p < p$, then $(-n) \text{ MOD } p = p - n \text{ MOD } p$ and $(-n) \text{ DIV } p = -(n \text{ DIV } p) - 1$.

This exercise shows that one can easily compute $n \text{ DIV } p$ and $n \text{ MOD } p$ for negative n using DIV p and MOD p functions that are only defined for $n \in \mathbb{N}$.

19. Show that q and r are unique in the Division Algorithm. That is, show that, if $p, q, r, q', r' \in \mathbb{Z}$ with $p \geq 1$ and if

$$q \cdot p + r = q' \cdot p + r', \quad 0 \leq r < p, \text{ and } 0 \leq r' < p,$$

then $q = q'$ and $r = r'$.

20. Let $p \geq 2$. Define the mapping $\theta \colon \mathbb{Z}(p) \to [\mathbb{Z}]_p$ by $\theta(m) = [m]_p$.

(a) Show that θ is a one-to-one correspondence of $\mathbb{Z}(p)$ onto $[\mathbb{Z}]_p$.

(b) Show that $\theta(m +_p n) = \theta(m) + \theta(n)$ and $\theta(m *_p n) = \theta(m) \cdot \theta(n)$, where the operations + and $\cdot$ on $[\mathbb{Z}]_p$ are as defined in Example 6(a).

21. (a) Verify the commutative law for $+_p$ in Theorem 4(a).

(b) Verify the associative lawfor $+_p$ in Theorem 4(b).

(c) Observe that the proofs of the commutative and associative laws for $*_p$ are almost identical to those for $+_p$.

22. (a) Show that $n^2 \not\equiv 2 \pmod 3$ for all n in $\mathbb{Z}$.

(b) Show that $n^2 \not\equiv 2 \pmod 5$ for all n in $\mathbb{Z}$.

Chapter Highlights

For the following items:

(a) Satisfy yourself that you can define and use each concept and notation.
(b) Give at least one reason why the item was included in this chapter.
(c) Think of at least one example of each concept and at least one situation in which each fact would be useful.

The purpose of this review is to tie each of the ideas to as many other ideas and to as many concrete examples as possible. That way, when an example or concept is brought to mind, everything tied to it will also be called up for possible use.

CONCEPTS AND NOTATION

binary relation on S or from S to T
- reflexive, antireflexive, symmetric, antisymmetric, transitive [for relations on S]
- converse relation, function as relation
- equivalence relation
  - equivalence class $[s]$, partition $[S]$
  - natural mapping ν of s to $[s]$
  - congruence mod p, $\equiv \pmod p$, $[n]_p$
- quotient, DIV p, remainder, MOD p
- $\mathbb{Z}(p)$, $+_p$, $*_p$
- digraph or graph [undirected]
  - vertex, edge, loop, parallel edges
  - initial, terminal vertex, endpoint of edge path
    - length
    - closed path, cycle
  - acyclic path, acyclic digraph
  - adjacency relation, reachable relation
  - picture
    - of a digraph or graph
    - of a relation

matrix
- transpose, sum, product, scalar multiple, negative, inverse
- square, symmetric
- special matrices **0**, **I**

adjacency matrix of a graph or digraph

matrix of a relation

FACTS

Equivalence relations and partitions are two views of the same concept.

Every equivalence relation $\sim$ is of form $a \sim b$ if and only if $f(a) = f(b)$ for some function f.

Matrix addition is associative and commutative.

Matrix multiplication is not commutative.

Definitions of functions on equivalence classes must be independent of choice of representatives.

Division Algorithm $[n = p \cdot q + r, 0 \le r < p]$.

The unique x in $\mathbb{Z}(p)$ with $x \equiv m \pmod{p}$ is m MOD p.

If $m \equiv m' \pmod{p}$ and $n \equiv n' \pmod{p}$, then $m + n \equiv m' + n' \pmod{p}$ and $m \cdot n \equiv m' \cdot n' \pmod{p}$.

Operations $+_p$ and $*_p$ on $\mathbb{Z}(p)$ mimic + and $\cdot$ on $\mathbb{Z}$ except for cancellation with respect to $*_p$.

Supplementary Exercises

1. A graph has the matrix

$$\begin{bmatrix} 2 & 1 & 0 & 1 \\ 1 & 0 & 1 & 2 \\ 0 & 1 & 1 & 0 \\ 1 & 2 & 0 & 0 \end{bmatrix}.$$

   (a) How many edges does the graph have?

   (b) Calculate the sum of the degrees of the vertices of the graph.

   (c) Draw a picture of the graph.

2. Let $\mathbf{A}$ and $\mathbf{B}$ be 3×3 matrices defined by $\mathbf{A}[i, j] = i + 2j$ and $\mathbf{B}[i, j] = i^2 + j$. Determine

   (a) $\mathbf{A}^T$

   (b) $\mathbf{A} + \mathbf{A}^T$

   (c) $\sum_{i=1}^{3} \mathbf{B}[i, i]$

   (d) $\sum_{i=1}^{3} \sum_{j=1}^{3} \mathbf{A}[i, j]$

   (e) $\sum_{j=1}^{3} \mathbf{A}[j, 4 - j]$

3. Which of the following are equivalence relations? For the equivalence relations, identify or describe the equivalence classes. For the other relations, specify which properties, reflexivity, symmetry, and transitivity, may fail.

   (a) For $m, n \in \mathbb{N}$, define $(m, n) \in R_1$ if $m + n$ is an even integer.

   (b) Let V be the set of vertices of a graph G, and for $u, v \in V$ define $(u, v) \in R_2$ if $u = v$ or there exists a path from u to v.

   (c) Let V be the set of vertices of a digraph D, and for $u, v \in V$ define $(u, v) \in R_3$ if $u = v$ or there exists a path from u to v.

4. Define the equivalence relation R on $\mathbb{Z}$ by $(m, n) \in R$ if $m \equiv n \pmod{7}$.

   (a) How many equivalence classes of R are there?

   (b) List three elements in each of the following equivalence classes: $[1], [-73], [73], [4], [2^{250}]$.

5. Let A be an $n \times n$ matrix. Show that $A + A^T$ must be a symmetric matrix.

6. Let $S = \{1, 2, 3, 4, 5, 6, 7\}$ and define $m \sim n$ if $m^2 \equiv n^2 \pmod{5}$.

   (a) Show that $\sim$ is an equivalence relation on S.

   (b) Find all the equivalence classes.

7. Let Σ be some alphabet and let Σ^* consist of all strings using letters from Σ. For strings s and t in Σ^*, we define $(s, t) \in R$ provided that length$(s) \le$ length(t). Is R reflexive? symmetric? transitive? an equivalence relation?

8. This exercise concerns the famous Fermat's Little Theorem, which states that if p is a prime, then

$$n^{p-1} \equiv 1 \pmod{p} \quad \text{unless} \quad p | n.$$

   This theorem will reappear in Exercise 62 on page 224.

   (a) State Fermat's Little Theorem for the case $p = 2$, and explain why this case is almost obvious.

   (b) State Fermat's Little Theorem for $p = 3$, and prove it. *Hint:* There are two cases, $n \equiv 1 \pmod{3}$ and $n \equiv -1 \pmod{3}$.

9. The set $\mathbb{Q}^+$ of positive rational numbers is the set $\{m/n : m, n \in \mathbb{N}\}$. Is $f(m/n) = m - n$ a well-defined function on $\mathbb{Q}^+$? Explain briefly.

10. Let R be the relation defined on $\mathbb{N} \times \mathbb{N}$ as follows: $((m, n), (p, q)) \in R$ if and only if $m \equiv p \pmod{3}$ or $n \equiv q \pmod{5}$.

    (a) Is R reflexive? If not, prove it is not.

    (b) Is R symmetric? If not, prove it is not.

    (c) Is R transitive? If not, prove it is not.

    (d) Is R an equivalence relation? If not, explain why not.

11. (a) Consider an equivalence relation $\approx$ on a set S. Let x, y, z be in S, where $x \approx y$ and $y \not\approx z$. Prove *carefully* that $x \not\approx z$.

    (b) Let m, n, r, p be integers with $p \ge 2$. Show that if $m \equiv n \pmod{p}$ and $n \not\equiv r \pmod{p}$, then $m \not\equiv r \pmod{p}$.

12. (a) Does the formula $f([n]_{10}) = [2n - 3]_{20}$ give a well-defined function $f\colon [\mathbb{Z}]_{10} \to [\mathbb{Z}]_{20}$? Justify your answer.

    (b) Does the formula $f([n]_{10}) = [5n - 3]_{20}$ give a well-defined function $f\colon [\mathbb{Z}]_{10} \to [\mathbb{Z}]_{20}$? Justify your answer.

Induction and Recursion

The theme of this chapter can be described by this statement: "If we start out right and if nothing can go wrong, then we will always be right." The context to which we apply this observation will be a list of propositions that we are proving, a succession of steps that we are executing in an algorithm, or a sequence of values that we are computing. In each case, we wish to be sure that the results we get are always correct.

In §4.1 we look at algorithm segments, called **while** loops, that repeat a sequence of steps as long as a specified condition is met. The next section introduces mathematical induction, a fundamental tool for proving sequences of propositions. Section 4.4 introduces the ideas of recursive definition of sequences and the calculation of terms from previous terms. Then §4.5 gives explicit methods that apply to sequences given by some common forms of recurrence equations. The principles of induction introduced in §4.2 apply only to a special kind of recursive formulation, so in §4.6 we extend those methods to more general recursive settings. The chapter concludes with a discussion of greatest common divisors and the Euclidean algorithm for computing them. Section 4.7 also includes applications to solving congruences.

4.1 Loop Invariants

In the last section of Chapter 3 we claimed that for every pair of integers m and n, with $m > 0$ and $n \geq 0$, there are integers q and r, the quotient and remainder, such that $n = m \cdot q + r$ and $0 \leq r < m$. We will prove that claim in this section by first giving an algorithm that allegedly constructs q and r and then proving that the algorithm really does what it should. The algorithm we present has a special feature, called a **while** loop, that makes the algorithm easy to understand and helps us to give a proof of its correctness. The main point of this section is to understand **while** loops and some associated notation. In the next section we will see how the logic associated with these loops can be used to develop mathematical induction, one of the most powerful proof techniques in mathematics.

Here is the basic outline of our algorithm for division.

Guess values of q and r so that $n = m \cdot q + r$.
Keep improving the guesses, without losing the property that $n = m \cdot q + r$, until eventually we stop, with $0 \leq r < m$.

Actually, our initial guess will not be very imaginative: $q = 0$ and $r = n$ certainly work. This q is probably too small and the r probably too large, but at any rate q is not too large, and this choice makes r easy to find. Once we are started, $n = m \cdot q + r$ will remain true if we increase q by 1 and decrease r by m, because $m \cdot q + r = m \cdot (q+1) + (r - m)$. If we make these changes enough times, we hope to get $r < m$, as desired.

We can refine this rough algorithm into one that will actually compute q and r with the required properties. Here is the final version, written out in the style of a computer program. Don't worry; it is not necessary to know programming to follow the ideas.

DivisionAlgorithm(integer, integer)

```
{Input: integers m > 0 and n ≥ 0.}
{Output: integers q and r with m · q + r = n and 0 ≤ r < m.}
begin
{First initialize.}
q := 0
r := n
{Then do the real work.}
while r ≥ m do
   q := q + 1
   r := r − m
return q and r
end ■
```

Here and from now on we will use the notational convention that statements in braces {· · ·} are comments, and not part of the algorithm itself. The sequence of lines in the algorithm is a recipe. It says:

1. Set the value of q to be 0. [The $\boldsymbol{a} := \boldsymbol{b}$ notation means "set, or define, the value of a to equal the value of b."]
2. Set the value of r to be n.
3. Check to see if $r \geq m$.
   If $r \geq m$, then:
      Increase the value of q by 1,
      Decrease the value of r by m and
      Go back to 3.
   Otherwise, i.e., if $r < m$,
      Go on to 4.
4. Stop.

The initial guesses for q and r are likely to be wrong, of course, but repetition of step 3 is meant to improve them until they are right.

The **while** loop here is step 3. In general, a **while loop** is a sequence of steps in an algorithm or program that has the form

```
while g do
   S
```

and is interpreted to mean

(∗) Check if g is true.
If g is true, then do whatever S says to do, and after that go back to (∗).
Otherwise, skip over S and go on to whatever follows the loop in the main program.

Here g is some proposition, for instance "$r \geq m$," called the **guard** of the loop, and S is a sequence of steps, called the **body** of the loop. We could let S contain instructions to jump out of the loop to some other place in the program, but there is no real loss of generality if we just think of S as some program segment that we execute, after which we go back to test g again. Of course, S may change the truth value of g; that's the whole idea. We call an execution of S in the loop a **pass** through the loop, or an **iteration**, and we say that the loop **terminates** or **exits** if at some stage g is false and the program continues past the loop.

EXAMPLE 1

What a loop actually does may depend on the status of various quantities at the time the program or algorithm arrives at the loop. Figure 1 shows some simple loops.

Figure 1 ▶

(a)	(b)	(c)
while $n < 5$ **do**	**while** $n \neq 8$ **do**	**while** $A \neq \emptyset$ **do**
$n := n + 1$	print n^2	choose x in A
print n^2	$n := n + 2$	remove x from A

If $n = 0$ when the program enters the loop in Figure 1(a), then the loop replaces n by 1, prints 1, replaces n by 2, prints $4, \ldots$, replaces n by 5, prints 25, observes that now the guard $n < 5$ is false, and goes on to the next part of the program, whatever that is. If $n = 4$ at the start of the loop, then the program just replaces n by 5 and prints 25 before exiting. If $n = 6$ initially, the program does nothing at all, since the guard $6 < 5$ is false immediately.

If $n = 0$ initially, the loop of Figure 1(b) prints 0, 4, 16, and 36 and then exits. With an input value of $n = 1$, however, this loop prints 1, 9, 25, 49, 81, 121, etc. It *never* exits because, although the values of n keep changing, the guard $n \neq 8$ is always true. Although termination is usually desirable, loops are not required to terminate, as this example shows. Figure 1(c) gives a nonnumeric example. If A is some set, for instance the set of edges of a graph, the loop just keeps throwing elements out of A until there are none left. If A is finite, the loop terminates; otherwise, it doesn't. ■

$m = 7, n = 17$	q	r	$r \geq m$
Initially	0	17	True
After the first pass	1	10	True
After the second pass	2	3	False

Figure 2 ▲

In the case of DivisionAlgorithm we set the initial values of q and r before we enter the loop. Then S instructs us to increase q by 1 and decrease r by m. Figure 2 shows the successive values of q, r and the proposition "$r \geq m$" during the execution of this algorithm with inputs $m = 7$ and $n = 17$.

The algorithm does not go through the **while** loop a third time, since the guard $r \geq m$ is false after the second pass through. The final values are $q = 2$, $r = 3$. Since $17 = 7 \cdot 2 + 3$, the algorithm produces the correct result for the given input data.

This may be the place to point out that the division algorithm we have given is much slower than the pencil-and-paper division method we learned as children. For example, it would take DivisionAlgorithm 117 passes through the **while** loop to divide 2002 by 17. One reason the childhood method and its computerized relatives are a lot faster is that they take advantage of the representation of the input data in decimal or binary form; some organizational work has already been done. The algorithm we have given could also be greatly speeded up by an improved strategy for guessing new values for q and r. Since we just want to know that the right q and r *can* be computed, though, we have deliberately kept the account uncomplicated.

For DivisionAlgorithm to be of any value, of course, we need to be sure that:

1. It stops after a while, and
2. When it stops, it gives the right answers.

Let us first see why the algorithm stops. The action is all in the **while** loop, since the initialization presents no problems. Each pass through the loop, i.e., each execution of the instructions $q := q+1$ and $r := r-m$, takes a fixed amount of time, so we just need to show that the algorithm only makes a finite number of passes through the loop. Since $m \cdot q + r = n$ will remain true, the key is what happens to r. At first, $r = n \geq 0$. If $n < m$, then the guard is false and the program makes no passes at all through the **while** loop. Otherwise, if $n \geq m$, then each execution of the body of the loop reduces the value of r by m, so the successive values of r are $n, n-m, n-2m$, etc. These integers keep decreasing, since $m > 0$. Sooner or later we reach a k with $r = n - k \cdot m < m$. Then the guard $r \geq m$ is false, we exit from the loop, and we quit.

How do we know we'll reach such a k? The set $\mathbb{N}$ of natural numbers has the following important property:

Well-Ordering Principle Every nonempty subset of $\mathbb{N}$ has a smallest element.

This principle implies, in particular, that decreasing sequences in $\mathbb{N}$ do not go on forever; every decreasing sequence $a > b > c > \cdots$ in $\mathbb{N}$ is finite in length, because no member of the sequence can come after the smallest member. Hence every infinite decreasing sequence in $\mathbb{Z}$ must eventually have negative terms.

In our case, the sequence $n > n - m > n - 2m > \cdots$ must eventually have negative terms. If $n-k\cdot m$ is the last member of the sequence in $\mathbb{N}$, then $n-k\cdot m \geq 0$ and $n - (k+1)\cdot m < 0$. Thus $n - k \cdot m < m$, and the algorithm stops after k steps. Figure 2 illustrates an example where $k = 2$.

We now know that DivisionAlgorithm stops, but why does it always give the right answers? Because the equation $m \cdot q + r = n$ stayed true and at the end $r = n - k \cdot m$ satisfied $0 \leq r < m$. To formalize this argument, we introduce a new idea. We say that a proposition p is an **invariant** of the loop

```
while g do
  S
```

in case it satisfies the following condition.

If p and g are true before we do S, then p is true after we finish S.

The following general theorem gives one reason why loop invariants are useful.

Loop Invariant Theorem Suppose that p is an invariant of the loop "**while** g **do** S" and that p is true on entry into the loop. Then p is true after each iteration of the loop. If the loop terminates, it does so with p true and g false.

Proof The theorem is really fairly obvious. By hypothesis, p is true at the beginning, i.e., after 0 iterations of the loop. Suppose, if possible, that p is false after some iteration. Then there is a first time that p is false, say after the kth pass. Now $k \geq 1$ and p was true at the end of the $(k-1)$th pass, and hence at the start of the kth pass. So was g, or we wouldn't even have made the kth iteration. Because p is a loop invariant, it must also be true after the kth pass, contrary to the choice of k.

This contradiction shows that it can never happen that p is false at the end of a pass through the loop, so it must always be true.

If the loop terminates [because g is false] it does so at the end of a pass, so p must be true at that time. ■

The idea that we used in this proof could be called the Principle of the Smallest Criminal—if there's a bad guy, then there's a smallest bad guy. This is really just another way of stating the Well-Ordering Principle. In our proof we looked at the set $\{k \in \mathbb{N} : p \text{ is false after the } k\text{th pass}\}$. Supposing that this set was nonempty made it have a smallest member and led eventually to a contradiction.

How does the Loop Invariant Theorem help us validate DivisionAlgorithm? Figure 3 shows the algorithm again, for reference, with a new comment added just before the **while** loop. We next show that the statement "$m \cdot q + r = n$ and $r \geq 0$" is a loop invariant. It is surely true before we enter the **while** loop, because we have set up q and r that way. Moreover, if $m \cdot q + r = n$ and $r \geq m$, and if we execute the body of the loop to get new values $q' = q + 1$ and $r' = r - m$, then

$$m \cdot q' + r' = m \cdot (q + 1) + (r - m) = m \cdot q + r = n$$

and

$$r' = r - m \geq 0,$$

so the statement remains true after the pass through the loop. We already know that the loop terminates. The Loop Invariant Theorem tells us that when it does stop $n = m \cdot q + r$ and $r \geq 0$ are true, and $r \geq m$ is false, i.e., $r < m$. These are exactly the conditions we want the output to satisfy. We have shown the following.

Figure 3 ▶

DivisionAlgorithm(integer, integer)

```
{Input: integers m > 0 and n ≥ 0.}
{Output: integers q and r with m · q + r = n and 0 ≤ r < m.}
begin
q := 0
r := n
{m · q + r = n and r ≥ 0.}
while r ≥ m do
   q := q + 1
   r := r − m
return q and r
end ■
```

Theorem Given integers m and n with $m > 0$ and $n \geq 0$, DivisionAlgorithm constructs integers q and r with $n = m \cdot q + r$ and $0 \leq r < m$.

Why did we choose the loop invariant we did? We wanted to have $n = m \cdot q + r$ and $0 \leq r < m$. Of course "$n = m \cdot q + r$" and "$0 \leq r$" are each invariants of our loop, but neither by itself is good enough. We want both, so we list both. We can't use "$r < m$" as an invariant for the loop we set up, but it's a natural negation for a guard "$r \geq m$," so it will hold when the guard fails on exit from the loop.

Loop invariants can be used to guide the development of algorithms, as well as to help verify their correctness. We illustrate the method by indicating how we might have developed DivisionAlgorithm. At the beginning of this section we first saw a rough outline of the algorithm. A slightly refined version might have looked like this.

```
begin
Initialize q and r so that n = m · q + r.
while r ≥ m do
```

```
    Make progress toward stopping, and keep n = m · q + r.
return the final values
end
```

The next version could be somewhat more explicit.

```
begin
Let q := 0.
Let r = n − m · q, so that n = m · q + r.
while r ≥ m do
    Increase q.
    Keep n = m · q + r {so r will decrease}.
return the final values of q and r
end
```

At this stage it is clear that r has to be n initially and that, if q is increased by 1 in the body of the loop, then r has to decrease by m in that pass. Finally, $r \geq 0$ comes from requiring $n \geq 0$ in the input, so $r \geq 0$ initially, and the fact that $r \geq m$ at the start of each pass through the loop, so $r \geq 0$ at the end of each pass. Developing the algorithm this way makes it easy to show, using the invariant, that the algorithm gives the correct result.

DivisionAlgorithm can provide q and r even if n is negative. In that case, DivisionAlgorithm$(m, -n)$ yields integers q' and r' with $-n = m \cdot q' + r'$ and $0 \leq r' < m$. If $r' = 0$, then $n = m \cdot (-q')$, so $n = m \cdot q + 0$ with $q = -q'$ and $r = 0$. If $r' > 0$, then $n = m \cdot (-q' - 1) + (m - r')$, so $n = m \cdot q + r$ with $q = -q' - 1$ and $r = m - r'$. Note that $0 < r < m$ in this case, since $0 < r' < m$. For more on this point, see Exercise 16 on page 180.

This argument and the verification we have given for DivisionAlgorithm give a complete proof of the assertion that we called The Division Algorithm on page 120 and illustrated in Example 1 of §3.5.

Of course, just listing a proposition p in braces doesn't make it an invariant, any more than writing down a program makes it compute what it's supposed to. In order to use the Loop Invariant Theorem, one needs to check that p does satisfy the definition of an invariant.

```
{Input: an integer n > 0.}
{Output: the integer n!.}
begin
m := 1
FACT := 1
{FACT = m!}
while m < n do
    m := m + 1
    FACT := FACT · m
return FACT
end
```

Figure 4 ▲

EXAMPLE 2

(a) We defined $n! = 1 \cdot 2 \cdots n$ in §1.6. If $n > 1$, then $n! = (n-1)! \cdot n$, a fact that lets us compute the value of $n!$ with a simple algorithm. Figure 4 shows one such algorithm.

The alleged loop invariant is "FACT $= m!$". This equation is certainly true initially, and if FACT $= m!$ at the start of the **while** loop, then

$$(\text{new FACT}) = (\text{old FACT}) \cdot (\text{new } m) = m! \cdot (m+1) = (m+1)! = (\text{new } m)!,$$

as required at the end of the loop. The loop exits with $m = n$ and hence with FACT $= n!$ Once we had the idea of using FACT $= m!$ as a loop invariant, this algorithm practically wrote itself.

With the convention that $0! = 1$, we could initialize the algorithm with $m := 0$ instead of $m := 1$ and still produce the correct output.

Notice one apparently unavoidable feature of this algorithm; even if we just want 150!, we must compute $1!, 2!, \ldots, 149!$ first.

(b) Figure 5 shows another algorithm for computing $n!$, this time building in the factors in reverse order, from the top down. To see how this algorithm works, write out its steps starting with a small input such as $n = 5$.

Again, the loop invariant is easy to check. Initially, it is obviously true, and if REV $\cdot\, m! = n!$ at the start of a pass through the loop, then

$$(\text{new REV}) \cdot (\text{new } m)! = (\text{REV} \cdot m) \cdot (m-1)! = \text{REV} \cdot m! = n!$$

at the end of the pass. The guard makes the exit at the right time, with $m = 1$, so the algorithm produces the correct output, even if $n = 0$.

```
{Input: an integer n ≥ 0.}
{Output: the integer n!.}
begin
REV := 1
m := n
{REV · m! = n!}
while m > 0 do
    REV := REV · m
    m := m − 1
return REV
end
```

Figure 5 ▲

This algorithm does not compute $1!, 2!, \ldots$ along the way to $n!$, but what it does is just as bad, or worse. We still must compute a lot of partial products, in this instance pretty big ones from the very beginning, before we are done. ■

The examples we have just seen begin to show the value of loop invariants. In the next section we will use the Loop Invariant Theorem to develop the method of proof by mathematical induction, one of the most important techniques in this book. Later chapters will show us how the use of loop invariants can help us to understand the workings of some fairly complicated algorithms, as well as to verify that they produce the correct results.

Both DivisionAlgorithm and Example 2(a), the first of our algorithms to compute $n!$, had variables that increased by 1 on each pass through the loop. In the case of DivisionAlgorithm, we did not know in advance how many passes we would make through the loop, but in Example 2(a) we knew that m would take the values 1, 2, 3, ... , n in that order and that the algorithm would then stop. There is a convenient notation to describe this kind of predictable incrementation.

The algorithm instruction "**for** $k = m$ **to** n **do** S" tells us to substitute m, $m+1, \ldots, n$, for k, in that order, and to do S each time. Thus the segment

```
FACT := 1
for k = 1 to n do
    FACT := FACT · k
```

produces FACT $= n!$ at the end.

Using a **while** loop, we could rewrite a segment "**for** $k = m$ to n **do** S" as

```
k := m
while k ≤ n do
    S
    k := k + 1
```

with one major and obvious warning: the segment S cannot change the value of k. If S changes k, then we should replace k in the algorithm by some other letter, not m or n, that S does not change. See Exercise 15 for what can go wrong otherwise.

Some programming languages contain a construction "**for** $k = n$ **downto** m **do** S," meaning substitute $n, n-1, \ldots, m$ in decreasing order for k, and do S each time. Our second factorial algorithm could have been written using this device.

EXAMPLE 3

(a) Here is an algorithm to compute the sum

$$\sum_{k=1}^{n} k^2 = 1 + 4 + 9 + \cdots + n^2,$$

where n is a given positive integer.

```
s := 0
for k = 1 to n do
    s := s + k^2
return s
```

(b) More generally, the algorithm

```
s := 0
for k = m to n do
    s := s + a_k
return s
```

computes $\sum_{k=m}^{n} a_k = a_m + \cdots + a_n$. ■

Exercises 4.1

1. (a) With input $m = 13$ and $n = 73$, give a table similar to Figure 2 showing successive values of q, r and "$r \geq m$" in DivisionAlgorithm on page 129.
   (b) Repeat part (a) with input $m = 13$ and $n = 39$.
   (c) Repeat part (a) with input $m = 73$ and $n = 13$.
2. (a) Write $-73 = 13 \cdot q + r$, where q and r are integers and $0 \leq r < 13$.
   (b) Write $-39 = 13 \cdot q + r$, where q and r are integers and $0 \leq r < 13$.
   (c) Write $-13 = 73 \cdot q + r$, where q and r are integers and $0 \leq r < 73$.
3. List the first five values of x for each of the following algorithm segments.

(a)
```
x := 0
while 0 ≤ x do
    x := 2x + 3
```
(b)
```
x := 1
while 0 ≤ x do
    x := 2x + 3
```
(c)
```
x := 1
while 0 ≤ x do
    x := 2x − 1
```

4. Find an integer b so that the last number printed is 6.

(a)
```
n := 0
while n < b do
    print n
    n := n + 1
```
(b)
```
n := 0
while n < b do
    n := n + 1
    print n
```

5. (a) List the values of m and n during the execution of the following algorithm, using the format of Figure 2.

```
begin
m := 0
n := 0
while n ≠ 4 do
    m := m + 2n + 1
    n := n + 1
end
```

   (b) Modify the algorithm of part (a) so that $m = 17^2$ when it stops.
6. Interchange the lines "$n := n + 1$" and "print n^2" in Figure 1(a).
   (a) List the values printed if $n = 0$ at the start of the loop.
   (b) Do the same for $n = 4$.
7. Interchange the lines "print n^2" and "$n := n + 2$" in Figure 1(b).
   (a) List the values printed if $n = 0$ at the start of the loop.
   (b) Do the same for $n = 1$.
8. (a) Write the segment

```
for i = 1 to 17 do
    k := k + 2i
```

   using a **while** loop.
   (b) Repeat part (a) for the segment

```
for k = 8 downto 1 do
    i := i + 2k
```

9. Show that the following are loop invariants for the loop

```
while 1 ≤ m do
    m := m + 1
    n := n + 1
```

   (a) $m + n$ is even.
   (b) $m + n$ is odd.
10. Show that the following are loop invariants for the loop

```
while 1 ≤ m do
    m := 2m
    n := 3n
```

   (a) $n^2 \geq m^3$
   (b) $2m^6 < n^4$
11. Consider the loop

```
while j ≥ n do
    i := i + 2
    j := j + 1
```

   where i and j are nonnegative integers.
   (a) Is $i < j^2$ a loop invariant if $n = 1$? Explain.
   (b) Is $0 \leq i < j^2$ a loop invariant if $n = 0$? Explain.
   (c) Is $i \leq j^2$ a loop invariant if $n = 0$? Explain.
   (d) Is $i \geq j^2$ a loop invariant if $n = 0$? Explain.
12. Consider the loop

```
while k ≥ 1 do
    k := 2k
```

   (a) Is $k^2 \equiv 1 \pmod 3$ a loop invariant? Explain.
   (b) Is $k^2 \equiv 1 \pmod 4$ a loop invariant? Explain.
13. For which values of the integer b do each of the following algorithms terminate?

(a)
```
begin
k := b
while k < 5 do
    k := 2k − 1
end
```
(b)
```
begin
k := b
while k ≠ 5 do
    k := 2k − 1
end
```

(c)
```
begin
k := b
while k < 5 do
    k := 2k + 1
end
```

14. Suppose that the loop body in Example 2(a) is changed to

```
FACT := FACT · m
m := m + 1.
```

How should the initialization and guard be changed in order to get the same output, FACT $= n!$? The loop invariant may change.

15. Do the two algorithms shown produce the same output? Explain.

Algorithm A

```
begin
k := 1
while k ≤ 4 do
    k := k^2
    print k
    k := k + 1
end
```

Algorithm B

```
begin
for k = 1 to 4 do
    k := k^2
    print k
end
```

16. Do the two algorithms shown produce the same output? Explain.

Algorithm C

```
begin
m := 1
n := 1
while 1 ≤ m ≤ 3 do
    while 1 ≤ n ≤ 3 do
        m := 2m
        n := n + 1
        print m
end
```

Algorithm D

```
begin
m := 1
n := 1
while 1 ≤ m ≤ 3
    and 1 ≤ n ≤ 3 do
        m := 2m
        n := n + 1
        print m
    end
```

17. Here's a primitive pseudorandom-number generator that generates numbers in $\{0, 1, 2, \dots, 72\}$, given a positive integer c:

```
begin
r := c
while r > 0 do
    r := 31 · r MOD 73
end
```

For example, if $c = 3$, we get 3, 20, 36, 21, 67, 33, ... Which of the following are loop invariants?

(a) $r < 73$ (b) $r \equiv 0 \pmod 5$ (c) $r = 0$

18. (a) Show that $S = I^2$ is an invariant of the loop

```
while 1 ≤ I do
    S := S + 2I + 1
    I := I + 1
```

(b) Show that $S = I^2 + 1$ is also an invariant of the loop in part (a).

(c) What is the 73rd number printed by the following? Explain.

```
begin
S := 1
I := 1
while 1 ≤ I do
    print S
    S := S + 2I + 1
    I := I + 1
end
```

(d) The same question as in part (c), but with $S := 2$ initially.

19. Which of these sets of integers does the Well-Ordering Principle say have smallest elements? Explain.

(a) $\mathbb{P}$ (b) $\mathbb{Z}$

(c) $\{n \in \mathbb{P} : n^2 > 17\}$ (d) $\{n \in \mathbb{P} : n^2 < 0\}$

(e) $\{n \in \mathbb{Z} : n^2 > 17\}$

(f) $\{n^2 : n \in \mathbb{P} \text{ and } n! > 2^{1000}\}$

20. (a) What would we get if we tried to apply the algorithm in Figure 4 with $n = -3$ or with $n = -73$? Do we obtain sensible outputs?

(b) Answer the same question for the algorithm in Figure 5.

21. Consider the following loop, with $a, b \in \mathbb{P}$.

```
while r > 0 do
    a := b
    b := r
    r := a MOD b
```

[Recall that $m = (m \text{ DIV } n) \cdot n + (m \text{ MOD } n)$ with $0 \le (m \text{ MOD } n) < n$.] Which of the following are invariants of the loop? Explain.

(a) a, b, and r are multiples of 5.

(b) a is a multiple of 5.

(c) $r < b$.

(d) $r \le 0$.

22. (a) Is $5^k < k!$ an invariant of the following loop?

```
while 4 ≤ k do
    k := k + 1
```

(b) Can you conclude that $5^k < k!$ for all $k \ge 4$?

23. Suppose that the propositions p and q are both invariants of the loop

```
while g do S
```

(a) Is $p \wedge q$ an invariant of the loop? Explain.

(b) Is $p \vee q$ an invariant of the loop? Explain.

(c) Is $\neg p$ an invariant of the loop? Explain.

(d) Is $p \rightarrow q$ an invariant of the loop? Explain.

(e) Is $g \rightarrow p$ an invariant of the loop? Explain.

24. Here is an algorithm that factors an integer as a product of an odd integer and a power of 2.

{Input: a positive integer n.}
{Output: nonnegative integers k and m with m odd and $m \cdot 2^k = n$.}
begin
$m := n$
$k := 0$
while m is even **do**
  $m := \frac{m}{2}$ and $k := k + 1$
return k and m
end

(a) Show that $m \cdot 2^k = n$ is a loop invariant.

(b) Explain why the algorithm terminates, and show that m is odd on exit from the loop.

25. The following algorithm gives a fast method for raising a number to a power.

{Input: a number a and a positive integer n.}
{Output: a^n.}
begin
$p := 1$
$q := a$
$i := n$
while $i > 0$ **do**
  if i is odd **then** $p := p \cdot q$
  {Do the next two steps whether or not i is odd.}
  $q := q \cdot q$
  $i := i \text{ DIV } 2 = \lfloor \frac{i}{2} \rfloor$
return p
end

(a) Give a table similar to Figure 2 showing successive values of p, q, and i with input $a = 2$ and $n = 11$.

(b) Verify that $q^i \cdot p = a^n$ is a loop invariant and that $p = a^n$ on exit from the loop.

26. (a) Consider the following algorithm with inputs $n \in \mathbb{P}$, $x \in \mathbb{R}$.

begin
$m := n$
$y := x$
while $m \neq 0$ **do**
  body
end

Suppose that "body" is chosen so that "$x^n = x^m \cdot y$ and $m \geq 0$" is an invariant of the **while** loop. What is the value of y at the time the algorithm terminates, if it ever does?

(b) Is it possible to choose z so that the loop

while $m \neq 0$ **do**
  $m := m - 1$
  $y := z$

has "$x^n = x^m \cdot y$ and $m \geq 0$" as an invariant? Explain.

27. Is the Well-Ordering Principle true if $\mathbb{N}$ is replaced by $\mathbb{R}$ or by $[0, \infty)$?

4.2 Mathematical Induction

This section develops a framework for proving lots of propositions all at once. We introduce the main idea, which follows from the Well-Ordering Principle for $\mathbb{N}$, as a natural consequence of the Loop Invariant Theorem on page 131.

EXAMPLE 1

We claim that $37^{500} - 37^{100}$ is a multiple of 10. One way to check this would be to compute $37^{500} - 37^{100}$ and, if the last digit is 0, to proclaim victory. The trouble with this plan is that 37^{500} is pretty big—it has over 780 decimal digits—so we might be a while computing it. Since we aren't claiming to know the exact value of $37^{500} - 37^{100}$, but just claiming that it's a multiple of 10, maybe there's a simpler way.

Since 37 is odd, so are 37^{500} and 37^{100}, so at least we know that $37^{500} - 37^{100}$ is a multiple of 2. To establish our claim, we just need to show that it's also a multiple of 5. We notice that $37^{500} = (37^{100})^5$, so we're interested in $(37^{100})^5 - 37^{100}$. Maybe $n^5 - n$ is *always* a multiple of 5 for $n \in \mathbb{P}$. If it is, then we can get our claim by letting $n = 37^{100}$, and we've learned a lot more than just the answer to the original question.

Experiments with small numbers look promising. For instance, we have $1^5 - 1 = 0$, $2^5 - 2 = 30$, $3^5 - 3 = 240$, and even $17^5 - 17 = 1{,}419{,}840$. So there is hope.

Here is our strategy. We construct a simple loop, in effect an elementary machine whose job it is to verify that $n^5 - n$ is a multiple of 5 for every n in $\mathbb{P}$, starting with $n = 1$ and running at least until $n = 37^{100}$. We will also give the machine

```
begin
  n := 1
  while n < 37^100 do
    if n^5 − n is a multiple of 5 then
      n := n + 1
end
```

Figure 1 ▲

some help. Figure 1 shows the loop. Its output is irrelevant; what will matter is its invariant.

The **while** loop simply checks that $n^5 - n$ is a multiple of 5 and, if that's true, it moves on to the next value of n. We claim that the algorithm terminates [with $n = 37^{100}$] and that "$n^5 - n$ is a multiple of 5" is a loop invariant. If these claims are true, then, since $1^5 - 1$ is a multiple of 5 when we arrive at the loop, the Loop Invariant Theorem will imply that $37^{500} - 37^{100}$ is a multiple of 5.

Consider a pass through the loop, say with $n = k < 37^{100}$. If $k^5 - k$ is *not* a multiple of 5, then the body of the loop does nothing, and the algorithm just repeats the loop forever with n stuck at k. On the other hand, if $k^5 - k$ *is* a multiple of 5, then the **if** condition is satisfied, the loop body increases the value of n to $k + 1$ and the algorithm goes back to check if the guard $n < 37^{100}$ is still true. To make sure that the algorithm terminates with $n = 37^{100}$, we want to be sure that at each iteration the **if** condition is true.

Here is where we give the algorithm some help. If it is making the $(k + 1)$st pass through the loop, then $k^5 - k$ must have been a multiple of 5. We give a little proof now to show that this fact forces $(k + 1)^5 - (k + 1)$ also to be a multiple of 5. The reason [using some algebra] is that

$$\begin{aligned}(k + 1)^5 - (k + 1) &= k^5 + 5k^4 + 10k^3 + 10k^2 + 5k + 1 - k - 1 \\ &= (k^5 - k) + 5(k^4 + 2k^3 + 2k^2 + k).\end{aligned}$$

If $k^5 - k$ is a multiple of 5, then, since the second term is obviously a multiple of 5, $(k + 1)^5 - (k + 1)$ must also be a multiple of 5. What this all means is that we can tell the algorithm not to bother checking the **if** condition each time; $n^5 - n$ will always be a multiple of 5, because it was the time before.

This argument is enough to show that the **if** condition forms a loop invariant. Moreover, n increases by 1 each time through the loop. Eventually, $n = 37^{100}$ and the loop terminates, as we claimed.

The loop we have devised may look like a stupid algorithm. To check that $37^{500} - 37^{100}$ is a multiple of 5, we seem to go to all the work of looking at $1^5 - 1$, $2^5 - 2$, $3^5 - 3$, ..., $(37^{100} - 1)^5 - (37^{100} - 1)$ first. That really *would* be stupid. But notice that, as a matter of fact, the only one we checked outright was $1^5 - 1$. For the rest, we just gave a short algebraic argument that *if* $k^5 - k$ is a multiple of 5, then so is $(k + 1)^5 - (k + 1)$. The algorithm was never intended to be run, but was just meant to give us a *proof* that $37^{500} - 37^{100}$ is a multiple of 5, without computing either of the huge powers. In fact, a tiny change in the algorithm shows, with the same proof, that 5 divides $73^{5000} - 73^{1000}$. Indeed, we can show that $n^5 - n$ is a multiple of 5 for *every* $n \in \mathbb{P}$ by applying exactly the same arguments to the loop in Figure 2, which does not terminate. ■

```
n := 1
while 1 ≤ n do
  if n^5 − n is a multiple of 5 then
    n := n + 1
```

Figure 2 ▲

Suppose, more generally, that we have a finite list of propositions, say $p(m)$, $p(m + 1), \ldots, p(n)$, where $m, m + 1, \ldots, n$ are successive integers. In Example 1, for instance, we had $p(k) =$ "$k^5 - k$ is a multiple of 5" for $k = 1, 2, \ldots, 37^{100}$. Suppose that we also know, as we did in Example 1, that

(B) $p(m)$ is true, and

(I) $p(k + 1)$ is true whenever $p(k)$ is true and $m \le k < n$.

We claim that all of the propositions $p(m), p(m + 1), \ldots, p(n)$ must then be true. The argument is just like the one we went through in Example 1. We construct the loop in Figure 3.

```
k := m
{p(k) is true}
while m ≤ k < n do
    if p(k) is true then
    k := k + 1
```

Figure 3 ▲

By (I), the statement "$p(k)$ is true" is a loop invariant, and the loop terminates with $k = n$. In fact, *we don't need to include* an "**if** $p(k)$ **then**" line, since $p(k)$ is guaranteed to be true on each pass through the loop. Of course, the condition that $p(k) \Longrightarrow p(k + 1)$ is the reason that $p(k)$ is true at the end of each pass. In terms of the Well-Ordering Principle, we are just observing that if $p(k)$ ever failed there would be a first time when it did, and the conditions (B) and (I) prevent every time from being that first bad time.

We have proved the following important fact.

Principle of Finite Mathematical Induction Let $p(m)$, $p(m+1), \ldots,$ $p(n)$ be a finite sequence of propositions. If

(B) $p(m)$ is true, and

(I) $p(k+1)$ is true whenever $p(k)$ is true and $m \le k < n$,

then all the propositions are true.

In many applications m will be 0 or 1. An infinite version is available, too. Just replace "**while** $m \le k < n$" by "**while** $m \le k$" to get the following.

Principle of Mathematical Induction Let $p(m)$, $p(m+1), \ldots$ be a sequence of propositions. If

(B) $p(m)$ is true, and

(I) $p(k+1)$ is true whenever $p(k)$ is true and $m \le k$,

then all the propositions are true.

Condition (B) in each of these principles of induction is called the **basis**, and (I) is the **inductive step**. Given a list of propositions, these principles help us to organize a proof that all of the propositions are true. The basis is usually easy to check; the inductive step is sometimes quite a bit more complicated to verify.

The principles tell us that *if* we can show (B) and (I), then we are done, but they do not help us show either condition. Of course, if the $p(k)$'s are not all true, the principles cannot show that they are. Either (B) or (I) must fail in such a case.

EXAMPLE 2

(a) For each positive integer n, let $p(n)$ be "$n! > 2^n$," a proposition that we claim is true for $n \ge 4$. To give a proof by induction, we verify $p(n)$ for $n = 4$, i.e., check that $4! > 2^4$, and then show

(I) If $4 \le k$ and if $k! > 2^k$, then $(k+1)! > 2^{k+1}$.

The proof of (I) is straightforward:

$$\begin{aligned}(k+1)! &= k! \cdot (k+1) \\ &> 2^k \cdot (k+1) && \text{[by the inductive assumption } k! > 2^k\text{]} \\ &\ge 2^k \cdot 2 && \text{[since } k+1 \ge 5 > 2\text{]} \\ &= 2^{k+1}.\end{aligned}$$

Since we have checked the basis and the inductive step, $p(n)$ is true for every integer $n \ge 4$ by induction.

(b) The fact that

$$1 + 2 + \cdots + n = \frac{n(n+1)}{2} \quad \text{for all } n \text{ in } \mathbb{P}$$

is useful to know. It can be proved by an averaging argument, but also by induction.

Let $p(n)$ be the proposition "$\sum_{i=1}^{n} i = n(n+1)/2$." Then $p(1)$ is the proposition "$\sum_{i=1}^{1} i = 1(1+1)/2$," which is true.

Assume inductively that $p(k)$ is true for some positive integer k, i.e., that $\sum_{i=1}^{k} i = k(k+1)/2$. We want to show that this assumption implies that $p(k+1)$

is true. Now

$$\sum_{i=1}^{k+1} i = \left(\sum_{i=1}^{k} i\right) + (k+1) \qquad \text{[definition of } \textstyle\sum \text{ notation]}$$

$$= \frac{k(k+1)}{2} + (k+1) \qquad \text{[assumption that } p(k) \text{ is true]}$$

$$= \left[\frac{k}{2} + 1\right](k+1) \qquad \text{[factor out } k+1\text{]}$$

$$= \left[\frac{k+2}{2}\right](k+1)$$

$$= \frac{((k+1)+1)(k+1)}{2},$$

so $p(k+1)$ holds. By induction, $p(n)$ is true for every n in $\mathbb{P}$.

(c) We can use induction to establish the formula for the sum of the terms in a geometric series:

$$\sum_{i=0}^{n} r^i = \frac{r^{n+1}-1}{r-1} \quad \text{if } r \neq 0, r \neq 1 \text{ and } n \in \mathbb{N}.$$

Let $p(n)$ be "$\sum_{i=0}^{n} r^i = \frac{r^{n+1}-1}{r-1}$." Then $p(0)$ is "$r^0 = \frac{r^1-1}{r-1}$," which is true because $r^0 = 1$. Thus the basis for induction is true.

We prove the inductive step $p(k) \Longrightarrow p(k+1)$ as follows:

$$\sum_{i=0}^{k+1} r^i = \left(\sum_{i=0}^{k} r^i\right) + r^{k+1}$$

$$= \frac{r^{k+1}-1}{r-1} + r^{k+1} \qquad \text{[by the inductive assumption } p(k)\text{]}$$

$$= \frac{r^{k+1}-1+r^{k+2}-r^{k+1}}{r-1} \qquad \text{[algebra]}$$

$$= \frac{r^{k+2}-1}{r-1} \qquad \text{[more algebra]}$$

$$= \frac{r^{(k+1)+1}-1}{r-1},$$

i.e., $p(k+1)$ holds. By induction, $p(n)$ is true for every $n \in \mathbb{N}$.

(d) We prove that all numbers of the form $8^n - 2^n$ are divisible by 6. More precisely, we show that $8^n - 2^n$ is divisible by 6 for each $n \in \mathbb{P}$. Our nth proposition is

$$p(n) = \text{"}8^n - 2^n \text{ is divisible by 6."}$$

The basis for the induction, $p(1)$, is clearly true, since $8^1 - 2^1 = 6$. For the inductive step, assume that $p(k)$ is true for some $k \geq 1$. Our task is to use this assumption somehow to establish $p(k+1)$:

$$8^{k+1} - 2^{k+1} \text{ is divisible by 6.}$$

Thus we would like to write $8^{k+1} - 2^{k+1}$ somehow in terms of $8^k - 2^k$, in such a way that any remaining terms are easily seen to be divisible by 6. A little trick is to write $8^{k+1} - 2^{k+1}$ as $8(8^k - 2^k)$ plus appropriate correction terms:

$$8^{k+1} - 2^{k+1} = 8(8^k - 2^k) + 8 \cdot 2^k - 2^{k+1}$$

$$= 8(8^k - 2^k) + 8 \cdot 2^k - 2 \cdot 2^k = 8(8^k - 2^k) + 6 \cdot 2^k.$$

Now $8^k - 2^k$ is divisible by 6 by assumption $p(k)$, and $6 \cdot 2^k$ is obviously a multiple of 6, so the same is true of $8^{k+1} - 2^{k+1}$. We have shown that the inductive step is valid, and so our proof is complete by the Principle of Mathematical Induction.

What if we hadn't thought of the "little trick"? Here's a straightforward approach. By assumption, $8^k - 2^k = 6m$ for some integer m, so

$$8^{k+1} = 8 \cdot 8^k = 8(2^k + 6m) = 2(2^k + 6m) + 6(2^k + 6m) = 2^{k+1} + 6p,$$

where $p = 2m + 2^k + 6m$. Thus $8^{k+1} - 2^{k+1}$ is divisible by 6. The idea is to try to express the thing you're looking for in terms of the thing you already know about, in this case to express 8^{k+1} in terms of 8^k, and then see what can be done. ■

It is worth emphasizing that, prior to the last sentence in each of these proofs, we did *not* prove "$p(k+1)$ is true." We merely proved an implication: "if $p(k)$ is true, then $p(k+1)$ is true." In a sense we proved an infinite number of assertions, namely: $p(1)$; if $p(1)$ is true, then $p(2)$ is true; if $p(2)$ is true, then $p(3)$ is true; if $p(3)$ is true, then $p(4)$ is true; etc. Then we applied mathematical induction to conclude: $p(1)$ is true; $p(2)$ is true; $p(3)$ is true; $p(4)$ is true; etc.

Note also that when we use the induction principles we don't need to write any loops. We used loops to justify the principles, but when we apply the principles we only need to verify conditions (B) and (I).

EXAMPLE 3

Here are some poor examples of the use of induction.

(a) For $n \in \mathbb{P}$, let $p(n)$ be "$n^2 \le 100$." Then we can check directly that $p(1)$, $p(2)$, ..., $p(10)$ are true. If we try to prove by induction that $p(n)$ is true for all $n \in \mathbb{P}$, or even for $1 \le n \le 20$, we will fail. We can show the basis (B), but the inductive step (I) must not be provable, because we know, for example, that $p(11)$ is false.

(b) For $n \in \mathbb{N}$, let $r(n) =$ "$n^3 - n + 1$ is a multiple of 3." We show that $r(k) \Longrightarrow r(k+1)$. Since

$$(k+1)^3 - (k+1) + 1 = k^3 + 3k^2 + 3k + 1 - k = k^3 - k + 1 + 3(k^2 + k),$$

since $k^3 - k + 1$ is a multiple of 3 by assumption, and since $3(k^2 + k)$ is obviously a multiple of 3, $(k+1)^3 - (k+1) + 1$ is also a multiple of 3. That is, $r(k) \Longrightarrow r(k+1)$ for $0 \le k$, and (I) holds. Does this imply that all $r(n)$ are true? No! We didn't check the basis $r(0) =$ "$0^3 - 0 + 1$ is a multiple of 3," which is false. In fact, all the propositions $r(n)$ are false.

As parts (a) and (b) illustrate, one must always check both the basis (B) and the inductive step (I) before applying a principle of induction.

(c) For $m \in \mathbb{P}$, let $s(m)$ be "$m \cdot (m+1)$ is even." We could certainly prove $s(m)$ true for all m in $\mathbb{P}$ using induction [think of how such a proof would go], but we don't need such an elaborate proof. Either m or $m+1$ is even, so their product is surely even in any case. Sometimes it is easier to prove directly that all propositions in a sequence are true than it is to invoke induction. ■

EXAMPLE 4

n	s_n
0	a
1	$2a + b$
2	$2(2a + b) + b = 4a + 3b$
3	$2(4a + 3b) + b = 8a + 7b$
4	$2(8a + 7b) + b = 16a + 15b$

Figure 4 ▲

(a) An application of induction often starts with a guessing game. Suppose, for instance, that the sequence $(s_0, s_1, s_2, \ldots)$ satisfies the conditions $s_0 = a$ and $s_n = 2s_{n-1} + b$ for some constants a and b and all $n \in \mathbb{P}$. Can we find a formula to describe s_n? We compute the first few terms of the sequence to see if there is a pattern. Figure 4 shows the results.

It looks as if we might have $s_n = 2^n a + (2^n - 1)b$ in general. [This kind of guessing a general fact from a few observations is what is called "inductive reasoning" in the sciences. The use of the word "inductive" in that context is quite different from the mathematical usage.] Now we have a list of propositions $p(n) =$ "$s_n = 2^n a + (2^n - 1)b$" that we suspect might be true for all n in $\mathbb{N}$. It turns out [Exercise 9] that we have made a good guess, and one can prove by *mathematical* induction that $p(n)$ is true for each n in $\mathbb{N}$.

(b) As we are looking for a pattern, it sometimes helps to examine closely how a particular case follows from the one before. Then we may be able to see how to construct an argument for the inductive step.

Let $\mathcal{P}(S)$ be the power set of some finite set S. If S has n elements, then $\mathcal{P}(S)$ has 2^n members. This proposition was shown to be plausible in Example 2 on page 19. We prove it now, by induction.

We verified this assertion earlier for $n = 0, 1, 2,$ and 3. In particular, the case $n = 0$ establishes the basis for induction. Before proving the inductive step, let's experiment a little and compare $\mathcal{P}(S)$ for $S = \{a, b\}$ and for $S = \{a, b, c\}$. Note that

$$\mathcal{P}(\{a, b, c\}) = \big\{\emptyset, \{a\}, \{b\}, \{a, b\}, \{c\}, \{a, c\}, \{b, c\}, \{a, b, c\}\big\}.$$

The first four sets make up $\mathcal{P}(\{a, b\})$; each of the remaining sets consists of a set in $\mathcal{P}(\{a, b\})$ with c added to it. This is why $\mathcal{P}(\{a, b, c\})$ has twice as many sets as $\mathcal{P}(\{a, b\})$. This argument looks as if it generalizes: every time an element is added to S, the size of $\mathcal{P}(S)$ doubles.

To prove the inductive step, we assume the proposition is valid for k. We consider a set S with $k+1$ elements; for convenience we use $S = \{1, 2, 3, \dots, k, k+1\}$. Let $T = \{1, 2, 3, \dots, k\}$. The sets in $\mathcal{P}(T)$ are simply the subsets of S that do not contain $k + 1$. By the assumption for k, $\mathcal{P}(T)$ contains exactly 2^k sets. Each remaining subset of S contains the number $k + 1$, so it is the union of a set in $\mathcal{P}(T)$ with the one-element set $\{k + 1\}$. That is, $\mathcal{P}(S)$ has another 2^k sets that are not subsets of T. It follows that $\mathcal{P}(S)$ has $2^k + 2^k = 2^{k+1}$ members. This completes the inductive step, and hence the proposition is true for all n by mathematical induction. ■

The method of mathematical induction applies to situations, such as the one in this example, in which

1. We know the answer in the beginning,
2. We know how to determine the answer at one stage from the answer at the previous stage.
3. We have a guess at the general answer.

Of course, if our guess is wrong, that's too bad; we won't be able to prove it is right with this method, or with any other. But if our guess is correct, then mathematical induction often gives us a framework for confirming the guess with a proof.

Sometimes it only makes sense to talk about a finite sequence of propositions $p(n)$. For instance, we might have an iterative algorithm that we know terminates after a while, and we might want to know that some condition holds while it is running. Verifying invariants for **while** loops is an example of this sort of problem. If we can write our condition as $p(k)$, in terms of some variable k that increases steadily during the execution, say $k = m, m + 1, \dots, N$, then we may be able to use the Principle of Finite Mathematical Induction to prove that $p(k)$ is true for each allowed value of k. Verifying an invariant of a segment "**for** $i = m$ to N **do** S" amounts to checking the inductive step for this principle.

This section has covered the basic ideas of mathematical induction and gives us the tools to deal with a great number of common situations. As we will see in §4.6, there are other forms of induction that we can use to handle many problems in which the principles from this section do not naturally apply.

Exercises 4.2

1. Explain why $n^5 - n$ is a multiple of 10 for all n in $\mathbb{P}$.
Hint: Most of the work was done in Example 1.

2. Write a loop in the style of Figure 2 that corresponds to the proof that $\sum_{i=1}^{n} i = \frac{n(n+1)}{2}$ in Example 2(b).

3. (a) Show that "$n^5 - n + 1$ is a multiple of 5" is an invariant of the loop in Figure 1.

(b) Is $n^5 - n + 1$ a multiple of 5 for all n in $\mathbb{P}$ with $n \le 37^{100}$?

4. (a) Show that $n^3 - n$ is a multiple of 6 for all n in $\mathbb{P}$.

(b) Use part (a) to give another proof of Example 2(d).

5. Prove

$$\sum_{i=1}^{n} i^2 = 1 + 4 + 9 + \cdots + n^2 = \frac{n(n+1)(2n+1)}{6} \quad \text{for} \quad n \in \mathbb{P}.$$

6. Prove

$$4 + 10 + 16 + \cdots + (6n - 2) = n(3n + 1) \text{ for all } n \in \mathbb{P}.$$

7. Show each of the following.

(a) $37^{100} - 37^{20}$ is a multiple of 10.

(b) $37^{20} - 37^4$ is a multiple of 10.

(c) $37^{500} - 37^4$ is a multiple of 10.

(d) $37^4 - 1$ is a multiple of 10.

(e) $37^{500} - 1$ is a multiple of 10.

8. Prove

$$\frac{1}{1 \cdot 5} + \frac{1}{5 \cdot 9} + \frac{1}{9 \cdot 13} + \cdots + \frac{1}{(4n-3)(4n+1)} = \frac{n}{4n+1} \quad \text{for} \quad n \in \mathbb{P}.$$

9. Show by induction that, if $s_0 = a$ and $s_n = 2s_{n-1} + b$ for $n \in \mathbb{P}$, then $s_n = 2^n a + (2^n - 1)b$ for every $n \in \mathbb{N}$.

10. Consider the following procedure.

```
begin
  S := 1
  while 1 ≤ S do
     print S
     S := S + 2√S + 1
```

(a) List the first four printed values of S.

(b) Use mathematical induction to show that the value of S is always an integer. [It is easier to prove the stronger statement that the value of S is always the square of an integer; in fact, $S = n^2$ at the start of the nth pass through the loop.]

11. Prove that $11^n - 4^n$ is divisible by 7 for all n in $\mathbb{P}$.

12. (a) Choose m and $p(k)$ in the segment

```
k := m
while m ≤ k do
   if p(k) is true then
      k := k + 1
```

so that proving $p(k)$ an invariant of the loop would show that $2^n < n!$ for all integers $n \ge 4$.

(b) Verify that your $p(k)$ in part (a) is an invariant of the loop.

(c) The proposition $p(k) =$ "$8^k < k!$" is an invariant of this loop. Does it follow that $8^n < n!$ for all $n \ge 4$? Explain.

13. (a) Show that $\sum_{i=0}^{k} 2^i = 2^{k+1} - 1$ is an invariant of the loop in the algorithm

```
begin
  k := 0
  while 0 ≤ k do
     k := k + 1
end
```

(b) Repeat part (a) for the invariant $\sum_{i=0}^{k} 2^i = 2^{k+1}$.

(c) Can you use part (a) to prove that $\sum_{i=0}^{k} 2^i = 2^{k+1} - 1$ for every k in $\mathbb{N}$? Explain.

(d) Can you use part (b) to prove that $\sum_{i=0}^{k} 2^i = 2^{k+1}$ for every k in $\mathbb{N}$? Explain.

14. Prove that $n^2 > n + 1$ for $n \ge 2$.

15. (a) Calculate $1 + 3 + \cdots + (2n - 1)$ for a few values of n, and then guess a general formula for this sum.

(b) Prove the formula obtained in part (a) by induction.

16. For which n in $\mathbb{P}$ does the inequality $4n \le n^2 - 7$ hold? Explain.

17. Consider the proposition $p(n) =$ "$n^2 + 5n + 1$ is even."

(a) Prove that $p(k) \Longrightarrow p(k+1)$ for all k in $\mathbb{P}$.

(b) For which values of n is $p(n)$ actually true? What is the moral of this exercise?

18. Prove $(2n + 1) + (2n + 3) + (2n + 5) + \cdots + (4n - 1) = 3n^2$ for all n in $\mathbb{P}$. The sum can also be written $\sum_{i=n}^{2n-1} (2i + 1)$.

19. Prove that $5^n - 4n - 1$ is divisible by 16 for n in $\mathbb{P}$.

20. Prove $1^3 + 2^3 + \cdots + n^3 = (1 + 2 + \cdots + n)^2$, i.e., $\sum_{i=1}^{n} i^3 = \left[\sum_{i=1}^{n} i\right]^2$ for all n in $\mathbb{P}$. *Hint:* Use the identity in Example 2(b).

21. Prove that

$$\frac{1}{n+1} + \frac{1}{n+2} + \cdots + \frac{1}{2n} = 1 - \frac{1}{2} + \frac{1}{3} - \frac{1}{4} + \cdots + \frac{1}{2n-1} - \frac{1}{2n}$$

for n in $\mathbb{P}$. For $n = 1$ this equation says that $\frac{1}{2} = 1 - \frac{1}{2}$, and for $n = 2$ it says that $\frac{1}{3} + \frac{1}{4} = 1 - \frac{1}{2} + \frac{1}{3} - \frac{1}{4}$.

22. For n in $\mathbb{P}$, prove

(a) $\sum_{i=1}^{n} \frac{1}{\sqrt{i}} \ge \sqrt{n}$ (b) $\sum_{i=1}^{n} \frac{1}{\sqrt{i}} \le 2\sqrt{n} - 1$

23. Prove that $5^{n+1} + 2 \cdot 3^n + 1$ is divisible by 8 for $n \in \mathbb{N}$.

24. Prove that $8^{n+2} + 9^{2n+1}$ is divisible by 73 for $n \in \mathbb{N}$.

25. This exercise requires a little knowledge of trigonometric identities. Prove that $|\sin nx| \le n|\sin x|$ for all x in $\mathbb{R}$ and all n in $\mathbb{P}$.

Office Hours 4.2

I'm having big trouble with induction. Are we really going to need it later?

That's a question that we can think about on several levels. There's the basic idea of induction and then there's the mechanical part of writing up proofs in good form. Right now, in Chapter 4, we're trying to understand the idea *and* work on form, so it's easy to get confused. As we go on, there will be less emphasis on form, but it will still be important to know how you *could* write up an induction proof if you had to. We'll also generalize these ideas, so be sure you understand basic induction now.

You'll see careful induction proofs in later chapters of the book, and I'll expect you to be able to understand them, but you'll also see the authors say things like, "An induction proof is hiding here" or "If those three dots make you nervous, give a proof by induction." They think you should be able to recognize where induction is needed to justify "and so on" claims. They don't necessarily expect you to stop and fill in the details, but just to think about how you *might* do it. You'll have very few exercises after this chapter that absolutely require induction proofs, although there will be some.

What's most important is that you understand the *idea* of induction, and the standard form for writing inductive proofs can help make the pattern clear. You need to make sure your claim is true to begin with, and then you need to check that if it's true for some value of n, then it's always true for the next value.

I sort of remember induction from high school. I thought the idea was to take a formula, change all the n's to $n + 1$, then assume it and prove it. There was a picture of dominoes in the book, but nothing about loop invariants.

Well, you can see now what they were trying to say. Maybe they just worked with formulas, but of course induction applies to any sequence of propositions. Let's take a formula, though, say the standard example

$$1 + 2 + \cdots + n = \frac{n(n+1)}{2}.$$

Actually, this isn't a formula, it's a *proposition* that says that the sum $1 + 2 + \cdots + n$ is given by the formula $n(n + 1)/2$. Anyhow, we start by checking it for $n = 1$, with only one term in the sum. Sure enough, $1 = 1 \cdot (1 + 1)/2$, so the formula is right so far. In other words, the proposition is true for $n = 1$.

Now here comes the "replace n by $n+1$" part. Let's call our proposition $P(n)$. Actually, to reduce confusion we've been using k, rather than n, in the inductive steps. Then the induction principle tells us that it will be enough to show that $P(k)$ always implies $P(k + 1)$. This means that we want to show that, no matter what value k has, if $P(k)$ is true, then so is $P(k + 1)$. So our argument should show that if $P(17)$ is true, then so is $P(18)$. If we somehow know that

$$1 + 2 + \cdots + 16 + 17 = \frac{17 \cdot (17 + 1)}{2},$$

then it should follow *from our argument* that

$$1 + 2 + \cdots + 16 + 17 + 18 = \frac{18 \cdot (18 + 1)}{2},$$

but of course we can't just replace 17 by 18 in the first equation and automatically expect the second equation to be true. And our argument has to work for every k, not just for $k = 17$.

In this case the argument is pretty easy. We do replace k by $k + 1$ everywhere, to get

$$1 + 2 + \cdots + (k + 1) = \frac{(k + 1)(k + 1 + 1)}{2},$$

but this equation is *what we have to prove*. It's not just automatically true. To prove it, we're *allowed to assume* that the equation

$$1 + 2 + \cdots + k = \frac{k(k+1)}{2}$$

is true. Not for *every* k, though, but just for the one we're considering! This means that the thing we care about, $1 + 2 + \cdots + k + (k + 1)$, can be rewritten as $\frac{k(k+1)}{2} + (k + 1)$, and then a little algebra lets us rewrite this as $(k + 1)(k + 2)/2$, which is what we want. So, yes, we changed all the k's to $k + 1$ to see the result $P(k + 1)$ we wanted to prove, but what we assumed was the original $P(k)$. There's some work involved—more than just writing down $P(k+1)$. This particular example was one where we just added the same thing to both sides of an equation, but of course that won't be appropriate for lots of induction problems.

The domino picture is meant to convey the idea that somehow $P(k)$ needs to push over $P(k + 1)$ and that somebody needs to push over the first domino to start the process. The book's approach, using loop invariants, gives another way of thinking about induction. We'll see a lot more loop invariants when we begin studying algorithms, and of course you'll see them in your computer science classes, too.

But back to induction. The method gives a framework for proofs, but you'll need to supply the details. Remind me to say a few words in class about how the induction framework is sort of like a format for a business letter. You still have to put in the addressee and the contents of the letter and the signature, but at least you know what pieces go where.

4.3 Big-Oh Notation

One way that sequences arise naturally in computer science is as lists of successive computed values. The sequences FACT and TWO in Example 2 on page 36 and the sequence s_n in Example 4 on page 141 are of this sort, and we have seen a number of others. Another important application of sequences, especially later when we analyze algorithms, is to the problem of estimating how long a computation will take for a given input.

For example, think of sorting a list of n given integers into increasing order. There are lots of algorithms available for doing this job; you can probably think of several different methods yourself. Some algorithms are faster than others, and all of them take more time as n gets larger. If n is small, it probably doesn't make much difference which method we choose, but for large values of n a good choice of algorithm may lead to a substantial saving in time. We need a way to describe the time behavior of our algorithms.

In our sorting example, say the sequence t measures the time a particular algorithm takes, so $t(n)$ is the time to sort a list of length n. On a faster computer we might cut all the values of $t(n)$ by a factor of 2 or 100 or even 1000. But then *all* of our algorithms would get faster, too. For choosing between methods, what really matters is some measure, not of the absolute size of $t(n)$, but of the rate at which $t(n)$ grows as n gets large. Does it grow like 2^n, n^2, n, or $\log_2 n$ or like some other function of n that we have looked at, for instance in §1.6?

The main point of this section is to develop notation to describe rates of growth. Before we do so, we study more closely the relationships among familiar sequences, such as $\log_2 n$, $\sqrt{n}$, n, n^2, and 2^n, and get some practice working with inequalities and basic properties of logarithms. Some of the results can be verified by induction, but we give direct proofs when they seem clearer. Several of the general results, but not the careful estimates, also follow easily from L'Hôspital's Rule in calculus.

EXAMPLE 1

(a) For all positive integers n, we have

$$\cdots \le \sqrt[4]{n} \le \sqrt[3]{n} \le \sqrt{n} \le n \le n^2 \le n^3 \le n^4 \le \cdots.$$

Of course, other exponents of n can be inserted into this string of inequalities. For example,

$$n \le n\sqrt{n} \le n^2 \quad \text{for all } n;$$

recall that $n\sqrt{n} = n^{3/2}$.

(b) We have $n < 2^n$ for all $n \in \mathbb{N}$. Actually, we have $n \le 2^{n-1}$ for all n; this is clear for small values of n like 1, 2, and 3. In general,

$$n = 2 \cdot \frac{3}{2} \cdot \frac{4}{3} \cdot \frac{5}{4} \cdots \frac{n-1}{n-2} \cdot \frac{n}{n-1}.$$

There are $n - 1$ positive factors on the right, none of them larger than 2, so $n \le 2^{n-1}$.

In fact, the estimates we have made here have been truly crude; 2^n is incredibly large compared with n for even moderate values of n. Already for $n = 10$, we have $2^n \approx 1000$, and each increase of n by 1 doubles 2^n, so that $2^{20} \approx 1{,}000{,}000$, $2^{40} \approx 1{,}100{,}000{,}000{,}000$, and 2^{100} has over 30 decimal digits, though 100 is still a fairly small number. For fun, try graphing $y = x$ and $y = 2^x$ on the same axes to get an idea of how much faster an exponential function grows than a linear one, or even than a polynomial one like x^{10}, for that matter, as we will see in Example 2(b).

(c) $n^2 \le \frac{9}{8} \cdot 2^n$ for $n \ge 1$. This is easily checked for $n = 1, 2, 3$, and 4. Note that we get equality with $n = 3$. For $n > 4$, observe that

$$n^2 = 4^2 \cdot \left(\frac{5}{4}\right)^2 \cdots \left(\frac{n-1}{n-2}\right)^2 \cdot \left(\frac{n}{n-1}\right)^2.$$

Each factor on the right except the first one is at most $\left(\frac{5}{4}\right)^2$, and there are $n-4$ such factors. Since $\left(\frac{5}{4}\right)^2 = 1.5625 < 2$,

$$n^2 < 4^2 \cdot 2^{n-4} = 2^4 \cdot 2^{n-4} = 2^n \quad \text{if} \quad n > 4.$$

An induction proof that $n^2 < 2^n$ for $n > 4$ is asked for in Exercise 15 on page 180. ■

EXAMPLE 2

(a) Since $n \le 2^{n-1}$ for all n in $\mathbb{N}$ by Example 1(b), we have

$$\log_2 n \le \log_2 2^{n-1} = n - 1 \quad \text{for} \quad n \ge 1.$$

We can use this fact to show that

$$\log_2 x < x \quad \text{for all real numbers} \quad x > 0;$$

see Figure 1. Indeed, if $n = \lceil x \rceil$, the smallest integer at least as big as x, then $n - 1 < x \le n$. Thus $\log_2 x \le \log_2 n$, and $\log_2 n \le n - 1$ from above, so we have $\log_2 x \le n - 1 < x$.

Figure 1 ▶

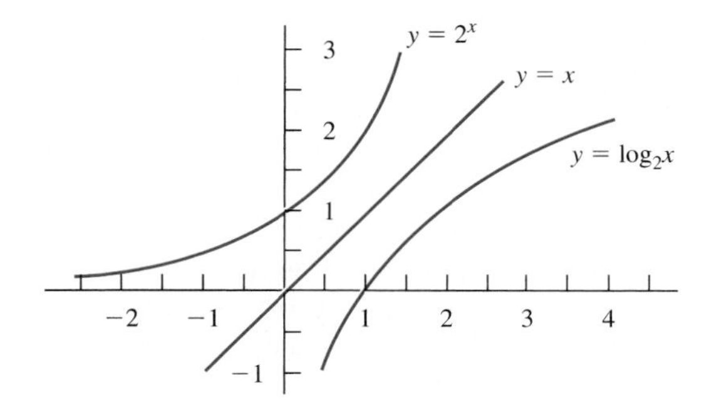

(b) In Example 1(c) we saw that $n^2 < 2^n$ for $n > 4$. In fact, given any fixed positive constant m, we have

$$n^m < 2^n \quad \text{for all sufficiently large } n.$$

This is a bit tricky to show. We first note that $\frac{1}{2}\log_2 n = \log_2 n^{1/2} = \log_2 \sqrt{n} < \sqrt{n}$ by part (a), so

$$\log_2 n^m = m \cdot \log_2 n < 2m \cdot \sqrt{n} = \frac{2m}{\sqrt{n}} \cdot n.$$

Now $2m/\sqrt{n} \le 1$ for $n \ge 4m^2$, so

$$\log_2 n^m < n \quad \text{for} \quad n \ge 4m^2.$$

Hence we have

$$n^m < 2^n \quad \text{for} \quad n \ge 4m^2.$$

Induction isn't an attractive approach to this problem. For one thing, we would need to do the analysis we just did to determine an appropriate basis, i.e., to determine how large n needs to be to start the induction.

(c) Again let m be a fixed positive integer. From part (b) we have

$$\log_2 n^m < n \quad \text{for} \quad n \ge 4m^2.$$

This inequality holds even if n is not an integer, so we can replace n by $\sqrt[m]{n}$ to obtain $\log_2(\sqrt[m]{n})^m < \sqrt[m]{n}$ for $\sqrt[m]{n} \ge 4m^2$; i.e.,

$$\log_2 n < \sqrt[m]{n} \quad \text{for} \quad n \ge (4m^2)^m.$$

Thus $\log_2 n < \sqrt[m]{n}$ for sufficiently large n.

This fact is a counterpart to our observation earlier that 2^n grows incredibly faster than n^k for any k. Now we see that $\log_2 n$ grows slower than any root of n that we care to choose. Compared with n, $\log_2 n$ is a real slowpoke. For instance, $\log_2 1{,}000{,}000$ is only about 20, $\log_2 2{,}000{,}000$ is only about 21, etc. ■

Before making these ideas about growth more precise, we observe some additional inequalities.

EXAMPLE 3

(a) We have

$$2^n < n! < n^n \quad \text{for} \quad n \ge 4.$$

We proved $2^n < n!$ using induction in Example 2(a) on page 139; see also Exercise 12 on page 143. Here is a direct argument. For $n = 4$, both inequalities are evident: $16 < 24 < 256$. For $n > 4$, we have $n! = (4!) \cdot 5 \cdot 6 \cdots (n-1) \cdot n$. The first factor $4!$ exceeds 2^4, and each of the remaining $n-4$ factors exceeds 2. So $n! > 2^4 \cdot 2^{n-4} = 2^n$.

The inequality $n! < n^n$ is true since $n!$ is a product of integers, all but one of which are less than n.

(b) Let's be greedy and claim that $40^n < n!$ for sufficiently large n. This will be a little trickier than $2^n < n!$ to verify. Observe that for $n > 80$ we can write

$$\begin{aligned} n! &> n(n-1)\cdots 81 && [n-80 \text{ factors}] \\ &> 80 \cdot 80 \cdots 80 && [n-80 \text{ factors}] \\ &= 80^{n-80} = 40^n\left\{2^n \cdot \frac{1}{80^{80}}\right\}, \end{aligned}$$

which will be greater than 40^n provided that $2^n > 80^{80}$ or $n > \log_2(80^{80}) = 80 \cdot \log_2 80 \approx 505.8$. This was a "crude" argument in the sense that we lost a lot [$80!$ in fact] when we wrote $n! > n(n-1)\cdots 81$. If we had wanted better information about where $n!$ actually overtakes 40^n, we could have made more careful estimates. Our work was enough, though, to show that $40^n < n!$ for all sufficiently large n. ■

To be more definite about what we mean when we say "$\cdots$ grows like $\cdots$ for large n," we need to develop some new notation, called the **big-oh notation**. Our main use for the notation will be to describe algorithm running times.

Given two sequences, say s and a, with nonnegative real values, the statement "$s(n) = O(a(n))$" [read "$s(n)$ is big oh of $a(n)$"] is intended to mean that as n gets large the values of s are no larger than some fixed constant multiple of the values of a. The equals sign $=$ is being used in a new way here, as just part of a complete notational package, so it has to be read and used carefully. For example, $O(a(n)) = s(n)$ is not a meaningful statement. Reading the equals sign as "is" gives the right idea. We might say "love is blind," or "$s(n)$ is $O(a(n))$," but not "blind is love" or "$O(a(n))$ is $s(n)$."

EXAMPLE 4

(a) Example 1(a) tells us that $\sqrt{n} = O(n)$, $n = O(n^2)$, etc. Example 1(b) tells us that $n = O(2^n)$, and Example 1(c) tells us that $n^2 = O(2^n)$. In fact, $n^m = O(2^n)$ for each m by Example 2(b). Example 3(a) shows that $2^n = O(n!)$ and that $n! = O(n^n)$. Also $2^n = O(n^n)$, and similar statements can be inferred from these examples.

(b) The dominant term of $6n^4 + 20n^2 + 2000$ is $6n^4$, since for large n the value of n^4 is much greater than n^2 or the constant 2000. We will write this observation as

$$6n^4 + 20n^2 + 2000 = O(n^4)$$

to indicate that the left expression "grows no worse than a multiple of n^4." It actually grows a little bit *faster* than $6n^4$. What matters, though, is that for large enough n the total quantity is no larger than some fixed multiple of n^4. [In this case, if $n \geq 8$, then $20n^2 + 2000 < n^4$, so $6n^4 + 20n^2 + 2000 < 7n^4$.] We could also say, of course, that it grows no faster than a multiple of n^5, but that's not as useful a piece of information. ■

Here is the precise definition. Let s be a sequence of real numbers, and let a be a sequence of positive real numbers. We write

$$\boldsymbol{s(n) = O(a(n))}$$

in case there is some positive constant C such that

$$|s(n)| \leq C \cdot a(n) \quad \text{for all sufficiently large values of } n.$$

Here we are willing to have the inequality fail for a few small values of n, perhaps because $s(n)$ or $a(n)$ fails to be defined for those values. All we really care about are the large values of n. In practice, $s(n)$ will represent some sequence of current interest [such as an upper bound on the time some algorithm will run], while $a(n)$ will be some simple sequence, like n, $\log_2 n$, n^3, etc., whose growth we understand. The next theorem lists some of what we have learned in Examples 1 to 3.

Theorem 1 Here is the hierarchy of several familiar sequences, in the sense that each sequence is big-oh of any sequence to its right:

$$1,\ \log_2 n,\ \ldots,\ \sqrt[4]{n},\ \sqrt[3]{n},\ \sqrt{n},\ n,\ n\log_2 n,\ n\sqrt{n},\ n^2,\ n^3,\ n^4,\ \ldots,\ 2^n,\ n!,\ n^n.$$

The constant sequence 1 in the theorem is defined by $1(n) = 1$ for all n. It doesn't grow at all. Of course, lots of sequences are not on this list, including some whose growth rates would put them between ones that are listed. We have just picked some popular examples.

EXAMPLE 5

(a) Suppose that $a(n) = n$ for all $n \in \mathbb{N}$. The statement $s(n) = O(n)$ means that $|s(n)|$ is bounded by a constant multiple of n; i.e., there is some $C > 0$ so that $|s(n)| \le Cn$ for all large enough $n \in \mathbb{N}$.

(b) Suppose $a(n) = 1$ for all $n \in \mathbb{N}$. We say that $s(n)$ is $O(1)$ if there is a constant C such that $|s(n)| \le C$ for all large n, that is, in case the values of $|s|$ are bounded above by some constant.

(c) The sequence s defined by $s_n = 3n^2 + 15n$ satisfies $s_n = O(n^2)$, because $n \le n^2$ for $n \ge 1$, and thus $|s_n| \le 3n^2 + 15n^2 = 18n^2$ for all large enough n.

(d) The sequence t given by $t_n = 3n^2 + (-1)^n 15n$ also satisfies $t_n = O(n^2)$. As in part (c), we have $|t_n| \le 3n^2 + 15n^2 = 18n^2$ for $n \ge 1$.

(e) We can generalize the examples in parts (c) and (d). Consider a polynomial $s(n) = a_m n^m + a_{m-1}n^{m-1} + \cdots + a_0$ in n of degree m with $a_m \ne 0$. Here n is the variable, and m and the coefficients $a_0, a_1, \ldots, a_m$ are constants. Since $|a_k n^k| \le |a_k| \cdot n^m$ for $k = 0, 1, \ldots, m-1$, we have

$$|s(n)| \le |a_m n^m| + |a_{m-1}n^{m-1}| + \cdots + |a_0|$$
$$\le (|a_m| + |a_{m-1}| + \cdots + |a_0|)n^m,$$

and hence $s(n) = O(n^m)$. The first inequality holds because

$$|x_1 + x_2 + \cdots + x_i| \le |x_1| + |x_2| + \cdots + |x_i|$$

for any finite sequence $x_1, x_2, \ldots, x_i$ in $\mathbb{R}$. ■

Because of Example 5(e), a sequence $s(n)$ is said to have **polynomial growth** if $s(n) = O(n^m)$ for some positive integer m. From a theoretical point of view, algorithms whose time behaviors have polynomial growth are regarded as manageable. They certainly are, compared to those with time behavior as bad as 2^n, say. In practice, efficient time behavior like $O(n)$ or $O(n \log_2 n)$ is most desirable.

People sometimes write "$s(n) = O(a(n))$" when what they intend to say is that $s(n)$ and $a(n)$ grow at essentially the same rate. The **big-theta** notation, defined by

$$\boldsymbol{s(n) = \Theta(a(n))}$$

in case both $s(n) = O(a(n))$ and $a(n) = O(s(n))$, expresses that meaning precisely. Here a and s are both sequences of positive numbers. While a sequence that is $O(n \log n)$ might actually be $O(n)$, a sequence that is $\Theta(n \log n)$ could not be $O(n)$; it must in fact grow as rapidly as a multiple of $n \log n$.

EXAMPLE 6

(a) We saw in Example 5(c) that $3n^2 + 15n = O(n^2)$. Since also $n^2 \le 3n^2 \le 3n^2 + 15n$, we have $n^2 = O(3n^2 + 15n)$, and hence $3n^2 + 15n = \Theta(n^2)$.

(b) More generally, if $s(n) = a_m n^m + a_{m-1}n^{m-1} + \cdots + a_0$ is a polynomial with $a_m > 0$ and $a_k \ge 0$ for $k = 0, \ldots, m-1$, then $s(n) = O(n^m)$ by Example 5(e), and $n^m = (1/a_m) \cdot a_m n^m \le (1/a_m)s(n)$, so $n^m = O(s(n))$. Thus $s(n) = \Theta(n^m)$. ■

The next example concerns a sequence that arises in estimating the time behavior of algorithms.

EXAMPLE 7

Let $s_n = 1 + \frac{1}{2} + \cdots + \frac{1}{n}$ for $n \ge 1$. Sequences defined by sums, such as this one, may be bounded or they may grow larger without bound, even if the terms being added get smaller and smaller. Calculus provides methods that can often determine what the behavior is. In the case of this sequence s_n, integral calculus yields the inequalities

$$\ln(n+1) < s_n < 1 + \ln n$$

for $n \in \mathbb{P}$, where ln is $\log_e$ with $e \approx 2.718$. Thus $s_n = O(\ln n)$. Also, we have $\ln n < s_{n-1} < s_n$ for $n \geq 2$, so $\ln n = O(s_n)$. Thus $s_n = \Theta(\ln n)$. Since $\ln n = \ln 2 \cdot \log_2 n$, we also have $s_n = \Theta(\log_2 n)$. ■

Our next theorem lists some general facts about big-oh notation. Remember that writing $s(n) = O(a(n))$ just signifies that $s(n)$ is some sequence that is $O(a(n))$.

Theorem 2 Let $a(n)$ and $b(n)$ be sequences of positive numbers.

(a) If $s(n) = O(a(n))$ and if c is a constant, then $c \cdot s(n) = O(a(n))$.
(b) If $s(n) = O(a(n))$ and $t(n) = O(a(n))$, then $s(n) + t(n) = O(a(n))$.
(c) If $s(n) = O(a(n))$ and $t(n) = O(b(n))$, then $s(n) + t(n) = O(\max\{a(n), b(n)\})$.
(d) If $s(n) = O(a(n))$ and $t(n) = O(b(n))$, then $s(n) \cdot t(n) = O(a(n) \cdot b(n))$.

Corresponding facts about big-theta can be obtained by using the theorem twice, once in each direction. You may wish to skip the theorem's proof on first reading and focus on the examples and discussion after the proof to see what the theorem is really saying.

Proof Parts (a) and (d) are left to Exercise 25.

(b) If $s(n) = O(a(n))$ and $t(n) = O(a(n))$, then there exist positive constants C and D such that

$$|s(n)| \leq C \cdot a(n) \quad \text{for all sufficiently large } n,$$

and

$$|t(n)| \leq D \cdot a(n) \quad \text{for all sufficiently large } n.$$

Since $|x + y| \leq |x| + |y|$ for $x, y \in \mathbb{R}$, we conclude that

$$|s(n) + t(n)| \leq |s(n)| + |t(n)| \leq (C + D) \cdot a(n)$$

for all sufficiently large n. Consequently, we have $s(n) + t(n) = O(a(n))$.

(c) Let $s(n) = O(a(n))$ and $t(n) = O(b(n))$. Then there exist positive constants C and D such that

$$|s(n)| \leq C \cdot a(n) \quad \text{and} \quad |t(n)| \leq D \cdot b(n) \quad \text{for sufficiently large } n.$$

Then we have

$$\begin{aligned} |s(n) + t(n)| &\leq |s(n)| + |t(n)| \leq C \cdot a(n) + D \cdot b(n) \\ &\leq C \cdot \max\{a(n), b(n)\} + D \cdot \max\{a(n), b(n)\} \\ &= (C + D) \cdot \max\{a(n), b(n)\} \end{aligned}$$

for sufficiently large n. Therefore, $s(n) + t(n) = O(\max\{a(n), b(n)\})$. ■

EXAMPLE 8

(a) Since $n^2 + 13n = O(n^2)$ and $(n + 1)^3 = O(n^3)$ [think about this],

$$(n^2 + 13n) + (n + 1)^3 = O(\max\{n^2, n^3\}) = O(n^3)$$

by Theorem 2(c), and

$$(n^2 + 13n)(n + 1)^3 = O(n^2 \cdot n^3) = O(n^5)$$

by Theorem 2(d).

(b) If $s(n) = O(n^4)$ and $t(n) = O(\log_2 n)$, then we have $s(n) \cdot t(n) = O(n^4 \log_2 n)$, $s(n)^2 = O(n^8)$, and $t(n)^2 = O(\log_2^2 n)$. Note that $\log_2^2 n$ denotes $(\log_2 n)^2$.

(c) The general principles in Theorem 2 would have shortened some arguments in Examples 5 and 7. For instance, we know that $n^k = O(n^m)$ if $k \le m$, so Theorem 2 gives

$$\begin{aligned} a_m n^m &+ a_{m-1}n^{m-1} + \cdots + a_1 n + a_0 \\ &= O(n^m) + O(n^{m-1}) + \cdots + O(n) + O(1) \quad \text{[by Theorem 2(a)]} \\ &= O(n^m) \quad \text{[by Theorem 2(c)].} \end{aligned}$$

We are able to use Theorem 2(c) here because the number of summands, $m+1$, does not depend on n. ■

Looking at Example 4(b) and Example 5, we see that it might be useful to describe a sequence by giving its dominant term plus terms that contribute less for large n. Thus we might write

$$6n^4 + 20n^2 + 1000 = 6n^4 + O(n^2),$$

where the big-oh term here stands for the expression $20n^2 + 1000$, which we know satisfies the condition $20n^2 + 1000 = O(n^2)$. This usage of the big-oh notation in expressions such as $s(n) + O(a(n))$ is slightly different in spirit from the usage in equations like $s(n) = O(a(n))$, but both are consistent with an interpretation that $O(a(n))$ represents "a sequence whose values are bounded by some constant multiple of $a(n)$ for all large enough n." It is this meaning that we shall use from now on.

Just as we made sense out of $s(n) + O(a(n))$, we can write $s(n) \cdot O(a(n))$ to represent a product sequence $s(n) \cdot t(n)$, where $t(n) = O(a(n))$.

EXAMPLE 9 As a consequence of Example 5(e), we can write

$$\sum_{k=0}^{m} a_k n^k = a_m n^m + O(n^{m-1}).$$ ■

Big-oh and big-theta notation let us work with rough estimates, but they are not a license to be sloppy. On the contrary, they provide careful, precise meaning to statements about comparative rates of growth, statements loose enough to omit irrelevant details but tight enough to capture the essentials.

Exercises 4.3

1. For each sequence below, find the smallest number k such that $s(n) = O(n^k)$. You need not prove that your k is the best possible.
(a) $s(n) = 13n^2 + 4n - 73$
(b) $s(n) = (n^2 + 1)(2n^4 + 3n - 8)$
(c) $s(n) = (n^3 + 3n - 1)^4$ (d) $\sqrt{n+1}$

2. Repeat Exercise 1 for
(a) $s(n) = (n^2 - 1)^7$ (b) $s(n) = \sqrt{n^2 - 1}$
(c) $s(n) = \sqrt{n^2 + 1}$
(d) $s(n) = (n^2 + n + 1)^2 \cdot (n^3 + 5)$

3. For each sequence below, give the sequence $a(n)$ in the hierarchy of Theorem 1 such that $s(n) = O(a(n))$ and $a(n)$ is as far to the left as possible in the hierarchy.
(a) $s(n) = 3^n$ (b) $s(n) = n^3 \log_2 n$
(c) $s(n) = \sqrt{\log_2 n}$

4. Repeat Exercise 3 for
(a) $s(n) = n + 3\log_2 n$ (b) $s(n) = (n \log_2 n + 1)^2$
(c) $s(n) = (n+1)!$

5. State whether each of the following is true or false. In each case give a reason for your answer.
(a) $2^{n+1} = O(2^n)$ (b) $(n+1)^2 = O(n^2)$
(c) $2^{2n} = O(2^n)$ (d) $(200n)^2 = O(n^2)$
(e) $2^{n+1} = \Theta(2^n)$ (f) $(n+1)^2 = \Theta(n^2)$
(g) $2^{2n} = \Theta(2^n)$ (h) $(200n)^2 = \Theta(n^2)$

6. Repeat Exercise 5 for
(a) $\log_2^{73} n = O(\sqrt{n})$ (b) $\log_2(n^{73}) = O(\log_2 n)$
(c) $\log_2 n^n = O(\log_2 n)$ (d) $(\sqrt{n} + 1)^4 = O(n^2)$

7. True or false. In each case give a reason.
(a) $40^n = O(2^n)$ (b) $(40n)^2 = O(n^2)$
(c) $(2n)! = O(n!)$ (d) $(n+1)^{40} = O(n^{40})$

8. Let $s(n)$ be a sequence of positive numbers.
(a) Show that $73 \cdot s(n) = \Theta(s(n))$.
(b) Show that if $s(n) \ge 1$ for all n, then $\lfloor s(n) \rfloor = \Theta(s(n))$.

9. For each of the following sequences $s(n)$, find a number k with $s(n) = \Theta(n^k)$.

(a) $s(n) = 20n^5 - 3n^3$

(b) $s(n) = (5n^6 + 2n^4)\sqrt{n}$

(c) $s(n) = (n^2 + 6n - 1)(n^4 + n^2)$

(d) $s(n) = (n^5 - 1)/(n - 1)$

(e) $s(n) = \sum_{i=1}^{n} i$

10. For each of the following sequences $s(n)$, find a number a with $s(n) = \Theta(a^n)$.

(a) $s(n) = 3^{n+5}$

(b) $s(n) = 2^{2n}$

(c) $s(n) = (\log_2 10)^{n-3}$

(d) $s(n) = 1^{n^2+200n+5}$

11. (a) Show that $t_n = O(n^2)$, where $t_n = \sum_{k=1}^{n} k = 1 + 2 + \cdots + n$.

(b) Show that if $s_n = \sum_{k=1}^{n} k^2$, then $s_n = O(n^3)$.

12. Let A be a positive constant. Show that $A^n < n!$ for sufficiently large n. *Hint:* Analyze Example 3(b).

13. Show that if $s(n) = 3n^4 + O(n)$ and $t(n) = 2n^3 + O(n)$, then

(a) $s(n) + t(n) = 3n^4 + O(n^3)$

(b) $s(n) \cdot t(n) = 6n^7 + O(n^5)$

14. Show that

(a) If $s(n) = O(n^2)$ and $t(n) = O(n^3)$, then $(5n^3 + s(n)) \cdot (3n^4 + t(n)) = 15n^7 + O(n^6)$.

(b) If $s(n) = O(n)$ and $t(n) = O(n^2)$, then $(5n^3 + s(n)) \cdot (3n^4 + t(n)) = 15n^7 + O(n^5)$.

15. This exercise shows that one must be careful using division in big-oh calculations. Big-theta notation works better for division: see Exercise 23.

(a) Let $s(n) = n^5$ and $t(n) = n$. Observe that $s(n) = O(n^5)$ and $t(n) = O(n^2)$, but that $s(n)/t(n)$ is not $O(n^3)$.

(b) Give examples of sequences $s(n)$ and $t(n)$ such that $s(n) = O(n^6)$ and $t(n) = O(n^2)$, but $s(n)/t(n)$ is not $O(n^4)$.

16. Show that $\log_{10} n = \Theta(\log_2 n)$.

17. With an input of size n, algorithm A requires $5n \log_2 n$ operations and algorithm B requires $80n$ operations.

(a) Which algorithm is more efficient—requires fewer operations for a given input size—in the long run, i.e., for large n? Explain.

(b) For which n is algorithm A at least twice as efficient as algorithm B?

(c) For which n is algorithm B at least twice as efficient as algorithm A?

18. With an input of size n, algorithm A requires $1000\sqrt{n}$ operations and algorithm B requires $5n$ operations.

(a) Which algorithm is more efficient in the long run, i.e., for large n? Explain.

(b) For which n is algorithm A at least twice as efficient as algorithm B?

(c) For which n is algorithm B at least twice as efficient as algorithm A?

19. For each $n \in \mathbb{P}$, let $\text{DIGIT}(n)$ be the number of digits in the decimal expansion of n.

(a) Show that $10^{\text{DIGIT}(n)-1} \le n < 10^{\text{DIGIT}(n)}$.

(b) Show that $\log_{10} n$ is $O(\text{DIGIT}(n))$.

(c) Show that $\text{DIGIT}(n)$ is $O(\log_{10} n)$.

(d) Observe that $\text{DIGIT}(n) = \Theta(\log_{10} n)$.

20. Let $\text{DIGIT2}(n)$ be the number of digits in the binary expansion of n. Show that $\text{DIGIT2}(n) = \Theta(\log_2 n)$ and $\text{DIGIT2}(n) = \Theta(\text{DIGIT}(n))$ [see Exercise 19].

21. (a) Here again is the algorithm for computing a^n in Exercise 25 on page 137.

```
{Input: a number a and a positive integer n.}
{Output: a^n.}
begin
p := 1
q := a
i := n
while i > 0 do
    if i is odd then p := p · q
    {Do the next two steps whether or not i is odd.}
    q := q · q
    i := i DIV 2 = ⌊i/2⌋
return p
end
```

Show that the algorithm makes $O(\log_2 n)$ passes through the **while** loop.

(b) How many passes does the algorithm make through the loop for $n = 3$, $n = 2^{100}$, and $n = 3 \cdot 2^{100}$?

22. (a) Show that the following algorithm for computing a^n makes $\Theta(n)$ passes through the **while** loop.

```
{Input: a number a and a positive integer n.}
{Output: a^n.}
begin
p := 1
q := a
i := n
while i > 0 do
    p := p · a
    i := i − 1
return p
end
```

(b) Is this algorithm faster than the one in Exercise 21 for large values of n?

23. Show that if $s(n) = \Theta(n^5)$ and $t(n) = \Theta(n^2)$, then $s(n)/t(n) = \Theta(n^3)$. [Compare this statement with Exercise 15.]

24. Explain why part (b) of Theorem 2 can be viewed as a special case of part (c).

25. Explain why parts (a) and (d) of Theorem 2 are true.

4.4 Recursive Definitions

The values of the terms in a sequence may be given explicitly by formulas such as $s_n = n^3 - 73n$ or by descriptions such as "let t_n be the weight of the nth edge on the path." Terms may also be defined by descriptions that involve other terms that come before them in the sequence.

We say that a sequence is defined **recursively** provided:

(B) Some finite set of values, usually the first one or first few, are specified.

(R) The remaining values of the sequence are defined in terms of previous values of the sequence. A formula that gives such a definition is called a **recurrence formula** or **relation**.

The requirement (B) gives the **basis** or starting point for the definition. The remainder of the sequence is determined by using the relation (R) repeatedly, i.e., recurrently.

EXAMPLE 1

(a) We can define the familiar sequence FACT recursively by

(B) $\text{FACT}(0) = 1$,

(R) $\text{FACT}(n+1) = (n+1) \cdot \text{FACT}(n)$ for $n \in \mathbb{N}$.

Condition (R) lets us calculate $\text{FACT}(1)$, then $\text{FACT}(2)$, then $\text{FACT}(3)$, etc., and we quickly see that $\text{FACT}(n) = n!$ for the first several values of n. This equation can be proved for all n by mathematical induction, as follows. By (B), $\text{FACT}(0) = 1 = 0!$ Assuming inductively that $\text{FACT}(m) = m!$ for some $m \in \mathbb{N}$ and using (R), we get $\text{FACT}(m+1) = (m+1) \cdot \text{FACT}(m) = (m+1) \cdot m! = (m+1)!$ for the inductive step. Since we understand the sequence $n!$ already, the recursive definition above may seem silly, but we will try to convince you in part (b) that recursive definitions of even simple sequences are useful.

(b) Consider the sequence $\text{SUM}(n) = \sum_{i=0}^{n} \frac{1}{i!}$. To write a computer program that calculates the values of SUM for large values of n, one could use the following recursive definition:

(B) $\text{SUM}(0) = 1$,

(R) $\text{SUM}(n+1) = \text{SUM}(n) + \dfrac{1}{(n+1)!}$.

The added term in (R) is the reciprocal of $(n+1)!$, so $\text{FACT}(n+1)$ will be needed as the program progresses. At each n, one could instruct the program to calculate $\text{FACT}(n+1)$ from scratch or one could store a large number of these values. Clearly, it would be more efficient to alternately calculate $\text{FACT}(n+1)$ and $\text{SUM}(n+1)$ using the recursive definition in part (a) for FACT and the recursive definition above for SUM. Figure 1 shows an algorithm designed that way.

SUMAlgorithm(integer, integer)

```
{Input: integer n ≥ 0.}
{Output: SUM, the sum Σ_{i=0}^{n} 1/i!.}
{Auxiliary variable: FACT.}
begin
i := 0
SUM := 0
FACT := 1
while i ≤ n do
    SUM := SUM + 1/FACT
    FACT := (i + 1) · FACT
    i := i + 1
return SUM
end
```

Figure 1 ▲

(c) Define the sequence SEQ as follows:

(B) $\text{SEQ}(0) = 1$,

(R) $\text{SEQ}(n+1) = (n+1)/\text{SEQ}(n)$ for $n \in \mathbb{N}$.

With $n = 0$, (R) gives $\text{SEQ}(1) = 1/1 = 1$. Then with $n = 1$, we find that $\text{SEQ}(2) = 2/1 = 2$. Continuing in this way, we see that the first few terms are 1, 1, 2, 3/2, 8/3, 15/8, 16/5, 35/16. It is by no means apparent what a general formula for $\text{SEQ}(n)$ might be. It is evident that $\text{SEQ}(73)$ exists, but it would take considerable calculation to find it. ■

In Example 1, how did we *know* that SEQ(73) exists? Our certainty is based on the belief that recursive definitions do indeed define sequences on all of $\mathbb{N}$, unless some step leads to an illegal computation such as division by 0. We could prove by induction that the recursive definition in Example 1 defines a sequence, but we will use the Well-Ordering Principle instead; see page 131. Let

$$S = \{n \in \mathbb{N} : \text{SEQ}(n) = 0 \text{ or } \text{SEQ}(n) \text{ is not defined}\}.$$

We want to show that S is empty. If not, it has a smallest member, say m. By (B) $m \neq 0$, so $m - 1 \in \mathbb{N}$. Since m is the smallest bad guy, $\text{SEQ}(m-1) \neq 0$ and $\text{SEQ}(m-1)$ is defined. But then (R) defines $\text{SEQ}(m) = m/\text{SEQ}(m-1) \neq 0$, contrary to $m \in S$. Thus S must be empty.

We have just shown that the two conditions

(B) $\quad \text{SEQ}(0) = 1,$

(R) $\quad \text{SEQ}(n+1) = (n+1)/\text{SEQ}(n) \quad$ for $n \in \mathbb{N}$

define SEQ, and we can just as easily show that SEQ is the only sequence $a_0, a_1, a_2, \ldots$ that satisfies the conditions

(B′) $\quad a_0 = 1,$

(R′) $\quad a_{n+1} = (n+1)/a_n \quad$ for $n \in \mathbb{N}$.

For suppose that (a_n) satisfies (B′) and (R′). Then $a_0 = 1 = \text{SEQ}(0)$, and if $a_n = \text{SEQ}(n)$, then

$$a_{n+1} = (n+1)/a_n = (n+1)/\text{SEQ}(n) = \text{SEQ}(n+1).$$

By the Principle of Mathematical Induction, $a_n = \text{SEQ}(n)$ for every n in $\mathbb{N}$. If we could somehow find a formula for a sequence (a_n) satisfying (B′) and (R′), then it would be a formula for SEQ.

The proof we just gave that $\text{SEQ}(n)$ is uniquely defined for each n uses the fact that $\text{SEQ}(n+1)$ only depends on $\text{SEQ}(n)$. Recursive definitions allow a term to depend on terms before it besides the one just preceding it. In such cases, either the Well-Ordering Principle or an enhanced version of the Principle of Mathematical Induction that we take up in §4.6 can be used to prove that the sequences are well defined and uniquely determined by the conditions. In particular, an extension of the argument above shows the following [see Exercise 19 on page 180].

Proposition If a sequence (s_n) is defined by a basis and a recurrence condition, then it is the only sequence that satisfies these conditions.

The values of the terms in a recursively defined sequence can be calculated in more than one way. An **iterative calculation** finds s_n by computing all the values $s_1, s_2, \ldots, s_{n-1}$ first so that they are available to use in computing s_n. We had in mind an iterative calculation of the sequences in Example 1. To calculate FACT(73), for instance, we would first calculate $\text{FACT}(k)$ for $k = 1, 2, \ldots, 72$, even though we might have no interest in these preliminary values themselves.

In the case of FACT, there really seems to be no better alternative. Sometimes, though, there is a more clever way to calculate a given value of s_n. A **recursive calculation** finds the value of s_n by looking to see which terms s_n depends on, then which terms those terms depend on, and so on. It may turn out that the value of s_n only depends on the values of a relatively small set of its predecessors, in which case the other previous terms can be ignored.

EXAMPLE 2 (a) Define the sequence T by

(B) $\quad T(1) = 1,$

(R) $\quad T(n) = 2 \cdot T(\lfloor n/2 \rfloor) \quad$ for $n \geq 2$.

Then

$$T(73) = 2 \cdot T(\lfloor 73/2 \rfloor) = 2 \cdot T(36) = 2 \cdot 2 \cdot T(18) = 2 \cdot 2 \cdot 2 \cdot T(9)$$
$$= 2 \cdot 2 \cdot 2 \cdot 2 \cdot T(4) = 2 \cdot 2 \cdot 2 \cdot 2 \cdot 2 \cdot T(2)$$
$$= 2 \cdot 2 \cdot 2 \cdot 2 \cdot 2 \cdot 2 \cdot T(1) = 2^6.$$

For this calculation we only need the values of $T(36)$, $T(18)$, $T(9)$, $T(4)$, $T(2)$, and $T(1)$, and we have no need to compute the other 66 values of $T(n)$ that precede $T(73)$.

This sequence can be described in another way as follows [Exercise 19 on page 171]:

$$T(n) \text{ is the largest integer } 2^k \text{ with } 2^k \leq n.$$

Using this description we could compute $T(73)$ by looking at the list of powers of 2 less than 73 and taking the largest one.

(b) A slight change in (R) from part (a) gives the sequence Q with

(B) $Q(1) = 1$,

(R) $Q(n) = 2 \cdot Q(\lfloor n/2 \rfloor) + n$ for $n \geq 2$.

Now the general term is not so clear, but we can still find $Q(73)$ recursively from $Q(36)$, $Q(18)$, ... , $Q(2)$, $Q(1)$. ■

EXAMPLE 3

(a) The **Fibonacci sequence** is defined as follows:

(B) $\text{FIB}(1) = \text{FIB}(2) = 1$,

(R) $\text{FIB}(n) = \text{FIB}(n-1) + \text{FIB}(n-2)$ for $n \geq 3$.

Note that the recurrence formula makes no sense for $n = 2$, so $\text{FIB}(2)$ had to be defined separately in the basis. The first few terms of this sequence are

$$1, 1, 2, 3, 5, 8, 13, 21, 34, 55, 89.$$

This sequence will reappear in various places in the book.

(b) Here is an easy way to define the sequence

$$0, 0, 1, 1, 2, 2, 3, 3, \ldots.$$

(B) $\text{SEQ}(0) = \text{SEQ}(1) = 0$,

(R) $\text{SEQ}(n) = 1 + \text{SEQ}(n-2)$ for $n \geq 2$.

Compare Exercise 11(a) on page 38. ■

EXAMPLE 4

Let $\Sigma = \{a, b\}$.

(a) We are interested in the number s_n of words of length n that do not contain the string aa. Let's write A_n for the set of words in Σ^n of this form. Then $A_0 = \{\lambda\}$, $A_1 = \Sigma$, and $A_2 = \Sigma^2 \setminus \{aa\}$, so $s_0 = 1$, $s_1 = 2$, and $s_2 = 4 - 1 = 3$. To get a recurrence formula for s_n, we consider $n \geq 2$ and count the number of words in A_n in terms of shorter words. If a word in A_n ends in b, then the b can be preceded by any word in A_{n-1}. So there are s_{n-1} words in A_n that end in b. If a word in A_n ends in a, then the last two letters must be ba since the string aa is not allowed, and this string can be preceded by any word in A_{n-2}. So there are s_{n-2} words in A_n that end in a. Thus $s_n = s_{n-1} + s_{n-2}$ for $n \geq 2$. This is the recurrence relation for the Fibonacci sequence, but note that the basis is different: $s_1 = 2$ and $s_2 = 3$, while $\text{FIB}(1) = 1$ and $\text{FIB}(2) = 1$. In fact, $s_n = \text{FIB}(n+2)$ for $n \in \mathbb{N}$ [Exercise 13].

(b) Since Σ^n has 2^n words in it, there are $2^n - s_n = 2^n - \text{FIB}(n+2)$ words of length n that do contain aa. ■

EXAMPLE 5

Let $\Sigma = \{a, b, c\}$, let B_n be the set of words in Σ^n with an even number of a's, and let t_n denote the number of words in B_n. Then $B_0 = \{\lambda\}$, $B_1 = \{b, c\}$, and $B_2 = \{aa, bb, bc, cb, cc\}$, so $t_0 = 1$, $t_1 = 2$, and $t_2 = 5$. We count the number of words in B_n by looking at the last letter. If a word in B_n ends in b, then that b can be preceded by any word in B_{n-1}. So t_{n-1} words in B_n end in b. Similarly, t_{n-1} words in B_n end in c. If a word in B_n ends in a, then the a must be preceded by a word in Σ^{n-1} with an *odd* number of a's. Since Σ^{n-1} has 3^{n-1} words, $3^{n-1} - t_{n-1}$ of them must have an odd number of a's. Hence $3^{n-1} - t_{n-1}$ words in B_n end in a. Thus

$$t_n = t_{n-1} + t_{n-1} + (3^{n-1} - t_{n-1}) = 3^{n-1} + t_{n-1}$$

for $n \geq 1$. Hence $t_3 = 3^2 + t_2 = 9 + 5 = 14$, $t_4 = 3^3 + t_3 = 27 + 14 = 41$, etc.

In this case, it's relatively easy to find an explicit formula for t_n. First note that

$$t_n = 3^{n-1} + t_{n-1} = 3^{n-1} + 3^{n-2} + t_{n-2} = \cdots$$
$$= 3^{n-1} + 3^{n-2} + \cdots + 3^0 + t_0 = 1 + \sum_{k=0}^{n-1} 3^k.$$

If the first set of three dots makes you nervous, you can supply a proof by induction. Now we apply Example 2(c) on page 139 to obtain

$$t_n = 1 + \frac{3^n - 1}{3 - 1} = 1 + \frac{3^n - 1}{2} = \frac{3^n + 1}{2}.$$

Our mental process works for $n \geq 1$, but the formula works for $n = 0$, too. The formula for t_n looks right: about half of the words in Σ^n use an even number of a's.

Notice the technique that we used to get the formula for t_n. We wrote t_n in terms of t_{n-1}, then t_{n-2}, then $t_{n-3}, \ldots$ and collected the leftover terms to see if they suggested a formula. They looked like $3^{n-1} + 3^{n-2} + 3^{n-3} + \cdots + 3^{n-k} + t_{n-k}$, so we guessed that carrying the process backward far enough would give $t_n = 3^{n-1} + \cdots + 3^0 + t_0$.

Another way to guess the formula would have been to start at the small end, but not to do the simplifying arithmetic. Here's how that process might go.

$$t_0 = 1$$
$$t_1 = 3^0 + 1$$
$$t_2 = 3^1 + (3^0 + 1)$$
$$t_3 = 3^2 + (3^1 + (3^0 + 1))$$
$$t_4 = 3^3 + (3^2 + (3^1 + (3^0 + 1)))$$
$$\vdots$$
$$t_n = 3^{n-1} + 3^{n-2} + \cdots + 3 + 1 + 1 = 1 + \sum_{k=0}^{n-1} 3^k.$$

We have illustrated two of the tricks of the trade, but it really doesn't matter how we arrive at the formula as long as we can show that it's correct. To prove that our guess is correct, it is enough to prove that it satisfies the recurrence conditions that define t_n: $t_0 = 1$ and $t_n = 3^{n-1} + t_{n-1}$. We simply check:

$$\frac{3^0 + 1}{2} = 1 \quad \text{and} \quad \frac{3^n + 1}{2} = 3^{n-1} + \frac{3^{n-1} + 1}{2} \quad \text{for } n \in \mathbb{P}.$$

This method of proof is legitimate because the recurrence conditions define the sequence t_n uniquely. ■

EXAMPLE 6

(a) Define the sequence S by

(B) $S(0) = 0, S(1) = 1,$

(R) $S(n) = S(\lfloor n/2 \rfloor) + S(\lfloor n/5 \rfloor)$ for $n \geq 2$.

It makes sense to calculate the values of S recursively, rather than iteratively. For instance,

$$\begin{aligned} S(73) &= S(36) + S(14) = [S(18) + S(7)] + [S(7) + S(2)] \\ &= S(18) + 2S(7) + S(2) \\ &= S(9) + S(3) + 2[S(3) + S(1)] + S(1) + S(0) \\ &= \cdots \\ &= 8S(1) + 6S(0) = 8. \end{aligned}$$

The calculation of $S(73)$ involves the values of $S(36)$, $S(18)$, $S(14)$, $S(9)$, $S(7)$, $S(4)$, $S(3)$, $S(2)$, $S(1)$, and $S(0)$, but that's still better than finding all values of $S(k)$ for $k = 1, \ldots, 72$.

(b) Recursive calculation requires storage space for the intermediate values that have been called for but not yet computed. It may be possible, though, to keep the number of storage slots fairly small. For example, the recursive calculation of $\text{FACT}(6)$ goes like this:

$$\text{FACT}(6) = 6 \cdot \text{FACT}(5) = 30 \cdot \text{FACT}(4) = 120 \cdot \text{FACT}(3) = \cdots .$$

Only one address is needed for the intermediate [unknown] value $\text{FACT}(k)$ for $k < 6$. Similarly,

$$\begin{aligned} \text{FIB}(7) &= \text{FIB}(6) + \text{FIB}(5) = (\text{FIB}(5) + \text{FIB}(4)) + \text{FIB}(5) \\ &= 2 \cdot \text{FIB}(5) + \text{FIB}(4) \\ &= 3 \cdot \text{FIB}(4) + 2 \cdot \text{FIB}(3) \\ &= 5 \cdot \text{FIB}(3) + 3 \cdot \text{FIB}(2) \\ &= 8 \cdot \text{FIB}(2) + 5 \cdot \text{FIB}(1) \end{aligned}$$

only requires two intermediate addresses at each step.

Beware, though. Even if you know how to do the calculation with only a few addresses, behind the scenes your software may be setting up not only an address, but even a complete recursion tree for each as-yet-uncomputed intermediate value. ■

Of course, we can give recursive definitions even if the values of the sequence are not real numbers.

EXAMPLE 7

Let S be a set and let f be a function from S into S. Our goal is to recursively define the composition $f^{(n)}$ of f with itself n times. It is convenient to define $f^{(0)}$ to be the identity function 1_S, i.e., the function given by $1_S(s) = s$ for all $s \in S$. Here is the recursive definition.

(B) $f^{(0)} = 1_S$ [the identity function on S],

(R) $f^{(n+1)} = f^{(n)} \circ f$.

Thus

$$f^{(1)} = f, \quad f^{(2)} = f \circ f, \quad f^{(3)} = f \circ f \circ f, \quad \text{etc.} \tag{1}$$

In other words,

$$f^{(n)} = f \circ f \circ \cdots \circ f \quad [n \text{ times}]. \tag{2}$$

The recursive definition is more precise than the "etc." in (1) or the three dots in (2). ■

As in Example 7, recursive definitions often give concise, unambiguous definitions of objects that we understand well already. They can also be especially useful for writing programs that deal with such objects.

EXAMPLE 8

Let a be a nonzero real number. Define the powers of a by

(B) $a^0 = 1$,

(R) $a^{n+1} = a^n \cdot a \quad$ for $n \in \mathbb{N}$.

Equivalently,

(B) $\text{POW}(0) = 1$,

(R) $\text{POW}(n+1) = \text{POW}(n) \cdot a \quad$ for $n \in \mathbb{N}$.

An algorithm for computing a^n based on this definition practically writes itself. [Unfortunately, it's incredibly slow compared with the algorithm in Exercise 25 on page 137. See Exercises 21 and 22 of §4.3.] ■

EXAMPLE 9

Let $(a_j)_{j \in \mathbb{P}}$ be a sequence of real numbers. We can define the product notation by

(B) $\displaystyle\prod_{j=1}^{1} a_j = a_1$,

(R) $\displaystyle\prod_{j=1}^{n+1} a_j = a_{n+1} \cdot \prod_{j=1}^{n} a_j \quad$ for $n \geq 1$.

Equivalently,

(B) $\text{PROD}(1) = a_1$,

(R) $\text{PROD}(n+1) = a_{n+1} \cdot \text{PROD}(n) \quad$ for $n \geq 1$.

These recursive definitions start at $n = 1$. An alternative is to define the "empty product" to be 1; i.e.,

(B) $\displaystyle\prod_{j=1}^{0} a_j = 1 \quad$ [which looks peculiar],

or

(B) $\text{PROD}(0) = 1$.

Then the same recursive relation (R) as before serves to define the remaining terms of the sequence. ■

Exercises 4.4

1. We recursively define $s_0 = 1$ and $s_{n+1} = 2/s_n$ for $n \in \mathbb{N}$.
   (a) List the first few terms of the sequence.
   (b) What is the *set* of values of s?
2. We recursively define $\text{SEQ}(0) = 0$ and $\text{SEQ}(n+1) = 1/[1 + \text{SEQ}(n)]$ for $n \in \mathbb{N}$. Calculate $\text{SEQ}(n)$ for $n = 1$, 2, 3, 4, and 6.
3. Consider the sequence $(1, 3, 9, 27, 81, \ldots)$.
   (a) Give a formula for the nth term $\text{SEQ}(n)$, where $\text{SEQ}(0) = 1$.
   (b) Give a recursive definition for the sequence SEQ.
4. (a) Give a recursive definition for the sequence $(2, 2^2, (2^2)^2, ((2^2)^2)^2, \ldots)$, i.e., $(2, 4, 16, 256, \ldots)$.
   (b) Give a recursive definition for the sequence $(2, 2^2, 2^{(2^2)}, 2^{(2^{(2^2)})}, \ldots)$, i.e., $(2, 4, 16, 65536, \ldots)$.

5. Is the following a recursive definition for a sequence SEQ? Explain.

(B) $\text{SEQ}(0) = 1$,

(R) $\text{SEQ}(n+1) = \text{SEQ}(n)/[100 - n]$.

6. (a) Calculate SEQ(9), where SEQ is as in Example 1(c).

(b) Calculate FIB(12), where FIB is as in Example 3(a).

(c) Calculate $Q(19)$, where Q is as in Example 2(b).

7. Let $\Sigma = \{a, b, c\}$ and let s_n denote the number of words of length n that do not contain the string aa.

(a) Calculate s_0, s_1, and s_2.

(b) Find a recurrence formula for s_n.

(c) Calculate s_3 and s_4.

8. Let $\Sigma = \{a, b\}$ and let s_n denote the number of words of length n that do not contain the string ab.

(a) Calculate s_0, s_1, s_2, and s_3.

(b) Find a formula for s_n and prove it is correct.

9. Let $\Sigma = \{a, b\}$ and let t_n denote the number of words of length n with an even number of a's.

(a) Calculate t_0, t_1, t_2, and t_3.

(b) Find a formula for t_n and prove it is correct.

(c) Does your formula for t_n work for $n = 0$?

10. Consider the sequence defined by

(B) $\text{SEQ}(0) = 1, \text{SEQ}(1) = 0$,

(R) $\text{SEQ}(n) = \text{SEQ}(n-2) \quad$ for $n \ge 2$.

(a) List the first few terms of this sequence.

(b) What is the set of values of this sequence?

11. We recursively define $a_0 = a_1 = 1$ and $a_n = a_{n-1} + 2a_{n-2}$ for $n \ge 2$.

(a) Calculate a_6 recursively.

(b) Prove that all the terms a_n are odd integers.

12. Calculate the following values of S in Example 6 recursively.

(a) $S(4)$ (b) $S(5)$ (c) $S(10)$

13. Show that, if the sequence s_n satisfies $s_1 = 2$, $s_2 = 3$, and $s_n = s_{n-1} + s_{n-2}$ for $n \ge 3$, then $s_n = \text{FIB}(n+2)$ for $n \in \mathbb{N}$. *Hint:* Apply the Well-Ordering Principle to the set $S = \{n \in \mathbb{P} : s_n \ne \text{FIB}(n+2)\}$.

14. Recursively define $b_0 = b_1 = 1$ and $b_n = 2b_{n-1} + b_{n-2}$ for $n \ge 2$.

(a) Calculate b_5 iteratively.

(b) Explain why all the terms b_n are odd. *Hint:* Consider the first one that is even.

15. Let $\text{SEQ}(0) = 1$ and $\text{SEQ}(n) = \sum_{i=0}^{n-1} \text{SEQ}(i)$ for $n \ge 1$. This is actually a simple, familiar sequence. What is it?

16. We recursively define $a_0 = 0$, $a_1 = 1$, $a_2 = 2$, and $a_n = a_{n-1} - a_{n-2} + a_{n-3}$ for $n \ge 3$.

(a) List the first few terms of the sequence until the pattern is clear.

(b) What is the set of values of the sequence?

17. The process of assigning n children to n classroom seats can be broken down into (1) choosing a child for the first seat and (2) assigning the other $n - 1$ children to the remaining seats. Let $A(n)$ be the number of different assignments of n children to n seats.

(a) Write a recursive definition of the sequence A.

(b) Calculate $A(6)$ recursively.

(c) Is the sequence A familiar?

18. Consider the process of assigning $2n$ children to n cars in a playground train so that two children go in each car. First choose two children for the front car [there are $2n(2n-1)/2$ ways to do this, as we will see in §5.1]. Then distribute the rest of the children to the remaining $n - 1$ cars. Let $B(n)$ be the number of ways to assign $2n$ children to n cars.

(a) Write a recursive definition of the sequence B.

(b) Calculate $B(3)$ recursively.

(c) Calculate $B(5)$ iteratively.

(d) Give an explicit formula for $B(n)$.

19. Consider the sequence FOO defined by

(B) $\text{FOO}(0) = 1, \text{FOO}(1) = 1$

(R) $\text{FOO}(n) = \dfrac{10 \cdot \text{FOO}(n-1) + 100}{\text{FOO}(n-2)} \quad$ for $n \ge 2$.

(a) What is the set of values of FOO?

(b) Repeat part (a) for the sequence GOO defined by

(B) $\text{GOO}(0) = 1, \text{GOO}(1) = 2$

(R) $\text{GOO}(n) = \dfrac{10 \cdot \text{GOO}(n-1) + 100}{\text{GOO}(n-2)} \quad$ for $n \ge 2$.

We learned about these entertaining sequences from our colleague Ivan Niven.

20. Let $(a_1, a_2, \ldots)$ be a sequence of real numbers.

(a) Give a recursive definition for $\text{SUM}(n) = \sum_{j=1}^{n} a_j$ for $n \ge 1$.

(b) Revise your recursive definition for SUM(n) by starting with $n = 0$. What is the "empty sum"?

21. Let $(A_1, A_2, \ldots)$ be a sequence of subsets of some set S.

(a) Give a recursive definition for $\bigcup_{j=1}^{n} A_j$.

(b) How would you define the "empty union"?

(c) Give a recursive definition for $\bigcap_{j=1}^{n} A_j$.

(d) How would you define the "empty intersection"?

22. Let $(A_1, A_2, \ldots)$ be a sequence of subsets of some set S. Define

(B) $\text{SYM}(1) = A_1$,

(R) $\text{SYM}(n+1) = A_{n+1} \oplus \text{SYM}(n) \quad$ for $n \ge 1$.

Recall that $\oplus$ denotes symmetric difference. It turns out that an element x in S belongs to SYM(n) if and only if the set $\{k : x \in A_k \text{ and } k \le n\}$ has an odd number of elements. Prove this by mathematical induction.

4.5 Recurrence Relations

Sequences that appear in mathematics and science are frequently defined by recurrences, rather than by formulas. A variety of techniques have been developed to obtain explicit formulas for the values in some cases. In this section we give the complete answer for sequences defined by recurrences of the form

$$s_n = as_{n-1} + bs_{n-2},$$

where a and b are constants, and we get substantial information about sequences that satisfy recurrences of the form

$$s_{2n} = 2 \cdot s_n + f(n)$$

for known functions f. We begin by considering

$$s_n = as_{n-1} + bs_{n-2},$$

assuming that the two initial values s_0 and s_1 are known. The cases where $a = 0$ or $b = 0$ are especially easy to deal with.

If $b = 0$, so that $s_n = as_{n-1}$ for $n \geq 1$, then $s_1 = as_0$, $s_2 = as_1 = a^2 s_0$, etc. A simple induction argument shows that $s_n = a^n s_0$ for all $n \in \mathbb{N}$.

Now suppose that $a = 0$. Then $s_2 = bs_0$, $s_4 = bs_2 = b^2 s_0$, etc., so that $s_{2n} = b^n s_0$ for all $n \in \mathbb{N}$. Similarly, $s_3 = bs_1$, $s_5 = b^2 s_1$, etc., so that $s_{2n+1} = b^n s_1$ for all $n \in \mathbb{N}$.

EXAMPLE 1

(a) Consider the recurrence relation $s_n = 3s_{n-1}$ with $s_0 = 5$. Here $a = 3$, so $s_n = 5 \cdot 3^n$ for $n \in \mathbb{N}$.

(b) Consider the recurrence relation $s_n = 3s_{n-2}$ with $s_0 = 5$ and $s_1 = 2$. Here $b = 3$, so $s_{2n} = 5 \cdot 3^n$ and $s_{2n+1} = 2 \cdot 3^n$ for $n \in \mathbb{N}$. ■

From now on we will assume that $a \neq 0$ and $b \neq 0$. It will be convenient to ignore the specified values of s_0 and s_1 until later. From the special cases we have examined it seems reasonable to hope that some solutions have the form $s_n = cr^n$ for a nonzero constant c. This hope, if true, would force

$$r^n = ar^{n-1} + br^{n-2}.$$

Dividing by r^{n-2} would then give $r^2 = ar + b$, or $r^2 - ar - b = 0$. In other words, if it turns out that $s_n = cr^n$ for all n, then r must be a solution of the quadratic equation $x^2 - ax - b = 0$, which is called the **characteristic equation** of the recurrence relation. This equation has either one or two solutions, and these solutions tell the story.

Theorem 1 Consider a recurrence relation of the form

$$s_n = as_{n-1} + bs_{n-2}$$

with characteristic equation

$$x^2 - ax - b = 0,$$

where a and b are nonzero constants.

(a) If the characteristic equation has two different solutions r_1 and r_2, then there are constants c_1 and c_2 such that

$$s_n = c_1 r_1^n + c_2 r_2^n \quad \text{for } n \in \mathbb{N}. \tag{1}$$

If s_0 and s_1 are specified, then the constants can be determined by setting $n = 0$ and $n = 1$ in (1) and solving the two equations for c_1 and c_2.

(b) If the characteristic equation has only one solution r, then there are constants c_1 and c_2 such that

$$s_n = c_1 r^n + c_2 \cdot n \cdot r^n \quad \text{for } n \in \mathbb{N}. \qquad \textbf{(2)}$$

As in part (a), c_1 and c_2 can be determined if s_0 and s_1 are specified.

Warning. This fine theorem only applies to recurrence relations of the form $s_n = as_{n-1} + bs_{n-2}$.

Even if a and b are real numbers, the solutions to the characteristic equation need not be real. The theorem is still true in the nonreal case, but you may wish to think in terms of real solutions in what follows. We will prove Theorem 1 later, but first let's look at some examples of its use.

EXAMPLE 2

Consider the recurrence relation $s_n = s_{n-1} + 2s_{n-2}$, where $s_0 = s_1 = 3$. Here $a = 1$ and $b = 2$. The characteristic equation $x^2 - x - 2 = 0$ has solutions $r_1 = 2$ and $r_2 = -1$, since $x^2 - x - 2 = (x-2)(x+1)$, so part (a) of the theorem applies. By Theorem 1,

$$s_n = c_1 \cdot 2^n + c_2 \cdot (-1)^n$$

for constants c_1 and c_2. By setting $n = 0$ and $n = 1$, we find that

$$s_0 = c_1 \cdot 2^0 + c_2 \cdot (-1)^0 \quad \text{and} \quad s_1 = c_1 \cdot 2^1 + c_2 \cdot (-1)^1;$$

i.e.,

$$3 = c_1 + c_2 \quad \text{and} \quad 3 = 2c_1 - c_2.$$

Solving this system of two equations gives $c_1 = 2$ and $c_2 = 1$. We conclude that

$$s_n = 2 \cdot 2^n + 1 \cdot (-1)^n = 2^{n+1} + (-1)^n \quad \text{for} \quad n \in \mathbb{N}.$$

It is easy to check that this answer satisfies the recurrence equation and the initial conditions [they're how we got c_1 and c_2], so it must be the correct answer. ■

EXAMPLE 3

Consider again the Fibonacci sequence in Example 3 on page 155. Let $s_0 = 0$ and $s_n = \text{FIB}(n)$ for $n \geq 1$. Then $s_n = s_{n-1} + s_{n-2}$ for $n \geq 2$. Here $a = b = 1$, so we solve $x^2 - x - 1 = 0$. By the quadratic formula, the equation has two solutions:

$$r_1 = \frac{1+\sqrt{5}}{2} \quad \text{and} \quad r_2 = \frac{1-\sqrt{5}}{2}.$$

Thus part (a) of the theorem applies, so

$$s_n = c_1 r_1^n + c_2 r_2^n = c_1 \left(\frac{1+\sqrt{5}}{2}\right)^n + c_2 \left(\frac{1-\sqrt{5}}{2}\right)^n \quad \text{for} \quad n \in \mathbb{N}.$$

While solving for c_1 and c_2, it's convenient to retain the notation r_1 and r_2. Setting $n = 0$ and $n = 1$ gives

$$0 = c_1 + c_2 \quad \text{and} \quad 1 = c_1 r_1 + c_2 r_2.$$

If we replace c_2 by $-c_1$ in the second equation, we get $1 = c_1 r_1 - c_1 r_2 = c_1(r_1 - r_2)$ and hence

$$c_1 = \frac{1}{r_1 - r_2}.$$

Since $r_1 - r_2 = \sqrt{5}$, we conclude that $c_1 = 1/\sqrt{5}$ and $c_2 = -1/\sqrt{5}$. Finally,

$$s_n = c_1 r_1^n + c_2 r_2^n = \frac{1}{\sqrt{5}} r_1^n - \frac{1}{\sqrt{5}} r_2^n = \frac{1}{\sqrt{5}}(r_1^n - r_2^n) \quad \text{for} \quad n \geq 0,$$

so

$$\text{FIB}(n) = \frac{1}{\sqrt{5}}\left[\left(\frac{1+\sqrt{5}}{2}\right)^n - \left(\frac{1-\sqrt{5}}{2}\right)^n\right] \quad \text{for} \quad n \geq 1.$$

This description for $\text{FIB}(n)$ is hardly what one could have expected when we started. $\text{FIB}(n)$ is an integer, of course, whereas the expression with the radicals certainly doesn't look much like an integer. The formula is correct, though. Try it with a calculator for a couple of values of n to be convinced.

By using the formula, we can compute any value of $\text{FIB}(n)$ without having to find the previous values, and we can also estimate the rate at which $\text{FIB}(n)$ grows as n increases. Since

$$\frac{\sqrt{5}-1}{2} < 0.62,$$

the absolute value of

$$\frac{1}{\sqrt{5}} \cdot \left(\frac{1-\sqrt{5}}{2}\right)^n$$

is less than 0.5 for every positive n, and thus

$$\text{FIB}(n) \text{ is the closest integer to } \tfrac{1}{\sqrt{5}} \cdot \left(\tfrac{1+\sqrt{5}}{2}\right)^n. \qquad \blacksquare$$

EXAMPLE 4

Consider the sequence (s_n) defined by $s_0 = 1, s_1 = -3.$ and $s_n = 6s_{n-1} - 9s_{n-2}$ for $n \geq 2$. Here the characteristic equation is $x^2 - 6x + 9 = 0$, which has exactly one solution, $r = 3$. By part (b) of the theorem,

$$s_n = c_1 \cdot 3^n + c_2 \cdot n \cdot 3^n \quad \text{for} \quad n \in \mathbb{N}.$$

Setting $n = 0$ and $n = 1$, we get

$$s_0 = c_1 \cdot 3^0 + 0 \quad \text{and} \quad s_1 = c_1 \cdot 3^1 + c_2 \cdot 3^1$$

or

$$1 = c_1 \quad \text{and} \quad -3 = 3c_1 + 3c_2.$$

So $c_1 = 1$ and $c_2 = -2$. Therefore,

$$s_n = 3^n - 2n3^n \quad \text{for} \quad n \in \mathbb{N},$$

a solution that is easily checked to be correct. ■

Proof of Theorem 1

(a) No matter what s_0 and s_1 are, the equations

$$s_0 = c_1 + c_2 \quad \text{and} \quad s_1 = c_1 r_1 + c_2 r_2$$

can be solved for c_1 and c_2, since $r_1 \neq r_2$. The original sequence (s_n) is determined by the values s_0 and s_1 and the recurrence condition $s_n = as_{n-1} + bs_{n-2}$, so it suffices to show that the sequence defined by (1) also satisfies this recurrence condition. Since $x = r_1$ satisfies $x^2 = ax + b$, we have $r_1^n = ar_1^{n-1} + br_1^{n-2}$, so the sequence (r_1^n) satisfies the condition $s_n = as_{n-1} + bs_{n-2}$. So does (r_2^n). It is now easy to check that the sequence defined by (1) also satisfies the recurrence condition:

$$\begin{aligned} as_{n-1} + bs_{n-2} &= a(c_1 r_1^{n-1} + c_2 r_2^{n-1}) + b(c_1 r_1^{n-2} + c_2 r_2^{n-2}) \\ &= c_1(ar_1^{n-1} + br_1^{n-2}) + c_2(ar_2^{n-1} + br_2^{n-2}) \\ &= c_1 r_1^n + c_2 r_2^n = s_n. \end{aligned}$$

(b) If r is the only solution of the characteristic equation, then the characteristic equation has the form $(x - r)^2 = 0$. Thus the polynomials $x^2 - 2rx + r^2$ and $x^2 - ax - b$ are equal and, equating their coefficients, we obtain $a = 2r$ and $b = -r^2$. The recurrence relation can now be written

$$s_n = 2rs_{n-1} - r^2 s_{n-2}.$$

Putting $n = 0$ and $n = 1$ in equation (2) gives the equations

$$s_0 = c_1 \quad \text{and} \quad s_1 = c_1 r + c_2 r.$$

Since $r \neq 0$, these equations have the solutions $c_1 = s_0$ and $c_2 = -s_0 + s_1/r$. As in our proof of part (a), it suffices to show that any sequence defined by (2) satisfies $s_n = 2rs_{n-1} - r^2 s_{n-2}$. But

$$\begin{aligned} 2rs_{n-1} &- r^2 s_{n-2} \\ &= 2r(c_1 r^{n-1} + c_2(n-1)r^{n-1}) - r^2(c_1 r^{n-2} + c_2(n-2)r^{n-2}) \\ &= 2c_1 r^n + 2c_2(n-1)r^n - c_1 r^n - c_2(n-2)r^n \\ &= c_1 r^n + c_2 \cdot n \cdot r^n = s_n. \end{aligned}$$

■

The proof of the theorem is still valid if the roots of the characteristic equation are not real, so the theorem is true in that case as well. Finding the values of the terms with formula (1) will then involve complex arithmetic, but the answers will, of course, all be real if a, b, s_0 and s_1 are real. This situation is analogous to the calculation of the Fibonacci numbers, which are integers, using $\sqrt{5}$, which is not.

Theorem 1 applies only to recurrences of the form $s_n = as_{n-1} + bs_{n-2}$, but it is a useful tool, since recurrences like this come up fairly frequently. There is a version of the theorem that applies to recurrences of the form

$$s_n = a_{n-1}s_{n-1} + \cdots + a_{n-k}s_{n-k}$$

with $k \geq 3$ and constants $a_{n-1}, \ldots, a_{n-k}$. The characteristic equation in this case is $x^k - a_{n-1}x^{k-1} - \cdots - a_{n-k} = 0$. Its roots can be found, at least approximately, using symbolic computation software, and appropriate constants $c_1, \ldots, c_k$ can then be obtained as solutions of a system of linear equations. In case there is one root r that is larger than all of the others, as there was in the Fibonacci example, the solution for large values of n will be approximately Cr^n for some constant C. The general theorem is somewhat complicated to state; see, for example, the text *Introductory Combinatorics* by Brualdi for more details.

Another important kind of recurrence arises from estimating running times of what are called **divide-and-conquer** algorithms. The general algorithm of this type splits its input problem into two or more pieces, solves the pieces separately, and then puts the results together for the final answer. A simple example of a divide-and-conquer algorithm is the process of finding the largest number in a set of numbers by breaking the set into two batches of approximately equal size, finding the largest number in each batch, and then comparing these two to get the largest number overall. Merge sorting of a set of numbers by sorting the two batches separately and then fitting the sorted lists together is another example. For more about this topic, see §7.5.

If $T(n)$ is the time such an algorithm takes to process an input of size n, then the structure of the algorithm leads to a recurrence of the form

$$T(n) = T\left(\frac{n}{2}\right) + T\left(\frac{n}{2}\right) + F(n),$$

where the two $T(n/2)$'s give the time it takes to process the two batches separately, and $F(n)$ is the time to fit the two results together. Of course, this equation only makes sense if n is even.

Our next theorem gives information about recurrences such as this, of the form

$$s_{2n} = 2 \cdot s_n + f(n).$$

We have stated the theorem in a fairly general form, with one very important special case highlighted. Exercise 19 gives two variations on the theme.

Theorem 2 Let (s_n) be a sequence that satisfies a recurrence relation of the form

$$s_{2n} = 2 \cdot s_n + f(n) \qquad \text{for} \qquad n \in \mathbb{P}.$$

Then

$$s_{2^m} = 2^m \cdot \left(s_1 + \frac{1}{2} \sum_{i=0}^{m-1} \frac{f(2^i)}{2^i} \right) \qquad \text{for} \qquad m \in \mathbb{N}.$$

In particular, if

$$s_{2n} = 2 \cdot s_n + A + B \cdot n$$

for constants A and B, then

$$s_{2^m} = 2^m \cdot s_1 + (2^m - 1) \cdot A + \frac{B}{2} \cdot 2^m \cdot m.$$

Thus, if $n = 2^m$ in this case, we have

$$s_n = ns_1 + (n-1)A + \frac{B}{2} \cdot n \cdot \log_2 n.$$

Before we discuss the proof, let's see how we can use the theorem.

EXAMPLE 5

(a) Finding the largest member of a set by dividing and conquering leads to the recurrence

$$T(2n) = 2T(n) + A,$$

where the constant A is the time it takes to compare the winners from the two halves of the set. According to Theorem 2 with $B = 0$,

$$T(2^m) = 2^m \cdot T(1) + (2^m - 1) \cdot A.$$

Here $T(1)$ is the time it takes to find the biggest element in a set with one element. It seems reasonable to regard $T(1)$ as an overhead or handling charge for examining each element, whereas A is the cost of an individual comparison. If $n = 2^m$, we get

$$T(n) = n \cdot T(1) + (n-1) \cdot A,$$

so T is big-oh of n; i.e., $T(n) = O(n)$. In this algorithm, we examine n elements and make $n - 1$ comparisons to find the largest one. Thus, dividing and conquering gives no essential improvement over simply running through the elements one at a time, at each step keeping the largest number seen so far.

(b) Sorting the set in part (a) by dividing and conquering gives a recurrence

$$T(2n) = 2 \cdot T(n) + B \cdot n,$$

since the time to fit together the two ordered halves into one set is proportional to the sizes of the two sets being merged. Theorem 2 with $A = 0$ says that

$$T(2^m) = 2^m \cdot T(1) + \frac{B}{2} \cdot 2^m \cdot m.$$

If $n = 2^m$, then

$$T(n) = n \cdot T(1) + \frac{B}{2} \cdot n \cdot \log_2 n,$$

so $T(n) = O(n \cdot \log_2 n)$. For large n, the biggest part of the cost comes from the merging, not from examining individual elements. This algorithm is actually a reasonably efficient one for many applications. ■

You may want to skip the following proof, which is just an algebraic verification.

Proof of Theorem 2 We verify that if

$$s_{2^m} = 2^m \cdot \left(s_1 + \frac{1}{2} \sum_{i=0}^{m-1} \frac{f(2^i)}{2^i} \right) \qquad (*)$$

for some $m \in \mathbb{N}$, then

$$s_{2^{m+1}} = 2^{m+1} \cdot \left(s_1 + \frac{1}{2} \sum_{i=0}^{m} \frac{f(2^i)}{2^i} \right).$$

Since $s_{2^0} = 2^0 \cdot s_1$, because there are no terms in the summation if $m = 0$, it will follow by induction on m that $(*)$ holds for every $m \in \mathbb{N}$. [Though it's not necessary to check the case $m = 1$ separately, the reader may find it comforting.] Here are the details:

$$\begin{aligned} s_{2^{m+1}} &= s_{2 \cdot 2^m} = 2 \cdot s_{2^m} + f(2^m) && \text{[recurrence]} \\ &= 2^{m+1} \cdot \left(s_1 + \frac{1}{2} \sum_{i=0}^{m-1} \frac{f(2^i)}{2^i} \right) + f(2^m) && \text{[by } (*)\text{]} \\ &= 2^{m+1} \cdot \left(s_1 + \frac{1}{2} \sum_{i=0}^{m-1} \frac{f(2^i)}{2^i} + \frac{1}{2 \cdot 2^m} f(2^m) \right) \\ &= 2^{m+1} \cdot \left(s_1 + \frac{1}{2} \sum_{i=0}^{m} \frac{f(2^i)}{2^i} \right). \end{aligned}$$

The special case $f(n) = A + Bn$ gives a summation that we can compute fairly easily, but it is just as simple to verify directly [Exercise 18] that

$$s_{2^m} = 2^m \cdot s_1 + (2^m - 1) \cdot A + \frac{B}{2} \cdot 2^m \cdot m$$

if (s_n) satisfies the recurrence $s_{2n} = 2 \cdot s_n + A + Bn$. ■

This proof is technically correct, but not illuminating. How did we get the formula for s_{2^m} in the first place? We just used the recurrence to calculate the first few terms s_2, s_4, s_8, s_{16} and then guessed the pattern.

The theorem only tells us the values of s_n in cases in which n is a power of 2. That leaves some enormous gaps in our knowledge. It is often the case, however, that the sequence s is **monotone**; that is, $s_k \le s_n$ whenever $k \le n$. Timing estimates for algorithms generally have this property, for example. If s is monotone and if we know a function g such that $s_n \le g(n)$ whenever n is a power of 2, then for a general integer k we have $2^{m-1} < k \le 2^m$ for some m, so $s_{2^{m-1}} \le s_k \le s_{2^m} \le g(2^m)$. Since $m = \lceil \log_2 k \rceil$, we have $s_k \le g(2^{\lceil \log_2 k \rceil})$ for all k. This inequality isn't beautiful, but it does show how g provides a bound of s_k for all k.

We often just want to bound the values of s_n, typically by concluding that $s_n = O(h(n))$ for some function h. Replacing "=" by "≤" in the proof of Theorem 2 shows that if

$$s_{2n} \le 2 \cdot s_n + f(n) \qquad \text{for} \qquad n \in \mathbb{P}$$

then

$$s_{2^m} \le 2^m \cdot \left(s_1 + \frac{1}{2} \sum_{i=0}^{m-1} \frac{f(2^i)}{2^i} \right) \qquad \text{for} \qquad m \in \mathbb{N}.$$

In particular, if $s_{2n} \le 2 \cdot s_n + A$, then

$$s_{2^m} \le 2^m \cdot s_1 + (2^m - 1) \cdot A = O(2^m).$$

If, in addition, s is monotone and $2^{m-1} < k \le 2^m$, then $s_k \le s_{2^m} \le C \cdot 2^m$ for some C [for instance, $C = s_1 + A$], so $s_k \le C \cdot 2 \cdot 2^{m-1} \le C \cdot 2 \cdot k$. That is, $s_k = O(k)$.

The inequality $s_{2n} \le 2 \cdot s_n + B \cdot n$ leads similarly to $s_{2^m} = O(2^m \cdot m)$ and to $s_k = O(k \cdot \log_2 k)$. We were actually a little imprecise when we estimated the time for the merge sort in Example 5(b). If one of the two lists is exhausted early, we can just toss the remainder of the other list in without further action. What we *really* have in that example is the *inequality* $T(2n) \le 2T(n) + Bn$, which of course still gives $T(n) = O(n \cdot \log_2 n)$.

Exercises 4.5

1. Give an explicit formula for s_n, where $s_0 = 3$ and $s_n = -2s_{n-1}$ for $n \ge 1$.

2. (a) Give an explicit formula for $s_n = 4s_{n-2}$, where $s_0 = s_1 = 1$.

(b) Repeat part (a) for $s_0 = 1$ and $s_1 = 2$.

3. Prove that if $s_n = as_{n-1}$ for $n \ge 1$ and $a \ne 0$, then $s_n = a^n \cdot s_0$ for $n \in \mathbb{N}$.

4. Verify that the sequence given by $s_n = 2^{n+1} + (-1)^n$ in Example 2 satisfies the conditions $s_0 = s_1 = 3$ and $s_n = s_{n-1} + 2s_{n-2}$ for $n \ge 2$.

5. Verify that the sequence $s_n = 3^n - 2 \cdot n \cdot 3^n$ in Example 4 satisfies the conditions $s_0 = 1$, $s_1 = -3$, and $s_n = 6s_{n-1} - 9s_{n-2}$ for $n \ge 2$.

6. Use the formula for FIB(n) in Example 3 and a hand calculator to verify that FIB(6) = 8.

7. Give an explicit formula for s_n, where $s_0 = 3$, $s_1 = 6$, and $s_n = s_{n-1} + 2s_{n-2}$ for $n \ge 2$. *Hint:* Imitate Example 2, but note that now $s_1 = 6$.

8. Repeat Exercise 7 with $s_0 = 3$ and $s_1 = -3$.

9. Give an explicit formula for the sequence in Example 4 on page 155: $s_0 = 1$, $s_1 = 2$, and $s_n = s_{n-1} + s_{n-2}$ for $n \ge 2$. *Hint:* Use Example 3 on page 161.

10. Consider the sequence s_n, where $s_0 = 2$, $s_1 = 1$, and $s_n = s_{n-1} + s_{n-2}$ for $n \ge 2$.

(a) Calculate s_n for $n = 2, 3, 4, 5$, and 6.

(b) Give an explicit formula for s_n.

(c) Check your answers in part (a) using the formula obtained in part (b).

11. In each of the following cases, give an explicit formula for s_n.

(a) $s_0 = 2$, $s_1 = -1$, and $s_n = -s_{n-1} + 6s_{n-2}$ for $n \ge 2$.

(b) $s_0 = 2$ and $s_n = 5 \cdot s_{n-1}$ for $n \ge 1$.

(c) $s_0 = 1$, $s_1 = 8$, and $s_n = 4s_{n-1} - 4s_{n-2}$ for $n \ge 2$.

(d) $s_0 = c$, $s_1 = d$, and $s_n = 5s_{n-1} - 6s_{n-2}$ for $n \ge 2$. Here c and d are unspecified constants.

(e) $s_0 = 1$, $s_1 = 4$, and $s_n = s_{n-2}$ for $n \ge 2$.

(f) $s_0 = 1$, $s_1 = 2$, and $s_n = 3 \cdot s_{n-2}$ for $n \ge 2$.

(g) $s_0 = 1$, $s_1 = -3$, and $s_n = -2s_{n-1} + 3s_{n-2}$ for $n \ge 2$.

(h) $s_0 = 1$, $s_1 = 2$, and $s_n = -2s_{n-1} + 3s_{n-2}$ for $n \ge 2$.

12. Give an explicit formula for s_n.

(a) $s_0 = 0$ and $s_n = 5s_{n-1}$ for $n \ge 1$.

(b) $s_0 = 0$, $s_1 = 0$, $s_2 = 0$, and $s_n = 17s_{n-2} - 73s_{n-3}$ for $n \ge 3$.

(c) $s_0 = 5$ and $s_n = s_{n-1}$ for $n \ge 1$.

(d) $s_0 = 3$, $s_1 = 5$, and $s_n = 2s_{n-1} - s_{n-2}$ for $n \ge 2$.

13. Suppose that the sequence $s_0, s_1, s_2, \ldots$ of positive numbers satisfies

$$s_n = as_{n-1} + bs_{n-2} \quad \text{for } n \ge 2$$

and that $x^2 - ax - b = (x - r_1)(x - r_2)$ for real numbers r_1 and r_2.

(a) Show that if $r_1 > r_2 > 0$ then $s_n = O(r_1^n)$.

(b) Give an example with $r_1 = r_2 = 2$ for which $s_n = O(2^n)$ is false.

(c) What big-oh estimate for s_n can you give in general if $r_1 = r_2 > 0$?

14. Recall that if $s_n = bs_{n-2}$ for $n \ge 2$, then $s_{2n} = b^n s_0$ and $s_{2n+1} = b^n s_1$ for $n \in \mathbb{N}$. Show that Theorem 1 holds for $a = 0$ and $b > 0$, and reconcile this assertion with the preceding sentence. That is, specify r_1, r_2, c_1, and c_2 in terms of b, s_0, and s_1.

15. In each of the following cases, give an explicit formula for s_{2^m}.

(a) $s_{2n} = 2s_n + 3$, $s_1 = 1$

(b) $s_{2n} = 2s_n$, $s_1 = 3$

(c) $s_{2n} = 2s_n + 5n$, $s_1 = 0$

(d) $s_{2n} = 2s_n + 3 + 5n$, $s_1 = 2$

(e) $s_{2n} = 2s_n - 7$, $s_1 = 1$

(f) $s_{2n} = 2s_n - 7$, $s_1 = 5$

(g) $s_{2n} = 2s_n - n$, $s_1 = 3$

(h) $s_{2n} = 2s_n + 5 - 7n$, $s_1 = 0$.

16. Suppose that the sequence (s_n) satisfies the given inequality and that $s_1 = 7$. Give your best estimate for how large s_{2^m} can be.

(a) $s_{2n} \le 2s_n + 1$

(b) $s_{2n} \le 2(s_n + n)$

17. Suppose that the sequence (s_n) satisfies the recurrence

$$s_{2n} = 2s_n + n^2.$$

Give a formula for s_{2^m} in terms of s_1, and give an argument that your formula is correct. *Suggestion:* Apply Theorem 2 or guess a formula and check that it is correct.

18. Verify the formula in Theorem 2 for s_{2^m} in the case $f(n) = A + B \cdot n$.

19. Theorem 2 does not specifically apply to the following two recurrences, but the ideas carry over.

(a) Given that $t_{2n} = b \cdot t_n + f(n)$ for some constant b and function f, find a formula for t_{2^m} in terms of b, t_1, and the values of f.

(b) Given that $t_{3n} = 3t_n + f(n)$, find a formula for t_{3^m}.

4.6 More Induction

The principle of mathematical induction studied in §4.2 is often called the First Principle of Mathematical Induction. We restate it here in a slightly different form. Convince yourself that there is no substantive change from the previous version.

First Principle of Mathematical Induction Let m be an integer and let $p(n)$ be a sequence of propositions defined on the set $\{n \in \mathbb{Z} : n \geq m\}$. If

(B) $p(m)$ is true and

(I) for $k > m$, $p(k)$ is true whenever $p(k-1)$ is true,

then $p(n)$ is true for every $n \geq m$.

In the inductive step (I), each proposition is true provided the proposition immediately preceding it is true. To use this principle for constructing a proof, we need to check that $p(m)$ is true and that each proposition is true *assuming that the proposition just before it is true.* It is this right to assume the immediately previous case that makes the method of proof by induction so powerful. It turns out that in fact we are permitted to assume *all* previous cases. This apparently stronger assertion is a consequence of the following principle, whose proof we discuss at the end of this section.

Second Principle of Mathematical Induction Let m be an integer and let $p(n)$ be a sequence of propositions defined on the set $\{n \in \mathbb{Z} : n \geq m\}$. If

(B) $p(m)$ is true and

(I) for $k > m$, $p(k)$ is true whenever $p(m), \ldots, p(k-1)$ are all true,

then $p(n)$ is true for every $n \geq m$.

To verify (I) for $k = m + 1$, one shows that $p(m)$ implies $p(m+1)$. To verify (I) for $k = m+2$, one shows that $p(m)$ and $p(m+1)$ together imply $p(m+2)$. And so on. To verify (I) in general, one considers a $k > m$, assumes that the propositions $p(n)$ are true for $m \leq n < k$, and shows that these propositions imply that $p(k)$ is true. The Second Principle of Mathematical Induction is the appropriate version to use when the truths of the propositions follow from predecessors other than the immediate predecessors.

Our first example begins to fill in a proof that we promised long ago, in §1.2.

EXAMPLE 1

We show that every integer $n \geq 2$ can be written as a product of primes, a fact that we stated without proof in Theorem 2 back on page 9. Note that if n is prime the "product of primes" is simply the number n by itself. For $n \geq 2$, let $p(n)$ be the proposition

"n can be written as a product of primes."

Observe that the First Principle of Mathematical Induction is really unsuitable here. The lone fact that the integer 1,311,819, say, happens to be a product of primes is of no help in showing that 1,311,820 is also a product of primes. We apply the Second Principle. Clearly, $p(2)$ is true, since 2 is a prime.

Consider $k > 2$ and assume that $p(n)$ is true for all n satisfying $2 \leq n < k$. We need to show that it follows from this assumption that $p(k)$ is true. If k is prime, then $p(k)$ is clearly true. Otherwise, k can be written as a product $i \cdot j$, where i and j are integers greater than 1. Thus $2 \leq i < k$ and $2 \leq j < k$. Since both $p(i)$ and $p(j)$ are assumed to be true, we can write i and j as products of primes. Then $k = i \cdot j$ is also a product of primes. We have checked the basis and induction step for the Second Principle of Mathematical Induction, and so we infer that all the propositions $p(n)$ are true.

In the next section, we will prove that the factorization of a positive integer into a product of primes is unique. ■

Often the general proof of the inductive step (I) does not work for the first few values of k. In such a case, these first few values need to be checked separately, and then they may serve as part of the basis. We restate the Second Principle of Mathematical Induction in a more general version that applies in such situations.

Second Principle of Mathematical Induction Let m be an integer, let $p(n)$ be a sequence of propositions defined on the set $\{n \in \mathbb{Z} : n \geq m\}$, and let l be a nonnegative integer. If

(B) $p(m), \ldots, p(m+l)$ are all true and

(I) for $k > m + l$, $p(k)$ is true whenever $p(m), \ldots, p(k-1)$ are true,

then $p(n)$ is true for all $n \geq m$.

If $l = 0$, this is our original version of the Second Principle.

In §4.4 we saw that many sequences are defined recursively using earlier terms other than the immediate predecessors. The Second Principle is the natural form of induction for proving results about such sequences.

EXAMPLE 2

(a) In Exercise 14 on page 159 we recursively defined $b_0 = b_1 = 1$ and $b_n = 2b_{n-1} + b_{n-2}$ for $n \geq 2$. In part (b), we asked for an explanation of why all b_n's are odd integers. Given what we had discussed up to that point, a proof would have required use of the Well-Ordering Principle. It is more straightforward to apply the Second Principle of Mathematical Induction, as follows.

The nth proposition $p(n)$ is "b_n is odd." In the inductive step we will use the relation $b_k = 2b_{k-1} + b_{k-2}$, and so we'll need $k \geq 2$. Hence we'll check the cases $n = 0$ and 1 separately. Thus we will use the Second Principle with $m = 0$ and $l = 1$.

(B) The propositions $p(0)$ and $p(1)$ are obviously true, since $b_0 = b_1 = 1$.

(I) Consider $k \geq 2$ and assume that b_n is odd for all n satisfying $0 \leq n < k$. In particular, b_{k-2} is odd. Clearly, $2b_{k-1}$ is even, and so $b_k = 2b_{k-1} + b_{k-2}$ is the sum of an even and an odd integer. Thus b_k is odd.

It follows from the Second Principle of Mathematical Induction that all b_n's are odd. Note that in this proof the oddness of b_k followed from the oddness of b_{k-2}. An application of Theorem 1 on page 160 gives us the formula

$$b_n = \frac{1}{2}(1 + \sqrt{2})^n + \frac{1}{2}(1 - \sqrt{2})^n,$$

but it is not at all obvious from this formula that each b_n is an odd integer.

(b) The first few values of the Fibonacci sequence are $\text{FIB}(1) = 1$, $\text{FIB}(2) = 1$, $\text{FIB}(3) = 2$, $\text{FIB}(4) = 3$, $\text{FIB}(5) = 5$, $\text{FIB}(6) = 8$, ... Most of these are odd numbers. In fact, the next even one is $\text{FIB}(9) = 34$. Let's see if we can show that $\text{FIB}(n)$ is even if and only if n is a multiple of 3.

This statement is true for $1 \le n \le 6$, as we have just seen. Consider $n > 6$, and assume that for $1 \le k < n$ the number $\text{FIB}(k)$ is even if and only if k is a multiple of 3. If n is a multiple of 3, then $n-1$ and $n-2$ are not, so $\text{FIB}(n-1)$ and $\text{FIB}(n-2)$ are both odd. Hence $\text{FIB}(n) = \text{FIB}(n-1) + \text{FIB}(n-2)$ is even in this case. If n is *not* a multiple of 3, then exactly one of $n-1$ and $n-2$ is, so just one of $\text{FIB}(n-1)$ and $\text{FIB}(n-2)$ is even and their sum, $\text{FIB}(n)$, must be odd. By the Second Principle of Mathematical Induction, $\text{FIB}(n)$ is even if and only if n in $\mathbb{P}$ is a multiple of 3. ■

EXAMPLE 3

We recursively define $a_0 = a_1 = a_2 = 1$ and $a_n = a_{n-2} + a_{n-3}$ for $n \ge 3$. The first few terms of the sequence are 1, 1, 1, 2, 2, 3, 4, 5, 7, 9, 12, 16, 21, 28, 37, 49. The generalized version of Theorem 1 that we mentioned briefly in §4.5 just after the proof of that theorem suggests that (a_n) grows at most exponentially, with $a_n \le r^n$ for some constant r. Can we guess some suitable r and then prove this inequality in general? A typical strategy is to try to prove the inequality by induction and see what r needs to be for such a proof to succeed.

Here is our scratchwork. The inductive step will go like this: Assume that $a_n \le r^n$ for $1 \le n < k$. Then

$$a_k = a_{k-2} + a_{k-3} \le r^{k-2} + r^{k-3} = (r+1)r^{k-3} \le r^k.$$

The last inequality here will only be true if $r + 1 \le r^3$. We experiment: $r = 1$ is too small, while $r = 2$ works but seems too large. A little further experimentation shows that $r = \frac{4}{3}$ works, since $\frac{4}{3} + 1 = \frac{7}{3} < \frac{64}{27} = \left(\frac{4}{3}\right)^3$. This guess is not the best possible, but we didn't ask for the best possible.

Our conjecture now is that $a_n \le \left(\frac{4}{3}\right)^n$ for $n \in \mathbb{N}$, which is clear for $n = 0$, 1, and 2, by inspection. These cases provide the basis for induction, and the scratchwork above suggests that we will be able to execute the inductive step. Here are the careful details. We consider $k \ge 3$ and assume that $a_n \le \left(\frac{4}{3}\right)^n$ for $0 \le n < k$. In particular, $a_{k-2} \le \left(\frac{4}{3}\right)^{k-2}$ and $a_{k-3} \le \left(\frac{4}{3}\right)^{k-3}$. Thus we have

$$a_k = a_{k-2} + a_{k-3} \le \left(\frac{4}{3}\right)^{k-2} + \left(\frac{4}{3}\right)^{k-3} = \left(\frac{4}{3}\right)^{k-3} \cdot \left(\frac{4}{3} + 1\right).$$

Since $\frac{4}{3} + 1 < \left(\frac{4}{3}\right)^3$, we conclude that $a_k \le \left(\frac{4}{3}\right)^k$. This establishes the inductive step. Hence we infer from the Second Principle of Mathematical Induction [with $m = 0$ and $l = 2$] that $a_n \le \left(\frac{4}{3}\right)^n$ for all $n \in \mathbb{N}$.

This proof worked because there actually *is* an r with $a_n \le r^n$ and because we were able to find a number r with $r + 1 \le r^3$ to fill in our inductive step. If we hadn't found such a number, then we would have had to find another proof, prove something else, or abandon the problem. If we had tried a different but suitable r, then the details of the argument would have changed slightly. Induction gives us a framework for proofs, but it doesn't provide the details, which are determined by the particular problem at hand. ■

We have already stated the First Principle of Mathematical Induction for finite sequences in §4.2. Both versions of the Second Principle can also be stated for finite sequences. The changes are simple. Suppose that the propositions $p(n)$ are defined for $m \le n \le m^*$. Then the first version of the Second Principle can be stated as follows. If

(B) $p(m)$ is true and

(I) for $m < k \le m^*$, $p(k)$ is true whenever $p(m), \dots, p(k-1)$ are all true,

then $p(n)$ is true for all n satisfying $m \le n \le m^*$.

We return to the infinite principles of induction and end this section by discussing the logical relationship between the two principles and explaining why we regard both as valid for constructing proofs.

It turns out that each of the two principles implies the other, in the sense that if we accept either as valid, then the other is also valid. It is clear that the Second Principle implies the First Principle, since, if we are allowed to assume all previous cases, then we are surely allowed to assume the immediately preceding case. A rigorous proof can be given by showing that (B) and (I) of the Second Principle are consequences of (B) and (I) of the First Principle.

It is perhaps more surprising that the First Principle implies the Second. A proof can be given by using the propositions

$$q(n) = p(m) \wedge \cdots \wedge p(n) \qquad \text{for} \qquad n \ge m$$

and showing that, if the sequence $p(n)$ satisfies (B) and (I) of the Second Principle, then the sequence $q(n)$ satisfies (B) and (I) of the First Principle. It then follows that every $q(n)$ is true by the First Principle, so every $p(n)$ is also true.

The equivalence of the two principles is of less concern to us than an assurance that they are valid rules. For this we rely on the Well-Ordering Principle stated on page 131.

Proof of the Second Principle Assume

(B) $p(m), \dots, p(m+l)$ are all true and

(I) for $k > m + l$, $p(k)$ is true whenever $p(m), \dots, p(k-1)$ are true,

but that $p(n)$ is false for some $n \ge m$. Then the set

$$S = \{n \in \mathbb{Z} : n \ge m \text{ and } p(n) \text{ is false}\}$$

is nonempty. By the Well-Ordering Principle, S has a smallest element n_0. In view of (B) we must have $n_0 > m + l$. Since $p(n)$ is true for $m \le n < n_0$, condition (I) implies that $p(n_0)$ is true also. This contradicts the fact that n_0 belongs to S. It follows that if (B) and (I) hold, then every $p(n)$ is true. ■

A similar proof can be given for the First Principle but, because the Second Principle clearly implies the First, it is not needed.

Since the Second Principle is true and lets us assume more, why don't we just use it every time and forget about the First Principle? The answer is that we *could*. The account that we have given included the First Principle because it ties in naturally with loop invariants and because many people find it easier to begin with than the Second Principle. The First Principle is a little more natural for some problems, too, but there's nothing wrong with using the Second Principle of Mathematical Induction whenever a proof by induction is called for.

Exercises 4.6

Some of the exercises for this section require only the First Principle of Mathematical Induction and are included to provide extra practice. Most of them deal with sequences. You will see a number of applications later in which sequences are not so obvious.

1. Prove $3 + 11 + \cdots + (8n - 5) = 4n^2 - n$ for $n \in \mathbb{P}$.

2. For $n \in \mathbb{P}$, prove

(a) $1 \cdot 2 + 2 \cdot 3 + \cdots + n(n+1) = \frac{1}{3}n(n+1)(n+2)$

(b) $\dfrac{1}{1 \cdot 2} + \dfrac{1}{2 \cdot 3} + \cdots + \dfrac{1}{n(n+1)} = \dfrac{n}{n+1}$

3. Prove that $n^5 - n$ is divisible by 10 for all $n \in \mathbb{P}$.

4. (a) Calculate b_6 for the sequence (b_n) in Example 2.
(b) Use the recursive definition of (a_n) in Example 3 to calculate a_9.

5. Is the First Principle of Mathematical Induction adequate to prove the fact in Exercise 11(b) on page 159? Explain.

6. Recursively define $a_0 = 1$, $a_1 = 2$, and $a_n = \frac{a_{n-1}^2}{a_{n-2}}$ for $n \geq 2$.
(a) Calculate the first few terms of the sequence.
(b) Using part (a), guess the general formula for a_n.
(c) Prove the guess in part (b).

7. Recursively define $a_0 = a_1 = 1$ and $a_n = \frac{a_{n-1}^2 + a_{n-2}}{a_{n-1} + a_{n-2}}$ for $n \geq 2$. Repeat Exercise 6 for this sequence.

8. Recursively define $a_0 = 1$, $a_1 = 2$, and $a_n = \frac{a_{n-1}^2 - 1}{a_{n-2}}$ for $n \geq 2$. Repeat Exercise 6 for this sequence.

9. Recursively define $a_0 = 0$, $a_1 = 1$, and $a_n = \frac{1}{4}(a_{n-1} - a_{n-2} + 3)^2$ for $n \geq 2$. Repeat Exercise 6 for this sequence.

10. Recursively define $a_0 = 1$, $a_1 = 2$, $a_2 = 3$, and $a_n = a_{n-2} + 2a_{n-3}$ for $n \geq 3$.
(a) Calculate a_n for $n = 3, 4, 5, 6, 7$.
(b) Prove that $a_n > \left(\frac{3}{2}\right)^n$ for all $n \geq 1$.

11. Recursively define $a_0 = a_1 = a_2 = 1$ and $a_n = a_{n-1} + a_{n-2} + a_{n-3}$ for $n \geq 3$.
(a) Calculate the first few terms of the sequence.
(b) Prove that all the a_n's are odd.
(c) Prove that $a_n \leq 2^{n-1}$ for all $n \geq 1$.

12. Recursively define $a_0 = 1$, $a_1 = 3$, $a_2 = 5$, and $a_n = 3a_{n-2} + 2a_{n-3}$ for $n \geq 3$.
(a) Calculate a_n for $n = 3, 4, 5, 6, 7$.
(b) Prove that $a_n > 2^n$ for $n \geq 1$.
(c) Prove that $a_n < 2^{n+1}$ for $n \geq 1$.
(d) Prove that $a_n = 2a_{n-1} + (-1)^{n-1}$ for $n \geq 1$.

13. Recursively define $b_0 = b_1 = b_2 = 1$ and $b_n = b_{n-1} + b_{n-3}$ for $n \geq 3$.
(a) Calculate b_n for $n = 3, 4, 5, 6$.
(b) Show that $b_n \geq 2b_{n-2}$ for $n \geq 3$.
(c) Prove the inequality $b_n \geq (\sqrt{2})^{n-2}$ for $n \geq 2$.

14. For the sequence in Exercise 13, show that $b_n \leq \left(\frac{3}{2}\right)^{n-1}$ for $n \geq 1$.

15. Recursively define $\text{SEQ}(0) = 0$, $\text{SEQ}(1) = 1$, and

$$\text{SEQ}(n) = \frac{1}{n} \cdot \text{SEQ}(n-1) + \frac{n-1}{n} \cdot \text{SEQ}(n-2)$$

for $n \geq 2$. Prove that $0 \leq \text{SEQ}(n) \leq 1$ for all $n \in \mathbb{N}$.

16. As in Exercise 15 on page 159, let $\text{SEQ}(0) = 1$ and $\text{SEQ}(n) = \sum_{i=0}^{n-1} \text{SEQ}(i)$ for $n \geq 1$. Prove that $\text{SEQ}(n) = 2^{n-1}$ for $n \geq 1$.

17. Recall the Fibonacci sequence in Example 2(b) defined by

(B) $\text{FIB}(1) = \text{FIB}(2) = 1$,
(R) $\text{FIB}(n) = \text{FIB}(n-1) + \text{FIB}(n-2)$ for $n \geq 3$.

Prove that

$$\text{FIB}(n) = 1 + \sum_{k=1}^{n-2} \text{FIB}(k) \qquad \text{for } n \geq 3.$$

18. The **Lucas sequence** is defined as follows:

(B) $\text{LUC}(1) = 1$ and $\text{LUC}(2) = 3$,
(R) $\text{LUC}(n) = \text{LUC}(n-1) + \text{LUC}(n-2)$ for $n \geq 3$.

(a) List the first eight terms of the Lucas sequence.
(b) Prove that $\text{LUC}(n) = \text{FIB}(n+1) + \text{FIB}(n-1)$ for $n \geq 2$, where FIB is the Fibonacci sequence defined in Exercise 17.

19. Let the sequence T be defined as in Example 2(a) on page 154 by

(B) $T(1) = 1$,
(R) $T(n) = 2 \cdot T(\lfloor n/2 \rfloor)$ for $n \geq 2$.

Show that $T(n)$ is the largest integer of the form 2^k with $2^k \leq n$. That is, $T(n) = 2^{\lfloor \log n \rfloor}$, where the logarithm is to the base 2.

20. (a) Show that if T is defined as in Exercise 19, then $T(n)$ is $O(n)$.
(b) Show that if the sequence Q is defined as in Example 2(b) on page 154 by

(B) $Q(1) = 1$,
(R) $Q(n) = 2 \cdot Q(\lfloor n/2 \rfloor) + n$ for $n \geq 2$,

then $Q(n)$ is $O(n^2)$.
(c) Show that, in fact, $Q(n)$ is $O(n \log_2 n)$ for Q as in part (b).

21. Show that if S is defined as in Example 6 on page 157 by

(B) $S(0) = 0$, $S(1) = 1$,
(R) $S(n) = S(\lfloor n/2 \rfloor) + S(\lfloor n/5 \rfloor)$ for $n \geq 2$,

then $S(n)$ is $O(n)$.

4.7 The Euclidean Algorithm

In §1.2 we introduced common divisors and greatest common divisors, and we briefly discussed the Euclidean algorithm, which computes greatest common divisors. In this section we develop this algorithm, and we show how to modify the algorithm to solve

$x \cdot n \equiv a \pmod{m}$ for the unknown x. We also use the algorithm to show that the factorization of positive integers into products of primes is (essentially) unique.

Recall from §1.2 that a **common divisor** of integers m and n is an integer that is a divisor of both m and n. Thus 1 is always a common divisor, and if m and n are not both 0, then they have only a finite number of common divisors. We define the largest of these to be the **greatest common divisor** or **gcd** of m and n, and we denote it by $\mathbf{gcd}(m, n)$. These definitions make sense even for negative integers, but since the divisors of an integer are the same as the divisors of its negative, and since a greatest common divisor will always be positive, it is no real restriction to consider just nonnegative m and n. The case in which one of the two integers is 0 will be an important one, though. Since 0 is a multiple of every integer, every integer is a divisor of 0, and the common divisors of m and 0 are simply the divisors of m. Thus, if m is a positive integer, then $\gcd(m, 0) = m$.

The Euclidean algorithm will produce $\gcd(m, n)$, and it also will produce integers s and t such that

$$\gcd(m, n) = sm + tn.$$

We claimed the existence of such integers s and t a long time ago, in our proofs of Theorems 3 and 4 on pages 12 and 13. The Euclidean algorithm will establish that claim constructively, but it is useful for much more than theoretical arguments.

As we will see, this algorithm is *fast*; it requires only $O(\log_2(\max\{m, n\}))$ arithmetic operations and is practical for numbers with hundreds of digits. But why would anyone care about finding gcd's or finding the values of s and t for large numbers anyway? Aside from purely mathematical problems, there are everyday applications to public-key cryptosystems for secure transmission of data and to fast implementation of computer arithmetic for very large numbers. The applications depend upon solving congruences of the form $n \cdot x \equiv a \pmod{m}$ for unknown x, given m, n, and a, a topic that we will return to later in this section. It turns out that the most important case is the one in which $\gcd(m, n) = 1$ and that finding t with $1 = sm + tn$ is the critical step.

The Euclidean algorithm will compute gcd's rapidly using just addition, subtraction, and the functions DIV and MOD from §3.5. The key to the algorithm is the following fact.

Proposition If m and n are integers with $n > 0$, then the common divisors of m and n are the same as the common divisors of n and $m \text{ MOD } n$. Hence

$$\gcd(m, n) = \gcd(n, m \text{ MOD } n).$$

Proof We have $m = n \cdot (m \text{ DIV } n) + m \text{ MOD } n$. This equation holds even if $n = 1$, since $m \text{ DIV } 1 = m$ and $m \text{ MOD } 1 = 0$. If n and $m \text{ MOD } n$ are multiples of d, then both $n \cdot (m \text{ DIV } n)$ and $m \text{ MOD } n$ are multiples of d, so their sum m is too. In the other direction, if m and n are multiples of d', then $n \cdot (m \text{ DIV } n)$ is also a multiple of d'; so the difference $m \text{ MOD } n = m - n \cdot (m \text{ DIV } n)$ is a multiple of d'.

Since the pairs (m, n) and $(n, m \text{ MOD } n)$ have the same sets of common divisors, the two gcd's must be the same. ■

EXAMPLE 1

(a) We have $\gcd(45, 12) = \gcd(12, 45 \text{ MOD } 12) = \gcd(12, 9) = \gcd(9, 12 \text{ MOD } 9) = \gcd(9, 3) = \gcd(3, 9 \text{ MOD } 3) = \gcd(3, 0) = 3$. Recall that $\gcd(m, 0) = m$ for all nonzero integers m.

(b) Similarly, we get the chain $\gcd(20, 63) = \gcd(63, 20 \text{ MOD } 63) = \gcd(63, 20) = \gcd(20, 63 \text{ MOD } 20) = \gcd(20, 3) = \gcd(3, 20 \text{ MOD } 3) = \gcd(3, 2) = \gcd(2, 3 \text{ MOD } 2) = \gcd(2, 1) = \gcd(1, 2 \text{ MOD } 1) = \gcd(1, 0) = 1$.

(c) As a third example, $\gcd(12, 6) = \gcd(6, 12 \text{ MOD } 6) = \gcd(6, 0) = 6$. ■

These examples suggest a general strategy for finding $\gcd(m, n)$: Replace m and n by n and $m \text{ MOD } n$ and try again. We hope to reduce in this way to a case where we know the answer. Here is the resulting algorithm, which was already known to Euclid over 2000 years ago and may be older than that.

AlgorithmGCD(integer, integer)

{Input: $m, n \in \mathbb{N}$, not both 0.}
{Output: $\gcd(m, n)$.}
{Auxiliary variables: integers a and b.}
$a := m$; $b := n$
{The pairs (a, b) and (m, n) have the same gcd .}
while $b \neq 0$ **do**
    $(a, b) := (b, a \text{ MOD } b)$
return a ■

EXAMPLE 2

The table in Figure 1 lists the successive values of (a, b) when AlgorithmGCD is applied to the numbers in Example 1. The output gcd's are shown in boldface. [At the end of the middle column, recall that $2 \text{ MOD } 1 = 0$.] ■

Figure 1 ▶

(a, b)	(a, b)	(a, b)
(45,12) (12,9) (9,3) (**3**,0)	(20,63) (63,20) (20,3) (3,2) (2,1) (**1**,0)	(12,6) (**6**,0)
$m = 45$, $n = 12$	$m = 20$, $n = 63$	$m = 12$, $n = 6$

Theorem 1 AlgorithmGCD computes the gcd of the input integers m and n.

Proof We need to check that the algorithm terminates and that, when it does, a is the gcd of m and n.

If $n = 0$, the algorithm sets $a := m$, $b := 0$ and does not execute the **while** loop at all, so it terminates with $a = m$. Since $\gcd(m, 0) = m$ for $m \in \mathbb{P}$, the algorithm terminates correctly in this case. Thus suppose that $n > 0$.

As long as b is positive, the **while** loop replaces b by $a \text{ MOD } b$. Then, since $0 \leq a \text{ MOD } b < b$, the new value of b is smaller than the old value. A decreasing sequence of nonnegative integers must terminate, so b is eventually 0 and the algorithm terminates.

The Proposition shows that the statement "$\gcd(a, b) = \gcd(m, n)$" is an invariant of the **while** loop. When the loop terminates, $a > b = 0$, and since $a = \gcd(a, 0)$ we have $a = \gcd(m, n)$, as required. ■

The argument that the algorithm terminates shows that it makes at most n passes through the **while** loop. In fact, the algorithm is much faster than that.

Theorem 2 For input integers $m > n \geq 0$, AlgorithmGCD makes at most $2 \log_2(m + n)$ passes through the loop.

Proof In the next paragraph, we will show in general that if $a \geq b$, then

$$b + a \text{ MOD } b < \frac{2}{3} \cdot (a + b). \qquad (*)$$

This inequality means that the value of $a+b$ decreases by a factor of at least 2/3 with each pass through the loop

while $b \neq 0$ **do**
  $(a, b) := (b, a \text{ MOD } b)$.

At first, $a+b=m+n$, and after k passes through the loop we have $a+b \leq \left(\frac{2}{3}\right)^k \cdot (m+n)$. Now $a+b \geq 1+0$ always, so $1 \leq \left(\frac{2}{3}\right)^l \cdot (m+n)$ after the last, lth, pass through the loop. Hence $m+n \geq \left(\frac{3}{2}\right)^l$, so $\log_2(m+n) \geq l \cdot \log_2\left(\frac{3}{2}\right) > \frac{1}{2}l$ and thus $l < 2 \cdot \log_2(m+n)$, as desired. [Note that $\log_2\left(\frac{3}{2}\right) > \frac{1}{2}$ follows from the inequality $\frac{3}{2} > 2^{1/2} \approx 1.414$.]

It only remains to prove $(*)$ as promised, which we rewrite as

$$3b + 3 \cdot (a \text{ MOD } b) < 2a + 2b \quad \text{or} \quad b + 3 \cdot (a \text{ MOD } b) < 2a.$$

Since $a = b \cdot (a \text{ DIV } b) + a \text{ MOD } b$, this is equivalent to

$$b + 3 \cdot (a \text{ MOD } b) < 2b \cdot (a \text{ DIV } b) + 2 \cdot (a \text{ MOD } b)$$

or

$$b + a \text{ MOD } b < 2b \cdot (a \text{ DIV } b). \qquad (\dagger)$$

Since $a \geq b$, we have $a \text{ DIV } b \geq 1$. Moreover, $a \text{ MOD } b < b$; thus

$$b + a \text{ MOD } b < 2b \leq 2b \cdot (a \text{ DIV } b),$$

proving $(\dagger)$. ■

In fact, AlgorithmGCD typically terminates after far fewer than $2\log_2(m+n)$ passes through the loop, though there are inputs m and n (Fibonacci numbers!) that do require at least $\log_2(m+n)$ passes [see Exercise 15]. Moreover, the numbers a and b get smaller with each pass, so even starting with very large inputs one quickly gets manageable numbers. Note also that the hypothesis $m > n$ is not a serious constraint; if $n > m$ as in Example 1(b), then AlgorithmGCD replaces (m, n) by $(n, m \text{ MOD } n) = (n, m)$ on the first pass.

We will now modify AlgorithmGCD so that it produces not only the integer $d = \gcd(m, n)$, but also integers s and t with $d = sm + tn$. To make the new algorithm clearer, we first slightly rewrite AlgorithmGCD, using DIV instead of MOD. Here is the result.

AlgorithmGCD$^+$(*integer, integer*)

{Input: $m, n \in \mathbb{N}$, not both 0.}
{Output: $d = \gcd(m, n)$.}
{Auxiliary variables: integers a, b and q.}
$a := m$; $b := n$
while $b \neq 0$ **do**
  $q := a \text{ DIV } b$
  $(a, b) := (b, a - qb)$
$d := a$
return d ■

a	$b = a_{\text{next}}$	q	$-qb$
135	40	3	−120
40	15	2	−30
15	10	1	−10
10	**5**	2	−10
5	0		

Figure 2 ▲

EXAMPLE 3

Figure 2 gives a table of successive values of a, b, q, and $-qb$ for $m = 135$ and $n = 40$. The output d is in boldface. We have labeled b as a_{next} to remind ourselves that each b becomes a in the next pass. In fact, since the b's just become a's, we could delete the b column from the table without losing information. ■

In the general case, just as in Example 3, if we set $a_0 = m$ and $b_0 = a_1 = n$, then AlgorithmGCD$^+$ builds the sequences $a_0, a_1, \ldots, a_l, a_{l+1} = 0$ and $b_0, b_1, \ldots, b_l$, as well as $q_1, q_2, \ldots, q_l$. If a_{i-1} and b_{i-1} are the values of a and b at the start of the ith pass through the loop and a_i and b_i the values at the end, then

$a_i = b_{i-1}$, $q_i = a_{i-1}$ DIV a_i, and $a_{i+1} = b_i = a_{i-1} - q_i b_{i-1} = a_{i-1} - q_i a_i$. That is, the finite sequences (a_i) and (q_i) satisfy

$$q_i = a_{i-1} \text{ DIV } a_i \quad \text{and} \quad a_{i+1} = a_{i-1} - q_i a_i$$

for $i = 1, \ldots, l$. Moreover, $a_l = \gcd(m, n)$.

Now we are ready to compute s and t. We will construct two sequences $s_0, s_1, \ldots, s_l$ and $t_0, t_1, \ldots, t_l$ such that

$$a_i = s_i m + t_i n \quad \text{for} \quad i = 0, 1, \ldots, l.$$

Taking $i = l$ will give $\gcd(m, n) = a_l = s_l m + t_l n$, which is what we want.

To start, we want $m = a_0 = s_0 m + t_0 n$; to accomplish this we set $s_0 = 1$ and $t_0 = 0$. Next we want $n = a_1 = s_1 m + t_1 n$, so we set $s_1 = 0$ and $t_1 = 1$. From now on we exploit the recurrence $a_{i+1} = a_{i-1} - q_i a_i$. If we already have

$$a_{i-1} = s_{i-1} m + t_{i-1} n \quad \text{and} \quad a_i = s_i m + t_i n,$$

then

$$a_{i+1} = a_{i-1} - q_i a_i = [s_{i-1} - q_i s_i] \cdot m + [t_{i-1} - q_i t_i] \cdot n.$$

Thus, if we set

$$s_{i+1} = s_{i-1} - q_i s_i \quad \text{and} \quad t_{i+1} = t_{i-1} - q_i t_i,$$

then we get $a_{i+1} = s_{i+1} m + t_{i+1} n$, as desired.

EXAMPLE 4

Figure 3 shows the successive values of a_i, q_i, s_i, and t_i for the case $m = 135$, $n = 40$ of Example 3. Sure enough, $5 = 3 \cdot 135 + (-10) \cdot 40$ as claimed. ■

i	a_i	q_i	s_i	t_i
0	135		1	0
1	40	3	0	1
2	15	2	1	−3
3	10	1	−2	7
4	**5**	2	**3**	**−10**
5	0			

Figure 3 ▲

To recast the recursive definition of the sequences (s_i) and (t_i) in forms suitable for a **while** loop, we introduce new variables s and t, corresponding to s_i and t_i, and new variables s' and t', corresponding to s_{i+1} and t_{i+1}. While we are at it, we let $a' = b$ as well. Here is the algorithm that results.

Euclidean Algorithm(*integer*, *integer*)

```
{Input: m, n ∈ ℕ, not both 0.}
{Output: d = gcd(m, n), integers s and t with d = sm + tn.}
{Auxiliary variables: integers q, a, a', s, s' , t, t'.}
a := m; a' := n; s := 1; s' := 0; t := 0; t' := 1
{a = sm + tn and a' = s'm + t'n}
while a' ≠ 0 do
    q := a DIV a'
    (a, a') := (a', a − qa')
    (s, s') := (s', s − qs')
    (t, t') := (t', t − qt')
d := a
return d, s and t  ■
```

We have simply added two more lines to the loop in AlgorithmGCD$^+$, without affecting the value of a, so this algorithm makes exactly as many passes through the loop as AlgorithmGCD did. The argument above shows that the equations $a = sm + tn$ and $a' = s'm + t'n$ are loop invariants, so $d = a = sm + tn$ when the algorithm terminates, which it does after at most $2\log_2(m + n)$ iterations. We have shown the following.

Theorem 3 For input integers $m > n \geq 0$, the Euclidean algorithm produces $d = \gcd(m, n)$ and integers s and t with $d = sm + tn$, using $O(\log_2 m)$ arithmetic operations of the form $-$, $\cdot$, and DIV.

As noted in §1.2, every common divisor of m and n will also divide $sm + tn = \gcd(m, n)$. We restate this fact as follows.

Corollary The greatest common divisor $\gcd(m, n)$ of m and n is a multiple of all the common divisors of m and n.

The actual running time for the Euclidean algorithm will depend on how fast the arithmetic operations are. For large m and n, the operation DIV is the bottleneck. In our applications to solving congruences $n \cdot x \equiv a \pmod{m}$, which we are about to describe, we will only need t and not s, so we can omit the line of the algorithm that calculates s, with corresponding savings in time. Moreover, since we will only need $t \text{ MOD } m$ in that case, we can replace $t - qt'$ by $(t - qt') \text{ MOD } m$ in the computation and keep all the numbers t and t' smaller than m.

We can apply the tools we have developed to the problem of solving congruences of the form

$$n \cdot x \equiv a \pmod{m}$$

for x, given a, m, and n, which we said earlier has important applications. If we could simply divide a by n, then $x = a/n$ would solve the congruence. Unfortunately, with typical integers n and a, it's not likely that a will be a multiple of n, so this method usually won't work. Fortunately, we don't need to have $n \cdot x = a$ anyway, since we just want $n \cdot x \equiv a \pmod{m}$, and there is another approach that does work. It is based on the properties of arithmetic mod m that we developed in Theorem 2 on page 122 and Theorem 3 on page 123.

Dividing by n is like multiplying by $1/n$. If we can somehow find an integer t with

$$n \cdot t \equiv 1 \pmod{m},$$

then t will act like $1/n \pmod{m}$, and we will have $n \cdot t \cdot a \equiv 1 \cdot a \equiv a \pmod{m}$, so $x = t \cdot a$ [which looks like $\frac{1}{n} \cdot a \pmod{m}$] will solve the congruence, no matter what a is. Finding t, if it exists, is the key. If $\gcd(m, n) = 1$, then we can use the t in the equation $1 = sm + tn$ that EuclideanAlgorithm gives us, and our problem is solved.

If $\gcd(m, n) \neq 1$, however, then it can happen that there is no solution to $n \cdot x \equiv a \pmod{m}$. Even if a solution exists for a particular lucky value of a, there will be no t with $n \cdot t \equiv 1 \pmod{m}$. For example, $4 \cdot x \equiv 2 \pmod{6}$ has the solution $x = 2$, but $4 \cdot t \equiv 1 \pmod{6}$ has no solution at all, because $4 \cdot t \equiv 1 \pmod{6}$ is another way of saying that $4t - 1$ is a multiple of 6, but $4t - 1$ is an odd number.

This last argument shows more generally that if $\gcd(m, n) = d > 1$, then the congruence $n \cdot x \equiv a \pmod{m}$ has no solution unless a is a multiple of d. [Exercise 13(c) shows that this necessary condition is also sufficient.] If we expect to be able to solve $n \cdot x \equiv a \pmod{m}$ for arbitrary values of a, then we must have $\gcd(m, n) = 1$; i.e., m and n need to be relatively prime.

EXAMPLE 5

(a) Since 5 is prime, each of the congruences $n \cdot x \equiv 1 \pmod{5}$, with $n = 1, 2, 3, 4$, has a solution. These congruences are equivalent to the equations $n *_5 x = 1$ in $\mathbb{Z}(5)$, which were solved in Exercise 11 on page 125 using a table. With such small numbers, the Euclidean algorithm was not needed. The solutions are $1 \cdot 1 \equiv 1 \pmod{5}$, $2 \cdot 3 \equiv 1 \pmod{5}$, $3 \cdot 2 \equiv 1 \pmod{5}$, and $4 \cdot 4 \equiv 1 \pmod{5}$.

(b) Let's solve $10x \equiv 1 \pmod{37}$. Since 37 is prime, we know there is a solution, but creating a table would be inconvenient. We apply the Euclidean algorithm with $m = 37$ and $n = 10$ to obtain $s = 3$ and $t = -11$, so $3 \cdot 37 + (-11) \cdot 10 = 1$. Then $10 \cdot (-11) \equiv 1 \pmod{37}$ and hence $x = -11$ is a solution of $10x \equiv 1 \pmod{37}$. Any integer y with $y \equiv -11 \pmod{37}$ is also a solution, so $-11 \text{ MOD } 37 = -11 + 37 = 26$ is a solution of $10 *_{37} x = 1$ in $\mathbb{Z}(37) = \{0, 1, 2, \dots, 36\}$. Notice that the value of s is completely irrelevant here.

(c) Consider the equation $n \cdot x \equiv 1 \pmod{15}$. We regard n as fixed and seek a solution x in $\mathbb{Z}(15)$. If n and 15 are not relatively prime, then there is no solution. In particular, there is no solution if n is a multiple of 3 or 5.

We can, however, solve $13x \equiv 1 \pmod{15}$, which we know has a solution, since $\gcd(13, 15) = 1$. We apply the Euclidean algorithm with $m = 15$ and

$n = 13$ to obtain $s = -6$ and $t = 7$, so $(-6) \cdot 15 + 7 \cdot 13 = 1$. [Again, we don't care about s.] Thus $13 \cdot 7 \equiv 1 \pmod{15}$; i.e., $x = 7$ is a solution of $13x \equiv 1 \pmod{15}$. ■

We made the comment in §3.5 that, although $+_p$ and $*_p$ behave on $\mathbb{Z}(p) = \{0, 1, \ldots, p-1\}$ like $+$ and $\cdot$ on $\mathbb{Z}$, one must be careful about cancellation in $\mathbb{Z}(p)$. We have seen in this section that if n and p are relatively prime, then we can find t in $\mathbb{Z}$ so that $n \cdot t \equiv 1 \pmod{p}$. If $n \in \mathbb{Z}(p)$, then by Theorem 3 on page 123 we have

$$n *_p (t \text{ MOD } p) = (n \text{ MOD } p) *_p (t \text{ MOD } p) = (n \cdot t) \text{ MOD } p = 1,$$

so $t \text{ MOD } p$ acts like $1/n$ in $\mathbb{Z}(p)$ and $a *_p (t \text{ MOD } p)$ is a/n in $\mathbb{Z}(p)$. Thus division by n is possible in $\mathbb{Z}(p)$ if $\gcd(n, p) = 1$. In case p is prime, $\gcd(n, p) = 1$ for every nonzero n in $\mathbb{Z}(p)$, so division by all nonzero members of $\mathbb{Z}(p)$ is legal.

Cancellation of a factor on both sides of an equation is the same as division of both sides by the canceled factor. Thus, if $\gcd(n, p) = 1$ and $a *_p n = b *_p n$ in $\mathbb{Z}(p)$, then $a = b$, because $a = a *_p n *_p (t \text{ MOD } p) = b *_p n *_p (t \text{ MOD } p) = b$, where $t \text{ MOD } p$ acts like $1/n$ in $\mathbb{Z}(p)$. In terms of congruences, if n and p are relatively prime and if $a \cdot n \equiv b \cdot n \pmod{p}$, then $a \equiv b \pmod{p}$. We collect some of these facts in the following theorem, which just gives a taste of the rich algebraic properties of $\mathbb{Z}(p)$.

Theorem 4 If p is a prime, then it is possible to divide by, and hence to cancel, every nonzero element in $\mathbb{Z}(p)$. More generally, if n in $\mathbb{Z}(m)$ is relatively prime to m, then it is possible to divide by and to cancel n in $\mathbb{Z}(m)$.

We promised in §§1.2 and 4.6 that we would prove that prime factorizations are unique. The Euclidean algorithm gives us the key lemma.

Lemma 1 Let $m, n \in \mathbb{P}$. If a prime p divides the product mn, then it divides m or n (or both).

Proof Let $d = \gcd(m, p)$. Then d divides the prime p, so $d = p$ or $d = 1$. If $d = p$, then p divides m since $p = \gcd(m, p)$ and we're done. If $d = 1$, then the Euclidean algorithm provides integers s and t such that $1 = sm + tp$. Therefore, $n = s \cdot mn + p \cdot tn$. Since p divides mn by hypothesis and p obviously divides $p \cdot tn$, p divides n. ■

An easy induction argument extends this lemma to multiple factors.

Lemma 2 If a prime p divides the product $m_1 \cdot m_2 \cdots m_k$ of positive integers, then it divides at least one of the factors.

We showed in Example 1 on page 167 that every positive integer can be written as a product of primes. We now prove that this factorization is unique.

Theorem 5 A positive integer n can be written as a product of primes in essentially only one way.

Proof We can write the integer n as a product of primes $p_1 \cdot p_2 \cdots p_j$, where each prime is listed according to the number of times that it appears in the factorization. For example, if $n = 168 = 2 \cdot 2 \cdot 2 \cdot 3 \cdot 7$, then $j = 5$, $p_1 = p_2 = p_3 = 2$, $p_4 = 3$, and $p_5 = 7$. The statement of this theorem means simply that, if our general n is also written as a product of primes $q_1 \cdot q_2 \cdots q_k$ in some way, then $j = k$ and, moreover, by relabeling the q-primes if necessary, we can arrange for $p_1 = q_1$, $p_2 = q_2, \ldots, p_j = q_j$. Thus, no matter how 168 is written as a product of primes, there will be five factors: three of them will be 2, one will be 3, and the other will be 7.

For the proof, note that the prime p_1 divides the product $q_1 \cdot q_2 \cdots q_k$, so Lemma 4.7 says that p_1 divides one of the prime q-factors. Since the q-factor is a prime, it must be equal to p_1. Let's relabel the q-primes so that $q_1 = p_1$. Now $p_2 \cdot p_3 \cdots p_j = q_2 \cdot q_3 \cdots q_k$. By the same argument, we can relabel the remaining q-primes so that $q_2 = p_2$, etc. To give a tighter argument, an application of the Second Principle of Mathematical Induction is called for here. Assume that the statement of the theorem is true for integers less than n. As above, arrange for $q_1 = p_1$. Then apply the inductive assumption to

$$\frac{n}{p_1} = p_2 \cdot p_3 \cdots p_j = q_2 \cdot q_3 \cdots q_k$$

to conclude that $j = k$ and that the remaining q-primes can be relabeled so that $p_2 = q_2, \ldots, p_j = q_j$. ■

Exercises 4.7

Notice that many of the answers are easy to check, once found.

1. Use any method to find $\gcd(m, n)$ for the following pairs.
   (a) $m = 20, n = 20$ (b) $m = 20, n = 10$
   (c) $m = 20, n = 1$ (d) $m = 20, n = 0$
   (e) $m = 20, n = 72$ (f) $m = 20, n = -20$
   (g) $m = 120, n = 162$ (h) $m = 20, n = 27$
2. Repeat Exercise 1 for the following pairs.
   (a) $m = 17, n = 34$ (b) $m = 17, n = 72$
   (c) $m = 17, n = 850$ (d) $m = 170, n = 850$
   (e) $m = 289, n = 850$ (f) $m = 2890, n = 850$
3. List the pairs (a, b) that arise when AlgorithmGCD is applied to the numbers m and n, and find $\gcd(m, n)$.
   (a) $m = 20, n = 14$ (b) $m = 20, n = 7$
   (c) $m = 20, n = 30$ (d) $m = 2000, n = 987$
4. Repeat Exercise 3 for the following.
   (a) $m = 30, n = 30$ (b) $m = 30, n = 10$
   (c) $m = 30, n = 60$ (d) $m = 3000, n = 999$
5. Use the Euclidean algorithm to find $\gcd(m, n)$ and integers s and t with $\gcd(m, n) = sm + tn$ for the following. *Suggestion:* Make tables like the one in Figure 3.
   (a) $m = 20, n = 14$ (b) $m = 72, n = 17$
   (c) $m = 20, n = 30$ (d) $m = 320, n = 30$
6. Repeat Exercise 5 for the following.
   (a) $m = 14{,}259, n = 3521$ (b) $m = 8359, n = 9373$
7. For each value of n solve $n \cdot x \equiv 1 \pmod{26}$ with $0 \le x < 26$, or explain why no solution exists.
   (a) $n = 5$ (b) $n = 11$
   (c) $n = 4$ (d) $n = 9$
   (e) $n = 17$ (f) $n = 13$
8. Repeat Exercise 7 for the congruence $n \cdot x \equiv 1 \pmod{24}$ with $0 \le x < 24$.
9. Solve the following congruence equations for x.
   (a) $8x \equiv 1 \pmod{13}$ (b) $8x \equiv 4 \pmod{13}$
   (c) $99x \equiv 1 \pmod{13}$ (d) $99x \equiv 5 \pmod{13}$
10. Solve for x.
    (a) $2000x \equiv 1 \pmod{643}$ (b) $643x \equiv 1 \pmod{2000}$
    (c) $1647x \equiv 1 \pmod{788}$ (d) $788x \equiv 24 \pmod{1647}$
11. Show that the equations $a = sm + tn$ and $a' = s'm + t'n$ are invariants for the loop in EuclideanAlgorithm no matter how q is defined in the loop. [Thus a mistake in calculating or guessing q does no permanent harm.]
12. Consider integers m and n, not both 0. Show that $\gcd(m, n)$ is the *smallest* positive integer that can be written as $am + bn$ for integers a and b.
13. Suppose that $d = \gcd(m, n) = sm + tn$ for some integers s and t and positive integers m and n.
    (a) Show that m/d and n/d are relatively prime.
    (b) Show that if $d = s'm + t'n$ for $s', t' \in \mathbb{Z}$, then $s' = s + k \cdot n/d$ for some $k \in \mathbb{Z}$.
    (c) Show that if a is a multiple of d, then $nx \equiv a \pmod{m}$ has a solution. *Hint:* $nt \equiv d \pmod{m}$.
14. Suppose that $d = \gcd(m, n) = sm + tn$ for some $s, t \in \mathbb{Z}$ and positive integers m and n.
    (a) Show that if $s' = s + k \cdot n/d$ and $t' = t - k \cdot m/d$, then $d = s'm + t'n$.
    (b) Show that there are $s', t' \in \mathbb{Z}$ such that $d = s'm + t'n$ and $0 \le s' < n/d$.
15. (a) Show that, with $m = \text{FIB}(l + 2)$ and $n = \text{FIB}(l + 1)$ as inputs and $l \ge 1$, AlgorithmGCD makes exactly l passes through the loop. The Fibonacci sequence is defined in Example 3 on page 155. *Hint:* Use induction on l.
    (b) Show that $k \ge \log_2 \text{FIB}(k + 3)$ for $k \ge 3$.
    (c) Show that, with inputs $m = \text{FIB}(l + 2)$ and $n = \text{FIB}(l + 1)$ and with $l \ge 2$, AlgorithmGCD makes at least $\log_2(m + n)$ passes through the loop.
16. Write out the induction proof for Lemma 4.7 on page 177.

Chapter Highlights

As usual:

(a) Satisfy yourself that you can define each concept and can describe each method.
(b) Give at least one reason why the item was included in this chapter.
(c) Think of at least one example of each concept and at least one situation in which each fact or method would be useful.

CONCEPTS AND NOTATION

while loop
guard, body, pass = iteration, terminate = exit

invariant

for loop, **for** ... downto loop

mathematical induction
basis, inductive step

recursive definition of sequence
basis, recurrence formula
iterative, recursive calculation
Fibonacci sequence

characteristic equation

divide-and-conquer recurrence

divisor, common divisor, gcd

relatively prime integers

FACTS AND PRINCIPLES

Loop Invariant Theorem

Well-Ordering Principle for $\mathbb{N}$

First and Second Principles of [Finite] Mathematical Induction
[These two principles are logically equivalent.]

Theorem 2 on page 164 on recurrences of form $s_{2n} = 2s_n + f(n)$

If p is prime, then division by nonzero elements is legal in $\mathbb{Z}(p)$.

Prime factorization is unique.

METHODS

Use of loop invariants to develop algorithms and verify correctness

Solution of $s_n = as_{n-1} + bs_{n-2}$ with characteristic equation

Euclidean algorithm to compute $\gcd(m, n)$ and integers s, t with $\gcd(m, n) = sm + tn$

Application of the Euclidean algorithm to congruences $m \cdot x \equiv a \pmod{n}$

Supplementary Exercises

1. Consider the loop "**while** $n < 200$ **do** $n := 10 - 3n$." Which of the following are invariants of the loop? Explain in each case.

(a) n is odd. (b) n is even.

2. (a) Is $3n < 2m$ an invariant of the loop "**while** $n < m$ **do** $n := 3n$ and $m := 2m$"? Explain.

(b) Is $3n < 2m$ an invariant of the loop "**while** $n > m$ **do** $n := 3n$ and $m := 2m$"? Explain.

3. For each statement, indicate whether it is true or false.

(a) $\sqrt{n^2 + n} = O(n)$ (b) $\sum_{k=1}^{n} k^2 = O(n^2)$
(c) $n! = O(2^n)$ (d) $2^n = O(n^2)$

4. Suppose that we are given that $f(n) = O(n^2)$ and that $g(n) = O(n^3)$. For each function below, indicate its growth $O(n^k)$ with the smallest k that you can be sure of.

(a) $f(n) + g(n)$ (b) $f(n) \cdot g(n)$ (c) $f(n)^2$
(d) $f(n)^2 + g(n)^2$

5. (a) Show that $3(n + 7)^2 = \Theta(n^2)$.

(b) Show that $2^{2n} = O(3^n)$ is false.

(c) Show that $\log(n!) = O(n \log n)$.

6. Show that $n^n = O(n!)$ is false.

7. The sequence $h_0, h_1, h_2, \ldots$ satisfies $h_0 = 0$, $h_1 = 1$ and $h_n = 2h_{n-1} + h_{n-2}$ for $n \geq 2$.

(a) Find h_3. (b) Find a formula for h_n.

(c) Verify that your formula gives the correct value of h_3.

8. Solve the recurrence relation $a_n = a_{n-1} + 6a_{n-2}$ in each of the following cases.

(a) $a_0 = 2$, $a_1 = 6$

(b) $a_0 = 1$, $a_2 = 4$ [Note: it's a_2, not a_1.]

(c) $a_0 = 0$, $a_1 = 5$

9. Define the sequence (b_n) by $b_0 = 5$ and $b_n = 2b_{n-1} - 3$ for $n \geq 1$. Show that $b_n > 8n$ for $n \geq 4$.

10. (a) What is the number a such that $(3n-1)^5 = \Theta(n^a)$?

(b) What is the smallest number a such that $4^{2n+1} + 5n = O(a^n)$?

11. Give examples of sequences $s(n)$ and $t(n)$ such that

(a) $s(n) = O(t(n))$ is true, but $t(n) = O(s(n))$ is false.

(b) $\log_2(s(n)) = O(\log_2(t(n)))$ is true, but $s(n) = O(t(n))$ is false.

12. Prove carefully that $3^n + 2n - 1$ is divisible by 4 for all positive integers n.

13. Prove that $(2n)! \geq 2^n \cdot n!$ for all $n \in \mathbb{N} = \{0, 1, 2, 3, \ldots\}$.

14. For each sequence below, give the sequence $a(n)$ in the hierarchy of Theorem 1 on page 148 such that $s(n) = O((a(n))$ and $a(n)$ is as far to the left as possible in the hierarchy.

(a) $3n\sqrt{n^6 + n^2 + 1}$ (b) $3n + \sqrt{n^6 + n^2 + 1}$

(c) 3^n (d) $\sqrt{3n}$

(e) $(3n+5)^4$ (f) $(n^2+7)^3$

15. (a) For which positive integers $n \leq 5$ is the proposition "$2^n > n^2$" true?

(b) Prove by induction that $2^n > n^2$ for all integers $n \geq 5$.

16. Revise DivisionAlgorithm to obtain an algorithm that finds q and r for all positive integers m and all integers n, positive, negative, or 0.

17. Suppose that we change DivisionAlgorithm's **while** loop to

while $r \geq m$ **do**
  $q := q + 2$
  $r := r - 2m$

Is $n = m \cdot q + r$ still a loop invariant? How about $0 \leq r$? Does the changed algorithm still produce the q and r we want? Does your answer fit the pattern of the "somewhat more explicit" algorithm on page 133?

18. Suppose that $1 = sm + tn$ for integers s, m, t, and n.

(a) Show that $\gcd(s, t)$ divides 1.

(b) Show that if m and n are positive, then they must be relatively prime.

(c) What can you say about $\gcd(s, n)$?

(d) Is your answer to part (c) correct if $s = 0$?

19. Prove the proposition on page 154. That is, show that if the sequence $s_0, s_1, s_2 \ldots$ is defined by a basis and recurrence condition and if the sequence $t_0, t_1, t_2, \ldots$ satisfies the same basis and recurrence condition, then $s_n = t_n$ for every n in $\mathbb{N}$.

20. Define the sequence $a_0, a_1, a_2, \ldots$ by $a_0 = 1$, $a_1 = 2$, and

$$a_n = 2a_{n-1} - a_{n-2} + 2$$

for $n \geq 2$. Use any method to prove that $a_n = n^2 + 1$ for all n in $\mathbb{P}$.

21. Use the *definition* of big-oh notation to show that if $a(n) = O(\log_2 n)$, then $n^2 \cdot a(n) = O(n^2 \log_2 n)$.

22. Prove the existence part of the Division Algorithm stated on page 13 directly from the Well-Ordering Principle. *Hint:* If the statement is true, then $r = n - m \cdot q$ is the smallest nonnegative integer in the set $S = \{n - m \cdot q' : q' \in \mathbb{Z}$ and $n - m \cdot q' \geq 0\}$. Show S is nonempty, and let r denote its smallest element.

23. (a) Show that there are an infinite number of positive integers that are not divisible by 2, 3, 5, 7, 11, 13, or 17.

(b) Prove that there is a smallest integer larger than 10^9 that is not divisible by 2, 3, 5, 7, 11, 13, or 17.

(c) Is your proof constructive? Explain.

24. Let $a_0 = 0$, $a_1 = 2$, $a_2 = 8$, and $a_n = 6a_{n-1} - 12a_{n-2} + 8a_{n-3}$ for $n \geq 3$.

(a) Calculate a_3.

(b) Prove carefully that $a_n = n \cdot 2^n$ for all positive integers n.

25. Recursively define $a_0 = a_1 = 1$ and $a_n = a_{n-1} + a_{n-2} + 1$ for $n \geq 2$.

(a) List the first few members of the sequence.

(b) Prove that all members of the sequence are odd integers.

(c) Prove that $a_n < 2^n$ for all $n > 0$.

26. Let (a_n) and (b_n) be sequences of positive numbers. Prove carefully, using the definition of big-oh notation, that if $a_n = O(b_n)$, then there is a constant, C, such that $a_n \leq Cb_n$ for $n = 1, 2, \ldots$.

27. Suppose that a sequence a_n is $O(n)$ and consider the sequence b_n defined by $b_1 = a_1$, $b_2 = a_1 + a_2$, $\ldots$, $b_n = a_1 + a_2 + \cdots + a_n$, etc. Prove carefully (using definitions) that the sequence b_n is $O(n^2)$.

28. Let (a_n) be a sequence of positive integers that is $O(n)$, and consider the sequence (c_n) defined by $c_0 = a_0$ and $c_n = c_{n-1} + a_{n-1}$ for $n \geq 1$. Prove carefully (using definitions) that the sequence c_n is $O(n^2)$.

29. Prove the Second Principle of Induction from the First Principle by using the propositions

$$q(n) = p(m) \wedge \cdots \wedge p(n) \qquad \text{for} \qquad n \geq m$$

and showing that, if the sequence $p(n)$ satisfies (B) and (I) of the Second Principle, then the sequence $q(n)$ satisfies (B) and (I) of the First Principle. Conclude that every $q(n)$ is true by the First Principle, so every $p(n)$ is also true.

Counting

One major goal of this chapter is to develop methods for counting large finite sets without actually listing their elements. The theory that results has applications to computing probabilities in finite sample spaces, which we introduce in §5.2. In the last section, we will see how an obvious principle, the Pigeon-Hole Principle, can be used to establish some nontrivial and interesting results.

5.1 Basic Counting Techniques

Imagine that you have developed a process, for example, a computer program, that takes 8 different quantities as inputs. Suppose that there are 20 different types of inputs and that you want to test your process by trying it on all possible choices of 8 of the 20 input types. Is this a realistic plan, or will there be too many possibilities to try? How can we tell? Is the question posed properly? Does it matter which order the inputs have? Questions such as these are the focus of this section. The answers to some of the ones above are these: if order doesn't matter, there are 125,970 choices, which may or may not be an impractical number to try; if order does matter, there are about 5,079,000,000 possibilities, which is a lot. By the end of this section you should be able to see where these numbers came from. See Exercise 5.

We start with some easier questions and work up, collecting tools as we go along.

EXAMPLE 1

(a) How many integers in $A = \{1, 2, 3, \ldots, 1000\}$ are divisible by 3? We answered questions like this in §1.1. There are $\lfloor 1000/3 \rfloor = 333$ such integers.

(b) How many members of A are *not* divisible by 3? That's easy. It's the difference $1000 - 333 = 667$.

(c) How many members of A are divisible by 3 or by 5 or by both 3 and 5? There are 333 members divisible by 3 and $\lfloor 1000/5 \rfloor = 200$ divisible by 5; but if we just add 333 and 200 to get 533, we will have too many; the multiples of 15 will have been counted twice. The correct answer is $\lfloor 1000/3 \rfloor + \lfloor 1000/5 \rfloor - \lfloor 1000/15 \rfloor = 467$. ■

This example reminds us of the following familiar properties of counting. [Here, as in §1.4, we write $|S|$ for the number of elements in the set S.]

Union Rules Let S and T be finite sets.

(a) If S and T are disjoint, i.e., if $S \cap T = \emptyset$, then $|S \cup T| = |S| + |T|$.
(b) In general, $|S \cup T| = |S| + |T| - |S \cap T|$.
(c) If $S \subseteq T$, then $|T \setminus S| = |T| - |S|$.

Union Rule (a) simply says that we can count a union of disjoint sets by counting the individual sets and adding the results. Union Rule (b) corrects for possible double-counting of the overlap $S \cap T$. We can deduce it from Rule (a) as follows, remembering that S and $T \setminus S$ are disjoint, as are $T \setminus S$ and $S \cap T$. By Rule (a) we have

$$|S \cup T| = |S| + |T \setminus S| \quad \text{and} \quad |T| = |T \setminus S| + |S \cap T|.$$

Thus

$$|S \cup T| + |S \cap T| = |S| + |T \setminus S| + |S \cap T| = |S| + |T|,$$

which implies (b).

Rule (c) also comes from Rule (a), since $T \setminus S$ and S are disjoint. It just says that one way to count a set is to count everything in sight and then subtract the number of elements you don't want.

There is a k-set version of Rule (a). We say that a collection of sets is **pairwise disjoint** if every two sets in it are disjoint, i.e., if no two overlap. If $\{S_1, \ldots, S_k\}$ is a *pairwise disjoint* collection of finite sets, then

$$|S_1 \cup \cdots \cup S_k| = |S_1| + \cdots + |S_k|.$$

This equation is false if the collection is not pairwise disjoint. The Inclusion–Exclusion Principle, a generalization of Rule (b) for counting possibly overlapping unions of more than two sets, appears in §5.3.

Another common type of problem arises when we are performing a succession of steps, each of which can be done in several ways, and we want to count the total number of different ways to perform the steps.

EXAMPLE 2

(a) How many strings, i.e., words, of length 4 can we make from the letters a, b, c, d, and e if repetitions are allowed? In the notation of §1.3, we want $|\Sigma^4|$, where $\Sigma = \{a, b, c, d, e\}$. We can also think of the strings as ordered 4-tuples, i.e., as members of $\Sigma \times \Sigma \times \Sigma \times \Sigma$. We imagine a step-by-step process for forming such a string. There are 5 choices for the first letter, then 5 choices for the second letter, then 5 choices for the third letter, and finally 5 choices for the last letter, so altogether there are $5 \cdot 5 \cdot 5 \cdot 5 = 5^4$ possible strings. We could list all 625 of them, but we don't need to if we just want to know how many there are.

(b) How many 2-digit numbers are there? We don't allow the first digit to be 0 (for example, 07 is not permitted), so we want $|\{10, 11, \ldots, 99\}|$. One easy solution is to use Union Rule (c) and compute $|\{00, \ldots, 99\}| - |\{00, \ldots, 09\}| = 100 - 10 = 90$. Alternatively, we could consider choosing the 1's digit in any of 10 ways and then the 10's digit in any of 9 ways, avoiding 0. There are then $10 \cdot 9$ different sequences of choices. Still 90. We could also pick the 10's digit first and then the 1's digit, to get $9 \cdot 10$.

(c) How many of these 2-digit numbers are odd? We could count the even ones, using the methods of §1.1, and subtract. We could also use Union Rule (a) in its generalized form. Break the set of 2-digit numbers into pairwise disjoint sets $S_0, S_1, S_2, \ldots, S_9$, where S_k is the set of numbers with 1's digit k. Then we want $|S_1 \cup S_3 \cup S_5 \cup S_7 \cup S_9| = |S_1| + |S_3| + |S_5| + |S_7| + |S_9|$. Since $|S_k| = 9$ for each k [why?], the answer is $5 \cdot 9 = 45$.

We could also have thought of constructing a number as a compound process in which the first step is to choose the 1's digit from $\{1, 3, 5, 7, 9\}$ in one of 5 ways and the second step is to choose the 10's digit in one of 9 ways. All told, there are $5 \cdot 9$ possible pairs of choices. Or reverse the order of the steps and get $9 \cdot 5$. Either way, half of the 90 numbers are odd.

(d) How many of these odd 2-digit numbers have their two digits different? The easy solution is to subtract from the total in part (c) the number that have both digits the same, i.e., $|\{11, 33, 55, 77, 99\}|$, to get $45 - 5 = 40$. A more complicated solution, but one that generalizes, is to choose the 1's digit in any of 5 ways and then the 10's digit in one of 8 ways, avoiding 0 and the digit chosen for the 1's digit. We get $5 \cdot 8$, which is still 40. Notice that it is not easy now to solve the problem by first picking the 10's digit in one of 9 ways and then picking the 1's digit, because the number of choices available for the 1's digit depends on whether the 10's digit chosen was even or odd. This example illustrates one of the reasons counting is often challenging. Among two or more approaches that initially look equally reasonable, often some will turn out to work more easily than others. One needs to be flexible in tackling such problems.

(e) How many odd 17-digit numbers have no two digits the same? That's easy. None, because there are only 10 digits available, so we'd have to repeat some digits. How many odd 7-digit numbers have no two digits the same? This question, which we leave as Exercise 13, yields to the second approach in part (d), though not to the easy first method. ■

We can state the idea that we have just been using as a general principle.

Product Rule Suppose that a set of ordered k-tuples $(s_1, s_2, \ldots, s_k)$ has the following structure. There are n_1 possible choices of s_1. Given an s_1, there are n_2 possible choices of s_2; given any s_1 and s_2, there are n_3 possible choices of s_3; and in general, given any $s_1, s_2, \ldots, s_{j-1}$, there are n_j choices of s_j. Then the set has $n_1 n_2 \cdots n_k$ elements.

In particular, for finite sets $S_1, S_2, \ldots, S_k$ we have

$$|S_1 \times S_2 \times \cdots \times S_k| = |S_1| \cdot |S_2| \cdots |S_k|.$$

If all of the S_i's are the same set S, then we have the nice formula $|S^k| = |S|^k$.

We will use the Product Rule often, but almost never with the forbidding formalism suggested in its statement. The way we have put it is meant to strongly encourage thinking of problems in terms of multistep processes; indeed, actually visualizing such a process is often a helpful problem-solving technique.

EXAMPLE 3

(a) Let $\Sigma = \{a, b, c, d, e, f, g\}$. The number of words in Σ^* having length 5 is $7^5 = 16,807$; i.e., $|\Sigma^5| = 16,807$, by the Product Rule. The number of words in Σ^5 that have no letters repeated is $7 \cdot 6 \cdot 5 \cdot 4 \cdot 3 = 2520$, because the first letter can be selected in 7 ways, then the second letter can be selected in 6 ways, etc.

(b) Let $\Sigma = \{a, b, c, d\}$. The number of words in Σ^2 without repetitions of letters is $4 \cdot 3 = 12$ by the Product Rule. We can illustrate these words by a picture

Figure 1 ▶

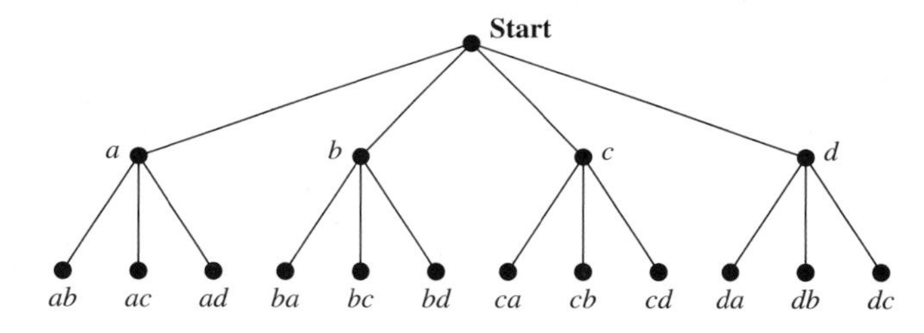

called a tree; see Figure 1. Each path from the start corresponds to a word in Σ^2 without repetitions. For example, the path going through b and ending at bc corresponds to the word bc. One can imagine a similar but very large tree for the computation in part (a). In fact, one can imagine such a tree for any situation to which the Product Rule applies. ■

EXAMPLE 4

The number of paths from s to f in Figure 2(a) is $3 \cdot 2 \cdot 2 \cdot 1 = 12$ since there are 3 choices for the first edge and then 2 each for the second and third edges. After we select the first three edges, the fourth edge is forced.

Figure 2 ▶

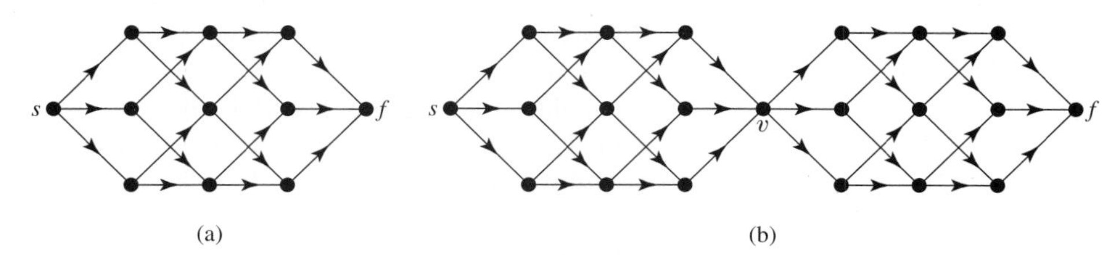

A similar computation can be used to count the number of paths from s to f in Figure 2(b). Alternatively, notice that from above there are 12 paths from s to v and 12 paths from v to f, so there are $12 \cdot 12 = 144$ paths from s to f. ■

EXAMPLE 5

Let S and T be finite sets. We will count the number of functions $f: S \to T$. Here it is convenient to write

$$S = \{s_1, s_2, \ldots, s_m\} \quad \text{and} \quad T = \{t_1, t_2, \ldots, t_n\}$$

so that $|S| = m$ and $|T| = n$. A function $f: S \to T$ is described by specifying $f(s_1)$ to be one of the n elements in T, then specifying $f(s_2)$ to be one of the n elements in T, etc. This process leads to $n \cdot n \cdots n = n^m$ different results, each of which specifies a different function. We conclude that there are n^m functions mapping S into T. I.e., $|\text{FUN}(S, T)| = n^m = |T|^{|S|}$, where we write **FUN(*S*, *T*)** for the set of all functions f from S into T. ■

As we hinted in Example 2(c), we can view the Product Rule as an extended version of Union Rule (a), in the same way that multiplication is just repeated addition. The n_1 different choices for s_1 yield n_1 pairwise disjoint sets of k-tuples. The number of k-tuples with a particular choice of s_1 is the number of $(k-1)$-tuples $(s_2, \ldots, s_k)$ possible for that, *or for any other*, choice of s_1. Say this number is N_1. By the generalized Union Rule (a) for pairwise disjoint sets, the total number of k-tuples is

$$\underbrace{N_1 + \cdots + N_1}_{n_1 \text{ times}} = n_1 \cdot N_1.$$

Now repeat this argument to get $n_1 \cdot n_2 \cdot N_2 = \cdots = n_1 \cdot n_2 \cdots n_k$. [A detailed proof would use induction, of course.]

A **permutation** of a finite set is an ordered list of its elements, with each element occurring exactly once. For instance, (b, a, d, c) and (d, c, a, b) are two different permutations of $\{a, b, c, d\}$. We can think of the permutations of $\{a, b, c, d\}$ as words of length $|\{a, b, c, d\}|$ with no repeated letters.

As in Example 2, we will often be interested in ordered lists of *some* of the elements in S. To deal with this more general situation, suppose that $|S| = n$ and consider a positive integer $r \leq n$. An ***r*-permutation** of S is a sequence, i.e., an ordered list, of r distinct elements of S. If $S = \{s_1, \ldots, s_n\}$, then one way to write an r-permutation of S is as $(s_{i_1}, \ldots, s_{i_r})$, where $i_1, \ldots, i_r$ are distinct members of $\{1, \ldots, n\}$. Alternatively, we can view an ***r*-permutation** as a one-to-one mapping σ [lowercase Greek sigma] of the set $\{1, 2, \ldots, r\}$ into S, with $\sigma(j) = s_{i_j}$ for $j = 1, \ldots, r$. An r-permutation σ is completely described by the ordered r-tuple

$(\sigma(1), \sigma(2), \dots, \sigma(r))$. An r-permutation can be obtained by assigning any of the n elements in S to 1, then any of the $n-1$ remaining elements to 2, etc. Hence, by the Product Rule, the number of r-permutations of S is $n(n-1)(n-2)\cdots$, where the product has exactly r factors. The last factor is $n-(r-1) = n-r+1$. We will denote this product by $\boldsymbol{P(n, r)}$. Thus S has exactly

$$P(n,r) = n(n-1)(n-2)\cdots(n-r+1) = \prod_{j=0}^{r-1}(n-j)$$

r-permutations. The n-permutations are simply called **permutations**. Thus an n-element set has exactly $P(n,n) = n!$ permutations.

Note that $P(n,r)\cdot(n-r)! = n!$, so

$$P(n,r) = \frac{n!}{(n-r)!} \quad \text{for} \quad 1 \le r \le n.$$

We could view the equation $P(n,r)\cdot(n-r)! = n!$ in terms of a multistep process. To build an n-permutation, build an r-permutation first and then finish the job by choosing a permutation of the remaining set of $n-r$ elements. The total number of ways to perform the process is the number of n-permutations, $n!$, but by the Product Rule it's also $P(n,r)\cdot(n-r)!$.

The trick we just used, applying the Product Rule when we know the product and want one of the factors, is worth remembering.

It is convenient to agree that $P(n,0) = 1$, the unique 0-permutation being the "empty permutation." Indeed, it's natural as well to define $P(n,r)$ to be 0 if $r < 0$ or $r > n$, since there are no r-permutations in these cases.

EXAMPLE 6

(a) There are $7! = 5040$ permutations of any 7-element set. In particular, there are $7!$ permutations of the alphabet Σ in Example 3(a). The words in Σ^5 that have no letters repeated are the 5-permutations of Σ. There are $P(7,5) = 7\cdot6\cdot5\cdot4\cdot3$ of them. Note that the empty word λ is the empty permutation of Σ.

(b) The 2-letter words without repetitions in the tree of Example 3(b) are 2-permutations of the 4-element set Σ. There are $P(4,2) = 4\cdot3 = 12$ of them. ■

EXAMPLE 7

(a) The number of different orderings of a deck of 52 cards is the number of permutations of the deck, i.e., $52! \approx 8.07\cdot10^{67}$, which is enormous. The act of shuffling a deck of cards is a way of creating various permutations of the deck with various probabilities. Some very interesting theorems about card shuffling were established as recently as +the 1980s. They deal with questions such as How many shuffles are needed to get a reasonably well mixed deck of cards? The methods are mostly theoretical; not even the largest computers can store 52! items and mindlessly verify results case by case.

(b) We calculate the number of ways of selecting 5 cards with replacement from a deck of 52 cards. Thus we are counting ordered 5-tuples consisting of cards from the deck. **With replacement** means that each card is returned to the deck before the next card is drawn. The set of ways of selecting 5 cards with replacement is in one-to-one correspondence with $D \times D \times D \times D \times D = D^5$, where D is the 52-element set of cards. The Product Rule applies, so there are $|D|^5 = 52^5$ ways of selecting 5 cards with replacement.

(c) Now we calculate the number of ways of selecting 5 cards without replacement from a deck of 52 cards. **Without replacement** means that, once a card is drawn, it is not returned to the deck. This time not all ordered 5-tuples in D^5 are allowed. Specifically, ordered 5-tuples with cards repeated are forbidden. But the Product Rule does apply, and the 5 cards can be selected without replacement in $52\cdot51\cdot50\cdot49\cdot48 = P(52,5)$ ways.

So far we have only counted *ordered* 5-tuples of cards, not 5-card subsets. We will return to the subset question in Example 10. ■

The previous examples show that r-permutations can be relevant in counting problems where order matters. Often order is irrelevant, in which case the ability to count *unordered* sets becomes important. We already know that a set S with n elements has 2^n subsets altogether. For $0 \le r \le n$, let $\binom{n}{r}$ be the number of r-element subsets of S. The number $\binom{n}{r}$, called a **binomial coefficient**, is read "n choose r" and is sometimes called the number of **combinations** of n things taken r at a time. Binomial coefficients get their name from the Binomial Theorem, which will be discussed in §5.3.

EXAMPLE 8

We count the strings of 0's and 1's of length 72 that contain exactly 17 1's. Each such string is determined by choosing some set of 17 positions for the 1's among the 72 possible, so the number of such strings is $\binom{72}{17}$. In general, the number of strings of length n with r 1's is $\binom{n}{r}$. To finish off this example, we need to know the value of $\binom{72}{17}$. ■

Theorem For $0 \le r \le n$, we have

$$\binom{n}{r} = \frac{n!}{(n-r)!r!} = \binom{n}{n-r}.$$

Proof Let S be a set with n elements. Consider the process of choosing an r-permutation in two steps: first choose an r-element subset of S in one of the $\binom{n}{r}$ ways, and then arrange it in order, in one of $r!$ ways. Altogether, there are $P(n, r)$ possible outcomes, so the Product Rule applies to give

$$P(n, r) = \binom{n}{r} \cdot r!.$$

Hence

$$\binom{n}{r} = \frac{P(n, r)}{r!} = \frac{n!}{(n-r)! \cdot r!}.$$

Since $n - (n - r) = r$, this formula is the same if we interchange r and $n - r$. Hence $\binom{n}{r} = \binom{n}{n-r}$. Or view choosing r elements as the same as choosing the $n - r$ *other* elements. ■

Did you notice that trick again, the one we said was worth remembering? By the way, now we know that

$$\binom{72}{17} = \frac{72!}{17! \cdot 55!} \approx 1.356 \times 10^{16}.$$

We wouldn't want to try listing all those strings of 0's and 1's, but at least we can count them.

EXAMPLE 9

Consider a graph with no loops that is **complete** in the sense that each pair of distinct vertices has exactly one edge connecting them. If the graph has n vertices, how many edges does it have? Let's assume $n \ge 2$. Each edge determines a 2-element subset of the set V of vertices and, conversely, each 2-element subset of V determines an edge. In other words, the set of edges is in one-to-one correspondence with the set of 2-element subsets of V. Hence there are

$$\binom{n}{2} = \frac{n!}{(n-2)!2!} = \frac{n(n-1)}{2}$$

edges of the graph. ■

An excellent way to illustrate the techniques of this section is to calculate the numbers of various kinds of poker hands. A deck of cards consists of 4 suits called clubs, diamonds, hearts and spades. Each suit consists of 13 cards with values A, 2, 3, 4, 5, 6, 7, 8, 9, 10, J, Q, K. Here A stands for ace, J for jack, Q for queen, and K for king. There are 4 cards of each value, one from each suit. A **poker hand** is a set of 5 cards from a 52-card deck of cards. The order in which the cards are chosen does not matter. A **straight** consists of 5 cards whose values form a consecutive sequence, such as 8, 9, 10, J, Q. The ace A can be at the bottom of a sequence A, 2, 3, 4, 5 or at the top of a sequence 10, J, Q, K, A. Poker hands are classified into pairwise disjoint sets as follows, listed in reverse order of their likelihood.

Royal flush 10, J, Q, K, A all in the same suit.

Straight flush A straight all in the same suit that is not a royal flush.

Four of a kind Four cards in the hand have the same value. For example, four 3's and a 9.

Full house Three cards of one value and two cards of another value. For example, three jacks and two 8's.

Flush Five cards all in the same suit, but not a royal or straight flush.

Straight A straight that is not a royal or straight flush.

Three of a kind Three cards of one value, a fourth card of a second value, and a fifth card of a third value.

Two pairs Two cards of one value, two more cards of a second value, and the remaining card a third value. For example, two queens, two 4's, and a 7.

One pair Two cards of one value, but not classified above. For example, two kings, a jack, a 9, and a 6.

Nothing None of the above.

EXAMPLE 10

(a) There are $\binom{52}{5}$ poker hands. Note that

$$\binom{52}{5} = \frac{52 \cdot 51 \cdot 50 \cdot 49 \cdot 48}{5 \cdot 4 \cdot 3 \cdot 2 \cdot 1} = 52 \cdot 17 \cdot 10 \cdot 49 \cdot 6 = 2{,}598{,}960.$$

(b) How many poker hands are full houses? Let's call a hand consisting of three jacks and two 8's a full house of type $(J, 8)$, with similar notation for other types of full houses. Order matters, since hands of type $(8, J)$ have three 8's and two jacks. Also, types like (J, J) and $(8, 8)$ are impossible. So types of full houses correspond to 2-permutations of the set of possible values of cards; hence there are $13 \cdot 12$ different types of full houses.

Now we count the number of full houses of each type, say type $(J, 8)$. There are $\binom{4}{3} = 4$ ways to choose three jacks from four jacks, and there are then $\binom{4}{2} = 6$ ways to select two 8's from four 8's. Thus there are $4 \cdot 6 = 24$ hands of type $(J, 8)$. This argument works for all $13 \cdot 12$ types of hands, and so there are $13 \cdot 12 \cdot 24 = 3744$ full houses.

(c) How many poker hands are two pairs? Let's say that a hand with two pairs is of type $\{Q, 4\}$ if it consists of two queens and two 4's and some card of a third value. This time we have used set notation because order does not matter: hands of type $\{4, Q\}$ are hands of type $\{Q, 4\}$ and we don't want to count them twice. There are exactly $\binom{13}{2}$ types of hands. For each type, say $\{Q, 4\}$, there are $\binom{4}{2}$ ways of choosing two queens, $\binom{4}{2}$ ways of choosing two 4's and $52 - 8 = 44$ ways of choosing the fifth card. Hence there are

$$\binom{13}{2} \cdot \binom{4}{2} \cdot \binom{4}{2} \cdot 44 = 123{,}552$$

poker hands consisting of two pairs.

(d) How many poker hands are straights? First we count all possible straights even if they are royal or straight flushes. Let's call a straight consisting of the values 8, 9, 10, J, Q a straight of type Q. In general, the type of a straight is the highest value in the straight. Since any of the values 5, 6, 7, 8, 9, 10, J, Q, K, A can be the highest value in a straight, there are 10 types of straights. Given a type of straight, there are 4 choices for each of the 5 values. So there are 4^5 straights of each type and $10 \cdot 4^5 = 10{,}240$ straights altogether. There are 4 royal flushes and 36 straight flushes, so there are 10,200 straights that are not of these exotic varieties.

(e) You are asked to count the remaining kinds of poker hands in Exercise 15, for which all answers are given. ■

Exercises 5.1

1. Calculate

(a) $\binom{8}{3}$ (b) $\binom{8}{0}$ (c) $\binom{8}{5}$

(d) $\binom{52}{50}$ (e) $\binom{52}{52}$ (f) $\binom{52}{1}$

2. (a) Give an example of a counting problem whose answer is $P(26, 10)$.

(b) Give an example of one whose answer is $\binom{26}{10}$.

3. Give the value of

(a) $P(10, 1)$

(b) $P(10, 0)$

(c) $P(10, -2) + P(10, 17)$

(d) $P(100, 40)/P(99, 39)$

(e) $P(1000, 350)/P(999, 350)$

(f) $P(100, 2)$

4. Let $A = \{1, 2, 3, 4, 5, 6, 7, 8, 9, 10\}$ and $B = \{2, 3, 5, 7, 11, 13, 17, 19\}$.

(a) Determine the sizes of the sets $A \cup B$, $A \cap B$, and $A \oplus B$.

(b) How many subsets of A are there?

(c) How many 4-element subsets of A are there?

(d) How many 4-element subsets of A consist of 3 even and 1 odd number?

5. Given 20 different types of inputs to a program, in how many ways can 8 of them be selected if

(a) order does not matter.

(b) order matters.

6. A certain class consists of 12 men and 16 women. How many committees can be chosen from this class consisting of

(a) 7 people?

(b) 3 men and four women?

(c) 7 women or seven men?

7. (a) How many committees consisting of 4 people can be chosen from 9 people?

(b) Redo part (a) if there are two people, Ann and Bob, who will not serve on the same committee.

8. How many committees consisting of 4 men and 4 women can be chosen from a group of 8 men and 6 women?

9. Let $S = \{a, b, c, d\}$ and $T = \{1, 2, 3, 4, 5, 6, 7\}$.

(a) How many one-to-one functions are there from T into S?

(b) How many one-to-one functions are there from S into T?

(c) How many functions are there from S into T?

10. Let $P = \{1, 2, 3, 4, 5, 6, 7, 8, 9\}$ and $Q = \{A, B, C, D, E\}$.

(a) How many 4-element subsets of P are there?

(b) How many permutations, i.e., 5-permutations, of Q are there?

(c) How many license plates are there consisting of 3 letters from Q followed by 2 numbers from P? Repetition is allowed; for example, DAD 88 is allowed.

11. Cards are drawn from a deck of 52 cards with replacement.

(a) In how many ways can ten cards be drawn so that the tenth card is not a repetition?

(b) In how many ways can ten cards be drawn so that the tenth card is a repetition?

12. Let Σ be the alphabet $\{a, b, c, d, e\}$ and let $\Sigma^k = \{w \in \Sigma^* : \text{length}(w) = k\}$. How many elements are there in each of the following sets?

(a) Σ^k, for each $k \in \mathbb{N}$

(b) $\{w \in \Sigma^3 :$ no letter in w is used more than once$\}$

(c) $\{w \in \Sigma^4 :$ the letter c occurs in w exactly once$\}$

(d) $\{w \in \Sigma^4 :$ the letter c occurs in w at least once$\}$

13. Let S be the set of 7-digit numbers $\{n \in \mathbb{N} : 10^6 \leq n < 10^7\}$.

(a) Find $|S|$.

(b) How many members of S are odd?

(c) How many are even?

(d) How many are multiples of 5?

(e) How many have no two digits the same?

(f) How many odd members of S have no two digits the same?

(g) How many even members of S have no two digits the same?

14. Let S be the set of integers between 1 and 10,000.

(a) How many members of S are multiples of 3 and also multiples of 7?

(b) How many members of S are multiples of 3 or of 7 or of both 3 and 7?

(c) How many members of S are not divisible by either 3 or 7?

(d) How many members of S are divisible by 3 or by 7 but not by both?

15. Count the number of poker hands of the following kinds:

(a) four of a kind

(b) flush [but not a straight or royal flush]

(c) three of a kind

(d) one pair

16. (a) In how many ways can the letters a, b, c, d, e, f be arranged so that the letters a and b are next to each other?

(b) In how many ways can the letters a, b, c, d, e, f be arranged so that the letters a and b are not next to each other?

(c) In how many ways can the letters a, b, c, d, e, f be arranged so that the letters a and b are next to each other, but a and c are not.

17. (a) Give the adjacency matrix for a complete graph with n vertices; see Example 9.

(b) Use the matrix in part (a) to count the number of edges of the graph. *Hint:* How many entries in the matrix are equal to 1?

18. Let A and B be the sets in Exercise 4. Calculate the sizes of $\text{FUN}(A, B)$ and $\text{FUN}(B, A)$. Which is bigger? *Hint:* See Example 5.

19. Consider a complete graph with n vertices, $n \geq 4$.

(a) Find the number of paths of length 3.

(b) Find the number of paths of length 3 whose vertex sequences consist of distinct vertices.

(c) Find the number of paths of length 3 consisting of distinct edges.

20. Use a two-step process and the Product Rule to show that

$$P(n, r) \cdot P(n - r, s) = P(n, r + s)$$

for $r + s \leq n$. Describe your process.

5.2 Elementary Probability

In the last section we calculated the number of poker hands and the numbers of various kinds of poker hands. These numbers may not be all that fascinating, but poker players *are* interested in the fraction of poker hands that are flushes, full houses, etc. Why? Because if all poker hands are equally likely, then these fractions represent the likelihood or probability of getting one of these good hands.

The underlying structure in probability is a set, called a **sample space**, consisting of **outcomes** that might result from an experiment, a game of chance, a survey, etc. It is traditional to denote a generic sample space by big omega $\boldsymbol{\Omega}$ [a sort of Greek O, for **o**utcome]. Generic possible outcomes in Ω are denoted by little omegas ω or other Greek letters. Subsets of Ω are called **events**.

A probability on Ω is a function P that assigns a number $P(E)$ to each event $E \subseteq \Omega$. The idea is that $\boldsymbol{P(E)}$, the **probability of** E, should be a measure of the likelihood or the chance that an outcome in E occurs. We want $P(E) = 0$ to mean that there is really no chance the outcome of our experiment will be in E and $P(E) > 0$ to mean that there is some chance that an outcome in E occurs.

There are certainly experiments with infinitely many possible outcomes, but for now we think of Ω as a finite set.

Our first example uses calculations from Example 10 on page 187.

EXAMPLE 1

(a) In poker the set Ω of all possible outcomes is the set of all poker hands. Thus $|\Omega| = 2{,}598{,}960$. For an event $E \subseteq \Omega$, $P(E)$ represents the probability that a dealt poker hand belongs to E. A typical event of interest is

$$H = \{\omega \in \Omega : \omega \text{ is a full house}\}.$$

The event "She was dealt a full house" in everyday language corresponds to the statement "Her poker hand belongs to H." We assume that there is no cheating, so all hands are equally likely, and the probability of being dealt a full house is

$$P(H) = \frac{|H|}{|\Omega|} = \frac{3744}{2{,}598{,}960} \approx 0.00144$$

or about 1 chance in 700. Real poker players see full houses much more often than this, because real poker games allow players to selectively exchange cards after the original deal or allow players to receive more than five cards. The value $P(H)$ above is the probability of being dealt a full house without taking advantage of the fancy rules.

(b) Another important event in poker is

$$F = \{\omega \in \Omega : \omega \text{ is a flush}\}.$$

We might abbreviate this by writing $F =$ "ω is a flush" or simply $F =$ "flush." From Exercise 15(b) on page 189, we have $|F| = 5108$, so

$$P(F) = \frac{|F|}{|\Omega|} = \frac{5108}{2{,}598{,}960} \approx 0.00197$$

or about 1 chance in 500.

(c) The probability of obtaining a full house or flush is $P(H \cup F)$. No poker hand is both a full house and a flush, so the events H and F are disjoint. In this case, it seems reasonable that $P(H \cup F) = P(H) + P(F)$, so that

$$P(H \cup F) = P(H) + P(F) \approx 0.00341.$$

In fact, this addition property of probabilities of disjoint events will be taken as an axiom for probability.

(d) What is $P(\Omega)$? The event Ω is the event that a poker hand is dealt, so $P(\Omega)$ is the probability that if a poker hand is dealt, then a poker hand is dealt. This is a sure thing. We surely want $P(\Omega) > 0$, and it is the standard convention to take $P(\Omega) = 1$. ■

The discussions in parts (c) and (d) of the preceding example suggest that a probability P should at least satisfy the following axioms:

(P_1) $P(\Omega) = 1$,

(P_2) $P(E \cup F) = P(E) + P(F)$ for disjoint events E and F.

After we look at some more examples, we will show that these simple conditions (P_1) and (P_2) by themselves are enough to give us a workable mathematical model of probability. Thus we make the following definition. A **probability on** Ω is a function P that assigns to each event $E \subseteq \Omega$ a number $P(E)$ in $[0, 1]$ and that satisfies axioms (P_1) and (P_2). Remember, for now we think of Ω as a finite set.

Often, as with poker hands, it is reasonable to assume that all the outcomes are **equally likely**. From (P_2) and (P_1) we would then have

$$\sum_{\omega \in \Omega} P(\{\omega\}) = P(\Omega) = 1,$$

so in this case $P(\{\omega\}) = \frac{1}{|\Omega|}$ for each ω in Ω. Using (P_2) again, we see that

$$P(E) = |E| \cdot \frac{1}{|\Omega|} = \frac{|E|}{|\Omega|} \quad \text{for } E \subseteq \Omega \quad \text{when outcomes are equally likely.}$$

This is the formula that we quietly used in our computations in parts (a) and (b) of Example 1.

Warning. This is a handy formula, but don't use it unless you have reason to believe that all the outcomes are equally likely.

Consider again any probability P on a sample space Ω. Since Ω consists of *all* possible outcomes, the probability that nothing happens surely should be zero. That is, $P(\emptyset) = 0$. We did not make this an axiom, because it follows easily from axiom (P_2):

$$P(\emptyset) = P(\emptyset \cup \emptyset) = P(\emptyset) + P(\emptyset), \quad \text{so} \quad P(\emptyset) = 0.$$

If E is an event, then, since E and its complement E^c are disjoint, we have

$$P(E) + P(E^c) = P(E \cup E^c) = P(\Omega) = 1,$$

so $P(E^c) = 1 - P(E)$ for $E \subseteq \Omega$. Thus the probability that an event fails is 1 minus the probability of the event. For example, the probability of *not* getting a full house is $1 - P(H) \approx 1 - 0.00144 = 0.99856$.

We have already established parts (a) and (b) of the next theorem. Note that in our analysis above, and in the proof below, we never use the equation $P(E) = |E|/|\Omega|$ because, in general, outcomes need not be equally likely. Consider, for example, the outcomes of winning or losing a lottery.

Theorem Let P be a probability on a sample space Ω.

(a) $P(\emptyset) = 0$.

(b) $P(E^c) = 1 - P(E)$ for events E.

(c) $P(E \cup F) = P(E) + P(F) - P(E \cap F)$ for events E and F, whether they are disjoint or not.

(d) If $E_1, E_2, \dots, E_m$ are pairwise disjoint events, then

$$P(E_1 \cup E_2 \cup \cdots \cup E_m) = \sum_{k=1}^{m} P(E_k) = P(E_1) + P(E_2) + \cdots + P(E_m).$$

By **pairwise disjoint** we mean here, as in §5.1, that $E_j \cap E_k = \emptyset$ for $j \neq k$; i.e., any two different sets in the sequence $E_1, E_2, \dots, E_m$ have no elements in common.

Proof

(c) The argument is essentially the same as for Union Rule (b) in §5.1. We have $E \cup F = E \cup (F \setminus E)$ and $F = (F \setminus E) \cup (E \cap F)$ with both unions disjoint; if this isn't obvious, draw Venn diagrams. Hence $P(E \cup F) = P(E) + P(F \setminus E)$ and $P(F) = P(F \setminus E) + P(E \cap F)$, so

$$P(E \cup F) = P(E) + P(F \setminus E) = P(E) + [P(F) - P(E \cap F)].$$

(d) This is an easy induction argument. The result is true for $m = 2$ by axiom (P_2). Assume that the identity holds for m sets and that $E_1, E_2, \dots, E_{m+1}$ are pairwise disjoint. Then $E_1 \cup E_2 \cup \cdots \cup E_m$ is disjoint from E_{m+1}, so

$$P(E_1 \cup E_2 \cup \cdots \cup E_{m+1}) = P(E_1 \cup E_2 \cup \cdots \cup E_m) + P(E_{m+1}).$$

Since we are assuming the result holds for m sets, we obtain

$$P(E_1 \cup E_2 \cup \cdots \cup E_{m+1}) = P(E_1) + P(E_2) + \cdots + P(E_m) + P(E_{m+1}). \quad \blacksquare$$

EXAMPLE 2

A number is selected at random from the set $\Omega = \{1, 2, 3, \dots, 100\}$. We calculate the probability that the number selected is divisible by 3 or by 5. That is, we calculate $P(D_3 \cup D_5)$, where

$$D_3 = \{3, 6, 9, \dots, 99\} \quad \text{and} \quad D_5 = \{5, 10, 15, \dots, 100\}.$$

Since the numbers in Ω have equal chances of being selected,

$$P(D_3) = \frac{|D_3|}{|\Omega|} = \frac{33}{100} = 0.33 \quad \text{and} \quad P(D_5) = \frac{|D_5|}{|\Omega|} = \frac{20}{100} = 0.20.$$

We want $P(D_3 \cup D_5)$, but it isn't easy to list $D_3 \cup D_5$ and determine its size. However, $D_3 \cap D_5$ is easily seen to consist of just multiples of 15; i.e., $D_3 \cap D_5 = \{15, 30, 45, 60, 75, 90\}$, so $P(D_3 \cap D_5) = \frac{|D_3 \cap D_5|}{|\Omega|} = \frac{6}{100} = 0.06$. By part (c) of the theorem, we conclude that

$$P(D_3 \cup D_5) = P(D_3) + P(D_5) - P(D_3 \cap D_5) = 0.33 + 0.20 - 0.06 = 0.47. \quad \blacksquare$$

EXAMPLE 3

In Example 4 on page 184 we counted 12 paths from s to f in Figure 2(a) on page 184. Note that each such path has length 4. Suppose a set of 4 edges is selected at random. What is the probability that the 4 edges are the edges of a path from s to f?

Here Ω consists of all the 4-element subsets of the set of all edges. There are 18 edges in all, so $|\Omega| = \binom{18}{4} = 3060$. Only 12 of these sets give paths from s to f, so the answer to the question is $12/3060 \approx 0.00392$. ■

Time for an easier example.

EXAMPLE 4

The usual die [plural is dice] has six sides with 1, 2, 3, 4, 5, and 6 dots on them, respectively. Figure 1 shows two dice. When a die is tossed, one of the six numbers appears at the top. If all six numbers are equally likely, we say the die is **fair**. We will assume that our dice are fair. Let $\Omega = \{1, 2, 3, 4, 5, 6\}$ be the set of possible outcomes. Since the die is fair, $P(k) = \frac{1}{6}$ for each k in Ω; here we are writing $P(k)$ in place of $P(\{k\})$. If E is the event "k is even," i.e., if $E = \{2, 4, 6\}$, then $P(E) = \frac{1}{6} + \frac{1}{6} + \frac{1}{6} = \frac{1}{2}$. If F is the event "4 or 5," i.e., if $F = \{4, 5\}$, then $P(F) = \frac{1}{3}$. Also

$$P(E \cup F) = P(\{2, 4, 5, 6\}) = \frac{2}{3} \quad \text{and} \quad P(E \cap F) = P(\{4\}) = \frac{1}{6}.$$

Figure 1 ▶

As always, we have $P(E \cup F) = P(E) + P(F) - P(E \cap F)$. ■

The last example was too easy. Here is one that is just right.

EXAMPLE 5

We now consider tossing two fair dice, one black and one red, as in Figure 1. Usually the outcome of interest is the sum of the two values appearing at the tops of the dice; for example, the sum is 9 for the dice in Figure 1. For the first time in this section it is not absolutely clear what the sample space Ω should be. In fact, the choice is up to us, but some choices may be easier to work with than others. Since the outcomes of interest are the sums, it is tempting to set Ω equal to the 11-element set consisting of 2, 3, 4, 5, 6, 7, 8, 9, 10, 11, and 12. The trouble is that these outcomes are not equally likely. While we could use their probabilities [determined below], it is easier and more informative to let Ω consist of all ordered pairs of values on the two dice:

$$\Omega = \{(k, l) : 1 \le k \le 6,\ 1 \le l \le 6\}.$$

Here the first entry, k, is the value on the black die and the second entry, l, is the value on the red die. The outcome shown in Figure 1 corresponds to the pair (5,4). It seems reasonable, and is justified in §9.1, that these 36 outcomes are equally likely. They are listed in Figure 2(a). The events of special interest are the ones "sum is k" for $k = 2, 3, 4, \ldots, 12$. For example,

$$P(\text{sum is } 9) = P(\{(3, 6), (4, 5), (5, 4), (6, 3)\}) = \frac{4}{36} = \frac{1}{9}.$$

This and the other values of $P(\text{sum is } k)$ are given in Figure 2(b). Note that the outcomes $(4, 5)$ and $(5, 4)$ are really different. On the other hand, as we can see from Figure 2(a), there is only one way to get two 5's, namely $(5, 5)$. Thus

$$P(\text{sum is } 10) = P(\{(4, 6), (5, 5), (6, 4)\}) = \frac{3}{36} = \frac{1}{12}.$$

Figure 2 ▶

(1, 1)	(1, 2)	(1, 3)	(1, 4)	(1, 5)	(1, 6)
(2, 1)	(2, 2)	(2, 3)	(2, 4)	(2, 5)	(2, 6)
(3, 1)	(3, 2)	(3, 3)	(3, 4)	(3, 5)	(3, 6)
(4, 1)	(4, 2)	(4, 3)	(4, 4)	(4, 5)	(4, 6)
(5, 1)	(5, 2)	(5, 3)	(5, 4)	(5, 5)	(5, 6)
(6, 1)	(6, 2)	(6, 3)	(6, 4)	(6, 5)	(6, 6)

(a)

Sum	P (Sum)	Sum
2	1/36	12
3	2/36 = 1/18	11
4	3/36 = 1/12	10
5	4/36 = 1/9	9
6	5/36	8
7	6/36 = 1/6	

(b)

(a) What is the probability that the sum of the values on the dice is greater than 7? This event consists of the ordered pairs below the dashed line in Figure 2(a). There are 15 such pairs, so the answer is $\frac{15}{36} = \frac{5}{12}$. We can also use the table in Figure 2(b) to get

$$P(\text{sum} > 7) = \sum_{k=8}^{12} P(\text{sum is } k) = \frac{5}{36} + \frac{4}{36} + \frac{3}{36} + \frac{2}{36} + \frac{1}{36} = \frac{15}{36}.$$

We have just illustrated the fact that either Figure 2(a) or 2(b) can be used to handle questions only involving sums. However, as we next show, we cannot use Figure 2(b) to solve all probabilistic questions about our two dice.

(b) What is the probability that the number on the black die divides the number on the red die? That is, what is $P(E)$, where $E = \{(k, l) : k|l\}$? This time Figure 2(b) is no help, but we can list the elements of E with or without using Figure 2(a):

$$E = \{(1, 1), (1, 2), (1, 3), (1, 4), (1, 5), (1, 6), (2, 2), (2, 4), (2, 6), (3, 3), (3, 6), (4, 4), (5, 5), (6, 6)\}.$$

Since $|E| = 14$, we have $P(E) = \frac{14}{36}$. ■

EXAMPLE 6

(a) A tossed coin is said to be **fair** if the probability of heads is $\frac{1}{2}$. If H signifies heads and T signifies tails, then $P(H) = P(T) = \frac{1}{2}$. We want to toss the coin several times, say n times. The set of possible outcomes corresponds to the set Ω of all n-tuples of H's and T's. For example, (T, H, H, T, H) corresponds to $n = 5$ tosses, where the first and fourth tosses are tails and the other three are heads. As in Example 5, it's reasonable to assume that all 2^n outcomes in Ω are equally likely.

For $r = 0, 1, \ldots, n$, we calculate $P(\text{exactly } r \text{ of the tosses are heads})$. The number of n-tuples with r heads is exactly the number of strings of 0's and 1's having length n in which exactly r entries are 1's. This is $\binom{n}{r}$, as shown in Example 8 on page 186. Hence

$$P(\text{exactly } r \text{ of the tosses are heads}) = \binom{n}{r} \cdot \frac{1}{2^n}.$$

(b) If we tossed a fair coin 10 times, we would be surprised if we got 8 or more heads or if we got 8 or more tails. Should we be? We'll decide after we find out how likely this event is. Thus we calculate

$$P(\text{number of heads is } \le 2 \text{ or } \ge 8)$$

$$= \frac{1}{2^{10}}\left[\binom{10}{0} + \binom{10}{1} + \binom{10}{2} + \binom{10}{8} + \binom{10}{9} + \binom{10}{10}\right]$$

$$= \frac{1}{1024}[1 + 10 + 45 + 45 + 10 + 1] = \frac{112}{1024} \approx 0.109.$$

Therefore, there is more than a 10 percent chance of getting 8 or more heads or 8 or more tails; we conclude that it is only mildly surprising when this happens.

Our use of the words "mildly" and "surprising" here was clearly subjective. However, in many statistical analyses, events with probabilities less than 0.05 are regarded as surprising or unexpected, while events with probabilities greater than 0.05 are not.

The probability $P(\text{number of heads is } \le 2 \text{ or } \ge 8)$ can also be calculated by first calculating the probability of the complementary event:

$$P(\text{number of heads is } \ge 3 \text{ and } \le 7) = \sum_{r=3}^{7} P(\text{number of heads is } r)$$

$$= \frac{1}{2^{10}}\left[\binom{10}{3} + \binom{10}{4} + \binom{10}{5} + \binom{10}{6} + \binom{10}{7}\right]$$

$$= \frac{1}{1024}[120 + 210 + 252 + 210 + 120] = \frac{912}{1024} \approx 0.891.$$

(c) The computation in part (b) would get out of hand if we tossed the coin 100 times and wanted to calculate, say,

$$P(\text{number of heads is } \le 35 \text{ or } \ge 65).$$

It turns out that such probabilities can be approximated by using the "Gaussian" or "normal" distribution that is studied in statistics and probability courses. This distribution is intimately connected with the famous bell curve. We will have more to say about all of this in Chapter 9. Incidentally, using the Gaussian distribution, we have

$$P(\text{number of heads is } \le 35 \text{ or } \ge 65) \approx 0.004,$$

so it really would be surprising if we obtained more than 64 or fewer than 36 heads in 100 tosses of a fair coin. ■

Sample spaces are often infinite. In this case the axioms need to be modified. First we give some examples.

EXAMPLE 7

(a) Consider the experiment of tossing a fair coin until a head is obtained. Our sample space will be $\Omega = \mathbb{P} = \{1, 2, 3, \dots\}$, where outcome k corresponds to k tosses, with only the last toss being a head. From Example 6(a) we see that

$$P(1) = \text{probability of head on first toss } = \frac{1}{2},$$

$$P(2) = \text{probability of a tail, then a head } = \frac{1}{2^2},$$

$$P(3) = \text{probability of 2 tails, then a head } = \frac{1}{2^3},$$

etc. In general, $P(k) = 1/2^k$. For finite $E \subseteq \Omega$, we define $P(E) = \sum_{k \in E} \frac{1}{2^k}$.

For example, the probability of getting a head by tossing the coin fewer than 6 times is

$$P(\{1, 2, 3, 4, 5\}) = \frac{1}{2} + \frac{1}{4} + \frac{1}{8} + \frac{1}{16} + \frac{1}{32} = \frac{31}{32} \approx 0.969.$$

We haven't verified that P satisfies the axioms for a probability, but it does. For example, if you've ever summed an infinite series you know that

$$P(\Omega) = \sum_{k \in \mathbb{P}} P(k) = \sum_{k=1}^{\infty} \frac{1}{2^k} = 1.$$

If you haven't, this is just shorthand for the statement

$$\sum_{k=1}^{n} \frac{1}{2^k} \text{ is very close to 1 for very large } n.$$

More generally, the "sum" $\sum_{k \in E} \frac{1}{2^k}$ makes sense for every $E \subseteq \Omega$, and we define $P(E)$ to be this sum. For example, the probability that an odd number of tosses are needed is

$$P(\{1, 3, 5, 7, 9, 11, 13, \ldots\}) = \sum_{k \text{ odd}} \frac{1}{2^k},$$

which turns out to be $\frac{2}{3}$.

(b) People talk about random numbers in the interval $[0, 1) = \Omega$. Computers claim that they can produce random numbers in $[0, 1)$. What do they mean? If we ignore the fact that people and computers *really* only work with finite sets, we have to admit that Ω has an infinite number of outcomes that we want to be equally likely. Then $P(\omega)$ must be 0 for all ω in $[0, 1)$, and the definition $P(E) = \sum_{\omega \in E} P(\omega)$ for $E \subseteq \Omega$ leads to nonsense.

Not all is lost, but we cannot base our probability P on outcomes ω alone. We'd like to have $P([0, \frac{1}{2})) = \frac{1}{2}$, $P([\frac{3}{4}, 1)) = \frac{1}{4}$, etc. It turns out that there is a probability P defined on *some* subsets of $[0, 1)$, called events, such that $P([a, b)) = b - a$ whenever $[a, b) \subseteq [0, 1)$. This P satisfies axioms (P_1) and (P_2) for a probability, and more:

$$P\left(\bigcup_{k=1}^{\infty} E_k\right) = \sum_{k=1}^{\infty} P(E_k)$$

for pairwise disjoint sequences of events $E_1, E_2, \ldots$ in $[0, 1)$. This is the probability P that underlies the concept of a random number in the interval $[0, 1)$.

(c) There are many useful probabilities P on $\Omega = [0, 1)$. The probability in part (b) is the one where $P([a, b)) = b - a$ for $[a, b) \subseteq [0, 1)$. There are also many probabilities on $\mathbb{P}$ in addition to the one in part (a). ■

Here is the modified definition for a probability P on an infinite sample space Ω. As hinted in Example 7(b), in general, only certain subsets of Ω are regarded as events. We retain axiom (P_1), that $P(\Omega) = 1$, and axiom (P_2) is strengthened to

$$(P_2') \quad P\left(\bigcup_{k=1}^{\infty} E_k\right) = \sum_{k=1}^{\infty} P(E_k) \quad \text{for pairwise disjoint sequences of events in } \Omega.$$

In Chapter 9, where we continue the discussion of probability, we will generally restrict our attention to "discrete" probabilities, i.e., to probabilities P for which the set of values can be listed as a sequence, which will certainly be the case if the sample space Ω itself can be listed as a sequence. All the examples of this section, except for parts (b) and (c) of Example 7, are discrete probabilities.

Exercises 5.2

Whenever choices or selections are "at random," the possible outcomes are assumed to be equally likely.

1. An integer in $\{1, 2, 3, \ldots, 25\}$ is selected at random. Find the probability that the number is

(a) divisible by 3 (b) divisible by 5

(c) a prime

2. A letter of the alphabet is selected at random. What is the probability that it is a vowel [$a, e, i, o,$ or u]?

3. A 4-letter word is selected at random from Σ^4, where $\Sigma = \{a, b, c, d, e\}$.

(a) What is the probability that the letters in the word are distinct?

(b) What is the probability that there are no vowels in the word?

(c) What is the probability that the word begins with a vowel?

4. A 5-letter word is selected at random from Σ^5, where $\Sigma = \{a, b, c\}$. Repeat Exercise 3.

5. An urn contains 3 red and 4 black balls. [Ball and urn probability problems have been around for ages, it seems.] A set of 3 balls is removed at random from the urn without replacement. Give the probabilities that the 3 balls are

(a) all red (b) all black

(c) 1 red and 2 black (d) 2 red and 1 black

(e) Sum the answers to parts (a)–(d).

6. An urn has 3 red and 2 black balls. Two balls are removed at random without replacement. What is the probability that the 2 balls are

(a) both red? (b) both black?

(c) different colors?

7. Suppose that an experiment leads to events A, B, and C with the following probabilities: $P(A) = 0.5$, $P(B) = 0.8$, $P(A \cap B) = 0.4$. Find

(a) $P(B^c)$ (b) $P(A \cup B)$ (c) $P(A^c \cup B^c)$

8. Suppose that an experiment leads to events A and B with the following probabilities: $P(A) = 0.6$ and $P(B) = 0.7$. Show that $P(A \cap B) \geq 0.3$.

9. A 5-card poker hand is dealt. Find the probability of getting

(a) four of a kind (b) three of a kind

(c) a [non-exotic] straight (d) two pairs

(e) one pair

Hint: Use Example 10 on page 187 and the answers to Exercise 15 on page 189.

10. A poker hand is dealt.

(a) What is the probability of getting a hand better than one pair? "Better" here means any other special hand listed in §5.1.

(b) What is the probability of getting a pair of jacks or better? The only pairs that are "better" than a pair of jacks are a pair of queens, a pair of kings, and a pair of aces.

11. A black die and a red die are tossed as in Example 5. What is the probability that

(a) the sum of the values is even?

(b) the number on the red die is bigger than the number on the black die?

(c) the number on the red die is twice the number on the black die?

12. Two dice are tossed as in Exercise 11. What is the probability that

(a) the maximum of the numbers on the dice is 4?

(b) the minimum of the numbers on the dice is 4?

(c) the product of the numbers on the dice is 4?

13. Let P be a probability on a sample space Ω. For events E_1, E_2, and E_3, show that

$$\begin{aligned} P(E_1 \cup E_2 \cup E_3) &= P(E_1) + P(E_2) + P(E_3) \\ &\quad - P(E_1 \cap E_2) - P(E_1 \cap E_3) \\ &\quad - P(E_2 \cap E_3) + P(E_1 \cap E_2 \cap E_3). \end{aligned}$$

14. Let P be a probability on a sample space Ω. Show that if E and F are events and $E \subseteq F$, then $P(E) \leq P(F)$.

15. A fair coin is tossed 6 times. Find the probabilities of getting

(a) no heads

(b) 1 head

(c) 2 heads

(d) 3 heads

(e) more than 3 heads

16. A fair coin is tossed until a head is obtained. What is the probability that the coin was tossed at least 4 times?

17. A fair coin is tossed n times. Show that the probability of getting an even number of heads is $\frac{1}{2}$.

18. The probability of my winning the first game of backgammon is 0.5, of my winning the second game is 0.4, and of my winning both games is 0.3. What is the probability that I will lose both games?

19. A set of 4 numbers is selected at random from $S = \{1, 2, 3, 4, 5, 6, 7, 8\}$ without replacement. What is the probability that

(a) exactly 2 of them are even?

(b) none of them is even?

(c) exactly 1 of them is even?

(d) exactly 3 of them are even?

(e) all of them are even?

20. (a) A student answers a 3-question true-false test at random. What is the probability that she will get at least two-thirds of the questions correct?

(b) Repeat part (a) for a 6-question test.

(c) Repeat part (a) for a 9-question test.

21. A computer program selects an integer in the set $\{k : 1 \leq k \leq 1{,}000{,}000\}$ at random and prints the result. This process is repeated 1 million times. What is the probability that the value $k = 1$ appears in the printout at least once? *Hints:*

(a) A number is selected at random from $\{1, 2, 3\}$ three times. What's the probability that 1 was selected at least once? [First find the probability that 1 is not selected.]

(b) A number is selected at random. What's the probability that 1 was selected at least once?

(c) A number is selected at random n times from the set $\{1, 2, \ldots, n\}$. What's the probability that 1 was selected at least once?

(d) Set $n = 1{,}000{,}000$.

22. The 26 letters A, B, ..., Z are arranged in a random order. [Equivalently, the letters are selected sequentially at random without replacement.]

(a) What is the probability that A comes before B in the random order?

(b) What is the probability that A comes before Z in the random order?

(c) What is the probability that A comes just before B in the random order?

5.3 Inclusion-Exclusion and Binomial Methods

This section contains extensions of the counting methods we introduced in §5.1. The Inclusion-Exclusion Principle generalizes Union Rule (b) to count unions of more than two sets. The Binomial Theorem, one of the basic facts of algebra, is closely related to counting. Our third topic, putting objects in boxes, shows yet another application for binomial coefficients.

It is often easy to count elements in an intersection of sets, where the key connective is "and." On the other hand, a direct count of the elements in a union of sets is often difficult. The Inclusion-Exclusion Principle will tell us the size of a union in terms of the sizes of various intersections.

Let $A_1, A_2, \ldots, A_n$ be finite sets. For $n = 2$, Union Rule (b) on page 182 states that

$$|A_1 \cup A_2| = |A_1| + |A_2| - |A_1 \cap A_2|.$$

For $n = 3$, the Inclusion-Exclusion Principle below will assert that

$$\begin{aligned}|A_1 \cup A_2 \cup A_3| = |A_1| &+ |A_2| + |A_3| \\ &-(|A_1 \cap A_2| + |A_1 \cap A_3| + |A_2 \cap A_3|) \\ &+|A_1 \cap A_2 \cap A_3|,\end{aligned}$$

and for $n = 4$ it will say that

$$\begin{aligned}&|A_1 \cup A_2 \cup A_3 \cup A_4| = |A_1| + |A_2| + |A_3| + |A_4| \\ &- (|A_1 \cap A_2| + |A_1 \cap A_3| + |A_1 \cap A_4| + |A_2 \cap A_3| + |A_2 \cap A_4| + |A_3 \cap A_4|) \\ &+ (|A_1 \cap A_2 \cap A_3| + |A_1 \cap A_2 \cap A_4| + |A_1 \cap A_3 \cap A_4| + |A_2 \cap A_3 \cap A_4|) \\ &- |A_1 \cap A_2 \cap A_3 \cap A_4|.\end{aligned}$$

Here is a statement of the general principle in words. A version with symbols is offered in Exercise 16.

Inclusion-Exclusion Principle To calculate the size of $A_1 \cup A_2 \cup \cdots \cup A_n$, calculate the sizes of all possible intersections of sets from $\{A_1, A_2, \ldots, A_n\}$, add the results obtained by intersecting an odd number of the sets, and then subtract the results obtained by intersecting an even number of the sets.

In terms of the phrase "inclusion-exclusion," include or add the sizes of the sets, then exclude or subtract the sizes of all intersections of two sets, then include or add the sizes of all intersections of three sets, etc.

EXAMPLE 1

(a) We count the number of integers in $S = \{1, 2, 3, \ldots, 2000\}$ that are divisible by 9 or 11. Let $D_k = \{n \in S : n \text{ is divisible by } k\}$ for each k in $\mathbb{P}$. We seek $|D_9 \cup D_{11}|$. Our first task is to find the sizes of the sets D_9 and D_{11}. This problem is just like Example 1 on page 181. We have $|D_k| = \lfloor 2000/k \rfloor$ in general, so $|D_9| = 222$, $|D_{11}| = 181$, and $|D_9 \cap D_{11}| = |D_{99}| = 20$. Thus, by Union Rule (b), i.e., the Inclusion-Exclusion Principle for $n = 2$, we find

$$|D_9 \cup D_{11}| = |D_9| + |D_{11}| - |D_9 \cap D_{11}| = 222 + 181 - 20 = 383.$$

(b) Only the arithmetic gets messier if we want the number of integers in S that are divisible by 9, 11, or 13, say. Here we will use the Inclusion-Exclusion Principle for $n = 3$:

$$\begin{aligned}|D_9 \cup D_{11} \cup D_{13}| = |D_9| &+ |D_{11}| + |D_{13}| - |D_9 \cap D_{11}| - |D_9 \cap D_{13}| \\ &-|D_{11} \cap D_{13}| + |D_9 \cap D_{11} \cap D_{13}|.\end{aligned}$$

Some of these numbers were obtained in part (a). In the same way we obtain

$$|D_{13}| = 153, \qquad |D_9 \cap D_{13}| = |D_{117}| = 17,$$
$$|D_{11} \cap D_{13}| = |D_{143}| = 13, \quad |D_9 \cap D_{11} \cap D_{13}| = |D_{1287}| = 1,$$

and hence

$$|D_9 \cup D_{11} \cup D_{13}| = 222 + 181 + 153 - 20 - 17 - 13 + 1 = 507.$$

(c) Let's tackle one more problem like this and count the number of integers in S that are divisible by 9, 11, 13, or 15. We will need to count intersections involving D_{15}, and we'll need to be careful. For example, $D_9 \cap D_{15} = D_{45}$ [not D_{135}] because an integer n is divisible by both 9 and 15 if and only if it is divisible by $\text{lcm}(9, 15) = 45$. Taking such care, we find

$$|D_{15}| = 133,$$
$$|D_9 \cap D_{15}| = |D_{45}| = 44, \qquad |D_{11} \cap D_{15}| = |D_{165}| = 12,$$
$$|D_{13} \cap D_{15}| = |D_{195}| = 10, \qquad |D_9 \cap D_{11} \cap D_{15}| = |D_{495}| = 4,$$
$$|D_9 \cap D_{13} \cap D_{15}| = |D_{585}| = 3, \qquad |D_{11} \cap D_{13} \cap D_{15}| = |D_{2145}| = 0,$$
$$|D_9 \cap D_{11} \cap D_{13} \cap D_{15}| = |D_{6435}| = 0.$$

Applying the Inclusion-Exclusion Principle for $n = 4$ to the sets D_9, D_{11}, D_{13}, and D_{15}, we obtain

$$|D_9 \cup D_{11} \cup D_{13} \cup D_{15}| = 222 + 181 + 153 + 133$$
$$- (20 + 17 + 44 + 13 + 12 + 10) + (1 + 4 + 3 + 0) - 0 = 581. \quad ■$$

The Inclusion-Exclusion Principle is ideally suited to situations in which

(a) we just want the size of $A_1 \cup \cdots \cup A_n$, not a listing of its elements,

and

(b) multiple intersections are fairly easy to count.

Example 1 illustrated several such problems.

If we want an actual list of the members of $A_1 \cup \cdots \cup A_n$, then we can use an iterative algorithm to list A_1 first, then $A_2 \setminus A_1$, then $A_3 \setminus (A_1 \cup A_2)$, and so on, or we can list A_1 and then recursively list $(A_2 \setminus A_1) \cup \cdots \cup (A_n \setminus A_1)$. Look again at Example 1 to see the sorts of calculations that either of these methods would require to produce a list of the elements of $D_9 \cup D_{11} \cup D_{13} \cup D_{15}$. One would certainly want machine help and a friendly data structure. If we just want a *count*, though, and not a list, the Inclusion-Exclusion Principle makes the job fairly painless.

The Inclusion-Exclusion Principle is also the tool to use if we want to find the size of an intersection of sets whose complements are relatively easy to work with.

EXAMPLE 2

Select a number at random from $T = \{1000, 1001, \ldots, 9999\}$. We find the probability that the number has at least one digit that is 0, at least one that is 1, and at least one that is 2. For example, 1072 and 2101 are two such numbers. It is easier to count numbers that exclude certain digits, and so we deal with complements. That is, for $k = 0$, 1, and 2, we let

$$A_k = \{n \in T : n \text{ has no digit equal to } k\}.$$

Then each A_k^c consists of those n in T that have at least one digit equal to k, so $A_0^c \cap A_1^c \cap A_2^c$ consists of those n in T that have at least one 0, one 1, and one 2 among their digits. This is exactly the set whose size we are after.

Since $A_0^c \cap A_1^c \cap A_2^c = (A_0 \cup A_1 \cup A_2)^c$ by De Morgan's law, we will first calculate $|A_0 \cup A_1 \cup A_2|$ using the Inclusion–Exclusion Principle. By the Product Rule, we have $|A_1| = 8 \cdot 9 \cdot 9 \cdot 9$, since there are 8 choices for the first digit, which cannot be 0 or 1, and 9 choices for each of the other digits, which cannot be 1.

Similar computations yield

$|A_0| = 9 \cdot 9 \cdot 9 \cdot 9 = 6561,$ $\qquad |A_1| = |A_2| = 8 \cdot 9 \cdot 9 \cdot 9 = 5832,$

$|A_0 \cap A_1| = |A_0 \cap A_2| = 8 \cdot 8 \cdot 8 \cdot 8 = 4096,$

$|A_1 \cap A_2| = 7 \cdot 8 \cdot 8 \cdot 8 = 3584,$

$|A_0 \cap A_1 \cap A_2| = 7 \cdot 7 \cdot 7 \cdot 7 = 2401.$

By the Inclusion-Exclusion Principle,

$$|A_0 \cup A_1 \cup A_2| = 6561 + 5832 + 5832 - (4096 + 4096 + 3584) + 2401 = 8850,$$

so

$$|(A_0 \cup A_1 \cup A_2)^c| = |T| - |A_0 \cup A_1 \cup A_2| = 9000 - 8850 = 150.$$

There are 150 integers in T whose digits include at least one 0, 1, and 2. Hence the probability of this event is $\frac{150}{|T|} = \frac{150}{9000} = \frac{1}{60}$. ■

EXAMPLE 3

If the digits $1, 2, \ldots, 9$ are listed in random order, what is the probability that no digit is in its natural place? Here 1 is naturally first, 2 second, etc. It's easier to find the probability of the complementary event, so we count the permutations of $\{1, \ldots, 9\}$ in which at least one digit *is* in its natural place. Let A_k be the set of permutations in which the kth digit is k. We want $|A_1 \cup \cdots \cup A_9|$.

Now $|A_1| = 8!$ [put 1 first and arrange the other 8 digits arbitrarily] and indeed $|A_i| = 8!$ for each i. Next $|A_1 \cap A_2| = 7!$ [put 1 and 2 in correct position and then distribute the 7 others] and similarly $|A_i \cap A_j| = 7!$ whenever $i < j$. In the same way we see that $|A_i \cap A_j \cap A_k| = 6!$ whenever $i < j < k$, and so on. There are 9 sets A_i, $\binom{9}{2}$ sets $A_i \cap A_j$ with $i < j$, $\binom{9}{3}$ sets $A_i \cap A_j \cap A_k$, etc. Hence the Inclusion-Exclusion Principle gives us

$$|A_1 \cup \cdots \cup A_9| = 9 \cdot 8! - \binom{9}{2} \cdot 7! + \binom{9}{3} \cdot 6! - \cdots - \binom{9}{8} \cdot 1! + \binom{9}{9} \cdot 0!.$$

Note that the last term $\binom{9}{9} \cdot 0! = 1$ counts the "identity" permutation, i.e., the permutation for which none of the digits is moved. Because $\binom{9}{m} \cdot (9 - m)! = \frac{9!}{m!}$ for $m = 0, \ldots, 9$, we have

$$|A_1 \cup \cdots \cup A_9| = 9! \left(1 - \frac{1}{2!} + \frac{1}{3!} - \cdots - \frac{1}{8!} + \frac{1}{9!}\right).$$

Since there are $9!$ permutations of $\{1, \ldots, 9\}$, the probability of $A_1 \cup \cdots \cup A_9$ is

$$1 - \frac{1}{2!} + \frac{1}{3!} - \cdots - \frac{1}{8!} + \frac{1}{9!}$$

and the probability of $(A_1 \cup \cdots \cup A_9)^c$ is

$$\frac{1}{2!} - \frac{1}{3!} + \cdots + \frac{1}{8!} - \frac{1}{9!}.$$

You may have recognized this number as $1 - \frac{1}{1!} + \frac{1}{2!} - \frac{1}{3!} + \cdots + \frac{1}{8!} - \frac{1}{9!}$, which is the first part of an infinite series for e^{-1}, where $e \approx 2.71828$. Thus our answer is approximately $1/e \approx 0.368$. ■

An Explanation of the Inclusion-Exclusion Principle. The main barrier to proving the general principle is the notation [cf. Exercise 16]. The principle can be proved by induction on n. We show how the result for $n = 2$ leads to the result for $n = 3$. Using the $n = 2$ case, we have

$$|A \cup B \cup C| = |A \cup B| + |C| - |(A \cup B) \cap C| \qquad \textbf{(1)}$$

and

$$|A \cup B| = |A| + |B| - |A \cap B|. \tag{2}$$

Applying the distributive law for unions and intersections [rule 3b in Table 1 on page 25], we also obtain

$$\begin{aligned} |(A \cup B) \cap C| &= |(A \cap C) \cup (B \cap C)| \\ &= |A \cap C| + |B \cap C| - |A \cap B \cap C|. \end{aligned} \tag{3}$$

Substitution of (2) and (3) into (1) yields

$$\begin{aligned} |A \cup B \cup C| = |A| + |B| - |A \cap B| + |C| - |A \cap C| \\ - |B \cap C| + |A \cap B \cap C|, \end{aligned}$$

which is the principle for $n = 3$. ■

The next theorem is probably familiar from algebra. It has many applications, and because $a+b$ is a binomial it explains why we called $\binom{n}{r}$ a "binomial coefficient."

Binomial Theorem For real numbers a and b and for $n \in \mathbb{N}$, we have

$$(a+b)^n = \sum_{r=0}^{n} \binom{n}{r} a^r b^{n-r}.$$

Proof The theorem can be proved by induction, using the recurrence relation

$$\binom{n+1}{r} = \binom{n}{r-1} + \binom{n}{r} \qquad \text{for} \qquad 1 \le r \le n;$$

see Exercise 10. This relation can in turn be proved by algebraic manipulation, but let us give a set-theoretic explanation instead, in the spirit of counting.

There are $\binom{n+1}{r}$ r-element subsets of $\{1, 2, \ldots, n, n+1\}$. We separate them into two classes. There are $\binom{n}{r}$ subsets that contain only members of $\{1, 2, \ldots, n\}$. Each remaining subset consists of the number $n+1$ and some $r-1$ members of $\{1, 2, \ldots, n\}$. Since there are $\binom{n}{r-1}$ ways to choose the elements that aren't $n+1$, there are $\binom{n}{r-1}$ subsets of this type. Hence there are exactly $\binom{n}{r} + \binom{n}{r-1}$ r-element subsets of $\{1, 2, \ldots, n, n+1\}$, so that

$$\binom{n}{r} + \binom{n}{r-1} = \binom{n+1}{r},$$

as claimed.

The *real* reason, though, that $\binom{n}{r}$ is the coefficient of $a^r b^{n-r}$ in $(a+b)^n$ is that it counts the terms with r a's in them that arise when we multiply out $(a+b)^n$. Look at

$$(a+b)^2 = (a+b)(a+b) = aa + (ab + ba) + bb \quad \text{and}$$

$$\begin{aligned} (a+b)^3 &= (a+b)(a+b)(a+b) \\ &= aaa + (aab + aba + baa) + (abb + bab + bba) + bbb. \end{aligned}$$

There are $3 = \binom{3}{2}$ terms with two a's in $(a+b)^3$, corresponding to the choices

$$(\underline{a}+b)(\underline{a}+b)(a+\underline{b}), \quad (\underline{a}+b)(a+\underline{b})(\underline{a}+b), \quad \text{and} \quad (a+\underline{b})(\underline{a}+b)(\underline{a}+b).$$

In the general case of $(a+b)^n$, each choice of r factors $(a+b)$ from which to take the a's produces a term with r a's, and so also with $n-r$ b's. There are $\binom{n}{r}$ possible choices, so there are $\binom{n}{r}$ such terms. ■

EXAMPLE 4

(a) We already know that $(1-1)^m = 0$ for every positive integer m, and we can use this fact to get an interesting formula. For every positive integer m,

$$0 = (-1+1)^m = \sum_{k=0}^{m} \binom{m}{k} \cdot (-1)^k \cdot (+1)^{m-k}$$

$$= 1 - \binom{m}{1} + \binom{m}{2} - \cdots + (-1)^{m-1}\binom{m}{m-1} + (-1)^m.$$

If m is odd, this equation is not too hard to see, because then the terms $\binom{m}{k}$ and $\binom{m}{m-k}$ have opposite signs and cancel each other out; but if m is even, the equation is not so obvious in general without the binomial theorem. For example, with $m = 5$ we get

$$0 = 1 - 5 + 10 - 10 + 5 - 1,$$

with lots of cancellation, but with $m = 6$ it's

$$0 = 1 - 6 + 15 - 20 + 15 - 6 + 1.$$

(b) The formula in part (a) gives another way to verify the Inclusion-Exclusion Principle. Our claim is that

$$|A_1 \cup A_2 \cup \cdots \cup A_n| = \sum_i |A_i| - \sum_{i<j} |A_i \cap A_j| + \sum_{i<j<k} |A_i \cap A_j \cap A_k|$$

$$+ \cdots + (-1)^n |A_1 \cap \cdots \cap A_n|.$$

Consider an element in the union. For some m it is in exactly m of the sets, say in $A_1, A_2, \ldots, A_m$. In the elaborate formula the element gets counted once for each A_i it's in, then minus once for each two of $A_1, A_2, \ldots, A_m$ it's in, then once for each three of them it's in, etc. Altogether, the count for this element is

$$\binom{m}{1} - \binom{m}{2} + \cdots + (-1)^m \binom{m}{m-1} + (-1)^{m+1},$$

which is 1 by part (a). Thus every element of $A_1 \cup A_2 \cup \cdots \cup A_n$ gets counted exactly once overall in the formula. ■

Sometimes the Binomial Theorem is useful for computing the value of a sum formula, such as $\sum_{r=0}^{n} (-1)^r \binom{n}{r}$ in Example 4, $\sum_{r=0}^{n} \binom{n}{r}$ [Exercise 11], or $\sum_{r=0}^{n} \binom{n}{r} 2^r$ [Exercise 12]. At other times we can use it to compute specific coefficients, often the first few, in powers that we don't want to write out in detail.

EXAMPLE 5

The first few terms of $(2+x)^{100}$ are

$$2^{100} + \binom{100}{1} 2^{99} x + \binom{100}{2} 2^{98} x^2 + \cdots = 2^{100} + 100 \cdot 2^{99} x + 4950 \cdot 2^{98} x^2 + \cdots.$$

If x is very near 0, it is tempting to say that, since $x^3, x^4, \ldots$ are tiny, we can pretty much ignore the terms after x^2. Beware! The coefficients in the middle, such as $\binom{100}{50}$, can get pretty big, and collectively the discarded terms can add up to a lot. In our case, if $x = 0.01$, we make about a 1.4 percent error in $(2.01)^{100}$ by dropping the higher powers, but with $x = 0.1$ our "approximate" answer would be less than 5 percent of the correct value. ■

Our next counting principle can be applied in a variety of settings. We offer it in a form that is easy to remember.

Placing Objects in Boxes There are $\binom{n+k-1}{k-1} = \binom{n+k-1}{n}$ ways to place n identical objects into k distinguishable boxes.

Proof The proof is both elegant and illuminating; we illustrate it for the case $n = 5$ and $k = 4$. We let five 0's represent the objects, and then we add three 1's to serve as dividers among the four boxes. We claim that there is a one-to-one correspondence between the strings consisting of five 0's and three 1's and the ways to place the five 0's into four boxes. Specifically, a given string corresponds to the placement of the 0's before the first 1 into the first box, the 0's between the first and second 1 into the second box, the 0's between the second and third 1 into the third box, and the 0's after the third 1 into the fourth box. For example,

$$0\,0\,1\,1\,0\,0\,0\,1 \longrightarrow 0\,0\,\big|\,\big|\,0\,0\,0\,\big| \longrightarrow \boxed{0\,0}\boxed{}\boxed{0\,0\,0}\boxed{}.$$

box 1	box 2	box 3	box 4
0 0		0 0 0	

In this instance, boxes 2 and 4 are empty, because there are no 0's between the first and second dividers and there are no 0's after the last divider. More examples:

$$1\,0\,0\,1\,0\,0\,1\,0 \longrightarrow \boxed{}\boxed{0\,0}\boxed{0\,0}\boxed{0};$$

$$0\,0\,0\,1\,1\,1\,0\,0 \longrightarrow \boxed{0\,0\,0}\boxed{}\boxed{}\boxed{0\,0}.$$

There are $\binom{8}{3}$ strings having five 0's and three 1's; just choose where to put the 1's. Since $\binom{8}{3} = \binom{5+4-1}{4-1}$, this establishes the result for $n = 5$ and $k = 4$.

In the general case, we consider strings of n 0's and $k - 1$ 1's. The 0's correspond to objects and the 1's to dividers. There are $\binom{n+k-1}{k-1}$ such strings and, as above, there is a one-to-one correspondence between these strings and the placing of n 0's into k boxes. The alternate formula $\binom{n+k-1}{n}$ comes from choosing where to put the 0's in the strings. ■

EXAMPLE 6

(a) In how many ways can 10 identical marbles be placed into 5 distinguishable bags? Here $n = 10$, $k = 5$, and the answer is

$$\binom{10+5-1}{5-1} = \binom{14}{4} = 1001.$$

(b) In how many ways can 15 indistinguishable fish be placed into 5 different ponds so that each pond contains at least 1 fish? Start by putting a fish in each pond. Now we can put the remaining 10 fish into the ponds in $\binom{14}{4}$ ways, just as in part (a), so the answer is again $\binom{14}{4}$.

(c) In how many ways can 10 tigers be placed into 5 *in*distinguishable bags? This problem is much harder. You should be aware that counting problems can get difficult quickly. Here one would like to apply part (a) somehow. However, even with the methods of the next section, there is no natural way to do so. We abandon this problem. Any solution we are aware of involves the consideration of several cases. ■

Sometimes problems need to be reformulated before it is clear how our principles apply.

EXAMPLE 7

How many numbers in $\{1, 2, 3, \ldots, 100{,}000\}$ have the property that the sum of their digits is 7? We can ignore the very last number, 100,000, and we can assume that all the numbers have 5 digits, by placing zeros in front if necessary. So, for example, we replace 1 by 00001 and 73 by 00073. Our question is now this: How many strings of 5 digits have the property that the sum of their digits is 7? We can associate each such string with the placement of 7 balls in 5 boxes; for example,

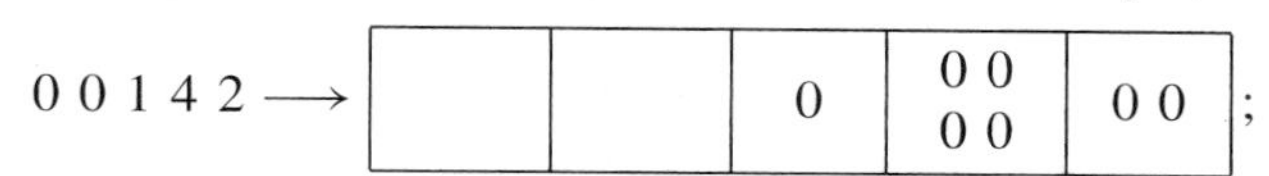

3 0 1 2 1 ⟶

0 0 0		0	0 0	0

There are $\binom{11}{4} = 330$ such placements, so there are 330 numbers with the desired property. ■

We can also take objects out of their boxes. Assume that each of k boxes contains an unlimited supply of objects labeled according to which box they are in. Applying the principle in reverse, we see that there are $\binom{n+k-1}{k-1}$ ways to remove n objects from the k boxes. In other words,

Fact The number of ways to select a set of n objects of k distinguishable types, allowing repetitions, is

$$\binom{n+k-1}{k-1} = \binom{n+k-1}{n}.$$

EXAMPLE 8 In how many ways can 10 coins be selected from an unlimited supply of pennies, nickels, dimes, and quarters? This problem is tailor-made for the principle just stated. Let $n = 10$ [for the 10 coins] and $k = 4$ [for the 4 types of coins]. Then the answer is

$$\binom{10+4-1}{4-1} = \binom{13}{3} = 286.$$

From another viewpoint, this problem is equivalent to counting ordered 4-tuples of nonnegative integers whose sum is 10. For example, (5, 3, 0, 2) corresponds to the selection of 5 pennies, 3 nickels, and 2 quarters. Counting these ordered 4-tuples is equivalent to counting the ways of placing 10 indistinguishable objects into 4 boxes, which can be done in $\binom{13}{3}$ ways. ■

Exercises 5.3

1. Among 200 people, 150 either swim or jog or both. If 85 swim and 60 swim and jog, how many jog?

2. Let $S = \{100, 101, 102, \ldots, 999\}$ so that $|S| = 900$.

   (a) How many numbers in S have at least one digit that is a 3 or a 7? Examples: 300, 707, 736, 103, 997.

   (b) How many numbers in S have at least one digit that is a 3 *and* at least one digit that is a 7? Examples: 736 and 377, but not 300, 707, 103, 997.

3. An integer is selected at random from $\{1, 2, 3, \ldots, 1000\}$. What is the probability that the integer is

   (a) divisible by 7? (b) divisible by 11?

   (c) not divisible by 7 or 11?

   (d) divisible by 7 or 11 but *not* both?

4. An investor has 7 \$1000 bills to distribute among 3 mutual funds.

   (a) In how many ways can she invest her money?

   (b) In how many ways can she invest her money if each fund must get at least \$1000?

5. An integer is selected at random from $\{1, 2, 3, \ldots, 1000\}$. What is the probability that it is divisible by at least one of the integers 4, 5, or 6?

6. Let $\Sigma = \{a, e, f, g, i\}$. Find the probability that a randomly selected 5-letter word will use all three vowels a, e, i. *Hint:* First find the number of words in Σ^5 that do not use all three vowels.

7. Twelve identical letters are to be placed into 4 mailboxes.

   (a) In how many ways can this be done?

   (b) How many ways are possible if each mailbox must receive at least 2 letters?

8. How many different mixes of candy are possible if a mix consists of 10 pieces of candy and 4 different kinds of candy are available in unlimited quantities?

9. Use the binomial theorem to expand the following:

(a) $(x + 2y)^4$ (b) $(x - y)^6$

(c) $(3x + 1)^4$ (d) $(x + 2)^5$

10. (a) Complete the proof of the Binomial Theorem by providing the induction proof. Use the recurrence relation provided, which is also given in part (b).

(b) Prove $\binom{n+1}{r} = \binom{n}{r-1} + \binom{n}{r}$ for $1 \le r \le n$ algebraically.

11. Prove that $2^n = \sum_{r=0}^{n} \binom{n}{r}$

(a) by setting $a = b = 1$ in the binomial theorem.

(b) by counting subsets of an n-element set.

(c) by induction using the recurrence relation in Exercise 10(b).

12. Prove that $\sum_{r=0}^{n} \binom{n}{r} 2^r = 3^n$ for $n \in \mathbb{P}$.

13. (a) Verify that $\sum_{k=m}^{n} \binom{k}{m} = \binom{n+1}{m+1}$ for some small values of m and n, such as $m = 3$ and $n = 5$.

(b) Prove the identity by induction on n for $n \ge m$.

(c) Prove the identity by counting the $(m+1)$-element subsets of the $(n+1)$-element set $\{1, 2, \ldots, n+1\}$. *Hint:* How many of these sets are there whose largest element is $k + 1$? What can k be?

14. In the proof of the "Placing Objects in Boxes" principle, we set up a one-to-one correspondence between strings of five 0's and three 1's and placements of five 0's in four boxes.

(a) Give the placements that correspond to the following strings:

1 0 1 0 1 0 0 0, 0 1 0 0 1 0 0 1,

1 0 0 0 0 0 1 1, 1 1 1 0 0 0 0 0.

(b) Give the strings that correspond to the following placements:

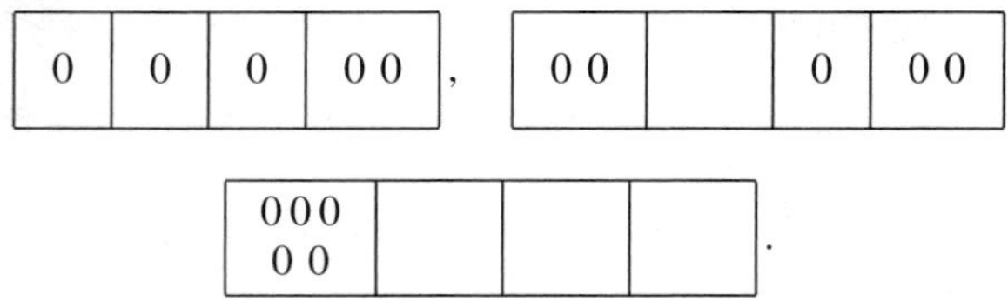

15. (a) How many ways are there to put 14 objects in 3 boxes with at least 8 objects in one box?

(b) How many ways are there to put 14 objects in 3 boxes with no more than 7 objects in any box?

(c) For how many numbers between 0 and 999 is the sum of their digits equal to 20? *Hint:* Each digit must be at least 2; part (b) applies.

16. Consider finite sets $\{A_1, A_2, \ldots, A_n\}$. Let $\mathcal{P}_+(n)$ be the set of nonempty subsets I of $\{1, 2, \ldots, n\}$. Show that the Inclusion-Exclusion Principle says that

$$\left|\bigcup_{i=1}^{n} A_i\right| = \sum_{I \in \mathcal{P}_+(n)} (-1)^{|I|+1} \cdot \left|\bigcap_{i \in I} A_i\right|.$$

17. (a) For how many integers between 1000 and 9999 is the sum of the digits exactly 9? Examples: 1431, 5121, 9000, 4320.

(b) How many of the integers counted in part (a) have all nonzero digits?

18. Six passengers on a small airplane are randomly assigned to the six seats on the plane. On the return trip they are again randomly assigned seats.

(a) What is the probability that every passenger has the same seat on both trips?

(b) What is the probability that exactly five passengers have the same seats on both trips?

(c) What is the probability that at least one passenger has the same seat on both trips?

19. How many sets of l integers from $\{1, \ldots, p\}$ contain no two consecutive integers? Here $1 \le l \le p$. *Suggestion:* Convert this question to one about sequences of 0's and 1's with l 1's, and then view it in terms of putting 0's in boxes.

20. This exercise is motivated by an article by David M. Berman, "Lottery drawings often have consecutive numbers," *College Math. J.* **25** (1994), 45–47. In the Louisiana lottery, 6 numbers are selected at random from a set of 44. People have been surprised that the winning selection often contains two consecutive numbers. Should they be?

(a) If l lucky numbers are selected randomly from $\{1, 2, \ldots, p\}$, show that the probability that no two are consecutive is $\dfrac{\binom{p-l+1}{l}}{\binom{p}{l}}$.

(b) Show that if $l = 6$ and $p = 44$, then the probability that the set of winning lucky numbers includes two consecutive numbers is approximately 0.538.

5.4 Counting and Partitions

Recall that a partition of a set S is a collection of pairwise disjoint nonempty subsets whose union is the set S itself. This section will focus on problems that are associated with partitions.

EXAMPLE 1

In how many ways can we form three committees of sizes 3, 5, and 7 from a group of 20 people if nobody can be on more than one committee? We may think of forming a fourth, leftover committee whose 5 members are the people *not* chosen for the real committees, though the 5-member real committee and the leftover committee are of course not interchangeable. The four committees together form a partition of the group of people, so our question can be rephrased: how many ways are there to partition a set of 20 elements into subsets of sizes 3, 5, 7, and 5, respectively? We will discover that the answer is

$$\frac{20!}{3! \cdot 5! \cdot 7! \cdot 5!} \approx 5.587 \cdot 10^9.$$

There are several ways to arrive at this answer. One way is to use the Product Rule. Visualize a selection process that consists of first choosing the 3-person committee, then the 5-person committee from the 17 people remaining, then the 7-person committee from the 12 still remaining. We can choose the leftover committee too, but of course by that time there is only one way to do so. According to the Product Rule, the total number of choices available is

$$\binom{20}{3} \cdot \binom{17}{5} \cdot \binom{12}{7} \cdot \binom{5}{5}.$$

If we rewrite this product in terms of factorials, we get

$$\frac{20!}{3! \cdot 17!} \cdot \frac{17!}{5! \cdot 12!} \cdot \frac{12!}{7! \cdot 5!} \cdot \frac{5!}{5! \cdot 0!},$$

which upon cancellation and noting that $0! = 1$ gives $\dfrac{20!}{3! \cdot 5! \cdot 7! \cdot 5!}$.

■

There is a general method here. An **ordered partition** of a set S is a sequence $(A_1, A_2, \ldots, A_k)$ whose members $A_1, A_2, \ldots, A_k$ form a partition of S. The A_i's themselves are not assumed to be ordered internally, but the order in which they appear in the list does matter. The committees in Example 1, including the leftover committee, form one ordered partition of the 20-person group.

EXAMPLE 2 Let $S = \{1, 2, 3, 4, 5, 6, 7, 8\}$. Here are some ordered partitions of S:

$$(\{1, 3, 5\}, \{2, 4, 6, 7, 8\}), \quad (\{2, 4, 6, 7, 8\}, \{1, 3, 5\}),$$
$$(\{3, 6\}, \{2, 5, 8\}, \{1, 4, 7\}), \quad (\{1\}, \{2, 4, 6, 8\}, \{3, 5, 7\}),$$
$$(\{1, 6\}, \{2, 5, 8\}, \{3, 4\}, \{7\}), \quad \text{and} \quad (\{6, 1\}, \{2, 5, 8\}, \{4, 3\}, \{7\}).$$

The last two are the same, since $\{1, 6\} = \{6, 1\}$ and $\{3, 4\} = \{4, 3\}$, but all of the others are distinct. ■

The argument in Example 1 generalizes to establish the following principle.

Counting Ordered Partitions If $n = n_1 + n_2 + \cdots + n_k$ and if a set has n elements, then there are

$$\frac{n!}{n_1! n_2! \cdots n_k!}$$

ordered partitions $(A_1, A_2, \ldots, A_k)$ of the set with $|A_i| = n_i$ for $i = 1, 2, \ldots, k$.

As illustrated in Example 1, the number given by the formula can also be written as

$$\binom{n}{n_1} \cdot \binom{n - n_1}{n_2} \cdots \binom{n - n_1 - \cdots - n_{k-1}}{n_k}.$$

The last factor is $\binom{n_k}{n_k}$, so it can be omitted. One possible advantage of the factored form is that, when n is very large and $n!$ is enormous, the binomial coefficient factors are much smaller and can perhaps be calculated more readily.

EXAMPLE 3

(a) A **bridge deal** is an ordered partition of 52 cards into four sets, one for each player, with 13 cards each. The order *within* each set is irrelevant, but which player gets which hand is crucial. There are

$$\frac{52!}{13!\,13!\,13!\,13!} = \frac{52!}{(13!)^4} \approx 5.3645 \cdot 10^{28}$$

bridge deals.

(b) Given a bridge deal, what is the probability that each hand of 13 cards contains one ace? First we deal out the aces, one to each player; this can be done in $4! = 24$ ways. Just as in part (a), the remaining cards can be partitioned in $48!/(12!)^4$ ways. So $24 \cdot 48!/(12!)^4$ of the bridge deals yield one ace in each hand. The probability of such a deal is

$$24\frac{48!}{(12!)^4} \cdot \frac{(13!)^4}{52!} = \frac{24 \cdot 13^4}{49 \cdot 50 \cdot 51 \cdot 52} \approx 0.1055.$$

(c) A single **bridge hand** consists of 13 cards drawn from a 52-card deck. There are

$$\binom{52}{13} \approx 6.394 \cdot 10^{11}$$

bridge hands. We say that a bridge hand has distribution n_1—n_2—n_3—n_4, where $n_1 \geq n_2 \geq n_3 \geq n_4$ and $n_1 + n_2 + n_3 + n_4 = 13$ if there are n_1 cards of some suit, n_2 cards of a second suit, n_3 cards of a third suit, and n_4 cards of the remaining suit. To illustrate, we count the number of bridge hands having 4—3—3—3 distribution. There are

$$\binom{13}{4}\binom{13}{3}\binom{13}{3}\binom{13}{3}$$

ways to choose 4 clubs and 3 of each of the other suits. We get the same result if we replace clubs by one of the other suits. So we conclude that there are

$$4\binom{13}{4}\binom{13}{3}^3 \approx 6.6906 \cdot 10^{10}$$

bridge hands having 4—3—3—3 distribution. The probability that a bridge hand will have 4—3—3—3 distribution is

$$\frac{6.6906}{63.94} \approx 0.1046.$$ ■

A fraction of the form $\frac{20!}{3!\cdot5!\cdot7!\cdot5!}$ is sometimes called a **multinomial coefficient** and written, in a style like that of a binomial coefficient, as $\binom{20}{3\ 5\ 7\ 5}$. Here's the reason. Imagine multiplying out $(a+b+c+d)^{20}$ and collecting terms that have the same powers of a, b, c and d. The coefficient of $a^3b^5c^7d^5$ would be the number of ways to get 3 a's, 5 b's, 7 c's and 5 d's by choosing one letter from each of the 20 factors $(a+b+c+d)$. Choosing the letters in this way is like selecting which 3 of the factors go in committee A, which 5 in committee B, etc.; so there are $\frac{20!}{3!\cdot5!\cdot7!\cdot5!}$ terms $a^3b^5c^7d^5$ in the multiplied-out product, and the coefficient of $a^3b^5c^7d^5$ is $\binom{20}{3\ 5\ 7\ 5}$.

The theorem to describe this situation has the Binomial Theorem as a special case.

Multinomial Theorem For real numbers $a_1, \ldots, a_k$ and for $n \in \mathbb{N}$, we have

$$(a_1 + \cdots + a_k)^n = \sum_{n_1+\cdots+n_k=n} \binom{n}{n_1\ \ldots\ n_k} a_1^{n_1} \cdots a_k^{n_k}.$$

Here $\binom{n}{n_1\ \ldots\ n_k}$ stands for $\frac{n!}{n_1!\cdots n_k!}$, and the sum is over all possible ways to write n as $n_1 + \cdots + n_k$.

EXAMPLE 4

An alternative approach to the committee problem of Example 1 consists of lining up the people in order and assigning the first 3 to the first committee, the next 5 to the second committee, and so on. There are 20! ways to line up the people, but different lineups can result in the same committees. Rearranging the first 3 people in line or the next 5 or the next 7 or the last 5 will not change the committee assignments, but any other rearrangements will cause a change. Therefore, for any given committee assignment there are $3! \cdot 5! \cdot 7! \cdot 5!$ ways to rearrange the people without changing the assignment. We conclude that the number of different committee assignments is the number of possible lineups divided by the number of lineups that produce any given assignment; i.e., it's

$$\frac{20!}{3! \cdot 5! \cdot 7! \cdot 5!},$$

just as before.

Still another approach, not very different from the last one, is to consider a two-step process for choosing an ordering, i.e., a permutation, of all 20 people. The first step is to choose the committees, say in one of x ways, and the second step is to arrange the people in order so that the members of the first committee come first, then the members of the second committee, etc. By the Product Rule, we have

$$x \cdot (\text{number of ways to do step 2}) = n!.$$

The number of arrangements possible in step 2 is $3! \cdot 5! \cdot 7! \cdot 5!$ by the Product Rule, so

$$x = \frac{20!}{3! \cdot 5! \cdot 7! \cdot 5!}$$

again. ■

This last approach used the trick we mentioned in §5.1 of applying the Product Rule when we know the product and want one of the factors. Here's another example. To count the number of sheep in a flock, simply count the legs and divide by 4. In function-theoretic terms, let L be the set of legs and S the set of sheep, and let $\psi: L \to S$ be the function that maps each leg to the sheep it belongs to. For each sheep s, the set

$$\psi^{\leftarrow}(s) = \{l \in L : \psi(l) = s\}$$

is the set of legs of s. We are given that $|\psi^{\leftarrow}(s)| = 4$ for every s, so $|L| = 4 \cdot |S|$ and $|S| = |L|/4$.

It's time to formalize this idea.

Counting Lemma If $\psi: A \to B$ maps the finite set A onto B and if all the sets

$$\psi^{\leftarrow}(b) = \{a \in A : \psi(a) = b\}$$

for b in B have the same number of elements, say r, then

$$|B| = \frac{|A|}{r}.$$

Proof The set A is the union of the pairwise disjoint sets $\psi^{\leftarrow}(b)$, so the Union Rule gives

$$|A| = \sum_{b \in B} |\psi^{\leftarrow}(b)| = \sum_{b \in B} r = r \cdot |B|.$$

Therefore, $|B| = |A|/r$.

Alternatively, think of choosing members of A in two stages, by first choosing $b \in B$ in one of $|B|$ ways and then choosing one of the r a's for which $\psi(a) = b$. Again, $|A| = |B| \cdot r$. ■

EXAMPLE 5

This lemma applies to Example 4 if we let B be the set of all 3—5—7—5 committee assignment partitions P and let A be the set of all permutations of the set of 20 people. For each permutation σ of A, let $\psi(\sigma)$ assign the first 3 people to the first committee, the next 5 to the second, etc. Then $\psi: A \to B$, and each set $\psi^{\leftarrow}(P)$ has $3! \cdot 5! \cdot 7! \cdot 5!$ permutations in it. Since $|A| = 20!$, we conclude yet again that $|B| = \frac{20!}{3!5!7!5!}$. ■

When we were putting objects in boxes in §5.3, we regarded the objects as indistinguishable, but we could tell the boxes apart. When we took the objects out of the boxes later, we could think of the objects as somewhat distinguishable, if we remembered which boxes they came from. A mixture of 3 apples, 5 oranges, 7 pears, and 5 turnips is different from a mixture of 2 apples, 6 oranges, 9 pears, and 3 turnips, even if all apples look alike, all oranges look alike, etc. The term **multiset** is sometimes used to describe a set with this sort of partial distinguishability, with notation $\{n_1 \cdot a_1, n_2 \cdot a_2, \ldots, n_k \cdot a_k\}$ to indicate that it contains n_i members of type a_i for $i = 1, \ldots, k$. Our first mixture would be described as

$$\{3 \cdot \text{apple}, 5 \cdot \text{orange}, 7 \cdot \text{pear}, 5 \cdot \text{turnip}\}$$

in this notation.

One can think of a multiset as an ordinary set by putting distinguishing labels on all the members, for instance, $\text{apple}_1, \text{apple}_2, \ldots, \text{apple}_s$. Then ignore the labels when thinking of the set as a multiset.

A **combination** of elements of a multiset is just a subset of its set of elements with the labels ignored. A **permutation** of a multiset is an ordered list of its elements with distinguishing labels ignored, as in the next example.

EXAMPLE 6

This problem begins with a hat containing slips of paper on which letters are written. The slips are drawn from the hat one at a time and placed in a row as they are drawn, and the question is how many different words can arise in this way. Here we regard two words as different if there is at least one position in which they have different letters.

If all the letters in the hat are different, the problem is easy. For example, if the hat contains the 10 letters I, M, P, O, R, T, A, N, C, E, then there are 10! different words that can be drawn, most of which are gibberish and only one of which is IMPORTANCE.

If the hat contains the 8 letters E, E, N, N, N, O, S, S, however, then there are 8! ways to pull them from the hat, but more than one way can lead to the same word. To analyze this, let's put subscripts on the duplicate letters, so that the hat contains E_1, E_2, N_1, N_2, N_3, O, S_1, and S_2. There are 8! permutations of these subscripted letters. If ψ is the function that maps each such permutation to itself but with the subscripts erased, then the image of ψ is the set of distinguishable words using E, E, N, N, N, O, S, S. For example,

$$\psi(N_3\, O\, N_1\, S_2\, E_1\, N_2\, S_1\, E_2) = \text{N O N S E N S E}.$$

How many of the permutations map onto NONSENSE? There are 3! different orders in which N_1, N_2, N_3 can appear, 2! different orders for S_1 and S_2, and 2! different orders for E_1 and E_2. So there are $3! \cdot 2! \cdot 2! = 24$ different permutations that give NONSENSE. That is, the inverse image $\psi^{\leftarrow}(\text{NONSENSE})$ has 24 elements. Similarly, there are 24 permutations that give SNEENONS, or any other word in these letters. The Counting Lemma says that the total number of distinguishable words is

$$\frac{8!}{3! \cdot 2! \cdot 2!} = 1680.$$

Thus, if we draw slips from the hat at random, the chance of getting NONSENSE is 1/1680. Note that $1680 = \binom{8}{3\ 2\ 2\ 1}$; the 1 term corresponds to the 1! order in which the letter O can appear. ■

The argument in this example can be used to prove the following general fact.

Counting Permutations of Multisets Suppose that a set of n objects is partitioned into subsets of k different types, with $n_1, n_2, \ldots, n_k$ members, so that $n = n_1 + n_2 + \cdots + n_k$. Regard two permutations of the set as distinguishable in case their entries in at least one position are of different types. Then there are

$$\frac{n!}{n_1!\,n_2!\cdots n_k!} = \binom{n}{n_1\ n_2\ \ldots\ n_k}$$

distinguishable permutations of the set.

EXAMPLE 7

Let $\Sigma = \{a, b, c\}$. The number of words in Σ^* having length 10 using 4 a's, 3 b's and 3 c's is

$$\frac{10!}{4!\,3!\,3!} = 4200.$$

The number of words using 5 a's, 3 b's, and 2 c's is

$$\frac{10!}{5!\,3!\,2!} = 2520,$$

and the number using 5 a's and 5 b's is

$$\frac{10!}{5!\,5!} = 252.$$

For comparison, note that Σ^{10} has $3^{10} = 59{,}049$ words. ■

EXAMPLE 8

We count the number of ordered partitions of the set $\{1, \ldots, 8\}$ of the form (A, B, C, D), where $|A| = 2$, $|B| = 3$, $|C| = 1$, and $|D| = 2$. This problem can be viewed as a question of counting permutations of an 8-element multiset with objects of 4 types, that is, as a letters-in-a-hat problem. Here's how.

Consider an ordered partition (A, B, C, D) of $\{1, 2, \ldots, 8\}$ into sets with 2, 3, 1, and 2 members, respectively. For instance, take $(\{5, 8\}, \{1, 3, 6\}, \{2\}, \{4, 7\})$. Then 5 and 8 are of type A, while 1, 3, and 6 are of type B, etc. If we list the types of $1, 2, 3, \ldots, 8$ in order, we get the sequence B, C, B, D, A, B, D, A, which completely describes the partition, as long as we know how to decode it. [If, instead of calling the blocks A, B, C, D, we label them with the letters E, N, O, and S, then $(\{5, 8\}, \{1, 3, 6\}, \{2\}, \{4, 7\})$ corresponds to our friend N O N S E N S E.]

This argument shows that the ordered partitions of $\{1, 2, \ldots, 8\}$ into blocks of sizes 2, 3, 1, and 2 correspond one-to-one with the distinguishable permutations of A, A, B, B, B, C, D, D, so the number of ordered partitions of type 2—3—1—2 is $\frac{8!}{2!\cdot 3!\cdot 1!\cdot 2!} = 1680$, which we already knew, of course, from the Counting Ordered Partitions formula. ■

Sometimes problems reduce to counting unordered partitions. In such cases, count ordered partitions first and then divide by suitable numbers to take into account the lack of order.

EXAMPLE 9

(a) In how many ways can 12 students be divided into three groups, with 4 students in each group, so that one group studies English, one studies history and one studies mathematics? Here order matters: if we permuted groups of students, the students would be studying different topics. So we count ordered partitions, of which there are

$$\frac{12!}{4!\,4!\,4!} = \binom{12}{4} \cdot \binom{8}{4} = 495 \cdot 70 = 34{,}650.$$

(b) In how many ways can 12 climbers be divided into three teams, with 4 climbers in each team, but with all teams climbing the same mountain? Now we wish to count unordered partitions, since we regard partitions like (A, B, C) and (B, A, C) as equivalent, because they correspond to the same partition of the 12 climbers into three groups. From part (a), there are 34,650 ordered partitions. If we map each ordered partition (A, B, C) to the unordered partition $\psi((A, B, C)) = \{A, B, C\}$, we find that $\psi^{\leftarrow}(\{A, B, C\})$ has $3! = 6$ elements, namely (A, B, C), (A, C, B), (B, A, C), (B, C, A), (C, A, B), and (C, B, A). So, by the Counting Lemma, there are 34,650/6 = 5775 unordered partitions of the desired type. Hence the answer to our question is 5775. ■

EXAMPLE 10

(a) In how many ways can 19 students be divided into five groups, two groups of 5 and three groups of 3, so that each group studies a different topic? As in Example 9(a), we count ordered partitions, of which there are

$$\frac{19!}{5!\,5!\,3!\,3!\,3!} \approx 3.911 \cdot 10^{10}.$$

(b) In how many ways can the students in part (a) be divided if all five groups are to study the same topic? In part (a) we counted ordered partitions (A, B, C, D, E), where $|A| = |B| = 5$ and $|C| = |D| = |E| = 3$. If A and B are permuted and C, D, E are permuted, we will get the same study groups, but we cannot permute groups of different sizes, such as A and D. To count unordered partitions, we let $\psi((A, B, C, D, E)) = (\{A, B\}, \{C, D, E\})$. Each of the inverse image sets $\psi^{\leftarrow}(\{A, B\}, \{C, D, E\})$ has $2! \cdot 3!$ elements [such as (B, A, C, E, D)], so, by the Counting Lemma, there are

$$\frac{19!}{5!\,5!\,3!\,3!\,3!} \cdot \frac{1}{2!\,3!} \approx 3.26 \cdot 10^{9}$$

unordered partitions of students into study groups. ■

EXAMPLE 11

In how many ways can we split a group of 12 contestants into 4 teams of 3 contestants each? More generally, what if we want to divide $3n$ contestants among n teams of 3? We are asking for unordered partitions. There are

$$\frac{(3n)!}{(3!)\cdots(3!)} = \frac{(3n)!}{6^n}$$

ordered partitions $(A_1, A_2, \ldots, A_n)$ for which each A_i has 3 members. Any permutation of the n sets gives the same unordered partition, and so there are $\frac{(3n)!}{6^n \cdot n!}$ unordered partitions of $3n$ elements into n sets with 3 elements each. Note that this number is

$$\frac{3n(3n-1)(3n-2)}{6n} \cdot \frac{(3n-3)(3n-4)(3n-5)}{6(n-1)} \cdots \frac{6 \cdot 5 \cdot 4}{6 \cdot 2} \cdot \frac{3 \cdot 2 \cdot 1}{6 \cdot 1}$$

$$= \frac{(3n-1)(3n-2)}{2} \cdot \frac{(3n-4)(3n-5)}{2} \cdots \frac{5 \cdot 4}{2} \cdot \frac{2 \cdot 1}{2}$$

$$= \binom{3n-1}{2} \cdot \binom{3n-4}{2} \cdots \binom{5}{2} \cdot \binom{2}{2}.$$

This factorization suggests another way to solve the problem. List the contestants in some order. Pick the first contestant and choose 2 more for her team, in one of $\binom{3n-1}{2}$ ways. Now pick the next unchosen contestant and choose 2 more to go with him, in one of $\binom{3n-4}{2}$ ways. And so on. [Or use induction.]

In case there are just 12 contestants, there are

$$\binom{11}{2} \cdot \binom{8}{2} \cdot \binom{5}{2} \cdot \binom{2}{2} = 15{,}400$$

ways to choose the teams. ■

The principles in this section are applicable in a variety of common situations, but they won't handle every problem that comes up. The thought processes that we demonstrated in the proofs of the principles are as valuable as the principles themselves, since the same sort of analysis can often be used on problems to which the ready-made tools don't apply.

Exercises 5.4

1. From a total of 15 people, 3 committees consisting of 3, 4, and 5 people, respectively, are to be chosen.

(a) How many such sets of committees are possible if no person may serve on more than one committee?

(b) How many such sets of committees are possible if there is no restriction on the number of committees on which a person may serve?

2. Compare.

(a) $\binom{7}{2} \cdot \binom{5}{2}$ and $\frac{7!}{2! \cdot 2! \cdot 3!}$

(b) $\binom{12}{3} \cdot \binom{9}{4}$ and $\frac{12!}{3! \cdot 4! \cdot 5!}$

(c) $\binom{n}{k} \cdot \binom{n-k}{r}$ and $\frac{n!}{k! \cdot r! \cdot (n-k-r)!}$

(d) $\binom{n}{r\ \ s}$ for $r + s = n$ and $\binom{n}{r}$

(e) $\binom{n}{1\ \ldots\ 1}$ and $n!$

3. Three pairwise disjoint teams are to be selected from a group of 13 students, with all three teams competing in the same programming contest. In how many ways can the teams be formed if they are to have

(a) 5, 3, and 2 students? (b) 4, 3, and 3 students?

(c) 3 students each?

4. How many different signals can be created by lining up 9 flags in a vertical column if 3 flags are white, 2 are red, and 4 are blue?

5. Let S be the set of all sequences of 0's, 1's, and 2's of length 10. For example, S contains 0 2 1 1 0 1 2 2 0 1.

(a) How many elements are in S?

(b) How many sequences in S have exactly five 0's and five 1's?

(c) How many sequences in S have exactly three 0's and seven 1's?

(d) How many sequences in S have exactly three 0's?

(e) How many sequences in S have exactly three 0's, four 1's, and three 2's?

(f) How many sequences in S have at least one 0, at least one 1, and at least one 2?

6. Find the number of permutations that can be formed from all the letters of the following words.

(a) FLORIDA (b) CALIFORNIA

(c) MISSISSIPPI (d) OHIO

7. (a) How many 4-digit numbers can be formed using only the digits 3, 4, 5, 6, and 7?

(b) How many of the numbers in part (a) have at least one digit appear more than once?

(c) How many of the numbers in part (a) are even?

(d) How many of the numbers in part (b) are bigger that 5000?

8. Let $\Sigma = \{a, b, c\}$. If $n_1 \geq n_2 \geq n_3$ and $n_1 + n_2 + n_3 = 6$, we call a word in Σ^6 of type n_1—n_2—n_3 if one of the letters appears in the word n_1 times, another letter appears n_2 times, and the other letter appears n_3 times. For example, $a\,c\,c\,a\,b\,c$ is of type 3—2—1 and $c\,a\,c\,c\,c\,a$ is of type 4—2—0. The number of words in Σ^6 of each type is

Type	6—0—0	5—1—0	4—2—0	4—1—1
Number	3	36	90	90

Type	3—3—0	3—2—1	2—2—2
Number	60	360	90

(a) Verify three of the numbers in the table above.

(b) Calculate the sum of the numbers in the table above. What is its significance?

9. In how many ways can $2n$ elements be partitioned into two sets with n elements each?

10. Consider finite sets satisfying $\chi_A + \chi_B = \chi_C + \chi_D$. Show that $|A| + |B| = |C| + |D|$.

11. Two of the 12 contestants in Example 11, Ann and Bob, want to be on the same team.

(a) How many ways are there to choose the teams so that Ann and Bob are together?

(b) If all ways of choosing teams are equally probable, what is the probability that Ann and Bob are chosen for the same team?

12. A basketball tournament has 16 teams. How many ways are there to match up the teams in 8 pairs for the first playoffs?

13. How many equivalence relations are there on $\{0, 1, 2, 3\}$? *Hint:* Count unordered partitions. Why does this solve the problem?

14. A hat contains the letters B, B, B, E, O, O, P, P. What is the probability that when they are drawn out of the hat one after another the letters will spell BEBOPBOP in that order?

15. Find

(a) the coefficient of $a^3b^2c^4$ in $(a + b + c)^9$.

(b) the coefficient of $x^2y^3z^4$ in $(x + y + z)^9$.

(c) the coefficient of $x^2y^3z^2$ in $(1 + x + y + z)^9$.

(d) the coefficient of $x^2y^3z^4$ in $(1 + x + y + z)^9$.

(e) the coefficient of $x^2y^3z^4$ in $(1 + 2x + 3y + 4z)^9$.

Office Hours 5.4

How can I tell what method to use on these counting problems? Some of them seem to be about permutations, and some about objects in boxes. Now we have partitions. There seem to be so many methods, and sometimes none of them look useful.

Right! Even with all the methods we've covered, it's easy to find problems we can't handle, at least not without breaking them down into pieces. The good news is that often we *can* recognize a familiar type.

Objects-in-boxes problems, for instance, are fairly easy to spot. Some people also call them "stars and bars" problems, thinking about the way we turned objects in boxes into strings of 0's and 1's. Once you get to that stage, you're counting possible places to put the 1's, so it's no surprise that the answer is a binomial coefficient. Understanding where the answer comes from will also help you see how to do problems when the formula doesn't quite apply.

Straight permutation problems are also pretty easy, if you think in terms of processes. Choose the first one, then the next one, etc., until you have all you need. "First" and "next" are the tip-off: order matters here. I can probably think of a tricky way to ask a question so that it seems that order matters when it doesn't or doesn't matter when it does, but if you concentrate on a *process* for making your choices, you'll see the right method.

A question that asks how many *sets* there are of such and such a type is likely to involve binomial coefficients, whereas one that asks about how many strings you can build is going to involve permutations, maybe like the ones in the NONSENSE example. Here's where things get trickier, and you'll want to look at the examples in the book for ideas.

As I see it, the authors are really just taking a few very basic principles, like the Union Rule, the Product Rule, and the Counting Lemma, and applying them in various orders to get what they call "methods." You can use their methods if they seem to apply, but you can also go back to basics if you want, at least to turn a problem into one that the methods will work on.

Let me point out one other possible source of confusion. There can be lots of different types of questions with the same setup. For instance, suppose we have a big pile of fish and some boxes. If we put the fish in the boxes, then we can ask how many different combinations of fish in boxes there are—like 5 fish in the first box, 8 in the second, etc. That's a pure objects-in-boxes question. We can also ask how many different ways there are to assign fish to the boxes so that the first box gets 5 fish, the second gets 8, and so on. That's an ordered partition problem, and probably easiest to think of in terms of a process that chooses the fish for the first box, then those for the second box, etc. This problem only makes sense if we can tell the fish apart and "assign" them. Think of committees of fish, so one group of 5 is different from another. Or we could put 5 fish in the first box, 8 in the second, and so on, and give them little tags to wear so that the ones in the first box have A's on them, the ones in the second box have B's on them, etc. Then we could dump all the fish out, line them up and ask how many different words the tags can spell. That kind of problem would be like the NONSENSE example, counting permutations of a multiset. Or how about assigning the fish to the boxes at random and asking what the probability is that the first box has exactly 5 fish? My point is, you can't always tell what the question will be just from the setup.

One other piece of general advice on this material:

Don't just guess.

Be sure you see a reason for any claim you make or method you choose. It's OK to memorize the formulas, but they're worthless if you don't use the right ones for the job. Look hard at the examples in the book, not to imitate them so much as to understand their overall plan of attack. As usual, ask yourself how the answers would change if you changed the examples a little. Your homework exercises will give you some practice, too.

Now give me an example of one you're still having trouble on, and we'll work a problem like it.

5.5 Pigeon-Hole Principle

The usual Pigeon-Hole Principle asserts that, if m objects are placed in k boxes or pigeon-holes and if $m > k$, then some box will receive more than one object. Here is a slight generalization of this obvious fact.

Pigeon-Hole Principle If a finite set S is partitioned into k sets, then at least one of the sets has $|S|/k$ or more elements.

Proof Say the sets in the partition are $A_1, \dots, A_k$. Then the average value of $|A_i|$ is $\frac{1}{k} \cdot (|A_1| + \cdots + |A_k|) = \frac{1}{k} \cdot |S|$, so the largest A_i has at least this many elements. ■

Note that $|S|/k$ need not be an integer. Since the sizes of the sets are integers, we can say that some set in the partition has at least $\lceil |S|/k \rceil$ members. The Pigeon-Hole Principle doesn't say which set is this large, and indeed there could be more than one of them.

We will often apply the principle when the partition is given by a function, in which case it can be stated as follows.

Pigeon-Hole Principle If S and T are finite sets satisfying $|S| > r \cdot |T|$ and if $f: S \to T$, then at least one of the sets $f^{\leftarrow}(t)$ has more than r elements.

Proof Let $\text{Im}(f)$ be the image of f, so that $\text{Im}(f) \subseteq T$, and let $k = |\text{Im}(f)|$. The family $\{f^{\leftarrow}(t) : t \in \text{Im}(f)\}$ partitions S into k sets with $k \le |T|$. By the principle just proved, some set $f^{\leftarrow}(t)$ has at least $|S|/k$ members. Since $|S|/k \ge |S|/|T| > r$ by hypothesis, such a set $f^{\leftarrow}(t)$ has more than r elements. ■

When $r = 1$, this principle tells us that if $f: S \to T$ and $|S| > |T|$, then at least one of the sets $f^{\leftarrow}(t)$ has more than one element. It is remarkable how often this simple observation is helpful in problem solving.

EXAMPLE 1

(a) If three distinct integers are given, then there must be two of them whose sum is even, because either two of the integers are even or else two of them are odd, and in either case their sum must be even.

Here is a tighter argument. Let S be the set of three integers. Since the function MOD 2 maps even integers to 0 and odd integers to 1, we have MOD 2: $S \to \{0, 1\}$ and, by the Pigeon-Hole Principle, either $(\text{MOD } 2)^{\leftarrow}(0)$ or $(\text{MOD } 2)^{\leftarrow}(1)$ has more than one element. That is, either S contains two [or more] even integers or else S contains two [or more] odd integers.

(b) Part (a) shows that, given three different integers, at least two of them must be congruent mod 2. More generally, if $n_1, n_2, \dots, n_{p+1}$ are any $p + 1$ integers, distinct or not, we claim that at least two of them are congruent mod p; i.e., $n_j \equiv n_k \pmod{p}$ for some $j \ne k$. If two of the integers are equal, they are certainly congruent mod p, so we may as well assume that the integers are distinct. To prove our claim, we set $S = \{n_1, \dots, n_{p+1}\}$ and apply the Pigeon-Hole Principle to the function MOD p: $S \to \mathbb{Z}(p)$. Since $|S| = p + 1 > p = |\mathbb{Z}(p)|$, two numbers n_j and n_k in S have the same image in $\mathbb{Z}(p)$. Thus we have $n_j \text{ MOD } p = n_k \text{ MOD } p$; hence $n_j \equiv n_k \pmod{p}$. ■

EXAMPLE 2

We show that if $a_1, a_2, \ldots, a_p$ are any p integers, not necessarily distinct or nonnegative, then some of them add up to a number that is a multiple of p. Consider the $p+1$ numbers

$$0, \ a_1, \ a_1+a_2, \ a_1+a_2+a_3, \ldots, \ a_1+a_2+a_3+\cdots+a_p.$$

By Example 1(b), some two of these are congruent mod p, so the difference between some two of them is a multiple of p. This difference has the form

$$a_s + a_{s+1} + \cdots + a_t,$$

which is a sum of integers from our list.

Note that we have proved more than we claimed; some *consecutive* batch of a_i's adds up to a multiple of p. The result just proved is sharp, in the sense that we can give integers $a_1, a_2, \ldots, a_{p-1}$ for which no nonempty subset has sum that is a multiple of p. For a simple example, let $a_j = 1$ for $j = 1, 2, \ldots, p-1$. ■

EXAMPLE 3

Let A be some fixed 10-element subset of $\{1, 2, 3, \ldots, 50\}$. We show that A possesses two different 5-element subsets, the sums of whose elements are equal. Let $\mathcal{S}$ be the family of 5-element subsets B of A. For each B in $\mathcal{S}$, let $f(B)$ be the sum of the numbers in B. Note that we must have both inequalities $f(B) \geq 1+2+3+4+5 = 15$ and $f(B) \leq 50+49+48+47+46 = 240$ so that $f\colon \mathcal{S} \to T$, where $T = \{15, 16, 17, \ldots, 240\}$. Since $|T| = 226$ and $|\mathcal{S}| = \binom{10}{5} = 252$, the Pigeon-Hole Principle shows that $\mathcal{S}$ contains different sets with the same image under f, i.e., different sets the sums of whose elements are equal. ■

EXAMPLE 4

You probably know that the decimal expansions of rational numbers repeat themselves, but you may never have seen a proof. For example,

$$\frac{29}{54} = 0.537037037037037\cdots;$$

check this *by long division* before proceeding further! The general fact can be seen as a consequence of the Pigeon-Hole Principle, as follows.

We may assume that the given rational number has the form m/n for integers m and n with $0 < m < n$. Let us analyze the steps in the algorithm for long division. When we divide m by n, we obtain $.d_1d_2d_3\cdots$, where

$$\begin{aligned} 10 \cdot m &= n \cdot d_1 + r_1, & 0 \leq r_1 < n, \\ 10 \cdot r_1 &= n \cdot d_2 + r_2, & 0 \leq r_2 < n, \\ 10 \cdot r_2 &= n \cdot d_3 + r_3, & 0 \leq r_3 < n, \end{aligned}$$

etc., so that $10 \cdot r_j = n \cdot d_{j+1} + r_{j+1}$, where $0 \leq r_{j+1} < n$. That is,

$$r_{j+1} = (10 \cdot r_j) \text{ MOD } n \quad \text{and} \quad d_{j+1} = (10 \cdot r_j) \text{ DIV } n.$$

Figure 1 shows the details for our sample division, where $m = 29$ and $n = 54$. The remainders r_j all take their values in $\{0, 1, 2, \ldots, n-1\} = \mathbb{Z}(n)$. By the Pigeon-Hole Principle, after a while the values must repeat. In fact, two of the numbers $r_1, r_2, \ldots, r_{n+1}$ must be equal. Hence there are k and l in $\{1, 2, \ldots, n+1\}$ with $k < l$ and $r_k = r_l$. Let $p = l - k$, so that $r_k = r_{k+p}$. [In our carefully selected example, k can be 1 and p can be 3.] We will show that the sequences of r_i's and d_i's repeat every p terms beginning with $i = k+1$.

First, we show by induction that the remainders repeat:

$$r_j = r_{j+p} \quad \text{for} \quad j \geq k. \tag{$*$}$$

We have $r_k = r_{k+p}$ by the choice of k and p. Assume inductively that $r_j = r_{j+p}$ for some j. Then $r_{j+1} = (10 \cdot r_j) \text{ MOD } n = (10 \cdot r_{j+p}) \text{ MOD } n = r_{j+p+1}$. Thus $(*)$ holds for $j+1$, and we conclude by induction that $(*)$ holds for $j \geq k$.

Now, for $j > k$, $(*)$ implies that

$$d_j = (10 \cdot r_{j-1}) \text{ DIV } n = (10 \cdot r_{j+p-1}) \text{ DIV } n = d_{j+p}.$$

Figure 1 ▶

$.d_1d_2d_3d_4d_5\ldots$	$= .53703\ldots$	
$54\,\overline{)290}$	$10\cdot 29 = 10m$	
270	$54\cdot 5 = nd_1$	
200	$10\cdot 20 = 10r_1$	$r_1 = 20$
162	$54\cdot 3 = nd_2$	
380	$10\cdot 38 = 10r_2$	$r_2 = 38$
378	$54\cdot 7 = nd_3$	
20	$10\cdot 2 = 10r_3$	$r_3 = 2$
0	$54\cdot 0 = nd_4$	
200	$10\cdot 20 = 10r_4$	$r_4 = 20$
162	$54\cdot 3 = nd_5$	
380	$10\cdot 38 = 10r_5$	$r_5 = 38$
378	etc.	

Thus the d_js also repeat in cycles of length p, and we have

$$d_{k+1} = d_{k+p+1}, \quad d_{k+2} = d_{k+p+2}, \quad \ldots, \quad d_{k+p} = d_{k+2p}$$

so that

$$d_{k+1}d_{k+2}\cdots d_{k+p} = d_{k+p+1}d_{k+p+2}\cdots d_{k+2p}.$$

In fact, this whole block $d_{k+1}d_{k+2}\cdots d_{k+p}$ repeats indefinitely. In other words, the decimal expansion of m/n is a repeating expansion. ■

Some applications of the Pigeon-Hole Principle require considerable ingenuity. The method is typically suitable when the goal is just to show that something must exist and when something else is not too hard to count. The idea is always to find one set to map into another.

EXAMPLE 5

Here we show that, if $a_1, a_2, \ldots, a_{n^2+1}$ is a sequence of n^2+1 distinct real numbers, then there is either a subsequence with $n+1$ terms that is increasing or one that is decreasing. This means that there exist subscripts $s(1) < s(2) < \cdots < s(n+1)$ so that either

$$a_{s(1)} < a_{s(2)} < \cdots < a_{s(n+1)}$$

or

$$a_{s(1)} > a_{s(2)} > \cdots > a_{s(n+1)}.$$

For each j in $\{1, 2, \ldots, n^2+1\}$, let $\text{INC}(j)$ be the length of the longest increasing subsequence stopping at a_j and $\text{DEC}(j)$ be the length of the longest decreasing subsequence stopping at a_j. Then define $f(j) = (\text{INC}(j), \text{DEC}(j))$. For example, suppose that $n = 3$ and the original sequence is given by

a_1	a_2	a_3	a_4	a_5	a_6	a_7	a_8	a_9	a_{10}
11	3	15	8	6	12	17	2	7	1.

Here $a_5 = 6$; $\text{INC}(5) = 2$ since a_2, a_5 is the longest increasing subsequence stopping at a_5, and $\text{DEC}(5) = 3$ since a_1, a_4, a_5 and a_3, a_4, a_5 [i.e., 11, 8, 6 and 15, 8, 6] are longest decreasing subsequences stopping at a_5. Similarly, $\text{INC}(6) = 3$ and $\text{DEC}(6) = 2$, so $f(5) = (2, 3)$ and $f(6) = (3, 2)$. Indeed, in this example

$$\begin{array}{lll} f(1) = (1,1) & f(2) = (1,2) & f(3) = (2,1) \\ f(4) = (2,2) & f(5) = (2,3) & f(6) = (3,2) \\ f(7) = (4,1) & f(8) = (1,4) & f(9) = (3,3) \\ f(10) = (1,5). & & \end{array}$$

This particular example has increasing subsequences of length 4, such as a_2, a_5, a_6, a_7 [note $\text{INC}(7) = 4$], and also decreasing subsequences of length 4, such as a_1, a_4, a_5, a_8 [note $\text{DEC}(8) = 4$]. Since $\text{DEC}(10) = 5$, it even has a decreasing subsequence of length 5. In this example f is one-to-one, so it cannot map the 10-element set $\{1, 2, 3, \ldots, 10\}$ into the 9-element set $\{1, 2, 3\} \times \{1, 2, 3\}$. In other words, the one-to-oneness of f alone forces at least one $\text{INC}(j)$ or $\text{DEC}(j)$ to exceed 3, and this in turn forces our sequence to have an increasing or decreasing subsequence of length 4.

To prove the general result, we first prove directly that f must *always* be one-to-one. Consider j, k in $\{1, 2, 3, \ldots, n^2 + 1\}$ with $j < k$. If $a_j < a_k$, then $\text{INC}(j) < \text{INC}(k)$, since a_k could be attached to the longest increasing sequence ending at a_j to get a longer increasing sequence ending at a_k. Similarly, if $a_j > a_k$, then $\text{DEC}(j) < \text{DEC}(k)$. In either case the *ordered pairs* $f(j)$ and $f(k)$ cannot be equal; i.e., $f(j) \neq f(k)$. Since f is one-to-one, the Pigeon-Hole Principle says that f cannot map $\{1, 2, 3, \ldots, n^2 + 1\}$ into $\{1, 2, \ldots, n\} \times \{1, 2, \ldots, n\}$, so there is either a j such that $\text{INC}(j) \geq n + 1$ or one such that $\text{DEC}(j) \geq n + 1$. Hence the original sequence has an increasing or decreasing subsequence with $n + 1$ terms. ■

The next example doesn't exactly apply the Pigeon-Hole Principle, but it is a pigeon-hole problem in spirit.

EXAMPLE 6

Consider nine nonnegative real numbers $a_1, a_2, a_3, \ldots, a_9$ with sum 90.

(a) We show that there must be three of the numbers having sum at least 30. This is easy because

$$90 = (a_1 + a_2 + a_3) + (a_4 + a_5 + a_6) + (a_7 + a_8 + a_9),$$

so at least one of the sums in parentheses must be at least 30.

(b) We show that there must be four of the numbers having sum at least 40. There are several ways to do this, but none of them is quite as simple as the method in part(a). Our first approach is to observe that the sum of all the numbers in Figure 2 is 360, since each row sums to 90. Hence one of the nine columns must have sum at least 360/9 = 40.

Figure 2 ▶

a_1	a_2	a_3	a_4	a_5	a_6	a_7	a_8	a_9
a_2	a_3	a_4	a_5	a_6	a_7	a_8	a_9	a_1
a_3	a_4	a_5	a_6	a_7	a_8	a_9	a_1	a_2
a_4	a_5	a_6	a_7	a_8	a_9	a_1	a_2	a_3

Our second approach is to use part (a) to select three of the numbers having sum $s \geq 30$. One of the remaining six numbers must have value at least $\frac{1}{6}$ of their sum, which is $90 - s$. Adding this number to the selected three gives four numbers with sum at least

$$s + \frac{1}{6}(90 - s) = 15 + \frac{5}{6}s \geq 15 + \frac{5}{6} \cdot 30 = 40.$$

Our third approach is to note that we may as well assume $a_1 \geq a_2 \geq \cdots \geq a_9$. Then it is clear that $a_1 + a_2 + a_3 + a_4$ is the largest sum using four of the numbers and our task is relatively concrete: to show that $a_1 + a_2 + a_3 + a_4 \geq 40$. Moreover, this rephrasing suggests showing that

$$a_1 + a_2 + \cdots + a_n \geq 10n \qquad (*)$$

for $1 \leq n \leq 9$. We accomplish this by doing a finite induction, i.e., by noting that $(*)$ holds for $n = 1$ [why?] and showing that if $(*)$ holds for n with $1 \leq n < 9$, then $(*)$ holds for $n + 1$. We will adapt the method of our second approach. Assume that $(*)$ holds for n, and let $s = a_1 + a_2 + \cdots + a_n$. Since a_{n+1} is the largest of the remaining $9 - n$ numbers, we have $a_{n+1} \geq (90 - s)/(9 - n)$. Hence

$$\begin{aligned} a_1 + a_2 + \cdots + a_n + a_{n+1} &= s + a_{n+1} \geq s + \frac{90 - s}{9 - n} \\ &= s + \frac{90}{9-n} - \frac{s}{9-n} = s\left(1 - \frac{1}{9-n}\right) + \frac{90}{9-n} \\ &\geq 10n\left(1 - \frac{1}{9-n}\right) + \frac{90}{9-n} \qquad \text{[inductive assumption]} \\ &= 10n + \frac{90 - 10n}{9 - n} = 10n + 10 = 10(n+1). \end{aligned}$$

This finite induction argument shows that $(*)$ holds for $1 \leq n \leq 9$.

For this particular problem, the first and last approaches are far superior because they generalize in an obvious way without further tricks. ■

The Pigeon-Hole Principle can be generalized to allow the sets A_i to overlap.

Generalized Pigeon-Hole Principle Let $A_1, \ldots, A_k$ be subsets of the finite set S such that each element of S is in at least t of the sets A_i. Then the average number of elements in the A_i's is at least $t \cdot |S|/k$.

Proof We use the interesting and powerful technique of counting a set of pairs in two ways. Let P be the set of all pairs (s, A_i) with $s \in A_i$. We can count P by counting the pairs for each s in S and then adding up the numbers. We get

$$\begin{aligned} |P| &= \sum_{s \in S} (\text{number of pairs } (s, A_i) \text{ with } s \in A_i) \\ &= \sum_{s \in S} (\text{number of } A_i\text{'s with } s \in A_i) \\ &\geq \sum_{s \in S} t \qquad \text{[by assumption]} \\ &= t \cdot |S|. \end{aligned}$$

We can also count P by counting the pairs for each A_i and adding the numbers. Thus

$$\begin{aligned} |P| &= \sum_{i=1}^{k} (\text{number of pairs } (s, A_i) \text{ with } s \in A_i) \\ &= \sum_{i=1}^{k} (\text{number of } s\text{'s with } s \in A_i) = \sum_{i=1}^{k} |A_i|. \end{aligned}$$

Putting these two results together, we get

$$\sum_{i=1}^{k} |A_i| \geq t \cdot |S|,$$

so the average of the $|A_i|$'s, namely

$$\frac{1}{k} \cdot \sum_{i=1}^{k} |A_i|,$$

is at least $t \cdot |S|/k$.

Our proof also shows that if each s is in exactly t of the sets A_i, then the average of the $|A_i|$'s is exactly $t \cdot |S|/k$. The ordinary Pigeon-Hole Principle is the special case $t = 1$. ■

EXAMPLE 7 A roulette wheel is divided into 36 sectors with numbers 1, 2, 3, ..., 36 in some unspecified order. [We are omitting sections with 0 and 00 that are included in Las Vegas and give the house the edge in gambling.] The average value of a sector is 18.5. If this isn't obvious, observe that $1 + 2 + \cdots + 36 = 666$ by Example 2(b) on page 139 and that $666/36 = 18.5$.

(a) There are 36 pairs of consecutive sectors. We show that the average of the sums of the numbers on these pairs of consecutive sectors is 37. Imagine that each sector is replaced by a bag containing as many marbles as the number on the sector. For each pair of consecutive sectors, consider the set of marbles from the two bags. Each marble belongs to exactly two of these 2-sector sets, and there are 36 sets in all. Since the total number of marbles is 666, the remark at the end of the proof of the Generalized Pigeon-Hole Principle shows that the average number of marbles in these sets is $2 \cdot 666/36 = 37$. The number of marbles in each set corresponds to the sum of the numbers on the two consecutive sectors.

The fact that each marble belongs to exactly two of the sets is essential to the argument above. Suppose, for example, that we omit one pair of consecutive sectors from consideration. If the numbers on these two sectors are, say, 5 and 17, then the average of the sums of the numbers on the remaining 35 pairs of sectors would be

$$\frac{2 \cdot 666 - 22}{35} \approx 37.43.$$

(b) Some consecutive pair of sectors must have sum at least 38. To see this, group the 36 sectors into 18 pairwise disjoint consecutive pairs. The average sum of the 18 pairs is 37. Some sum exceeds 37 or else each pair has sum exactly 37. In the latter case, shift clockwise one sector, which changes each sum. Some new sum will have to be greater than 37.

(c) Part (a) easily generalizes. Let t be an integer with $2 \le t \le 36$. There are 36 blocks of t consecutive sectors, and the average of the sums of the numbers on these blocks is $18.5 \cdot t$. Imagine the marbles in part (a) again. Each block corresponds to the set of marbles associated with the t sectors of the block. Since each marble belongs to exactly t of these sets, the average number of marbles in these sets is

$$\frac{t \cdot 666}{36} = 18.5 \cdot t.$$

As before, the number of marbles in each set is the sum of the numbers on the sectors in the corresponding block. ■

Exercises 5.5

Most, but not all, of the following exercises involve the Pigeon-Hole Principle. They also provide more practice using the techniques from §§5.1 to 5.4. The exercises are not all equally difficult, and some may require extra ingenuity.

1. Explain why, in any set of 11 positive integers, some two integers must have the same 1's digit.

2. (a) A sack contains 50 marbles of 4 different colors. Explain why there are at least 13 marbles of the same color.

(b) If exactly 8 of the marbles are red, explain why there are at least 14 of the same color.

3. Suppose that 73 marbles are placed in 8 boxes.

(a) Show that some box contains at least 10 marbles.

(b) Show that if two of the boxes are empty, then some box contains at least 13 marbles.

4. (a) Let B be a 12-element subset of $\{1, 2, 3, 4, 5, 6\} \times \{1, 2, 3, 4, 5, 6\}$. Show that B contains two different ordered pairs, the sums of whose entries are equal.

(b) How many times can a pair of dice be tossed without obtaining the same sum twice?

5. Let A be a 10-element subset of $\{1, 2, 3, \ldots, 50\}$. Show that A possesses two different 4-element subsets, the sums of whose elements are equal.

6. Let S be a 3-element set of integers. Show that S has two different nonempty subsets such that the sums of the numbers in each of the subsets are congruent mod 6.

7. Let A be a subset of $\{1, 2, 3, \ldots, 149, 150\}$ consisting of 25 numbers. Show that there are two disjoint pairs of numbers from A having the same sum [for example, $\{3,89\}$ and $\{41,51\}$ have the same sum, 92].

8. For the following sequences, first apply the result in Example 5 to decide whether they *must* have an increasing or decreasing subsequence of length 5. Then find an increasing or decreasing subsequence of length 5 if you can.

(a) 4, 3, 2, 1, 8, 7, 6, 5, 12, 11, 10, 9, 16, 15, 14, 13

(b) 17, 13, 14, 15, 16, 9, 10, 11, 12, 5, 6, 7, 8, 1, 2, 3, 4

(c) 10, 6, 2, 14, 3, 17, 12, 8, 7, 16, 13, 11, 9, 15, 4, 1, 5

9. Find the decimal expansions for 1/7, 2/7, 3/7, 4/7, 5/7, and 6/7 and compare them.

10. (a) Show that if 10 nonnegative integers have sum 101, there must be 3 with sum at least 31.

(b) Prove a generalization of part (a): If $1 \le k \le n$ and if n nonnegative integers have sum m, there must be k with sum at least ________ .

11. In this problem the 24 numbers 1, 2, 3, 4, ... , 24 are permuted in some way, say $(n_1, n_2, n_3, n_4, \dots, n_{24})$.

(a) Show that there must be 4 consecutive numbers in the permutation that are less than 20, i.e., at most 19.

(b) Show that $n_1 + n_2 + n_3 + \cdots + n_{24} = 300$.

(c) Show that there must be 3 consecutive numbers in the permutation with sum at least 38.

(d) Show that there must be 5 consecutive numbers in the permutation with sum at least 61.

12. Consider the roulette wheel in Example 7.

(a) Use Example 7(c) to show that there are 4 consecutive sectors with sum at least 74.

(b) Show that, in fact, 74 in part (a) can be improved to 75.

(c) Use Example 7(c) to show that there are 5 consecutive sectors with sum at least 93.

(d) Show that, in fact, 93 in part (c) can be improved to 95.

13. Let n_1, n_2, and n_3 be distinct positive integers. Show that at least one of n_1, n_2, n_3, $n_1 + n_2$, $n_2 + n_3$ or $n_1 + n_2 + n_3$ is divisible by 3. *Hint:* Map the set $\{n_1, n_1 + n_2, n_1 + n_2 + n_3\}$ to $\mathbb{Z}(3)$ by $f = \text{MOD } 3$.

14. A club has 6 men and 9 women members. A committee of 5 is selected at random. Find the probability that

(a) there are 2 men and 3 women on the committee.

(b) there are at least 1 man and at least 1 woman on the committee.

(c) the committee consists of only men or else consists of only women.

15. Six-digit numbers are to be formed using only the integers in the set

$$A = \{1, 2, 3, 4, 5, 6, 7, 8\}.$$

(a) How many such numbers can be formed if repetitions of the digits are allowed?

(b) In part (a), how many of the numbers contain at least one 3 and at least one 5?

(c) How many 6-digit numbers can be formed if each digit in A can be used at most once?

(d) How many 6-digit numbers can be formed that consist of one 2, two 4's, and three 5's?

16. How many divisors are there of 6000? *Hint:* $6000 = 2^4 \cdot 3 \cdot 5^3$ and every divisor has the form $2^m 3^n 5^r$, where $m \le 4$, $n \le 1$, and $r \le 3$.

17. Consider n in $\mathbb{P}$, and let S be a subset of $\{1, 2, \dots, 2n\}$ consisting of $n + 1$ numbers.

(a) Show that S contains two numbers that are relatively prime. *Suggestion:* Consider pairs $(2k - 1, 2k)$.

(b) Show that S contains two numbers such that one of them divides the other.

(c) Show that part (a) can fail if S has only n elements.

(d) Show that part (b) can fail if S has only n elements.

18. (a) Consider a subset A of $\{0, 1, 2, \dots, p\}$ such that $|A| > \frac{1}{2}p + 1$. Show that A contains two different numbers whose sum is p.

(b) For $p = 6$, find A with $|A| = \frac{1}{2}p + 1$ not satisfying the conclusion in part (a).

(c) For $p = 7$, find A with $|A| = \frac{1}{2}(p - 1) + 1$ not satisfying the conclusion in part (a).

19. A class of 21 students wants to form 7 study groups so that each student belongs to exactly 2 study groups.

(a) Show that the average size of the study groups would have to be 6.

(b) Indicate how the students might be assigned to the study groups so that each group has exactly 6 students.

20. The English alphabet consists of 21 consonants and 5 vowels. The vowels are a, e, i, o, and u.

(a) Prove that no matter how the letters of the English alphabet are listed in order [e.g., z u v a r q l g h ⋯] there must be 4 consecutive consonants.

(b) Give a list to show that there need not be 5 consecutive consonants.

(c) Suppose now that the letters of the English alphabet are put in a circular array, for example, as shown in Figure 3. Prove that there must be 5 consecutive consonants in such an array.

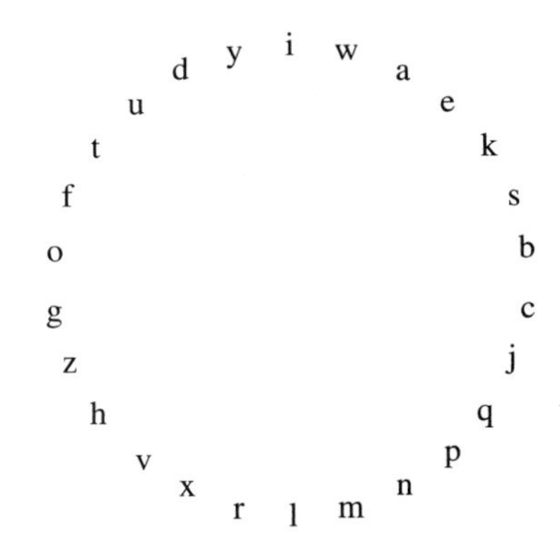

Figure 3 ▲

Chapter Highlights

To check your understanding of the material in this chapter, follow our usual suggestions for review. Think always of examples.

CONCEPTS AND NOTATION

selection with/without replacement
r-permutation, permutation, $P(n, r)$, combination
ordered partition
multiset
sample space, event, outcome, probability [on Ω], $P(E)$

FACTS

Union Rules, $|S \cup T| = |S| + |T| - |S \cap T|$ for finite sets

Product Rule, $|S_1 \times S_2 \times \cdots \times S_n| = \prod_{k=1}^{n} |S_k|, \quad |S^n| = |S|^n$

$P(E^c) = 1 - P(E)$

$P(E \cup F) = P(E) + P(F) - P(E \cap F)$

$\binom{n}{r} = \frac{n!}{(n-r)!\,r!}, \quad \binom{n}{r_1 \ \ldots \ r_k} = \frac{n!}{r_1! \cdots r_k!}$

Binomial Theorem: $(a + b)^n = \sum_{r=0}^{n} \binom{n}{r} a^r b^{n-r}$

Multinomial Theorem: $(a_1 + \cdots + a_k)^n = \sum_{n_1+\cdots+n_k=n} \binom{n}{n_1 \ \ldots \ n_k} a_1^{n_1} \cdots a_k^{n_k}$

Counting Lemma

Formula $\frac{n!}{n_1!\,n_2! \cdots n_k!}$ for counting multiset permutations or ordered partitions

Formula $\binom{n+k-1}{k-1}$ for ways to place n objects in k boxes

Inclusion–Exclusion Principle

Pigeon-Hole Principle, Generalized Pigeon-Hole Principle

METHODS

Visualizing a process as a sequence of steps.
Counting a set of pairs in two different ways.
Numerous clever ideas illustrated in the examples.

Supplementary Exercises

1. How many strings of 7 English letters read the same backward as forward?

2. Show that if 63 animals out of a set of 100 are woolly and 50 animals are fierce, then there must be some fierce, woolly animals. What's the smallest possible number of fierce, woolly ones?

3. In how many ways can the 26 letters $A, B, \ldots, Z$ of the English alphabet be listed so that the letters A and Z are *not* next to each other? Explain.

4. How many *odd* numbers between 100,000 and 1,000,000 have no two digits the same? Explain.

5. How many different fruit assortments can be made from oranges, limes, and mangos satisfying the following conditions?

   (a) There are 8 pieces of fruit altogether.

   (b) There are an even number of pieces of each type, and 16 pieces altogether.

   (c) There are more limes than oranges and 8 pieces total.

6. How many ways are there to break a class of 15 distinct students into

   (a) a group of 7 and a group of 8?

   (b) three groups of 5?

7. (a) How many numbers between 100,000 and 1,000,000 have no two successive digits the same?

   (b) How many odd numbers between 100,000 and 1,000,000 have no two digits the same? [Think of a good order in which to choose the digits.]

8. A pit contains 14 snakes, 3 of which are poisonous.

(a) How many ways are there to select 5 snakes from the pit without replacement?

(b) How many of these ways leave at least 1 poisonous snake in the pit?

(c) How many ways select at least 2 poisonous snakes?

9. How many ways are there to put 3 distinct flags on 2 identical poles if each pole gets at least one flag?

10. The English alphabet Σ consists of 26 letters.

(a) What is the number of sequences of 8 different letters from Σ?

(b) What is the number of ways to choose a *set* of 8 different letters from Σ with exactly 3 of them in $\{A, B, C, D, E\}$?

(c) What is the number of sequences of 9 letters from Σ with repetitions allowed that contain exactly 3 A's?

(d) Suppose that the set $\{F, I, S, H, W, R, A, P\}$ has been selected. What is the number of ways these letters can be arranged in sequence starting with $SRAF$?

11. A box contains tickets of 4 colors: red, blue, yellow, and green. Each ticket has a number from $\{0, 1, \ldots, 9\}$ written on it.

(a) What is the largest number of tickets the box can contain without having at least 21 tickets of the same color?

(b) What is the smallest number of tickets the box must contain to be sure that it contains at least two tickets of the same color with the same number on them?

12. A teacher with a class of 20 students decides, instead of grading their midterm exams, just to divide the exams randomly into 5 piles, give the ones in the first pile A's, the ones in the second pile B's, and so on.

(a) What is the number of different ways to put the papers in piles with 3 A's, 6 B's, 7 C's, 1 D, and 3 F's?

(b) What is the number of ways to divide the set of papers into 2 indistinguishable subsets with 10 papers in each?

(c) What is the number of ways to divide the set of papers as in part (b) so that all 6 B papers are in the same set?

13. Let S be the set of integers between 10,000 and 19,999.

(a) What is the number of members of S that have at least one digit that is a 3?

(b) What is the number of members of S that have at least one digit that is a 3 and at least one digit that is a 4?

14. Twenty-one numbers are drawn at random, without replacement, from the set $\{100, 101, \ldots, 998, 999\}$. What is the probability that at least three of the numbers have the same last digit? Explain.

15. Let $\Sigma = \{a, b, c, d, e\}$.

(a) How many words [i.e., strings] of length 5 in Σ^* contain each letter in Σ exactly once?

(b) How many words of length 6 in Σ^* contain the letter d?

(c) How many words of length 7 in Σ^* begin with the string ded?

16. A dessert is formed by choosing 2 different flavors of ice cream and 3 different toppings. If 31 flavors of ice cream are available and 18 different toppings, how many different desserts are possible?

17. Given three sets A, B, and C, each with 800 elements, such that

$$|A \cap B| = |A \cap C| = |B \cap C| = 250 \text{ and } |A \cap B \cap C| = 50,$$

evaluate $|A \cup B \cup C|$.

18. Half of the 40 frogs in a bucket are male and half are female. If we draw 8 frogs at random from the bucket without replacement, what is the probability that 4 will be male and 4 will be female?

19. We have a standard deck of 52 cards.

(a) Draw a card at random. What's the probability that it's the ace of spades?

(b) Draw 2 cards with replacement. What's the probability that both cards are the same?

(c) Draw 2 cards without replacement. What's the probability both are the same?

(d) Draw 6 cards with replacement. What's the probability no two are the same, i.e., all 6 are different?

(e) Draw 6 cards without replacement. What's the probability no two are the same?

(f) Draw 6 cards with replacement. What's the probability at least 2 are the same?

(g) Draw 6 cards with replacement. What's the probability that exactly 2 are the same?

(h) Draw 6 cards with replacement. What's the probability that exactly 4 of them are the ace of spades?

(i) Draw 6 cards with replacement. What's the probability that at most 4 are the ace of spades?

(j) Draw 6 cards with replacement. What's the probability that at least 3 are the ace of spades?

20. *There was an old man with a beard*
Who said, "It is just as I feared.
Two owls and a hen,
Four larks and a wren
Have all built their nests in my beard."
— Edgar Lear

(a) A bird is chosen at random from the beard. What is the probability that it is a wren?

(b) Each bird has a little tag with a letter on it, O for owl, H for hen, L for lark, and W for wren. One morning the birds are called out from the beard at random and lined up in a row from left to right. What is the probability that their tags spell out $WOLHOLLL$?

(c) How many teams of 4 birds can be chosen from the beard dwellers if each team must include at least 1 lark?

(d) At least 2 larks?

(e) What's the least number of birds that can be chosen from the beard to be sure that at least 1 is a lark?

(f) What's the largest number that can be chosen without having any larks?

(g) What's the least number that can be chosen so that at least 2 birds must be of the same kind?

(h) Another old man with a beard has an unlimited supply of owls, hens, larks, and wrens and wishes to choose 8 birds to nest in his beard. How many different distributions, i.e., assortments, such as 3 owls, 2 hens, no larks, and 3 wrens, can he choose?

(i) Does it make sense to ask what the probability is that the two old men have the same beard-bird assortments?

21. How many positive divisors does $2^3 \cdot 5^4 \cdot 7$ have?

22. (a) Show how to color the edges of K_5 with red and green so that there is no red triangle and no green triangle.

(b) Consider coloring the edges of K_6 with two colors, red and green. Show that if there is no red triangle, then there must be a green triangle. *Hint:* Pick a vertex, argue that at least 3 edges to it must be the same color, and go from there.

23. We have $m + 1$ people with integer ages between 1 and $2m$. Show that there must be at least 2 people whose ages are within a year of each other.

24. Eight students are seated in the front row at the game.

(a) In how many different ways can they be seated?

(b) In how many different ways can they be seated if Bob and Ann refuse to sit side by side?

(c) In how many ways can they be seated if Bob, Chris, and Ann insist on sitting together, with Chris in the middle?

(d) Four of the students are selected to come early and hold seats for the whole group. In how many ways can these 4 be chosen if Pat and Chris insist on coming at the same time?

(e) After the game the students walk over to a restaurant in two groups of 4. In how many ways can they split up?

(f) What if Pat and Chris insist on being in the same group?

25. (a) In how many ways can 10 frogs, 10 gerbils, and 10 snakes be placed in 40 cages if each cage gets no more than one animal? The animals of each type are indistinguishable, but we can tell the cages apart.

(b) In how many ways can these animals be placed in 40 cages if there is no restriction on how many animals each cage gets?

(c) What if both the cages and the animals are distinguishable, but there's no limit on the number of animals per cage?

(d) We plan to paint the 40 cages with 4 colors, fuchsia, green, salmon, and puce, so that every cage gets painted just one color. How many ways are there to do this if the cages are now otherwise indistinguishable?

(e) How many ways are there if the cages are distinguishable?

26. A 100-element set S has subsets A, B, and C of sizes 50, 70, and 65, respectively.

(a) What is the smallest $|A \cup B \cup C|$ could be?

(b) What is the largest $|A \cup B \cup C|$ could be?

(c) What is the smallest $|B \cap C|$ could be?

(d) What is the smallest $|A \cap B \cap C|$ could be? *Hint:* How large can $(A \cap B \cap C)^c$ be?

(e) What is the smallest $|(A \cap B) \cup (A \cap C)|$ could be?

27. Consider words of length 12 made from letters in the alphabet $\Sigma = \{a, b, c, d, e, x\}$.

(a) If a word is chosen at random, with all words equally likely, what is the probability that it does not contain the letter x?

(b) If a word is chosen at random, what is the probability that it does not contain x but does contain a?

(c) How many of the words do not contain x or do contain b?

(d) How many of the words contain c exactly 4 times?

(e) How many of the words contain a and b exactly 3 times each and contain d exactly 6 times?

(f) How many of the words contain 3 different letters exactly 4 times each?

(g) How many of the words contain exactly 3 different letters?

28. (a) How many ways are there to distribute 6 indistinguishable fish among 3 indistinguishable bowls if each bowl must receive at least 1 fish?

(b) How many ways are there to distribute 6 distinguishable people into 3 distinguishable cars if each car must get at least 1 person?

29. How many ordered strings of length 7 can be made from a set of 3 N's, one O, 2 S's and 2 E's? [Note that one letter will be left out in each string.] Explain.

30. How many ways are there to arrange m distinguishable flags on n distinguishable poles? *Hint:* Think of putting the flags on one long pole and then sawing the pole into n pieces.

31. (a) How many ordered strings (i.e., "words") of length 8 can be made using only the letters a, b, and c, each an unlimited number of times?

(b) How many such strings have at most one a in them?

(c) How many such strings are there in which each of the three letters occurs at least 2 times?

32. Let w_n count the n-letter words in A, B, C, and D with an odd number of B's. Find a recurrence relation satisfied by w_n for all $n \geq 2$, and calculate w_0, w_1, w_2, w_3, and w_4.

33. A city contains a large number of men, women, and children. If 16 people are chosen from its population, how many different distributions, such as 9 men, no women, and 7 children, are possible?

34. We have observed a correspondence between questions about arrangements of strings of letters and questions about placing objects in boxes. Rephrase the following as a question about objects and boxes, and then answer it. How many ways are there to rearrange the word $R\,E\,M\,E\,M\,B\,E\,R$?

35. We have 30 distinct blood samples.

(a) How many ways are there to break up the samples into 6 distinguishable groups of 5 samples each?

(b) How many ways are there to put the samples into 3 distinguishable boxes, allowing empty boxes?

36. Rewrite the following question as a balls-in-boxes problem, and then give the answer. What is the coefficient of $A^3B^4C^2$ in $(A+B+C)^9$?

37. We have an unlimited number of red, green, and blue marbles, and we wish to make an assortment of 20 marbles, such as 5 red, 12 green, and 3 blue. All marbles of the same color are indistinguishable.

(a) How many different assortments are possible?

(b) How many different assortments are possible that contain at least 2 red marbles and at least 5 green ones?

(c) How many different assortments are possible that contain at most 5 blue marbles?

38. A list contains 100 problems, of which 75 are easy and 40 are important.

(a) What is the smallest possible number of problems that are both easy and important?

(b) Suppose that n different problems are chosen at random from the list, with all problems equally likely to be chosen. What is the probability that all n are important?

(c) How many problems must be chosen in order to be sure that at least 5 of them are easy?

39. Let $D = \{1, 2, 3, 4, 5\}$ and let $L = \{a, b, c\}$.

(a) How many functions are there from the set D to the set L?

(b) How many of these functions map D *onto* L?

40. A pen contains 100 animals, 70 of which are large, 45 of which are hairy, and 10 of which are neither large nor hairy.

(a) An animal is selected at random from the pen, with all animals equally likely to be selected. What is the probability that the animal is large?

(b) Two animals are selected at random from the pen, without replacement. What is the probability that both are hairy?

(c) Two animals are selected as in part (b). What is the probability that exactly one of them is large?

(d) A team of 50 animals is selected and removed from the pen. What is the number of ways this can be done so that at least one member of the team is hairy?

41. A pit contains 25 snakes, 5 of which are poisonous and 20 of which are harmless [= not poisonous]. Twenty-one snakes are drawn at random from the pit, without replacement.

(a) What is the probability that the first snake drawn is poisonous?

(b) What is the probability that the first 3 snakes drawn are all harmless?

(c) What is the probability that all 21 snakes drawn are harmless?

42. We wish to form a committee of 10 members from a group of 50 Democrats, 40 Republicans, and 20 snakes.

(a) How many different distributions, such as 5 Democrats, 4 Republicans, and 1 snake, are possible?

(b) How many different distributions are possible with no snakes?

(c) How many are possible with at least 3 snakes?

43. Let S be the set of all sequences of 0's, 1's, and 2's of length 8. For example, S contains 2 0 0 1 0 2 1 1.

(a) How many elements are in S?

(b) How many sequences in S have exactly five 0's and three 1's?

(c) How many sequences in S have exactly three 0's, three 1's, and two 2's?

(d) How many sequences in S have exactly three 0's?

(e) For how many 8-bit strings is either the second or fourth bit a 1 (or both)?

(f) How many sequences in S have at least one 0, at least one 1, and at least one 2?

44. (a) How many integers between 1 and 1000 (inclusive) are not divisible by 2, 5, or 17?

(b) How many are divisible by 2, but not by 5, or 17?

45. We're going to make a team of 6 people from a set of 8 men and 9 women.

(a) How many ways are there to choose the team? [People are distinguishable. Otherwise, there are just 7 distributions. Why?]

(b) How many ways are there to choose the team with no men, with 1 man, with at most 3 men?

(c) What's $\displaystyle\sum_{i=0}^{t}\binom{m}{i}\binom{w}{t-i}$ in general? Could you prove it?

(d) Now add 5 children. How many 6-person teams can be formed? How many with at most 2 children?

46. Three dice are rolled, and their top spots are added. Possible sums are $3, 4, \ldots, 18$.

(a) Are all these sums equally likely? Explain.

(b) How many ways are there to roll a sum of 1, of 2, of 3, of 4, of 5, of 6?

(c) How many ways are there to distribute 6 objects into 3 boxes if each box gets at least 1 object and at most 6 objects?

(d) Are the answers to parts (b) and (c) related?

(e) How many ways are there to roll a sum of 8 with 3 dice?

47. How many functions are there from the English alphabet $\Sigma = \{a, b, \ldots, z\}$ onto $D = \{0, 1, 2, \ldots, 9\}$, the set of decimal digits? Give a formula, not a numerical answer.

48. (a) How many ordered triples (m, n, r) of nonnegative integers satisfy $m + n + r = 10$?

(b) How many ordered triples (m, n, r) of positive integers satisfy $m + n + r = 10$?

(c) How many ways are there to break a 10-element set of indistinguishable objects into 3 disjoint nonempty subsets? [This is harder.]

49. Let s and m be positive integers.

(a) How many ways are there to choose $s + 1$ different numbers from $\{1, 2, \ldots, k\}$ so that k is the largest number chosen?

(b) Write $\sum_{i=s}^{m} \binom{i}{s}$ as a single binomial coefficient, and explain your answer.

50. What is the probability that if we rearrange the letters A, A, B, B, C randomly the resulting word has two identical letters in succession?

51. (a) Find the number of words of length 5 in $\{A, B, C, D, E\}$ in which the string ABC does not occur. [Note that not all letters are required to occur.]

(b) Find the number of such words in which the strings ABC and BCD do not occur.

52. We deal a standard deck of 52 cards randomly to 4 players so that each player gets 13 cards. What is the probability that a particular player, say Sam, gets all 4 aces?

53. (a) How many different words of length 3 can be made from the 26 letters $\{A, B, \ldots, Z\}$, allowing repetitions?

(b) How many of the words in part (a) have letters that occur more than once?

(c) How many of the words in part (a) have letters that occur two or more times in a row?

54. Let w_n be the number of words of length n in A, B, C that have two A's in a row.

(a) Give the values of w_1, w_2, w_3, and w_4.

(b) Give a recurrence relation satisfied by w_n for $n \geq 3$.

55. (a) How many different words, i.e., strings, can be made using all of the letters $I, I, I, I, M, P, P, S, S, S, S$?

(b) How many of the words in part (a) have all the I's together?

(c) How many of the words in part (a) read the same backward and forward?

(d) If the letters are arranged in random order, what is the probability that they spell MISSISSIPPI?

56. (a) Generalize Example 3 and Exercise 18 in §5.3 as follows. If an ordered list $s_1, \ldots, s_n$ is randomized, what is the probability that no object will still be in its original place? For large n, what is the approximate value of this probability?

(b) If a phone company sends out 100 bills at random, what is the probability that at least one bill will go to the correct customer?

57. In the notation of the Inclusion-Exclusion Principle, let S_1 be the sum of the sizes of the sets A_i, let S_2 be the sum of the sizes of their intersections taken 2 at a time, and, in general, let S_k for $k \geq 3$ be the sum of the sizes of their intersections taken k at a time. Finally, let S_0 be the size of a set U that contains all of the sets A_i. Write a formula for $|(A_i \cup \cdots \cup A_n)^c|$ in terms of $S_0, S_1, \ldots,$ where complementation is with respect to the set U.

58. (a) Find the number of 5-letter multisets of A, B, C in which some letter occurs at least 3 times.

(b) Find the number of 5-letter multisets of A, B, C in which no letter occurs more than twice.

(c) Find the number of 5-letter *words* in A, B, C in which some letter occurs at least 3 times.

(d) Find the number of 5-letter words in A, B, C in which no letter occurs more than twice.

(e) What complications would arise if you were to look at 7-letter multisets and 7-letter words?

59. A collection of 14 distinct sociologists is to be divided up into 4 nonoverlapping committees, each of which will have a chair, a vice-chair and a secretary. All committees will be given the same task. How many different ways are there to create the committees and assign the 3 titles to 3 different members of each committee?

60. Rearrangements of ordered lists with no entry in its original position are called **derangements**.

(a) How many derangements are there of the letters in the word *peach*? That is, how many words can be made from those letters with no letter in the same positions as it is in *peach*? *Hint:* See Example 3 on page 199.

(b) Suppose we try to ask the same question about the letters in the word *apple*. What should the question be, and what is its answer?

(c) Repeat part (b) for the word *acacia*.

(d) Repeat part (b) for the word *poppy*.

61. (a) Show that if p is a prime, then $\binom{p}{r}$ is a multiple of p for $0 < r < p$. *Hint:* $p! = \binom{p}{r} \cdot r! \cdot (p - r)!$.

(b) Show that if p is a prime, then $(a + b)^p \equiv a^p + b^p \pmod{p}$ for all integers a and b, just as some algebra students wish.

62. Recall from Exercise 8 on page 127 that if p is a prime, then $n^{p-1} \equiv 1 \pmod{p}$ unless $p|n$. This is Fermat's Little Theorem,.

(a) Prove that $\binom{p}{n} \equiv 0 \pmod{p}$ for $0 < n < p$.

(b) Use induction to prove that $n^p \equiv n \pmod{p}$ for $n \in \mathbb{N}$.

(c) Show that $n^p \equiv n \pmod{p}$ for $n \in \mathbb{Z}$.

(d) Prove Fermat's Little Theorem.

Introduction to Graphs and Trees

We have already seen graphs and digraphs in Chapter 3, where they gave us ways to picture relations. In this chapter we develop the fundamental theory of the graphs and digraphs themselves. Section 6.1 introduces the main ideas and terminology, and §§6.2 and 6.5 discuss paths with special properties. The remainder of the chapter is devoted to trees, which we will study further in Chapter 7. Trees are graphs that can also be thought of as digraphs in a natural way. Section 6.4 is devoted to rooted trees, which arise commonly as data structures. The final section of the chapter treats graphs whose edges have weights and contains two algorithms for constructing spanning trees of smallest weight.

6.1 Graphs

We introduced the concepts of digraph and graph in §3.2, and now we begin to look at graphs in more detail. Recall from §3.2 that a **path** of **length** n in a graph is a sequence $e_1\, e_2 \cdots e_n$ of edges together with a sequence $v_1\, v_2 \cdots v_{n+1}$ of vertices, where v_i and v_{i+1} are the endpoints of e_i for $i = 1, \ldots, n$. The function γ specifies the vertices of each edge by $\gamma(e_i) = \{v_i, v_{i+1}\}$. The vertices v_i and v_{i+1} can be equal, in which case e_i is a **loop**. If $v_{n+1} = v_1$, the path is said to be **closed**.

A closed path can consist just of going out and back along the same sequence of edges, for example, $e\, f\, g\, g\, f\, e$. The most important closed paths in what follows, though, will be the ones in which no edge is repeated. A path is called **simple** if all of its edges are different. Thus a simple path cannot use any edge twice, though it may go through the same vertex more than once. A closed simple path with vertex sequence $x_1 \cdots x_n\, x_1$ for $n \geq 1$ is called a **cycle** if the vertices $x_1, \ldots, x_n$ are distinct. A graph that contains no cycles is called **acyclic**. We will see soon that a graph is acyclic if and only if it contains no simple closed paths.

A path itself is **acyclic** if the subgraph consisting of the vertices and edges of the path is acyclic. In general, a graph H is a **subgraph** of a graph G if $V(H) \subseteq V(G)$, $E(H) \subseteq E(G)$, and the function γ for G defined on $E(G)$ agrees with the γ for H on $E(H)$. If G has no parallel edges and if we think of $E(G)$ as a set of one- or two-element subsets of $V(G)$, then the agreement condition on γ is a consequence of $E(H) \subseteq E(G)$. It follows from the definitions of acyclicity and subgraph that if H is a subgraph of G and if G is acyclic, then so is H.

EXAMPLE 1

(a) The path $e\,f$ with vertex sequence $x_1\,x_2\,x_1$ in the graph of Figure 1(b) is a cycle. So is the path $e\,f$ with vertex sequence $x_2\,x_1\,x_2$.

(b) The path $e\,e$ with vertex sequence $x_1\,x_2\,x_1$ in the graph of Figure 1(a) is a closed path, but it is not a cycle because it is not simple. Neither is the path $e\,e$ with vertex sequence $x_2\,x_1\,x_2$. This graph is acyclic.

(c) In the graph of Figure 1(c) the path $e\,f\,h\,i\,k\,g$ of length 6 with vertex sequence $u\,v\,w\,x\,y\,w\,u$ is closed and simple, but it is not a cycle because the first six vertices u, v, w, x, y and w are not all different. The path with vertex sequence $u\,w\,v\,w\,u\,v\,u$ is a closed path that is not simple and also fails to be a cycle. The graph as a whole is not acyclic, and neither are these two paths, since $u\,v\,w\,u$ is a cycle in both of their subgraphs. ■

Figure 1 ▶

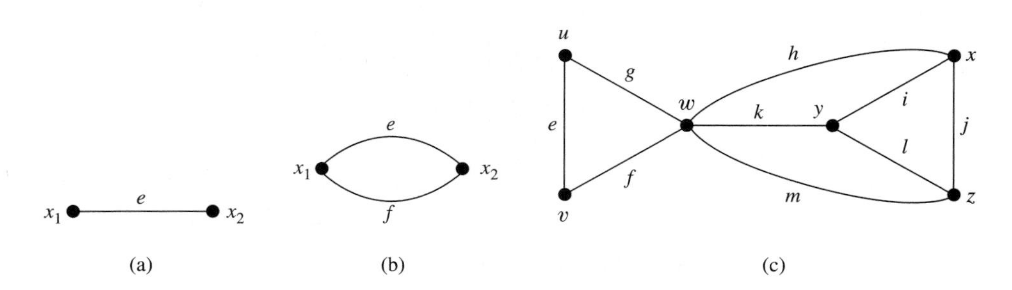

A path $e_1 \cdots e_n$ in G with vertex sequence $x_1 \cdots x_{n+1}$ consisting of distinct vertices must surely be simple, since no two edges in it can have the same set of endpoints. Example 1(b) shows, however, that a closed path $x_1 \cdots x_n\,x_{n+1}$ with $x_1, \ldots, x_n$ distinct need not be simple. That bad example, which is a closed path of length 2, is essentially the only one there is, as the following shows.

Proposition 1 Every closed path $e_1 \cdots e_n$ of length at least 3 with $x_1, \ldots, x_n$ distinct is a cycle.

Proof We only need to show that the edges $e_1, \ldots, e_n$ are different. Since $x_1, \ldots, x_n$ are distinct, the path $e_1 \cdots e_{n-1}$ is simple. That is, $e_1, \ldots, e_{n-1}$ are all different. But $\gamma(e_n) = \{x_n, x_1\}$ and $\gamma(e_i) = \{x_i, x_{i+1}\}$ for $i < n$. Since $\{x_n, x_1\}$ could equal $\{x_i, x_{i+1}\}$ only if $i = 1$ and $n = 2$, we must have $e_n \neq e_i$ for $i < n$, so the path is simple. ■

For paths that are not closed, distinctness of vertices can be characterized another way.

Proposition 2 A path has all of its vertices distinct if and only if it is simple and acyclic.

Proof Consider first a path with distinct vertices. It must be simple, as we observed just before Proposition 1. A picture of the subgraph consisting of the vertices and edges of the path could be drawn as a straight line with vertices on it, which is evidently acyclic.

For the converse, suppose a simple path has vertex sequence $x_1 \cdots x_{n+1}$ with some repeated vertices. Consider two such vertices, say x_i and x_j with $i < j$ and with the difference $j - i$ as small as possible. Then $x_i, x_{i+1}, \ldots, x_{j-1}$ are all distinct, so the simple path $x_i\,x_{i+1} \cdots x_{j-1}\,x_j$ is a cycle, and the original path contains a cycle. Note that if $j = i + 1$, then the cycle is a loop, and if $j = i + 2$, we get a cycle like the graph in Figure 1(b). ■

It follows from the proof of Proposition 2 that every simple closed path contains a cycle and, in fact, such a cycle can be built out of successive edges in the path. Moreover, if a graph is acyclic, then it contains no cycles, so it contains no simple closed paths. We will look more closely at such graphs in the later sections on trees.

In the language of §3.2, the next theorem says that the presence of cycles has no influence on the reachable relation.

Theorem 1 If u and v are distinct vertices of a graph G and if there is a path in G from u to v, then there is a simple acyclic path from u to v. Indeed, every shortest path from u to v is simple and acyclic.

Proof Among all paths from u to v in G, choose one of smallest length; say its vertex sequence is $x_1 \cdots x_{n+1}$, with $x_1 = u$ and $x_{n+1} = v$. As we noted in Proposition 2, this path is simple and acyclic provided the vertices $x_1, \ldots, x_{n+1}$ are distinct. But, if they were not, we would have $x_i = x_j$ for some i and j with $1 \le i < j \le n+1$. Then the path $x_i\, x_{i+1} \cdots x_j$ from x_i to x_j would be closed [see Figure 2 for an illustration], and the shorter path $x_1 \cdots x_i\, x_{j+1} \cdots x_{n+1}$ obtained by omitting this part would still go from u to v. Since $x_1 \cdots x_n\, x_{n+1}$ has smallest length, a shorter path is impossible. So the vertices are distinct and every shortest path from u to v is simple and acyclic. ■

Figure 2 ▶

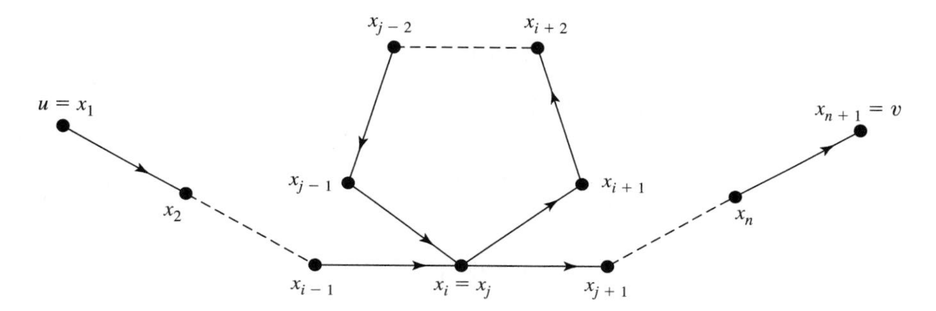

In the next proof, we will remove an edge e from a graph G to get a new graph $G \setminus \{e\}$. The graph $G \setminus \{e\}$ is the subgraph of G with the same vertex set as G, i.e., $V(G \setminus \{e\}) = V(G)$, and with the edge e removed, i.e., $E(G \setminus \{e\}) = E(G) \setminus \{e\}$. We will have other occasions, especially in §6.3, to remove edges from graphs.

Corollary If e is an edge in a simple closed path in G, then e belongs to some cycle.

Proof If e is a loop, we are done, so suppose e joins distinct vertices u and v. Remove e from G. Since the given path through u and v is closed, even with e gone there is a path from u to v; just go the long way around. This is a path in $G \setminus \{e\}$, since the original closed path was simple and hence does not repeat the edge e. By Theorem 1, applied to $G \setminus \{e\}$, there is a simple acyclic path from u to v not including e. Joining the ends of such a path with e creates a cycle using e. ■

We will need the following fact when we study trees in §6.3.

Theorem 2 If u and v are distinct vertices of the acyclic graph G, then there is at most one simple path in G from u to v.

Proof We suppose the theorem is false, and among pairs of vertices joined by two different simple paths, choose a pair (u, v) with the shortest possible path length from u to v.

Consider two simple paths from u to v, one of which is as short as possible. If the two paths had no vertices but u and v in common, then going from u to v along one path and returning from v to u along the other would yield a cycle, contrary to the fact that G is acyclic. Thus the paths must both go through at least one other vertex, say w. Then parts of the paths from u to v form two different simple paths from u to w or else parts form two different simple paths from w to v. But w is closer to u and to v than they are to each other, contradicting the choice of the pair (u, v). ■

EXAMPLE 2

This example illustrates the idea in the proof of Theorem 2. Consider the graph in Figure 1(c). This graph is *not* acyclic. We will exhibit alternative simple paths in it and cycles derived from them. The path $x\ w\ y\ z\ x$ formed by going out from x to w on one path and backward to x on the other is a cycle. So is $w\ u\ v\ w$, formed by going on from w to u and then back to w another way.

The simple paths $u\ w\ y\ x\ w\ v$ and $u\ w\ y\ z\ w\ v$ go from u to v. The cycle $y\ x\ w\ z\ y$ is made out of parts of these paths, one part from y to w and the other back from w to y. ■

It frequently happens that two graphs are "essentially the same," even though they differ in the names of their edges and vertices. General graph-theoretic statements that can be made about one of the graphs are then equally true of the other. To make these ideas mathematically precise, we introduce the concept of isomorphism. Generally speaking, two sets with some mathematical structure are said to be **isomorphic** [pronounced eye-so-MOR-fik] if there is a one-to-one correspondence between them that preserves [i.e., is compatible with] the structure. In the case of graphs, the vertices and edges of one graph should match up with the vertices and edges of the other, and the edge matching should be consistent with the vertex matching. We can state this idea precisely. Suppose first that G and H are graphs without parallel edges, so we can consider their edges to be one- or two-element sets of vertices. An **isomorphism** of G onto H in this case is a one-to-one correspondence $\alpha\colon V(G) \to V(H)$ matching up the vertices so that $\{u, v\}$ is an edge of G if and only if $\{\alpha(u), \alpha(v)\}$ is an edge of H. Two graphs G and H are **isomorphic**, written $G \simeq H$, if there is an isomorphism α of one onto the other. In this case, the inverse correspondence α^{-1} is also an isomorphism.

For graphs with parallel edges the situation is slightly more complicated: we require two one-to-one correspondences $\alpha\colon V(G) \to V(H)$ and $\beta\colon E(G) \to E(H)$ such that an edge e of $E(G)$ joins vertices u and v in $V(G)$ if and only if the corresponding edge $\beta(e)$ joins $\alpha(u)$ and $\alpha(v)$. Thus two graphs are isomorphic if and only if they have the same picture except for the labels on edges and vertices. This observation is mostly useful as a way of verifying an alleged isomorphism by actually drawing matching pictures of the two graphs.

EXAMPLE 3

The correspondence α with $\alpha(t) = t'$, $\alpha(u) = u', \ldots, \alpha(z) = z'$ is an isomorphism between the graphs pictured in Figures 3(a) and 3(b). The graphs shown in Figures 3(c) and 3(d) are also isomorphic to each other [see Exercise 17], but not to the graphs in parts (a) and (b). ■

To tell the graphs of Figures 3(a) and 3(c) apart, we can simply count vertices. Isomorphic graphs must have the same number of vertices and the same number of edges. These two numbers are examples of **isomorphism invariants** for graphs. Other examples include the number of loops and number of simple paths of a given length.

It is often useful to count the number of edges attached to a particular vertex. To get the right count, we need to treat loops differently from edges with two distinct vertices. We define $\mathbf{deg}(v)$, the **degree** of the vertex v, to be the number of 2-vertex edges with v as a vertex plus twice the number of loops with v as vertex. If you think of a picture of G as being like a road map, then the degree of v is simply the number of roads you can take to leave v, with each loop counting as two roads.

Figure 3 ▶

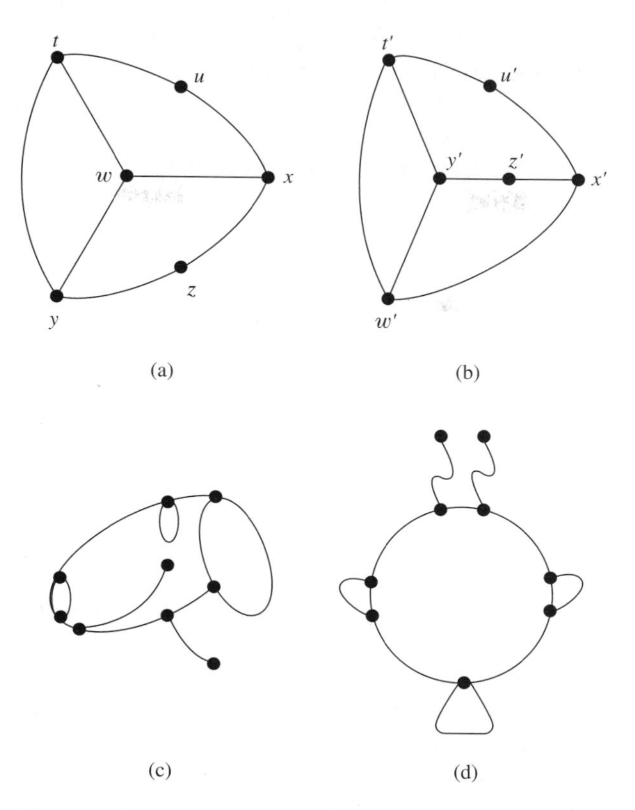

The number $D_k(G)$ of vertices of degree k in G is an isomorphism invariant, as is the **degree sequence** $(D_0(G), D_1(G), D_2(G), \ldots)$.

EXAMPLE 4

(a) The graphs shown in Figures 3(a) and 3(b) each have degree sequence $(0, 0, 2, 4, 0, 0, \ldots)$. Those in Figures 3(c) and 3(d) have degree sequence $(0, 2, 0, 6, 1, 0, 0, \ldots)$. From this information we can conclude that neither of the graphs in Figures 3(a) and 3(b) can be isomorphic to either of the graphs in Figures 3(c) and 3(d). As we noted in Example 3, the graphs in Figures 3(a) and 3(b) turn out to be isomorphic graphs, but we *cannot* draw this conclusion just from the fact that their degree sequences are the same; see part (b).

(b) All four of the graphs in Figure 4 have eight vertices of degree 3 and no others. It turns out that $H_1 \simeq H_2 \simeq H_3$, but that none of these three is isomorphic to H_4. See Exercise 18. Having the same degree sequence does not guarantee isomorphism. ■

Figure 4 ▶

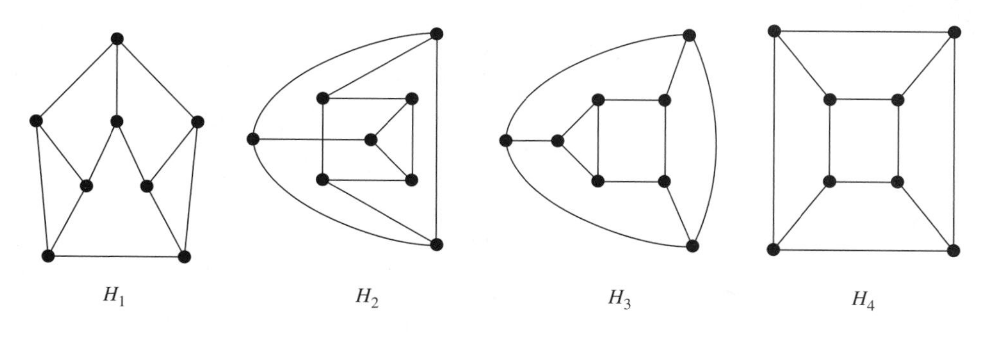

A graph in which all vertices have the same degree, as in Figure 4, is called a **regular** graph. As the example shows, regular graphs with the same number of vertices need not be isomorphic. Graphs without loops or multiple edges and in which every vertex is joined to every other by an edge are called **complete** graphs. A complete graph with n vertices has all vertices of degree $n - 1$, so such a graph is regular. All complete graphs with n vertices are isomorphic to each other, so we use the symbol $\boldsymbol{K_n}$ for any of them.

EXAMPLE 5

Figure 5(a) shows the first five complete graphs. The graph in Figure 5(b) has four vertices, each of degree 3, but it is not complete. ■

Figure 5 ▶

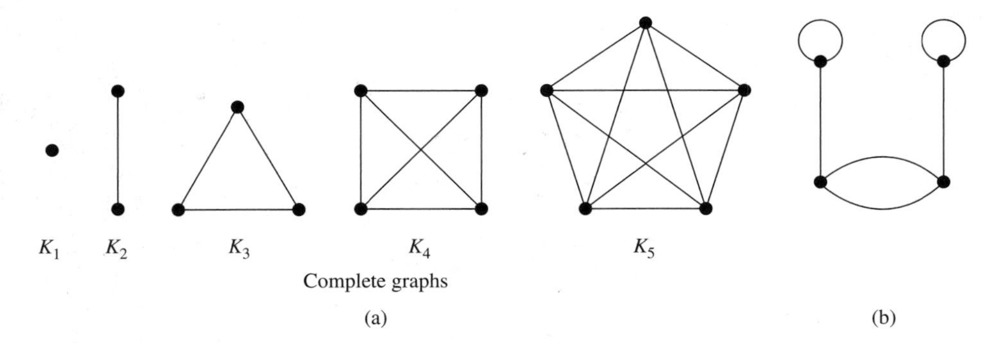

A complete graph K_n contains subgraphs isomorphic to the graphs K_m for $m = 1, 2, \ldots, n$. Such a subgraph can be obtained by selecting any m of the n vertices and using all the edges in K_n joining them. Thus K_5 contains $\binom{5}{2} = 10$ subgraphs isomorphic to K_2, $\binom{5}{3} = 10$ subgraphs isomorphic to K_3 [i.e., triangles], and $\binom{5}{4} = 5$ subgraphs isomorphic to K_4. In fact, every graph with n or fewer vertices and with no loops or parallel edges is isomorphic to a subgraph of K_n; just delete the unneeded edges from K_n.

Complete graphs have a high degree of symmetry. Each permutation α of the vertices of a complete graph gives an isomorphism of the graph onto itself, since both $\{u, v\}$ and $\{\alpha(u), \alpha(v)\}$ are edges whenever $u \neq v$.

The next theorem relates the degrees of vertices to the number of edges of the graph.

Theorem 3

(a) The sum of the degrees of the vertices of a graph is twice the number of edges. That is,

$$\sum_{v \in V(G)} \deg(v) = 2 \cdot |E(G)|.$$

(b) $D_1(G) + 2D_2(G) + 3D_3(G) + 4D_4(G) + \cdots = 2 \cdot |E(G)|$.

Proof

(a) Each edge, whether a loop or not, contributes 2 to the degree sum. This is a place where our convention that each loop contributes 2 to the degree of a vertex pays off.

(b) The total degree sum contribution from the $D_k(G)$ vertices of degree k is $k \cdot D_k(G)$. ■

EXAMPLE 6

(a) The graph of Figure 3(c) [as well as the isomorphic graph in Figure 3(d)] has vertices of degrees 1, 1, 3, 3, 3, 3, 3, 3, and 4, and has twelve edges. The degree sequence is $(0, 2, 0, 6, 1, 0, 0, 0, \ldots)$. Sure enough,

$$1+1+3+3+3+3+3+3+4 = 2 \cdot 12 = 2 + 2 \cdot 0 + 3 \cdot 6 + 4 \cdot 1.$$

(b) The graphs in Figure 4 have twelve edges and eight vertices of degree 3. Their degree sequence is $(0, 0, 0, 8, 0, 0, 0, \ldots)$, and

$$2 \cdot 12 = 0 + 2 \cdot 0 + 3 \cdot 8$$

confirms Theorem 3(b) in this case.

(c) The complete graph K_n has n vertices, each of degree $n-1$, and has $n(n-1)/2$ edges. Thus the sum of the degrees of the vertices is $n(n-1)$, which is twice the number of edges. Also, $D_{n-1}(K_n) = n$ and the other $D_k(K_n)$'s equal 0, so the sum in Theorem 3(b) is $(n-1) \cdot D_{n-1}(K_n) = (n-1) \cdot n$. ■

Exercises 6.1

1. For the graph in Figure 6(a), give the vertex sequence of a shortest path connecting the following pairs of vertices, and give its length.

   (a) s and v (b) s and z (c) u and y (d) v and w

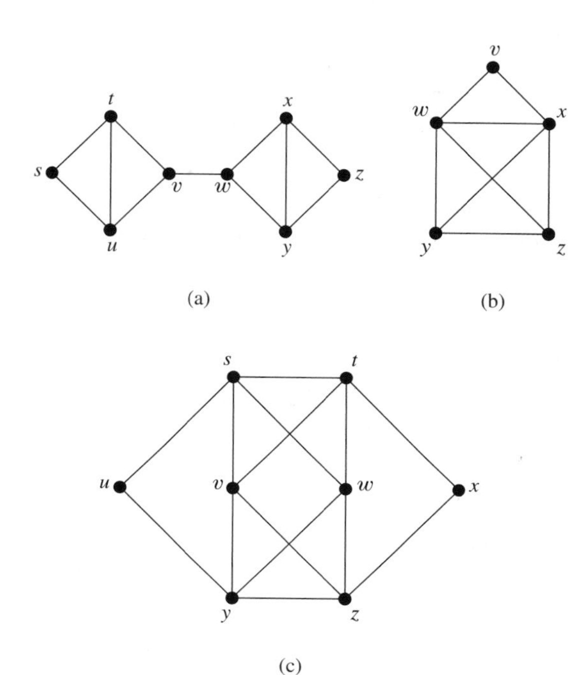

Figure 6 ▲

2. For each pair of vertices in Exercise 1, give the vertex sequence of a longest path connecting them that repeats no edges. Is there a longest path connecting them if edges can be repeated?
3. True or False. "True" means "true in all circumstances under consideration." Consider a graph.
   (a) If there is an edge from a vertex u to a vertex v, then there is an edge from v to u.
   (b) If there is an edge from a vertex u to a vertex v and an edge from v to a vertex w, then there is an edge from u to w.
4. Repeat Exercise 3 with the word "edge" replaced by "path" everywhere.
5. Repeat Exercise 3 with the word "edge" replaced by "path of even length" everywhere.
6. Confirm Theorem 3 for each graph in Figure 6 by calculating
   (i) the sum of the degrees of all of the vertices;
   (ii) the number of edges;
   (iii) the sum $D_1(G) + 2D_2(G) + 3D_3(G) + \cdots$.
7. Give an example of a graph with vertices x, y, and z that has all three of the following properties:
   (i) there is a cycle using vertices x and y;
   (ii) there is a cycle using vertices y and z;
   (iii) no cycle uses vertices x and z.
8. Can a graph have an odd number of vertices of odd degree? Explain.
9. (a) Give a table of the function γ for the graph G pictured in Figure 7.
   (b) List the edges of this graph, considered as subsets of $V(G)$. For example, $a = \{w, x\}$.

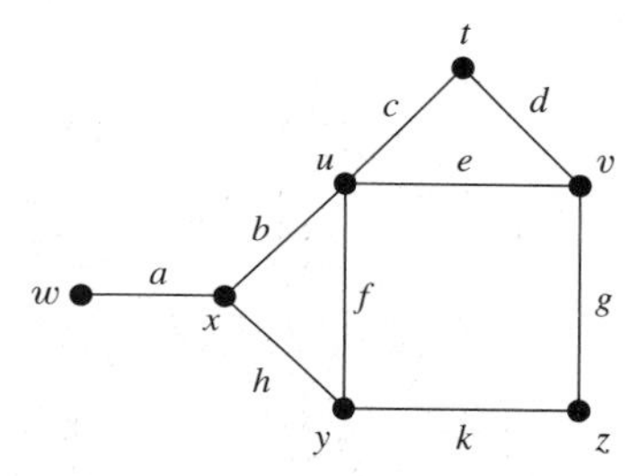

Figure 7 ▲

10. Draw a picture of the graph G with $V(G) = \{x, y, z, w\}$, $E(G) = \{a, b, c, d, f, g, h\}$, and γ as given by the table

e	a	b	c	d	f	g	h
$\gamma(e)$	$\{x, y\}$	$\{x, y\}$	$\{w, x\}$	$\{w, y\}$	$\{y, z\}$	$\{y, z\}$	$\{w, z\}$

11. In each part of this exercise, two paths are given that join a pair of points in the graph of Figure 7. Use the idea of the proof of Theorem 2 to construct cycles out of sequences of edges in the two paths.
    (a) $e\,b\,a$ and $g\,k\,h\,a$ (b) $a\,b\,c\,d\,g\,k$ and $a\,h$
    (c) $e\,b\,a$ and $d\,c\,f\,h\,a$
12. Suppose that a graph H is isomorphic to the graph G of Figure 7.
    (a) How many vertices of degree 1 does H have?
    (b) Give the degree sequence of H.
    (c) How many different isomorphisms are there of G onto G? Explain.
    (d) How many isomorphisms are there of G onto H?
13. (a) Draw pictures of all five of the regular graphs that have four vertices, each vertex of degree 2. "All" here means that every regular graph with four vertices and each vertex of degree 2 is isomorphic to one of the five, and no two of the five are isomorphic to each other.

(b) Draw pictures of all regular graphs with four vertices, each of degree 3, and with no loops or parallel edges.

(c) Draw pictures of all regular graphs with five vertices, each of degree 3.

14. (a) Draw pictures of all 14 graphs with three vertices and three edges. [See Exercise 13 for the meaning of "all" here.]

(b) Draw pictures of the two graphs with four vertices and four edges that have no loops or parallel edges.

(c) List the four graphs in parts (a) and (b) that are regular.

15. Which, if any, of the pairs of graphs shown in Figure 8 are isomorphic? Justify your answer by describing an isomorphism or explaining why one does not exist.

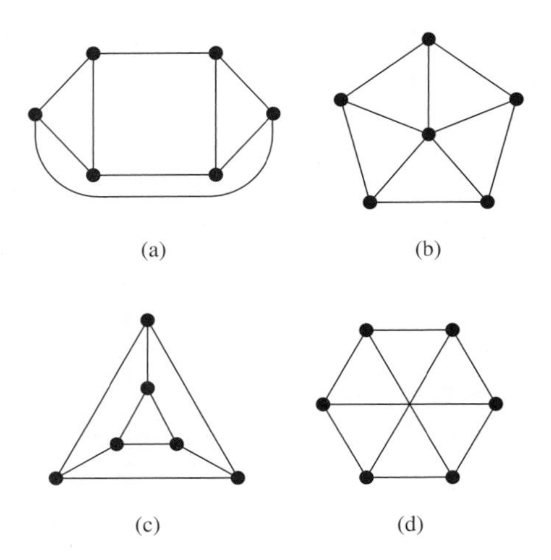

Figure 8 ▲

16. Describe an isomorphism between the graphs shown in Figure 9.

17. Describe an isomorphism between the graphs in (c) and (d) of Figure 3 on page 229.

18. Verify the claims in Example 4(b).

19. Consider the complete graph K_8 with vertices $v_1, v_2, \ldots, v_8$.

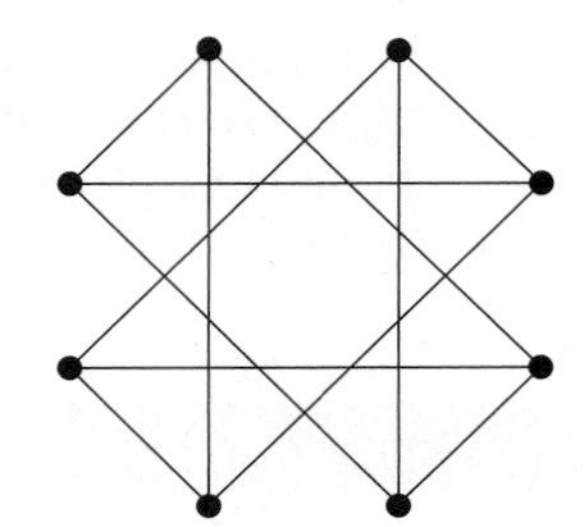
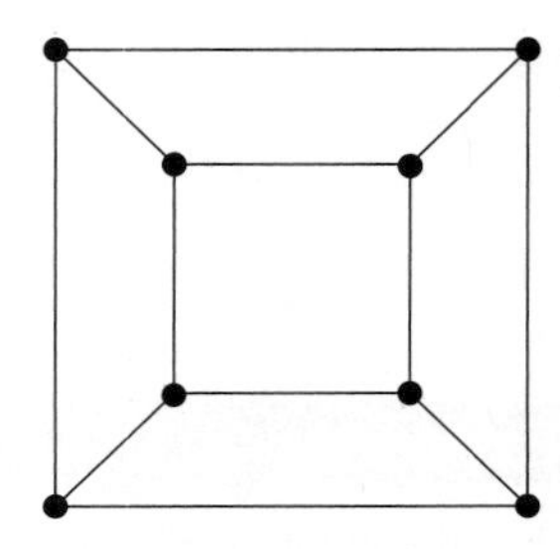

Figure 9 ▲

(a) How many subgraphs of K_8 are isomorphic to K_5?

(b) How many simple paths with three or fewer edges are there from v_1 to v_2?

(c) How many simple paths with three or fewer edges are there altogether in K_8?

20. (a) A graph with 21 edges has seven vertices of degree 1, three of degree 2, seven of degree 3 and the rest of degree 4. How many vertices does it have?

(b) How would your answer to part (a) change if the graph also had six vertices of degree 0?

21. Which of the following are degree sequences of graphs? In each case, either draw a graph with the given degree sequence or explain why no such graph exists.

(a) $(1, 1, 0, 3, 1, 0, 0, \ldots)$ (b) $(4, 1, 0, 3, 1, 0, 0, \ldots)$

(c) $(0, 1, 0, 2, 1, 0, 0, \ldots)$ (d) $(0, 0, 2, 2, 1, 0, 0, \ldots)$

(e) $(0, 0, 1, 2, 1, 0, 0, \ldots)$ (f) $(0, 1, 0, 1, 1, 1, 0, \ldots)$

(g) $(0, 0, 0, 4, 0, 0, 0, \ldots)$ (h) $(0, 0, 0, 0, 5, 0, 0, \ldots)$

22. (a) Let $\mathcal{S}$ be a set of graphs. Show that isomorphism $\simeq$ is an equivalence relation on $\mathcal{S}$.

(b) How many equivalence classes are there if $\mathcal{S}$ consists of the four graphs in Figure 4?

23. Show that every finite graph in which each vertex has degree at least 2 contains a cycle.

24. Show that every graph with n vertices and at least n edges contains a cycle. *Hint:* Use induction on n and Exercise 23.

25. Show that

$$2|E(G)| - |V(G)| = -D_0(G) + D_2(G) + 2D_3(G) + \cdots + (k-1)D_k(G) + \cdots.$$

6.2 Edge Traversal Problems

One of the oldest problems involving graphs is the Königsberg bridge problem: Is it possible to take a walk in the town shown in Figure 1(a) crossing each bridge exactly once and returning home? The Swiss mathematician Leonhard Euler [pronounced OIL-er] solved this problem in 1736. He constructed the graph shown in Figure 1(b), replacing the land areas by vertices and the bridges joining them by edges. The question then became this: Is there a closed path in this graph that uses each edge exactly once? We call such a path an **Euler circuit** of the graph. More generally, a simple path that contains all edges of a graph G is called an **Euler path** of G. Euler showed that there is no Euler circuit for the Königsberg bridge graph,

Figure 1 ▶

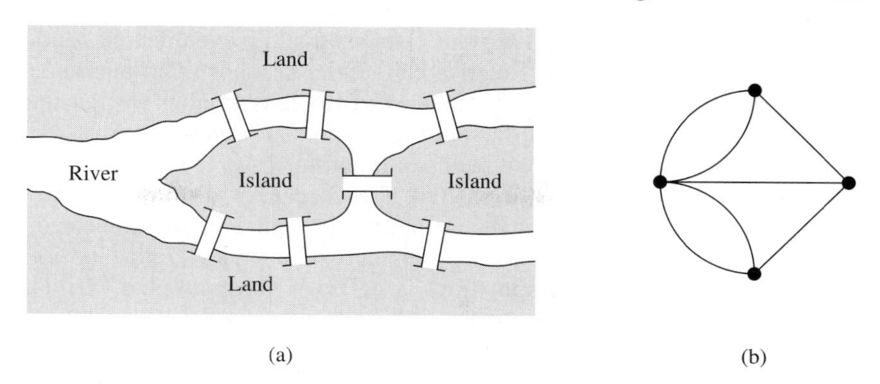

(a) (b)

in which all vertices have odd degree, by establishing the following elementary fact.

Theorem 1 A graph that has an Euler circuit must have all vertices of even degree.

Proof Start at some vertex on the circuit and follow the circuit from vertex to vertex, erasing each edge as you go along it. When you go through a vertex, you erase one edge going in and one going out, or else you erase a loop. Either way, the erasure reduces the degree of the vertex by 2. Eventually, every edge gets erased and all vertices have degree 0. So all vertices must have had even degree to begin with. ■

Again, the convention concerning degrees of vertices with loops is crucial. Without it, this proof would not work and the theorem would be false.

Corollary A graph G that has an Euler path has either two vertices of odd degree or no vertices of odd degree.

Proof Suppose that G has an Euler path starting at u and ending at v. If $u = v$, the path is closed and Theorem 1 says that all vertices have even degree. If $u \neq v$, create a new edge e joining u and v. The new graph $G \cup \{e\}$ has an Euler circuit consisting of the Euler path for G followed by e, so all vertices of $G \cup \{e\}$ have even degree. Remove e. Then u and v are the only vertices of $G = (G \cup \{e\}) \setminus \{e\}$ of odd degree. ■

EXAMPLE 1

The graph shown in Figure 2(a) has no Euler circuit, since u and v have odd degree, but the path $b\ a\ c\ d\ g\ f\ e$ is an Euler path. The graph in Figure 2(b) has all vertices of even degree and in fact has an Euler circuit. The graph in Figure 2(c) has all vertices of even degree but has no Euler circuit, for the obvious reason that the graph is broken into two subgraphs that are not connected to each other. Each of the subgraphs, however, has its own Euler circuit. ■

Figure 2 ▶

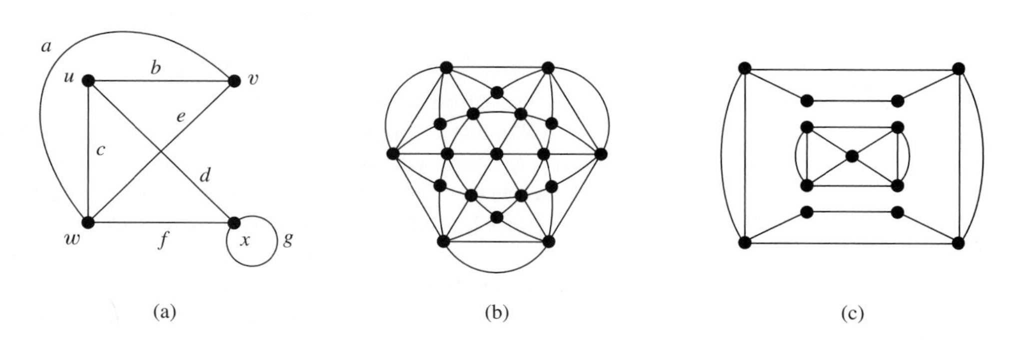

(a) (b) (c)

Theorem 1 shows that the even-degree condition is necessary for the existence of an Euler circuit. Euler's major contribution to the problem was his proof that, except for the sort of obvious trouble we ran into in Figure 2(c), the condition is also sufficient to guarantee an Euler circuit.

We need some terminology to describe the exceptional cases. A graph is said to be **connected** if each pair of distinct vertices is joined by a path in the graph. The graphs in Figures 2(a)and 2(b) are connected, but the one in Figure 2(c) is not. A connected subgraph of a graph G that is not contained in a larger connected subgraph of G is called a **component** of G. The component containing a given vertex v consists of v together with all vertices and edges on paths starting at v.

EXAMPLE 2

(a) The graphs of Figures 2(a) and 2(b) are connected. In these cases the graph has just one component, the graph itself.

(b) The graph of Figure 2(c) has two components, the one drawn on the outside and the one on the inside. Another picture of this graph is shown in Figure 3(a). In this picture there is no "inside" component, but of course there are still two components.

(c) The graph of Figure 3(b) has seven components, two of which are isolated vertices, i.e., vertices which cannot be reached from any other vertices. ■

Figure 3 ▶

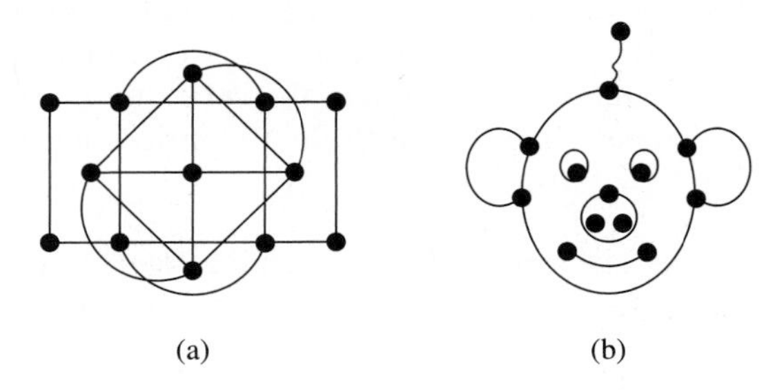

Here's what Euler showed and a consequence of it.

Theorem 2 (Euler's theorem) A finite connected graph in which every vertex has even degree has an Euler circuit.

Corollary A finite connected graph that has exactly two vertices of odd degree has an Euler path.

Proof of the Corollary Say u and v have odd degree. Create a new edge e joining them. Then $G \cup \{e\}$ has all vertices of even degree, so it has an Euler circuit by Theorem 2. Remove e again. What remains of the circuit is an Euler path for G. ■

One way to prove this theorem is to consider a simple path of greatest possible length and show that every such path must be an Euler circuit. A proof along these lines does not lead to a practical algorithm, since for even moderate-sized graphs it is unrealistic to search for a longest simple path by examining paths one by one. To really understand this theorem, we would like to find a proof that leads to an algorithm or procedure that will always produce an Euler circuit. A full understanding of a proof often leads to an algorithm, and behind every algorithm there's a proof.

Here is a simple explanation of Euler's theorem, which we illustrate using Figure 4. Start with any vertex, say w, and any edge connected to it, say a. The other vertex, x in this case, has even degree and has been used an odd number of times [once], so there is an unused edge leaving x. Pick one, say b. Continue in this way. The process won't stop until the starting vertex w is reached since, whenever any other vertex is reached, only an odd number of its edges have been used. In our example, this algorithm might start out with edges $a\,b\,e$ and vertices $w\,x\,w\,y$. At y

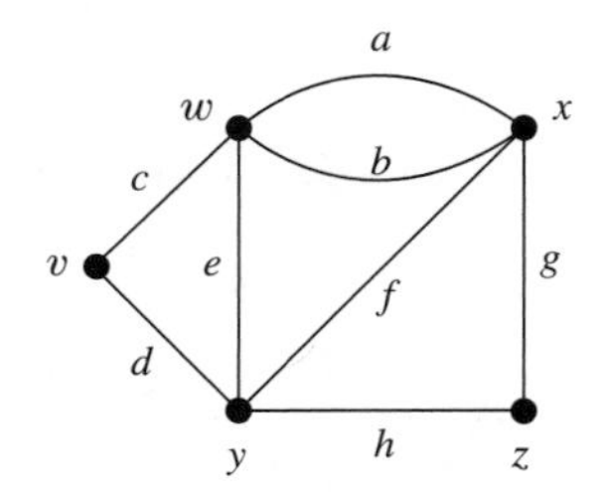

Figure 4 ▲

we can choose any of three edges: d, f, or h. If we select f, the rest of the process is determined. We end up with the Euler circuit $a\ b\ e\ f\ g\ h\ d\ c$ with vertex sequence $w\ x\ w\ y\ x\ z\ y\ v\ w$.

Simple, wasn't it? Well, it is too simple. What would have happened if, when we first reached vertex y, we had chosen edge d? After choosing edge c, we'd have been trapped at vertex w and our path $a\ b\ e\ d\ c$ would have missed edges f, g, and h. Our explanation and our algorithm must be too simple. In our example it is clear that edge d should have been avoided when we first reached vertex y, but why? What general principle should have warned us to avoid this choice? Think about it.

Let's take another look at the Euler circuit that we found for Figure 4, which we reproduce in Figure 5(a). As we select edges, let's remove them from the graph and consider the subgraphs we obtain. Our path started out with edges $a\ b\ e$; Figure 5(b) shows the graph with these edges removed. In our successful search for an Euler circuit, we next selected f, and we noted that if we had selected d we were doomed. Figure 5(c) shows the graph if f is also removed, while Figure 5(d) shows the graph if instead d is removed. There is a difference: removal of d disconnects the graph, and it's pretty clear we're trapped in a component of the graph that is left. On the other hand, removal of f does not disconnect the graph.

Figure 5 ▶

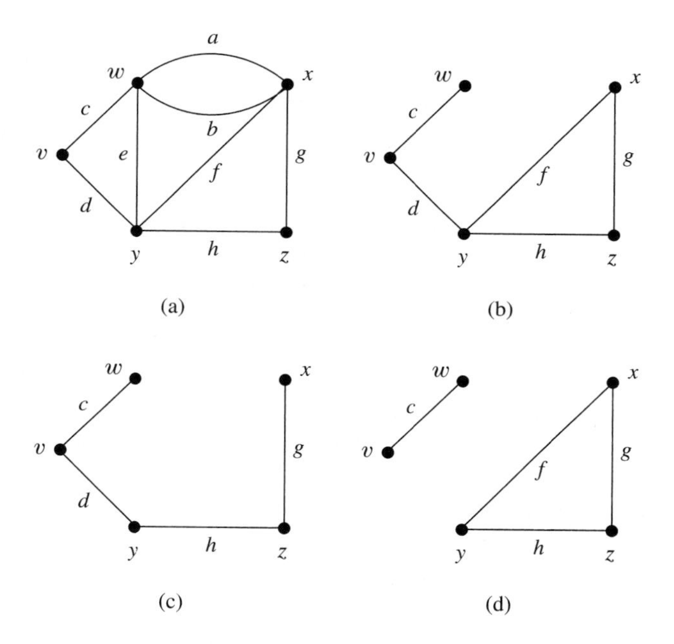

It turns out that it is always possible to select an edge whose removal will not disconnect the graph, unless there's only one edge to choose, in which case we choose that edge and delete both it and the vertex it led from. This fact, which is not trivial to prove, is the basis for Fleury's algorithm to construct Euler paths. For details on this algorithm and for a proof of Euler's theorem based on longest paths, see either the first or third edition of this book. One critical drawback of Fleury's algorithm is that it requires repeatedly examining subgraphs of G to see if they are connected, a time-consuming operation. Even using a fairly fast connectivity check, we know of no way to have Fleury's algorithm take fewer than $\Theta(|V(G)|^2 \cdot |E(G)|)$ steps. [Big-theta Θ and big-oh O notation were explained in §4.3.]

There is another approach that is conceptually fairly simple and that leads to an $O(|E(G)|)$ algorithm, which is much faster than Fleury's algorithm for large graphs. Instead of looking for a longest closed path, let's take any closed path we can get and then attach others to it until we've used up all the edges. Figure 6 shows the idea.

Figure 6 ▶

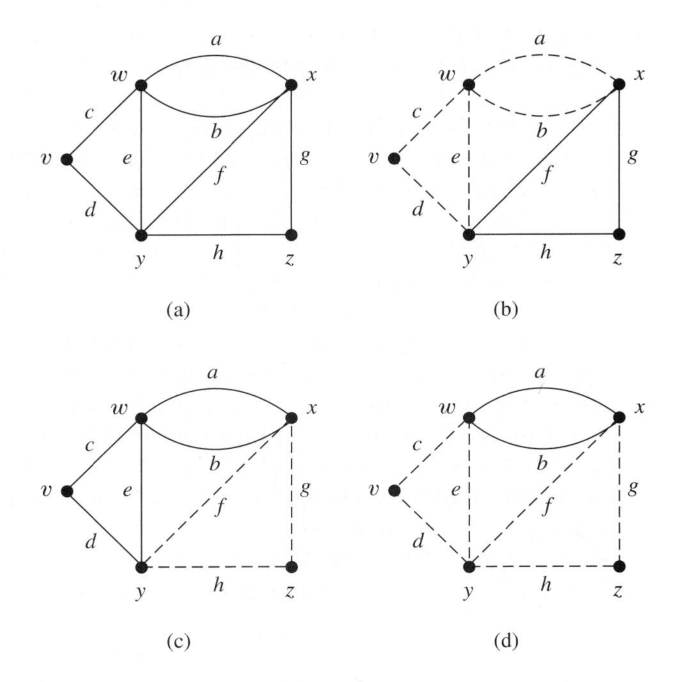

Figure 6(a) shows our initial graph again. In Figure 6(b) we start by choosing the dashed closed path $a\,b\,e\,d\,c$, the one that got us stuck at w earlier. The remaining edges form the closed path $f\,g\,h$, which we can attach to $a\,b\,e\,d\,c$, say at x, to get the Euler circuit $a\,g\,h\,f\,b\,e\,d\,c$. We could have attached the path at y as well.

In Figure 6(c) we have chosen the closed path $f\,g\,h$ initially. What's left is a closed path, if we traverse the edges as $a\,b\,c\,d\,e$. We can attach this path to $f\,g\,h$ at x or at y to get an Euler circuit. Suppose that we had not seen the full path $a\,b\,c\,d\,e$, but only $c\,d\,e$. We could still have attached $c\,d\,e$ to $f\,g\,h$ at y to get the dashed closed path $f\,g\,h\,c\,d\,e$ in Figure 6(d) and then attached the remaining path $a\,b$ at w or x.

The idea is simply to keep on attaching closed paths until all the edges are used up. Here is the algorithm. The notation $G \setminus C$ denotes the graph with vertex set $V(G)$ and edge set $E(G) \setminus C$.

EulerCircuit(graph)

{Input: A connected graph G with all vertices of positive even degree}
{Output: An Euler circuit C of G}
Choose $v \in V(G)$.
Construct a simple closed path C in G with v as one vertex.
{C is a closed path in G with no repeated edges.}
while length(C) $< |E(G)|$ **do**
  Choose a vertex w on C of positive degree in $G \setminus C$.
  Construct a simple closed path in $G \setminus C$ through w.
  Attach this path to C at w to obtain a longer simple closed path C.
return C ■

Proof That EulerCircuit Works If the statement "C is a closed path in G with no repeated edges" is a loop invariant and if the algorithm does not break down somewhere, then this algorithm will produce an Euler circuit for G, because the path C will be closed at the end of each pass through the loop, the number of edges remaining will keep going down, and the loop will terminate with all edges of G in C.

If C is a simple closed path, if w is a vertex of C, and if D is a simple closed path through w whose edges are not in C, then attaching D to C at w yields a simple closed path: follow C to w, take a trip around D and back to w, then continue with C to its starting point. Thus the statement in braces is a loop invariant.

Will there always be a place to attach another closed path to C, i.e., a vertex on C of positive degree in $G \setminus C$? That is, can the instruction "Choose a vertex w on C of positive degree in $G \setminus C$" be executed? Yes, unless C contains all the edges of G, in which case the algorithm stops. Here's why. Suppose that e is an edge not in C and that u is a vertex of e. If C goes through u, then u itself has positive degree in $G \setminus C$, and we can attach at u. So suppose that u is not on C. Since G is connected, there is a path in G from u to the vertex v on C. [Here's where we need connectedness.] Let w be the first vertex in such a path that is on C [then $w \neq u$, but possibly $w = v$]. Then the edges of the part of the path from u to w don't belong to C. In particular, the last one [the one to w] does not belong to C. So w is on C and has positive degree in $G \setminus C$.

We also need to be sure that the instruction "Construct a simple closed path in $G \setminus C$ through w" can be executed. Thus the proof will be complete once we show that the following algorithm works to construct the necessary paths. ■

ClosedPath(graph, vertex)

{Input: A graph H in which every vertex has even degree,
  and a vertex v of positive degree}
{Output: A simple closed path P through v}
Choose an edge e of H with endpoint v.
Let $P := e$ and remove e from $E(H)$.
while there is an edge at the terminal vertex of P **do**
  Choose such an edge, add it to the end of P and remove it from $E(H)$.
return P ■

Proof That ClosedPath Works We want to show that the algorithm produces a simple closed path from v to v. Simplicity is automatic, because the algorithm deletes edges from further consideration as it adds them to the path P.

Since v has positive degree initially, there is an edge e at v to start with. Could the algorithm get stuck someplace and not get back to v? When P passes through a vertex w other than v, it reduces the degree of w by 2—it removes an edge leading into w and one leading away—so the degree of w stays an even number. [Here's where we use the hypothesis about degrees.] Hence, whenever we have chosen an edge leading into a w, there's always another edge leading away to continue P. The path must end somewhere, since no edges are used twice, but it cannot end at any vertex other than v. ■

Here's EulerCircuit again with the calls to ClosedPath shown explicitly.

EulerCircuit(graph)

{Input: A connected graph G with all vertices of positive even degree}
{Output: An Euler circuit C of G}
Choose $v \in V(G)$.
Let $C :=$ ClosedPath(G, v).
while length$(C) < |E(G)|$ **do**
  Choose a vertex w on C of positive degree in $G \setminus C$.
  Attach ClosedPath$(G \setminus C, w)$ to C at w to obtain a longer simple
    closed path C.
return C ■

The operations in EulerCircuit and its subroutine ClosedPath consist of adding or removing edges. Each operation takes a fixed amount of time, and each edge only gets handled once; so with a suitable data structure the total time to construct an Euler circuit is $O(|E(G)|)$.

Exercises 6.2

1. Which of the graphs in Figure 7 have Euler circuits? Give the vertex sequence of an Euler circuit in each case in which one exists.

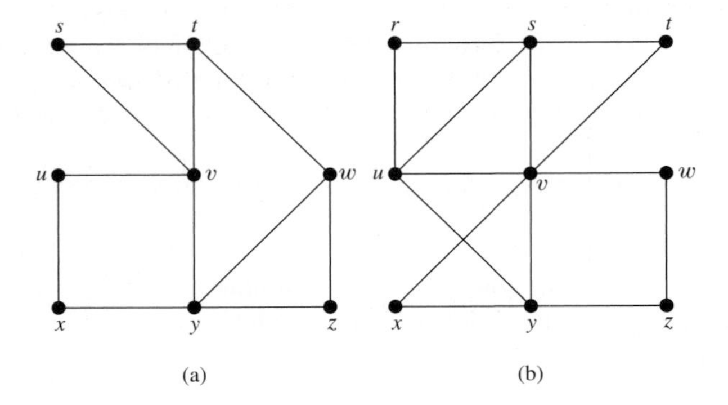

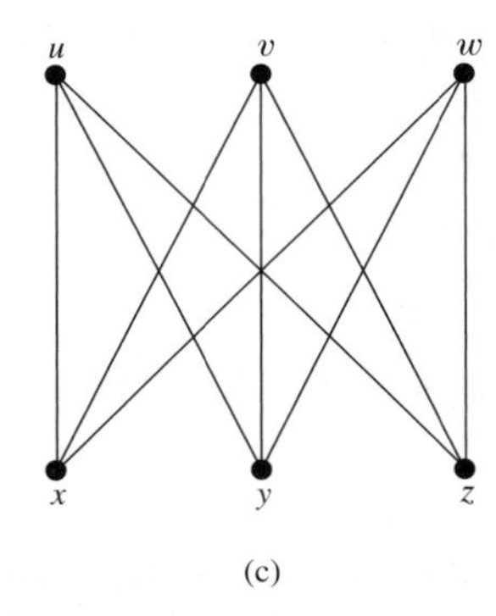

Figure 7 ▲

2. (a) Give the vertex sequence of a simple closed path of largest possible length in the graph of Figure 7(a).
   (b) Is your path in part (a) an Euler circuit for the graph?
3. Apply the algorithm EulerCircuit to the graph in Figure 7(b).
4. Draw K_5 as in Figure 5 on page 230 and label the vertices. Then apply algorithm EulerCircuit to it.
5. Apply the algorithm EulerCircuit to the graph of Figure 7(c) until something goes wrong. What's the trouble?
6. Is it possible for an insect to crawl along the edges of a cube so as to travel along each edge exactly once? Explain.
7. Consider the graph shown in Figure 8(a).
   (a) Describe an Euler path for this graph or explain why there isn't one.
   (b) Describe an Euler circuit for this graph or explain why there isn't one.
8. Repeat Exercise 7 for the graph of Figure 8(b).
9. (a) Explain how to modify the algorithm EulerCircuit(G) to obtain an Euler path for a connected graph with exactly two vertices of odd degree.
   (b) What will the algorithm EulerCircuit(G) return if the input graph G is not connected?
10. Apply a modification of the EulerCircuit algorithm as suggested in Exercise 9(a) to get an Euler path for the

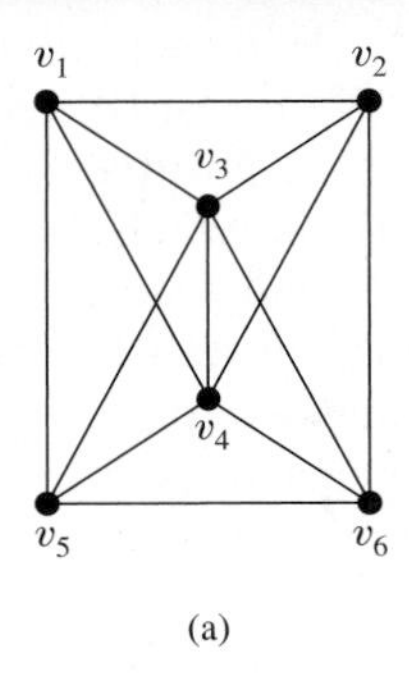

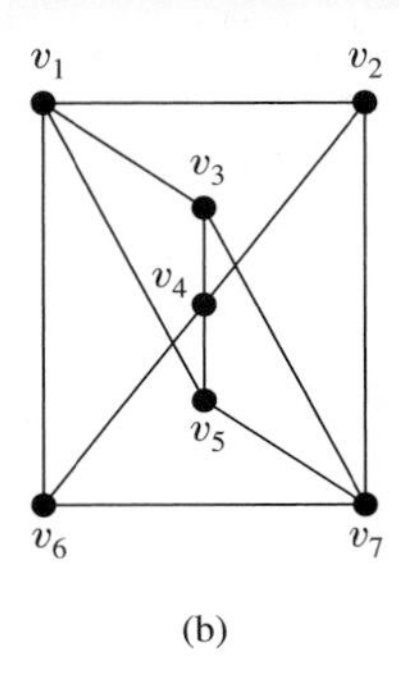

Figure 8 ▲

graph of Figure 2(a). Describe the steps in your construction.

11. Construct a graph with vertex set $\{0, 1\}^3$, i.e., the set $\{0, 1\} \times \{0, 1\} \times \{0, 1\}$, and with an edge between vertices if they differ in exactly two coordinates.
    (a) How many components does the graph have?
    (b) How many vertices does the graph have of each degree?
    (c) Does the graph have an Euler circuit?
12. Answer the same questions as in Exercise 11 for the graph with vertex set $\{0, 1\}^3$ and with an edge between vertices if they differ in two or three coordinates.
13. (a) Show that if a connected graph G has exactly $2k$ vertices of odd degree and $k \geq 2$, then $E(G)$ is the disjoint union of the edge sets of k simple paths. *Hint:* Add more edges, as in the proof of the corollary to Theorem 2.
    (b) Find two disjoint simple paths whose edge set union is $E(G)$ for the Königsberg graph in Figure 1.
    (c) Repeat part (b) for the graph in Figure 8(b).
14. Which complete graphs K_n have Euler circuits?
15. An old puzzle presents a house with 5 rooms and 16 doors, as shown in Figure 9. The problem is to figure out how to walk around and through the house so as to go through each door exactly once.
    (a) Is such a walk possible? Explain.
    (b) How does your answer change if the door adjoining the two large rooms is sealed shut?

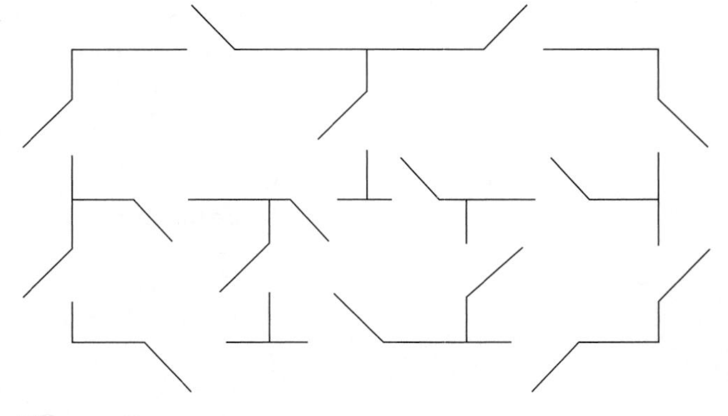

Figure 9 ▲

6.3 Trees

Trees are probably the graphs that people are most familiar with, especially in connection with data structures. Figure 1(d) on page 100, Figure 1 on page 183, and Figures 1, 2, and 3 in §6.4 coming up all illustrate ways in which tree structures can be used as conceptual frameworks for arranging data. We will see other important examples later as well. Trees also arise as subgraphs of other graphs, with the most familiar examples probably related to network designs in which a number of nodes must be connected, with exactly one path between any two nodes. This section introduces trees in general, with subgraph applications in mind. In §6.4 we will return to network questions.

Each of the figures we have just mentioned shows a connected graph that contains no cycles. That's the definition: a **tree** is a connected acyclic graph. Since they are acyclic, trees have no parallel edges or loops. Here are some more examples.

Figure 1 ▶

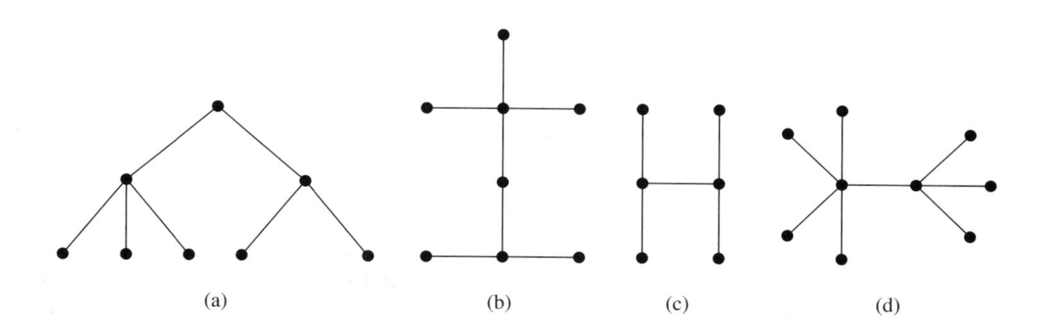

EXAMPLE 1

Figure 1 contains pictures of some trees. The ones in Figures 1(a) and 1(b) are isomorphic. Their pictures are different, but the essential structures [vertices and edges] are the same. They share all graph-theoretic properties, such as the numbers of vertices and edges and the number of vertices of each degree. To make this clear, we have redrawn them in Figures 2(a) and 2(b) and labeled corresponding vertices. The two trees in Figures 2(c) and 2(d) are also isomorphic. ■

Figure 2 ▶

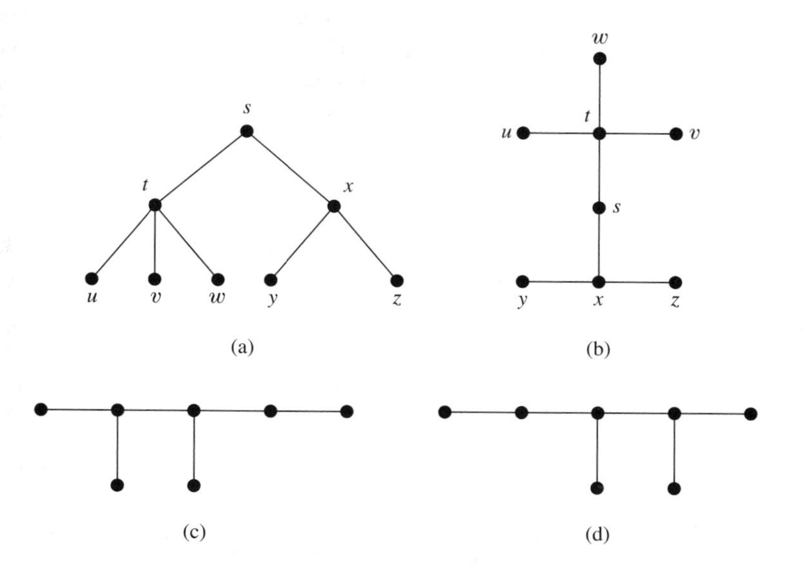

EXAMPLE 2

The eleven trees with seven vertices are pictured in Figure 3. In other words, every tree with seven vertices is isomorphic to one drawn in Figure 3, and no two of the trees in Figure 3 are isomorphic to each other. ■

Figure 3 ▶

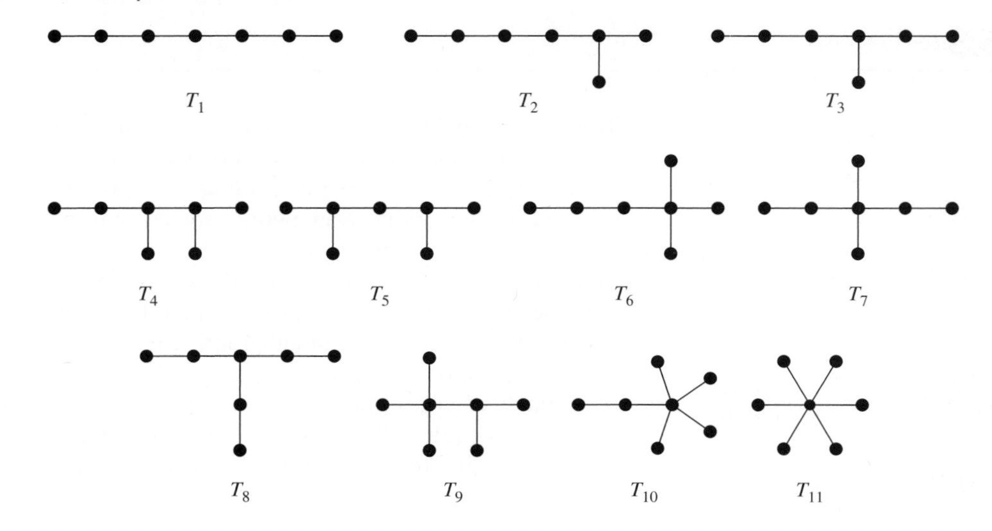

Trees are connected and acyclic, so to tell whether a connected graph is a tree, we need to be able to detect cycles. The next theorem lists some properties of edges that can help us decide acyclicity questions.

Theorem 1 Let e be an edge of a connected graph G. The following statements are equivalent:

(a) $G \setminus \{e\}$ is connected.
(b) e is an edge of some cycle in G.
(c) e is an edge of some simple closed path in G.

Proof First note that if e is a loop, then $G \setminus \{e\}$ is connected, while e is a cycle all by itself. Since cycles are simple closed paths, the theorem holds in this case, and we may assume that e is not a loop. Thus e connects distinct vertices u and v. If f is another edge connecting u and v, then clearly $G \setminus \{e\}$ is connected and $e\,f$ is a cycle containing e. So the theorem also holds in this case. Hence we may assume that e is the unique edge connecting u and v.

(a) $\Longrightarrow$ (b). Suppose that $G \setminus \{e\}$ is connected. By Theorem 1 on page 227 there is a simple acyclic path $x_1 x_2 \cdots x_m$ with $u = x_1$ and $x_m = v$. Since there is no edge from u to v in $G \setminus \{e\}$, we have $x_2 \neq v$, so $m \geq 3$. As noted in Proposition 2 on page 226, the vertices $x_1, x_2, \ldots, x_m$ are distinct, so $x_1 x_2 \cdots x_m u$ is a cycle in G containing the edge e.

(b) $\Longleftrightarrow$ (c). Obviously, (b) $\Longrightarrow$ (c) since cycles are simple closed paths, and (c) $\Longrightarrow$ (b) even if G is not connected, by the Corollary to Theorem 1 on page 227.

(b) $\Longrightarrow$ (a). The edge e is one path between u and v, while the rest of the cycle containing e gives an alternative route. It is still possible to get from any vertex of G to any other, even if the bridge is out at e, just by replacing e by the alternative if necessary. ■

EXAMPLE 3

We illustrate Theorem 1 using the connected graph in Figure 4(a). Note that e_1 does not belong to a cycle and that $G \setminus \{e_1\}$ is disconnected, since no path in $G \setminus \{e_1\}$ connects v to the other vertices. Likewise, e_5 belongs to no cycle and $G \setminus \{e_5\}$ is disconnected. The remaining edges belong to cycles. Removal of any *one* of them will not disconnect G. ■

Given a connected graph G, we are interested in minimal subgraphs that connect all the vertices. Such a subgraph must be acyclic, since by Theorem 1 one edge of any cycle could be removed without losing the connectedness property. In other words, such a subgraph T is a **spanning tree**: it's a tree that includes every vertex

Figure 4 ▶

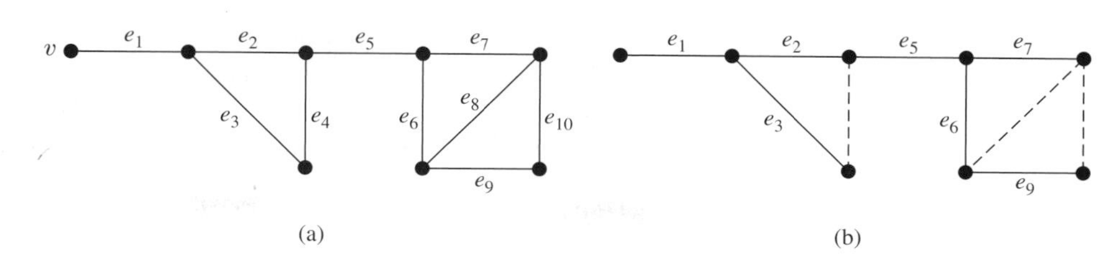

of G, i.e., $V(T) = V(G)$. Thus T is a tree obtained from G by removing some of the edges, perhaps, but keeping all of the vertices.

EXAMPLE 4

The graph H in Figure 5(a) has over 300 spanning trees, of which 4 have been sketched. They all have 6 edges. ■

Figure 5 ▶

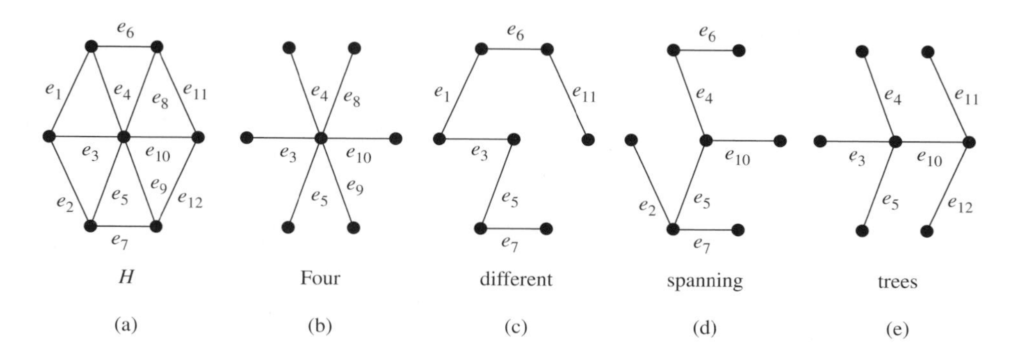

The next theorem guarantees that spanning trees exist for all finite connected graphs. In §6.6 we will see how to construct them.

Theorem 2 Every finite connected graph G has a spanning tree.

Proof Consider a connected subgraph G' of G that uses all the vertices of G and has as few edges as possible. Suppose that G' contains a cycle, say one involving the edge e. By Theorem 1, $G' \setminus \{e\}$ is a connected subgraph of G that has fewer edges than G' has, contradicting the choice of G'. So G' has no cycles. Since it's connected, G' is a tree. ■

EXAMPLE 5

We illustrate Theorem 2 using the connected graph in Figure 5(a). Note that $G \setminus \{e_{10}\}$ is still connected but has cycles. If we also remove e_8, the resulting graph still has the cycle $e_2\, e_3\, e_4$. But if we then remove one of the edges in the cycle, say e_4, we obtain an acyclic connected subgraph, i.e., a spanning tree. See Figure 5(b). Clearly, several different spanning trees can be obtained in this way. ■

In characterizing trees we lose nothing by restricting our attention to graphs with no loops or parallel edges. Our first characterizations hold even if the graph is infinite.

Theorem 3 Let G be a graph with more than one vertex, no loops, and no parallel edges. The following are equivalent:

(a) G is a tree.
(b) Each pair of distinct vertices is connected by exactly one simple path.
(c) G is connected, but will not be if any edge is removed.
(d) G is acyclic, but will not be if any edge is added.

Proof This proof consists of four short proofs.

(a) $\Longrightarrow$ (b). Suppose that G is a tree, so that G is connected and acyclic. By Theorem 1 on page 227, each pair of vertices is connected by at least one simple path, and by Theorem 2 on page 227, there is just one simple path.

(b) $\Longrightarrow$ (c). If (b) holds, then G is clearly connected. Let $e = \{u, v\}$ be an edge of G and assume that $G \setminus \{e\}$ is still connected. Note that $u \neq v$, since G has no loops. By Theorem 1 on page 227, there is a simple path in $G \setminus \{e\}$ from u to v. Since this path and the one-edge path e are different simple paths in G from u to v, we contradict (b).

(c) $\Longrightarrow$ (d). Suppose that (c) holds. If G had a cycle, then we could remove an edge in the cycle from G and retain connectedness by Theorem 1. So G is acyclic. Now consider an edge e not in the graph G, and let G' denote the graph G with this new edge adjoined. Since $G' \setminus \{e\} = G$ is connected, we apply Theorem 1 to G' to conclude that e belongs to some cycle of G'. In other words, adding e to G destroys acyclicity.

(d) $\Longrightarrow$ (a). If (d) holds and G is not a tree, then G is not connected. Then there exist distinct vertices u and v that are not connected by any paths in G. Consider the new edge $e = \{u, v\}$. According to our assumption (d), $G \cup \{e\}$ has a cycle and e must be part of it. The rest of the cycle is a path in G that connects u and v. This contradicts our choice of u and v. Hence G is connected and G is a tree. ■

To appreciate Theorem 3, draw a tree or look at a tree in one of the Figures 1 through 4 and observe that it possesses all the properties (a)–(d) in Theorem 3. Then draw or look at a nontree and observe that it possesses none of the properties (a)–(d).

Vertices of degree 1 in a tree are called **leaves** [the singular is **leaf**].

EXAMPLE 6

Of the trees in Figure 3, T_1 has two leaves; T_2, T_3, and T_8 have three leaves; T_4, T_5, T_6, and T_7 have four leaves; T_9 and T_{10} have five leaves; and T_{11} has six leaves. ■

The next two lemmas help us characterize finite trees.

Lemma 1 A finite tree with at least one edge has at least two leaves.

Proof Consider a longest simple acyclic path, say $v_1 v_2 \cdots v_n$. Because the path is acyclic, its vertices are distinct, and both vertices v_1 and v_n are leaves, since otherwise the path could be made longer. ■

Lemma 2 A tree with n vertices has exactly $n - 1$ edges.

Proof We apply induction. For $n = 1$ or $n = 2$, the lemma is clear. Assume the statement is true for some $n \geq 2$, and consider a tree T with $n + 1$ vertices. By Lemma 1, T has a leaf v_0. Let T_0 be the graph obtained by removing v_0 and the edge attached to v_0. Then T_0 is a tree, as is easily checked, and it has n vertices. By the inductive assumption, T_0 has $n - 1$ edges, so T has n edges. ■

Theorem 4 Let G be a finite graph with n vertices, no loops, and no parallel edges. The following are equivalent:

(a) G is a tree.
(b) G is acyclic and has $n - 1$ edges.
(c) G is connected and has $n - 1$ edges.

In other words, any two of the properties "connectedness," "acyclicity," and "having $n - 1$ edges" imply the third.

Proof The theorem is obvious for $n = 1$ and $n = 2$, so we assume $n \geq 3$. Both (a) $\Longrightarrow$ (b) and (a) $\Longrightarrow$ (c) follow from Lemma 2.

(b) $\Longrightarrow$ (a). Assume that (b) holds but that G is not a tree. Then (d) of Theorem 3 cannot hold. Since G is acyclic, we can evidently add some edge and retain acyclicity. Now add as many edges as possible and still retain acyclicity. The graph G' so obtained will satisfy Theorem 3(d), so G' will be a tree. Since G' has n vertices and at least n edges, this contradicts Lemma 2. Thus our assumption that G is not a tree is wrong, and G is a tree.

(c) $\Longrightarrow$ (a). Assume that (c) holds but that G is not a tree. By Theorem 2, G has a spanning tree T, which must have fewer than $n - 1$ edges. This contradicts Lemma 2, so G is a tree. ■

As with Theorem 3, to appreciate Theorem 4, draw some trees and observe that all the properties in Theorem 4 hold, and then draw some nontrees and observe that none of the properties holds.

An acyclic graph, whether or not it is connected, is sometimes called a **forest**. The connected components of a forest are acyclic, so they are trees. It is possible to generalize Theorem 4 to characterize forests as well as trees; see Exercise 11.

Exercises 6.3

1. Find all trees with six vertices.
2. The trees in Figure 6 have seven vertices. Specify which tree in Figure 3 each is isomorphic to.

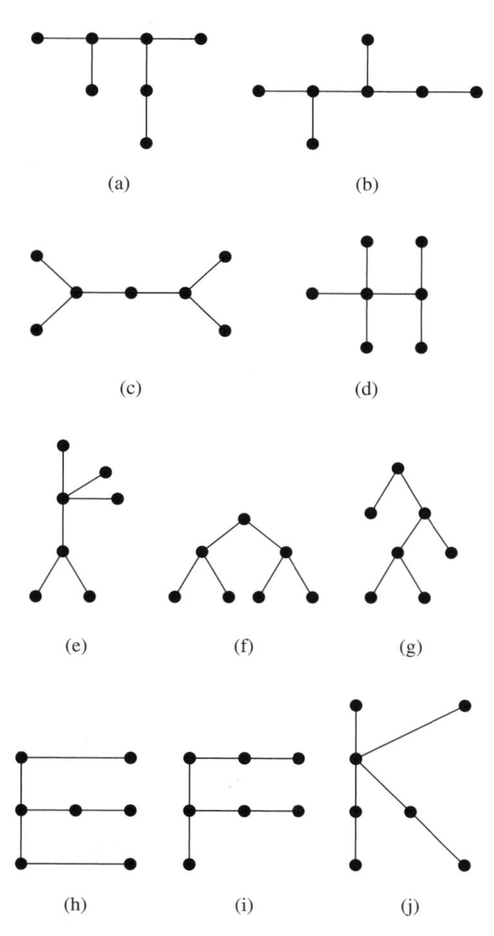

Figure 6 ▲

3. Count the number of spanning trees in the graphs of Figure 7.

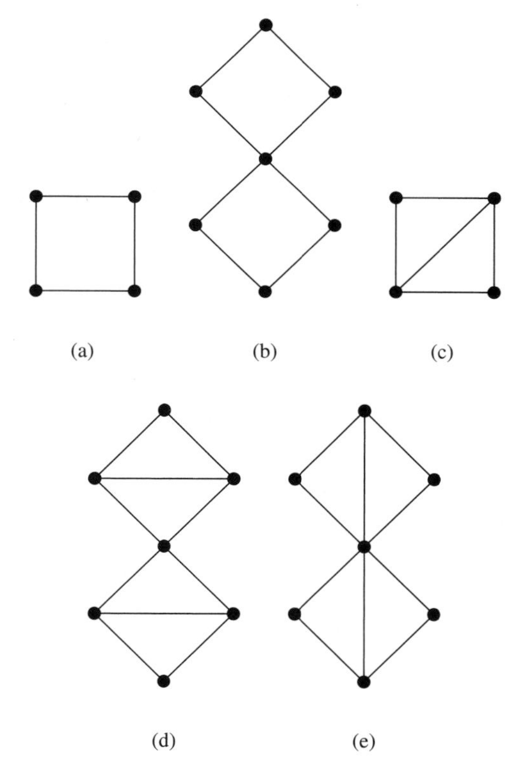

Figure 7 ▲

4. Repeat Exercise 3 for Figure 5(a).
5. For which edges in Figure 5(a) do (a)–(c) of Theorem 1 hold?
6. Find two nonisomorphic spanning trees of $K_{3,3}$ drawn in Figure 7 on page 256.

7. Consider a tree with n vertices. It has exactly $n-1$ edges [Lemma 2], so the sum of the degrees of its vertices is $2n-2$ [Theorem 3 on page 230].

   (a) A certain tree has two vertices of degree 4, one vertex of degree 3, and one vertex of degree 2. If the other vertices have degree 1, how many vertices are there in the graph? *Hint:* If the tree has n vertices, $n-4$ of them will have to have degree 1.

   (b) Draw a tree as described in part (a).

8. Repeat Exercise 7 for a tree with two vertices of degree 5, three of degree 3, two of degree 2, and the rest of degree 1.

9. (a) Show that there is a tree with six vertices of degree 1, one vertex of degree 2, one vertex of degree 3, one vertex of degree 5, and no others.

   (b) For $n \geq 2$, consider n positive integers $d_1, \ldots, d_n$ whose sum is $2n-2$. Show that there is a tree with n vertices whose vertices have degrees $d_1, \ldots, d_n$.

   (c) Show that part (a) illustrates part (b), where $n = 9$.

10. Draw pictures of the nine connected graphs with four edges and four vertices. Don't forget loops and parallel edges.

11. (a) Show that a forest with n vertices and m components has $n-m$ edges.

    (b) Show that a graph with n vertices, m components, and $n-m$ edges must be a forest.

12. Figure 4 illustrates four spanning trees for the graph in Figure 4(a). They have 6, 2, 4, and 5 leaves, respectively. For each part below, either draw a spanning tree with the property, or explain why you cannot do so.

    (a) The tree has 3 leaves. (b) The tree has 1 leaf.

    (c) The central vertex is a leaf.

13. Sketch a tree with at least one edge and no leaves. *Hint:* See Lemma 1 to Theorem 4.

14. Show that a connected graph with n vertices has at least $n-1$ edges.

6.4 Rooted Trees

A **rooted tree** is a tree with one vertex, called its **root**, singled out. This concept is simple, but amazingly useful. Besides having important applications to data structures, rooted trees also help us to organize and visualize relationships in a wide variety of settings. In this section we look at several examples of rooted trees and at some of their properties.

Rooted trees are commonly drawn with their roots at the top, just upside down from the trees in the woods. Our first examples include some typical applications, with accompanying diagrams.

EXAMPLE 1

(a) An ordered set such as a list of numbers or an alphabetized file can be conveniently organized by a special type of rooted tree, called a **binary search tree**, in which each vertex corresponds to an element of the list and there are at most two branches, a left one and a right one, below each vertex. If a vertex has branches below it, then all elements obtained by following the left branch come before the element associated with the vertex itself, which in turn comes before all elements obtained by following the right branch. Figure 1 shows an example of such a tree that holds a set of client records, organized in alphabetical order.

Figure 1 ▶

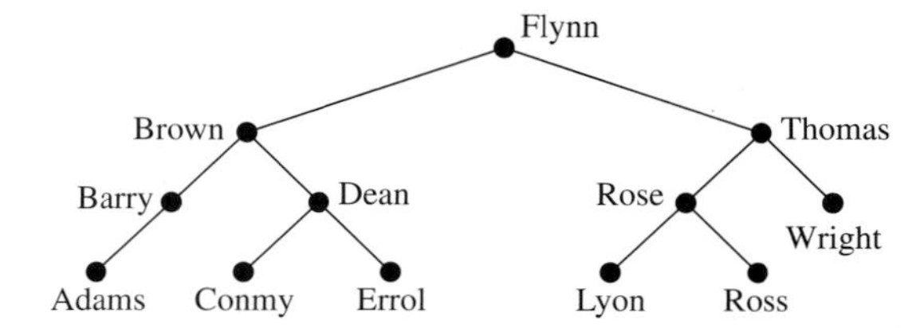

To locate the records for Conmy, say, we compare Conmy alphabetically with the root, Flynn. Conmy comes first, so we take the left branch and compare Conmy with Brown. Conmy comes second, so we take the right branch. Then go left at Dean to find Conmy. If we had wanted the records for Dean, we would have stopped when we got to Dean's vertex. What makes the search procedure work and gives the tree its "binary" name is the fact that at each vertex [or **node**, as they are frequently called in this setting] there are at most two edges downward, at most one to the left and at most one to the right. It is easy to see where the records for a client named Lobell should go: to the right of Flynn, left of Thomas, left of Rose and left of Lyon, and hence at a new vertex below and to the left of Lyon.

The chief advantage to organizing data with a binary search tree is that one only needs to make a few comparisons in order to locate the correct address, even when the total number of records is large. The idea as we have described it here is quite primitive. The method is so important that a number of schemes have been devised for creating and updating search trees to keep the average length of a search path relatively small, on the order of $\log_2 n$, where n is the number of vertices in the tree.

(b) Data structures for diagnosis or identification can frequently be viewed as rooted trees. Figure 2 shows the idea. To use such a data structure, we start at the top and proceed from vertex to vertex, taking an appropriate branch in each case to match the symptoms of the patient. The final leaf on the path gives the name of the most likely condition or conditions for the given symptoms. The same sort of rooted tree structure is the basis for the key system used in field guides for identifying mushrooms, birds, wildflowers, and the like. In the case of the client records in part (a), there was a natural left-to-right way to arrange the labels on the nodes. In this example and the next, the order in which the vertices are listed has no special significance.

Figure 2 ▶

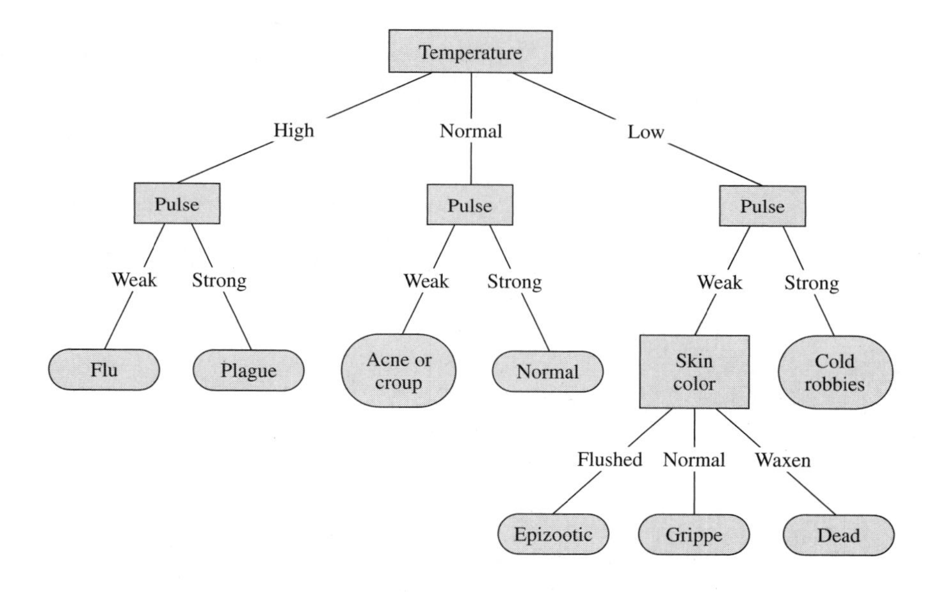

(c) The chains of command of an organization can often be represented by a rooted tree. Part of the hierarchy of a university is indicated in Figure 3.

Figure 3 ▶

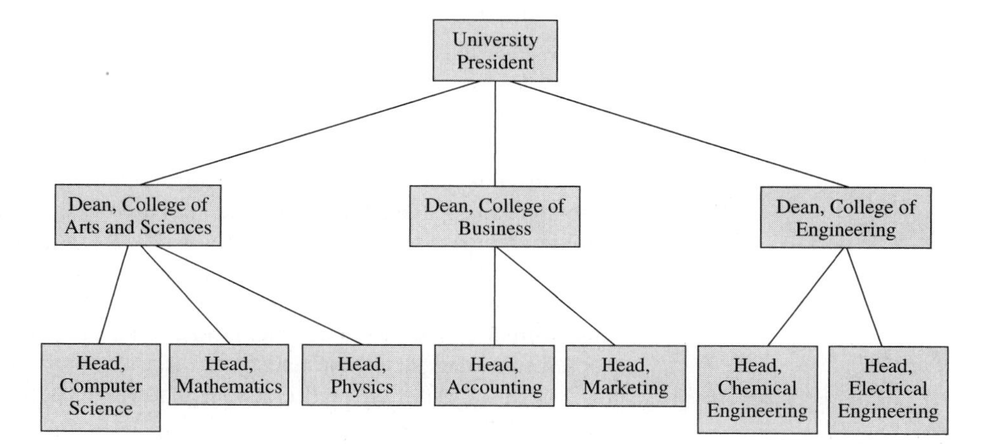

(d) Directory trees provide a way to conceptualize how files are related to each other on a computer and provide the user with a schematic picture in which it is easy to add or delete pieces. For example, in the UNIX model one can imagine file **foo** as being inside directory **goo/**, which is itself inside directory **moo/**, which lives just under the root **/**. Then **foo** is addressed by its path **/moo/goo/foo** from the root directory. Starting at the root, take the **moo** trunk and then the **goo** branch to the **foo** leaf. To get from **foo** to the file **/moo/zip/tip**, one backs up the tree toward the root, through **/moo/goo/** to **/moo/** and then heads down through **/moo/zip/** to **/moo/zip/tip**. The computer scientists who designed UNIX had a tree structure in mind. The idea of documents within folders within folders in other operating systems also is clearly based on trees. ■

In these examples we have given the vertices of the trees fancy names or labels, such as "Ross" or "epizootic," rather than u, v, w, etc., to convey additional information. In fact, the *name* of a vertex and its *label* do not have to be the same, and one could quite reasonably have several vertices with different names but the same label. For instance, the label of a vertex might be a dollar figure associated with it. When we speak of a **labeled tree**, we will mean a tree that has some additional information attached to its vertices. In practical applications we can think of the label of a vertex as being the information that is stored at the vertex. We will see shortly that it may also be possible to choose the vertex names themselves to convey information about the locations of the vertices in the graph.

There is a natural way to view a rooted tree as a digraph. Simply pick up the tree by its root and let gravity direct the edges downward. Figure 4(a) shows a tree with root r directed in this way. It is common to leave the arrow heads off the edges, with the agreement that the edges are directed downward. We can think of the rooted tree in Figure 4(b) as either an undirected tree with distinguished vertex r or as the digraph in Figure 4(a); there is no essential difference between the two points of view.

Figure 4 ▶

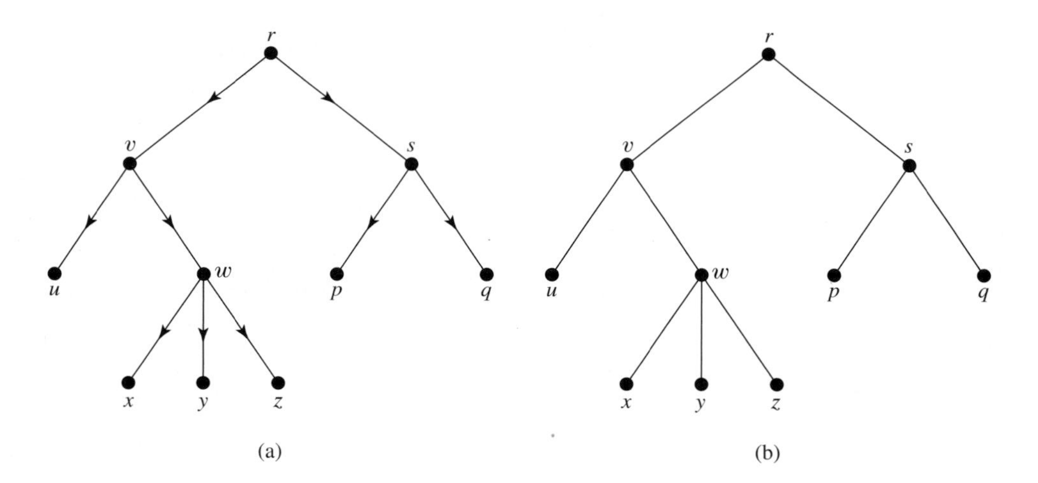

To make the gravitational definition mathematically respectable, we recall from Theorem 3 on page 241 that, whenever r and v are vertices of a tree, there is a unique simple path joining r to v. If r is the root of T and if e is an edge of T with vertices u and w, then either u is on the unique simple path from r to w or w is on the simple path from r to u. In the first case, direct e from u to w, and in the second case, direct e from w to u. This is just what gravity does. We will denote by $\boldsymbol{T_r}$ the rooted tree made from the undirected tree T by choosing the root r. When we think of T_r as a digraph, we will consider it to have the natural digraph structure that we have just described.

EXAMPLE 2

Consider the [undirected] tree in Figure 5(a). If we select v, x, and z to be the roots, we obtain the three rooted trees illustrated in Figures 5(b), (c), and (d). The exact placement of the vertices is unimportant; Figures 5(b) and 5(b′) represent the same rooted tree.

Figure 5 ▶

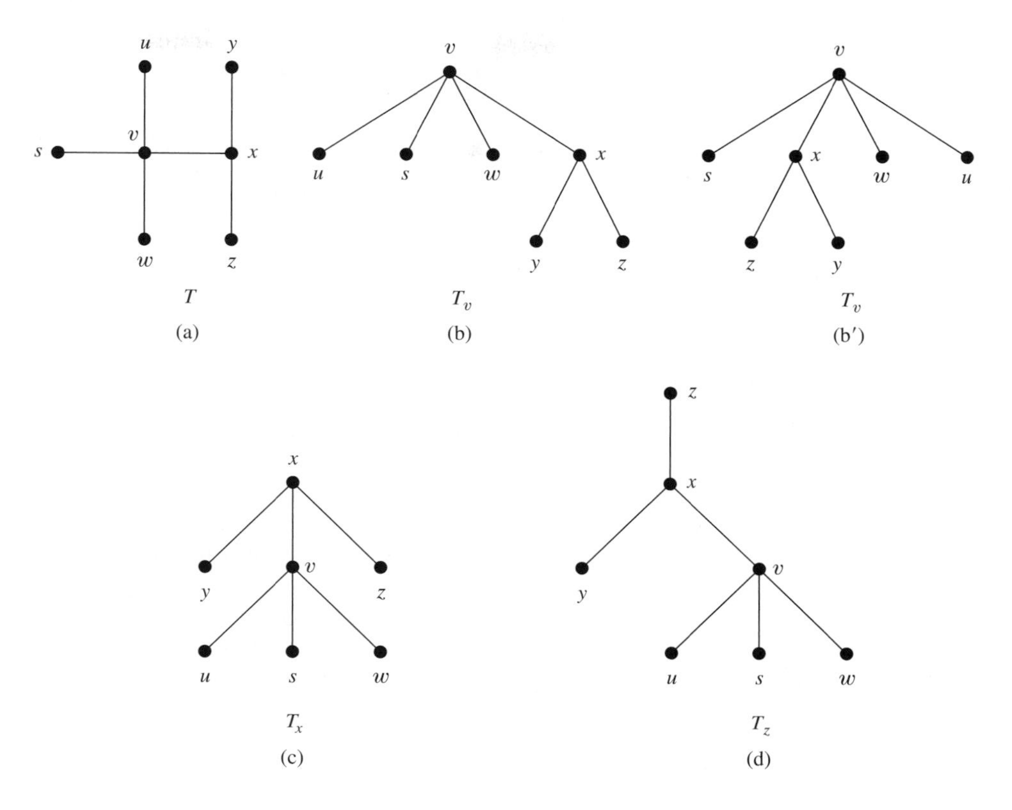

Note that (v, w) is a directed edge of the rooted tree T_z in Figure 5(d), because $\{v, w\}$ is an edge of the unique simple path in Figure 5(a) from z to w. On the other hand, (w, v) is not a directed edge of T_z; even though $\{w, v\}$ is an edge of the original tree, it is not an edge of the unique simple path from z to v. Similar remarks apply to all of the other edges. ■

The terms used to describe various parts of a tree are a curious mixture derived both from the trees in the woods and from family trees. As before, the vertices of degree 1 are called **leaves**; there is one exception: occasionally [as in Figure 5(d)], the root will have degree 1, but we will not call it a leaf. Viewing a tree as a digraph, we see that its root is the only vertex that is not a terminal vertex of an edge, while the leaves are the ones that are not initial vertices. The remaining vertices are sometimes called "branch nodes" or "interior nodes," and the leaves are sometimes called "terminal nodes." We adopt the convention that if (v, w) is a directed edge of a rooted tree, then v is the **parent** of w, and w is a **child** of v. Every vertex except the root has exactly one parent. A parent may have several **children**. More generally, w is a **descendant** of v provided $w \neq v$ and v is a vertex of the unique simple path from r to w. Finally, for any vertex v the **subtree with root** v is precisely the tree $\boldsymbol{T_v}$ consisting of v, all its descendants, and all the directed edges connecting them. Whenever v is a leaf, the subtree with root v is a trivial one-vertex tree.

EXAMPLE 3

Consider the rooted tree in Figure 4, which is redrawn in Figure 6. There are six leaves. The parent v has two children, u and w, and five descendants: u, w, x, y, and z. All the vertices except r itself are descendants of r. The whole tree itself is clearly a subtree rooted at r, and there are six trivial subtrees consisting of leaves. The interesting subtrees are given in Figure 6. ■

Figure 6 ▶

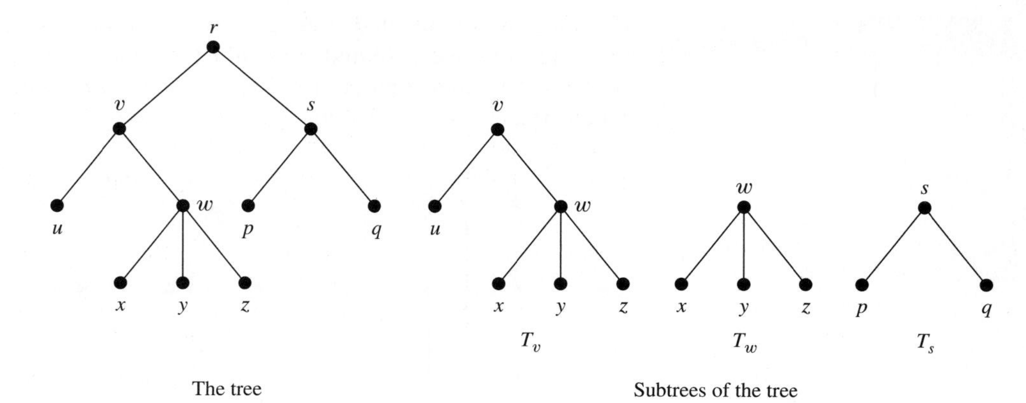

A rooted tree is a **binary tree** in case each node has at most two children: a left child, a right child, both a left child and a right child, or no children at all. This sort of right–left distinction was important in the binary search tree of Example 1(a). Similarly, for $m > 2$ an **m-ary tree** is one in which the children of each parent are labeled with distinct members of $\{1, 2, \ldots, m\}$. A parent is not required to have a complete set of m children; we say that the ith child is **absent** in case no child is labeled with i. In the digraph terminology of §8.1, an m-ary tree has $\text{outdeg}(v) \le m$ for all vertices v. An m-ary tree [or binary tree] is a **regular m-ary tree** if $\text{outdeg}(v) = m$ or 0 for all vertices v, i.e., if each vertex that has any children has exactly m of them.

The **level number** of a vertex v is the length of the unique simple path from the root to v. In particular, the root itself has level number 0. The **height** of a rooted tree is the largest level number of a vertex. Only leaves can have their level numbers equal to the height of the tree. A regular m-ary tree is said to be a **full m-ary tree** if all the leaves have the same level number, namely the height of the tree.

EXAMPLE 4

(a) The rooted tree in Figures 4 and 6 is a 3-ary tree and, in fact, is an m-ary tree for $m \ge 3$. It is not a regular 3-ary tree, since vertices r, v, and s have outdegree 2. Vertices v and s have level number 1; vertices u, w, p, and q have level number 2; and the leaves x, y, and z have level number 3. The height of the tree is 3.

(b) The labeled tree in Figure 8(a) on page 250 is a full regular binary tree of height 3. The labeled tree in Figure 8(b) is a regular 3-ary tree of height 3. It is not a full 3-ary tree, since one leaf has level number 1 and five leaves have level number 2. ■

EXAMPLE 5

Consider a full m-ary tree of height h. There are m vertices at level 1. Each parent at level 1 has m children, so there are m^2 vertices at level 2. A simple induction shows that, because the tree is full, it has m^l vertices at level l for each $l \le h$. Thus it has $1 + m + m^2 + \cdots + m^h$ vertices in all. Since

$$(m-1)(1 + m + m^2 + \cdots + m^h) = m^{h+1} - 1,$$

as one can check by multiplying and canceling, we have

$$1 + m + m^2 + \cdots + m^h = \frac{m^{h+1} - 1}{m - 1}.$$

[An inductive proof of this formula is given in Example 2(c) of Section 4.1.] Note that the same tree has $p = (m^h - 1)/(m - 1)$ parents and $t = m^h$ leaves. ■

The binary search tree in Example 1(a) depended on having alphabetical ordering to compare client names. The same idea can be applied more generally; all we need is some ordering or listing to tell us which of two given objects comes first. [The technical term for this kind of listing, which we will discuss further in §§11.1

and 11.2, is **linear order**.] Whenever we want to, we can decide on an order in which to list the children of a given parent in a rooted tree. If we order the children of *every* parent in the tree, we obtain what is called an **ordered rooted tree**. When we draw such a tree, we draw the children in order from left to right. It is convenient to use the notation $\boldsymbol{v < w}$ to mean that v comes before w in the order, even in cases in which v and w are not numbers.

EXAMPLE 6

(a) If we view Figure 5(b) as an ordered rooted tree, then the children of v are ordered: $u < s < w < x$. The children of x are ordered: $y < z$. Figure 5(b′) is the picture of a different *ordered* rooted tree, since $s < x < w < u$ and $z < y$.

(b) As soon as we draw a rooted tree, it looks like an ordered rooted tree, even if we do not care about the order structure. For example, the important structure in Figure 3 is the rooted tree structure. The order of the "children" is not important. The head of the computer science department precedes the head of the mathematics department simply because we chose to list the departments in alphabetical order.

(c) A binary tree or, more generally, an m-ary tree is in a natural way an ordered tree, but there is a difference between the two ideas. In a binary tree the right child will be the first child if there is no left child. In the 3-ary tree of Figure 6, the children w, s, and q each may have the label 3, even though their parents only have two children. Also, the children v, u, and p each may have label 1 or 2. ■

EXAMPLE 7

(a) Consider an alphabet Σ ordered in some way. We make Σ^* into a rooted tree as follows. The empty word λ will serve as the root. For any word w in Σ^*, its set of children is

$$\{wx : x \in \Sigma\}.$$

Since Σ is ordered, we can order each set of children to obtain an ordered rooted tree Σ^*_λ by letting $wx < wy$ in case $x < y$ in Σ.

(b) Suppose that $\Sigma = \{a, b\}$, where $a < b$. Each vertex has two children. For instance, the children of $a\,b\,b\,a$ are $a\,b\,b\,a\,a$ and $a\,b\,b\,a\,b$. Part of the infinite ordered rooted tree Σ^*_λ is drawn in Figure 7.

(c) Figure 1 on page 183 showed part of the tree $\{a, b, c, d\}^*_\lambda$ with the natural ordering determined by $a < b < c < d$. ■

Figure 7 ▶

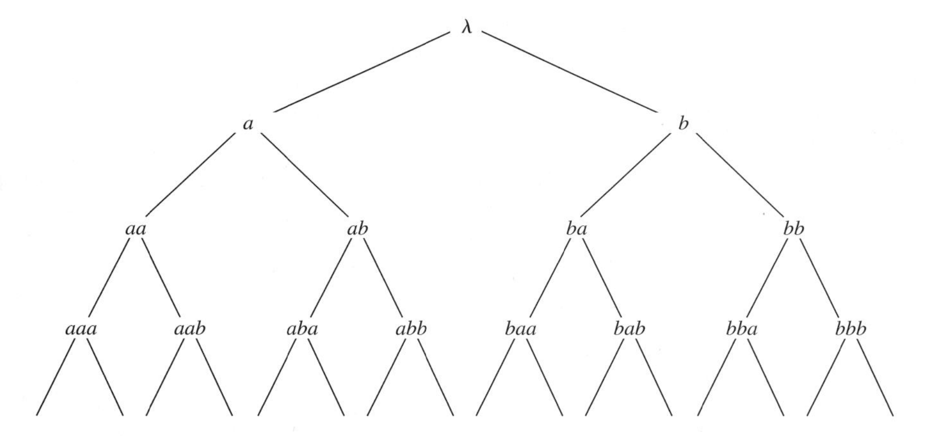

There are a variety of ways to name [or label] the vertices of an ordered rooted tree so that the names describe their locations. One such scheme resembles Example 7: vertices of an m-ary tree can be named using words from Σ^*, where $\Sigma = \mathbb{Z}(m) = \{0, 1, \ldots, m-1\}$. The ordered children of the root have names from $\{0, 1, 2, \ldots, m-1\}$. If a vertex is named by the word w, then its ordered children are named using $w0$, $w1$, $w2$, etc. The name of a vertex tells us the exact location of the vertex in the tree. For example, a vertex named 1021 would be the number two child of the vertex named 102, which, in turn, would be the number three child of the vertex named 10, etc. The level of the vertex is the length of its name; a vertex named 1021 is at level 4.

EXAMPLE 8

All of the vertices except the root in Figure 8(a) are named in this way. In Figure 8(b) we have only named the leaves. The names of the other vertices should be clear. ■

Figure 8 ▶

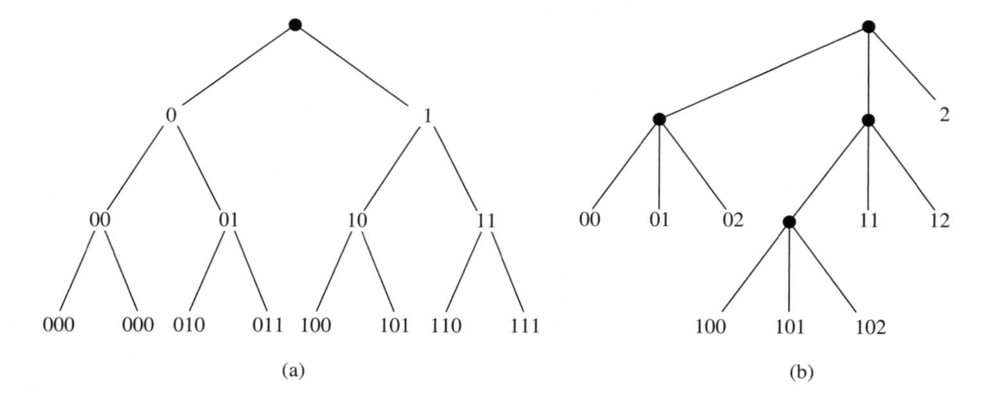

Trees have been used for a long time to organize or store information. In old mathematics textbooks the individual paragraphs were frequently labeled by using numbers, i.e., words in Σ^*, where $\Sigma = \{1, 2, 3, \dots\}$. Decimal points were used to set off the letters of the words in Σ^*. Thus 3.4.1.2 would refer to the second paragraph of the first subsection of the fourth section of Chapter 3, while 3.4.12 would refer to the twelfth subsection of the fourth section of Chapter 3. This scheme is not very pretty, so modern authors usually avoid it, but it has some real advantages that carry over to present-day uses of trees. One can always insert new paragraphs or sections without disrupting the numbering system. With a little care, paragraphs and sections can be deleted without causing trouble, especially if one doesn't mind gaps. Also a label, such as 3.4.12, specifies exactly where the subsection or paragraph fits into the book.

The book you are reading now has been produced with the typesetting language LaTeX, which uses the magic of computers to effortlessly keep track of chapters, sections, and subsections, renumbering them as required. Although the book itself is printed linearly as one long string of characters, the numbering scheme lets us visualize a tree structure associated with it. In contrast, one famous mathematics book has theorems numbered from 1 to 460. All the label "Theorem 303" tells us is that this theorem probably appears about two-thirds of the way through the book.

Of course, being able to locate a node in a tree quickly is most useful if we know which node we want to reach. If we are storing patient records with a tree indexed by last names and if we need to look at every node, for example, to determine which patients are taking a particular drug, the tree structure is not much help. In §7.3 we will investigate search schemes for systematically going through all the nodes in a rooted tree.

Exercises 6.4

1. (a) For the tree in Figure 4, draw a rooted tree with new root v.
   (b) What is the level number of the vertex r?
   (c) What is the height of the tree?
2. Create a binary search tree of height 4 for the English alphabet $\{a, b, \dots, z\}$ with its usual order.
3. (a) For each rooted tree in Figure 5, give the level numbers of the vertices and the height of the tree.
   (b) Which of the trees in Figure 5 are regular m-ary for some m?
4. Discuss why ordinary family trees are not rooted trees.

5. (a) There are seven different types of rooted trees of height 2 in which each node has at most two children. Draw one tree of each type.
   (b) To which of the types in part (a) do the regular binary trees of height 2 belong?
   (c) Which of the trees in part (a) are full binary trees?
   (d) How many different types of binary trees are there of height 2?
6. (a) Repeat Exercise 5(a) for the seven types of rooted trees of height 3 in which each node that is not a leaf has two children.
   (b) How many different types of regular binary trees are there of height 3?
7. For each n specified below, draw a binary search tree with nodes $1, 2, 3, \ldots, n$ and with height as small as possible.
   (a) $n = 7$ (b) $n = 15$
   (c) $n = 4$ (d) $n = 6$
8. A **2-3 tree** is a rooted tree such that each interior node, including the root if the height is 2 or more, has either two or three children and all paths from the root to the leaves have the same length. There are seven different types of 2-3 trees of height 2. Draw one tree of each type.
9. (a) Draw full m-ary trees of height h for $m = 2$, $h = 2$; $m = 2$, $h = 3$; and $m = 3$, $h = 2$.
   (b) Which trees in part (a) have m^h leaves?
10. Consider a full binary tree T of height h.
   (a) How many leaves does T have?
   (b) How many vertices does T have?
11. Consider a full m-ary tree with p parents and t leaves. Show that $t = (m - 1)p + 1$ no matter what the height is.
12. Give some real-life examples of information storage that can be viewed as labeled trees.
13. Let $\Sigma = \{a, b\}$ and consider the rooted tree Σ_λ^*; see Example 7. Describe the set of vertices at level k. How big is this set?
14. Draw part of the rooted tree Σ_λ^*, where $\Sigma = \{a, b, c\}$ and $a < b < c$ as usual.

The next two exercises illustrate some of the problems associated with updating binary search trees.

15. (a) Suppose that in Example 1(a) client Rose moves away. How might we naturally rearrange the tree to delete the records of Rose without disturbing the rest of the tree too much?
   (b) Repeat part (a) with Brown moving instead of Rose.
16. (a) Suppose that in Example 1(a) a new client, Smith, must be added to the binary search tree. Show how to do so without increasing the height of the tree. Try not to disturb the rest of the tree more than necessary.
   (b) Suppose that three new clients, Smith1, Smith2, and Smith3, must be added to the tree in Example 1(a). Show how to do so without increasing the height of the binary search tree.
   (c) What happens in part (b) if there are four new clients?

6.5 Vertex Traversal Problems

Euler's Theorem on page 234 tells us which graphs have closed paths that use each edge exactly once, and the algorithm EulerCircuit on page 236 gives a way to construct the paths when they exist. In contrast, much less is known about graphs with paths that use each vertex exactly once. The Irish mathematician Sir William Hamilton was one of the first to study such graphs and at one time even marketed a puzzle based on the problem.

A path is called a **Hamilton path** if it visits every vertex of the graph exactly once. A closed path that visits every vertex of the graph exactly once, except for the last vertex, which duplicates the first one, is called a **Hamilton circuit**. A graph with a Hamilton circuit is called a **Hamiltonian graph**. A Hamilton path must be simple [why?], and by Proposition 1 on page 226, if G has at least three vertices, then a Hamilton circuit of G must be a cycle.

EXAMPLE 1

(a) The graph shown in Figure 1(a) has Hamilton circuit $v\,w\,x\,y\,z\,v$.

(b) Adding more edges can't hurt, so the graph K_5 of Figure 1(b) is also Hamiltonian. In fact, every complete graph K_n for $n \geq 3$ is Hamiltonian; we can go from vertex to vertex in any order we please.

(c) The graph of Figure 1(c) has the Hamilton path $v\,w\,x\,y\,z$, but has no Hamilton circuit, since no cycle goes through v.

(d) The graph of Figure 1(d) has no Hamilton path, so it certainly has no Hamilton circuit. ■

Figure 1 ▶

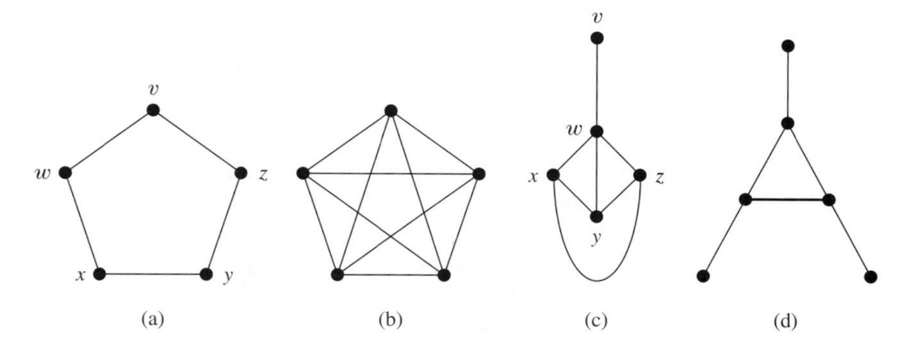

With Euler's theorem, the theory of Euler circuits is nice and complete. What can be proved about Hamilton circuits? Under certain conditions, graphs will have so many edges compared to the number of vertices that they *must* have Hamilton circuits. But the graph in Figure 1(a) has very few edges, and yet it has a Hamilton circuit. And the graph in Figure 1(c) has lots of edges but no Hamilton circuit. It turns out that there is no known simple characterization of those connected graphs possessing Hamilton circuits. The concept of Hamilton circuit seems very close to that of Euler circuit, yet the theory of Hamilton circuits is vastly more complicated. In particular, no efficient algorithm is known for finding Hamilton circuits. The problem is a special case of the Traveling Salesperson Problem, in which one begins with a graph whose edges are assigned **weights** that may represent mileage, cost, computer time, or some other quantity that we wish to minimize. In Figure 2 the weights might represent mileage between cities on a traveling salesperson's route. The goal is to find the shortest round trip that visits each city exactly once. That is, the goal is to find a Hamilton circuit minimizing the sum of the weights of the edges. An algorithm solving this problem would also be able to find Hamilton circuits in an unweighted graph, since we could always assign weight 1 to each edge.

Figure 2 ▶

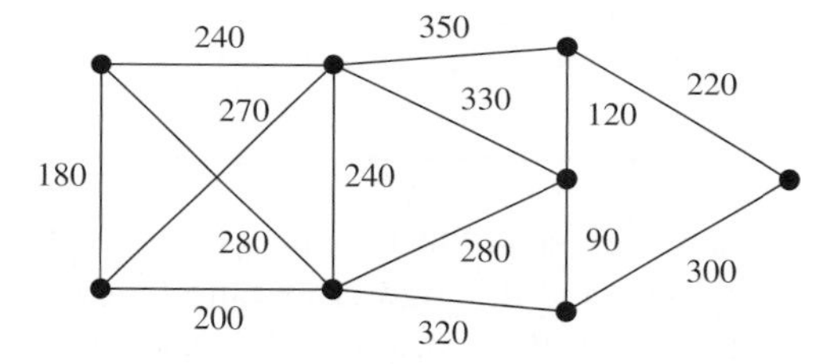

Obviously, a Hamiltonian graph with n vertices must have at least n edges. This necessary condition may not be sufficient, as Figures 1(c) and 1(d) illustrate. Of course, loops and parallel edges are of no help. The next three theorems give conditions assuring that a graph is Hamiltonian. We will illustrate the ideas with small graphs, where one can easily check the Hamiltonian property without the theory, but the theorems are powerful for large graphs. The first theorem gives a simple sufficient condition.

Theorem 1 If the graph G has no loops or parallel edges, if $|V(G)| = n \geq 3$, and if $\deg(v) \geq n/2$ for each vertex v of G, then G is Hamiltonian.

EXAMPLE 2

(a) The graph K_5 in Figure 1(b) has $\deg(v) = 4$ for each v and has $|V(G)| = 5$, so it satisfies the condition of Theorem 1. None of the other graphs in Figure 1 satisfies the condition of Theorem 1, because in each case $|V(G)|/2 = 5/2$ or 3, while the graph has at least one vertex of degree 1 or 2. The graph in (a) is Hamiltonian, but the two in (c) and (d) are not. Failure of the condition does not tell us whether a graph is Hamiltonian or not.

(b) Each of the graphs in Figure 3 has $|V(G)|/2 = 5/2$ and has a vertex of degree 2. These graphs do not satisfy the hypotheses of Theorem 1, but are nevertheless Hamiltonian. ■

Figure 3 ▶

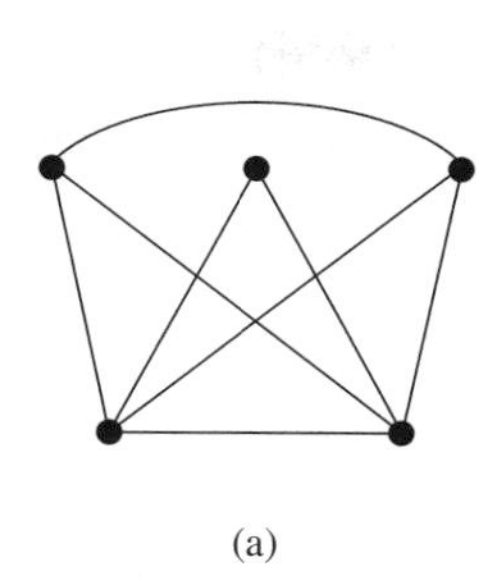

(a)

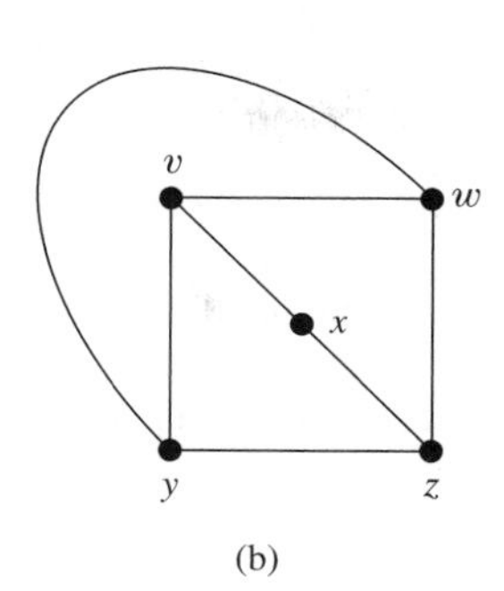

(b)

Theorem 1 imposes a uniform condition: there are many (at least $n/2$) edges at each vertex. Our next theorem requires only that there be enough edges somewhere in the graph. We will establish both of these sufficient conditions as consequences of Theorem 3, which gives a criterion in terms of degrees of pairs of vertices.

Theorem 2 Every graph with n vertices that has no loops or parallel edges and that has at least $\frac{1}{2}(n-1)(n-2)+2$ edges is Hamiltonian.

EXAMPLE 3

(a) The Hamiltonian graph of Figure 3(a) has $n = 5$, so $\frac{1}{2}(n-1)(n-2)+2 = 8$. It has 8 edges, so it satisfies the hypotheses and the conclusion of Theorem 2.

(b) The Hamiltonian graph of Figure 3(b) also has $n = 5$, and so again we have $\frac{1}{2}(n-1)(n-2)+2 = 8$, but the graph has only 7 edges. It fails to satisfy the hypotheses of Theorem 2 as well as Theorem 1. If there were no vertex in the middle, we would have K_4 with $n = 4$, so $\frac{1}{2}(n-1)(n-2)+2 = 5$, and the 6 edges would be more than enough. As it stands, the graph satisfies the hypotheses of the next theorem. ■

Theorem 3 Suppose that the graph G has no loops or parallel edges and that $|V(G)| = n \geq 3$. If

$$\deg(v) + \deg(w) \geq n$$

for each pair of vertices v and w that are not connected by an edge, then G is Hamiltonian.

EXAMPLE 4

(a) For the graph in Figure 3(b), $n = 5$. There are three pairs of distinct vertices that are not connected by an edge. We verify the hypotheses of Theorem 3 by examining them:

$$\begin{aligned} &\{v, z\}, \quad \deg(v) + \deg(z) = 3 + 3 = 6 \geq 5; \\ &\{w, x\}, \quad \deg(w) + \deg(x) = 3 + 2 = 5 \geq 5; \\ &\{x, y\}, \quad \deg(x) + \deg(y) = 2 + 3 = 5 \geq 5. \end{aligned}$$

(b) For the graph in Figure 1(a), $n = 5$ and every vertex has degree 2. This graph doesn't satisfy the hypotheses of Theorem 3 [or Theorems 1 and 2], but it is Hamiltonian anyway. None of the theorems characterizes Hamiltonian graphs, so none of them can be used to show that a graph is not Hamiltonian! ■

Proof of Theorem 3 Suppose the theorem is false for some n, and let G be a counterexample with $V(G) = n$ and with $|E(G)|$ as large as possible. Since G has no loops or parallel edges, G is a subgraph of the Hamiltonian graph K_n. Adjoining to G an edge from K_n would give a graph that still satisfies the degree condition, but has more than $|E(G)|$ edges. By the choice of G, any such graph would have a

Hamilton circuit. This implies that G must already have a Hamilton *path*, say with vertex sequence $v_1\, v_2 \cdots v_n$. Since G has no Hamilton circuit, v_1 and v_n are not connected by an edge in G, so $\deg(v_1) + \deg(v_n) \ge n$ by hypothesis.

Define subsets S_1 and S_n of $\{2, \dots, n\}$ by

$$S_1 = \{i : \{v_1, v_i\} \in E(G)\} \quad \text{and} \quad S_n = \{i : \{v_{i-1}, v_n\} \in E(G)\}.$$

Then we have $S_1 \cup S_2 \subseteq \{2, 3, \dots, n\}$, $|S_1| = \deg(v_1)$, and $|S_n| = \deg(v_n)$. Since $|S_1| + |S_n| \ge n$ and $S_1 \cup S_n$ has at most $n - 1$ elements, $S_1 \cap S_n$ must be nonempty. Thus there is an i for which both $\{v_1, v_i\}$ and $\{v_{i-1}, v_n\}$ are edges of G. Then [see Figure 4] the path $v_1 \cdots v_{i-1}\, v_n \cdots v_i\, v_1$ is a Hamilton circuit in G, contradicting the choice of G as a counterexample. ∎

Figure 4 ▶

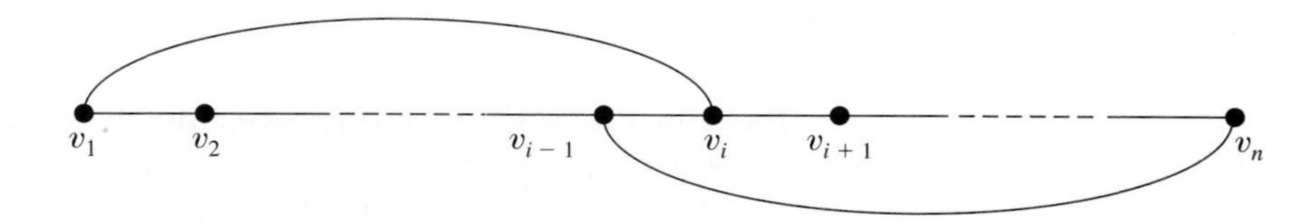

Our first two theorems follow easily from Theorem 3.

Proofs of Theorems 1 and 2 Suppose that G has no loops or parallel edges and that $|V(G)| = n \ge 3$.

If $\deg(v) \ge n/2$ for each v, then $\deg(v) + \deg(w) \ge n$ for every v and w, whether joined by an edge or not, so the hypothesis of Theorem 3 is satisfied and G is Hamiltonian.

Suppose that

$$|E(G)| \ge \frac{1}{2}(n-1)(n-2) + 2 = \binom{n-1}{2} + 2,$$

and consider vertices u and v with $\{u, v\} \notin E(G)$. Remove from G the vertices u and v and all edges with u or v as a vertex. Since $\{u, v\} \notin E(G)$, we have removed $\deg(u) + \deg(v)$ edges and 2 vertices. The graph G' that is left is a subgraph of K_{n-2}, so

$$\binom{n-2}{2} = |E(K_{n-2})| \ge |E(G')| \ge \binom{n-1}{2} + 2 - \deg(u) - \deg(v).$$

Hence

$$\begin{aligned}
\deg(u) + \deg(v) &\ge \binom{n-1}{2} - \binom{n-2}{2} + 2 \\
&= \frac{1}{2}(n-1)(n-2) - \frac{1}{2}(n-2)(n-3) + 2 \\
&= \frac{1}{2}(n-2)[(n-1) - (n-3)] + 2 \\
&= \frac{1}{2}(n-2)[2] + 2 = n.
\end{aligned}$$

Again, G satisfies the hypothesis of Theorem 3, so G is Hamiltonian. ∎

Theorems 1, 2, and 3 are somewhat unsatisfactory in two ways. Not only are their sufficient conditions not necessary, but the theorems also give no guidance for finding a Hamilton circuit when one is guaranteed to exist. As we mentioned earlier, as of this writing no efficient algorithm is known for finding Hamilton paths or circuits. On the positive side, a Hamiltonian graph must certainly be connected, so all three theorems give sufficient conditions for a graph to be connected.

EXAMPLE 5

A **Gray code** of length n is a list of all 2^n distinct strings of n binary digits such that adjacent strings differ in exactly one digit and the last string differs from the first string in exactly one digit. For example, 00, 01, 11, 10 is a Gray code of length 2.

We can view the construction of a Gray code as a graph-theoretic problem. Let $V(G)$ be the set $\{0, 1\}^n$ of binary n-tuples, and join n-tuples by an edge if they differ in exactly one digit. As we will illustrate, a Gray code of length n is, in effect, a Hamilton circuit of the graph G.

Figure 5(a) shows the graph G for $n = 2$. Figure 5(b) shows the same graph redrawn. This graph has eight Hamilton circuits, four in each direction, which shows that there are eight Gray codes of length 2. We can regard two Hamilton circuits as "equivalent" in case they really have the same path, but just start at different points along the way. From this point of view, this graph has only two equivalence classes of Hamilton circuits. If we then regard two Gray codes as equivalent in case their Hamilton circuits are equivalent, there are just two equivalence classes of Gray codes of length 2.

Figure 5 ▶

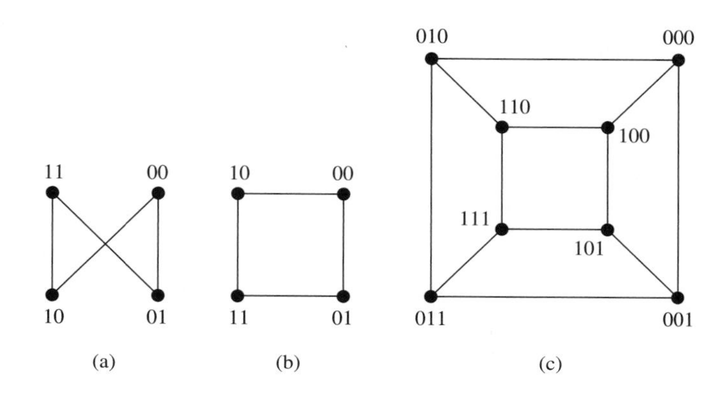

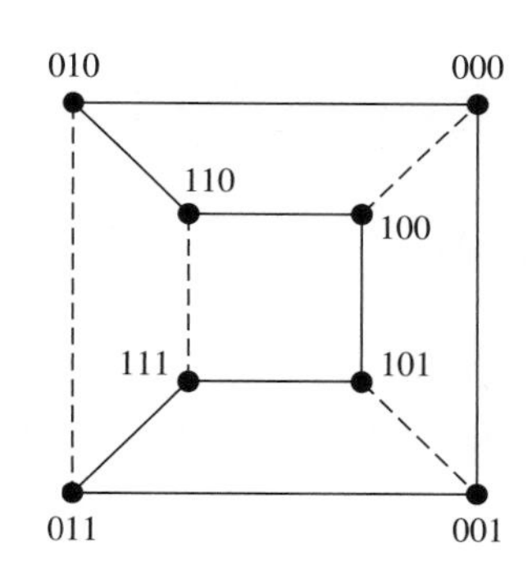

Figure 6 ▲

Figure 5(c) shows the graph for $n = 3$. There are 12 equivalence classes of Gray codes of length 3. Figure 6 indicates the Hamilton circuit corresponding to one such code. Thus

$$000,\quad 001,\quad 011,\quad 111,\quad 101,\quad 100,\quad 110,\quad 010$$

is a Gray code of length 3.

Gray codes can be used to label the individual processors in a hypercube array. [The square and cube in Figure 5 are hypercubes of dimensions 2 and 3.] Using such a labeling scheme, two processors are connected if and only if their labels differ in just 1 bit. ■

The vertices in the graphs we constructed in Example 5 can be partitioned into two sets, those with an even number of 1's and those with an odd number, so that each edge joins a member of one set to a member of the other. We conclude this section with some observations about Hamilton circuits in graphs with this sort of partition.

A graph G is called **bipartite** if $V(G)$ is the union of two disjoint nonempty subsets V_1 and V_2 such that every edge of G joins a vertex of V_1 to a vertex of V_2. A graph is called a **complete bipartite** graph if, in addition, every vertex of V_1 is joined to every vertex of V_2 by a unique edge.

EXAMPLE 6

The graphs shown in Figure 7 are all bipartite. All but the one in Figure 7(b) are complete bipartite graphs. ■

Given m and n, the complete bipartite graphs with $|V_1| = m$ and $|V_2| = n$ are all isomorphic to each other; we denote them by $\boldsymbol{K_{m,n}}$. Note that $K_{m,n}$ and $K_{n,m}$ are isomorphic.

Figure 7 ▶

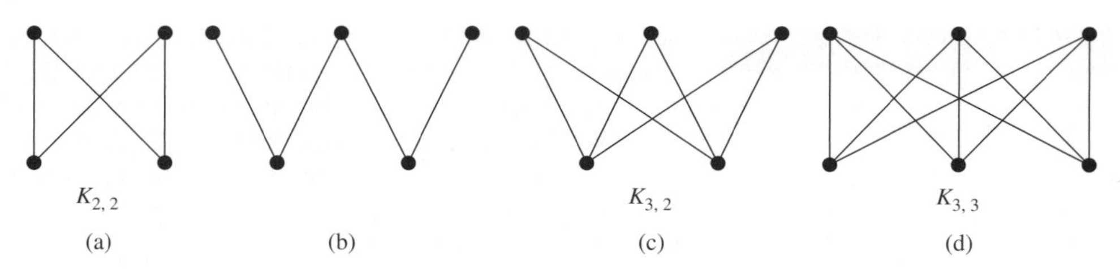

Theorem 4 Let G be a bipartite graph with partition $V(G) = V_1 \cup V_2$. If G has a Hamilton circuit, then $|V_1| = |V_2|$. If G has a Hamilton path, then the numbers $|V_1|$ and $|V_2|$ differ by at most 1. For complete bipartite graphs with at least three vertices the converse statements are also true.

Proof The vertices on a path in G alternately belong to V_1 and V_2. If the closed path $x_1 x_2 \cdots x_n x_1$ goes through each vertex once, then $x_1, x_3, x_5, \ldots$ must belong to one of the two sets, say V_1. Since $\{x_n, x_1\}$ is an edge, n must be even and $x_2, x_4, \ldots, x_n$ all belong to V_2. So $|V_1| = |V_2|$. Similar remarks apply to a nonclosed Hamilton path $x_1 x_2 \cdots x_n$, except that n might be odd, in which case one of V_1 and V_2 will have an extra vertex.

Now suppose $G = K_{m,n}$. If $m = n$, we can simply go back and forth from V_1 to V_2, since edges exist to take us wherever we want. If $|V_1|$ and $|V_2|$ differ by 1, we should start in the larger of V_1 and V_2 to get a Hamilton path. ■

A computer scientist we know tells the story of how he once spent over two weeks on a computer searching for a Hamilton path in a bipartite graph with 42 vertices before he realized that the graph violated the condition of Theorem 4. The story has two messages: (1) people do have practical applications for bipartite graphs and Hamilton paths, and (2) *thought should precede computation.*

Exercises 6.5

1. (a) Explain why the graph in Figure 1(c) has no Hamilton circuit. *Warning:* None of Theorems 1–3 can be used to show that a graph has no Hamilton circuit. An analysis of the particular graph is needed.

   (b) Explain why the graph in Figure 1(d) has no Hamilton path.

2. For each graph in Figure 8, give a Hamilton circuit or explain why none exists.

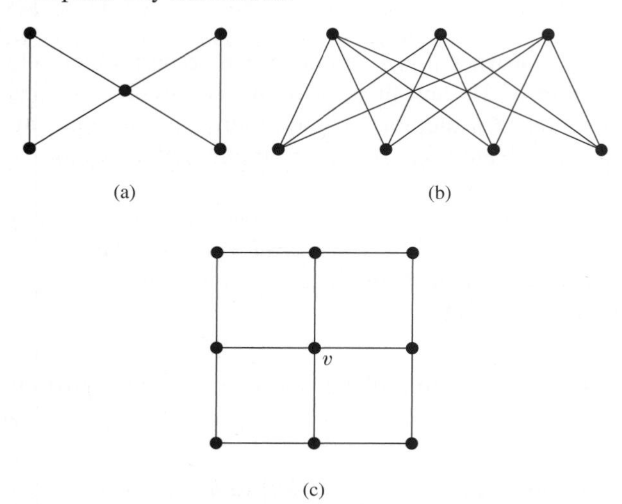

Figure 8 ▲

3. Consider the graph shown in Figure 9(a).

   (a) Is this a Hamiltonian graph?

   (b) Is this a complete graph?

   (c) Is this a bipartite graph?

   (d) Is this a complete bipartite graph?

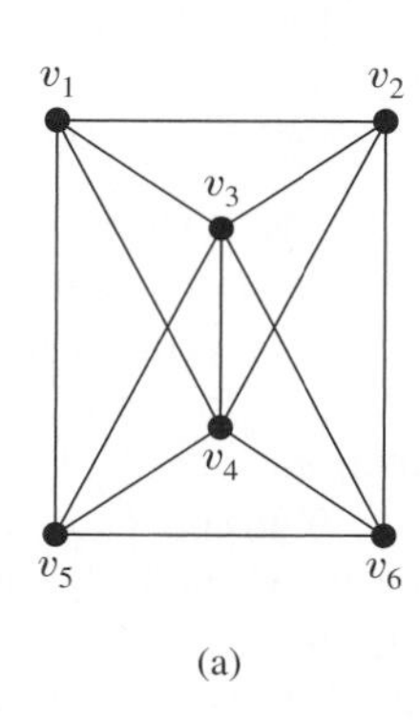

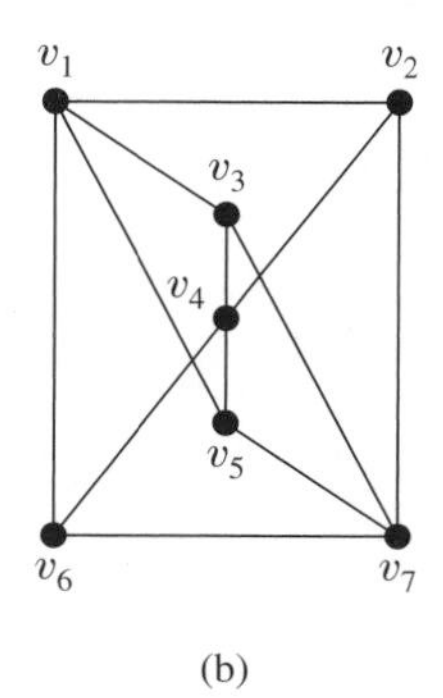

Figure 9 ▲

4. Answer the same questions as in Exercise 3 for the graph in Figure 9(b).

5. (a) How many Hamilton circuits does the graph $K_{n,n}$ have for $n \geq 2$? [Count circuits as different if they have different starting points or different vertex sequences.]

(b) How many Hamilton paths does $K_{n,n-1}$ have for $n \geq 2$?

(c) Which complete bipartite graphs $K_{m,n}$ have Euler paths?

6. Redraw the graphs in Figure 5 and mark each of the subsets V_1 and V_2 of the bipartite partition of $V(G)$. Which of the graphs are complete bipartite graphs?

7. Arrange eight 0's and 1's in a circle so that each 3-digit binary number occurs as a string of three consecutive symbols somewhere in the circle. *Hint:* Find a Hamilton circuit in the graph with vertex set $\{0, 1\}^3$ and with an edge between vertices (v_1, v_2, v_3) and (w_1, w_2, w_3) whenever $(v_1, v_2) = (w_2, w_3)$ or $(v_2, v_3) = (w_1, w_2)$.

8. Give two examples of Gray codes of length 3 that are not equivalent to the one in Example 5.

9. Consider the graph that has vertex set $\{0, 1\}^3$ and an edge between vertices whenever they differ in two coordinates. Does the graph have a Hamilton circuit? Does it have a Hamilton path?

10. Repeat Exercise 9 for the graph that has vertex set $\{0, 1\}^3$ and an edge between vertices if they differ in two or three coordinates.

11. For $n \geq 4$, build the graph K_n^+ from the complete graph K_{n-1} by adding one more vertex in the middle of an edge of K_{n-1}. [Figure 3(b) shows K_5^+.]

(a) Show that K_n^+ does not satisfy the condition of Theorem 2.

(b) Use Theorem 3 to show that K_n^+ is nevertheless Hamiltonian.

12. For $n \geq 3$, build a graph K_n^{++} from the complete graph K_{n-1} by adding one more vertex and an edge from the new vertex to a vertex of K_{n-1}. [Figure 1(c) shows K_5^{++}.] Show that K_n^{++} is not Hamiltonian. Observe that K_n^{++} has n vertices and $\frac{1}{2}(n-1)(n-2)+1$ edges. This example shows that the number of edges required in Theorem 2 cannot be decreased.

13. The **complement** of a graph G is the graph with vertex set $V(G)$ and with an edge between distinct vertices v and w if G does *not* have an edge joining v and w.

(a) Draw the complement of the graph of Figure 3(b).

(b) How many components does the complement in part (a) have?

(c) Show that if G is not connected, then its complement is connected.

(d) Give an example of a graph that is isomorphic to its complement.

(e) Is the converse to the statement in part (c) true?

14. Suppose that the graph G is regular of degree $k \geq 1$ [i.e., each vertex has degree k] and has at least $2k+2$ vertices. Show that the complement of G is Hamiltonian. *Hint:* Use Theorem 1.

15. Show that Gray codes of length n always exist. *Hint:* Use induction on n and consider the graph G_n, in which a Hamilton circuit corresponds to a Gray code of length n, as described in Example 5.

16. None of the theorems in this section can be used to solve Exercise 15. Why?

6.6 Minimum Spanning Trees

The theorems that characterize trees suggest two methods for finding a spanning tree of a finite connected graph. Using the idea in the proof of Theorem 2 on page 241, we could just remove edges one after another without destroying connectedness, i.e., remove edges that belong to cycles, until we are forced to stop. This will work, but if G has n vertices and more than $2n$ edges, this procedure will examine and throw out more than half of the edges. It might be faster, if we could do it, to build up a spanning tree by choosing its $n-1$ edges one at a time so that at each stage the subgraph of chosen edges is acyclic. The algorithms in this section all build trees in this way.

Our first algorithm starts from an initially chosen vertex v. If the given graph is connected, then the algorithm produces a spanning tree for it. Otherwise, the algorithm gives a spanning tree for the connected component of the graph that contains v.

Tree(vertex)

```
{Input: A vertex v of the finite graph G}
{Output: A set E of edges of a spanning tree for the component of G
        that contains v}
Let V := {v} and E := Ø.
        {V is a list of visited vertices.}
while there are edges of G joining vertices in V to vertices that are
        not in V do
    Choose such an edge {u, w} with u in V and w not in V.
    Put w in V and put the edge {u, w} in E.
return E  ■
```

To get a spanning forest for G, we just keep growing trees.

Forest(graph)

{Input: A finite graph G}
{Output: A set E^* of edges of a spanning forest for G}
Set $V^* := \varnothing$ and $E^* := \varnothing$.
while $V^* \neq V(G)$ **do**
  Choose $v \in V(G) \setminus V^*$.
  Let E be the edge set Tree(v) for a tree spanning the
    component of G containing v, and let V be its vertex set.
  Put the members of V in V^* and put the members of E in E^*.
return E^* ■

EXAMPLE 1

We can illustrate the operations of Tree and Forest on the graph shown in Figure 1(a). Figure 2 shows the steps in finding Tree(1) and Tree(2); then Forest puts them together. Where choices are available, we have chosen in increasing numerical order and exhausted all edges from a given vertex before going on to the next vertex. Other choice schemes are possible, of course. Figure 1(b) shows the two trees that are grown in the spanning forest, with vertices labeled according to whether they are in the component of 1 or the component of 2. ■

Figure 1 ▶

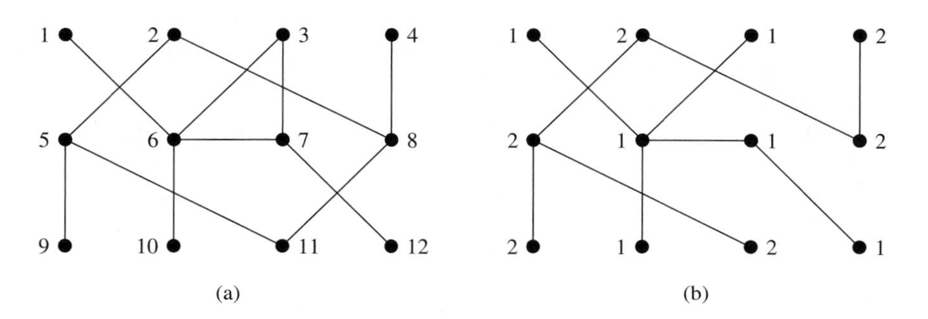

Figure 2 ▶

Tree(1)
$V := \{1\}; E := \varnothing$.
Choose edge $\{1, 6\}$.
$V := \{1, 6\}; E := \{\{1, 6\}\}$.
Choose edges $\{6, 3\}$, $\{6, 7\}$, $\{6, 10\}$ [order doesn't matter here].
$V := \{1, 6, 3, 7, 10\}; E := \{\{1, 6\}, \{6, 3\}, \{6, 7\}, \{6, 10\}\}$.
Choose edge $\{7, 12\}$ [not $\{3, 7\}$ since 3 and 7 are both in V].
$V := \{1, 6, 3, 7, 10, 12\}; E := \{\{1, 6\}, \{6, 3\}, \{6, 7\}, \{6, 10\}, \{7, 12\}\}$.

Tree(2)
$V := \{2\}; E := \varnothing$.
Choose edges $\{2, 5\}$, $\{2, 8\}$.
$V := \{2, 5, 8\}; E := \{\{2, 5\}, \{2, 8\}\}$.
Choose edges $\{5, 9\}$, $\{5, 11\}$.
$V := \{2, 5, 8, 9, 11\}; E := \{\{2, 5\}, \{2, 8\}, \{5, 9\}, \{5, 11\}\}$.
Choose edge $\{8, 4\}$.
$V := \{2, 5, 8, 9, 11, 4\}; E := \{\{2, 5\}, \{2, 8\}, \{5, 9\}, \{5, 11\}, \{8, 4\}\}$.

Forest
Put together the two lists E from Tree(1) and Tree(2) to form E^*.
$E^* := \{\{1, 6\}, \{6, 3\}, \{6, 7\}, \{6, 10\}, \{7, 12\}, \{2, 5\}, \{2, 8\}, \{5, 9\}, \{5, 11\}, \{8, 4\}\}$.

Theorem 1 Tree(v) is a spanning tree for the component of G containing v. Hence Forest produces a spanning forest for G.

Proof Here is Tree again for reference.

Tree(vertex)

```
Let V := {v} and E := Ø.
while there are edges of G joining vertices in V to vertices that are
        not in V do
    Choose such an edge {u, w} with u in V and w not in V.
    Put w in V and put {u, w} in E.
return E
```

Each pass through the **while** loop increases the size of V, so the algorithm must stop eventually. The statement "V and E are the vertices and edges of a tree with v as vertex" is clearly true when we first enter the loop. We show that it is an invariant of the loop, so when the algorithm stops, it stops with a tree containing v. Theorem 4 on page 242 implies that attaching a new vertex to a tree with a new edge yields a tree. Since this is precisely what the **while** loop does, the statement is a loop invariant.

It remains to show that the algorithm does not stop as long as there are vertices in the component V' containing v. Suppose that there is a vertex v' in $V' \setminus V$ at the end of a **while** loop. There will be a path in V' from v to v', and along this path there will be a first edge $\{u, w\}$ with initial vertex u in V and terminal vertex w in $V' \setminus V$. [u might be v, and w might be v'.] Hence the guard on the **while** loop is still true, and the algorithm does not stop.

We conclude that Tree(v) is as advertised. ■

The time that Tree takes depends on the scheme for making choices and on how the list of available edges is maintained. If we mark each vertex when we choose it and, at the same time, tell its neighbors that it's marked, then each vertex and each edge are handled only once. If G is connected, then Tree builds a spanning tree in time $O(|V(G)| + |E(G)|)$. The same argument shows that, in the general case, Forest also runs in time $O(|V(G)| + |E(G)|)$.

The number of components of G is the number of trees in a spanning forest. It is easy to keep track of the components as we build trees with Forest, by using a function C that assigns the value u to each vertex w in the list V determined from Tree(u). Then each pass through the **while** loop in Forest produces a different value of C, which is shared by all the vertices in the component for that pass. The labels 1 and 2 in Figure 1(b) were assigned using such a function C. To modify the algorithm, simply set $C(v) := v$ at the start of Tree and add the line "Set $C(w) := C(u)$" after putting the edge $\{u, w\}$ in E.

Forest can be used to test connectivity of a graph; just check whether the algorithm produces more than one tree. Forest can also be used to give a relatively fast test for the presence of cycles in a graph. If G is acyclic, then the spanning forest produced is just G itself; otherwise, $|E^*| < |E(G)|$ at the conclusion of the algorithm.

The question of finding spanning trees for connected graphs is especially interesting if the edges are **weighted**, i.e., if each edge e of G is assigned a nonnegative number $W(e)$. The **weight** $W(H)$ of a subgraph H of G is simply the sum of the weights of the edges of H. The problem is to find a **minimum spanning tree**, i.e., a spanning tree whose weight is less than or equal to that of any other. If a graph G is not weighted and if we assign each edge the weight 1, then all spanning trees are minimum, since they all have weight $|V(G)| - 1$ [by Lemma 2 on page 242].

Our next algorithm builds a minimum spanning tree for a weighted graph G whose edges $e_1, \ldots, e_m$ have been initially sorted so that

$$W(e_1) \le W(e_2) \le \cdots \le W(e_m).$$

The algorithm proceeds one by one through the list of edges of G, beginning with the smallest weights, choosing edges that do not introduce cycles. When the algorithm stops, the set E is supposed to be the set of edges in a minimum spanning tree for G. The notation $E \cup \{e_j\}$ in the statement of the algorithm stands for the subgraph whose edge set is $E \cup \{e_j\}$ and whose vertex set is $V(G)$.

Kruskal's Algorithm(weighted graph)

{Input: A finite weighted connected graph G with edges listed in order of increasing weight}
{Output: A set E of edges of a minimum spanning tree for G}
Set $E := \emptyset$.
for $j = 1$ to $|E(G)|$ **do**
  if $E \cup \{e_j\}$ is acyclic **then**
    Put e_j in E.
return E ■

EXAMPLE 2 Figure 3(a) shows a weighted graph with the weights indicated next to the edges. Figure 3(b) shows one possible way to number the edges of the graph so that the weights form a nondecreasing sequence, i.e., with $W(e_i) \le W(e_j)$ whenever $i < j$.

Figure 3 ▶

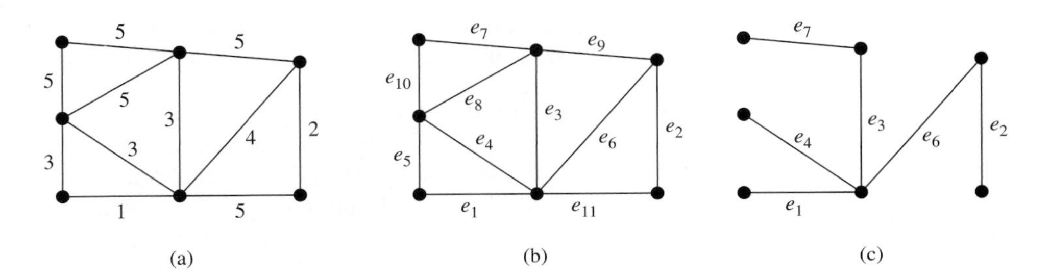

Applying Kruskal's algorithm to this weighted graph gives the spanning tree T with edges e_1, e_2, e_3, e_4, e_6, e_7 sketched in Figure 3(c). Edges e_1, e_2, e_3 and e_4 were put in E, because they created no cycles. Edge e_5 was rejected because $e_1\, e_4\, e_5$ would form a cycle. Edges e_8 through e_{11} were rejected for similar reasons. The spanning tree T has weight 18. ■

Theorem 2 Kruskal's algorithm produces a minimum spanning tree.

Proof We show first that the statement "E is contained in a minimum spanning tree of G" is a loop invariant. This statement is clearly true initially when E is empty. Suppose it is true at the start of the jth pass through the `for` loop, so that E is contained in some minimum spanning tree T [about which we only know that it exists]. If $E \cup \{e_j\}$ is not acyclic, then E doesn't change, so we may suppose that $E \cup \{e_j\}$ is acyclic. We want to find a minimum spanning tree T^* that contains $E \cup \{e_j\}$. If e_j is in T, then we can take $T^* = T$. Thus suppose that e_j is not in T.

Then e_j belongs to some cycle C in $T \cup \{e_j\}$ by Theorem 4 on page 242. Since $E \cup \{e_j\}$ is acyclic, the cycle C must contain some edge f in T with f in $T \setminus (E \cup \{e_j\})$. Form $T^* := (T \cup C) \setminus \{f\} = (T \cup \{e_j\}) \setminus \{f\}$. Then T^* is connected, spans G, and has $|V(G)| - 1$ edges. Hence, by Theorem 4 on page 242, T^* is a spanning tree for G containing $E \cup \{e_j\}$. Because $E \cup \{f\} \subseteq T$, $E \cup \{f\}$ is acyclic. Since f has not yet been picked to be adjoined to E, it must be that e_j has first chance; i.e., $W(e_j) \le W(f)$. Since $W(T^*) = W(T) + W(e_j) - W(f) \le W(T)$, and T is a minimum spanning tree, in fact we have $W(T^*) = W(T)$. Thus T^* is a minimum spanning tree, as desired.

Since E is always contained in a minimum spanning tree, it only remains to show that the graph with edge set E and vertex set $V(G)$ is connected when the algorithm stops. Let u and v be two vertices of G. Since G is connected, there is a path from u to v in G. If some edge f on that path is not in E, then $E \cup \{f\}$ contains a cycle [else f would have been chosen in its turn], so f can be replaced in the path by the part of the cycle that's in E. Making necessary replacements in this way, we obtain a path from u to v lying entirely in E. ■

Note, by the way, that Kruskal's algorithm works even if G has loops or parallel edges. It never chooses loops, and it will select the first edge listed in a collection of parallel edges. It is not even necessary for G to be connected in order to apply Kruskal's algorithm. In the general case the algorithm produces a **minimum spanning forest** made up of minimum spanning trees for the various components of G.

If G has n vertices, Kruskal's algorithm can't produce more than $n - 1$ edges in E. The algorithm could be programmed to stop when $|E| = n - 1$, but it might still need to examine every edge in G before it stopped.

Each edge examined requires a test to see if e_j belongs to a cycle. The algorithm Forest can be applied to the graph $E \cup \{e_j\}$ to test whether it contains a cycle. The idea in Forest can be applied in another way to give an acyclicity check, using the observations in Theorem 3 on page 241. Suppose that G' is the graph with $V(G') = V(G)$ and $E(G') = E$ when the algorithm is examining e_j. If we know in which components of G' the endpoints of e_j lie, then we can add e_j to E if they lie in different components and reject e_j otherwise.

This test is quick, provided we keep track of the components. At the start, each component consists of a single vertex, and it's easy to update the component list after accepting e_j; the components of the two endpoints of e_j just merge into a single component. The resulting version of Kruskal's algorithm runs in time $O(|E(G)| \cdot \log_2 |E(G)|)$, including the time that it takes to sort $E(G)$ initially. For complete details, see the account of Kruskal's algorithm in a standard text.[1]

In the case of the graph in Example 2, it would have been quicker to delete a few bad edges from G than it was to build T up one edge at a time. There is a general algorithm that works by deleting edges: Given a connected graph with the edges listed in increasing order of weight, go through the list starting at the big end, throwing out an edge if and only if it belongs to a cycle in the current subgraph of G. The subgraphs that arise during the operation of this algorithm all have vertex set $V(G)$ and are all connected. The algorithm only stops when it reaches an acyclic graph, so the final result is a spanning tree for G. It is in fact a minimum spanning tree [Exercise 16]. Indeed, it's the same tree that Kruskal's algorithm produces. If $|E(G)| < 2|V(G)| - 1$, this procedure may take less time than Kruskal's algorithm, but, of course, if G has so few edges, both algorithms work quite quickly.

Kruskal's algorithm makes sure that the subgraph being built is always acyclic, while the deletion procedure we have just described keeps all subgraphs connected. Both algorithms are greedy, in the sense that they always choose the smallest edge to add or the largest to delete. Greed does not always pay—consider trying to get 40 cents out of a pile of dimes and quarters by picking a quarter first—but it pays off this time.

The algorithm we next describe is doubly greedy; it makes minimum choices while simultaneously keeping the subgraph both acyclic and connected. Moreover, it does not require the edges of G to be sorted initially. The procedure works just like Tree, but it takes weights into account. It grows a tree T inside G, with $V(T) = V$ and $E(T) = E$. At each stage the algorithm looks for an edge of smallest weight that joins a vertex in T to some new vertex outside T. Then it adds such an edge and vertex to T and repeats the process.

---

[1] See, for example, *Data Structures and Algorithms* by Aho, Hopcroft, and Ullman, *Introduction to Algorithms* by Cormen, Leiserson, and Rivest, or *Data Structures and Network Algorithms* by Tarjan.

Prim's Algorithm(weighted graph)

{Input: A finite weighted connected graph G [with edges listed in any order]}
{Output: A set E of edges of a minimum spanning tree for G}
Set $E := \emptyset$.
Choose w in $V(G)$ and set $V := \{w\}$.
while $V \neq V(G)$ **do**
  Choose an edge $\{u, v\}$ in $E(G)$ of smallest possible weight with $u \in V$ and $v \in V(G) \setminus V$.
  Put $\{u, v\}$ in E and put v in V.
return E ■

EXAMPLE 3

We apply Prim's algorithm to the weighted graph shown in Figure 4(a). Since choices are possible at several stages of the execution, the resulting tree is not uniquely determined. The solid edges in Figure 4(b) show one possible outcome, where the first vertex chosen is marked "Start." The first edge selected is labeled a. The second edge selected is b, but the dashed edge b' is an alternative choice. Then edges c, d and e are chosen in that order, with the dashed edge d' as an alternative choice to d. Note that Kruskal's algorithm would have chosen edges c and e before edge b. ■

Figure 4 ▶

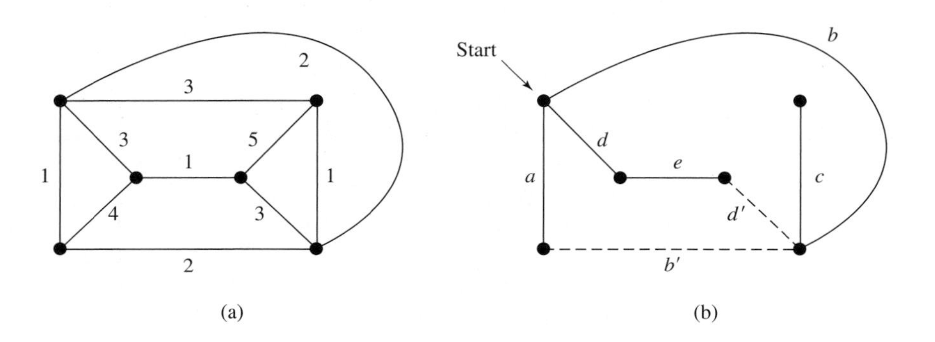

Theorem 3 Prim's algorithm produces a minimum spanning tree for a connected weighted graph.

Proof The proof of Theorem 1, that Tree works, shows that Prim's algorithm terminates and yields a spanning tree for G; all that is new here is greed. The statement "T is contained in a minimum spanning tree of G" is surely true at the start, when T is just a single vertex. We claim that this statement is an invariant of the **while** loop.

Suppose that, at the beginning of some pass through the **while** loop, T is contained in the minimum spanning tree T^* of G. Suppose that the algorithm now chooses the edge $\{u, v\}$. If $\{u, v\} \in E(T^*)$, then the new T is still contained in T^*, which is wonderful. Suppose not. Because T^* is a spanning tree, there is a path in T^* from u to v. Since $u \in V$ and $v \notin V$, there must be some edge in the path that joins a vertex z in V to a vertex w in $V(G) \setminus V$. Since Prim's algorithm chose $\{u, v\}$ instead of $\{z, w\}$, we have $W(u, v) \leq W(z, w)$. Take $\{z, w\}$ out of $E(T^*)$ and replace it with $\{u, v\}$. The new graph T^{**} is still connected, so it's a tree by Theorem 4 on page 242. Since $W(T^{**}) \leq W(T^*)$, T^{**} is also a minimum spanning tree, and T^{**} contains the new T. At the end of the loop, T is still contained in some minimum spanning tree, as we wanted to show. ■

Prim's algorithm makes $n - 1$ passes through the **while** loop for a graph G with n vertices. Each pass involves choosing a smallest edge subject to a specified condition. A stupid implementation could require looking through all the edges of

$E(G)$ to find the right edge. A more clever implementation would keep a record for each vertex x in $V(G) \setminus V$ of the vertex u in V with smallest value $W(u, x)$ and would also store the corresponding value of $W(u, x)$. The algorithm could then simply run through the list of vertices x in $V(G) \setminus V$, find the smallest $W(u, x)$, add $\{u, x\}$ to E, and add x to V. Then it could check, for each y in $V(G) \setminus V$, whether x is now the vertex in V closest to y and, if so, update the record for y. The time to find the closest x to V and then update records is just $O(n)$, so Prim's algorithm with the implementation we have just described runs in time $O(n^2)$.

Prim's algorithm can easily be modified [Exercise 15] to produce a minimum spanning forest for a graph, whether or not the graph is connected. The algorithm as it stands will break down if the given graph is not connected.

As a final note, we observe that the weight of a minimum spanning tree helps provide a lower bound for the Traveling Salesperson Problem we mentioned in §6.5. Suppose that the path $C = e_1 e_2 \cdots e_n$ is a solution for the Traveling Salesperson Problem on G, i.e., a Hamilton circuit of smallest possible weight. Then $e_2 \cdots e_n$ visits each vertex just once, so it's a spanning tree for G. If M is the weight of a minimum spanning tree for G [a number that we can compute using Kruskal's or Prim's algorithm], then

$$M \leq W(e_2) + \cdots + W(e_n) = W(C) - W(e_1).$$

Hence $W(C) \geq M$ + (smallest edge weight in G). If we can find, by any method, some Hamilton circuit C of G with weight close to M + (smallest edge weight), then we should probably take it and not spend time trying to do better.

Exercises 6.6

1. (a) Find Tree(1) for the graph in Figure 5(a). Draw a picture of the tree that results, and label the edges with a, b, c, d, e, f in the order in which they are chosen. Use the choice scheme of Example 1. Repeat part (a) for the graph in Figure 5(b).

(b) Repeat part (a) for the graph in Figure 5(c).

(d) Repeat part (a) for the graph in Figure 5(d).

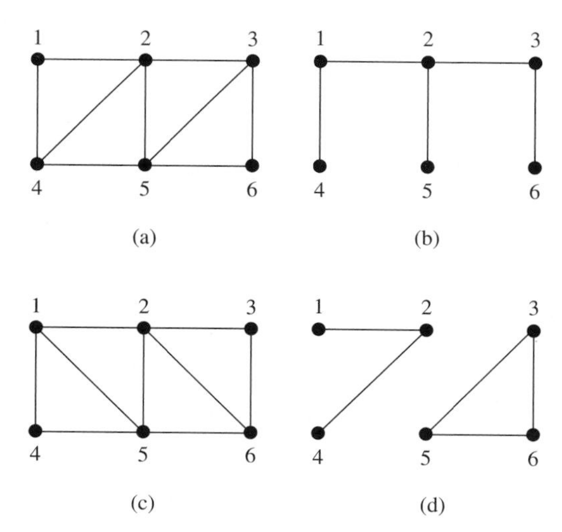

Figure 5 ▲

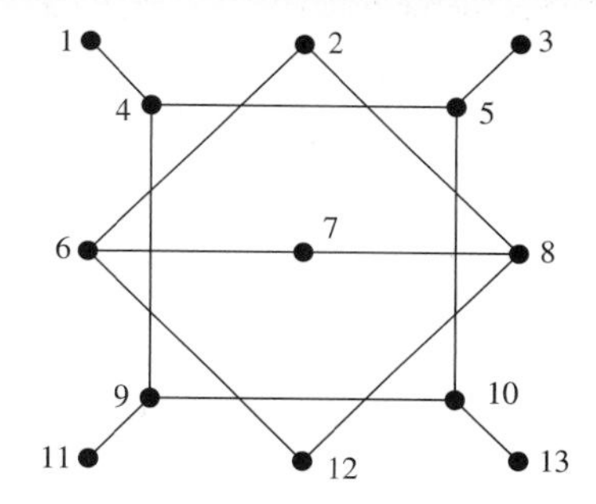

Figure 6 ▲

2. Apply Forest to the graph in Figure 6.

3. Figure 7(a) shows a weighted graph, and Figures 7(b) and 7(c) show two different ways to label its edges with $a, b, \ldots, n$ in order of increasing weight.

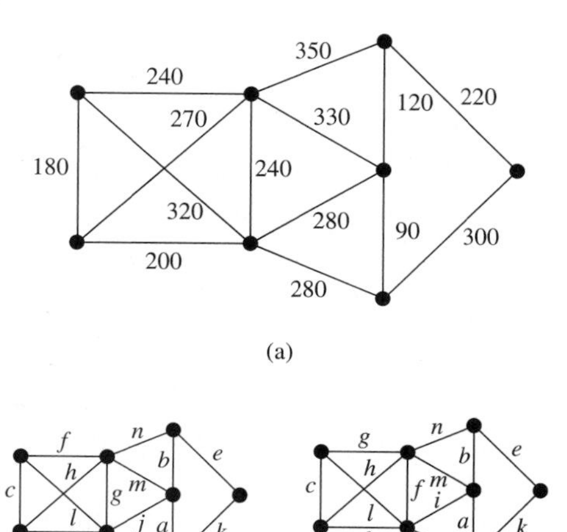

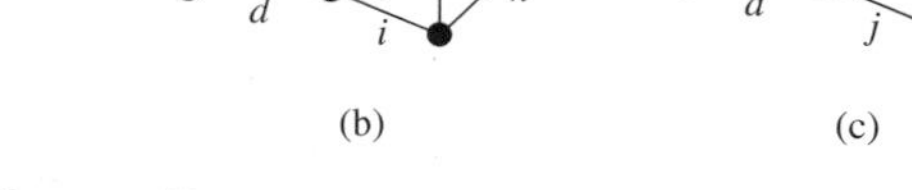

Figure 7 ▲

(a) Apply Kruskal's algorithm to the graph with the edges ordered as in Figure 7(b). Draw the resulting minimum spanning tree and give its weight.

(b) Repeat part (a) with the ordering in Figure 7(c).

4. Suppose that the graph in Figure 7(b) is weighted so that

$$W(a) \geq W(b) \geq \cdots \geq W(n).$$

Draw a minimum spanning tree for the graph.

5. Apply Prim's algorithm to the graph in Figure 4(a), starting at one of the

(a) central vertices. (b) vertices on the right.

6. Suppose that the graph in Figure 3(b) is weighted so that $W(e_i) \leq W(e_j)$ when $i > j$, so e_1 is the heaviest edge. Draw a minimum spanning tree for the graph.

7. (a) Apply Prim's algorithm to the graph in Figure 7(a), starting at the vertex of degree 4. Draw the resulting minimum spanning tree and label the edges alphabetically in the order chosen.

(b) What is the weight of a minimum spanning tree for this graph?

(c) How many different answers are there to part (a)?

8. Suppose that the graph in Figure 5(a) on page 241 is weighted so that

$$W(e_1) > W(e_2) > \cdots > W(e_{10}).$$

(a) List the edges in a minimum spanning tree for this graph in the order in which Kruskal's algorithm would choose them.

(b) Repeat part (a) for Prim's algorithm, starting at the upper right vertex.

9. Repeat Exercise 8 with weights satisfying

$$W(e_1) < W(e_2) < \cdots < W(e_{10}).$$

10. (a) Use Kruskal's algorithm to find a minimum spanning tree of the graph in Figure 8(a). Label the edges in alphabetical order as you choose them. Give the weight of the minimum spanning tree.

(b) Repeat part (a) with Prim's algorithm, starting at the lower middle vertex.

11. (a) Repeat Exercise 10(a) for the graph in Figure 8(b).

(b) Repeat Exercise 10(b) for the graph in Figure 8(b), starting at the top vertex.

12. (a) Find all spanning trees of the graph in Figure 9.

(b) Which edges belong to every spanning tree?

(c) For a general finite connected graph, characterize the edges that belong to every spanning tree. Prove your assertion.

13. An oil company wants to connect the cities in the mileage chart in Figure 10 by pipelines going directly between cities. What is the minimum number of miles of pipeline needed?

14. Does every edge of a finite connected graph with no loops belong to some spanning tree? Justify your answer. [For this exercise, the edges are not weighted.]

15. (a) Where does Prim's algorithm break down if G is not connected?

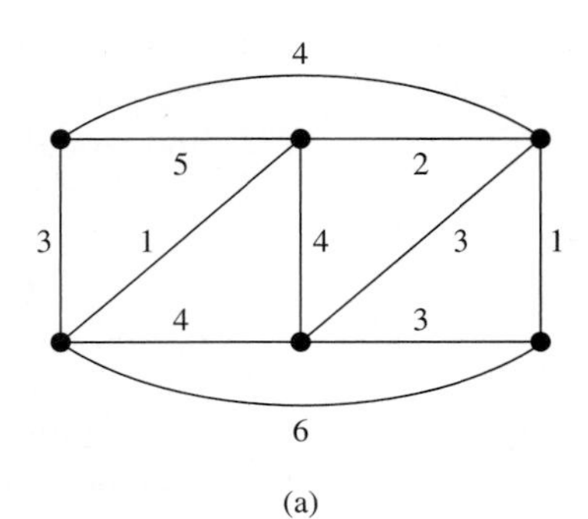

(a)

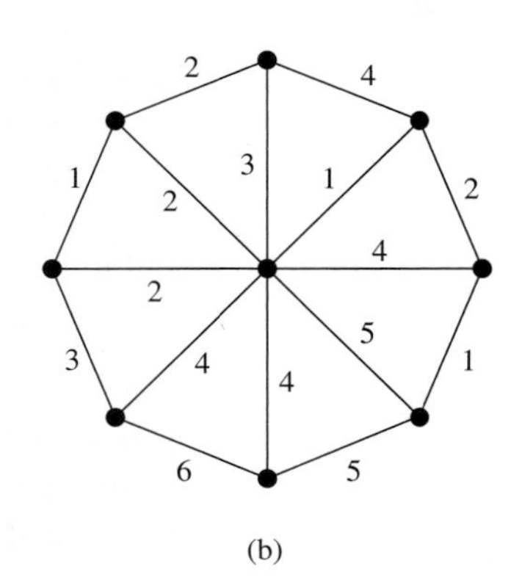

(b)

Figure 8 ▲

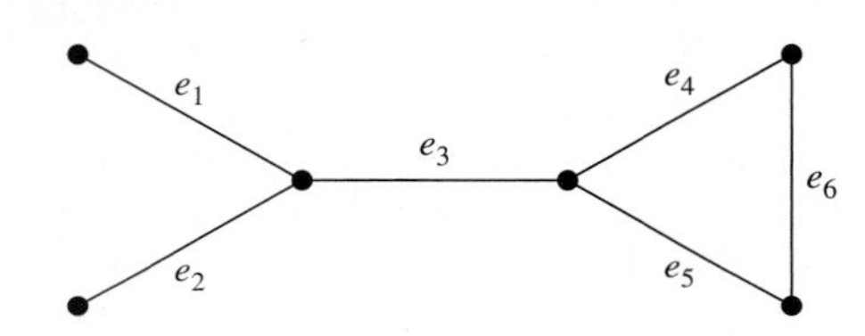

Figure 9 ▲

	Des Moines	Milwaukee	Minneapolis	Omaha	Pierre	Winnipeg
Bismarck	670	758	427	581	211	369
Des Moines		361	252	132	492	680
Milwaukee			332	493	690	759
Minneapolis				357	394	431
Omaha					391	650
Pierre						521

Figure 10 ▲

(b) Modify Prim's algorithm to produce a minimum spanning forest.

16. (a) Show that if H is a subgraph of the weighted graph G that contains a minimum spanning tree of G, then every minimum spanning tree of H is a minimum spanning tree of G.

(b) Show that the edge-deletion algorithm described after Kruskal's algorithm produces a minimum spanning tree. *Hint:* Show that the set of edges remaining after each deletion contains a minimum spanning tree for G. *Outline:* Suppose that the set E of edges remaining when e is about to be deleted contains a minimum spanning tree for G. Apply part (a) to the minimum spanning tree K *produced for* E by Kruskal's algorithm. Let C be a cycle in E containing e. Argue that every edge f in $C \setminus (K \cup \{e\})$ precedes e on the edge list and forms a cycle with some edges in K that also precede e. Get a path in $K \setminus \{e\}$ that joins the ends of e and conclude that $e \notin K$.

17. Let G be a finite weighted connected graph in which different edges have different weights. Show that G has exactly one minimum spanning tree. *Hint:* Assume that G has more than one minimum spanning tree. Consider the edge of smallest weight that belongs to some but not all minimum spanning trees.

Chapter Highlights

As always, one of the best ways to use this material for review is to follow the suggestions at the end of Chapter 1. Ask yourself: What does it mean? Why is it here? How can I use it? Keep thinking of examples. Though there are lots of items on the lists below, there are not really as many new ideas to master as the lists would suggest. Have courage.

CONCEPTS

path
- closed, simple, cycle, acyclic
- Euler path, Euler circuit
- Hamilton path, Hamilton circuit, Gray code

isomorphism, invariant

degree, degree sequence

graph
- regular, complete, bipartite, complete bipartite

connected, component

tree, leaf, forest

spanning tree, spanning forest, minimum spanning tree [forest]

rooted tree
- root, parent, child, descendant, subtree with root v
- binary search tree, labeled tree, ordered rooted tree
- binary [m-ary] rooted tree
  - regular, full
- level number, height

weight of edge, path, subgraph

FACTS

A path has all vertices distinct if and only if it is simple and acyclic.

If there is a path between distinct vertices, then there is a simple acyclic path between them.

There is at most one simple path between two vertices in an acyclic graph or digraph. There is exactly one simple path between two vertices in a tree.

If e is an edge of a connected graph G, then e belongs to some cycle if and only if $G \setminus \{e\}$ is connected. Thus an algorithm that checks connectedness can test for cycles.

The following statements are equivalent for a graph G with $n \geq 1$ vertices and no loops:

(a) G is a tree.
(b) G is connected, but won't be if an edge is removed.
(c) G is acyclic, but won't be if an edge is added.
(d) G is acyclic and has $n - 1$ edges [as many as possible].
(e) G is connected and has $n - 1$ edges [as few as possible].

Choosing a root gives a tree a natural directed structure.

$$\sum_{v \in V(G)} \deg(v) = 2 \cdot |E(G)|.$$

A graph has an Euler circuit if and only if it is connected and all vertices have even degree. Euler paths exist if at most two vertices have odd degree.

If a graph has no loops or parallel edges, and if $|V(G)| = n \geq 3$, then G is Hamiltonian if any of the following is true:

(a) $\deg(v) \ge n/2$ for each vertex v [high degrees].
(b) $|E(G)| \ge \frac{1}{2}(n-1)(n-2)+2$ [lots of edges].
(c) $\deg(v)+\deg(w) \ge n$ whenever v and w are not connected by an edge.

Theorem 4 on page 256 gives information on Hamilton paths in bipartite graphs.

ALGORITHMS

EulerCircuit and ClosedPath algorithms to construct an Euler circuit of a graph.

Tree and Forest algorithms to build a spanning forest or find components of a graph in time $O(|V(G)|+|E(G)|)$.

Kruskal's and Prim's algorithms to construct minimum spanning trees [or forests] for weighted graphs.

Supplementary Exercises

1. All parts of this question refer to the following two graphs G and H.

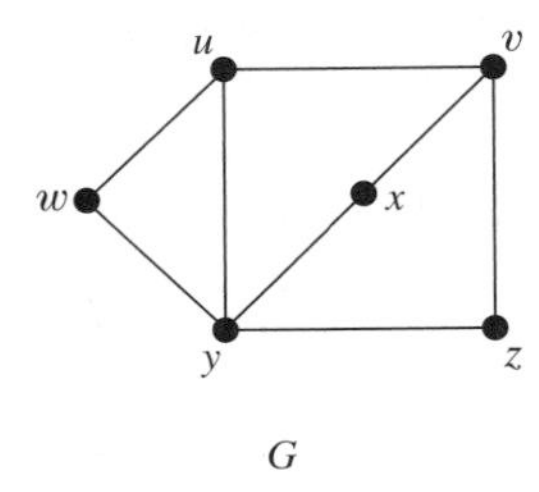

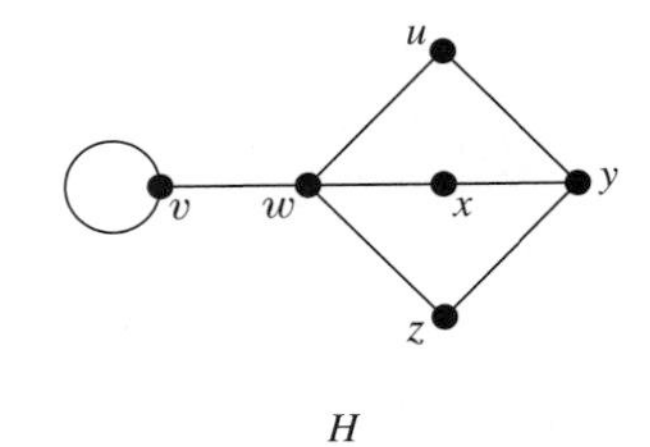

(a) Does the graph G have an Euler circuit? Justify your answer.

(b) Give the vertex sequence of an Euler path for the graph H.

(c) Give the vertex sequence of a cycle in H.

(d) Is every simple closed path in G a cycle? Explain.

2. This exercise refers to the graphs G and H in Exercise 1.

(a) How many edges can a spanning tree of G have?

(b) Does G have a Hamiltonian path? Explain.

(c) How many edges can a spanning tree of H have?

(d) Does H have a Hamiltonian path? Explain.

3. We are given a tree with n vertices, $n \ge 3$. For each part of this question, decide whether the statement must be true, must be false, or could be either true or fâlse.

(a) The tree has n edges.

(b) There is at least one vertex of degree 2.

(c) There are at least two vertices of degree 1.

(d) Given two distinct vertices of the tree, there is exactly one path connecting them.

4. (a) Does the complete bipartite graph $K_{2,7}$ have an Euler circuit? Explain.

(b) Does $K_{2,15}$ have an Euler path? Explain.

(c) Does $K_{4,6}$ have a Hamilton path? Explain.

5. (a) Draw a binary search tree of height as small as possible for the letters a, b, c, d, e, f in their usual (alphabetical) order.

(b) What are the possible labels for the root in a tree like the one in part (a)?

6. Suppose that a connected graph without loops or parallel edges has 11 vertices, each of degree 6.

(a) Must the graph have an Euler circuit? Explain.

(b) Must the graph have a Hamilton circuit? Explain.

(c) If the graph *does* have an Euler circuit, how many edges does the circuit contain?

(d) If the graph *does* have a Hamilton circuit, what is its length?

7. (a) Find a minimal spanning tree for the following graph.

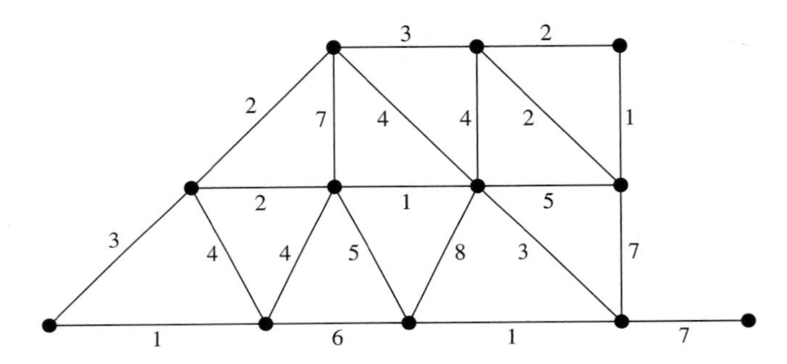

(b) What is the weight of the minimal spanning tree?

(c) Does the graph have an Euler path or circuit? Explain.

(d) Does the graph have a Hamilton circuit? Explain.

8. The figure below shows a weighted graph G, with weights given next to the edges. The thickly drawn edges have already been chosen as part of a minimum spanning tree for G.

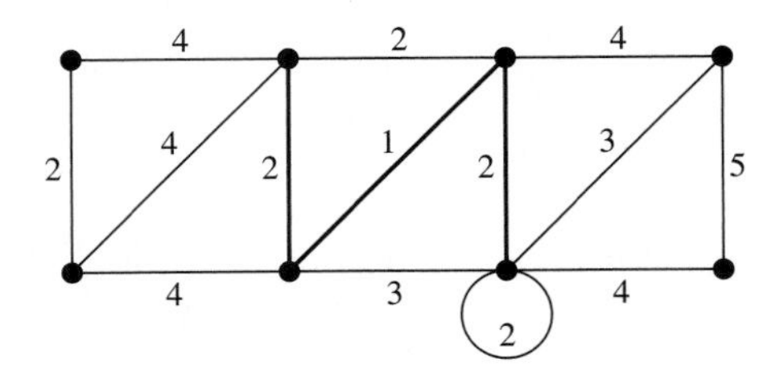

(a) What is the next edge that Prim's algorithm would choose in this case?

(b) What is the next edge that Kruskal's algorithm would choose in this case?

(c) What is the weight of a minimal spanning tree for this graph?

(d) What is the smallest weight of a Hamilton circuit for this graph?

(e) What is the smallest weight of a Hamilton path for this graph?

9. Consider the following weighted graph, where $W(e_1) < W(e_2) < \cdots < W(e_{11}) < W(e_{12})$.

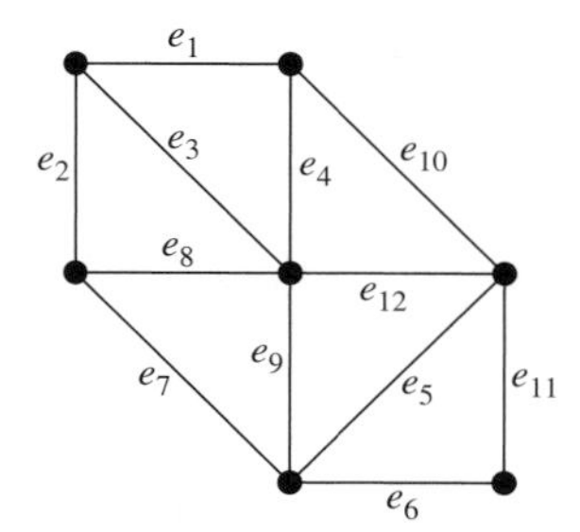

(a) Use Kruskal's algorithm to obtain the minimal spanning tree. List the edges in the order in which Kruskal's algorithm will select them.

(b) Use Prim's algorithm, starting at vertex u, to obtain the minimal spanning tree. List the edges in the order in which Prim's algorithm will select them.

10. The drawing shows a binary search tree containing five records, organized alphabetically from left to right.

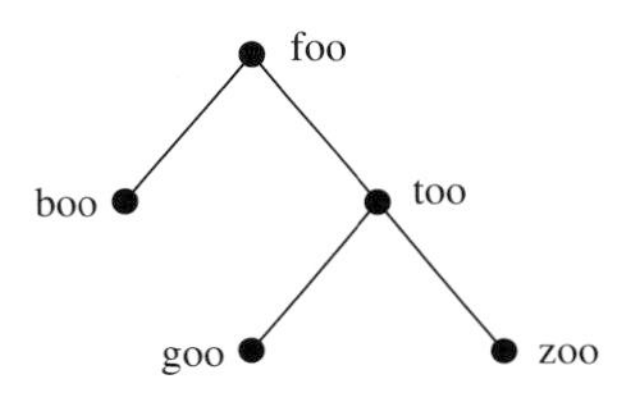

(a) Describe how one would search this tree for the record **goo**.

(b) Describe how one would search this tree for the record **poo**.

(c) Show how to modify this tree to a binary search tree of the same height that contains all of the given records *and* the record **poo**.

11. (a) How many essentially different, i.e., nonisomorphic, rooted trees have four vertices?

(b) How many essentially different *ordered* rooted trees have four vertices?

12. We are given a full binary (rooted) tree with n vertices and t leaves. Let h be the height of the rooted tree. For each part of this question, indicate whether the statement must hold, cannot hold, or might or might not hold.

(a) If a vertex is at level 5, then there is a path of length 5 from it to the root.

(b) $h \geq \log_2 t$.

(c) $h > \log_2 n - 1$.

(d) If $n \geq 3$, then there are two vertices at level 1.

13. Ann claims that every tree with more than one vertex is a bipartite graph. She shows the following tree as an example.

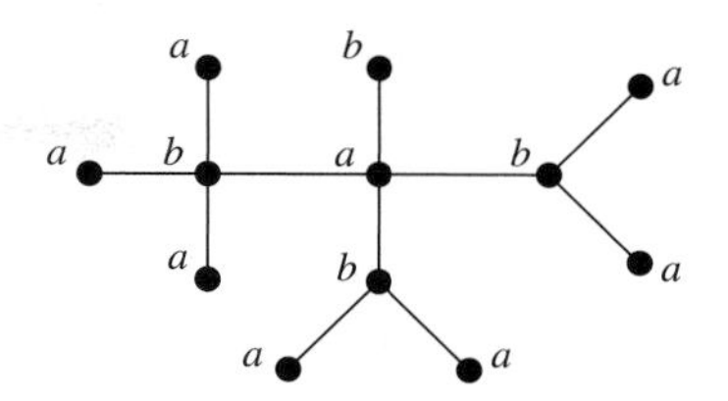

Bob says that, despite this example, Ann is wrong in general. Who is correct?

14. Find all complete bipartite graphs $K_{m,n}$ that have a closed path that is both an Euler circuit and a Hamilton circuit. Justify your claim.

15. Let K_n be the complete graph with n vertices ($n \geq 3$).

(a) Show that if we remove at most $n - 3$ edges, then the subgraph obtained is Hamiltonian.

(b) Show that $n - 2$ edges can be removed in such a way that the subgraph obtained is not Hamiltonian.

16. Let G be a graph with $n \geq 2$ vertices, no loops, and no parallel edges. Prove that G has two different vertices that have the same degree. *Hint:* How big can the degree of a vertex be? If some vertex actually has this largest possible degree, can there be a vertex of degree 0?

17. Prove that the number of vertices of a graph is at most the sum of the number of its edges and the number of its connected components.

18. (a) Show that the edges of a Hamilton path H for a graph G form a spanning tree for G.

(b) Show that if G is a weighted graph with weight function W and if H is a Hamilton circuit for G, then $W(H) \geq W(T) + W(e)$ for every minimal spanning tree T of G and edge e of H.

19. (a) Bob has drawn a complicated graph with ten vertices, each of degree at least 5. Ann asks Bob if his graph has a Hamilton circuit. After a long time, he says that it does not. Ann tells him that his graph must have a loop or parallel edges. Explain why Ann must be correct if Bob's statement about Hamilton circuits is correct.

(b) Is there a Hamiltonian graph with ten vertices in which every vertex has degree less than 5? Explain.

20. Let $T_{m,h}$ denote a full m-ary tree of height h. A node is chosen at random in $T_{2,3}$ with all nodes equally likely to be chosen.

(a) What is the probability that the node is at level 2?

(b) At level 3?

(c) What is the most likely level for the node?

The **expected level** $L(T)$ for the tree T is defined to be

$$\sum_{k=0}^{\infty} k \cdot (\text{probability that a random node is at level } k).$$

[For large k the probabilities here are of course 0, so it's not really an infinite sum.]

(d) Find $L(T_{2,3})$.

(e) Find $L(T_{3,2})$.

21. Let S be a set of records of the form (*name*, *number*), and let A be the set $\{1, \ldots, 1000\}$ of memory addresses. A **hash function** from S to A randomly assigns to each member of S some address in A, unless two members of S are assigned the same address, in which case we say there is a **collision** and the function does something else. Imagine that S is being built up one record at a time.

(a) What is the probability that the second member of S does not produce a collision?

(b) What is the probability that the first 20 records do not produce a collision?

(c) Suppose that the first 20 records in S have been assigned 20 different addresses. What is the probability that the 21st member of S yields a collision?

(d) If the records can contain only 100 possible different numbers, how many names are possible before a collision is certain?

22. Ann looks at the graph shown and observes that it doesn't have an Euler cycle. She says that if she adds just one more vertex and adds an edge from the new vertex to each of the other vertices in the graph, then the resulting graph will have an Euler cycle. Bob says that if she just wants an Euler cycle she can do it by adding only three new edges and no new vertices. What's the story?

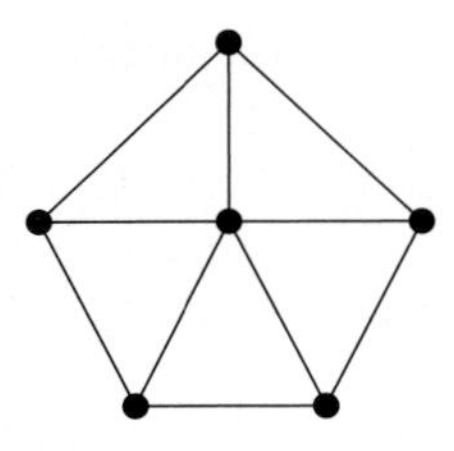

23. (a) Find the degree sequence of each of the graphs shown below.

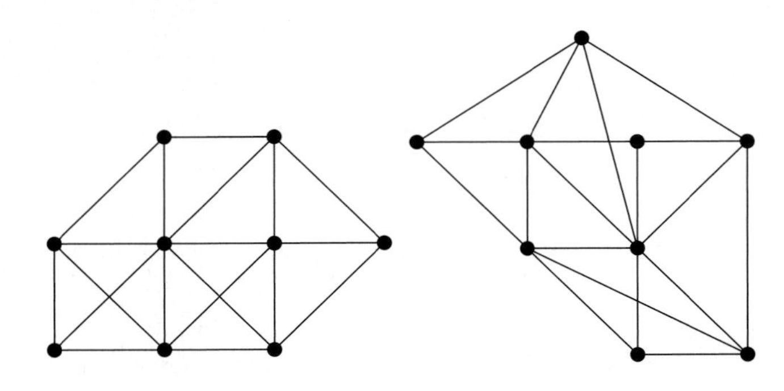

(b) Are these two graphs isomorphic? Explain.

(c) Are these graphs Hamiltonian? Explain.

(d) Do these graphs satisfy the hypotheses of any of the theorems in §6.5? Explain.

Recursion, Trees, and Algorithms

One of the main goals of this chapter is an understanding of how recursive algorithms work and how they can be verified. Before we discuss algorithms, we look at the general topic of recursive definition, which we touched on briefly for sequences in Chapter 4, and we consider a recursive generalization of mathematical induction. Recursive algorithms can often be thought of as working their way downward through some sort of tree. To give emphasis to this view, we have chosen examples of algorithms that explicitly examine trees or that naturally give rise to trees. The idea of a sorted labeling of a digraph, which our algorithms produce in §7.3, will play a role later in Chapter 8. The last two sections of this chapter present applications of recursive methods to algebraic notation and to prefix codes, which are used for file compression and for designing data structures.

7.1 General Recursion

The recursive definitions in §4.4 allowed us to define sequences by giving their first few terms, together with a recipe for getting later terms from earlier ones. In our treatment of tree algorithms, it will be convenient to have a more general version of recursion, which we consider now. There are three main themes to this section: recursively defined sets, a generalization of induction, and recursively defined functions. These topics are closely linked, as we will see.

Roughly speaking, a set of objects is defined recursively in case it is built up by a process in which some elements are put in at the beginning and others are added later because of elements that are already members of the club. Of course, such a description is too fuzzy to be useful. We will be more precise in a moment, but first here are some examples of recursively defined sets.

EXAMPLE 1

(a) We build $\mathbb{N}$ recursively by

(B) $0 \in \mathbb{N}$;

(R) If $n \in \mathbb{N}$, then $n + 1 \in \mathbb{N}$.

Condition (B) puts 0 in $\mathbb{N}$, and (R) gives a recipe for generating new members of $\mathbb{N}$ from old ones. By invoking (B) and then repeatedly using (R), we get $0, 1, 2, 3, \ldots$ in $\mathbb{N}$.

(b) The conditions

(B) $1 \in S$;

(R) If $n \in S$, then $2n \in S$

give members of a subset S of $\mathbb{P}$ that contains 1 [by (B)], 2 [by (R), since $1 \in S$], 4 [since $2 \in S$], 8, 16, etc. It is not hard to see, by induction, that S contains $\{2^m : m \in \mathbb{N}\}$. Moreover, the numbers of the form 2^m are the only ones that are forced to belong to S by (B) and (R) [see Exercise 1]. It seems reasonable to say that (B) and (R) define the set $\{2^m : m \in \mathbb{N}\}$. ■

EXAMPLE 2

Consider an alphabet Σ. The rules

(B) $\lambda \in \Sigma^*$;

(R) If $w \in \Sigma^*$ and $x \in \Sigma$, then $wx \in \Sigma^*$

give a recursive description of Σ^*. The empty word λ is in Σ^* by decree, and repeated applications of (R) let us build longer words. For instance, if Σ is the English alphabet, then the words λ, b, bi, big, $\ldots$, $bigwor$, $bigword$ are in the set described by (B) and (R). ■

A **recursive definition of a set** S consists of two parts: a **basis** of the form

(B) $X \subseteq S$,

where X is some specified set, and a **recursive clause**

(R) If s is determined from members of S by following certain rules, then s is in S.

The particular rules in (R) will be specified, as they were in Examples 1 and 2, and they can be anything that makes sense. As in Example 2, they can refer to objects that are not in S itself.

Our implicit understanding is always that an element belongs to S only if it is required to by (B) and (R). Thus in Example 1(b) only the powers of 2 are *required* to be in S, so S consists only of powers of 2.

The conditions (B) and (R) allow us to build S up by spreading out from X in layers. Define the chain of sets $S_0 \subseteq S_1 \subseteq S_2 \subseteq \cdots$ by

$S_0 = X$, and

S_{n+1} = everything in S_n or constructed from members of S_n by using a rule in (R).

Then S_n consists of the members of S that can be built from X by n or fewer applications of rules in (R), and $S = \bigcup_{n=0}^{\infty} S_n$.

EXAMPLE 3

(a) For $S = \mathbb{N}$ as in Example 1(a), $S_0 = \{0\}$, $S_1 = \{0, 1\}$, $S_2 = \{0, 1, 2\}, \ldots$, $S_n = \{0, 1, \ldots, n\}, \ldots$.

(b) For $S = \{2^m : m \in \mathbb{N}\}$ as defined in Example 1(b), $S_0 = \{1\}$, $S_1 = \{1, 2\}$, $S_2 = \{1, 2, 4\}, \ldots, S_n = \{1, 2, 4, \ldots, 2^n\}$.

(c) For $\Sigma = \{a, b\}$, the definition of Σ^* in Example 2 leads to $\Sigma^*{}_0 = \{\lambda\}$, $\Sigma^*{}_1 = \{\lambda, a, b\}$, $\Sigma^*{}_2 = \{\lambda, a, b, aa, ab, ba, bb\}$, etc., so $\Sigma^*{}_n = \{\lambda\} \cup \Sigma \cup \Sigma^2 \cup \cdots \cup \Sigma^n$ in our old notation.

(d) Let $\Sigma = \{a, b\}$ and define S by

(B) $\Sigma \subseteq S$, and

(R) If $w \in S$, then $awb \in S$.

Then $S_0 = \Sigma = \{a, b\}$, $S_1 = \{a, b, aab, abb\}$, $S_2 = S_1 \cup \{aaabb, aabbb\}$, etc., and S itself consists of the words in Σ^* of the form $a \cdots ab \cdots b$ with one more or one fewer a than the number of b's. ■

EXAMPLE 4

We can define the class of finite trees recursively. For convenience, let us say that the graph G' is **obtained from** the graph G **by attaching** v **as a leaf** in case

(a) $V(G') = V(G) \cup \{v\} \neq V(G)$, and

(b) $E(G') = E(G) \cup \{e\}$, where the edge e joins v to a vertex in $V(G)$.

Then the class of trees is defined by

(B) Every graph with one vertex and no edges is a [trivial] tree, and

(R) If T is a tree and if T' is obtained from T by attaching a leaf, then T' is a tree.

We can think of this recursive definition as building up trees by adding leaves. Figure 1 shows a typical construction sequence.

Figure 1 ▶

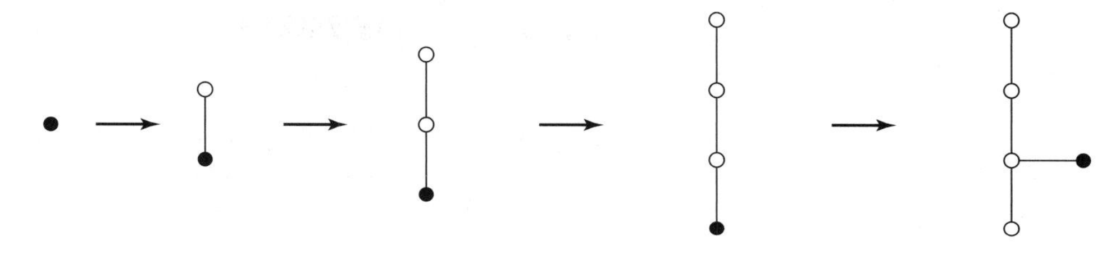

We previously defined trees to be connected acyclic graphs. To show that the new recursive definition coincides with the old definition, we would need to check that

(1) (B) and (R) only produce connected acyclic graphs, and

(2) Every finite connected acyclic graph can be constructed using only (B) and (R).

It is fairly easy to see that (1) is true, because (B) gives trees, and if T is connected and acyclic, then so is any T' constructed from T by (R). In effect, we are arguing by induction that if S_n consists of trees, then so does S_{n+1}.

To show (2) is a little harder. Imagine that there are [finite] trees that we can't construct with (B) and (R), and suppose that T is such a tree with as few vertices as possible. By (B), T has more than one vertex. Then T has at least two leaves by Lemma 1 on page 242. Prune a leaf of T to get a new tree T''. Then T'' is constructible from (B) and (R), by the minimal choice of T. But T is obtainable from T'' by (R); just attach the leaf again. Hence T itself is constructible, which is a contradiction to the way it was chosen. ■

EXAMPLE 5

(a) We can mimic the recursive definition in Example 4 to obtain the class of [finite] rooted trees.

(B) A graph with one vertex v and no edges is a [trivial] rooted tree with root v;

(R) If T is a rooted tree with root r, and T' is obtained by attaching a leaf to T, then T' is a rooted tree with root r.

As in Example 4, we see that this definition gives nothing but rooted trees. The second argument in Example 4 can be adapted to show that every rooted tree is constructible using (B) and (R); since every nontrivial tree has at least two leaves, we can always prune a leaf that's not the root.

(b) Here is another way to describe the class of rooted trees recursively. We will define a class $\mathcal{R}$ of ordered pairs (T, r) in which T is a tree and r is a vertex of T, called the **root** of the tree. For convenience, say that (T_1, r_1) and (T_2, r_2) are **disjoint** in case T_1 and T_2 have no vertices in common. If the pairs $(T_1, r_1), \ldots, (T_k, r_k)$ are disjoint, then we will say that T is **obtained by hanging** $(T_1, r_1), \ldots, (T_k, r_k)$ **from** r in case

(1) r is not a vertex of any T_i,
(2) $V(T) = V(T_1) \cup \cdots \cup V(T_k) \cup \{r\}$, and
(3) $E(T) = E(T_1) \cup \cdots \cup E(T_k) \cup \{e_1, \ldots, e_k\}$, where each edge e_i joins r to r_i.

Figure 2 shows the tree obtained by hanging (T_1, r_1), (T_2, r_2), and (T_3, r_3) from the root r.

Figure 2 ▶

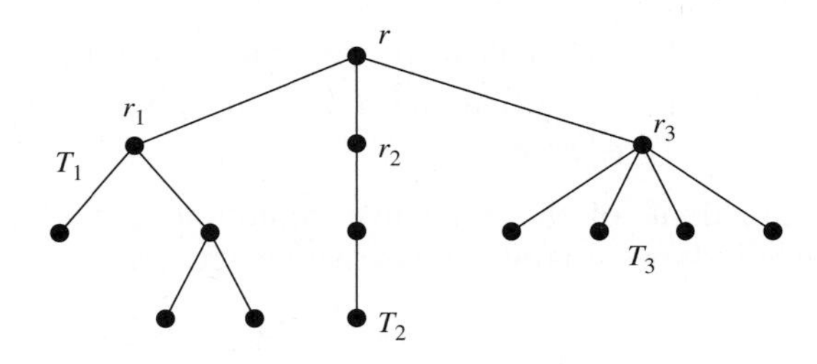

Here is the definition of $\mathcal{R}$:

(B) If T is a graph with one vertex v and no edges, then $(T, v) \in \mathcal{R}$;

(R) If $(T_1, r_1), \ldots, (T_k, r_k)$ are disjoint members of $\mathcal{R}$ and if (T, r) is obtained by hanging $(T_1, r_1), \ldots, (T_k, r_k)$ from r, then $(T, r) \in \mathcal{R}$.

As in part (a), we have no difficulty in showing that $\mathcal{R}$ consists of pairs (T, r) in which T is a tree with vertex r, that is, a tree rooted at r. To see that every rooted tree arises from (B) and (R), imagine a rooted tree T that doesn't, with $|V(T)|$ as small as possible. Because T doesn't come from (B), it has more than one vertex. Consider the children of the root r of T. The subtrees that have these children as their roots are smaller than T, so they're obtainable by (B) and (R). Since T is the result of hanging these subtrees from r, (T, r) is in $\mathcal{R}$ after all, contrary to assumption. ■

In our examples so far we know which sets we are trying to define, and the recursive definitions are aimed directly at them. One can also give perfectly acceptable recursive definitions without knowing for sure what they define, as we now illustrate.

EXAMPLE 6

(a) We define a set A of integers by

(B) $1 \in A$;

(R) If $n \in A$, then $3n \in A$, and if $2n + 1 \in A$, then $n \in A$.

These conditions describe A unambiguously, but it is not completely clear which integers are in A. A little experimentation suggests that A contains no n satisfying $n \equiv 2 \pmod 3$, and in fact [Exercise 9] this is the case. More experimentation suggests that perhaps $A = \{n \in \mathbb{P} : n \not\equiv 2 \pmod 3\}$, i.e., $A = \{1, 3, 4, 6, 7, 9, 10, 12, 13, \ldots\}$. It is not clear how we could verify such a guess. The chain 1, 3, 9, 27, 13, 39, 19, 57, 171, 85, 255, 127, 63, 31, 15, 7, which one can show is the shortest route to showing that 7 is in A, gives no hint at a general argument. Several hours of study have not completely settled the question. We do not know whether our guess is correct, although experts on recursive functions could perhaps tell us the answer.

(b) Consider the set S defined by

(B) $1 \in S$;

(R) If $n \in S$, then $2n \in S$, and if $3n + 1 \in S$ with n odd, then $n \in S$.

As of this writing, *nobody* knows whether $S = \mathbb{P}$, though many people have looked at the question. ■

Recursively defined classes support a kind of generalized induction. Suppose that the class S is defined recursively by a set X, given in (B), and a set of rules for producing new members, given in (R). Suppose also that each s in S has an associated proposition $p(s)$.

Generalized Principle of Induction In this setting, if

(b) $p(s)$ is true for every s in X, and

(r) $p(s)$ is true whenever s is produced from members t of S for which $p(t)$ is true,

then $p(s)$ is true for every s in S.

Proof Let $T = \{s \in S : p(s) \text{ is true}\}$, and recall the sequence $X = S_0 \subseteq S_1 \subseteq S_2 \subseteq \cdots$ defined earlier. By (b), $X = S_0 \subseteq T$. If $S \not\subseteq T$, then there is a smallest $n \in \mathbb{P}$ for which $S_n \not\subseteq T$, and there is some $s \in S_n \setminus T$. Because $S_{n-1} \subseteq T$, $s \notin S_{n-1}$, so s must be constructed from members of S_{n-1} using rules in (R). But then (r) says that $p(s)$ is true, contrary to the choice of s. Thus $S \subseteq T$, as claimed. ■

EXAMPLE 7

(a) If S is $\mathbb{N}$, given by

(B) $0 \in S$;

(R) $n \in S$ implies $n + 1 \in S$,

then (b) and (r) become

(b) $p(0)$ is true;

(r) $p(n + 1)$ is true whenever $p(n)$ is true.

So the generalized principle in this case is just the ordinary principle of mathematical induction.

(b) For the class $\mathcal{R}$ of finite rooted trees, defined as in Example 5(b), we get the following principle: Suppose that

(b) $p(T)$ is true for every 1-vertex rooted tree T, and

(r) $p(T)$ is true whenever p is true for every subtree hanging from the root of T.

Then $p(T)$ is true for every finite rooted tree T.

(c) The recursive definition of the class of trees in Example 4 yields the following: Suppose that $p(T)$ is true for every 1-vertex graph T with no edges, and suppose that $p(T')$ is true whenever $p(T)$ is true and T' is obtained by attaching a leaf to T. Then $p(T)$ is true for every finite tree T.

(d) We apply the principle in part (c) to give another proof that a tree with n vertices has $n - 1$ edges; compare Lemma 2 on page 242. Let $p(T)$ be the proposition "$|E(T)| = |V(T)| - 1$." If T has one vertex and no edges, then $p(T)$ is true. If $p(T)$ is true for some tree T and we attach a leaf to T to obtain a tree T', then both sides of the equation increase by 1, so $p(T')$ is still true. The generalized principle implies that $p(T)$ is true for every finite tree T. ■

In general, if the class S is recursively defined, then it can happen that a member of S_n can be constructed from members of S_{n-1} in more than one way. There are often advantages to working with recursive definitions in which new members of S are produced from other members in only one way. We will call such recursive definitions **uniquely determined**.

EXAMPLE 8

Most of the recursive definitions given so far have been uniquely determined. Here we discuss the exceptions.

(a) The recursive definition of the class of finite trees in Example 4 is not uniquely determined. In fact, the tree constructed in Figure 1 can also be constructed as indicated in Figure 3.

Figure 3 ▶

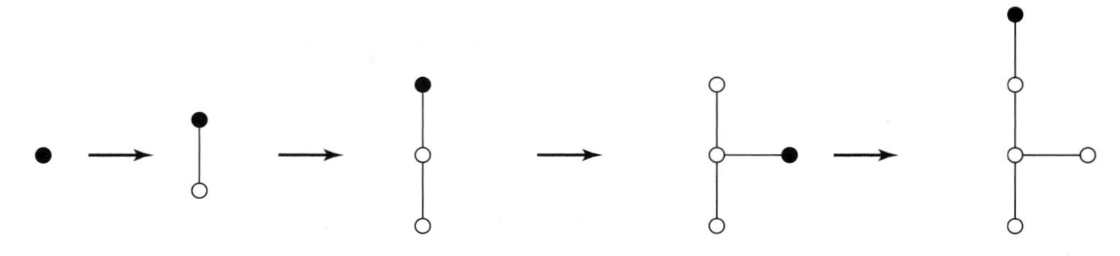

(b) The first recursive definition of the class of finite rooted trees in Example 5 is also not uniquely determined. See Exercise 7. On the other hand, the "hanging from the root" definition in Example 5(b) is uniquely determined. A rooted tree is uniquely determined by the subtrees that hang from its root, each of the subtrees is uniquely determined in the same way, and so on.

(c) The definitions in Example 6 are not uniquely determined. For example, 1 is in A by (B), and also by using (R) twice to get 3 and then 1 in A. Similarly, 1 reappears in S of Example 6(b) via the sequence 1, 2, 4, 1. ■

If the set S is recursively defined by (B) and (R), then we may be able to define functions f on S recursively, using the following procedure:

(1) Define $f(s)$ explicitly for each s in the set X described in (B);

(2) For the other elements s in S, give a recipe for defining $f(s)$ in terms of values of $f(t)$ for t's in S that come before s.

To say what we mean by "come before," we need to recall our chain $X = S_0 \subseteq S_1 \subseteq S_2 \subseteq \cdots$ for building up S. For each s in S there is a first value of n for which $s \in S_n$. We view all the members of S_{n-1} as coming before such an s, though of course the rules in (R) may produce s from just a few of them. If the recursive definition of S is uniquely determined, then the recipe in step (2) will be unambiguous. Otherwise, one must be sure that the definition is well defined. That is, one must check that the value of $f(s)$ does not depend on how s is constructed from t's that come before s.

EXAMPLE 9

(a) Let $\mathbb{N}$ be given by

(B) $0 \in \mathbb{N}$;

(R) $n \in \mathbb{N}$ implies $n + 1 \in \mathbb{N}$.

As we saw earlier, $S_m = \{0, \ldots, m\}$, so n appears for the first time in S_n and the numbers that come before n are $0, \ldots, n-1$. To define a function f on $\mathbb{N}$, we can define $f(0)$ and then, for each $n \geq 1$, explain how to get $f(n)$ from $f(0), \ldots, f(n-1)$.

(b) Define $\mathbb{P}$ recursively by

(B) $\{1, 2\} \subseteq \mathbb{P}$;

(R) $n \in \mathbb{P}$ implies $n + 1 \in \mathbb{P}$.

We can define the function f on $\mathbb{P}$ as follows: $f(1) = 1$, $f(2) = 1$, and $f(n) = f(n-1) + f(n-2)$ for $n \geq 3$. Then f describes the Fibonacci sequence. With this particular recursive definition of $\mathbb{P}$, we have

$$\{1, 2\} = S_0 \subseteq \{1, 2, 3\} = S_1 \subseteq \{1, 2, 3, 4\} = S_2 \subseteq \cdots,$$

so whenever $n \in S_m$, we have $n - 1$ and $n - 2$ in S_{m-1}. ■

EXAMPLE 10

(a) The set Σ^* was defined recursively in Example 2. Define the **length** function l on Σ^* by

(1) $l(\lambda) = 0$;
(2) If $w \in \Sigma^*$ and $x \in \Sigma$, then $l(wx) = l(w) + 1$.

Then $l(w)$ is the number of letters in w.

(b) We could also have defined Σ^* as follows:

(B) $\{\lambda\} \cup \Sigma \subseteq \Sigma^*$;
(R) If $w, u \in \Sigma^*$, then $wu \in \Sigma^*$.

Then length could have been defined by

(1) $l'(\lambda) = 0$, $l'(x) = 1$ for $x \in \Sigma$;
(2) If $w, u \in \Sigma^*$, then $l'(wu) = l'(w) + l'(u)$.

Or could it? The potential difficulty comes because this new recursive definition of Σ^* is not uniquely determined. It may be possible to construct a given word as wu in more than one way. For instance, if $\Sigma = \{a, b\}$, then a, ab, $(ab)a$, $((ab)a)b$ is one path to $abab$, but a, ab, $(ab)(ab)$ is another, and a, b, ba, $(ba)b$, $a((ba)b)$ is still another. How do we know that our "definition" of l' gives the same value of $l'(abab)$ for all of these paths?

It turns out [Exercise 13] that l' *is* well defined, and in fact it is the function l that we saw in (a) above.

(c) Let's try to define a **depth** function d on Σ^* by

(1) $d(\lambda) = 0$, $d(x) = 1$ for $x \in \Sigma$;
(2) If $w, u \in \Sigma^*$, then $d(wu) = \max\{d(w), d(u)\} + 1$.

This time we fail. If d were well defined, we would have $d(a) = 1$, $d(ab) = 1 + 1 = 2$, $d((ab)a) = 2 + 1 = 3$, and $d(((ab)a)b) = 3 + 1 = 4$, but also $d((ab)(ab)) = 2 + 1 = 3$. Thus the value of $d(abab)$ is ambiguous, and d is not well defined. ■

Examples 10(b) and 10(c) illustrate the potential difficulty in defining functions recursively when the recursive definition is not uniquely determined. A uniquely determined recursive definition is preferable for defining functions recursively. Thus the definition of Σ^* that we used in Example 10(a) is preferable to the one in Example 10(b). Likewise, the definition of rooted trees in Example 5(b) has advantages over that in Example 5(a). Sometimes we don't have a choice: We know of no uniquely determined recursive definition for finite trees, but the definition in Example 4 is still useful.

EXAMPLE 11

We wish to recursively define the height of finite rooted trees so that the height is the length of a longest simple path from the root to a leaf. For this purpose, our recursive definition of rooted trees in Example 5(a) is of little use, since the addition of a leaf need not increase the height of the tree. [If this isn't clear, think about adding leaves to the tree in Figure 4 on page 246.] However, the recursive definition in Example 5(b) is ideal for the task.

(a) Recall that the class $\mathcal{R}$ of finite rooted trees is defined by

(B) If T is a graph with one vertex v and no edges, then $(T, v) \in \mathcal{R}$;

(R) If $(T_1, r_1), \ldots, (T_k, r_k)$ are disjoint members of $\mathcal{R}$, and if (T, r) is obtained by hanging $(T_1, r_1), \ldots, (T_k, r_k)$ from r, then $(T, r) \in \mathcal{R}$.

The height of a member of $\mathcal{R}$ is defined by

(1) Trivial one-vertex rooted trees have height 0;
(2) If (T, r) is defined as in (R), and if the trees $(T_1, r_1), \ldots, (T_k, r_k)$ have heights $h_1, \ldots, h_k$, then the height of (T, r) is $1 + \max\{h_1, \ldots, h_k\}$.

(b) Other concepts for finite rooted trees can be defined recursively. For example, for an integer m greater than 1, the class of **m-ary rooted trees** is defined by

(B) A trivial one-vertex rooted tree is an m-ary rooted tree;

(R) A rooted tree obtained by hanging at most m m-ary rooted trees from the root is an m-ary rooted tree.

(c) We use these recursive definitions to prove that an m-ary tree of height h has at most m^h leaves. This statement is clear for a trivial tree, since $m^0 = 1$, and also for $h = 1$, since in that case the leaves are the children of the root. Consider an m-ary tree (T, r), defined as in part (b), with the property that each subtree (T_i, r_i) has height h_i. Since (T, r) is an m-ary tree, each (T_i, r_i) is an m-ary tree, and so it has at most m^{h_i} leaves. Let $h^* = \max\{h_1, \ldots, h_k\}$. Then the number of leaves of (T, r) is bounded by

$$m^{h_1} + \cdots + m^{h_k} \le m^{h^*} + \cdots + m^{h^*} = k \cdot m^{h^*}.$$

Since the root has at most m children, $k \le m$, and so (T, r) has at most $m \cdot m^{h^*} = m^{h^*+1}$ leaves. Since $h^* + 1$ equals the height h of (T, r), we are done. ■

Exercises 7.1

1. Let S be the recursively defined set in Example 1(b).

(a) Show that $2^m \in S$ for all $m \in \mathbb{N}$.

(b) Show that if n is in S, then n has the form 2^m for some $m \in \mathbb{N}$. *Hint:* Use the Generalized Principle of Induction; i.e., verify $p(1)$ and verify that $p(n)$ implies $p(2n)$, where $p(n) =$ "n has the form 2^m for some $m \in \mathbb{N}$."

(c) Conclude that $S = \{2^m : m \in \mathbb{N}\}$.

2. Use the definition in Example 2 to show that the following objects are in Σ^*, where Σ is the usual English alphabet.

(a) *cat* (b) *math* (c) *zzpq* (d) *aint*

3. (a) Describe the subset S of $\mathbb{N} \times \mathbb{N}$ recursively defined by

(B) $(0, 0) \in S$;

(R) If $(m, n) \in S$ and $m < n$, then $(m + 1, n) \in S$, and if $(m, m) \in S$, then $(0, m + 1) \in S$.

(b) Use the recursive definition to show that $(1, 2) \in S$.

(c) Is the recursive definition uniquely determined?

4. (a) Describe the subset T of $\mathbb{N} \times \mathbb{N}$ recursively defined by

(B) $(0, 0) \in T$;

(R) If $(m, n) \in T$, then $(m, n + 1) \in T$ and $(m + 1, n + 1) \in T$.

(b) Use the recursive definition to show that $(3, 5) \in T$.

(c) Is the recursive definition uniquely determined?

5. Let $\Sigma = \{a, b\}$ and let S be the set of words in Σ^* in which all the a's precede all the b's. For example, *aab*, *abbb*, *a*, *b*, and the empty word λ belong to S, but *bab* and *ba* do not.

(a) Give a recursive definition for the set S.

(b) Use your recursive definition to show that $abbb \in S$.

(c) Use your recursive definition to show that $aab \in S$.

(d) Is your recursive definition uniquely determined?

6. Let $\Sigma = \{a, b\}$ and let T be the set of words in Σ^* that have exactly one a.

(a) Give a recursive definition for the set T.

(b) Use your recursive definition to show that $bbab \in T$.

(c) Is your recursive definition uniquely determined?

7. (a) Describe two distinct constructions of the rooted tree in Figure 4, using the recursive definition in Example 5(a).

(b) Describe the construction of the rooted tree in Figure 4, using the uniquely determined definition in Example 5(b).

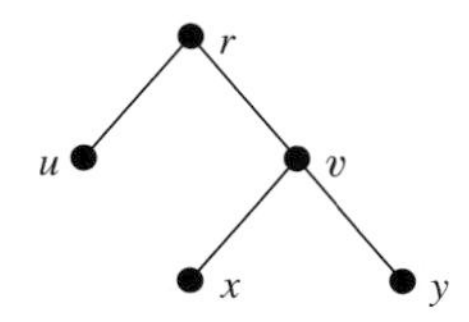

Figure 4 ▲

8. Verify (2) of Example 4 by showing that S_n contains all trees with at most $n+1$ vertices.

9. Let A be the recursively defined set in Example 6(a). Show that if n is in A, then $n \equiv 0 \pmod 3$ or $n \equiv 1 \pmod 3$.

10. Show that 4, 6, 10, and 12 are in the set A defined in Example 6(a).

11. (a) Show that the set S defined in Example 6(b) includes 1, 2, 3, 4, 5, and 6.

(b) Show that 7 is in S.

12. (a) Give a recursive definition for the class of regular m-ary trees.

(b) Do the same for full m-ary trees.

13. Define Σ^* as in Example 10(b), and let l' be as in that example. Show that $l'(w)$ is the number of letters in w for every $w \in \Sigma^*$. *Hint:* Use generalized induction with $X = \{\lambda\} \cup \Sigma$.

14. Let Σ^* be as defined in Example 2. We define the **reversal** $\overleftarrow{w}$ of a word w in Σ^* recursively as follows:

(B) $\overleftarrow{\lambda} = \lambda$,

(R) If $\overleftarrow{w}$ has been defined and $x \in \Sigma$, then $\overleftarrow{wx} = x\overleftarrow{w}$.

This is another well-defined definition.

(a) Prove that $\overleftarrow{x} = x$ for all $x \in \Sigma$.

(b) Use this definition to find the reversal of cab.

(c) Use this definition to find the reversal of $abbaa$.

(d) If w_1 and w_2 are in Σ^*, what is $\overleftarrow{w_1 w_2}$ in terms of $\overleftarrow{w_1}$ and $\overleftarrow{w_2}$? What is $\overleftarrow{\overleftarrow{w_1}}$?

15. Here is a recursive definition for a subset S of $\mathbb{N} \times \mathbb{N}$:

(B) $(0, 0) \in S$,

(R) If $(m, n) \in S$, then $(m+2, n+3) \in S$.

(a) List four members of S.

(b) Prove that if $(m, n) \in S$, then 5 divides $m+n$.

(c) Is the converse to the assertion in part (b) true for pairs (m, n) in $\mathbb{N} \times \mathbb{N}$?

16. Here is a recursive definition for another subset T of $\mathbb{N} \times \mathbb{N}$:

(B) $(0, 0) \in T$,

(R) If $(m, n) \in T$, then each of $(m+1, n)$, $(m+1, n+1)$, and $(m+1, n+2)$ is in T.

(a) List six members of T.

(b) Prove that $2m \geq n$ for all $(m, n) \in T$.

(c) Is this recursive definition uniquely determined?

17. Consider the following recursive definition for a subset A of $\mathbb{N} \times \mathbb{N}$:

(B) $(0, 0) \in A$,

(R) If $(m, n) \in A$, then $(m+1, n)$ and $(m, n+1)$ are in A.

(a) Show that $A = \mathbb{N} \times \mathbb{N}$.

(b) Let $p(m, n)$ be a proposition-valued function on $\mathbb{N} \times \mathbb{N}$. Use part (a) to devise a general recursive procedure for proving $p(m, n)$ true for all m and n in $\mathbb{N}$.

18. Let $\Sigma = \{a, b\}$ and let B be the subset of Σ^* defined recursively as follows:

(B) a and b are in B,

(R) If $w \in B$, then abw and baw are in B.

(a) List six members of B.

(b) Prove that if $w \in B$, then length(w) is odd.

(c) Is the converse to the assertion in part (b) true?

(d) Is this recursive definition uniquely determined?

19. Let Σ be a finite alphabet. For words w in Σ^*, the reversal $\overleftarrow{w}$ is defined in Exercise 14.

(a) Prove that length(w) = length($\overleftarrow{w}$) for all $w \in \Sigma^*$.

(b) Prove that $\overleftarrow{w_1 w_2} = \overleftarrow{w_2}\,\overleftarrow{w_1}$ for all $w_1, w_2 \in \Sigma^*$. *Hint:* Fix w_1, say, and prove that the propositions $p(w) =$ "$\overleftarrow{w_1 w} = \overleftarrow{w}\,\overleftarrow{w_1}$" are all true.

7.2 Recursive Algorithms

This section describes the structure of a general recursive algorithm. It also discusses how to prove that a recursive algorithm does what it is claimed to do. The presentation is a continuation of the account of recursion begun in the last section.

For both generalized induction and the recursive definition of functions, we start out with a recursively defined set S. Our goal is to show that the subset of S for which $p(s)$ is true or $f(s)$ is well defined is the whole set S. Another important question we can ask about a recursively defined set S is this: How can we tell whether something is a member of S?

EXAMPLE 1

As in Example 1(b) on page 269, define $S = \{2^m : m \in \mathbb{N}\}$ recursively by

(B) $1 \in S$;

(R) $2k \in S$ whenever $k \in S$.

Then $S \subseteq \mathbb{P}$. Given $n \in \mathbb{P}$, we wish to test whether n is in S, and if it turns out that n is in S, then we would like to find m with $n = 2^m$.

Suppose that n is odd. If $n = 1$, then $n \in S$ by (B), and $n = 2^0$. If $n > 1$, then (R) can't force n to be in S, so $n \notin S$. We know the answer quickly in the odd case.

If n is even, say $n = 2k$, then $n \in S$ if $k \in S$, by (R), but otherwise (R) won't put n in S. So we need to test $k = n/2$ to see if $k \in S$. If $k \in S$ with $k = 2^a$, then n will be in S with $n = 2^{a+1}$, and if $k \notin S$, then $n \notin S$.

This analysis leads us to design the following algorithm, in which $-\infty$ is just a symbol, but we define $-\infty + 1$ to be $-\infty$.

Test(integer)

```
{Input: A positive integer n}
{Output: A pair (b, m), where b is true if n ∈ S and false if n ∉ S, and
    m = log₂ n if n ∈ S, −∞ if n ∉ S}
if n is odd then
   if n = 1 then
      b := true; m := 0
   else
      b := false; m := −∞
else {so n is even}
   (b′, m′) := Test(n/2)
      {b′ is true if and only if n/2 ∈ S, in which case n/2 = 2^m′}
   b := b′; m := m′ + 1.
return (b, m)  ■
```

As usual, the statements in braces { . . . } are comments and not part of the algorithm itself.

The way that Test works is that if the input n is 1, then it reports success; if n is odd and not 1, it reports failure; and if n is even, it checks out $n/2$. For example, with input $n = 4$, Test(4) calls for Test(2), which calls for Test(1). Since Test(1) = (true, 0), we have Test(2) = (true, 1), and hence Test(4) = (true, 2). These outputs reflect the facts that $1 = 2^0 \in S$, $2 = 2^1 \in S$, and $4 = 2^2 \in S$. With input $n = 6$, Test checks Test(3), which reports failure. Thus Test(3) = (false, $-\infty$) leads to Test(6) = (false, $-\infty$). ■

The membership-testing algorithm Test in Example 1 has the interesting feature that if the original input n is even, then Test asks itself about $n/2$. Algorithms that call themselves are termed **recursive algorithms**.

Algorithm Test doesn't stop calling itself until the recursive calls reach an odd number. Test only determines that 4 is in S by computing Test(2), which gets its value from Test(1). The value b = true is then passed back through the calling chain. Here are the executions of Test for inputs 4 and 6.

```
Test(4)                          Test(6)
  Test(2)                          Test(3) = (false, −∞)
    Test(1) = (true, 0)          = (false, −∞ + 1 = −∞)
  = (true, 0 + 1 = 1)
= (true, 1 + 1 = 2)
```

The notation we use here has an = sign wherever a quantity is returned. Either the = appears on the same line as the call that returns the value, for example, in the line Test(1) = (true, 0), or it appears below it with the same indentation as the call, as when Test(4) returns (true, 2).

EXAMPLE 2

The class of finite trees was defined in Example 4 on page 271 by

(B) Trivial [1-vertex] graphs are trees;

(R) If T is obtained by attaching a leaf to a tree, then T is a tree.

We can devise a recursive algorithm TestTree for testing graphs to see if they are trees.

TestTree(graph)

```
{Input: A finite graph G}
{Output: True if G is a tree, and false if it is not}
if G has one vertex and no edges then
   return true
else
   if G has a leaf then
      Prune a leaf and its associated edge, to get G'.
      return TestTree(G')   {= true if G' is a tree, false if it's not}
   else
      return false   ■
```

If G really *is* a tree, this algorithm will prune edges and vertices off it until all that's left is a single vertex, and then it will report success, i.e., "true," all the way back up the recursive ladder. Figure 1 shows a chain G, G', G'', ... of inputs considered in one such recursion. When TestTree(G'''') is called, the value "true" is returned and then this value is passed to TestTree(G'''), TestTree(G''), TestTree(G') and back to TestTree(G).

Figure 1 ▶

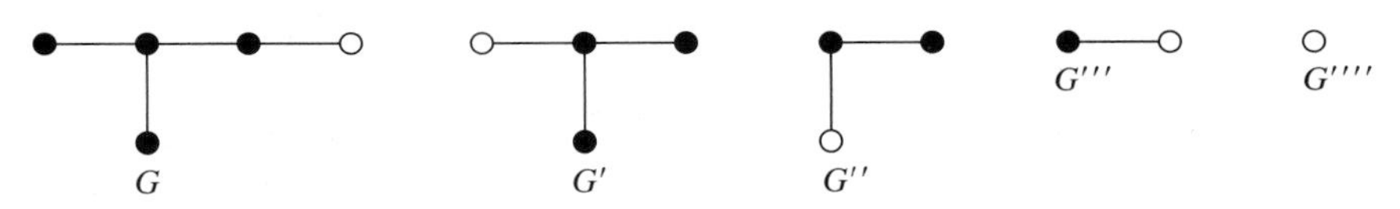

Figure 2 shows a chain for an input H that is not a tree. In this case the computation of TestTree(H''') cannot either apply the case for a trivial tree or recur to a smaller graph. It simply reports "false," and then TestTree(H''), TestTree(H'), and TestTree(H) accept that verdict. ■

Figure 2 ▶

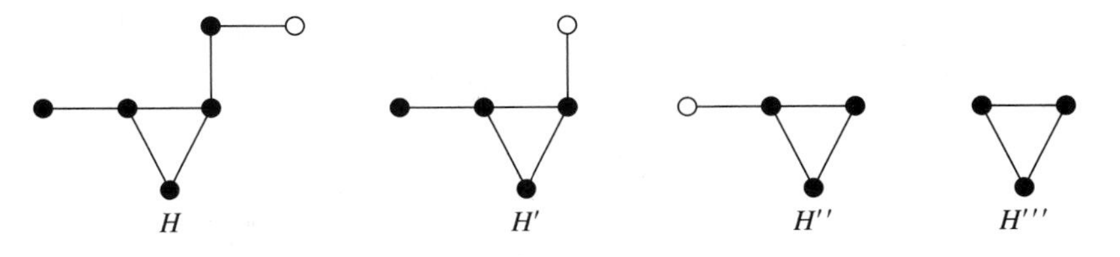

To help motivate the next example, consider the following questions that probably did not arise in algebra: What is an acceptable formula? What is it that makes $(x + y)(x - y)$ look good and makes $(x + -(^4/y$ look worthless? Why can a computer make sense of the first expression but not the second? An answer leads to the notion of a **well-formed formula** or **wff**.

EXAMPLE 3

(a) Here is a definition of wff's for algebra.

(B) Numerical constants and variables, such as x, y, and z, are wff's.

(R) If f and g are wff's, so are $(f+g)$, $(f-g)$, (fg), (f/g), and (f^g).

Being variables, x and y are wff's. Therefore, both $(x+y)$ and $(x-y)$ are wff's. We can hence conclude that $((x+y)(x-y))$ is a wff. The definition isn't entirely satisfactory, since the outside parentheses here seem unneeded. However, without them the square $(((x+y)(x-y))^2)$ and the expression $((x+y)(x-y)^2)$ would look the same, and these two expressions have different meanings. The problem is that in algebra we traditionally allow the omission of parentheses in some circumstances. Taking all the exceptional cases into account would lead us to a very complicated definition. Note also that our definition does not exclude division by 0. Thus (0/0) is a wff even though we would not assign a numerical value to this expression.

(b) In computer science, the symbol $*$ is often used for multiplication and $\wedge$ is used for exponentiation [$a \wedge b$ means a^b]. With this notation, the definition of wff's can be rewritten as

(B) Numerical constants and variables are wff's.

(R) If f and g are wff's, so are $(f+g)$, $(f-g)$, $(f*g)$, (f/g), and $(f \wedge g)$.

For example,

$$(((((X+Y) \wedge 2) - (2*(X*Y))) - (X \wedge 2)) - (Y \wedge 2))$$

is a wff.

(c) In §7.4 we will discuss Polish notation, which is a parenthesis-free notation. The preceding examples and related exercises may help you appreciate its value.

(d) How can we test a string of symbols to see if it's a wff? The recursive definition gives us an idea for an algorithm. Suppose that we have some way to recognize numerical constants and variables and that our input string is made out of allowable symbols. The question is whether the symbols are put together in a meaningful way. Here's an outline of a test algorithm.

TestWFF(string)

```
{Input: A string w of allowable symbols}
{Output: True if w is a wff, false otherwise}
if w is a numerical constant or a variable then
   return true
else
   if w is of the form (f + g), (f − g), (f ∗ g), (f/g) or (f ^ g) then
      return TestWFF(f) ∧ TestWFF(g)
         {= true if and only if both f and g are wff's}
   else
      return false   ■
```

Recall from Chapter 2 that $P \wedge Q$ is true if and only if both P and Q are true. Condition (R) only allows each new wff to be constructed in one way. A formula $(f*g)$ cannot also be of the form (h/k) for wff's h and k or of the form $(f'*g')$ for wff's f' and g' with $f' \neq f$ or $g' \neq g$, so a given w can lead to at most one pair f and g. Writing out the details for the main **else** branch would be tedious, but would yield no surprises. It is easy to believe that one could write a computer program to recognize wff's. ■

A computer needs to be able to read strings of input to see if they contain instructions or other meaningful expressions. Designing a compiler amounts to designing an instruction language and a recognition algorithm so that the computer can "parse," i.e., make sense out of, input strings. The recent development of the theory of formal languages has been greatly motivated by questions about parsing.

Recursive algorithms are useful for much more than just testing membership in sets. We will look at some other illustrations in this section, and wewill have important examples in this and later chapters as well. The basic feature of a recursive algorithm is that it calls itself, but of course that may not be much help unless the recursive calls are somehow closer to termination than the original call. To make sure that an algorithm terminates, we must know that chains of recursive calls cannot go on forever. We need some **base cases** for which the result is produced without further recursive calls, and we need to guarantee that the inputs for the recursive calls always get closer to the base cases somehow. A typical recursive algorithm with a chance of terminating has the following rough form.

```
Recur(input)
{Input: I; Output: O}
if I is a base case then
    O := whatever it should be.
else
    Find some I_k's closer to base cases than I is.
    For each k let O_k := Recur(I_k).
    Use the O_k's somehow to determine O.
return O
```

EXAMPLE 4

Figure 3 shows six very simple algorithms. All of them look alike for the first few lines. In each instance the input n is assumed to be a positive integer, and the base case is $n = 1$. Do these algorithms terminate, and if they do, how is the output r related to the input n? We can experiment a little. With $n = 1$, FOO returns 1. With $n = 2$ as input, it finds FOO(1) to get 1, so FOO(2) $= 1 + 1 = 2$. With $n = 3$ as input, FOO finds FOO(2) to get 2, as we just saw, so FOO(3) $= 2+1 = 3$. In fact, it looks as if FOO(n) $= n$ every time. In any case, since the chain $n, n-1, n-2, \ldots$ can't go on forever without reaching the base case, FOO must terminate. We will come back in a moment to the question of how one might prove that FOO(n) $= n$ for every n.

Figure 3 ▶

```
FOO(n)                      GOO(n)
if n = 1 then               if n = 1 then
   return 1                    return 1
else                        else
   return FOO(n − 1) + 1       return GOO(n − 1) * n

BOO(n)                      MOO(n)
if n = 1 then               if n = 1 then
   return 1                    return 1
else                        else
   return BOO(n + 1) + 1       return MOO(n/2) + 1

TOO(n)                      ZOO(n)
if n = 1 then               if n = 1 then
   return 1                    return 1
else                        else
   return TOO(⌊n/2⌋) + 1       return ZOO(⌊n/2⌋) * 2
```

Algorithm GOO also terminates, for the same reason that FOO does. Here are its first few values: GOO(1) $=$ 1; GOO(2) $=$ GOO(1) $* 2 = 1 * 2 = 2$; GOO(3) $=$ GOO(2) $* 3 = 2 * 3 = 6$. In fact, it looks as if we'll always get GOO(n) $= n!$, so GOO gives us a recursive way to compute factorials.

Algorithm BOO never terminates. For instance, BOO(2) calls BOO(3), which calls BOO(4), etc. The recursive calls don't get closer to the base case $n = 1$.

TABLE 1

n	FOO	GOO	BOO	MOO	TOO	ZOO
1	1	1	1	1	1	1
2	2	2		2	2	2
3	3	6			2	2
4	4	24		3	3	4
5	5	120			3	4
6	6	720			3	4
7	7	5040			3	4

Algorithm MOO fails for a different reason. Even though $n/2$ is closer to 1 than n is, $n/2$ may not be an integer, so MOO($n/2$) may not make sense. If we decide to allow noninteger inputs, then each MOO($n/2$) makes sense, but MOO need not terminate. For instance, the chain 6, 3, 3/2, 3/4, 3/8, ... goes on forever.

The MOO problem is fixed in TOO, since $\lfloor n/2 \rfloor$ is surely an integer, and if $n > 1$, then $n > \lfloor n/2 \rfloor \geq 1$. Table 1 shows the first few values returned by TOO, as well as by the other algorithms in Figure 3.

Algorithm ZOO terminates, as TOO does, but produces a different output. ■

Once we know that a recursive algorithm terminates, we still want to know what its output values are. If the algorithm is intended to produce a particular result, we must verify that it does what it is supposed to do. In the notation of our generic algorithm Recur(I), we need to check that

(b) If I is a base case, then O has the correct value;

(r) If each O_k has the correct value in the recursive calls Recur(I_k), then O has the correct value for I.

These conditions look like the ones we must check in a proof by generalized induction, and of course there is a connection, as we now explain. Recursively define the set S of allowable inputs for Recur as follows. For (B), let X be the set of base case inputs. Agree that I is **obtained from** $\{I_1, \ldots, I_m\}$ in case the **else** branch of Recur(I) can compute O from the outputs $O_1, \ldots, O_m$ of Recur(I_1), ..., Recur(I_m). Let the recursive clause be

(R) If I is obtained from $I_1, \ldots, I_m$ in S, then $I \in S$.

Then S is precisely the set of inputs for which Recur can compute outputs. Setting $p(I)$ = "Recur gives the correct value for input I" defines a set of propositions on S, and (b) and (r) become

(b′) $p(I)$ is true for all $I \in X$;

(r′) $p(I)$ is true whenever I is obtained from members I_k of S for which $p(I_k)$ is true.

The Generalized Principle of Induction on page 273 says that these conditions guarantee that $p(I)$ is true for every I in S.

EXAMPLE 5

(a) To show that GOO produces $n!$, we simply need to verify that

(b) $1! = 1$, and

(r) $n! = (n-1)! * n$.

These equations are true by definition of $n!$.

(b) We claim that the algorithm Test of Example 1 produces (true, $\log_2 n$) when the input n is a power of 2 and gives (false, $-\infty$) otherwise. Here is the algorithm again, for reference.

Test(integer)

```
if n is odd then
   if n = 1 then
      b := true; m := 0
   else
      b := false; m := −∞
else
   (b′, m′) := Test(n/2)
   b := b′; m := m′ + 1
return (b, m)     ■
```

What are the base cases and the allowable inputs? We want the set S of allowable inputs to be $\mathbb{P}$. Certainly, $n = 1$ is a base case. Since (R) in Example 1 just lets $2k$ be obtained from k, if we only permitted $n = 1$ as a base case, then S would only consist of the powers of 2. Instead, the algorithm treats all odd numbers in $\mathbb{P}$ as base cases.

To check that the algorithm is correct, i.e., that it gives the answer we claim, we just need to verify

(b) If $n = 1$, then $n = 2^0$, and if n is odd and greater than 1, then n is not a power of 2;

(r) If $n/2 = 2^{m'}$, then $n = 2^{m'+1}$.

Both (b) and (r) are clearly true.

While we are at it, we can estimate the time, $T(n)$, that Test takes for an input n. The base case verification and execution in the odd case take some constant time, say C. If n is even, the **else** branch takes time $T(n/2)$ to find Test($n/2$), plus some time D to compute b and m from b' and m'. Thus

$$T(n) \leq C + T(n/2) + D. \qquad (*)$$

Now $n = 2^m \cdot k$ with k odd and $m > 0$. Hence

$$\begin{aligned} T(n) = T(2^m \cdot k) &\leq T(2^{m-1} \cdot k) + C + D \leq T(2^{m-2} \cdot k) + C + D + C + D \\ &\leq \cdots \\ &\leq T(k) + m \cdot (C + D) \leq C + m \cdot (C + D) \\ &\leq 2m \cdot (C + D). \end{aligned}$$

Since $\dfrac{n}{k} = 2^m$, we have

$$m = \log_2\left(\frac{n}{k}\right) \leq \log_2 n, \qquad \text{so} \quad T(n) \leq 2(C + D) \cdot \log_2 n.$$

Thus $T(n) = O(\log_2 n)$. The three dots in the argument are harmless; once we guess the inequality $T(n) \leq 2(C + D) \cdot \log_2 n$, we can prove it inductively using $(*)$.

This algorithm is more efficient for large n than any of the divide-and-conquer algorithms covered by Theorem 2 on page 164. Note that $(*)$ gives

$$T(2n) \leq T(n) + (C + D) \qquad \text{for all } n.$$

This is a sharper estimate than any inequality of the form

$$s_{2n} \leq \mathbf{2}s_n + f(n)$$

because of the coefficient 2. However, an inequality version of Exercise 19 on page 167, with $b = 1$ and $f(n) = C + D$, could be applied to obtain

$$T(2^m) \leq C + m \cdot (C + D) \qquad \text{for} \quad m > 0.$$

(c) The base cases for TestTree in Example 2 include not only the graphs with one vertex, but also all finite graphs without leaves. For any such graph, the algorithm returns false without recurring further, unless the graph is the trivial 1-vertex graph, in which case the algorithm returns true.

(d) Similarly, the base cases for TestWFF in Example 3 include all strings that are *not* of form $(f + g)$, $(f - g)$, $(f * g)$, (f/g), or $(f \,\hat{}\, g)$. ■

At the expense of added storage for intermediate results, we can convert **while** loops to recursive procedures. For instance, the recursive segment

```
Recur( )
if g is false then
    do nothing
else
    S
    Recur( )
```

produces the same effect as the loop

```
while g do
    S.
```

Since the main drawback to recursive algorithms is that they tend to gobble memory space, for a really big computation one might be more inclined to convert a recursive program to an iterative one, rather than the other way around. From a conceptual viewpoint, however, recursion—reducing to a simpler problem—is often a clearer approach than iteration.

EXAMPLE 6

We presented the Euclidean algorithm on page 175 using a **while** loop, but recursion seems more natural. The key observation that $\gcd(m, n) = \gcd(n, m \text{ MOD } n)$ if $n \neq 0$ gives the following recursive algorithm.

Euclid(integer, integer)

```
{Input: m, n ∈ ℕ, not both 0}
{Output: gcd(m, n)}
if n = 0 then
    return m
else
    return Euclid(n, m MOD n)    ■
```

A minor addition gives an algorithm that computes integers s and t with $\gcd(m, n) = sm + tn$.

Euclid$^+$(integer, integer)

```
{Input: m, n ∈ ℕ, not both 0}
{Output: (d, s, t), where d = gcd(m, n) and integers s and t
        satisfy d = sm + tn}
if n = 0 then
    return (m, 1, 0)
else
    (d', s', t') := Euclid⁺(n, m MOD n)
    return (d', t', s' − t' · (m DIV n)).    ■
```

The running times for these algorithms are essentially the same as for the iterative versions in §4.7. ■

For reference, we repeat the conditions a recursive algorithm must satisfy.

VERIFICATION CONDITIONS FOR RECURSIVE ALGORITHMS

(a) The algorithm must terminate. In particular, recursive calls must make progress toward base cases.

(b) The algorithm must give the correct results in all base cases.

(c) The algorithm must give the correct result if all recursive calls produce correct results.

Exercises 7.2

1. (a) Illustrate the execution of the algorithm Test for input 20.

   (b) Do the same for input 8.

2. Illustrate the algorithm TestTree for the graphs in Figure 4. You may draw pictures like those in Figures 1 and 2.

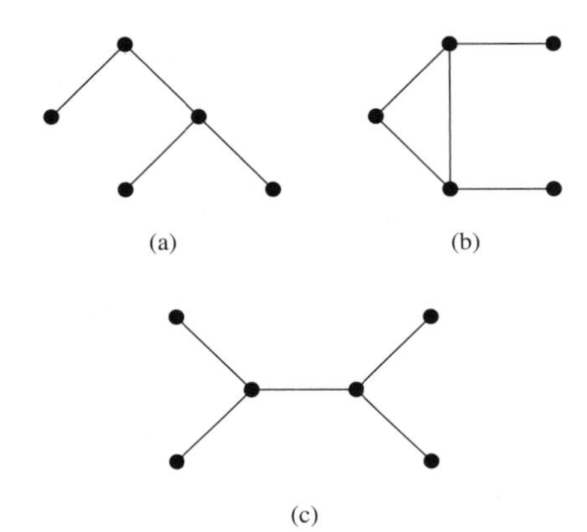

Figure 4 ▲

3. Add enough parentheses to the following algebraic expressions so that they are wff's as defined in Example 3.

   (a) $x + y + z$ (b) $x + y/z$

   (c) xyz (d) $(x + y)^{x+y}$

4. Add enough parentheses to the following algebraic expressions so that they are wff's as defined in Example 3.

   (a) $X + Y + Z$ (b) $X * (Y + Z)$

   (c) $X \wedge 2 + 2 * X + 1$ (d) $X + Y/Z - Z * X$

5. Use the recursive definition of wff in Example 3 to show that the following are wff's.

   (a) $((x^2) + (y^2))$ (b) $(((X \wedge 2) + (Y \wedge 2)) \wedge 2)$

   (c) $((X + Y) * (X - Y))$

6. Recall the algorithm GOO of Example 4:

```
GOO(n)
if n = 1 then
  return 1
else
  return GOO(n − 1) ∗ n
```

   What happens if the input is a positive integer and

   (a) The line "**return** 1" is replaced by "**return** 0"?

   (b) The last line is replaced by the line "**return** GOO$(n - 1) * (n - 1)$"?

   (c) The last line is replaced by the line "**return** GOO$(n - 2) * n$"?

7. Supply the lines for $n = 8$ and $n = 9$ in Table 1.

8. Verify that the algorithm TOO produces $k + 1$ whenever $2^k \le n < 2^{k+1}$. *Hint:* Use induction on k.

9. Verify that the algorithm ZOO produces 2^k whenever $2^k \le n < 2^{k+1}$.

10. Show how the algorithm Euclid$^+$ computes $d = \gcd(80, 35)$ and finds s and t satisfying $d = 80s + 35t$.

11. Repeat Exercise 10 for $d = \gcd(108, 30)$.

12. Repeat Exercise 10 for $d = \gcd(56, 21)$.

13. (a) Verify that the algorithm Euclid terminates and that it produces the correct output.

    (b) Do the same for the algorithm Euclid$^+$.

14. We recursively define the **depth** of a wff in algebra as follows; see Example 3.

    (B) Numerical constants and variables have depth 0;

    (R) If depth(f) and depth(g) have been defined, then each of $(f + g)$, $(f - g)$, $(f * g)$, (f/g), and $(f \wedge g)$ has depth equal to $1 + \max\{\text{depth}(f), \text{depth}(g)\}$.

    This turns out to be a well-defined definition; compare with Example 10 on page 275. Calculate the depths of the following algebraic expressions:

    (a) $((x \wedge 2) + (y \wedge 2))$

    (b) $(((X \wedge 2) + (Y \wedge 2)) \wedge 2)$

    (c) $((X + Y) * (X - Y))$

    (d) $(((((X + Y) \wedge 2) - (2 * (X * Y))) - (X \wedge 2)) - (Y \wedge 2))$

    (e) $(((x + (x + y)) + z) - y)$

    (f) $(((X * Y)/X) - (Y \wedge 4))$

15. The following is a uniquely determined recursive definition of wff's for the propositional calculus.

    (B) Variables, such as p, q, and r, are wff's;

    (R) If P and Q are wff's, so are $(P \vee Q)$, $(P \wedge Q)$, $(P \to Q)$, $(P \leftrightarrow Q)$, and $\neg P$.

    Note that we do not require parentheses when we negate a proposition. Consequently, the negation symbol $\neg$ always negates the shortest subexpression following it that is a wff. In practice, we tend to omit the outside parentheses and, for the sake of readability, brackets [ ] and braces { } may be used for parentheses. Show that the following are wff's in the propositional calculus.

    (a) $\neg(p \vee q)$

(b) $(\neg p \wedge \neg q)$

(c) $((p \leftrightarrow q) \rightarrow ((r \rightarrow p) \vee q))$

16. Modify the definition in Exercise 15 so that the "exclusive or" connective $\oplus$ is allowable.

17. Throughout this exercise, let p and q be *fixed* propositions. We recursively define the family $\mathcal{F}$ of compound propositions using only p, q, $\wedge$, and $\vee$ as follows:

(B) $p, q \in \mathcal{F}$;

(R) If $P, Q \in \mathcal{F}$, then $(P \wedge Q)$ and $(P \vee Q)$ are in $\mathcal{F}$.

(a) Use this definition to verify that $(p \wedge (p \vee q))$ is in $\mathcal{F}$.

(b) Prove that if p and q are false, then all the propositions in $\mathcal{F}$ are false. *Hint:* Apply the general principle of induction to the proposition-valued function r on $\mathcal{F}$, where $r(P)$ = "if p and q are false, then P is false." Show (B): $r(P)$ and $r(Q)$ are true; and (I): if $r(P)$ and $r(Q)$ are true, so are $r((P \wedge Q))$ and $r((P \vee Q))$.

(c) Show that $p \rightarrow q$ is not logically equivalent to any proposition in $\mathcal{F}$. This verifies the unproved claim in the answer to Exercise 17(c) on page 85; see page 544.

18. The following algorithm returns the smallest number in a list of numbers.

SmallestEntry(list)

{Input: A list L of numbers}
{Output: The smallest entry in L}
if L has only 1 entry **then**
  return that entry
else
  Cut L into two lists M and N of approximately equal sizes.
  return min{SmallestEntry(M), SmallestEntry(N)} ■

Illustrate the operation of this algorithm for each of the following input lists. Break lists so that length(M) = $\lceil$length(L)/2$\rceil$.

(a) 1 2 3 4 5 6

(b) 5 2 8 4 2

(c) 5 4 8 4 2 5 3

19. The following algorithm computes the digits of the binary [i.e., base 2] representation of a positive integer n.

ConvertToBinary(integer)

{Input: A positive integer n}
{Output: The string of binary coefficients of n}
if $n = 1$ **then**
  return n
else
  return ConvertToBinary(n DIV 2) followed by n MOD 2 ■

(a) Illustrate the execution of this algorithm to give the binary representation of 25.

(b) Illustrate the execution to give the binary representation of 16.

(c) How could you modify the algorithm to yield hexadecimal [i.e., base 16] representations?

7.3 Depth-First Search Algorithms

A tree-traversal algorithm is an algorithm for listing [or visiting or searching] all the vertices of a finite ordered rooted tree. There are various reasons why one might wish to traverse a tree. A binary search tree, such as the client record tree of Example 1 on page 244, provides a quick way to arrive at or store each individual record; but to generate a list of all clients who have been contacted within the past year, one would need to examine *all* the records. A traversal algorithm provides a systematic way to do so, even if all that is desired is a list of the records at the leaves. We will see another less obvious application of tree traversal later in this section when we discuss labeling digraphs.

The essential idea of tree traversal is to move along the edges of the tree in a path that visits each vertex, but that visits no vertex more times than necessary. Depending on the purpose of the traversal, one might wish to take some action when visiting the vertices—perhaps labeling them, perhaps adding their existing labels to a list, perhaps examining the records associated with them, and so on—but not to take an action twice for the same vertex. The general structure of a tree traversal algorithm is independent of what actions, if any, are to take place at the vertices. Our discussion will focus on listing and labeling, but the ideas are more widely applicable.

Three common tree-traversal algorithms provide preorder listing, inorder listing [for binary trees *only*], and postorder listing. All three are recursive algorithms.

In the **preorder listing**, the root is listed first and then the subtrees T_w rooted at its children are listed in order of their roots. Because of the way we draw ordered rooted trees, we will refer to the order as left to right. In this algorithm the root gets listed first, and its children get the job of listing the descendants.

Preorder(rooted tree)

{Input: A finite ordered rooted tree with root v}
{Output: A list of all the vertices of the tree, in which parents always come before their children}
Put v on the list $L(v)$.
for each child w of v, taken from left to right, **do**
  Attach Preorder(T_w) {a list of w and its descendants} to the end of the list $L(v)$.
return $L(v)$ ■

A base case occurs when the root v has no children. In this case, the first step adds the root to the list and the **for** loop is vacuously completed, so the list from a base case has only one entry, v.

In the **postorder listing**, the subtrees are listed in order first, and then the root is listed at the end. Again a base case occurs when the tree has just one vertex, v. In this case, the first step is vacuously completed, and again the list has only one entry, v.

Postorder(rooted tree)

{Input: A finite ordered rooted tree with root v}
{Output: A list of all the vertices of the tree, in which parents always appear after their children}
Begin with an empty list; $L(v) := \lambda$.
for each child w of v, taken from left to right, **do**
  Attach Postorder(T_w) {a list of w and its descendants} to the end of the list $L(v)$ obtained so far.
Put v at the end of $L(v)$.
return $L(v)$ ■

EXAMPLE 1

We apply the algorithms Preorder and Postorder to the tree $T = T_r$ in Figure 1. Since an understanding of all the algorithms in this section depends on a solid understanding of these simple ones, we will explain every step in this example.

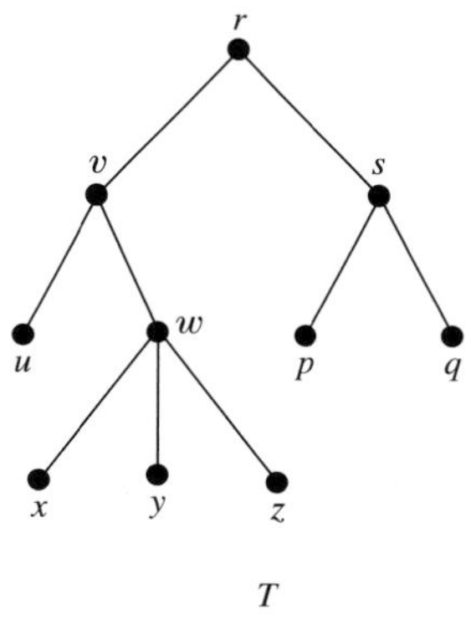

r
v
s
u
w
p
q
x
y
z
T

Figure 1 ▲

(a) The Preorder algorithm proceeds as follows, using the general trace format that we introduced for recursive algorithms in §7.2:

Preorder(T_r); $L(r) = r$
  Preorder(T_v); $L(v) = v$ {v is r's first child}
    Preorder(T_u) $= u$ {v's first child u has no children}
    $L(v) = v$ Preorder(T_u) $= v\,u$
    Preorder(T_w); $L(w) = w$ {w is v's next child}
      Preorder(T_x) $= x$ {no children for w's first child x}
      $L(w) = w$ Preorder(T_x) $= w\,x$
      Preorder(T_y) $= y$ {y is also a leaf}
      $L(w) = w\,x$ Preorder(T_y) $= w\,x\,y$
      Preorder(T_z) $= z$ {z is another leaf}
      $L(w) = w\,x\,y$ Preorder(T_z) $= w\,x\,y\,z$
    $= w\,x\,y\,z$ {no more children for w}
    $L(v) = v\,u$ Preorder(T_w) $= v\,u\,w\,x\,y\,z$
  $= v\,u\,w\,x\,y\,z$ {no more children for v}
  $L(r) = r$ Preorder(T_v) $= r\,v\,u\,w\,x\,y\,z$
  Preorder(T_s); $L(s) = s$ {s is r's second child}
    Preorder(T_p) $= p$ {another leaf}
    $L(s) = s$ Preorder(T_p) $= s\,p$
    Preorder(T_q) $= q$ {another leaf}
    $L(s) = s\,p$ Preorder(T_q) $= s\,p\,q$
  $= s\,p\,q$ {s has no more children}
  $L(r) = r\,v\,u\,w\,x\,y\,z$ Preorder(T_s) $= r\,v\,u\,w\,x\,y\,z\,s\,p\,q$
$= r\,v\,u\,w\,x\,y\,z\,s\,p\,q$ {done}.

In the Preorder algorithm, each vertex is listed when it is first visited, so this listing can be obtained from the picture of T as illustrated in Figure 2. Just follow the dashed line, listing each vertex the first time it is reached.

Figure 2 ▶

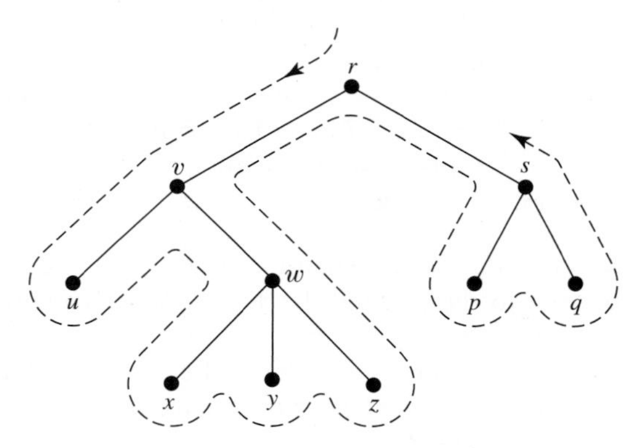

(b) The Postorder algorithm does the listing of the parents after their children are listed:

Postorder(T_r); $L(r) = \lambda$
  Postorder(T_v); $L(v) = \lambda$ {v is still r's first child}
    Postorder(T_u) $= u$ {v's first child u has no children}
    $L(v) = \lambda$ Postorder(T_u) $= u$
    Postorder(T_w); $L(w) = \lambda$ {w is v's next child}
      Postorder(T_x) $= x$ {leaf}
      $L(w) = \lambda$ Postorder(T_x) $= x$
      Postorder(T_y) $= y$ {also a leaf}
      $L(w) = x$ Postorder(T_y) $= x\ y$
      Postorder(T_z) $= z$ {another leaf}
      $L(w) = x\ y$ Postorder(T_z) $= x\ y\ z$
    $= x\ y\ z\ w$ {no more children for w}
    $L(v) = u$ Postorder(T_w) $= u\ x\ y\ z\ w$
  $= u\ x\ y\ z\ w\ v$ {done with v}
  $L(r) = \lambda$ Postorder(T_v) $= u\ x\ y\ z\ w\ v$
  Postorder(T_s); $L(s) = \lambda$ {s is r's second child}
    Postorder(T_p) $= p$ {another leaf}
    $L(s) = \lambda$ Postorder(T_p) $= p$
    Postorder(T_q) $= q$ {another leaf}
    $L(s) = p$ Postorder(T_q) $= p\ q$
  $= p\ q\ s$ {done with s}
  $L(r) = u\ x\ y\ z\ w\ v$ Postorder(T_s) $= u\ x\ y\ z\ w\ v\ p\ q\ s$
$= u\ x\ y\ z\ w\ v\ p\ q\ s\ r$ {all finished}.

In the Postorder algorithm, each vertex is listed when it is last visited, so this listing can also be obtained from the picture of T in Figure 2, provided each vertex is listed the *last* time it is visited. ■

For *binary* ordered rooted trees, a third kind of listing is available. The **inorder listing** puts each vertex in between the lists for the subtrees rooted at its left and right children.

Inorder(rooted tree)

{Input: A finite ordered binary tree with root v}
{Output: A list of the vertices of the tree, in which left children appear before their parents, and right children appear after them}
Begin with an empty list; $L(v) = \lambda$.
if v has a left child w **then**

Attach Inorder(T_w) {a list of w and its descendants}
to the end of $L(v)$.
Attach v at the end of $L(v)$.
if v has a right child u **then**
Attach Inorder(T_u) to the end of $L(v)$.
return $L(v)$ ■

If the input tree is not regular, then some vertices have just one child. Since each such child must be designated as either a left or a right child, the algorithm still works. As with Preorder and Postorder, the list for a childless vertex consists of just the vertex itself.

EXAMPLE 2

(a) The Inorder algorithm proceeds as follows for the binary tree in Figure 3.

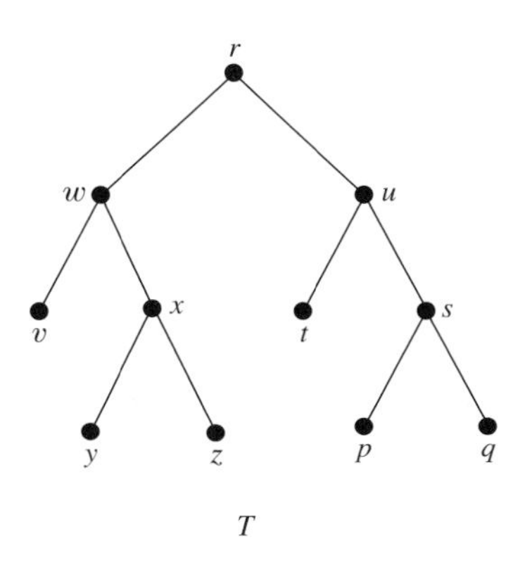

Figure 3 ▲

Inorder(T_r); $L(r) = \lambda$
Inorder(T_w); $L(w) = \lambda$ {w is r's left child}
Inorder(T_v) $= v$ {v, a leaf, is w's left child}
$L(w) =$ Inorder(T_v) $w = v\,w$ {back to w}
Inorder(T_x); $L(x) = \lambda$ {x is w's right child}
Inorder(T_y) $= y$ {y, a leaf, is x's left child}
$L(x) =$ Inorder(T_y) $x = y\,x$
Inorder(T_z) $= z$ {z is x's right child}
$= y\,x$ Inorder(T_z) $= y\,x\,z$
$= v\,w$ Inorder(T_x) $= v\,w\,y\,x\,z$
$L(r) =$ Inorder(T_w) $r = v\,w\,y\,x\,z\,r$
Inorder(T_u); $L(u) = \lambda$ {u is r's right child}
Inorder(T_t) $= t$ {and so on}
$L(u) =$ Inorder(T_t) $u = t\,u$
Inorder(T_s); $L(s) = \lambda$
Inorder(T_p) $= p$
$L(s) =$ Inorder(T_p) $s = p\,s$
Inorder(T_q) $= q$
$= p\,s$ Inorder(T_q) $= p\,s\,q$
$= t\,u$ Inorder(T_s) $= t\,u\,p\,s\,q$
$= v\,w\,y\,x\,z\,r$ Inorder(T_u) $= v\,w\,y\,x\,z\,r\,t\,u\,p\,s\,q$

In this example each parent has both left and right children. If some of the children had been missing, then the algorithm still would have performed successfully. For example, if y had been missing and if z had been designated as x's right child, then the list Inorder(T_x) would have been just $x\,z$. If the whole right branch x, y, z had been missing, Inorder(T_w) would have been just $v\,w$.

(b) Consider the labeled tree in Figure 8(a) on page 250. The inorder listing of the subtree with root 00 is 000, 00, 001; the other subtrees are also easy to list. The inorder listing of the entire tree is

$$000, 00, 001, 0, 010, 01, 011, \text{ root}, 100, 10, 101, 1, 110, 11, 111.$$ ■

If we are given a listing of a tree, can we reconstruct the tree from the listing? In general, the answer is no. Different trees can have the same listing.

EXAMPLE 3

Consider again the tree T in Figure 3. Figure 4 gives two more binary trees having the same inorder listing. The binary tree in Figure 5(a) has the same preorder listing as T, and the tree in Figure 5(b) has the same postorder listing as T. Exercise 8 asks you to verify these assertions. ■

In spite of these examples, there are important situations under which trees can be recovered from their listings. Since our applications will be to Polish notation and

Figure 4 ▶

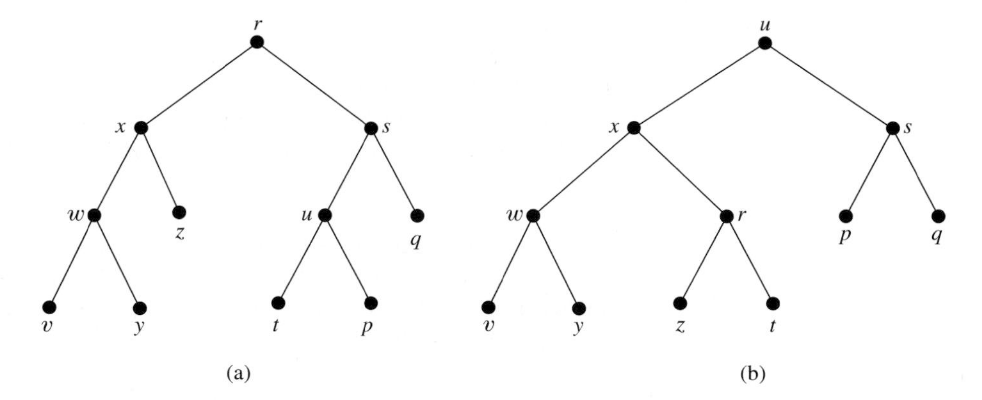

Figure 5 ▶

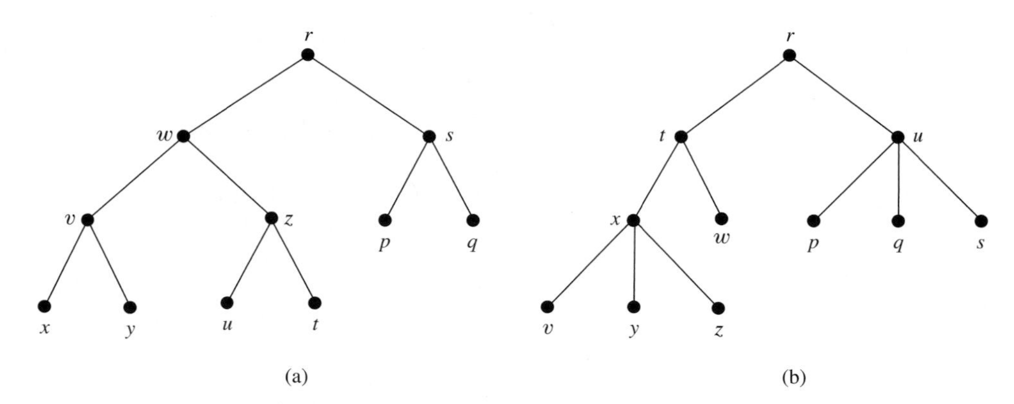

since the ideas are easier to grasp in that setting, we will return to this matter at the end of the next section.

We can analyze how long our three listing-labeling algorithms take by using a method of **charges**. Let t_1 be the time that it takes to label a vertex, and let t_2 be the time that it takes to attach a list of vertices to another list. [We can assume that the attachment time is constant by using a linked-list representation for our lists of vertices.] The idea now is to think of the total time for the various operations in the execution of the algorithm as being charged to the individual vertices and edges of the tree. We charge a vertex v with the time t_1 that it takes to label it, and we charge an edge (v, w) with the time t_2 that it takes to attach the list $L(w)$ obtained by the call to w to make it part of the list $L(v)$. Then the time for every operation gets charged somewhere, every vertex and every edge gets charged, and nobody gets charged twice. By adding up the charges, we find that the total time is $t_1|V(T)| + t_2|E(T)| = t_1|V(T)| + t_2(|V(T)| - 1) = O(|V(T)|)$.

The three listing algorithms above are all examples of what are called **depth-first search** or **backtrack** algorithms. Each goes as far as it can by following the edges of the tree away from the root, then it backs up a bit and again goes as far as it can, and so on. The same idea can be useful in settings in which we have no obvious tree to start with. We will illustrate with an algorithm that labels the vertices of a digraph G with integers in a special way. Before we get into the details of the general case, we look at how such an algorithm might work on an ordered rooted tree.

The list of vertices given by Postorder provides a natural way to label the vertices of the tree; simply give the kth vertex on the list the label k. This labeling has the property that parents have larger labels than their children and hence have larger labels than any of their descendants. We call such a labeling a **sorted labeling**, but some people also call it a **topological sort**. For such a labeling the root has the largest label. The natural labeling produced by Preorder—label the kth vertex on the list with k—gives children labels that are larger than the labels of their parents. For many purposes, either of these two natural labelings is as good as the other. Because of our application to digraphs later in this section, however, we will concentrate on the Postorder version. Exercise 14 illustrates one trouble with Preorder.

Observe that if we wish parents to have labels larger than their children, then we are forced to use an ordering like that produced by Postorder. The root r has to get the biggest label, so it's labeled last after we traverse the whole tree and find out how many vertices there are. For the same reason, each vertex v has to be labeled after all of its descendants, i.e., labeled at the end of the traversal of the subtree T_v of which it is the root. When we label T_v, we need to know which labels have already been used up. The Postorder listing provides this information automatically. Here is the modified recursive algorithm.

LabelTree(rooted tree, integer)

{Input: A finite rooted tree with root v, and a positive integer k}
{Gives the tree a sorted labeling with labels starting at k}
$n := k$
for w a child of v (children taken in some definite order) **do**
  LabelTree(T_w, n) {labels T_w with $n, \ldots$}
  $n := 1 + $ label of w {sets the starting point for the next labels}
Label v with n.
return ■

In the base case in which the root v is childless, the **for** loop is not executed, so LabelTree(T_v, k) labels v with k. To label the whole tree rooted at r, starting with the label 1, we call LabelTree(T_r, 1).

If we only want a sorted labeling, then it doesn't matter in which order the children do their labeling tasks, as long as no two children try to use the same labels.

EXAMPLE 4 Here is how LabelTree would deal with our tree of Example 1.

LabelTree(T_r, 1); $n = 1$
  LabelTree(T_v, 1)
    LabelTree(T_u, 1); label u with 1 {u is childless}; $n = 2$
    LabelTree(T_w, 2)
      LabelTree(T_x, 2); label x with 2; $n = 3$
      LabelTree(T_y, 3); label y with 3; $n = 4$
      LabelTree(T_z, 4); label z with 4; $n = 5$
    Label w with 5; $n = 6$
  Label v with 6; $n = 7$
  LabelTree(T_s, 7)
    LabelTree(T_p, 7); label p with 7; $n = 8$
    LabelTree(T_q, 8); label q with 8; $n = 9$
  Label s with 9; $n = 10$
Label r with 10 ■

Another way to produce a sorted labeling recursively is to find some leaf to label 1, remove the leaf, and recursively label what's left with 2, 3, ...

LabelTree2(rooted tree, integer)

{Input: A finite rooted tree with root v, and a positive integer k}
{Gives the tree a sorted labeling with labels starting at k}
if T_v has only one vertex **then**
  Label it with k.
else
  Find a leaf of T_v and label it with k.
  Remove the leaf and its edge from T_v.
  LabelTree2(T_v, $k+1$) {gives a sorted labeling to what's left of T_v}
return ■

This procedure labels every vertex—sooner or later every vertex gets to be a leaf—and never gives a vertex a label smaller than any of its descendants, which have all been pruned away by the time the vertex gets its turn to be a leaf. Finding a leaf with each recursion might be time consuming, however, depending on the data structure used for storing information about the tree.

Sorted labelings can be defined and studied in any digraph. Consider a finite digraph G and a subset L of $V(G)$. A **sorted labeling** of L is a numbering of its vertices with $1, 2, \ldots, |L|$ such that $i > j$ whenever there is a path in G from vertex i to vertex j.

EXAMPLE 5 Figure 6 shows two digraphs whose vertex sets have sorted labelings. Notice that if $i > j$, then there does not have to be a path from i to j; we simply require that $i > j$ if there *is* such a path. ■

Figure 6 ▶

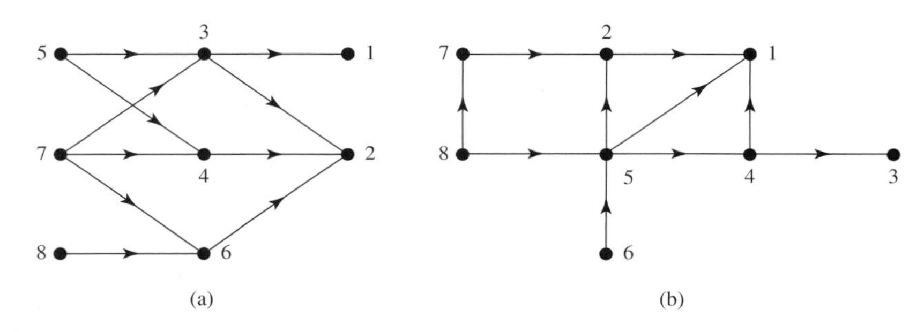

A digraph whose vertex set has a sorted labeling cannot possibly contain a cycle, for if the vertices of a cycle were labeled $i\,j \ldots k\,i$, then we would have $i > j > \cdots > k > i$. If we restrict ourselves to acyclic digraphs, i.e., to ones that contain no cycles, then a sorted labeling of $V(G)$ is always possible. After some preparation, we will prove this fact by exhibiting an algorithm, Label, that constructs one. The "child" and "descendant" terminology for the tree case gets changed into the language of "successors" and "accessible" for a general digraph, but the idea of the algorithm is the same.

We use the notation $\textbf{SUCC}(v)$ for the set of immediate successors of v, those vertices w for which there is an edge from v to w. We also let $\textbf{ACC}(v)$ be the set of vertices that are accessible from v, i.e., the set consisting of v itself and those vertices w for which there is a path from v to w. More generally, if $L \subseteq V(G)$, we let $\textbf{ACC}(L)$ be the union of the sets $\text{ACC}(v)$ for v in L. Observe that $\text{SUCC}(v) \subseteq \text{ACC}(v)$ for all vertices v and that $L \subseteq \text{ACC}(L)$ for all subsets L of $V(G)$. Moreover,

$$\text{ACC}(v) = \{v\} \cup \text{ACC}(\text{SUCC}(v)) \quad \text{and} \quad \text{ACC}(\text{ACC}(v)) = \text{ACC}(v).$$

EXAMPLE 6 The successor and accessible sets for the graph in Figure 6(a) are given in Figure 7. ■

There are two reasons why algorithm Label will be more complicated than LabelTree. In a rooted tree, viewed as a digraph, all vertices are accessible from the

Figure 7 ▶

Vertex	1	2	3	4	5	6	7	8
SUCC( )	∅	∅	{1,2}	{2}	{3,4}	{2}	{3,4,6}	{6}
ACC( )	{1}	{2}	{1,2,3}	{2,4}	{1,2,3,4,5}	{2,6}	{1,2,3,4,6,7}	{2,6,8}

root, so the algorithm starts at the root and works down. However, there may not be a unique vertex in an acyclic digraph from which all other vertices are accessible. For example, see either of the digraphs in Figure 6. Therefore, Label will need subroutines that label sets like $\text{ACC}(v)$. The other difference is that the children of a parent in a rooted tree have disjoint sets of descendants, so none of the children ever tries to label some other child's descendant. On the other hand, a vertex in an acyclic digraph can be accessible from different successors of the same vertex. For example, vertex 2 in Figure 6(a) is accessible from all three successors of 7. Thus the algorithm will need to keep track of vertices already labeled and avoid trying to label them again.

The algorithm Label first chooses a vertex v, labels all the vertices accessible from v [including v], and then calls itself recursively to handle the vertices that don't have labels yet. For clarity, we give the procedure for labeling the vertices accessible from v as a separate algorithm, which we call TreeSort for a reason that will become clear.

TreeSort is also recursive. In case we apply TreeSort to a rooted tree, starting at the root r, it asks a child of r to label its descendants and then it recursively labels the remaining descendants of r and r itself. So in this case TreeSort works rather like LabelTree. The order of traversal for the two algorithms is exactly the same.

When either Label or TreeSort is called recursively, some of the vertices may already have been assigned labels. The algorithms will have to be passed that information so that they won't try to label those vertices again. For convenience in verifying the algorithms' correctness, we make one more definition. Say that a subset L of $V(G)$ is **well-labeled** in case $L = \text{ACC}(L)$ and L has a sorted labeling. Here is the first algorithm. We will prove that it works after we describe its operation and give an example.

TreeSort(digraph, set of vertices, vertex)

{Input: A finite acyclic digraph G, a well-labeled subset L of $V(G)$,
    and a vertex $v \in V(G) \setminus L$}
{Makes $L \cup \text{ACC}(v)$ well-labeled with a sorted labeling that
    agrees with the given labeling on L}
if $\text{SUCC}(v) \subseteq L$ **then**
    Label L as before and label v with $|L| + 1$.
else
  Choose $w \in \text{SUCC}(v) \setminus L$.
  TreeSort(G, L, w)
    {well-labels $L \cup \text{ACC}(w)$ to agree with the labeling on L}
  TreeSort($G, L \cup \text{ACC}(w), v$)
    {well-labels $L \cup \text{ACC}(v)$ to agree with the labeling on $L \cup \text{ACC}(w)$}
return ■

The first instruction takes care of the base case in which all the vertices in $\text{ACC}(v)$ except v are already labeled. Note that this base case holds vacuously if v has no successors at all. As with Postorder [where successors are called children], the else instruction refers the labeling task to the successors, but with an extra complication because already labeled vertices need to be dealt with. After the first chosen successor w_1 and the vertices accessible from w_1 are all labeled, TreeSort($G, L \cup \text{ACC}(w_1), v$) leads to one of two possibilities. If all of v's successors are in $L \cup \text{ACC}(w_1)$, then v is the only unlabeled vertex in $L \cup \text{ACC}(v)$, so it gets labeled. Otherwise, another successor w_2 of v is chosen; w_2 and those vertices accessible from w_2 that were not previously labeled are given labels. And so on.

EXAMPLE 7

We illustrate TreeSort on the digraph G of Figure 6(a), which is redrawn in Figure 8(a) with letter names for the vertices. As usual, arbitrary choices are made in alphabetical order.

Figure 8 ▶

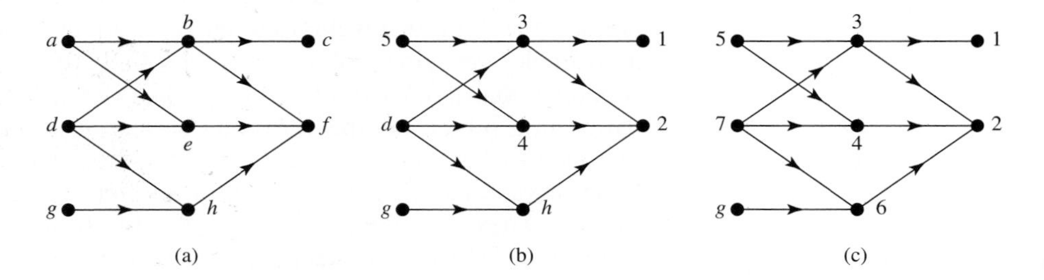

(a) First we use TreeSort$(G, \emptyset, a)$ with $v = a$ and $L = \emptyset$ to label $\text{ACC}(a)$.

Start with a, choose its successor b, choose its successor c, and label c with 1.
Return to b, choose its successor f, and label f with 2.
Return to b and label b with 3.
Return to a, choose its successor e, which has no unlabeled successors, and label e with 4.
Return to a and label a with 5.

Figure 8(b) shows the corresponding labeling of $\text{ACC}(a)$. The sequence just given describes an execution of TreeSort, and here's the detailed explanation specifying the subroutines.

TreeSort$(G, \emptyset, a)$
  Choose successor b of a {since a has unlabeled successors}
  TreeSort$(G, \emptyset, b)$
    Choose successor c of b {since b has unlabeled successors}
    TreeSort$(G, \emptyset, c)$
      Label c with 1 {since c has no successors at all}
    TreeSort$(G, \{c\}, b)$
      Choose successor f of b {f is still unlabeled}
      TreeSort$(G, \{c\}, f)$
        Label f with 2 {since f has no successors}
      TreeSort$(G, \{c, f\}, b)$
        Label b with 3 {since b now has no unlabeled successors}
          {$\text{ACC}(b) = \{c, f, b\}$ is now labeled}
  TreeSort$(G, \{c, f, b\}, a)$
    Choose successor e of a {a still has an unlabeled successor}
    TreeSort$(G, \{c, f, b\}, e)$
      Label e with 4 {the successor f is labeled already}
    TreeSort$(G, \{c, f, b, e\}, a)$
      Label a with 5 {a now has no unlabeled successors}

(b) We next use TreeSort$(G, \{a, b, c, e, f\}, d)$ to label $\{a, b, c, e, f\} \cup \text{ACC}(d)$. See Figure 8(b).

Start with d, choose its unlabeled successor h, then label h with 6, since h has no unlabeled successors.
Return to d and label d with 7, since d now has no unlabeled successors.

Figure 8(c) shows the labeling of $\{a, b, c, e, f\} \cup \text{ACC}(d)$.

(c) Finally, we use TreeSort(G, L, g) with $L = \{a, b, c, d, e, f, h\}$ to label $V(G) = L \cup \text{ACC}(g)$. See Figure 8(c). This will be easy.

Start with g and label g with 8, since g has no unlabeled successors.

Figure 6(a) shows the labeling of G. We would, of course, obtain different labelings if we used different alphabetic preferences. See Exercise 13. ■

In Example 7 the algorithm performed depth-first searches following the paths indicated by the dashed curves in Figure 9. Each vertex gets labeled the last time it is visited. In each search, the set of vertices and edges visited forms a tree if the directions on the edges are ignored. Thus the algorithm sorts within a tree that it finds inside G; this is why we called it TreeSort. The algorithm Tree on page 257 worked in a similar way to build spanning trees inside connected graphs.

Figure 9 ▶

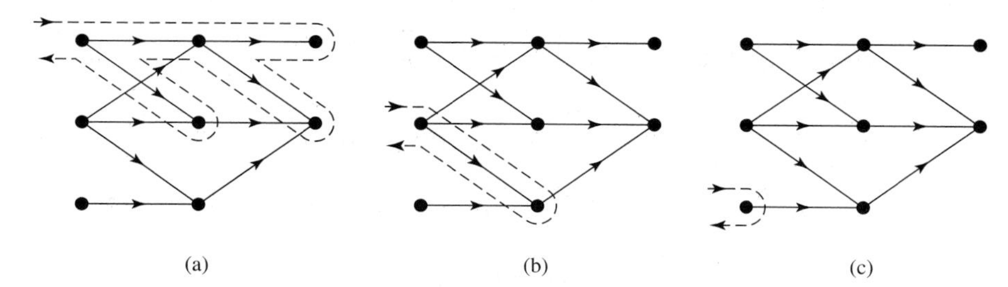

Proof That TreeSort Works. Here is the algorithm again, for reference.

TreeSort(digraph, set of vertices, vertex)

```
{Input: A finite acyclic digraph G, a well-labeled subset L of V(G),
        and a vertex v ∈ V(G) \ L}
{Makes L ∪ ACC(v) well-labeled with a sorted labeling that
        agrees with the given labeling on L}
if SUCC(v) ⊆ L then
        Label L as before and label v with |L| + 1.
else
    Choose w ∈ SUCC(v) \ L.
    TreeSort(G, L, w)
        {well-labels L ∪ ACC(w) to agree with the labeling on L}
    TreeSort(G, L ∪ ACC(w), v)
        {well-labels L ∪ ACC(v) to agree with the labeling on L ∪ ACC(w)}
return
```

Since L is well-labeled, $L = \text{ACC}(L)$, so

$$\text{ACC}(L \cup \text{ACC}(v)) = \text{ACC}(L) \cup \text{ACC}(\text{ACC}(v)) = L \cup \text{ACC}(v).$$

Hence we only need to verify that TreeSort(G, L, v) labels $L \cup \text{ACC}(v)$ correctly. For this we need to check that it gives the right result in the base case $\text{SUCC}(v) \subseteq L$ and that, if its recursive calls TreeSort(G, L, w) and TreeSort$(G, L \cup \text{ACC}(w), v)$ give the right results, then so does TreeSort(G, L, v). We should also check that the recursive calls are valid and are to cases closer to the base case.

Suppose that $\text{SUCC}(v) \subseteq L$. Then TreeSort labels L as before with $1, \ldots, |L|$ and labels v with $|L| + 1$; i.e., it gives $L \cup \{v\}$ a sorted labeling. Because L is well-labeled and $\text{SUCC}(v) \subseteq L$, we have

$$\text{ACC}(v) = \{v\} \cup \text{ACC}(\text{SUCC}(v)) \subseteq \{v\} \cup \text{ACC}(L) = \{v\} \cup L,$$

and hence

$$\text{ACC}(L \cup \{v\}) = \text{ACC}(L) \cup \text{ACC}(v) = L \cup \{v\}$$

and

$$L \cup \text{ACC}(v) = L \cup \{v\}.$$

Thus TreeSort well-labels $L \cup \text{ACC}(v)$ in this case.

Now consider what happens in the **else** case. Since $w \in \text{SUCC}(v) \setminus L$ and L is well-labeled, TreeSort(G, L, w) is meaningful. Assume that TreeSort(G, L, w) well-labels $L \cup \text{ACC}(w)$ to agree with the labeling on L. To make sense of the recursive call TreeSort$(G, L \cup \text{ACC}(w), v)$, we must also have $v \notin L \cup \text{ACC}(w)$. We are given that $v \notin L$. Since $w \in \text{SUCC}(v)$ and G is acyclic, we also have $v \notin \text{ACC}(w)$, so $v \notin L \cup \text{ACC}(w)$. Thus TreeSort$(G, L \cup \text{ACC}(w), v)$ is meaningful. Assuming that it well-labels $(L \cup \text{ACC}(w)) \cup \text{ACC}(v) = L \cup \text{ACC}(v)$ to agree with the labeling on $L \cup \text{ACC}(w)$, and hence with the labeling on L, we get a correct labeling for $L \cup \text{ACC}(v)$ in the **else** case.

It remains to show that the recursive calls approach base cases. We can use $|\text{ACC}(v) \setminus L|$, the number of unlabeled vertices in $\text{ACC}(v)$, as a measure of distance from a base case, for which this number is 0. Suppose that $w \in \text{SUCC}(v) \setminus L$. Since $\text{ACC}(w) \subseteq \text{ACC}(v)$ and since v belongs to $\text{ACC}(v) \setminus L$, but not to $\text{ACC}(w) \setminus L$, we have $|\text{ACC}(v) \setminus L| > |\text{ACC}(w) \setminus L|$; and so TreeSort$(G, L, w)$ is closer to a base case than TreeSort(G, L, v) is. Similarly, w belongs to $\text{ACC}(v) \setminus L$, but not to $\text{ACC}(v) \setminus (L \cup \text{ACC}(w))$. Hence $|\text{ACC}(v) \setminus L| > |\text{ACC}(v) \setminus (L \cup \text{ACC}(w))|$ and TreeSort$(G, L \cup \text{ACC}(w), v)$ is closer to a base case than TreeSort(G, L, v) is. Thus both recursive calls are closer to base cases than TreeSort(G, L, v) is, and TreeSort does perform as claimed. ■

By the time we finished Example 7 we had labeled the entire acyclic digraph G. We repeatedly applied TreeSort to unlabeled vertices until all the vertices were labeled; i.e., we executed algorithm Label(G, L) below with $L = \emptyset$.

Label(digraph, set of vertices)

{Input: A finite acyclic digraph G and a well-labeled subset L of $V(G)$}
{Gives $V(G)$ a sorted labeling that agrees with that on L}
if $L = V(G)$ **then**
  Do nothing.
else
  Choose $v \in V(G) \setminus L$.
  TreeSort(G, L, v) {well-labels $L \cup \text{ACC}(v)$ to agree with L labels}
  Label$(G, L \cup \text{ACC}(v))$ {labels what's left of $V(G)$, and retains the labels on $L \cup \text{ACC}(v)$}.
return ■

The verification that Label performs correctly is similar to the argument for TreeSort. The appropriate measure of distance from the base case $L = V(G)$ is $|V(G) \setminus L|$, the number of vertices not yet labeled.

EXAMPLE 8

As already noted, if we apply algorithm Label$(G, \emptyset)$ to the digraph in Figure 8(a), we obtain the sorted labeling in Figure 6(a). We summarize.

Choose a
Label $\text{ACC}(a) = \{c, b, f, e, a\}$ with TreeSort$(G, \emptyset, a)$.   {Example 7(a)}
Recur to $L := \{c, b, f, e, a\}$.
  Choose d.
  Label $L \cup \text{ACC}(d) = L \cup \{h, d\}$ with TreeSort(G, L, d).   {Example 7(b)}
  Recur to $L := \{c, b, f, e, a, h, d\}$.
    Choose g.
    Label $L \cup \text{ACC}(g) = L \cup \{g\} = V(G)$.   {Example 7(c)}
    Recur to $L := V(G)$.
      Stop. ■

Although our examples have all been connected digraphs, algorithms TreeSort and Label would have worked as well for digraphs with more than one component.

The method of charges gives an estimate for how long Label takes. Charge each vertex of G with the time it takes to label it. Assign charges to the edges of G in two different ways: if $w \in \text{SUCC}(v)$, but w is already in L when TreeSort looks at successors of v, then charge the edge (v, w) from v to w with the time it takes to remove w from the list of available choices in $\text{SUCC}(v)$; if w *is* chosen in $\text{SUCC}(v)$, then charge (v, w) with the time it takes to recur to w and later get back to v. Then all activity gets charged somewhere, and thus the time to label $V(G)$ starting with $L = \emptyset$ is $O(|V(G)| + |E(G)|)$.

One might object and say that, since only the edges (v, w) for $w \in \text{SUCC}(v) \setminus L$ get used, they are the only ones that take time. In fact, though, each edge in G has to be considered, if only briefly, to determine whether it is available for use. For example, in the graph of Figure 8(a) the edge (e, f) is not followed; i.e., TreeSort does not recur from e to f, because f is already labeled when the algorithm gets to e. On the other hand, we see that the dashed curve in Figure 9 goes down and up the edge (a, e)—down for the call to TreeSort at e and back up to report the result to TreeSort at a.

Exercises 7.3

1. Use the "dashed-line" idea of Figure 2 to give the preorder and postorder listings of the vertices of the tree in Figure 4(a).
2. Repeat Exercise 1 for Figure 5(a).
3. Repeat Exercise 1 for Figure 5(b).
4. Give the inorder listing of the vertices of the labeled tree in Figure 10.

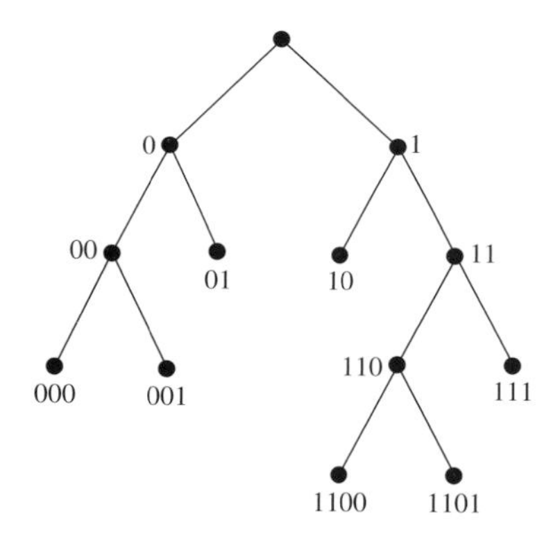

Figure 10 ▲

5. (a) Use the Preorder algorithm to list the vertices of the tree T in Figure 3. Give the values of $L(w)$ and $L(u)$ that the algorithm produces.
   (b) Repeat part (a) using the Postorder algorithm.
6. List the successors and accessible sets for the graph in Figure 6(b); use the format of Figure 7.
7. (a) Use the Postorder algorithm to list the vertices in Figure 4(b).
   (b) Repeat part (a) using the Preorder algorithm.
   (c) Repeat part (a) using the Inorder algorithm.
8. Verify the statements in Example 3.
9. (a) Apply the algorithm TreeSort$(G, \emptyset, a)$ to the digraph G of Figure 11(a). At each choice point in the algorithm, use alphabetical order to choose successors of the vertex. Draw a dashed curve to show the search pattern, and show the labels on the vertices, as in Figure 8(c).
   (b) List the steps for part (a) in the less detailed manner of Example 7(a).
   (c) Apply the algorithm Label$(G, \emptyset)$ to the digraph G of Figure 11(a) and draw a picture that shows the resulting labels on the vertices.

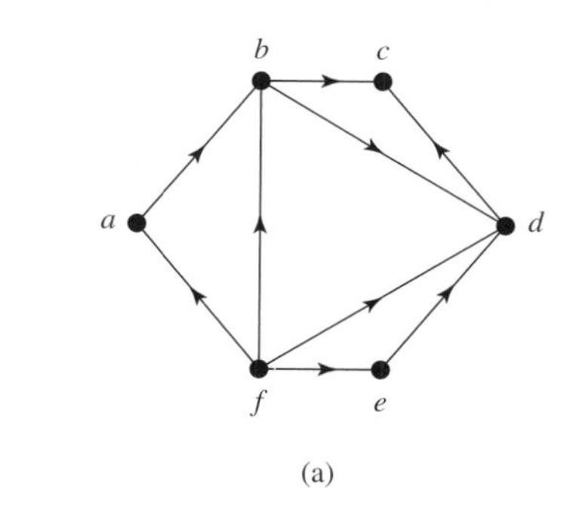

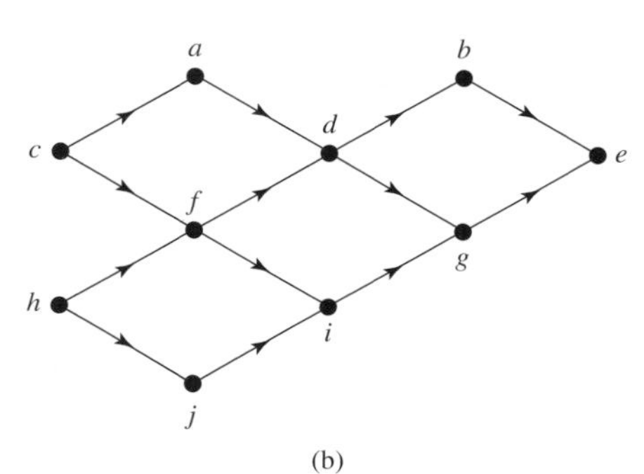

Figure 11 ▲

10. (a) Repeat Exercise 9(a) for the digraph G of Figure 11(b).

(b) Repeat Exercise 9(c) for this digraph.

11. Apply TreeSort(G, L, v) to the digraph G in Figure 12 with

(a) $v = a$, $L = \emptyset$ (b) $v = d$, $L = \{a, b, c\}$

(c) $v = h$, $L = \{a, b, c, d, e, f, g\}$

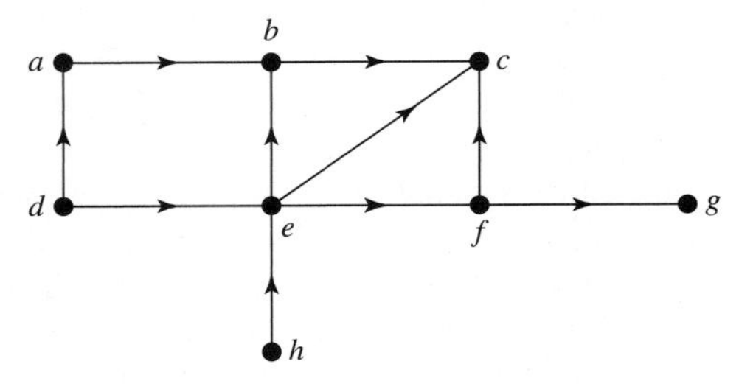

Figure 12 ▲

12. Use Exercise 1 and algorithm Label to give a sorted labeling of the digraph in Figure 12.

13. (a) The digraph in Figure 8(a) is redrawn in Figure 13(a) with different letter names on the vertices. Use the algorithm Label and the usual alphabetic ordering to give a sorted labeling of the digraph.

(b) Repeat part (a) using the digraph of Figure 13(b).

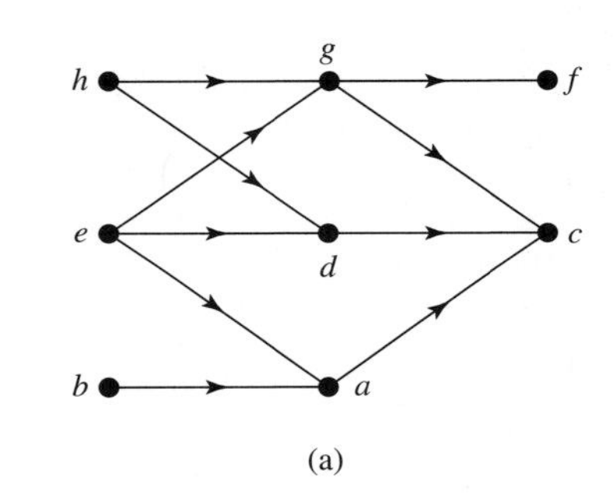

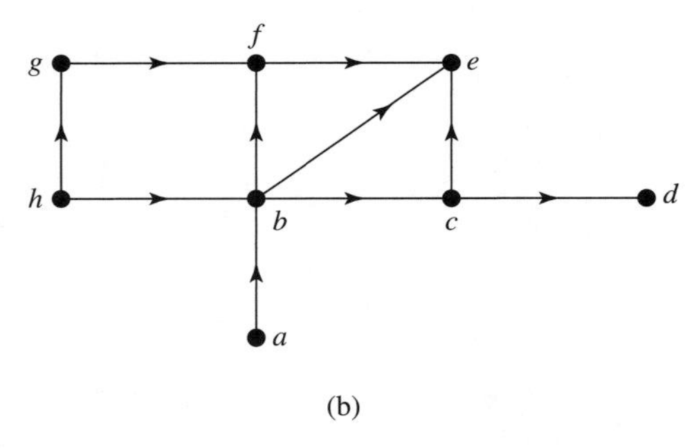

Figure 13 ▲

14. The algorithm TreeSort resembles Postorder in that it labels each vertex the last time it is visited. Here is an algorithm BushSort that tries to imitate Preorder.

BushSort(digraph, set of vertices, vertex)

{Input: A finite acyclic digraph G, a subset L of $V(G)$ labeled somehow, and a vertex $v \in V(G) \setminus L$}
{Labels $L \cup \text{ACC}(v)$ somehow to agree with the labeling on L}
Label v with $|L| + 1$ {when we first come to v}.
Replace L by $L \cup \{v\}$.
for $w \in \text{SUCC}(v) \setminus L$ (successors taken in some order) **do**
  BushSort(G, L, w) {labels $L \cup \text{ACC}(w)$ somehow}
  Replace L by $L \cup \text{ACC}(w)$. ■

(a) Apply BushSort$(G, \emptyset, a)$ to label the digraph G in Figure 14.

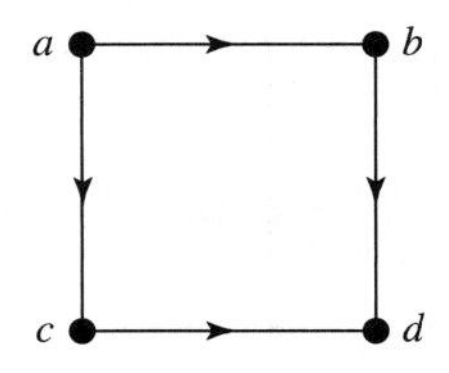

Figure 14 ▲

(b) Is your labeling in part (a) a sorted labeling for the digraph? Is it the reverse of a sorted labeling; i.e., do vertices have smaller numbers than their successors? Explain.

15. Give a measure of the distance from a base case for the algorithm LabelTree, and show that the recursive calls at (T_w, n) in the algorithm are closer to base cases than is the case (T_v, k) that calls them.

16. (a) Give an example of an acyclic digraph without parallel edges that has 4 vertices and $4(4-1)/2 = 6$ edges. How many sorted labelings does your example have?

(b) Show that every acyclic digraph without parallel edges that has n vertices and $n(n-1)/2$ edges has exactly one sorted labeling.

17. Show that an acyclic digraph without parallel edges that has n vertices cannot have more than $n(n-1)/2$ edges. [Hence Label takes time no more than $O(n^2)$ for such digraphs.]

18. (a) Modify the algorithm Preorder to obtain an algorithm LeafList that returns a list of the leaves of T_v ordered from left to right.

(b) Illustrate your algorithm LeafList on the tree of Figure 1.

7.4 Polish Notation

Preorder, postorder, and inorder listings give ways to list the vertices of an ordered rooted tree. If the vertices have labels such as numbers, addition signs, multiplication signs, and the like, then the list itself may have a meaningful interpretation. For instance, using ordinary algebraic notation, the list $4*3\div 2$ determines the number 6. The list $4 + 3 * 2$ seems ambiguous; is it the number 14 or 10? Polish notation, which we describe below, is a method for defining algebraic expressions without parentheses, using lists obtained from trees. It is important that the lists completely determine the corresponding labeled trees and their associated expressions. After we discuss Polish notation, we prove that under quite general conditions the lists do determine the trees uniquely.

Polish notation can be used to write expressions that involve objects from some system [of numbers, or matrices, or propositions in the propositional calculus, etc.] and certain operations on the objects. The operations are usually, but not always, **binary**, i.e., ones that combine two objects; or **unary**, i.e., ones that act on only one object. Examples of binary operations are $+$, $*$, $\wedge$, and $\rightarrow$. The operations $\neg$ and ^{c} are unary, and the symbol $-$ is used for both a binary operation and a unary one. The ordered rooted trees that correspond to expressions have leaves labeled by objects from the system [such as numbers] or by variables representing objects from the system [such as x]. The other vertices are labeled by the operations.

EXAMPLE 1

The algebraic expression

$$((x - 4)\,^\wedge\, 2) * ((y + 2)/3)$$

is represented by the tree in Figure 1(a). This expression uses several familiar binary operations on $\mathbb{R}$: $+, -, *, /, ^\wedge$. Recall that $*$ represents multiplication and that $^\wedge$ represents exponentiation: $a\,^\wedge\, b$ means a^b. Thus our expression is equivalent to

$$(x - 4)^2 \left(\frac{y + 2}{3}\right).$$

Note that the tree is an *ordered* tree; if x and 4 were interchanged, for example, the tree would represent a different algebraic expression.

Figure 1 ▶

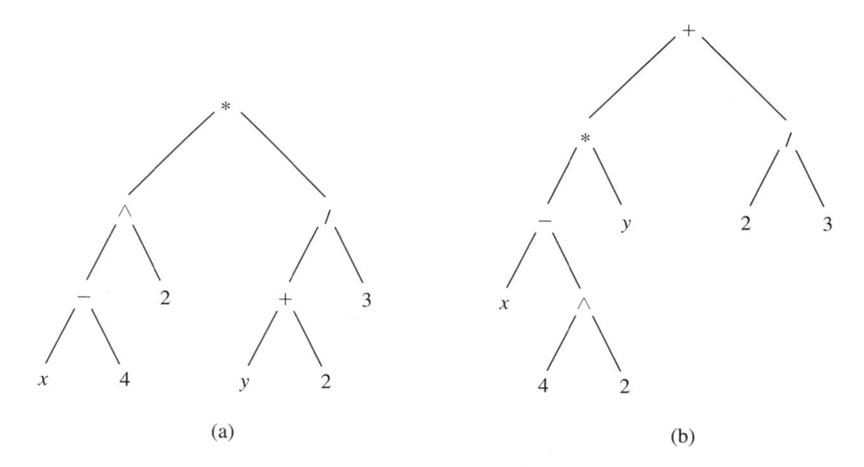

It is clear that the ordered rooted tree determines the algebraic expression. Note that the inorder listing of the vertices yields $x - 4\,^\wedge 2 * y + 2\,/\,3$, and this is exactly the original expression *except for the parentheses*. Moreover, the parentheses are crucial, since this expression determines neither the tree nor the original algebraic expression. This listing could just as well come from the algebraic expression

$$((x - (4\,^\wedge\, 2)) * y) + (2\,/\,3)$$

whose tree is drawn in Figure 1(b). This algebraic expression is equivalent to $(x - 16)y + \frac{2}{3}$, a far cry from

$$(x - 4)^2 \left(\frac{y + 2}{3}\right).$$

Let us return to our original algebraic expression given by Figure 1(a). Preorder listing yields

$$* \wedge - x\, 4\, 2 \,/\, + y\, 2\, 3$$

and postorder listing yields

$$x\, 4 - 2 \wedge y\, 2 + 3 \,/\, *.$$

It turns out that each of these listings uniquely determines the tree and hence the original algebraic expression. Thus these expressions are unambiguous *without parentheses*. This extremely useful observation was made by the Polish logician Łukasiewicz. The preorder listing is known now as **Polish notation** or **prefix notation**. The postorder listing is known as **reverse Polish notation** or **postfix notation**. Our usual algebraic notation, with the necessary parentheses, is known as **infix** notation. ■

EXAMPLE 2

Consider the compound proposition $(p \to q) \vee (\neg\, p)$. We treat the binary operations $\to$ and $\vee$ as before. However, $\neg$ is a 1-ary or unary operation. We decree that its child is a right child, since the operation precedes the proposition that it operates on. The corresponding binary tree in Figure 2(a) can be traversed and listed in all three ways. The preorder listing is $\vee \to p\, q \neg\, p$, and the postorder listing is $p\, q \to p \neg \vee$. The inorder list is the original expression *without parentheses*. Another tree with the same inorder expression is drawn in Figure 2(b). In this case, as in Example 1, the preorder and postorder listings determine the tree and hence the original compound proposition. ■

Figure 2 ▶

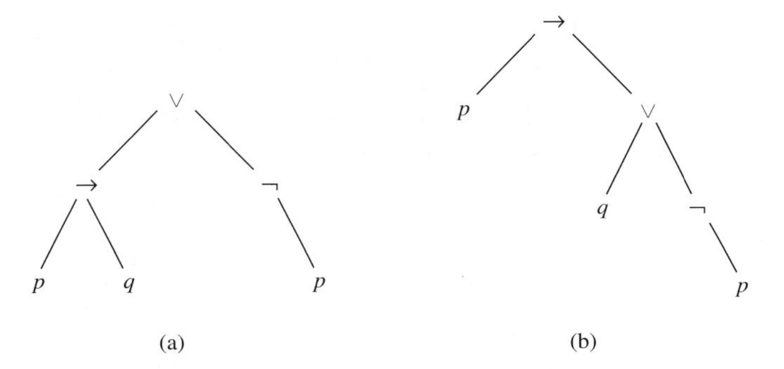

As we illustrated in Example 3 on page 289, in general the preorder or postorder list of vertices will not determine a tree, even in the case of a regular binary tree. More information is needed. It turns out that if we are provided with the level of each vertex, then the preorder or postorder list determines the tree. We will not pursue this. It also turns out that if we know how many children each vertex has, then the tree is determined by its preorder or postorder list.

EXAMPLE 3

We illustrate the last sentence by beginning with the expression

$$x\, 4 - 2 \wedge y\, 2 + 3 \,/\, *$$

in postfix notation. Each vertex $*$, $\wedge$, $-$, $/$, and $+$ represents a binary operation and so has two children. The other vertices have no children, so we know exactly how many children each vertex has.

We now recover the tree, but instead of drawing subtrees we'll determine their corresponding subexpressions. To reconstruct the tree, we recall that in the postorder listing each vertex is immediately preceded by lists of the ordered subtrees rooted at its children. Our task is to recognize these subtree lists in the given expression. To do so, we will borrow the idea of a "stack" from computer science.

Starting from the left, we read x and 4, which have no children of their own. Let's put them in a stack $\begin{matrix}4\\x\end{matrix}$, which we can write on one line as $[x, 4]$, with the most recently read entry on the right. Next we read $-$, which must have two children, namely the leaves x and 4. The subtree rooted at $-$ has postfix listing $x\,4\,-$, corresponding to the infix expression $x-4$. We replace $x\,4\,-$ by $(x-4)$ to get the modified sequence $(x-4)\,2\,{}^{\wedge}\,y\,2\,+\,3\,/\,*$, with $(x-4)$ acting like a leaf. Then we read 2. It also has no children, so we put it on top of the stack. At this point we have two unclaimed children in the stack $\begin{smallmatrix}2\\(x-4)\end{smallmatrix}$, which we can write as $[(x-4), 2]$. We next read the binary operation ${}^{\wedge}$, which has two children, $(x-4)$ and 2, corresponding to the two subtrees hanging from it. We take these children off the stack and replace them by $((x-4)^2)$, which becomes an unclaimed child, and consider the modified sequence $((x-4)^2)\,y\,2\,+\,3\,/\,*$. Now y and 2 have no children, so we build the stack $[((x-4)^2), y, 2]$. The next symbol, $+$, picks up the top, i.e., rightmost, two unclaimed children on the stack, the leaves y and 2, building the subtree rooted at $+$, corresponding to $(y+2)$. We get the new stack $[((x-4)^2), (y+2)]$ and corresponding modified sequence $((x-4)^2)\,(y+2)\,3\,/\,*$. Since 3 has no children, we add it to the stack to get $[((x-4)^2, (y+2), 3]$. Then $/$ takes $(y+2)$ and 3 off the stack to give $(\frac{(y+2)}{3})$, corresponding to the subtree rooted at $/$, and yields the sequence $((x-4)^2)\,(\frac{(y+2)}{3})\,*$. Finally, $*$ takes the two remaining children off the stack and multiplies them to give $((x-4)^2)(\frac{(y+2)}{3})$, which is the infix algebraic expression that originally gave rise to the tree in Example 1. We have recovered the tree and its associated value from its postfix listing.

We briefly summarize the procedure that we just went through. Starting with an empty stack, we work from left to right on the input string. When we encounter a symbol with no children, we add it to the top of the stack. When we encounter a symbol with children, we take a suitable number of children from the top of the stack, combine them as instructed by the symbol, and put the result on top of the stack. When we reach the end, the single element remaining in the stack is the result. In our example, each symbol with children was a binary operation, so it had two children; but in general, for example as in Figure 2(a), a symbol might have just one child or more than two.

We can also describe the tree we construct by giving a table such as the following:

Vertex	x	4	$-$	2	$\wedge$	y	2	$+$	3	$/$	$*$	Result
Stack		x	4	$-$	2	$\wedge$	y	2	$+$	3	$/$	$*$
			x		$-$		$\wedge$	y	$\wedge$	$+$	$\wedge$	
								$\wedge$		$\wedge$		

Here the stack listed below a vertex is the stack available at the time the vertex is reached. The children of a vertex can be read from the table. For example, the children of $+$ are y and 2, and the children of $*$ are ${}^{\wedge}$ and $/$. To use such a table, we need to know how many children each vertex has.

In a hand-held calculator that uses reverse Polish logic, each application of a binary operation key actually pops the top two entries off a stack, combines them, and pushes the result back on top of the stack. ■

EXAMPLE 4

The same method works for a compound proposition in reverse Polish notation, but when we meet the unary operation $\neg$, it acts on just the immediately preceding subexpression. For example, starting with $p\,q \to\ p\,q \wedge \neg \vee$, we get

$$\boldsymbol{p\,q \to}\ p\,q \wedge \neg \vee$$
$$(p \to q)\ \boldsymbol{p\,q \wedge} \neg \vee$$
$$(p \to q)\boldsymbol{(p \wedge q)\neg} \vee \qquad [\neg \text{ just acts on } (p \wedge q)]$$
$$\boldsymbol{(p \to q)(\neg(p \wedge q)) \vee}$$
$$(p \to q) \vee (\neg(p \wedge q)).$$

The reader can draw a tree representing this compound proposition. Alternatively, we can pick off the children by working from left to right as in Example 3. The sequence of stacks is given in the table

Vertex	p	q	$\to$	p	q	$\wedge$	$\neg$	$\vee$	Result
Stack		p	q p	$\to$	p $\to$	q p $\to$	$\wedge$ $\to$	$\neg$ $\to$	$\vee$

or, in terms of values of subtrees, in the following more complicated table.

Vertex	p	q	$\to$	p	q	$\wedge$	$\neg$	$\vee$	Result
Stack		p	q p	$(p \to q)$	p $(p \to q)$	q p $(p \to q)$	$(p \wedge q)$ $(p \to q)$	$(\neg(p \wedge q))$ $(p \to q)$	$(p \to q) \vee (\neg(p \wedge q))$

■

Not all strings of operations and symbols lead to meaningful expressions.

EXAMPLE 5

Suppose it is alleged that

$$y + 2\,x * \,{}^\wedge 4 \quad \text{and} \quad q \neg p\, q \vee \wedge \to$$

are in reverse Polish notation. The first one is hopeless right away, since $+$ is not preceded by two expressions. The second one breaks down when we attempt to decode [i.e., parse] it as in Example 4:

$$\boldsymbol{q \neg}\, p\, q \vee \wedge \to$$

$$(\neg q)\, \boldsymbol{p\, q \vee} \wedge \to$$

$$\boldsymbol{(\neg q)(p \vee q) \wedge} \to$$

$$((\neg q) \wedge (p \vee q)) \to .$$

Unfortunately, the operation $\to$ has only one subexpression preceding it. We conclude that neither of the strings of symbols given above represents a meaningful expression. ■

Just as we did with ordinary algebraic expressions in §7.2, we can recursively define what we mean by well-formed formulas [wff's] for Polish and reverse Polish notation. We give one example by defining **wff's for reverse Polish notation** of algebraic expressions:

(B) Numerical constants and variables are wff's.

(R) If f and g are wff's, so are $f\,g\,+$, $f\,g\,-$, $f\,g\,*$, $f\,g\,/$, and $f\,g\,{}^\wedge$.

This definition omits the unary $-$ operation, which slightly complicates matters [see Exercises 21 and 22].

EXAMPLE 6

We show that $x\,2\,{}^\wedge\,y - x\,y * /$ is a wff. All the variables and constants are wff's by (B). Then $x\,2\,{}^\wedge$ is a wff by (R). Hence $x\,2\,{}^\wedge\,y\,-$ is a wff, where we use (R) with $f = x\,2\,{}^\wedge$ and $g = y$. Similarly, $x\,y\,*$ is a wff by (R). Finally, the entire expression is a wff by (R), since it has the form $f\,g\,/$, where $f = x\,2\,{}^\wedge\,y\,-$ and $g = x\,y\,*$. ■

We end this section by showing that an expression in Polish or reverse Polish notation uniquely determines its tree, and hence determines the original expression. The proof is based on the second algorithm illustrated in Examples 3 and 4.

Theorem Let T be a finite ordered rooted tree whose vertices have been listed by a preorder listing or a postorder listing. Suppose that the number of children of each vertex is known. Then the tree can be recovered from the listing.

Proof We consider only the case of postorder listing. We are given the postordered list $v_1\, v_2 \cdots v_n$ of T and the numbers $c_1, \ldots, c_n$ of children of $v_1, \ldots, v_n$. We'll show that, for each vertex v_m of T, the set $C(v_m)$ of children of v_m and its order are uniquely determined.

Consider some vertex v_m. Then v_m is the root of the subtree T_m consisting of v_m and its descendants. When algorithm Postorder on page 287 lists v_m, the subtrees of T_m have already been listed and their lists immediately precede v_m, in the order determined by the order of $C(v_m)$. Moreover, the algorithm does not insert later entries into the list of T_m, so the list of T_m appears in the complete list $v_1\, v_2 \cdots v_n$ of T as an unbroken string with v_m at the right end.

Since v_1 has no predecessors, v_1 is a leaf. Thus $C(v_1) = \varnothing$, and its order is vacuously determined. Assume inductively that for each k with $k < m$ the set $C(v_k)$ and its order are determined, and consider v_m. [We are using the second principle of induction here on the set $\{1, 2, \ldots, n\}$.] Then the stack set $S(m) = \{v_k : k < m\} \setminus \bigcup_{k<m} C(v_k)$ is determined. It consists of the vertices v_k to the left of v_m whose parents are not to the left of v_m, i.e., the children as yet unclaimed by their parents. The children of v_m are the members of $S(m)$ that are in the tree T_m. Since the T_m list is immediately to the left of v_m, the children of v_m are the c_m members of $S(m)$ farthest to the right. Since $S(m)$ is determined and c_m is given, the set $C(v_m)$ of children of v_m is determined. Moreover, its order is the order of appearance in the list $v_1\, v_2 \cdots v_n$.

By induction, each ordered set $C(v_m)$ of children is determined by the postordered list and the sequence $c_1, c_2, \ldots, c_n$. The root of the tree is, of course, the last vertex v_n. Thus the complete structure of the ordered rooted tree is determined. ■

EXAMPLE 7

We illustrate the proof of the theorem. The list $u\, x\, y\, z\, w\, v\, p\, q\, s\, r$ and sequence 0, 0, 0, 0, 3, 2, 0, 0, 2, 2 give the sets shown in the following table.

v_k	u	x	y	z	w	v	p	q	s	r
$S(k)$		u	x u	y x u	z y x u	w u	v	p v	q p v	s v
$C(v_k)$	$\varnothing$	$\varnothing$	$\varnothing$	$\varnothing$	$\{x, y, z\}$	$\{u, w\}$	$\varnothing$	$\varnothing$	$\{p, q\}$	$\{v, s\}$

The ordered sets $C(v_k)$ can be assembled recursively into the tree T of Figure 1 on page 287. To see this, it's probably easiest to assemble in the order: $C(r)$, $C(v)$, $C(s)$, $C(u)$, $C(w)$, $C(p)$, $C(q)$, etc. ■

Exercises 7.4

1. Write the algebraic expression given by Figure 1(b) in reverse Polish and in Polish notation.

2. For the ordered rooted tree in Figure 3(a), write the corresponding algebraic expression in reverse Polish notation and also in the usual infix algebraic notation.

3. (a) For the ordered rooted tree in Figure 3(b), write the corresponding algebraic expression in Polish notation and also in the usual infix algebraic notation.

   (b) Simplify the algebraic expression obtained in part (a) and then draw the corresponding tree.

4. Calculate the following expressions given in reverse Polish notation.

   (a) $3\,3\,4\,5\,1 - * + +$ (b) $3\,3 + 4 + 5 * 1 -$

   (c) $3\,3\,4 + 5 * 1 - +$

5. Calculate the following expressions given in reverse Polish notation.

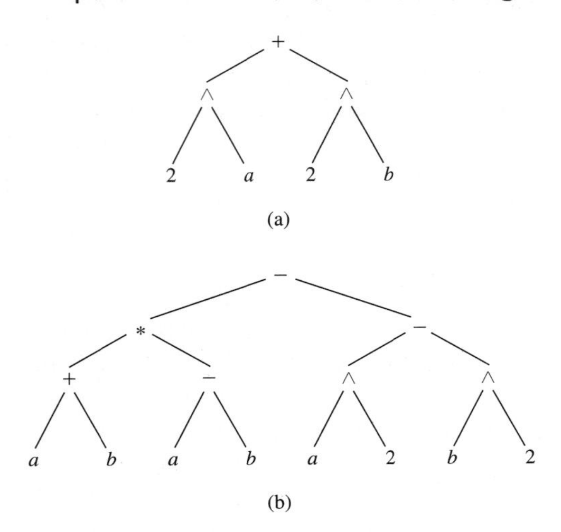

Figure 3 ▲

(a) $6\,3\,/\,3+7\,3-*$

(b) $3\,2\,{}^\wedge 4\,2\,{}^\wedge +5\,/\,2\,*$

6. Calculate the following expressions in Polish notation.

(a) $-*3\,{}^\wedge 5\,2\,2$ (b) ${}^\wedge *3\,5-2\,2$

(c) $-{}^\wedge *3\,5\,2\,2$ (d) $/*2+2\,5\,{}^\wedge +3\,4\,2$

(e) $*+/6\,3\,3-7\,3$

7. Write the following algebraic expressions in reverse Polish notation.

(a) $(3x-4)^2$ (b) $(a+2b)/(a-2b)$

(c) $x-x^2+x^3-x^4$

8. Write the algebraic expressions in Exercise 7 in Polish notation.

9. (a) Write the algebraic expressions $a(bc)$ and $(ab)c$ in reverse Polish notation.

(b) Do the same for $a(b+c)$ and $ab+ac$.

(c) What do the associative and distributive laws look like in reverse Polish notation?

10. Write the expression $x\,y+2\,{}^\wedge x\,y-2\,{}^\wedge -x\,y*/$ in the usual infix algebraic notation and simplify.

11. Consider the compound proposition represented by Figure 2(b).

(a) Write the proposition in the usual infix notation [with parentheses].

(b) Write the proposition in reverse Polish and in Polish notation.

12. The following compound propositions are given in Polish notation. Draw the corresponding rooted trees, and rewrite the expressions in the usual infix notation.

(a) $\leftrightarrow \neg \wedge \neg p \neg q \vee p q$ (b) $\leftrightarrow \wedge p q \neg \rightarrow p \neg q$

[These are laws from Table 1 on page 62.]

13. Repeat Exercise 12 for the following.

(a) $\rightarrow \wedge p \rightarrow p q q$

(b) $\rightarrow \wedge \wedge \rightarrow p q \rightarrow r s \vee p r \vee q s$

14. Write the following compound propositions in reverse Polish notation.

(a) $[(p \rightarrow q) \wedge (q \rightarrow r)] \rightarrow (p \rightarrow r)$

(b) $[(p \vee q) \wedge \neg p] \rightarrow q$

15. Illustrate the ambiguity of "parenthesis-free infix notation" by writing the following pairs of expressions in infix notation without parentheses.

(a) $(a/b)+c$ and $a/(b+c)$

(b) $a+(b^3+c)$ and $(a+b)^3+c$

16. Use the recursive definition for wff's for reverse Polish notation to show that the following are wff's.

(a) $3\,x\,2\,{}^\wedge *$ (b) $x\,y+1\,x\,/\,1\,y\,/+*$

(c) $4\,x\,2\,{}^\wedge y\,z+2\,{}^\wedge/-$

17. (a) Define wff's for Polish notation for algebraic expressions.

(b) Use the definition in part (a) to show that ${}^\wedge +x\,/\,4\,x\,2$ is a wff.

18. Let $S_1 = x_1\,2\,{}^\wedge$ and $S_{n+1} = S_n\,x_{n+1}\,2\,{}^\wedge +$ for $n \geq 1$. Here $x_1, x_2, \ldots$ represent variables.

(a) Show that each S_n is a wff for reverse Polish notation. *Suggestion:* Use induction.

(b) What does S_n look like in the usual infix notation?

19. (a) Define wff's for reverse Polish notation for the propositional calculus; see the definition just before Example 6.

(b) Use the definition in part (a) to show that $p q \neg \wedge \neg p q \neg \rightarrow \vee$ is a wff.

(c) Define wff's for Polish notation for the propositional calculus.

(d) Use the definition in part (c) to show that $\vee \neg \wedge p \neg q \rightarrow p \neg q$ is a wff.

20. (a) Draw the tree with postorder vertex sequence $s\,t\,v\,y\,r\,z\,w\,u\,x\,q$ and number of children sequence 0, 0, 0, 2, 2, 0, 0, 0, 2, 3.

(b) Is there a tree with $s\,t\,v\,y\,r\,z\,w\,u\,x\,q$ as preorder vertex sequence and number of children sequence 0, 0, 0, 2, 2, 0, 0, 0, 2, 3? Explain.

21. Explain why a hand-held calculator needs different keys for the operation that takes x to $-x$ and the one that takes (x, y) to $x-y$.

22. Let $\ominus$ be the unary negation operation that takes x to $-x$.

(a) Give the infix expression corresponding to the reverse Polish expression $4\,x\,y\,+\,-\,z\,\ominus\,*\ominus$.

(b) How many children does each $\ominus$ node have in the corresponding tree, and where are they located?

7.5 Weighted Trees

A **weighted tree** is a finite rooted tree in which each leaf is assigned a nonnegative real number, called the **weight** of the leaf. In this section we discuss general weighted trees, and we give applications to prefix codes and sorted lists.

To establish some notation, we assume that our weighted tree T has t leaves, whose weights are $w_1, w_2, \ldots, w_t$. We lose no generality if we also assume that $w_1 \le w_2 \le \cdots \le w_t$. It will be convenient to label the leaves by their weights, and we will often refer to a leaf by referring to its weight. Let $l_1, l_2, \ldots, l_t$ denote the corresponding level numbers of the leaves, so l_i is the length of the path from the root to the leaf w_i. The **weight** of the tree T is the number

$$\boldsymbol{W(T)} = \sum_{i=1}^{t} w_i l_i,$$

in which the weights of the leaves are multiplied by their level numbers.

EXAMPLE 1

(a) The six leaves of the weighted tree in Figure 1(a) have weights 2, 4, 6, 7, 7, and 9. Thus $w_1 = 2$, $w_2 = 4$, $w_3 = 6$, $w_4 = 7$, $w_5 = 7$, and $w_6 = 9$. There are two leaves labeled 7, and it does not matter which we regard as w_4 and which we regard as w_5. For definiteness, we let w_4 represent the leaf labeled 7 at level 2. Then the level numbers are $l_1 = 3$, $l_2 = 1$, $l_3 = 3$, $l_4 = 2$, $l_5 = 1$, and $l_6 = 2$. Hence

$$W(T) = \sum_{i=1}^{6} w_i l_i = 2 \cdot 3 + 4 \cdot 1 + 6 \cdot 3 + 7 \cdot 2 + 7 \cdot 1 + 9 \cdot 2 = 67.$$

Figure 1 ▶

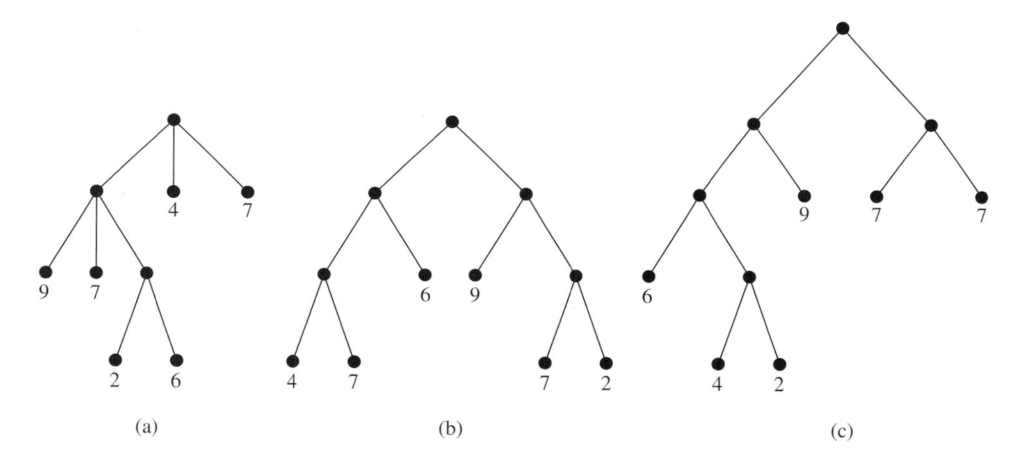

(b) The same six weights can be placed on a binary tree, as in Figure 1(b) for instance. Now the level numbers are $l_1 = 3$, $l_2 = 3$, $l_3 = 2$, $l_4 = l_5 = 3$, and $l_6 = 2$, so

$$W(T) = 2 \cdot 3 + 4 \cdot 3 + 6 \cdot 2 + 7 \cdot 3 + 7 \cdot 3 + 9 \cdot 2 = 90.$$

(c) Figure 1(c) shows another binary tree with these weights. Its weight is

$$W(T) = 2 \cdot 4 + 4 \cdot 4 + 6 \cdot 3 + 7 \cdot 2 + 7 \cdot 2 + 9 \cdot 2 = 88.$$

The total weight is less than that in part (b), because the heavier leaves are near the root and the lighter ones are farther away. Later in this section we will discuss an algorithm for obtaining a binary tree with minimum weight for any specified sequence of weights $w_1, w_2, \ldots, w_t$.

(d) As noted in part (c), we are often interested in binary trees with minimum weight. If we omit the binary requirement and allow many weights near the root, as in part (a), then we can get the lowest possible weight by placing *all* the weights on leaves at level 1 so that $W(T) = \sum_{i=1}^{t} w_i$. For weights 2, 4, 6, 7, 7, and 9, this gives a tree of weight 35. Such weighted trees are boring, and they won't help us solve any interesting problems. ■

EXAMPLE 2

(a) As we will explain later in this section, certain sets of binary numbers can serve as codes. One such set is {00, 01, 100, 1010, 1011, 11}. These numbers are the labels of the leaves in the binary tree of Figure 2(a), in which the nodes are labeled as in Example 8 on page 250. This set could serve as a code for the letters in an alphabet Σ that has six letters. Suppose that we know how frequently each letter in Σ is used in sending messages. In Figure 2(b) we have placed a weight at each leaf that signifies the percentage of code symbols using that leaf. For example, the letter coded 00 appears 25 percent of the time, the letter coded 1010 appears 20 percent of the time, etc. Since the length of each code symbol as a word in 0's and 1's is exactly its level in the binary tree, the average length of a code message using 100 letters from Σ will just be the weight of the weighted tree, in this case

$$25 \cdot 2 + 10 \cdot 2 + 10 \cdot 3 + 20 \cdot 4 + 15 \cdot 4 + 20 \cdot 2 = 280.$$

This weight measures the efficiency of the code. As we will see in Example 7, there are more efficient codes for this example, i.e., for the set of frequencies 10, 10, 15, 20, 20, 25.

Figure 2 ▶

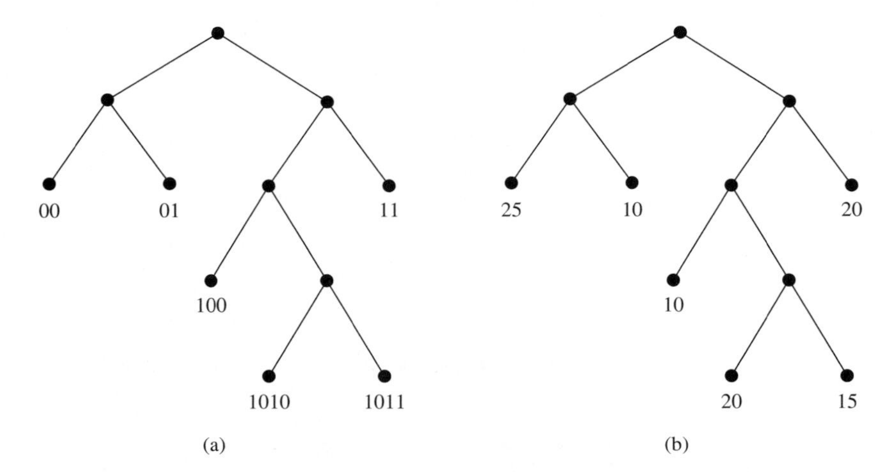

(b) Instead of standing for letters in the alphabet Σ, the code words in part (a) can be thought of as addresses for the leaves in the binary search tree in Figure 2(a). Suppose that the time it takes to reach a leaf and look up the record stored there is proportional to the number of decision [branch] nodes on the path to the leaf. Then, to minimize the average lookup time, the most frequently looked up records should be placed at leaves near the root, with the less popular ones farther away. If the weights in Figure 2(b) show the percentages of lookups of records in the corresponding locations, then the average lookup time is proportional to the weight of the weighted tree. ■

EXAMPLE 3

Consider a collection of sorted lists, say $L_1, L_2, \ldots, L_n$. For example, each L_i could be an alphabetically sorted mailing list of clients or a pile of exam papers arranged in increasing order of grades. To illustrate the ideas involved, let's suppose that each list is a set of real numbers arranged by the usual order $\leq$. Suppose that we can merge lists together two at a time to produce new lists. Our problem is to determine how to merge the n lists most efficiently to produce a single sorted list.

Two lists are merged by comparing the first numbers of both lists and selecting the smaller of the two [either one if they are equal]. The selected number is removed and becomes the first member of the merged list, and the process is repeated for the two lists that remain. The next number selected is placed second on the merged list, and so on. The process ends when one of the remaining lists is empty.

For instance, to merge 4, 8, 9 and 3, 6, 10, 11: compare 3 and 4, choose **3**, and reduce to lists 4, 8, 9 and 6, 10, 11; compare 4 and 6, choose **4**, and reduce to 8, 9 and 6, 10, 11; compare 8 and 6, choose **6**, and reduce to 8, 9 and 10, 11;

compare 8 and 10, choose **8**, and reduce to 9 and 10, 11; compare 9 and 10, choose **9**, and reduce to the empty list and 10, 11; choose **10**, then **11**. The merged list is 3, 4, 6, 8, 9, 10, 11. In this example five comparisons were required. If the lists contain j and k elements, respectively, then in general the process must end after $j + k - 1$ or fewer comparisons. The goal is to merge $L_1, L_2, \ldots, L_n$ by repeated two-list merges while minimizing the worst-case number of comparisons involved.

Figure 3 ▶

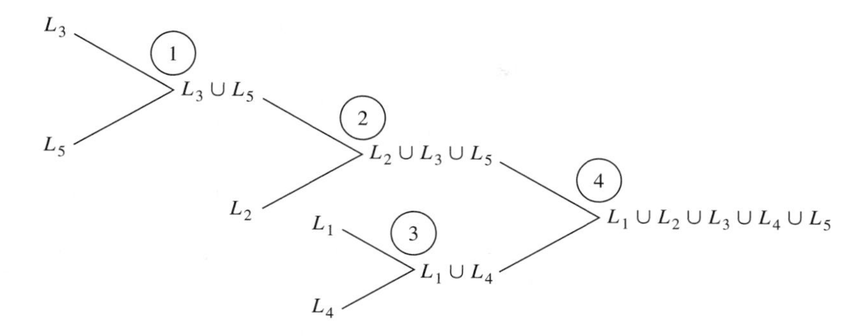

Suppose, for example, that we have five lists L_1, L_2, L_3, L_4, and L_5 with 15, 22, 31, 34, and 42 items and suppose that they are merged as indicated in Figure 3. There are four merges, indicated by the circled numbers. The first merge involves at most $|L_3| + |L_5| - 1 = 72$ comparisons. The second merge involves at most $|L_2| + |L_3| + |L_5| - 1 = 94$ comparisons. The third and fourth merges involve at most $|L_1| + |L_4| - 1 = 48$ and $|L_1| + |L_2| + |L_3| + |L_4| + |L_5| - 1 = 143$ comparisons. The entire process involves at most 357 comparisons. This number isn't very illuminating by itself, but note that

$$357 = 2 \cdot |L_1| + 2 \cdot |L_2| + 3 \cdot |L_3| + 2 \cdot |L_4| + 3 \cdot |L_5| - 4.$$

This is just 4 less than the weight of the tree in Figure 4. Note the intimate connection between Figures 3 and 4. No matter how we merge the five lists in pairs, there will be four merges. A computation like the one above shows that the merge will involve at most $W(T) - 4$ comparisons, where T is the tree corresponding to the merge. So finding a merge that minimizes the worst-case number of comparisons is equivalent to finding a binary tree with weights 15, 22, 31, 34, and 42 having minimal weight. We return to this problem in Example 8.

Figure 4 ▶

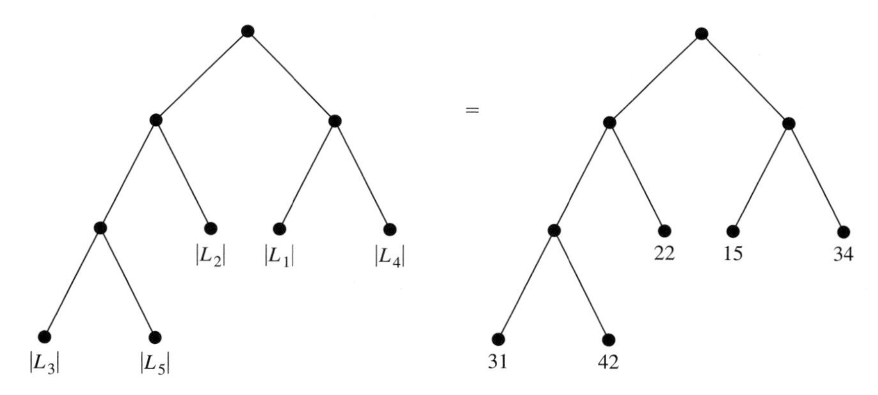

The merging of n lists in pairs involves $n - 1$ merges. In general, a merge of n lists will involve at most $W(T) - (n - 1)$ comparisons, where T is the weighted tree corresponding to the merge. ■

Examples 2 and 3 suggest the following general problem. We are given a list $L = (w_1, \ldots, w_t)$ of at least two nonnegative numbers, and we want to construct a binary weighted tree T with the members of L as weights so that $W(T)$ is as small as possible. We call such a tree T an **optimal binary tree** for the weights

$w_1, \ldots, w_t$. The following recursive algorithm solves the problem by producing an optimal binary tree $T(L)$.

Huffman(list)

```
{Input: A list L = (w_1, ..., w_t) of nonnegative numbers, t ≥ 2}
{Output: An optimal binary tree T(L) for L}
if t = 2 then
   Let T(L) be a tree with 2 leaves, of weights w_1 and w_2.
else
   Find the two smallest members of L, say u and v.
   Let L' be the list obtained from L by removing u and v and
      inserting u + v.
   Recur to L' and form T(L) from Huffman(L') by replacing a
      leaf of weight u + v in Huffman(L') by a subtree with two leaves,
      of weights u and v.
return T(L)  ■
```

This algorithm ultimately reduces the problem to the base case of finding optimal binary trees with two leaves, which is trivial to solve. We will show shortly that Huffman's algorithm always produces an optimal binary tree. First we look at some examples of how it works, and we apply the algorithm to the problems that originally motivated our looking at optimal trees.

EXAMPLE 4

Consider weights 2, 4, 6, 7, 7, and 9. First the algorithm repeatedly combines the smallest two weights to obtain shorter and shorter weight sequences. Here is the recursive chain:

Huffman(**2**, **4**, 6, 7, 7, 9) replaces 2 and 4 by 2 + 4 and calls for
  Huffman(**6**, **6**, 7, 7, 9) which replaces 6 and 6 by 12 and calls for
    Huffman(**7**, **7**, 9, 12) which calls for
      Huffman(**9**, **12**, 14) which calls for
        Huffman(14, 21) which builds the first tree in Figure 5.

Figure 5 ►

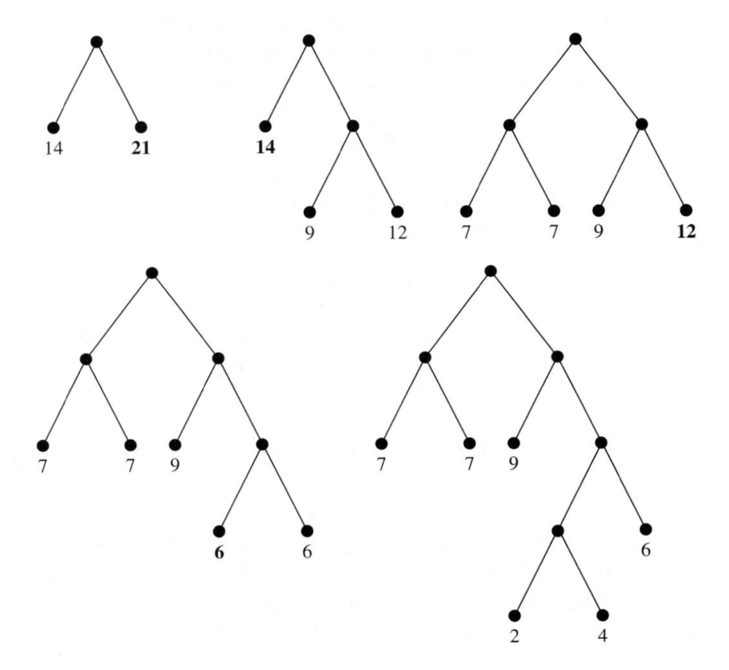

Now each of the previous recursive calls constructs its tree and passes it back up the calling chain. Figure 5 shows the full sequence of trees Huffman(14, 21),

Huffman(9, 12, 14), ..., Huffman(2, 4, 6, 7, 7, 9). For example, we get the third tree Huffman(7, 7, 9, 12) from the second tree Huffman(9, 12, 14) by replacing the leaf of weight **14** = 7 + 7 with a subtree with two leaves of weight 7 each. The final weighted tree is essentially the tree drawn in Figure 1(c), which must be an optimal binary tree. As noted in Example 1(c), it has weight 88. ■

EXAMPLE 5

Let's find an optimal binary tree with weights 2, 3, 5, 7, 10, 13, and 19. We repeatedly combine the smallest two weights to obtain the weight sequences

$$\mathbf{2}, \mathbf{3}, 5, 7, 10, 13, 19 \to \mathbf{5}, \mathbf{5}, 7, 10, 13, 19$$

$$\to \mathbf{7}, \mathbf{10}, 10, 13, 19 \to \mathbf{10}, \mathbf{13}, 17, 19 \to \mathbf{17}, \mathbf{19}, 23 \to 23, 36.$$

Then we use Huffman's algorithm to build the optimal binary trees in Figure 6. After the fourth tree is obtained, either leaf of weight 10 could have been replaced by the subtree with weights 5 and 5. Thus the last two trees also could have been as drawn in Figure 7. Either way, the final tree has weight 150 [Exercise 2]. Note that the optimal tree is by no means unique; the one in Figure 6 has height 4, while the one in Figure 7 has height 5. ■

Figure 6 ▶

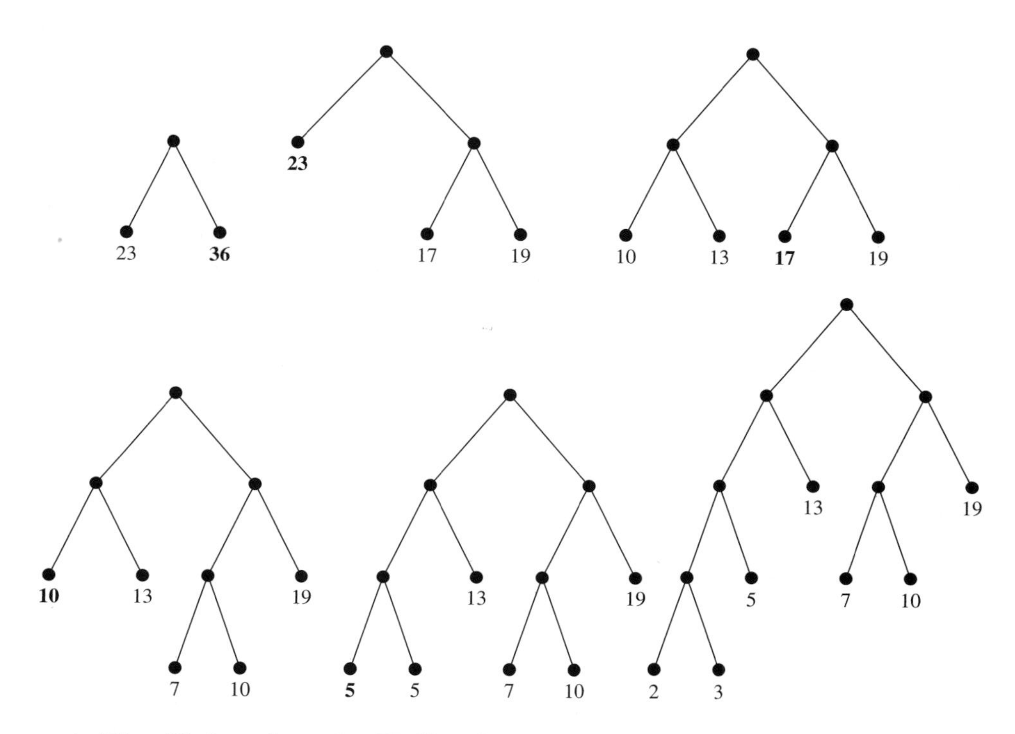

We will show later that Huffman's algorithm always works. First, though, we look at how it applies to our motivating examples, starting with codes using binary numbers. To avoid ambiguity, we must impose some restrictions; for example, a code should not use all three of 10, 01, and 0110, since a string such as 011010 would be ambiguous. If we think of strings of 0's and 1's as labels for vertices in a binary tree, then each label gives instructions for finding its vertex from the root: go left on 0, right on 1. Motivated in part by the search tree in Example 2(b), we impose the condition that no code symbol can be the initial string of another code symbol. Thus, for instance, 01 and 0110 cannot both be code symbols. In terms of the binary tree, this condition means that no vertex labeled with a symbol can lie below another such vertex in the tree. We also impose one more condition on the code: The binary tree whose leaves are labeled by the code words must be regular so that every nonleaf has two children. A code meeting these two conditions is called a **prefix code**.

Figure 7 ▶

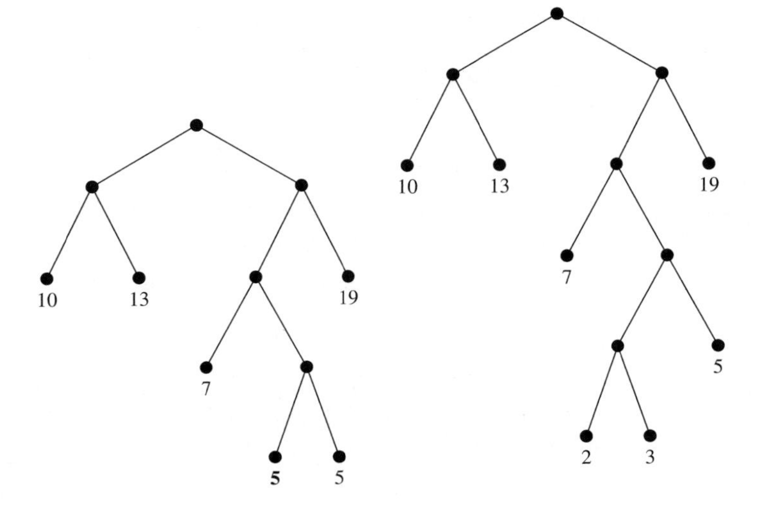

EXAMPLE 6

The set {00, 01, 100, 1010, 1011, 11} is a prefix code. It is the set of leaves for the labeled binary tree in Figure 2(a), which we have redrawn in Figure 8(a) with all vertices but the root labeled. Every string of 0's and 1's of length 4 begins with one of these code symbols, since every path of length 4 from the root in the full binary tree in Figure 8(b) runs into one of the code vertices. This means that we can attempt to decode any string of 0's and 1's by proceeding from left to right in the string, finding the first substring that is a code symbol, then the next substring after that, etc. This procedure either uses up the whole string or it leaves at most three 0's and 1's undecoded at the end.

Figure 8 ▶

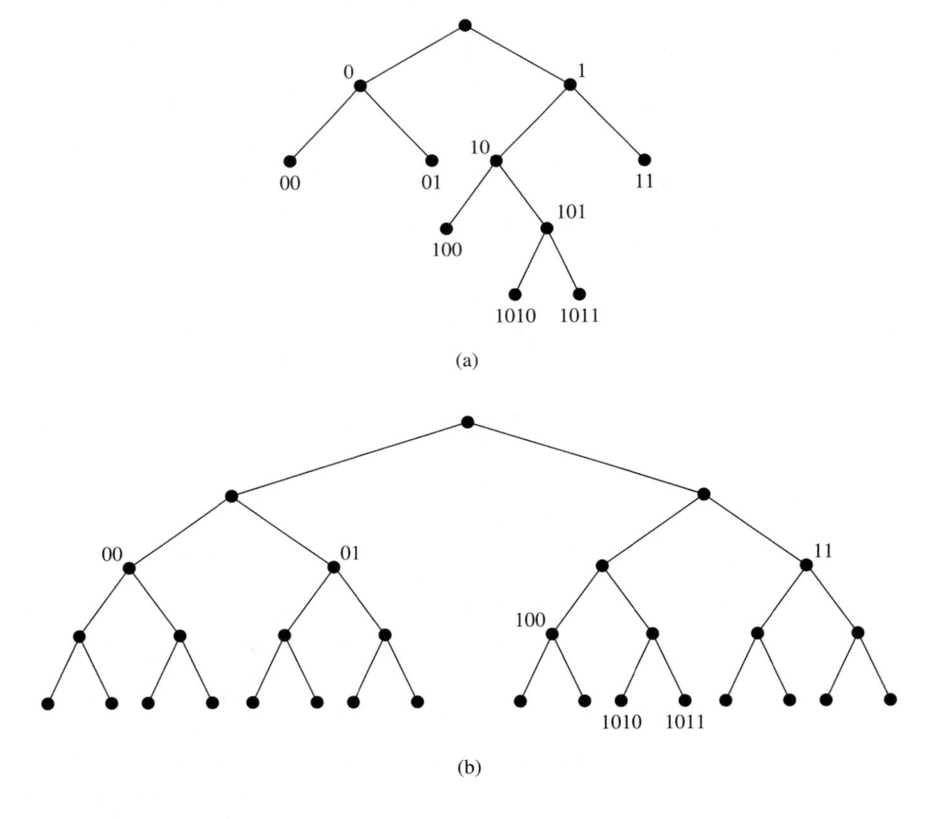

For example, consider the string

$$1\,1\,1\,0\,1\,0\,1\,1\,0\,1\,1\,0\,0\,0\,1\,0\,0\,1\,1\,1\,1\,1\,0\,0\,1\,0.$$

We visit vertex 1, then vertex 11. Since vertex 11 is a leaf, we record 11 and return to the root. We next visit vertices 1, 10, 101, and 1010. Since 1010 is a leaf, we record 1010 and return again to the root. Proceeding in this way, we obtain the sequence of code symbols

$$11, \quad 1010, \quad 11, \quad 01, \quad 100, \quad 01, \quad 00, \quad 11, \quad 11, \quad 100 \quad STOP$$

and have 10 left over. This scheme for decoding arbitrary strings of 0's and 1's will work for any code with the property that every path from the root in a full binary tree runs into a unique code vertex. Prefix codes have this property by their definition. ■

EXAMPLE 7

(a) We now solve the problem suggested by Example 2(a); that is, we find a prefix code for the set of frequencies 10, 10, 15, 20, 20, and 25 that is as efficient as possible. We want to minimize the average length of a code message using 100 letters from Σ. Thus all we need is an optimal binary tree for these weights. Using the procedure illustrated in Examples 4 and 5, we obtain the weighted tree in Figure 9(a). We label this tree with binary digits in Figure 9(b). Then {00, 01, 10, 110, 1110, 1111} will be a most efficient code for Σ, provided we match the letters of Σ to code symbols so that the frequencies of the letters are given by Figure 9(a). With this code, the average length of a code message using 100 letters from Σ is

$$20 \cdot 2 + 20 \cdot 2 + 25 \cdot 2 + 15 \cdot 3 + 10 \cdot 4 + 10 \cdot 4 = 255,$$

an improvement over the average length 280 obtained in Example 2.

(b) The solution we have just obtained also gives a most efficient binary search tree for looking up records whose lookup frequencies are 10, 10, 15, 20, 20, and 25 as in Example 2(b). ■

Figure 9 ▶

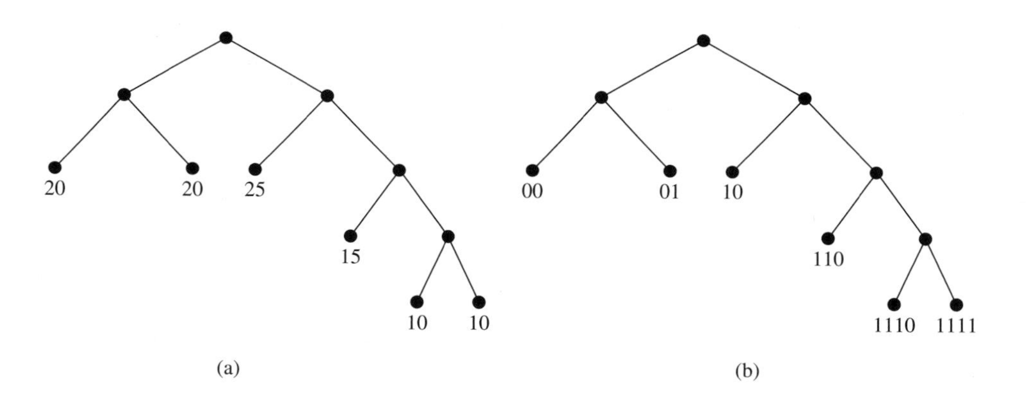

EXAMPLE 8

We complete the discussion on sorted lists begun in Example 3. There we saw that we needed an optimal binary tree with weights 15, 22, 31, 34, and 42. Using the procedure in Examples 4 and 5, we obtain the tree in Figure 10. This tree has weight 325. The corresponding merge in pairs given in Figure 11 will require at most $325 - 4 = 321$ comparisons. ■

To show that Huffman's algorithm works, we first prove a lemma that tells us that in optimal binary trees the heavy leaves are near the root. The lemma and its corollary are quite straightforward if all the weights are distinct [Exercise 14]. However, we need the more general case. Even if we begin with distinct weights, Huffman's algorithm may lead to the case where the weights are not all distinct, as occurred in Example 5.

Figure 10 ▶

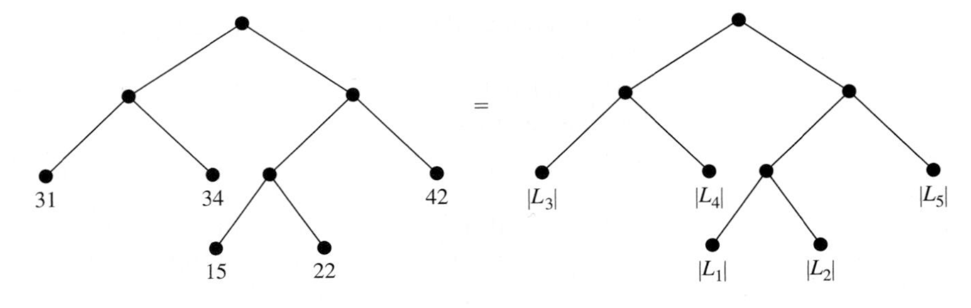

Figure 11 ▶

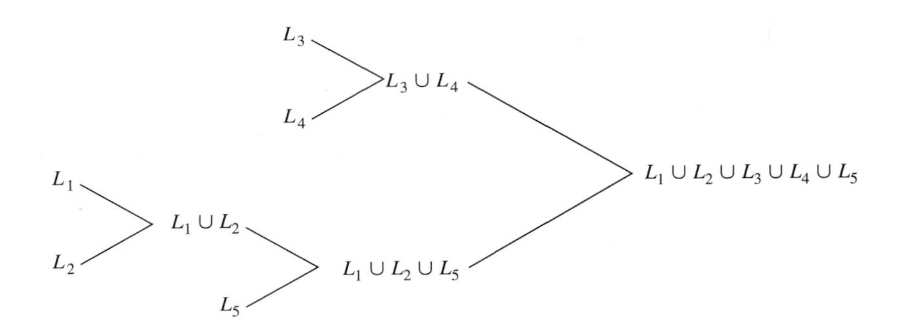

Lemma Let T be an optimal binary tree for weights $w_1, w_2, \ldots, w_t$. For $i = 1, 2, \ldots, t$, let l_i denote the level of w_i. If $w_j < w_k$, then $l_j \geq l_k$.

Proof Assume that $w_j < w_k$ and $l_j < l_k$ for some j and k. Let T' be the tree obtained by interchanging the weights w_j and w_k. In calculating $W(T)$, the leaves w_j and w_k contribute $w_j l_j + w_k l_k$, while in calculating $W(T')$, they contribute $w_j l_k + w_k l_j$. Since the other leaves contribute the same to both $W(T)$ and $W(T')$, we have

$$W(T) - W(T') = w_j l_j + w_k l_k - w_j l_k - w_k l_j = (w_k - w_j)(l_k - l_j) > 0.$$

Hence $W(T') < W(T)$ and T is not an optimal binary tree, contrary to our hypothesis. ■

Corollary Suppose that $0 \leq w_1 \leq w_2 \leq \cdots \leq w_t$. There is an optimal binary tree T for which the two smallest weights w_1 and w_2 are both at the lowest level l, i.e., farthest from the root.

Proof There are at least two leaves at the lowest level, say w_j and w_k. If $w_1 < w_j$, then $l_1 \geq l_j = l$ by the lemma, so $l_1 = l$ and w_1 is at level l. If $w_1 = w_j$, then conceivably $l_1 < l_j$, but we can interchange w_1 and w_j without changing the total weight of T. Similarly, by interchanging w_2 and w_k if necessary, we may suppose that w_2 also is at level l. ■

The following result shows that Huffman's algorithm works.

Theorem Suppose that $0 \leq w_1 \leq w_2 \leq \cdots \leq w_t$. Let T' be an optimal binary tree with weights $w_1+w_2, w_3, \ldots, w_t$, and let T be the weighted binary tree obtained from T' by replacing a leaf of weight $w_1 + w_2$ by a subtree with two leaves having weights w_1 and w_2. Then T is an optimal binary tree with weights $w_1, w_2, \ldots, w_t$.

Proof Since there are only finitely many binary trees with t leaves, there must be an optimal binary tree T_0 with weights $w_1, w_2, \ldots, w_t$. Our task is to show that $W(T) = W(T_0)$. By the corollary of the lemma, we may suppose that the weights w_1 and w_2 for T_0 are at the same level. The total weight of T_0 won't change if weights at the same level are interchanged. Thus we may assume that w_1 and w_2

are children of the same parent p. These three vertices form a little subtree T_p with p as root.

Now let T_0' be the tree with weights $w_1 + w_2, w_3, \dots, w_t$ obtained from T_0 by replacing the subtree T_p by a leaf $\overline{p}$ of weight $w_1 + w_2$. Let l be the level of the vertex p, and observe that in calculating $W(T_0)$ the subtree T_p contributes $w_1(l+1) + w_2(l+1)$, while in calculating $W(T_0')$ the vertex $\overline{p}$ with weight $w_1 + w_2$ contributes $(w_1 + w_2)l$. Thus

$$W(T_0) = W(T_0') + w_1 + w_2.$$

The same argument shows that

$$W(T) = W(T') + w_1 + w_2.$$

Since T' is optimal for the weights $w_1 + w_2, w_3, \dots, w_t$, we have $W(T') \leq W(T_0')$, so

$$W(T) = W(T') + w_1 + w_2 \leq W(T_0') + w_1 + w_2 = W(T_0).$$

Of course, $W(T_0) \leq W(T)$, since T_0 is optimal for the weights $w_1, w_2, \dots, w_t$, so $W(T) = W(T_0)$, as desired. That is, T is an optimal binary tree with weights $w_1, w_2, \dots, w_t$. ■

Huffman's algorithm, applied to the list $L = (w_1, w_2, \dots, w_t)$, leads recursively to $t - 1$ choices of parents in $T(L)$. Each choice requires a search for the two smallest members of the current list, which can be done in time $O(t)$ by just running through the list. Thus the total operation of the algorithm takes time $O(t^2)$.

There are at least two ways to speed up the algorithm. It is possible to find the two smallest members in time $O(\log_2 t)$, using a binary tree as a data structure for the list L. Alternatively, there are algorithms that can initially sort L into nondecreasing order in time $O(t \log_2 t)$. Then the smallest elements are simply the first two on the list, and after we remove them we can maintain the nondecreasing order on the new list by inserting their sum at the appropriate place, just as we did in Examples 4 and 5. The correct insertion point can be found in time $O(\log_2 t)$, so this scheme, too, works in time $O(t \log_2 t)$.

Minimum-weight prefix codes, called **Huffman codes**, are of considerable practical interest because of their applications to efficient data transmission or encoding and to the design of search tree data structures, as described in Examples 2(a) and (b). In both of these settings the actual frequencies of message symbols or record lookups are determined by experience and may change over time. In the search tree setting, the number of code symbols may even change as records are added or deleted. The problem of dynamically modifying a Huffman code to reflect changing circumstances is challenging and interesting, but unfortunately beyond the scope of this book.

Exercises 7.5

1. (a) Calculate the weights of all the trees in Figure 5.

(b) Calculate the weights of all the trees in Figure 6.

2. Calculate and compare the weights of the two trees in Figures 6 and 7 with weights 2, 3, 5, 7, 10, 13, and 19.

3. Construct an optimal binary tree for the following sets of weights and compute the weight of the optimal tree.

(a) $\{1, 3, 4, 6, 9, 13\}$

(b) $\{1, 3, 5, 6, 10, 13, 16\}$

(c) $\{2, 4, 5, 8, 13, 15, 18, 25\}$

(d) $\{1, 1, 2, 3, 5, 8, 13, 21, 34\}$

4. Find an optimal binary tree for the weights 10, 10, 15, 20, 20, and 25 and compare your answer with Figure 9(a).

5. Which of the following sets of sequences are prefix codes? If the set is a prefix code, construct a binary tree whose leaves represent this binary code. Otherwise, explain why the set is not a prefix code.

(a) $\{000, 001, 01, 10, 11\}$

(b) $\{00, 01, 110, 101, 0111\}$

(c) $\{00, 0100, 0101, 011, 100, 101, 11\}$

6. Here is a prefix code: {00, 010, 0110, 0111, 10, 11}.

(a) Construct a binary tree whose leaves represent this binary code.

(b) Decode the string

0 0 1 0 0 0 0 1 1 0 0 1 0 0 0 1 0 0 1 1 1 1 1 0 1 1 0

if $00 = A$, $10 = D$, $11 = E$, $010 = H$, $0110 = M$, and 0111 represents the apostrophe '. You will obtain the very short poem titled "Fleas."

(c) Decode 0 1 0 1 1 0 1 1 0 0 0 1 0 1 1 0 1 1 0 1 1 0 1 1 0 0 0 1 0.

(d) Decode the following soap opera. 1 0 0 0 1 0 0 1 0 0 0 1 0 0 1 1 0 0 0 1 0 0 0 0 1 1 0. 0 1 0 1 1 0 1 1 0 0 0 1 0 1 1 0 1 1 0 0 0 0 1 1 0 0 0 1 0. 0 1 1 0 0 0 0 1 1 0 0 0 1 0 1 1 1 0 0 0 1 0 1 0 1 1 0 0 1 0.

7. Suppose we are given an alphabet Σ of seven letters a, b, c, d, e, f and g with the following frequencies per 100 letters: a–11, b–20, c–4, d–22, e–14, f–8, and g–21.

(a) Design an optimal binary prefix code for this alphabet.

(b) What is the average length of a code message using 100 letters from Σ?

8. Repeat Exercise 7 for the frequencies a–25, b–2, c–15, d–10, e–38, f–4, and g–6.

9. (a) Show that the code {000, 001, 10, 110, 111} satisfies all the requirements of a prefix code, except that the corresponding binary tree is not regular.

(b) Show that some strings of binary digits are meaningless for this code.

(c) Show that {00, 01, 10, 110, 111} is a prefix code, and compare its binary tree with that of part (a).

10. Repeat Exercise 7 for the frequencies a–31, d–31, e–12, h–6, and m–20.

11. Let L_1, L_2, L_3, and L_4 be sorted lists having 23, 31, 61 and 73 elements, respectively. How many comparisons at most are needed if the lists are merged as indicated?

(a)

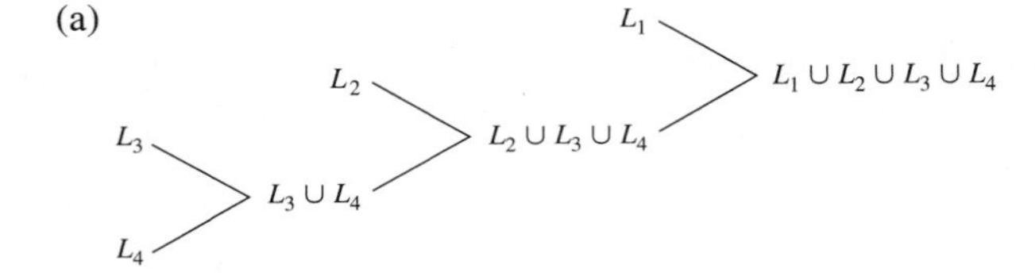

(b)

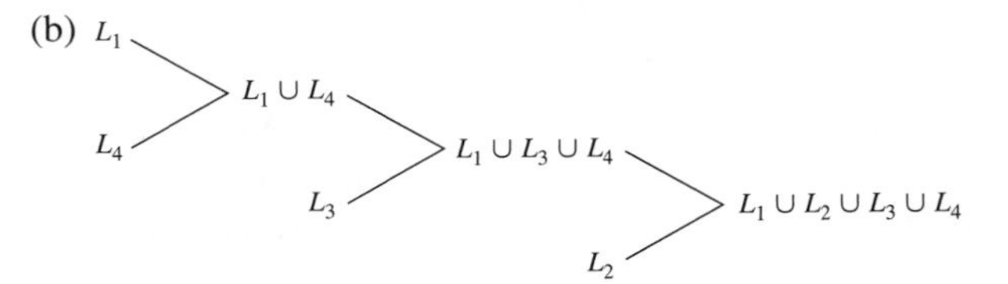

(c)

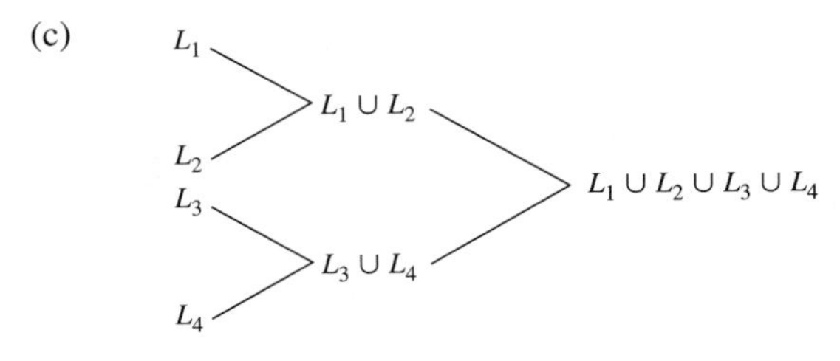

(d)

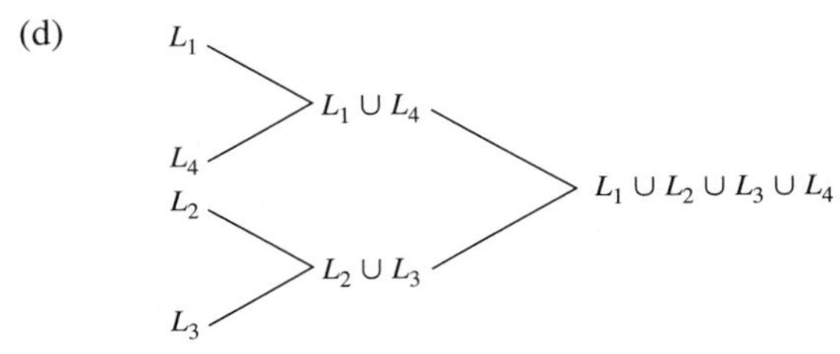

(e) How should the four lists be merged so that the total number of comparisons is a minimum? It is not sufficient simply to examine parts (a)–(d), since there are other ways to merge the lists.

12. Let L_1, L_2, L_3, L_4, L_5, and L_6 be sorted lists having 5, 6, 9, 22, 29, and 34 elements, respectively.

(a) Show how the six lists should be merged so that the total number of comparisons can be kept to a minimum.

(b) How many comparisons might be needed in your procedure?

13. Repeat Exercise 12 for seven lists having 2, 5, 8, 12, 16, 22, and 24 elements, respectively.

14. Let T be an optimal binary tree whose weights satisfy $w_1 < w_2 < \cdots < w_t$. Show that the corresponding level numbers satisfy $l_1 \geq l_2 \geq l_3 \geq \cdots \geq l_t$.

15. Look at Exercise 1 again, and note that, whenever a vertex of weight $w_1 + w_2$ in a tree T' is replaced by a subtree with weights w_1 and w_2, then the weight increases by $w_1 + w_2$. That is, the new tree T has weight $W(T') + w_1 + w_2$, just as in the proof of the theorem.

Chapter Highlights

For suggestions on how to use this material, see the highlights at the end of Chapter 1.

CONCEPTS AND NOTATION

recursive definition of a set, of a function
  basis, recursive clause
  uniquely determined
wff for algebra, for reverse Polish notation
recursive algorithm

base case
input obtained from other inputs
preorder, postorder, inorder listing
sorted labeling of a rooted tree, of an acyclic digraph
SUCC(v), ACC(v)
Polish, reverse Polish, infix notation
binary, unary operation
weighted tree [weights at leaves], weight of a tree
optimal binary tree
merge of lists
prefix code
Huffman code = minimum-weight prefix code

FACTS

The classes of finite trees and finite rooted trees can be defined recursively.

Recursive algorithms can test membership in sets defined recursively and compute values of functions defined recursively.

Generalized Principle of Induction

Verification Conditions for Recursive Algorithms

An ordered rooted tree cannot always be recovered from its preorder, inorder, or postorder listing.

But it *can* be recovered from its preorder or postorder listing given knowledge of how many children each vertex has.

METHODS AND ALGORITHMS

Depth-first search to traverse a rooted tree.

Preorder, Postorder and Inorder to list vertices of an ordered rooted tree.

Method of charges for estimating running time of an algorithm.

LabelTree to give a sorted labeling of a rooted tree.

TreeSort and Label to give a sorted labeling of an acyclic digraph in time $O(|V(G)| + |E(G)|)$.

Use of binary weighted trees to determine efficient merging patterns and efficient prefix codes.

Huffman's algorithm to find an optimal binary tree with given weights.

Supplementary Exercises

1. (a) Find the value of the expression $1\ 2\ +\ 3\ 4\ *\ 5\ -\ *$ given in reverse Polish notation.

(b) Draw the tree associated with this expression.

2. Let $\Sigma = \{a, b\}$. Recursively define the subset S of Σ^* by

(B) The empty word λ is in S.

(R) If $w \in S$, then $awb \in S$ and $bwa \in S$.

(a) Which of the following words are in S?

λ, aa, aba, $abba$, $bbaa$, $baba$, $abbb$, $babbaba$

(b) What can you say about the lengths of words in S? Explain.

(c) Is $abbaab$ in S? Explain.

(d) How are the numbers of a's and b's related for words in S. Explain.

3. The sketch shows a rooted tree ordered from left to right.

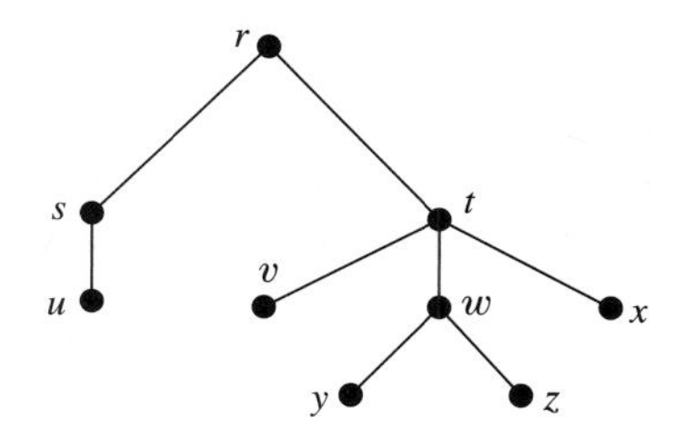

(a) Put labels on the tree as the algorithm LABELTREE would assign them, starting at 1.

(b) Give the preorder listing of the vertices of the tree.

4. (a) Draw an optimal binary tree for the list $L = (3, 3, 4, 5, 6, 8)$.

(b) What is the weight of the optimal tree for this list?

5. Suppose that a formula has reverse Polish notation $2\ 3 + 4\ 7 + * 9\ 5\ 8 * + /$. Write the formula in Polish notation, and evaluate the formula to get a number.

6. Consider the following tree.

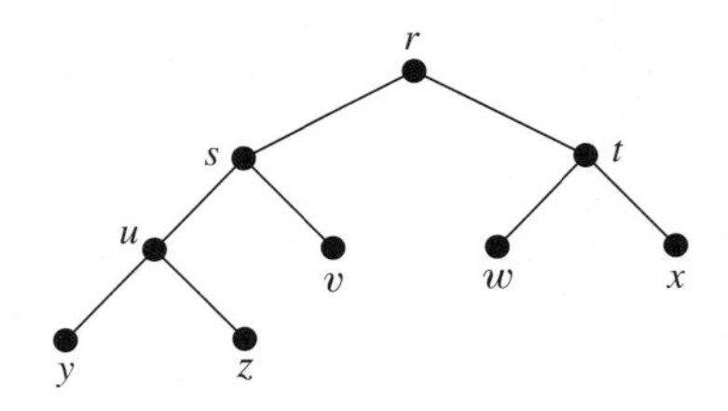

(a) Is this a binary tree? a regular binary tree? a full binary tree?

(b) What is the height of the tree?

(c) How many leaves are at level 2?

(d) Give the postorder listing of the vertices of the tree.

(e) Give the inorder listing of the vertices of the tree.

(f) The rooted tree is a digraph with arrows pointing downward. Label the vertices so that if (a, b) is an edge, then $L(a) < L(b)$.

7. Suppose that we want to encode messages using the letters C, E, L, S, U and Y with frequencies proportional to 7, 31, 20, 24, 12, and 6, respectively, by converting each letter to a binary word.

(a) Draw a tree that shows the most efficient encoding for these letters.

(b) Encode the message $CLUE$ with the efficient encoding.

8. Let L_1, L_2, L_3, L_4, L_5, and L_6 be sorted lists having 3, 5, 7, 8, 14, and 20 elements, respectively.

(a) Show how the lists should be merged so that the total number of comparisons can be kept to a minimum.

(b) How many comparisons might be needed in your procedure?

9. Let $\{a, b\}^*$ be the set of all strings of a's and b's. Recursively define the subset S of $\{a, b\}^*$ by

(B) The empty string and b are in S;

(R) If w and u are strings in S, then awa and wu are in S.

Show that every member of S contains an even number of a's.

10. (a) Give a recursive definition of wff's in Polish notation for algebraic expressions.

(b) Using your definition from part (a), which of the following are wff's?

$$* + * 3\, x\, 1 * 2\, x$$
$$+ - * 3 \wedge x\, 2 * 5\, x\, 4$$
$$- * 3 \wedge x + 2 * 5\, x$$

(c) For each of the expressions in part (b) that is a wff, write the expression in ordinary algebraic notation with parentheses and also in reverse Polish notation.

11. SUMAlgorithm on page 153 uses a **while** loop and an auxiliary variable FACT to compute $\text{SUM}(n) = \sum_{i=0}^{n} \frac{1}{i!}$.

Write a recursive algorithm $\text{SUM}^+(integer)$ that returns the ordered pair

$$(\text{SUM}_1^+(n), \text{SUM}_2^+(n)) = (\text{SUM}(n), n!)$$

for input n in $\mathbb{N}$.

12. Verify that the algorithm Label on page 296 performs as claimed and that Label$(G, \emptyset)$ produces a sorted labeling of a finite acyclic digraph G.

13. Let $\Sigma = \{a, b\}$, and let B be the subset of Σ^* defined recursively as follows:

(B) The empty word λ is in B;

(R) If w is in B, then aw and wb are in B.

(a) List seven members of B.

(b) List three words in Σ^* that are not in B.

(c) Give a simple description of all the words in B.

(d) Is this recursive definition uniquely determined?

14. Suppose that we are given an alphabet Σ of six letters a, b, c, d, e, and f with the following frequences per 100 letters: a–30, b–4, c–20, d–5, e–33, and f–8.

(a) Design an optimal binary prefix code for the alphabet.

(b) What is the average length of a code message using 100 letters from Σ?

(c) Use your code to encode *Dead Beef Cafe*.

15. The following recursive algorithm ALGO accepts as input a positive integer n.

```
ALGO(n)
if n = 1 then
    return 17
else
    return 3*ALGO(n − 1)
```

(a) What value does ALGO return if $n = 4$?

(b) What value does ALGO return for a general positive integer n?

(c) Give a careful, complete verification of your statement in part (b).

(d) How would your answer to part (b) change if $n - 1$ were replaced by $n - 2$ in the recursive call?

16. (a) Consider the postorder listing $L\ P\ M\ N\ R\ Q\ E\ D\ C\ B\ G\ H\ T\ J\ K\ F\ A$. The table below shows the number of children of each entry on the list. Reconstruct the tree.

Vertex	L	P	M	N	R	Q	E	D	C
Children	0	0	0	0	2	0	4	0	1

Vertex	B	G	H	T	J	K	F	A
Children	1	0	0	0	0	3	2	3

(b) A computer program to reconstruct the tree would use a stack. Vertically list the elements of the stack just after the element R has been added to the top. Also vertically list the elements of the stack just after the element G has been added to the top.

17. Let $\Sigma = \{a, b, c\}$, and let B be the subset of Σ^* defined recursively as follows:

(B) The empty word λ is in B;

(R) If w is in B, then awb and cwa are in B.

(a) List six members of B.

(b) Prove carefully that length(w) is even for all $w \in B$.

(c) Does every word of even length belong to B? Explain.

(d) Is this recursive definition uniquely determined?

18. Let $\Sigma = \{a, b\}$, and define a subset S of Σ^* as follows:

(B) The empty word λ is in S.

(R) If $w_1, w_2 \in S$, then aw_1b, bw_1a and w_1w_2 are in S.

For this exercise, let's say a word is "balanced" if it has the same number of a's as b's. Then clearly every word in S is balanced. Show that every balanced word is in S.

19. The recursive algorithm LabelTree puts labels on the vertices of a rooted tree so that parents have larger labels than their children. Can you think of a nonrecursive algorithm to do the same job?

Digraphs

This chapter is devoted to directed graphs, which were introduced in Chapter 3. The first section is primarily concerned with acyclic digraphs. It contains an account of sorted labelings that is independent of the treatment in Chapter 7, and it includes a digraph version of Euler's theorem. The remainder of the chapter deals with weighted digraphs, the primary focus being on paths of smallest weight between pairs of vertices. Section 8.2 introduces weighted digraphs and also contains a discussion of scheduling networks, in which the important paths are the ones of largest weight. Although this chapter contains a few references to sorted labelings and to §7.3, it can be read independently of Chapter 7.

8.1 Digraphs Revisited

We first looked at directed graphs, i.e., digraphs, in Chapter 3, where we linked them to matrices and relations. Later, in §6.4, we pointed out the natural way to turn a rooted tree into a digraph by making its edges all lead away from the root. Although trees will come up again in this chapter, the questions we ask now will have quite a different flavor from those we considered in Chapter 6.

Imagine a large and complicated digraph. One example could be a diagram whose vertices represent possible states a processor can be in, with an edge from state s to state t in case the processor can change from s to t in response to some input. Another example could be a complicated street map, perhaps with some one-way streets, in which vertices correspond to intersections, and we view a two-way street as a pair of edges, one in each direction. We can ask a variety of interesting questions. Are there vertices, for example, processor states or weird dead-end intersections, that have no edges leading away from them? If there are, how can we locate at least one of them? Can we get from every vertex to every other vertex by following directed paths? If not, what vertices *can* we get to from a given v? If there is a cost associated with each edge, how can we find paths of minimum cost? In this chapter we'll explore these and other questions about digraphs and will consider several interesting applications. To begin with, we need to collect some basic facts.

As we saw in the undirected case, cycles in paths are redundant.

Theorem 1 If u and v are different vertices of a digraph G, and if there is a path in G from u to v, then there is an acyclic path from u to v.

The proof is exactly the same as that of Theorem 1 on page 227. The only difference now is that the edges of the paths are directed.

Corollary 1 If there is a closed path from v to v, then there is a cycle from v to v.

Proof If there is an edge e of the graph from v to itself, then the one-element sequence e is a cycle from v to v. Otherwise, there is a closed path from v to v having the form $v\,x_2 \cdots x_n\,v$, where $x_n \neq v$. Then, by Theorem 1, there is an acyclic path from v to x_n. Tacking on the last edge from x_n to v gives the desired cycle. ■

Corollary 2 A path is acyclic if and only if all its vertices are distinct.

Proof If a path has no repeated vertex, then it is surely acyclic. If a path has a repeated vertex, then it contains a closed path, so by Corollary 1 it contains a cycle. ■

Two special kinds of vertices are more interesting for digraphs than they are for undirected graphs. A vertex that is not an initial vertex of any edge, i.e., one with no arrows leading away from it, is called a **sink**. A **source** is a vertex that is not a terminal vertex of any edge. We have encountered sources and sinks before, although we didn't give them these names, when we were looking at sorted labelings in §7.3.

EXAMPLE 1

In the digraph drawn in Figure 1, the vertices v and y are sinks, while t and z are sources. This digraph is acyclic. The next theorem shows that it is not an accident that there is at least one sink and at least one source. ■

Figure 1 ▶

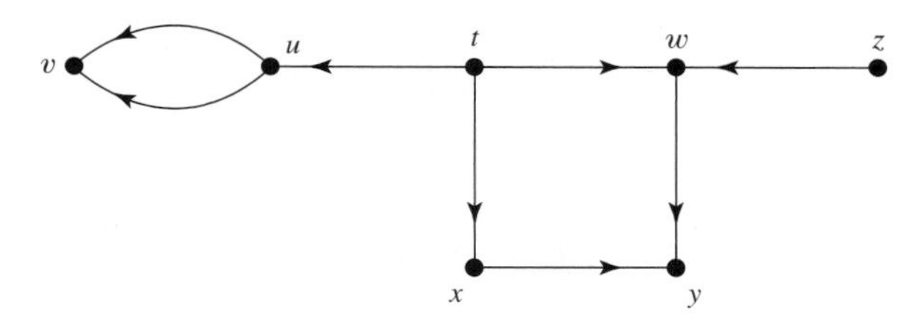

Theorem 2 Every finite acyclic digraph has at least one sink and at least one source.

We give three proofs for sinks; analogous facts about sources can be verified by reversing all the arrows. See Exercise 13.

First Proof Since the digraph is acyclic, every path in it is acyclic. Since the digraph is finite, the path lengths are bounded and there must be a path of largest length, say $v_1\,v_2 \cdots v_n$. Then v_n must be a sink. [Of course, if the digraph has no edges at all, every vertex is a sink.] ■

This proof is short and elegant, but it doesn't tell us how to find v_n or any other sink. Our next argument is constructive.

Second Proof Choose any vertex v_1. If v_1 is a sink, we are done. If not, there is an edge from v_1 to some v_2. If v_2 is a sink, we are done. If not, etc. We obtain in this way a sequence $v_1, v_2, v_3, \ldots$ such that $v_1\,v_2 \cdots v_k$ is a path for each k. As in the first proof, such paths cannot be arbitrarily long, so at some stage we reach a sink. ■

Third Proof Omit this proof unless you have studied §7.3. The algorithm Label on page 296 provides a sorted labeling for a finite acyclic digraph. This means that if there is a path from a vertex labeled i to one labeled j, then $i > j$. In particular, the vertex with the smallest label cannot be the initial vertex of any edge, so it must be a sink. ■

We will give an algorithm, based on the construction in the second proof, that returns a sink when it is applied to a finite acyclic digraph G. The algorithm uses **immediate successor sets** $\text{SUCC}(v)$, defined by $\textbf{succ}(\boldsymbol{v}) = \{u \in V(G) : \text{there is an edge from } v \text{ to } u\}$. These sets also appeared in the algorithm TreeSort on page 293, which is a subroutine for the algorithm Label mentioned in the third proof above. Note that a vertex v is a sink if and only if $\text{SUCC}(v)$ is the empty set. These data sets $\text{SUCC}(v)$ would be supplied in the description of the digraph G when the algorithm is carried out.

Sink(digraph)

```
{Input: A finite acyclic digraph G}
{Output: A sink S of G}
Choose a vertex v in V(G).
Let S := v.
while SUCC(v) ≠ Ø do
    Choose u in SUCC(v).
    Let S := u.
    Let v := u.
return S ■
```

EXAMPLE 2

(a) Consider the acyclic digraph G shown in Figure 1. The immediate successor sets are $\text{SUCC}(t) = \{u, w, x\}$, $\text{SUCC}(u) = \{v\}$, $\text{SUCC}(v) = \emptyset$, $\text{SUCC}(w) = \{y\}$, $\text{SUCC}(x) = \{y\}$, $\text{SUCC}(y) = \emptyset$, and $\text{SUCC}(z) = \{w\}$. One possible sequence of choices using the algorithm Sink on G is t, w, y. Others starting with t are t, x, y and t, u, v. A different first choice could lead to z, w, y. We could even get lucky and choose a sink the first time. In any case Sink(G) is either v or y.

(b) There is only one sink, v_n, in the digraph sketched in Figure 2. The immediate successor sets are $\text{SUCC}(v_k) = \{v_{k+1}\}$ for $1 \leq k \leq n - 1$. If algorithm Sink started with vertex v_1, it would take $n - 1$ passes through the **while** loop before reaching the sink v_n. ■

Figure 2 ▶

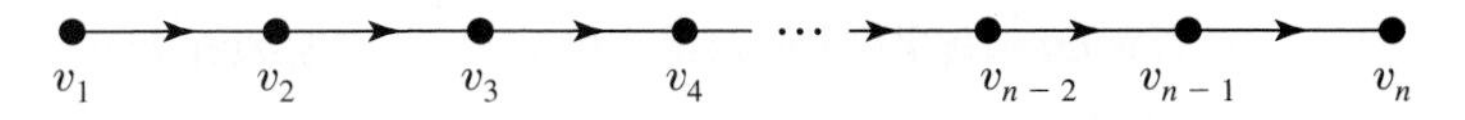

The time Sink takes is proportional to the number of vertices it chooses before it gets to a sink, so for a digraph with n vertices the algorithm runs in time $O(n)$.

If a digraph with n vertices can have its vertices numbered from 1 to n in such a way that $i > j$ whenever there is a path from vertex i to vertex j, then such a labeling of its vertices is called a **sorted labeling**. The algorithm Label on page 296 gives a sorted labeling to any acyclic digraph. We now give an independent proof that such a labeling is possible using sinks. This proof will lead to another algorithm.

Theorem 3 Every finite acyclic digraph has a sorted labeling.

Proof We use induction on the number n of vertices and note that the assertion is obvious for $n = 1$. Assume inductively that acyclic digraphs with fewer than n vertices can be labeled as described, and consider an acyclic digraph G with n vertices. By Theorem 2, G has a sink, say s. Give s the number 1. Form a new graph H with $V(H) = V(G) \setminus \{s\}$ and with all the edges of G that do not have s as a vertex. Since G has no cycles, H has no cycles. Since H has only $n - 1$ vertices, it has a sorted labeling by the inductive assumption. Let's increase the value of each label by 1 so that the vertices of H are numbered $2, 3, \ldots, n$. The vertex s is numbered 1, so every vertex in $V(G)$ has a number between 1 and n.

Now suppose there is a path in G from vertex i to vertex j. If the path lies entirely in H, then $i > j$, since H is properly numbered. Otherwise, some vertex along the path is s, and since s is a sink, it must be the last vertex, vertex j. But then $j = 1$, so $i > j$ in this case too. Hence G, with n vertices, has a sorted labeling. The Principle of Mathematical Induction now shows that the theorem holds for all n. ■

The idea in the proof of Theorem 3 can be developed into a procedure for constructing sorted labelings for acyclic digraphs.

NumberVertices(digraph)

```
{Input: A finite acyclic digraph G with n vertices}
{Gives V(G) a sorted labeling}
Let V := V(G) and E := E(G).
while V ≠ Ø do
   Let H be the digraph with vertex set V and edge set E.
   Apply Sink to H        {to get a sink of H}.
   Label Sink(H) with n − |V| + 1.
   Remove Sink(H) from V and all edges attached to it from E.
return ■
```

Each pass through the **while** loop removes from further consideration one vertex from the original set of vertices $V(G)$, so the algorithm must stop, and when it stops each vertex is labeled. The labels used are $1, 2, \ldots, n$. As written, the algorithm calls Sink as a subroutine. You may find it instructive to apply this algorithm to number the graph of Figure 1. Also see Exercise 14 for a procedure that begins numbering with n.

If we estimate the running time of NumberVertices, assuming that it calls Sink to find sinks as needed, the best we can do is $O(n^2)$, where n is the number of vertices of G. This algorithm is inefficient because the Sink subroutine may examine the same vertex over and over. Algorithm Label on page 296 is more efficient, although harder to execute by hand.

EXAMPLE 3

Consider again the digraph in Figure 2. The algorithm NumberVertices might begin with vertex v_1 every time it calls Sink. It would find the sink v_n the first time, label it and remove it. The next time it would find v_{n-1} and label and remove it. And so on. The number of vertices that it would examine is

$$n + (n-1) + \cdots + 2 + 1 = \frac{1}{2}n(n+1) > \frac{1}{2}n^2.$$

In contrast, suppose that algorithm Label began with vertex v_1. Instead of starting over each time, once it reached v_n it would backtrack to v_{n-1}, label it, backtrack to v_{n-2}, label it, etc., for a total of $2n$ vertex examinations. ■

The second proof of Theorem 2 consisted of constructing a path from a given vertex v to a sink. We call a vertex u **reachable from** v in G if there is a path [of length at least 1] in G from v to u, and we define

$$\boldsymbol{R(v)} = \{u \in V(G) : u \text{ is reachable from } v\}.$$

Then $R(v) = \emptyset$ if and only if v is a sink, and the theorem's second proof showed that each nonempty set $R(v)$ in an acyclic digraph contains at least one sink. The notion of reachability has already appeared in a slightly different context. A pair (v, w) of vertices belongs to the reachable relation, as defined in §3.2, if and only if the vertex w is reachable from v. Also, in §7.3 we used the set $\text{ACC}(v)$ of vertices accessible from v, i.e., the set $R(v)$ together with the vertex v itself.

EXAMPLE 4

(a) For the digraph in Figure 2, we have

$$R(v_k) = \{v_j : k < j \leq n\} \qquad \text{for} \quad k = 1, 2, \ldots, n.$$

Equality holds here for $k = n$, since both sets are empty in that case.

(b) For the digraph in Figure 1, we have $R(t) = \{u, v, w, x, y\}$, $R(u) = \{v\}$, $R(v) = \emptyset$, $R(w) = \{y\}$, $R(x) = \{y\}$, $R(y) = \emptyset$, and $R(z) = \{w, y\}$. ■

Even if G is not acyclic, the sets $R(v)$ may be important. As we shall see in §11.5, determining all sets $R(v)$ amounts to finding the transitive closure of a certain relation. In §8.3 we will study algorithms that can find the $R(v)$'s as well as answer other graph-theoretic questions.

Euler's theorem on page 234 has a digraph version. Since a digraph can be viewed as a graph, the notion of **degree** of a vertex still makes sense. To take into account which way the edges are directed, we refine the idea. For a vertex v of a digraph G, the **indegree** of v is the number of edges of G with v as terminal vertex, and the **outdegree** of v is the number with v as initial vertex. With obvious notation, we have

$$\text{indeg}(v) + \text{outdeg}(v) = \deg(v)$$

for all vertices v. A loop gets counted twice here, once going out and once coming in.

EXAMPLE 5

(a) Consider again the acyclic digraph in Figure 1. The sinks v and y have outdegree 0, while the sources t and z have indegree 0. Other in- and outdegrees are easy to read off. For example, $\text{indeg}(u) = 1$ and $\text{outdeg}(u) = 2$. The sum of all indegrees is just the number of edges 8, which is also the sum of the outdegrees.

(b) The digraphs in Figure 3 have the special property that at each vertex the indegree and the outdegree are equal, which is the digraph analog of having even undirected degree at every vertex. ■

Figure 3 ▶

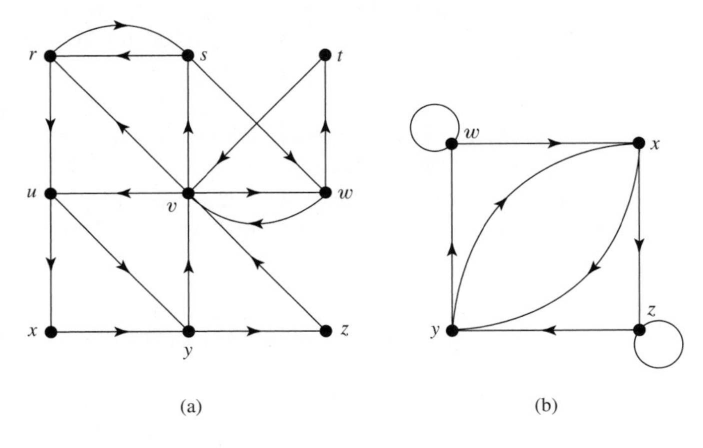

The Euler story for digraphs is very like the one for undirected graphs as presented in §6.2.

Theorem 4 (Euler's Theorem for Digraphs) Suppose G is a finite digraph that is connected when viewed as a graph. There is a closed [directed] path in G using all the edges of G if and only if $\text{indeg}(v) = \text{outdeg}(v)$ for every vertex v.

Proof Suppose that such a closed path exists. We start at some vertex and follow the path, erasing each edge in order. As we go through a vertex, we erase one edge going in and one going out, or else we erase a loop. Either way, the erasure reduces both the indegree and the outdegree by 1. Eventually, all edges are erased, so all indegrees and outdegrees are equal to 0. Hence, at each vertex the indegree and outdegree must have been equal at the start.

As in the graph case, the converse implication is harder to prove. Suppose that the indegree and outdegree are equal at each vertex. It is easy to modify the algorithms EulerCircuit and ClosedPath on page 237 to work for digraphs and so prove the implication constructively. We can also give a less constructive proof, which highlights where the connectedness of G comes in. Let P be a simple path in G of greatest possible length, say from u to v, and follow it, starting at u and throwing away edges as we take them. When P goes through any vertex w other than u, it reduces both its indegree and outdegree by 1, so $\text{indeg}(w)$ and $\text{outdeg}(w)$ remain equal. If P ever comes back to w again, it must be that $\text{indeg}(w) > 0$, so $\text{outdeg}(w) > 0$, and there is an edge available for P to take to leave w. Thus P cannot end at such a w, so v must be u and P must be closed.

We next claim that P visits every vertex. If not, then, since G is connected as an undirected graph, there is an *undirected* path from some unvisited vertex to a vertex on P. Some edge, say e, on this path joins an unvisited vertex v and a vertex w on P. [Draw a sketch.] Depending on the direction of e, we can either go from v to w on e and then back around on P to w or else go around P from w to w and then take e to v. In either case, we get a simple path longer than P, which is impossible.

Finally, we claim that P uses every edge of G. If e were an edge not in P, then, since both of its endpoints must lie on P, we could attach it at one end to P as in the last paragraph to get a simple path longer than P, again a contradiction. ■

Note that if a finite digraph is not connected but satisfies the degree conditions of Euler's theorem, the theorem still applies to its connected components.

EXAMPLE 6

(a) Each digraph in Figure 3 has a closed path using all of its edges [Exercise 9].

(b) A **de Bruijn sequence of order n** is a circular arrangement of 0's and 1's so that each string of length n appears exactly once as n consecutive digits. Figure 4(a) is such an arrangement for $n = 4$. Since there are 2^n such strings and each digit in the sequence is the first digit of one of them, the de Bruijn sequence has length 2^n.

Figure 4 ▶

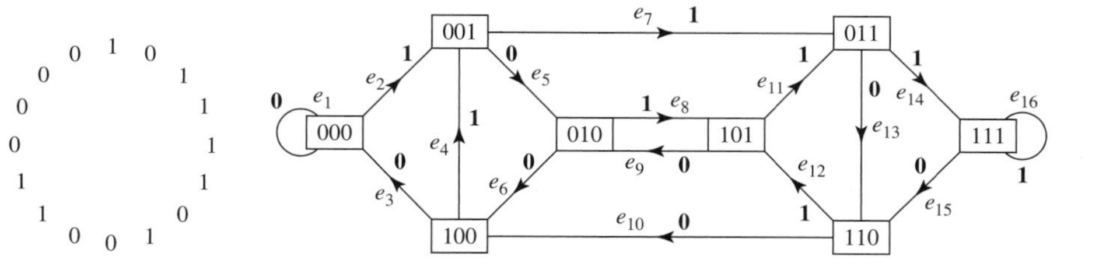

We obtained the arrangement in Figure 4(a) using the digraph in Figure 4(b), which needs some explanation. The set of $2^{n-1} = 8$ vertices of the digraph consists of all the strings of 0's and 1's of length $n - 1 = 3$. A directed edge connects two such strings provided the last two digits of the initial vertex agree with the first two digits of the terminal vertex. We label each edge with the last digit of the terminal vertex. Thus edges e_1, e_3, e_5, e_6, etc., are labeled 0, and e_2, e_4, etc., are labeled 1. Put differently, the label of the edge of a path of length 1 gives the last digit of the terminal vertex of the path. It follows that the labels of the two edges of a path of length 2 give the last two digits of the terminal vertex of the path in the same order. Similarly, the labels of the edges of a path of length 3 give all the digits of the terminal vertex in the same order.

Now observe that each vertex of the digraph has indegree 2 and outdegree 2. Since the digraph is connected, Euler's theorem guarantees that there is a closed path that uses each edge exactly once. We claim that the labels of these $2^n = 16$ edges provide a de Bruijn sequence of order $n = 4$. For example, the labels of the edges of the closed path

$$e_2, e_5, e_8, e_{11}, e_{14}, e_{16}, e_{15}, e_{12}, e_9, e_6, e_4, e_7, e_{13}, e_{10}, e_3, e_1$$

give the circular array in Figure 4(a) starting at the top and proceeding clockwise. Since there are only 16 different consecutive sequences of digits in the circular array, it suffices to show that any sequence $d_1\,d_2\,d_3\,d_4$ of 0's and 1's does appear in the circular arrangement. Since $d_1\,d_2\,d_3$ is the initial vertex for an edge labeled 0 and also for an edge labeled 1, some edge in the path has initial vertex $d_1\,d_2\,d_3$ and is labeled d_4. As noted in the last paragraph, the three edges preceding it are labeled d_1, d_2, and d_3, in that order. So the labels of the four consecutive edges are d_1, d_2, d_3, and d_4, as claimed.

Persi Diaconis has shown us a clever card trick based on a de Bruijn sequence of length 32. After finding out the colors of the cards five audience members hold, the performer is magically able to tell them what rank and suits their cards are. Do you see how that might be done? [Hint: He is not playing with a full 52-card deck.] ■

Exercises 8.1

1. Find the sinks and sources for the digraph in Figure 5.

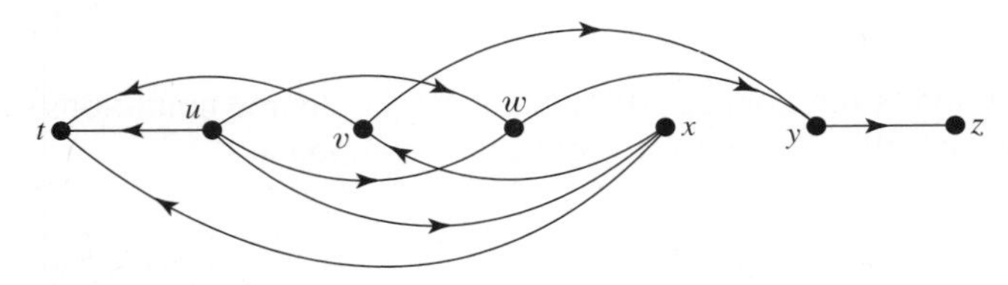

Figure 5 ▲

2. (a) Give the immediate successor sets SUCC(v) for all vertices in the digraph shown in Figure 5.
   (b) What is Sink (G) for an initial choice of vertex w?
   (c) What sinks of G are in $R(x)$?
3. Consider the digraph G pictured in Figure 6.
   (a) Find $R(v)$ for each vertex v in $V(G)$.

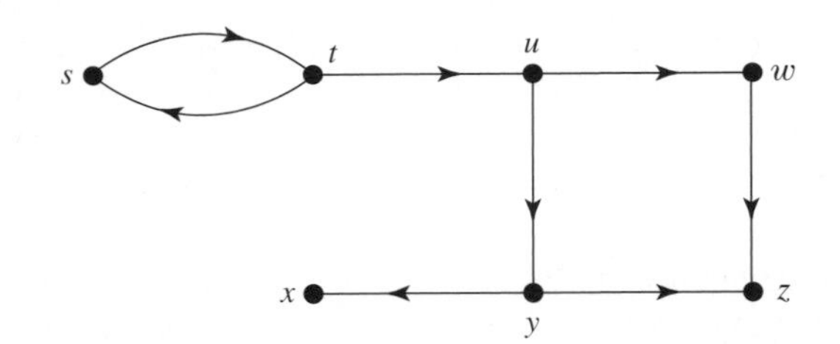

Figure 6 ▲

   (b) Find all sinks of G.
   (c) Is G acyclic?
4. Does the algorithm Sink work on digraphs that have cycles? Explain.
5. Consider a digraph G with the following immediate successor sets: SUCC(r) $= \{s, u\}$, SUCC(s) $= \emptyset$, SUCC(t) $= \{r, w\}$, SUCC(u) $= \emptyset$, SUCC(w) $= \{r, t, x, y\}$, SUCC(x) $= \emptyset$, SUCC(y) $= \{w, z\}$, and SUCC(z) $= \emptyset$.
   (a) Draw a picture of such a digraph.
   (b) Do these sets SUCC(v) determine $E(G)$ uniquely? Explain.
   (c) Find all sinks in G.
   (d) Find paths from the vertex w to three different sinks in the digraph.
6. Give two sorted labelings for the acyclic digraph in Figure 1.
7. Give two sorted labelings for the acyclic digraph in Figure 5.
8. A **tournament** is a digraph in which every two vertices have exactly one edge between them. [Think of (x, y) as an edge provided that x defeats y.]
   (a) Give an example of a tournament with four vertices.
   (b) Show that a tournament cannot have two sinks.
   (c) Can a tournament with a cycle have a sink? Explain.
   (d) Would you like to be the sink of a tournament?
9. For each digraph in Figure 3, give a closed path using all of its edges.
10. Use the digraph in Figure 4(b) to give two essentially different de Bruijn sequences that are also essentially different from the one in Figure 4(a). [We regard two circular sequences as essentially the same if one can be obtained by rotating or reversing the other.]
11. (a) Create a digraph, similar to the one in Figure 4(b), that can be used to find de Bruijn sequences of order 3.
    (b) Use your digraph to draw two de Bruijn sequences of order 3.
12. (a) Explain how the discussion in Example 6(b) needs to be modified to show that de Bruijn sequences exist of all orders $n \geq 3$.
    (b) It is easy to draw a de Bruijn sequence of order 2. Do so.

13. The **reverse** of a digraph G is the digraph $\overleftarrow{G}$ obtained by reversing all the arrows of G. That is, $V(\overleftarrow{G}) = V(G)$, $E(\overleftarrow{G}) = E(G)$, and, if $\gamma(e) = (x, y)$, then $\overleftarrow{\gamma}(e) = (y, x)$. Use $\overleftarrow{G}$ and Theorem 2 to show that if G is acyclic [and finite], then G has a source.

14. (a) Use algorithm Sink and $\overleftarrow{G}$ [see Exercise 13] to give an algorithm Source that produces a source in any finite acyclic digraph.

(b) Modify the algorithm NumberVertices by using sources instead of sinks to produce an algorithm that numbers $V(G)$ in decreasing order.

(c) Use your algorithm to number the digraph of Figure 1.

15. (a) Suppose that a finite acyclic digraph has just one sink. Show that there is a path to the sink from each vertex.

(b) What is the corresponding statement for sources?

16. Let G be a digraph and define the relation $\sim$ on $V(G)$ by $x \sim y$ if $x = y$ or if x is reachable from y and y is reachable from x.

(a) Show that $\sim$ is an equivalence relation.

(b) Find the equivalence classes for the digraph pictured in Figure 6.

(c) Describe the relation $\sim$ in the case that G is acyclic.

17. (a) Show that in Theorem 1 and Corollary 1 the path without repeated vertices can be constructed from edges of the given path. Thus every closed path contains at least one cycle.

(b) Show that if u and v are vertices of a digraph and if there is a path from u to v, then there is a path from u to v in which no edge is repeated. [Consider the case $u = v$, as well as $u \neq v$.]

18. Let G be a digraph.

(a) Show that if u is reachable from v, then $R(u) \subseteq R(v)$.

(b) Give an alternative proof of Theorem 2 by choosing v in $V(G)$ with $|R(v)|$ as small as possible.

(c) Does your proof in part (b) lead to a useful constructive procedure? Explain.

19. Show that a path in a graph G is a cycle if and only if it is possible to assign directions to the edges of G so that the path is a [directed] cycle in the resulting digraph.

8.2 Weighted Digraphs and Scheduling Networks

In many applications of digraphs one wants to know if a given vertex v is reachable from another vertex u, that is, if it is possible to get to v from u by following arrows. For instance, suppose that each vertex represents a state a machine can be in such as **fetch**, **defer**, or **execute**, and there is an edge from s to t whenever the machine can change from state s to state t in response to some input. If the machine is in state u, can it later be in state v? The answer is yes if and only if the digraph contains a path from u to v.

Now suppose there is a cost associated with each transition from one state to another, i.e., with each edge in the digraph. Such a cost might be monetary, might be a measure of the time involved to carry out the change, or might have some other meaning. We could now ask for a path from u to v with the smallest total associated cost obtained by adding all the costs for the edges in the path.

If all edges cost the same amount, then the cheapest path is simply the shortest. In general, however, edge costs might differ. A digraph with no parallel edges is called **weighted** if each edge has an associated number, called its **weight**. In a given application it might be called "cost" or "length" or "capacity" or have some other interpretation. The exclusion of parallel edges is not particularly restrictive, since, in practice, if a digraph does have parallel edges, one can generally eliminate all but one of them from consideration as part of a winning solution.

We can describe the weighting of a digraph G using a function W from $E(G)$ to $\mathbb{R}$, where $\boldsymbol{W(e)}$ is the weight of the edge e. The **weight** of a path $e_1 e_2 \cdots e_m$ in G is then the sum $\sum_{i=1}^{m} W(e_i)$ of the weights of its edges. Because of our assumption that weighted digraphs have no parallel edges, we may suppose that $E(G) \subseteq V(G) \times V(G)$ and write $\boldsymbol{W(u, v)}$ for the weight of the edge (u, v) from u to v, if there is one. Weights are normally assumed to be nonnegative, but many of the results we will obtain for weighted digraphs are true without such a limitation.

EXAMPLE 1

(a) The digraph shown in Figure 1 is stolen from Figure 1 on page 100, where it described a rat and some cages. It could just as well describe a machine, perhaps a robotic tool, with states A, B, C, and D, and the number next to an

Figure 1 ▶

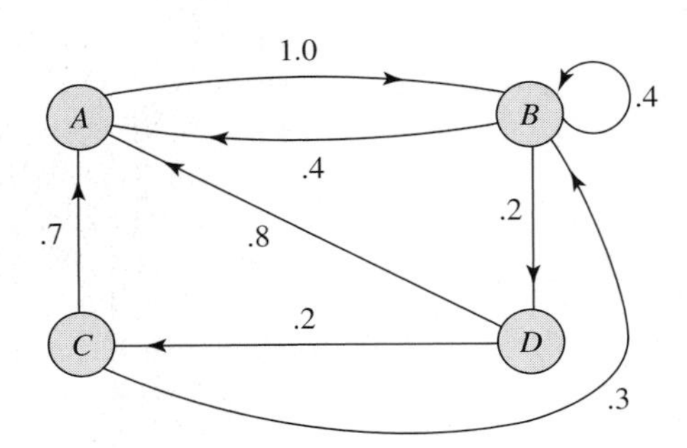

arrow [the weight of the edge] could be the number of microseconds necessary to get from its initial state to its terminal state. With that interpretation it takes 0.3 microsecond to go from state C to state B, 0.9 microsecond to go from D to A by way of C, 0.8 microsecond to go directly from D to A, and 0.4 microsecond to stay in state B in response to an input. What is the shortest time to get from D to B?

(b) Figure 2 shows a more complicated example. Now the shortest paths $s\,v\,x\,f$ and $s\,w\,x\,f$ from s to f have weights $6 + 7 + 4 = 17$ and $3 + 7 + 4 = 14$, respectively, but the longer path $s\,w\,v\,y\,x\,z\,f$ has weight $3 + 2 + 1 + 3 + 1 + 3 = 13$, which is less than either of these. Thus length is not directly related to weight. This example also shows a path $s\,w\,v$ from s to v that has smaller weight than the edge from s to v.

Figure 2 ▶

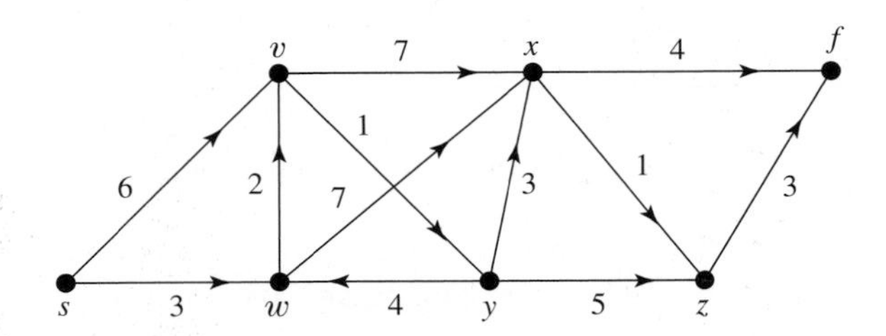

This digraph has a cycle $w\,v\,y\,w$. Clearly, the whole cycle cannot be part of a path of minimum weight, but pieces of it can be. For instance, $w\,v\,y$ is the path of smallest weight from w to y, and the edge $y\,w$ is the path of smallest weight from y to w. ■

If we wish, we can display the weight function W in a tabular form by labeling rows and columns of an array with the members of $V(G)$ and entering the value of $W(u, v)$ at the intersection of row u and column v.

EXAMPLE 2

The array for the digraph in Figure 2 is given in Figure 3(a). The numbers appear in locations corresponding to edges of the digraph. The table in Figure 3(a) contains enough information to let us reconstruct the weighted digraph, since from the table we know just where the edges go and what their weights are. Figure 3(b) is a tabulation of the weight function W^*, where $W^*(u, v)$ is the smallest weight of a path from u to v if such a path exists. ■

Figure 3 ▶

W	s	v	w	x	y	z	f
s		6	3				
v				7	1		
w		2		7			
x						1	4
y			4	3		5	
z							3
f							

(a)

W^*	s	v	w	x	y	z	f
s		5	3	9	6	10	13
v		7	5	4	1	5	8
w		2	7	6	3	7	10
x						1	4
y		6	4	3	7	4	7
z							3
f							

(b)

We will call the smallest weight of a path [of length at least 1] from u to v the **min-weight** from u to v, and we will generally denote it by $\boldsymbol{W^*(u, v)}$, as in Example 2. We also call a path from u to v that has this weight a **min-path**.

It is no real restriction to suppose that our weighted digraphs have no loops [why?], so we could just decide not to worry about $W(u, u)$, or we might define $W(u, u) := 0$ for all vertices u. It turns out, though, that we will learn more useful information by another choice, based on the following idea.

Consider vertices u and v with no edge from u to v in G. We can create a fictitious new edge from u to v of enormous weight, so big that the edge would never get chosen in a min-path if a real path from u to v were available. Suppose we create such fictitious edges wherever edges are missing in G. If we ever find that the weight of a path in the enlarged graph is enormous, then we will know that the path includes at least one fictitious edge, so it can't be a path in the original digraph G. A convenient notation is to write $W(u, v) = \infty$ if there is no edge from u to v in G; $W^*(u, v) = \infty$ means that there is no path from u to v. The operating rules for the symbol ∞ are $\infty + x = x + \infty = \infty$ for every x and $a < \infty$ for every real number a.

With this notation, we will write $W(u, u) = a$ if there is a loop at u of weight a and write $W(u, u) = \infty$ otherwise. Then $W^*(u, u) < \infty$ means that there is a path [of length at least 1] from u to itself in G, while $W^*(u, u) = \infty$ means that there is no such path. The digraph G is acyclic if and only if $W^*(v, v) = \infty$ for every vertex v.

EXAMPLE 3

Using this notation, we would fill in all the blanks in the tables of Figures 3(a) and (b) with ∞'s. ■

In Example 2 we simply announced the values of W^*, and the example is small enough that it is easy to check the correctness of the values given. For more complicated digraphs, the determination of W^* and of the min-paths can be nontrivial problems. In §8.3 we will describe algorithms for finding both W^* and the corresponding min-paths, but until then we will stare at the picture until the answer is clear. For small digraphs this method is as good as any.

In many situations described by weighted graphs, the min-weights and min-paths are the important concerns. However, one class of problems for which weighted digraphs are useful, but where min-paths are irrelevant, is the scheduling of processes that involve a number of steps. Such scheduling problems are of considerable importance in business and manufacturing. The following example illustrates the sort of situation that arises.

EXAMPLE 4

Consider a cook preparing a simple meal of curry and rice. The curry recipe calls for the following steps.

(a) Cut up meatlike substance—about 10 minutes.
(b) Grate onion—about 2 minutes with a food processor.
(c) Peel and quarter potatoes—about 5 minutes.
(d) Marinate substance, onions, and spices—about 30 minutes.
(e) Heat oil—4 minutes. Fry potatoes—15 minutes.
Fry cumin seed—2 minutes.
(f) Fry marinated substance—4 minutes.
(g) Bake fried substance and potatoes—60 minutes.

In addition, there is

(h) Cook rice—20 minutes.

We have grouped three steps together in (e), since they must be done in sequence. Some of the other steps can be done simultaneously if enough help is available. We suppose our cook has all the help needed.

Figure 4(a) gives a digraph that shows the sequence of steps and the possibilities for parallel processing. Cutting, grating, peeling, and rice cooking can all go on at once. The dashed arrows after cutting and peeling indicate that frying and marinating cannot begin until cutting and peeling are completed. The other two dashed arrows have similar meanings. The picture has been redrawn in Figure 4(b) with weights on the edges to indicate the time involved. [Ignore the numbers at the vertices for the moment.] The vertices denote stages of partial completion of the overall process, starting at the left and finishing at the right. In this case the min-path from left to right has weight 20, but there is much more total time required to prepare the meal than just the 20 minutes to cook the rice. The min-weight is no help. The important question here is this: What is the smallest total time required to complete all steps in the process?

Figure 4 ▶

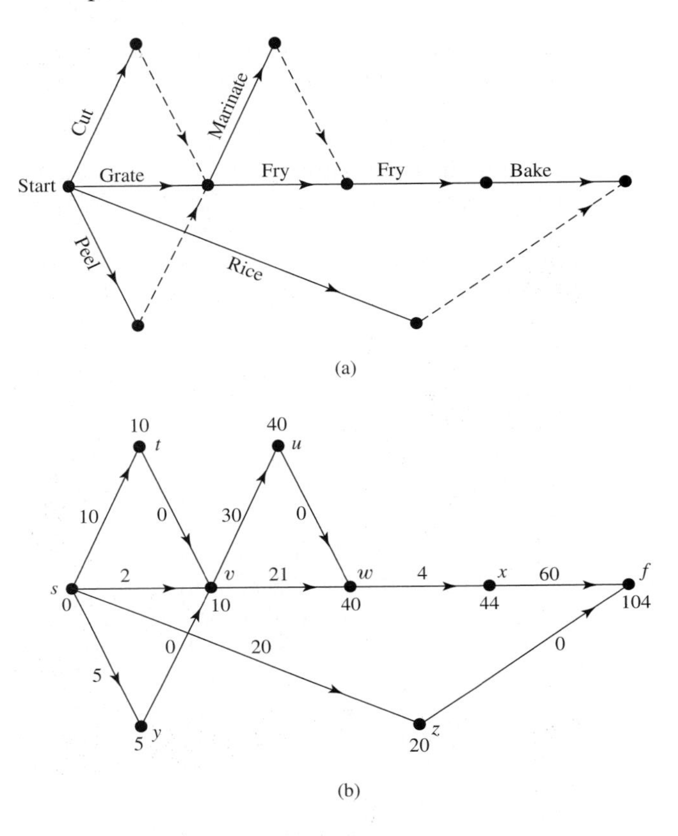

To answer the question, we first examine vertices from left to right. Suppose we start at s at time 0. What is the earliest time we could finish cutting, grating, and peeling and arrive at vertex v? Clearly, 10 minutes, since we must wait for the cutting no matter how soon we start grating or peeling. In fact, 10 is the *largest* weight of a path from s to v. Now what is the earliest time we can arrive at vertex w? The shortest time to get from v to w is 30 minutes [the largest weight of a path from v to w], so the shortest time to get from s to w is $10 + 30 = 40$ minutes. Similarly, the earliest time by which we can complete the whole process and arrive at the vertex f is $40 + 4 + 60 = 104$ minutes after we start.

In each instance, the smallest time to arrive at a given vertex is the largest weight of a path from s to that vertex. The numbers beside the vertices in Figure 4(b) give these smallest times. ■

An acyclic digraph with nonnegative weights and unique source and sink, such as the digraph in Figure 4, is called a **scheduling network**. For the rest of this section we suppose that we are dealing with a scheduling network G with source s [start] and sink f [finish]. For vertices u and v of G a **max-path** from u to v is a path of largest weight, and its weight is the **max-weight** from u to v, which we denote by

$\boldsymbol{M(u, v)}$. Later we will need to define $M(u, v)$ when there is *no* path from u to v, in which case we will set $M(u, v) = -\infty$.

Max-weights and max-paths can be analyzed in much the same way as min-weights and min-paths. In §8.3 we will describe how to modify an algorithm for W^* to get one for M. For now we determine M by staring at the digraph. A max-path from s to f is called a **critical path**, and an edge belonging to such a path is a **critical edge**.

We introduce three functions on $V(G)$ that will help us understand scheduling networks. If we start from s at time 0, the earliest time we can arrive at a vertex v, having completed all tasks preceding v, is $M(s, v)$. We denote this earliest arrival time by $\boldsymbol{A(v)}$. In particular, $A(f) = M(s, f)$, the time in which the whole process can be completed. Let $\boldsymbol{L(v)} = M(s, f) - M(v, f) = A(f) - M(v, f)$. Since $M(v, f)$ represents the shortest time required to complete all steps from v to f, $L(v)$ is the latest time we can leave v and still complete all remaining steps by time $A(f)$. To calculate $L(v)$, we may work backward from f. Example 5(a) below gives an illustration.

The **slack time** $\boldsymbol{S(v)}$ of a vertex v is defined by $\boldsymbol{S(v)} = L(v) - A(v)$. This is the maximum time that all tasks starting at v could be idle without delaying the entire process. So, of course, $S(v) \geq 0$ for all v. We can prove this formally by noting that

$$S(v) = L(v) - A(v) = M(s, f) - M(v, f) - M(s, v) \geq 0,$$

since $M(s, v) + M(v, f) \leq M(s, f)$. The last inequality holds because we can join max-paths from s to v and from v to f to obtain a path from s to f having weight $M(s, v) + M(v, f)$.

EXAMPLE 5

(a) The scheduling network in Figure 4(b) has only one critical path: $s\,t\,v\,u\,w\,x\,f$. Thus the steps (a), (d), (f), and (g) are critical. The network is redrawn in Figure 5 with the critical path highlighted. Note that, if the weights of the non-critical edges were decreased by improving efficiency, then the entire process would still take 104 minutes. Even eliminating the noncritical tasks altogether would not speed up the process.

Figure 5 ▶

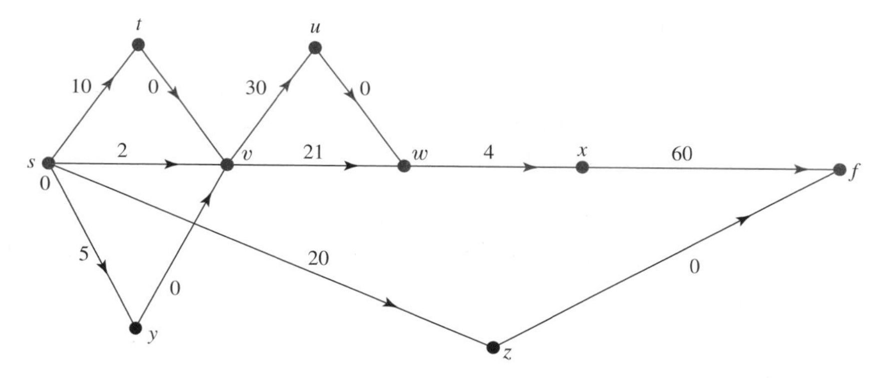

The functions A, L, and S are given in the following table. The values of A are written next to the vertices in Figure 4(b).

	s	t	u	v	w	x	y	z	f
A	0	10	40	10	40	44	5	20	104
L	0	10	40	10	40	44	10	104	104
S	0	0	0	0	0	0	5	84	0

Here are some sample calculations. Given that $A(u) = 40$, $W(u, v) = 0$, $A(v) = 10$, and $W(v, w) = 21$, we have

$$A(w) = \max\{40 + 0, 10 + 21\} = 40.$$

Given that $L(w) = 40$, $W(v, w) = 21$, $L(u) = 40$, and $W(v, u) = 30$, we have

$$L(v) = \min\{40 - 21, 40 - 30\} = 10.$$

This calculation illustrates what we meant earlier when we spoke of working backward from f to compute $L(v)$. The entries in the $A(v)$ row are computed from left to right, and then, once $A(f)$ is known, the entries in the $L(v)$ row are computed from right to left. Exercise 21 asks for proofs that these methods for calculating $A(v)$ and $L(v)$ are correct in general.

Observe that the slack time is 0 at each vertex on the critical path. This always happens [Exercise 18]. Since $S(y) = 5$, the task of peeling potatoes can be delayed 5 minutes without delaying dinner. Since $S(z) = 84$, one could delay the rice 84 minutes; in fact, one would normally wait about 74 minutes before starting the rice to allow it to "rest" 10 minutes at the end.

A glance at edge (v, w) shows that we could delay this task 9 [= 30 − 21] minutes, but we cannot deduce this information from the slack times. From $S(v) = 0$, we can only infer that we cannot delay *all* tasks starting at v. For this reason, we will introduce below a useful function on the set $E(G)$ of edges, called the float time.

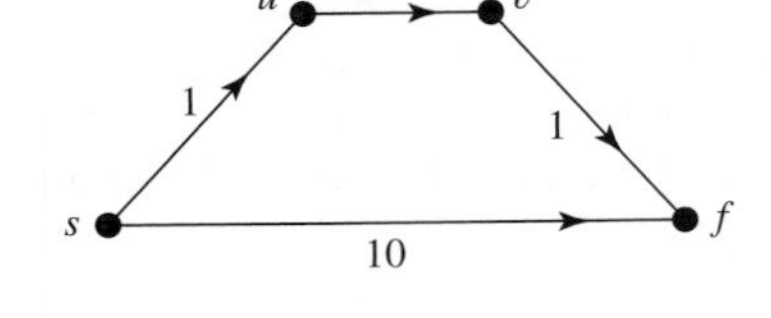

Figure 6 ▲

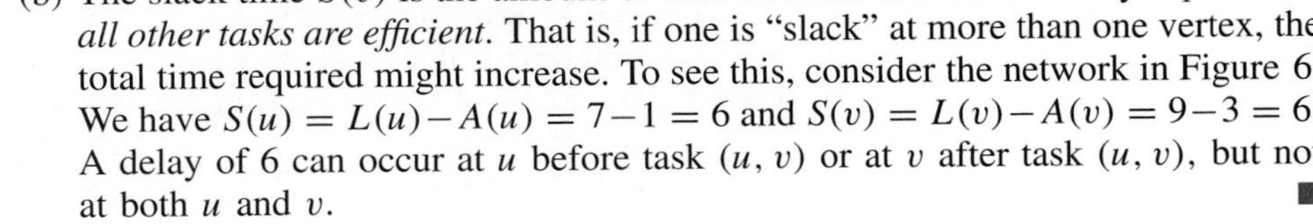

(b) The slack time $S(v)$ is the amount of time all tasks at v can be delayed *provided all other tasks are efficient*. That is, if one is "slack" at more than one vertex, the total time required might increase. To see this, consider the network in Figure 6. We have $S(u) = L(u) - A(u) = 7 - 1 = 6$ and $S(v) = L(v) - A(v) = 9 - 3 = 6$. A delay of 6 can occur at u before task (u, v) or at v after task (u, v), but not at both u and v. ■

If (u, v) is an edge of a scheduling network, then

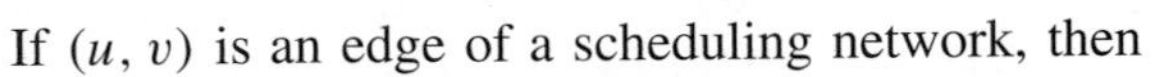

$$M(s, u) + W(u, v) + M(v, f) \le M(s, f),$$

since a path $s \cdots u\, v \cdots f$ certainly has no greater weight than a critical path from s to f. Thus

$$W(u, v) \le [M(s, f) - M(v, f)] - M(s, u) = L(v) - A(u).$$

This inequality is not a surprise, since $L(v) - A(u)$ is the maximum time the task at edge (u, v) could take without increasing the total time. We define the **float time** $\boldsymbol{F(u, v)}$ of the edge (u, v) by

$$\boldsymbol{F(u, v)} = L(v) - A(u) - W(u, v).$$

This is the maximum delay possible for the task corresponding to the edge (u, v). The next theorem will give a connection between float times for edges and slack times for vertices.

EXAMPLE 6

The critical edges in Figure 5 all have float time 0. This statement can be verified directly, but the next theorem shows that it always holds. Also,

$$F(s, v) = L(v) - A(s) - W(s, v) = 10 - 0 - 2 = 8;$$

$$F(v, w) = L(w) - A(v) - W(v, w) = 40 - 10 - 21 = 9;$$

$$F(s, y) = L(y) - A(s) - W(s, y) = 10 - 0 - 5 = 5;$$

$$F(y, v) = L(v) - A(y) - W(y, v) = 10 - 5 - 0 = 5; \text{ etc.}$$ ■

Theorem Consider a scheduling network.

(a) The float time $F(u, v)$ is 0 if and only if (u, v) is a critical edge.

(b) $F(u, v) \ge \max\{S(u), S(v)\}$ for all edges of the network.

Proof

(a) Suppose that $F(u, v) = 0$; then

$$[M(s, f) - M(v, f)] - M(s, u) - W(u, v) = 0,$$

so

$$M(s, f) = M(s, u) + W(u, v) + M(v, f).$$

This equation implies that, if we attach the edge (u, v) to max-paths from s to u and from v to f, we will obtain a critical path for the network. Hence (u, v) is a critical edge.

If (u, v) is an edge of a critical path $s \cdots u\, v \cdots f$, then $s \cdots u$ must have weight $M(s, u)$, since otherwise $s \cdots u$ could be replaced by a heavier path. Similarly, $v \cdots f$ has weight $M(v, f)$, so

$$M(s, u) + W(u, v) + M(v, f) = M(s, f).$$

Therefore, we have $F(u, v) = 0$.

(b) The inequality $S(u) \le F(u, v)$ is equivalent to each of the following:

$$L(u) - A(u) \le L(v) - A(u) - W(u, v);$$

$$W(u, v) \le L(v) - L(u);$$

$$W(u, v) \le [M(s, f) - M(v, f)] - [M(s, f) - M(u, f)];$$

$$W(u, v) \le M(u, f) - M(v, f);$$

$$W(u, v) + M(v, f) \le M(u, f).$$

The last inequality is clear, since the edge (u, v) can be attached to a max-path from v to f to obtain a path from u to f of weight $W(u, v) + M(v, f)$.

The inequality $S(v) \le F(u, v)$ is slightly easier to verify [Exercise 18]. ■

As we have seen, shortening the time required for a noncritical edge does not decrease the total time $M(s, f)$ required for the process. Identification of critical edges focuses attention on those steps in a process where improvement may make a difference and where delays will surely be costly. Since its introduction in the 1950s, the method of critical path analysis, sometimes called PERT for Program Evaluation and Review Technique, has been a popular way of dealing with industrial management scheduling problems.

Exercises 8.2

Use the ∞ notation in all tables of W and W^*.

1. Give tables of W and W^* for the digraph of Figure 1 with the loop at B removed.
2. Give a table of W^* for the digraph of Figure 7(a).
3. Give tables of W and W^* for the digraph of Figure 7(b).
4. The path $s\,w\,v\,y\,x\,z\,f$ is a min-path from s to f in the digraph of Figure 2. Find another min-path from s to f in that digraph.
5. Figure 8 shows a weighted digraph. The directions and weights have been left off the edges, but the number at each vertex v is $W^*(s, v)$.
   (a) Give three different weight functions W that yield these values of $W^*(s, v)$. [An answer could consist of three pictures with appropriate numbers on the edges.]
   (b) Do the different weight assignments yield different min-paths between points? Explain.

Figure 7 ▶

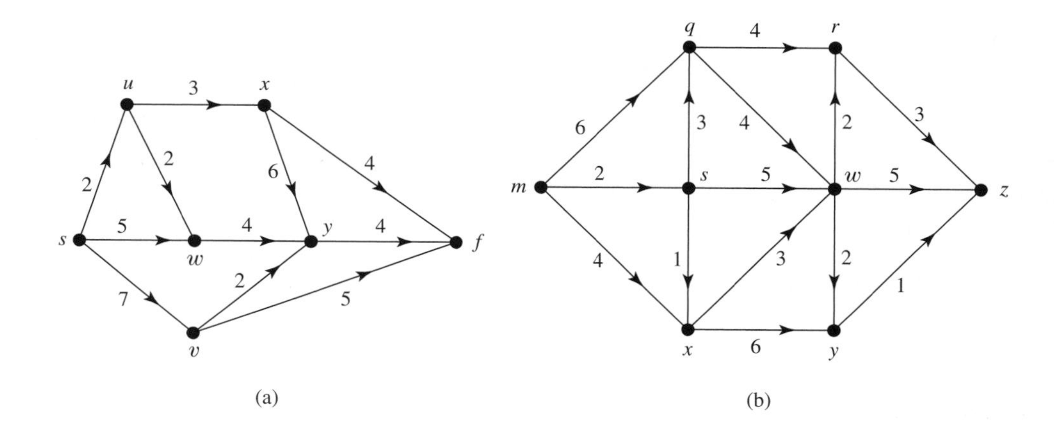

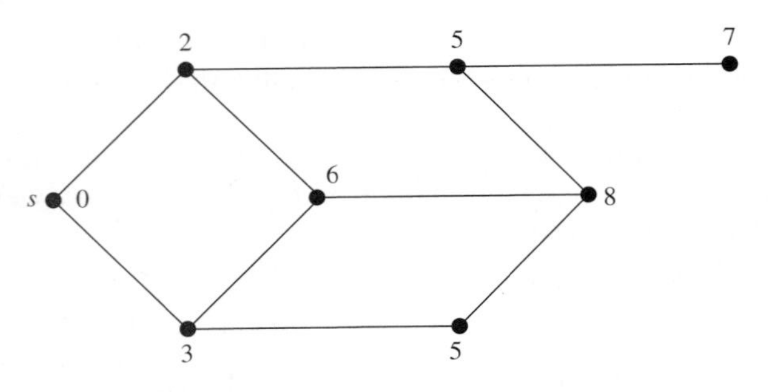

Figure 8 ▲

6. Suppose that u, v, and w are vertices of a weighted digraph with min-weight function W^* and that $W^*(u, v) + W^*(v, w) = W^*(u, w)$. Explain why there is a min-path from u to w through v.

7. (a) Find a critical path for the digraph of Figure 2 with the edge from y to w removed.
   (b) Why does the critical path method apply only to acyclic digraphs?

8. Calculate the float times for the edges (s, z) and (z, f) in Figure 5. *Hint:* See Example 5(a) for some useful numbers.

9. (a) Give a table of A and L for the network in Figure 7(a).
   (b) Find the slack times for the vertices of this network.
   (c) Find the critical paths for this network.
   (d) Find the float times for the edges of this network.

10. Repeat Exercise 9 for the digraph in Figure 7(a) with each edge of weight 1.

11. Repeat Exercise 9 for the digraph in Figure 7(b).

12. (a) Calculate the float times for the edges in the network in Figure 6.
    (b) Can each task in a network be delayed by its float time without delaying the entire process? Explain.

13. Consider the network in Figure 9.
    (a) How many critical paths does this digraph have?
    (b) What is the largest float time for an edge in this digraph?
    (c) Which edges have the largest float time?

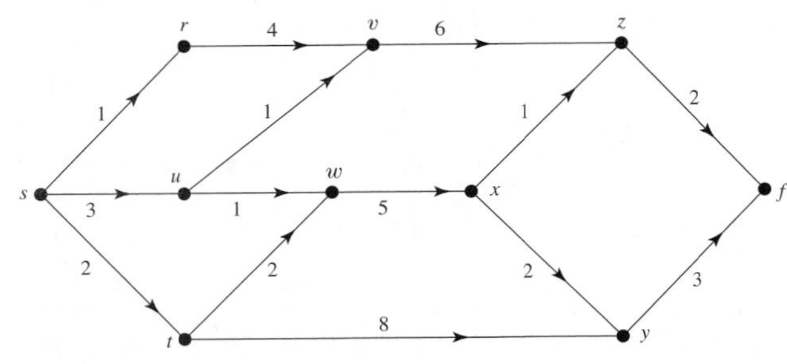

Figure 9 ▲

14. Find the slack times for the vertices of the network in Figure 9.

15. In Example 4 we used edges of weight 0 as a device to indicate that some steps could not start until others were finished.
    (a) Explain how to avoid such 0-edges if parallel edges are allowed.
    (b) Draw a digraph for the process of Figure 4 to illustrate your answer.

16. If the cook in Example 4 has no helpers, then steps (a), (b), and (c) must be done one after the other, but otherwise the situation is as in the example.
    (a) Draw a scheduling network for the no-helper process.
    (b) Find a critical path for this process.
    (c) Which steps in the process are not critical?

17. (a) Give tables of W and W^* for the digraph of Figure 10.

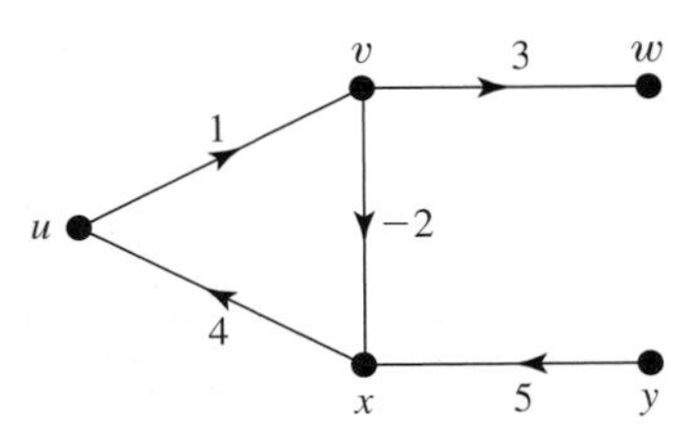

Figure 10 ▲

    (b) Explain how to tell just from the table for W^* whether this digraph is acyclic.
    (c) How would your answer to part (a) change if the edge of weight -2 had weight -6 instead?
    (d) Explain how to tell which are the sources and sinks of this digraph from the table for W.

18. (a) Complete the proof of the theorem by showing $S(v) \le F(u, v)$.
    (b) Show that if (u, v) is a critical edge, then $S(u) = S(v) = 0$.
    (c) If (u, v) is an edge with $S(u) = S(v) = 0$, must (u, v) be critical? Explain.

19. The float time $F(u, v)$ can be thought of as the amount of time we can delay starting along the edge (u, v) without delaying completion of the entire process. Define the **free-float time** $FF(u, v)$ to be the amount of time we can delay starting along (u, v) without increasing $A(v)$.
    (a) Find an expression for $FF(u, v)$ in terms of the functions A and W.
    (b) Find $FF(u, v)$ for all edges of the digraph in Figure 7(a).
    (c) What is the difference between $F(u, v)$ and $FF(u, v)$?

20. (a) Show that increasing the weight of a critical edge in a scheduling network increases the max-weight from the source to the sink.
    (b) Is there any circumstance in which reducing the amount of time for a critical step in a process does not reduce the total time required for the process? Explain.

21. (a) Show that $A(u) = \max\{A(w) + W(w, u) : (w, u) \in E(G)\}$.
    (b) Show that $L(u) = \min\{L(v) - W(u, v) : (u, v) \in E(G)\}$.

Office Hours 8.2

I was looking ahead at the next section, like you told us to, and I can't imagine ever memorizing all those algorithms. It would take the whole hour to apply one on the exam. What do you really expect us to know about them? Will they be on the exam?

Count on it. But you're right, I won't ask you to carry out a whole execution of Warshall's algorithm on the exam. More likely, I'll ask you to explain in a general way how it works—maybe give an explanation entirely in words—or I'll start you at some point partway along in the execution and ask you to take the next step. Dijkstra's algorithm is different. I could very well ask you for a full run on some fairly small example.

The way to learn these algorithms and others, such as the ones we saw for spanning trees, is not to memorize their programs but to think about what they produce and how they work. The book proves that they *do* work, which is important, of course, but to make sense out of the algorithms, you need to play with them enough to see why they make the choices they do. Both Dijkstra's and Warshall's algorithms depend on building a bigger and bigger set of allowable vertices on paths. Both make changes if new information comes in about paths with smaller weights. So what's the difference between the two algorithms? Asking questions like this will help you understand the algorithms, and once you understand them they're not so hard to remember.

Describing paths with pointers is fairly natural, especially if the arrows point forward. Backward pointers in Dijkstra's algorithm tell us how to head back to our starting point from wherever we are. When pointers change in the algorithms, it's because some new, better way has been found to get from one place to another. All of this is pretty straightforward.

The rest of the material in the section is about modifying Warshall's algorithm. You should be sure you understand how to get reachable sets this way and also max-weights. The MaxWeight algorithm is really just a stripped-down version of Warshall's algorithm. You could maybe use it to find critical paths for fixing dinner, but it's probably not worth memorizing.

In the long run, of course, you may not even remember how these algorithms work, but at least you'll remember that they exist, and you'll remember what kinds of problems they solve. For the short run, expect to see some questions about them on the exam.

By the way, have you tried programming any of these algorithms? They're not hard to program, and just getting the input and output set up will help you nail down what the algorithms are good for.

8.3 Digraph Algorithms

Our study of digraphs and graphs has led us to a number of concrete questions. Given a digraph, what is the length of a shortest path from one vertex to another? If the digraph is weighted, what is the minimum or maximum weight of such a path? How can we efficiently find such paths? Is there any path at all?

This section describes some algorithms for answering these questions and others, algorithms that can be implemented on computers as well as used to organize hand computations. The algorithms we have chosen are reasonably fast, and their workings are comparatively easy to follow. For a more complete discussion we refer the reader to books on the subject, such as *Data Structures and Algorithms* by Aho, Hopcroft and Ullman or *Introduction to Algorithms* by Cormen, Leiserson and Rivest.

Of course min-weight problems can be important for graphs with parallel edges—think of airline scheduling, for example—but if two parallel edges have different weights, then the one with the higher weight will never appear in a min-path. Neither will a loop. [Why?] For these reasons, and to keep the account as uncluttered as possible, we will assume here that our digraphs have no loops or parallel edges. Hence $E(G) \subseteq V(G) \times V(G)$, and we can describe the digraph with a table of the edge-weight function $W(u, v)$, as we did in §8.2. Our goal will be to obtain some or all of the min-weights $W^*(u, v)$, with corresponding min-paths.

The min-weight algorithms that we will consider all begin by looking at paths of length 1, i.e., single edges, and then systematically consider longer and longer paths between vertices. As they proceed, the algorithms find smaller and smaller path weights between vertices, and when the algorithms terminate, the weights are the best possible.

Our first algorithm, due to Dijkstra, finds min-paths from a selected vertex s to all other vertices, given that all edge weights are nonnegative. The algorithm as presented in its final form on page 338 looks somewhat complex and unmotivated. To see where it comes from and why it works, we will develop it slowly, using an example to illustrate the ideas. The essence of the algorithm is to maintain a growing set M of "marked vertices" and at each stage to determine paths from s of smallest weight whose intermediate vertices are all marked. Such paths will be referred to as "marked paths." Eventually, all vertices will be marked, so every path will be a marked path. Here is the outline.

```
Mark s.
while there are unmarked vertices do
    Choose one and mark it.
    for each vertex v do
        Find a path of smallest weight from s to v that
        only goes through marked vertices.
```

Our aim, of course, is to find actual paths that have weights as small as possible. The vertex s itself is a special case. We don't care about paths from s to s or about paths with s as intermediate vertex, since coming back through s would just make paths weigh more. Remember, edge weights are nonnegative. To keep from having to say "except for s" again and again, we have marked s, and we will also agree that $W(s, s) = 0$.

At the start, the set M of marked vertices will be $\{s\}$, so the only paths from s using marked vertices are the ones with no intermediate vertices at all, i.e., the edges (s, v) with s as initial vertex. Their weights will be $W(s, v)$.

Now suppose that we have marked some set M of vertices. Call a path an ***M*-path** in case all its vertices, except perhaps the first and last, are in M. If there are any M-paths from s to v, we write $W_M(s, v)$ for the weight of an M-path of smallest weight from s to v, which we call an ***M*-min-path**. If there are no M-min-paths from s to v, we set $W_M(s, v) = \infty$.

The key observation is that, when the marked set M is just $\{s\}$, the input provides us with $W_M(s, v) = W(s, v)$ for $v \neq s$, and when M is the set $V(G)$ of all vertices, $W_M(s, v) = W^*(s, v)$ for $v \neq s$, as desired. Dijkstra's algorithm starts with $M = \{s\}$ and marks one vertex at a time, i.e., executes the **while** loop repeatedly, until M is equal to $V(G)$.

EXAMPLE 1

Consider the weighted digraph shown in Figure 1. As we will explain after this example, Dijkstra's algorithm chooses the vertices in a particular order that allows the algorithm to efficiently keep track of M-min-paths as well as values of $W_M(s, v)$. In this first example, we will mark the vertices, starting with $s = v_2$, in the order given by Dijkstra's algorithm *without explanation* and will focus on seeing how the algorithm gets us from the edge-weights $W(s, v)$ to the min-weights $W^*(s, v)$. We will illustrate the algorithm by marking vertices in the following order: $v_2, v_5, v_3, v_4, v_6, v_7, v_1$. Figure 2 indicates the values of $W_M(s, v)$ as M grows. The

Figure 1 ▶

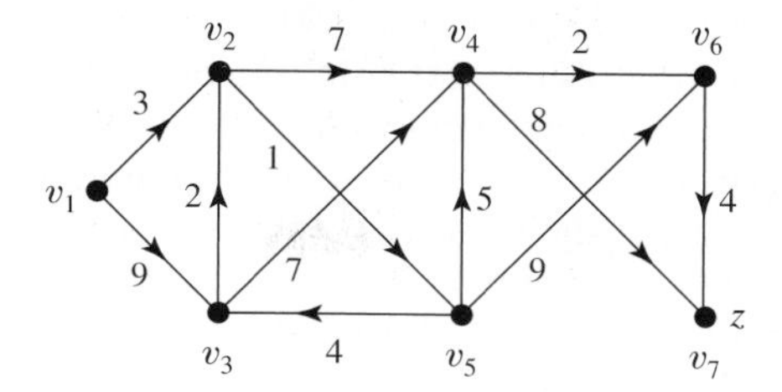

Figure 2 ▶

M	$W_M(s, v_j) = W_M(v_2, v_j)$					
	v_1	v_3	v_4	v_5	v_6	v_7
$\{v_2\}$	∞	∞	7	1	∞	∞
$\{v_2, v_5\}$	∞	5	6	1	10	∞
$\{v_2, v_5, v_3\}$	∞	5	6	1	10	∞
$\{v_2, v_5, v_3, v_4\}$	∞	5	6	1	8	14
$\{v_2, v_5, v_3, v_4, v_6\}$	∞	5	6	1	8	12
$\{v_2, v_5, v_3, v_4, v_6, v_7\}$	∞	5	6	1	8	12
$\{v_2, v_5, v_3, v_4, v_6, v_7, v_1\}$	∞	5	6	1	8	12

Figure 3 ▶

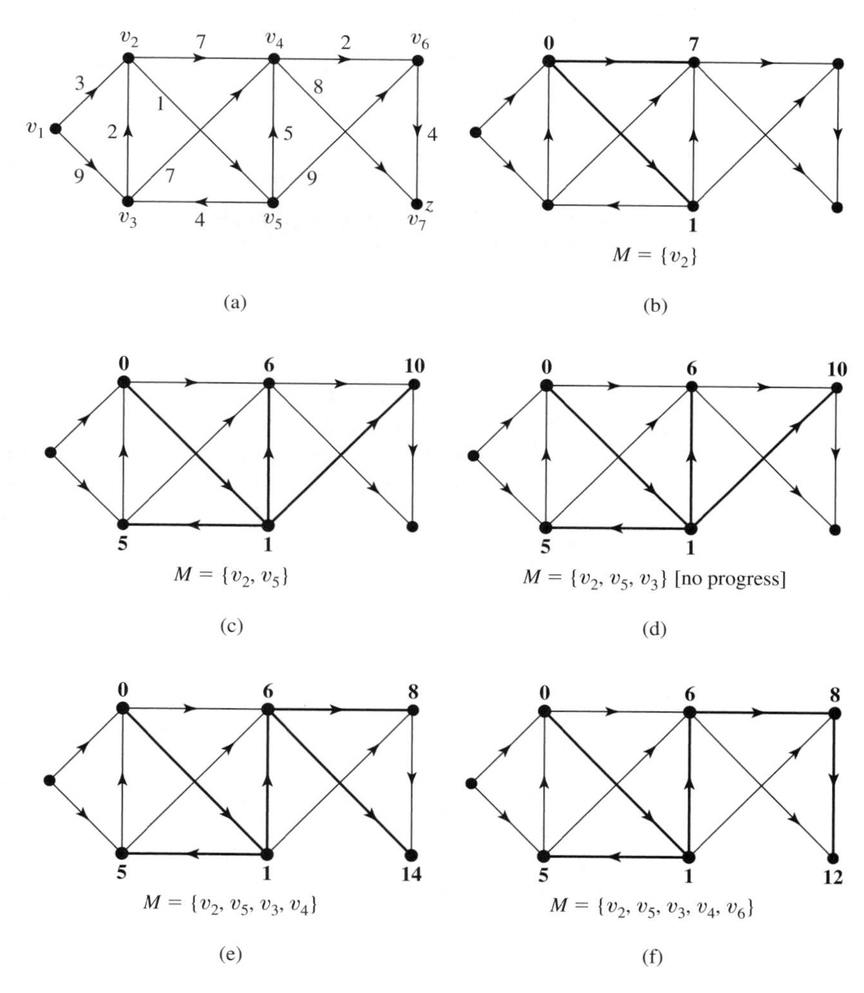

In the next two steps mark v_7 and v_1, with no other changes.

values in the first row are just the values $W(s, v)$, and the values in the last row are the min-weights $W^*(s, v)$.

The successive pictures in Figure 3 indicate how the algorithm proceeds and will help explain how we got the entries $W_M(s, v_j)$ in Figure 2 and why they are correct. We have shown the marked vertices as heavier dots and some of the paths through them with heavier lines. In addition, we have put bold numbers by the vertices to indicate the minimal weights of heavy paths to them from s. Notice that, as the sequence progresses, the bold numbers on some of the vertices get smaller, reflecting the fact that better marked paths to them have been found. For example, the values of $W_M(s, v_j)$ in the third row of Figure 2 can be verified by staring at Figure 3(d) to see the shortest paths with intermediate vertices in $\{v_3, v_5\}$. At the end, in Figure 3(f), the bold number at each vertex v is the min-weight $W^*(s, v)$, and the heavy path from s to v is an associated min-path. There is no bold number at vertex v_1, because there are no paths at all from s to v_1. Look at the figure to convince yourself of these facts. As promised, we will explain below how we decided to select vertices to mark in this example. It may be helpful to refer to Figure 3 as you read the following discussion. ■

Let's look at the algorithm again.

Mark s.
while there are unmarked vertices **do**
    Choose one and mark it.
    for each vertex v **do**
        Find a path of smallest weight from s to v that
        only goes through marked vertices.

The loop must terminate, because the set of unmarked vertices keeps decreasing. When the loop does terminate, every vertex will be marked, so for each vertex v we will have found a min-path from s to v if there are any paths at all from s to v. It looks as if the hard part of the job will be the update step—finding the new smallest-weight paths when a new vertex has been marked. Dijkstra's algorithm has a clever way of choosing a new vertex w so that this update is easy. The following account suggests one way to arrive at this choice and also shows that it is correct.

At the start, the only marked paths from s are the ones with no intermediate vertices at all, i.e., the edges with s as initial vertex, and their weights will be $W(s, v)$. There will never be marked paths to vertices that are unreachable from s. If there is no actual marked path from s to v, let's say that the marked path of smallest weight from s to v is a "virtual" path, of weight ∞. Virtual paths in this sense are purely for notational convenience and are not paths of any real length or weight. With this convention, when s is the only marked vertex, the smallest weight of a marked path from s to v is $W(s, v)$, whether $s\,v$ is a path or not. To summarize, when s is the only marked vertex, the marked path of smallest weight from s to any vertex v is $s\,v$ or a virtual path, and in either case its weight is $W(s, v)$. Our next task is to see how to mark the other vertices and continue to find marked min-paths.

Suppose that we have marked some set M of vertices. We want to choose a new vertex w to mark so that we can get $M \cup \{w\}$-min-paths in some easy way. Let's look at what happens when we choose w. Consider a particular vertex v. One possibility is that there is no actual $M \cup \{w\}$-path from s to v. Then there was no M-path either, and we have $W_{M\cup\{w\}}(s, v) = W_M(s, v) = \infty$. Another possibility is that there is an $M \cup \{w\}$-min-path from s to v that doesn't go through w. In this case the path is also an M-min-path, so it has the same weight as our known M-min-path to v, and we can just keep the known one. This case is also easy.

The remaining possibility is that there is at least one $M \cup \{w\}$-min-path from s to v and that w is on every such path. Then there is some path from s to v through members of M *and* through w that has weight less than the weight of every M-path from s to v. We want to find such a path P. Its weight will be the weight of the part from s to w plus the weight of the part from w to v. We know an M-min-path from s to w, which we might as well use for the part of P from s to w. Its weight is $W_M(s, w)$.

How about the part from w to v? If we have chosen w so that this part can't go through any vertices of M, then it will have to consist simply of an edge from w to v, and we can attach the edge to the end of our known M-min-path from s to w to get an $M \cup \{w\}$-min-path from s to v. Its weight will be the sum $W_M(s, w) + W(w, v)$ of the M-min-weight of w and the edge weight. The update will be easy.

Can we choose w so that there are no marked vertices on the part of the $M \cup \{w\}$-min-path from w to v? Suppose we have set things up so that $W_M(s, m) \le W_M(s, w)$ for every m in M. If there were an m in M on the piece of the $M \cup \{w\}$-min-path from w to v, then we could get an $M \cup \{w\}$-path from s to v by simply going directly from s to m on an M-min-path and going from there to v. Such a path would bypass w entirely, and because $W_M(s, m) \le W_M(s, w)$, it would contradict the assumption that every $M \cup \{w\}$-min-path from s to v goes through w. So if $W_M(s, m) \le W_M(s, w)$ for every m in M, then the part of P from w to v can't contain vertices in M and must be a single edge.

One way to be sure that $W_M(s, m) \le W_M(s, w)$ for every m in M, no matter how we choose w, is to arrange that

$$W_M(s, m) \le W_M(s, v) \text{ for every } m \text{ in } M \text{ and every } v \text{ in } V(G) \setminus M. \qquad \textbf{(1)}$$

Condition (1) holds at the start of the algorithm, since at that stage $M = \{s\}$ and $W_M(s, s) = 0 \le W_M(s, v)$ for every v in $V(G)$, by assumption. Thus it will be enough to have condition (1) be an invariant of the **`while`** loop. Assuming that (1) holds for some M, we want to choose w so that for every m in M

$$W_{M\cup\{w\}}(s, m) \le W_{M\cup\{w\}}(s, v) \qquad \text{for every } v \text{ in } V(G) \setminus (M \cup \{w\}) \qquad \textbf{(2)}$$

and also

$$W_{M\cup\{w\}}(s, w) \le W_{M\cup\{w\}}(s, v) \qquad \text{for every } v \text{ in } V(G) \setminus (M \cup \{w\}). \qquad \textbf{(3)}$$

Since condition (1) holds, for every m in M and v in $V(G) \setminus M \cup \{w\}$, we have

$$W_{M\cup\{w\}}(s, m) \le W_M(s, m) \le \min\{W_M(s, v), W_M(s, w)\}. \qquad \textbf{(4)}$$

If $W_M(s, v) = W_{M\cup\{w\}}(s, v)$, then

$$\min\{W_M(s, v), W_M(s, w)\} \le W_{M\cup\{w\}}(s, v) \qquad \textbf{(5)}$$

by the definition of minimum. If $W_M(s, v) > W_{M\cup\{w\}}(s, v)$, then $W_{M\cup\{w\}}(s, v)$ must be the weight of a path from s to v through w, so $W_{M\cup\{w\}}(s, v) \ge W_M(s, w)$ [using the fact that W is nonnegative] and (5) is still true. Hence, by (4) and (5),

$$W_{M\cup\{w\}}(s, m) \le W_{M\cup\{w\}}(s, v)$$

for every v in $V(G) \setminus (M \cup \{w\})$, so (2) holds for *every* choice of w in $V(G) \setminus M$.

The key condition, therefore, is (3). Since $W_{M\cup\{w\}}(s, w) = W_M(s, w)$ and $W_{M\cup\{w\}}(s, v) \le W_M(s, v)$ for all v, condition (3) forces us to choose w with

$$W_M(s, w) \le W_M(s, v) \qquad \text{for every } v \text{ in } V(G) \setminus M.$$

If we do so, then $W_M(s, w) = \min\{W_M(s, v), W_M(s, w)\} \le W_{M\cup\{w\}}(s, v)$ by (5). Hence

$$W_{M\cup\{w\}}(s, w) \le W_M(s, w) \le W_{M\cup\{w\}}(s, v),$$

i.e., (3) holds. Thus, if at each stage we choose w in $V(G) \setminus M$ with $W_M(s, w)$ as small as possible, then (1) will be an invariant of the **`while`** loop and the rest of the algorithm, including the update step, almost takes care of itself.

Before we give the final version of the algorithm, we need to say more about paths. One way to describe a path is to associate with each vertex a pointer to the next vertex along the path until we come to the end. Think, perhaps, of a traveler in a strange city asking for directions to the railway station, going for a block, and then asking again, etc. We can describe the path $x\,y\,z\,w$ by a pointer from x to y, one from y to z, and one from z to w, i.e., by a function P such that $P(x) = y$, $P(y) = z$, and $P(z) = w$.

We can describe a path just as well with a function that points backward. For example, $Q(w) = z$, $Q(z) = y$, $Q(y) = x$ gives the backward path $w\,z\,y\,x$, and we can simply reverse its vertex sequence to get $x\,y\,z\,w$. For Dijkstra's algorithm, the backward pointers seem most natural.

Here is the official version of the algorithm, which we have just shown yields $W^*(s, v)$ for each v.

Dijkstra's Algorithm (digraph, vertex, weight function)

```
{Input: A digraph G without loops or parallel edges, a selected vertex s
    of G, and a nonnegative edge-weight function W for G}
{Output: The min-weights W*(s, v) for v ∈ V(G) \ {s}, and backward
    pointers P(v) that describe min-paths from s}
{Auxiliary variables: A set M of marked vertices and a "distance"
    function D with D(v) = W_M(s, v)}
Set M := {s}.
for v in V(G) do        {set things up}
    Set D(v) := W(s, v).
    if W(s, v) ≠ ∞ then        {there is an edge from s to v}
        Set P(v) := s.        {point back to s}
while V(G) \ M ≠ Ø do        {there are unmarked vertices}
    Choose w in V(G) \ M with D(w) as small as possible.
    Put w in M.
    for v in V(G) \ M do        {update D and P}
        if D(v) > D(w) + W(w, v) then
            Set D(v) := D(w) + W(w, v).        {less weight through w}
            Set P(v) := w.        {point back to w}
for v ∈ V(G) do
    Set W*(s, v) := D(v).
return the weights W*(s, v) and pointers P(v) for v ∈ V(G) \ {s}.    ■
```

We still need to verify that the pointers $P(v)$ returned by the algorithm describe min-paths from s. If $P(s)$ is never defined, then there must be no edge from s to v, and $D(v)$ must never change in an update step. In this case, $W^*(s, v) = W(s, v) = \infty$; i.e., there is no path at all from s to v. If $P(v)$ is defined initially to be s and then never changes, then again $W^*(s, v) = W(s, v)$, the edge from s to v is a min-path, and $P(v)$ correctly points backward to s. If $P(v)$ is defined and then changes during update steps, consider the last such update, say when w is chosen for M. From that point on, $D(v)$ never changes, so $W^*(s, v) = D(v) = D(w) + W(w, v)$, the edge from w to v is the final edge of a min-path from s to v, and $P(v)$ correctly points backward to w.

We could stop the algorithm if $D(w)$ is ever ∞, since all the vertices reachable from s will have been marked by then, and no values will change later. Note also that, after a vertex v goes into the marked set M, the values $D(v)$ and $P(v)$ never change. Of course, if we just want to know min-weights, then we can omit all parts of the algorithm that refer to P.

EXAMPLE 2

Consider Example 1 again. The drawings in Figure 3 show the situation at the beginning and after each pass through the `while` loop in the execution of Dijkstra's algorithm. The bold numbers in Figure 3 are the finite values of $D(v)$.

Figure 4 gives the values of M, $D(v)$, and $P(v)$ in a table. Each row in the table corresponds to one pass through the `while` loop. The minimum weights of paths from $s = v_2$ are the D-values in the last row. For example, min-paths from v_2 to v_6 have weight 8. To find one, use the backward pointer's values in the last row: $P(v_6) = v_4$, $P(v_4) = v_5$, and $P(v_5) = v_2$ give us the min-path $v_2\, v_5\, v_4\, v_6$. This is the only min-path from v_2 to v_6 in this example, but, in general, there can be several min-paths connecting two vertices. The algorithm will always produce just one, though.

Figure 4 ▶

M	$D(v_j)$						$P(v_j)$					
	v_1	v_3	v_4	v_5	v_6	v_7	v_1	v_3	v_4	v_5	v_6	v_7
$\{v_2\}$	∞	∞	7	**1**	∞	∞			v_2	$\mathbf{v_2}$		
$\{v_2, v_5\}$	∞	**5**	6	1	10	∞		$\mathbf{v_5}$	v_5	v_2	v_5	
$\{v_2, v_5, v_3\}$	∞	5	**6**	1	10	∞		v_5	$\mathbf{v_5}$	v_2	v_5	
$\{v_2, v_5, v_3, v_4\}$	∞	5	6	1	**8**	14		v_5	v_5	v_2	$\mathbf{v_4}$	v_4
$\{v_2, v_5, v_3, v_4, v_6\}$	∞	5	6	1	8	**12**		v_5	v_5	v_2	v_4	$\mathbf{v_6}$
$\{v_2, v_5, v_3, v_4, v_6, v_7\}$	∞	5	6	1	8	12		v_5	v_5	v_2	v_4	v_6

Since $D(v_1) = \infty$ at the end, v_1 is not reachable from v_2, which is clear from the digraph, of course. ■

Dijkstra's algorithm spends the largest part of its time going through the `while` loop, removing one of the original n vertices at each pass. The time to find the smallest $D(w)$ is $O(n)$ if we simply examine the vertices in $V(G) \setminus M$ one by one, but in fact there are sorting algorithms that will locate w faster. For each chosen vertex w, there are at most n comparisons and replacements, so the total time for one pass through the loop is $O(n)$. All told, the algorithm makes n passes, so it takes total time $O(n^2)$.

If the digraph is presented in terms of successor lists, then the algorithm can be rewritten so that the replacement–update step only looks at successors of w. During the total operation, each edge is then considered just once in an update step. Such a modification speeds up the overall performance if $|E(G)|$ is much less than n^2.

Dijkstra's algorithm finds the weights of min-paths from a given vertex. To find $W^*(v_i, v_j)$ for all pairs of vertices v_i and v_j, we could just apply the algorithm n times, starting from each of the n vertices. There is another algorithm, originally due to Warshall and refined by Floyd, that produces all of the values $W^*(v_i, v_j)$ and min-paths between vertices and that is easy to program. Like Dijkstra's algorithm, it builds an expanding list of examined vertices and looks at paths through vertices on the list.

The idea is again to keep enlarging the list of allowed intermediate vertices on paths, but in a different way. At first consider only paths with no intermediate vertices at all, i.e., edges. Next allow paths that go through v_1, then ones that go through v_1 or v_2 [or both], etc. Eventually, all vertices are allowed along the way, so all paths are considered.

Suppose that $V(G) = \{v_1, \ldots, v_n\}$. Warshall's algorithm works with an $n \times n$ matrix $\mathbf{W}$, which at the beginning is the edge-weight matrix $\mathbf{W}_0$ with $\mathbf{W}_0[i, j] = W(v_i, v_j)$ for all i and j and at the end is the min-weight matrix $\mathbf{W}_n = \mathbf{W}^*$ with $\mathbf{W}^*[i, j] = W^*(v_i, v_j)$.

The min-paths produced by Warshall's algorithm also can be described by a pointer function P. In this case, it is just as easy to have the pointers go in the forward direction so that at all times $(v_i, v_{P(i,j)})$ is the first edge in a path of smallest known weight from v_i to v_j, if such a path has been found. When the algorithm stops, $P(i, j) = 0$ if there is no path from v_i to v_j. Otherwise, the sequence

$$i, \quad P(i, j), \quad P(P(i, j), j), \quad P(P(P(i, j), j), j), \quad \ldots$$

lists the indices of vertices on a min-path from v_i to v_j. Since each subscript k on the list is followed by $P(k, j)$, this sequence is easy to define recursively using the function P.

We can describe P by an $n \times n$ matrix $\mathbf{P}$ with entries in $\{0, \ldots, n\}$. At the start, let $\mathbf{P}[i, j] = j$ if there is an edge from v_i to v_j, and let $\mathbf{P}[i, j] = 0$ otherwise. Whenever the algorithm discovers a path from v_i to v_j allowing $v_1, \ldots, v_k$ along the way that is better than the best path allowing $v_1, \ldots, v_{k-1}$, the pointer value $\mathbf{P}[i, j]$ is set equal to $\mathbf{P}[i, k]$; a better way to head from v_i to v_j is to head for v_k first. Of course, if we only want $\mathbf{W}^*$, then we can forget about $\mathbf{P}$. Here is the algorithm.

Warshall's Algorithm (edge-weight matrix)

```
{Input: A nonnegative edge-weight matrix W0 of a weighted digraph
        without loops or parallel edges}
{Output: The corresponding min-weight matrix W* and a pointer
        matrix P* giving min-paths for the digraph}
{Intermediate variables: Matrices W and P}
Set W := W0.
for i = 1 to n do        {this loop just initializes P}
   for j = 1 to n do
      if there is an edge from v_i to v_j then
         Set P[i, j] := j.
      else
         Set P[i, j] := 0.
for k = 1 to n do        {this loop does the work}
   for i = 1 to n do
      for j = 1 to n do
         if W[i, j] > W[i, k] + W[k, j] then
            Replace W[i, j] by W[i, k] + W[k, j].
            Replace P[i, j] by P[i, k].
Set W* := W and P*: = P.
return W* and P*.   ■
```

EXAMPLE 3

We apply Warshall's algorithm to the digraph shown in Figure 5. Hand calculations with Warshall's algorithm lead to n new matrices, one for each value of k. For this example, the matrices are the following.

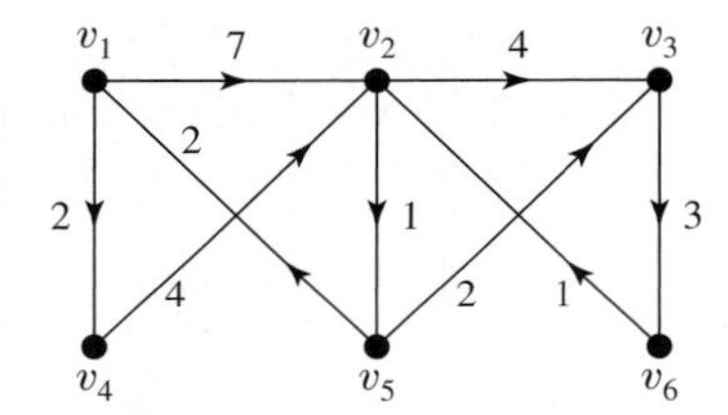

Figure 5 ▲

$$\mathbf{W} = \mathbf{W}_0 = \begin{bmatrix} \infty & 7 & \infty & 2 & \infty & \infty \\ \infty & \infty & 4 & \infty & 1 & \infty \\ \infty & \infty & \infty & \infty & \infty & 3 \\ \infty & 4 & \infty & \infty & \infty & \infty \\ 2 & \infty & 2 & \infty & \infty & \infty \\ \infty & 1 & \infty & \infty & \infty & \infty \end{bmatrix}, \quad \mathbf{W}_1 = \begin{bmatrix} \infty & 7 & \infty & 2 & \infty & \infty \\ \infty & \infty & 4 & \infty & 1 & \infty \\ \infty & \infty & \infty & \infty & \infty & 3 \\ \infty & 4 & \infty & \infty & \infty & \infty \\ 2 & 9 & 2 & 4 & \infty & \infty \\ \infty & 1 & \infty & \infty & \infty & \infty \end{bmatrix},$$

$$\mathbf{W}_2 = \begin{bmatrix} \infty & 7 & 11 & 2 & 8 & \infty \\ \infty & \infty & 4 & \infty & 1 & \infty \\ \infty & \infty & \infty & \infty & \infty & 3 \\ \infty & 4 & 8 & \infty & 5 & \infty \\ 2 & 9 & 2 & 4 & 10 & \infty \\ \infty & 1 & 5 & \infty & 2 & \infty \end{bmatrix}, \quad \mathbf{W}_3 = \begin{bmatrix} \infty & 7 & 11 & 2 & 8 & 14 \\ \infty & \infty & 4 & \infty & 1 & 7 \\ \infty & \infty & \infty & \infty & \infty & 3 \\ \infty & 4 & 8 & \infty & 5 & 11 \\ 2 & 9 & 2 & 4 & 10 & 5 \\ \infty & 1 & 5 & \infty & 2 & 8 \end{bmatrix},$$

$$\mathbf{W}_4 = \begin{bmatrix} \infty & 6 & 10 & 2 & 7 & 13 \\ \infty & \infty & 4 & \infty & 1 & 7 \\ \infty & \infty & \infty & \infty & \infty & 3 \\ \infty & 4 & 8 & \infty & 5 & 11 \\ 2 & 8 & 2 & 4 & 9 & 5 \\ \infty & 1 & 5 & \infty & 2 & 8 \end{bmatrix}, \quad \mathbf{W}_5 = \begin{bmatrix} 9 & 6 & 9 & 2 & 7 & 12 \\ 3 & 9 & 3 & 5 & 1 & 6 \\ \infty & \infty & \infty & \infty & \infty & 3 \\ 7 & 4 & 7 & 9 & 5 & 10 \\ 2 & 8 & 2 & 4 & 9 & 5 \\ 4 & 1 & 4 & 6 & 2 & 7 \end{bmatrix},$$

$$\mathbf{W}^* = \mathbf{W}_6 = \begin{bmatrix} 9 & 6 & 9 & 2 & 7 & 12 \\ 3 & 7 & 3 & 5 & 1 & 6 \\ 7 & 4 & 7 & 9 & 5 & 3 \\ 7 & 4 & 7 & 9 & 5 & 10 \\ 2 & 6 & 2 & 4 & 7 & 5 \\ 4 & 1 & 4 & 6 & 2 & 7 \end{bmatrix},$$

and

$$\mathbf{P}_0 = \begin{bmatrix} 0 & 2 & 0 & 4 & 0 & 0 \\ 0 & 0 & 3 & 0 & 5 & 0 \\ 0 & 0 & 0 & 0 & 0 & 6 \\ 0 & 2 & 0 & 0 & 0 & 0 \\ 1 & 0 & 3 & 0 & 0 & 0 \\ 0 & 2 & 0 & 0 & 0 & 0 \end{bmatrix}, \quad \mathbf{P}_1 = \begin{bmatrix} 0 & 2 & 0 & 4 & 0 & 0 \\ 0 & 0 & 3 & 0 & 5 & 0 \\ 0 & 0 & 0 & 0 & 0 & 6 \\ 0 & 2 & 0 & 0 & 0 & 0 \\ 1 & 1 & 3 & 1 & 0 & 0 \\ 0 & 2 & 0 & 0 & 0 & 0 \end{bmatrix},$$

$$\mathbf{P}_2 = \begin{bmatrix} 0 & 2 & 2 & 4 & 2 & 0 \\ 0 & 0 & 3 & 0 & 5 & 0 \\ 0 & 0 & 0 & 0 & 0 & 6 \\ 0 & 2 & 2 & 0 & 2 & 0 \\ 1 & 1 & 3 & 1 & 1 & 0 \\ 0 & 2 & 2 & 0 & 2 & 0 \end{bmatrix}, \quad \mathbf{P}_3 = \begin{bmatrix} 0 & 2 & 2 & 4 & 2 & 2 \\ 0 & 0 & 3 & 0 & 5 & 3 \\ 0 & 0 & 0 & 0 & 0 & 6 \\ 0 & 2 & 2 & 0 & 2 & 2 \\ 1 & 1 & 3 & 1 & 1 & 3 \\ 0 & 2 & 2 & 0 & 2 & 2 \end{bmatrix},$$

$$\mathbf{P}_4 = \begin{bmatrix} 0 & 4 & 4 & 4 & 4 & 4 \\ 0 & 0 & 3 & 0 & 5 & 3 \\ 0 & 0 & 0 & 0 & 0 & 6 \\ 0 & 2 & 2 & 0 & 2 & 2 \\ 1 & 1 & 3 & 1 & 1 & 3 \\ 0 & 2 & 2 & 0 & 2 & 2 \end{bmatrix}, \quad \mathbf{P}_5 = \begin{bmatrix} 4 & 4 & 4 & 4 & 4 & 4 \\ 5 & 5 & 5 & 5 & 5 & 5 \\ 0 & 0 & 0 & 0 & 0 & 6 \\ 2 & 2 & 2 & 2 & 2 & 2 \\ 1 & 1 & 3 & 1 & 1 & 3 \\ 2 & 2 & 2 & 2 & 2 & 2 \end{bmatrix},$$

$$\mathbf{P}^* = \mathbf{P}_6 = \begin{bmatrix} 4 & 4 & 4 & 4 & 4 & 4 \\ 5 & 5 & 5 & 5 & 5 & 5 \\ 6 & 6 & 6 & 6 & 6 & 6 \\ 2 & 2 & 2 & 2 & 2 & 2 \\ 1 & 3 & 3 & 1 & 3 & 3 \\ 2 & 2 & 2 & 2 & 2 & 2 \end{bmatrix}.$$

We illustrate the computations by calculating $\mathbf{W}_4[5, 2]$ and $\mathbf{P}_4[5, 2]$. Here $k = 4$. The entry $\mathbf{W}_3[5, 2]$ is 9, and $\mathbf{P}_3[5, 2] = 1$, corresponding to the shortest path $v_5\, v_1\, v_2$ with intermediate vertices from $\{v_1, v_2, v_3\}$. To find $\mathbf{W}_4[5, 2]$, we look at $\mathbf{W}_3[5, 4] + \mathbf{W}_3[4, 2]$, which is $4 + 4 = 8$, corresponding to the pair of paths $v_5\, v_1\, v_4$ and $v_4\, v_2$. Since $9 > 8$, we replace $\mathbf{W}_3[5, 2]$ by $\mathbf{W}_3[5, 4] + \mathbf{W}_3[4, 2] = 8$, corresponding to $v_5\, v_1\, v_4\, v_2$. At the same time, we replace $\mathbf{P}_3[5, 2]$ by $\mathbf{P}_3[5, 4] = 1$, which in this case is no change; the improvement in the path from v_5 to v_2 came in the part from v_1 onward. As you can perhaps see from this example, the computations involved in Warshall's algorithm, while not difficult, are not well suited to hand calculation. A given entry, such as $\mathbf{W}[5, 2]$, may change several times during the calculations as k runs through all possible values, and the intermediate values of $\mathbf{W}$ and $\mathbf{P}$ are not particularly informative.

We can check the values returned by the algorithm for this example, since only a few vertices are involved. Most of the rows of $\mathbf{P}_6$ are clearly correct. For instance, *every* path from v_4 starts by going to v_2. Look at the entries in the fifth row and convince yourself that they are correct.

The matrix $\mathbf{W}^*$ provides us with the weights of min-paths, and the matrix $\mathbf{P}^*$ of pointers provides us with min-paths between vertices. For example, the min-paths from v_1 to v_3 have weight 9 because $\mathbf{W}^*[1, 3] = 9$. Since $\mathbf{P}^*[1, 3] = 4$, $\mathbf{P}^*[4, 3] = 2$, $\mathbf{P}^*[2, 3] = 5$, and $\mathbf{P}^*[5, 3] = 3$, the path $v_1\, v_4\, v_2\, v_5\, v_3$ is a min-path from v_1 to v_3.

The diagonal entries $\mathbf{W}^*[i, i]$ and $\mathbf{P}^*[i, i]$ in this example give some information about cycles. For instance, $\mathbf{W}^*[2, 2] = 7$ indicates that there is a cycle through v_2 of weight 7 whose vertex sequence $v_2\, v_5\, v_3\, v_6\, v_2$ can be deduced from $\mathbf{P}^*[2, 2] = 5$, $\mathbf{P}^*[5, 2] = 3$, $\mathbf{P}^*[3, 2] = 6$, and $\mathbf{P}^*[6, 2] = 2$. ■

Now that we have seen an example of the algorithm's operation, let's look at why it works.

Theorem 1 Warshall's algorithm produces the min-weight matrix $\mathbf{W}^*$.

Proof We have written the algorithm as a nest of **for** loops. We can also present it in the following form, which looks more like Dijkstra's algorithm.

> Set $\mathbf{W} := \mathbf{W}_0$, $L := \emptyset$, $V := \{1, \ldots, n\}$.
> **for** each i and j in V **do**
>   Set $\mathbf{P}[i, j] := j$ if $\mathbf{W}[i, j] \neq \infty$ and $\mathbf{P}[i, j] := 0$ otherwise.
> **while** $V \setminus L \neq \emptyset$ **do**
>   Choose $k \in V \setminus L$ and put k in L.
>   **for** each i, j in V **do**

Set $\mathbf{W}[i, j] := \min\{\mathbf{W}[i, j], \mathbf{W}[i, k] + \mathbf{W}[k, j]\}$, and if this is a change then set $\mathbf{P}[i, j] := k$.

return $\mathbf{W}$ and $\mathbf{P}$

The assignment $\mathbf{W}[i, j] := \min\{\mathbf{W}[i, j], \mathbf{W}[i, k] + \mathbf{W}[k, j]\}$ here has the same effect as our previous "**if** ... **then** Replace ... " segment; the value of $\mathbf{W}[i, j]$ changes only if $\mathbf{W}[i, j] > \mathbf{W}[i, k] + \mathbf{W}[k, j]$, i.e., only if things get better when we are allowed to go through v_k.

Similarly to our usage for Dijkstra's algorithm, we will call a path with intermediate vertices v_i for i in L an L-path and will call an L-path of minimal weight an L-min-path. We will show that the following statement is an invariant of the **while** loop.

For all $i, j \in \{1, \dots, n\}$, if $\mathbf{W}[i, j] \neq \infty$, then the number $\mathbf{W}[i, j]$ is the weight of an L-min-path from v_i to v_j that begins with the edge from v_i to $v_{\mathbf{P}[i,j]}$.

We will also show that the statement holds at the start of the algorithm, and it will follow that it holds on exit from the loop when $L = \{1, \dots, n\}$, which is what we want.

At the start L is empty, so an L-path from v_i to v_j is simply the edge from v_i to v_j, if there is one, or is a virtual path if there is no edge. In either case it is an L-min-path, its weight is $\mathbf{W_0}[i, j]$, and, if it's not virtual, then $\mathbf{P_0}[i, j] = j$.

Assume now that the statement is true at the start of the pass in which k is chosen. The smallest weight of an $L \cup \{k\}$-path from i to j that goes through k is $\mathbf{W}[i, k] + \mathbf{W}[k, j]$. If this number is less than $\mathbf{W}[i, j]$, then it is used as the new value of $\mathbf{W}[i, j]$. Otherwise, we can do at least as well with an $L \cup \{k\}$-path that misses k, i.e., with an L-path; so the current value of $\mathbf{W}[i, j]$, which is not changed, is the correct weight of an $L \cup \{k\}$-min-path from i to j. If $\mathbf{W}[i, j]$ changes, then the value of $\mathbf{P}[i, j]$ is set to $\mathbf{P}[i, k]$, so the first edge on the $L \cup \{k\}$-min-path from v_i to v_j is the first edge on the L-min-path to v_k. This argument proves that the statement is a loop invariant, as claimed. ■

It is easy to analyze how long Warshall's algorithm takes. The comparison–replacement step inside the j-loop takes at most some fixed amount of time, say t. The step is done exactly n^3 times, once for each possible choice of the triple (k, i, j), so the total time to execute the algorithm is $n^3 t$, which is $O(n^3)$.

The comparison–replacement step for Warshall's algorithm is the same as the one in Dijkstra's algorithm, which includes other sorts of steps as well. Since Dijkstra's algorithm can be done in $O(n^2)$ time, doing it once for each of the n vertices gives an $O(n^3)$ algorithm to find all min-weights. The time constants involved in this multiple Dijkstra's algorithm are different from the ones for Warshall's algorithm, however, so the choice of which algorithm to use may depend on the computer implementation available. If $|E(G)|$ is small compared with n^2, a successor list presentation of the digraph favors choosing Dijkstra's algorithm.

This is probably the time to confess that we are deliberately ignoring one possible complication. Even if G has only a handful of vertices, the weights $W(i, j)$ could still be so large that it would take years to write them down, so both Dijkstra's and Warshall's algorithms might take a long time. In practical situations, though, the numbers that come up in the applications of these two algorithms are of manageable size. In any case, given the same set of weights, our comparison of the relative run times for the algorithms is still valid.

The nonnegativity of edge weights gets used in a subtle way in the proof of Theorem 1. For example, if $\mathbf{W}[k, k] < 0$ for some k, then $\mathbf{W}[k, k] + \mathbf{W}[k, j] < \mathbf{W}[k, j]$, but there is no smallest $L \cup \{k\}$-path weight from k to j. We can go around some path of negative weight from k to itself as many times as we like before setting out for j. One way to rule out this possibility is to take nonnegative edge weights to begin with, as we have done, so that every replacement value is also nonnegative. One can permit negative input weights for Warshall's algorithm by requiring the digraph to be acyclic so that looping back through k cannot occur.

Warshall's algorithm can be adapted to find max-weights in an acyclic digraph. Replace all the ∞'s by $-\infty$'s, with $-\infty + x = -\infty = x + (-\infty)$ for all x, and $-\infty < a$ for all real numbers a. Change the inequality in the replacement step to $\mathbf{W}[i, j] < \mathbf{W}[i, k] + \mathbf{W}[k, j]$. The resulting algorithm computes $\mathbf{W}_n[i, j] = M(v_i, v_j)$, where M is the max-weight function of §8.2.

If we just want max-weights from a single source, as we might for a scheduling network, we can simplify the algorithm somewhat, at the cost of first relabeling the digraph, using an algorithm such as NumberVertices on page 321 or Label on page 296. It seems natural in this setting to use a **reverse sorted labeling**, in which the labels are arranged so that $i < j$ if there is a path from v_i to v_j. Such a labeling is easy to obtain from an ordinary sorted labeling; for $i = 1, \ldots, n$, simply relabel v_i as v_{n+1-i}.

To find max-weights from a given vertex v_s, for instance from v_1, just fix $i = s$ in the max-modified Warshall's algorithm. The reverse sorted labeling also means that $\mathbf{W}[k, j] = -\infty$ if $j \le k$, so the j-loop does not need to go all the way from 1 to n. Here is the resulting algorithm, with $i = s = 1$. A pointer function can be added if desired [Exercise 15(a)].

MaxWeight(edge-weight matrix)

{Input: An edge-weight matrix $\mathbf{W}_0$ of an acyclic digraph with a reverse sorted labeling}
{Output: The max-weights $M(1, j)$ for $j = 2, \ldots, n$}
for $k = 2$ to $n - 1$ **do**
  for $j = k + 1$ to n **do**
    if $\mathbf{W}[1, j] < \mathbf{W}[1, k] + \mathbf{W}[k, j]$ **then**
      Replace $\mathbf{W}[1, j]$ by $\mathbf{W}[1, k] + \mathbf{W}[k, j]$.
return $\mathbf{W}(1, 2), \ldots, \mathbf{W}(1, n)$ ■

We emphasize that this algorithm is only meant for acyclic digraphs with reverse sorted labelings. The proof that it works in time $O(n^2)$ [Exercise 16] is similar to the argument for Warshall's algorithm. The algorithm can be speeded up somewhat if the digraph is given by successor lists and if we just consider j in SUCC(k). Then each edge gets examined exactly once to see if it enlarges max-weights. The algorithm then runs in a time of $O(\max\{|V(G)|, |E(G)|\})$, which is comparable to the time it takes to sort the vertices initially using the algorithm Label.

Dijkstra's algorithm does not work with negative weights [Exercise 10]. Moreover [Exercise 11], there seems to be no natural way to modify it to find max-weights.

EXAMPLE 4

We apply MaxWeight to the digraph of Figure 6(a). Figure 6(b) gives the initial matrix $\mathbf{W}_0$. For convenience, we use a row matrix $\mathbf{D}$ with $\mathbf{D}[j] = \mathbf{W}[1, j]$ for $j = 1, \ldots, 6$. The sequence of matrices is as follows.

$$\mathbf{D}_0 = \mathbf{D}_1 = [\ -\infty \quad 1 \quad 2 \quad -\infty \quad -\infty \quad -\infty\ ],$$

$$\mathbf{D}_2 = [\ -\infty \quad 1 \quad 2 \quad 3 \quad 4 \quad -\infty\ ],$$

$$\mathbf{D}_3 = [\ -\infty \quad 1 \quad 2 \quad 7 \quad 4 \quad -\infty\ ],$$

$$\mathbf{D}_4 = \mathbf{D}_5 = [\ -\infty \quad 1 \quad 2 \quad 7 \quad 4 \quad 9\ ].$$

Figure 6 ▶

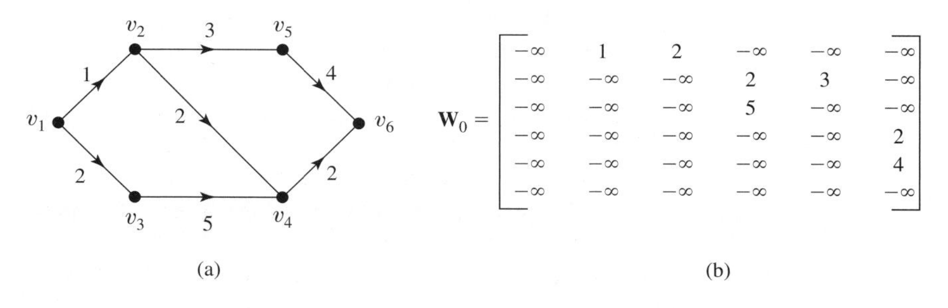

(a) (b)

As an illustration, we compute $\mathbf{D}_4$, assuming that $\mathbf{D}_3$ gives the right values of $\mathbf{W}[1, j]$ for $k = 3$. The j-loop for $k = 4$ is

```
for j = 5 to 6 do
    if D[j] < D[4] + W[4, j] then
        Replace D[j] by D[4] + W[4, j].
```

Since $\mathbf{D}_3[5] = 4 > -\infty = 7 + (-\infty) = \mathbf{D}_3[4] + \mathbf{W}[4, 5]$, we make no replacement and get $\mathbf{D}_4[5] = \mathbf{D}_3[5] = 4$. Since $\mathbf{D}_3[6] = -\infty < 9 = 7 + 2 = \mathbf{D}_3[4] + \mathbf{W}[4, 6]$, we make a replacement and obtain $\mathbf{D}_4[6] = 9$. The values of $\mathbf{D}_3[1], \ldots, \mathbf{D}_3[4]$ are, of course, unchanged in $\mathbf{D}_4$. ■

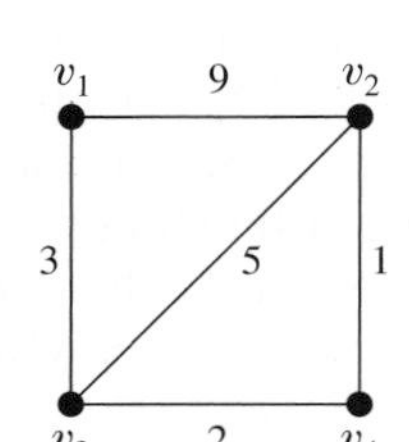

Figure 7 ▲

Either Dijkstra's or Warshall's algorithm can be used on undirected graphs, in effect by replacing each undirected edge $\{u, v\}$ such that $u \neq v$ by two directed edges (u, v) and (v, u). If the undirected edge has weight w, assign each of the directed edges the same weight w. If the graph is unweighted, assign weight 1 to all edges. Loops are really irrelevant in the applications of min-paths to undirected graphs, so it is convenient when using the algorithms to set $W(i, i) = 0$ for all i.

EXAMPLE 5

Consider the weighted graph shown in Figure 7. We use Warshall's algorithm to find min-weights and min-paths between vertices, allowing travel in either direction along the edges. Figure 8 gives the successive values of $\mathbf{W}$ and $\mathbf{P}$. As a sample, we calculate $\mathbf{P}_4[2, 1]$. Since $\mathbf{W}_3[2, 4] + \mathbf{W}_3[4, 1] = 1 + 5 = 6 < 8 = \mathbf{W}_3[2, 1]$, we have $\mathbf{W}_4[2, 1] = 6$ and also $\mathbf{P}_4[2, 1] = \mathbf{P}_3[2, 4] = 4$. The min-path from v_2 to v_1 is described by the sequence 2, $\mathbf{P}^*[2, 1] = 4$, $\mathbf{P}^*[4, 1] = 3$, $\mathbf{P}^*[3, 1] = 1$; i.e., the min-path is $v_2\, v_4\, v_3\, v_1$.

Figure 8 ▶

$$\mathbf{W}_0 = \mathbf{W}_1 = \begin{bmatrix} 0 & 9 & 3 & \infty \\ 9 & 0 & 5 & 1 \\ 3 & 5 & 0 & 2 \\ \infty & 1 & 2 & 0 \end{bmatrix} \qquad \mathbf{P}_0 = \mathbf{P}_1 = \begin{bmatrix} 0 & 2 & 3 & 0 \\ 1 & 0 & 3 & 4 \\ 1 & 2 & 0 & 4 \\ 0 & 2 & 3 & 0 \end{bmatrix}$$

$$\mathbf{W}_2 = \begin{bmatrix} 0 & 9 & 3 & 10 \\ 9 & 0 & 5 & 1 \\ 3 & 5 & 0 & 2 \\ 10 & 1 & 2 & 0 \end{bmatrix} \qquad \mathbf{P}_2 = \begin{bmatrix} 0 & 2 & 3 & 2 \\ 1 & 0 & 3 & 4 \\ 1 & 2 & 0 & 4 \\ 2 & 2 & 3 & 0 \end{bmatrix}$$

$$\mathbf{W}_3 = \begin{bmatrix} 0 & 8 & 3 & 5 \\ 8 & 0 & 5 & 1 \\ 3 & 5 & 0 & 2 \\ 5 & 1 & 2 & 0 \end{bmatrix} \qquad \mathbf{P}_3 = \begin{bmatrix} 0 & 3 & 3 & 3 \\ 3 & 0 & 3 & 4 \\ 1 & 2 & 0 & 4 \\ 3 & 2 & 3 & 0 \end{bmatrix}$$

$$\mathbf{W}^* = \mathbf{W}_4 = \begin{bmatrix} 0 & 6 & 3 & 5 \\ 6 & 0 & 3 & 1 \\ 3 & 3 & 0 & 2 \\ 5 & 1 & 2 & 0 \end{bmatrix} \qquad \mathbf{P}^* = \mathbf{P}_4 = \begin{bmatrix} 0 & 3 & 3 & 3 \\ 4 & 0 & 4 & 4 \\ 1 & 4 & 0 & 4 \\ 3 & 2 & 3 & 0 \end{bmatrix}$$

The diagonal entries of $\mathbf{P}$ in this example have no significance. ■

Warshall's algorithm can be used to find reachable sets $R(v)$ for a digraph. Simply give all edges, including loops, weight 1. The final value $\mathbf{W}^*[i, j]$ is ∞ if there is no path from v_i to v_j, and $\mathbf{W}^*[i, j]$ is a positive integer if a path exists. In particular, $\mathbf{W}^*[i, i] < \infty$ if and only if v_i is a vertex of a cycle in G. We can test whether a digraph G is acyclic by applying Warshall's algorithm and looking at the diagonal entries of $\mathbf{W}^*$. In §6.6 we saw acyclicity checks based on Kruskal's algorithm and Forest that are even faster.

Exercises 8.3

1. (a) Give the min-weight matrix $\mathbf{W}^*$ for the digraph shown in Figure 9(a). Any method is allowed, including staring at the picture.
   (b) Repeat part (a) for the digraph in Figure 9(b).
   (c) Give the initial and final min-path pointer matrices $\mathbf{P}$ for Warshall's algorithm applied to the digraph in Figure 9(a). Again any method is allowed.
   (d) Repeat part (c) for the digraph in Figure 9(b).

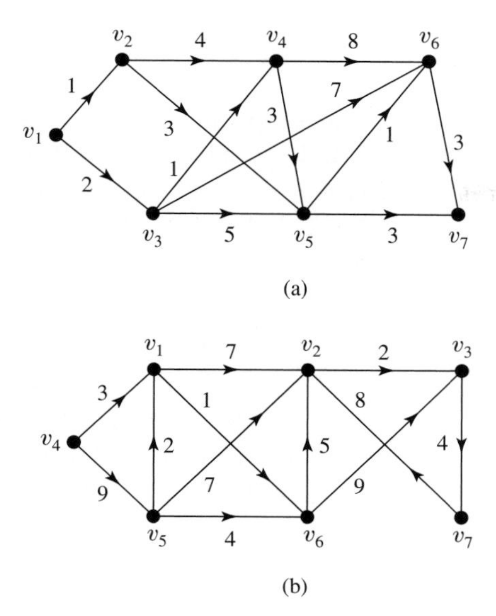

Figure 9 ▲

2. Find the max-weight matrix for the digraph of Figure 9(a).
3. (a) Apply Dijkstra's algorithm to the digraph of Figure 9(a). Start at v_1 and use the format of Figure 4.
   (b) Repeat part (a) for the digraph of Figure 9(b).
   (c) Repeat part (a) for the digraph of Figure 10(a).
   (d) Repeat part (a) for the digraph of Figure 10(b).

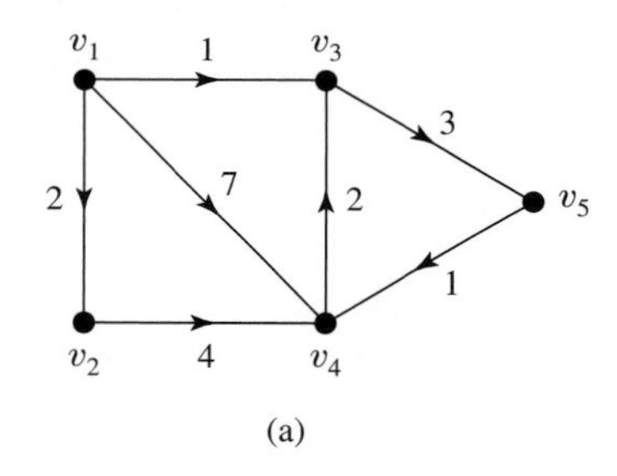

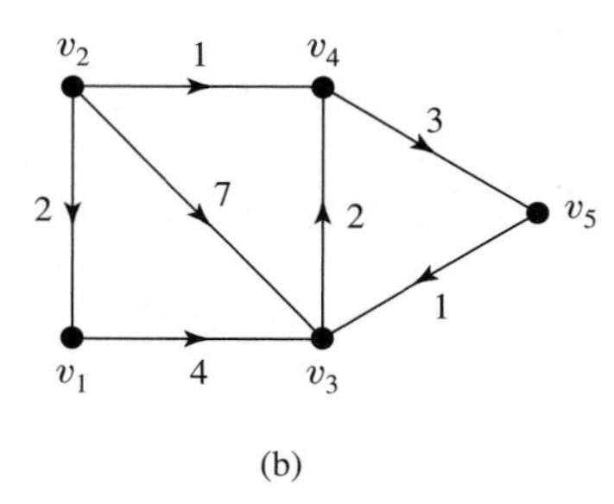

Figure 10 ▲

4. Apply Dijkstra's algorithm to the digraph of Figure 5, starting at v_1. Compare your answer with the answer obtained by Warshall's algorithm in Example 3.
5. (a) Use Warshall's algorithm to find minimum path lengths in the digraph of Figure 11. Give the matrix for $\mathbf{W}$ at the start of each k-loop.

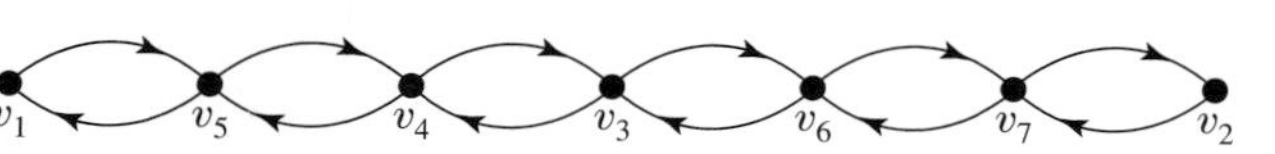

Figure 11 ▲

   (b) Use Dijkstra's algorithm on this digraph to find minimum path lengths from v_1. Write your answer in the format of Figure 4.
6. (a) Use Warshall's algorithm to find $\mathbf{W}^*$ for the digraph of Figure 6(a).
   (b) Find max-weights for the same digraph using the modified Warshall's algorithm.
7. (a) Use Warshall's algorithm to find $\mathbf{P}^*$ for the digraph of Figure 12(a).
   (b) Repeat part (a) for max-paths instead of min-paths.

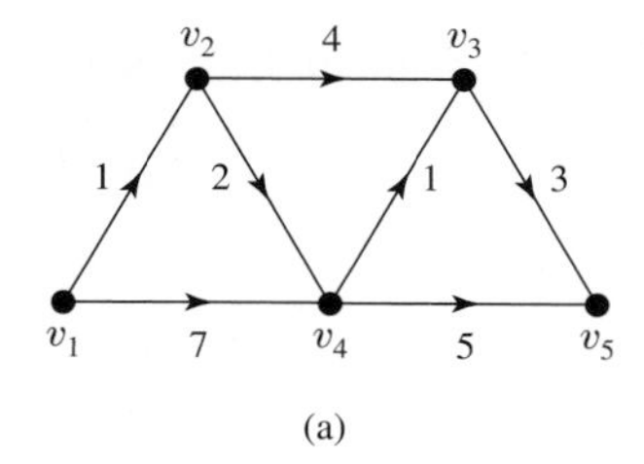

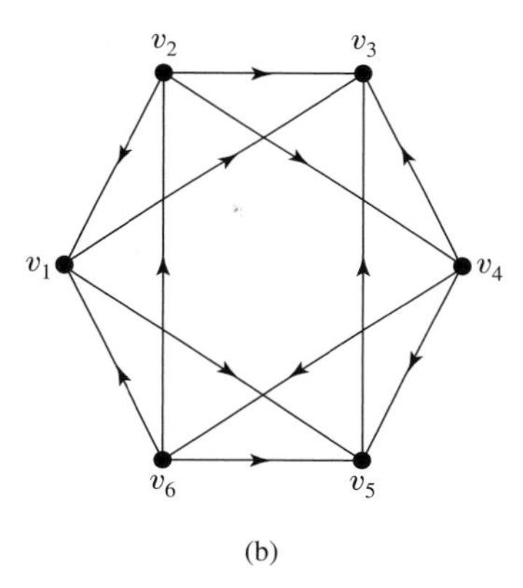

Figure 12 ▲

8. Apply Dijkstra's algorithm to find min-weights from v_3 in the digraph of Figure 9(a).
9. (a) Use the algorithm MaxWeight to find max-weights from v_1 to the other vertices in the digraph of Figure 9(a).
   (b) Use MaxWeight to find max-weights from s to the other vertices in the digraph of Figure 13. [Start by sorting the digraph.]

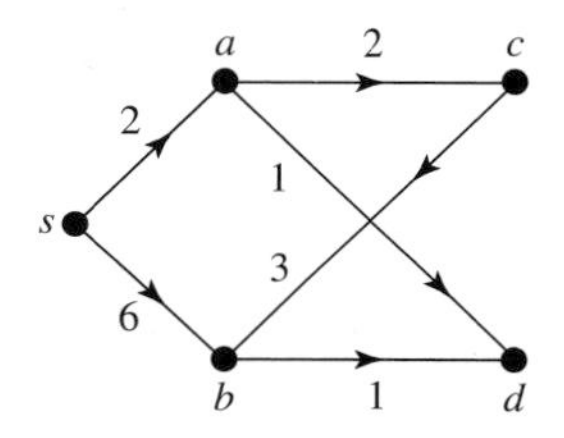

Figure 13 ▲

10. (a) Show that Dijkstra's algorithm does not produce correct min-weights from v_1 for the acyclic digraph shown in Figure 14. [So negative weights can cause trouble.]

(b) Would Dijkstra's algorithm give the correct min-weights from v_1 for this digraph if the "**for** v in $V(G) \setminus M$ **do**" line of the algorithm were replaced by "**for** v in V **do**"? Explain.

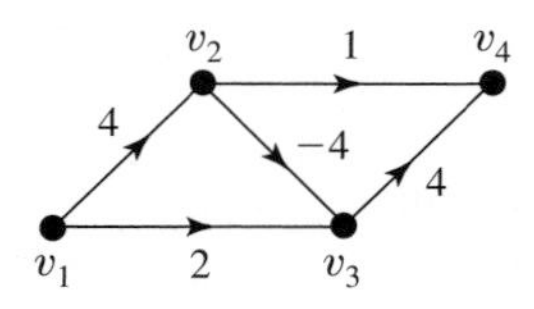

Figure 14 ▲

11. This exercise shows some of the difficulties in trying to modify Dijkstra's algorithm to get max-weights. Change the replacement step in Dijkstra's algorithm to the following:

if $D(v) < D(w) + W(w, v)$ **then**
  Replace $D(v)$ by $D(w) + W(w, v)$.

(a) Suppose that this modified algorithm chooses w in $V(G) \setminus M$ with $D(w)$ as large as possible. Show that the new algorithm fails to give the right answer for the digraph in Figure 15(a). [Start at v_1.]

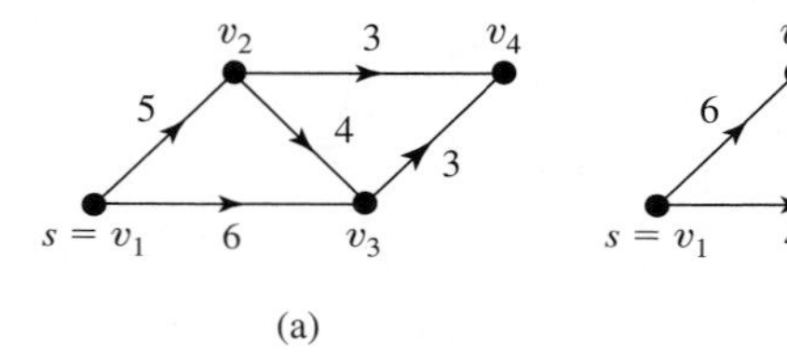

(a) (b)

Figure 15 ▲

(b) Suppose that the modified algorithm instead chooses w in $V(G) \setminus M$ with $D(w)$ nonnegative, but as small as possible. Show that the new algorithm fails for the digraph in Figure 15(b).

(c) Would it help in either part (a) or (b) if "**for** v in $V(G) \setminus M$ **do**" were replaced by "**for** v in $V(G)$ **do**"?

12. The **reachability matrix** $\mathbf{M}_R$ for a digraph is defined by $\mathbf{M}_R[i, j] = 1$ if $v_j \in R(v_i)$ and $\mathbf{M}_R[i, j] = 0$ otherwise.

(a) Find the reachability matrix $\mathbf{M}_R$ for the digraph of Figure 12(b) by using Warshall's algorithm.

(b) Is this digraph acyclic?

13. (a) Draw a picture of the digraph with weight matrix

$$\mathbf{W}_0 = \begin{bmatrix} 0 & 0 & 1 & 0 & 0 & 0 \\ 0 & 0 & 0 & 1 & 0 & 0 \\ 1 & 0 & 0 & 0 & 1 & 0 \\ 0 & 1 & 0 & 0 & 0 & 1 \\ 0 & 0 & 1 & 0 & 0 & 0 \\ 0 & 0 & 0 & 1 & 0 & 0 \end{bmatrix}.$$

(b) Is this digraph acyclic?

(c) Find the reachability matrix $\mathbf{M}_R$ [see Exercise 12] for this digraph.

14. Repeat Exercise 13 for the digraph with weight matrix

$$\mathbf{W}_0 = \begin{bmatrix} 0 & 0 & 0 & 0 & 0 & 0 \\ 1 & 0 & 0 & 0 & 0 & 0 \\ 0 & 1 & 0 & 0 & 0 & 0 \\ 0 & 0 & 1 & 0 & 0 & 0 \\ 0 & 0 & 0 & 1 & 0 & 0 \\ 0 & 0 & 0 & 0 & 1 & 0 \end{bmatrix}.$$

15. (a) Modify the algorithm MaxWeight to get an algorithm that finds pointers along max-paths from a single vertex.

(b) Apply your algorithm from part (a) to find max-path pointers from v_1 in the digraph of Figure 6(a).

16. (a) Show that MaxWeight finds max-weights from v_1 to other vertices. *Suggestion:* Show by induction on k that MaxWeight and the max-modified Warshall's algorithm produce the same values for $\mathbf{W}_k[1, j]$ at the end of the kth pass through their respective loops.

(b) Show that MaxWeight operates in time $O(n^2)$.

17. Modify Warshall's algorithm so that $\mathbf{W}[i, j] = 0$ if no path has been discovered from v_i to v_j and $\mathbf{W}[i, j] = 1$ if a path is known. *Hint:* Initialize suitably and use $\min\{\mathbf{W}[i, k], \mathbf{W}[k, j]\}$.

Chapter Highlights

As usual: What does it mean? Why is it here? How can I use it? Think of examples.

CONCEPTS AND NOTATION

sink, source
sorted labeling, reverse sorted labeling
indegree, outdegree
weighted digraph
  min-weight, min-path, $W^*(u, v)$
  max-weight, max-path, $M(u, v)$
scheduling network
  critical path, edge
  $A(v)$, $L(v)$

slack time $S(v)$ of a vertex
float time $F(u, v)$ of an edge
pointer

FACTS

A path in a digraph is acyclic if and only if its vertices are distinct.

Every finite acyclic digraph has a sorted labeling [also known from Chapter 7].

A digraph has a closed path using all edges if and only if it is connected as a graph and every vertex has the same indegree as outdegree.

ALGORITHMS

Sink to find a sink in a finite acyclic digraph.

NumberVertices to give a sorted labeling of an acyclic digraph in time $O(|V(G)|^2)$.

Dijkstra's algorithm to compute min-weights [-paths] from a selected vertex in time $O(|V(G)|^2)$ [or better], given a sorted labeling.

Warshall's algorithm to compute min-weights [-paths] or max-weights [-paths] between all pairs of vertices, as well as to determine reachability, in time $O(|V(G)|^3)$.

MaxWeight to find max-weights [-paths] from a selected vertex in time $O(|V(G)|^2)$ or $O(|V(G)| + |E(G)|)$, given a reverse sorted labeling.

Dijkstra's and Warshall's algorithms to give min-weights [-paths] for undirected graphs [with each $W(i, i)$ initialized to 0].

Supplementary Exercises

1. The sketch shows a digraph.

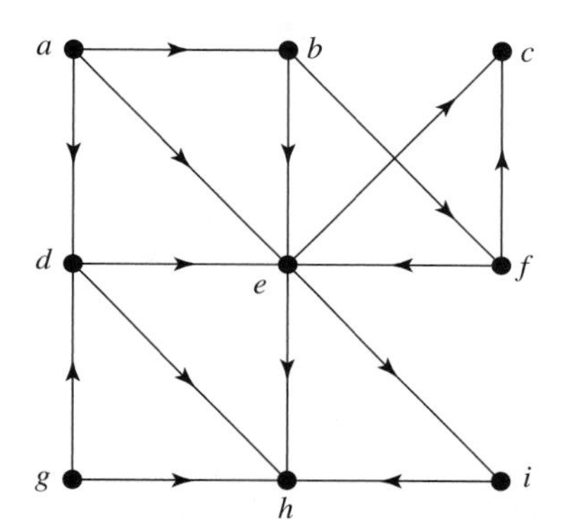

(a) Is this digraph acyclic?

(b) Does this digraph have an Euler path?

(c) List all the sinks of this digraph.

(d) List all the sources of this digraph.

(e) Give a sorted labeling of this digraph.

2. A digraph has vertices a, b, c, d, e, f, g with successor sets $\text{SUCC}(a) = \emptyset$, $\text{SUCC}(b) = \{c, d, e\}$, $\text{SUCC}(c) = \{e, f\}$, $\text{SUCC}(d) = \emptyset$, $\text{SUCC}(e) = \{d\}$, $\text{SUCC}(f) = \{f\}$, and $\text{SUCC}(g) = \{d, e, f\}$.

(a) Is this digraph acyclic? Explain.

(b) Without drawing the digraph, determine the sources and sinks.

(c) Draw a picture of the digraph.

(d) Is it possible to give a sorted labeling to this digraph? Explain.

3. Consider the scheduling network shown.

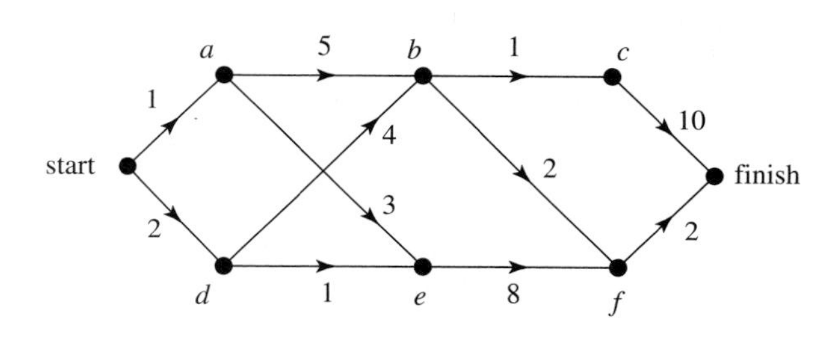

(a) Find $A(v)$, $L(v)$, and $S(v)$ for each vertex v.

(b) Give the float time of each edge for the network.

(c) This network has two critical paths. Give their vertex sequences.

4. (a) Give the table of W^* for the digraph shown.

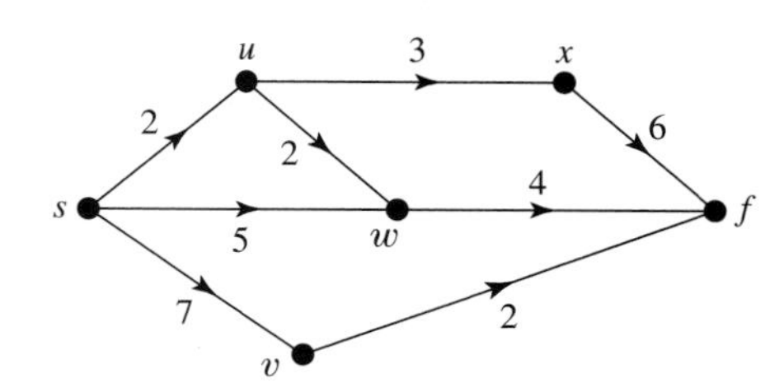

(b) List the edges in a critical path from s to f for this digraph viewed as a scheduling network.

(c) Give the slack times at u and at v.

5. The following digraph is acyclic.

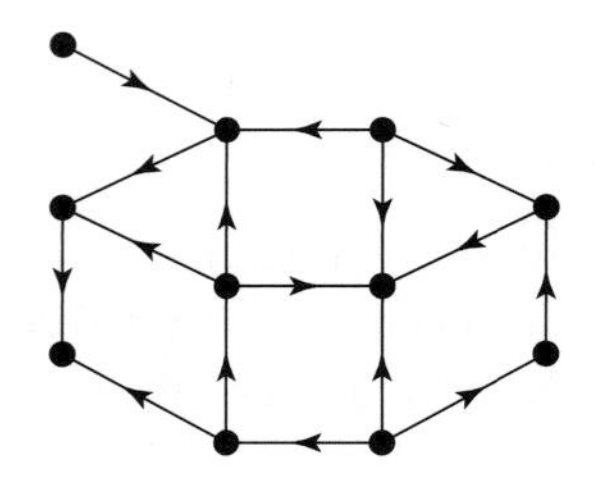

(a) Number its vertices so that $i > j$ if there is a path from vertex i to vertex j.

(b) List the sinks and sources of this digraph.

(c) Is every sink reachable from every source? Explain.

6. Consider the scheduling network shown. Determine the arrival and leave time for each vertex and the float time of each edge. Find all critical paths.

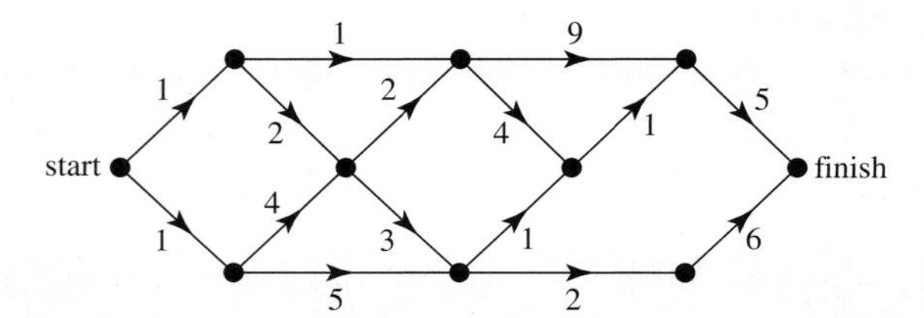

7. Ann says that reducing the time for an edge in a critical path will reduce the overall time from start to finish for a scheduling network. Bob says that may be true sometimes, but not always. Who is right?

8. Ann says that since she can find all the critical paths in a scheduling network by just finding the vertices v with $S(v) = 0$, there's no point in studying float times of edges. Bob, who has worked Exercise 18 on page 332, says she's not quite right, that $S(v) = 0$ if and only if v is on a critical path, but that you can't always figure out what the path is from that information. Who is right?

9. Consider the digraph shown.

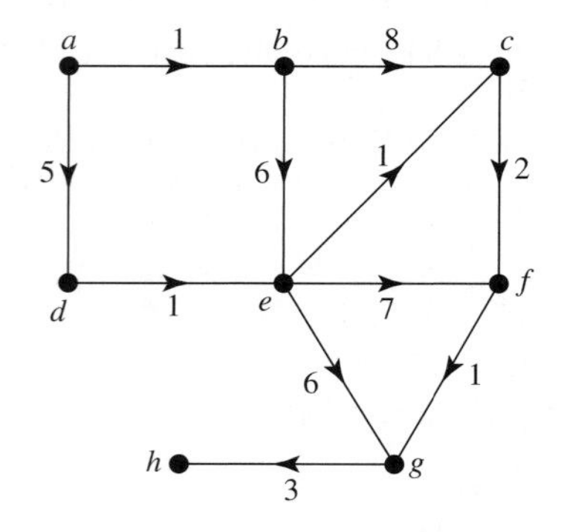

(a) Apply Dijkstra's algorithm to this digraph, starting with vertex a. Using your answer, give the total weight of the minimal path from a to h, and find this path.

(b) Repeat part (a), starting from vertex b.

10. (a) Suppose that we want to use Warshall's algorithm to solve the two problems in Exercise 9. Must we use two starting matrix pairs, or is one enough?

(b) Give starting matrices $\mathbf{W}_0$ and $\mathbf{P}_0$ for Warshall's algorithm on the digraph of Exercise 9.

(c) The final matrices $\mathbf{W}^*$ and $\mathbf{P}^*$ for this digraph are given below. Use these matrices to solve Exercises 9(a) and (b) again.

$$\mathbf{W}^* = \begin{bmatrix} \infty & 1 & 7 & 5 & 6 & 9 & 10 & 13 \\ \infty & \infty & 7 & \infty & 6 & 9 & 10 & 13 \\ \infty & \infty & \infty & \infty & \infty & 2 & 3 & 6 \\ \infty & \infty & 2 & \infty & 1 & 4 & 5 & 8 \\ \infty & \infty & 1 & \infty & \infty & 3 & 4 & 7 \\ \infty & \infty & \infty & \infty & \infty & \infty & 1 & 4 \\ \infty & \infty & \infty & \infty & \infty & \infty & \infty & 3 \\ \infty & \infty & \infty & \infty & \infty & \infty & \infty & \infty \end{bmatrix},$$

$$\mathbf{P}^* = \begin{bmatrix} 0 & 2 & 4 & 4 & 4 & 4 & 4 & 4 \\ 0 & 0 & 5 & 0 & 5 & 5 & 5 & 5 \\ 0 & 0 & 0 & 0 & 0 & 6 & 6 & 6 \\ 0 & 0 & 5 & 0 & 5 & 5 & 5 & 5 \\ 0 & 0 & 3 & 0 & 0 & 3 & 3 & 3 \\ 0 & 0 & 0 & 0 & 0 & 0 & 7 & 7 \\ 0 & 0 & 0 & 0 & 0 & 0 & 0 & 8 \\ 0 & 0 & 0 & 0 & 0 & 0 & 0 & 0 \end{bmatrix}.$$

(d) Viewing the digraph as a scheduling network, give the vertex sequence for a critical path.

11. Consider the digraph shown.

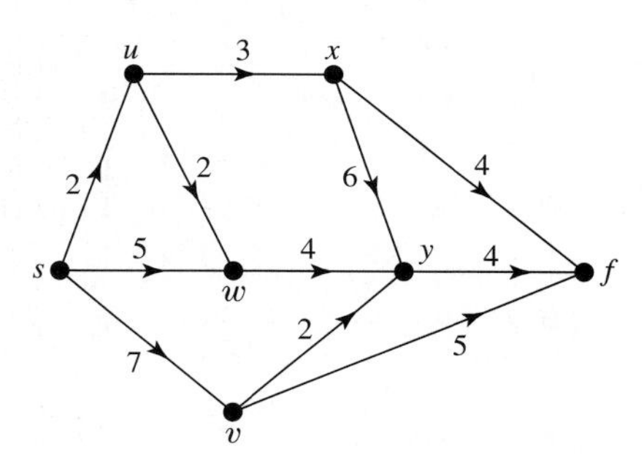

(a) Viewing this digraph as a scheduling network, use any method to find and mark a critical path for the network.

(b) Which vertices in this network have slack time 0?

(c) Give the float times of the edges from u to w and from x to f.

(d) Now view the digraph as an ordinary weighted digraph. Suppose that we are using Dijkstra's algorithm to calculate min-weights from the vertex s and that we have already marked vertices s and u to arrive at the table below. Fill in the next two rows of the table.

M	$D(u)$	$D(v)$	$D(w)$	$D(x)$	$D(y)$	$D(f)$
$\emptyset$	2	7	5	∞	∞	∞
$\{u\}$	2	7	4	5	∞	∞

12. Find a de Bruijn sequence for strings of length 2 formed of the digits 1, 2, and 3. To obtain it, modify the technique given for binary strings of length n.

Discrete Probability

This chapter expands on the brief introduction to probability in §5.2. The first section develops the ideas of independence and conditional probability of events. In §9.2 the emphasis shifts from sample spaces to discrete random variables and includes a discussion of independence of random variables. Section 9.3 gives some of the basic properties of expectation and variance for discrete random variables. In §9.4 we introduce cumulative distribution functions, including the binomial distribution. We then show how the normal distributions arise, both in approximating the binomial distribution and in practical measurement problems.

9.1 Independence in Probability

Our setting for probability is a sample space Ω and a probability P. If E is an event, i.e., a subset of Ω, then $P(E)$ is a number that represents the probability of E. As we said in §5.2, this means that $P(\Omega) = 1$, that $0 \leq P(E) \leq 1$ for every event E, and that $P(E \cup F) = P(E) + P(F)$ whenever E and F are disjoint. Given a sample space Ω, we want the probability P to match our estimates of the likelihoods of events.

Our estimates of these likelihoods may change if more information becomes available, for example, if we learn that the outcome belongs to some particular subset of Ω. For instance, if I happen to see that my opponent in poker has an ace, then I will think it's more likely that she has two aces than I would have thought before I got the information. On the other hand, if I learn that she has just won the lottery, this fact seems to be unrelated to the likelihood that she has two aces. Our goal in this section is to give mathematical meaning to statements such as "the probability of A given B" and "A is independent of B."

EXAMPLE 1

Figure 1 lists the set Ω of 36 equally likely outcomes when two fair dice are tossed, one black and one red. This situation is discussed in Example 5 on page 192. Consider the events B = "the value on the black die is ≤ 3," R = "the value on the red die is ≥ 5," and S = "the sum of the values is ≥ 8." The outcomes in B are in the top three rows in Figure 1, those in R are in the last two columns, and the red ones below the dashed line are in S.

Figure 1 ▶

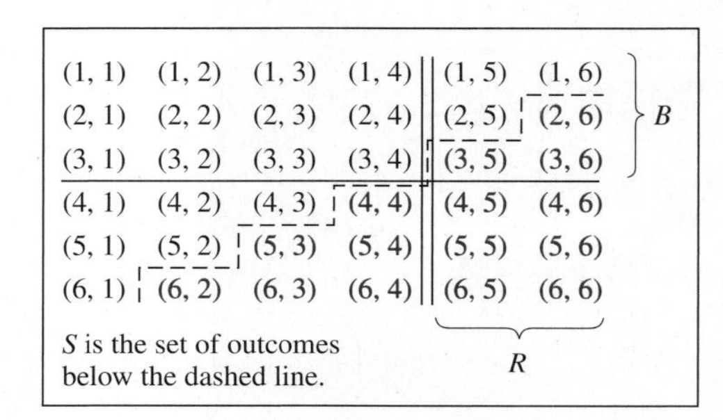

(a) Let's have a friend toss our dice, but keep the result hidden. Observe that $P(R) = \frac{1}{3}$; i.e., we think there's one chance in three that the outcome is in R. Now suppose we learn from our friend that the sum of the values on the two dice is ≥ 8. Then the 36 outcomes in Ω are no longer equally likely. In fact, the ones with sum ≤ 7 cannot have occurred, and we now think that each outcome with sum ≥ 8 is more likely than we thought it was before we got the new information.

How likely? Well, the set S consists of the 15 outcomes shown in red in Figure 1, and the outcomes in it still seem as equally likely as they were before, so each of them now should have probability $\frac{1}{15}$. Since the red die has value ≥ 5 in nine of these outcomes, i.e., $|R \cap S| = 9$, we estimate the probability of R now, given S, to be $\frac{9}{15} = 0.6$, rather than $\frac{1}{3}$.

(b) Originally, we estimated $P(B) = \frac{1}{2}$, but if we know the outcome is in S, then our estimate of the probability that the black die has value ≤ 3 becomes $|B \cap S|/|S| = \frac{3}{15} = 0.2$. Just as in part (a), we switch from using our original probability to a new one that matches the likelihoods of outcomes given that S has occurred.

(c) How would our estimate of the likelihood of B change if we knew that R had occurred? That is, what's the probability that the black die is ≤ 3 if we know the red die is ≥ 5? Twelve outcomes in Figure 1 lie in R and six of them also lie in B, so the probability is still $\frac{1}{2}$. This is reasonable; we don't expect knowledge of the red die to tell us about the black die. The fact that we know the outcome is in R makes the set of possibilities smaller, but it does not change the likelihood that the outcome is in B.

The preceding paragraph is slightly misleading. It looks as if we've used probability to show that knowledge about the red die doesn't affect probabilities involving the black die. The truth is the other way around. We *believe* that the red die's outcome doesn't change the probabilities involving the black die. So, as we will explain in Example 7, we *arranged* for the probability P to reflect this fact. In other words, we view the 36 outcomes in Figure 1 as equally likely *because* this probability has the property that knowledge of the red die doesn't affect the probabilities of events involving only the black die, and vice versa. ■

We formalize the idea discussed in Example 1. Suppose that we know that an outcome ω is in some event $S \subseteq \Omega$. Then the outcome is in an event E if and only if it's in $E \cap S$; so the probability that ω is in E, given that it's in S, should depend only on the probability of $E \cap S$. If all outcomes in S are equally likely, then the probability of E given S, i.e., the probability of $E \cap S$ given S, is just the fraction of the outcomes in S that are in $E \cap S$, so it is $\frac{|E \cap S|}{|S|}$. More generally, the probability of $E \cap S$ given S should be the fraction of $P(S)$ associated with $E \cap S$. For $P(S) > 0$ we define the **conditional probability** $\boldsymbol{P(E|S)}$, read "the probability of E given S," by

$$\boldsymbol{P(E|S)} = \frac{P(E \cap S)}{P(S)} \quad \text{for} \quad E \subseteq \Omega.$$

Then $P(\Omega|S) = \frac{P(\Omega \cap S)}{P(S)} = \frac{P(S)}{P(S)} = 1$, and it is not hard to see that the function $E \to P(E|S)$ satisfies the conditions that define a probability on Ω. Note that we have $P(E|S) = P(E \cap S|S)$ for $E \subseteq \Omega$.

Since $P(S|S) = 1$, one could also think of S as the sample space for the conditional probability, instead of Ω, in which case we would only define $P(E|S)$ for $E \subseteq S$. Figure 2 suggests how we can view S as a cut-down version of Ω, with $S \cap E$ as the event in S corresponding to the event E in Ω. If $E \subseteq S$ and $P(S) < 1$, then $P(E|S) = \frac{P(E)}{P(S)} > P(E)$, as expected; the fact that S has occurred makes such an event E more likely than it would have been without the extra information. In general, for an event E not contained in S, $P(E|S)$ can be greater than, equal to, or less than $P(E)$, depending on the circumstances.

Figure 2 ▶

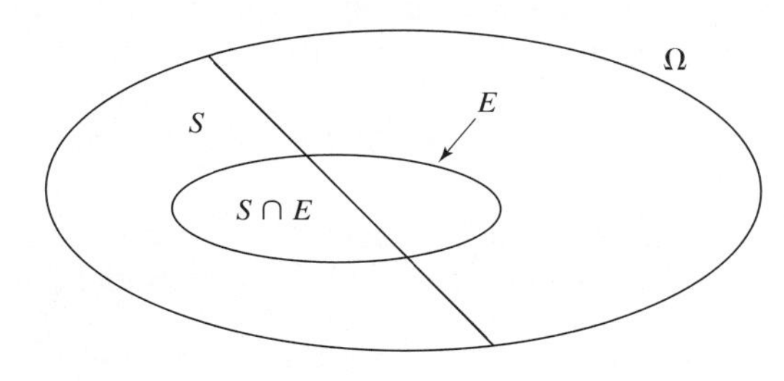

EXAMPLE 2

We return to the two-dice problem, so Ω, B, R, and S are as in Example 1 and Figure 1.

(a) With our new notation, the assertion in Example 1(a) is that $P(R|S) = \frac{9}{15}$. Indeed,

$$P(R|S) = \frac{P(R \cap S)}{P(S)} = \frac{9/36}{15/36} = \frac{9}{15}.$$

In Example 1(b) we observed that $P(B|S) = \frac{3}{15}$; this agrees with

$$P(B|S) = \frac{P(B \cap S)}{P(S)} = \frac{3/36}{15/36} = \frac{3}{15}.$$

It seems that our new definition of $P(E|S)$ has only complicated matters by introducing a bunch of 36's that get canceled anyway. But the new definition makes good sense even when the outcomes are not equally likely, in which case counting possible outcomes no longer helps.

(b) In Example 1(c) we saw that $P(B|R) = \frac{1}{2} = P(B)$. Similarly,

$$P(R|B) = \frac{P(R \cap B)}{P(B)} = \frac{6/36}{18/36} = \frac{1}{3} = P(R).$$

We repeat: the knowledge that R has occurred does not affect the probability of B, and vice versa. ■

One of the most important questions in probability and statistics is this: Given the occurrence of one event, does this change the likelihood of the other? If not, the events are viewed as probabilistically independent. Otherwise, the one event has a probabilistic effect on the other and, if the problem is important, one may look for causes for this effect.

EXAMPLE 3

Let Ω be the set of all adult American males. We assume that all men in Ω are equally likely to be selected in any study. If S is the event "he has smoked at least 10 years" and C = "he has lung cancer," then studies show that $P(C|S)$ is a lot larger than $P(C)$. There seems to be a cause and effect here. Smoking increases the probability of getting lung cancer even though not every smoker gets lung cancer. Lung cancer doesn't "depend" on smoking, in the sense that you must smoke to get lung cancer, but smoking does increase your chances. Lung cancer "depends" on smoking in the probabilistic sense.

Let M = "he studied college mathematics." So far as we know, $P(C|M) = P(C)$. If this is true, getting lung cancer is probabilistically independent of studying mathematics. Briefly, the events C and M are [probabilistically] independent.

Suppose that it was discovered that $P(C|M) > P(C)$. Common sense tells us that mathematics doesn't directly cause cancer. So we would look for more reasonable explanations. Perhaps college-educated men lead more stressful lives and cancer thrives in such people. And so on. ■

For events A and B with $P(B) > 0$, we say that A and B are **independent** if $P(A|B) = P(A)$. This condition is equivalent to $\frac{P(A\cap B)}{P(B)} = P(A)$ and to $P(A \cap B) = P(A) \cdot P(B)$ and, if $P(A) > 0$, also to

$$P(B|A) = \frac{P(A \cap B)}{P(A)} = P(B).$$

Thus, if A and B are independent, then B and A are independent; i.e., independence is a symmetric relation. The equivalent formulation

(I) $\qquad P(A \cap B) = P(A) \cdot P(B) \qquad$ for independent events A and B

is useful in computations. Also, it makes sense even if $P(A)$ or $P(B)$ is 0, so we will adopt (I) as the definition of **independence**. Following tradition, we will refer to pairs of independent events, but we will write $\{A, B\}$ to stress that order doesn't matter.

EXAMPLE 4

Back to the two-dice problem in Examples 1 and 2. The events R and S are not independent, since $P(R) = \frac{1}{3}$ while $P(R|S) = \frac{9}{15} = \frac{3}{5}$. Also

$$P(S) = \frac{15}{36} = \frac{5}{12} \quad \text{while} \quad P(S|R) = \frac{9}{12} = \frac{3}{4}.$$

Similarly, B and S are not independent [check this].

However, B and R are independent:

$$P(B|R) = \frac{1}{2} = P(B) \quad \text{and} \quad P(R|B) = \frac{1}{3} = P(R).$$

Knowing that one of these events has occurred does not change the likelihood that the other event has occurred. ■

If A and B are independent, then so are A and the complementary event $B^c = \Omega \setminus B$, since

$$\begin{aligned} P(A \cap B^c) &= P(A) - P(A \cap B) \\ &= P(A) - P(A)P(B) \\ &= P(A)(1 - P(B)) \\ &= P(A)P(B^c). \end{aligned}$$

Thus—no surprise—if knowing that B has occurred doesn't change our estimate of $P(A)$, then neither does knowing that B has *not* occurred. See also Exercise 20.

People sometimes refer to events as independent if they cannot both occur, i.e., if they are disjoint. This is ***not*** compatible with our usage and should be avoided. If $A \cap B = \emptyset$, then knowledge that B occurs *does* tell us something about A, namely that A does *not* occur: $P(A|B) = 0$. Disjoint events are not independent unless $P(A)$ or $P(B)$ is 0.

What should it mean to say that *three* events, A, B and C, are independent? The idea we're trying to capture is that our estimate of the likelihood of any of the events isn't affected by knowing whether one or more of the others have occurred. For the event A, what we require mathematically is

$$P(A) = P(A|B) = P(A|C) = P(A|B \cap C);$$

i.e.,

$$P(A \cap B) = P(A)P(B),$$
$$P(A \cap C) = P(A)P(C),$$
$$P(A \cap B \cap C) = P(A)P(B \cap C),$$

and similarly, for B and C,

$$P(B \cap A) = P(B)P(A),$$
$$P(B \cap C) = P(B)P(C),$$
$$P(B \cap A \cap C) = P(B)P(A \cap C),$$
$$P(C \cap A) = P(C)P(A),$$
$$P(C \cap B) = P(C)P(B),$$
$$P(C \cap A \cap B) = P(C)P(A \cap B).$$

Some of these equations repeat others. In fact, the three equations for $P(A \cap B \cap C)$ all say the same thing, that

$$P(A \cap B \cap C) = P(A)P(B)P(C),$$

because of the conditions on $P(A \cap B)$, $P(A \cap C)$, and $P(B \cap C)$.

Altogether then, when duplicates are removed, the conditions become

$$P(A \cap B) = P(A)P(B),$$
$$P(A \cap C) = P(A)P(C),$$
$$P(B \cap C) = P(B)P(C),$$
$$P(A \cap B \cap C) = P(A)P(B)P(C).$$

Could we leave out the last equation here? The next example shows that we can't.

EXAMPLE 5

We again consider the two-dice problem. Let B_o = "the black die's value is odd," R_o = "the red die's value is odd," and E = "the sum of the values is even." It is easy to check that the pairs $\{B_o, R_o\}$, $\{B_o, E\}$, and $\{R_o, E\}$ are all independent [Exercise 1]. On the other hand, if both B_o and R_o occur, then E is a certainty, since the sum of two odd numbers is even. In symbols,

$$P(E|B_o \cap R_o) = 1 \neq P(E) = \frac{1}{2}$$

or, equivalently,

$$P(E \cap B_o \cap R_o) = P(B_o \cap R_o) = \frac{1}{4} \neq \frac{1}{8} = P(E)P(B_o)P(R_o).$$

Thus we would not regard B_o, R_o, E as independent even though any two of them are independent. ■

The right definition, which works in general, is to say that the events A_1, $A_2, \ldots, A_n$ are **independent** if

$$\text{(I}'\text{)} \qquad P\left(\bigcap_{i \in J} A_i\right) = \prod_{i \in J} P(A_i)$$

for all nonempty subsets J of $\{1, 2, \ldots, n\}$. Before going on, let's get comfortable with this intimidating notation. Suppose that $n = 3$ again. For $J = \{1, 2, 3\}$, the index i in (I′) takes the values 1, 2, 3, so condition (I′) asserts that

$$P(A_1 \cap A_2 \cap A_3) = P(A_1) \cdot P(A_2) \cdot P(A_3),$$

the requirement we ran into before Example 5. For the three two-element subsets of $\{1, 2, 3\}$, condition (I′) gives the conditions for pairwise independence: $P(A_1 \cap A_2) = P(A_1) \cdot P(A_2)$, etc. And for the three one-element subsets of $\{1, 2, 3\}$, condition (I′) just states the obvious:

$$P(A_1) = P(A_1), \quad P(A_2) = P(A_2), \quad \text{and} \quad P(A_3) = P(A_3).$$

For $n = 4$, there are 15 nonempty subsets of $\{1, 2, 3, 4\}$, so the shorthand (I′) is really worthwhile. You should write out a few examples, perhaps using J equal to $\{1, 4\}$, $\{2, 3\}$, $\{1, 3, 4\}$, and $\{1, 2, 3, 4\}$, until you are sure what condition (I′) is saying.

EXAMPLE 6

A fair coin is tossed n times, as in Example 6 on page 193. For $k = 1, 2, \ldots, n$, let E_k be the event "the kth toss is a head." We assume that these events are independent, since knowing what happened on the first and fifth tosses, say, shouldn't affect the probability of a head on the second or seventh tosses. Of course, $P(E_k) = \frac{1}{2}$ for each k, since the coin is fair. From independence we obtain

$$P\left(\bigcap_{k=1}^{n} E_k\right) = \prod_{k=1}^{n} P(E_k) = \frac{1}{2} \cdot \frac{1}{2} \cdot \frac{1}{2} \cdots \frac{1}{2} = \frac{1}{2^n}.$$

Since $\bigcap_{k=1}^{n} E_k$ consists of the single event that "all n tosses are heads," this equation tells us that the probability of getting all heads is $1/2^n$. Similar reasoning shows that the probability of any particular sequence of n heads and tails is $1/2^n$. Thus the assumption that the outcomes of different tosses are independent implies that each n-tuple should have probability $1/2^n$, as in Example 6 on page 193. ■

EXAMPLE 7

We now show that the probability we've used for the two-dice problem is correct under the following assumptions: The probability of each value on the black die is $\frac{1}{6}$, the probability of each value on the red die is $\frac{1}{6}$, and the outcomes on the black and red dice are independent. Let's focus on the outcome (4, 5). This is the single outcome in $B_4 \cap R_5$, where B_4 = "the black die is 4" and R_5 = "the red die is 5." Then $P(B_4) = P(R_5) = \frac{1}{6}$ and, by independence,

$$P(4, 5) = P(B_4 \cap R_5) = P(B_4) \cdot P(R_5) = \frac{1}{36}.$$

The same thinking shows that $P(k, l) = \frac{1}{36}$ for all outcomes (k, l). This is why we used $P(k, l) = \frac{1}{36}$ for all (k, l) drawn in Figure 1. ■

If A and B are not independent, the equation $P(A \cap B) = P(A) \cdot P(B)$ *is not valid.* If $P(A) \neq 0$ and $P(B) \neq 0$, though, we always have the equations

$$P(A \cap B) = P(A) \cdot P(B|A) \quad \text{and} \quad P(A \cap B) = P(B) \cdot P(A|B),$$

which are useful when one of $P(B|A)$ or $P(A|B)$ is easy to determine. As we'll illustrate, often some of these values are a lot easier to calculate than others.

EXAMPLE 8

Two cards are drawn at random from a deck of cards, and we are interested in whether they are both aces. The relevant events are A_1 = "the first card is an ace" and A_2 = "the second card is an ace."

(a) What is the probability that both cards are aces if the first card is put back into the deck [which is then shuffled] before the second card is drawn, i.e., with replacement? In this case, A_1 and A_2 are independent events, so

$$P(A_1 \cap A_2) = P(A_1) \cdot P(A_2) = \frac{4}{52} \cdot \frac{4}{52} = \frac{1}{13} \cdot \frac{1}{13} = \frac{1}{169}.$$

(b) What is the probability that both cards are aces if the first card is not put back into the deck [drawing without replacement]? The events are no longer independent: if A_1 occurs, there is one less ace in the remaining deck, so we'd expect the probability of A_2, given A_1, to be smaller than $P(A_2)$. In fact, we have $P(A_2|A_1) = \frac{3}{51}$, which is indeed less than $P(A_2) = \frac{1}{13}$. The probability that both cards are aces in this case is

$$P(A_1 \cap A_2) = P(A_1) \cdot P(A_2|A_1) = \frac{4}{52} \cdot \frac{3}{51} = \frac{1}{221}.$$

We can also view this problem as one of selecting a random 2-element set from the 52-card deck, assuming that all 2-card sets are equally likely. The probability of getting two aces is then

$$\frac{\binom{4}{2}}{\binom{52}{2}} = \frac{\frac{4 \cdot 3}{2 \cdot 1}}{\frac{52 \cdot 51}{2 \cdot 1}} = \frac{1}{221}.$$

Getting the same answer again supports our assumption that all 2-elements sets *are* equally likely. One can prove by induction that all k-element subsets are equally likely to be selected if, after some elements have been selected, all remaining elements are equally likely to be drawn. ■

EXAMPLE 9 A company purchases cables from three firms and keeps a record of how many are defective. The facts are summarized in Figure 3. Thus 30 percent of the cables are purchased from firm C and 2 percent of them are defective. In terms of probability, if a purchased cable is selected at random, then

$$P(A) = 0.50, \qquad P(B) = 0.20, \qquad P(C) = 0.30,$$

where A = "the cable came from firm A," etc. If D = "the cable is defective," then we are also given

$$P(D|A) = 0.01, \qquad P(D|B) = 0.04, \qquad P(D|C) = 0.02.$$

Figure 3 ▶

Firm	A	B	C
Fraction of cables purchased	0.50	0.20	0.30
Fraction of defective cables	0.01	0.04	0.02

(a) The probability that a cable was purchased from firm A and was defective is $P(A \cap D) = P(A) \cdot P(D|A) = 0.50 \times 0.01 = 0.005$. Similarly

$$P(B \cap D) = P(B) \cdot P(D|B) = 0.20 \times 0.04 = 0.008$$

and

$$P(C \cap D) = P(C) \cdot P(D|C) = 0.30 \times 0.02 = 0.006.$$

(b) What is the probability that a random cable is defective? Since D is the pairwise disjoint union of $A \cap D$, $B \cap D$, and $C \cap D$, the answer is

$$P(D) = P(A) \cdot P(D|A) + P(B) \cdot P(D|B) + P(C) \cdot P(D|C)$$
$$= 0.005 + 0.008 + 0.006 = 0.019.$$

The tree in Figure 4 may be a helpful picture of what's going on. ■

The calculation of $P(D)$ in Example 9(b) illustrates the next general observation with A, B, C being the partition.

Figure 4 ▶

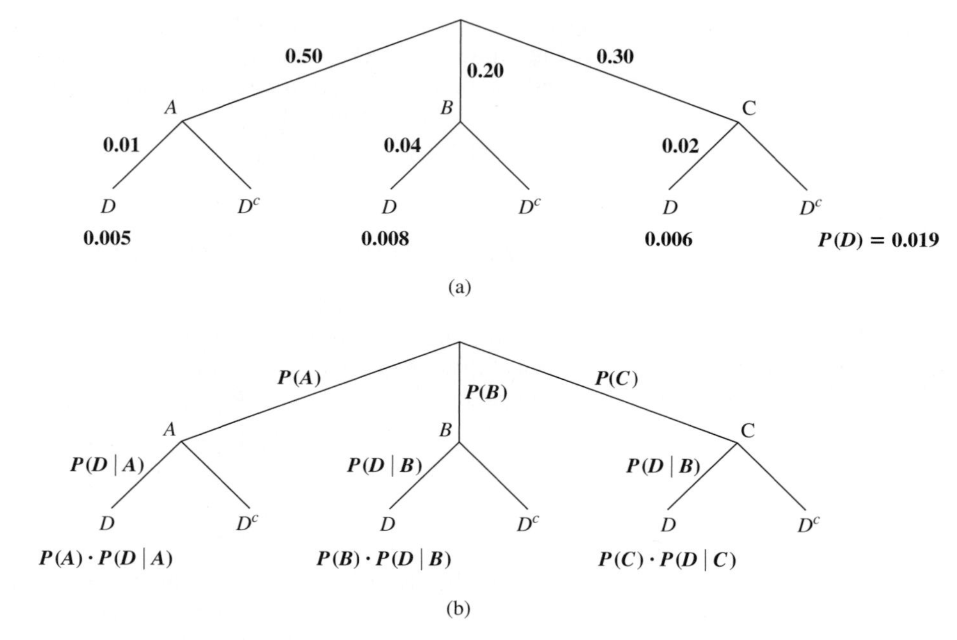

Total Probability Formula If events $A_1, A_2, \ldots, A_k$ partition the sample space Ω and if $P(A_j) > 0$ for each j, then for any event B we have

$$P(B) = P(A_1) \cdot P(B|A_1) + P(A_2) \cdot P(B|A_2) + \cdots + P(A_k) \cdot P(B|A_k)$$
$$= \sum_{j=1}^{k} P(A_j) \cdot P(B|A_j).$$

This formula is true because B is the disjoint union of the sets $A_j \cap B$, and the probability of each of these sets is $P(A_j) \cdot P(B|A_j)$.

Sometimes we know that event B has occurred and want to know the likelihood that each A_j has occurred. That is, we want $P(A_1|B), P(A_2|B), \ldots, P(A_k|B)$, numbers that are not obvious in the beginning. Since

$$P(A_j|B) = \frac{P(A_j \cap B)}{P(B)} = \frac{P(A_j) \cdot P(B|A_j)}{P(B)},$$

the Total Probability Formula implies the following.

Bayes' Formula Suppose that the events $A_1, \ldots, A_k$ partition the set Ω, where $P(A_j) > 0$ for each j, and suppose that B is any event for which $P(B) > 0$. Then

$$P(A_j|B) = \frac{P(A_j) \cdot P(B|A_j)}{P(B)}$$

for each j, where

$$P(B) = P(A_1) \cdot P(B|A_1) + P(A_2) \cdot P(B|A_2) + \cdots + P(A_k) \cdot P(B|A_k).$$

EXAMPLE 10

The randomly selected cable in Example 9 is found to be defective. What is the probability that it came from firm A? We want $P(A|D)$, and Bayes' Formula gives

$$P(A|D) = \frac{P(A) \cdot P(D|A)}{P(D)} = \frac{0.50 \times 0.01}{0.019} = \frac{0.005}{0.019} \approx 0.263.$$

Even though half the cables are purchased from firm A, only about a quarter of the defective cables come from this firm. Similar computations yield

$$P(B|D) = \frac{0.008}{0.019} \approx 0.421 \quad \text{and} \quad P(C|D) = \frac{0.006}{0.019} \approx 0.316.$$

Note how these fractions can be read from the top drawing in Figure 4. Note also that the three probabilities add to 1. When a defective cable is selected, we're sure that it came from exactly one of the firms A, B, or C. ■

The formula $P(A \cap B) = P(A) \cdot P(B|A)$ generalizes as follows:

$$P(A_1 \cap A_2 \cap \cdots \cap A_n)$$
$$= P(A_1) \cdot P(A_2|A_1) \cdot P(A_3|A_1 \cap A_2) \cdots P(A_n|A_1 \cap A_2 \cap \cdots \cap A_{n-1}).$$

Exercise 26 asks for a proof.

EXAMPLE 11 Four cards are drawn from an ordinary 52-card deck without replacement. What is the probability that the first two drawn will be aces and the second two will be kings? Using suggestive notation, we obtain

$$P(A_1 \cap A_2 \cap K_3 \cap K_4)$$
$$= P(A_1) \cdot P(A_2|A_1) \cdot P(K_3|A_1 \cap A_2) \cdot P(K_4|A_1 \cap A_2 \cap K_3)$$
$$= \frac{4}{52} \cdot \frac{3}{51} \cdot \frac{4}{50} \cdot \frac{3}{49} \approx 0.000022.$$ ■

Exercises 9.1

Wherever conditional probabilities such as $P(A|B)$ occur, assume that $P(B) > 0$.

1. (a) Show that the events B_o, R_o, and E in Example 5 are pairwise independent.

(b) Determine $P(B_o|E \cap R_o)$ and compare with $P(B_o)$.

2. Two dice, one black and one red, are tossed. Consider the events

S = "the sum is ≥ 8,"
L = "the value on the black die is less than the value on the red die,"
E = "the values on the two dice are equal,"
G = "the value on the black die is greater than the value on the red die."

Which of the following pairs of events are independent? $\{S, L\}$, $\{S, E\}$, $\{L, E\}$, $\{L, G\}$. Don't calculate unless necessary [but see Exercise 3].

3. By calculating suitable probabilities, determine which pairs of events in Exercise 2 are independent.

4. The two dice in Exercise 2 are tossed again.

(a) Find the probability that the value on the red die is ≥ 5, given that the sum is 9.

(b) Do the same, given that the sum is ≥ 9.

5. A fair coin is tossed four times. Consider the events A = "exactly one of the first two tosses is heads" and B = "exactly two of the four tosses are heads." Are the events A and B independent? Justify your answer.

6. A fair coin is tossed four times.

(a) What is the probability of getting at least two consecutive heads?

(b) What is the probability of getting consecutive heads, given that at least two of the tosses are heads?

7. Suppose that an experiment leads to three events A, B, and C with $P(A) = 0.3$, $P(B) = 0.4$, $P(A \cap B) = 0.1$, and $P(C) = 0.8$.

(a) Find $P(A|B)$.

(b) Find $P(A^c)$.

(c) Are A and B independent? Explain.

(d) Are A^c and B independent?

8. Given independent events A and B with $P(A) = 0.4$ and $P(B) = 0.6$, find

(a) $P(A|B)$ (b) $P(A \cup B)$ (c) $P(A^c \cap B)$

9. A box has three red marbles and eight black marbles. Two marbles are drawn at random without replacement. Find the probability that

(a) they are both red.

(b) they are both black.

(c) one is red and one is black.

10. Repeat Exercise 9 if one marble is drawn, observed, and then replaced before the second is drawn [i.e., drawn with replacement]. Compare the answers to those in Exercise 9.

11. Three cards are drawn from an ordinary 52-card deck without replacement. Determine the probability that

(a) three aces are drawn.

(b) an ace, king and queen are drawn in that order.

(c) at least one ace is drawn.

12. Recall that the probability of a poker hand being a flush is about 0.00197 [Example 1 on page 189]. What is the probability of a flush, given that all five of the cards are red?

13. Three urns contain marbles as indicated in Figure 5. An urn is selected at random and then a marble is selected at random from the urn.

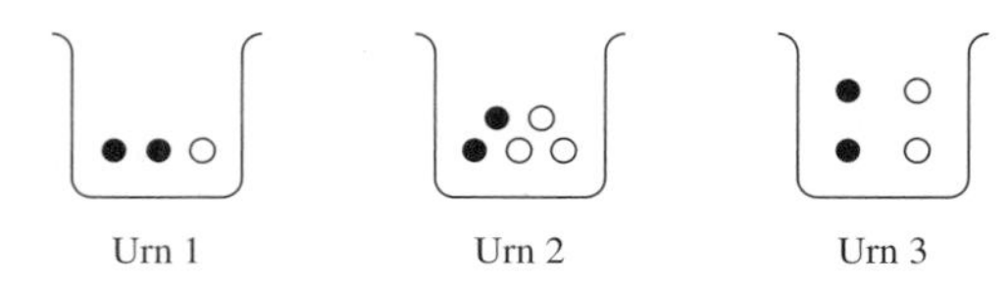

Figure 5 ▲

(a) What is the probability that a black marble was selected?

(b) Given that a black marble was selected, what is the probability that it was selected from urn 1? urn 2? urn 3?

(c) What is the probability that a black marble was selected from urn 1?

14. An urn is selected at random from Figure 5 and then two marbles are selected at random [no replacement]. Find the probability that

(a) both marbles are white. (b) both marbles are black.

15. (a) Given that the two marbles selected in Exercise 14 are both black, find the probability that urn 1 was selected.

(b) Repeat if both of the marbles are white.

16. One marble is selected at random from each urn in Figure 5. What is the probability that all three marbles are white? that all three are black?

17. Ellen, Frank, and Gayle handle all the orders at Burger Queen. Figure 6 indicates the fraction of orders handled by each and, for each employee, the fraction of complaints received.

Employee	Ellen	Frank	Gayle
Fraction of orders	0.25	0.35	0.40
Fraction of complaints	0.04	0.06	0.03

Figure 6 ▲

(a) For a randomly selected order, what is the probability that there was a complaint?

(b) Given an order with a complaint, what is the probability that it was prepared by Ellen? Frank? Gayle?

(c) What is the sum of the answers to part (b)?

18. A box contains two pennies, one fair and one with two heads. A coin is chosen at random and tossed.

(a) What is the probability that it comes up heads?

(b) Given that it comes up heads, what is the probability that the coin is the two-headed coin?

(c) If the selected coin is tossed twice, what is the probability that it comes up heads both times?

19. The following is observed for a test of a rare disease, where D = "the subject has the disease," N = "the subject tests negative," and N^c = "the subject tests positive": $P(N^c \cap D) = 0.004$, $P(N \cap D) = 0.0001$, $P(N^c) = 0.044$. Verify the following assertions.

(a) The test is over 97.5 percent accurate on diseased subjects. *Hint:* Find $P(N^c|D)$.

(b) The test is nearly 96 percent accurate on the general population. *Hint:* Find $P((N^c \cap D) \cup (N \cap D^c))$.

(c) Yet it is misleading over 90 percent of the time in cases where the test is positive! *Hint:* Find $P(D|N^c)$. *Moral:* One must be very careful interpreting tests for rare diseases.

20. Show that if events A and B are independent, so are the events A^c and B^c.

21. An electronic device has n components. Each component has probability q of failure before the warranty is up. [Or think of light bulbs in your living unit.]

(a) What is the probability that some component will fail before the warranty is up? What assumptions are you making?

(b) What if $n = 100$ and $q = 0.01$?

(c) What if $n = 100$ and $q = 0.001$?

(d) What if $n = 100$ and $q = 0.1$?

22. Prove that if $P(A|B) > P(A)$, then $P(B|A) > P(B)$.

23. A box of 20 items is inspected by checking a sample of 5 items. If none of the items has a defect, then the box is accepted.

(a) What is the probability that a box with exactly two defective items will be accepted?

(b) Repeat part (a) if 10 items are checked instead of 5.

24. Prove that if $P(A) = P(B) = \frac{2}{3}$, then $P(A|B) \geq \frac{1}{2}$. *Hint:* First show that $P(A \cap B) \geq \frac{1}{3}$.

25. Suppose that A, B, and C are independent events. Must $A \cap B$ and $A \cap C$ be independent? Explain.

26. Prove that

$$P(A_1 \cap A_2 \cap \cdots \cap A_n) = P(A_1) \cdot P(A_2|A_1) \cdot P(A_3|A_1 \cap A_2) \cdots P(A_n|A_1 \cap A_2 \cap \cdots \cap A_{n-1}).$$

If the induction step is too complicated, verify the result for $n = 3$ and $n = 4$ instead.

27. Prove or disprove:

(a) If A and B are independent and if B and C are independent, then A and C are independent.

(b) Every event A is independent of itself.

(c) If A and B are disjoint events, then they are independent.

28. It is, of course, false that $(b \to a) \wedge a \Longrightarrow b$. [Consider the case where a is true and b is false.] On the other hand, if $b \to a$ and a hold, then b is more likely than it was before, in the following sense.

(a) Show that if $P(A|B) = 1$ and $P(A) < 1$, then $P(B|A) > P(B)$.

(b) What if $P(A) = 1$?

29. Let S be an event in Ω with $P(S) > 0$.

(a) Show that if P^* is defined by $P^*(E) = P(E|S)$ for $E \subseteq \Omega$, then P^* preserves ratios as follows:

$$\frac{P^*(E)}{P^*(F)} = \frac{P(E \cap S)}{P(F \cap S)} \quad \text{whenever } P(F \cap S) \neq 0.$$

(b) Show that if P^* is a probability on Ω that preserves ratios, as in part (a), then $P^*(E) = P(E|S)$ for all $E \subseteq \Omega$.

9.2 Random Variables

In this section we shift the point of view slightly. Rather than focus on events, i.e., subsets of the sample space Ω, we look at numerical-valued functions defined on Ω. Since events E can be studied by using their characteristic functions $\chi_E : \Omega \to \{0, 1\}$, we lose nothing by this change of perspective and will gain a great deal.

Here is an illustration based on an example at the end of this section. Suppose that a device has 17 components, each of which has a probability 0.02 of failure within a year. What's the probability that at most two components fail within a year, assuming that different component failures are independent of each other? If we use a sample space Ω consisting of 17-tuples of F's and S's, where F represents failure and S stands for survival, then we want the probability of the event consisting of all 17-tuples with at most two F's. Since the outcomes in Ω are not all equally likely, the problem looks harder than the ones we've been doing. The methods of this section will give us another way to attack such questions and many others as well. In this instance, we'll consider the function that counts the number of failures in a year and ask for the probability that it has the value 0, 1, or 2.

A function from Ω into $\mathbb{R}$ is called a **random variable**. The terminology comes historically from the fact that the values of such a function vary as we go from one random outcome ω to another. It is traditional to use capital letters near the end of the alphabet for random variables, with generic ones named X, Y, or Z.

To avoid complications, we will restrict our attention to **discrete random variables**, i.e., to random variables X whose set of values

$$X(\Omega) = \{X(\omega) : \omega \in \Omega\}$$

can be listed as a sequence. By making this restriction, we will be able to do our calculations with sums, whereas, in the general case, integrals may be needed from calculus, and other complications can arise. The set $X(\Omega)$ is called the **value set** of X. Often it will be a set of integers.

EXAMPLE 1

(a) A natural random variable on the sample space of outcomes when two dice are tossed is the one that gives the sum of the values shown on the two dice; that is,

$$X_s(k, l) = k + l \quad \text{for} \quad (k, l) \in \Omega.$$

The value set of X_s is $\{2, 3, 4, 5, 6, 7, 8, 9, 10, 11, 12\}$.

(b) If we toss a coin 5 times, one natural random variable Y_h to associate with the set Ω of outcomes is the one that counts the heads that come up. Thus, for instance,

$$Y_h(HTHTT) = 2.$$

The value set of Y_h is $\{0, 1, 2, 3, 4, 5\}$.

(c) A natural random variable W to associate with the sample space Ω of strings $H, TH, TTH, TTTH, \ldots$ is length. Thus $W(TT \cdots TH) = 1 +$ the number of T's in the string. Such a random variable would be suited to an experiment consisting of tossing a coin until it comes up heads, in which case we could say that W counts the number of tosses. The value set of W is $\mathbb{P}$. ■

Given a random variable X on Ω and a condition C that some of its values may satisfy, we use the notation **$\{X$ satisfies $C\}$** as an abbreviation for the event $\{\omega \in \Omega : X(\omega) \text{ satisfies } C\}$. For example,

$$\begin{aligned}
&\{5 < X \le 7\} \quad \text{represents} \quad \{\omega \in \Omega : 5 < X(\omega) \le 7\},\\
&\{X = a \text{ or } X^2 = b\} \quad \text{represents} \quad \{\omega \in \Omega : X(\omega) = a \text{ or } X(\omega)^2 = b\},\\
&\{|X - 4| < 3\} \quad \text{represents} \quad \{\omega \in \Omega : |X(\omega) - 4| < 3\}, \text{ etc.}
\end{aligned}$$

When writing probabilities of these events, we will usually drop the braces { and }. So, for example,

$$P(X = 2) \quad \text{means} \quad P(\{X = 2\}), \quad \text{i.e.,} \quad P(\{\omega \in \Omega : X(\omega) = 2\}).$$

EXAMPLE 2

We return again to the sample space Ω consisting of the 36 equally likely outcomes when two fair dice are tossed.

(a) As in Example 1, let X_s give the sum of the values of the two dice. All the numbers $P(X_s = k)$ are given in Figure 1(a), which contains the same information as Figure 2(b) on page 193. With these values, we can calculate all the probabilities of interest that involve the random variable X_s. For example,

$$P(X_s \leq 5) = \frac{1}{36} + \frac{2}{36} + \frac{3}{36} + \frac{4}{36} = \frac{10}{36}$$

and

$$P(4 < X_s < 10) = \frac{4}{36} + \frac{5}{36} + \frac{6}{36} + \frac{5}{36} + \frac{4}{36} = \frac{2}{3}.$$

Figure 1 ▶

k	2	3	4	5	6	7	8	9	10	11	12
$P(X_s = k)$	$\frac{1}{36}$	$\frac{2}{36}$	$\frac{3}{36}$	$\frac{4}{36}$	$\frac{5}{36}$	$\frac{6}{36}$	$\frac{5}{36}$	$\frac{4}{36}$	$\frac{3}{36}$	$\frac{2}{36}$	$\frac{1}{36}$

(a)

k	1	2	3	4	5	6
$P(X_b = k) = P(X_r = k)$	$\frac{1}{6}$	$\frac{1}{6}$	$\frac{1}{6}$	$\frac{1}{6}$	$\frac{1}{6}$	$\frac{1}{6}$

(b)

If E is the set of even integers, then

$$P(X_s \text{ is even}) = P(X_s = 2) + P(X_s = 4) + \cdots + P(X_s = 12)$$
$$= \frac{1}{36}(1 + 3 + 5 + 5 + 3 + 1) = \frac{1}{2}.$$

(b) Let X_b be the value on the black die and X_r the value of the red die, so for each outcome (k, l) we have $X_b(k, l) = k$ and $X_r(k, l) = l$. The random variables X_b and X_r each have value set $\{1, 2, 3, 4, 5, 6\}$. Their possible values are equally likely; see Figure 1(b). As an example, we calculate $P(X_b = 3)$. The set $\{X_b = 3\} = \{(k, l) \in \Omega : X_b(k, l) = 3\}$ is

$$\{(k, l) \in \Omega : k = 3\} = \{(3, 1), (3, 2), (3, 3), (3, 4), (3, 5), (3.6)\},$$

so $P(X_b = 3) = \frac{6}{36} = \frac{1}{6}$. The random variables X_b and X_r have the property that they take the same values with the same probability.

(c) The random variable X_s is the sum of the random variables X_b and X_r, since $X_s(k, l) = k+l = X_b(k, l)+X_r(k, l)$ for all $(k, l) \in \Omega$. The events in Example 1 on page 349 can be written in terms of these random variables:

$$B = \text{"value on the black die is } \leq 3\text{"} = \{X_b \leq 3\},$$
$$R = \text{"value on the red die is } \geq 5\text{"} = \{X_r \geq 5\},$$
$$S = \text{"sum is } \geq 8\text{"} = \{X_s \geq 8\}.$$

Calculations in Examples 1 and 2 on pages 349 and 351 can now be written as

$$P(X_r \geq 5) = \frac{1}{3}, \qquad P(X_r \geq 5 | X_s \geq 8) = \frac{9}{15},$$

$$P(X_b \leq 3) = \frac{1}{2}, \qquad P(X_b \leq 3 | X_s \geq 8) = \frac{3}{15},$$

$$P(X_b \leq 3 | X_r \geq 5) = \frac{1}{2} = P(X_b \leq 3),$$

$$P(X_r \geq 5 | X_b \leq 3) = \frac{1}{3} = P(X_r \geq 5).$$

The last two equations imply that the events $\{X_r \geq 5\}$ and $\{X_b \leq 3\}$ are independent. This fact is not surprising, and there is nothing special about the values 5 and 3 nor the choices of the inequalities $\geq$ and $\leq$. We expect the random variables X_r and X_b to be independent; i.e., knowledge of X_r doesn't change probabilistic knowledge of X_b, and vice versa. ■

The preceding discussion suggests that we define random variables X and Y on a sample space Ω to be **independent** provided that the events

$$\{\omega \in \Omega : X(\omega) \text{ is in } I\} \quad \text{and} \quad \{\omega \in \Omega : Y(\omega) \text{ is in } J\}$$

are independent for all choices of intervals I and J in $\mathbb{R}$. Knowing that $X(\omega)$ belongs to some interval I shouldn't affect the probability that $Y(\omega)$ belongs to some interval J. For discrete random variables, this requirement is equivalent to requiring that the events $\{X = k\}$ and $\{Y = l\}$ be independent for all k in the value set of X and all l in the value set of Y. In other words, X and Y are independent if

(I) $$P(X = k \text{ and } Y = l) = P(X = k) \cdot P(Y = l)$$

for all k in the value set of X and all l in the value set of Y.

EXAMPLE 3

(a) The random variables X_b and X_r that give the values on the black die and red die are independent. Indeed,

$$P(X_b = k \text{ and } X_r = l) = \frac{1}{36} = P(X_b = k) \cdot P(X_r = l) \quad \text{for all } (k, l).$$

(b) As before, let X_s be the sum of the values on the two dice. Then X_s and X_b are not independent; for example, if X_b is big, X_s is more likely to be big. This intuitive explanation is *not* a proof and needs mathematical reinforcement. Rather than test condition (I) mindlessly, we'll use this comment as a guide. If X_b is big, like 6, the sum X_s surely can't be small; indeed, it can't be less than 7. So, for instance,

$$P(X_b = 6 \text{ and } X_s = 2) = 0 \neq P(X_b = 6) \cdot P(X_s = 2) = \frac{1}{6} \cdot \frac{1}{36}.$$

This single example is enough to *prove* that X_b and X_s are not independent. ■

A sequence $X_1, X_2, \ldots, X_n$ of random variables on a sample space Ω is **independent** if

$$P\{X_i(\omega) \text{ is in } J_i \text{ for } i = 1, 2, \ldots, n\} = \prod_{i=1}^{n} P\{X_i(\omega) \text{ is in } J_i\}$$

for all intervals $J_1, J_2, \ldots, J_n$ in $\mathbb{R}$. Our random variables are discrete, so this condition is equivalent to

(I′) $$P(X_i = k_i \text{ for } i = 1, 2, \ldots, n) = \prod_{i=1}^{n} P(X_i = k_i)$$

whenever each k_i is in the value set of the random variable X_i.

EXAMPLE 4

(a) A fair coin is tossed n times, as in Example 6 on page 354 [and Example 6 on page 193]. The sample space consists of n-tuples of H's and T's. For $i = 1, 2, \ldots, n$, let $X_i = 1$ if the ith toss is a head and $X_i = 0$ otherwise. Thus X_i is the characteristic function of the event E_i = "ith toss is a head." In Example 6 on page 354 we assumed that the outcomes of the different tosses were independent. Equivalently, for each sequence $k_1, k_2, \ldots, k_n$ of 0's and 1's, we assumed that the events $\{X_i = k_i\}$ were independent, so that

$$P(X_i = k_i \text{ for } i = 1, 2, \ldots, n) = \prod_{i=1}^{n} P(X_i = k_i) = \frac{1}{2^n}.$$

In other words, we assumed that the random variables $X_1, X_2, \ldots, X_n$ were independent.

(b) The sum $S_n = X_1 + X_2 + \cdots + X_n$ is a useful random variable. Since a member of the sample space is an n-tuple of H's and T's and since $X_i = 1$ if the ith entry is H and $X_i = 0$ otherwise, S_n counts the number of heads in the n tosses. The value set of S_n is $\{0, 1, 2, \ldots, n\}$. As we explained in Example 6 on page 193,

$$P(S_n = k) = \frac{1}{2^n} \cdot \binom{n}{k} \quad \text{for} \quad k \in \{0, 1, 2, \ldots, n\}.$$

(c) In Example 7(a) on page 194 and Example 1(c), a fair coin is tossed until a head is obtained. Let W be the random variable, the "waiting time," that counts the number of tosses needed to get a head. Then

$$P(W = k) = \frac{1}{2^k} \quad \text{for} \quad k = 1, 2, 3, \ldots .$$

■

So far we've assumed that the coins we've tossed have been fair coins. A coin is said to be **unfair** or **biased** if the probability p of a head is different from $\frac{1}{2}$. Most U.S. coins are slightly biased; nickels are the worst. We can still ask for the probability of k heads in n tosses, but the formula in Example 4(b) no longer applies. It is useful to consider the following more general model, which has numerous applications.

We imagine an experiment with one possible outcome of interest, traditionally called **success**; the complementary event is called **failure**. We assume that $P(\text{success}) = \boldsymbol{p}$) for some p, $0 < p < 1$. We set $\boldsymbol{q} = P(\text{failure})$, so that $p + q = 1$. We further assume that the experiment is repeated several times, say n times, and that the outcomes of the different experiments are independent: a successful first experiment does not change the likelihood that the second experiment will be successful, and so on.

EXAMPLE 5

(a) The tossing of a fair coin n times can be viewed as an experiment in which we regard heads as a success and tails as a failure. Then $p = P(\text{success}) = \frac{1}{2}$. We already know that

$$P(k \text{ heads in } n \text{ tosses}) = \frac{1}{2^n} \cdot \binom{n}{k}$$

for $k = 0, 1, 2, \ldots, n$. In general,

$$P(k \text{ successes in } n \text{ experiments}) = \frac{1}{2^n} \cdot \binom{n}{k}$$

provided that $p = \frac{1}{2}$.

(b) Suppose that an electronic device has n components and that the probability for each component to fail before the warranty is up is q. A component is successful, then, provided it is still working when the warranty is up. [If it fails the next day, we will still regard this a success!] With $p = 1 - q$, this situation fits our general scheme even though the n "experiments" will be running simultaneously. As in Exercise 21 on page 358, we are assuming that the survivals of the components are independent, which might or might not be a reasonable assumption. Unless $p = \frac{1}{2}$, we do not yet have a formula for

$$P(k \text{ successes in } n \text{ experiments}),$$

i.e., for

$$P(\text{exactly } k \text{ components are working when the warranty is up}).$$

(c) If the probability of curing a certain disease is p and if the cure is applied to n people with the disease, then the collection of "experiments" fits our general model. Again, we are assuming that the success rates for various patients are independent.

■

Given p and n, we now calculate the probability of exactly k successes in n independent experiments, where p is the probability of success for each experiment. In terms of random variables, we want $P(S_n = k)$, where S_n counts the number of successes in n experiments. The sample space Ω consists of all n-tuples of S's and F's [for successes and failures]. There are 2^n such n-tuples but they are *not* equally likely. Compare the likelihood of all S's to that of all F's if $p = 0.001$. For a fixed k, there are $\binom{n}{k}$ n-tuples with exactly k S's, and these turn out to be equally likely. To illustrate the idea, we consider a special case.

EXAMPLE 6

Given $n = 5$ and any p, we find the probability of exactly 3 successes. First fix a particular 5-tuple with 3 successes, say (S, S, F, S, F). The probability of this particular outcome is

$$P(\text{first is S}) \cdot P(\text{second is S}) \cdot P(\text{third is F}) \cdot P(\text{fourth is S}) \cdot P(\text{fifth is F})$$
$$= p \cdot p \cdot q \cdot p \cdot q = p^3q^2.$$

For another such 5-tuple, the order of factors will be different, but the result p^3q^2 will be the same, since each of the 3 successes has probability p and each of the 2 failures has probability q. Without actually listing all such outcomes, we know that there are $\binom{5}{3}$ of them, each with probability p^3q^2, and we conclude that

$$P(S_5 = 3) = P(3 \text{ successes in 5 experiments}) = \binom{5}{3}p^3q^2.$$ ■

Exactly the same argument works in the general case; hence

$$P(S_n = k) = P(k \text{ successes in } n \text{ experiments}) = \binom{n}{k}p^kq^{n-k}.$$

Note that these probabilities must sum to 1; i.e.,

$$\sum_{k=0}^{n}\binom{n}{k}p^kq^{n-k} = 1.$$

This equation is true by the binomial theorem on page 200, which tells us that the sum on the left is equal to $(p+q)^n$, which equals 1 since $p+q = 1$. Because of this intimate connection between S_n and the binomial theorem, S_n is called a **binomial random variable**. Such a random variable can be used whenever one has a sequence of independent experiments, each with probability p of success. In applications we can specify "success" to be anything we want.

EXAMPLE 7

(a) A fair coin is tossed n times. The experiment is the tossing of the coin, and we choose to view heads as success. Here $p = q = \frac{1}{2}$. Then the probability $P(S_n = k)$ of k heads in n tosses, i.e., k successes in n experiments, is equal to

$$\binom{n}{k}p^kq^{n-k} = \binom{n}{k}\left(\frac{1}{2}\right)^k\left(\frac{1}{2}\right)^{n-k} = \binom{n}{k}\left(\frac{1}{2}\right)^n = \frac{1}{2^n}\binom{n}{k},$$

which fortunately agrees with our earlier formula. For $n = 10$, the $p = \frac{1}{2}$ row of Table 1 contains the numerical values of these probabilities (accurate to three decimal places), and Figure 2(a) gives a graph of them.

TABLE 1 $P(S_n = k)$ for Binomial S_n and $n = 10$

k	0	1	2	3	4	5	6	7	8	9	10
$p = 1/2$	0.001	0.010	0.044	0.117	0.205	0.246	0.205	0.117	0.044	0.010	0.001
$p = 1/3$	0.017	0.087	0.195	0.260	0.228	0.136	0.057	0.017	0.003	0	0
$p = 1/10$	0.349	0.387	0.194	0.057	0.011	0.001	0	0	0	0	0

Figure 2 ▶

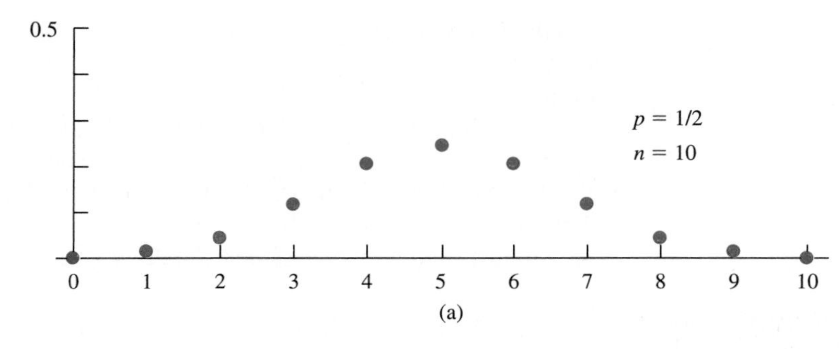

(a)

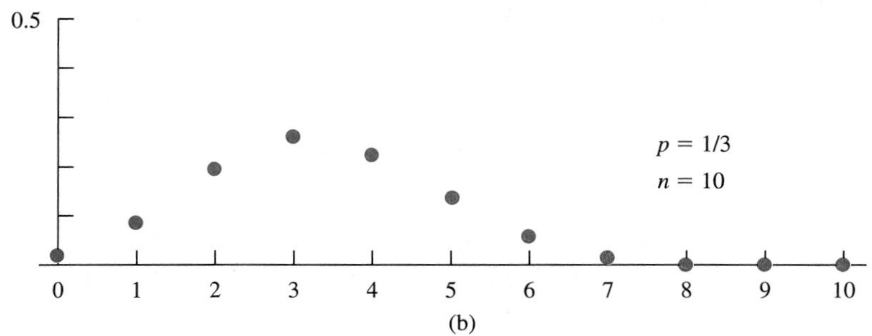

(b)

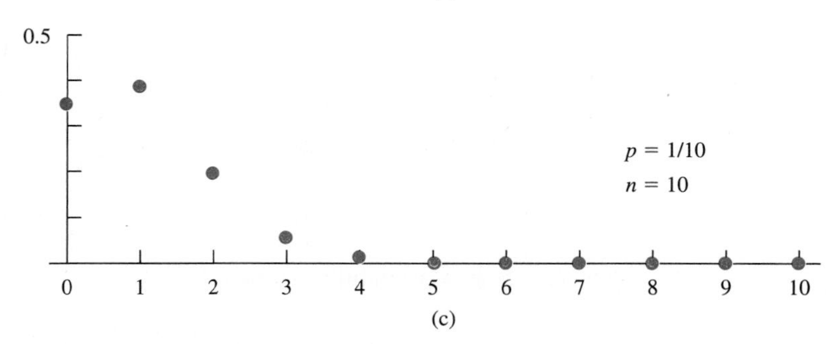

(c)

(b) A rather biased coin might have the probability $p = \frac{1}{3}$ of a head on each toss. Then we would have

$$P(k \text{ heads in } n \text{ tosses}) = \binom{n}{k} \cdot \left(\frac{1}{3}\right)^k \cdot \left(\frac{2}{3}\right)^{n-k}.$$

The $p = \frac{1}{3}$ row of Table 1 and Figure 2(b) contain the numerical values and a graph of these probabilities for $n = 10$.

(c) Even if our original "experiment" has several outcomes of interest, we can focus on some particular event and define it to be a "success." For example, suppose that a fair die is tossed n times and we are interested in how often the value is 1 or 2. We define this event to be success, so $P(\text{success}) = \frac{1}{3} = p$. Thus

$$P(k \text{ successes in } n \text{ tosses}) = \binom{n}{k} \cdot \left(\frac{1}{3}\right)^k \cdot \left(\frac{2}{3}\right)^{n-k}.$$

If the die is tossed 10 times, then we can use Table 1. For instance, the probability of getting a 1 or 2 at most two times in 10 tosses is

$$P(S_{10} = 0) + P(S_{10} = 1) + P(S_{10} = 2) \approx 0.017 + 0.087 + 0.195 = 0.299.$$

(d) Slim Hulk is a basketball player who makes $\frac{2}{3}$ of his free throws. If his successes are independent, what is the probability that he will make at least 7 of the next 10 free throws? Here $n = 10$ and $p = \frac{2}{3}$. We can't apply Table 1 directly, but we can if [to Slim's dismay] we call a miss a success and calculate $P(\text{number of misses} \leq 3)$. Now $P(\text{miss}) = P(\text{success}) = \frac{1}{3}$; so we use Table 1 with $p = \frac{1}{3}$ to obtain

$$P(\text{at least 7 free throws}) = P(\text{number of misses } \leq 3)$$
$$= P(S_{10} \leq 3) = P(S_{10} \leq 2) + P(S_{10} = 3) \approx 0.299 + 0.260 = 0.559,$$

where we used the calculation in part (c) to estimate $P(S_{10} \leq 2)$. Even though Slim only makes $\frac{2}{3}$ of his free throws, he has better than a 50 percent chance of making 7 or more in the next 10 tries! ■

EXAMPLE 8

We return to the n components, each of which has probability q of failure before the warranty is up. The probability that at least one component fails is

$$1 - P(n \text{ successes}) = 1 - \binom{n}{n} \cdot p^n = 1 - p^n = 1 - (1-q)^n.$$

This result agrees with the answer to Exercise 21, §9.1.

The probability of at most 2 failures is

$$P(n-2 \text{ successes}) + P(n-1 \text{ successes}) + P(n \text{ successes})$$
$$= \binom{n}{n-2} \cdot p^{n-2}q^2 + \binom{n}{n-1} \cdot p^{n-1}q + \binom{n}{n} \cdot p^n$$
$$= \frac{n}{2}(n-1) \cdot p^{n-2}q^2 + n \cdot p^{n-1}q + p^n.$$

If we interchange the roles of success and failure, we obtain

$$P(2 \text{ failures}) + P(1 \text{ failure}) + P(0 \text{ failures})$$
$$= \binom{n}{2} \cdot q^2p^{n-2} + \binom{n}{1} \cdot qp^{n-1} + \binom{n}{0} \cdot p^n,$$

which is the same value, of course. ■

Did you notice how the sample spaces slowly disappeared from view as this section progressed? They are still there, of course, because random variables *are* defined on sample spaces. But one of the early triumphs in probability theory was the realization that practically all of probability involves random variables, which often can be studied without reference to sample spaces.

Exercises 9.2

1. (a) Three fair dice are tossed. What is the value set for the random variable that sums the values on the dice?

(b) Repeat part (a) for n dice.

2. Two fair dice are tossed. Find

(a) the probability that the sum is less than or equal to 7.

(b) $P(5 \leq \text{sum} \leq 10)$.

(c) the probability that the sum is a multiple of 3.

3. Suppose that the independent random variables X and Y have value set $\{0, 1, 2\}$, that $P(X = 0) = P(X = 1) = P(Y = 0) = P(Y = 1) = \frac{1}{4}$, and that $P(X = 2) = P(Y = 2) = \frac{1}{2}$. Calculate

(a) $P(X = 0 \text{ and } Y = 2)$

(b) $P(X = 0 \text{ or } Y = 2)$

(c) $P(X \leq 1 \text{ and } Y \geq 1)$

4. (a) For X as in Exercise 3, find the value set for the random variable $3X + 2$.

(b) Calculate $P(3X + 2 = k)$ for all k in the value set of $3X + 2$.

5. (a) Give the value set for the sum $X + Y$ of the random variables in Exercise 3.

(b) Calculate $P(X + Y = 2)$.

(c) Calculate $P(X + Y = k)$ for all k in the value set of $X + Y$.

6. Here is a random variable, T, on the set Ω of 36 equally likely outcomes when two fair dice are tossed: $T(k, l) = k \cdot l$ [k times l].

(a) Give the value set for T.

(b) Calculate $P(T \leq 2)$.

(c) Calculate $P(T = 12)$.

7. For Ω in Exercise 6, let D and M be the random variables defined by $D(k, l) = |k - l|$ and $M(k, l) = \max\{k, l\}$.

(a) Give the value sets for D and M.

(b) Make a table of probability values for D and M as in Figure 1(a) on page 360. *Hint:* Use Figure 1 on page 350.

(c) Calculate $P(D \le 1)$, $P(M \le 3)$, and $P(D \le 1$ and $M \le 3)$.

(d) Are the random variables D and M independent? Explain.

8. An urn has five red marbles and five blue marbles. Four marbles are selected at random [without replacement]. Determine the value set for the random variable X that counts the number of red marbles selected, and calculate $P(X = k)$ for k in the value set.

9. Repeat Exercise 8 if seven marbles are selected at random.

10. A fair die is tossed and a player wins \$5 if a 5 appears and loses \$1 otherwise. Find the value set for the random variable W that records the player's winnings, and calculate $P(W = k)$ for k in the value set. [A loss is a negative win.]

11. A fair die is tossed. Let W be the random variable that counts the number of tosses until the first 6 appears. What is the value set for W? Give the probabilities $P(W = k)$ for k in the value set.

12. A fair die is tossed n times.

(a) Give the probability that a 4, 5, or 6 appears at each toss.

(b) Give the probability that a 5 or 6 appears at each toss.

(c) Give the probability that a 6 appears at each toss.

13. An electronic device has 10 components. Each has probability 0.10 of burning out in the next year. Assume that the components burn out independently.

(a) What is the probability that none of the components burn out next year?

(b) What is the probability that at most two of the components burn out in the next year? *Hint:* Use Table 1.

14. Half of Burger Queen customers order French fries. Assume that they order independently.

(a) What is the probability that exactly five of the next ten customers will order French fries?

(b) What is the probability that at least three of the next ten customers will order French fries?

15. An electronic device has 50 components, each of which has probability 0.02 of failure before the warranty is up. Use Example 8 to find the

(a) probability that at least one component fails.

(b) probability that at most two components fail.

16. Where is independence used in Example 6?

17. Show that two events E and F are independent if and only if their characteristic functions χ_E and χ_F are independent random variables. You may use the result of Exercise 20 on page 358.

9.3 Expectation and Standard Deviation

Experience suggests that, if we toss a fair die lots of times, then the various possible outcomes 1, 2, 3, 4, 5, and 6 will each happen about the same number of times, and the average value of the outcomes will be about the average of 1, 2, 3, 4, 5, and 6; that is $(1 + 2 + 3 + 4 + 5 + 6)/6 = 3.5$. More generally, if X is a random variable on a finite sample space Ω with all outcomes equally likely, then the average value,

$$A = \frac{1}{|\Omega|} \sum_{\omega \in \Omega} X(\omega),$$

of X on Ω has a probabilistic interpretation: If elements ω of Ω are selected at random many times and the values $X(\omega)$ are recorded, then the average of the selected values will probably be close to A. This statement is actually a theorem that needs proof, but we hope you will accept it as reasonably intuitive.

When the outcomes are not equally likely, we are still interested in the probabilistic average, so we will weight the values of the random variable accordingly. Given a random variable X on a finite sample space Ω, its **expectation**, **expected value**, or **mean** is defined by

$$E(X) = \mu = \sum_{\omega \in \Omega} X(\omega) \cdot P(\{\omega\}).$$

If all outcomes are equally likely, then $P(\{\omega\}) = 1/|\Omega|$ for all ω in Ω, so $E(X)$ is exactly the unweighted average A discussed above. Note that we use two notations and terms for the same concept: the expectation $E(X)$ of X and the mean μ [lowercase Greek mu], or μ_X if we need to specify the random variable.

EXAMPLE 1

An amateur dart player knows from experience that each time he throws a dart at the dart board in Figure 1 he will hit the regions with the probabilities indicated. Thus $P(A) = 0.02$, $P(B) = 0.06$, etc. A concessionaire owns the dart board and offers to pay \$10 whenever the dart hits region A, \$3 whenever it hits region B, and \$1 whenever it hits region C. What is the amateur's expected average income per throw? It is certainly not the average $\frac{1}{4}(10 + 3 + 1 + 0) = \$3.50$.

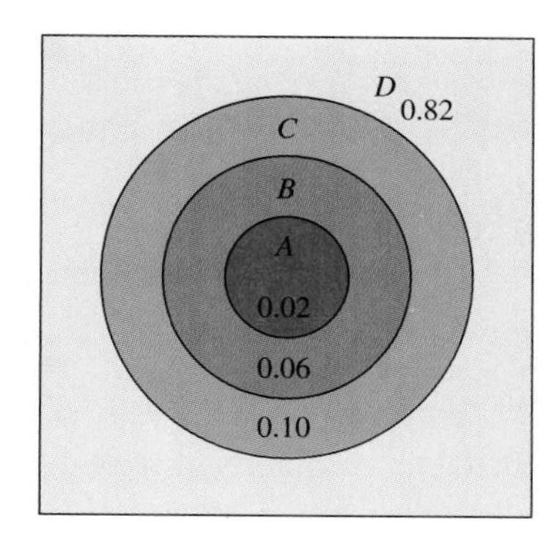

Figure 1 ▲

If he made, say, 100 throws, then he would expect to make \$10 about 2 times, i.e., make about \$20 from region A, and similarly about $\$3 \times 6 = \$18$ from region B, $\$1 \times 10 = \$10$ from region C, and $\$0 \times 82 = \$0$ from region D. All told, that would give him \$48 for his 100 throws or just \$0.48 per throw. That is, his probabilistic average or mean payoff per throw is

$$\mu = \$10 \times 0.02 + \$3 \times 0.06 + \$1 \times 0.10 + \$0 \times 0.82 = \$0.48.$$

We tried to make this expected value seem reasonable by imagining 100 throws, but it really has nothing to do with the number of throws. How likely it is that the player's actual average will be close to \$0.48 does depend on how many throws he makes and also on what we mean by "close," but that's another story. The value μ is $E(X)$ for the random variable X defined on the sample space $\Omega = \{A, B, C, D\}$, by $X(A) = 10$, $X(B) = 3$, $X(C) = 1$, and $X(D) = 0$.

Normally, concessionaires charge for the privilege of playing games. A charge of 48 cents per try would be "fair" for this player, in the sense that the expected gain would then be 0. Since the concessionaire owns the dart board, it would be reasonable if the charge were somewhat more than 48 cents per try.

Let's clarify why the expected gain would be 0 if the charge were μ per try. The new random variable representing the player's net gain would be $X - \mu$. See Figure 2. For this example, one can check directly that $E(X - \mu) = 0$, but in fact this equation always holds, as we will see after the next theorem. ■

Figure 2 ▶

Region	X [no charge]	$X - \mu$ [charging μ]
A	10.00	9.52
B	3.00	2.52
C	1.00	0.52
D	0.00	−0.48

Since random variables are functions with values in $\mathbb{R}$, we can add, subtract, and multiply them just as one does in algebra and calculus. The only difference is that the domain is now a sample space Ω instead of a set of real numbers. For example, given random variables X and Y on Ω, the new random variable $X + Y$ is defined by the equation

$$(X + Y)(\omega) = X(\omega) + Y(\omega) \quad \text{for all} \quad \omega \in \Omega.$$

For a real number a, the random variable aX is defined by $(aX)(\omega) = a \cdot X(\omega)$ for $\omega \in \Omega$. Also, for a real number c, the constant function defined by $X(\omega) = c$ for all $\omega \in \Omega$ is a random variable, which we'll simply call the "constant random variable c."

Theorem 1 Let X and Y be random variables on a finite sample space Ω. Then

(a) $E(X + Y) = E(X) + E(Y)$.

(b) $E(aX) = a \cdot E(X)$ for real numbers a in $\mathbb{R}$.

(c) $E(c) = c$ for any constant random variable c on Ω.

Proof We have

$$E(X + Y) = \sum_{\omega \in \Omega} (X + Y)(\omega) \cdot P(\{\omega\}) = \sum_{\omega \in \Omega} [X(\omega) + Y(\omega)] \cdot P(\{\omega\}).$$

We now use the distributive law for real numbers to rearrange terms:

$$E(X+Y) = \sum_{\omega\in\Omega} X(\omega)\cdot P(\{\omega\}) + \sum_{\omega\in\Omega} Y(\omega)\cdot P(\{\omega\}) = E(X)+E(Y).$$

Thus part (a) holds. We omit the proof of part (b), which is even easier. Finally,

$$E(c) = \sum_{\omega\in\Omega} c\cdot P(\{\omega\}) = c\cdot \sum_{\omega\in\Omega} P(\{\omega\}) = c\cdot P(\Omega) = c. \qquad \blacksquare$$

Corollary For any random variable X, we have $E(X-\mu)=0$.

Proof Here μ is the expectation $E(X)$ of X. Thus

$$E(X-\mu) = E(X) - E(\mu) = E(X) - \mu = 0. \qquad \blacksquare$$

One can sum more than two random variables. In fact, we quietly discussed such a sum in Example 4(b) on page 361. A simple induction, using Theorem 1(a), shows that

$$E(X_1+X_2+\cdots+X_n) = E(X_1)+E(X_2)+\cdots+E(X_n)$$

for random variables $X_1, X_2, \ldots, X_n$. The next theorem gives the expectation in terms of the value set

$$X(\Omega) = \{k\in\mathbb{R} : k = X(\omega) \text{ for some } \omega\in\Omega\}.$$

Theorem 2 For a random variable X on a finite sample space Ω, we have

$$E(X) = \sum_{k\in X(\Omega)} k\cdot P(X=k).$$

Proof The set Ω is a disjoint union of the sets $\{X=k\}$, $k\in X(\Omega)$, on which X takes the constant value k. So we can break the sum defining $E(X)$ into sums over these sets:

$$E(X) = \sum_{\omega\in\Omega} X(\omega)\cdot P(\{\omega\}) = \sum_{k\in X(\Omega)} \left\{ \sum_{\omega\in\{X=k\}} X(\omega)\cdot P(\{\omega\}) \right\}.$$

Since $X(\omega)=k$ for ω in $\{X=k\}$, the inside sum is equal to

$$\sum_{\omega\in\{X=k\}} X(\omega)\cdot P(\{\omega\}) = k\cdot \sum_{\omega\in\{X=k\}} P(\{\omega\}) = k\cdot P(X=k).$$

Consequently, we have

$$E(X) = \sum_{k\in X(\Omega)} k\cdot P(X=k). \qquad \blacksquare$$

The formula in Theorem 2 could have been taken as the definition of the expectation for discrete random variables. Generalizations of Theorem 2 hold for many other types of random variables as well.

EXAMPLE 2 We illustrate the previous theorems by again considering the random variables X_s, X_b, and X_r on the space Ω modeling the two-dice problem [Examples 2 and 3 on pages 360 and 361].

(a) To calculate $E(X_s)$ from the definition, we have to sum 36 numbers:

$$E(X_s) = \sum_{(k,l)\in\Omega} X_s(k,l)\cdot P(k,l) = \frac{1}{36}(2+3+3+4+\cdots+11+12).$$

You may finish this, but we'd rather show how the theorems make this easier.

(b) We calculate $E(X_s)$ using Theorem 2 and the values $P(X_s = k)$ in Figure 1(a) on page 360:

$$E(X_s) = 2 \cdot \frac{1}{36} + 3 \cdot \frac{2}{36} + 4 \cdot \frac{3}{36} + 5 \cdot \frac{4}{36} + 6 \cdot \frac{5}{36} + 7 \cdot \frac{6}{36} + 8 \cdot \frac{5}{36}$$
$$+ 9 \cdot \frac{4}{36} + 10 \cdot \frac{3}{36} + 11 \cdot \frac{2}{36} + 12 \cdot \frac{1}{36}$$
$$= \frac{1}{36}[2 + 6 + 12 + 20 + 30 + 42 + 40 + 36 + 30 + 22 + 12]$$
$$= \frac{252}{36} = 7.$$

With such a nice answer, one can't help wondering if there's an even better way to see this result. There is and we give it in part (c).

(c) Recall that $X_s = X_b + X_r$. From the discussion at the beginning of this section, we have $E(X_b) = E(X_r) = 3.5$, so

$$E(X_s) = E(X_b) + E(X_r) = 7.$$

In words, the expected value on the black die is 3.5 and the expected value on the red die is 3.5, so the expected sum is 7. ■

EXAMPLE 3

A fair coin is tossed n times. What is the expected number of heads? If your intuition gives $n/2$, you are right. But please read on.

(a) The question is this: What is $E(S_n)$, where S_n is the random variable that counts heads? Using Theorem 2, we find

$$E(S_n) = \sum_{k=0}^{n} k \cdot P(S_n = k) = \sum_{k=0}^{n} k \cdot \frac{1}{2^n}\binom{n}{k}.$$

This expression is too complicated, but we can learn from our experience in Example 2. The random variable S_n is the sum $X_1 + X_2 + \cdots + X_n$, where X_i is 1 if the ith toss is a head and 0 otherwise. Thus $P(X_i = 1) = P(X_i = 0) = \frac{1}{2}$. Then

$$E(X_i) = 1 \cdot P(X_i = 1) + 0 \cdot P(X_i = 0) = \frac{1}{2}$$

for each i, and therefore

$$E(S_n) = \sum_{i=1}^{n} E(X_i) = \frac{1}{2} + \frac{1}{2} + \cdots + \frac{1}{2} = \frac{n}{2},$$

as we expected.

(b) You may be puzzled by the first sum for $E(S_n)$ in part (a). It's correct, though, so what we have is an interesting proof of a "binomial identity":

$$\sum_{k=0}^{n} k \cdot \binom{n}{k} = n \cdot 2^{n-1}.$$

Many similar intriguing identities are used and studied in probability and combinatorics. ■

EXAMPLE 4

Let W be the random variable that counts the number of tosses of a fair coin needed to get a head, i.e., the "waiting time" for the first head; see Example 4(c) on page 361. What's the expected average wait? You might experiment to see. The answer is given by the infinite sum

$$\sum_{k=1}^{\infty} k \cdot P(W = k) = \sum_{k=1}^{\infty} k \cdot \frac{1}{2^k},$$

which turns out to be 2.

Is 2 the intuitively correct answer? We imagine that it is to some people, but not to most. Here's an argument that appears to be nonsense, though it isn't totally nonsense: You expect half a head in one toss, so you ought to get a whole head in two tosses! ■

The expectation of a random variable X gives us its probabilistic average. However, it doesn't tell us how close to the average we are likely to be. We need another measurement. A natural choice is the probabilistic average distance of X from its mean μ. This is the "mean deviation" $E(|X - \mu|)$, i.e., the mean of all the deviations $|X(\omega) - \mu|$, $\omega \in \Omega$. While this measure is sometimes used, it turns out that a similar measure, called the standard deviation, is much more manageable and useful. The **standard deviation** $\boldsymbol{\sigma}$ [or $\boldsymbol{\sigma_X}$] is the square root of $E((X - \mu)^2)$. Like the mean deviation, it lies between the smallest deviations $|X(\omega) - \mu|$ and the largest deviations $|X(\omega) - \mu|$ and serves as sort of an average of the deviations.

The primary difficulty with $E(|X - \mu|)$ is that the absolute value function $|x|$ is troublesome, whereas x^2 is not. Students of calculus may recall that $|x|$ is difficult to handle mainly because of its abrupt behavior at 0, where its graph takes a right-angle turn.

For a random variable X, with mean μ, the square of the standard deviation σ is called the **variance** of X and written $\boldsymbol{V(X)}$. Thus

$$V(X) = \sigma^2 = \sigma_X^2 = E((X - \mu)^2).$$

The variance and standard deviation measure how spread out the values of X are. The smaller $V(X)$ is, the more confident we can be that $X(\omega)$ is close to μ for a randomly selected ω.

To develop formulas to help us calculate $V(X)$, we start with a lemma.

Lemma For a random variable X on a finite sample space Ω, we have

$$E((X - c)^2) = \sum_{k \in X(\Omega)} (k - c)^2 \cdot P(X = k).$$

As with Theorem 2, this lemma makes sense for all discrete random variables.

Proof This proof is very like the proof of Theorem 2. Let Y be the random variable $(X - c)^2$; i.e., $Y(\omega) = (X(\omega) - c)^2$ for all $\omega \in \Omega$. We need to show that $E(Y) = \sum_{k \in X(\Omega)} (k - c)^2 \cdot P(X = k)$.

The set Ω is a disjoint union of the sets $\{X = k\}$, $k \in X(\Omega)$, on which Y takes the constant value $(k - c)^2$. Then

$$E(Y) = \sum_{\omega \in \Omega} Y(\omega) \cdot P(\{\omega\}) = \sum_{k \in X(\Omega)} \left\{ \sum_{\omega \in \{X = k\}} Y(\omega) \cdot P(\{\omega\}) \right\},$$

so

$$E(Y) = \sum_{k \in X(\Omega)} (k - c)^2 \left\{ \sum_{\omega \in \{X = k\}} P(\{\omega\}) \right\} = \sum_{k \in X(\Omega)} (k - c)^2 \cdot P(X = k). \quad ■$$

The next theorem follows by setting $c = \mu$ in the Lemma.

Theorem 3 For a discrete random variable X with mean μ, we have

$$V(X) = \sum_{k \in X(\Omega)} (k - \mu)^2 \cdot P(X = k).$$

Note that this theorem implies that $V(X) = 0$ if and only if X is the constant random variable μ.

EXAMPLE 5

(a) Let X be the random variable that records the value of a tossed fair die. Since $\mu = 3.5$ from our discussion at the start of this section, we have

$$V(X) = \sum_{k=1}^{6} (k-\mu)^2 \cdot \frac{1}{6}$$
$$= \frac{1}{6}[(1-3.5)^2 + (2-3.5)^2 + (3-3.5)^2 + (4-3.5)^2$$
$$+ (5-3.5)^2 + (6-3.5)^2] = \frac{1}{6}[17.5] = \frac{35}{12}.$$

The standard deviation is $\sigma = \sqrt{\frac{35}{12}} \approx 1.71$.

(b) Let X be the coin-tossing random variable so that $P(X=0) = P(X=1) = \frac{1}{2}$. Then we have

$$\mu = \frac{1}{2}, \quad V(X) = \left(0 - \frac{1}{2}\right)^2 \cdot \frac{1}{2} + \left(1 - \frac{1}{2}\right)^2 \cdot \frac{1}{2} = \frac{1}{4},$$

$$\text{and} \quad \sigma = \sqrt{V(X)} = \frac{1}{2}.$$ ■

The next formula for $V(X)$ makes the computations a little easier because the mean μ only has to be handled once.

Theorem 4 For a discrete random variable X with mean μ, we have

$$V(X) = E(X^2) - \mu^2,$$

where

$$E(X^2) = \sum_{k \in X(\Omega)} k^2 \cdot P(X=k).$$

Proof Since $(X-\mu)^2 = X^2 - 2\mu X + \mu^2$, parts (a) and (b) of Theorem 1 imply that

$$V(X) = E(X^2) - 2\mu \cdot E(X) + \mu^2.$$

But $E(X) = \mu$, so

$$V(X) = E(X^2) - 2\mu^2 + \mu^2 = E(X^2) - \mu^2.$$

Finally, the formula for $E(X^2)$ is the special case of the Lemma with $c = 0$. ■

EXAMPLE 6

Here we redo the computations in Example 5.

(a) We have

$$E(X^2) = \sum_{k=1}^{6} k^2 \cdot \frac{1}{6} = \frac{1}{6}[1+4+9+16+25+36] = \frac{91}{6},$$

so

$$V(X) = E(X^2) - \mu^2 = \frac{91}{6} - \left(\frac{7}{2}\right)^2 = \frac{35}{12}.$$

(b) We have $E(X^2) = 1 \cdot \frac{1}{2} = \frac{1}{2}$, so

$$V(X) = E(X^2) - \mu^2 = \frac{1}{2} - \left(\frac{1}{2}\right)^2 = \frac{1}{4}.$$ ■

Mean and variance have especially good properties for independent random variables.

Theorem 5 If X and Y are independent random variables, then

$$E(XY) = E(X) \cdot E(Y).$$

Proof The statement is true in general, but our proof only works for discrete random variables. To avoid infinite sums, we'll assume that X and Y have finite value sets. The sample space Ω is a disjoint union of the sets $\{X = k\} \cap \{Y = l\}$, where k and l range over $X(\Omega)$ and $Y(\Omega)$, respectively. By a now familiar argument, we conclude that

$$E(XY) = \sum_{k \in X(\Omega)} \sum_{l \in Y(\Omega)} k \cdot l \cdot P(\{X = k\} \cap \{Y = l\}).$$

Since X and Y are independent, $P(\{X = k\} \cap \{Y = l\}) = P(X = k) \cdot P(Y = l)$ by the condition (I) on page 361. Therefore,

$$E(XY) = \sum_{k \in X(\Omega)} \sum_{l \in Y(\Omega)} k \cdot l \cdot P(X = k) \cdot P(Y = l).$$

Since these are finite sums, we can use the distributive law for real numbers to rewrite the equation as

$$E(XY) = \left\{ \sum_{k \in X(\Omega)} k \cdot P(X = k) \right\} \cdot \left\{ \sum_{l \in Y(\Omega)} l \cdot P(Y = l) \right\} = E(X) \cdot E(Y). \quad \blacksquare$$

In general, the formula for the variance $V(X + Y)$ is complicated, but for independent random variables we have the following.

Theorem 6 If $X_1, X_2, \ldots, X_n$ are independent random variables, then

$$V(X_1 + X_2 + \cdots + X_n) = V(X_1) + V(X_2) + \cdots + V(X_n).$$

Proof We will only give a proof for two independent random variables X and Y. And, as before, we will assume that the value sets of X and Y are finite. Our task, then, is to show that $V(X + Y) = V(X) + V(Y)$.

We write μ_X and μ_Y for the means of X and Y; then $\mu_X + \mu_Y$ is the mean for $X + Y$, by Theorem 1(a). By Theorem 4, we have

$$\begin{aligned} V(X + Y) &= E((X + Y)^2) - (\mu_X + \mu_Y)^2 \\ &= E(X^2 + 2XY + Y^2) - (\mu_X^2 + 2\mu_X \cdot \mu_Y + \mu_Y^2) \\ &= E(X^2) + 2E(XY) + E(Y^2) - \mu_X^2 - 2\mu_X \cdot \mu_Y - \mu_Y^2. \end{aligned}$$

Since $V(X) = E(X^2) - \mu_X^2$ and $V(Y) = E(Y^2) - \mu_Y^2$, we can rewrite this equation as

$$V(X + Y) = V(X) + V(Y) + 2[E(XY) - \mu_X \cdot \mu_Y].$$

Since X and Y are independent, $E(XY) = E(X) \cdot E(Y) = \mu_X \cdot \mu_Y$ by Theorem 5, so $V(X + Y) = V(X) + V(Y)$.

Essentially the same argument, but with more terms in the sums, works for $n > 2$. ■

EXAMPLE 7

(a) What is the variance of the random variable X_s that adds the values on two fair dice? A direct assault using the definition or Theorem 4 would be hard work. But we know that $X_s = X_b + X_r$, where X_b and X_r are independent. We also know that $V(X_b) = V(X_r) = \frac{35}{12}$ from Example 5. Therefore, $V(X_s) = \frac{35}{6}$ by Theorem 6, and the standard deviation of X_s is $\sqrt{\frac{35}{6}} \approx 2.42$.

(b) To obtain the variance of the random variable S_n that counts the heads from n tosses of a fair coin, we recall that S_n is the sum of the independent random variables $X_1, X_2, \ldots, X_n$, where $P(X_i = 0) = P(X_i = 1) = \frac{1}{2}$. From Example 5, $V(X_i) = \frac{1}{4}$ for each i. Theorem 6 now implies that

$$V(S_n) = \sum_{i=1}^{n} V(X_i) = \sum_{i=1}^{n} \frac{1}{4} = \frac{n}{4}.$$

The standard deviation is $\frac{1}{2}\sqrt{n}$, which grows as n does, but much more slowly for large n. ■

We introduced binomial random variables S_n in §9.2 starting after Example 4. These random variables model the repetition of n independent experiments, where p and q are the probabilities of "success" and "failure," respectively. The expectation calculated in the next theorem should come as no surprise.

Theorem 7 If S_n is a binomial random variable for some n and p, then

$$E(S_n) = np \quad \text{and} \quad V(S_n) = npq.$$

The standard deviation is $\sqrt{npq}$.

Proof Just as with the special case $p = \frac{1}{2}$, i.e., the case of tossing fair coins, a direct assault on $E(S_n)$ and $V(S_n)$ would be unwise.

For $i = 1, 2, \ldots, n$, define $X_i = 1$ if the ith experiment is a success, and define $X_i = 0$ otherwise. Then $S_n = X_1 + X_2 + \cdots + X_n$; first we find the expectation and variance for each X_i. Since $P(X_i = 1) = p$ and $P(X_i = 0) = q$, we have

$$E(X_i) = 1 \cdot P(X_i = 1) + 0 \cdot P(X_i = 0) = p.$$

Since $X_i^2 = X_i$, we have $E(X_i^2) = p$ and

$$V(X_i) = E(X_i^2) - [E(X_i)]^2 = p - p^2 = p(1 - p) = pq.$$

Since $S_n = X_1 + X_2 + \cdots + X_n$, we conclude that

$$E(S_n) = \sum_{i=1}^{n} E(X_i) = \sum_{i=1}^{n} p = np.$$

Since the X_i are independent, Theorem 6 implies that

$$V(S_n) = \sum_{i=1}^{n} V(X_i) = \sum_{i=1}^{n} pq = npq.$$ ■

EXAMPLE 8

Table 1 on page 363 gives us probabilities of S_{10} for $p = \frac{1}{2}, \frac{1}{3}$, and $\frac{1}{10}$. The corresponding values of $E(S_{10})$ are 5, $\frac{10}{3}$, and 1, and the values of σ are ≈ 1.58, ≈ 1.49, and ≈ 0.95. The values of σ approach 0 as the probability p approaches 0. See Figure 2 on page 364. The same phenomenon occurs for any fixed n when p goes to 0, and a similar phenomenon occurs when p goes to 1. ■

Exercises 9.3

1. Let X be a random variable with $P(X = -1) = P(X = 0) = P(X = 1) = \frac{1}{5}$ and $P(X = 2) = \frac{2}{5}$. Find the expectation of
   (a) X (b) X^2 (c) $3X + 2$
2. Find the standard deviation for the random variable X in Exercise 1.
3. Find the standard deviation for the random variable X^2 in Exercise 1.
4. Suppose that certain lottery tickets cost \$1.00 each and that the only prize is \$1,000,000. If each ticket has probability 0.0000006 of winning, what is the expected gain when one ticket is purchased?

5. Consider independent random variables X and Y with

$$P(X = 0) = P(X = 1) = P(Y = 0) = P(Y = 1) = \frac{1}{4}$$

and $P(X = 2) = P(Y = 2) = \frac{1}{2}$. Find the mean and the standard deviation for X, Y and $X + Y$.

6. Find the mean and standard deviation for the random variables in Exercise 7 on page 365, defined on the set Ω of 36 equally likely outcomes when two fair dice are tossed: $D(k, l) = |k - l|$ and $M(k, l) = \max\{k, l\}$.

7. As in Exercise 9 on page 366, seven marbles are selected at random [without replacement] from an urn containing five red marbles and five blue marbles.

(a) What is the expected number of red marbles selected?

(b) What is the standard deviation?

8. If the concessionnaire in Example 1 charged \$1 for the privilege of throwing a dart at her dartboard, what would the player's expected average income be? Would playing this game be a good investment for the player? Discuss.

9. Calculate the mean deviations for the random variables in Example 5 and compare with the standard deviations.

10. A newspaper article states that the average family has 2.1 children and 1.8 automobiles. How many average families are there? Discuss.

11. What is the expected number of aces in a five-card poker hand? *Hint:* The direct assault using Theorem 2 can be avoided.

12. A fair die is tossed until the first 4 appears. Use the "nonsensical" argument in Example 4 to determine what the expected waiting time is for the first 4.

13. An urn has four red marbles and one blue marble.

(a) Marbles are drawn from the urn without replacement until the blue marble is obtained. What is the expected waiting time for getting the blue marble?

(b) Repeat part (a) if each marble is replaced after each drawing. *Hint:* See Exercise 12.

(c) Compare your answers to parts (a) and (b) and comment.

14. Show that $\sum_{k=1}^{n} k^2 \cdot \binom{n}{k} = n(n+1) \cdot 2^{n-2}$ for $n \geq 0$.

Hint: Let S_n be the random variable in Example 7, so $V(S_n) = n/4$ and $\mu = E(S_n) = n/2$. Calculate $E(S_n^2)$ two ways, using the two equations in Theorem 4.

15. Would you rather have a 50 percent chance of winning \$1,000,000 or a 20 percent chance of winning \$3,000,000? Discuss.

16. A baseball player has batting average 0.333. That is, he is successful in $\frac{1}{3}$ of his official attempts at hits. We assume that the attempts are independent.

(a) What is the expected number of hits in three official attempts?

(b) What is the probability that he will get at least one hit in his next three official attempts?

17. Suppose that the baseball player in Exercise 16 has ten official attempts.

(a) What is the expected number of hits?

(b) What is the probability that he will get at most three hits? *Hint:* Use Table 1 on page 363.

(c) What is the probability that he will get at least three hits?

18. Show that $E(XY)$ does not always equal $E(X) \cdot E(Y)$.

19. Show that $V(X + a) = V(X)$ and $V(aX) = a^2 \cdot V(X)$ for a in $\mathbb{R}$ and a random variable X with finite value set.

20. Let X be a random variable with mean μ and standard deviation σ. Give the mean and the standard deviation for $-X$.

21. (a) Suppose that $X_1, X_2, \ldots, X_n$ are independent random variables, each with mean μ and standard deviation σ. Find the mean and the standard deviation of $S = X_1 + X_2 + \cdots + X_n$.

(b) Do the same for the average $\frac{1}{n}S = \frac{1}{n}(X_1 + X_2 + \cdots + X_n)$.

22. (a) Show that if random variables X and Y on a finite sample space Ω satisfy $X(\omega) \leq Y(\omega)$ for all $\omega \in \Omega$, then $E(X) \leq E(Y)$.

(b) Show that if $|X(\omega) - \mu| \leq |X(\omega_0) - \mu|$ for all $\omega \in \Omega$, then $|X(\omega_0) - \mu| \geq \sigma_X$. That is, the largest deviation of $|X - \mu|$ is greater than or equal to the standard deviation of X.

9.4 Probability Distributions

In this section, we will study random variables using their probability distributions. At the end, we will be led to the famous bell curve and see why it plays a central role in probability and statistics.

Consider a typical binomial experiment consisting of 300 independent trials, each with probability $\frac{1}{3}$ of success. We know the expected number of successes, $300 \cdot \frac{1}{3} = 100$, and the standard deviation, $\sqrt{300 \cdot \frac{1}{3} \cdot \frac{2}{3}} \approx 8.16$; but what's the probability that the number of successes is between 90 and 110, say? That is, what's $P(90 \leq S_{300} \leq 110)$ for $p = \frac{1}{3}$? We have the formula

$$P(S_{300} = k) = \binom{300}{k} \cdot \left(\frac{1}{3}\right)^k \cdot \left(\frac{2}{3}\right)^{300-k}$$

from page 363, so couldn't we just ask a computer or programmable calculator to find the value of the sum

$$\sum_{k=90}^{110} \binom{300}{k} \cdot \frac{2^{300-k}}{3^{300}} ?$$

One trouble with this method is that numerical problems with overflow and round-off might give us an answer we can't trust. The binomial coefficients here are enormous—$\binom{300}{90}$ has 79 decimal digits—but they're being multiplied by incredibly tiny numbers, too. Plainly, a calculator won't be up to the job. And this was just a simple example. What if we had 3000 trials, or $p = 0.423$, or wanted the probability of a larger range of values?

One of our goals in this section is to develop a method to handle problems such as this one with ease, using a single standard function available on many calculators and spreadsheet computer applications. Along the way we will see some other important probabilistic ideas.

Each random variable X has a function F_X, defined on the real line $\mathbb{R}$, that encodes the probabilistic information about X. So far, with discrete random variables, we've focused on the values $P(X = k)$ and used appropriate sums of these values to answer probabilistic questions about X. In fact, most questions about X involve quantities such as $P(X \leq 5)$, $P(X < 5)$, $P(X > 5)$, $P(X \geq 5)$, and $P(5 \leq X \leq 10)$. All these quantities can be written in terms of probabilities of the form $P(X \leq y)$ and perhaps also of the form $P(X = k)$. Of course, $P(X \leq 5)$ already has this form. Since $P(X \leq 5) = P(X < 5) + P(X = 5)$, we can write $P(X < 5) = P(X \leq 5) - P(X = 5)$. Similarly, $P(X > 5) = 1 - P(X \leq 5)$, $P(X \geq 5) = 1 - P(X < 5)$, and also $P(5 \leq X \leq 10) = P(X \leq 10) - P(X < 5)$. The probabilities of the form $P(X \leq y)$ contain key information about X. Their values vary with y in $\mathbb{R}$, and it is useful to formalize how they do.

For any random variable X on any sample space, the **cumulative distribution function** or **cdf** of X is the function $\boldsymbol{F_X}$ defined on $\mathbb{R}$ by

$$F_X(y) = P(X \leq y) \quad \text{for} \quad y \in \mathbb{R}.$$

This function accumulates or collects the values of $P(X)$ up to and including the value for $X = y$. If X is a discrete random variable, then F_X *sums* the values:

$$F_X(y) = \sum_{k \leq y} P(X = k) \quad \text{for discrete random variables.}$$

If X is understood, we will sometimes write F instead of F_X.

EXAMPLE 1

Consider the random variables X_s, X_b, and X_r with the probabilities specified in Figure 1 on page 360. The cdf for X_b and X_r is F, where

$$F(y) = \begin{cases} 0 & \text{for } y < 1 \\ 1/6 & \text{for } 1 \leq y < 2 \\ 2/6 & \text{for } 2 \leq y < 3 \\ 3/6 & \text{for } 3 \leq y < 4 \\ 4/6 & \text{for } 4 \leq y < 5 \\ 5/6 & \text{for } 5 \leq y < 6 \\ 1 & \text{for } y \geq 6 \end{cases} .$$

The cdf F_s for X_s is given by

$$F_s(y) = \sum_{k \leq y} P(X_s = k).$$

These cdf's are shown in Figure 1. ■

Figure 1 ▶

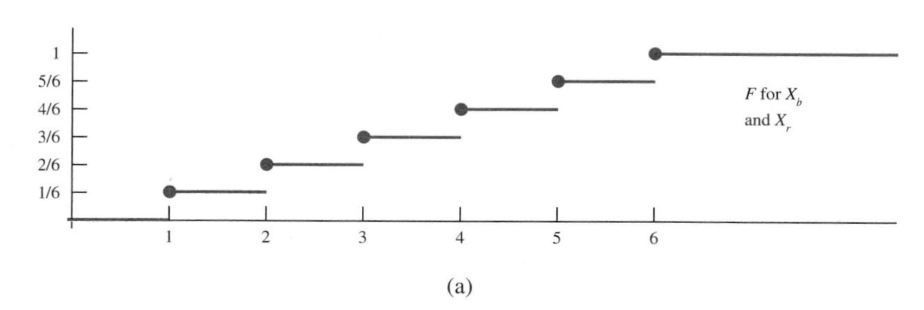

(a)

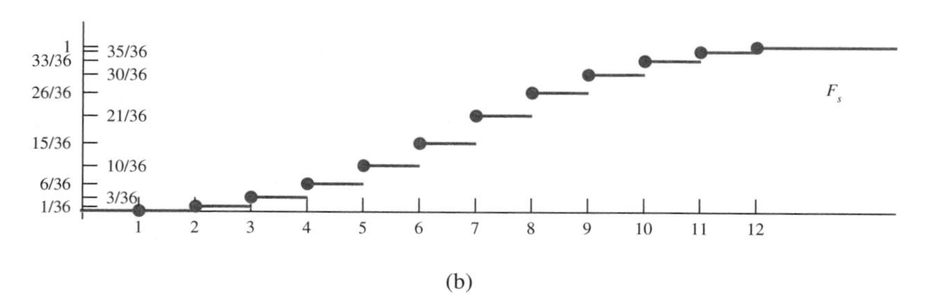

(b)

EXAMPLE 2

We return to the binomial random variable S_n that counts the number of "successes" in n independent experiments, with probability p of success for each experiment. We saw in §9.2 that $P(S_n = k) = \binom{n}{k} p^k q^{n-k}$ for $k = 0, 1, 2, \dots, n$, where $q = 1 - p$ is the probability of "failure"; so the cdf F for S_n is

$$F(y) = \sum_{k \le y} \binom{n}{k} p^k q^{n-k} \quad \text{for} \quad y \in \mathbb{R}.$$

This cdf is called the **cumulative binomial distribution**. Note that it is a step function that is constant between integers and has jumps at points in the value set, a property that holds for all discrete random variables. See Figure 2 for a picture of the cdf for S_5 when $p = 0.5$. Table 1 gives the values of the cumulative binomial distribution in case $n = 10$ and p is $\frac{1}{2}$, $\frac{1}{3}$, or $\frac{1}{10}$. Much more complete tables can be found in probability and statistics books, and many calculators can compute values of cdf's. ■

Figure 2 ▶

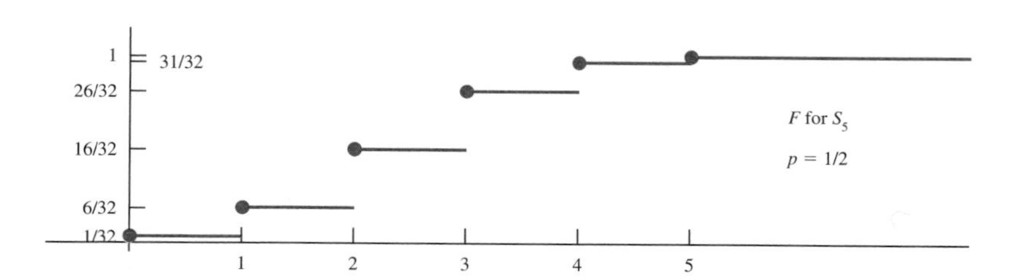

TABLE 1 Cumulative Binomial Distribution for $n = 10$

k	0	1	2	3	4	5	6	7	8	9	10
$p = 1/2$	0.001	0.011	0.055	0.172	0.377	0.623	0.828	0.945	0.989	0.999	1.00
$p = 1/3$	0.017	0.104	0.299	0.559	0.787	0.923	0.980	0.997	1.00	1.00	1.00
$p = 1/10$	0.349	0.736	0.930	0.998	1.00	1.00	1.00	1.00	1.00	1.00	1.00

EXAMPLE 3

We illustrate the power of cdf's by using Table 1 to calculate some probabilities. Note that any of the calculations could be made by summing entries in Table 1 on page 363, but this could get tedious.

(a) A fair coin is tossed ten times. We can calculate many probabilities quickly:

$$\begin{aligned} P(\text{at most 5 heads}) &= P(S_{10} \le 5) = F(5) \approx 0.623, \\ P(\text{at least 4 heads}) &= P(S_{10} \le 10) - P(S_{10} \le 3) \\ &= F(10) - F(3) \approx 1 - 0.172 = 0.828, \\ P(3 \le \text{number of heads} \le 7) &= F(7) - F(2) \approx 0.945 - 0.055 = 0.890. \end{aligned}$$

Note that we can use the table to calculate any of the values $P(S_{10} = k)$. For example, $P(S_{10} = 5) = F(5) - F(4) \approx 0.623 - 0.377 = 0.246$, which agrees with the corresponding entry in Table 1 on page 363.

(b) With a biased coin the probability of a head on each toss is $\frac{1}{3}$. Then

$$\begin{aligned} &P(\text{at most 5 heads}) = F(5) \approx 0.923, \\ &P(\text{at least 4 heads}) = F(10) - F(3) \approx 1 - 0.559 = 0.441 \\ &P(3 \le \text{number of heads} \le 7) = F(7) - F(2) \approx 0.997 - 0.299 = 0.698. \end{aligned}$$ ■

The advantage of a cdf, as illustrated in Example 3, would be even more apparent if we wanted to calculate, say, the probability of obtaining between 30 and 70 heads in 100 tosses of a fair coin. Then we'd want $P(30 \le S_{100} \le 70)$. If we had a table of *values* of $P(S_{100} = k)$, we would have to sum $\sum_{k=30}^{70} P(S_{100} = k)$, a sum of 41 numbers! If we were given a table for the *cdf*, we would only need to calculate the difference $F(70) - F(29)$. But even if this idea works for $n = 100$ and $p = \frac{1}{2}$, how about our example at the start of this section, with $n = 300$ and $p = \frac{1}{3}$? Or how about $p = 0.315$? We won't find tables of cdf's for all interesting values of n and p.

The miraculous fact is that we won't need those tables to solve such problems for large n, which is where the troubles come up. Instead, we can use a special distribution function, called the "Gaussian distribution." It's not a distribution of a discrete random variable, though, so we'll need to extend our horizons to consider nondiscrete random variables.

EXAMPLE 4

The simplest examples of nondiscrete random variables are the uniform ones, based on the probability introduced briefly in Example 7(b) on page 194.

When people talk about choosing a random number in the interval $[0, 1)$, they don't mean that all numbers in $[0, 1)$ are equally likely to be chosen. That would be a true statement, but a useless one, since the probability of choosing any given number, such as $\frac{1}{2}$ or $\sqrt{3}/4$, is 0. What they mean instead is that the probability of choosing a number in any given subinterval $[a, b)$ is proportional to the length of the subinterval. The probability of choosing the number in $[0, 1)$ is 1, so the probability of choosing it in $[a, b)$ is $b - a$. In particular, if U is the random variable on $[0, 1)$ that gives the value of the number chosen, then $P(U \text{ in } [0, x)) = x - 0 = x$ for $0 \le x < 1$. Since $P(U = x) = 0$ for every x, we also have $P(0 \le U \le x) = P(0 \le U < x)$. Thus the cdf F_U for U, called the **uniform distribution** on $[0, 1)$ or on $[0, 1]$, is given by

$$F_U(y) = \begin{cases} 0 & \text{for } y < 0 \\ y & \text{for } 0 \le y < 1 \\ 1 & \text{for } 1 \le y \end{cases}.$$

The random variable U itself is called a **uniform random variable**. Its value set is $[0, 1)$, since every value in $[0, 1)$ is *possible*, even though $P(U = x) = 0$ for each x in $[0, 1)$. For this random variable, probabilities of intervals are useful, but probabilities of single points are not.

The cdf's for discrete random variables have jumps at points in the value set. In contrast, the cdf F_U is a nice continuous function with no jumps. See Figure 3. ■

Figure 3 ▶

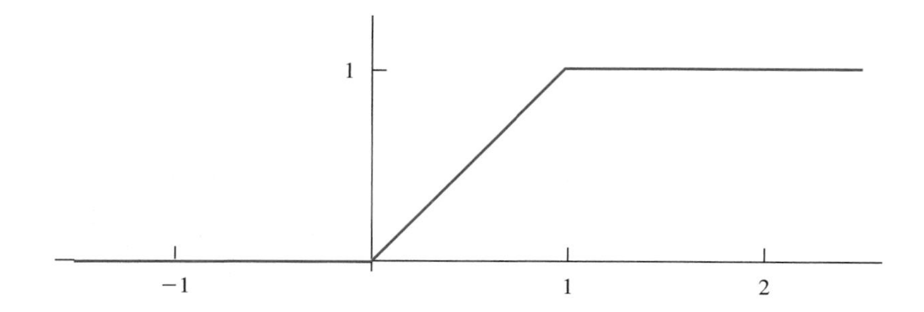

We next show a couple of examples of how cdf's are related to areas under the graphs of certain functions whose graphs lie on or above the x-axis.

EXAMPLE 5

(a) Consider the random variable X that records the value obtained when a single fair die is tossed. Thus $P(X = k) = \frac{1}{6}$ for $k = 1, 2, 3, 4, 5, 6$. The cdf for X is the cdf F in Example 1 and Figure 1. We create a function f on $\mathbb{R}$ in stages. First we define $f(k) = P(X = k) = \frac{1}{6}$ for $k = 1, 2, 3, 4, 5, 6$; see Figure 4(a). Next we replace each of the six points by a rectangle of width 1; see Figure 4(b). We extended the graph to the left of each point, but it would be as reasonable to extend it to the right of each point, and one could argue that it would more fair to arrange for the rectangle to extend $\frac{1}{2}$ to each side of the points. Finally, we define the function f to be 0 elsewhere; see Figure 4(c).

Figure 4 ▶

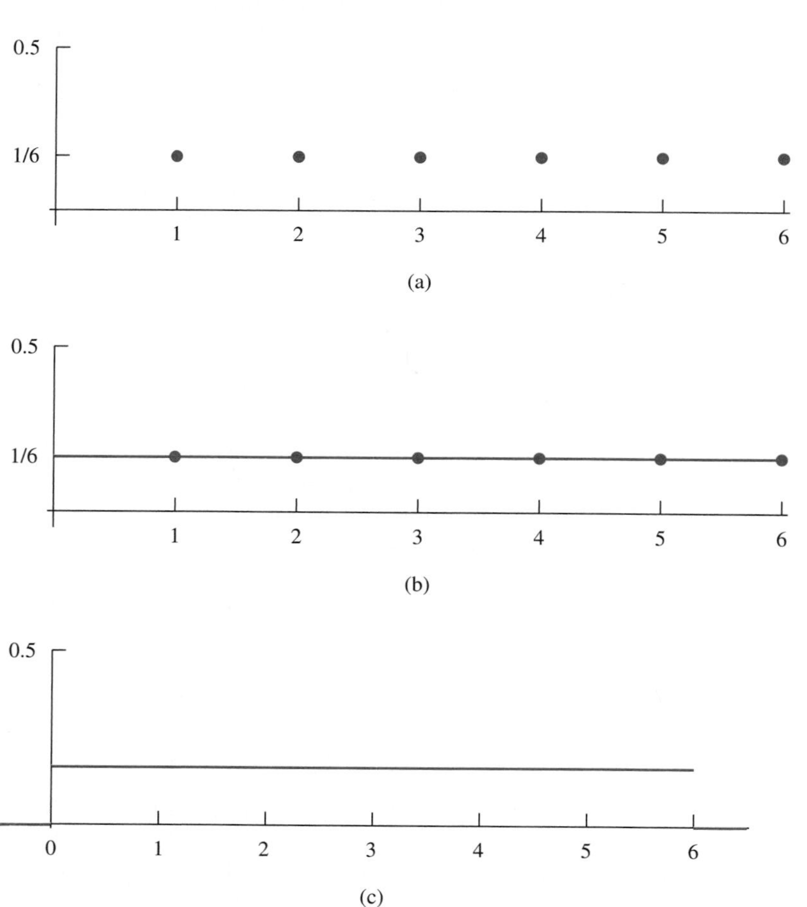

Note that the area of each little rectangle is $\frac{1}{6}$ and that the total area under the graph of f is equal to 1. Finally, and this is the connection between F and areas under f, note that

$$F(k) = P(X \leq k) = \text{area under } f \text{ on } (-\infty, k]$$

for $k = 1, 2, 3, 4, 5, 6$. Remember, $(-\infty, k] = \{y \in \mathbb{R} : y \leq k\}$. [This area formula doesn't work for $F(y)$ if y is a noninteger in $[0, 6]$, but please don't worry about this.]

(b) Now consider the binomial cdf in Example 2 and the picture for $n = 5$ and $p = \frac{1}{2}$ in Figure 2. Just as in part (a), we create a function f on $\mathbb{R}$ starting with $f(k) = P(S_5 = k) = \frac{1}{2^5}\binom{5}{k} = \frac{1}{32}\binom{5}{k}$ for $k = 0, 1, 2, 3, 4, 5$. See the development in Figure 5. Again the areas of the little rectangles are $P(S_5 = 0), P(S_5 = 1), \ldots, P(S_5 = 5)$, and the total area under the graph of f is 1. And, as in part (a),

$$F(k) = P(S_5 \leq k) = \text{area under } f \text{ on } (-\infty, k]$$

for $k = 0, 1, 2, 3, 4, 5$.

Figure 5 ▶

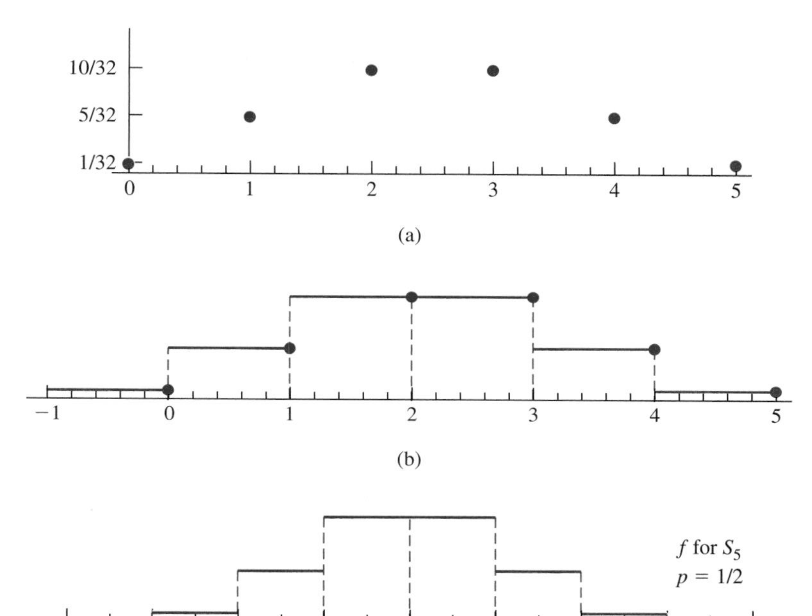

(c) Now we return to the uniform distribution F_U in Example 4 and Figure 3. This time we pull a nice f_U for F_U out of the hat, though, if you know some calculus, note that $f_U(x)$ is just the derivative $F_U'(x)$ [except for $x = 0$ and $x = 1$]. Here's

$$f_U(x) = \begin{cases} 0 & \text{for} \quad x < 0 \\ 1 & \text{for} \quad 0 \leq x < 1 \, ; \\ 0 & \text{for} \quad x \geq 1 \end{cases}$$

see Figure 6. Again $f_U(x) \geq 0$ for all $x \in \mathbb{R}$, and the total area under f_U is obviously 1. Moreover, by mentally checking the three cases $x < 0$, $0 \leq x < 1$, and $x \geq 1$, one can easily see that

$$F_U(y) = P(U \leq y) = \text{area under } f_U \text{ on } (-\infty, y].$$

This holds *for all* y in $\mathbb{R}$; in this respect, the random variable U is superior to discrete random variables. ■

Figure 6 ▶

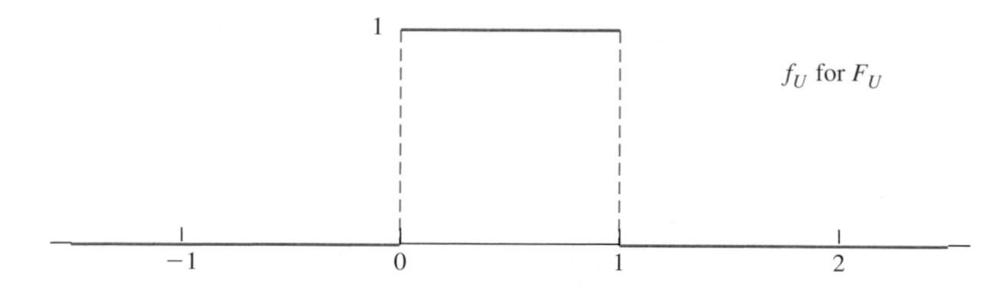

Note the key features of each f and f_U in the last example:

$f(x) \geq 0$ for all $x \in \mathbb{R}$.
Total area under f is equal to 1.
$F(y) = P(X \leq y) =$ area under f on $(-\infty, y]$, at least for some values of y.

Before we give more illustrations of the connection between cdf's and areas under special functions f, we introduce a way of "normalizing" random variables.

Why would anyone want to do that? Sometimes, even though we might wish otherwise, nice finite discrete objects lead to more complicated theoretical objects. Such is the case with binomial distributions. In applications, the number n of experiments or samples is often large, and the computations of the binomial random variable get difficult. Clearly, it would be nice if, for large n, the cdf's F_n of the binomial distributions were close to a single cdf of some well-behaved random variable. [We will clarify what we mean by "close" later.] As stated, this can't be true. The means and standard deviations [np and $\sqrt{npq}$, respectively, by Theorem 7 on page 373] get large as n gets large, so the cdf's F_n drift off to the right on $\mathbb{R}$. For example, if n = 1,000,000 and $p = \frac{1}{2}$, the mean of the binomial distribution is 500,000, and the most probable values are around 500,000. The corresponding cdf doesn't reach $\frac{1}{2}$ until y = 500,000.

On the other hand, except for the horizontal stretching, the graphs of these cdf's all have similar general shapes. They go along near 0 for a while, swing up near the mean, at $x = \mu$, and then flatten out again to approach 1. If we could shrink and shift them appropriately, maybe the altered versions would all look pretty much alike. This is the idea behind "normalization," which shifts the x-axis to put the mean at 0 and then changes the scale to make the standard deviation equal to 1. As we will see, normalizing binomial distributions is just the right trick.

Given any random variable X, with mean μ and standard deviation $\sigma > 0$, the **normalization of** X is $\widetilde{X} = \frac{X-\mu}{\sigma}$. This new random variable only differs from X by a constant multiple $1/\sigma$ and an additive constant, so knowledge of $\widetilde{X}$ is easily transferred back to X. The nice properties of $\widetilde{X}$ are stated in part (a) of the next theorem. The rest of the theorem shows the simple connection between the cdf's of X and $\widetilde{X}$; if you know one, then you can easily compute the other.

Theorem 1 Let X be a random variable with mean μ, standard deviation $\sigma > 0$, and cdf F. Let $\widetilde{X}$ be the normalization of X and let $\widetilde{F}$ be its cdf.

(a) $E(\widetilde{X}) = 0, \quad V(\widetilde{X}) = 1$ and $\sigma_{\widetilde{X}} = 1$.
(b) $F(y) = \widetilde{F}\left(\frac{y-\mu}{\sigma}\right)$ for $y \in \mathbb{R}$.
(c) $\widetilde{F}(y) = F(\sigma y + \mu)$ for $y \in \mathbb{R}$.

Proof

(a) Using Theorem 1 on page 367, we obtain

$$E(\widetilde{X}) = \frac{1}{\sigma}E(X - \mu) = \frac{1}{\sigma}[E(X) - \mu] = \frac{1}{\sigma}[\mu - \mu] = 0.$$

Exercise 19 on page 374 shows that $V(X+a) = V(X)$ and $V(aX) = a^2V(X)$ in general. Hence

$$V(\widetilde{X}) = V\left(\frac{X}{\sigma}\right) = \frac{1}{\sigma^2}V(X),$$

and since $V(X) = \sigma^2$, we obtain $V(\widetilde{X}) = 1$.

(b) By definition, $F(y) = P(X \le y)$. Also, for ω in the sample space of X,

$$X(\omega) \le y \iff X(\omega) - \mu \le y - \mu \iff \frac{X(\omega)-\mu}{\sigma} \le \frac{y-\mu}{\sigma}.$$

So

$$\{X \le y\} = \left\{\frac{X-\mu}{\sigma} \le \frac{y-\mu}{\sigma}\right\} = \left\{\widetilde{X} \le \frac{y-\mu}{\sigma}\right\}.$$

It follows that

$$F(y) = P(X \le y) = P\left(\widetilde{X} \le \frac{y-\mu}{\sigma}\right) = \widetilde{F}\left(\frac{y-\mu}{\sigma}\right).$$

(c) Replace y by $\sigma y + \mu$ in part (b). ■

EXAMPLE 6

Let S_n be a random variable with binomial distribution for n and some fixed p, $0 < p < 1$. The corresponding normalized random variable is

$$\widetilde{S_n} = \frac{S_n - np}{\sigma} = \frac{S_n - np}{\sqrt{npq}}.$$

We know the value set of S_n is $\{0, 1, 2, \dots, n\}$. The value set of $\widetilde{S_n}$ is more complicated:

$$\left\{\frac{-np}{\sigma}, \frac{-np+1}{\sigma}, \frac{-np+2}{\sigma}, \dots, \frac{n-np}{\sigma}\right\}.$$

Note that the last term is $\frac{nq}{\sigma}$ since $1-p=q$. For $n=5$ and $p=\frac{1}{2}$, the value set of $\widetilde{S_5}$ is

$$\left\{\frac{-2.5}{\sqrt{5}/2}, \frac{-1.5}{\sqrt{5}/2}, \frac{-0.5}{\sqrt{5}/2}, \frac{0.5}{\sqrt{5}/2}, \frac{1.5}{\sqrt{5}/2}, \frac{2.5}{\sqrt{5}/2}\right\};$$

see Exercise 4. This value set is $\approx \{-2.236, -1.342, -0.447, 0.447, 1.342, 2.236\}$.

We write F_n and $\widetilde{F_n}$ for the cdf's of S_n and $\widetilde{S_n}$, respectively. For $n=5$ and $p=\frac{1}{2}$, Figure 5 shows the picture of a function f with the property

$$F_5(k) = P(S_5 \le k) = \text{area under } f \text{ on } (-\infty, k]$$

for k in the value set of S_5. Figure 7 shows the picture of the corresponding function for $\widetilde{S_5}$, which satisfies

$$\widetilde{F_5}(y) = P(\widetilde{S_5} \le y) = \text{area under } f \text{ on } (-\infty, y]$$

for y in the value set of $\widetilde{S_5}$. Since the little rectangles don't have width 1 anymore, the heights of the rectangles have been adjusted so that the total area under f is 1. In fact, the heights are σ times the heights in Figure 5.

Figure 7 ▶

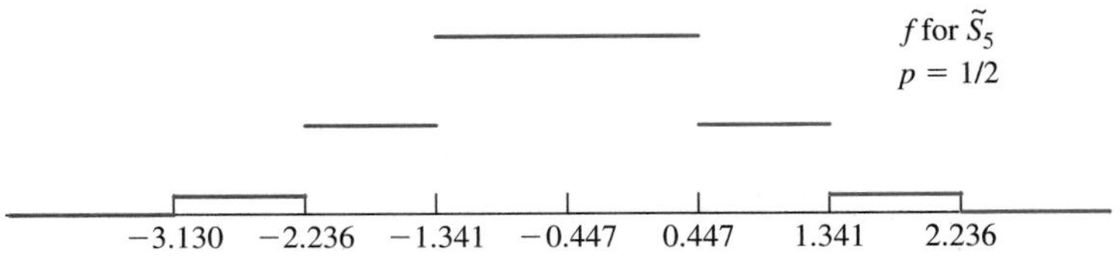

When n is large, the corresponding function f satisfying

$$\widetilde{F_n}(y) = P(\widetilde{S_n} \le y) = \text{area under } f \text{ on } (-\infty, y]$$

for y in the value set of $\widetilde{S}_n$ looks something like the function sketched in Figure 8. [The corresponding graph for $n = 100$ and $p = \frac{1}{2}$ would have the peak slightly more to the right.] ∎

Figure 8 ▶

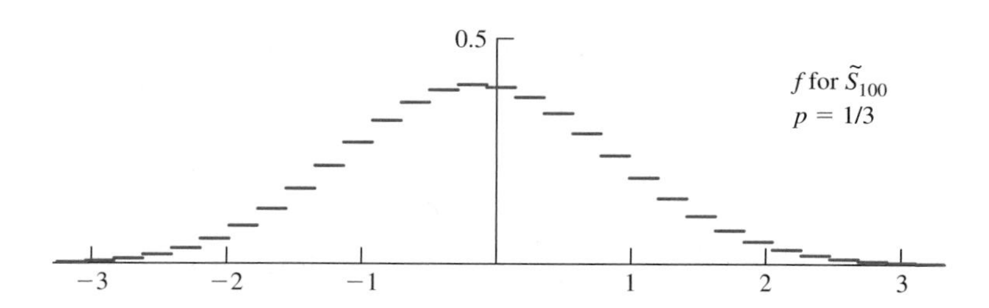

We're finally ready to put things together. Since the random variables $\widetilde{S}_n$ all have the same mean and standard deviation, it is reasonable to hope that their cdf's $\widetilde{F}_n$ are close to some fixed cdf when n is large. From Figure 8 it is even reasonable to hope that such a cdf would be based on the area under the graph of a function that looks like a bell curve; see Figure 9(a). Such a cdf won't be the cdf for any discrete random variable, and it won't be the uniform distribution, so it must be some new sort of cdf. A central result in probability theory is that there does exist a cdf $\mathbf{\Phi}$, called the **Gaussian** or **standard normal distribution**, such that

$$\widetilde{F}_n(y) \approx \Phi(y) \qquad \text{for large } n \text{ and for } y \in \mathbb{R}. \qquad (*)$$

Thus, for large n, the values of $P(a < \widetilde{S}_n \leq b) = \widetilde{F}_n(b) - \widetilde{F}_n(a)$ will be close to $\Phi(b) - \Phi(a)$. The distribution Φ does not depend on p; for any fixed p, the cdf's $\widetilde{F}_n$ are close to Φ for large n. This extraordinary result and some powerful generalizations of it are known as the "central limit theorem." The proof is given in more advanced texts.

Figure 9 ▶

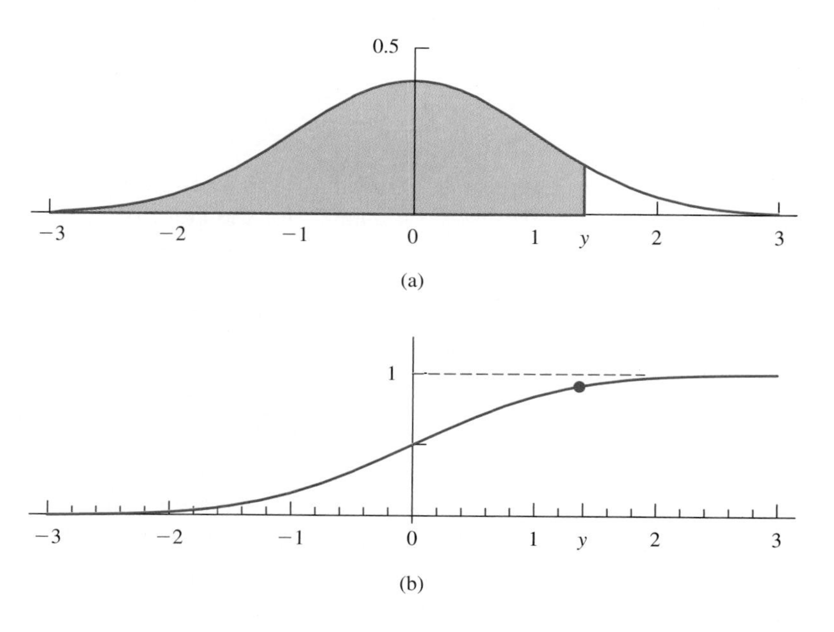

We will call Φ the **Gaussian distribution**. It is a special case of a more general class of distributions, which we will introduce later and refer to as "normal distributions."

Before illustrating the uses of this powerful tool, we clarify exactly what Φ is. Figure 9(a) is a picture of the standard bell curve. Its formula is

$$\phi(x) = \frac{1}{\sqrt{2\pi}} e^{-x^2/2}$$

TABLE 2 The Gaussian Distribution Φ

y	$\Phi(y)$	y	$\Phi(y)$	y	$\Phi(y)$	y	$\Phi(y)$	y	$\Phi(y)$	y	$\Phi(y)$
−3.0	0.001	−2.0	0.023	−1.0	0.159	0.0	0.500	1.0	0.841	2.0	0.977
−2.9	0.002	−1.9	0.029	−0.9	0.184	0.1	0.540	1.1	0.864	2.1	0.982
−2.8	0.003	−1.8	0.036	−0.8	0.212	0.2	0.579	1.2	0.885	2.2	0.986
−2.7	0.003	−1.7	0.045	−0.7	0.242	0.3	0.618	1.3	0.903	2.3	0.989
−2.6	0.005	−1.6	0.055	−0.6	0.274	0.4	0.655	1.4	0.919	2.4	0.992
−2.5	0.006	−1.5	0.067	−0.5	0.309	0.5	0.691	1.5	0.933	2.5	0.994
−2.4	0.008	−1.4	0.081	−0.4	0.345	0.6	0.726	1.6	0.945	2.6	0.995
−2.3	0.011	−1.3	0.097	−0.3	0.382	0.7	0.758	1.7	0.955	2.7	0.996
−2.2	0.014	−1.2	0.115	−0.2	0.421	0.8	0.788	1.8	0.964	2.8	0.997
−2.1	0.018	−1.1	0.136	−0.1	0.460	0.9	0.816	1.9	0.971	2.9	0.998

where $e \approx 2.71828$. It is interesting to see the two most interesting irrational numbers, π and e, appear in this formula. The total area under the bell curve is 1. For $y \in \mathbb{R}$,

$$\Phi(y) = \text{ area under } \phi \text{ to the left of } y;$$

see Figure 9(b). For the y indicated in Figure 9(a), this is the shaded area. Students of calculus would write

$$\Phi(y) = \frac{1}{\sqrt{2\pi}} \int_{-\infty}^{y} e^{-x^2/2} dx,$$

and might hope that there is a simple formula for this integral. Alas, there is no simple formula for $\Phi(y)$, but the values of Φ can be approximated as closely as desired and can be found in tables and on calculators. A short table is given in Table 2. By the way, there is one value that is clear: $\Phi(0) = 0.500$ since half the area under ϕ lies to the left of 0.

EXAMPLE 7

Suppose that an experiment is repeated $n = 10{,}000$ times with $P(\text{success}) = 0.1 = p$ each time. The expected number of successes is $\mu = np = 1000$, and we want to know

$$P(950 \le \text{number of successes} \le 1050).$$

This is $F_n(1050) - F_n(949)$. We convert this into a difference involving the normalized random variable $\widetilde{F}_n$ and use Theorem 1. Since

$$\sigma = \sqrt{npq} = \sqrt{10,000 \cdot \frac{1}{10} \cdot \frac{9}{10}} = 30,$$

we have

$$F_n(1050) = \widetilde{F}_n\left(\frac{1050 - 1000}{30}\right) \approx \Phi(1.7) \approx 0.955$$

and

$$F_n(949) = \widetilde{F}_n\left(\frac{949 - 1000}{30}\right) \approx \Phi(-1.7) \approx 0.045.$$

We conclude that

$$P(950 \le \text{number of successes} \le 1050) \approx 0.910.$$

The approximations for the values of Φ were obtained using Table 2. Use of a better table, or a calculator, gives the answer 0.901. ■

Example 7 illustrates the power of the central limit theorem. For any p and large n, we only need to understand *one* cdf. At the computational level, this means that we only need one table of values or one function key on a calculator. In this brief introduction, we haven't faced the question of how big n needs to be to get a good approximation, or even what we mean by a good approximation. It turns out that the approximation will be useful for most purposes provided that $np \ge 10$ and $n(1 - p) \ge 10$.

The Gaussian distribution is a special case of a more general class of distributions. Any random variable X whose normalized random variable $\widetilde{X}$ has distribution Φ will be called a **normal random variable**, and its distribution will be called a **normal distribution**. Normal distribution cdf's look a lot like Φ. In fact, their graphs are just stretched or shrunk shifts of Φ; see Exercise 21. A random variable whose distribution function is close to a normal distribution will be said to be "approximately normally distributed."

Normal distributions arise in probability and statistics in many ways. It turns out that observed measured data, such as heights of all adult women or SAT exam scores, are approximately normally distributed; i.e., their normalized random variables have approximately the Gaussian distribution.

EXAMPLE 8

A random sample is taken of the heights of some adult women. Thus we have a random variable X = height defined on the given set of women. Our statistician friends look at the data and tell us that they are approximately normally distributed, with mean $\mu = 66$ inches and standard deviation $\sigma = 2.5$ inches. We estimate the probability that an adult woman has height between 61 and 71 inches. We will ignore those with height exactly 61 inches and estimate $P(61 < X \le 71)$. This is $F(71) - F(61)$, where F is the cdf of X. If $\widetilde{F}$ is the cdf for the normalized random variable $\widetilde{X}$, then Theorem 1(b) shows that

$$F(71) - F(61) = \widetilde{F}\left(\frac{71-66}{2.5}\right) - \widetilde{F}\left(\frac{61-66}{2.5}\right) = \widetilde{F}(2) - \widetilde{F}(-2).$$

Since the cdf $\widetilde{F}$ is assumed to be approximately the Gaussian distribution Φ,

$$P(61 < X \le 71) \approx \Phi(2) - \Phi(-2).$$

From the table, this value is approximately $0.977 - 0.023 = 0.954$. ■

The fact that we were supplied with μ and σ made estimating the probability in this example fairly painless, compared with going through the data set and actually counting the women with heights between 61 and 71 inches. It would have been just as easy to use Φ to find the proportion of women whose heights fall into any other given range, too. Moreover, if our sample set of women has been sufficiently randomly drawn from, say, the population of all U.S. women, then the probability that a woman drawn from the total population has height between 61 and 71 inches is approximately the same as the probability for women in our sample, about 0.95. The justification for this last statement lies in the field of statistics, which we will not pursue.

The computation in the example applies in considerable generality. Let X be any random variable with a normal distribution, so that Φ is the distribution function of $\widetilde{X}$. If μ and σ are the mean and the standard deviation for X, then

$$\begin{aligned} P(\mu - 2\sigma < X \le \mu + 2\sigma) &= F(\mu + 2\sigma) - F(\mu - 2\sigma) \\ &\approx \Phi\left(\frac{\mu + 2\sigma - \mu}{\sigma}\right) - \Phi\left(\frac{\mu - 2\sigma - \mu}{\sigma}\right) = \Phi(2) - \Phi(-2) \approx 0.954. \end{aligned}$$

In the example, $\mu = 66$ and $\sigma = 2.5$. Probabilists would say "the probability that a normally distributed random variable is within 2 standard deviations of its mean is approximately 0.95."

The probability that a normally distributed random variable is within 1 standard deviation of its mean is approximately

$$\Phi(1) - \Phi(-1) \approx 0.841 - 0.159 = 0.682.$$

There's more than a two-thirds chance that a randomly selected value from a normally distributed random variable is within 1 standard deviation of the mean. Similarly, such a random variable is within 3 standard deviations of its mean with probability approximately equal to

$$\Phi(3) - \Phi(-3) \approx 0.9987 - 0.0013 = 0.9974.$$

For this approximation, we used a better table than Table 2. Our conclusion is that a randomly selected value from a normally distributed random variable is almost surely within 3 standard deviations of the mean.

Exercises 9.4

1. Let F_s be the cdf for the random variable giving the sum of two fair dice, as in Example 1 on page 375. Calculate the values $F_s(k)$ for k in the value set of F_s, i.e., for $k \in \{2, 3, 4, \ldots, 12\}$. Sketch F_s. *Hint:* Use Figure 1 on page 360.

2. Consider the random variable D on the set Ω of 36 equally outcomes when two fair dice are tossed, where $D(k, l) = |k - l|$ for $(k, l) \in \Omega$. Also, let F be the cdf for D. Calculate the values $F(k)$ for k in the value set of D, and sketch F. *Hint:* Use the answers for Exercise 7 on page 365.

3. Repeat Exercise 2 for the cdf F of the random variable M, where $M(k, l) = \max\{k, l\}$ for $(k, l) \in \Omega$. The same hint applies.

4. The value set of $\widetilde{S}_5$, with $p = \frac{1}{2}$, is given in Example 6. Verify that the result is correct.

5. Let F be the cdf for the random variable W in Exercise 11 on page 366 that counts the number of tosses of a fair die until the first 6 appears. Give the values of $F(k)$ for $k = 1, 2, 3, 4$.

6. Draw the cdf for a random variable X satisfying $P(X = 0) = P(X = 1) = \frac{1}{2}$. This will be the cdf for the random variable that counts the number of heads when a fair coin is tossed *once*. It is also the cdf for the random variables $X_1, \ldots, X_n$ in Example 4(b) on page 361.

7. Let F be the cdf for some random variable X. Verify

(a) $0 \le F(y) \le 1$ for all $y \in \mathbb{R}$.

(b) F is nondecreasing; i.e., $y_1 < y_2$ implies that $F(y_1) \le F(y_2)$.

8. A random sample is taken of the heights of college men. The mean μ is 69 inches and the standard deviation σ is 3 inches. Estimate the probability that a random college man will have height

(a) between 66 and 72 inches.

(b) between 63 and 75 inches.

(c) between 60 and 78 inches.

9. What is the probability that a random college man in Exercise 8 will have height

(a) between 64 and 69 inches?

(b) between 70.5 and 74.5 inches?

(c) of at least 72 inches tall?

(d) less than 63 inches tall?

10. An experiment is repeated 30,000 times with probability of success $\frac{1}{4}$ each time.

(a) Show that the expected number μ of successes is 7500.

(b) Show that the standard deviation σ is 75.

11. This exercise concerns the experiment in Exercise 10. Estimate the probability that the number of successes

(a) will be less than 7700.

(b) will exceed 7550.

(c) will lie in the interval [7400,7600].

12. This exercise concerns the experiment in Exercise 10.

(a) What is the probability that the number of failures will exceed 7400?

(b) Explain why the probability that the number of successes is less than 7200 is very close to 0.

(c) Explain why the probability that the number of successes is less than 7900 is very close to 1.

13. An experiment is repeated 1800 times with $\frac{1}{3}$ probability of success for each experiment.

(a) What is the expected number of successes?

(b) What is the standard deviation?

(c) Recall from Example 8 that $\Phi(2) - \Phi(-2) \approx 0.954$. Give an interval so that we are about 95 percent sure that the number of successes will lie in that interval.

14. For a large dinner party, 1000 invitations are sent out. The host estimates that each invitee has a 60 percent chance of attending the party, and he believes that their decisions are independent. How many people should he plan for if he wants to be about 97 percent sure that he has enough place settings?

15. A fair coin is tossed many times. Estimate the probability that the fraction of heads is between 49 and 51 percent if the coin is tossed

(a) 1000 times. (b) 10,000 times.

(c) 1,000,000 times.

16. On a national test taken by 1,000,000 students, only 51 percent got the right answer on a particular true–false question. Did any of the students understand the question? Or was it just luck that 51 percent were successful?

17. A sequence of three independent experiments is performed. The probability of success on the first experiment is $\frac{1}{2}$, the probability of success on the second is $\frac{1}{3}$, and the probability of success on the third is only $\frac{1}{4}$.

(a) Find the expectation of the number X of successes.

(b) Find the standard deviation.

(c) Is X a binomial random variable? Explain.

18. Let X_1 and X_2 be random variables with means μ_1, μ_2 and standard deviations σ_1, σ_2, respectively. Suppose that X_1 and X_2 both have the same normalization $\widetilde{X}$. Give a formula for the cdf F_2 in terms of $F_1, \mu_1, \mu_2, \sigma_1$ and σ_2.

19. Let Φ be the Gaussian distribution.

(a) Explain why $\Phi(y) + \Phi(-y) = 1$ for $y \in \mathbb{R}$.

(b) Show that $\Phi(y) - \Phi(-y) = 2 \cdot \Phi(y) - 1$ for $y \in \mathbb{R}$.

20. Let X be a normally distributed random variable with mean μ and standard deviation σ. Recall from Example 8 that $P(\mu - 2\sigma < X \leq \mu + 2\sigma) = \Phi(2) - \Phi(-2)$. For any $c > 0$, give $P(\mu - c\sigma < X \leq \mu + c\sigma)$ in terms of Φ, and verify your formula.

21. (a) Show that if X is a normal random variable with mean μ and standard deviation $\sigma > 0$, then its cdf F, which is a normal distribution, is given by $F(y) = \Phi(\frac{y-\mu}{\sigma})$ for $y \in \mathbb{R}$.

(b) When is the normal distribution F in part (a) equal to the Gaussian distribution Φ?

Chapter Highlights

To check your understanding of the material in this chapter, follow our usual suggestions for review. Think always of examples.

CONCEPTS AND NOTATION

conditional probability, $P(E|S)$
independent events
random variable
  value set
  discrete
  independent random variables
  expectation = expected value = mean, $E(X) = \mu$
  standard deviation, variance, $V(X) = \sigma^2$
cumulative distribution function = cdf, F_X
  binomial distribution, S_n
    success, p, failure, q
  uniform distribution
normalization, $\widetilde{X}$, $\widetilde{F}$
Gaussian distribution, Φ
normal random variables and normal distributions

FACTS

Total Probability Formula for events that partition Ω.
Bayes' Formula:

$$P(A_j|B) = \frac{P(A_j) \cdot P(B|A_j)}{P(B)} \text{ with } P(B) = \sum_i P(A_i) \cdot P(B|A_i).$$

If all outcomes are equally likely, then $E(X)$ is the average value of X.
$E(X + Y) = E(X) + E(Y)$, $E(aX) = aE(X)$, and $E(c) = c$ for constant random variables.
$V(X) = E(X^2) - \mu^2$.
If X and Y are independent random variables, then $E(XY) = E(X) \cdot E(Y)$.
If $X_1, \ldots, X_n$ are independent, then

$$V(X_1 + \cdots + X_n) = V(X_1) + \cdots + V(X_n).$$

If S_n has a binomial distribution, then $E(S_n) = np$ and $V(S_n) = npq$.
The normalization of X has mean 0 and variance 1; its distribution function $\widetilde{F}$ is related to the distribution function F of X by

$$F(y) = \widetilde{F}\left(\frac{y-\mu}{\sigma}\right) \text{ and } \widetilde{F}(y) = F(\sigma y + \mu) \quad \text{for} \quad y \in \mathbb{R}.$$

The cdf of the normalization of a random variable with a binomial distribution is approximately the normal cdf Φ for large n.

METHODS

Applying the binomial random variable.
Use of tables of cdf's to calculate probabilities of events.
Use of a table of Φ to estimate $P(a \leq S_n \leq b)$ for binomially distributed random variables S_n.
Recognizing that measurements are often approximately normally distributed.

Supplementary Exercises

1. An experiment consists of tossing a fair coin three times. Let A be the event that the first toss is a head, B the event that the first two tosses are heads, and C the event that the total number of heads tossed is even.

(a) Calculate $P(A|B)$, $P(C|B)$, and $P(A \cap C)$.

(b) Are the events A and C independent? Explain.

(c) Describe a random variable Y for which B is the event $\{Y = 2\}$.

2. A bag contains three red marbles and one black one. A box contains two red marbles and two black ones. A marble is drawn at random from the bag and placed in the box.

(a) What is the probability that the box now contains three red marbles?

(b) A marble is now drawn at random from the box. It is red. What is the probability that the black marble was initially drawn from the bag? Show your method. [To fix notation, let B be the event that the black marble was drawn from the bag and R the event that a red marble was drawn from the box.]

3. Suppose that an experiment leads to events A and B such that $P(A) = 0.7$, $P(B) = 0.4$, and $P(A|B) = 0.3$.

(a) Calculate $P(A \cap B)$ and $P(A \cap B^c)$.

(b) Are the events A and B independent? Explain.

4. An urn contains one white marble and two black marbles. A second urn contains three white marbles and two black marbles. An urn is selected at random, and then a marble is selected at random.

(a) What is the probability that the marble selected is white?

(b) Given that the marble selected is white, what is the probability that it came from the first urn?

5. We are given events A and B in some sample space Ω, with $P(A) = 1/4$ and $P(B) = 1/3$. For each of the following statements, determine whether the statement must be true, might be true or might be false, or must be false.

(a) $P(A \cap B) = 1/12$ (b) $P(A \cup B) = 7/12$

(c) $P(B|A) = P(B)/P(A)$ (d) $P(A|B) \geq P(A)$

(e) $P(A^c) = 3/4$

(f) $P(A) = P(B)P(A|B) + P(B^c)P(A|B^c)$

6. A bag contains twelve red marbles and one black one. A box contains three red marbles. A marble is drawn at random from the bag and placed in the box, and then a marble is drawn at random from the box. It is red. What is the probability that the black marble was drawn from the bag? Explain. [For notation, let B be the event that the black marble was drawn from the bag and R the event that a red marble was drawn from the box.]

7. An experiment consists of drawing three marbles at random (without replacement) from a jar containing five red and three green marbles. The random variable X counts the number of red marbles selected.

(a) Show that $P(X = 0) = 1/56$, $P(X = 1) = 15/56$, and $P(X = 2) = 30/56$.

(b) Find $P(X = 3)$.

(c) Calculate $E(X)$, the expected number of red marbles selected.

(d) Calculate the variance $V(X)$ for X.

(e) Let Y be the random variable that counts the number of green marbles. Are X and Y independent random variables? Explain why this particular pair of random variables is or is not independent.

8. A TV dealer received two shipments of ten TV sets each. The dealer learned that all the sets in one shipment were in good shape, but two of the sets in the other shipment were defective (bad). She selected one shipment at random and randomly tested five sets. None were defective. What is the probability that she opened the "all good" shipment?

9. A basketball player attempts 1200 free throws, with probability 3/4 of success on each independent attempt. The random variable X that counts the total number of successes has mean 900 and standard deviation 15.

(a) Let $\widetilde{X}$ be the normalization of X. Find a number L such that

$$P(900 \leq X \leq 915) = P(0 \leq \widetilde{X} \leq L).$$

(b) Estimate $P(900 \leq X \leq 915)$.

10. Two independent experiments are performed. The probability of success for the first experiment is 0.7 and for the second is 0.2. Let X be the random variable that counts the total number of successes for the two experiments.

(a) What is the expected value of X?

(b) What is the probability of exactly one success?

(c) What is the probability of at most one success?

(d) Sketch the cumulative density function (cdf) F_X.

(e) What is the variance of X?

(f) How have you used the independence of the two experiments in your answers to parts (a) and (b)?

11. A basketball player generally makes $3/4$ of her free throws. That is, her probability of succeeding on any given attempt is 0.75. [Assume that different attempts are independent.]

(a) If she makes four attempts in a game, what is the probability that exactly three of them are successes?

(b) Let Y be the random variable that counts her successes on a single attempt. Calculate $V(Y)$.

Suppose now that she makes 400 attempts, and let X count the number of her successes.

(c) What is the expected value of X?

(d) Write, but do not try to evaluate, a sum whose value is $P(290 \le X \le 310)$.

(e) Give a convincing argument that the standard deviation σ_X is $5\sqrt{3}$ [≈ 8.66].

(f) Let $\widetilde{X}$ be the normalization of X. Find a number L such that

$$P(290 \le X \le 310) = P(-L \le \widetilde{X} \le L).$$

(g) Estimate $P(290 \le X \le 310)$.

12. Consider independent random variables X and Y with the same probabilities: $P(X = -1) = P(Y = -1) = P(X = 2) = P(Y = 2) = 1/8$, $P(X = 0) = P(Y = 0) = P(X = 1) = P(Y = 1) = 3/8$, and $P(X = a) = P(Y = a) = 0$ for all other a. Let F be the cdf of X and Y.
Calculate the following, indicating whenever you have used independence.

(a) $P(X \le 0)$
(b) $P(X < 0)$
(c) $P(Y > 0)$
(d) $F(0)$
(e) $P(X \le 0$ and $Y < 0)$
(f) $P(X \le 0$ or $Y < 0)$
(g) $P(X + Y \le 3)$
(h) $E(X)$
(i) $E(X - 3Y)$
(j) $E(XY)$
(k) $E(|X|)$
(l) $V(X)$
(m) $V(X - 3Y)$
(n) standard deviation of $X + Y$

13. The game of Russian roulette consists of repeated independent trials until the first failure. The probability of success for each trial is 5/6.

(a) What is the probability that failure occurs on the fourth trial?

(b) What is the expected number of trials, i.e., the expected length of the game?

14. Four nickels and one dime are tossed. [All coins are fair.] Let N be the random variable that counts the number of nickels that come up heads, and D the random variable that counts the number of dimes that come up heads.

(a) What is the value set of N?

(b) Sketch the cumulative distribution function (cdf) for $2D$.

(c) The random variables N and D are independent. What does this statement *mean* in mathematical terms?

(d) What is the expectation $E(N + D)$?

(e) What is the probability $P(N = E(N))$?

15. A fair die is tossed 18,000 times and we are interested in the random variable S that counts the number of sixes that appear. [The probability of a six at each toss is, of course, 1/6.]

(a) What is the expected number of sixes, i.e., the mean of S?

(b) Calculate the standard deviation for the random variable S.

(c) Estimate $P(17{,}900 < S \le 18{,}200)$.

(d) What unstated assumption is being used in this exercise?

16. Often the sample space Ω for a probability P can be viewed as a product set, say $\Omega_1 \times \Omega_2$, where the "events" in Ω_1 and the "events" in Ω_2 are independent. What we really mean is that, for $E \subseteq \Omega_1$ and $F \subseteq \Omega_2$, the events $E \times \Omega_2$ and $\Omega_1 \times F$ in Ω are independent.

(a) With the setup above, show that $P(E \times F) = P(E \times \Omega_2) \cdot P(\Omega_1 \times F)$.

(b) The experiment of drawing a single card from a deck of cards can be viewed this way, where Ω_1 consists of the 4 possible suits (spades, hearts, diamonds, clubs) and Ω_2 consists of the 13 possible values. For example, the event "four of clubs" corresponds to the element (club, four) in $\Omega_1 \times \Omega_2$. Let $C = \{\text{club}\} \subseteq \Omega_1$ and $F = \{\text{four}\} \subseteq \Omega_2$. Describe in words the events $C \times \Omega_2$, $\Omega_1 \times F$ and $C \times F$. Calculate and compare $P(C \times F)$ and $P(C \times \Omega_2) \cdot P(\Omega_1 \times F)$.

(c) Show that Example 1 on page 349 involving two fair dice can also be viewed in this way by stating what Ω_1 and Ω_2 would be; see Figure 1 on page 350. Show that the events B and R have the form $B_1 \times \Omega_2$ and $\Omega_1 \times R_2$, respectively. Calculate and compare the values $P(B \cap R)$ and $P(B_1 \times \Omega_2) \cdot P(\Omega_1 \times R_2)$.

CHAPTER 10

Boolean Algebra

The term Boolean algebra has two distinct but related uses. On the one hand, it means the kind of symbolic arithmetic first developed by George Boole in the nineteenth century to manipulate logical truth values in an algebraic way, a kind of algebra well suited to describe two-valued computer logic. In this broad sense, Boolean algebra can be taken to include all sorts of mathematical methods that describe the operation of logical circuits.

Boolean algebra is also the name for a specific kind of algebraic structure whose operations satisfy an explicit set of rules. The rules and operations are chosen to provide concrete models for logical arithmetic.

This chapter begins with an account of Boolean algebras as structures, develops the link between the algebras and their logical interpretations, applies the resulting theory to logical networks, and describes one method for simplifying complex logical expressions. It concludes with a characterization of finite Boolean algebras.

10.1 Boolean Algebras

This section contains the definitions and basic properties of Boolean algebras and introduces some important examples. The motivation comes from our hope to apply the symbolic logic that we saw in Chapter 2 in a systematic way to circuit design. The key new objects will be the atoms, which play the role of building blocks for many Boolean algebras, including all finite ones. This fact will be exploited in §10.5, where we will show that every finite Boolean algebra is essentially the algebra of subsets of a nonempty finite set, the kind of algebra we visualized with Venn diagrams back in §1.4.

The section ends with an introduction to the algebras BOOL(n) of Boolean functions, whose connection with Boolean expressions will be explained in §10.2 and whose atoms will correspond to elementary Boolean expressions called minterms. The analysis of Boolean expressions is the key to analyzing and simplifying logic networks in §10.3.

First we will build the theory on its own foundation, but the motivation will come from the rules of set theory, on the one hand, and from truth-table arithmetic on the other. A good way to study this section would be to omit the proofs on the first reading and focus on the concepts and the statements of the theorems. Before we proceed with the formal definition, here are some examples of Boolean algebras.

EXAMPLE 1

(a) Consider any nonempty set S. The set $\mathcal{P}(S)$ of all subsets of the set S has the familiar operations $\cup$ and $\cap$. These operations combine two members A and B of $\mathcal{P}(S)$, i.e., two subsets of S, to give members $A \cup B$ and $A \cap B$ of $\mathcal{P}(S)$. The complementation operation ^{c} takes A to its complement $A^c = S \setminus A$, another member of $\mathcal{P}(S)$. The operations $\cup$, $\cap$, and ^{c} enjoy a variety of properties, some of which are listed in Table 1 on page 25. The sets S and $\emptyset$ have special properties, such as $A \cap S = A$ and $A \cap \emptyset = \emptyset$ for all members A of $\mathcal{P}(S)$.

$\vee$	0	1
0	0	1
1	1	1

and

$\wedge$	0	1
0	0	0
1	0	1

Figure 1 ▲

(b) The set $\mathbb{B} = \{0, 1\}$ has the usual logical operations $\vee$ and $\wedge$, together with the operation $'$ defined by $0' = 1$ and $1' = 0$. Here 0 and 1 have the interpretations "false" and "true," respectively, and $'$ corresponds to negation $\neg$. Figure 1 describes the operations $\vee$ and $\wedge$. In terms of ordinary integer arithmetic, $a \vee b = \max\{a, b\}$, $a \wedge b = \min\{a, b\} = a \cdot b$, and $a' = 1 - a$ for $a, b \in \mathbb{B}$. ■

EXAMPLE 2

(a) The set $\mathbb{B}^n = \mathbb{B} \times \cdots \times \mathbb{B}$ [with n factors] of n-tuples of 0's and 1s has Boolean operations $\vee$, $\wedge$, and $'$ obtained by applying the corresponding operations on $\mathbb{B}$ to each coordinate. Thus

$$(a_1, a_2, \ldots, a_n) \vee (b_1, b_2, \ldots, b_n) = (a_1 \vee b_1, a_2 \vee b_2, \ldots, a_n \vee b_n),$$
$$(a_1, a_2, \ldots, a_n) \wedge (b_1, b_2, \ldots, b_n) = (a_1 \wedge b_1, a_2 \wedge b_2, \ldots, a_n \wedge b_n),$$
$$(a_1, a_2, \ldots, a_n)' = (a_1', a_2', \ldots, a_n').$$

Here are some sample Boolean computations in $\mathbb{B}^4$:

$$(1,0,0,1) \vee (1,1,0,0) = (1,1,0,1), \qquad (1,0,0,1) \wedge (1,1,0,0) = (1,0,0,0),$$
$$(1,0,0,1)' = (0,1,1,0), \qquad (1,1,0,0)' = (0,0,1.1).$$

(b) Let S be any nonempty set. As in Section 5.1, we write $\text{FUN}(S, \mathbb{B})$ for the set of all functions from S to $\mathbb{B}$. We provide $\text{FUN}(S, \mathbb{B})$ with Boolean operations $\vee$, $\wedge$, and $'$ by defining them elementwise. That is, for $f, g \in \text{FUN}(S, \mathbb{B})$, we define new functions $f \vee g$, $f \wedge g$, and f' in $\text{FUN}(S, \mathbb{B})$ by the rules

$$(f \vee g)(s) = f(s) \vee g(s) \quad \text{for all} \quad s \in S,$$
$$(f \wedge g)(s) = f(s) \wedge g(s) \quad \text{for all} \quad s \in S,$$
$$f'(s) = f(s)' \quad \text{for all} \quad s \in S.$$

For example, suppose that $S = \{a, b, c, d\}$ and that Figure 2 gives the values of the functions f and g in $\text{FUN}(S, \mathbb{B})$. The columns for $(f \vee g)(x)$, $(f \wedge g)(x)$, $f'(x)$, and $g'(x)$ give the values of the functions $f \vee g$, $f \wedge g$, f', and g'. To calculate $(f \wedge g)(d)$, for example, we use $(f \wedge g)(d) = f(d) \wedge g(d) = 1 \wedge 0 = 0$.

Figure 2 ▶

x	$f(x)$	$g(x)$	$(f \vee g)(x)$	$(f \wedge g)(x)$	$f'(x)$	$g'(x)$
a	1	1	1	1	0	0
b	0	1	1	0	1	0
c	0	0	0	0	1	1
d	1	0	1	0	0	1

If you compare these calculations with those in part (a), you may suspect that the two examples are closely related. They are, and we will make the relationship more precise in the next section.

Notice also that if the members of S had been 0 0, 0 1, 1 0, and 1 1, instead of a, b, c, and d, then Figure 2 would have looked just like a truth table.

(c) There is a natural link between the operations $\vee$ and $\wedge$ for $\text{FUN}(S, \mathbb{B})$ and the operations $\cup$ and $\cap$ on S. If A is the set of elements s of S for which $f(s) = 1$ and B is the set for which $g(s) = 1$, then $A \cup B$ is the set of elements s for which $(f \vee g)(s) = 1$. Similarly, we have $A \cap B = \{s \in S : (f \wedge g)(s) = 1\}$ and $A^c = \{s \in S : f(s) = 0\}$. In the terminology of characteristic functions, introduced in §1.5, this means

$$\chi_A \vee \chi_B = \chi_{A \cup B}, \qquad \chi_A \wedge \chi_B = \chi_{A \cap B}, \qquad (\chi_A)' = \chi_{A^c}. \qquad ■$$

In general, we define a **Boolean algebra** to be a set with two binary operations $\vee$ and $\wedge$, a unary operation $'$, and distinct elements 0 and 1 satisfying the following laws.

1Ba.	$x \vee y = y \vee x$	commutative laws
b.	$x \wedge y = y \wedge x$	
2Ba.	$(x \vee y) \vee z = x \vee (y \vee z)$	associative laws
b.	$(x \wedge y) \wedge z = x \wedge (y \wedge z)$	
3Ba.	$x \vee (y \wedge z) = (x \vee y) \wedge (x \vee z)$	distributive laws
b.	$x \wedge (y \vee z) = (x \wedge y) \vee (x \wedge z)$	
4Ba.	$x \vee 0 = x$	identity laws
b.	$x \wedge 1 = x$	
5Ba.	$x \vee x' = 1$	complementation laws
b.	$x \wedge x' = 0$	

The operation $\vee$ is called **join**, $\wedge$ is called **meet**, and the unary operation $'$ is called **complementation**. A particular Boolean algebra might have different notations for its operations and distinguished elements, but they would still satisfy these laws. Table 1 summarizes the notations for the Boolean algebras in Examples 1 and 2.

TABLE 1 Some Boolean Algebras

The Boolean Algebra	0	1	join	meet	$'$
$\mathcal{P}(S)$	$\varnothing$	S	$\cup$	$\cap$	c
$\mathbb{B}$	0	1	$\vee$	$\wedge$	$'$
$\mathbb{B}^n$	$(0, \ldots, 0)$	$(1, \ldots, 1)$	$\vee$, $\wedge$ coordinatewise		$'$coordinatewise
$\text{FUN}(S, \mathbb{B})$	$\chi_\varnothing$	χ_S	$\vee$, $\wedge$ elementwise		$(\chi_A)' = \chi_{A^c}$

If we interchange $\vee$ and $\wedge$ in law 3Ba, we obtain law 3Bb. In fact, if we interchange $\vee$ and $\wedge$ in all the laws defining Boolean algebras and at the same time switch 0 with 1, we get all the same laws back again. In each case, laws a and b are interchanged. In particular, the properties of complementation are unchanged. The properties that hold in every Boolean algebra are the ones that are consequences of the defining laws, so we have the fundamental **duality principle**:

> If $\wedge$ is interchanged with $\vee$ and 0 with 1 everywhere in a formula valid for all Boolean algebras, then the resulting formula is also valid for all Boolean algebras.

Here are some consequences of the defining laws. Their proofs illustrate typical algebraic manipulations in Boolean algebra arguments.

Theorem 1 The following properties hold in every Boolean algebra:

$$\left.\begin{array}{ll} \text{6Ba.} & x \vee x = x \\ \text{b.} & x \wedge x = x \end{array}\right\} \quad \text{idempotent laws}$$

$$\left.\begin{array}{ll} \text{7Ba.} & x \vee 1 = 1 \\ \text{b.} & x \wedge 0 = 0 \end{array}\right\} \quad \text{more identity laws}$$

$$\left.\begin{array}{ll} \text{8Ba.} & (x \wedge y) \vee x = x \\ \text{b.} & (x \vee y) \wedge x = x \end{array}\right\} \quad \text{absorption laws}$$

Proof Here is a derivation of 6Ba:

$$\begin{aligned} x \vee x &= (x \vee x) \wedge 1 && \text{identity law 4Bb} \\ &= (x \vee x) \wedge (x \vee x') && \text{complementation law 5Ba} \\ &= x \vee (x \wedge x') && \text{distributive law 3Ba} \\ &= x \vee 0 && \text{complementation law 5Bb} \\ &= x && \text{identity law 4Ba.} \end{aligned}$$

For 7Ba, observe that

$$\begin{aligned} x \vee 1 &= x \vee (x \vee x') && \text{complementation law 5Ba} \\ &= (x \vee x) \vee x' && \text{associative law 2Ba} \\ &= x \vee x' && \text{idempotent law 6Ba just proved} \\ &= 1 && \text{complementation law 5Ba.} \end{aligned}$$

And for 8Ba, we have

$$\begin{aligned} (x \wedge y) \vee x &= (x \wedge y) \vee (x \wedge 1) && \text{identity law 4Bb} \\ &= x \wedge (y \vee 1) && \text{distributive law 3Bb} \\ &= x \wedge 1 && \text{identity law 7Ba just proved} \\ &= x && \text{identity law 4Bb.} \end{aligned}$$

Now 6Bb, 7Bb, and 8Bb follow by the duality principle. ■

It turns out that the associative laws are consequences of the other defining laws for a Boolean algebra, so they are redundant. In fact, Theorem 1 can be proved without using the associative laws. Proofs of these facts are tedious and not very informative, so we omit them.

The next lemma shows that if an element z in a Boolean algebra *acts like* the complement of an element w, then it *is* the complement.

Lemma 1 In a Boolean algebra,

$$\text{if} \quad w \vee z = 1 \quad \text{and} \quad w \wedge z = 0, \quad \text{then} \quad z = w'.$$

Proof

$$\begin{aligned} z &= z \vee 0 && \text{identity law 4Ba} \\ &= z \vee (w \wedge w') && \text{complementation law 5Bb} \\ &= (z \vee w) \wedge (z \vee w') && \text{distributive law 3Ba} \\ &= (w \vee z) \wedge (w' \vee z) && \text{commutative law 1Ba used twice} \\ &= 1 \wedge (w' \vee z) && \text{hypothesis} \\ &= (w \vee w') \wedge (w' \vee z) && \text{complementation law 5Ba} \end{aligned}$$

$$
\begin{aligned}
&= (w' \vee w) \wedge (w' \vee z) && \text{commutative law 1Ba} \\
&= w' \vee (w \wedge z) && \text{distributive law 3Ba} \\
&= w' \vee 0 && \text{hypothesis} \\
&= w' && \text{identity law 4Ba.}
\end{aligned}
$$
■

Corollary In a Boolean algebra, $(z')' = z$ for all z.

Proof Let $w = z'$. Our claim is that $w' = z$, which follows from the lemma since $w \vee z = z' \vee z = 1$ and $w \wedge z = z' \wedge z = 0$. ■

Theorem 2 Every Boolean algebra satisfies the De Morgan laws

$$
\left.
\begin{aligned}
&\text{9Ba.} \quad (x \vee y)' = x' \wedge y' \\
&\text{b.} \quad (x \wedge y)' = x' \vee y'
\end{aligned}
\right\}.
$$

Proof By Lemma 1, law 9Ba will hold if we prove $(x \vee y) \vee (x' \wedge y') = 1$ and $(x \vee y) \wedge (x' \wedge y') = 0$. We have

$$
\begin{aligned}
(x \vee y) \vee (x' \wedge y') &= [(x \vee y) \vee x'] \wedge [(x \vee y) \vee y'] && \text{distributivity} \\
&= [y \vee (x \vee x')] \wedge [x \vee (y \vee y')] && \text{associativity and commutativity} \\
&= [y \vee 1] \wedge [x \vee 1] = 1 \wedge 1 = 1.
\end{aligned}
$$

Similarly, $(x \vee y) \wedge (x' \wedge y') = 0$, and so 9Ba holds. Formula 9Bb follows by duality. ■

EXAMPLE 3 It is not hard to verify that the sets and operations in Examples 1 and 2 form Boolean algebras. By Theorem 2, they must also satisfy the De Morgan laws. In the context of the algebra $\mathcal{P}(S)$ with operations $\cup$, $\cap$ and ^{c}, the De Morgan laws are our old friends

$$(A \cup B)^c = A^c \cap B^c \quad \text{and} \quad (A \cap B)^c = A^c \cup B^c$$

from Chapter 1. The laws for the two-element Boolean algebra $\mathbb{B}$ and for $\mathbb{B}^n$ correspond to the familiar logical De Morgan laws

$$\neg(p \vee q) \Longleftrightarrow (\neg p) \wedge (\neg q) \quad \text{and} \quad \neg(p \wedge q) \Longleftrightarrow (\neg p) \vee (\neg q).$$
■

It should be emphasized that our axiomatic development of Boolean algebras does not allow the taking of infinite joins and meets, even though one can take infinite unions and intersections in the Boolean algebra $\mathcal{P}(S)$. Boolean algebras allowing infinite joins and meets are studied and even have a name. They are called **complete** Boolean algebras.

The Boolean algebra $\mathcal{P}(S)$ has a relation $\subseteq$ that links some of its members. That is, the inclusion $A \subseteq B$ holds for some pairs of subsets A and B of S. The relation can be expressed in terms of $\cup$, since $A \subseteq B$ if and only if $A \cup B = B$. This fact suggests the following definition in the general case.

Define the relation $\leq$ on a Boolean algebra by

$$x \leq y \quad \text{if and only if} \quad x \vee y = y.$$

It follows from the idempotent law $x \vee x = x$ that $x \leq x$ for every x. We also define $x < y$ to mean $x \leq y$ but $x \neq y$, while $x \geq y$ means $y \leq x$ and $x > y$ means $y < x$.

EXAMPLE 4

(a) For the Boolean algebra $\mathcal{P}(S)$, the relations $\leq$, $<$, $\geq$, and $>$ are the relations $\subseteq$, $\subset$, $\supseteq$, and $\supset$, respectively.

(b) In the Boolean algebra $\mathbb{B}$, we have $0 \leq 0$, $0 \leq 1$, $1 \leq 1$, and $0 < 1$, so also $0 \geq 0$, $1 \geq 0$, $1 \geq 1$, and $1 > 0$. No surprises here!

(c) In $\mathbb{B}^n$ we have $(a_1, \ldots, a_n) \leq (b_1, \ldots, b_n)$ if and only if

$$(a_1 \vee b_1, \ldots, a_n \vee b_n) = (a_1, \ldots, a_n) \vee (b_1, \ldots, b_n) = (b_1, \ldots, b_n),$$

i.e., if and only if $a_k \vee b_k = b_k$ for each k. Thus $(a_1, \ldots, a_n) \leq (b_1, \ldots, b_n)$ if and only if $a_k \leq b_k$ for each k. ■

Although we defined the relation $\leq$ in terms of the operation $\vee$, we could just as easily have used $\wedge$, as the next lemma shows.

Lemma 2 In a Boolean algebra,

$$x \vee y = y \quad \text{if and only if} \quad x \wedge y = x.$$

Proof If $x \vee y = y$, then

$$\begin{aligned} x &= (x \vee y) \wedge x && \text{absorption law 8Bb} \\ &= y \wedge x && \text{assumption} \\ &= x \wedge y && \text{commutative law 1Bb.} \end{aligned}$$

If $x \wedge y = x$, then, using laws 8Ba and 1Ba, we obtain $y = (y \wedge x) \vee y = (x \wedge y) \vee y = x \vee y$. ■

It may be a bit of a surprise to find that the relations $\leq$ and $<$ that we have defined abstractly in terms of $\vee$ or $\wedge$ satisfy some familiar properties of the relations $\leq$ and $<$ for numbers.

Lemma 3 In a Boolean algebra,

(a) If $x \leq y$ and $y \leq z$, then $x \leq z$;
(b) If $x \leq y$ and $y \leq x$, then $x = y$;
(c) If $x < y$ and $y < z$, then $x < z$.

Properties (a) and (b) say, in the terminology of §11.1, that the relation $\leq$ is a partial order relation. Property (a), **transitivity**, is familiar from our study of equivalence relations in §3.4, but $\leq$ is quite different from an equivalence relation.

Proof of Lemma 3

(a) If $x \leq y$ and $y \leq z$, then

$$\begin{aligned} z &= y \vee z && \text{since } y \leq z \\ &= (x \vee y) \vee z && \text{since } x \leq y \\ &= x \vee (y \vee z) && \text{associative law 2Ba} \\ &= x \vee z && \text{since } y \leq z. \end{aligned}$$

Thus $x \leq z$.

(b) If $x \leq y$ and $y \leq x$, then $x \vee y = y$ and $y \vee x = x$. By the commutative law 1Ba, $x = y$.

(c) If $x < y$ and $y < z$, then $x \leq z$ by part (a). The case $x = z$ is impossible, since it would give $x \leq y$ and $y \leq z = x$ but $x \neq y$, contrary to part (b). ■

The relation $\leq$ is also linked to the operations $\vee$ and $\wedge$ and to the special elements 0 and 1 in ways that parallel the structure of $\mathcal{P}(S)$.

Lemma 4 In a Boolean algebra,

(a) $x \wedge y \leq x \leq x \vee y$ for every x and y;
(b) $0 \leq x \leq 1$ for every x.

Proof

(a) Since $(x \wedge y) \vee x = x$ by rule 8Ba, we have $x \wedge y \leq x$. Also, $x \vee (x \vee y) = (x \vee x) \vee y = x \vee y$ by rules 2Ba and 6Ba; so $x \leq x \vee y$.
(b) Rules 4Ba and 7Ba (and commutativity) give $0 \vee x = x$ and $x \vee 1 = 1$. Thus $0 \leq x$ and $x \leq 1$ by the definition of $\leq$. ■

Finite sets can be built up as unions of 1-element subsets. Indecomposable building blocks play an important role in analyzing more general Boolean algebras as well. An **atom** of a Boolean algebra is a nonzero element a that cannot be written in the form $a = b \vee c$ with $a \neq b$ and $a \neq c$; i.e., a cannot be written as a join of two elements different from itself.

EXAMPLE 5

(a) The atoms of $\mathcal{P}(S)$ are the 1-element sets $\{s\}$; every A in $\mathcal{P}(S)$ with more than one member can be decomposed as $(A \setminus \{s\}) \cup \{s\}$ for $s \in A$.
(b) The only atom of $\mathbb{B}$ is 1. ■

EXAMPLE 6

(a) The atoms of $\mathbb{B}^n$ are the n-tuples with exactly one entry 1 and the rest of the entries 0. The atoms of $\mathbb{B}^3$ are $(1, 0, 0)$, $(0, 1, 0)$, and $(0, 0, 1)$.
(b) For the Boolean algebra $\text{FUN}(S, \mathbb{B})$ in Example 2, the atoms are the functions taking the value 1 at one element of S and the value 0 elsewhere. In other words, the atoms are the characteristic functions of the one-element subsets of S. ■

Atoms are like prime numbers in the sense that they have no nontrivial decompositions. We will see that we can also use them like primes to build up other elements of our Boolean algebras. First, we give an alternative characterization of atoms as minimal nonzero elements.

Proposition A nonzero element a of a Boolean algebra is an atom if and only if there is no element x with $0 < x < a$.

Proof Suppose that a is an atom and that $x < a$, so that $x \vee a = a$ and $x \wedge a = x$ by Lemma 2. Then $a = a \wedge 1 = (x \vee a) \wedge (x \vee x') = x \vee (a \wedge x')$ by distributivity. Since a is an atom, one of the terms x or $a \wedge x'$ must be a itself. Since $x \neq a$ by assumption, $a \wedge x' = a$. But then $x = a \wedge x = (a \wedge x') \wedge x = a \wedge (x' \wedge x) = a \wedge 0 = 0$.

On the other hand, if a is not an atom, then $a = x \vee y$ for some x and y not equal to a. Then $0 \leq x \leq a$ by parts (b) and (a) of Lemma 4. We have $x \neq 0$, since otherwise $a = 0 \vee y = y$. Also, $x \neq a$, so $0 < x < a$. [The same argument shows that $0 < y < a$, but we don't need this observation here.] ■

Corollary If a and b are atoms in a Boolean algebra and if $a \wedge b \neq 0$, then $a = b$. Thus, if $a \neq b$, then $a \wedge b = 0$.

Proof By Lemma 4, $0 < a \wedge b \leq a$. By the Proposition, we must have $a \wedge b = a$. Similarly, $a \wedge b = b$, so $a = b$. ■

We are now ready to describe how the atoms form the basic building blocks of a finite Boolean algebra. First, let's look at some examples.

EXAMPLE 7

(a) Consider a nonempty finite set S. The atoms of $\mathcal{P}(S)$ are the 1-element sets $\{s\}$. If $T = \{t_1, \ldots, t_m\}$ is any m-element subset of S, then the atoms $\{s\}$ of $\mathcal{P}(S)$ that satisfy $\{s\} \subseteq T$ are the atoms $\{t_1\}, \ldots, \{t_m\}$, and T is their union $\{t_1\} \cup \cdots \cup \{t_m\}$.

(b) The atoms of $\mathbb{B}^n$ are the n-tuples in $\mathbb{B}^n$ with exactly one 1. Say $\boldsymbol{a_i}$ is the atom with its 1 in the ith position. From Example 4(c), if $\boldsymbol{x} = (x_1, \ldots, x_n)$ is in $\mathbb{B}^n$, then $\boldsymbol{a_i} \leq \boldsymbol{x}$ if and only if $x_i = 1$, and $\boldsymbol{x}$ is the join of the atoms $\boldsymbol{a_i}$ for which $x_i = 1$. For instance, if $\boldsymbol{x} = (1, 1, 0, 1, 0)$ in $\mathbb{B}^5$, then

$$\boldsymbol{x} = (1, 0, 0, 0, 0) \vee (0, 1, 0, 0, 0) \vee (0, 0, 0, 1, 0) = \boldsymbol{a_1} \vee \boldsymbol{a_2} \vee \boldsymbol{a_4}. \quad \blacksquare$$

The next theorem shows that what occurred in these examples happens in general. Compare it with the theorem about products of primes that we first discussed on page 9. Note that, while the same prime can occur more than once in a factorization, there is no reason to list the same atom twice, since $a \vee a = a$.

Theorem 3 Let B be a finite Boolean algebra with set of atoms $A = \{a_1, \ldots, a_n\}$. Each nonzero x in B can be written as a join of distinct atoms:

$$x = a_{i_1} \vee \cdots \vee a_{i_k}.$$

Moreover, such an expression is unique, except for the order of the atoms.

Proof We first show that every nonzero element can be written in the form shown, where the atoms themselves are written as joins with only one term. Suppose not, and let S be the set of nonzero members of B that are not joins of atoms. If $x \in S$, then x is not itself an atom, so, just as in the second part of the proof of the Proposition above, $x = y \vee z$ with $0 < y < x$ and $0 < z < x$. At least one of y and z is also in S, since otherwise y and z would be joins of atoms and x, the join of y and z, would be too. Thus, for each x in S, there is some other element w of S with $x > w$. It follows that, starting from any x in S, there is a chain $x = x_0 > x_1 > x_2 > \cdots$ all of whose elements are in S. Since B is finite, the elements $x_0, x_1, x_2, \ldots$ cannot all be different; sooner or later $x_k = x_m$ with $k < m$. Then transitivity and $x_k > \cdots > x_m$ imply that $x_k > x_m$ by Lemma 3(c), contradicting $x_k = x_m$. [A little induction is hiding here.] Assuming that S is nonempty leads to a contradiction, so every nonzero x must be a join of atoms. [A recursive algorithm for finding an expression as a join of atoms can be based on this argument.]

To show uniqueness, suppose that x can be written as the join of atoms in two ways:

$$x = a_1 \vee a_2 \vee \cdots \vee a_j = b_1 \vee b_2 \vee \cdots \vee b_k$$

where $a_1, a_2, \ldots, a_j$ are distinct atoms and $b_1, b_2, \ldots, b_k$ are distinct atoms. We will show that

$$\{a_1, a_2, \ldots, a_j\} = \{b_1, b_2, \ldots, b_k\}; \quad \textbf{(1)}$$

this is what we mean by "unique, except for the order of the atoms." Note that (1) implies that $j = k$. Since $a_1 \leq x$, we have

$$a_1 = a_1 \wedge x = a_1 \wedge (b_1 \vee b_2 \vee \cdots \vee b_k) = (a_1 \wedge b_1) \vee \cdots \vee (a_1 \vee b_k).$$

Since $a_1 \neq 0$, we must have $a_1 \wedge b_s \neq 0$ for some s. By the last corollary, $a_1 = b_s$. In other words, a_1 belongs to the set $\{b_1, b_2, \ldots, b_k\}$. This argument works for all a_i, and so we have $\{a_1, a_2, \ldots, a_j\} \subseteq \{b_1, b_2, \ldots, b_k\}$. The reversed inclusion has a similar proof, so (1) holds. ■

Corollary In a finite Boolean algebra B, 1 is the join of all the atoms of B.

Proof We know that 1 is the join of some of the atoms. Since $1 \vee a = 1$ for all a, sticking on more atoms can't hurt. [In fact, by the uniqueness property, we know that we *need* all the atoms.] ■

As we have already indicated, finite Boolean algebras often arise in contexts different from $\mathcal{P}(S)$. Again let $\mathbb{B} = \{0, 1\}$ and for $n = 1, 2, \ldots$, let $\mathbb{B}^n$ be the Boolean algebra of Example 2. An **n-variable Boolean function** is a function

$$f: \mathbb{B}^n \to \mathbb{B}.$$

We write $\text{BOOL}(n)$ for the set of all n-variable Boolean functions. If $f \in \text{BOOL}(n)$ and $(x_1, \ldots, x_n) \in \mathbb{B}^n$, we write $f(x_1, \ldots, x_n)$ for the value $f((x_1, \ldots, x_n))$.

EXAMPLE 8

A 3-variable Boolean function is an f such that $f(x, y, z) = 0$ or 1 for each of the 2^3 choices of x, y, and z in $\mathbb{B} \times \mathbb{B} \times \mathbb{B}$. We could think of the input variables x, y, and z as amounting to three switches, each in one of two positions. Then f behaves like a black box that produces an output of 0 or 1 depending on the settings of the x, y, z switches and the internal structure of the box. Since there are 8 ways to set the switches and since each setting can lead to either of 2 outputs, depending on the function, there are $2^8 = 256$ different 3-variable Boolean functions. That is, $|\text{BOOL}(3)| = 256$.

A 3-variable Boolean function f can also be viewed as a column in a truth table. For example, Figure 3 describes a unique function f, which is just one of the $2^8 = 256$ possible functions, since the column can contain any arrangement of eight 0's and 1's. Each of the columns x, y, and z can be viewed as a function in $\text{BOOL}(3)$, i.e., a function from $\mathbb{B}^3$ to $\mathbb{B}$. For instance, column x describes the function h such that $h(a, b, c) = a$ for $(a, b, c) \in \mathbb{B}^3$. ■

x	y	z	f
0	0	0	1
0	0	1	1
0	1	0	0
0	1	1	1
1	0	0	0
1	0	1	0
1	1	0	0
1	1	1	1

Figure 3 ▲

The counting argument in Example 8 works in general to give $|\text{BOOL}(n)| = 2^{(2^n)}$, a very big number unless n is very small. As in Example 2(b), $\text{BOOL}(n)$ is a Boolean algebra with the Boolean operations defined coordinatewise.

EXAMPLE 9

Figure 4 illustrates the Boolean operations in $\text{BOOL}(3)$ with a truth table for the indicated functions f and g. For this illustration, we selected the functions f and g and the examples $f \vee g$, $f \wedge g$, f', and $f' \wedge g$ quite arbitrarily. Note that $f \wedge g$ is an atom of $\text{BOOL}(3)$, since it takes the value 1 at exactly one member of $\mathbb{B}^3$. There are seven other atoms in $\text{BOOL}(3)$. In §10.2 we will show how to write any member of a finite Boolean algebra, in particular any member of $\text{BOOL}(n)$, as a join of atoms. ■

Figure 4 ▶

x	y	z	f	g	$f \vee g$	$f \wedge g$	f'	$f' \wedge g$
0	0	0	1	0	1	0	0	0
0	0	1	1	1	1	1	0	0
0	1	0	0	0	0	0	1	0
0	1	1	1	0	1	0	0	0
1	0	0	0	1	1	0	1	1
1	0	1	0	0	0	0	1	0
1	1	0	0	1	1	0	1	1
1	1	1	1	0	1	0	0	0

Exercises 10.1

1. (a) Verify that $\mathbb{B} = \{0, 1\}$ in Example 1(b) is a Boolean algebra by checking some of the laws 1Ba through 5Bb.

   (b) Do the same for $\text{FUN}(S, \mathbb{B})$ in Example 1(b).

2. Verify the sample computations in Example 2(a).

3. Complete the proof of Theorem 2 by showing that $(x \vee y) \wedge (x' \wedge y') = 0$.

4. Provide reasons for all the equalities

$$a = a \wedge 1 = (x \vee a) \wedge (x \vee x') = x \vee (a \wedge x'),$$

$$x = a \wedge x = (a \wedge x') \wedge x = a \wedge (x' \wedge x) = a \wedge 0 = 0$$

   in the proof of the Proposition on page 395.

5. (a) Let $S = \{a, b, c, d, e\}$ and write $\{a, c, d\}$ as a join of atoms in $\mathcal{P}(S)$.

   (b) Write $(1, 0, 1, 1, 0)$ as a join of atoms in $\mathbb{B}^5$.

(c) Let f be the function in $\text{FUN}(S, \mathbb{B})$ that maps a, c, and d to 1 and maps b and e to 0. Write f as a join of atoms in $\text{FUN}(S, \mathbb{B})$.

6. Describe the atoms of $\text{FUN}(S, \mathbb{B})$ in Example 2(b). Is your description valid even if S is infinite?

7. (a) Give tables for the atoms of the Boolean algebra $\text{BOOL}(2)$.

(b) Write the function $g: \mathbb{B}^2 \to \mathbb{B}$ defined by $g(x, y) = x$ as a join of atoms in $\text{BOOL}(2)$.

(c) Write the function $h: \mathbb{B}^2 \to \mathbb{B}$ defined by $h(x, y) = x' \vee y$ as a join of atoms in $\text{BOOL}(2)$.

8. (a) How many atoms are there in $\text{BOOL}(4)$?

(b) Consider a function in $\text{BOOL}(4)$ described by a column that has five 1's and the rest of its entries 0. How many atoms appear in its representation as a join of atoms?

(c) How many different elements of $\text{BOOL}(4)$ are joins of five atoms?

9. There is a natural way to draw pictures of finite Boolean algebras. For x and y in the Boolean algebra B, we say that x **covers** y in case $x > y$ and there are no elements z with $x > z > y$. [Thus atoms are the elements that cover 0.] A **Hasse diagram** of B is a picture of the digraph whose vertices are the members of B and that has an edge from x to y if and only if x covers y.

Draw Hasse diagrams of the following Boolean algebras. Draw the element 1 at the top, and direct the edges generally downward.

(a) $\mathcal{P}(\{1, 2\})$ (b) $\mathbb{B}$ (c) $\mathbb{B}^2$ (d) $\mathbb{B}^3$

10. (a) Describe the atoms of the Boolean algebra $\mathcal{P}(\mathbb{N})$.

(b) Is every nonempty member of $\mathcal{P}(\mathbb{N})$ a join of atoms? Discuss.

11. Let x and y be elements of a Boolean algebra, and let a be an atom.

(a) Show that $a \leq x \vee y$ if and only if $a \leq x$ or $a \leq y$.

(b) Show that $a \leq x \wedge y$ if and only if $a \leq x$ and $a \leq y$.

(c) Show that either $a \leq x$ or $a \leq x'$, but not both.

12. Let x and y be elements of a finite Boolean algebra, each written as a join of atoms:

$$x = a_1 \vee \cdots \vee a_n \quad \text{and} \quad y = b_1 \vee \cdots \vee b_m.$$

(a) Explain how to write $x \vee y$ and $x \wedge y$ as joins of distinct atoms. Illustrate with examples.

(b) How would you write x' as the join of distinct atoms?

10.2 Boolean Expressions

To see where this is all going, imagine that we want to build [or maybe simulate] a logic circuit with three inputs, x, y, and z, each of which can be 0 or 1, and with one output, which is 1 if y and z are not both 1 but either x is 1 or else y is 1 and z is 0, and which is 0 otherwise. Suppose also that we want to make this circuit as simple as possible, perhaps subject to some added conditions. Our first task is to figure out some sensible mathematical way to describe the problem. Not surprisingly, we will use Boolean algebra. After that, we'll need to think about how to build circuits, and after that we can work on simplifying them. This section and the following two will fill in the details of this program.

First, we develop terminology to describe and work with problems such as our circuit example. We can think of the inputs, x, y, and z for our circuits as variables that can take on the values 0 and 1 in $\mathbb{B}$. When we work with real number variables, we form new expressions such as $-x$, $x+y$, $5+x \cdot y$, and $(-(y \cdot z)) \cdot (x + (y \cdot (-z)))$, and we can do the same sort of thing with Boolean variables to form expressions such as x', $x \vee y$, $0 \vee (x \wedge y)$, and $(y \wedge z)' \wedge (x \vee (y \wedge z'))$. Here is the definition we need.

A Boolean expression is a string of symbols involving the constants 0 and 1, some variables, and the Boolean operations. To be more precise, we define **Boolean expressions in** n **variables** $x_1, x_2, \ldots, x_n$ recursively as follows:

(B) The symbols 0, 1 and $x_1, x_2, \ldots, x_n$ are Boolean expressions in $x_1, \ldots, x_n$;

(R) If E_1 and E_2 are Boolean expressions in $x_1, x_2, \ldots, x_n$, so are $(E_1 \vee E_2)$, $(E_1 \wedge E_2)$, and E_1'.

As usual, in practice we will normally omit the outside parentheses and will freely use the associative laws.

EXAMPLE 1

(a) Here are four Boolean expressions in the three variables x, y, z:

$$(x \vee y) \wedge (x' \vee z) \wedge 1, \qquad (x' \wedge z) \vee (x' \wedge y) \vee z', \qquad x \vee y, \qquad z.$$

The first two obviously involve all three variables. The last two don't. Whether we regard $x \vee y$ as an expression in two, three, or more variables often doesn't matter. When it does matter and the context doesn't make the variables clear, we will be careful to say how we are viewing the expression.

The Boolean expressions 0 and 1 can be viewed as expressions in any number of variables, just as constant functions can be viewed as functions of one or of several variables.

(b) The expression

$$(x_1 \wedge x_2 \wedge \cdots \wedge x_n) \vee (x_1' \wedge x_2 \wedge \cdots \wedge x_n) \vee (x_1 \wedge x_2' \wedge x_3 \wedge \cdots \wedge x_n)$$

is an example of a Boolean expression in n variables. ■

The usage of both symbols $\vee$ and $\wedge$ leads to bulky and awkward Boolean expressions, so we will usually replace the connective $\wedge$ by a dot or by no symbol at all.

EXAMPLE 2

(a) With this new convention for $\wedge$, the first two Boolean expressions in Example 1(a) can be written as

$$(x \vee y) \cdot (x' \vee z) \cdot 1 \qquad \text{and} \qquad (x'z) \vee (x'y) \vee z'$$

or, more simply, as

$$(x \vee y)(x' \vee z)1 \qquad \text{and} \qquad x'z \vee x'y \vee z';$$

just as in ordinary algebra, the "product" $\wedge$ or $\cdot$ takes precedence over the "sum" $\vee$.

(b) The Boolean expression in Example 1(b) is

$$x_1x_2 \cdots x_n \vee x_1'x_2 \cdots x_n \vee x_1x_2'x_3 \cdots x_n.$$

(c) The expression $xyz \vee xy'z \vee x'z$ is shorthand for

$$(x \wedge y \wedge z) \vee (x \wedge y' \wedge z) \vee (x' \wedge z).$$ ■

If we substitute 0 or 1 for each occurrence of each variable in a Boolean expression, then we get an expression involving 0, 1, $\vee$, $\wedge$, and $'$ that has a meaning as a member of the Boolean algebra $\mathbb{B} = \{0, 1\}$. For example, replacing x by 0, y by 1, and z by 1 in the Boolean expression $x'z \vee x'y \vee z'$ gives

$$0'1 \vee 0'1 \vee 1' = (1 \wedge 1) \vee (1 \wedge 1) \vee 0 = 1 \vee 1 \vee 0 = 1.$$

In general, if E is a Boolean expression in the n variables $x_1, x_2, \ldots, x_n$, then E defines a **Boolean function** mapping $\mathbb{B}^n$ into $\mathbb{B}$ whose function value at the n-tuple $(a_1, a_2, \ldots, a_n)$ is the element of $\mathbb{B}$ obtained by replacing x_1 by a_1, x_2 by a_2, ..., and x_n by a_n in E.

EXAMPLE 3

(a) The Boolean function mapping $\mathbb{B}^3$ into $\mathbb{B}$ that corresponds to $x'z \vee x'y \vee z'$ is given in the following table. Just as with truth tables for propositions, we first calculate the Boolean functions for some of the subexpressions. The fourth entry in the last column is the value that we calculated a moment ago. Note that the Boolean expression z' corresponds to the function on $\mathbb{B}^3$ that maps each triple (a, b, c) to c', where $a, b, c \in \{0, 1\}$. Similarly, z corresponds to the function that maps each (a, b, c) to c.

x	y	z	$x'z$	$x'y$	z'	$x'z \vee x'y \vee z'$
0	0	0	0	0	1	1
0	0	1	1	0	0	1
0	1	0	0	1	1	1
0	1	1	1	1	0	1
1	0	0	0	0	1	1
1	0	1	0	0	0	0
1	1	0	0	0	1	1
1	1	1	0	0	0	0

(b) The Boolean expression $(x \vee yz')(yz)'$ yields the Boolean function mapping $\mathbb{B}^3$ into $\mathbb{B}$ given by the last column in the next table. The circuit that we imagined ourselves building earlier would produce the output described by this function: 1 just in case y and z are not both 1 and either $x = 1$ or else $y = 1$ and $z = 0$. Since there are infinitely many 3-variable Boolean expressions, but only finitely many functions from $\mathbb{B}^3$ to $\mathbb{B}$, we can hope to find some Boolean expression nicer than $(x \vee yz')(yz)'$ to determine this same function. We will see this function again in Example 2(a) on page 406 and in Exercise 11 on page 417. ■

x	y	z	yz'	$x \vee yz'$	$(yz)'$	$(x \vee yz')(yz)'$
0	0	0	0	0	1	0
0	0	1	0	0	1	0
0	1	0	1	1	1	1
0	1	1	0	0	0	0
1	0	0	0	1	1	1
1	0	1	0	1	1	1
1	1	0	1	1	1	1
1	1	1	0	1	0	0

We will regard two Boolean expressions as **equivalent** provided that their corresponding Boolean functions are the same. For instance, $x(y \vee z)$ and $(xy) \vee (xz)$ are equivalent, since each corresponds to the function with value 1 at $(1, 1, 0)$, $(1, 0, 1)$, and $(1, 1, 1)$ and value 0 otherwise. We will write $x(y \vee z) = (xy) \vee (xz)$ and, in general, we will write $E = F$ if the two Boolean expressions E and F are equivalent. The usage of "=" to denote this equivalence relation is customary and seems to cause no confusion.

The use of notation in this way is familiar from our experience with algebraic expressions and algebraic functions on $\mathbb{R}$. Technically, the algebraic expressions $(x+1)(x-1)$ and x^2-1 are different [because they *look* different], but the functions f and g on $\mathbb{R}$ defined by

$$f(x) = (x + 1)(x - 1) \qquad \text{and} \qquad g(x) = x^2 - 1$$

are equal. We regard the two expressions as equivalent and commonly use either $(x + 1)(x - 1)$ or $x^2 - 1$ as a name for the function they define. Similarly, we will often use Boolean expressions as names for the Boolean functions they define.

EXAMPLE 4

The function in BOOL(3) named xy is defined by $xy(a, b, c) = ab$ for all (a, b, c) in $\mathbb{B}^3$, so

$$xy(a, b, c) = \begin{cases} 1 & \text{if } a = b = 1. \\ 0 & \text{otherwise} \end{cases}$$

Similarly, the functions named $x \vee z'$ and $xy'z$ satisfy

$$(x \vee z')(a, b, c) = a \vee c' = \begin{cases} 1 & \text{if } a = 1 \text{ or } c = 0 \\ 0 & \text{otherwise} \end{cases}$$

and

$$xy'z(a, b, c) = ab'c = \begin{cases} 1 & \text{if } a = 1,\ b = 0,\ c = 1. \\ 0 & \text{otherwise} \end{cases}$$

Since $xy'z$ takes the value 1 at exactly one point in $\mathbb{B}^3$, it is an atom of the Boolean algebra BOOL(3). The other seven atoms in BOOL(3) are

$$xyz, \quad xyz', \quad xy'z', \quad x'yz, \quad x'yz', \quad x'y'z, \quad \text{and} \quad x'y'z'.$$

■

Suppose that E_1, E_2, and E_3 are Boolean expressions in n variables. Since BOOL(n) is a Boolean algebra, both of the Boolean expressions $E_1(E_2 \vee E_3)$ and $(E_1E_2) \vee (E_1E_3)$ define the same function. Thus the two expressions are equivalent, and we can write the distributive law

$$E_1(E_2 \vee E_3) = (E_1E_2) \vee (E_1E_3).$$

In the same way, Boolean expressions also satisfy all the other laws of a Boolean algebra, as long as we are willing to write equivalences as if they were equations.

Boolean expressions consisting of a single variable or its complement, such as x or y', are called **literals**. The functions that correspond to them have the value 1 at half of the elements of $\mathbb{B}^n$. For example, the literal y' for $n = 3$ corresponds to the function with value 1 at the four elements $(a, 0, c)$ in $\mathbb{B}^3$ and value 0 at the four elements $(a, 1, c)$.

Just as in Example 9 on page 397, the atoms of BOOL(n) are the functions that have the value 1 at exactly one member of $\mathbb{B}^n$. Each atom corresponds to a Boolean expression of a special form, called a minterm. A **minterm** in n variables is a meet [i.e., product] of exactly n literals, each involving a different variable.

EXAMPLE 5

(a) The expressions $xy'z'$ and $x'yz'$ are minterms in the three variables x, y, z. The corresponding functions in BOOL(3) have the value 1 only at $(1, 0, 0)$ and $(0, 1, 0)$, respectively.

(b) The expression xz' is a minterm in the two variables x, z. It is *not* a minterm in the three variables x, y, z; the corresponding function in BOOL(3) has value 1 at both $(1, 0, 0)$ and $(1, 1, 0)$.

(c) The expression $xyx'z$ is not a minterm, since it involves the variable x in more than one literal. In fact, this expression is equivalent to 0. The expression $xy'zx$ is not a minterm either, because it involves the variable x twice, though it is equivalent to the minterm $xy'z$ in x, y, and z.

(d) In the following table we list the eight elements of $\mathbb{B}^3$ and the corresponding minterms that take the value 1 at the indicated elements. Note that the literals corresponding to 0 entries are complemented, while the other literals are not.

■

(a, b, c)	Minterm with value 1 at (a, b, c)
$(0, 0, 0)$	$x'y'z'$
$(0, 0, 1)$	$x'y'z$
$(0, 1, 0)$	$x'yz'$
$(0, 1, 1)$	$x'yz$
$(1, 0, 0)$	$xy'z'$
$(1, 0, 1)$	$xy'z$
$(1, 1, 0)$	xyz'
$(1, 1, 1)$	xyz

According to Theorem 3 on page 396, every member of BOOL(n) can be written as a join of atoms. Since atoms in BOOL(n) correspond to minterms, every Boolean expression in n variables is equivalent to a join of distinct minterms. Moreover, such a representation as a join is unique, apart from the order in which the minterms are written. We call the join of minterms that is equivalent to a given Boolean expression E the **minterm canonical form** of E. [Another popular term, which we will not use, is **disjunctive normal form**, or **DNF**.] Parts (b) and (c) of the next example illustrate two different procedures for finding minterm canonical forms.

EXAMPLE 6

(a) The Boolean expression

$$x'yz' \vee xy'z' \vee xy'z \vee xyz'$$

is a join of minterms in x, y, z as it stands, so this expression is its own minterm canonical form. The corresponding Boolean function has the values shown in the right-hand column of the table. The 1's in the column tell which atoms in BOOL(3) are involved, and hence determine the corresponding minterms. For instance, the 1 in the $(1, 1, 0)$ row corresponds to the minterm xyz'.

x	y	z	$x'yz'$	$xy'z'$	$xy'z$	xyz'	$x'yz' \vee xy'z' \vee xy'z \vee xyz'$
0	0	0	0	0	0	0	0
0	0	1	0	0	0	0	0
0	1	0	1	0	0	0	1
0	1	1	0	0	0	0	0
1	0	0	0	1	0	0	1
1	0	1	0	0	1	0	1
1	1	0	0	0	0	1	1
1	1	1	0	0	0	0	0

(b) The Boolean expression $(x \vee yz')(yz)'$, which matches the circuit we are thinking of building, is not written as a join of minterms. To get its minterm canonical form, we can calculate the values of the corresponding Boolean function, as we did in Example 3(b). When we calculate all eight values of the function, we get the right-hand column in the table in part (a). Thus $(x \vee yz')(yz)'$ is equivalent to the join of minterms in part (a); i.e., its minterm canonical form is $x'yz' \vee xy'z' \vee xy'z \vee xyz'$.

(c) We can attack $(x \vee yz')(yz)'$ directly and try to convert it into a join of minterms using Boolean algebra laws. Recall that we write $E = F$ in case the Boolean expressions E and F are equivalent. By the Boolean algebra laws,

$$\begin{aligned}
(x \vee yz')(yz)' &= (x \vee yz')(y' \vee z') && \text{De Morgan law} \\
&= (x(y' \vee z')) \vee ((yz')(y' \vee z')) && \text{distributive law} \\
&= (xy' \vee xz') \vee (yz'y' \vee yz'z') && \text{distributive law twice} \\
&= (xy' \vee xz') \vee (0 \vee yz') && yy' = 0,\ z'z' = z' \\
&= xy' \vee xz' \vee yz' && \text{associative law and property of 0.}
\end{aligned}$$

We first applied the De Morgan laws to get all complementation down to the level of the literals. Then we distributed $\vee$ across meets as far as possible.

Now we have an expression as a join of meets of literals, but not as a join of minterms in x, y, z. Consider the subexpression xy', which is missing the variable z. Since $z \vee z' = 1$, we have $xy' = xy'1 = xy'(z \vee z') = xy'z \vee xy'z'$, which is a join of minterms. We can do the same sort of thing to the other two terms and get

$$xy' \vee xz' \vee yz' = (xy'z \vee xy'z') \vee (xyz' \vee xy'z') \vee (xyz' \vee x'yz'),$$

which is a join of minterms. Deleting repetitions gives the minterm canonical form

$$xy'z \vee xy'z' \vee xyz' \vee x'yz'$$

for the expression $(x \vee yz')(yz)'$ we started with. This is of course the same as the answer obtained in part (b). ■

The methods illustrated in this example work in general. Given a Boolean expression, we can calculate the values of the Boolean function it defines—in effect, find its truth table. Then each value of 1 corresponds to a minterm in the canonical form of the expression. This is the method of Example 6(b). From this point of view, the minterm canonical form is just another way of looking at the Boolean function.

Alternatively, we can obtain the minterm canonical form as in Example 6(c). First use the De Morgan laws to move all complementation to the literals. Then distribute $\vee$ over products wherever possible. Then replace xx by x and xx' by 0 as necessary and insert missing variables using $x \vee x' = 1$. Finally, eliminate duplicates.

It is not always clear which technique is preferable for a given Boolean expression. One would not want to do a lot of calculations by hand using either method. Fortunately, the minterm canonical form is primarily useful as a theoretical tool, and when calculations do arise in practice, they can be performed by machine using simple algorithms.

From a theoretical point of view, the minterm canonical form of a Boolean expression is highly valuable, since it gives the expression in terms of its basic parts, the minterms or atoms. As we will illustrate in §10.3, Boolean expressions can be realized as electronic circuits, and equivalent Boolean expressions correspond to electronic circuits that perform identically, i.e., that give the same outputs for given inputs. Hence it is of interest to "simplify" Boolean expressions to get corresponding "simplified" electronic circuits.

There are various ways to measure the simplification. It would be impossible to describe here all methods that have practical importance, but we can at least discuss one simple criterion. Let's say that a join of products [i.e., meets] of literals is **optimal** if there is no equivalent Boolean expression that is a join of fewer products and if, among all equivalent joins of the same number of products, there are none with fewer literals. Our task is to find an optimal join of products equivalent to a given Boolean expression. We may suppose that we have already found *one* particular equivalent join of products, the minterm canonical form.

EXAMPLE 7

(a) Consider the expression $(xy)'z$. The table shows the values of the Boolean function it defines. It follows that the minterm canonical form is $x'y'z \vee x'yz \vee xy'z$. This expression is not optimal. By the Boolean algebra laws, $(xy)'z = (x' \vee y')z = x'z \vee y'z$, which is a join of only two terms with four literals. We will show in Example 2(d) on page 413 that $x'z \vee y'z$ is optimal [or see Exercise 13].

x	y	z	xy	$(xy)'$	$(xy)'z$
0	0	0	0	1	0
0	0	1	0	1	1
0	1	0	0	1	0
0	1	1	0	1	1
1	0	0	0	1	0
1	0	1	0	1	1
1	1	0	1	0	0
1	1	1	1	0	0

This example illustrates a problem that can arise in practice. It seems plausible that a circuit to produce $x'z \vee y'z$ might be simpler than one to produce the expression $x'y'z \vee x'yz \vee xy'z$; but perhaps a circuit to produce the original expression $(xy)'z$ would be simplest of all. We return to this point in §10.3.

(b) Consider the join of products $E = x'z' \vee x'y \vee xy' \vee xz$. Is it optimal? We use Boolean algebra calculations, including the $x \vee x' = 1$ trick, to find its minterm canonical form:

$$E = x'yz' \vee x'y'z' \vee x'yz \vee x'yz' \vee xy'z \vee xy'z' \vee xyz \vee xy'z$$
$$= x'yz' \vee x'y'z' \vee x'yz \vee xy'z \vee xy'z' \vee xyz.$$

This has just made matters worse—more products and more literals. We want to repackage the expression in some clever way. To do so, we reverse the $x \vee x' = 1$ trick. Observe that we can group the six minterms together in pairs $x'yz'$ and $x'y'z'$, $x'yz$ and xyz, and $xy'z$ and $xy'z'$ such that two minterms in the same pair differ in exactly one literal. Since

$$x'yz' \vee x'y'z' = x'(y \vee y')z' = x'z', \quad x'yz \vee xyz = yz,$$
$$\text{and} \quad xy'z \vee xy'z' = xy',$$

we have $E = x'z' \vee yz \vee xy'$. A different grouping gives

$$x'yz' \vee x'yz = x'y, \quad x'y'z' \vee xy'z' = y'z', \quad \text{and} \quad xy'z \vee xyz = xz,$$

so $E = x'y \vee y'z' \vee xz$. Each of these joins of products $x'z' \vee yz \vee xy'$ and $x'y \vee y'z' \vee xz$ will be shown to be optimal in Example 2(c) on page 413. Thus no join of products that is equivalent to E has fewer than three products, and no join with three products has fewer than six literals. Whether or not we believe these claims now, the two expressions look simpler than the join of four products that we started with. ■

There is a method, called the **Quine–McCluskey procedure**, that builds optimal expressions by systematically grouping together products that differ in only one literal. The algorithm is tedious to use by hand, but is readily programmed for computer calculation. Among other references, the textbooks *Applications-Oriented Algebra* by J. L. Fisher and *Modern Applied Algebra* by G. Birkhoff and T. C. Bartee contain readable accounts of the method.

Another procedure for finding optimal expressions, the method of **Karnaugh maps**, has a resemblance to Venn diagrams. The method works reasonably well for Boolean expressions in three or four variables, where the problems are fairly simple anyway, but is less useful for more than four variables. The textbook *Computer Hardware and Organization* by M. E. Sloan devotes several sections to Karnaugh maps and discusses their advantages and disadvantages in applications. We will look at the method in §10.4, after we have described the elements of logical circuitry.

Exercises 10.2

1. Let $f: \mathbb{B}^3 \to \mathbb{B}$ be the Boolean function such that $f(0,0,0) = f(0,0,1) = f(1,1,0) = 1$ and $f(a,b,c) = 0$ for all other $(a,b,c) \in \mathbb{B}^3$. Write the corresponding Boolean expression in minterm canonical form.

2. Give the Boolean function corresponding to the Boolean expression in Example 7(b).

3. For each of the following Boolean expressions in x, y, and z, describe the corresponding Boolean function and write the minterm canonical form.

   (a) xy (b) z' (c) $xy \vee z'$ (d) 1

4. Consider the Boolean expression $x \vee yz$ in x, y, and z.

   (a) Give a table for the corresponding Boolean function $f: \mathbb{B}^3 \to \mathbb{B}$.

   (b) Write the expression in minterm canonical form.

5. Find the minterm canonical form for the 4-variable Boolean expressions

   (a) $(x_1x_2x_3') \vee (x_1'x_2x_3x_4')$ (b) $(x_1 \vee x_2)x_3'x_4$

6. Use the method of Example 6(c) to find the minterm canonical form of the 3-variable Boolean expression $((x \vee y)' \vee z)'$.

7. (a) Find a join of products involving a total of three literals that is equivalent to the expression

   $$xz \vee (y' \vee y'z) \vee xy'z'.$$

   (b) Repeat part (a) for $((xy \vee xyz) \vee xz) \vee z$.

8. The Boolean function $f: \mathbb{B}^3 \to \mathbb{B}$ is given by $f(a,b,c) = a +_2 b +_2 c$ for $(a,b,c) \in \mathbb{B}^3$. Recall that $+_2$ refers to addition modulo 2, which is defined in §3.5.

   (a) Determine a Boolean expression corresponding to f.

   (b) Write the expression in minterm canonical form with variables x, y, and z.

9. Find an optimal expression equivalent to

   $$(x \vee y)' \vee z \vee x(yz \vee y'z').$$

10. Group the three minterms in $xyz \vee xyz' \vee xy'z$ in two pairs to obtain an equivalent expression as a join of two products with two literals each.

11. There is a notion of maxterm dual to the notion of minterm. A **maxterm** in $x_1, \ldots, x_n$ is a join of n literals, each involving a different one of $x_1, \ldots, x_n$.
   (a) Use De Morgan laws to show that every Boolean expression in variables $x_1, \ldots, x_n$ is equivalent to a product of maxterms.
   (b) Write $xy' \vee x'y$ as a product of maxterms in x and y.
12. Consider a product E of k literals chosen from among $x_1, x_1', \ldots, x_n, x_n'$ and involving k different variables x_i. Show that E determines a function in BOOL(n) that takes the value 1 on a subset of $\mathbb{B}^n$ having 2^{n-k} elements.
   Hint: Get a minterm canonical expression for E by using the $x \vee x'$ trick of Example 6(c) with the variables $w_1, \ldots, w_{n-k}$ that are not involved in E.
13. (a) Show that $x'z \vee y'z$ is not equivalent to a product of literals. *Hint:* Use Exercise 12.
   (b) Show that $x'z \vee y'z$ is not equivalent to a join of products of literals in which one "product" is a single literal. [Parts (a) and (b) together show that $x'z \vee y'z$ is optimal.]
14. Prove that if E_1 and E_2 are Boolean expressions in $x_1, \ldots, x_n$, then $E_1 \vee E_2$ and $E_2 \vee E_1$ are equivalent.

10.3 Logic Networks

Computer science at the hardware level includes the design of devices to produce appropriate outputs from given inputs. For inputs and outputs that are 0's and 1's, the problem is to design circuitry that transforms input data according to the rules for Boolean functions. The basic building blocks of logic networks are small units, called **gates**, that correspond to simple Boolean functions. Hardware versions of these units are available from manufacturers, packaged in a wide variety of configurations. Gates also appear combined in the circuitry of logic chips designed by sophisticated software. In this section we will only be able to touch on a few of the simplest ways in which Boolean algebra methods can be applied to logical design. Some of the methods in this section and the next also have applications to software logic for parallel processors.

Figure 1 ▶

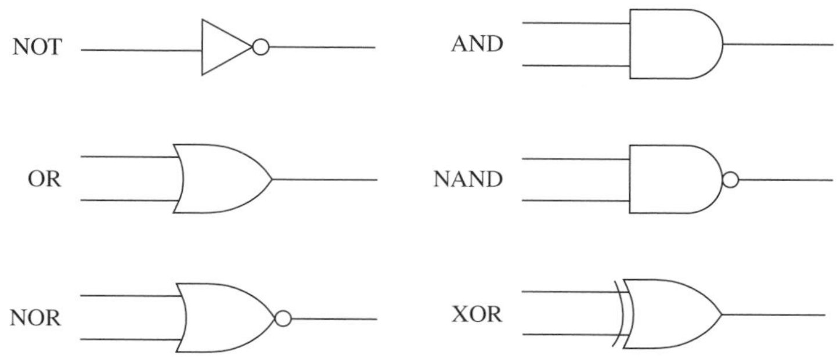

NOT AND

OR NAND

NOR XOR

Figure 1 shows the standard ANSI/IEEE symbols for the six most elementary gates. We use the convention that the lines entering the symbol from the left are input lines, and the line on the right is the output line. Placing a small circle on an input or output line complements the signal on that line. The following table shows the Boolean function values associated with these six gates and gives the corresponding Boolean function names for inputs x and y. Gates for AND, OR, NAND, and NOR are also available with more than two input lines.

		x'	$x \vee y$	$(x \vee y)'$	xy	$(xy)'$	$x \oplus y$
x	y	NOT	OR	NOR	AND	NAND	XOR
0	0	1	0	1	0	1	0
0	1	1	1	0	0	1	1
1	0	0	1	0	0	1	1
1	1	0	1	0	1	0	0

EXAMPLE 1

(a) The gate shown in Figure 2(a) corresponds to the Boolean function $(x \vee y')'$ or, equivalently, $x'y$.

(b) The 3-input AND gate in Figure 2(b) goes with the function $x'yz$.

(c) The gate in Figure 2(c) gives $(x'y')'$ or $x \vee y$, so it acts like an OR gate. ■

Figure 2 ▶

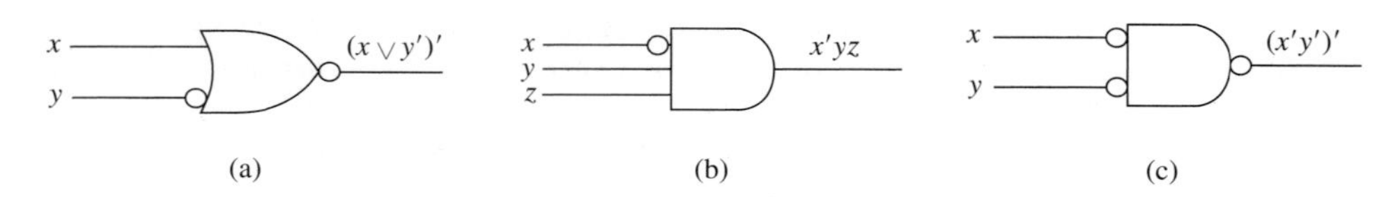

We consider the problem of designing a network of gates to produce a given complicated Boolean function of several variables. One major consideration is to keep the number of gates small. Another is to keep the length of the longest chain of gates small. Still other criteria arise in concrete practical applications.

EXAMPLE 2

(a) The circuit that we have been trying to build since the beginning of §10.2 corresponds to the Boolean expression $(x \vee yz')(yz)'$, so one possible answer to the problem is the circuit in Figure 3. [Small solid dots indicate points where input lines divide.] Perhaps, though, we can get the same function with fewer gates. Exercise 11 on page 417 deals with this question.

Figure 3 ▶

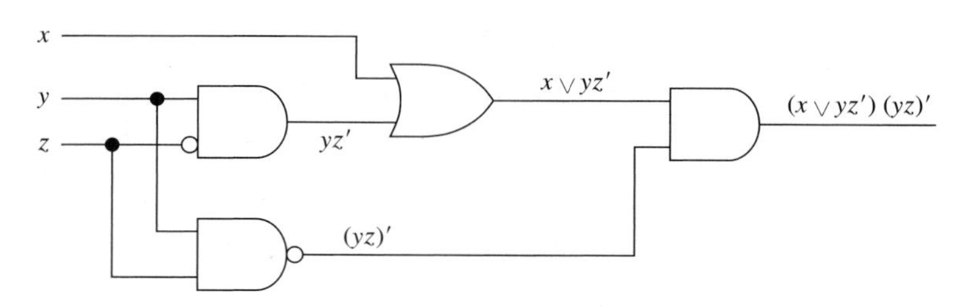

(b) Consider the foolishly designed network shown in Figure 4(a). There are four gates in the network. Reading from left to right there are two chains that are three gates long. We calculate the Boolean functions at A, B, C, and D:

Figure 4 ▶

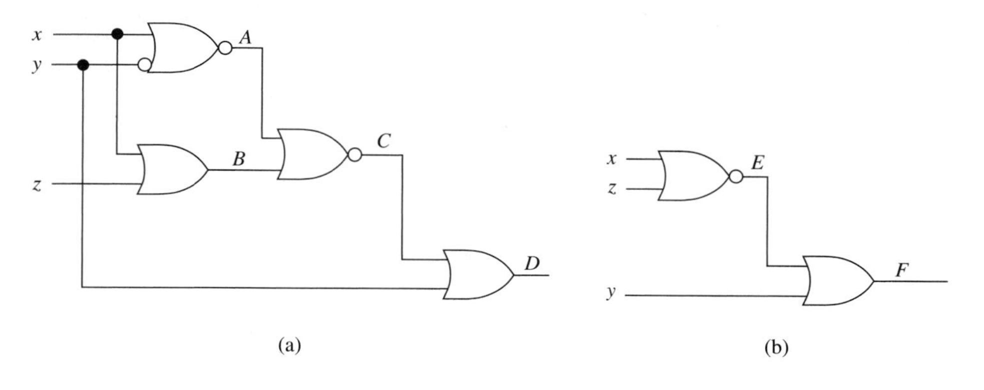

$$A = (x \vee y')'; \qquad B = x \vee z;$$
$$C = (A \vee B)' = ((x \vee y')' \vee (x \vee z))';$$
$$D = C \vee y = ((x \vee y')' \vee (x \vee z))' \vee y.$$

Boolean algebra laws give

$$\begin{aligned} D &= ((x \vee y')(x \vee z)') \vee y = ((x \vee y')x'z') \vee y \\ &= (xx'z' \vee y'x'z') \vee y = y'x'z' \vee y \\ &= y'x'z' \vee yx'z' \vee y = (y' \vee y)x'z' \vee y = x'z' \vee y. \end{aligned}$$

The network shown in Figure 4(b) produces the same output, since

$$E = (x \vee z)' = x'z' \qquad \text{and} \qquad F = E \vee y = x'z' \vee y. \qquad \blacksquare$$

This simple example shows how it is sometimes possible to redesign a complicated network into one that uses fewer gates. One reason for trying to reduce the lengths of chains of gates is that in many situations, including programmed simulations of hard-wired circuits, the operation of each gate takes a fixed basic unit of time, and the gates in a chain must operate one after the other. Long chains mean slow operation.

The expression $x'z' \vee y$ that we obtained for the complicated expression D in the last example happens to be an optimal expression for D in the sense of §10.2. Optimal expressions do not always give the simplest networks. For example, one can show [Exercise 7(a) on page 416] that $xz \vee yz$ is an optimal expression in x, y, and z. Now $xz \vee yz = (x \vee y)z$, which can be implemented with an OR gate and an AND gate, whereas to implement $xz \vee yz$ directly would require two AND gates to form xz and yz and an OR gate to finish the job. In practical situations our definition of "optimal" should change to match the hardware available.

In some settings it is desirable to have all gates be of the same type or be of at most two types. It turns out that we can do everything just with NAND or just with NOR. Which of these two types of gates is more convenient to use may depend on the particular technology being employed. Figure 5(a) shows how to write NOT, OR, and AND in terms of NAND. These equivalences also answer Exercise 18 on page 85, since NAND is another name for the Sheffer stroke referred to in that exercise. Figure 5(b) shows the corresponding networks. Exercise 2 asks for a corresponding table and figure for NOR. The network for OR in Figure 5 could also have been written as a single NAND gate with both inputs complemented. Complementation may or may not require separate gates in a particular application, depending on the technology involved and the source of the inputs. In most of our discussion we proceed as if complementation can be done at no cost.

Figure 5 ▶

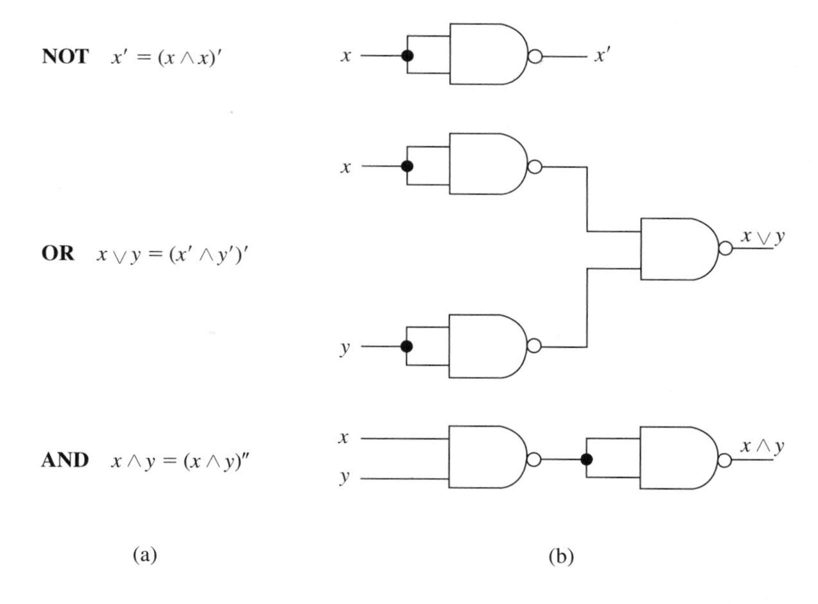

Combinations of AND and OR such as those that arise in joins of products can easily be done entirely with NAND's.

EXAMPLE 3

Figure 6 shows a simple illustration. Just replace all AND's and OR's by NAND's in an AND–OR 2-stage network to get an equivalent network. An OR–AND 2-stage network can be replaced by a NOR network in a similar way [Exercise 4]. ■

Logic networks can be viewed as acyclic digraphs with the sources labeled by variables $x_1, x_2, \ldots$, the other vertices labeled with $\vee$, $\wedge$, and $\oplus$, and some edges labeled $\neg$ for complementation. Each vertex then has an associated Boolean expression in the variables that label the sources.

Figure 6 ▶

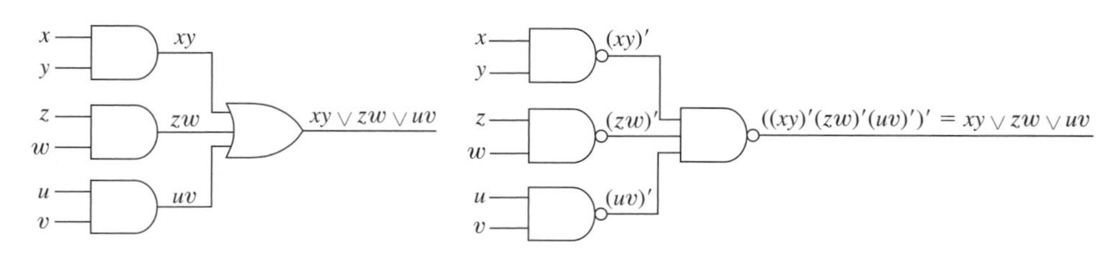

EXAMPLE 4

The network of Figure 7(a) yields the digraph of Figure 7(b), with all edges directed from left to right. If we insert a 1-input $\wedge$-vertex in the middle of the edge from z to $(x \wedge y)' \vee z \vee (x \wedge z' \wedge w)$, we don't change the logic, and we get a digraph in which the vertices appear in columns—first a variable column, then an $\wedge$ column, then an $\vee$ column—and in which edges go only from one column to the next. ■

Figure 7 ▶

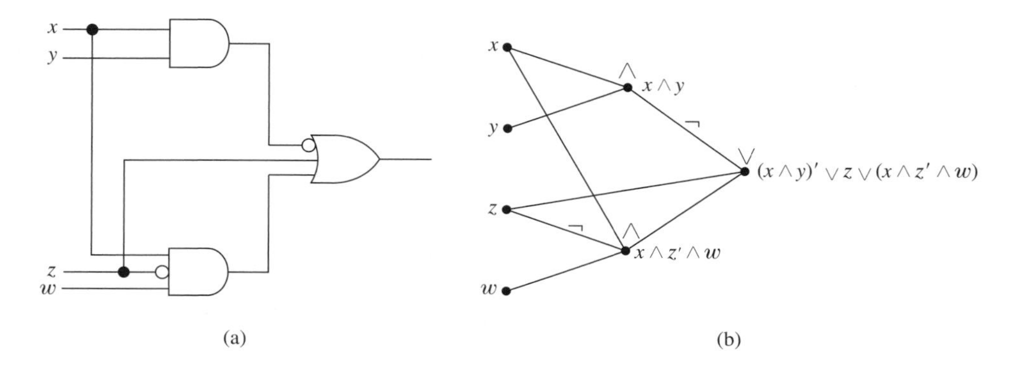

(a) (b)

The labeled digraph in Example 4 describes a computation of the Boolean function $(x \wedge y)' \vee z \vee (x \wedge z' \wedge w)$. In a similar way, any such labeled digraph describes computations for the Boolean functions that are associated with its sinks, i.e., the output vertices at the right. Since every Boolean expression can be written as a join of products of literals, every Boolean function has a computation that can be described by a digraph like the one in Example 4, with a variable column, an $\wedge$ column, and an $\vee$ column [consisting of a single vertex]. Indeed, as we saw in Example 3, all corresponding gates can be made NAND gates, so the $\vee$ vertex can be made an $\wedge$ vertex.

In a digraph of this sort, no path has length greater than 2. One interpretation is that the associated computation takes just 2 units of time. The price we pay may be an enormous number of gates.

EXAMPLE 5

Consider the Boolean expression $E = x_1 \oplus x_2 \oplus \cdots \oplus x_n$ in n variables. The corresponding Boolean function on $\mathbb{B}^n$ takes the value 1 at $(a_1, a_2, \ldots, a_n)$ if and only if an odd number of the entries $a_1, a_2, \ldots, a_n$ are 1. [See Exercise 15.] The corresponding minterms are the ones with an odd number of uncomplemented literals. Hence the minimal canonical form for E uses half of all the possible minterms and is a join of 2^{n-1} terms.

We next show that the minterm canonical form for this E is optimal; i.e., whenever E is written as a join of products of literals, the products that appear must be the minterms mentioned in the last paragraph. Otherwise, some term would be a product of fewer than n literals, say with x_k and x_k' both missing. Some choice of values of $a_1, a_2, \ldots, a_n$ makes this term have value 1, and an odd number of such values $a_1, a_2, \ldots, a_n$ must be 1. If we change a_k from 0 to 1 or from 1 to 0, the term will still have value 1, but an even number of the values $a_1, a_2, \ldots, a_n$ will be 1. No term of E can have this property, so each term for E must involve all n variables.

The observations of the last two paragraphs show that a length 2 digraph associated with $E = x_1 \oplus x_2 \oplus \cdots \oplus x_n$ must have at least $2^{n-1} + 1$ $\wedge$ and $\vee$ vertices, a number that grows exponentially with n. ■

If we are willing to let the paths grow in length, we can divide and conquer to keep the total number of vertices manageable. Figure 8 shows the idea for the expression $x_1 \oplus x_2 \oplus x_3 \oplus x_4$. This digraph has 9 $\wedge$ and $\vee$ vertices. So does the digraph associated with the join-of-products computation of $x_1 \oplus x_2 \oplus x_3 \oplus x_4$, since $2^3 + 1 = 9$; we have made no improvement. But how about $x_1 \oplus x_2 \oplus \cdots \oplus x_8$? The join-of-products digraph has $2^7 + 1 = 129$ $\wedge$ and $\vee$ vertices, while the analog of the Figure 8 digraph only has 9 + 9 + 2 + 1 = 21 $\wedge$ and $\vee$ vertices [Exercise 11].

Figure 8 ▶

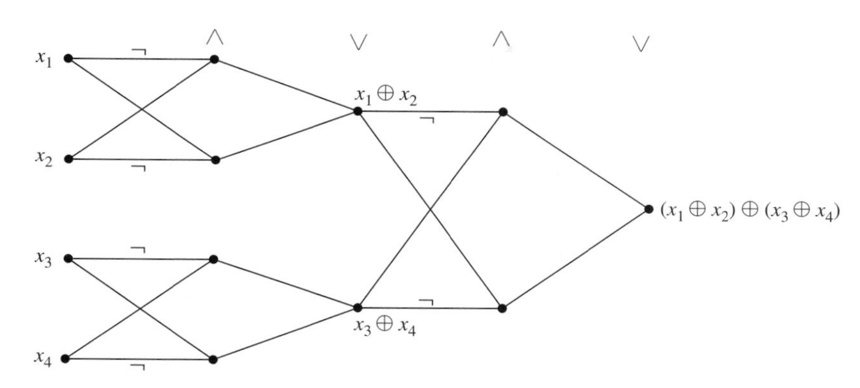

For $x_1 \oplus x_2 \oplus \cdots \oplus x_n$ in general, the comparison is $2^{n-1} + 1$ gates for the 2-stage computation versus only $3(n - 1)$ gates for the divide-and-conquer scheme, while the maximum path length increases from 2 to at most $2 \log_2 n$ [Exercise 13]. Thus doubling the number of inputs increases path length by at most 2.

EXAMPLE 6

Circuits to perform operations in binary arithmetic, for example to add two integers, are an important class of logic networks. We illustrate some of the methods that arise by adding the two integers 25 and 13, written in binary form as 11001 and 1101, respectively. This representation just means that

$$25 = 16 + 8 + 1 = \mathbf{1} \cdot 2^4 + \mathbf{1} \cdot 2^3 + \mathbf{0} \cdot 2^2 + \mathbf{0} \cdot 2 + \mathbf{1} \cdot 1 \qquad \text{and}$$
$$13 = 8 + 4 + 1 = \mathbf{1} \cdot 2^3 + \mathbf{1} \cdot 2^2 + \mathbf{0} \cdot 2 + \mathbf{1} \cdot 1,$$

so our problem looks like

$$\begin{array}{r} 11001 \\ +\quad 1101 \\ \hline ? \end{array}$$

Carry digits ⟶	11001
Numbers being added ⟶	11001
⟶	1101
Answer ⟶	100110

Figure 9 ▲

Binary addition is similar to ordinary decimal addition. Working from right to left, we can add the digits in each column. If the sum is 0 or 1, we write the sum in the answer line and carry a digit 0 one column to the left. If the sum is 2 or 3 [i.e., 10 or 11 in binary], we write 0 or 1, respectively, and go to the next column with a carry digit 1. Figure 9 gives the details for our illustration, with the top row inserted to show the carry digits. The answer represents $1 \cdot 2^5 + 0 \cdot 2^4 + 0 \cdot 2^3 + 1 \cdot 2^2 + 1 \cdot 2 + 0 \cdot 1 = 38$, as it should.

The rightmost column contains only two digits x and y [in our illustration $x = y = 1$]. The answer digit in this column is $(x + y)$ MOD 2, i.e., $x \oplus y$, and the carry digit for the next column is $(x + y)$ DIV 2, which is xy. The simple logic network shown in Figure 10, called a **half-adder**, produces the two outputs $S = x \oplus y$ and $C = xy$ from inputs x and y. Here S signifies "sum" and C signifies "carry."

Figure 10 ▶

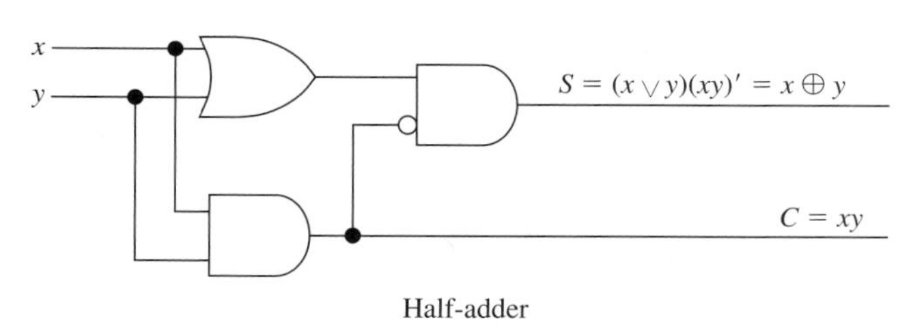

Half-adder

For the more general case with a carry input, C_I, as well as a carry output, C_O, we can combine two half-adders and an OR gate to get the network of Figure 11, called a **full-adder**. Here C_O is 1 if and only if at least two of x, y, and C_I are 1, i.e., if and only if either x and y are both 1 or exactly one of them is 1 and C_I is also 1.

Figure 11 ▶

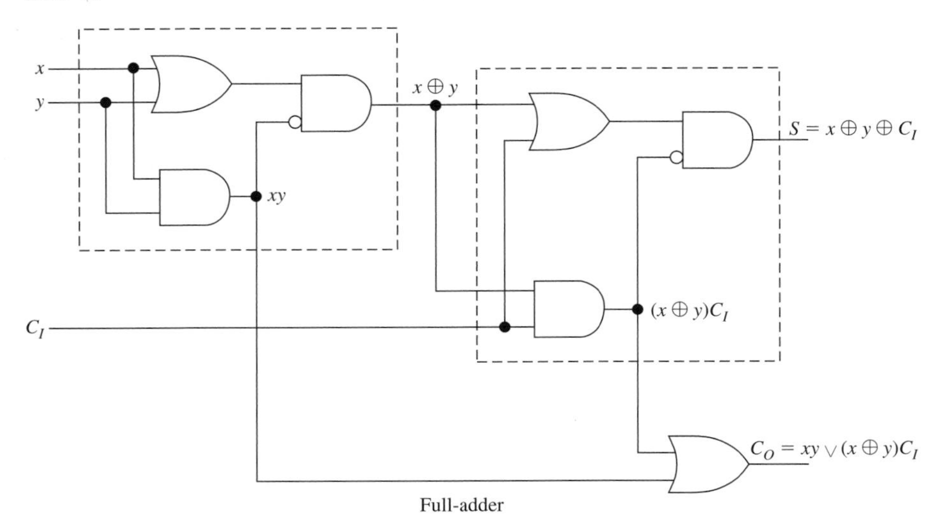

Full-adder

Several full-adders can be combined into a network for adding n-digit binary numbers, or a single full-adder can be used repeatedly with suitable delay devices to feed the input data bits in sequentially. In practice, each of these two schemes is slower than necessary. Fancy networks have been designed to add more rapidly and to perform other arithmetic operations. ■

Our purpose in including Example 6 was to illustrate the use of logic networks in hardware design and also to suggest how partial results from parallel processes can be combined. The full-adder shows how networks to implement two or more Boolean functions can be blended together. The minterm canonical form of $S = x \oplus y \oplus C_I$ is

$$xyC_I \vee xy'C_I' \vee x'yC_I' \vee x'y'C_I$$

which turns out to be optimal [see Example 5, with $n = 3$, or Exercise 7(c) on page 416]. It can be implemented with a logic network using four AND gates and one OR gate if we allow four input lines. The optimal join-of-products expression for C_O is $xy \vee xC_I \vee yC_I$, which can be produced with three AND gates and one OR gate. To produce S and C_O separately would appear to require $4 + 1 + 3 + 1 = 9$ gates, yet Figure 11 shows that we can get by with 7 gates if we want both S and C_O at once. Moreover, each gate in Figure 11 has only two input lines. As this discussion suggests, the design of economical logic networks is not an easy problem.

Exercises 10.3

Note. In these exercises, inputs may be complemented unless otherwise specified.

1. (a) Describe the Boolean function that corresponds to the logic network shown in Figure 12.

   (b) Sketch an equivalent network consisting of two 2-input gates.

2. Write logical equations and sketch networks as in Figure 5 that show how to express NOT, OR, and AND in terms of NOR without complementation of inputs.

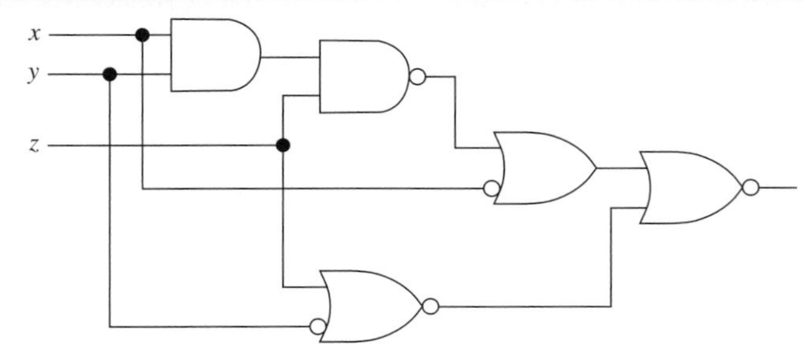

Figure 12 ▲

3. Sketch logic networks equivalent to those in Figure 13, but composed entirely of NAND gates.

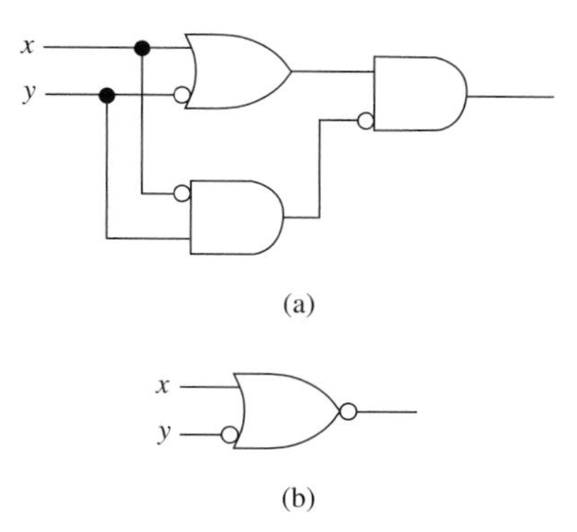

Figure 13 ▲

4. Sketch logic networks equivalent to those in Figure 14, but composed entirely of NOR gates.

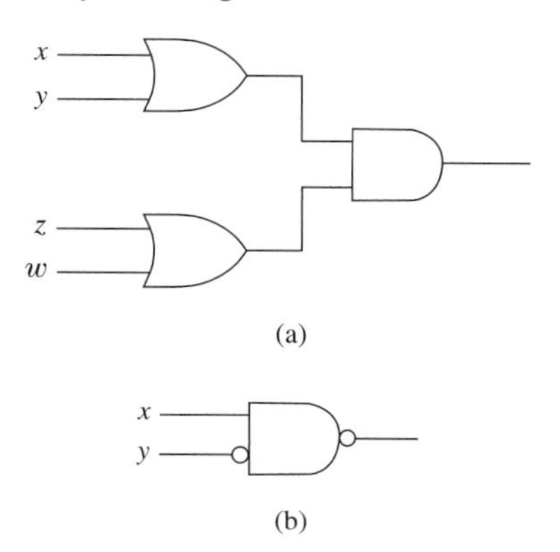

Figure 14 ▲

5. Sketch a logic network for the function XOR using
   (a) two AND gates and one OR gate.
   (b) two OR gates and one AND gate.
6. Sketch a logic network that has output 1 if and only if
   (a) exactly one of the inputs x, y, z has the value 1.
   (b) at least two of the inputs x, y, z, w have value 1.
7. Calculate the values of S and C_O for a full-adder with the given input values.
   (a) $x = 1$, $y = 0$, $C_I = 0$ (b) $x = 1$, $y = 1$, $C_I = 0$
   (c) $x = 0$, $y = 1$, $C_I = 1$ (d) $x = 1$, $y = 1$, $C_I = 1$
8. Find all values of x, y, and C_I that produce the following outputs from a full-adder.
   (a) $S = 0$, $C_O = 0$ (b) $S = 0$, $C_O = 1$
   (c) $S = 1$, $C_O = 1$
9. Consider the "triangle" and "circle" gates whose outputs are as shown in Figure 15. Show how to make a logic network from these two types of gates without complementation on input or output lines to produce the Boolean function.
   (a) x' (b) xy (c) $x \vee y$

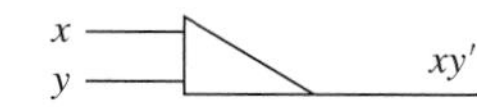

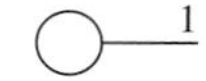

Figure 15 ▲

10. AND-OR-INVERT gates that produce the same effect as the logic network shown in Figure 16 are available commercially. What inputs should be used to make such a gate into an XOR gate?

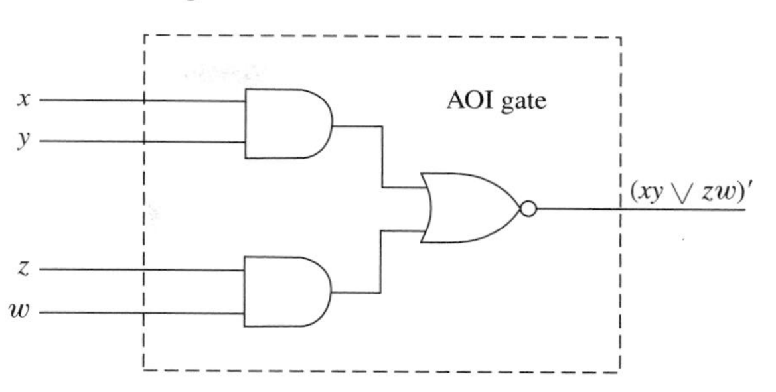

Figure 16 ▲

11. Draw a digraph like the one in Figure 8 for a divide-and-conquer computation of $x_1 \oplus x_2 \oplus \cdots \oplus x_8$.
12. (a) Draw the digraph for the 2-stage join-of-products computation of the expression $x_1 \oplus x_2 \oplus x_3 \oplus x_4$. How many $\wedge$ vertices are there in the digraph of the join-of-products computation of $x_1 \oplus x_2 \oplus \cdots \oplus x_8$?
    (b) Would you like to draw the digraph in part (b)?
13. Show by induction that for $n \geq 2$ there is a digraph for the computation of $x_1 \oplus x_2 \oplus \cdots \oplus x_n$ that has $3(n-1)$ $\wedge$ and $\vee$ vertices and is such that if $2^m \geq n$ then every path has length at most $2m$. *Suggestion:* Consider k with $2^{k-1} < n \leq 2^k$ and combine digraphs for 2^{k-1} and $n - 2^{k-1}$ variables.
14. Draw a digraph for the computation of $x_1 \oplus \cdots \oplus x_6$ with 15 $\wedge$ and $\vee$ vertices and all paths of length at most 6. *Suggestion:* See Exercise 13.
15. (a) Let E_1 and E_2 be Boolean expressions in n variables, with Boolean functions f_1 and f_2. Show that the Boolean function for $E_1 \oplus E_2$ takes the value 1 at a member of $\mathbb{B}^n$ when exactly one of f_1 and f_2 does.
    (b) Consider $1 \leq m \leq n$. Show that the Boolean expression $x_1 \oplus x_2 \oplus \cdots \oplus x_m$ takes the value 1 at $(a_1, a_2, \ldots, a_n)$ in $\mathbb{B}^n$ if and only if an odd number of the entries $a_1, a_2, \ldots, a_m$ are 1. *Hint:* Use [finite] induction on m.
16. The digraph in Figure 17, directed from left to right, has two sinks.
    (a) What are the two corresponding Boolean functions?
    (b) Would a Boolean 2-stage network associated with this digraph produce both functions at once?

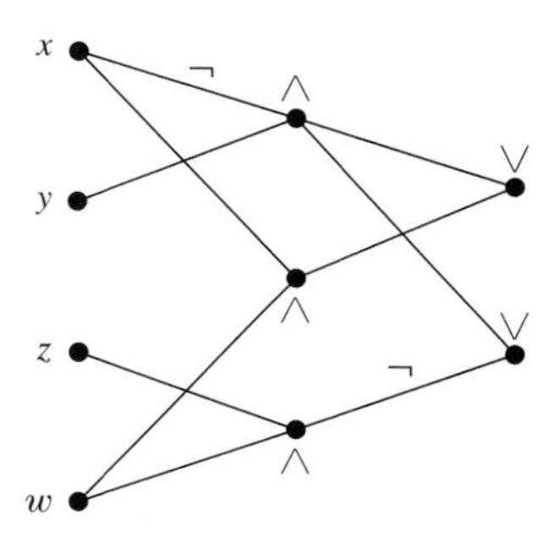

Figure 17 ▲

10.4 Karnaugh Maps

Instead of trying to find the most economical or "best" logic network possible, we may decide to settle for a solution that just seems reasonably good. Optimal solutions in the sense of §10.2 can be considered to be approximately best, so a technique for finding optimal solutions is worth having. The method of **Karnaugh maps**, which we now discuss briefly, is such a scheme. We can think of it as a sort of Boolean algebra mixture of the Venn diagrams and truth tables that we used earlier to visualize relationships between sets and between propositions.

We consider first the case of a 3-variable Boolean function in x, y, and z. The Karnaugh map of such a function is a 2×4 table, such as the ones in Figure 1. Each of the eight squares in the table corresponds to a minterm. The plus marks indicate which minterms are involved in the function described by the table. The columns of a Karnaugh map are arranged so that neighboring columns differ in just one literal. If we wrap the table around and sew the left edge to the right edge, then we get a cylinder whose columns still have this property.

Figure 1 ▶

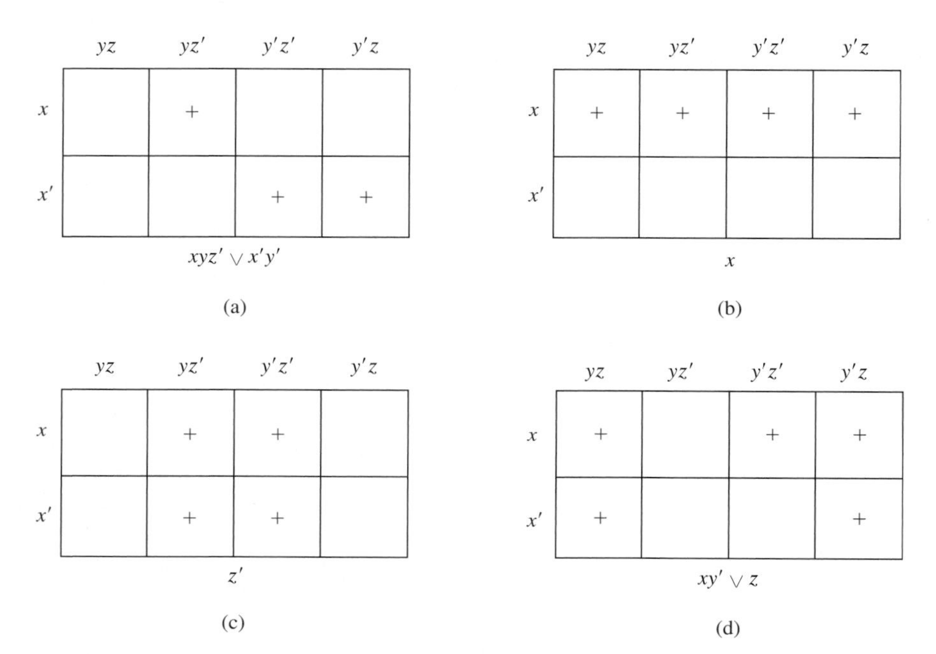

EXAMPLE 1

(a) In Figure 1(a) the minterm canonical form is $xyz' \vee x'y'z' \vee x'y'z$. Since $x'y'z' \vee x'y'z = x'y'(z' \vee z) = x'y'$, the function can also be written as $xyz' \vee x'y'$.

(b) The Karnaugh maps for literals are particularly simple. The map for x, shown in Figure 1(b), has the whole first row marked; x' has the whole second row marked. The map for y has the left 2×2 block marked and the one for y' has the right 2×2 block marked. The map for z' in Figure 1(c) has just the entries in the middle 2×2 block marked. If we sew the left edge to the right edge, then the columns involving z also form a 2×2 block.

(c) The map in Figure 1(d) describes the function $xy'z' \vee z$. Since both xy' boxes are marked, the function can also be written as $xy' \vee z$. ■

We now have a cylindrical map on which the literals x and x' correspond to 1×4 blocks, the literals y, z, y', and z' correspond to 2×2 blocks, products of two literals correspond to 1×2 or 2×1 blocks, and products of three literals correspond to 1×1 blocks.

To find an optimal expression for a given Boolean function in x, y, and z, we outline blocks corresponding to products by performing the following steps:

Step 1. Mark the squares on the Karnaugh map corresponding to the function.

Step 2. (a) Outline each marked block with 8 squares. [If all 8 boxes are marked, the Boolean function is 1, and we're done.]

(b) Outline each marked block with 4 squares that is not contained in a larger outlined block.

(c) Outline each marked block with 2 squares that is not contained in a larger outlined block.

(d) Outline each marked square that is not contained in a larger outlined block.

Step 3. Select a set of outlined blocks that

(a) has every marked square in at least one selected block,

(b) has as few blocks as possible, and

(c) among all sets satisfying (b) gives an expression with as few literals as possible.

We will say more about how to satisfy (b) and (c) in step 3 after we consider some examples.

EXAMPLE 2

(a) Consider the Boolean function with Karnaugh map in Figure 2(a). The "rounded rectangles" outline three blocks, one with four squares, corresponding to y, and two with two squares, corresponding to xz' and $x'z$. The $x'z$ block is made from squares on the two sides of the seam where we sewed the left and right edges together. Since it takes all three outlined blocks to cover all marked squares, we must use all three blocks in step 3. The resulting optimal expression is $y \vee xz' \vee x'z$.

(b) The Boolean function $(x'y'z)'$ is mapped in Figure 2(b). Here the outlined blocks go with x, y, and z'. Again, it takes all three to cover the marked squares, so the optimal expression is $x \vee y \vee z'$.

(c) The Karnaugh map in Figure 2(c) has six outlined blocks, each with two squares. The marked squares can be covered with either of two sets of three blocks, corresponding to

$$x'y \vee xz \vee y'z' \quad \text{and} \quad x'z' \vee yz \vee xy'.$$

Since no fewer than three of the blocks can cover six squares, both of these expressions are optimal. We saw this Boolean function in Example 7(b) on page 403.

Figure 2 ▶

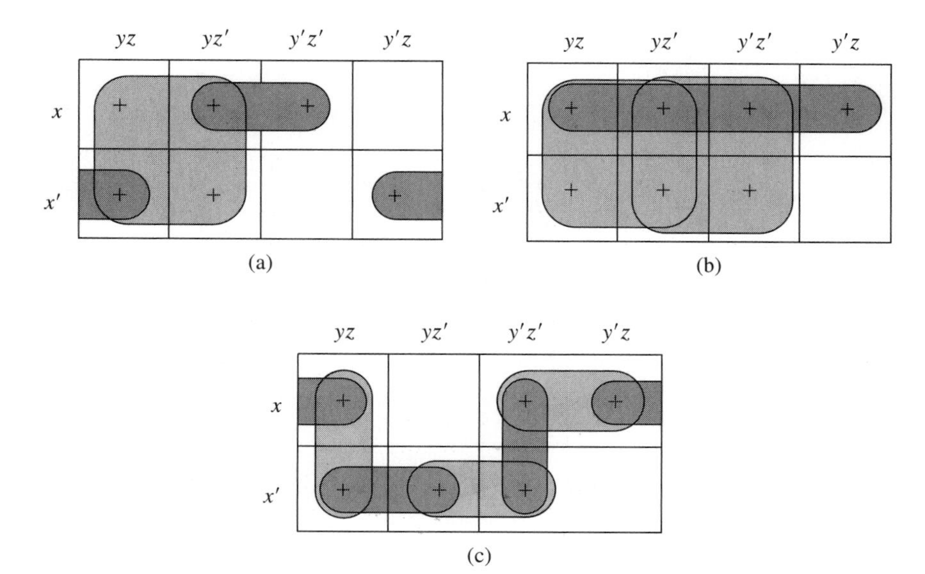

(d) Draw the Karnaugh map for the Boolean function $x'z \vee y'z$ and outline blocks following steps 1 to 3. Only the $x'z$ and $y'z$ blocks will be marked. Both will be needed to cover the marked squares, so $x'z \vee y'z$ is its own optimal Boolean expression. ■

Example 2(c) shows a situation in which more than one choice is possible. To illustrate the problems that choices may cause in selecting the blocks in step 3, we increase the number of variables to four, say w, x, y, and z. Now the map is a 4×4 table, such as the ones in Figure 3, and we can think of sewing the top and bottom edges together to form a tube and then the left and right edges together to form a doughnut-shaped surface. The three-step procedure is the same as before, except that in step 2 we start by looking for blocks with 16 squares.

EXAMPLE 3

(a) The map in Figure 3(a) has four outlined blocks, three with four squares corresponding to wy, yz', and $w'z'$ and one with two squares corresponding to $wx'z$. The two-square block is the only one containing the marked $wx'y'z$ square, and the blocks for wy and $w'z'$ are the only ones containing the squares for $wxyz$ and $w'x'y'z'$, respectively, so these three blocks must be used. Since they cover all the marked squares, they meet the conditions of step 3. The optimal expression is $wx'z \vee wy \vee w'z'$.

(b) The checkerboard pattern in Figure 3(b) describes the symmetrical Boolean function $w \oplus x \oplus y \oplus z$ in w, x, y, z. In this case all eight blocks are 1×1 and the optimal expression is just the minterm canonical form. A similar conclusion holds for $x_1 \oplus x_2 \oplus \cdots \oplus x_n$ in general, as noted in Example 5 on page 408.

(c) The map in Figure 3(c) has five blocks. Each two-square block is essential, since each is the only block containing one of the marked squares. The big four-square wx' block is superfluous, since its squares are already covered by

Figure 3 ▶

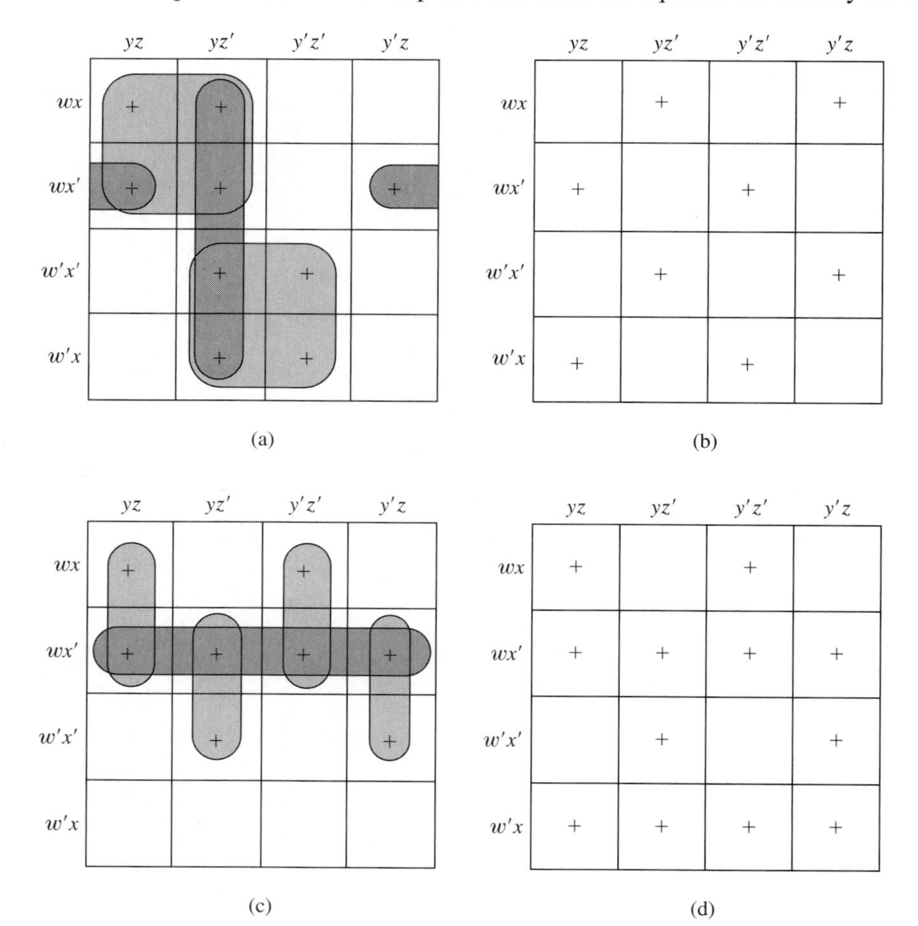

the other blocks. The optimal expression is $wyz \vee x'yz' \vee wy'z' \vee x'y'z$. Greed does not pay here, since any rule that selected the biggest blocks first would pick up the superfluous four-square block.

(d) Greed in part (c) would have been foolish; one should always select essential blocks first. How about the less foolish greedy rule "Choose the essential blocks first, then choose the largest remaining blocks to cover"? Let's fill in the bottom row of Figure 3(c) to get Figure 3(d). There are 2 four-square and 8 two-square blocks to outline. [What are they?] Now there are no essential blocks; i.e., every marked square is in at least two outlined blocks. If we choose the new 1×4 block $w'x$, then we are left with the example in part (c) and should not choose the four-square block wx'. Similarly, if we choose the block wx', then we should not take $w'x$. Even the less foolish rule does not lead to either of the two optimal expressions $wyz \vee x'yz' \vee wy'z' \vee x'y'z \vee w'x$ or $xyz \vee w'yz' \vee xy'z' \vee w'y'z \vee wx'$. ■

EXAMPLE 4

The maps in Figures 3(a), (b) and (c) offered no real choices; the essential blocks already covered all marked squares. The map of Figure 4(a), like the one in Figure 3(d), offers the opposite extreme. Every marked square is in at least two blocks. Clearly, we must use at least one two-square block to cover $wx'yz'$. Suppose that we choose the $wx'z'$ block. We can finish the job by choosing the four additional blocks shown in Figure 4(b). The resulting expression is

$$wx'z' \vee wy' \vee w'y \vee w'z \vee w'x.$$

Figure 4(c) shows another choice of blocks from Figure 4(a), this time with only four blocks altogether. The corresponding expression is

$$wx'z' \vee w'y \vee xy' \vee y'z.$$

Figure 4 ▶

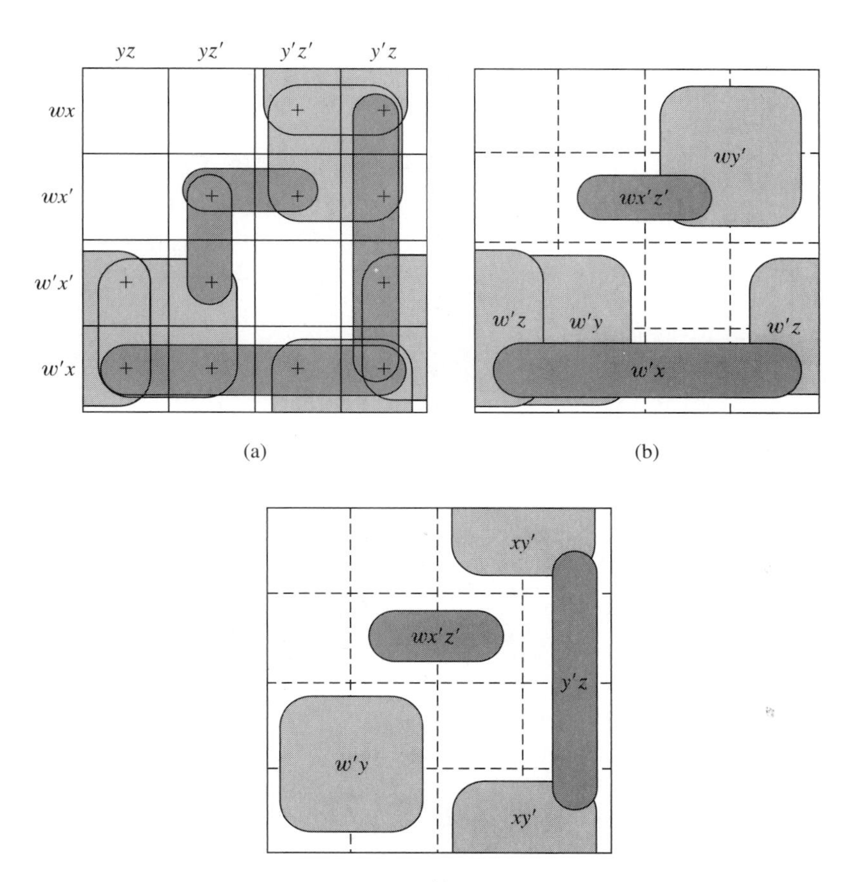

(a) (b) (c)

This expression is better, but is it optimal? The only possible improvement would be to reduce to one two-square and two four-square blocks. Since there are twelve squares to cover, no such improvement is possible, and so the expression we have found is optimal. ■

The rules for deciding which blocks to choose in situations like this are fairly complicated. It is not enough simply to choose the essential blocks, because we are forced to, and then cover the remaining squares with the largest blocks possible. For hand calculations, the tried-and-true method is to stare at the picture until the answer becomes clear. For machine calculations, which are necessary for more than five variables in any case, the Karnaugh map procedure is logically the same as the Quine–McCluskey method, for which software exists.

We close this discussion by emphasizing once again that the technical term "optimal" refers only to the complexity of a particular type of expression, as a join of products of literals. It is not synonymous with "best." An optimal expression for a Boolean function gives a way to construct a corresponding two-stage AND–OR network with as few gates as possible, but other kinds of networks may be cheaper to build.

Exercises 10.4

For each of the Karnaugh maps in Exercises 1 through 4, give the corresponding minterm canonical form and an optimal expression.

1.

	yz	yz'	$y'z'$	$y'z$
x	+	+	+	+
x'	+			+

2.

	yz	yz'	$y'z'$	$y'z$
x	+	+		+
x'	+		+	+

3.

	yz	yz'	$y'z'$	$y'z$
x	+	+		+
x'			+	+

4.

	yz	yz'	$y'z'$	$y'z$
x	+			+
x'			+	

5. Draw the Karnaugh maps and outline the blocks for the given Boolean functions of x, y, and z using the three-step procedure.

(a) $x \vee x'yz$ (b) $(x \vee yz)'$

(c) $y'z \vee xyz$ (d) $y \vee z$

6. Suppose the Boolean functions E and F each have Karnaugh maps consisting of a single block, and suppose the block for E contains the block for F.

(a) How are the optimal expressions for E and F related?

(b) Give examples of E and F related in this way.

7. Draw the Karnaugh map of each of the following Boolean expressions in x, y, and z, and show that the expression is optimal.

(a) $xz \vee yz$

(b) $xy \vee xz \vee yz$

(c) $xyz \vee xy'z' \vee x'yz' \vee x'y'z$

8. Repeat Exercise 7 for the following expressions in x, y, z, and w.

(a) $x' \vee yzw$

(b) $x'z' \vee xy'z \vee w'xy$

(c) $wxz \vee wx'z' \vee w'x'z \vee w'xz'$

9. Find optimal expressions for the Boolean functions with these Karnaugh maps.

(a)

	yz	yz'	$y'z'$	$y'z$
wx	+	+	+	
wx'		+	+	+
$w'x'$	+	+	+	+
$w'x$	+	+	+	

(b)

	yz	yz'	$y'z'$	$y'z$
wx	+			
wx'				
$w'x'$	+	+	+	+
$w'x$		+	+	+

(c)

	yz	yz'	$y'z'$	$y'z$
wx	+	+	+	
wx'	+		+	+
$w'x'$		+	+	
$w'x$			+	+

(d)

	yz	yz'	$y'z'$	$y'z$
wx	+			+
wx'	+			+
$w'x'$		+	+	
$w'x$	+			+

10. (a) Find an optimal expression for the Boolean function E of x, y, z, and w that has the value 1 if and only if at least two of x, y, z, and w have the value 1.

(b) Give a Boolean expression for the function E of part (a) that has eight $\vee$ and $\wedge$ operations. [Hence the optimal expression in part (a) does not minimize the number of gates in a logic network for E.]

11. Our favorite Boolean function $(x \vee yz')(yz)'$ from Example 3 on page 399, Example 6 on page 402, and Example 2(a) on page 406 can be produced by a logic network that uses four gates. By finding an optimal expression, show that it can also be produced by a network with just three gates [allowing complementation on inputs and outputs].

10.5 Isomorphisms of Boolean Algebras

What does it mean for two Boolean algebras to be essentially the same—or essentially different? For instance, Examples 1 and 2 on page 390 described two Boolean algebras $\mathcal{P}(\{a, b\})$ and $\mathbb{B}^2$, each with four members. Are these Boolean algebras really the same, or not? Figures 1 and 2 describe their structures. If we replace Ø by 00, $\{a\}$ by 10, $\{b\}$ by 01, and $\{a, b\}$ by 11 in Figure 1, then we get Figure 2; so it seems that $\mathbb{B}^2$ is really just $\mathcal{P}(\{a, b\})$ with the names changed. They do have essentially the same structure.

Figure 1 ▶

∪	Ø	$\{a\}$	$\{b\}$	$\{a,b\}$
Ø	Ø	$\{a\}$	$\{b\}$	$\{a,b\}$
$\{a\}$	$\{a\}$	$\{a\}$	$\{a,b\}$	$\{a,b\}$
$\{b\}$	$\{b\}$	$\{a,b\}$	$\{b\}$	$\{a,b\}$
$\{a,b\}$	$\{a,b\}$	$\{a,b\}$	$\{a,b\}$	$\{a,b\}$

∩	Ø	$\{a\}$	$\{b\}$	$\{a,b\}$
Ø	Ø	Ø	Ø	Ø
$\{a\}$	Ø	$\{a\}$	Ø	$\{a\}$
$\{b\}$	Ø	Ø	$\{b\}$	$\{b\}$
$\{a,b\}$	Ø	$\{a\}$	$\{b\}$	$\{a,b\}$

′	
Ø	$\{a,b\}$
$\{a\}$	$\{b\}$
$\{b\}$	$\{a\}$
$\{a,b\}$	Ø

Figure 2 ▶

∪	00	01	10	11
00	00	01	10	11
01	01	01	11	11
10	10	11	10	11
11	11	11	11	11

∩	00	01	10	11
00	00	00	00	00
01	00	01	00	01
10	00	00	10	10
11	00	01	10	11

′	
00	11
01	10
10	01
11	00

In general, we will want to think of two Boolean algebras as the same if we can match up their members with a one-to-one correspondence that preserves, i.e., is consistent with, the Boolean algebra structure, as given by the operations $\vee$, $\wedge$, and $'$.

Here is the formal definition. A **Boolean algebra isomorphism** is a one-to-one correspondence ϕ between Boolean algebras B_1 and B_2 that satisfies

$$\phi(x \vee y) = \phi(x) \vee \phi(y), \tag{1}$$

$$\phi(x \wedge y) = \phi(x) \wedge \phi(y), \tag{2}$$

and

$$\phi(x') = \phi(x)' \tag{3}$$

for all $x, y \in B_1$. Note that the Boolean operations on the left sides of equations (1)–(3) are operations in B_1, while the operations on the right sides are in B_2. Two Boolean algebras are said to be **isomorphic** if there is an isomorphism between them. In this case their algebraic structures are essentially the same.

We'll begin by finding some isomorphisms between Boolean algebras, such as those in Figures 1 and 2, that at first sight may seem quite different. Then we'll look at the big theorem of this section, which says that all finite Boolean algebras of any given size are isomorphic, so we should not be surprised by the isomorphisms in our examples. We should be surprised and amazed instead by the *theorem*, which even gives us a familiar concrete example of each type of finite Boolean algebra.

EXAMPLE 1

We can say more about the Boolean algebras of Examples 1 and 2 on page 390.

(a) For a fixed integer n, let $S = \{1, 2, \dots, n\}$, and consider the three Boolean algebras $\mathcal{P}(S)$, $\mathbb{B}^n$, and $\text{FUN}(S, \mathbb{B})$. First notice that each of these Boolean algebras has 2^n elements, so they have a chance to be isomorphic. In fact, they are, as we will now show.

The characteristic function χ_A of a set A in $\mathcal{P}(S)$ belongs to $\text{FUN}(S, \mathbb{B})$. If we define $\phi_1 \colon \mathcal{P}(S) \to \text{FUN}(S, \mathbb{B})$ by

$$\phi_1(A) = \chi_A,$$

then ϕ_1 is a one-to-one correspondence of $\mathcal{P}(S)$ onto $\text{FUN}(S, \mathbb{B})$ [why?]. The equations in Example 2(c) on page 390 say that this correspondence is also a Boolean algebra isomorphism. For example,

$$\phi_1(A \cup B) = \chi_{A \cup B} = \chi_A \vee \chi_B = \phi_1(A) \vee \phi_1(B),$$

so ϕ_1 preserves the join operation.

Example 2(b) on page 390 hinted at the existence of an isomorphism $\phi_2 \colon \text{FUN}(S, \mathbb{B}) \to \mathbb{B}^n$. Given f in $\text{FUN}(S, \mathbb{B})$, the n-tuple $(f(1), \dots, f(n))$ belongs to $\mathbb{B}^n$. The rule

$$\phi_2(f) = (f(1), \dots, f(n))$$

defines a one-to-one correspondence of $\text{FUN}(S, \mathbb{B})$ onto $\mathbb{B}^n$ [check this]. The correspondence is again a Boolean algebra isomorphism; for example,

$$\begin{aligned} \phi_2(f \wedge g) &= ((f \wedge g)(1), \dots, (f \wedge g)(n)) && \text{definition of } \phi_2 \\ &= (f(1) \wedge g(1), \dots, f(n) \wedge g(n)) && \text{definition of } f \wedge g \\ &= (f(1), \dots, f(n)) \wedge (g(1), \dots, g(n)) && \text{definition of } \wedge \text{ in } \mathbb{B}^n \\ &= \phi_2(f) \wedge \phi_2(g). && \text{definition of } \phi_2 \end{aligned}$$

The composite mapping $\phi_2 \circ \phi_1$ is a Boolean algebra isomorphism of $\mathcal{P}(S)$ onto $\mathbb{B}^n$ [see Exercise 5], so all three Boolean algebras $\mathcal{P}(S)$, $\mathbb{B}^n$, and $\text{FUN}(S, \mathbb{B})$ are isomorphic to each other.

(b) Note that each of the Boolean algebras in part (a) has n atoms. Moreover, our Boolean algebra isomorphisms map atoms to atoms. For example, for s in $S = \{1, 2, \dots, n\}$, we have $\phi_1(\{s\}) = \chi_{\{s\}}$. The one-element sets $\{s\}$ are the atoms of $\mathcal{P}(S)$, and the functions $\chi_{\{s\}}$ that take the value 1 at exactly one element of S are the atoms of $\text{FUN}(S, \mathbb{B})$. ■

The last example suggests that if $\phi \colon B_1 \to B_2$ is a Boolean algebra isomorphism, then a is an atom of B_1 if and only if $\phi(a)$ is an atom in B_2. In fact, since a Boolean algebra isomorphism preserves meets, joins, and complements, it must preserve any Boolean algebra structure that we can build from them, such as the order relation $\leq$ or the set of atoms [see Exercise 9]. This observation about atoms will enable us to characterize finite Boolean algebras. The next example illustrates the idea.

EXAMPLE 2

Let A be the set of atoms in $\mathbb{B}^3$, and set $(1, 0, 0) = a_1$, $(0, 1, 0) = a_2$, and $(0, 0, 1) = a_3$; so $A = \{a_1, a_2, a_3\}$. We claim that the mapping ϕ defined below is a Boolean algebra isomorphism of $\mathcal{P}(A)$ onto $\mathbb{B}^3$:

$$\begin{aligned} &\phi(\emptyset) = (0, 0, 0) = 0 && \phi(\{a_1, a_2\}) = (1, 1, 0) = a_1 \vee a_2 \\ &\phi(\{a_1\}) = (1, 0, 0) = a_1 && \phi(\{a_1, a_3\}) = (1, 0, 1) = a_1 \vee a_3 \\ &\phi(\{a_2\}) = (0, 1, 0) = a_2 && \phi(\{a_2, a_3\}) = (0, 1, 1) = a_2 \vee a_3 \\ &\phi(\{a_3\}) = (0, 0, 1) = a_3 && \phi(\{a_1, a_2, a_3\}) = (1, 1, 1) = a_1 \vee a_2 \vee a_3 = 1. \end{aligned}$$

Note that for each subset C of A the image $\phi(C)$ is the join of the members of the set C. For the empty set, we agree that the "empty join" is 0.

We can check the isomorphism conditions for a few cases, such as

$$\phi(\{a_1, a_2\} \cup \{a_1, a_3\}) = \phi(\{a_1, a_2\}) \vee \phi(\{a_1, a_3\}),$$

$$\phi(\{a_1, a_2\} \cap \{a_1, a_3\}) = \phi(\{a_1, a_2\}) \wedge \phi(\{a_1, a_3\}), \quad \text{and}$$

$$\phi(\{a_1, a_2\}^c) = \phi(\{a_1, a_2\})'$$

to see that the claim is plausible; see Exercise 2. Note that the operations, as well as the elements, in the two Boolean algebras happen to be different. It would be incredibly tedious to check conditions (1), (2), and (3) in the definition of a Boolean algebra isomorphism for all values of x and y in $\mathcal{P}(A)$. Is there a better way to verify isomorphism? The next theorem says that there is. ■

The theorem will tell us that every finite Boolean algebra B looks like $\mathcal{P}(S)$ for some set S. Moreover, if B and $\mathcal{P}(S)$ are isomorphic, then they must have the same number of atoms, $|S|$. Since $\mathcal{P}(S)$ is isomorphic to $\mathcal{P}(\{1, 2, \ldots, |S|\})$ [see Exercise 8], B must look like $\mathcal{P}(\{1, 2, \ldots, |S|\})$. It follows that the Boolean algebras $\mathcal{P}(\{1, \ldots, n\})$ for $n = 1, 2, \ldots$ form a complete set of examples of finite Boolean algebras. The Boolean algebra $\mathbb{B}^n$ also has n atoms, so it's isomorphic to $\mathcal{P}(\{1, \ldots, n\})$ as well, and hence the algebras $\mathbb{B}, \mathbb{B}^2, \mathbb{B}^3, \ldots$ form another complete list of examples, up to isomorphism.

The idea in the theorem is to start with B and its subset A of atoms, use A to build the new Boolean algebra $\mathcal{P}(A)$, and concoct a mapping from $\mathcal{P}(A)$ onto B, using the facts that $\mathcal{P}(A)$ and B have the same number of atoms and that everything can be written in terms of atoms. The isomorphism will have to map atoms to atoms, and it will also have to map joins of atoms to joins of the corresponding atoms. It turns out that preserving joins can tell us the whole story. Exercise 13 shows that every join-preserving one-to-one correspondence between Boolean algebras is an isomorphism, even if the algebras are not finite. Atoms make it easy to describe joins in finite Boolean algebras, which is one reason why they are so important.

Theorem Let B be a finite Boolean algebra and A its set of atoms. Then there is a Boolean algebra isomorphism of $\mathcal{P}(A)$ onto B. In particular, if A has n elements, then B has 2^n elements.

Proof Our isomorphism ϕ is a generalization of the one in Example 2. We define $\phi(\emptyset) = 0$, the 0 of the Boolean algebra B, and for nonempty $C \subseteq A$, we define

$$\phi(C) = \text{ the join of the atoms in } C.$$

In particular, $\phi(\{a\}) = a$ for each $a \in A$, and ϕ maps the join [i.e., union] of atoms $\{a\}$ in $\mathcal{P}(A)$ to the join of atoms a in B. Theorem 3 on page 396 tells us that every member of B is the image of ϕ; i.e., ϕ maps $\mathcal{P}(A)$ *onto* B. The uniqueness part of that theorem tells us that ϕ is *one-to-one*. So ϕ is a one-to-one correspondence, and hence $|B| = |\mathcal{P}(A)| = 2^n$.

We need to check that ϕ is really a Boolean algebra isomorphism. Suppose that $C = \{c_1, \ldots, c_m\}$ and $D = \{d_1 \ldots, d_n\}$ are subsets of A, i.e., members of $\mathcal{P}(A)$. Then $\phi(C) = c_1 \vee \cdots \vee c_m$ and $\phi(D) = d_1 \vee \cdots \vee d_n$, so $\phi(C) \vee \phi(D) = c_1 \vee \cdots \vee c_m \vee d_1 \vee \cdots \vee d_n$. If $c_i = d_j$ for some i and j, then $c_i \vee d_j = c_i$, so we can drop the duplicate, d_j, out of the expression for $\phi(C) \vee \phi(D)$. After we remove all duplicates, we're left with $\phi(C \cup D)$ [why?], so $\phi(C \cup D) = \phi(C) \vee \phi(D)$; i.e., ϕ preserves joins.

To show that ϕ preserves meets, we want to prove that

$$\phi(C \cap D) = (c_1 \vee \cdots \vee c_m) \wedge (d_1 \vee \cdots \vee d_n).$$

It seems easiest to look at the right-hand expression here first. If we expand out $(c_1 \vee \cdots \vee c_m) \wedge (d_1 \vee \cdots \vee d_n)$ using the distributive law $(a \vee b) \wedge c = (a \wedge c) \vee (b \wedge c)$

again and again and then $d \wedge (e \vee f) = (d \wedge e) \vee (d \wedge f)$ repeatedly, we end up with the join of all terms of form $c_i \wedge d_j$ for $i = 1, \ldots, m$ and $j = 1, \ldots, n$. Try it with $m = n = 2$ to see the idea. In fancy notation we would write

$$\left(\bigvee_{i=1}^{m} c_i\right) \wedge \left(\bigvee_{j=1}^{n} d_j\right) = \bigvee_{i=1}^{m}\left(c_i \wedge \left(\bigvee_{j=1}^{n} d_j\right)\right) = \bigvee_{i=1}^{m}\left(\bigvee_{j=1}^{n}(c_i \wedge d_j)\right).$$

Now c_i and d_j are atoms, so by the corollary on page 395 either $c_i \wedge d_j = 0$ or $c_i = c_i \wedge d_j = d_j$. In the join of all terms $c_i \wedge d_j$, the only nonzero terms are the atoms $c_i = d_j$ that are in *both* $\{c_1, \ldots, c_m\}$ and $\{d_1, \ldots, d_n\}$. We have thus shown that $(c_1 \vee \cdots \vee c_m) \wedge (d_1 \vee \cdots \vee d_n)$ is the join of all the atoms in $\{c_1, \ldots, c_m\} \cap \{d_1, \ldots, d_j\}$, as claimed, so ϕ preserves meets.

We still must show that $\phi(C^c) = \phi(C)'$. By Lemma 1 on page 392, with $w = \phi(C)$ and $z = \phi(C^c)$, it suffices to show that

$$\phi(C) \vee \phi(C^c) = 1 \quad \text{and} \quad \phi(C) \wedge \phi(C^c) = 0.$$

Because ϕ preserves meets, we have

$$\phi(C) \wedge \phi(C^c) = \phi(C \cap C^c) = \phi(\emptyset) = 0.$$

Since $\phi(A) = 1$ by the corollary on page 396, we also have

$$\phi(C) \vee \phi(C^c) = \phi(C \cup C^c) = \phi(A) = 1.$$

This completes the proof that ϕ is a Boolean algebra isomorphism. ■

Corollary A finite Boolean algebra has 2^n elements for some n in $\mathbb{P}$, and a Boolean algebra with 2^n elements has n atoms.

Remark It follows from the preceding theorem that if the finite Boolean algebras B_1 and B_2 have the same number n of atoms, then they are isomorphic Boolean algebras. Here's why. Let A_1 and A_2 be the sets of atoms of B_1 and B_2, respectively. As noted in Exercise 8, any one-to-one correspondence between A_1 and A_2 gives a Boolean algebra isomorphism $\phi\colon \mathcal{P}(A_1) \to \mathcal{P}(A_2)$. The theorem provides two Boolean algebra isomorphisms

$$\phi_1\colon \mathcal{P}(A_1) \to B_1 \quad \text{and} \quad \phi_2\colon \mathcal{P}(A_2) \to B_2.$$

One can show that the inverse of a Boolean algebra isomorphism is again a Boolean algebra isomorphism and that the composition of Boolean algebra isomorphisms is also a Boolean algebra isomorphism [see Exercise 5]; so

$$\phi_2 \circ \phi \circ \phi_1^{-1}\colon B_1 \to B_2$$

gives a Boolean algebra isomorphism of B_1 onto B_2.

In earlier editions of this book we proved directly that any two Boolean algebras of the same size must be isomorphic. This time we decided to consider B and $\mathcal{P}(A)$ first, because there's a natural one-to-one correspondence between their atoms: $\{a\}$ is an atom of $\mathcal{P}(A)$ for every atom a of B, and vice versa.

The situation for infinite Boolean algebras is quite complicated. In particular, there is no simple analogue of the theorem.

EXAMPLE 3

(a) Consider the Boolean algebra $\mathcal{P}(\mathbb{N})$. Each one-element set $\{n\}$ is an atom, but $\mathbb{N}$ is not a join of atoms [recall that only finite joins are allowed], so the corollary on page 396 does not carry over to infinite Boolean algebras.

(b) There is an infinite Boolean algebra $\mathcal{A}$ that doesn't have any atoms at all; see Exercise 11.

(c) It can be shown that there is a one-to-one correspondence between the Boolean algebra $\mathcal{P}(\mathbb{N})$ in part (a) and the Boolean algebra $\mathcal{A}$ in part (b). So, in a sense [elaborated on in §13.3], these Boolean algebras are the "same size." Nevertheless, these Boolean algebras are *not* isomorphic. Even though finite Boolean algebras of the same size *are* isomorphic [by the theorem], the natural generalization to infinite Boolean algebras does not hold. ■

Exercises 10.5

1. Find a set S so that $\mathcal{P}(S)$ and $\mathbb{B}^5$ are isomorphic Boolean algebras. Exhibit a Boolean algebra isomorphism from $\mathbb{B}^5$ to $\mathcal{P}(S)$.

2. Verify the identities suggested in Example 2(a).

3. (a) Is there a Boolean algebra with 6 elements? Explain.

(b) Is every finite Boolean algebra isomorphic to a Boolean algebra BOOL(n) of Boolean functions? Explain.

4. Verify the following special cases of calculations in the proof of the theorem, for distinct atoms a_1, a_2, a_3, a_4, a_5:

$$\phi(\{a_1, a_2, a_3\} \cup \{a_3, a_5\}) = \phi(\{a_1, a_2, a_3\}) \vee \phi(\{a_3, a_5\})$$

and

$$(a_1 \vee a_2 \vee a_3) \wedge (a_3 \vee a_5) = a_3.$$

5. (a) Show that the inverse of a Boolean algebra isomorphism $\phi\colon B_1 \to B_2$ is again a Boolean algebra isomorphism.

(b) Show that the composition $\phi_2 \circ \phi_1$ of two Boolean algebra isomorphisms $\phi_1\colon B_1 \to B_2$ and $\phi_2\colon B_2 \to B_3$ is a Boolean algebra isomorphism.

6. For an integer n greater than 1, let D_n be the set of divisors of n. Define $\vee$, $\wedge$, and $'$ on D_n by $a \vee b = \text{lcm}(a, b)$ [see §1.2], $a \wedge b = \gcd(a, b)$, and $a' = n/a$.

(a) The set $D_6 = \{1, 2, 3, 6\}$ with these operations $\vee$, $\wedge$, and $'$ is a Boolean algebra. What are its 0 and 1 elements?

(b) What are the atoms of D_6?

(c) Find a set S so that D_6 and $\mathcal{P}(S)$ are isomorphic, and exhibit an isomorphism between them.

7. Let D_n be as defined in Exercise 6.

(a) Show that D_7 is a Boolean algebra that is isomorphic to $\mathbb{B}$.

(b) Show that D_4 with these operations is not a Boolean algebra. *Hint:* See the corollary on page 420.

(c) Show that D_8 with these operations is not a Boolean algebra.

(d) Give an example of an integer n such that D_n is a Boolean algebra isomorphic to $\mathbb{B}^3$.

8. Let S and T be finite sets with the same number of elements. Show directly that $\mathcal{P}(S)$ and $\mathcal{P}(T)$ are isomorphic Boolean algebras. *Hint:* If $f\colon S \to T$ is a one-to-one correspondence, then $\phi(C) = f(C)$ defines a one-to-one correspondence $\phi\colon \mathcal{P}(S) \to \mathcal{P}(T)$.

9. Let $\phi\colon B_1 \to B_2$ be a Boolean algebra isomorphism between Boolean algebras B_1 and B_2.

(a) For $x, y \in B_1$, show that $x \le y$ if and only if $\phi(x) \le \phi(y)$.

(b) For $a \in B_1$, show that a is an atom of B_1 if and only if $\phi(a)$ is an atom of B_2. *Hint:* The result of Exercise 5(a) is useful here.

10. Explain why the Boolean algebras $\mathcal{P}(\mathbb{N})$ and $\mathcal{A}$ in Example 3 are not isomorphic.

11. Let $S = [0, 1)$ and let $\mathcal{A}$ consist of the empty set $\emptyset$ and all subsets of S that can be written as finite unions of intervals of the form $[a, b)$. *Warning:* While the ideas aren't that hard, writing out careful proofs of the statements below is somewhat challenging.

(a) Show that each member of $\mathcal{A}$ can be written as a finite *disjoint* union of intervals of the form $[a, b)$.

(b) Show that $\mathcal{A}$ is a Boolean algebra with respect to the operations $\cup$, $\cap$, and complementation.

(c) Show that $\mathcal{A}$ has no atoms whatever.

12. Let $B = \mathbb{B}^2$, with set of atoms $A = \{(1, 0), (0, 1)\}$. Describe the isomorphism between $\mathcal{P}(A)$ and B constructed in the proof of the theorem, by listing the images under ϕ of all the elements of $\mathcal{P}(A)$.

13. This exercise gives some additional facts about Boolean algebra isomorphisms and yields an alternative proof of the theorem that avoids massive use of the distributive law. Suppose that θ is a one-to-one correspondence of the Boolean algebra B_1 onto the Boolean algebra B_2 such that θ preserves joins.

(a) Show that θ^{-1} also preserves joins; i.e., if $x, y \in B_2$ and $a, b \in B_1$ with $\theta(a) = x$ and $\theta(b) = y$, then

$$\theta^{-1}(x \vee y) = a \vee b = \theta^{-1}(x) \vee \theta^{-1}(y).$$

(b) Is θ^{-1} also a one-to-one correspondence? Explain.

(c) Show that θ preserves the order relation $\le$; i.e., if $c \le d$ in B_1, then $\theta(c) \le \theta(d)$ in B_2. *Suggestion:* Rewrite $\le$ using $\vee$.

(d) Show that θ^{-1} also preserves the order relation.

(e) Show that if 0_1 and 0_2 are the 0 elements of B_1 and B_2, respectively, then $\theta(0_1) = 0_2$.

(f) Show that θ maps the 1 element of B_1 to the 1 element of B_2.

(g) Show that $\theta(g) \wedge \theta(g') = 0$ for each g in B_1. [Be careful: We don't know yet that θ preserves meets, though it does preserve order.]

(h) Show that $\theta(g) \vee \theta(g') = 1$ for each g in B_1.

(i) Show that $\theta(g') = \theta(g)'$ for each g in B_1. *Hint:* See Lemma 1 on page 392.

(j) Use De Morgan laws in B_1 and B_2 to show that θ preserves meets. Use parts (i) and (j) to conclude that θ is a Boolean algebra isomorphism.

Chapter Highlights

As usual: What does it mean? Why is it here? How can I use it? Think of examples.

CONCEPTS AND NOTATION

Boolean algebra
- join, meet, complement, atom
- duality principle
- $\leq$ order
- isomorphism

$\mathbb{B}$, $\mathbb{B}^n$

Boolean function, BOOL(n)

Boolean expression
- equivalent expressions
- minterm, minterm canonical form
- optimal join of products of literals

logic network
- NOT, AND, OR, NAND, NOR, XOR gates
- equivalent networks

Karnaugh map, block

FACTS

Boolean algebra laws in Theorems 1 and 2 on pages 392 and 393.

Properties of $\leq$ given in Lemmas 2, 3, and 4 on pages 394, 394 and 395.

Nonzero elements of finite Boolean algebras are uniquely expressible as joins of atoms.

Every logic network is equivalent to one using just NAND gates or just NOR gates.

Boolean expressions and logic networks correspond to labeled acyclic digraphs [§10.3].

Optimal Boolean expressions may not correspond to simplest networks.

Choosing essential blocks first in a Karnaugh map and then greedily choosing the largest remaining blocks to cover may not give an optimal expression.

Any two Boolean algebras with n atoms are isomorphic.

METHODS

Determination of minterm canonical form by calculating the corresponding Boolean function or by using Boolean algebra laws.

Use of a Karnaugh map to find all optimal expressions equivalent to a given Boolean expression.

Supplementary Exercises

1. Find the values of the following in $\mathbb{B}^4$.
(a) $(0, 1, 1, 0) \wedge (1, 1, 0, 0)$ (b) $(1, 1, 0, 1)'$
(c) $(1, 0, 1, 0) \vee (0, 1, 1, 0)$

2. Write the minterm canonical form of the 3-variable Boolean expression $(x \vee z)y'$.

3. Write each of the following as a join of atoms in the Boolean algebra indicated.
(a) $(1, 0, 0, 1)$ in $\mathbb{B}^4$
(b) $(0, 1, 0)'$ in $\mathbb{B}^3$
(c) $(0, 1, 1, 1, 0) \wedge (1, 0, 1, 1, 0)$ in $\mathbb{B}^5$
(d) $\{1, 3\}$ in $\mathcal{P}(\{1, 2, 3, 4, 5\})$
(e) $xy'z$ in BOOL(3)

4. (a) Give the minterm canonical form of the 3-variable Boolean expression $(x \vee z)(y' \vee (xz))$.
(b) Mark the squares on a Karnaugh map that correspond to the expression in part (a).

5. List the elements x of $\mathbb{B}^3$ with $x \leq (1, 0, 1)$.

6. (a) Give the minterm canonical form of the 3-variable Boolean expression $(xy \vee z')(y \vee z)$.
(b) Mark the squares on a Karnaugh map that correspond to the expression in part (a).
(c) Use any method to find an optimal expression for the Boolean function whose Karnaugh map is shown.

	yz	yz'	$y'z'$	$y'z$
wx	+	+		+
wx'	+	+	+	+
$w'x'$			+	+
$w'x$	+			+

7. (a) Mark the squares on a Karnaugh map corresponding to the Boolean function $(x \vee z)y' \vee x'yz$.

(b) Find an optimal expression for this Boolean function.

8. (a) Give an example of an atom in BOOL(4).

(b) How many atoms does BOOL(4) have?

(c) Is BOOL(4) isomorphic to $\mathbb{B}^n$ for some integer n? Explain.

9. One Boolean algebra isomorphism φ from $\mathbb{B}^2$ to $\mathcal{P}(\{a, b\})$ satisfies

$$\varphi((1, 0)) = \{a\} \quad \text{and} \quad \varphi((0, 1)) = \{b\}.$$

(a) Complete the description of φ by giving its values at the other members of $\mathbb{B}^2$.

(b) Give another example of a Boolean algebra isomorphism from $\mathbb{B}^2$ to $\mathcal{P}(\{a, b\})$.

(c) How many Boolean algebra isomorphisms of $\mathbb{B}^2$ onto $\mathcal{P}(\{a, b\})$ are there?

10. The logic network shown produces a Boolean function f of the three variables x, y, and z.

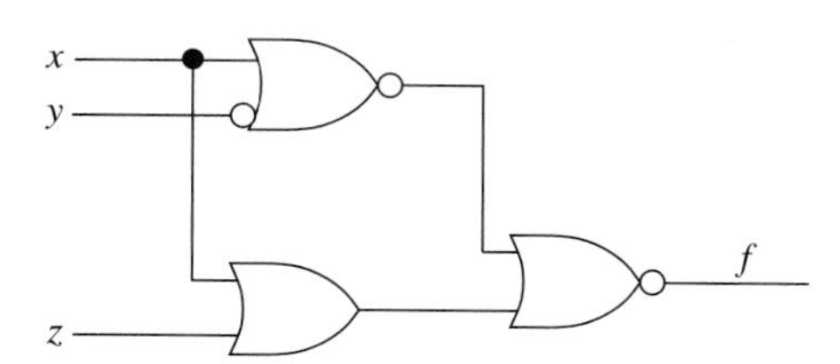

Draw a network with just two 2-input gates that produces the same function f. [You may complement inputs or outputs of your gates if you wish.]

11. Use any method to find an optimal expression for the Boolean function whose Karnaugh map is shown.

	yz	yz'	$y'z'$	$y'z$
wx	+	+		
wx'	+	+		+
$w'x'$		+	+	
$w'x$		+	+	

12. Consider the following Karnaugh map.

	yz	yz'	$y'z'$	$y'z$
wx	+		+	+
wx'		+	+	+
$w'x'$				+
$w'x$			+	

(a) Follow the Karnaugh procedure in §10.4 for finding a Boolean expression, but replace step 3 by the following greedy subalgorithm: Select the largest outlined blocks possible and continue until every marked square is in at least one selected block.

(b) Find an optimal Boolean expression following the Karnaugh procedure in the text.

(c) What did you learn from parts (a) and (b)? Does greed pay? Compare with Example 3 on page 414.

13. (a) Draw a logic network with exactly three 2-input gates that produces the Boolean function $(y(x \vee z)) \vee xyz$.

(b) Is there a logic network with *two* 2-input gates that produces this same Boolean function? Explain.

14. (a) The Boolean algebra $\mathbb{B}^5$ has five atoms. Give another example of a Boolean algebra with five atoms.

(b) How many isomorphisms φ are there from $\mathbb{B}^3$ to $\mathcal{P}(\{a, b, c\})$?

15. How many (nonisomorphic) Boolean algebras with six elements are there?

16. Describe a Boolean algebra isomorphism φ from $\mathcal{P}(\{a, b\})$ to $\mathbb{B}^2$ by giving the values $\varphi(S)$ for all sets S in $\mathcal{P}(\{a, b\})$.

More on Relations

This chapter continues the study of relations that we began long ago in Chapter 3. It may be wise to review that account quickly for terminology. The first two sections of this chapter discuss relations that order the elements of a set, beginning with general partial orderings and then turning to specific order relations on the sets $S_1 \times \cdots \times S_n$ and Σ^*. Section 11.4 discusses composition of relations in general and develops matrix analogs of statements about relations. Before that, though, we must revisit matrix multiplication. Section 11.3 develops this machinery and introduces the Boolean matrices that naturally describe relations. The final section determines the smallest relations with various properties that contain a given relation R on a set S. In particular, it describes the smallest equivalence relation containing R.

Sections 11.3 to 11.5 are independent of the first two sections of the chapter and may be studied separately.

11.1 Partially Ordered Sets

In this section we look at sets whose members can be compared with each other in some way. In typical instances we will think of one element as being smaller than another or as coming before another in some sort of order. The most familiar order is $\leq$ on $\mathbb{R}$, which has the following properties:

(R) $x \leq x$ for all x,

(AS) $x \leq y$ and $y \leq x$ imply that $x = y$,

(T) $x \leq y$ and $y \leq z$ imply that $x \leq z$,

(L) given x and y, either $x \leq y$ or $y \leq x$, with both true if $x = y$.

Except for property (L), these properties were discussed in Example 4 on page 96. Property (L) assures us that every two elements are comparable. A relation $\leq$ on a set that satisfies the properties listed above is called a **total order** or **linear order**; the label (L) refers to "linear." The term suggests that the elements can be listed in a line.

Any set S whose elements can be listed, perhaps using subscripts from $\mathbb{N}$, can be given an order satisfying the properties (R), (AS), (T), and (L) by agreeing that a member s precedes another member t and writing $s \leq t$, if s appears in the list before t or if $s = t$.

EXAMPLE 1

(a) Let S be the set of all students in some class. Listing all the students by the alphabetical order of their last names, with ties broken using first names or other data, gives a natural linear order of the class. Other linear orders of the class could be obtained based on the student ID numbers or the heights or ages of the students.

(b) The input for Kruskal's algorithm on page 260 is a finite weighted connected graph with edges $e_1, e_2, \ldots, e_m$ listed in order of increasing [more precisely, nondecreasing] weight. Such a linear order of the edges can be read off from the subscripts: edge e_j comes before [or equals] edge e_k if and only if $j \leq k$. ■

Linear orders, for which all pairs of elements are comparable, are common and will be studied further in the next section. Note, however, that in many sets that arise naturally we know how to compare some elements with others, but also have pairs that are not comparable. In such a case, the linear requirement (L) no longer holds, but the remaining properties (R), (AS), and (T) often do. A set whose members can be compared in such a way is said to be **ordered**, and the specification of how its members compare with each other is called an **order relation** on the set. We will make these definitions more precise after the next example.

EXAMPLE 2

(a) If we try to compare makes of automobiles, we can perhaps agree that Make RR is better than Make H because it is better in every respect, but we may not be able to say that either Make F or Make C is better than the other, since each may be superior in some ways.

(b) We can agree to compare two numbers in $\{1, 2, 3, \ldots, 73\}$ if one is a divisor of the other. Then 6 and 72 are comparable and so are 6 and 3. But 6 and 8 are not, since neither 6 nor 8 divides the other.

(c) We can compare two subsets of a set S [i.e., members of $\mathcal{P}(S)$] if one is a subset of the other. If S has more than one member, then it has some incomparable subsets. For example, if $s_1 \neq s_2$ and s_1 and s_2 belong to S, then the sets $\{s_1\}$ and $\{s_2\}$ are incomparable.

(d) We can compare functions pointwise. For instance, if f and g are defined on the set S and have values in $\{0, 1\}$, then we could consider f to be less than or equal to g in case $f(s) \leq g(s)$ for every $s \in S$. This is essentially the order we gave $\mathbb{B}^n$ in Example 4 on page 394. ■

Sets with comparison relations that allow the possibility of incomparable elements, such as those in Example 2, are said to be partially ordered.

Recall that a relation R on a set S is a subset of $S \times S$. A **partial order** on a set S is a relation R that is reflexive, antisymmetric, and transitive. These conditions mean that if we write $\boldsymbol{x \preceq y}$ as an alternative notation for $(x, y) \in R$, then a partial order satisfies

(R) $s \preceq s$ for every s in S;

(AS) $s \preceq t$ and $t \preceq s$ imply $s = t$;

(T) $s \preceq t$ and $t \preceq u$ imply $s \preceq u$.

If $\preceq$ is a partial order on S, the pair $(S, \preceq)$ is called a **partially ordered set**, or **poset** for short. We use the notation $\preceq$ as a general-purpose, generic name for a partial order. If there is already a notation, such as $\leq$ or $\subseteq$, for a particular partial order, then we will generally use it in preference to $\preceq$.

In Example 2 the understood relations were "is not as good as," "is a divisor of," "is a subset of," and "is never bigger than." We could just as well have considered the relations "is as good as," "is a multiple of," "contains," and "is always at least as big as," since these relations convey the same comparative information as the chosen ones. In general, each partial order on a set determines such a **converse** relation, in which x and y are related if and only if y and x are related in the original way. The converse of a partial order $\preceq$ is usually denoted by $\succeq$. Thus $x \succeq y$ means the same as $y \preceq x$. The converse relation is also a partial order [Exercise 7(a)]. If we view $\preceq$ on S as a subset R of $S \times S$, then $\succeq$ corresponds to the converse relation $R^{\leftarrow}$ defined in §3.1.

Given a partial order $\preceq$ on a set S, we can define another relation $\prec$ on S by

$$\boldsymbol{x \prec y} \qquad \text{if and only if} \qquad x \preceq y \text{ and } x \neq y.$$

For example, if $\preceq$ is set inclusion $\subseteq$, then $A \prec B$ means A is a proper subset of B, i.e., $A \subset B$. The relation $\prec$ is antireflexive and transitive:

(AR) $s \prec s$ is false for all s in S;

(T) $s \prec t$ and $t \prec u$ imply $s \prec u$.

Property (AR) for $\prec$ just comes from its definition, since $x \neq x$ is false. Transitivity is almost, but not quite, obvious. If $s \prec t$ and $t \prec u$, then $s \preceq t$ and $t \preceq u$, so $s \preceq u$, since $\preceq$ is transitive. To show that $s \prec u$, we still need to show that $s \neq u$. But if $s = u$, then $u \preceq t$. Since $t \preceq u$, (AS) for $\preceq$ would make $t = u$, contrary to $t \prec u$.

We call an antireflexive transitive relation a **quasi-order**. Each partial order on S yields a quasi-order and, conversely, if $\prec$ is a quasi-order on S, then the relation $\preceq$ defined by

$$x \preceq y \quad \text{if and only if} \quad x \prec y \text{ or } x = y$$

is a partial order on S [Exercise 7(b)]. Whether one chooses a partial order or its associated quasi-order to describe comparisons between members of a poset depends on the particular problem at hand. We will generally use the partial order, but will switch back and forth as convenient.

EXAMPLE 3

(a) A natural way to order the vertices of an acyclic digraph G is to say that vertex u precedes vertex v, written $u \prec v$, if there is a path from u to v. Then $\prec$ is a quasi-order on the set $V(G)$ of vertices of G. Property (AR), holds because $u \prec u$ would imply that there was a path from u to u, contradicting acyclicity of the digraph. Property (T) holds, because if $u \prec v$ and $v \prec w$, then there are paths from u to v and from v to w. Joining the paths yields a path from u to w, so $u \prec w$.

(b) In §7.3 we discussed sorted labelings for finite acyclic digraphs. A sorted labeling can be viewed as a function $l\colon V(G) \to \mathbb{N}$ that is consistent with the quasi-order $\prec$ in part (a), in the sense that

$$u \prec v \quad \text{implies} \quad l(u) > l(v).$$

Thus, if there is a path from vertex u to vertex v, then the label on u is larger than the label on v. This relationship is illustrated in Figure 6 on page 292.

(c) If we define $u \preceq v$ to mean that $u \prec v$ or $u = v$, then, as noted above, $\preceq$ is a partial order on $V(G)$. In other words, $u \preceq v$ if $u = v$ or if there is a path from u to v. This last sentence could have been used as the definition of $\preceq$ and then $\prec$ could have been defined in terms of $\preceq$, but in this setting the order relation $\prec$ seems to be the more natural one. The relation $\preceq$ is the reachable relation defined in §3.2 and again in §8.1. ■

It is possible, at least in principle, to draw a diagram that shows at a glance the order relation on a finite poset. Given a partial order $\preceq$ on S, we say the element t **covers** the element s in case $s \prec t$ and there is no u in S with $s \prec u \prec t$. A **Hasse** [pronounced HAH-suh] **diagram** of the poset $(S, \preceq)$ is a picture of the digraph whose vertices are the members of S, with an edge from t to s if and only if t covers s. Hasse diagrams, like rooted trees, are generally drawn with their edges directed downward and with the arrowheads left off.

EXAMPLE 4

(a) Let $S = \{1, 2, 3, 4, 5, 6\}$. As usual, we write $m|n$ in case m divides n. The diagram in Figure 1 is a Hasse diagram of the poset $(S, |)$. There is no edge between 1 and 6 because 6 does not cover 1. We can see from the diagram, though, that $1|6$, because the relation is transitive and there is a chain of edges corresponding to $1|2$ and $2|6$. Similarly, we can see that $1|4$ from the path $1|2|4$. Note that, in general, strings of comparisons for transitive relations can be run together without causing confusion: $x \preceq y \preceq z$ means $x \preceq y$, $y \preceq z$, *and* $x \preceq z$.

Figure 1 ▶

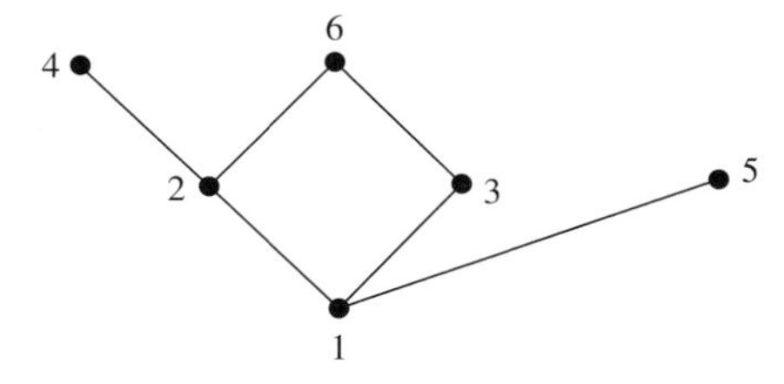

(b) Consider the power set $\mathcal{P}(\{a, b, c\})$ with $\subseteq$ as partial order. Figure 2 shows a Hasse diagram of $(\mathcal{P}(\{a, b, c\}), \subseteq)$. Note that the line joining $\{a, c\}$ to $\{a\}$ happens to cross the line joining $\{a, b\}$ to $\{b\}$, but this crossing is simply a feature of the drawing and has no significance as far as the partial order is concerned. In particular, the intersection of the two lines does *not* represent an element of the poset.

Figure 2 ▶

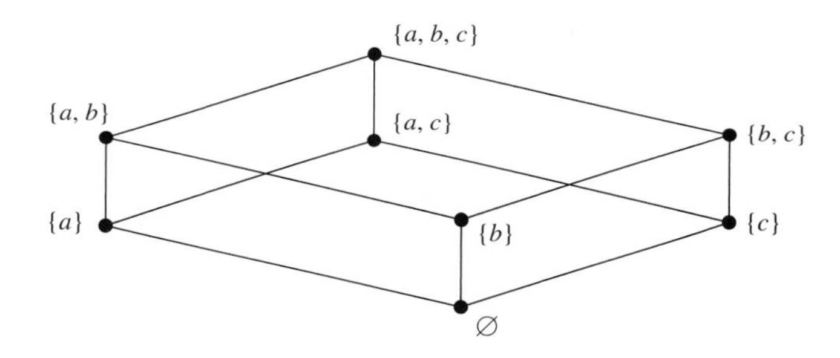

(c) The diagram in Figure 3 is not a Hasse diagram, because u cannot cover x if u covers y and y covers x. If any of the three edges connecting u, x, and y were removed, then the figure would be a Hasse diagram. The three Hasse diagrams obtained in this way would represent three different partial orders for the set $\{u, v, x, y, z\}$.

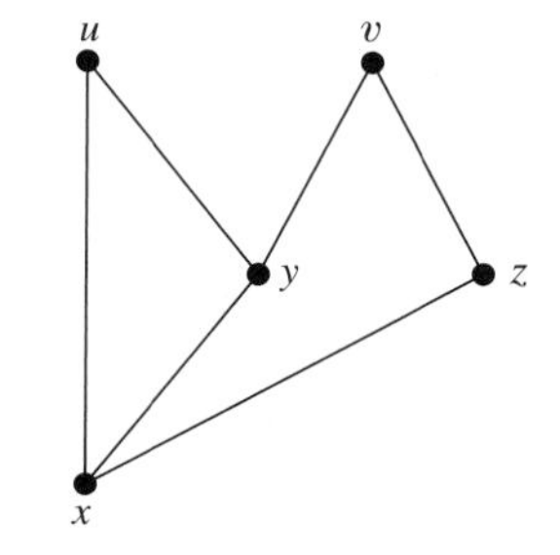

Figure 3 ▲

(d) The diagrams in Figure 4 are Hasse diagrams of posets whose order relations can be read off directly from the diagrams. All elements are related to themselves.

In addition:

For $S = \{a, b, c, d, e, f\}$, we have $a \preceq b$, $a \preceq c$, $a \preceq d$, $a \preceq e$, $a \preceq f$, $b \preceq e$, $b \preceq f$, and $c \preceq f$. We saw this picture before in part (a) of this example.

For $T = \{x, y, z, w\}$, we have $x \preceq y$, $x \preceq z$, $x \preceq w$, $y \preceq z$, $y \preceq w$, and $z \preceq w$. This is the picture we would get for divisors of 8 or of 27 or of 125 with the divisor order relation $|$.

For $U = \{A, B, C, D, E\}$, we have $A \preceq B$, $A \preceq C$, $A \preceq D$, $A \preceq E$, $B \preceq E$, $C \preceq E$, and $D \preceq E$. This picture is the Hasse diagram of the poset consisting of the sets $\{1\}$, $\{1, 2\}$, $\{1, 3\}$, $\{1, 4\}$, and $\{1, 2, 3, 4\}$ with set inclusion as order relation.

(e) The relation $<$ that we defined on a Boolean algebra in §10.1 is a quasi-order. The atoms are the elements of the algebra that cover the element 0. The poset $(\mathcal{P}(\{a, b, c\}), \subseteq)$ in part (b) is an example of a Boolean algebra viewed as a poset. ■

Figure 4 ▶

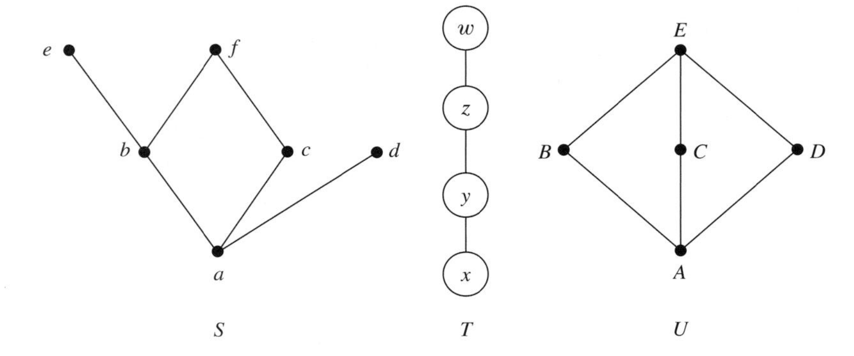

In general, given a Hasse diagram for a poset, we see that $s \preceq t$ in case either $s = t$ or there is a [downward] path from t to s. The reflexive and transitive laws are understood, and the covering information tells us the rest.

The fact that every finite poset has a Hasse diagram may be intuitively obvious, but we will provide a proof anyway, using properties of acyclic digraphs.

Theorem Every finite poset has a Hasse diagram.

Proof Given the poset $(P, \preceq)$, let H be the digraph with vertex set P and with an edge from x to y whenever x covers y. A typical path in H has a vertex sequence $x_1 x_2 \cdots x_{n+1}$ in which x_1 covers x_2, x_2 covers x_3, etc.; so $x_1 \succ x_2 \succ \cdots \succ x_{n+1}$. By transitivity of $\succ$, we have $x_1 \succ x_{n+1}$; in particular, $x_1 \neq x_{n+1}$ and the path is not closed. Hence H is an acyclic digraph. We showed in §7.3 and in Theorem 3 on page 320 that every finite acyclic digraph has a sorted labeling. By giving the digraph H such a labeling and drawing its picture so that the vertices with larger numbers are higher, we obtain a Hasse diagram for $(P, \preceq)$. ■

The proof of the preceding theorem suggests a formal algorithm for drawing a Hasse diagram for a finite poset. However, we can draw a finite Hasse diagram without labeling vertices. First draw elements that cover no other elements, then draw elements that cover elements already drawn, with edges to the elements they cover, and continue in this way.

Some infinite posets also have Hasse diagrams. A Hasse diagram of $\mathbb{Z}$ with the usual order $\leq$ is a vertical line with dots spaced along it. On the other hand, no real number covers any other in the usual $\leq$ order, so $(\mathbb{R}, \leq)$ has no Hasse diagram.

EXAMPLE 5

(a) Starting with an alphabet Σ, we can make the set Σ^* of all words using letters from Σ into an infinite poset as follows. For words w_1, w_2, in Σ^*, define $w_1 \preceq w_2$ if w_1 is an **initial segment** of w_2, i.e., if there is a word w in Σ^* with $w_1 w = w_2$. For example, we have $ab \preceq abbaa$, since $w_1 w = w_2$ with $w_1 = ab$, $w = baa$, and $w_2 = abbaa$. Also, $\lambda \preceq w$ for *all* words because $\lambda w = w$. Note that $abbaa$ does not cover ab, since $u = abb$ and $u = abba$ both satisfy $ab \prec u \prec abbaa$. However, $abbaa$ covers $abba$, $abba$ covers abb, and abb covers ab. In general, if w_2 covers w_1, then $\text{length}(w_2) = 1 + \text{length}(w_1)$.

For $\Sigma = \{a, b\}$, part of the Hasse diagram for $(\Sigma^*, \preceq)$ is drawn in Figure 5. This Hasse diagram is a tree. In §6.4 [Figure 7 on page 249] we viewed the diagram as a rooted tree, at which point tradition forced us to draw it upside down.

(b) A finite rooted tree T has a natural order in which the root r is the largest element. Define the relation $\preceq$ on the set V of vertices of T by saying that $v \preceq w$ in case $v = w$ or w is on the [unique] path from r to v. As in Example 3, $\preceq$ is a partial order on V. One Hasse diagram for $(V, \preceq)$ is the original tree T, drawn as usual with the root at the top and branches going downward. Pictures of rooted trees can be thought of as pictures of rather special posets.

Figure 5 ▶

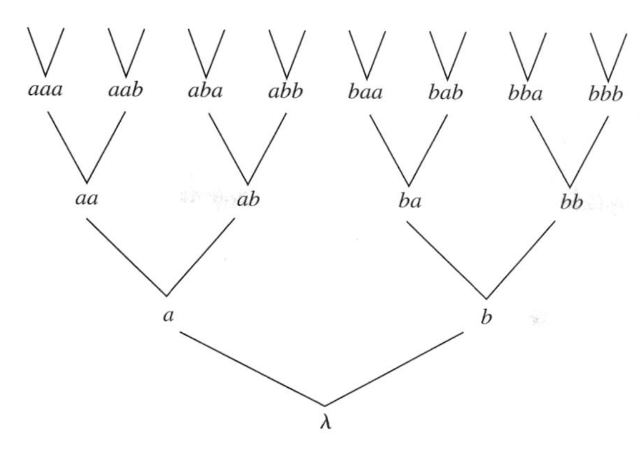

(c) The examples in parts (a) and (b) described trees with their roots at the bottom and at the top. There is a natural connection between the two sorts of presentation. A Hasse diagram determined by the converse relation $\succeq$ to the relation $\preceq$ on a poset S is simply a Hasse diagram for $(S, \preceq)$ turned top to bottom. The reason is that y covers x in the original relation if and only if x covers y in the converse relation; so the edges for a diagram of $(S, \succeq)$ are the edges for a diagram of $(S, \preceq)$ with the directions of all the arrows reversed. ■

The elements corresponding to points near the top or bottom of a Hasse diagram often turn out to be important. If $(P, \preceq)$ is a poset, we call an element x of P **maximal** in case there is no y in P with $x \prec y$ and call x **minimal** if there is no y in P with $y \prec x$. In the posets with Hasse diagrams shown in Figure 4, the elements d, e, f, w, and E are maximal, while a, x, and A are minimal. The infinite poset in Figure 5 has no maximal elements; the empty word λ is its only minimal element.

A subset S of a poset P inherits the partial order on P and is itself a poset, since the laws (R), (AS), and (T) apply to all members of P. We call S a **subposet** of P.

EXAMPLE 6

(a) The sets $\{2, 3, 4, 5, 6\}$ and $\{1, 2, 3, 6\}$ are subposets of the poset $\{1, 2, 3, 4, 5, 6\}$ given in Example 4(a), with Hasse diagrams shown in Figure 6. [Notice the placement of the primes in Figure 6(a).]

Figure 6 ▶

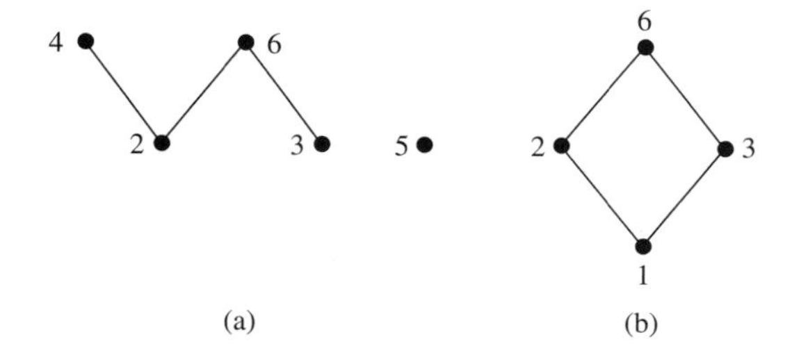

(b) The set of nonempty proper subsets of $\{a, b, c\}$ is a subposet of $\mathcal{P}(\{a, b, c\})$ with partial order $\subseteq$. Figure 7 shows a Hasse diagram for it. Compare with Figure 2. ■

Figure 7 ▶

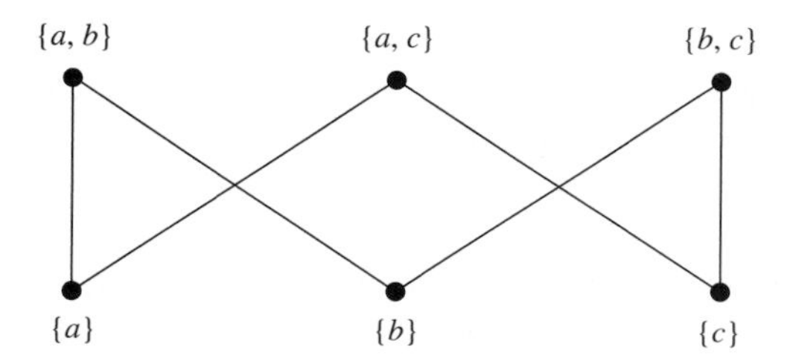

If S is a subposet of a poset $(P, \preceq)$, then it may happen that S has a member M such that $s \preceq M$ for every s in S. In Figure 6(b), $s \preceq 6$ for every s, while no such element M exists in Figure 6(a) or 7. An element M with this property is called the **largest member** of S or the **maximum** of S and denoted **max(S)**. [There is at most one such M; why?] This notation is consistent with our usage of $\max\{m, n\}$ to denote the larger of the two numbers m and n. Similarly, if S has a member m such that $m \preceq s$ for every s in S, then m is called the **smallest member** of S or the **minimum** of S and is denoted **min(S)**.

EXAMPLE 7

(a) Consider again the poset $(\{1, 2, 3, 4, 5, 6\}, |)$ illustrated in Figure 1. This poset has no largest member or maximum even though 4, 6, and 5 are all maximal elements. The element 1 is a minimum of the poset and is the only minimal element. The subset $\{2, 3\}$ has no largest member; 3 is larger than 2 in the usual $\leq$ order, but not in the order under discussion.

(b) If a Boolean algebra is viewed as a poset, then its largest element is 1 and its smallest element is 0. ■

Whether or not a subposet S of a poset $(P, \preceq)$ has a largest member, there may be elements x in the larger set P such that $s \preceq x$ for every s in S. [For example, both elements in the set $\{2, 3\}$ of Example 7 divide 6.] Such an element x is called an **upper bound** for S in P. If x is an upper bound for S in P and is such that $x \preceq y$ for every upper bound y for S in P, then x is called a **least upper bound** of S in P, and we write $x =$ **lub(S)**. Similarly, an element z in P such that $z \preceq s$ for all s in S is a **lower bound** for S in P. A lower bound z such that $w \preceq z$ for every lower bound w is called a **greatest lower bound** of S in P and is denoted by **glb(S)**. By the antisymmetric law (AS), a subset of P cannot have two different least upper bounds or two different greatest lower bounds.

EXAMPLE 8

(a) Consider the poset $(\mathbb{P}, |)$ where, as usual, $m|n$ if and only if m divides n. An upper bound for $\{m, n\}$ is an integer k in $\mathbb{P}$ such that m divides k and n divides k, i.e., a common multiple of m and n. The least upper bound $\text{lub}\{m, n\}$ is the **least common multiple** of m and n that we studied in §1.2. Similarly, the greatest lower bound $\text{glb}\{m, n\}$ is the **greatest common divisor** of m and n, the largest positive integer that divides both m and n. As we noted in the proof of Theorem 3 on page 12, the numbers $\text{lub}\{m, n\}$ and $\text{glb}\{m, n\}$ can be determined from the factorizations of m and n into products of primes. The primes themselves are the minimal members of the subposet $\mathbb{P} \setminus \{1\}$; i.e., they are the numbers that cover 1 in $\mathbb{P}$.

(b) In the poset $(\{1, 2, 3, 4, 5, 6\}, |)$, the subset $\{2, 3\}$ has exactly one upper bound, namely 6, so $\text{lub}\{2, 3\} = 6$. Similarly, $\text{glb}\{2, 3\} = 1$. The subset $\{4, 6\}$ has no upper bounds in the poset; 2 and 1 are both lower bounds, so $\text{glb}\{4, 6\} = 2$. The subset $\{3, 6\}$ has 6 as an upper bound, and has 3 and 1 as lower bounds; hence $\text{lub}\{3, 6\} = 6$ and $\text{glb}\{3, 6\} = 3$. Thus least upper bounds and greatest lower bounds for a subset may or may not exist, and if they do exist, they may or may not belong to the subset. For this particular poset, greatest lower bounds are greatest common divisors, and least upper bounds, when they exist in the poset, are least common multiples.

(c) In the poset P shown in Figure 8, the subset $\{b, c\}$ has d, e, g, and h as upper bounds in P, and h is an upper bound for $\{d, f\}$. The set $\{b, c\}$ has no least upper bound in P [why?], but $h = \text{lub}\{d, f\}$. The elements a and c are lower bounds for $\{d, e, f\}$, which has no greatest lower bound because a and c are not comparable. Element a is the greatest lower bound of $\{b, d, e, f\}$. ■

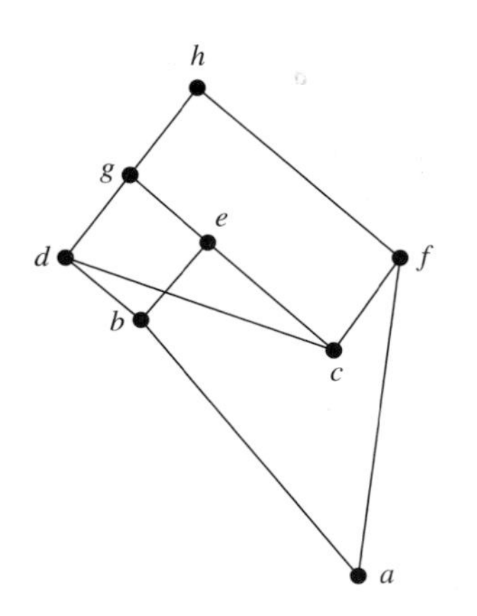

Figure 8 ▲

Many of the posets that come up in practice have the property that every 2-element subset has both a least upper bound and a greatest lower bound. A **lattice** is a poset in which lub$\{x, y\}$ and glb$\{x, y\}$ exist for every x and y. In a lattice $(P, \preceq)$, the equations

$$\boldsymbol{x} \vee \boldsymbol{y} = \text{lub}\{x, y\} \quad \text{and} \quad \boldsymbol{x} \wedge \boldsymbol{y} = \text{glb}\{x, y\}$$

define binary operations $\vee$ and $\wedge$ on P. As we will see in the next example, this usage is consistent with that introduced for Boolean algebras in §10.1. Note that glb$\{x, y\} = x \wedge y = x$ if and only if $x \preceq y$, which is true if and only if lub$\{x, y\} = x \vee y = y$. In particular, we can recover the order relation $\preceq$ if we know either binary operation $\wedge$ or $\vee$; see Exercise 11. One can show by induction [Exercise 19(b)] that every finite subset of a lattice has both a least upper bound and a greatest lower bound.

EXAMPLE 9

(a) The poset $(\mathcal{P}(\{a, b, c\}), \subseteq)$ shown in Figure 2 is a lattice. For instance,

$$\text{lub}(\{a\}, \{c\}) = \{a\} \vee \{c\} = \{a, c\},$$
$$\text{lub}(\{a, b\}, \{a, c\}) = \{a, b\} \vee \{a, c\} = \{a, b, c\},$$
$$\text{glb}(\{a, b\}, \{c\}) = \{a, b\} \wedge \{c\} = \emptyset,$$

and

$$\text{glb}(\{a, b\}, \{b, c\}) = \{a, b\} \wedge \{b, c\} = \{b\}.$$

In general, for any set S whatever, $(\mathcal{P}(S), \subseteq)$ is a lattice with lub$(A, B) = A \cup B$ and glb$(A, B) = A \cap B$ so that

$$\text{lub}\{A, B, \ldots, Z\} = A \cup B \cup \cdots \cup Z$$

and

$$\text{glb}\{A, B, \ldots, Z\} = A \cap B \cap \cdots \cap Z.$$

The poset shown in Figure 7 is not a lattice; for example, $\{a, b\}$ and $\{a, c\}$ have no least upper bound in this poset. In fact, they have no upper bounds at all.

(b) The full poset $(\mathbb{P}, |)$ discussed in Example 8(a) is a lattice. However, the subposet $S = \{1, 2, 3, 4, 5, 6\}$ of $\mathbb{P}$, shown in Figure 1, is not a lattice, since $\{3, 4\}$ has no upper bound in S.

(c) Consider the set $\text{FUN}(\{a, b, c\}, \{0, 1\})$ of all functions from the 3-element set $\{a, b, c\}$ to $\{0, 1\}$. As in Example 2(d), we obtain a partial order $\leq$ on this set by defining

$$f \leq g \quad \text{if and only if} \quad f(x) \leq g(x) \text{ for } x = a, b, c.$$

It is convenient to label the eight functions in this poset with subscripts, such as 101, that list the values the functions take at a, b, and c, respectively. For example, f_{101} represents the function such that $f_{101}(a) = 1$, $f_{101}(b) = 0$, and $f_{101}(c) = 1$. The Hasse diagram for the poset $(\text{FUN}(\{a, b, c\}, \{0, 1\}), \leq)$ is given in Figure 9. This poset is a lattice with the same structure as the lattice $\mathcal{P}(\{a, b, c\})$ in Figure 2. It is essentially the Boolean algebra BOOL(3) of Boolean functions of three variables.

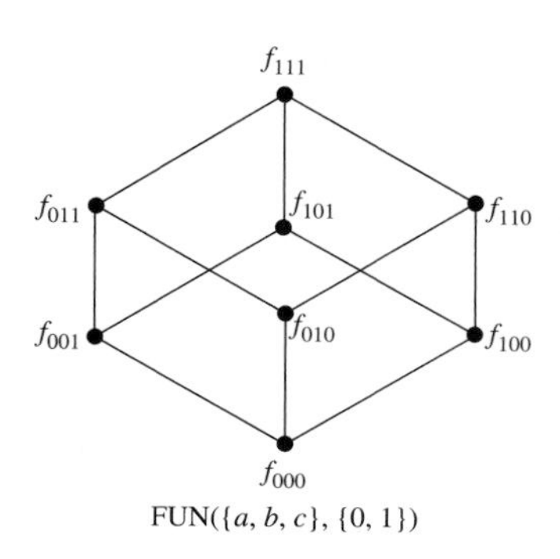

Figure 9 ▲

(d) In §10.1 we started out with a Boolean algebra $(B, \vee, \wedge, ')$ and created the relation $\leq$ by defining $x \leq y$ if and only if $x \vee y = y$ if and only if $x \wedge y = x$. Lemma 3 on page 394 shows that $\leq$ is a partial order. We now show that, with respect to the order $\leq$, the element $a \wedge b$ is the greatest lower bound of $\{a, b\}$; similarly, $a \vee b$ is lub$\{a, b\}$. We will use here just the algebraic properties of $\wedge$ and $\vee$.

First, $a \wedge b \leq a$ because $(a \wedge b) \wedge a = a \wedge b$, and similarly $a \wedge b \leq b$. Hence $a \wedge b$ is a lower bound for $\{a, b\}$. If $c \leq a$ and $c \leq b$ too, then $c \wedge a = c$ and $c \wedge b = c$. It follows that $c \wedge (a \wedge b) = (c \wedge a) \wedge b = c \wedge b = c$, and so $c \leq a \wedge b$. Thus $a \wedge b$ is the greatest lower bound of $\{a, b\}$. ■

Posets arise in a variety of ways, and in many cases the fact that there are pairs of elements that cannot be compared is an essential feature. Indeed, most of the posets that we have looked at in this section have had pairs of incomparable elements. Rooted trees, which can be thought of as Hasse diagrams of posets, are useful data structures, even though they have incomparable elements, because it is possible to start at the root and follow the ordering to get to any element fairly quickly.

In the next section we will return to those posets in which every element is related to every other element. We will also discuss how to use orders on relatively simple sets to produce orders for more complicated ones.

Exercises 11.1

1. Draw Hasse diagrams for the following posets.
   (a) $(\{1, 2, 3, 4, 6, 8, 12, 24\}, |)$ where $m|n$ means m divides n.
   (b) The set of subsets of $\{3, 7\}$ with $\subseteq$ as partial order.
2. (a) Give examples of two posets that come from everyday life or from other courses.
   (b) Do your examples have maximal or minimal elements? If so, what are they?
   (c) What are the converses of the partial orders in your examples?
3. Figure 10 shows the Hasse diagrams of three posets.
   (a) What are the maximal members of these posets?
   (b) Which of these posets have minimal elements?
   (c) Which of these posets have smallest members?
   (d) Which elements cover the element e?
   (e) Find each of the following if it exists.
   $$\text{lub}\{d, c\}, \quad \text{lub}\{w, y, v\}, \quad \text{lub}\{p, m\}, \quad \text{glb}\{a, g\}.$$
   (f) Which of these posets are lattices?
4. Find the maximal proper subsets of the 3-element set $\{a, b, c\}$. That is, find the maximal members of the subposet of $\mathcal{P}(\{a, b, c\})$ consisting of proper subsets of $\{a, b, c\}$.
5. Consider $\mathbb{R}$ with the usual order $\leq$.
   (a) Is $\mathbb{R}$ a lattice? If it is, what are the meanings of $a \vee b$ and $a \wedge b$ in $\mathbb{R}$?
   (b) Give an example of a nonempty subset of $\mathbb{R}$ that has no least upper bound.
   (c) Find $\text{lub}\{x \in \mathbb{R} : x < 73\}$.
   (d) Find $\text{lub}\{x \in \mathbb{R} : x \leq 73\}$.
   (e) Find $\text{lub}\{x \in \mathbb{R} : x^2 < 73\}$.
   (f) Find $\text{glb}\{x \in \mathbb{R} : x^2 < 73\}$.
6. Let S be a set of subroutines of a computer program. For A and B in S, write $A \prec B$ if A must be completed before B can be completed. What sort of restriction must be placed on subroutine calls in the program to make $\prec$ a quasi-order on S?
7. (a) Show that if $\preceq$ is a partial order on a set S, then so is its converse relation $\succeq$.
   (b) Show that if $\prec$ is a quasi-order on a set S, then the relation $\preceq$ defined by
   $$x \preceq y \quad \text{if and only if} \quad x \prec y \text{ or } x = y$$
   is a partial order on S.

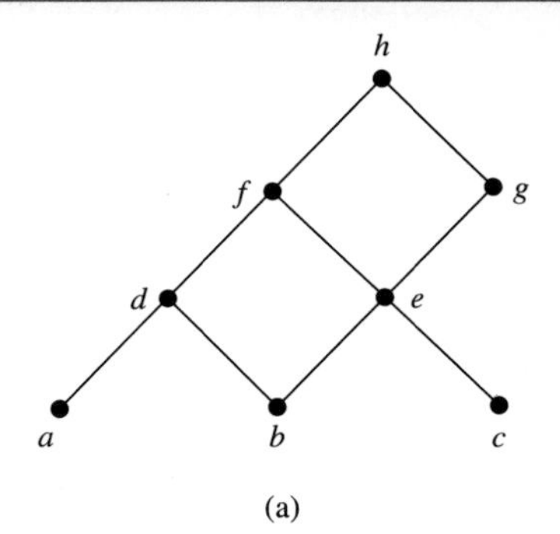

(a)

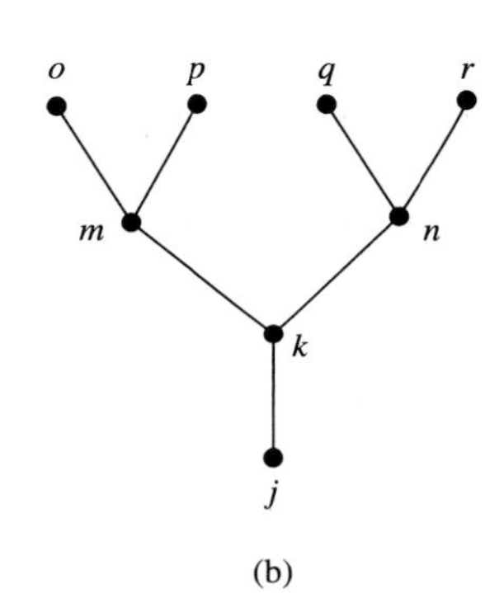

(b)

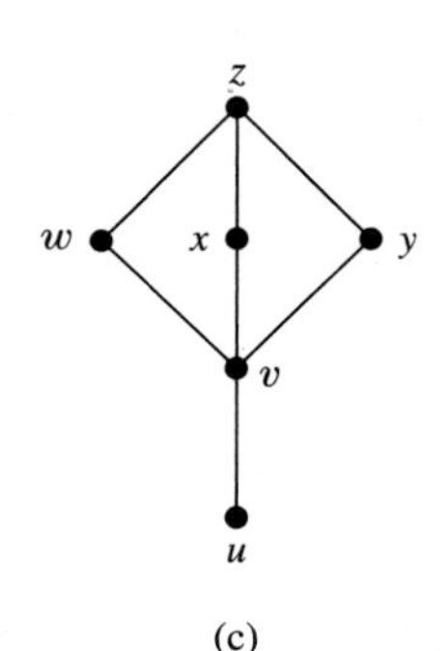

(c)

Figure 10 ▲

8. Let Σ be an alphabet. For $w_1, w_2 \in \Sigma^*$, define $w_1 \preceq w_2$ if there are w and w' in Σ^* with $w_2 = ww_1w'$. Is the relation $\preceq$ a partial order on Σ^*? Explain.
9. Let Σ be an alphabet. For $w_1, w_2 \in \Sigma^*$, let $w_1 \preceq w_2$ mean $\text{length}(w_1) \leq \text{length}(w_2)$. Is $\preceq$ a partial order on Σ^*? Explain.

10. Verify that the partial order $\preceq$ on Σ^* in Example 5(a) is reflexive and transitive.

11. The table in Figure 11 has been partially filled in. It gives the values of $x \vee y$ for x and y in a certain lattice $(L, \preceq)$. For example, $b \vee c = d$.

$\vee$	a	b	c	d	e	f
a		e	a	e	e	a
b			d	d	e	b
c				d	e	c
d					e	d
e						e
f						

Figure 11 ▲

(a) Fill in the rest of the table. *Hint:* Start with the diagonal and then use symmetry.
(b) Which are the largest and smallest elements of L?

12. Let $(L, \preceq)$ be the lattice in Exercise 11.
(a) Show that $f \preceq c \preceq d \preceq e$.
(b) Draw a Hasse diagram for L.

13. Let $\mathcal{F}(\mathbb{N})$ be the collection of all *finite* subsets of $\mathbb{N}$. Then $(\mathcal{F}(\mathbb{N}), \subseteq)$ is a poset.
(a) Does $\mathcal{F}(\mathbb{N})$ have a maximal element? If yes, give one. If no, explain.
(b) Does $\mathcal{F}(\mathbb{N})$ have a minimal element? If yes, give one. If no, explain.
(c) Given A, B in $\mathcal{F}(\mathbb{N})$, does $\{A, B\}$ have a least upper bound in $\mathcal{F}(\mathbb{N})$? If yes, specify it. If no, provide a specific counterexample.
(d) Given A, B in $\mathcal{F}(\mathbb{N})$, does $\{A, B\}$ have a greatest lower bound in $\mathcal{F}(\mathbb{N})$? If yes, specify it. If no, provide a specific counterexample.
(e) Is $\mathcal{F}(\mathbb{N})$ a lattice? Explain.

14. Repeat Exercise 13 for the collection $\mathcal{I}(\mathbb{N})$ of all *infinite* subsets of $\mathbb{N}$.

15. Define the relations $<$, $\leq$, and $\preceq$ on the plane $\mathbb{R} \times \mathbb{R}$ by

$$\begin{aligned} (x, y) < (z, w) &\quad \text{if } x^2 + y^2 < z^2 + w^2, \\ (x, y) \leq (z, w) &\quad \text{if } (x, y) < (z, w) \text{ or } (x, y) = (z, w), \\ (x, y) \preceq (z, w) &\quad \text{if } x^2 + y^2 \leq z^2 + w^2. \end{aligned}$$

(a) Which of these relations are partial orders? Explain.
(b) Which are quasi-orders? Explain.
(c) Draw a sketch of $\{(x, y) : (x, y) \leq (3, 4)\}$.
(d) Draw a sketch of $\{(x, y) : (x, y) \preceq (3, 4)\}$.

16. Let $\mathcal{E}(\mathbb{N})$ be the set of all finite subsets of $\mathbb{N}$ that have an even number of elements, with partial order $\subseteq$.
(a) Let $A = \{1, 2\}$ and $B = \{1, 3\}$. Find four upper bounds for $\{A, B\}$.
(b) Does $\{A, B\}$ have a least upper bound in $\mathcal{E}(\mathbb{N})$? Explain.
(c) Is $\mathcal{E}(\mathbb{N})$ a lattice?

17. Is every subposet of a lattice a lattice? Explain.

18. (a) Show that every nonempty finite poset has a minimal element. *Hint:* Use induction.
(b) Give an example of a poset with a maximal element but with no minimal element.

19. (a) Consider elements x, y, z in a poset. Show that if $\text{lub}\{x, y\} = a$ and $\text{lub}\{a, z\} = b$, then $\text{lub}\{x, y, z\} = b$.
(b) Show that every finite subset of a lattice has a least upper bound.
(c) Show that if x, y, and z are members of a lattice, then $(x \vee y) \vee z = x \vee (y \vee z)$.

20. Consider the poset C whose Hasse diagram is shown in Figure 10. Show that $w \vee (x \wedge y) \neq (w \vee x) \wedge (w \vee y)$ and $w \wedge (x \vee y) \neq (w \wedge x) \vee (w \wedge y)$. This example shows that lattices need not satisfy "distributive" laws for $\vee$ and $\wedge$.

11.2 Special Orderings

This section is about sets with partial orderings in which every two elements are comparable, that is, the ones that satisfy

(L) for each choice of s and t in S either $s \preceq t$ or $t \preceq s$ [or both].

A set with such a linear, or total, ordering is called, suggestively, a **chain**.

EXAMPLE 1

(a) The poset of Figure 1(b) is a chain, but the other posets in Figure 1 are not.
(b) The set $\mathbb{R}$ with the usual order $\leq$ is a chain.
(c) The lists of names in a phone book or words in a dictionary are chains if we define $w_1 \preceq w_2$ to mean that $w_1 = w_2$ or that w_1 comes before w_2. ■

Every subposet of a chain is itself a chain. For example, the posets $(\mathbb{Z}, \leq)$ and $(\mathbb{Q}, \leq)$ are subposets of $(\mathbb{R}, \leq)$ and are linearly ordered by the orders they inherit from $\mathbb{R}$. The words in the dictionary between "start" and "stop" form a subchain of the chain of all words in Example 1(c).

Every poset, whether or not it is itself a chain, will have subposets that are chains. It is often useful to know something about such subposets.

Figure 1 ▶

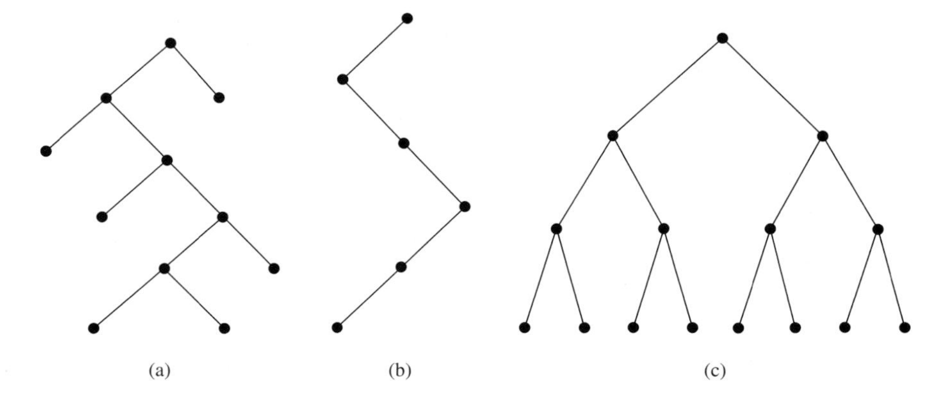

EXAMPLE 2

(a) Let S be the set of all people at some family reunion, and write $m \prec n$ in case m is a descendant of n. Then $\prec$ is a quasi-order that gives a partial order $\preceq$ by defining $m \preceq n$ if and only if $m \prec n$ or $m = n$. A chain in the poset $(S, \preceq)$ is a set of the form $\{m, n, p, \dots, r\}$ in which m is a descendant of n, n is a descendant of p, and so on. It would be unusual for such a chain to have more than five members, though the set S itself might be quite large.

(b) The Hasse diagram shown in Figure 2 describes a poset with a number of subchains [49 if we count the 1-element chains, but not the empty chain]. ■

Figure 2 ▶

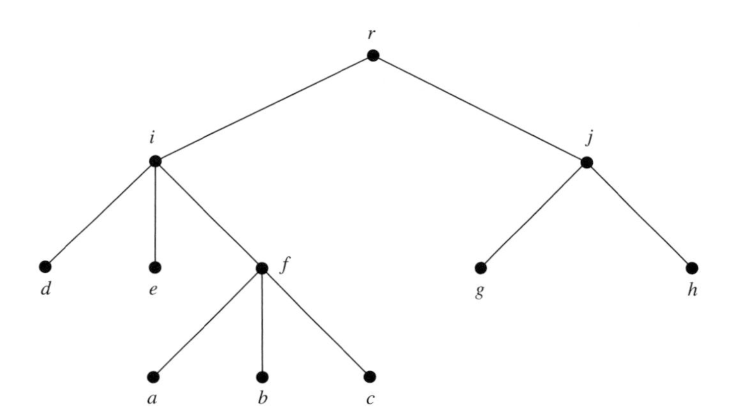

A finite chain must have a smallest member, and so must each of its nonempty subchains. Infinite chains, on the other hand, can exhibit a variety of behaviors. The infinite chains $(\mathbb{R}, \leq)$ and $(\mathbb{Z}, \leq)$ with their usual orders do not have smallest members. The chain $(\{x \in \mathbb{R} : 0 \leq x\}, \leq)$ has a smallest member, 0, but has subsets such as $\{x \in \mathbb{R} : 1 < x\}$ without smallest members. The infinite chain $(\mathbb{N}, \leq)$ has a smallest member, and every nonempty subset of $\mathbb{N}$ does too, by the well-ordering property of $\mathbb{N}$ stated on page 131. We used this property when we observed that every decreasing chain in $\mathbb{N}$ is finite, and hence that the Division Algorithm terminates.

We say that a chain C is **well-ordered** in case each nonempty subset of C has a smallest member. If C is well-ordered and if, for each c in C, we have a statement $p(c)$, then we can hope to prove that all the statements $p(c)$ are true by supposing that $\{c \in C : p(c) \text{ is false}\}$ is a nonempty subset of C, considering the smallest c for which $p(c)$ is false and deriving a contradiction. This was the idea behind our explanation of the principles of induction in §§4.2 and 4.6.

We devote most of this section to studying how to build new partial orders from known ones, with a special emphasis on chains. Suppose first that $(S, \preceq)$ is a given poset and that T is a nonempty set. We can define a partial order, which we also denote by $\preceq$, on the set of functions from T to S by defining

$$f \preceq g \quad \text{if} \quad f(t) \preceq g(t) \text{ for all } t \text{ in } T.$$

This is the partial order we used in Example 2(d) on page 425 and Example 9(c) on page 431. The verification that this new relation is a partial order on FUN(T, S) is straightforward [Exercise 13(a)]. If the order on S is $\leq$, we will write $\leq$ in place of $\preceq$ for the order relation on the set of functions.

EXAMPLE 3

(a) If $S = T = \mathbb{R}$ with the usual order, then $f \leq g$ means that the graph of f lies on or below the graph of g, as in Figure 3. Note that the graphs touch at one point; i.e., $f(x) = g(x)$ for one value of x in R.

Figure 3 ▶

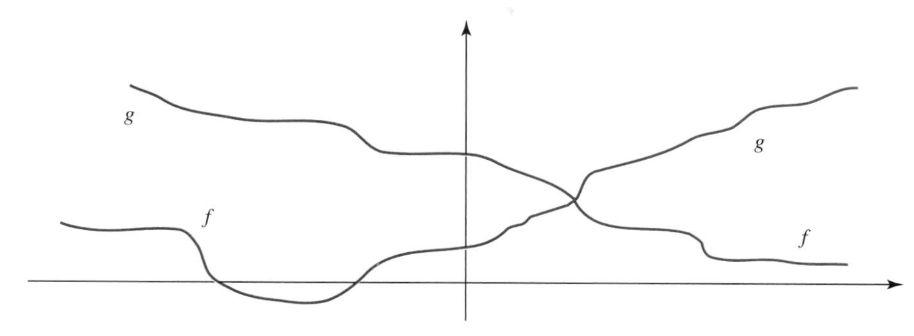

(b) Consider $S = \{0, 1\}$ with $0 < 1$. The functions in FUN$(T, \{0, 1\})$ are the characteristic functions of subsets of T. Each subset A of T has a corresponding function χ_A in FUN$(T, \{0, 1\})$, with $\chi_A(x) = 1$ if $x \in A$ and 0 if $x \notin A$. Then $\chi_A \leq \chi_B$ if and only if $x \in B$ whenever $x \in A$, i.e., if and only if $A \subseteq B$. Thus Hasse diagrams for (FUN$(T, \{0, 1\}), \leq)$ and $(\mathcal{P}(T), \subseteq)$ look alike. Figure 2 on page 427 and Figure 9 on page 431 show the diagrams for $T = \{a, b, c\}$. These two posets are in fact isomorphic Boolean algebras. ■

Example 3(b) shows that (FUN$(T, S), \preceq)$ need not be a chain even if S is a chain. The poset (FUN$(T, S), \preceq)$ does inherit some properties from S, however. If S has largest or smallest elements, then so does FUN(T, S); and if S is a lattice, so is FUN(T, S). For more in this vein see Exercise 13.

Another way to combine two sets into a new one is to form their product. Suppose that $(S, \preceq_1)$ and $(T, \preceq_2)$ are posets, where we use the subscripts 1 and 2 to keep track of which partial order is which. There is more than one natural way to make $S \times T$ into a poset. Our preference will depend on the problem at hand.

The first partial order we will describe for $S \times T$ is called the **product order**. For $s, s' \in S$ and $t, t' \in T$ define

$$(s, t) \preceq (s', t') \quad \text{if} \quad s \preceq_1 s' \text{ and } t \preceq_2 t'.$$

EXAMPLE 4

Let $S = T = \mathbb{N}$ with the usual order $\leq$ in each case. Then $(2, 5) \preceq (3, 7)$, since $2 \leq 3$ and $5 \leq 7$. Also, $(2, 5) \preceq (3, 5)$, since $2 \leq 3$ and $5 \leq 5$. But the pairs $(2, 7)$ and $(3, 5)$ are not comparable, since $(2, 7) \preceq (3, 5)$ would mean $2 \leq 3$ and $7 \leq 5$, while $(3, 5) \preceq (2, 7)$ would mean $3 \leq 2$ and $5 \leq 7$. Figure 4 indicates the pairs (m, n) in $S \times T = \mathbb{N} \times \mathbb{N}$ with $(2, 1) \preceq (m, n) \preceq (3, 4)$. ■

	⋮	⋮	⋮	⋮	⋮	⋮	⋮	
5	*	*	*	*	*	*	*	⋯
4	*	*	●	●	*	*	*	⋯
3	*	*	●	●	*	*	*	⋯
2	*	*	●	●	*	*	*	⋯
1	*	*	●	●	*	*	*	⋯
0	*	*	*	*	*	*	*	⋯
	0	1	2	3	4	5	6	

Figure 4 ▲

Consider again two posets $(S, \preceq_1)$ and $(T, \preceq_2)$. The fact that the product order $\preceq$ is a partial order on $S \times T$ is almost immediate from the definition. For example, if $(s, t) \preceq (s', t')$ and $(s', t') \preceq (s, t)$, then $s \preceq_1 s'$, $t \preceq_2 t'$, $s' \preceq_1 s$, and $t' \preceq_2 t$. Since $\preceq_1$ and $\preceq_2$ are antisymmetric, $s = s'$ and $t = t'$. So $(s, t) = (s', t')$. Thus $\preceq$ is antisymmetric.

There is no difficulty in extending this idea to define a partial order on the product $S_1 \times S_2 \times \cdots \times S_n$ of a finite number of posets. We define

$$(s_1, s_2, \ldots, s_n) \preceq (s'_1, s'_2, \ldots, s'_n) \quad \text{if} \quad s_i \preceq s'_i \text{ for all } i.$$

For many purposes, in particular, for algorithms that run through sets of input elements, it is desirable to be able to arrange the elements of a set as an ordered list, i.e., as a chain. In Example 4, $\mathbb{N}$ is a chain, but $\mathbb{N} \times \mathbb{N}$ is not; for instance, $(2, 4)$ and $(3, 1)$ are not related in the product order. In fact, the product order is almost never a total order [Exercise 4]. On the other hand, if $S_1, S_2, \ldots, S_n$ are chains, then there is another natural order on $S_1 \times S_2 \times \cdots \times S_n$ that makes it a chain, which we now illustrate.

EXAMPLE 5

(a) If S is the set $\{0, 1, 2, \ldots, 9\}$ with the usual order, then the set $S \times S$ consists of the pairs (m, n) of digits, which we can identify with the integers $00, 01, 02, \ldots, 98, 99$ from 0 to 99. To make $S \times S$ a chain, we can simply define $(m, n) \prec (m', n')$ for pairs if their corresponding integers are so related, i.e., if $m < m'$ or if $m = m'$ and $n < n'$. For example, this order makes $(5, 7) \prec (6, 3)$ since $57 < 63$, and $(3, 5) \prec (3, 7)$ since $35 < 37$.

In a similar way we can identify the set $S \times S \times S$ with the set of integers from 0 to 999 and make $S \times S \times S$ into a chain by defining $(m, n, p) \prec (m', n', p')$ if $m < m'$ or if $m = m'$ and $n < n'$ or if $m = m'$, $n = n'$, and $p < p'$.

(b) We can do this with letters, too. For instance, let Σ be the English alphabet in the usual order. If we identify k-tuples $(a_1, \ldots, a_k)$ in the product Σ^k with words $a_1 \cdots a_k$ of length k in Σ^*, then Σ^k is the set of strings of letters of length k, which we can order with the usual alphabetic, or dictionary, order. Thus, if $k = 3$, we have

$$fed \preceq few \preceq one \preceq six \preceq ten \preceq two \preceq won.$$

(c) In fact, the dictionary order works to compare words of differing lengths. To find words in a dictionary with the ordinary alphabetical order, one scans words from left to right looking for differences and ignoring the lengths of the words. Thus "aardvark" is listed before "axe" and "break" precedes "breakfast." We will examine this order in more detail shortly. ■

EXAMPLE 6

Imagine a device that has subassemblies, labeled by letters, each of which has at most 10 parts. It would be reasonable to label spare parts with letter–number combinations and to file them in bins arranged according to the filing order on $S \times T$, where $S = \{a, b, c, \ldots, z\}$ and $T = \{0, 1, \ldots, 9\}$ with the usual orders. We can identify the members of $S \times T$ with 2-symbol strings such as $a5$ and $x3$. Then bin $a5$ would come before bin $x3$ because a precedes x, but $a5$ would come after $a3$ since $3 < 5$. Note that all the parts for subassembly a are in the first bins $a0, a1, a2, \ldots, a9$. ■

The idea of this example works in general. If $(S_1 \preceq_1), \ldots, (S_n, \preceq_n)$ are posets, then we can define a relation $\prec$ on $S_1 \times \cdots \times S_n$ by

> $(s_1, s_2, \ldots, s_n) \prec (t_1, t_2, \ldots, t_n)$ if $s_1 \prec_1 t_1$ or if there is an r in $\{2, \ldots, n\}$ such that $s_1 = t_1, \ldots, s_{r-1} = t_{r-1}$ and $s_r \prec_r t_r$.

Then $\prec$ is a quasi-order [Exercise 19] that induces a partial order $\preceq$ on the product set $S_1 \times S_2 \times \cdots \times S_n$, which we call the **filing order**.

The filing order is primarily useful if each S_i is a chain.

Theorem 1 Let $(S_1, \preceq_1), \ldots, (S_n, \preceq_n)$ be chains. Then the filing order makes the product $S_1 \times \cdots \times S_n$ a chain.

Proof We already know that $\preceq$ is a partial order on $S_1 \times \cdots \times S_n$. Let $s = (s_1, \ldots, s_n)$ and $t = (t_1, \ldots, t_n)$ be distinct elements in $S_1 \times \cdots \times S_n$. Since $s \neq t$, there is a first value of r for which $s_r \neq t_r$. Since $(S_r, \preceq_r)$ is a chain, either $s_r \prec_r t_r$ or $t_r \prec_r s_r$. In the first case $s \prec t$; in the second, $t \prec s$. In either case, the two elements of $S_1 \times \cdots \times S_n$ are comparable. ■

The orders that we saw in Examples 5(a) and 5(b) can be viewed as special cases of the filing order in which all the posets $(S_i, \preceq_i)$ are the same. We now examine such orders in more detail, starting with a finite alphabet Σ equipped with an order $\preceq$ that makes it a chain. As in Example 5(b), we identify k-tuples with words of length k in Σ^*, i.e., words in Σ^k. The filing order gives us a natural order $\preceq^k$ on Σ^k that makes it a chain, by Theorem 1. Thus

$a_1 \cdots a_k \preceq^k b_1 \cdots b_k$ if the two words are the same or if $a_r \prec b_r$ for the first r at which a_r and b_r differ.

This order gives us a fine way to compare words of the same length. We can extend the idea to the whole set Σ^* of all words in letters from Σ by defining the **lexicographic** or **dictionary order** $\preceq_L$ on Σ^* as follows. Let $a_1 \cdots a_m$ and $b_1 \cdots b_n$ be in Σ^* with, say, $m \leq n$. Then

$$a_1 \cdots a_m \preceq_L b_1 \cdots b_n \quad \text{in case} \quad a_1 \cdots a_m \preceq^m b_1 \cdots b_m,$$

i.e., in case either $a_1 \cdots a_m$ is an initial segment of $b_1 \cdots b_n$ or $a_r \prec b_r$ for the first r for which $a_r \neq b_r$; similarly,

$$b_1 \cdots b_n \prec_L a_1 \cdots a_m \quad \text{in case} \quad b_1 \cdots b_m \prec^m a_1 \cdots a_m$$

EXAMPLE 7

For the words in Example 5(c) we have

aardvark	$\prec_L$ *axe*	[since $a \prec x$, so *aar* $\prec^3$ *axe*]
	$\prec_L$ *break*	[since $a \prec b$, so *axe* $\prec^3$ *bre*]
	$\prec_L$ *breakfast*	[since *break* is an initial segment]. ■

We can describe $\preceq_L$ in another way. For words w and z in Σ^*, we have $w \preceq_L z$ if and only if either

(a) $w = xu$ and $z = xv$ for words x, u, and v in Σ^* such that the first letter of u precedes the first letter of v in the ordering of Σ, or

(b) w is an initial segment of z, i.e., $z = wu$ for some $u \in \Sigma^*$.

Note that x in (a) can be any word, possibly the empty word λ.

Since every two words in Σ^* are comparable under $\preceq_L$, lexicographic order makes Σ^* into a chain, but a chain with some bizarre properties, as the next example shows.

EXAMPLE 8

Let $\Sigma = \{a, b\}$ with $a \prec b$. The first few terms of Σ^* in the lexicographic order are

$$\lambda, a, aa, aaa, aaaa, aaaaa, \ldots .$$

Any word using the letter b is preceded by an *infinite* number of words, including all the words using only the letter a. Moreover, Σ^* contains infinite decreasing sequences of words; for example,

$$b \succ_L ab \succ_L aab \succ_L aaab \succ_L \cdots .$$

The chain $(\Sigma^*, \preceq_L)$ is *not* well-ordered, since the set $\{b, ab, aab, aaab, aaaab, \ldots\}$ here has no smallest member. We see also that there are infinitely many words between any two members of this sequence; for example, the words

$$aaab, aaaba, aaabaa, aaabaaa, aaabaaaa, \ldots$$

all follow aaa and precede aab. Thus the lexicographic order on the infinite set Σ^* is very complicated and is difficult to visualize. ■

If we apply lexicographic order to some chosen finite subset of Σ^*, such as the set of all words in a particular dictionary, then the infinite problems in Example 8 disappear. But suppose that we are designing a search algorithm and want to list the elements of Σ^* in such a way that every element eventually shows up. Then lexicographic order won't work. What people often use instead is the **lenlex** order, $\preceq_{LL}$, in which short words come before longer ones, and words of the same length are ordered lexicographically. ["Lenlex" comes from "length" plus "lexicographic."] More precisely,

$$w_1 \preceq_{LL} w_2 \quad \text{if} \quad \begin{cases} w_1 \in \Sigma^k \text{ and } w_2 \in \Sigma^r \text{ with } k < r \text{ or} \\ w_1 \in \Sigma^k \text{ and } w_2 \in \Sigma^k \text{ for the same } k \text{ and } w_1 \preceq^k w_2 \end{cases}.$$

EXAMPLE 9

(a) If Σ is the English alphabet in the usual order, then the first few terms of Σ^* in the lenlex order are

$$\lambda, a, b, \ldots, z, aa, ab, \ldots, az, ba, bb, \ldots, bz, ca, cb, \ldots, cz, \\ da, db, \ldots, dz, \ldots, za, zb, \ldots, zz, aaa, aab, aac, \ldots.$$

(b) Let $\Sigma = \{0, 1\}$ with $0 < 1$. The first few terms of Σ^* in the lenlex order are

$$\lambda, 0, 1, 00, 01, 10, 11, 000, 001, 010, 011, \\ 100, 101, 110, 111, 0000, 0001, 0010, \ldots.$$

■

Note that if a dictionary were constructed using the lenlex order in Example 9(a), then all the short words would be at the beginning of the dictionary, and to find a word it would be essential to know its exact length. In fact, some dictionaries designed for crossword-puzzle solvers are arranged this way.

The lenlex order has the property we want.

Theorem 2 If $(\Sigma, \preceq)$ is a finite chain, then $(\Sigma^*, \preceq_{LL})$ is a well-ordered chain.

Proof We know from Theorem 1 that each $(\Sigma^k, \preceq^k)$ is a chain. The lenlex order $\preceq_{LL}$ simply links these chains end to end for $k = 0, 1, 2, \ldots$, so $(\Sigma^*, \preceq_{LL})$ is a chain. To check that $\preceq_{LL}$ well-orders Σ^*, consider a nonempty subset A of Σ^*. Let k be the shortest length of a word in A. Since $A \cap \Sigma^k$ is nonempty and finite, $A \cap \Sigma^k$ possesses a smallest element w_0 in $(\Sigma^k, \preceq^k)$. Then $w_0 \preceq_{LL} w$ for all w in A, so w_0 is the smallest member of A. ■

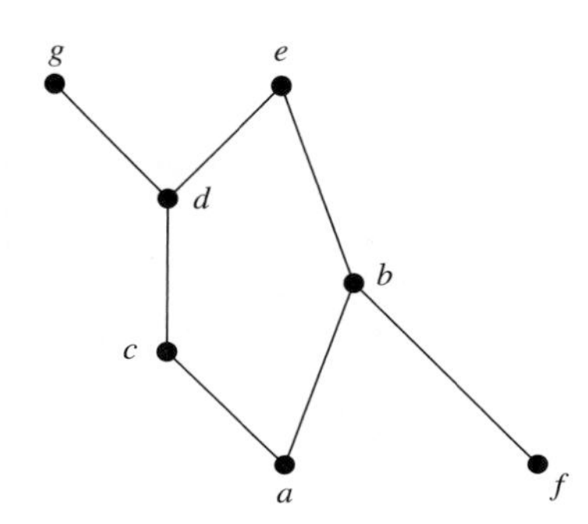

Figure 5 ▲

We are sometimes interested in chains in posets that cannot be made any larger. A **maximal chain** in a poset $(S, \preceq)$ is defined to be a chain that is not properly contained in another chain. Observe that if $\mathcal{C}(S)$ is the set of chains in S, then $(\mathcal{C}(S), \subseteq)$ is also a poset, and a maximal chain in S is simply a maximal member of $\mathcal{C}(S)$.

EXAMPLE 10

(a) In the poset of Figure 2 the maximal chains are $\{a, f, i, r\}$, $\{b, f, i, r\}$, $\{c, f, i, r\}$, $\{d, i, r\}$, $\{e, i, r\}$, $\{g, j, r\}$, and $\{h, j, r\}$. Notice that the maximal chains are not all of the same size.

(b) In the poset shown in Figure 5 the two maximal chains $\{a, c, d, e\}$ and $\{a, b, e\}$ containing a and e have different numbers of elements. This poset has four maximal chains in all.

(c) Paths from the root to the leaves in a rooted tree correspond to maximal chains with respect to the usual partial ordering of the tree. ■

Exercises 11.2

1. Let $A = \{1, 2, 3, 4\}$ with the usual order, and let $S = A \times A$ with the product order.
   (a) Find a chain in S with seven members.
   (b) Can a chain in S have eight members? Explain.
2. Let $\mathbb{N} \times \mathbb{N}$ have the product order. Draw a sketch like the one in Figure 4 that shows the set $\{(m, n) : (m, n) \preceq (5, 2)\}$.
3. Let $S = \{0, 1, 2\}$ and $T = \{3, 4\}$ with both sets given the usual order. List the members of the following sets in increasing filing order.
   (a) $S \times S$ (b) $S \times T$ (c) $T \times S$
4. Suppose that $(S, \preceq_1)$ and $(T, \preceq_2)$ are posets, each with more than one element. Show that $S \times T$ with the product order is not a chain.
5. Let $\mathbb{B} = \{0, 1\}$ with the usual order. List the elements 101, 010, 11, 000, 10, 0010, 1000 of $\mathbb{B}^*$ in increasing order
   (a) for the lexicographic order.
   (b) for the lenlex order.

6. Let Σ be the English alphabet with the usual order.
   (a) List the words of this sentence in increasing lenlex order.
   (b) List the words of this sentence in increasing lexicographic order.
7. Let $(\Sigma, \preceq)$ be a nonempty chain.
   (a) Does $(\Sigma^*, \preceq_{LL})$ have a maximal member? Explain.
   (b) Does $(\Sigma^*, \preceq_L)$ have a maximal member? Explain.
8. Under what conditions on Σ are the lexicographic order and lenlex order on Σ^* the same?
9. Let $S = \{0, 1, 2\}$ with the usual order, and let $T = \{a, b\}$ with $a < b$.
   (a) Draw a Hasse diagram for the poset $(\text{FUN}(T, S), \preceq)$ with the order $\preceq$ described just prior to Example 3. *Hint:* See Example 9(c) on page 431.
   (b) Draw a Hasse diagram for the poset $(S \times S, \preceq)$ with the product order.
   (c) Draw a Hasse diagram for $S \times T$ with the product order.
10. Suppose that $(S, \preceq_1)$ and $(T, \preceq_2)$ are posets and we define $\preceq$ on $S \times T$ by

$$(s, t) \preceq (s', t') \quad \text{if} \quad s \preceq_1 s' \quad \text{or} \quad t \preceq_2 t'.$$

    Note the word "or." Is $\preceq$ a partial order? Explain.
11. Is every chain a lattice? Explain.
12. Let $(C_1, \preceq_1), (C_2, \preceq_2), \ldots, (C_n, \preceq_n)$ be a set of disjoint chains. Describe a way to make $C_1 \cup C_2 \cup \cdots \cup C_n$ into a chain.
13. Let $(S, \preceq)$ be a poset and T a set. Define the relation $\preceq$ on $\text{FUN}(T, S)$ by

$$f \preceq g \quad \text{if} \quad f(t) \preceq g(t) \text{ for all } t \text{ in } T.$$

    (a) Show that $\preceq$ is a partial order.
    (b) Show that if m is a maximal element in S, then the function f_m defined by $f_m(t) = m$ for all $t \in T$ is a maximal element of $\text{FUN}(T, S)$.
    (c) Show that if S is a lattice and if $f, g \in \text{FUN}(T, S)$, then the function h defined by $h(t) = f(t) \vee g(t)$ for $t \in T$ is the least upper bound of $\{f, g\}$.
    (d) Describe the quasi-order $\prec$ associated with $\preceq$.
14. Let P be the set of all subsets of $\{1, 2, 3, 4, 5\}$.
    (a) Give two examples of maximal chains in $(P, \subseteq)$.
    (b) How many maximal chains are there in $(P, \subseteq)$?
15. Consider the partially ordered set $(S, |)$, where $S = \{2, 3, 4, \ldots, 999, 1000\}$.
    (a) There are exactly 500 maximal elements of $(S, |)$. What are they?
    (b) Give two examples of maximal chains in $(S, |)$.
    (c) Does every maximal chain contain a minimal element of S? Explain.
16. (a) Suppose that no chain in the poset $(S, \preceq)$ has more than 73 members. Must a chain in S with 73 members be a maximal chain? Explain.
    (b) Give an example of a poset that has two maximal chains with four members and four maximal chains with two members.
17. Show that in a finite poset $(S, \preceq)$ every maximal chain contains a minimal element of S.
18. Let $(S, \preceq_1)$ and $(T, \preceq_2)$ be posets, and give $S \times T$ the filing order.
    (a) Show that if m_1 is maximal in S and m_2 is maximal in T, then (m_1, m_2) is maximal in $S \times T$.
    (b) Does $S \times T$ have other maximal elements besides the ones described in part (a)? Explain.
    (c) Suppose $S \times T$ has a largest element. Must S or T have a largest element? Explain.
19. Let $(S_1, \preceq_1), \ldots, (S_n, \preceq_n)$ be posets and define $\prec$ on $S_1 \times \cdots \times S_n$ by

    $(s_1, \ldots, s_n) \prec (t_1, \ldots, t_n)$ if $s_1 \prec_1 t_1$ or if there is an r in $\{2, \ldots, n\}$ such that $s_1 = t_1, \ldots, s_{r-1} = t_{r-1}$ and $s_r \prec_r t_r$.

    Show that $\prec$ is a quasi-order.

11.3 Multiplication of Matrices

As we saw in §3.3, addition and scalar multiplication of matrices are straightforward, but the definition of the product of two matrices is more elaborate. Here it is again, for reference.

If $\mathbf{A}$ is an $m \times n$ matrix and $\mathbf{B}$ is an $n \times p$ matrix, then the **product AB** is the $m \times p$ matrix $\mathbf{C}$ defined by

$$c_{ik} = \sum_{j=1}^{n} a_{ij} b_{jk} \quad \text{for} \quad 1 \le i \le m \quad \text{and} \quad 1 \le k \le p.$$

Schematically, the (i, k)-entry of $\mathbf{AB}$ is obtained by multiplying terms of the ith row of $\mathbf{A}$ by corresponding terms of the kth column of $\mathbf{B}$ and summing. See Figure 1. One can calculate c_{ik} by mentally lifting the ith row of $\mathbf{A}$, rotating it clockwise by 90°, placing it on top of the kth column of $\mathbf{B}$, and then summing the products of the corresponding terms:

$$c_{ik} = a_{i1}b_{1k} + a_{i2}b_{2k} + \cdots + a_{in}b_{nk}.$$

Alternatively, train one eye to scan across rows of **A** while the other eye runs down columns of **B**. For this calculation to make sense, the rows of **A** must have the same number of entries as the columns of **B**. If **A** is $m \times n$ and **B** is $r \times p$, then the matrix product **AB** is only defined if $n = r$, in which case **AB** is an $m \times p$ matrix.

Figure 1 ▶

$$\underset{\mathbf{A}}{\begin{bmatrix} a_{11} & a_{12} & \cdots & a_{1n} \\ & \vdots & & \\ a_{i1} & a_{i2} & \cdots & a_{in} \\ & \vdots & & \\ a_{m1} & a_{m2} & \cdots & a_{mn} \end{bmatrix}} \underset{\mathbf{B}}{\begin{bmatrix} b_{11} & b_{12} & \cdots & b_{1k} & \cdots & b_{1p} \\ b_{21} & b_{22} & \cdots & b_{2k} & \cdots & b_{2p} \\ \vdots & \vdots & & \vdots & & \vdots \\ b_{n1} & b_{n2} & \cdots & b_{nk} & \cdots & b_{np} \end{bmatrix}} = \underset{\mathbf{AB} = \mathbf{C}}{\begin{bmatrix} c_{11} & c_{12} & \cdots & c_{1p} \\ c_{21} & c_{22} & \cdots & c_{2p} \\ \vdots & \vdots & & \vdots \\ & & c_{ik} & \\ c_{m1} & c_{m2} & \cdots & c_{mp} \end{bmatrix}}$$

The standard linear algebra explanation for choosing this definition is that matrices correspond to certain functions, called linear transformations, and then multiplication of matrices corresponds to composition of the linear transformations. A motivation can also be given in terms of systems of linear equations. A treatment along either of these lines would take us too far into linear algebra, so we will draw our motivation instead from graphs and digraphs, where we have applications.

EXAMPLE 1

Consider the digraph in Figure 2(a). Its adjacency matrix is

$$\mathbf{M} = \begin{bmatrix} 1 & 2 & 1 & 1 \\ 0 & 2 & 0 & 0 \\ 1 & 0 & 0 & 1 \\ 0 & 1 & 0 & 0 \end{bmatrix}.$$

Observe that the (i, j)-entry of **M** gives the number of paths of length 1 from v_i to v_j. Let's count the number of paths of length 2 in the digraph. The edges are labeled in Figure 2(b). By inspection we see that the paths of length 2 from v_1 to v_2 are $a\,b$, $a\,c$, $b\,d$, $b\,e$, $c\,d$, $c\,e$, and $h\,j$. So there are seven such paths. A similar inspection could be applied to find the number of paths of length 2 from any v_i to any v_j.

Figure 2 ▶

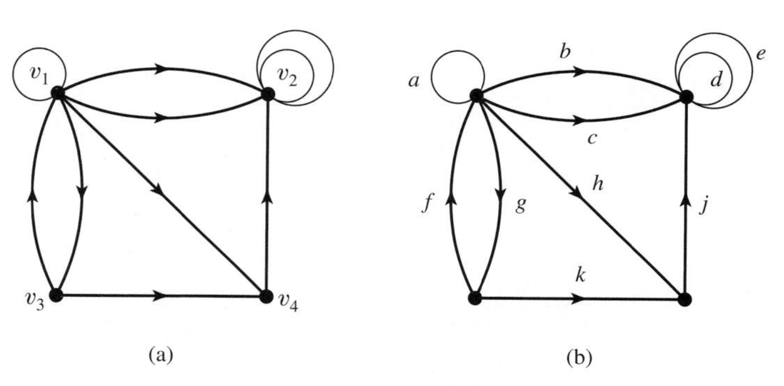

Is there a better method than inspection for counting paths, especially when we deal with larger digraphs? Let's count again the paths of length 2 from v_1 to v_2. Such a path passes through v_1, v_2, v_3, or v_4 on the way, so we can count the number of paths from v_1 to v_2 through v_1, the number through v_2, through v_3, and through v_4 and add these numbers to get the total. Now, to get the number of paths with vertex sequence $v_1\,v_1\,v_2$, for example, we count the edges from v_1 to v_1 [i.e., loops] and the edges from v_1 to v_2 and then multiply these numbers to get $1 \cdot 2 = 2$. The corresponding paths are $a\,b$ and $a\,c$. The numbers we multiply, $\mathbf{M}[1, 1]$ and $\mathbf{M}[1, 2]$, are given in the matrix **M**. As one more illustration, to get the number of paths with vertex sequence $v_1\,v_2\,v_2$, we count the edges from v_1 to v_2 and from v_2 to v_2 and multiply to get $\mathbf{M}[1, 2] \cdot \mathbf{M}[2, 2] = 2 \cdot 2 = 4$. The corresponding paths are $b\,d$, $b\,e$, $c\,d$, and $c\,e$. The general situation for paths from v_1 to v_2 of length 2 is illustrated in Table 1. The total number of such paths is thus

$$\mathbf{M}[1, 1] \cdot \mathbf{M}[1, 2] + \mathbf{M}[1, 2] \cdot \mathbf{M}[2, 2] + \mathbf{M}[1, 3] \cdot \mathbf{M}[3, 2] + \mathbf{M}[1, 4] \cdot \mathbf{M}[4, 2]$$
$$= 2 + 4 + 0 + 1 = 7.$$

TABLE 1

Vertex v_i	Number of edges from v_1 to v_i	Number of edges from v_i to v_2	Number of paths with vertex sequence $v_1\ v_i\ v_2$
v_1	$\mathbf{M}[1,1] = 1$	$\mathbf{M}[1,2] = 2$	$\mathbf{M}[1,1] \cdot \mathbf{M}[1,2] = 1 \cdot 2 = 2$
v_2	$\mathbf{M}[1,2] = 2$	$\mathbf{M}[2,2] = 2$	$\mathbf{M}[1,2] \cdot \mathbf{M}[2,2] = 2 \cdot 2 = 4$
v_3	$\mathbf{M}[1,3] = 1$	$\mathbf{M}[3,2] = 0$	$\mathbf{M}[1,3] \cdot \mathbf{M}[3,2] = 1 \cdot 0 = 0$
v_4	$\mathbf{M}[1,4] = 1$	$\mathbf{M}[4,2] = 1$	$\mathbf{M}[1,4] \cdot \mathbf{M}[4,2] = 1 \cdot 1 = 1$

This number turns out to be the entry in the first row and second column of the product matrix $\mathbf{MM} = \mathbf{M}^2$. It is the sum of products of entries from the first row of $\mathbf{M}$ and the second column of $\mathbf{M}$.

In general, the (i, j)-entry of the matrix $\mathbf{M}^2$ is the number of paths of length 2 from v_i to v_j. It turns out for our example that

$$\mathbf{M}^2 = \begin{bmatrix} 2 & 7 & 1 & 2 \\ 0 & 4 & 0 & 0 \\ 1 & 3 & 1 & 1 \\ 0 & 2 & 0 & 0 \end{bmatrix}.$$

The entries in the product $\mathbf{M}^2\mathbf{M} = \mathbf{M}^3$ tell us, similarly, the number of paths of length 3 connecting vertices, etc. These matrices are easy to compute [see Exercises 11 and 13] using matrix multiplication. ■

EXAMPLE 2 Let's look now at some matrices not associated with digraphs. Consider the matrices

$$\mathbf{A} = \begin{bmatrix} 3 & -1 \\ -2 & 4 \end{bmatrix} \quad \text{and} \quad \mathbf{B} = \begin{bmatrix} -1 & 0 & 3 \\ 2 & 1 & -5 \end{bmatrix}.$$

(a) To calculate $\mathbf{AB}$, we begin by mentally placing the first row of $\mathbf{A}$ over the first, second, and third columns of $\mathbf{B}$ in turn. These three computations give the entries in the first row of $\mathbf{AB}$:

$$\mathbf{AB} = \begin{bmatrix} -3-2 & 0-1 & 9+5 \\ & & \end{bmatrix} = \begin{bmatrix} -5 & -1 & 14 \\ & & \end{bmatrix}.$$

Using the second row of $\mathbf{A}$ in the same way, we obtain the second row of $\mathbf{AB}$:

$$\mathbf{AB} = \begin{bmatrix} -5 & -1 & 14 \\ 10 & 4 & -26 \end{bmatrix}.$$

(b) The product $\mathbf{BA}$ is not defined, since $\mathbf{B}$ is a 2×3 matrix, $\mathbf{A}$ is a 2×2 matrix, and $3 \neq 2$. Furthermore, our schematic procedure breaks down, since the rows of $\mathbf{B}$ have three terms and the columns of $\mathbf{A}$ have two terms; so it is not clear how we would mentally place the rows of $\mathbf{B}$ on top of the columns of $\mathbf{A}$.

(c) We have

$$\mathbf{A}^2 = \mathbf{AA} = \begin{bmatrix} 3 & -1 \\ -2 & 4 \end{bmatrix}\begin{bmatrix} 3 & -1 \\ -2 & 4 \end{bmatrix} = \begin{bmatrix} 11 & -7 \\ -14 & 18 \end{bmatrix}.$$

(d) We have

$$\mathbf{B}^T\mathbf{A}^T = \begin{bmatrix} -1 & 2 \\ 0 & 1 \\ 3 & -5 \end{bmatrix}\begin{bmatrix} 3 & -2 \\ -1 & 4 \end{bmatrix} = \begin{bmatrix} -5 & 10 \\ -1 & 4 \\ 14 & -26 \end{bmatrix},$$

so $\mathbf{B}^T\mathbf{A}^T = (\mathbf{AB})^T$ by part (a). This is not an accident; see Exercise 19. ■

Just as in Example 1, powers of adjacency matrices can be used to count paths in [undirected] graphs.

EXAMPLE 3

Consider the graph with adjacency matrix

$$\mathbf{M} = \begin{bmatrix} 1 & 2 & 2 & 1 \\ 2 & 2 & 0 & 1 \\ 2 & 0 & 0 & 1 \\ 1 & 1 & 1 & 0 \end{bmatrix}.$$

The graph can be obtained from Figure 2 by removing the arrowheads. We obtain

$$\mathbf{M}^2 = \begin{bmatrix} 10 & 7 & 3 & 5 \\ 7 & 9 & 5 & 4 \\ 3 & 5 & 5 & 2 \\ 5 & 4 & 2 & 3 \end{bmatrix}.$$

Thus we see that there are 10 paths of length 2 from v_1 to itself, 7 paths of length 2 from v_1 to v_2, etc. To get the number of paths of length 3, we would use

$$\mathbf{M}^3 = \mathbf{M}^2\mathbf{M} = \begin{bmatrix} 35 & 39 & 25 & 20 \\ 39 & 36 & 18 & 21 \\ 25 & 18 & 8 & 13 \\ 20 & 21 & 13 & 11 \end{bmatrix}.$$

It is clear that counting paths of length 3 in this graph by inspection would be tedious and that errors would be hard to avoid. ■

EXAMPLE 4

Consider

$$\mathbf{A} = \begin{bmatrix} 1 & 3 \\ 3 & 9 \end{bmatrix}, \quad \mathbf{B} = \begin{bmatrix} 3 & -1 \\ -6 & 2 \end{bmatrix}.$$

We have

$$\mathbf{AB} = \begin{bmatrix} 1 & 3 \\ 3 & 9 \end{bmatrix}\begin{bmatrix} 3 & -1 \\ -6 & 2 \end{bmatrix} = \begin{bmatrix} -15 & 5 \\ -45 & 15 \end{bmatrix}$$

and

$$\mathbf{BA} = \begin{bmatrix} 3 & -1 \\ -6 & 2 \end{bmatrix}\begin{bmatrix} 1 & 3 \\ 3 & 9 \end{bmatrix} = \begin{bmatrix} 0 & 0 \\ 0 & 0 \end{bmatrix}.$$

This example shows that multiplication of matrices is not commutative! Even when both products **AB** and **BA** are defined, they may or may not be equal. Also the product of two nonzero matrices can be a zero matrix. ■

Multiplication of matrices is associative, i.e., $(\mathbf{AB})\mathbf{C} = \mathbf{A}(\mathbf{BC})$ whenever either side makes sense. At the computational level, this fact is rather mysterious [see Exercise 22], though we could give an argument for adjacency matrices of graphs based on counting paths. The mystery vanishes in the context of linear transformations, because composition of transformations is associative. Thus, rather than give a complicated argument based on counting paths, we simply state the general associative law here without proof.

Associative Law for Matrices If **A** is an $m \times n$ matrix, **B** is an $n \times p$ matrix, and **C** is a $p \times q$ matrix, then $(\mathbf{AB})\mathbf{C} = \mathbf{A}(\mathbf{BC})$.

Since multiplication of matrices is associative, we can write **ABC** without ambiguity. Also, powers such as $\mathbf{A}^3 = \mathbf{AAA}$ are unambiguous. We stress again that although *multiplication of matrices is associative* it is *not commutative*. **AB** need not equal **BA** even if both products are defined, as we saw in Example 4. Some other laws of arithmetic not satisfied by matrices appear in Exercises 18 and 20(b).

EXAMPLE 5

The special $n \times n$ matrix

$$\mathbf{I}_n = \begin{bmatrix} 1 & 0 & 0 & \cdots & 0 \\ 0 & 1 & 0 & \cdots & 0 \\ 0 & 0 & 1 & \cdots & 0 \\ \vdots & \vdots & \vdots & & \vdots \\ 0 & 0 & 0 & \cdots & 1 \end{bmatrix}$$

with $\mathbf{I}_n[i, i] = 1$ for $i = 1, 2, \ldots, n$ and $\mathbf{I}_n[i, j] = 0$ for $i \neq j$ is called the $n \times n$ **identity matrix**. For example,

$$\mathbf{I}_2 = \begin{bmatrix} 1 & 0 \\ 0 & 1 \end{bmatrix} \quad \text{and} \quad \mathbf{I}_4 = \begin{bmatrix} 1 & 0 & 0 & 0 \\ 0 & 1 & 0 & 0 \\ 0 & 0 & 1 & 0 \\ 0 & 0 & 0 & 1 \end{bmatrix}.$$

Identity matrices play roles in matrix algebra analogous to the role of 1 in the algebra of numbers. In particular, we have [see Exercise 24]

$$\mathbf{A}\mathbf{I}_n = \mathbf{A} \quad \text{for all} \quad \mathbf{A} \in \mathfrak{M}_{m,n}$$

and

$$\mathbf{I}_n\mathbf{B} = \mathbf{B} \quad \text{for all} \quad \mathbf{B} \in \mathfrak{M}_{n,p}.$$

Both assertions apply to $n \times n$ matrices, so

$$\mathbf{A}\mathbf{I}_n = \mathbf{I}_n\mathbf{A} = \mathbf{A} \quad \text{for all} \quad \mathbf{A} \in \mathfrak{M}_{n,n}.$$

■

Two $n \times n$ matrices $\mathbf{A}$ and $\mathbf{B}$ are said to be **inverses** to each other provided $\mathbf{AB} = \mathbf{BA} = \mathbf{I}_n$. A matrix $\mathbf{A}$ can have only one inverse. To see this, suppose that $\mathbf{B}$ and $\mathbf{C}$ are both inverses to $\mathbf{A}$. Then $\mathbf{BA} = \mathbf{I}_n$ and $\mathbf{AC} = \mathbf{I}_n$, so $\mathbf{B} = \mathbf{BI}_n = \mathbf{B}(\mathbf{AC}) = (\mathbf{BA})\mathbf{C} = \mathbf{I}_n\mathbf{C} = \mathbf{C}$. A matrix $\mathbf{A}$ that has an inverse is called **invertible**, and its unique **inverse** is written $\mathbf{A}^{-1}$. There are several techniques for finding inverses of matrices by hand, and many calculators can find them too. We will not pursue this topic, though Exercise 16 gives the formula for 2×2 matrices.

As we saw above, matrix multiplication can be used to count the number of paths of any given length between vertices of a graph or digraph. What if we don't want to know how many paths there are, but simply want to know if there are any paths at all from v_i to v_j? One way, of course, would be to compute powers of the adjacency matrix. If any of the powers have a nonzero entry in the (i, j) position, then there's at least one path from v_i to v_j, and otherwise there isn't. Since we only care whether entries are 0 or not, we can also use a Boolean approach, which we now describe and which we will apply in §11.4.

Instead of taking numbers for our matrix entries, let's take entries from the set $\mathbb{B} = \{0, 1\}$, which has **Boolean operations** $\vee$ and $\wedge$ defined by Figure 3. These operations are the same ones we have seen in our study of truth tables in Chapter 2 and Boolean algebras in Chapter 10. Note that $m \vee n = \max\{m, n\}$ and $m \wedge n = \min\{m, n\}$.

Figure 3 ▶

∨	0	1
0	0	1
1	1	1

∧	0	1
0	0	0
1	0	1

Matrices with entries in $\mathbb{B}$ are called **Boolean matrices**. We define $\mathbf{A}_1 * \mathbf{A}_2$, the **Boolean product** of the $m \times n$ Boolean matrix $\mathbf{A}_1$ and the $n \times p$ Boolean matrix $\mathbf{A}_2$, by using the ordinary definition of matrix product, but with the operations addition and multiplication replaced by $\vee$ and $\wedge$, respectively. That is, the (i, k)-entry of

$\mathbf{A}_1 * \mathbf{A}_2$ is

$$(\mathbf{A}_1 * \mathbf{A}_2)[i, k]$$
$$= (\mathbf{A}_1[i, 1] \wedge \mathbf{A}_2[1, k]) \vee (\mathbf{A}_1[i, 2] \wedge \mathbf{A}_2[2, k]) \vee \cdots \vee (\mathbf{A}_1[i, n] \wedge \mathbf{A}_2[n, k]),$$

which can be written more compactly as

$$\bigvee_{j=1}^{n} (\mathbf{A}_1[i, j] \wedge \mathbf{A}_2[j, k]).$$

By the definitions of $\vee$ and $\wedge$, the (i, k)-entry in $\mathbf{A}_1 * \mathbf{A}_2$ is 1 if and only if at least one of the terms $\mathbf{A}_1[i, j] \wedge \mathbf{A}_2[j, k]$ is 1, and such a term is 1 if and only if both $\mathbf{A}_1[i, j]$ and $\mathbf{A}_2[j, k]$ are 1.

EXAMPLE 6

The Boolean product of the matrices $\mathbf{A}_1$ and $\mathbf{A}_2$ in Figure 4 is the matrix $\mathbf{A}_1 * \mathbf{A}_2$ in the figure. For example, the (3, 3)-entry of $\mathbf{A}_1 * \mathbf{A}_2$ is $(1 \wedge 1) \vee (1 \wedge 1) \vee (0 \wedge 1) = 1 \vee 1 \vee 0 = 1$. The (5, 1)-entry is $(0 \wedge 1) \vee (1 \wedge 0) \vee (0 \wedge 1) = 0 \vee 0 \vee 0 = 0$. Compare the values in $\mathbf{A}_1 * \mathbf{A}_2$ with those in the ordinary matrix product $\mathbf{A}_1\mathbf{A}_2$, viewing 0 and 1 as integers. ■

Figure 4 ▶

$$\begin{bmatrix} 1 & 0 & 0 \\ 1 & 0 & 1 \\ 1 & 1 & 0 \\ 1 & 1 & 1 \\ 0 & 1 & 0 \end{bmatrix} \qquad \begin{bmatrix} 1 & 0 & 1 & 0 \\ 0 & 1 & 1 & 1 \\ 1 & 0 & 1 & 1 \end{bmatrix}$$

$\mathbf{A}_1$ $\qquad$ $\mathbf{A}_2$

$$\begin{bmatrix} 1 & 0 & 1 & 0 \\ 1 & 0 & 1 & 1 \\ 1 & 1 & 1 & 1 \\ 1 & 1 & 1 & 1 \\ 0 & 1 & 1 & 1 \end{bmatrix} \qquad \begin{bmatrix} 1 & 0 & 1 & 0 \\ 2 & 0 & 2 & 1 \\ 1 & 1 & 2 & 1 \\ 2 & 1 & 3 & 2 \\ 0 & 1 & 1 & 1 \end{bmatrix}$$

$\mathbf{A}_1 * \mathbf{A}_2$ $\qquad$ $\mathbf{A}_1\mathbf{A}_2$

There is a graph-theoretic interpretation for Boolean matrix products whenever they are defined. In Example 6 the matrix $\mathbf{A}_1$ could describe a set of edges joining vertices 1, 2, 3, 4, 5 to vertices a, b, c, while $\mathbf{A}_2$ could describe a set of edges joining a, b, c to vertices E, F, G, H, with the interpretation that entries of 1 indicate the existence of edges and 0's their nonexistence. Then $\mathbf{A}_1 * \mathbf{A}_2$ would tell us about paths from 1, 2, 3, 4, 5 to E, F, G, H.

EXAMPLE 7

Consider a graph with no parallel edges and with adjacency matrix $\mathbf{A}$. The Boolean power

$$\mathbf{A} * \mathbf{A} * \cdots * \mathbf{A} \qquad [n \text{ times}]$$

tells us exactly which pairs of vertices are connected by paths of length n. Figure 5 shows an example. For this simple example we have

Figure 5 ▶

$$\begin{bmatrix} 0 & 1 & 1 & 1 \\ 1 & 0 & 0 & 0 \\ 1 & 0 & 1 & 1 \\ 1 & 0 & 1 & 1 \end{bmatrix}$$

$\mathbf{A}$ = adjacency matrix

Graph

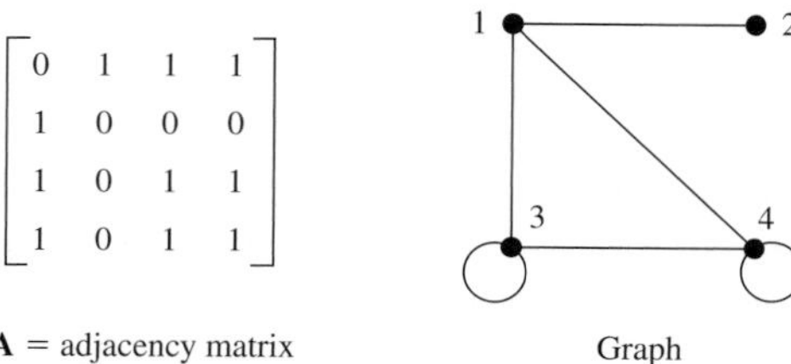

$$\mathbf{A} = \begin{bmatrix} 0 & 1 & 1 & 1 \\ 1 & 0 & 0 & 0 \\ 1 & 0 & 1 & 1 \\ 1 & 0 & 1 & 1 \end{bmatrix}, \qquad \mathbf{A} * \mathbf{A} = \begin{bmatrix} 1 & 0 & 1 & 1 \\ 0 & 1 & 1 & 1 \\ 1 & 1 & 1 & 1 \\ 1 & 1 & 1 & 1 \end{bmatrix},$$

$$\mathbf{A} * \mathbf{A} * \mathbf{A} = \begin{bmatrix} 1 & 1 & 1 & 1 \\ 1 & 1 & 1 & 1 \\ 1 & 1 & 1 & 1 \\ 1 & 1 & 1 & 1 \end{bmatrix},$$

so any two vertices are connected by a path of length 3—in fact, also by a path of length 1 *or* a path of length 2. ■

Exercises 11.3

1. Let

$$\mathbf{A} = \begin{bmatrix} 1 & 2 & 4 \\ 3 & 0 & 2 \end{bmatrix} \quad \text{and} \quad \mathbf{B} = \begin{bmatrix} 2 & 1 \\ -1 & 0 \\ -2 & 3 \end{bmatrix}.$$

Find the following when they exist.

(a) $\mathbf{AB}$ (b) $\mathbf{BA}$ (c) $\mathbf{ABA}$
(d) $\mathbf{A} + \mathbf{B}^T$ (e) $3\mathbf{A}^T - 2\mathbf{B}$ (f) $(\mathbf{AB})^2$

2. Let

$$\mathbf{C} = \begin{bmatrix} 1 \\ 0 \\ 1 \end{bmatrix}$$

and let $\mathbf{A}$ and $\mathbf{B}$ be as in Exercise 1. Find the following when they exist.

(a) $\mathbf{AC}$ (b) $\mathbf{BC}$ (c) $\mathbf{C}^2$
(d) $\mathbf{C}^T\mathbf{C}$ (e) $\mathbf{CC}^T$ (f) $73\mathbf{C}$

3. Let

$$\mathbf{A} = \begin{bmatrix} 3 & -4 & 3 & 1 \\ 2 & 0 & 1 & -2 \\ -1 & 1 & 2 & 0 \end{bmatrix}$$

$$\text{and} \quad \mathbf{B} = \begin{bmatrix} -1 & 1 & 0 \\ 1 & 2 & 1 \\ 0 & 1 & -1 \end{bmatrix}.$$

Find the following when they exist.

(a) $\mathbf{A}^2$ (b) $\mathbf{B}^2$ (c) $\mathbf{AB}$ (d) $\mathbf{BA}$

4. Let $\mathbf{A}$ and $\mathbf{B}$ be as in Exercise 3. Find the following when they exist.

(a) $\mathbf{BA}^T$ (b) $\mathbf{A}^T\mathbf{B}$
(c) $5(\mathbf{AB})^T - 3\mathbf{B}^T\mathbf{A}^T$

5. (a) Calculate both $(\mathbf{AB})\mathbf{C}$ and $\mathbf{A}(\mathbf{BC})$ for

$$\mathbf{A} = \begin{bmatrix} -1 & 4 \\ 2 & 5 \end{bmatrix}, \quad \mathbf{B} = \begin{bmatrix} 1 & 1 \\ 0 & 1 \end{bmatrix},$$

$$\text{and} \quad \mathbf{C} = \begin{bmatrix} 2 & -1 \\ 1 & 3 \end{bmatrix}.$$

(b) Calculate both $\mathbf{B}(\mathbf{AC})$ and $(\mathbf{BA})\mathbf{C}$.

6. Let $\mathbf{A}$, $\mathbf{B}$, and $\mathbf{C}$ be as in Exercise 5. Calculate:

(a) both $\mathbf{AB}$ and $\mathbf{BA}$
(b) both $\mathbf{AC}$ and $\mathbf{CA}$
(c) $\mathbf{A}^2$

7. Let

$$\mathbf{A} = \begin{bmatrix} 3 & -1 \\ 2 & 1 \\ -2 & 4 \end{bmatrix}, \quad \mathbf{B} = \begin{bmatrix} 1 & 2 \\ 0 & 1 \end{bmatrix},$$

$$\text{and} \quad \mathbf{C} = \begin{bmatrix} -1 & 3 \\ 2 & 1 \end{bmatrix}.$$

(a) Calculate $\mathbf{A}(\mathbf{BC})$ and $(\mathbf{AB})\mathbf{C}$.
(b) Calculate $\mathbf{A}(\mathbf{B}^2)$ and $(\mathbf{AB})\mathbf{B}$.

8. Let $\mathbf{A} = \begin{bmatrix} 1 & 0 \\ 1 & 1 \end{bmatrix}$. Calculate

(a) $\mathbf{A} * \mathbf{A}$ (b) $\mathbf{A} * \mathbf{A} * \mathbf{A}$
(c) $\mathbf{A} * \mathbf{A} * \cdots * \mathbf{A}$ for 17 factors

9. Let $\mathbf{A} = \begin{bmatrix} 0 & 1 & 0 \\ 0 & 0 & 1 \\ 1 & 0 & 0 \end{bmatrix}$. Calculate

(a) $\mathbf{A} * \mathbf{A}$ (b) $\mathbf{A} * \mathbf{A} * \mathbf{A}$
(c) $\mathbf{A} * \mathbf{A} * \cdots * \mathbf{A}$ for 72 factors

10. Draw a digraph associated with the matrix $\mathbf{A}$ of

(a) Exercise 8. (b) Exercise 9.

11. Let $\mathbf{M}$ be the adjacency matrix for the digraph in Example 1. One can check that $\mathbf{M}^2 = \begin{bmatrix} 2 & 7 & 1 & 2 \\ 0 & 4 & 0 & 0 \\ 1 & 3 & 1 & 1 \\ 0 & 2 & 0 & 0 \end{bmatrix}$.

Use $\mathbf{M}^2$ to find the number of paths of length 2

(a) from v_1 to itself. (b) from v_1 to v_3.
(c) from v_1 to v_4. (d) from v_2 to v_1.

12. Let $\mathbf{A}$ be the Boolean matrix for the digraph in Example 1 and Exercise 11. Use $\mathbf{A} * \mathbf{A}$ to determine whether there are or are not paths of length 2

(a) from v_1 to itself. (b) from v_1 to v_3.
(c) from v_1 to v_4. (d) from v_2 to v_1.

13. (a) Calculate $\mathbf{M}^3$ for the adjacency matrix in Example 1.

(b) Find the number of paths of length 3 from v_3 to v_2.

(c) List the paths of length 3 from v_3 to v_2 using the labeling of Figure 2(b).

14. This exercise refers to the graph in Example 3.

(a) Draw the graph; just remove the arrowheads from Figure 2(a). Label the edges as in Figure 2(b).

(b) How many paths of length 2 are there from v_3 to itself?

(c) List the paths from v_3 to itself of length 2.

(d) How many paths of length 3 are there from v_3 to itself?

(e) List the paths from v_3 to itself of length 3.

15. Repeat parts (a) to (d) of Exercise 14 for the vertex v_2.

16. Show that a 2×2 matrix $\mathbf{A} = \begin{bmatrix} a & b \\ c & d \end{bmatrix}$ has an inverse if and only if $ad - bc \neq 0$, in which case the inverse is

$$\mathbf{A}^{-1} = \frac{1}{ad - bc} \begin{bmatrix} d & -b \\ -c & a \end{bmatrix}.$$

Hint: Try to solve $\begin{bmatrix} a & b \\ c & d \end{bmatrix}\begin{bmatrix} x & y \\ z & w \end{bmatrix} = \begin{bmatrix} 1 & 0 \\ 0 & 1 \end{bmatrix}$ for x, y, z, and w.

17. Use Exercise 16 to determine which of the following matrices have inverses. Find the inverses when they exist and check your answers.

(a) $\mathbf{I} = \begin{bmatrix} 1 & 0 \\ 0 & 1 \end{bmatrix}$ (b) $\mathbf{A} = \begin{bmatrix} 1 & 1 \\ 0 & 1 \end{bmatrix}$

(c) $\mathbf{B} = \begin{bmatrix} 1 & 1 \\ 1 & 1 \end{bmatrix}$ (d) $\mathbf{C} = \begin{bmatrix} 2 & -3 \\ 5 & 8 \end{bmatrix}$

(e) $\mathbf{D} = \begin{bmatrix} 0 & 1 \\ 1 & 0 \end{bmatrix}$

18. Find 2×2 matrices that show that $(\mathbf{A} + \mathbf{B})(\mathbf{A} - \mathbf{B}) = \mathbf{A}^2 - \mathbf{B}^2$ does not generally hold.

19. Show that if $\mathbf{A}$ is an $m \times n$ matrix and $\mathbf{B}$ is an $n \times p$ matrix, then $(\mathbf{AB})^T = \mathbf{B}^T\mathbf{A}^T$. Note that both sides of the equality represent $p \times m$ matrices.

20. (a) Prove the cancellation law for $\mathfrak{M}_{m,n}$ under addition; i.e., prove that if $\mathbf{A}$, $\mathbf{B}$, $\mathbf{C}$ are in $\mathfrak{M}_{m,n}$ and $\mathbf{A} + \mathbf{C} = \mathbf{B} + \mathbf{C}$, then $\mathbf{A} = \mathbf{B}$.

(b) Show that the cancellation law for $\mathfrak{M}_{n,n}$ under multiplication fails; i.e., show that $\mathbf{AC} = \mathbf{BC}$ need not imply $\mathbf{A} = \mathbf{B}$ even when $\mathbf{C} \neq \mathbf{0}$.

21. (a) Let $\mathbf{A} = \begin{bmatrix} a & 0 \\ 0 & a \end{bmatrix}$ for some fixed a in $\mathbb{R}$. Show that $\mathbf{AB} = \mathbf{BA}$ for all $\mathbf{B}$ in $\mathfrak{M}_{2,2}$.

(b) Consider a fixed matrix $\mathbf{A}$ in $\mathfrak{M}_{2,2}$ that satisfies $\mathbf{AB} = \mathbf{BA}$ for all $\mathbf{B}$ in $\mathfrak{M}_{2,2}$. Show that

$$\mathbf{A} = \begin{bmatrix} a & 0 \\ 0 & a \end{bmatrix} \quad \text{for some} \quad a \in \mathbb{R}.$$

Hint: Write $\mathbf{A} = \begin{bmatrix} a & b \\ c & d \end{bmatrix}$ and try $\mathbf{B} = \begin{bmatrix} 1 & 0 \\ 0 & 0 \end{bmatrix}$ and $\begin{bmatrix} 0 & 1 \\ 0 & 0 \end{bmatrix}$.

22. (a) Show directly that $\mathbf{A}(\mathbf{BC}) = (\mathbf{AB})\mathbf{C}$ for matrices $\mathbf{A}$, $\mathbf{B}$, and $\mathbf{C}$ in $\mathfrak{M}_{2,2}$.

(b) Did you enjoy part (a)? If yes, give a direct proof of the general associative law for matrices.

23. (a) Let $\mathbf{A}$ and $\mathbf{B}$ be $m \times n$ matrices and let $\mathbf{C}$ be an $n \times p$ matrix. Show that the distributive law holds: $(\mathbf{A} + \mathbf{B})\mathbf{C} = \mathbf{AC} + \mathbf{BC}$.

(b) Verify the distributive law $\mathbf{A}(\mathbf{B} + \mathbf{C}) = \mathbf{AB} + \mathbf{AC}$. First specify the sizes of the matrices for which this makes sense.

24. Show that if $\mathbf{A}$ is an $m \times n$ matrix, then

(a) $\mathbf{I}_m\mathbf{A} = \mathbf{A}$. (b) $\mathbf{AI}_n = \mathbf{A}$.

11.4 Properties of General Relations

In §3.4 we studied equivalence relations, and in §§11.1 and 11.2 we discussed partial orders. Both of these types of relations are reflexive and transitive, but the difference between symmetry and antisymmetry gives the two subjects entirely different feelings. Partial orders have a sense of direction from small to large, whereas equivalence relations partition sets into unrelated blocks whose members are lumped together because they have something in common.

Each of these two types of binary relation provides us with some organized structure on a set S. In contrast, the theory of general binary relations, i.e., subsets of $S \times S$, is so abstract and nonspecific that there is really no structure to be developed. In this section we will be focusing on statements that are true for all relations. Not surprisingly, we will be limited to broad generalities. We begin by discussing how to compose two relations to get a third one. We then translate our account into matrix terms and revisit the links between relations, matrices, and digraphs that we first discussed in Chapter 3.

Functions can be viewed as relations, as we saw in §3.1. We identify the function $f: S \to T$ with the relation R_f from S to T defined by

$$R_f = \{(s, t) \in S \times T : f(s) = t\}.$$

If $g: T \to U$ is also a function, then the composite function $g \circ f: S \to U$ yields the relation

$$R_{g \circ f} = \{(s, u) \in S \times U : (g \circ f)(s) = g(f(s)) = u\},$$

which we can think of as the composite of R_f and R_g. The pair (s, u) is in $R_{g \circ f}$ if and only if $u = g(t)$, where $t = f(s) \in T$, so

$$R_{g \circ f} = \{(s, u) \in S \times U : \text{ there is a } t \in T \text{ with } (s, t) \in R_f \text{ and } (t, u) \in R_g\}.$$

We generalize this fact about relations associated with functions to define the composition of two arbitrary relations. If R_1 is a relation from S to T and R_2 is a relation from T to U, then the **composite** of R_1 and R_2 is the relation $\boldsymbol{R_2 \circ R_1}$ from S to U defined by

$$\boldsymbol{R_2 \circ R_1} = \{(s, u) \in S \times U : \text{ there is a } t \in T \text{ with } (s, t) \in R_1 \text{ and } (t, u) \in R_2\}.$$

For relations determined by functions, this definition gives $R_g \circ R_f = R_{g \circ f}$.

Since we think of R_1 as the first relation and R_2 as the second, on esthetic grounds alone this general definition seems backward. It will also turn out to be backward when we observe the connection between composition of relations and products of their matrices; so many people use $R_1 \circ R_2$ as a name for what we have called $R_2 \circ R_1$, even though such usage is inconsistent with the notation for functional composition and is a source of confusion. To avoid misunderstanding, we will write $\boldsymbol{R_1 R_2}$ for $R_2 \circ R_1$, rather than $R_1 \circ R_2$. To summarize:

Given relations R_1 from S to T and R_2 from T to U, the **composite relation**

$$\{(s, u) \in S \times U : (s, t) \in R_1 \text{ and } (t, u) \in R_2 \text{ for some } t \in T\}$$

will be denoted by either $R_1 R_2$ or $R_2 \circ R_1$. Thus (s, u) is in $R_1 R_2$ precisely if there is a t in T such that $(s, t) \in R_1$ and $(t, u) \in R_2$.

EXAMPLE 1

(a) Example 2 on page 95 concerns university students, courses, and departments. The relations are

$$R_1 = \{(s, c) \in S \times C : s \text{ is enrolled in } c\}$$

and

$$R_2 = \{(c, d) \in C \times D : c \text{ is required by } d\}.$$

Thus

$$R_1 R_2 = \{(s, d) \in S \times D : (s, c) \in R_1 \text{ and } (c, d) \in R_2 \text{ for some } c \in C\},$$

so (s, d) belongs to $R_1 R_2$ provided student s is taking some course that is required by department d.

Note that $R_2 R_1$ makes no sense, because the second entries of R_2 lie in the set D, while the first entries of R_1 lie in S; it could not happen that $(c, t) \in R_2$ and $(t, c') \in R_1$. Of course, $R_2 \circ R_1$ makes sense; this is just another name for $R_1 R_2$.

(b) Let $S = \{1, 2, 3, 4, 5\}$, $T = \{a, b, c\}$, and $U = \{e, f, g, h\}$. Consider the relations

$$R_1 = \{(1, a), (2, a), (2, c), (3, a), (3, b), (4, a), (4, b), (4, c), (5, b)\},$$

$$R_2 = \{(a, e), (a, g), (b, f), (b, g), (b, h), (c, e), (c, g), (c, h)\}.$$

The digraph in Figure 1 illustrates R_1, R_2 and $R_1 R_2$, where the edges from S to T are in R_1 and the edges from T to U are in R_2. Thus a pair (s, u) in $S \times U$ is in $R_1 R_2$ if and only if there is a path from s to u in the digraph. ■

EXAMPLE 2

Consider relations R_1 and R_2 from S to T and relations R_3 and R_4 from T to U.

(a) If $R_1 \subseteq R_2$ and $R_3 \subseteq R_4$, then $R_1 R_3 \subseteq R_2 R_4$. To see this, consider (s, u) in $R_1 R_3$. Then, for some $t \in T$, we have $(s, t) \in R_1$ and $(t, u) \in R_3$. Since $R_1 \subseteq R_2$ and $R_3 \subseteq R_4$, we also have $(s, t) \in R_2$ and $(t, u) \in R_4$. So $(s, u) \in R_2 R_4$. This argument shows that $R_1 R_3 \subseteq R_2 R_4$.

Figure 1 ▶

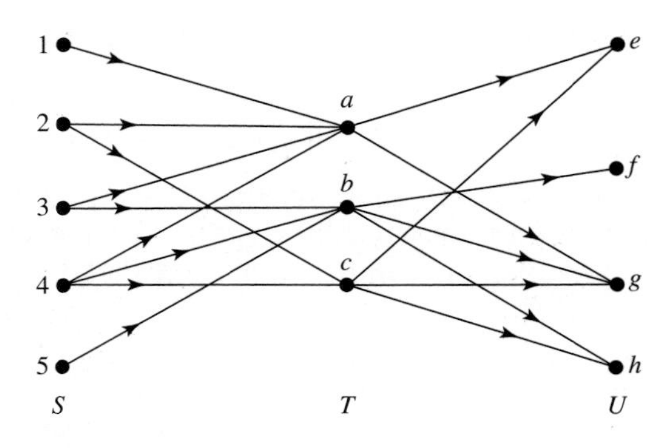

(b) The union of two relations from A to B, i.e., two subsets of $A \times B$, is a relation from A to B. We show that, for R_1, R_2, and R_3 as above,

$$(R_1 \cup R_2)R_3 = R_1R_3 \cup R_2R_3.$$

Since $R_1 \subseteq R_1 \cup R_2$, we have $R_1R_3 \subseteq (R_1 \cup R_2)R_3$ from part (a); similarly, $R_2R_3 \subseteq (R_1 \cup R_2)R_3$, so

$$R_1R_3 \cup R_2R_3 \subseteq (R_1 \cup R_2)R_3.$$

To check the reverse inclusion, consider $(s, u) \in (R_1 \cup R_2)R_3$. For some $t \in T$, we have $(s, t) \in R_1 \cup R_2$ and $(t, u) \in R_3$. Then either $(s, t) \in R_1$, in which case $(s, u) \in R_1R_3$, or else $(s, t) \in R_2$, in which case $(s, u) \in R_2R_3$. Either way, $(s, u) \in R_1R_3 \cup R_2R_3$ and hence

$$(R_1 \cup R_2)R_3 \subseteq R_1R_3 \cup R_2R_3.$$ ■

On page 33 we observed that composition of functions is associative. So is composition of relations.

Associative Law for Relations If R_1 is a relation from S to T, R_2 is a relation from T to U, and R_3 is a relation from U to V, then

$$(R_1R_2)R_3 = R_1(R_2R_3).$$

Proof We show that an ordered pair (s, v) in $S \times V$ belongs to $(R_1R_2)R_3$ if and only if

for some $t \in T$ and $u \in U$ we have $(s, t) \in R_1$, $(t, u) \in R_2$, and $(u, v) \in R_3$. $(*)$

A similar argument shows that (s, v) belongs to $R_1(R_2R_3)$ if and only if $(*)$ holds.

Consider (s, v) in $(R_1R_2)R_3$. Since R_1R_2 is a relation from S to U, this means that there exists $u \in U$ such that $(s, u) \in R_1R_2$ and $(u, v) \in R_3$. Since (s, u) is in R_1R_2, there exists $t \in T$ such that $(s, t) \in R_1$ and $(t, u) \in R_2$. Thus $(*)$ holds.

Now suppose that $(*)$ holds for an element (s, v) in $S \times V$. Then $(s, t) \in R_1$ and $(t, u) \in R_2$, so $(s, u) \in R_1R_2$. Since also $(u, v) \in R_3$, we conclude that (s, v) is in $(R_1R_2)R_3$. ■

In view of the associative law, we may write $R_1R_2R_3$ for either $(R_1R_2)R_3$ or $R_1(R_2R_3)$. As shown in the proof, (s, v) belongs to $R_1R_2R_3$ provided there exist $t \in T$ and $u \in U$ such that $(s, t) \in R_1$, $(t, u) \in R_2$, and $(u, v) \in R_3$.

The sets and relations that we have been considering so far could have been finite or infinite. We now restrict ourselves to finite sets S, T, and U and consider relations R_1 from S to T and R_2 from T to U. Just as we did in §3.3, we can associate matrices of 0's and 1's with these relations. We list the members of each of the sets S, T and U in some order. Then the (s, t)-entry of the matrix $\mathbf{A}_1$ corresponding to the relation R_1 is 1 if $(s, t) \in R_1$ and is 0 otherwise. The matrix $\mathbf{A}_2$ for R_2 is defined similarly. Given $\mathbf{A}_1$ and $\mathbf{A}_2$, we want to find the matrix for the composite relation R_1R_2.

It turns out that it's the Boolean product $\mathbf{A}_1 * \mathbf{A}_2$ as defined in §11.3. For $s \in S$ and $u \in U$, the (s, u)-entry of $\mathbf{A}_1 * \mathbf{A}_2$ is

$$\bigvee_{t \in T} (\mathbf{A}_1[s, t] \wedge \mathbf{A}_2[t, u]).$$

The value of this expression is 1 if and only if at least one term $\mathbf{A}_1[s, t] \wedge \mathbf{A}_2[t, u]$ is 1. A term $\mathbf{A}_1[s, t] \wedge \mathbf{A}_2[t, u]$ is 0 unless both $\mathbf{A}_1[s, t]$ and $\mathbf{A}_2[t, u]$ are 1, in which case we have $(s, t) \in R_1$ and $(t, u) \in R_2$. Thus the sum is 0 if there is no such t, i.e., if $(s, u) \notin R_1 R_2$, and is 1 if $(s, u) \in R_1 R_2$.

We have shown the following result.

Theorem 1 Consider relations R_1 from S to T and R_2 from T to U, where S, T, and U are finite. If $\mathbf{A}_1$ and $\mathbf{A}_2$ are the corresponding Boolean matrices, then the Boolean product $\mathbf{A}_1 * \mathbf{A}_2$ is the matrix for the composite relation $R_1 R_2$.

The Boolean product operation $*$ on Boolean matrices is associative. This fact can be shown directly or by using Theorem 1 and the associativity of the corresponding relations, as suggested in Exercise 17.

EXAMPLE 3 In Example 1(b) we had sets $S = \{1, 2, 3, 4, 5\}$, $T = \{a, b, c\}$, and $U = \{e, f, g, h\}$ with relations

$$R_1 = \{(1, a), (2, a), (2, c), (3, a), (3, b), (4, a), (4, b), (4, c), (5, b)\},$$

$$R_2 = \{(a, e), (a, g), (b, f), (b, g), (b, h), (c, e), (c, g), (c, h)\},$$

so

$$R_1 R_2 = \{(1, e), (1, g), (2, e), (2, g), (2, h), (3, e), (3, f), (3, g), (3, h), (4, e), (4, f), (4, g), (4, h), (5, f), (5, g), (5, h)\}.$$

The matrices $\mathbf{A}_1$, $\mathbf{A}_2$ and $\mathbf{A}_1 * \mathbf{A}_2$ for these relations are given in Figure 2. We saw them before in Example 6 on page 444, where we computed $\mathbf{A}_1 * \mathbf{A}_2$. ■

Figure 2 ▶

$$\underset{\mathbf{A}_1}{\begin{bmatrix} 1 & 0 & 0 \\ 1 & 0 & 1 \\ 1 & 1 & 0 \\ 1 & 1 & 1 \\ 0 & 1 & 0 \end{bmatrix}} \qquad \underset{\mathbf{A}_2}{\begin{bmatrix} 1 & 0 & 1 & 0 \\ 0 & 1 & 1 & 1 \\ 1 & 0 & 1 & 1 \end{bmatrix}} \qquad \underset{\mathbf{A}_1 * \mathbf{A}_2}{\begin{bmatrix} 1 & 0 & 1 & 0 \\ 1 & 0 & 1 & 1 \\ 1 & 1 & 1 & 1 \\ 1 & 1 & 1 & 1 \\ 0 & 1 & 1 & 1 \end{bmatrix}}$$

For the remainder of this section we consider relations on a single set S. Since relations on S are simply subsets of $S \times S$, the collection of all relations on S is $\mathcal{P}(S \times S)$.

Theorem 2 The composition operation on $\mathcal{P}(S \times S)$ is associative, and $\mathcal{P}(S \times S)$ has an identity E, i.e., a member E such that $RE = ER = R$ for all $R \in \mathcal{P}(S \times S)$.

Proof First note that if R_1 and R_2 are in $\mathcal{P}(S \times S)$, then so is their composite, $R_1 R_2$. Associativity of composition has already been verified. The identity is the **equality relation**

$$E = \{(s, s) \in S \times S : s \in S\}.$$ ■

The usual notational conventions for associative operations apply. Thus, if R is a relation on S, then $R^0 = E$ and, for n in $\mathbb{P}$, R^n is the composition of R with itself n times. Note that if $n > 1$, then (s, u) belongs to R^n provided there exist

$t_1, t_2, \ldots, t_{n-1}$ in S such that $(s, t_1), (t_1, t_2), \ldots, (t_{n-1}, u)$ are all in R. In other words, (s, u) is in R^n if s and u are R-related through a chain of length n or, equivalently, if the (s, u)-entry of the n-factor Boolean product matrix $\mathbf{A} * \cdots * \mathbf{A}$ is 1, where $\mathbf{A}$ is a Boolean matrix that describes R.

Once we have agreed on an order in which to list the members of a finite set S, every relation R on S has its associated Boolean matrix $\mathbf{A}$. This matrix can be viewed as the adjacency matrix for a digraph whose vertices are the members of S, with an edge from s to t if and only if $\mathbf{A}$ has a 1 as its (s, t)-entry. Hence the digraph has an edge from s to t if and only if $(s, t) \in R$; i.e., R is the adjacency relation for the digraph. We will call a picture of this digraph a **picture of** R. Note that, although $\mathbf{A}$ depends on the order in which we list the members of S, the digraph does not.

If R is symmetric, then the pairs of oppositely directed edges in the associated digraph can be combined into edges of an undirected graph for which R is the adjacency relation.

The next theorem shows the connection between transitivity and the composition of relations on S.

Theorem 3 If R is a relation on a set S, then R is transitive if and only if $R^2 \subseteq R$.

Proof Suppose first that R is transitive, and consider $(s, u) \in R^2$. By the definition of R^2, there is a t in S such that $(s, t) \in R$ and $(t, u) \in R$. Since R is transitive, (s, u) is also in R. We have shown that every (s, u) in R^2 is in R; i.e., $R^2 \subseteq R$.

For the converse, suppose that $R^2 \subseteq R$. Consider (s, t) and (t, u) in R. Then (s, u) is in R^2 and hence in R. This proves that R is transitive. ■

For $m \times n$ Boolean matrices $\mathbf{A}_1$ and $\mathbf{A}_2$, we write $\mathbf{A}_1 \le \mathbf{A}_2$ if every entry of $\mathbf{A}_1$ is less than or equal to the corresponding entry of $\mathbf{A}_2$, i.e., in case

$$\mathbf{A}_1[i, j] \le \mathbf{A}_2[i, j] \quad \text{for} \quad 1 \le i \le m \quad \text{and} \quad 1 \le j \le n.$$

If R_1 and R_2 are relations from S to T, with matrices $\mathbf{A}_1$ and $\mathbf{A}_2$, then

$$R_1 \subseteq R_2 \quad \text{if and only if} \quad \mathbf{A}_1 \le \mathbf{A}_2;$$

think about where the 1's and 0's are in $\mathbf{A}_1$ and $\mathbf{A}_2$ [Exercise 16(a)]. It follows that $\le$ is a partial order on the set of Boolean matrices. Moreover, a relation R on a set S satisfies $R^2 \subseteq R$ if and only if its matrix $\mathbf{A}$ satisfies $\mathbf{A} * \mathbf{A} \le \mathbf{A}$. Thus R is transitive if and only if $\mathbf{A} * \mathbf{A} \le \mathbf{A}$.

EXAMPLE 4 Consider the relation R on $\{1, 2, 3\}$ with matrix

$$\mathbf{A} = \begin{bmatrix} 1 & 0 & 0 \\ 1 & 0 & 0 \\ 1 & 1 & 0 \end{bmatrix}.$$

Since

$$\mathbf{A} * \mathbf{A} = \begin{bmatrix} 1 & 0 & 0 \\ 1 & 0 & 0 \\ 1 & 1 & 0 \end{bmatrix} * \begin{bmatrix} 1 & 0 & 0 \\ 1 & 0 & 0 \\ 1 & 1 & 0 \end{bmatrix} = \begin{bmatrix} 1 & 0 & 0 \\ 1 & 0 & 0 \\ 1 & 0 & 0 \end{bmatrix} \le \begin{bmatrix} 1 & 0 & 0 \\ 1 & 0 & 0 \\ 1 & 1 & 0 \end{bmatrix} = \mathbf{A},$$

R is transitive. The transitivity of R can also be seen from its corresponding digraph in Figure 3. Whenever a path of length 2 connects two vertices, a single edge also does. For example, 3 2 1 is a path and there is an edge from 3 to 1. ■

Since relations on finite sets correspond to Boolean matrices, properties of relations can be described in matrix terms. To list some of the more important matrix equivalents, we introduce a bit more notation. For Boolean $m \times n$ matrices $\mathbf{A}_1$ and $\mathbf{A}_2$, we define $\mathbf{A}_1 \vee \mathbf{A}_2$ by

$$(\mathbf{A}_1 \vee \mathbf{A}_2)[i, j] = \mathbf{A}_1[i, j] \vee \mathbf{A}_2[i, j] \quad \text{for} \quad 1 \le i \le m \quad \text{and} \quad 1 \le j \le n.$$

Figure 3 ▶

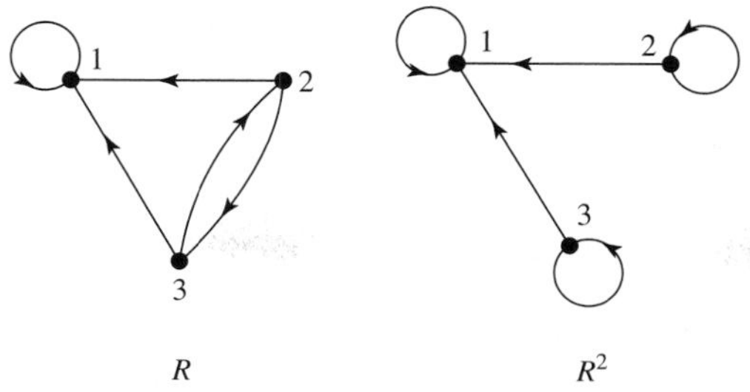

The matrix $\mathbf{A}_1 \wedge \mathbf{A}_2$ has a similar definition. For example, if

$$\mathbf{A}_1 = \begin{bmatrix} 1 & 0 & 1 & 1 \\ 0 & 1 & 1 & 0 \\ 1 & 1 & 0 & 1 \end{bmatrix} \quad \text{and} \quad \mathbf{A}_2 = \begin{bmatrix} 1 & 1 & 1 & 0 \\ 0 & 1 & 1 & 1 \\ 1 & 0 & 1 & 0 \end{bmatrix},$$

then

$$\mathbf{A}_1 \vee \mathbf{A}_2 = \begin{bmatrix} 1 & 1 & 1 & 1 \\ 0 & 1 & 1 & 1 \\ 1 & 1 & 1 & 1 \end{bmatrix} \quad \text{and} \quad \mathbf{A}_1 \wedge \mathbf{A}_2 = \begin{bmatrix} 1 & 0 & 1 & 0 \\ 0 & 1 & 1 & 0 \\ 1 & 0 & 0 & 0 \end{bmatrix}.$$

The summary below lists the matrix versions of some familiar properties that relations may have. Recall that $\mathbf{A}^T$ denotes the transpose of the matrix $\mathbf{A}$. The equivalences are straightforward to verify [see Exercises 15 and 16].

Summary. Let R be a relation on a finite set S with Boolean matrix $\mathbf{A}$. Then

(R) R is reflexive if and only if all the diagonal entries of $\mathbf{A}$ are 1;
(AR) R is antireflexive if and only if all the diagonal entries of $\mathbf{A}$ are 0;
(S) R is symmetric if and only if $\mathbf{A} = \mathbf{A}^T$;
(AS) R is antisymmetric if and only if $\mathbf{A} \wedge \mathbf{A}^T \leq \mathbf{I}$, where $\mathbf{I}$ is the identity matrix;
(T) R is transitive if and only if $\mathbf{A} * \mathbf{A} \leq \mathbf{A}$.

Let R_1 and R_2 be relations from a finite set S to a finite set T with Boolean matrices $\mathbf{A}_1$ and $\mathbf{A}_2$. Then

(a) $R_1 \subseteq R_2$ if and only if $\mathbf{A}_1 \leq \mathbf{A}_2$;
(b) $R_1 \cup R_2$ has Boolean matrix $\mathbf{A}_1 \vee \mathbf{A}_2$;
(c) $R_1 \cap R_2$ has Boolean matrix $\mathbf{A}_1 \wedge \mathbf{A}_2$.

Finally, composition of relations corresponds to the Boolean product of their associated matrices, as explained in Theorem 1.

Exercises 11.4

1. For each of the following Boolean matrices, consider the corresponding relation R on $\{1, 2, 3\}$. Find the Boolean matrix for R^2 and determine whether R is transitive.

(a) $\begin{bmatrix} 1 & 1 & 0 \\ 0 & 1 & 1 \\ 1 & 0 & 1 \end{bmatrix}$ (b) $\begin{bmatrix} 1 & 0 & 1 \\ 0 & 1 & 0 \\ 1 & 0 & 1 \end{bmatrix}$

(c) $\begin{bmatrix} 0 & 0 & 1 \\ 0 & 1 & 0 \\ 1 & 0 & 0 \end{bmatrix}$

2. Draw pictures of the relations in Exercise 1.

3. Let $S = \{1, 2, 3\}$ and $R = \{(2, 1), (2, 3), (3, 2)\}$.

(a) Find the matrices for R and R^2.

(b) Draw pictures of the relations in part (a).

(c) Is R transitive?

(d) Is R^2 transitive?

(e) Is $R \cup R^2$ transitive?

4. Let $S = \{1, 2, 3\}$ and $R = \{(1, 1), (1, 2), (1, 3), (3, 2)\}$.

(a) Find the matrices for R and R^2.

(b) Draw pictures of the relations in part (a).

(c) Show that R is transitive, i.e., $R^2 \subseteq R$, but that $R^2 \neq R$.

(d) Find R^n for all $n = 2, 3, \ldots$.

5. Let R be the relation on {1, 2, 3} with Boolean matrix

$$\mathbf{A} = \begin{bmatrix} 1 & 0 & 0 \\ 0 & 1 & 1 \\ 1 & 0 & 1 \end{bmatrix}.$$

(a) Find the Boolean matrix for R^n for $n \geq 0$.

(b) Is R reflexive? symmetric? transitive?

6. Repeat Exercise 5 for

$$\mathbf{A} = \begin{bmatrix} 0 & 1 & 0 \\ 1 & 1 & 1 \\ 0 & 1 & 0 \end{bmatrix}.$$

7. Consider the functions f and g from {1, 2, 3, 4} to itself defined by $f(m) = \max\{2, 4 - m\}$ and $g(m) = 5 - m$.

(a) Find Boolean matrices $\mathbf{A}_f$ and $\mathbf{A}_g$ for the relations R_f and R_g corresponding to f and g.

(b) Find Boolean matrices for $R_f R_g$ and $R_{f \circ g}$ and compare.

(c) Find Boolean matrices for the converse relations $R_f^{\leftarrow}$ and $R_g^{\leftarrow}$. Do these relations correspond to functions?

8. Give Boolean matrices for the following relations on $S = \{0, 1, 2, 3\}$.

(a) $(m, n) \in R_1$ if $m + n = 3$

(b) $(m, n) \in R_2$ if $m \equiv n \pmod 2$

(c) $(m, n) \in R_3$ if $m \leq n$

(d) $(m, n) \in R_4$ if $m + n \leq 4$

(e) $(m, n) \in R_5$ if $\max\{m, n\} = 3$

9. For each relation in Exercise 8, specify which of the properties (R), (AR), (S), (AS) and (T) the relation satisfies.

10. (a) Which of the relations in Exercise 8 are partial orders?

(b) Which of the relations in Exercise 8 are equivalence relations?

11. Let R_1 and R_2 be relations on a set S. Prove or disprove.

(a) If R_1 and R_2 are reflexive, so is $R_1 R_2$.

(b) If R_1 and R_2 are transitive, so is $R_1 R_2$.

(c) If R_1 and R_2 are symmetric, so is $R_1 R_2$.

12. What is the Boolean matrix for the equality relation E on a finite set S? Is this relation reflexive? symmetric? transitive?

13. Consider relations R_1 and R_2 from S to T and relations R_3 and R_4 from T to U.

(a) Show that $R_1(R_3 \cup R_4) = R_1 R_3 \cup R_1 R_4$.

(b) Show that $(R_1 \cap R_2)R_3 \subseteq R_1 R_3 \cap R_2 R_3$ and that equality need not hold.

(c) How are the relations $R_1(R_3 \cap R_4)$ and $R_1 R_3 \cap R_1 R_4$ related?

14. Let R_1 be a relation from S to T and R_2 be a relation from T to U. Show that the converse of $R_1 R_2$ is $R_2^{\leftarrow} R_1^{\leftarrow}$.

15. Verify statements (R), (AR), (S), (AS), and (T) in the summary at the end of this section.

16. Verify (a), (b), and (c) in the summary at the end of this section.

17. Use the associative law for relations to prove that the Boolean product is an associative operation.

18. Let R be a relation from a set S to a set T.

(a) Prove that $RR^{\leftarrow}$ is a symmetric relation on S. Don't use Boolean matrices, since S or T might be infinite.

(b) Use the fact proved in part (a) to quickly infer that $R^{\leftarrow}R$ is a symmetric relation on T.

(c) Under what conditions is $RR^{\leftarrow}$ reflexive?

19. Let R be an antisymmetric and transitive relation on a set S and, as usual, let E be the equality relation.

(a) Prove that $R \cup E$ is a partial order on S.

(b) Prove that $R \setminus E$ is a quasi-order on S.

11.5 Closures of Relations

Sometimes we may want to form new relations out of those that we already have. For example, we may have two equivalence relations, i.e., reflexive, symmetric, and transitive relations, R_1 and R_2 on S, and want to find an equivalence relation containing them both. Since R_1 and R_2 are subsets of $S \times S$, an obvious candidate is $R_1 \cup R_2$. Unfortunately, $R_1 \cup R_2$ need not be an equivalence relation; the trouble is that $R_1 \cup R_2$ need not be transitive. Well then, what *is* the smallest transitive relation containing $R_1 \cup R_2$? This turns out to be a loaded question. How do we know there is such a relation? We will see in what follows that if R is a relation on a set S, then there is always a smallest transitive relation containing R, which we will denote by $\boldsymbol{t(R)}$, and we will learn how to find it. There are also smallest relations containing R that are reflexive and symmetric; we'll denote them by $\boldsymbol{r(R)}$ and $\boldsymbol{s(R)}$.

EXAMPLE 1

Consider the relation R on {1, 2, 3, 4} whose Boolean matrix is

$$\mathbf{A} = \begin{bmatrix} 0 & 0 & 1 & 1 \\ 0 & 1 & 0 & 0 \\ 0 & 0 & 1 & 0 \\ 1 & 0 & 0 & 0 \end{bmatrix}.$$

Figure 1 gives a picture of R.

Figure 1 ▶

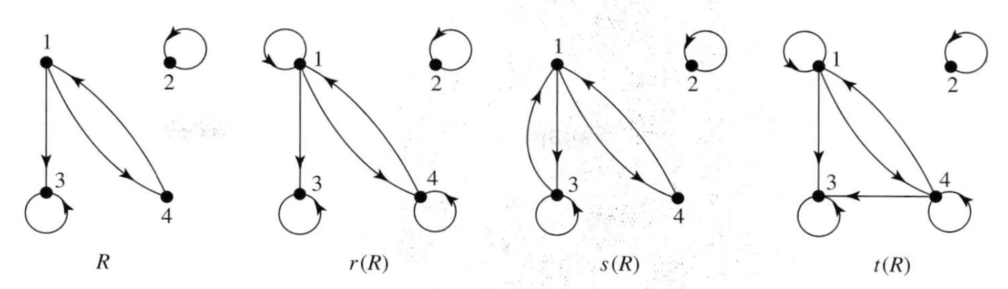

(a) The relation R is not reflexive, since neither 1 nor 4 is related to itself. To obtain the reflexive relation $r(R)$, we just need to add the two ordered pairs $(1, 1)$ and $(4, 4)$. The Boolean matrix $\mathbf{r}(\mathbf{A})$ of $r(R)$ is simply the matrix $\mathbf{A}$ with all the diagonal entries set equal to 1:

$$\mathbf{r}(\mathbf{A}) = \begin{bmatrix} 1 & 0 & 1 & 1 \\ 0 & 1 & 0 & 0 \\ 0 & 0 & 1 & 0 \\ 1 & 0 & 0 & 1 \end{bmatrix}.$$

To get the picture for $r(R)$ in Figure 1, we add the missing arrows from points to themselves.

(b) The relation R is not symmetric, since $(1, 3) \in R$, but $(3, 1) \notin R$. If we add the ordered pair $(3, 1)$ to R, we get the symmetric relation $s(R)$. Its Boolean matrix is

$$\mathbf{s}(\mathbf{A}) = \begin{bmatrix} 0 & 0 & 1 & 1 \\ 0 & 1 & 0 & 0 \\ 1 & 0 & 1 & 0 \\ 1 & 0 & 0 & 0 \end{bmatrix}.$$

To get the picture of $s(R)$ from the picture of R, we added the missing reverses of all the arrows.

(c) The relation R isn't transitive either. For example, we have $(4, 1) \in R$ and $(1, 3) \in R$, but $(4, 3) \notin R$. The scheme for finding $t(R)$ [or its Boolean matrix $\mathbf{t}(\mathbf{A})$] is not so simple as the methods for $r(R)$ and $s(R)$. Since $(4, 1) \in R$ and $(1, 3) \in R$, $t(R)$ will also contain $(4, 1)$ and $(1, 3)$. Since $t(R)$ must be transitive, $t(R)$ must contain $(4, 3)$, so we have to put $(4, 3)$ in $t(R)$. In general, if there is a path from x to y in the digraph of R, i.e., if there are points $x_1, x_2, \ldots, x_{m-1}$ so that $(x, x_1), (x_1, x_2), \ldots, (x_{m-1}, y)$ are all in R, then (x, y) must be in $t(R)$. If there is a path from x to y and also one from y to z, then there is one from x to z. So the set of all pairs (x, y) connected by paths in the digraph is a transitive relation and is the smallest transitive relation $t(R)$ containing R. In the terminology of §3.2, this relation is the reachable relation for the digraph.

To get the picture of $t(R)$ in Figure 1 from the picture of R, we added an edge from a point x to a point y whenever some path in R connected x to y and there wasn't already an edge from x to y. For example, we added the edge $(4, 3)$ because of the path 4 1 3 in R, and we added the loop $(1, 1)$ because of the path 1 4 1 in R. ■

The next proposition is nearly obvious. Think about why it's true before you read the proof.

Proposition Let R be a relation. Then $R = r(R)$ if and only if R is reflexive, $R = s(R)$ if and only if R is symmetric, and $R = t(R)$ if and only if R is transitive. Moreover,

$$r(r(R)) = r(R), \quad s(s(R)) = s(R), \quad \text{and} \quad t(t(R)) = t(R).$$

Proof If R is reflexive, then R is clearly the smallest reflexive relation containing R; i.e., $R = r(R)$. Conversely, if $R = r(R)$, then R is reflexive, since $r(R)$ is. Since $r(R)$ is reflexive, $r(R) = r(r(R))$ by what we have just shown.

The proofs for $s(R)$ and $t(R)$ are similar. ■

We can think of r, s, and t as functions that map relations to relations; for instance, s maps R to $s(R)$. Sometimes functions such as these are called operators. The proposition shows that repeating any of these three operators gives nothing new; operators with this property are called **closure operators**. On the other hand, combining two or more of these operators can lead to new relations.

EXAMPLE 2

For the relation R in Example 1 we obtained

$$\mathbf{r(A)} = \begin{bmatrix} 1 & 0 & 1 & 1 \\ 0 & 1 & 0 & 0 \\ 0 & 0 & 1 & 0 \\ 1 & 0 & 0 & 1 \end{bmatrix} \quad \text{and} \quad \mathbf{s(A)} = \begin{bmatrix} 0 & 0 & 1 & 1 \\ 0 & 1 & 0 & 0 \\ 1 & 0 & 1 & 0 \\ 1 & 0 & 0 & 0 \end{bmatrix}.$$

The Boolean matrix for $sr(R) = s(r(R))$ is

$$\mathbf{sr(A)} = \begin{bmatrix} 1 & 0 & 1 & 1 \\ 0 & 1 & 0 & 0 \\ 1 & 0 & 1 & 0 \\ 1 & 0 & 0 & 1 \end{bmatrix}.$$

This is also the matrix $\mathbf{rs(A)}$ for $rs(R)$, so $rs(R) = sr(R)$. Equality here is not an accident [Exercise 11(b)]. The relation $rs(R) = sr(R)$ is different from both of the relations $s(R)$ and $r(R)$ in this example. Combining two or more of the operators can indeed lead to new relations. ■

The next theorem gives explicit descriptions of the relations $r(R)$, $s(R)$, and $t(R)$, which are called the **reflexive, symmetric**, and **transitive closures** of R.

Theorem 1 If R is a relation on a set S and if $E = \{(x, x) : x \in S\}$, then

(r) $r(R) = R \cup E$;

(s) $s(R) = R \cup R^{\leftarrow}$;

(t) $t(R) = \bigcup_{k=1}^{\infty} R^k$.

Proof

(r) A relation on S is reflexive if and only if it contains E. Hence $R \cup E$ is reflexive, and every reflexive relation that contains R must contain $R \cup E$. So $R \cup E$ is the smallest reflexive relation containing R. This argument shows that $r(R) = R \cup E$.

(s) A relation is symmetric if and only if it is its own converse [Exercise 15 on page 100]. If $(x, y) \in R \cup R^{\leftarrow}$, then $(y, x) \in R^{\leftarrow} \cup R = R \cup R^{\leftarrow}$; thus $R \cup R^{\leftarrow}$ is symmetric. Consider a symmetric relation R' that contains R. If $(x, y) \in R^{\leftarrow}$, then $(y, x) \in R \subseteq R'$ and, since R' is symmetric, $(x, y) \in R'$. This argument shows that $R^{\leftarrow} \subseteq R'$. Since we also have $R \subseteq R'$, we conclude that $R \cup R^{\leftarrow} \subseteq R'$. Hence $R \cup R^{\leftarrow}$ is the smallest symmetric relation containing R, so $s(R) = R \cup R^{\leftarrow}$.

(t) First we show that the union $U = \bigcup_{k=1}^{\infty} R^k$ is transitive. Consider x, y, z in S such that $(x, y) \in U$ and $(y, z) \in U$. Then we must have $(x, y) \in R^k$ and $(y, z) \in R^j$ for some k and j in $\mathbb{P}$. Hence (x, z) belongs to $R^k R^j = R^{k+j}$, so that $(x, z) \in U$. Thus U is a transitive relation containing R.

Now consider any transitive relation R^* containing R. To show $U \subseteq R^*$, we prove by induction that $R^k \subseteq R^*$ for every $k \in \mathbb{P}$. The inclusion is given for $k = 1$. If the inclusion holds for some k, then

$$R^{k+1} = R^k R \subseteq R^* R \subseteq R^* R^* \subseteq R^*;$$

the last inclusion is valid because R^* is transitive [Theorem 3 on page 450]. By induction, $R^k \subseteq R^*$ for all $k \in \mathbb{P}$, so $U \subseteq R^*$. Thus U is the smallest transitive relation containing R, and

$$t(R) = U = \bigcup_{k=1}^{\infty} R^k. \quad \blacksquare$$

If S is finite, then we can improve on the description of $t(R)$ in Theorem 1.

Theorem 2 If R is a relation on a nonempty set S with n elements, then

$$t(R) = \bigcup_{k=1}^{n} R^k.$$

Proof Think of the digraph of R. The pair (x, y) is in $t(R)$ if and only if there is a path from x to y in the digraph. If there is such a path, there is one that doesn't visit the same vertex twice, unless $x = y$. It can't involve more than n vertices, so it can't have length more than n. Thus $(x, y) \in R^k$ for some $k \le n$. [This argument is essentially the proof of Theorem 1 on page 227.] ■

EXAMPLE 3

(a) Suppose that R is a relation on a set S with n elements and that $\mathbf{A}$ is a Boolean matrix for R. Theorem 2 and the matrix equivalents summarized in §11.4 show that the Boolean matrices of $t(R)$, $s(R)$, and $r(R)$ are

$$\mathbf{t(A)} = \mathbf{A} \vee \mathbf{A}^2 \vee \cdots \vee \mathbf{A}^n, \quad \mathbf{s(A)} = \mathbf{A} \vee \mathbf{A}^T, \quad \text{and} \quad \mathbf{r(A)} = \mathbf{A} \vee \mathbf{I},$$

where $\mathbf{I}$ is the $n \times n$ identity matrix and the powers $\mathbf{A}^k$ are Boolean powers.

(b) For the relation R back in Example 1, it is easy to see that $\mathbf{s(A)} = \mathbf{A} \vee \mathbf{A}^T$ and $\mathbf{r(A)} = \mathbf{A} \vee \mathbf{I}$, where $\mathbf{I}$ is the 4×4 identity matrix. One can also verify that

$$\mathbf{t(A)} = \mathbf{A} \vee \mathbf{A}^2 = \mathbf{A} \vee \mathbf{A}^2 \vee \mathbf{A}^3 \vee \mathbf{A}^4 = \begin{bmatrix} 1 & 0 & 1 & 1 \\ 0 & 1 & 0 & 0 \\ 0 & 0 & 1 & 0 \\ 1 & 0 & 1 & 1 \end{bmatrix}.$$

Of course, for such a simple relation it's easier to find $t(R)$ just by using the picture of R. ■

Given $\mathbf{A}$, it is easy to construct $\mathbf{s(A)}$ and $\mathbf{r(A)}$ using the formulas in Example 3(a). We can also use the formula in the example to compute $\mathbf{t(A)}$ by forming Boolean matrix powers $\mathbf{A}^k = \mathbf{A} * \cdots * \mathbf{A}$ for $k = 1, 2, \ldots, n$ and then combining the results. For large values of n, such a computation involves a great number of products of large matrices and is quite slow.

We can find $t(R)$ more quickly if S is large by thinking of the digraph for R. We observed in §3.2 and in Example 1 that $t(R)$ is the reachable relation for the digraph of R, whose adjacency relation is R. With each edge given weight 1, Warshall's algorithm on page 340 computes a matrix $\mathbf{W}^*$ whose (i, j)-entry is a positive integer if there is a path in the digraph from v_i to v_j and is ∞ otherwise. Replacing the ∞'s by 0's and the integers by 1's in the output produces the Boolean matrix $\mathbf{t(A)}$. In fact [Exercise 18], minor changes in the algorithm itself allow us to compute $\mathbf{t(A)}$ using Boolean entries and Boolean operations without the need for ∞'s or large integer calculations.

EXAMPLE 4

Again, let R be the relation in Example 1. We found in Example 2 that $rs(R) = sr(R)$ and that the Boolean matrix is

$$\mathbf{rs(A)} = \mathbf{sr(A)} = \begin{bmatrix} 1 & 0 & 1 & 1 \\ 0 & 1 & 0 & 0 \\ 1 & 0 & 1 & 0 \\ 1 & 0 & 0 & 1 \end{bmatrix}.$$

The transitive closure of $sr(R) = rs(R)$ turns out to have the matrix

$$\mathbf{tsr(A)} = \mathbf{trs(A)} = \begin{bmatrix} 1 & 0 & 1 & 1 \\ 0 & 1 & 0 & 0 \\ 1 & 0 & 1 & 1 \\ 1 & 0 & 1 & 1 \end{bmatrix},$$

which is also the matrix of the equivalence relation on $\{1, 2, 3, 4\}$ whose equivalence classes are $\{2\}$ and $\{1, 3, 4\}$. Thus $tsr(R)$ is transitive, symmetric, and reflexive. This fact may seem obvious from the notation tsr, but we must be careful here. It is conceivable, for example, that the transitive closure of a symmetric relation might not itself be symmetric. Our next example will show that the symmetric closure of a transitive relation need not be transitive, so we must be cautious about jumping to conclusions on the basis of notation. ■

EXAMPLE 5

Let R be the relation on $\{1, 2, 3\}$ with Boolean matrix

$$\mathbf{A} = \begin{bmatrix} 1 & 1 & 1 \\ 0 & 1 & 0 \\ 0 & 0 & 1 \end{bmatrix}.$$

Then R is reflexive [since $\mathbf{I} \leq \mathbf{A}$] and transitive [since $\mathbf{A} * \mathbf{A} = \mathbf{A}$], so $\mathbf{A} = \mathbf{r(A)} = \mathbf{t(A)} = \mathbf{tr(A)} = \mathbf{rt(A)}$. The relation $s(R)$ has matrix

$$\mathbf{s(A)} = \begin{bmatrix} 1 & 1 & 1 \\ 1 & 1 & 0 \\ 1 & 0 & 1 \end{bmatrix}.$$

Since $(2, 1)$ and $(1, 3)$ are in $s(R)$, but $(2, 3)$ is not, $s(R)$ is not transitive. We have

$$\mathbf{st(A)} = \mathbf{str(A)} = \mathbf{srt(A)} = \mathbf{s(A)} \neq \mathbf{ts(A)} = \mathbf{tsr(A)} = \mathbf{trs(A)}.$$

The order in which we form closures does matter. ■

The next lemma shows that the trouble we encountered in Example 5 is the only kind there is. Except in the case of symmetric closures, which can destroy transitivity as we just saw, forming closures preserves the properties that we already have.

Lemma

(a) If R is reflexive, then so are $s(R)$ and $t(R)$.
(b) If R is symmetric, then so are $r(R)$ and $t(R)$.
(c) If R is transitive, then so is $r(R)$.

Proof

(a) This is obvious, because if $E \subseteq R$, then $E \subseteq s(R)$ and $E \subseteq t(R)$.
(b) See Exercise 10.
(c) Suppose that R is transitive, and consider (x, y) and (y, z) in $r(R) = R \cup E$. If $(x, y) \in E$, then $x = y$, so $(x, z) = (y, z)$ is in $R \cup E$. If $(y, z) \in E$, then $y = z$, so $(x, z) = (x, y)$ is in $R \cup E$. If neither (x, y) nor (y, z) is in E, then both are in R, so $(x, z) \in R \subseteq R \cup E$ by the transitivity of R. Hence $(x, z) \in R \cup E$ in all cases. ■

The next theorem gives an answer to the basic question with which we began this section.

Theorem 3 For any relation R on a set S, $tsr(R)$ is the smallest equivalence relation containing R.

Proof Since $r(R)$ is reflexive, two applications of (a) of the lemma show that $tsr(R)$ is reflexive. Since $sr(R)$ is automatically symmetric, one application of (b) of the lemma shows that $tsr(R)$ is symmetric. Finally, $tsr(R)$ is automatically transitive, so $tsr(R)$ is an equivalence relation.

Now consider any equivalence relation R' such that $R \subseteq R'$. Then $r(R) \subseteq r(R') = R'$; hence $sr(R) \subseteq s(R') = R'$ and thus $tsr(R) \subseteq t(R') = R'$. Therefore, $tsr(R)$ is the smallest equivalence relation containing R. ■

EXAMPLE 6

(a) In Example 4, $tsr(R)$ was shown to be the equivalence relation with equivalence classes $\{2\}$ and $\{1, 3, 4\}$.

(b) Let R be the relation on $\{1, 2, 3\}$ in Example 5. Then

$$\mathbf{r(A)} = \begin{bmatrix} 1 & 1 & 1 \\ 0 & 1 & 0 \\ 0 & 0 & 1 \end{bmatrix}, \quad \mathbf{sr(A)} = \begin{bmatrix} 1 & 1 & 1 \\ 1 & 1 & 0 \\ 1 & 0 & 1 \end{bmatrix}, \quad \mathbf{tsr(A)} = \begin{bmatrix} 1 & 1 & 1 \\ 1 & 1 & 1 \\ 1 & 1 & 1 \end{bmatrix}.$$

The smallest equivalence relation containing R is $\{1, 2, 3\} \times \{1, 2, 3\}$, the universal relation. These computations can be double-checked by drawing pictures for the corresponding relations. ■

Theorem 3 has a graph-theoretic interpretation. Starting with a digraph whose adjacency relation is R, we form the digraph for $r(R)$ by putting loops at all vertices. Then we form the digraph for $sr(R)$ by adding edges, if necessary, so that whenever there is an edge from x to y, then there is also one from y to x. The result is essentially an undirected graph with a loop at each vertex. Now the graph for $tsr(R)$ has edges joining pairs (x, y) that are joined by paths in the graph for $sr(R)$. Thus (x, y) is in the relation $tsr(R)$ if and only if x and y are in the same connected component of the graph for $sr(R)$. Any algorithm that computes connected components, such as Forest on page 258, can compute $tsr(R)$ as well.

Exercises 11.5

1. Consider the relation R on $\{1, 2, 3\}$ with Boolean matrix $\mathbf{A} = \begin{bmatrix} 0 & 1 & 0 \\ 0 & 0 & 0 \\ 0 & 0 & 1 \end{bmatrix}$. Find the Boolean matrices for

(a) $r(R)$ (b) $s(R)$ (c) $rs(R)$
(d) $sr(R)$ (e) $tsr(R)$

2. Repeat Exercise 1 with $\mathbf{A} = \begin{bmatrix} 0 & 1 & 1 \\ 0 & 0 & 1 \\ 0 & 0 & 0 \end{bmatrix}$.

3. For R as in Exercise 1, list the equivalence classes of $tsr(R)$.

4. For R as in Exercise 2, list the equivalence classes of $tsr(R)$.

5. Repeat Exercise 1 for the relation R on $\{1, 2, 3, 4, 5\}$ with Boolean matrix

$$\mathbf{A} = \begin{bmatrix} 0 & 1 & 0 & 0 & 0 \\ 0 & 1 & 0 & 1 & 0 \\ 0 & 0 & 0 & 0 & 1 \\ 0 & 1 & 0 & 0 & 0 \\ 0 & 0 & 0 & 0 & 0 \end{bmatrix}.$$

6. List the equivalence classes of $tsr(R)$ for the relation R of Exercise 5.

7. Let R be the usual quasi-order relation on $\mathbb{P}$; i.e., $(m, n) \in R$ in case $m < n$. Find or describe

(a) $r(R)$ (b) $sr(R)$ (c) $rs(R)$
(d) $tsr(R)$ (e) $t(R)$ (f) $st(R)$

8. Repeat Exercise 7 where now $(m, n) \in R$ means that m divides n.

9. The Fraternal Order of Hostile Hermits is an interesting organization. Hermits know themselves. In addition, everyone knows the High Hermit, but neither he nor any of the other members knows any other member. Define the relation R on the F.O.H.H. by $(h_1, h_2) \in R$ if h_1 knows h_2. Determine $st(R)$ and $ts(R)$ and compare. [The High Hermit acts a little like a file server in a computer network.]

10. (a) Show that if $R_1, R_2, \ldots$ is a sequence of symmetric relations on a set S, then the union $\bigcup_{k=1}^{\infty} R_k$ is symmetric.

(b) Let R be a symmetric relation on S. Show that R^n is symmetric for each n in $\mathbb{P}$.

(c) Show that if R is symmetric, then so are $r(R)$ and $t(R)$.

11. Consider a relation R on a set S.

(a) Show that $tr(R) = rt(R)$.

(b) Show that $sr(R) = rs(R)$.

12. Let R_1 and R_2 be binary relations on the set S.

(a) Show that $r(R_1 \cup R_2) = r(R_1) \cup r(R_2)$.

(b) Show that $s(R_1 \cup R_2) = s(R_1) \cup s(R_2)$.

(c) Is $r(R_1 \cap R_2) = r(R_1) \cap r(R_2)$ always true? Explain.

(d) Is $s(R_1 \cap R_2) = s(R_1) \cap s(R_2)$ always true? Explain.

13. Consider two equivalence relations R_1 and R_2 on a set S.

(a) Show that $t(R_1 \cup R_2)$ is the smallest equivalence relation that contains both R_1 and R_2. *Hint:* Use Exercise 12, parts (a) and (b).

(b) Describe the largest equivalence relation contained in both R_1 and R_2.

14. Show that $st(R) \neq ts(R)$ for the relation R in Example 5.

15. Show that there does not exist a smallest antireflexive relation containing the relation R on $\{1, 2\}$ whose Boolean matrix is $\begin{bmatrix} 1 & 0 \\ 1 & 0 \end{bmatrix}$.

16. We say that a relation R on a set S is an **onto relation** if, for every $y \in S$, there is an x in S such that (x, y) is in R. Show that there does not exist a smallest onto relation containing the relation R on $\{1, 2\}$ specified in Exercise 15.

17. Suppose that a property p of relations on a nonempty set S satisfies the following:

(i) the universal relation $S \times S$ has property p;

(ii) p is **closed under intersections**; i.e., if $\{R_i : i \in I\}$ is a nonempty family of relations on S possessing property p, then the intersection $\bigcap_{i \in I} R_i$ also possesses property p.

(a) Prove that for every relation R there is a smallest relation that contains R and has property p.

(b) Observe that the properties reflexivity, symmetry, and transitivity satisfy both (i) and (ii).

(c) In view of Exercise 15, antireflexivity cannot satisfy (i) and (ii). Which of (i) and (ii) fail?

(d) Which of (i) and (ii) does the property "onto relation" in Exercise 16 fail to satisfy?

18. Give a modified version of Warshall's algorithm to compute $\mathbf{t}(\mathbf{A})$ directly from $\mathbf{A}$ using Boolean operations.

Chapter Highlights

As usual: What does it mean? Why is it here? How can I use it? Think of examples.

CONCEPTS AND NOTATION

partial order, $\preceq$, poset, subposet
- converse, $\succeq$
- quasi-order, $\prec$
- Hasse diagram
- maximal, minimal, largest, smallest elements
- upper, lower bound
  - least upper bound, $x \vee y = \text{lub}\{x, y\}$
  - greatest lower bound, $x \wedge y = \text{glb}\{x, y\}$
- lattice
- chain = totally ordered set = linearly ordered set, maximal chain
  - well-ordered set
- product order on $S_1 \times \cdots \times S_n$
- filing order on $S_1 \times \cdots \times S_n$, $\preceq^k$ on S^k
- lexicographic = dictionary order $\preceq_L$ on Σ^*
- lenlex order $\preceq_{LL}$ on Σ^*

composite relation $R_2 \circ R_1 = R_1 R_2$

equality relation E

Boolean matrix, Boolean product $*$

FACTS

Every finite poset has a Hasse diagram.

Filing order on $S_1 \times \cdots \times S_n$ is linear if each S_i is a chain.

If Σ is a chain, then lenlex order on Σ^* is a well-ordering, and lexicographic order is linear but not a well-ordering.

Composition of relations is associative.

Multiplication of matrices is associative.

The matrix of the composite R_1R_2 of relations is the Boolean product $\mathbf{A}_1 * \mathbf{A}_2$ of their matrices.

The relation R on S is transitive if and only if $R^2 \subseteq R$ if and only if $\mathbf{A} * \mathbf{A} \leq \mathbf{A}$.

Matrix analogs of some common relation statements are summarized at the end of §11.4.

The operators r, s, and t given by $r(R) = R \cup E$, $s(R) = R \cup R^{\leftarrow}$, and $t(R) = \bigcup_{k=1}^{\infty} R^k$ are closure operators on the class of all relations.

$t(R) = \bigcup_{k=1}^{n} R^k$ if $|S| = n$.

The relation $st(R)$ may not be transitive.

The smallest equivalence relation containing R is $tsr(R)$.

METHODS

Warshall's algorithm with all edge weights 1 can compute $t(R)$.

The algorithm Forest can compute $tsr(R)$ by computing connected components of the graph of $sr(R)$.

Supplementary Exercises

1. Draw the Hasse diagram for the poset $(P, \preceq)$ of positive integer divisors of 18, with $m \preceq n$ in case $m|n$.

2. Consider the poset $(\{1, 2, 3, 4, 5, 6, 8, 12\}, |)$, where $m|n$ means that m divides n; i.e., that n/m is an integer.

(a) Draw a Hasse diagram for the poset.

(b) Is there a largest element (i.e., maximum)? If yes, give it.

(c) Is there a smallest element (i.e., minimum)? If yes, give it.

(d) List all maximal elements.

(e) List all minimal elements.

(f) List all maximal chains.

(g) Is the poset a lattice? Explain briefly.

3. (a) Give an example of a maximal chain in $\mathbb{B}^4$.

(b) Give an example of a chain of length 4 in BOOL(2) or explain why no such chain exists.

4. (a) List four members of $\{a, b\}^*$ in increasing lexicographic order, given that $a < b$.

(b) List four members of $\{0, 1\}^2$ in increasing filing order.

(c) List four members of a chain in $\mathbb{B}^3$ in increasing $\leq$ order.

5. Give an example of each of the following.

(a) A poset that is not a chain.

(b) A chain with infinitely many members.

(c) The matrix of a transitive relation that is not reflexive.

(d) The digraph of a reflexive, symmetric relation that is not transitive.

6. For each part, indicate whether the statement is true or false.

(a) If a set Σ is totally ordered, then the corresponding lexicographic partial order on Σ^* also must be totally ordered.

(b) If a set Σ is totally ordered, then the corresponding lenex order on Σ^* also must be totally ordered.

(c) Every finite partially ordered set has a Hasse diagram.

(d) Every finite partially ordered set has a topological sorting.

(e) Every finite partially ordered set has a smallest element.

(f) Every finite totally ordered set has a largest element.

(g) An infinite partially ordered set cannot have a largest element.

7. (a) List the members 001, 00010, 11, and 101 of $\{0, 1\}^*$ in increasing lenlex order and in increasing lexicographic order.

(b) List the members of $\{0, 1\}^3$ between 001 and 101 in the filing order.

(c) List the members of $\{0, 1\}^*$ between 01 and 0100 in lexicographic order.

(d) How many members of $\{0, 1\}^*$ are there between 001 and 01 in lexicographic order?

8. Let $\mathbf{A}$ be the Boolean matrix

$$\mathbf{A} = \begin{bmatrix} 1 & 0 & 0 & 1 \\ 0 & 0 & 0 & 1 \\ 1 & 1 & 0 & 1 \\ 0 & 0 & 0 & 1 \end{bmatrix}.$$

(a) Calculate $\mathbf{A} * \mathbf{A}$.

(b) Is the relation corresponding to $\mathbf{A}$ transitive? Explain.

9. Let $\Sigma = \{a, b, c, e, r\}$ with the usual order. List the words *career*, *cab*, *car*, *crab*, *bare*, *arc*, λ, *bear*, *crabber*, *beer*, *care*, *brace*, *bar* in Σ^* in increasing order

(a) Using the lenlex order.

(b) Using the lexicographic order.

10. The relation R on $\{1, 2, 3, 4\}$ has corresponding matrix

$$\mathbf{A} = \begin{bmatrix} 0 & 1 & 0 & 0 \\ 0 & 0 & 0 & 0 \\ 1 & 0 & 1 & 1 \\ 0 & 1 & 0 & 1 \end{bmatrix}.$$

(a) Draw the digraph that corresponds to R and $\mathbf{A}$.
Find each of the following.

(b) $\mathbf{A} * \mathbf{A}$ (c) $s(\mathbf{A})$ (d) $t(\mathbf{A})$

(e) Is the symmetric closure of R an equivalence relation?

11. Suppose that $\{1, 2, 3, 4, 5\}$ and $\{15, 16, 17\}$ both have the natural partial order $\leq$ inherited from $\mathbb{Z}$.

(a) Explain why $\{1, 2, 3, 4, 5\} \times \{15, 16, 17\}$ with the product order is *not* a chain.

(b) Describe a way to use the order $\leq$ to make $\{1, 2, 3, 4, 5\} \times \{15, 16, 17\}$ into a chain. Give the details—don't just give a name.

12. Draw a Hasse diagram for a poset that has exactly five members, two of which are maximal and one of which is the poset's minimum. Label the maximal and minimum members on your diagram.

13. The set $\mathbb{B} = \{0, 1\}$ has the natural order $\leq$ with $0 \leq 0$, $0 \leq 1$ and $1 \leq 1$. Is $\mathbb{B}^{17}$ with the filing order derived from this order a chain? Explain.

14. Consider the poset P with Hasse diagram in Figure 2.

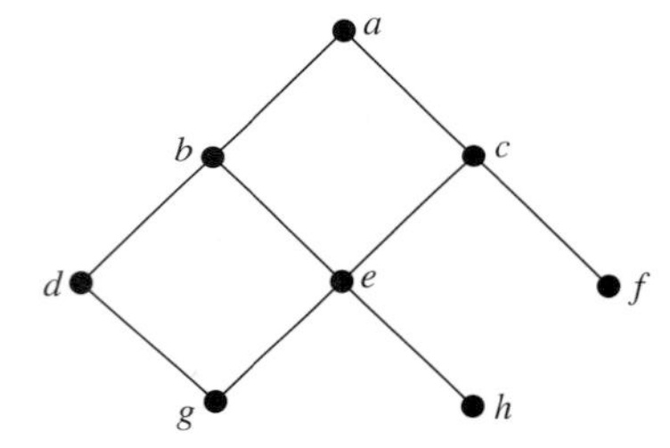

Figure 2 ▲

(a) List the maximal members of P.

(b) Is there a largest member of P? If so, what is it?

(c) List the minimal members of P.

(d) Is there a smallest member of P? If so, what is it?

(e) Determine the following when they exist: $\text{lub}\{b, c\}$, $\text{glb}\{b, c\}$, $\text{lub}\{b, f\}$, $\text{glb}\{b, f\}$, $\text{lub}\{g, c\}$, and $\text{glb}\{g, c\}$.

15. Consider the lattices in Figure 3.

(a) Calculate $D \vee (E \wedge A)$ and $(D \vee E) \wedge (D \vee A)$. What do you notice?

(b) Calculate $a \vee (b \wedge c)$ and $(a \vee b) \wedge (a \vee c)$. What do you notice?

(c) Calculate $A \wedge (B \vee C)$, $(A \wedge B) \vee C$, $x \wedge (z \vee y)$, and $(x \wedge z) \vee y$.

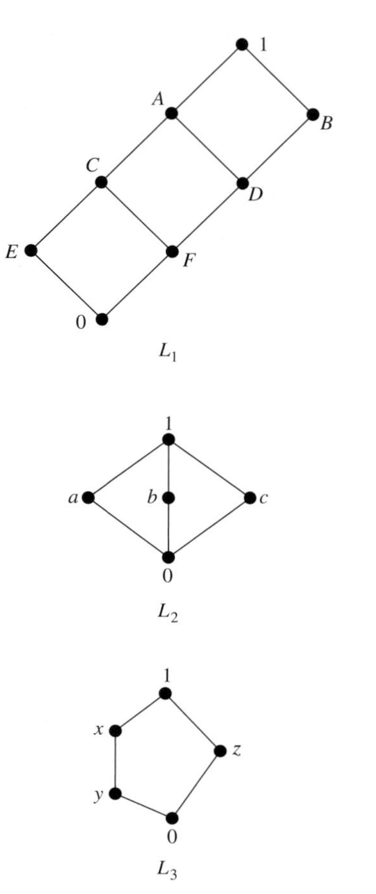

Figure 3 ▲

16. Consider the following relations on $\mathbb{P} = \{1, 2, 3, \ldots\}$: $(m, n) \in R_1$ if $m|n$; $(m, n) \in R_2$ if $|m - n| \leq 2$; $(m, n) \in R_3$ if $m + n$ is even; and $(m, n) \in R_4$ if 3 divides $m + n$. For each relation, determine whether it is reflexive, symmetric, anti-symmetric, or transitive. Also, which relations are partial orders, and which are equivalence relations?

17. Let R_1 and R_2 be the relations on $\{1, 2, 3\}$ with Boolean matrices $\mathbf{A}_1 = \begin{bmatrix} 1 & 1 & 1 \\ 0 & 1 & 0 \\ 0 & 1 & 1 \end{bmatrix}$ and $\mathbf{A}_2 = \begin{bmatrix} 0 & 0 & 1 \\ 1 & 0 & 0 \\ 0 & 0 & 1 \end{bmatrix}$.

(a) Find the Boolean matrices for R_1^2, R_1R_2, and the converse relation $\overleftarrow{R_1}$.

(b) Is the relation R_1 reflexive? symmetric? transitive?

(c) Repeat part (b) for R_2.

18. Let $(S, |)$ be the set $\{2, 3, 4, 5, 6, \ldots, 500\}$, with the partial order $|$, where $m|n$ means m divides n.

(a) There are 250 maximal elements. What are they?

(b) Describe the minimal elements and give four examples.

(c) Give two examples of maximal chains in $(S, |)$.

(d) Determine the following when they exist in S: $\text{lub}\{24, 54\}$, $\text{glb}\{24, 54\}$, $\text{lub}\{3, 5\}$, $\text{glb}\{3, 5\}$, and $\text{lub}\{2, 73\}$.

19. Let $S = \{2, 3, 4, 5, 6, 7, 8, 9, 10, 11, 12\}$, and define the relation R by $(m, n) \in R$ if and only if $m|n$.

(a) Does $r(R) = R$? If not, give an ordered pair in $r(R)$ that is not in R.

(b) Does $s(R) = R$? If not, give an ordered pair in $s(R)$ that is not in R.

(c) Does $t(R) = R$? If not, give an ordered pair in $t(R)$ that is not in R.

(d) For the equivalence relation $tsr(R)$, find the equivalence class containing 12.

20. Let Σ be the ordinary English alphabet. Then lexicographic order makes Σ^* into a chain. Let Δ be the set of words in Σ^* that appear in this book. If we order Δ with lexicographic order, is it also a chain? Explain.

21. (a) What should it mean for two lattices to be isomorphic?

(b) As usual, D_n is the lattice of all divisors of n with the ordering $j|k$. Which of the lattices D_{24}, D_{54}, D_{128}, and D_{30} are isomorphic to a lattice in Exercise 15?

CHAPTER 12

Algebraic Structures

This chapter contains an introduction to topics from the area of mathematics known as abstract algebra. The first three sections show how the study of groups acting on sets applies to answer some counting questions, and §12.4 presents the Fundamental Homomorphism Theorem as a tool for recognizing when two groups are essentially the same. Section 12.5 looks at semigroups and monoids, illustrating the ideas with familiar examples. The final section presents enough of the theory of rings and fields to indicate the connection between the Fundamental Homomorphism Theorem and the Chinese Remainder Theorem, a standard tool in computational algebra.

12.1 Groups Acting on Sets

Imagine being given the problem of manufacturing enough different logical circuits with four inputs and one output to compute all the different Boolean functions of four variables. On the one hand, since there are $2^4 = 16$ rows in a truth table for such a function, there are $2^{16} = 65{,}536$ such functions. On the other hand, just by switching the input connections around, maybe we can make one circuit compute various different functions, so we won't need as many different circuits as there are different functions. If we are willing to use external hardware to complement inputs and outputs, then we need to manufacture still fewer circuits. How many can we get by with? If we think of two functions as equivalent in case they can be computed by the same circuit, allowing complements, then we're really asking how many equivalence classes of functions there are.

Recall from §3.4 that these classes partition the set of functions. If all the classes had the same number of functions in them, say N, then the number of classes would be just $2^{16}/N$. In our case, though, the equivalence classes aren't all the same size. Think about this. If a function has three 1's in its truth table and thirteen 0's, then it's equivalent to at least one function with thirteen 1's and three 0's—just complement the output of the circuit—but it can't be equivalent to a function with four 1's and twelve 0's. Or can it? And how many different functions with three 0's and thirteen 1's is it equivalent to? The problem is not so easy as it might seem. Somehow we want to take advantage of the symmetry of the situation, but it's not clear how to do so. We will return to this problem at the end of §12.3.

EXAMPLE 1

Here is a simpler question on which we can use symmetry. How many essentially different ways are there to color the faces of a cube with six different given colors? We first need to clarify what we mean by "essentially different" colorings. Let's say two ways of coloring are equivalent in case it's possible to rotate the cube so that one coloring becomes the other. Not all colorings are equivalent. For example, if two colorings are equivalent and one of them has exactly one face colored green, then so must the other, and the face opposite the face colored green must have the same color for both colorings. We want to know how many equivalence classes of colorings there are. We could surely answer this question with a case-by-case count, but there's a more systematic way that will also give us the answer for other numbers of colors besides six.

First we tackle the following question: How many colorings are there in a given equivalence class? Since we are using six different colors, distinct rotations give distinct colorings. So there are as many colorings in an equivalence class as there are rotations that send the cube back onto itself. Figure 1(a) shows examples of the various sorts of rotations allowed, including the "identity" rotation, which does nothing at all. Figure 1(b) lists the numbers of each type. For instance, consider the $90°$ rotation of type b. Each such rotation has an axis through the centers of two opposite faces. There are 6 faces, so there are 3 opposite pairs and hence 3 such axes. Each axis gives two $90°$ rotations, one in each direction, so there are 6 rotations of type b. Let's also count rotations of type a. Each such rotation has an axis through opposite parallel edges of the cube. There are 12 edges of a cube and 6 pairs of opposite parallel edges. Each axis gives only one rotation, so there are 6 rotations of type a. The remaining values in Figure 1(b) have been determined by similar reasoning.

Figure 1 ▶

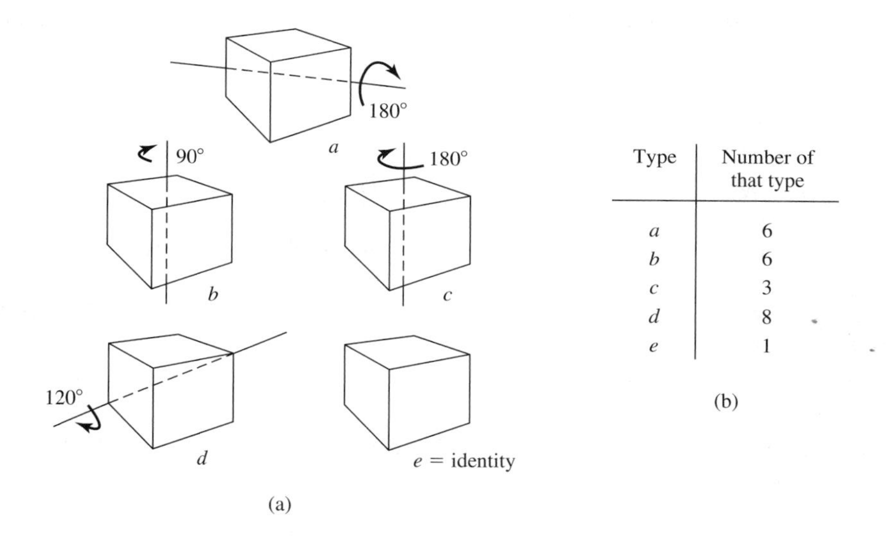

Type	Number of that type
a	6
b	6
c	3
d	8
e	1

Summing up, we see that there are 24 rotations of the cube. Since there are 6! ways to color the 6 faces of a cube, and since each equivalence class contains 24 such ways, there are $\frac{6!}{24} = 30$ essentially different ways to color the cube with 6 different colors. This problem was relatively easy, because all the equivalence classes are the same size. ■

EXAMPLE 2

A somewhat similar problem is that of counting the different ways to color the faces of a cube, with only the three colors red, blue, and green allowed. If one face is red and the rest are blue, it doesn't really matter which face is red, since we can rotate the cube to put the red face wherever we want. Given two ways of coloring the faces so that one is red, three are blue, and two are green, it may or may not be possible to rotate the cube so that the two colorings are really the same. How many essentially different ways are there to color the cube? We need to agree on what a coloring is and what it means for two colorings to be "the same," i.e., equivalent, and then we need to count the equivalence classes.

When we color a cube we are simply assigning to each face some color. That is, we can think of a **coloring** as being a function from the set X of faces of the cube to the set $C = \{\text{red}, \text{blue}, \text{green}\}$ of allowable colors. Thus the set of all allowable colorings is $\text{FUN}(X, C)$.

As in Example 1, we'll regard two colorings as equivalent in case it's possible to rotate the cube in space so that one coloring becomes the other. That is, the coloring φ is equivalent to the coloring θ in case we can color the cube with φ and then rotate the cube so that it looks as if it was colored with θ. The equivalence class of φ is the set of all colorings we can obtain by starting with φ and rotating the cube. This time not all classes have the same size. For example, a solid red coloring is not equivalent to any other coloring, but there are six equivalent colorings with one red face and five blue ones. [Why six?]

What's involved here is the symmetry of the cube. To answer our question, it's natural to consider again the group R of all rotations of space that send the cube back onto itself, since this group describes the rotational symmetry of the cube. If g and h are in R, then so is the rotation $g \circ h$ that we get by first rotating by h and then by g, and so also is g^{-1}, which simply undoes g. These two properties of R justify our using the technical term "group," which we will define later; but for now one can just think of R as a set of things that we can combine by composition.

The group R acts on the set X of faces of the cube, in the sense that if g is a rotation and x is a face of the cube, then $g(x)$, the image of the face x after the rotation, is also a face. The group R of rotations also acts on the set of colorings of the faces. The equivalence class of a coloring φ is just the set of colorings we can get from φ by acting on it with the rotations in R. ■

At this point we could continue to analyze our special cube-coloring problem, but the methods we would use are actually quite general. It makes sense to prove general results while we are at it and then see what they mean in our particular case. We will return to Example 2 in Example 9 on page 483.

In general, a set G of things that can be combined by a binary operation, say $*$, is called a **group** in case it satisfies the following four conditions.

(i) $x * y \in G$ whenever x and y are members of G.

(ii) $*$ is associative, i.e., $(x * y) * z = x * (y * z)$ for all $x, y, z \in G$.

(iii) G contains an **identity** element, say e, such that $e * x = x * e = x$ for every $x \in G$.

(iv) For every element x of G there is an **inverse**, x^{-1}, also in G, such that $x * x^{-1} = x^{-1} * x = e$.

We will often refer to the group $(G, *)$ so that we can specify the binary operation $*$. A group $(G, *)$ is **commutative** if it also satisfies the condition

$$x * y = y * x \text{ for all } x, y \in G.$$

Inverses are unique in a strong sense. If $x * y = e$, then

$$y = e * y = (x^{-1} * x) * y = x^{-1} * (x * y) = x^{-1} * e = x^{-1}.$$

Similarly, the equation $y * x = e$ forces y to be x^{-1}. To check that y is x^{-1}, then, it is enough to verify *either* of the conditions $x * y = e$ or $y * x = e$. The other will follow.

EXAMPLE 3

(a) The set R of rotations that sent the cube to itself in Example 2 is a group with binary operation composition, i.e., $\circ$. You can check (i) by holding onto the cube as you rotate it; following one rotation by another always results in a rotation. Since composition of functions is associative, (ii) holds. The identity rotation is simply the "rotation" consisting of doing nothing at all, and the inverse of a rotation, as we already observed, just undoes the original rotation.

(b) The set $\mathbb{Z}$ of integers is a commutative group with the operation $+$. The identity element is 0, and the inverse of n is $-n$. The set $\mathbb{Z}$ is *not* a group under multiplication as operation. Multiplication is associative and commutative, and the identity is 1, but the element 0 has no inverse in $\mathbb{Z}$.

$+_3$	0	1	2
0	0	1	2
1	1	2	0
2	2	0	1

(a)

$*_3$	0	1	2
0	0	0	0
1	0	1	2
2	0	2	1

(b)

Figure 2 ▲

(c) Similarly, the set $\mathbb{Z}(n)$ of integers mod n is a commutative group under the operation $+_n$, but not under multiplication, $*_n$. Figure 2 gives the tables for $+_3$ and $*_3$. The inverse of 2 under $+_3$ is 1, since $2 +_3 1 = 0 = 1 +_3 2$. Since $2 *_3 2 = 1$, 2 is its own inverse under $*_3$; but 0 has no inverse under $*_3$, as we see from the fact that the first row and column of the table contain no 1's.

(d) The set S_n of all permutations of the set $\{1, 2, \ldots, n\}$ is a group under composition as operation. Recall that a permutation of a set X is a one-to-one function from X onto X. We saw a long time ago that if f and g are permutations, then so is $f \circ g$ and so are f^{-1} and g^{-1}. In fact, Exercises 15 and 14 on page 45 assert that the composition of one-to-one correspondences is a one-to-one correspondence and the inverse of a one-to-one correspondence is also a one-to-one correspondence. In particular, $f \circ f^{-1} = f^{-1} \circ f = e$ for all $f \in S_n$, where e is the **identity permutation**: $e(k) = k$ for $k \in \{1, 2, \ldots, n\}$. The group S_n, with the operation understood to be $\circ$, is called the **symmetric group of degree n**.

For $n \geq 3$, S_n is *not* commutative; i.e., $f \circ g$ need not equal $g \circ f$. In fact, for $n \geq 3$ there are permutations f, g in S_n such that $f(1) = 1$, $f(2) = 3$, $g(1) = 2$, and $g(2) = 1$. Then $f \circ g(1) = 3 \neq 2 = g \circ f(1)$, so $f \circ g \neq g \circ f$. The group S_n has $n!$ elements, a number that gets large fast as n increases.

(e) More generally, just as in part (d), if X is any set whatever the collection PERM(X) of all permutations of X is a group under $\circ$ as operation. ■

Groups such as $(\mathbb{Z}, +)$ and $(\mathbb{Z}(n), +_n)$ in Examples 3(b) and (c) are interesting in their own right, but for the counting applications that we will study we will concentrate on groups of functions, such as those in Examples 3(a), (d), and (e). In such cases, and in general when the operation is not addition, we will frequently drop the $\circ$ and simply write fg for the product $f \circ g$.

Sometimes a group can be viewed as a group of permutations of a set, even though it was not originally defined that way. We say that the group $(G, \cdot)$ **acts on** the set X in case, for each $g \in G$, there is a permutation, say g^*, of X so that $(g \cdot h)^* = g^* \circ h^*$ for all $g, h \in G$. Usually, the action will be quite natural, and in fact we will often just write g instead of g^* when the action is obvious, so the condition for action becomes simply

$$(g \cdot h)(x) = g(h(x))$$

for all g and h in G and x in X.

EXAMPLE 4

(a) In Example 2 we observed that the group R of rotations that sends the cube back onto itself acts on the set of faces of the cube. Each rotation in R produces a permutation of the set of faces of the cube, and the permutation produced by the composition of two rotations is also a rotation.

(b) This same group also acts on the set of vertices of the cube. Each rotation sends vertices to vertices.

(c) This group of rotations also acts on the set of colorings of the faces of the cube. Rotating the cube turns one coloring into another [or maybe sometimes into the same coloring again]. If φ is a coloring and g is a rotation, then the coloring $g^*(\varphi)$ is the one we get by first coloring the cube with φ and then rotating it by g. To figure out what color $g^*(\varphi)$ gives a face x, we can rotate backward by g^{-1} and see what color φ gave $g^{-1}(x)$; i.e., $(g^*(\varphi))(x) = \varphi(g^{-1}(x))$.

If h is another rotation in R, then $(gh)^*(\varphi)$ is what we get by coloring by φ, then rotating by gh, i.e., rotating by h and then by g. This is the same as the coloring $g^*(h^*(\varphi)) = (g^* \circ h^*)(\varphi)$. Since $(gh)^*(\varphi) = (g^* \circ h^*)(\varphi)$ for every φ, we have $(gh)^* = g^* \circ h^*$, so this definition of g^* meets our condition to define an action of R.

(d) The symmetric group S_n also acts on the set $\{1, 2, \ldots, n\} \times \{1, 2, \ldots, n\}$ of ordered pairs (i, j) by letting $g^*((i, j)) = (g(i), g(j))$. For example, if $g(1) = 5$ and $g(3) = 2$, then $g^*((1, 3)) = (5, 2)$. ■

We will be most interested in cases in which G is a group of permutations associated with the symmetry of some object, such as the cube or a graph or digraph. Consider, for example, a digraph D with no parallel edges. A permutation g of the set $V(D)$ of vertices is a **digraph automorphism** provided it preserves the edge structure, that is, provided (x, y) is an edge if and only if $(g(x), g(y))$ is an edge. We can think of a digraph automorphism as moving the vertex labels around without essentially changing the structure of the digraph. We write **AUT(D)** for the set of automorphisms of D. It is easy to verify that AUT(D) is a group of permutations acting on $V(D)$ [Exercise 13]. Very roughly speaking, the larger AUT(D) is the more symmetry D has.

EXAMPLE 5

The digraph D drawn in Figure 3(a) has two sources and two sinks. Digraph automorphisms must send sources to sources and sinks to sinks. [Why?] One can verify that there are four automorphisms: the identity permutation e, which moves nothing; the permutation f that interchanges q and s; the permutation g that interchanges p and r; and the permutation fg that switches both the pair q and s and the pair p and r. For convenience, their function values are listed in Figure 3(b). Notice that g, for example, switches the labels p and r, but we could have done the same thing by just flipping the digraph over, without changing its structure. The multiplication table is in Figure 3(c). It is easily seen that the group AUT(D) is commutative. ■

Figure 3 ▶

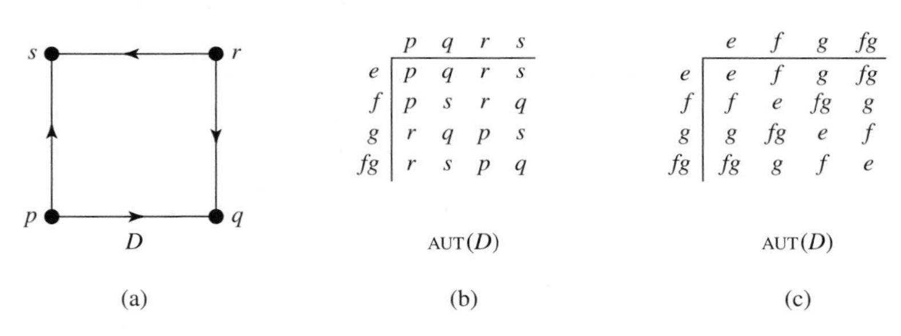

	p	q	r	s
e	p	q	r	s
f	p	s	r	q
g	r	q	p	s
fg	r	s	p	q

AUT(D)

(b)

	e	f	g	fg
e	e	f	g	fg
f	f	e	fg	g
g	g	fg	e	f
fg	fg	g	f	e

AUT(D)

(c)

EXAMPLE 6

(a) The digraph E in Figure 4(a) has another kind of symmetry and a different automorphism group. Every automorphism of E must send x to itself, but the remaining vertices can be permuted in any way. So AUT(E) is essentially PERM($\{y, z, w\}$) or S_3. Figure 4(b) gives the function values of the automorphisms, where g is the "rotation" that sends y to z, z to w, and w to y, and h is the "flip" that interchanges z and w. Figure 4(c) gives the multiplication table of AUT(E).

(b) There is another way to see AUT(E) as a group of permutations of three objects. In general, the group of automorphisms of a digraph, which we have defined to be a group of permutations of vertices, also acts on the set of edges of the digraph. If an edge goes from u to v, then an automorphism g takes it to the

Figure 4 ▶

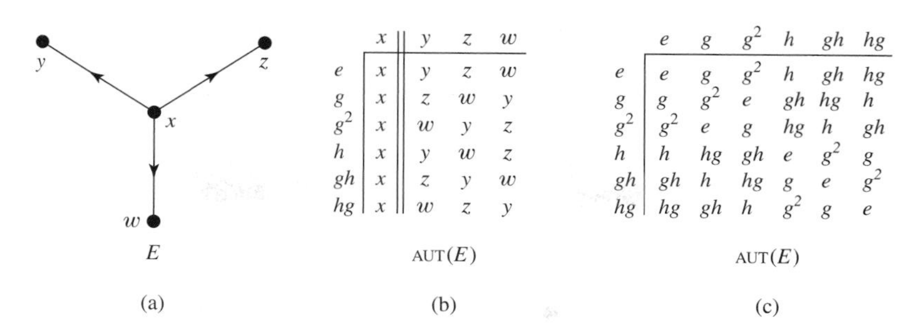

	x	y	z	w
e	x	y	z	w
g	x	z	w	y
g^2	x	w	y	z
h	x	y	w	z
gh	x	z	y	w
hg	x	w	z	y

AUT(E)

	e	g	g^2	h	gh	hg
e	e	g	g^2	h	gh	hg
g	g	g^2	e	gh	hg	h
g^2	g^2	e	g	hg	h	gh
h	h	hg	gh	e	g^2	g
gh	gh	h	hg	g	e	g^2
hg	hg	gh	h	g^2	g	e

AUT(E)

(a) (b) (c)

edge from $g(u)$ to $g(v)$, by the definition of automorphism. In the present case, AUT(E) acts on the three-element set of edges of E. ■

Automorphism groups of undirected graphs can be studied in the same way. For a graph H without parallel edges, a permutation g of $V(H)$ is a **graph automorphism** in case $\{x, y\}$ is an edge if and only if $\{g(x), g(y)\}$ is an edge. The group of all such automorphisms is denoted by AUT(H).

EXAMPLE 7 The symmetry of the graph H shown in Figure 5(a) is described by its group AUT(H), which is listed in the table in Figure 5(b). Note that g interchanges w and x, h interchanges y and z, whereas f flips the graph horizontally. ■

Figure 5 ▶

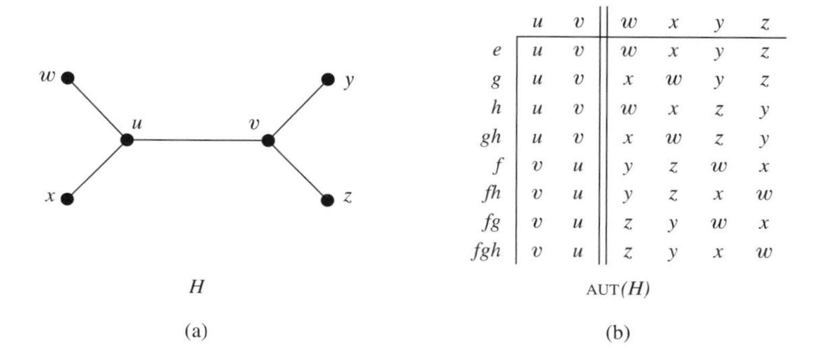

	u	v	w	x	y	z
e	u	v	w	x	y	z
g	u	v	x	w	y	z
h	u	v	w	x	z	y
gh	u	v	x	w	z	y
f	v	u	y	z	w	x
fh	v	u	y	z	x	w
fg	v	u	z	y	w	x
fgh	v	u	z	y	x	w

AUT(H)

(a) (b)

Now we come to the idea that links groups with our coloring and circuit-counting problems. Recall that we want to think of two colorings as equivalent in case we can turn one into the other by rotating the cube. Here's the general situation.

Proposition 3 Let G be a group acting on a set X. For $x, y \in X$, define $x \sim y$ if $y = g(x)$ for some $g \in G$. Then $\sim$ is an equivalence relation on X. For $x \in X$, the equivalence class containing x is

$$\boldsymbol{G(x)} = \{g(x) : g \in G\}.$$

[For general equivalence relations, we have usually written $[x]$ for equivalence classes, but $G(x)$ is a more natural notation here.]

Proof It suffices to verify that $\sim$ is reflexive, symmetric, and transitive.

(R) Let x be in X. Since $e \in G$ and $x = e(x)$, we have $x \sim x$.

(S) Suppose that $x \sim y$, where $x, y \in X$. This means that $y = g(x)$ for some $g \in G$. Since $g^{-1} \in G$ and $x = g^{-1}(y)$, we have $y \sim x$.

(T) Suppose that $x \sim y$ and $y \sim z$. Then $y = g(x)$ and $z = h(y)$ for some g and h in G. Since $hg \in G$ and $z = (h(g(x)) = (hg)(x)$, we conclude that $x \sim z$. ■

These special equivalence classes $G(x)$ are called **orbits of** G **on** X, or G-**orbits**. Equivalence relations determine partitions into equivalence classes [Theorem 1 on page 116], so we get the following.

Corollary If G acts on X, then the G-orbits partition X.

EXAMPLE 8

(a) Two colorings of the cube are equivalent in our original sense of Example 2 if and only if they are in the same orbit of the group R of rotations of the cube. Thus the number of essentially different colorings with a given set C of colors is the number of orbits of R acting on the set $\text{FUN}(X, C)$ of colorings.

(b) If $G = \text{PERM}(X)$, then every element of X is equivalent to every other element of X since, given any x and y in X, some permutation will map x to y. Hence there is only one big orbit, X itself.

(c) Consider again the digraph D in Figure 3. To find the orbits under $\text{AUT}(D)$ in a systematic way, we start with p and find

$$\text{AUT}(D)(p) = \{e(p), f(p), g(p), fg(p)\} = \{p, p, r, r\} = \{p, r\};$$

see the first column of Figure 3(b). Now we consider a vertex not yet found and repeat the process. We find

$$\text{AUT}(D)(q) = \{q, s, q, s\} = \{q, s\}$$

from the second column of Figure 3(b). We conclude that there are two orbits, $\{p, r\}$ and $\{q, s\}$.

(d) The orbits of vertices of the digraph E of Figure 4 under $\text{AUT}(E)$ can be read off from Figure 4(b) or from the picture. They are $\{x\}$ and $\{y, z, w\}$.

(e) One should also be able to see the orbits of vertices of the graph H in Figure 5 under $\text{AUT}(H)$ by staring at the figure and visualizing the graph automorphisms. [See Exercise 8.] The orbits are $\{u, v\}$ and $\{w, x, y, z\}$; see Figure 5(b). Notice that some of the automorphisms are not just the result of rotating or flipping the graph as a whole. ■

In §12.3 we will look at methods for counting how many orbits a group has on a set. Our methods will not involve actually finding the orbits and will apply no matter how complicated the situation is. For now, though, the only counting method we know is to determine the orbits directly somehow and then see how many we get.

Exercises 12.1

1. Consider the graph in Figure 6(a), and let G be its group of graph automorphisms.
   (a) Color the vertices by coloring w and x black and y and z red. Give all the equivalent colorings.
   (b) Repeat, where w is black and all the rest of the vertices are red.
   (c) Convince yourself that $G = \{e, g, h, gh\}$, where these automorphisms are given in Figure 6(b).
   (d) Give the multiplication table of G.
   (e) Can you guess how many different colorings of the vertices of this graph use the colors puce, magenta, or both?

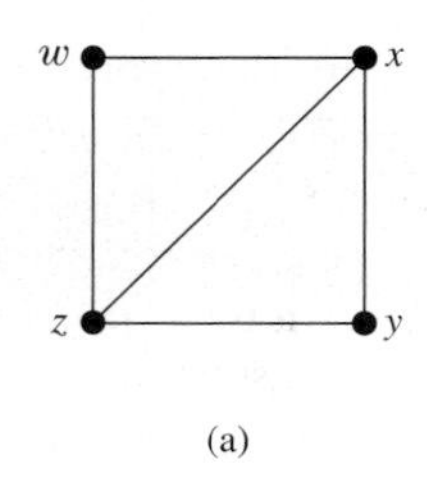

(a)

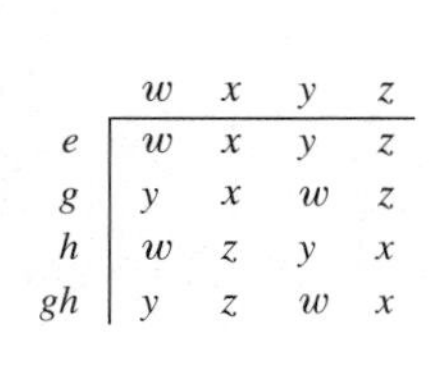

	w	x	y	z
e	w	x	y	z
g	y	x	w	z
h	w	z	y	x
gh	y	z	w	x

(b)

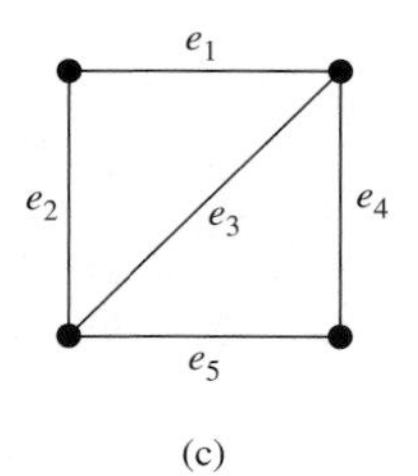

(c)

Figure 6 ▲

2. (a) Color the edges in Figure 6(c) by coloring e_1, e_3, e_5 red and e_2, e_4 black. Give all equivalent colorings. [Observe that the automorphism group here acts on the set of edges as well as on the set of vertices.]
   (b) Repeat, where e_1, e_2, e_3 are red and e_4, e_5 are black.

3. Consider the problem of coloring the four triangular faces of a square pyramid [like the ones in Egypt]. Suppose that we consider two colorings to be equivalent [think: "essentially the same"] if one can be turned into the other by rotating the pyramid.
   (a) What is a natural choice of a group R for this problem?
   (b) How many colorings are equivalent to a coloring with three red faces and one white face?
   (c) How many colorings are in the R-orbit of a coloring with one red face and three white faces?
   (d) How many colorings are equivalent to a coloring with two red and two white faces? Does it matter whether the red faces touch each other?

(e) How many ways are there to color the faces with red, white, blue, and green if each color must be used? [Try to avoid listing all the colorings.]

4. Consider the problem of assigning the numbers $1, 2, 3, 4$ to the triangular faces of the pyramid in Exercise 3, with two assignments equivalent in case one can be turned into the other by a rotation of the pyramid.

(a) Following the ideas in Example 1, how many numberings are in each equivalence class?

(b) How many numberings are there altogether?

(c) How many equivalence classes are there?

(d) What group is associated with this problem?

5. How many essentially different ways are there to color the vertices of the digraph D of Figure 3(a) using the given colors? [This exercise and Exercise 6 illustrate how the orbits get into the act.]

(a) just red

(b) both red and blue, each at least once

(c) red, white, and blue, each at least once

(d) all red, all blue, or some red and some blue

6. How many essentially different ways are there to color the edges of the digraph E of Figure 4(a) using the given colors?

(a) just red

(b) both red and blue, each at least once

(c) red, white, and blue, each at least once

(d) all red, all blue, or some red and some blue

7. Let F be the digraph obtained by reversing the direction of one edge, i.e., the edge from x to y in the digraph E of Figure 4(a).

(a) List the members of $\text{AUT}(F)$.

(b) List the orbits of $\text{AUT}(F)$ acting on the set $\{w, x, y, z\}$ of vertices.

(c) List the orbits of $\text{AUT}(F)$ acting on the set of edges of E.

8. List the elements of each of the following orbits, where H is the graph in Figure 5(a).

(a) $\text{AUT}(H)(w)$ (b) $\text{AUT}(H)(x)$ (c) $\text{AUT}(H)(u)$

(d) $\text{AUT}(H)(a)$, where a is the edge from x to u and $\text{AUT}(H)$ is acting on the set of edges of H

9. (a) How many orbits does S_3 have in its action on $\{1, 2, 3\} \times \{1, 2, 3\}$ as described in Example 4(d)?

(b) How many orbits does S_2 have on $\{1, 2\} \times \{1, 2\}$?

10. (a) Determine the products hg, $(hg)(fg)$, and $(fgh)(fgh)$ of the multiplication table for the group H in Example 7 and Figure 5. *Hint:* It is enough to see what happens to x and y and use the table in Figure 5. Do you see why?

(b) Is the group commutative?

11. Consider the graph in Figure 7(a).

(a) Convince yourself that the automorphism group for this graph has the eight permutations in Figure 7(b). Note that r signifies "**r**otation," h "**h**orizontal flip," v "**v**ertical flip," d "**d**iagonal flip," and f "other diagonal **f**lip."

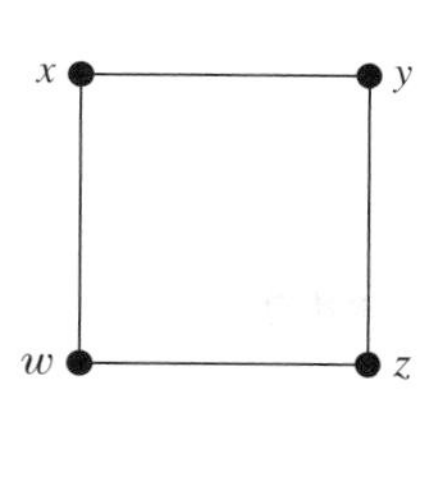

(a)

	x	y	z	w
e	x	y	z	w
r	y	z	w	x
r^2	z	w	x	y
r^3	w	x	y	z
h	w	z	y	x
v	y	x	w	z
d	x	w	z	y
f	z	y	x	w

(b)

Figure 7 ▲

(b) Give two members of this group that do not commute with each other.

(c) Color the vertices so that x, y are red and w, z are black. Give all equivalent colorings.

(d) Do the same if x, z are red and w, y are black.

(e) Do the same if x is red and the other vertices are black.

(f) Can you guess how many essentially different colorings use red, black, or both?

12. (a) Verify the rows for gh and hg in the function table for $\text{AUT}(E)$ in Figure 4(b).

(b) Is the composition operation of the group $\text{AUT}(E)$ commutative? Explain.

(c) Verify the entries $g(hg) = h$, $(hg)g = gh$, and $(gh)(gh) = e$ in the multiplication table of Figure 4(c) by checking what the products on the left-hand sides of these equations do to the vertices of E.

13. Let D be a digraph.

(a) Show that $\text{AUT}(D)$ is a group, with composition as operation.

(b) Digraph automorphisms preserve digraph properties. As examples, show that automorphisms must take sources to sources and sinks to sinks.

14. A group G acting on a set X is said to act **transitively** on X if $X = G(a)$ for some a in X. Suppose G acts transitively on X.

(a) Show that $X = G(x)$ for every x in X.

(b) How many orbits does G have on X?

15. Let H be a connected graph with no parallel edges, V its set of vertices, and E its set of edges. Also let G be the group of graph automorphisms acting on V.

(a) Show that if H has exactly two vertices and one edge, then two different automorphisms of G act the same way on the edge.

(b) Show that if H has more than one edge and if g and h are two different automorphisms of H, then there is at least one edge $\{x, y\}$ in E for which $g^*(\{x, y\}) \neq h^*(\{x, y\})$. [Hence specifying the action of an automorphism on the set of edges completely determines the automorphism.]

16. (a) Show how to construct, for each n in $\mathbb{P}$, a tree such that $|\text{AUT}(T)| = 2^n$. *Hint:* Keep attaching suitable graphs with two automorphisms.

(b) Show how to construct, for each n in $\mathbb{P}$, a digraph such that $|\text{AUT}(D)| = n$.

12.2 Fixed Points and Subgroups

In §12.1 we saw that it would be desirable to have a way of counting orbits of a group acting on a set. In this section we'll see that we can get information about how the group moves elements from information about how the group does *not* move elements, which may be easier to obtain.

EXAMPLE 1

Consider again the digraph E of Figure 4 on page 467, redrawn in Figure 1, and let $G = \text{AUT}(E)$.

Figure 1 ▶

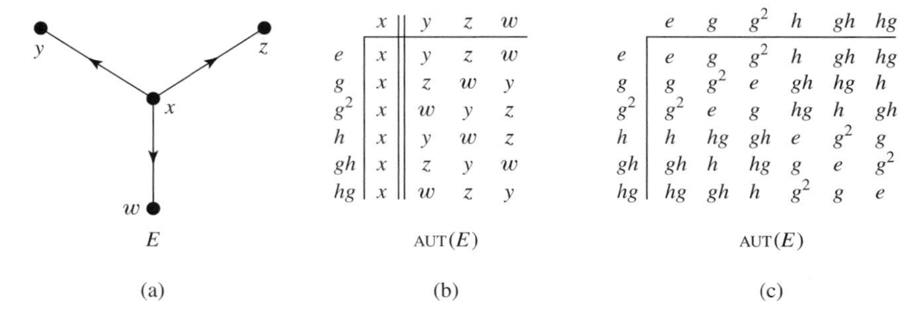

	x	y	z	w
e	x	y	z	w
g	x	z	w	y
g^2	x	w	y	z
h	x	y	w	z
gh	x	z	y	w
hg	x	w	z	y

AUT(E)

	e	g	g^2	h	gh	hg
e	e	g	g^2	h	gh	hg
g	g	g^2	e	gh	hg	h
g^2	g^2	e	g	hg	h	gh
h	h	hg	gh	e	g^2	g
gh	gh	h	hg	g	e	g^2
hg	hg	gh	h	g^2	g	e

AUT(E)

(a) (b) (c)

We saw in Example 8(d) on page 468 that G has two orbits on the set X of vertices of E: $\{y, z, w\}$ and $\{x\}$. The orbit $G(y) = \{y, z, w\}$ consists of the points $e(y)$, $g(y)$, $g^2(y)$, $h(y)$, $gh(y)$, and $hg(y)$, which are not all different. Indeed, $e(y) = h(y) = y$, $g(y) = gh(y) = z$, and $g^2(y) = hg(y) = w$, so $G(y) = \{e(y), g(y), g^2(y)\}$. Moreover, G breaks up into three sets, one for each member of the orbit of y. We have the members of G that send y to itself, i.e., $\{e, h\}$; the ones that take y to z, i.e., $\{g, gh\}$; and the ones that take y to w, i.e., $\{g^2, hg\}$. Since $g = ge$, we can consider $\{g, gh\}$ to be $\{ge, gh\}$, which we will write as $g\{e, h\}$. Similarly, since $hg = g^2h$, we have $\{g^2, hg\} = \{g^2e, g^2h\} = g^2\{e, h\}$, so

$$G = \{e, h\} \cup g\{e, h\} \cup g^2\{e, h\}.$$

The number of elements in G is the product $3 \cdot 2$ of the number of elements in the orbit of y and the number of elements of G that send y to itself.

If we try the same thing with the orbit of x, we see that the size of the orbit is 1, but the number of elements of G that send x to itself is 6, so the product $1 \cdot 6$ is still the number of elements of G. ■

Example 1 shows some specific cases of general facts that we can state with a bit more terminology and notation. If G is a group acting on the set X and if $x \in X$, then we say that the element g of G **fixes** x in case $g(x) = x$, and we define $\textbf{FIX}_G(x)$, the **subgroup fixing** $\boldsymbol{x}$, to be the set of all such g's that fix x. ["Fix" here means "keep in the same place," not "repair."] In symbols,

$$\textbf{FIX}_G(x) = \{g \in G : g(x) = x\}.$$

When only one group G is being considered, we may write $\textbf{FIX}(x)$ in place of $\text{FIX}_G(x)$.

EXAMPLE 2

(a) For the group AUT(D) of automorphisms of the digraph D in Figure 3 on page 466, the fixing subgroups are FIX$(p) = \{e, f\}$, FIX$(q) = \{e, g\}$, FIX$(r) = \{e, f\}$, and FIX$(s) = \{e, g\}$. These sets can be read from Figure 3(b) by finding the p's in the first column, the q's in the second column, etc.

(b) For the group AUT(H) in Figure 5 on page 467, the fixing subgroups are FIX$(u) =$ FIX$(v) = \{e, g, h, gh\}$, FIX$(w) =$ FIX$(x) = \{e, h\}$, and FIX$(y) =$ FIX$(z) = \{e, g\}$. Note that FIX(u) has 4 elements and the orbit AUT$(H)(u) = \{u, v\}$ has 2 elements, whereas FIX(w) has 2 elements and the orbit AUT$(H)(w) = \{w, x, y, z\}$ has 4 elements. Note also that $2 \cdot 4 = 4 \cdot 2 = 8 = |\text{AUT}(H)|$. ■

If you're still not sure what we're getting at in Examples 1 and 2, peek ahead at Theorem 1, which is our next goal. Our first fact characterizes the sets $g\text{FIX}_G(x)$.

Proposition 1 Let G be a group acting on the set X, and let $x \in X$. For each g in G, we have

$$g\text{FIX}_G(x) = \{gf : f \in \text{FIX}_G(x)\} = \{h \in G : h(x) = g(x)\}.$$

Proof The first equality is just the definition of $g\text{FIX}_G(x)$.

If $f \in \text{FIX}_G(x)$, then $(gf)(x) = g(f(x)) = g(x)$, so we have $g\text{FIX}_G(x) \subseteq \{h \in G : h(x) = g(x)\}$.

In the other direction, if $h \in G$ with $h(x) = g(x)$, then $x = g^{-1}(g(x)) = g^{-1}(h(x)) = (g^{-1}h)(x)$, so $g^{-1}h \in \text{FIX}_G(x)$. Hence $h = gg^{-1}h \in g\text{FIX}_G(x)$. It follows that $\{h \in G : h(x) = g(x)\} \subseteq g\text{FIX}_G(x)$. ■

We have used the term "subgroup" as part of the phrase "subgroup fixing x." In general a **subgroup** of a group G is a subset of G that is a group in its own right, i.e., a nonempty subset H of G such that, if f and g are in H, then so are f^{-1} and fg, where the inverse and product are the ones in G. Since the product of an element and its inverse is the identity element e, it follows that e belongs to every subgroup of G.

EXAMPLE 3

(a) The group G is always a subgroup of itself. The subset $\{e\}$ is also always a subgroup of G. These two subgroups, G and $\{e\}$, are sometimes called the **improper subgroups** of G; all other subgroups are called **proper**. The subgroup $\{e\}$ is also called the **trivial subgroup** of G.

(b) If G acts on the set X and if $x \in X$, then $\text{FIX}_G(x)$ is a subgroup of G. To see this, suppose that g and h fix x. Then $(gh)(x) = g(h(x)) = g(x) = x$, so $gh \in \text{FIX}_G(x)$, and $g^{-1}(x) = g^{-1}(g(x)) = x$, so $g^{-1} \in \text{FIX}_G(x)$.

More generally, if G acts on X and if Y is a subset of X, then the set $\{g \in G : g(Y) = Y\}$ is a subgroup of G, sometimes called the **stabilizer in *G* of *Y***. Exercise 14 asks for a careful proof of this fact. ■

The subsets $g\text{FIX}_G(x)$ of G that we saw in Proposition 1 are of a special sort that shows up frequently in the study of groups. In general, if H is a subgroup of the group G, then a **left coset** of H in G is a subset of the form

$$\boldsymbol{gH} = \{gh : h \in H\}$$

for some $g \in G$. The coset eH is H itself, and indeed $hH = H$ for every h in H [Exercise 27]. Since $e \in H$, the coset gH contains $ge = g$. Thus every g in G belongs to at least one left coset. In fact, each g belongs to just one left coset.

Proposition 2 The left cosets of a subgroup of a group form a partition of the group.

Proof We just showed that G is the union of the various cosets gH, so we only need to show that overlapping left cosets are identical. First we show that if $k \in gH$, then $kH = gH$. Say $k = gh'$ with $h' \in H$. Then, since H is a group, for every h in H we have $h'h \in H$ and hence $kh = gh'h \in gH$. So $kH \subseteq gH$. Since $g = kh'^{-1} \in kH$, similar reasoning shows that $gH \subseteq kH$.

Now suppose that gH and $g'H$ overlap; say $k \in gH \cap g'H$. Then, by what we have just shown, $gH = kH = g'H$. ■

We discuss an alternative proof of Proposition 2 at the very end of this section.

Instead of the left cosets gH that we have been considering, we could just as easily have looked at **right cosets**, sets of the form $\boldsymbol{Hg} = \{hg : h \in H\}$.

If G is commutative, then $gH = Hg$, in which case we just refer to **cosets**. In general, the left coset gH and the right coset Hg are different subsets of G. The proof of Proposition 2 is still valid for right cosets, with the obvious left–right switches.

EXAMPLE 4

(a) In Example 1 the subgroup $\{e, h\}$ fixing y has the three left cosets $\{e, h\}$, $g\{e, h\} = \{g, gh\}$, and $g^2\{e, h\} = \{g^2, hg\}$ in $\text{AUT}(E)$. It has the right cosets $\{e, h\}$, $\{e, h\}g = \{g, hg\}$, and $\{e, h\}g^2 = \{g^2, gh\}$, two of which are not left cosets. The set of right cosets forms a partition of $\text{AUT}(E) = \{e, g, g^2, h, gh, hg\}$, just as the set of left cosets does.

(b) Consider the subgroup $3\mathbb{Z}$ of the group $(\mathbb{Z}, +)$. The cosets are the sets of the form $3\mathbb{Z} + r = \{3k + r : k \in \mathbb{Z}\}$. There are just three of them, the congruence classes $[0]_3 = 3\mathbb{Z} = 3\mathbb{Z} + 0$, $[1]_3 = 3\mathbb{Z} + 1$, and $[2]_3 = 3\mathbb{Z} + 2$. Every n in $\mathbb{Z}$ can be written as $n = 3q + r$ with $r \in \{0, 1, 2\}$ and $q \in \mathbb{Z}$, so $\mathbb{Z}$ is the union of these three disjoint sets. Similarly, for every $p \in \mathbb{P}$ the cosets of $p\mathbb{Z}$ in $(\mathbb{Z}, +)$ are the sets $[r]_p = p\mathbb{Z} + r$, where $r = 0, 1, 2, \ldots, p - 1$. ■

Not only do the cosets of H in G partition G, but they also all have the same size, the size of H.

Proposition 3 Let H be a subgroup of the group G and let $g \in G$. The function $h \to gh$ is a one-to-one correspondence of H onto the coset gH. Hence $|gH| = |H|$ if H is finite.

Proof This function certainly maps H onto gH, and it is one-to-one because if $gh = gh'$, then $h = g^{-1}gh = g^{-1}gh' = h'$. ■

When we apply Propositions 1, 2, and 3 to the case $H = \text{FIX}_G(x)$, we get the following.

Theorem 1 Let G be a finite group acting on a set X, and let $x \in X$. The number of elements in G is the product of the number of elements in the orbit $G(x)$ by the number of elements in the fixing subgroup $\text{FIX}_G(x)$. In symbols,

$$|G| = |G(x)| \cdot |\text{FIX}_G(x)|.$$

In particular, the size of the orbit $G(x)$ divides the size of G.

Proof The distinct cosets $g\text{FIX}_G(x)$ partition G by Proposition 2, and by Proposition 3 they all have size $|\text{FIX}_G(x)|$. By Proposition 1 there is one coset for each $g(x)$ in the orbit of x, so the number of cosets is $|G(x)|$. ■

Theorem 1 can be viewed as a consequence of one of the basic facts of finite group theory.

Lagrange's Theorem Let H be a subgroup of the finite group G, and let $\boldsymbol{G/H}$ be the set of left cosets of H in G. Then

$$|G| = |G/H| \cdot |H|.$$

In particular, $|H|$ and $|G/H|$ divide $|G|$.

Proof The left cosets of H partition G, by Proposition 2. Each coset has $|H|$ members, by Proposition 3, and there are $|G/H|$ of them. ■

Proposition 3 and Lagrange's Theorem have valid analogues for right cosets, so G has the same number of right cosets of H as left cosets. Exercise 21 asks for a direct proof of this fact.

EXAMPLE 5

If G is a group and $g \in G$, then the set $\langle g \rangle = \{g^n : n \in \mathbb{Z}\}$ is a subgroup of G, called the **subgroup generated by g**. Here by g^0 we mean e. For positive n the symbol g^n represents the product of g by itself n times, and g^{-n} is $(g^{-1})^n$. With these agreements it can be shown [Exercise 12] that the familiar rule $g^{m+n} = g^m \cdot g^n$ holds for all $m, n \in \mathbb{Z}$. It follows that $\langle g \rangle$ really is a subgroup of G. ■

Groups of the form $\langle g \rangle$ are called **cyclic groups**. The idea is that the powers of g "cycle through" all the elements of the group $\langle g \rangle$. Since $g^m \cdot g^n = g^{m+n} = g^{n+m} = g^n \cdot g^m$, cyclic groups are commutative. A group element g is said to have [finite] **order** n if $\langle g \rangle$ has n elements and is said to have **infinite order** if it does not have finite order. We will see in Example 7 that if g has order n, then $g^n = e$, and n is the smallest positive integer with this property. In this case, $\langle g \rangle = \{g, g^2, \dots, g^n\}$.

EXAMPLE 6

(a) The subgroups $\langle h \rangle = \{e, h\}$ and $\langle g \rangle = \{e, g\}$ that we saw in Example 2(b) are cyclic. The elements g and h have order 2. The subgroup $\text{FIX}(u) = \{e, g, h, gh\}$ of that example is not cyclic. The powers of g are just e and g. Similarly, the subgroups of powers of h and of gh are just $\{e, h\}$ and $\{e, gh\}$, respectively, so this group does not consist of the powers of a single one of its members.

(b) If the operation for the group is $+$, then the identity element is usually denoted by 0, and instead of g^n we write ng. Thus $\langle g \rangle = \{ng : n \in \mathbb{Z}\} = \{\dots, -2g, -g, 0, g, 2g, \dots\}$. Of course, this set may have only finitely many members. If g has order n, then $ng = 0$ and $\langle g \rangle = \{g, 2g, \dots, ng\}$. As an example, the group $\mathbb{Z}(5)$ with operation $+_5$ is a cyclic group generated by any of its nonzero members; for instance, $\langle 2 \rangle = \{2, 4, 1, 3, 0\}$, since $1 \cdot 2 = 2$, $2 \cdot 2 = 4$, $3 \cdot 2 = 6 \equiv 1$, etc.

The infinite additive group $(\mathbb{Z}, +)$ is cyclic, generated by either 1 or -1. All the nonzero elements of $\mathbb{Z}$ have infinite order. The set $2\mathbb{Z}$ of even integers is a cyclic subgroup, generated by 2 or -2 and, more generally, $n\mathbb{Z} = \langle n \rangle = \langle -n \rangle$ for every integer n. Theorem 2 will show that these are the only subgroups that $\mathbb{Z}$ has. ■

All subgroups of $(\mathbb{Z}, +)$ turn out to be cyclic, which is the simplest possible situation.

Theorem 2 Every subgroup of $(\mathbb{Z}, +)$ is of the form $n\mathbb{Z}$ for some $n \in \mathbb{N}$.

Proof Consider a subgroup H of $(\mathbb{Z}, +)$. If $H = \{0\}$, then $H = 0\mathbb{Z}$, which is of the required form. Suppose $H \neq \{0\}$. If $0 \neq m \in H$, then also $-m \in H$. Thus $H \cap \mathbb{P}$ is nonempty, so the Well-Ordering Principle says that it has a smallest element, say n. We show that $H = n\mathbb{Z}$. Since $n \in H$ and H is a subgroup, we have $n\mathbb{Z} = \{kn : k \in \mathbb{Z}\} \subseteq H$. Consider an element k of H, which we want to show must be in $n\mathbb{Z}$. By the Division Algorithm, $k = nq + r$ with $0 \le r < n$. Since $n\mathbb{Z} \subseteq H$, we have $nq \in H$ and thus $r = k - nq \in H$. Since $r < n$ and n is the smallest positive member of H, we must have $r = 0$. That is, $k = nq \in n\mathbb{Z}$. Since k was arbitrary in H, $H \subseteq n\mathbb{Z}$ as claimed. ■

Subgroups can be generated by subsets with more than one element. Let A be a nonempty subset of G. The **subgroup generated by A** is the set $\langle A \rangle$ of all products of arbitrarily many elements from $A \cup A^{-1}$, where $A^{-1} = \{g^{-1} : g \in A\}$. It can be shown [Exercise 15(c)] that this set is a subgroup of G and that it is the smallest subgroup of G that contains A. The set $\langle A \rangle$ can also be defined recursively by

(B) $A \subseteq \langle A \rangle$.

(R_1) If $g, h \in \langle A \rangle$, then $g \cdot h \in \langle A \rangle$.

(R_2) If $g \in \langle A \rangle$, then $g^{-1} \in \langle A \rangle$.

Conditions (R_1) and (R_2) show that $\langle A\rangle$ is closed under the operations $\cdot$ and inversion. Consider g in A. Then $g \in \langle A\rangle$ by (B), so $g^{-1} \in \langle A\rangle$ by (R_2), and thus the identity $e = g \cdot g^{-1}$ also belongs to $\langle A\rangle$ by (R_1). Hence $\langle A\rangle$ is a subgroup of G.

The union of a collection of subgroups of a group is not, in general, a subgroup. The problem is that there is no reason why the product of two elements taken from different subgroups should lie in one of the chosen subgroups. On the other hand [Exercise 15(a)], the intersection of any collection of subgroups of a given group G is a subgroup of G. This fact gives us another way to view the subgroup $\langle A\rangle$; it's the intersection of all of the subgroups of G, including G itself, that contain A. Exercise 15(c) asks for a proof of this statement.

If G acts on a set X, then so do the subgroups of G, and it makes sense to consider their orbits. We know from the Corollary on page 468 that the $\langle g\rangle$-orbits partition X, so to describe how g acts on X, it will be enough to describe how it acts on each of its orbits. Consider an element x of X and some g in G. Since $\langle g\rangle = \{g^n : n \in \mathbb{Z}\}$, the orbit of x under $\langle g\rangle$ is

$$\langle g\rangle(x) = \{g^n(x) : n \in \mathbb{Z}\}.$$

First suppose that all the elements $g^n(x)$ are different. Then the orbit $\langle g\rangle(x)$ is simply $\{\dots, g^{-2}(x), g^{-1}(x), x, g(x), g^2(x), \dots\}$, and g just moves each element on the list along to the next one to the right. Nothing exciting here. Note that, in this case, $\langle g\rangle(x)$ is infinite, so of course X must be infinite.

The elements $g^n(x)$ need not all be different. For instance, if g fixes x, i.e., if $g(x) = x$, then $\langle g\rangle(x) = \{x\}$, no matter how large $\langle g\rangle$ is. Suppose it happens that $g^k(x) = g^\ell(x)$ for some $k > \ell$. Then $g^{k-\ell}(x) = g^{-\ell}(g^k(x)) = g^{-\ell}(g^\ell(x)) = x$. Thus the set $\{m \in \mathbb{P} : g^m(x) = x\}$ is nonempty. Let n be the smallest member of this set. We claim that

$$\langle g\rangle(x) = \{x, g(x), \dots, g^{n-1}(x)\}$$

and that the n elements listed in this set are distinct.

It's clear that $\{x, g(x), \dots, g^{n-1}(x)\}$ is a subset of $\langle g\rangle(x)$. To show the reverse containment, consider an element $g^s(x)$ in $\langle g\rangle(x)$. The integer s can be written as $s = r + qn$, where $r \in \{0, 1, \dots, n-1\}$. Since $g^{nq}(x) = g^n(g^n(\cdots(g^n(x))) = x$, we have $g^s(x) = g^{r+qn}(x) = g^r(g^{nq}(x)) = g^r(x)$. Since $r \in \{0, 1, \dots, n-1\}$, it follows that $g^s(x) = g^r(x) \in \{x, g(x), \dots, g^{n-1}(x)\}$. Thus every element of $\langle g\rangle(x)$ is in $\{x, g(x), \dots, g^{n-1}(x)\}$. The elements $x, g(x), \dots, g^{n-1}(x)$ must be distinct, because if $g^k(x) = g^\ell(x)$ with $0 \le \ell < k < n$ then, as above, $g^{k-\ell}(x) = x$ with $0 \le k-\ell < n$, contrary to the choice of n as the least positive integer such that $g^n(x) = x$.

We summarize what we have learned.

Proposition 4 Let G be a group acting on the set X, and let $g \in G$ and $x \in X$. If $|\langle g\rangle(x)| = n$, then $\langle g\rangle(x) = \{x, g(x), \dots, g^{n-1}(x)\}$, and n is the smallest positive integer such that $g^n(x) = x$.

EXAMPLE 7

(a) When we defined the order of an element g of G to be the size of the subgroup $\langle g\rangle$, we claimed that if g had order n, then n was the smallest positive integer for which $g^n = e$. Proposition 4 will let us justify this statement.

First observe that the group G can *act on itself* by letting $g(x) = gx$ for all $g, x \in G$. To check this assertion, we first must know that $x \to gx$ is a permutation of G, which follows from the fact that $x \to g^{-1}x$ is its inverse map. The other condition for an action is that $(gh)(x) = g(h(x))$ for every $g, h, x \in G$, which we verify by

$$\begin{aligned}(gh)(x) &= (gh)x && \text{[definition of the action]}\\ &= g(hx) && \text{[associativity of the group operation]}\\ &= g(h(x)) && \text{[definition of the action]}\end{aligned}$$

for every $g, h, x \in G$.

Now take $X = G$ and $x = e$ in Proposition 4. The proposition tells us that since $\langle g \rangle(e) = \{g^k \cdot e : k \in \mathbb{Z}\} = \{g^k : k \in \mathbb{Z}\} = \langle g \rangle$, if $|\langle g \rangle| = n$, i.e., if the order of g is n, then $\langle g \rangle = \{e, g, \ldots, g^{n-1}\}$, and n is the smallest positive integer such that $g^n = e$.

(b) A subgroup H of G also acts on G. The orbit under H of an element g in G [*the set on which H acts*] is

$$\{h(g) : h \in H\} = \{hg : h \in H\} = Hg.$$

In other words, the orbits under H are just the right cosets of H in G. The corollary to Proposition 3 on page 467 tells us that the H-orbits partition G. So the right cosets of H partition G, which establishes the right-coset version of Proposition 2. Similarly, $h^*(g) = g \cdot h^{-1}$ defines another action of H on G for which the orbits are just the left cosets [Exercise 22]. So the left cosets of H also partition G; this provides another confirmation of Proposition 2. ■

Exercises 12.2

1. Describe each of the following subgroups of $(\mathbb{Z}, +)$.

(a) $\langle 1 \rangle$ (b) $\langle 0 \rangle$ (c) $\langle\{-1, 2\}\rangle$

(d) $\langle \mathbb{Z} \rangle$ (e) $\langle\{2, 3\}\rangle$ (f) $\langle 6 \rangle \cap \langle 9 \rangle$

2. Which of the subgroups in Exercise 1 are cyclic groups? Justify your answers.

3. Which of the following subsets of $\mathbb{Z}$ are subgroups of $\mathbb{Z}$? Write the subgroups in the form $n\mathbb{Z}$; see Theorem 2.

(a) $\{0, -3, -6, -9, \ldots\}$ (b) $\{0, 5, 10, 15, 20, \ldots\}$

(c) $\{0, -2, -4, -8, -16, \ldots\}$

(d) $\{k \in \mathbb{Z} : k \text{ is a multiple of } 4\}$

(e) $\mathbb{N} \cup (-\mathbb{N})$

4. Which of the following subsets of $\mathbb{R}$ are subgroups of $\mathbb{R}$?

(a) $\mathbb{Z}$ (b) $\mathbb{N}$ (c) $\mathbb{Q}$

(d) $\{n\sqrt{2} : n \in \mathbb{Z}\}$ (e) $\{m + n\sqrt{2} : m, n \in \mathbb{Z}\}$

5. Consider $\text{AUT}(D)$ in Figure 3 on page 466. The fixing subgroups are given in Example 2(a) on page 470. Confirm the equality in Theorem 1 in these cases.

6. Repeat Exercise 5 for $\text{AUT}(E)$ in Figure 1 on page 470. The orbits are given in Example 1 on page 470.

7. (a) Find all generators of the group $(\mathbb{Z}(5), +_5)$.

(b) Find all generators of the group $(\mathbb{Z}(6), +_6)$.

8. (a) Find the intersection of all the subgroups $n\mathbb{Z}$ of $(\mathbb{Z}, +)$, where $n \in \mathbb{P}$.

(b) Is the intersection in part (a) cyclic? Explain.

9. (a) Give an example of a one-to-one correspondence between $4\mathbb{Z}$ and the coset $4\mathbb{Z} + 3$ in $(\mathbb{Z}, +)$.

(b) Give another example.

10. Consider the group $(\mathbb{Z}, +)$. Write $\mathbb{Z}$ as a disjoint union of five cosets of a subgroup.

11. (a) Prove that $(g_1 g_2)^{-1} = g_2{}^{-1} g_1{}^{-1}$ for g_1, g_2 in a group.

(b) Prove that $(g_1 g_2 g_3)^{-1} = g_3{}^{-1} g_2{}^{-1} g_1{}^{-1}$.

(c) Prove a generalization for $(g_1 g_2 \cdots g_n)^{-1}$.

12. For an element g in a group G, the powers g^n were defined for all $n \in \mathbb{Z}$ in Example 5. Prove that

(a) $(g^{-1})^{-k} = g^k$ for all $g \in G$ and $k \in \mathbb{Z}$.

(b) $g^m \cdot g^1 = g^{m+1}$ for all $g \in G$ and $m \in \mathbb{Z}$. [The case $m \in \mathbb{N}$ is easy.]

(c) $g^m \cdot g^n = g^{m+n}$ for all $g \in G$, $m \in \mathbb{Z}$, and $n \in \mathbb{N}$.

(d) $g^m \cdot g^n = g^{m+n}$ for all $g \in G$, $m \in \mathbb{Z}$, and $n \in \mathbb{Z}$. [*Hint:* You can use part (c) with g^{-1} instead of g.]

13. Let G be a group acting on a set X. Define

$$R = \{(x, y) \in X \times X : g(x) = y \text{ for some } g \in G\}.$$

(a) Show that R is an equivalence relation on X.

(b) Describe the partition of X corresponding to R.

14. Let G be a group acting on a set X, and for each subset Y of X let $\textbf{FIX}(\boldsymbol{Y}) = \{g \in G : g(Y) = Y\}$. Show that $\text{FIX}(Y)$ is a subgroup of G.
Note: Some authors call this subgroup the **set stabilizer** of Y in G and use the notation $\text{FIX}(Y)$ for another subgroup, the **pointwise stabilizer** $\{g \in G : g(y) = y \text{ for all } y \in Y\}$.

15. (a) Show that the intersection of any family of subgroups of a group G is again a subgroup of G.

(b) Give an example of a group G and subgroups H and K such that $H \cup K$ is not a subgroup of G.

(c) Show that if A is a subset of the group G, then

$$\langle A \rangle = \bigcap \{H : H \text{ is a subgroup of } G \text{ and } A \subseteq H\}.$$

(d) What is the subgroup $\langle A \rangle$ if A is empty?

16. Suppose the group G acts on the set X and K is a subgroup of G. Then K also acts on X. Show that each orbit of G on X is a union of orbits of K.

17. Suppose that G acts on X. Show that if $|G|$ is 2^k for some k and $|X|$ is odd, then some member of X must be fixed by *all* members of G. *Hint:* Since the orbits of G partition X, we can choose one element from each orbit; let $x_1, \ldots, x_m$ be such a collection. Then $|X| = \sum_{j=1}^{m} |G(x_j)|$. Apply Theorem 1 to each x_j.

18. What can you say about the sizes of the orbits of a group with 27 members?

19. Let $G = \text{AUT}(H)$ from Figure 5 on page 467. For subsets Y of H, define $\text{FIX}(Y)$ as in Exercise 14. Determine $\text{FIX}(\{w, y\})$. Does $\text{FIX}(w) \cap \text{FIX}(y) = \text{FIX}(\{w, y\})$?

20. Show that if H is a subgroup of $(G, \cdot)$, then $g \cdot H \cdot g^{-1}$ is also a subgroup for each g in G.

21. Let H be a subgroup of the group $(G, \cdot)$.

(a) Show that for each g in G the right coset $H \cdot g^{-1}$ consists of the inverses of the elements in the left coset $g \cdot H$.

(b) Describe a one-to-one correspondence between the set of left cosets of H in G and the set of right cosets.

22. Let H be a subgroup of a group G. For $h \in H$, define $h^*: G \to G$ by $h^*(g) = g \cdot h^{-1}$ for $g \in G$.

(a) Prove that this defines an action of H on G.

(b) Show that $h \to h^*$ is one-to-one.

23. Let G be a group and let H be a *finite*, nonempty subset of G. Show that if H is closed under the group operation, i.e., if $gh \in H$ whenever $g, h \in H$, then H is a subgroup of G. [This fact can save some work when dealing with finite groups.]

24. The table describes a binary operation $\bullet$ on the set $G = \{a, b, c, d, e\}$ with e as identity element.

$\bullet$	e	a	b	c	d
e	e	a	b	c	d
a	a	e	c	d	b
b	b	d	a	e	c
c	c	b	d	a	e
d	d	c	e	b	a

(a) Convince yourself that the set $\{e, a\}$ is a group under $\bullet$ as operation.

(b) Without doing any calculations, use the result of part (a) and Lagrange's Theorem to conclude that $(G, \bullet)$ is not a group.

25. The following table gives the binary operation for a group $(G, \bullet)$ with elements $a, b, c, d, e,$ and f.

$\bullet$	e	a	b	c	d	f
e	e	a	b	c	d	f
a	a	b	e	d	f	c
b	b	e	a	f	c	d
c	c	f	d	e	b	a
d	d	c	f	a	e	b
f	f	d	c	b	a	e

(a) List the members of the subgroup $\langle a \rangle$.

(b) Show that $\langle a \rangle \bullet c = c \bullet \langle a \rangle$.

(c) Find all the subgroups with two members.

(d) Find $|G/\langle d \rangle|$.

(e) Describe the right cosets of $\langle d \rangle$.

26. Repeat Exercise 25 for the group with the following table.

$\bullet$	e	a	b	c	d	f
e	e	a	b	c	d	f
a	a	b	e	d	f	c
b	b	e	a	f	c	d
c	c	d	f	a	b	e
d	d	f	c	b	e	a
f	f	c	d	e	a	b

27. If H is a subgroup of a group $(G, \cdot)$ and if $g \in G$, then $g \cdot H = H$ if and only if $g \in H$. Try to be clever and use Proposition 2.

28. Consider a finite group $(G, \cdot)$ with a subgroup H such that $|G| = 2|H|$. Show that $g \cdot H = H \cdot g$ for every g in G. [*Suggestion:* Consider the two cases $g \in H$ and $g \notin H$ separately.]

29. Let H be a subgroup of a group $(G, \cdot)$ and, for $g_1, g_2 \in G$, define $g_1 \sim g_2$ if $g_2^{-1} \cdot g_1 \in H$.

(a) Show that $\sim$ is an equivalence relation on G.

(b) Show that the partition in Proposition 2 is precisely the partition of equivalence classes for $\sim$ described in Theorem 1 on page 116.

12.3 Counting Orbits

We saw in §12.1 that to count the number of different ways to color something, or the number of inequivalent circuits of some sort, it would be enough to count the orbits of a suitable group G acting on a suitable set X. In this section we will see how to count such orbits without actually listing them. The key fact here will be quite surprising; we can count these orbits if we just know how many points in X each element of G fixes.

In §12.2 we looked at the group $\text{FIX}_G(x)$ of permutations in G that fixed a given element x of X. Now we change our point of view and, for each g in G, we look at the set

$$\text{FIX}_X(g) = \{x \in X : g(x) = x\}$$

consisting of all the members of X that g fixes. The subscript on FIX_X should help us remember that this is a subset of X; likewise, our old friends FIX_G are subsets of G. Here is the new fact.

Theorem 1 Let G be a finite group acting on a finite set X. The number of orbits of G on X equals

$$\frac{1}{|G|} \cdot \sum_{g \in G} |\text{FIX}_X(g)|.$$

The sum here has one term for each g in G. If you are moderately comfortable with Theorem 1 on page 472, read the next proof and then go on to see what this new theorem says in particular cases. Otherwise, we advise you to skip the proof now and return to it after studying the examples.

Proof We use an idea from the proof of the Generalized Pigeon-Hole Principle on page 217: count a set of pairs in two different ways. Finding a set to count can be tricky sometimes, but in our case the natural choice is

$$S = \{(g, x) \in G \times X : g(x) = x\}.$$

First, for each g in G, there are $|\text{FIX}_X(g)|$ pairs (g, x) with $g(x) = x$; so when we add these numbers we get

$$|S| = \sum_{g \in G} |\text{FIX}_X(g)|. \qquad \textbf{(1)}$$

We can also count the members of S by counting, for each x in X, the pairs (g, x) with $g(x) = x$; we get

$$|S| = \sum_{x \in X} |\text{FIX}_G(x)|. \qquad \textbf{(2)}$$

For each x in X, we have $|\text{FIX}_G(x)| = \dfrac{|G|}{|G(x)|}$ by Theorem 1 on page 472, so

$$|S| = \sum_{x \in X} \frac{|G|}{|G(x)|} = |G| \sum_{x \in X} \frac{1}{|G(x)|}.$$

Now consider the terms in this sum that come from a given orbit, say $G(x_0)$. Since $G(x) = G(x_0)$ for each x in the orbit $G(x_0)$, if we add these terms, we get

$$\sum_{x \in G(x_0)} \frac{1}{|G(x)|} = \sum_{x \in G(x_0)} \frac{1}{|G(x_0)|} = 1.$$

That is, the orbit contributes a total value of 1 to the sum $\sum_{x \in X} \dfrac{1}{|G(x)|}$. Thus, if there are m orbits, then

$$\sum_{x \in X} \frac{1}{|G(x)|} = 1 + 1 + \cdots + 1 = m.$$

It follows that $|S| = |G| \cdot m$, so by (1) we have

$$m = \frac{1}{|G|} \cdot |S| = \frac{1}{|G|} \cdot \sum_{g \in G} |\text{FIX}_X(g)|,$$

as claimed in the theorem. ■

In our first examples it will be easy to evaluate both sides of the equality in Theorem 1, so you might wonder about the value of the theorem. In nontrivial applications the number of orbits can be difficult to calculate directly, whereas the numbers $|\text{FIX}_X(g)|$ may be relatively easy to determine.

EXAMPLE 1

We return to the group $G = \text{AUT}(H)$ of automorphisms of the graph H in Figure 5 on page 467. Here is its picture again, in Figure 1(a). Ignore the Orbits column for now.

Figure 1 ▶

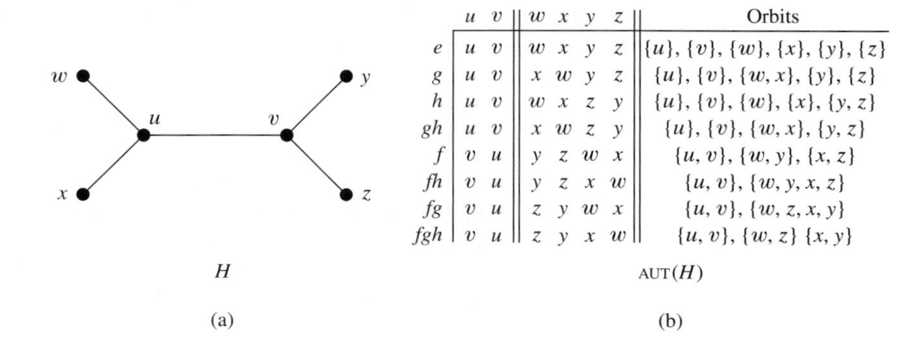

	u	v	w	x	y	z	Orbits
e	u	v	w	x	y	z	$\{u\}, \{v\}, \{w\}, \{x\}, \{y\}, \{z\}$
g	u	v	x	w	y	z	$\{u\}, \{v\}, \{w, x\}, \{y\}, \{z\}$
h	u	v	w	x	z	y	$\{u\}, \{v\}, \{w\}, \{x\}, \{y, z\}$
gh	u	v	x	w	z	y	$\{u\}, \{v\}, \{w, x\}, \{y, z\}$
f	v	u	y	z	w	x	$\{u, v\}, \{w, y\}, \{x, z\}$
fh	v	u	y	z	x	w	$\{u, v\}, \{w, y, x, z\}$
fg	v	u	z	y	w	x	$\{u, v\}, \{w, z, x, y\}$
fgh	v	u	z	y	x	w	$\{u, v\}, \{w, z\}\ \{x, y\}$

AUT(H)

(a) (b)

This group acts on the set $V = \{u, v, w, x, y, z\}$ of vertices of H. To see which members of V each automorphism fixes, we look in the table of Figure 1. We find $\text{FIX}_V(e) = \{u, v, w, x, y, z\}$, $\text{FIX}_V(g) = \{u, v, y, z\}$, $\text{FIX}_V(h) = \{u, v, w, x\}$, $\text{FIX}_V(gh) = \{u, v\}$, and $\text{FIX}_V(f) = \text{FIX}_V(fh) = \text{FIX}_V(fg) = \text{FIX}_V(fgh) = \emptyset$. These sets have 6, 4, 4, 2, 0, 0, 0, and 0 vertices, respectively. Also, $|G| = 8$, so Theorem 1 asserts that G has

$$\frac{1}{8}(6 + 4 + 4 + 2 + 0 + 0 + 0 + 0) = 2$$

orbits. Indeed, from the picture of H or the table, we see that there are exactly two orbits under G, $\{u, v\}$ and $\{w, x, y, z\}$. ■

Consider any graph H with vertex set V, and assume that H has no parallel edges. We have usually regarded $G = \text{AUT}(H)$ as acting on V, i.e., we have thought of G as a subset of $\text{PERM}(V)$; but we can also view G as acting on the set E of edges of the graph as in Example 6(b) on page 466. For each g in $G \subseteq \text{PERM}(V)$, define g^* in $\text{PERM}(E)$ by setting $g^*(\{u, v\}) = \{g(u), g(v)\}$ for each edge $\{u, v\}$. If $f, g \in G$, then

$$(f \circ g)^* = f^* \circ g^* \qquad (*)$$

meets the requirement to define an action of G. The composition on the left in $(*)$ is in $\text{PERM}(V)$, and the one on the right is in $\text{PERM}(E)$. Property $(*)$ holds because

$$(f \circ g)^*(\{u, v\}) = \{f \circ g(u), f \circ g(v)\}$$

and

$$f^* \circ g^*(\{u, v\}) = f^*(\{g(u), g(v)\}) = \{f(g(u)), f(g(v))\}.$$

EXAMPLE 2

(a) In Figure 2(a) we redraw the graph H yet again, but now the edges are also labeled. The table in Figure 2(b) shows how G acts on E. To show how the table was created, let's check that f^* is correct: f is the automorphism that reflects the graph about a vertical line through e_3, and so f^* interchanges e_1

Figure 2 ▶

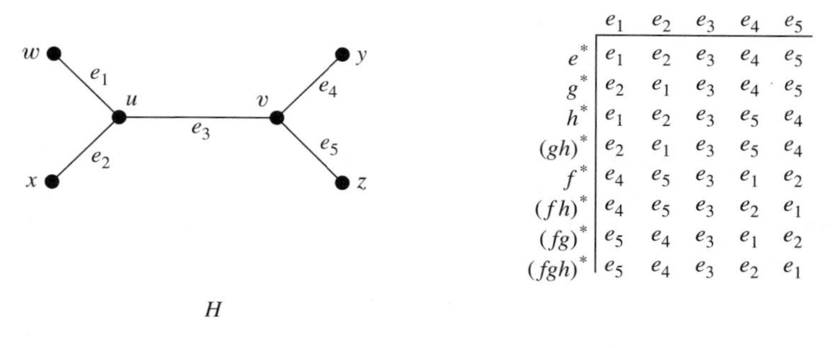

	e_1	e_2	e_3	e_4	e_5
e^*	e_1	e_2	e_3	e_4	e_5
g^*	e_2	e_1	e_3	e_4	e_5
h^*	e_1	e_2	e_3	e_5	e_4
$(gh)^*$	e_2	e_1	e_3	e_5	e_4
f^*	e_4	e_5	e_3	e_1	e_2
$(fh)^*$	e_4	e_5	e_3	e_2	e_1
$(fg)^*$	e_5	e_4	e_3	e_1	e_2
$(fgh)^*$	e_5	e_4	e_3	e_2	e_1

(a) (b)

and e_4, interchanges e_2 and e_5, and leaves e_3 fixed. We can check these claims formally by calculating

$$f^*(e_1) = f^*(\{u, w\}) = \{f(u), f(w)\} = \{v, y\} = e_4,$$
$$f^*(e_3) = f^*(\{u, v\}) = \{f(u), f(v)\} = \{v, u\} = e_3,$$

etc., but this shouldn't be necessary for graphs that we can draw.

Now from Figure 2(b) we find $\text{FIX}_E(e^*) = \{e_1, e_2, e_3, e_4, e_5\}$, $\text{FIX}_E(g^*) = \{e_3, e_4, e_5\}$, $\text{FIX}_E(h^*) = \{e_1, e_2, e_3\}$, and $\text{FIX}_E(\alpha^*) = \{e_3\}$ for the other five automorphisms α. These sets have 5, 3, 3, 1, 1, 1, 1, and 1 edges, respectively. According to Theorem 1, G has

$$\frac{1}{8}(5 + 3 + 3 + 1 + 1 + 1 + 1 + 1) = 2$$

orbits on E. Sure enough, the orbits are $\{e_3\}$ and $\{e_1, e_2, e_4, e_5\}$.

(b) The group G also acts on the set T of all two-element subsets of V, using the same definition for g^* as in (a). Note that T contains the set E of edges and a lot more. In fact, T has $\binom{6}{2} = 15$ elements. This time the situation is more abstract and less intuitive than when we viewed G as acting on vertices or edges. We don't have a useful picture of T, and a table like that in Figure 2(b) would be very cumbersome, so we will work algebraically. As always, $\text{FIX}_T(e^*)$ is the entire set on which G acts, T in this case. Now g maps each member of $\{u, v, y, z\}$ to itself, and so g^* sends each of the 2-element subsets of $\{u, v, y, z\}$ back to itself. It also fixes $\{w, x\}$, even though g moves w and x, since $g^*(\{w, x\}) = \{g(w), g(x)\} = \{x, w\}$. The permutation g^* doesn't fix any other two-element subsets [why not?], so $\text{FIX}_T(g^*)$ has $\binom{4}{2} + 1 = 7$ elements. Similarly, $|\text{FIX}_T(h^*)| = 7$. In addition, $\text{FIX}_T((gh)^*) = \{\{u, v\}, \{w, x\}, \{y, z\}\}$, $\text{FIX}_T(f^*) = \{\{u, v\}, \{w, y\}, \{x, z\}\}$, $\text{FIX}_T((fh)^*) = \text{FIX}_T((fg)^*) = \{\{u, v\}\}$, and $\text{FIX}_T((fgh)^*) = \{\{u, v\}, \{w, z\}, \{x, y\}\}$. So the eight sets $\text{FIX}_T(\quad)$ have 15, 7, 7, 3, 3, 1, 1, and 3 elements.

Theorem 1 then says that T consists of

$$\frac{1}{8}(15 + 7 + 7 + 3 + 3 + 1 + 1 + 3) = 5$$

orbits under G. This was not so obvious to begin with. We can observe that $\{w, u\}$, $\{w, v\}$, $\{w, x\}$, $\{w, y\}$, and $\{u, v\}$ belong to five different orbits. [Look at Figure 1 or 2 to see that none of these subsets can be mapped to another by automorphisms of the graph.] If we had been asked originally to find a representative of each orbit, we would know we were done when we had exhibited these five subsets. ■

Consider again a group G acting on a set X. We sometimes want to restrict the action to a subset Y of X by just ignoring what the members of G do to elements outside Y. For this plan to work, however, we need to be sure that the members of G map elements of Y into Y. We'll need Y to be an orbit of G or a union of several G-orbits. For each g in G, we define $g^*\colon Y \to Y$ by $\mathbf{g^*(y)} = g(y)$. This map g^*, called the **restriction** of g to Y, acts exactly as g does, but only on Y. Since Y is finite and g^* is a one-to-one map of Y into Y, g^* maps Y *onto* Y. Therefore, each g^* is a permutation of Y. Note also that

$$(f \cdot g)^* = f^* \circ g^* \qquad (*)$$

holds for all $f, g \in G$. In other words, G acts on Y.

EXAMPLE 3

(a) We return to the example in Figure 1 and restrict the action of $G = \text{AUT}(H)$ to the orbit $\{u, v\}$. To see the table of values of the restrictions to $\{u, v\}$, simply ignore the last four columns in Figure 1(b) and pretend that each permutation e, g, h, etc., has an asterisk * on it. We have eight different names for the restricted permutations, but there are only two genuinely different ones: e^* and f^*. The correspondence $g \to g^*$ is not one-to-one; for example, $e^* = g^* = h^* = (gh)^*$. Still, G acts on the orbit $\{u, v\}$.

(b) We use the same group G, but restrict to the other orbit $\{w, x, y, z\}$. The table of values is evident from Figure 1(b) by ignoring the u- and v-columns. Note that the eight restricted permutations are all different, so $g \to g^*$ is one-to-one this time. ■

Let's return to Theorem 1. Since there are $|G|$ terms in the sum

$$\sum_{g \in G} |\text{FIX}_X(g)|,$$

when we divide the sum by $|G|$, we obtain the average value of $|\text{FIX}_X(g)|$ over all members g of G. If some values are larger than average, then some must be smaller. This observation leads to the following somewhat surprising corollaries of Theorem 1.

Corollary 1 If X is the only orbit of G on X and if $|X| > 1$, then there exists an element g in G such that $g(x) \neq x$ for all $x \in X$.

Proof By Theorem 1, the average value of $|\text{FIX}_X(g)|$ is 1, since there is just one orbit. Moreover, $|\text{FIX}_X(e)| = |X| > 1$ and so $|\text{FIX}_X(g)| < 1$ for at least one g. For such a g, the set $\text{FIX}_X(g) = \{x \in X : g(x) = x\}$ must be the empty set. ■

Exercise 14 on page 469 concerns actions like the ones in Corollary 1.

Corollary 2 If G acts on X and if Y is an orbit of G on X with $|Y| > 1$, then G contains an element that moves every member of Y.

Proof The set Y is the only orbit under the group of restrictions g^* to Y for $g \in G$. By Corollary 1, there is a g^* such that $g^*(x) \neq x$ for all $x \in Y$. Thus $g(x) \neq x$ for all $x \in Y$. ■

EXAMPLE 4

We refer back to Figure 1. According to Corollary 2, some automorphism must move every member of the orbit $\{u, v\}$. In fact, the automorphisms f, fh, fg, and fgh all have this property.

The same corollary assures us that some automorphism moves every element in the orbit $\{w, x, y, z\}$. A glance at Figure 1(b) shows that gh, f, fh, fg, and fgh all have this property. ■

Sometimes, as we saw in §12.1, groups act on sets of functions that are natural and useful. Suppose that G acts on a set X. Let $\text{FUN}(X, C)$ be the set of all functions from X into some finite set, which we have called C because we are thinking of applications to colorings. As we saw in Example 4 on page 465, if we define $g^*(\varphi) = \varphi \circ g^{-1}$ for φ in $\text{FUN}(X, C)$, then

$$(fg)^* = f^* \circ g^* \qquad \text{for} \quad f, g \in G,$$

so G acts on $\text{FUN}(X, C)$.

The G-orbit of a function φ in $\text{FUN}(X, C)$ is $G(\varphi) = \{g^*(\varphi) : g \in G\}$, i.e., $\{\varphi \circ g^{-1} : g \in G\}$. Since every h in G is an inverse of something, namely $h = (h^{-1})^{-1}$, the set $\{g^{-1} : g \in G\}$ is G itself, and the orbit of φ is just $\{\varphi \circ g : g \in G\}$, which we will denote by $\varphi \circ G$.

Now let's go back to coloring problems.

EXAMPLE 5

In how many ways can we color the vertices of our favorite graph H in Figure 1 using red, black, or both colors?

Each of the six vertices can be colored red or black, so there are $2^6 = 64$ possible ways to assign colors to the six vertices. This count ignores the graph structure, though. We regard two colorings as equivalent if a graph automorphism will move one to the other. The colored graphs shown in Figure 3 correspond to four different ways of choosing the colors for the vertices, but we can see from the pictures that we can get from any one of these colored graphs to any other by applying a suitable graph automorphism. For instance, the automorphism h in the table of Figure 1 switches the upper and lower edges on the right and converts (a) into (b), and rotating the picture a half turn by fgh converts (a) into (d) and (b) into (c).

Figure 3 ▶

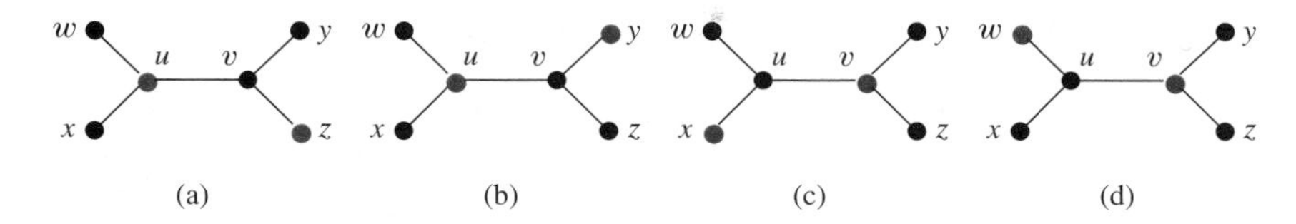

Each coloring of a graph in Figure 3 corresponds to a function $\varphi\colon V \to C$, where V is the set of vertices and $C = \{\text{Black, Red}\}$ is the set of colors; let φ_a, φ_b, φ_c, and φ_d be the names for these four colorings. The colorings φ_a and φ_b are really equivalent, since $\varphi_b = \varphi_a \circ h$; i.e., we can get the (b) coloring by first performing h, to switch vertices, and then doing the (a) coloring. Similarly, $\varphi_a \circ f = \varphi_c$ and $\varphi_a \circ fgh = \varphi_d$. These colorings all belong to the same orbit $\varphi \circ \text{AUT}(H)$ of $\text{AUT}(H)$ acting on $\text{FUN}(V, C)$. ■

The number of essentially different colorings in Example 5 is the number of $\text{AUT}(H)$-orbits in $\text{FUN}(V, C)$. The answer, 21, is *not* obvious. To find this number, we will apply Theorem 1 to $X = \text{FUN}(V, C)$, so we will need to be able to calculate numbers like $|\text{FIX}_{\text{FUN}(V,C)}(g^*)|$.

EXAMPLE 6

The automorphism fh of the graph in Example 5 has the $\langle fh\rangle$-orbits $\{u, v\}$ and $\{w, x, y, z\}$. [Here's where we use the Orbits column in Figure 1.] In order for $(fh)^*$ to fix a coloring, the vertices in each orbit must all be the same color; otherwise, applying fh would take at least one vertex to a vertex of a different color and would make a noticeable difference in the coloring. There are four colorings that meet this condition: all black, all red, u and v black but w, x, y, z all red, and u, v red but w, x, y, z black. These four colorings are the members of $\text{FIX}_{\text{FUN}(V,C)}((fh)^*)$. ■

The following theorem generalizes this example.

Theorem 2 Suppose that G acts on a set X and hence on $\text{FUN}(X, C)$. If g is in G, then $\text{FIX}_{\text{FUN}(X,C)}(g^*)$ consists of all the functions $\varphi\colon X \to C$ that are constant on the $\langle g\rangle$-orbits. If G, X, and C are finite, then the number of such functions is $|C|^m$, where m is the number of $\langle g\rangle$-orbits on X.

Proof We want to show that $g^*(\varphi) = \varphi$ if and only if φ is constant on each $\langle g\rangle$-orbit. Now $g^*(\varphi) = \varphi \circ g^{-1}$, so $g^*(\varphi) = \varphi$ if and only if $\varphi(g^{-1}(x)) = \varphi(x)$ for every $x \in X$. If φ is constant on each orbit $\langle g\rangle(y) = \{g^n(y) : n \in \mathbb{Z}\}$, then certainly $\varphi(g^{-1}(x)) = \varphi(x)$ for all $x \in X$; so $g^*(\varphi) = \varphi$. Now suppose that $g^*(\varphi) = \varphi$. Then, for all $x \in X$, we have $\varphi(x) = \varphi(g^{-1}(x)) = \varphi(g^{-2}(x)) = \cdots$; i.e., $\varphi(g^{-n}(x)) = \varphi(x)$ for $n \in \mathbb{N}$. Replacing x by $g^n(x)$, we also get $\varphi(x) = \varphi(g^n(x))$ for all $n \in \mathbb{N}$ and $x \in X$. Thus φ is constant on each orbit $\langle g\rangle(y) = \{g^n(y) : n \in \mathbb{Z}\}$ under $\langle g\rangle$.

A function φ that is constant on $\langle g\rangle$-orbits is completely determined by specifying its value on each orbit. Since the number of choices for each orbit value is $|C|$, there are $|C|^m$ such functions. ■

Theorems 1 and 2 provide us with the following answer to our coloring and circuit-counting questions.

Theorem 3 Consider a finite group G acting on a set X. The number $C(k)$ of G-equivalence classes of colorings of X using some or all of a set of k colors is given by the formula

$$C(k) = \frac{1}{|G|} \sum_{g \in G} k^{m(g)},$$

where $m(g)$ is the number of orbits of $\langle g \rangle$ on X.

EXAMPLE 7

Let's color the vertices of our running example, the graph H in Figure 1, using at most k colors. We have $|G| = |\text{AUT}(H)| = 8$. The numbers of orbits, $m(g)$, are given in Table 1. By Theorem 3, we have

$$\begin{aligned} C(k) &= \frac{1}{8}(k^6 + k^5 + k^5 + k^4 + k^3 + k^2 + k^2 + k^3) \\ &= \frac{1}{8}(k^6 + 2k^5 + k^4 + 2k^3 + 2k^2). \end{aligned}$$

TABLE 1

Subgroup	Its Orbits	Number of Orbits
$\langle e \rangle$	$\{u\}, \{v\}, \{w\}, \{x\}, \{y\}, \{z\}$	6
$\langle g \rangle$	$\{u\}, \{v\}, \{w, x\}, \{y\}, \{z\}$	5
$\langle h \rangle$	$\{u\}, \{v\}, \{w\}, \{x\}, \{y, z\}$	5
$\langle gh \rangle$	$\{u\}, \{v\}, \{w, x\}, \{y, z\}$	4
$\langle f \rangle$	$\{u, v\}, \{w, y\}, \{x, z\}$	3
$\langle fh \rangle$	$\{u, v\}, \{w, x, y, z\}$	2
$\langle fg \rangle$	$\{u, v\}, \{w, x, y, z\}$	2
$\langle fgh \rangle$	$\{u, v\}, \{w, z\}, \{x, y\}$	3

k	$C(k)$
1	1
2	21
3	171
4	820
5	2,850
6	8,001
7	19,306

Figure 4 ▲

The first few values of $C(k)$ are given in Figure 4. This problem for $k = 2$ was discussed in Example 5, where it appeared moderately difficult to solve directly. The problem for $k \geq 3$ looks hopeless without the theory we have developed. ■

EXAMPLE 8

Before we get back to the cube, we color the vertices of the square with k colors; see Exercise 11 on page 469. We regard two colorings as the same if we can turn one into the other by a suitable rotation of the square or by flipping it over. Figure 5(a) shows the square, and Figure 5(b) lists the relevant group of permutations of its vertex set. The table in Figure 6 lists the orbits of the cyclic subgroups of the group. We check the last line of Figure 6. Since $f^2 = e$, we have $\langle f \rangle = \{e, f\}$. Thus the orbits of $\langle f \rangle$ are the sets $\{e(v), f(v)\}$ for vertices v in $\{x, y, z, w\}$, so they are $\{e(x), f(x)\} = \{x, z\}$, $\{e(y), f(y)\} = \{y\}$, $\{e(z), f(z)\} = \{z, x\}$ and $\{e(w), f(w)\} = \{w\}$. There are just three different orbits; thus $m(f) = 3$. The other lines of Figure 6 are verified using $\langle e \rangle = \{e\}$, $\langle r \rangle = \langle r^3 \rangle = \{e, r, r^2, r^3\}$, $\langle r^2 \rangle = \{e, r^2\}$, $\langle h \rangle = \{e, h\}$, $\langle v \rangle = \{e, v\}$, and $\langle d \rangle = \{e, d\}$.

Figure 5 ▶

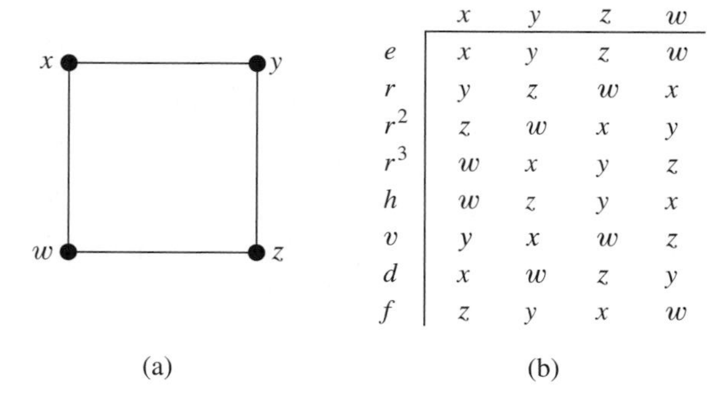

	x	y	z	w
e	x	y	z	w
r	y	z	w	x
r^2	z	w	x	y
r^3	w	x	y	z
h	w	z	y	x
v	y	x	w	z
d	x	w	z	y
f	z	y	x	w

(a) (b)

$\langle e\rangle$	$\{x\}, \{y\}, \{z\}, \{w\}$	$m(e) = 4$
$\langle r\rangle$	$\{x, y, z, w\}$	$m(r) = 1$
$\langle r^2\rangle$	$\{x, z\}, \{y, w\}$	$m(r^2) = 2$
$\langle r^3\rangle$	$\{x, y, z, w\}$	$m(r^3) = 1$
$\langle h\rangle$	$\{x, w\}, \{y, z\}$	$m(h) = 2$
$\langle v\rangle$	$\{x, y\}, \{z, w\}$	$m(v) = 2$
$\langle d\rangle$	$\{x\}, \{z\}, \{y, w\}$	$m(d) = 3$
$\langle f\rangle$	$\{x, z\}, \{y\}, \{w\}$	$m(f) = 3$

Figure 6 ▲

According to Theorem 3, there are

$$C(k) = \frac{1}{8}(k^4 + k + k^2 + k + k^2 + k^2 + k^3 + k^3)$$
$$= \frac{1}{8}(k^4 + 2k^3 + 3k^2 + 2k)$$

different ways to color the vertices of the square with k colors. For $k = 1$, this number is, of course, 1. The table in Figure 7 gives the numbers of colorings possible for the first few values of k. Figure 7(b) indicates the six different possibilities for two colors, including both of the one-color colorings. ■

Figure 7 ▶

k	Number of Ways to Color
1	1
2	6
3	21
4	55
5	120
6	231
7	406

(a)

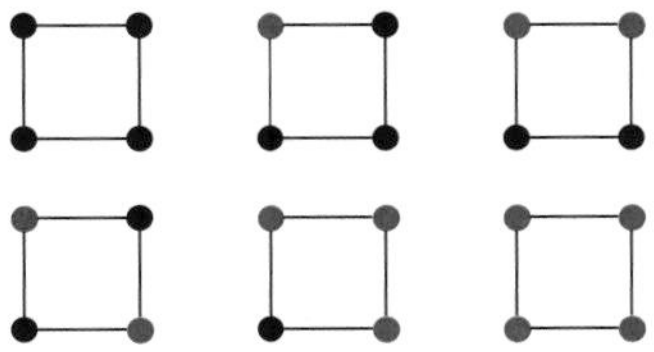

(b)

EXAMPLE 9

Now let us color the faces of the cube and answer our question from §12.1. There are 24 rotations that send the cube back to itself. To list them all would take a fair amount of work; but in fact we only need to know their orbit sizes, and for that we can just count the rotations of the five types illustrated in Figure 8(a). The table in Figure 8(b) tells how many there are of each type. It also gives the number $m(g)$ of orbits of their cyclic groups acting on the set of faces of the cube. For instance, the 90° rotation of type b has three $\langle g\rangle$-orbits: the faces the axis goes through form orbits of size 1 and the other four faces form an orbit of size 4. Thus $m(g) = 3$. The remaining values of $m(g)$ in Figure 8(b) were determined by similar reasoning. Theorem 3 gives the formula

$$C(k) = \frac{1}{24}(6k^3 + 6k^3 + 3k^4 + 8k^2 + k^6)$$

for the number of colorings of the faces with k colors. Figure 8(c) lists the first few values of $C(k)$. The value $C(3) = 57$ answers the question posed in Example 2 on page 463. ■

Figure 8 ▶

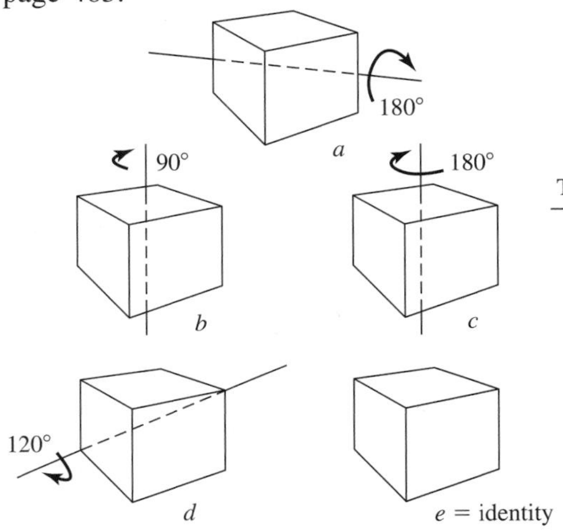

(a)

Type	Number of that Type	$m(g)$
a	6	3
b	6	3
c	3	4
d	8	2
e	1	6

(b)

k	$C(k)$
1	1
2	10
3	57
4	240
5	800
6	2226

(c)

Theorem 3 yields the number $C(k)$ of colorings using at most k colors. In some applications, one wants to know the number of colorings using *exactly* k colors. For this the Inclusion–Exclusion Principle on page 197 is useful.

EXAMPLE 10

(a) We can calculate the number of [inequivalent] colorings of the vertices of the square in Example 8 using exactly four colors, say red, blue, green, and yellow. Figure 7 gives us the number $C(k)$ of ways to color using at most k colors. For $i = 1, 2, 3, 4$, let A_i be the set of colorings that do not use the ith color. Then $A_1 \cup A_2 \cup A_3 \cup A_4$ is the set of colorings using three or fewer of the colors, and the answer we seek is $C(4) - |A_1 \cup A_2 \cup A_3 \cup A_4|$. Now, by the Inclusion–Exclusion Principle,

$$\begin{aligned} |A_1 \cup A_2 \cup A_3 \cup A_4| &= |A_1| + |A_2| + |A_3| + |A_4| - \{|A_1 \cap A_2| + |A_1 \cap A_3| \\ &\quad + |A_1 \cap A_4| + |A_2 \cap A_3| + |A_2 \cap A_4| + |A_3 \cap A_4|\} \\ &\quad + \{|A_1 \cap A_2 \cap A_3| + |A_1 \cap A_2 \cap A_4| + |A_1 \cap A_3 \cap A_4| + |A_2 \cap A_3 \cap A_4|\} \\ &\quad - |A_1 \cap A_2 \cap A_3 \cap A_4|. \end{aligned}$$

The set A_1 consists of all colorings using blue, green, or yellow, so $|A_1|$ is $C(3)$. Similarly, for A_2, A_3, and A_4. The set $A_1 \cap A_2$ consists of the colorings using green or yellow, so it has $C(2)$ elements; a similar observation applies to each intersection of two sets. Intersections like $A_1 \cap A_2 \cap A_3$ have only one coloring, i.e., they have $C(1)$ elements. Finally, $A_1 \cap A_2 \cap A_3 \cap A_4$ is the empty set. We conclude that

$$|A_1 \cup A_2 \cup A_3 \cup A_4| = 4C(3) - 6C(2) + 4C(1),$$

so the number of colorings using exactly four colors is

$$C(4) - 4C(3) + 6C(2) - 4C(1) = 55 - 4 \cdot 21 + 6 \cdot 6 - 4 \cdot 1 = 3.$$

Now that we know the answer, we can easily illustrate the different colorings in Figure 9.

Figure 9 ▶

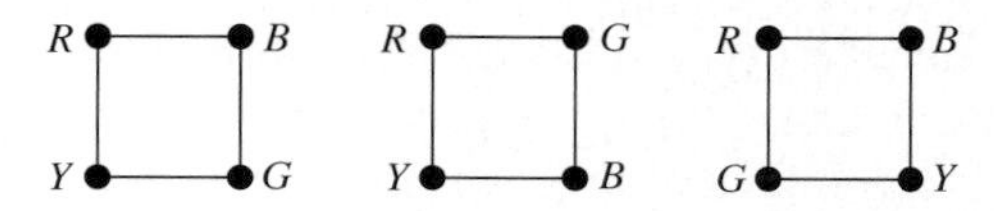

The same sort of reasoning would show that the number of colorings using exactly five colors is $C(5) - \binom{5}{4}C(4) + \binom{5}{3}C(3) - \binom{5}{2}C(2) + \binom{5}{1}C(1)$, the number using exactly six colors is $C(6) - \binom{6}{5}C(5) + \binom{6}{4}C(4) - \binom{6}{3}C(3) + \binom{6}{2}C(2) - \binom{6}{1}C(1)$, etc. Of course, such expressions $C(k)$ for $k > 4$ all evaluate to 0 for this example, since only four vertices are available to color.

(b) We calculate the number of [inequivalent] colorings of the cube using six different colors, where we regard two colorings as equivalent if one is the rotation of the other. As in part (a), this is $C(6) - 6C(5) + 15C(4) - 20C(3) + 15C(2) - 6C(1)$. Substituting the values $C(k)$ from Figure 8 and simplifying gives 30, a complicated confirmation of our work in Example 1 on page 463. ■

Now let us return to the problem of building logical circuits, at first for just two inputs. There are $2^4 = 16$ Boolean functions of two variables, namely the members of $\text{FUN}(\mathbb{B} \times \mathbb{B}, \mathbb{B})$, where $\mathbb{B} = \{0, 1\}$. We may think of a circuit as a black box with two input wires, one for x_1 and one for x_2, and one output wire. Each element (a_1, a_2) in $\mathbb{B} \times \mathbb{B}$ corresponds to a choice of values a_1 for x_1 and a_2 for x_2.

Interchanging the connections for x_1 and x_2 amounts to replacing (a_1, a_2) by (a_2, a_1) and corresponds to the permutation s of $\mathbb{B} \times \mathbb{B}$ that interchanges $(0, 1)$ and $(1, 0)$ and fixes $(0, 0)$ and $(1, 1)$. We want to regard two black boxes as equivalent if one will produce the same results as the other or will produce the same results if we interchange its input wires. That is, two boxes are equivalent if their Boolean functions f and f' are either the same or satisfy $f' = f \circ s$. Since $|\mathbb{B}| = 2$, the problem looks just like the two-color question for a four-element set, $\mathbb{B} \times \mathbb{B}$, with two elements that are interchangeable. We apply Theorem 3 with $X = \mathbb{B} \times \mathbb{B}$, $C = \mathbb{B}$, and $G = \langle s \rangle = \{e, s\}$. Here $m(e) = 4$ and $m(s) = 3$, so the number of G-orbits is

$$\frac{1}{2}(2^4 + 2^3) = 12.$$

We can confirm this result using the table in Figure 10, which lists all 16 Boolean functions from $\mathbb{B} \times \mathbb{B}$ to $\mathbb{B}$. Functions 2 and 4 can be performed with the same black box, as can functions 3 and 5, 10 and 12, and 11 and 13, so the number of orbits is $16 - 4 = 12$.

Figure 10 ▶

Function numbers

	0	1	2	3	4	5	6	7	8	9	10	11	12	13	14	15
(0, 0)	0	1	0	1	0	1	0	1	0	1	0	1	0	1	0	1
(0, 1)	0	0	1	1	0	0	1	1	0	0	1	1	0	0	1	1
(1, 0)	0	0	0	0	1	1	1	1	0	0	0	0	1	1	1	1
(1, 1)	0	0	0	0	0	0	0	0	1	1	1	1	1	1	1	1

Now suppose that we allow ourselves to complement inputs. Complementing the value on the first wire corresponds to interchanging (0, 0) with (1, 0) and (0, 1) with (1, 1). We denote this permutation of $\mathbb{B} \times \mathbb{B}$ by c_1 and the permutation that corresponds to complementing the second input by c_2. Altogether the permutations s, c_1, and c_2 generate the group G of permutations of $\mathbb{B} \times \mathbb{B}$ described in Figure 11(a). [This fact is not expected to be obvious; take our word for it.] This group acts on the 4-element set $\mathbb{B} \times \mathbb{B}$ in the same way that the group in Example 8 acts on the vertices of the square, as we see by comparing Figure 11(a) with Figure 11(b), which is just Figure 5(b) rewritten with some rows and columns interchanged. The correspondence $(0,0) \to x$, $(0,1) \to y$, $(1,0) \to w$, $(1,1) \to z$ converts one table into the other. From Figure 7 we know that there are $C(2) = 6$ ways to color the square with two colors, so there are six orbits of Boolean functions under the action of the group G, i.e., six essentially different black boxes. Using the function numbers from Figure 10, the orbits in $\text{FUN}(\mathbb{B} \times \mathbb{B}, \mathbb{B})$ are

$$\{0\}, \quad \{1, 2, 4, 8\}, \quad \{3, 5, 10, 12\}, \quad \{6, 9\}, \quad \{7, 11, 13, 14\}, \quad \{15\}.$$

To build circuits, it would be enough to have a circuit to compute one function from each orbit, say the functions 0, 1, 3, 6, 7, and 15.

Figure 11 ▶

	(0, 0)	(0, 1)	(1, 0)	(1, 1)
e	(0, 0)	(0, 1)	(1, 0)	(1, 1)
c_1	(1, 0)	(1, 1)	(0, 0)	(0, 1)
c_2	(0, 1)	(0, 0)	(1, 1)	(1, 0)
$c_1 \circ c_2$	(1, 1)	(1, 0)	(0, 1)	(0, 0)
s	(0, 0)	(1, 0)	(0, 1)	(1, 1)
$c_1 \circ s$	(1, 0)	(0, 0)	(1, 1)	(0, 1)
$c_2 \circ s$	(0, 1)	(1, 1)	(0, 0)	(1.0)
$c_1 \circ c_2 \circ s$	(1, 1)	(0, 1)	(1, 0)	(0, 0)

(*a*)

	x	y	w	z
e	x	y	w	z
h	w	z	x	y
v	y	x	z	w
r^2	z	w	y	x
d	x	w	y	z
r^3	w	x	z	y
r	y	z	x	w
f	z	y	w	x

(*b*)

If we also allow ourselves to complement the output of a circuit, then a circuit that computes the function numbered n will also compute $15 - n$, and we need even fewer black boxes. A circuit for 0 will also compute 15. One for 1 will compute 14 and hence also 7, 11, or 13. A circuit for 3 will also compute 12, which we already knew, and similarly a circuit for 6 will compute 9. The classes of functions are now

$$\{0, 15\}, \quad \{1, 2, 4, 8, 7, 11, 13, 14\}, \quad \{3, 5, 10, 12\}, \quad \text{and} \quad \{6, 9\}.$$

It still requires four different circuits to compute all 2-variable Boolean functions, allowing complementation on both input and output wires.

Our methods generalize, in theory, to count the number of black boxes needed for n-input Boolean functions. In practice, the detailed determination of orbits for all elements of G gets exceedingly complicated. For 4-input functions the answer is that 222 different circuits are required, even if we allow free complementation on inputs and outputs. This number is considerably smaller than $2^{16} = 65{,}536$. Knowing how many circuits there are does not help us find representative circuits, but it does tell us when we have found enough.

Our methods have not taken systematic advantage of the symmetry of the group G itself. By using such symmetry, one can obtain a formula for the number of G-orbits in $\text{FUN}(\mathbb{B}^n, \mathbb{B})$ whose members have exactly k of their values equal to 0 for $k = 1, 2, \ldots$. There is an extensive literature on the subject of using groups to count. For more in the spirit of this section, look in books on applied algebra, watching for the names Polyá and Burnside.

Exercises 12.3

1. Consider the group $G = \text{AUT}(H)$ of Example 1 acting on the set V of vertices of H.

(a) Find $|\text{FIX}_V(a)|$ for each automorphism a in G and add the results.

(b) Find $|\text{FIX}_G(p)|$ for each vertex p of H and add the results.

(c) Do the sums in parts (a) and (b) agree? Discuss.

2. Consider the square graph in Figure 7 on page 469. Confirm Theorem 1 for this example.

3. Consider the graph in Figure 12, and let G be the group of graph automorphisms acting on $\{w, x, y, z\}$. Confirm Theorem 1 for this example.

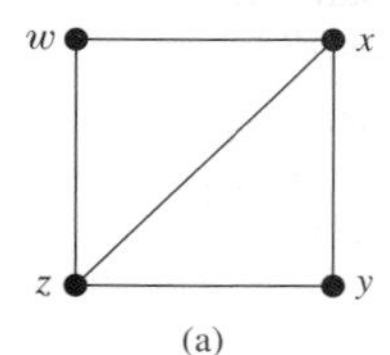

(a)

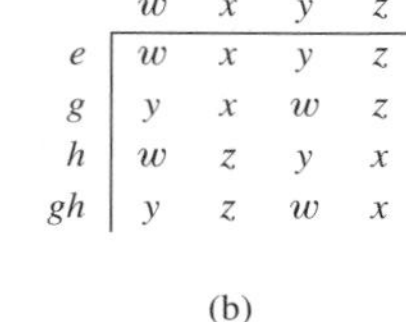

	w	x	y	z
e	w	x	y	z
g	y	x	w	z
h	w	z	y	x
gh	y	z	w	x

(b)

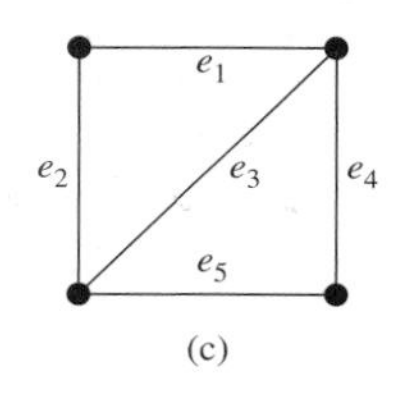

(c)

Figure 12 ▲

4. Let the group G in Exercise 3 act on the 5-element set E of edges. See Figure 12(c).

(a) Give the table of values, as in Figure 2(b).

(b) Confirm Theorem 1 for this example.

5. Show directly that Corollary 2 to Theorem 1 is true for each orbit of the group in Exercise 3.

6. Show directly that Corollary 2 is true for each orbit of the group in Exercise 4.

7. Verify that $|\text{FIX}_T(h^*)| = 7$ in Example 2(b).

8. Consider a finite group G acting on an n-element set X. Show that if $|\text{FIX}_X(g)| \geq 1$ for each $g \in G$, then G has at least $1 + \frac{n-1}{|G|}$ orbits in X. For $n > 1$, this implies Corollary 1 to Theorem 1. *Hint:* Treat the element e of G separately from the others.

9. The graph in Figure 13(a) has two automorphisms, which are described in Figure 13(b).

(a) What is the average number of vertices fixed by the automorphisms of this graph?

(b) Which of the automorphisms of this graph fix the average number of vertices?

(c) Find the number of ways to color the vertices of this graph with k colors.

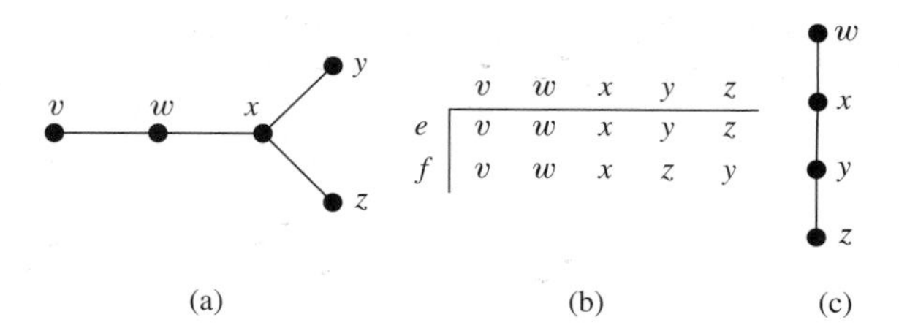

	v	w	x	y	z
e	v	w	x	y	z
f	v	w	x	z	y

(a) (b) (c)

Figure 13 ▲

10. Verify Theorem 1 for

(a) the group of automorphisms of the graph in Figure 13(a) acting on the set of vertices of the graph.

(b) the group in part (a) acting on the set of edges of the graph in Figure 13(a).

(c) the group of rotations in Example 9 acting on the set of faces of the cube.

11. The graph in Figure 13(c) has two automorphisms.

(a) How many ways are there to color the vertices of this graph with k colors?

(b) How many ways are there to label the vertices of this graph with four different labels?

12. How many different circular necklaces can be made from five beads of k different colors? Consider two necklaces to be the same if one looks just like the other when it is rotated or flipped over. *Hint:* The group here consists of e, four nontrivial rotations, and five flips. See Example 8 for the four-bead case.

13. The graph in Figure 14(a) has six automorphisms, which are described in Figure 14(b).

(a) Find the number of ways to color the vertices of this graph with k colors.

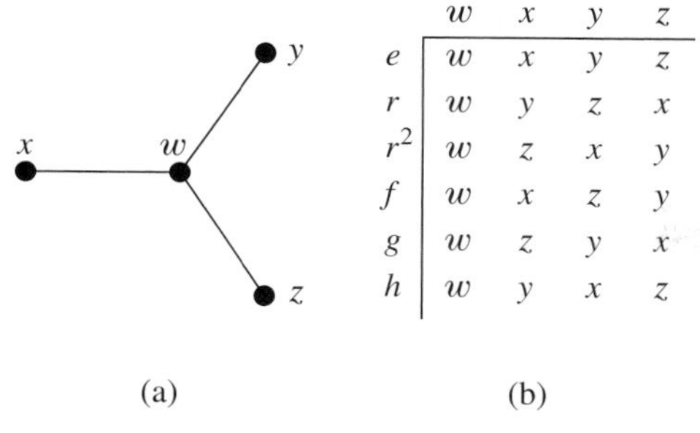

	w	x	y	z
e	w	x	y	z
r	w	y	z	x
r^2	w	z	x	y
f	w	x	z	y
g	w	z	y	x
h	w	y	x	z

Figure 14 ▲

(b) Find the number of ways to color the edges of this graph with k colors.

14. (a) Use the Inclusion–Exclusion Principle and the answer to Exercise 13(a) to find the number of ways to color the vertices of the graph in Figure 14(a) with exactly three colors.

(b) Describe all the different colorings in part (a), using the colors red, blue, and green.

(c) Find the number of ways to color the vertices of this graph with exactly four colors.

15. Find the number of colorings of the vertices of the square in Example 8 using

(a) exactly three colors,

(b) exactly two colors,

(c) exactly five colors.

16. Draw representatives of each coloring of the types counted in Exercise 15.

17. Consider the graph of Figure 12 and the group G acting on $V = \{w, x, y, z\}$.

(a) For each member g of G, find the number of $\langle g \rangle$-orbits in V.

(b) Find the number of ways to color the vertices of this graph with at most k colors.

(c) Use your answer to part (b) to calculate the number of ways to color the vertices using red or black or both. Draw a picture of a representative of each coloring.

18. The group G in Exercise 17 also acts on the set $E = \{e_1, e_2, e_3, e_4, e_5\}$ of edges; see Exercise 4.

(a) For each member g of G, find the number of $\langle g \rangle$-orbits in E.

(b) Find the number of ways to color the edges of this graph with at most k colors.

(c) How many ways are there to color the edges of this graph with exactly 2 colors?

19. (a) How many ways are there to color the vertices of a cube with k colors?

(b) How many ways are there to color the edges of a cube with k colors?

Two colorings in parts (a) and (b) are considered the same if one can be turned into the other by a rotation of the cube. *Suggestion:* Use Figure 8(a) to create new tables like Figure 8(b) for the actions on vertices and edges.

20. Consider the problem of coloring the faces of a cube using crayons that are red, green, and blue, as in Example 9.

(a) How many colorings use exactly two of the three colors?

(b) Use any method to find how many colorings have four red faces and two blue faces.

(c) How many colorings have exactly four red faces?

(d) Would you like to use the method of inspection to find all colorings with exactly two faces of each color?

21. (a) How many different 2-input logical circuits are there if we only regard two circuits as the same if they have the same or complementary outputs?

(b) How many are there if we also consider two circuits to be the same if they produce the same function when the input wires on one are interchanged?

12.4 Group Homomorphisms

One aim of this section is to show that every cyclic group $\langle g \rangle$ looks just like $(\mathbb{Z}, +)$ or a group $(\mathbb{Z}(p), +_p)$ for some positive integer p. To say what we mean by "looks just like," we will use the concept of isomorphism between groups. We have already encountered isomorphisms between various objects; they are the one-to-one correspondences that preserve whatever structure is under consideration. Thus a **group isomorphism** from one group $(G, \cdot)$ to another group $(G_0, \bullet)$ is a one-to-one correspondence $\varphi\colon G \to G_0$ satisfying $\varphi(g \cdot h) = \varphi(g) \bullet \varphi(h)$ for all $g, h \in G$. Isomorphisms preserve the group structure in a one-to-one way. Two groups are said to be **isomorphic** if there is a group isomorphism of one onto the other, and this is what we mean when we say informally that they "look just alike."

To show that every cyclic group is isomorphic to $\mathbb{Z}$ or to one of the groups $\mathbb{Z}(p)$, we will use mappings that preserve group structure, as isomorphisms do, but that are not necessarily either one-to-one or onto. Mappings of this sort are of fundamental importance in the study of groups and, indeed, of almost all sorts of algebraic structures.

A **homomorphism** from the group $(G, \cdot)$ to the group $(G_0, \bullet)$ is a function $\varphi\colon G \to G_0$ satisfying $\varphi(g \cdot h) = \varphi(g) \bullet \varphi(h)$ for all $g, h \in G$. Since the identity and inverses are part of the group structure, it appears that we should also require that φ preserve the identity and inverses. Fortunately, these properties are automatic consequences of the definition.

Proposition Let $\varphi\colon G \to G_0$ be a homomorphism.

(a) If e is the identity in G, then $\varphi(e)$ is the identity, e_0, of G_0.
(b) $\varphi(g^{-1}) = \varphi(g)^{-1}$ for all $g \in G$.
(c) $\varphi(G)$ is a subgroup of G_0.

Proof

(a) We have $\varphi(e) = \varphi(e) \bullet [\varphi(e) \bullet \varphi(e)^{-1}] = [\varphi(e) \bullet \varphi(e)] \bullet \varphi(e)^{-1} = \varphi(e \cdot e) \bullet \varphi(e)^{-1} = \varphi(e) \bullet \varphi(e)^{-1} = e_0$.
(b) Note that the first inverse is taken in the group G, while the second is taken in G_0. Since $\varphi(g) \bullet \varphi(g^{-1}) = \varphi(g \cdot g^{-1}) = \varphi(e) = e_0$ and inverses are unique, $\varphi(g^{-1}) = \varphi(g)^{-1}$.
(c) Since φ takes inverses to inverses, $\varphi(G)$ contains the inverses of all its elements. Since $\varphi(G)$ is also closed under products, it is a subgroup of G_0. ■

EXAMPLE 1

(a) Let $(G, \cdot)$ and $(G_0, \bullet)$ both be $(\mathbb{Z}, +)$. The homomorphism condition is

$$\varphi(m+n) = \varphi(m) + \varphi(n) \quad \text{for} \quad m, n \in \mathbb{Z}.$$

The function φ defined by $\varphi(n) = 5n$ for all n is an example of a homomorphism, since

$$\varphi(m+n) = 5 \cdot (m+n) = 5m + 5n = \varphi(m) + \varphi(n).$$

There is nothing special about 5; any other integer, including 0, would define a homomorphism from $(\mathbb{Z}, +)$ to $(\mathbb{Z}, +)$ in the same way.
(b) Let $(G, \cdot)$ be $(\mathbb{Z}, +)$, let $(G_0, \bullet)$ be $(\mathbb{R} \setminus \{0\}, \cdot)$, and let $\varphi(m) = 2^m$ for m in $\mathbb{Z}$. Since

$$\varphi(m+n) = 2^{m+n} = 2^m \cdot 2^n = \varphi(m) \cdot \varphi(n),$$

φ is a homomorphism of $(\mathbb{Z}, +)$ into $(\mathbb{R} \setminus \{0\}, \cdot)$. Note that $\varphi(0) = 1$, the identity of $(\mathbb{R} \setminus \{0\}, \cdot)$.
(c) Recall that $a = \log_2 b$ if and only if $b = 2^a$. The function φ given by $\varphi(x) = \log_2 x$ from $(\{x \in \mathbb{R} : x > 0\}, \cdot)$ to $(\mathbb{R}, +)$ is a group homomorphism, since $\log_2(xy) = \log_2(x) + \log_2(y)$.
(d) We saw the equations

$$(f \circ g)^* = f^* \circ g^* \quad \text{and} \quad (fg)^* = f^* \circ g^*$$

when we were looking at graph automorphisms in §12.3. In each case there was a homomorphism in the background. The first time $f \to f^*$ was a mapping of $\textsc{aut}(H)$ into the group of permutations of the edge set of the graph H, determined by $f^*(\{u, v\}) = \{f(u), f(v)\}$. Restriction of the members of $\textsc{aut}(H)$ to some union of orbits for $\textsc{aut}(H)$ gave another homomorphism $f \to f^*$ in Example 3 on page 479.

To say that a group G "acts on" a set X in the sense defined in §12.1 is simply to say that we have a homomorphism from G into $\textsc{perm}(X)$, i.e., a structure-preserving way of associating permutations with members of G. ■

EXAMPLE 2

(a) Consider a positive integer $p \geq 2$. As in §3.5, let $n \text{ MOD } p$ be the remainder when n is divided by p. Recall that $\text{MOD } p$ is a function from $\mathbb{Z}$ into $\mathbb{Z}(p)$. In fact, MOD p is a homomorphism of $(\mathbb{Z}, +)$ onto $(\mathbb{Z}(p), +_p)$, by Theorem 3 on page 123. That is,

$$(m+n) \text{ MOD } p = (m \text{ MOD } p) +_p (n \text{ MOD } p) \quad \text{for} \quad m, n \in \mathbb{Z}.$$

(b) The theorem also states that

$$(m \cdot n) \text{ MOD } p = (m \text{ MOD } p) *_p (n \text{ MOD } p).$$

It is tempting to conclude that MOD p is also a homomorphism of $(\mathbb{Z}, \cdot)$ onto $(\mathbb{Z}(p), *_p)$. But neither $(\mathbb{Z}, \cdot)$ nor $(\mathbb{Z}(p), *_p)$ is a group! Hence such a statement makes no sense in this setting. ■

EXAMPLE 3

Let $(G, \cdot)$ be a group and let $g \in G$. Exercise 12 on page 475 asserts that $g^{m+n} = g^m \cdot g^n$ for all $m, n \in \mathbb{Z}$. That is, the function φ from $(\mathbb{Z}, +)$ to $(G, \cdot)$ given by $\varphi(n) = g^n$ is a homomorphism. Its image $\varphi(\mathbb{Z})$ is the cyclic subgroup $\langle g \rangle$ of G. ■

Each homomorphism φ defined on G has associated with it a special subgroup of G, called its **kernel**, which is defined by

$$\text{kernel of } \varphi = \{g \in G : \varphi(g) = \varphi(e)\}.$$

In a sense, the kernel tells us what we are ignoring when we pass from G to $\varphi(G)$. Conclusion (d) of the next theorem can be viewed as saying that two members of G have the same image under φ if and only if they differ by a member of the kernel of φ.

Theorem 1 Let $(G, \cdot)$ and $(G_0, \bullet)$ be groups, with identities e and e_0, respectively. Let $\varphi: G \to G_0$ be a homomorphism with kernel K. Then

(a) K is a subgroup of G.
(b) $g \cdot K = K \cdot g$ for each g in G.
(c) $g \cdot K = \{h \in G : \varphi(h) = \varphi(g)\}$ for each g in G.
(d) $\varphi(g) = \varphi(h)$ if and only if $g \cdot h^{-1} \in K$.

Proof

(a) Since $e \in K$, the set K is nonempty. If $g, h \in K$, then $\varphi(g \cdot h) = \varphi(g) \bullet \varphi(h) = \varphi(e) \bullet \varphi(e) = e_0 \bullet e_0 = e_0$; so $g \cdot h$ is in K. Moreover, $\varphi(g^{-1}) = \varphi(g)^{-1} = e_0{}^{-1} = e_0$, so g^{-1} is in K. Thus K is closed under products and inverses; so it is a subgroup of G.

(b) We show first that $K \cdot g \subseteq g \cdot K$. It is enough to consider $k \in K$ and show that $k \cdot g \in g \cdot K$. Since $k \cdot g = g \cdot (g^{-1} \cdot k \cdot g)$, it suffices to show that $g^{-1} \cdot k \cdot g$ is in K. But $\varphi(g^{-1} \cdot k \cdot g) = \varphi(g^{-1}) \bullet \varphi(k) \bullet \varphi(g) = \varphi(g)^{-1} \bullet \varphi(e) \bullet \varphi(g) = \varphi(g^{-1} \cdot e \cdot g) = \varphi(e)$. A similar argument shows that $g \cdot K \subseteq K \cdot g$.

(c) For $k \in K$, $\varphi(g \cdot k) = \varphi(g) \bullet \varphi(k) = \varphi(g) \bullet e_0 = \varphi(g)$, and so $g \cdot K \subseteq \{h \in G : \varphi(h) = \varphi(g)\}$. On the other hand, if $\varphi(h) = \varphi(g)$, then $\varphi(g^{-1} \cdot h) = \varphi(g)^{-1} \bullet \varphi(h) = e_0$, which means $g^{-1} \cdot h \in K$; so $h = g \cdot (g^{-1} \cdot h) \in g \cdot K$.

(d) By parts (c) and (b), $\varphi(g) = \varphi(h) \iff g \cdot K = h \cdot K \iff K \cdot g = K \cdot h \iff K \cdot g \cdot h^{-1} = K \iff g \cdot h^{-1} \in K$. ■

Corollary 1 In the setting of Theorem 1, the cosets $g \cdot K$ are the equivalence classes for the equivalence relation that φ defines by $g \sim h$ if and only if $\varphi(g) = \varphi(h)$.

Proof This is just a restatement of (c). ■

We saw in Proposition 3 on page 472 that all cosets $g \cdot K$ have the same number of elements, $|K|$, and this fact gives us a useful test for one-to-oneness of homomorphisms.

Corollary 2 A homomorphism is one-to-one if and only if its kernel contains just the identity element.

Proof Clearly, K contains e. Moreover, by Theorem 1(c), φ is one-to-one if and only if all cosets $g \cdot K$ have exactly one element, which is true if and only if K itself has just one member. ■

EXAMPLE 4

(a) The homomorphism φ from $(\mathbb{Z}, +)$ to $(\mathbb{Z}, +)$ defined by $\varphi(n) = 5n$ is one-to-one, and its kernel is $\{n \in \mathbb{Z} : 5n = 0\} = \{0\}$. The homomorphism in part (b) of Example 1 is also one-to-one, and its kernel is $\{n \in \mathbb{Z} : 2^n = 1\} = \{0\}$. The homomorphism $\varphi(x) = \log_2 x$ in Example 1(c) is one-to-one, and its kernel is $\{x \in \mathbb{R} : \log_2 x = 0\} = \{1\}$.

(b) The homomorphism $\text{MOD}\, p$ [Example 2(a)] from $(\mathbb{Z}, +)$ to $(\mathbb{Z}(p), +_p)$ is not one-to-one. Its kernel is $\{n \in \mathbb{Z} : n \,\text{MOD}\, p = 0\} = \{n \in \mathbb{Z} : p \text{ divides } n\} = p\mathbb{Z}$. Two integers are in the same coset of the kernel if and only if their difference is a multiple of p, and the cosets are just the congruence classes $[r]_p = p\mathbb{Z} + r$.■

EXAMPLE 5

The homomorphism $n \to g^n$ from $(\mathbb{Z}, +)$ into $(\langle g \rangle, \cdot)$ that we saw in Example 3 is the key to understanding cyclic groups. Its kernel is $\{n \in \mathbb{Z} : g^n = e\}$. This subgroup of $\mathbb{Z}$ is just $\{0\}$ if $\langle g \rangle$ is infinite, but is $p\mathbb{Z}$ for some nonzero p otherwise. In the first case, the homomorphism is one-to-one, but in the second case it is not. We will exploit these facts in Example 8 after we have built up a little more machinery. ■

A subgroup K of a group G with the property that $gK = Kg$ for every g in G is called a **normal subgroup** of G. Theorem 1(b) shows that kernels of homomorphisms are normal. If G is commutative, then every subgroup is normal; but in the general case there will be nonnormal subgroups. When K is normal, its left and right cosets coincide, so we simply refer to its **cosets**.

If K is a normal subgroup of the group G, then the set G/K of cosets gK has a natural and useful group operation $*$ of its own. The members of G/K that we multiply are sets themselves [indeed, cosets], and the product of two cosets will be another such coset. This is not quite a new idea. We are used to making new sets out of two old ones, for instance by taking A and B and forming $A \cap B$ and $A \cup B$, and we have added and multiplied congruence classes in Example 6(a) on page 124. Our product of cosets generalizes that example.

The natural way to try to multiply two cosets gK and hK is to define

$$(\boldsymbol{gK}) * (\boldsymbol{hK}) = gKhK = \{gk_1hk_2 : k_1, k_2 \in K\}.$$

For an arbitrary subgroup K, this product may not be a coset, but if K is normal, then it must be one.

Theorem 2 Let K be a normal subgroup of the group G. Then

(a) $gKhK = (gh)K$ for all $g, h \in G$.

(b) The set G/K of cosets of K in G is a group under the operation $*$ defined by

$$(gK) * (hK) = ghK.$$

(c) The function $\boldsymbol{\nu} : G \to G/K$ defined by $\boldsymbol{\nu(g)} = gK$ is a homomorphism with kernel K. [ν is a Greek nu.]

Proof

(a) Since K is a subgroup of G, $K = Ke \subseteq KK \subseteq K$, so $K = KK$. Since K is normal, we have $hK = Kh$, so

$$gKhK = ghKK = ghK \in G/K.$$

(b) According to part (a), $(gK) * (hK)$ is the set $(gK)(hK)$, and thus $*$ is a well-defined binary operation on G/K. It is easy to check that it is associative, that K is the identity, and that $(gK)^{-1} = g^{-1}K$.

(c) We have $\nu(gh) = ghK = (gK) * (hK) = \nu(g) * \nu(h)$ by definition of $*$, so ν is a homomorphism. If $\nu(g) = \nu(e)$, then $g \in gK = eK = K$, and if $g \in K$, then $\nu(g) = gK = K = \nu(e)$ [Exercise 27 on page 476]. Thus K is the kernel of ν. ■

The mapping ν is called the **natural homomorphism** of G onto G/K. Naming it with the Greek letter nu is supposed to help us remember that it's **n**atural. Theorems 1 and 2 together tell us that kernels of homomorphisms are normal subgroups and, conversely, every normal subgroup is the kernel of some homomorphism. If K is the kernel of φ, then the natural mapping ν in Theorem 2 is the same one, $s \to [s]$, that we saw in the proof of Theorem 2 on page 117, since gK can be viewed as the equivalence class $[g]$ of g, by Corollary 1 of Theorem 1.

EXAMPLE 6

(a) Consider again the graph H in Figure 1 on page 478. The set $\{u, v\}$ is an orbit of $\text{AUT}(H)$, so restriction to this orbit [see page 479] gives a homomorphism

$$\varphi\colon \text{AUT}(H) \to \text{PERM}(\{u, v\}) = \{e, s\},$$

where e fixes u and v, and s switches them. The kernel of φ is the subgroup $K = \{e, g, h, gh\}$ of automorphisms that fix u and v, and φ maps each member of the coset $fK = \{f, fg, fh, fgh\}$ to s. The natural homomorphism for this example is the mapping ν from $\text{AUT}(H)$ to $\text{AUT}(H)/K = \{K, fK\}$ given by $\nu(e) = \nu(g) = \nu(h) = \nu(gh) = K$ and $\nu(f) = \nu(fg) = \nu(fh) = \nu(fgh) = fK$.

(b) Let $(G, \cdot) = (\mathbb{Z}, +)$ and let $K = 6\mathbb{Z}$. Theorem 2(b) tells us that $\mathbb{Z}/6\mathbb{Z}$ is a group under the operation

$$(k + 6\mathbb{Z}) * (m + 6\mathbb{Z}) = k + m + 6\mathbb{Z}.$$

The identity of $\mathbb{Z}/6\mathbb{Z}$ is $6\mathbb{Z}$.

Theorem 2(c) tells us that if ν is defined by $\nu(k) = k + 6\mathbb{Z}$, then ν maps $\mathbb{Z}$ onto $\mathbb{Z}/6\mathbb{Z}$ and the kernel of ν is $6\mathbb{Z}$; i.e.,

$$\{k \in \mathbb{Z} : \nu(k) = 6\mathbb{Z}\} = \{k \in \mathbb{Z} : k + 6\mathbb{Z} = 6\mathbb{Z}\} = 6\mathbb{Z}.$$

In Example 4(b) we observed that $6\mathbb{Z}$ is also the kernel of $\varphi = \text{MOD}\,6$, which maps $(\mathbb{Z}, +)$ onto $(\mathbb{Z}(6), +_6)$. If we define

$$\varphi^*(k + 6\mathbb{Z}) = k \quad \text{for} \quad k \in \{0, 1, 2, 3, 4, 5\} = \mathbb{Z}(6),$$

then φ^* is a one-to-one correspondence of $\mathbb{Z}/6\mathbb{Z}$ onto $\mathbb{Z}(6)$. The mapping φ^* is an isomorphism because

$$\begin{aligned}\varphi^*((k + 6\mathbb{Z}) * (m + 6\mathbb{Z})) &= \varphi^*(k + m + 6\mathbb{Z}) = \varphi^*(k +_6 m + 6\mathbb{Z})\\ &= k +_6 m = \varphi^*(k + 6\mathbb{Z}) +_6 \varphi^*(m + 6\mathbb{Z}).\end{aligned}$$

Thus the groups $\mathbb{Z}/6\mathbb{Z}$ and $\mathbb{Z}(6)$ are isomorphic. Using our G and K notation, this says that G/K and $\varphi(G)$ are isomorphic and illustrates the next theorem. ■

The Fundamental Homomorphism Theorem Let φ be a homomorphism from the group $(G, \cdot)$ to the group $(G_0, \bullet)$, with kernel K. Then G/K is isomorphic to $\varphi(G)$ under the isomorphism φ^* defined by

$$\varphi^*(g \cdot K) = \varphi(g).$$

Proof Theorem 1(c) says that $g \cdot K$ is the set of all h in G for which $\varphi(h) = \varphi(g)$. Thus $\varphi^*(g \cdot K)$ is the common value that φ has on all members of $g \cdot K$. The function φ has different values on different cosets, so φ^* is one-to-one. Its image is clearly

$\varphi(G)$, so we just need to observe that φ^* is a homomorphism:

$$\begin{aligned}
&\varphi^*((g \cdot K) * (h \cdot K)) \\
&= \varphi^*(g \cdot h \cdot K) && \text{[definition of } *\text{]} \\
&= \varphi(g \cdot h) && \text{[definition of } \varphi^*\text{]} \\
&= \varphi(g) \bullet \varphi(h) && \text{[}\varphi \text{ is a homomorphism]} \\
&= \varphi^*(g \cdot K) \bullet \varphi^*(h \cdot K) && \text{[definition of } \varphi^* \text{ again]}
\end{aligned}$$

■

The diagrams in Figure 1 show the Fundamental Homomorphism Theorem schematically. Figure 1(a) shows the groups and mappings. The theorem says that the homomorphism $\varphi: G \to G_0$ induces an isomorphism $\varphi^*: G/K \to \varphi(G)$ such that $\varphi = \varphi^* \circ \nu$. Here ν is the natural homomorphism of G onto G/K in Theorem 2. Figure 1(b) shows the images of the elements.

Figure 1 ▸

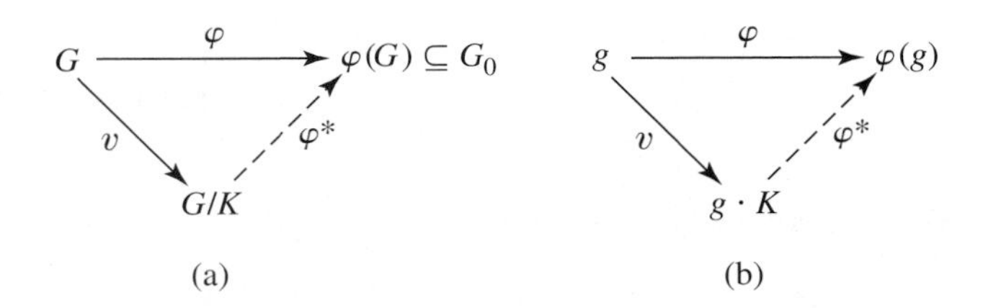

EXAMPLE 7

Example 6(a) described a homomorphism φ from $\text{AUT}(H)$ to $\text{PERM}(u, v) = \{e, s\}$ with kernel K, such that $\text{AUT}(H)/K = \{K, fK\}$. The corresponding isomorphism is the mapping $\varphi^*: \text{AUT}(H)/K \to \{e, s\}$ with $\varphi^*(K) = e$ and $\varphi^*(fK) = \varphi(f) = s$. ■

Generally, a theorem that says "A is isomorphic to B" is most useful if A and B look apparently quite different or are defined in quite different ways. The Fundamental Homomorphism Theorem is especially powerful because it says that the group $\varphi(G)$, which is a subgroup of G_0, looks just like the group G/K built in a standard way out of cosets of a normal subgroup of G. The next example shows how we can use this sort of information.

EXAMPLE 8

We show that every cyclic group is either isomorphic to a group $(\mathbb{Z}(p), +_p)$ for some $p \in \mathbb{P}$ or to the group $(\mathbb{Z}, +)$.

Given the cyclic group $\langle g \rangle = \{g^n : n \in \mathbb{Z}\}$, define $\varphi: \mathbb{Z} \to \langle g \rangle$ by $\varphi(n) = g^n$. Since $\varphi(m + n) = g^{m+n} = g^m \cdot g^n$ [Exercise 12 on page 475], φ is a homomorphism of $(\mathbb{Z}, +)$ onto $(\langle g \rangle, \cdot)$. So φ has a kernel, say K, and by the Fundamental Homomorphism Theorem the image group $\langle g \rangle$ is isomorphic to $\mathbb{Z}/K$ under $\varphi^*(n + K) = g^n$.

Since K is a subgroup of $\mathbb{Z}$, by Theorem 2 on page 473 we have $K = p\mathbb{Z}$ for some $p \in \mathbb{N}$. Thus $\langle g \rangle$ is isomorphic to $\mathbb{Z}/p\mathbb{Z}$.

If $p = 0$, then the kernel of φ is $\{0\}$, so φ is one-to-one [Corollary 2 of Theorem 1] and $\mathbb{Z}$ is isomorphic to $\langle g \rangle$ under $\varphi(n) = g^n$.

If $p > 0$, then $|\mathbb{Z}/p\mathbb{Z}| = |\{p\mathbb{Z}, 1 + p\mathbb{Z}, \ldots, (p-1) + p\mathbb{Z}\}| = p$, so $|\langle g \rangle| = p$. The mapping $\text{MOD}\, p: \mathbb{Z} \to \mathbb{Z}(p)$ is also a homomorphism with kernel $p\mathbb{Z}$, so $\mathbb{Z}/p\mathbb{Z}$ is also isomorphic to $\mathbb{Z}(p)$. Hence the composite mapping $k \to k + p\mathbb{Z} \to g^k$ is an isomorphism of $(\mathbb{Z}(p), +)$ onto $(\langle g \rangle, \cdot)$. [See Example 6 for the case $p = 6$.] ■

If G is finite, then Lagrange's Theorem on page 472 says that $|G/K| = |G|/|K|$. Since $|\varphi(G)| = |G/K|$ by the Fundamental Homomorphism Theorem, we have the following.

Corollary Let φ be a homomorphism defined on a finite group G, with kernel K. Then $|\varphi(G)| = |G|/|K|$. In particular, $|\varphi(G)|$ divides $|G|$.

EXAMPLE 9

(a) If $(G, \cdot)$ and $(H, \bullet)$ are groups, we can make the set $G \times H$ into a group, called the **direct product** of G and H, by defining

$$(g_1, h_1) * (g_2, h_2) = (g_1 \cdot g_2,\ h_1 \bullet h_2).$$

Exercise 9 asks for a proof that this operation makes $G \times H$ into a group with identity (e, e_0) and inverse $(g, h)^{-1} = (g^{-1}, h^{-1})$, where e and e_0 are the identity elements of G and H, respectively, and the inverses g^{-1} and h^{-1} are taken in the appropriate groups.

The **projection mapping** $\pi_1\colon G \times H \to G$ defined by $\pi_1(g, h) = g$ is a homomorphism onto G, since

$$\pi_1((g, h) * (g', h')) = \pi_1(g \cdot g', h \bullet h') = g \cdot g' = \pi_1(g, h) \cdot \pi_1(g', h')$$

and, similarly, $\pi_2\colon G \times H \to H$ given by $\pi_2(g, h) = h$ is a homomorphism onto H.

The kernel $\widetilde{H}$ of π_1 is $\{(g, h) : g = e, h \in H\}$, i.e., $\widetilde{H} = \{(e, h) : h \in H\}$. This subgroup of $G \times H$ is itself isomorphic to H, by the correspondence $(e, h) \to h$. The last corollary tells us that $|G| = |\pi_1(G \times H)| = |G \times H|/|\widetilde{H}| = |G \times H|/|H|$, which is no surprise.

(b) Define $\varphi\colon \mathbb{Z}(6) \to \mathbb{Z}(3) \times \mathbb{Z}(2)$ by $\varphi(n) = (n \text{ MOD } 3, n \text{ MOD } 2)$. To check that φ is a homomorphism, notice that throwing away 6's doesn't hurt (mod 2) or (mod 3), so

$$(m +_6 n) \text{ MOD } 2 = (m + n) \text{ MOD } 2 = (m \text{ MOD } 2) +_2 (n \text{ MOD } 2)$$

and, similarly,

$$(m +_6 n) \text{ MOD } 3 = (m + n) \text{ MOD } 3 = (m \text{ MOD } 3) +_3 (n \text{ MOD } 3).$$

Thus

$$\begin{aligned}\varphi(m +_6 n) &= ((m +_6 n) \text{ MOD } 3, (m +_6 n) \text{ MOD } 2)\\ &= ((m \text{ MOD } 3) +_3 (n \text{ MOD } 3), (m \text{ MOD } 2) +_2 (n \text{ MOD } 2))\\ &= (m \text{ MOD } 3, m \text{ MOD } 2) * (n \text{ MOD } 3, n \text{ MOD } 2)\\ &= \varphi(m) * \varphi(n).\end{aligned}$$

We have

$$\begin{aligned}&\varphi(0) = (0, 0) = \text{ the identity}, && \varphi(3) = (0, 1),\\ &\varphi(1) = (1, 1), && \varphi(4) = (1, 0),\\ &\varphi(2) = (2, 0), && \varphi(5) = (2, 1).\end{aligned}$$

The kernel of φ is just $\{0\}$, φ is one-to-one, and φ is onto $\mathbb{Z}(3) \times \mathbb{Z}(2)$, so φ is an isomorphism. Now $\mathbb{Z}(6)$ is cyclic, generated by 1. Hence $\varphi(\mathbb{Z}(6))$ is generated by $\varphi(1) = (1, 1)$. Thus $\mathbb{Z}(3) \times \mathbb{Z}(2)$ is also cyclic, though it might not have appeared so at first sight. ■

One way to think about homomorphisms and their images is in terms of photography. Different homomorphic images are like pictures taken from different angles. Each picture may lose some aspects of the original object, but enough pictures taken together can help us to reconstruct the object in three dimensions. This is the idea behind the Chinese Remainder Theorem, which we will see in the next section.

The message of the Fundamental Homomorphism Theorem is that to study the homomorphic images of G it is enough to look at the groups G/K that are made from cosets of certain subgroups K of G. The message of Theorem 2 is that we can identify the interesting subgroups; they are the ones satisfying the equations $gK = Kg$ for all g in G. If we fully understand the group and its normal subgroups, then we fully understand all its homomorphic images.

Exercises 12.4

1. Which of the following functions φ from $(\mathbb{Z}, +)$ to $(\mathbb{Z}, +)$ are homomorphisms?
   (a) $\varphi(n) = 6n$ (b) $\varphi(n) = n + 1$
   (c) $\varphi(n) = -n$ (d) $\varphi(n) = n^2$
2. Which of the following functions φ are homomorphisms from $(\mathbb{Z}, +)$ to the group $(\mathbb{R} \setminus \{0\}, \cdot)$?
   (a) $\varphi(n) = 6^n$ (b) $\varphi(n) = n$
   (c) $\varphi(n) = (-6)^n$ (d) $\varphi(n) = n^2 + 1$
   (e) $\varphi(n) = 2^{n+1}$
3. Which of the homomorphisms in Exercise 1 are isomorphisms? Explain briefly.
4. Which of the homomorphisms in (a), (b), and (c) of Example 1 on page 488 are isomorphisms? Explain briefly.
5. Let $F = \text{FUN}(\mathbb{R}, \mathbb{R})$, and define $+$ on F by
   $$(f + g)(x) = f(x) + g(x) \quad \text{for all} \quad x \in \mathbb{R}.$$
   (a) Show that $(F, +)$ is a group.
   (b) Is the group F commutative?
   (c) Define φ from F to $\mathbb{R}$ by $\varphi(f) = f(73)$. Show that φ is a homomorphism of $(F, +)$ onto $(\mathbb{R}, +)$.
6. Let $(G, \cdot)$ and $(G_0, \bullet)$ be groups with identity elements e and e_0, respectively.
   (a) Show that the mapping $\theta: g \to e_0$ is a homomorphism from G to G_0.
   (b) Find the kernel of θ.
   (c) Show that G is a normal subgroup of itself.
7. Find the kernel of the group homomorphism φ
   (a) from $(\mathbb{Z}, +)$ to $(\mathbb{Z}, +)$, defined by $\varphi(n) = 73n$.
   (b) from $(\mathbb{Z}, +)$ to $(\mathbb{Z}, +)$, defined by $\varphi(n) = 0$ for all n.
   (c) from $(\mathbb{Z}, +)$ to $(\mathbb{Z}(5), +_5)$, defined by $\varphi(n) = n \text{ MOD } 5$.
   (d) from $(\mathbb{Z}, +)$ to $(\mathbb{Z}, +)$, defined by $\varphi(n) = n$.
8. For each homomorphism in Exercise 7, describe the coset of the kernel that contains 73.
9. Let $(G, \cdot)$ and $(G_0, \bullet)$ be groups, with identity elements e and e_0, respectively.
   (a) Show that (e, e_0) is the identity element of $G \times H$ and that the inverse of an element (g, h) in $G \times H$ is (g^{-1}, h^{-1}).
   (b) Verify that the mapping $\pi_2: G \times H \to H$ defined by $\pi_2(g, h) = h$ is a homomorphism.
   (c) Find the kernel of the homomorphism π_2 in part (b).
   (d) Find a normal subgroup of $G \times H$ that is isomorphic to G.
   (e) Find a homomorphism $\psi: H \to G \times H$ such that $\pi_2(\psi(h)) = h$ for every $h \in H$.
10. Define φ from $(\mathbb{Z}, +)$ to $\mathbb{Z}(2) \times \mathbb{Z}(2)$ by $\varphi(k) = (k \text{ MOD } 2, k \text{ MOD } 2)$.
   (a) Verify that φ is a homomorphism.
   (b) Verify that $\varphi(\mathbb{Z})$ is a subgroup of $\mathbb{Z}(2) \times \mathbb{Z}(2)$.
   (c) Find the kernel of φ.
   (d) Is $\mathbb{Z}(2) \times \mathbb{Z}(2)$ isomorphic to $\mathbb{Z}(2)$? Explain.
   (e) Is $\mathbb{Z}(2) \times \mathbb{Z}(2)$ isomorphic to $\mathbb{Z}(4)$? Explain.
11. Suppose that φ is a homomorphism defined on a group G and that $|G| = 12$ and $|\varphi(G)| = 3$.
   (a) Find $|K|$, where K is the kernel of φ.
   (b) How many members of G does φ map onto each member of $\varphi(G)$?
   (c) What is $|G/K|$?
12. An **antihomomorphism** from $(G, \cdot)$ to $(G_0, \bullet)$ is a function ψ such that
   $$\psi(g \cdot h) = \psi(h) \bullet \psi(g) \quad \text{for all} \quad g, h \in G.$$
   (a) Show that the mapping $g \to g^{-1}$ is always an antihomomorphism of a group onto itself. [This ψ could be called an **anti-isomorphism** since it is one-to-one and onto.]
   (b) When is the anti-isomorphism in part (a) also an isomorphism?
   (c) Show that if ψ_1 and ψ_2 are antihomomorphisms for which the composition $\psi_1 \circ \psi_2$ is defined, then $\psi_1 \circ \psi_2$ is a homomorphism.
13. Show that if φ is a homomorphism defined on the group G and if $\varphi(g)$ has just one pre-image under φ for some g in G, then φ is one-to-one.
14. Consider a finite group G with a subgroup H such that $|G| = 2|H|$. Show that H must be a normal subgroup of G. *Suggestion:* See Exercise 28 on page 476.
15. Let H be a subgroup of G.
   (a) Show that $\{g \in G : gHg^{-1} = H\}$ is a subgroup of G.
   (b) Show that H is normal if and only if $gHg^{-1} = H$ for every $g \in G$.
   (c) Conclude that if G is generated by a subset A and if $gHg^{-1} = H$ for all $g \in A$, then H is a normal subgroup.
16. Show that if J and K are normal subgroups of a group $(G, \cdot)$, then $J \cap K$ is a normal subgroup. *Suggestion:* Consider $g \cdot (J \cap K) \cdot g^{-1}$.

17. Let H be the set of 2×2 matrices of the form $\begin{bmatrix} 1 & x \\ 0 & 1 \end{bmatrix}$ with matrix multiplication as operation.

(a) Verify that H is a group, with

$$\begin{bmatrix} 1 & x \\ 0 & 1 \end{bmatrix}^{-1} = \begin{bmatrix} 1 & -x \\ 0 & 1 \end{bmatrix}.$$

(b) Verify that

$$\begin{bmatrix} 0 & 1 \\ 1 & 0 \end{bmatrix} \cdot H \neq H \cdot \begin{bmatrix} 0 & 1 \\ 1 & 0 \end{bmatrix}.$$

This shows that H is not a normal subgroup of the multiplicative group G of all 2×2 matrices that have inverses.

(c) Show that H is a normal subgroup of the group T of all 2×2 matrices of the form

$$\begin{bmatrix} y & z \\ 0 & 1/y \end{bmatrix}, \qquad y \neq 0,$$

with multiplication as operation.

(d) The mapping φ from T to $(\mathbb{R} \setminus \{0\}, \cdot)$ defined by

$$\varphi\left(\begin{bmatrix} y & z \\ 0 & 1/y \end{bmatrix}\right) = y$$

is a group homomorphism. Find its kernel.

(e) Show that T/H is isomorphic to the group of nonzero real numbers under multiplication.

12.5 Semigroups

One frequently encounters structures with some but not all of the properties of groups. For instance, the associative law may hold even though elements don't have inverses. There might not even be an identity. To discuss such structures, we will use a little square $\Box$ as a symbol for a generic binary operation. This neutral symbol is not meant to suggest $+$ or $\cdot$ or any other familiar operation such as union or intersection, though in a particular example it might be any one of these.

A nonempty set S with a binary associative operation $\Box$ is called a **semigroup**. Thus $(S, \Box)$ is a semigroup in case $s_1 \Box s_2$ is in S for all $s_1, s_2 \in S$ and

$$s_1 \Box (s_2 \Box s_3) = (s_1 \Box s_2) \Box s_3 \qquad \text{for all} \quad s_1, s_2, s_3 \in S.$$

In view of associativity, expressions such as $s_1 \Box s_2 \Box s_3$ are unambiguous in a semigroup. Wherever we have encountered the associative law, there has always been a semigroup nearby. All that's new now is the emphasis on the overall structure.

As with groups, if the semigroup is **commutative** [i.e., if $s_1 \Box s_2 = s_2 \Box s_1$ for all $s_1, s_2 \in S$], we sometimes use additive notation $+$ instead of $\Box$. An element $\boldsymbol{e}$ in S is an **identity for** S provided

$$s \Box e = e \Box s = s \qquad \text{for all} \quad s \in S.$$

A semigroup with an identity is called a **monoid**. A **group**, then, is a monoid in which every element has an inverse. In this section we focus primarily on semigroups that are not groups. A **subsemigroup** of a semigroup is a subset that is closed under the operation.

EXAMPLE 1

(a) $(\mathbb{N}, +)$ is a semigroup, since the sum of two numbers in $\mathbb{N}$ is in $\mathbb{N}$ and since addition is associative. Its identity is 0, since $n + 0 = 0 + n = n$ for all $n \in \mathbb{N}$, and so it is a monoid. None of the positive numbers in $\mathbb{N}$ have additive inverses [i.e., negatives] *in* $\mathbb{N}$, so $(\mathbb{N}, +)$ is not a group. The monoid $(\mathbb{N}, +)$ is commutative.

(b) $(\mathbb{P}, +)$ is also a commutative semigroup, indeed, a subsemigroup of $(\mathbb{N}, +)$, but it has no identity. It makes no sense to inquire further about inverses. ■

EXAMPLE 2

(a) Multiplication $\cdot$ in $\mathbb{R}$ is associative and commutative, so any subset of $\mathbb{R}$ that is closed under multiplication is a semigroup. Thus $(\mathbb{R}, \cdot)$, $(\mathbb{Q}, \cdot)$, $(\mathbb{Z}, \cdot)$, and $(\mathbb{P}, \cdot)$ are all commutative semigroups. They all include the multiplicative identity 1, so all are monoids. None of these semigroups is a group: $(\mathbb{R}, \cdot)$, $(\mathbb{Q}, \cdot)$, and $(\mathbb{Z}, \cdot)$ all contain 0, and 0 has no multiplicative inverse [i.e., reciprocal]. The semigroup $(\mathbb{P}, \cdot)$ contains 2 but not its inverse $\frac{1}{2}$.

(b) Since the product of nonzero real numbers is nonzero, $\cdot$ is also a binary operation on $\mathbb{R} \setminus \{0\}$. Moreover, $(\mathbb{R} \setminus \{0\}, \cdot)$ is a bona fide commutative group.

(c) The set $\mathbb{Z}(p)$ is a commutative semigroup under $*_p$. This fact is proved in Theorem 4 on page 124. We already observed in Example 3(c) on page 464 that $\mathbb{Z}(p)$ is a group under $+_p$. ■

EXAMPLE 3

Let Σ be an alphabet. The set Σ^* of all words in the letters of Σ is informally defined in §1.5 and is recursively defined in Example 2 on page 270. Two words w_1 and w_2 in Σ^* are multiplied by **concatenation**; i.e., $w_1 w_2$ is the word obtained by placing the string w_2 right after the string w_1. Thus, if $w_1 = a_1 a_2 \cdots a_m$ and $w_2 = b_1 b_2 \cdots b_n$ with the a_j's and b_k's in Σ, then $w_1 w_2 = a_1 a_2 \cdots a_m b_1 b_2 \cdots b_n$. For example, if $w_1 = cat$ and $w_2 = nip$, then $w_1 w_2 = catnip$ and $w_2 w_1 = nipcat$. Multiplication by the empty word λ leaves a word unchanged:

$$w\lambda = \lambda w = w \qquad \text{for all} \quad w \in \Sigma^*.$$

It is evident from the definition that concatenation is an associative binary operation. Since the empty word λ serves as an identity for Σ^*, Σ^* is a monoid. It is certainly not a group; only the empty word has an inverse. ■

EXAMPLE 4

(a) Let $\mathfrak{M}_{m,n}$ be the set of all $m \times n$ matrices. Matrix addition $+$ is a binary operation on $\mathfrak{M}_{m,n}$ since

$$\mathbf{A}, \mathbf{B} \in \mathfrak{M}_{m,n} \qquad \text{implies} \qquad \mathbf{A} + \mathbf{B} \in \mathfrak{M}_{m,n}.$$

The set $\mathfrak{M}_{m,n}$ of all $m \times n$ matrices is a commutative group under addition. This fact is spelled out in the theorem on page 108.

(b) Matrix multiplication is also a binary operation on the set $\mathfrak{M}_{n,n}$ of $n \times n$ matrices. Under multiplication $\mathfrak{M}_{n,n}$ is a monoid with identity $\mathbf{I}_n$. The associative law is discussed on page 442. Except for the trivial case $n = 1$, this monoid is not commutative: **AB** does not necessarily equal **BA**. Some, but not all, matrices in $\mathfrak{M}_{n,n}$ have inverses. See Exercises 16 and 17 on page 446. The subsemigroup of invertible matrices is a group in its own right. ■

EXAMPLE 5

(a) Let $\mathcal{P}(U)$ be the set of all subsets of some set U. With the operation $\cup$, $\mathcal{P}(U)$ is a commutative semigroup with identity $\emptyset$; see laws 1a, 2a, and 5a in Table 1 on page 25. Only the empty set itself has an inverse, since $A \cup B \neq \emptyset$ whenever $A \neq \emptyset$.

(b) $\mathcal{P}(U)$ is a commutative semigroup under the operation $\cap$ with identity U; see laws 1b, 2b, and 5d in Table 1 on page 25.

(c) $\mathcal{P}(U)$ is also a semigroup using symmetric difference $\oplus$; see Exercise 2. ■

EXAMPLE 6

Consider a set $T = \{a, b, \ldots\}$ with at least two members. The set FUN(T, T) of all functions mapping T into T is a semigroup under composition; the associative law is discussed on page 33. The identity function 1_T on T is the identity for this semigroup, since

$$(1_T \circ f)(t) = 1_T(f(t)) = f(t) \qquad \text{for all} \quad t \in T,$$

so $1_T \circ f = f$ and, similarly, $f \circ 1_T = f$. This monoid is not commutative. For example, let f and g be the constant functions defined by $f(t) = a$ and $g(t) = b$ for all t in T. Then $(f \circ g)(t) = f(g(t)) = a$ for all t, and $(g \circ f)(t) = b$ for all t. That is, $f \circ g = f \neq g = g \circ f$.

The group PERM(T) of all permutations of T is an important subsemigroup of FUN(T, T). ■

EXAMPLE 7

(a) Let T be a nonempty set. The set FUN$(T, \mathbb{N})$ of functions mapping T into $\mathbb{N}$ is a semigroup under the operation $+$ defined by

$$(f + g)(t) = f(t) + g(t) \qquad \text{for all} \quad t \in T.$$

[Observe, for instance, that $(f+g)+h$ is defined by $((f+g)+h)(t) = (f+g)(t)+h(t) = (f(t)+g(t))+h(t)$ for all $t \in T$.] The identity element of $\text{FUN}(T, \mathbb{N})$ is the constant function z defined by $z(t) = 0$ for all $t \in T$. An inverse for f would be a function g with $f(t)+g(t) = 0$ for all t, i.e., with $g(t) = -f(t)$ for all t. Since f and g have values in $\mathbb{N}$, no such g exists if $f(t) > 0$ for some t.

(b) Replace $\mathbb{N}$ by $\mathbb{Z}$ in (a). Now inverses exist; the inverse of f is the function $-f$ defined by $(-f)(t) = -f(t)$ for all $t \in T$. Thus $\text{FUN}(T, \mathbb{Z})$ is a group under $+$.

(c) We can also make $\text{FUN}(T, \mathbb{N})$ into a semigroup under an operation $*$ defined by $(f * g)(t) = f(t) \cdot g(t)$. Associativity of $*$ comes from associativity of $\cdot$ on $\mathbb{N}$. The identity element is the constant function e with $e(t) = 1$ for all t.

(d) The usual definitions for adding and multiplying functions, that is, defining $(f+g)(x) = f(x)+g(x)$ and $(f \cdot g)(x) = f(x) \cdot g(x)$ for all x, make the set $\text{FUN}(\mathbb{R}, \mathbb{R})$ into a semigroup in two different ways. ■

Consider once again an arbitrary semigroup $(S, \Box)$. For $s \in S$ and $n \in \mathbb{P}$, we continue a familiar convention and write s^n for the $\Box$-product of s with itself n times. If S has an identity, we also write s^0 for e. The next theorem is obvious, though a formal induction proof can be given based on the recursive definition in which $s^{n+1} = s^n \Box s$.

Theorem 1 Let $(S, \Box)$ be a semigroup. For $s \in S$ and $m, n \in \mathbb{P}$, we have

(a) $s^m \Box s^n = s^{m+n}$

(b) $(s^m)^n = s^{mn}$.

If $(S, \Box)$ is a monoid, these formulas hold for $m, n \in \mathbb{N}$.

We have treated the generic operation $\Box$ the same way we would treat multiplication. Let's see how the foregoing would look when $\Box$ is replaced by $+$. The general product s^n is replaced by the general sum $ns = s+s+\cdots+s$ and, for $s \in S$ and $m, n \in \mathbb{P}$, the theorem now says

(a) $ms + ns = (m+n)s$

(b) $n(ms) = (mn)s$.

Starting with a nonempty subset A of S, we define the set A^+ **generated by** A recursively as follows.

(B) $A \subseteq A^+$.

(R) If $s, t \in A^+$ then $s \Box t \in A^+$.

Theorem 3 will tell us that A^+ consists of all products of elements from A. We say that A **generates** S if $A^+ = S$. Whether or not A generates S, A^+ is a subsemigroup of S by the following fact.

Theorem 2 Let A be a nonempty subset of the semigroup $(S, \Box)$. Then A^+ is the unique smallest subsemigroup of S that contains A.

Proof By (B), A^+ contains A. By (R), A^+ is closed under the operation $\Box$, so by definition A^+ is a subsemigroup of S.

Now consider an arbitrary subsemigroup T of S that contains A. We'll show that $A^+ \subseteq T$, from which it will follow that A^+ is the unique smallest subsemigroup containing A. We want to show that $s \in T$ for every s in A^+. Since A^+ is defined recursively, it's natural to use the Generalized Principle of Mathematical Induction introduced on page 273. Let $p(s)$ be the proposition "$s \in T$." To show that $p(s)$ is true for all s in A^+, we must establish two properties:

(B′) $p(s)$ is true for the members of A^+ specified in the basis.

(I) If a member u of A^+ is specified by (R) in terms of previously defined members, say $u = s\square t$, then $p(s) \wedge p(t) \Longrightarrow p(u)$.

Now (B′) is true, since $A \subseteq T$ by the choice of T. Condition (I) says "if $u = s\square t$ and if $s \in T$ and $t \in T$ then $u \in T$," which is true because T is a subsemigroup. The conditions for the Generalized Principle are met, so $s \in T$ for all s in A^+. ■

Theorem 2 helps us describe the members of A^+ without recursion, as follows.

Theorem 3 Let A be a nonempty subset of the semigroup $(S, \square)$. Then A^+ consists of all elements of S of the form $a_1\square\cdots\square a_n$ for $n \in \mathbb{P}$ and $a_1, \ldots, a_n \in A$.

Proof Let X be the set of all products $a_1\square\cdots\square a_n$. We want to show that $A^+ = X$. If $a_1, \ldots, a_n, b_1, \ldots, b_m$ are in A, then $a_1\square\cdots\square a_n\square b_1\square\cdots\square b_m$ is a product of this same form. Thus X is closed under $\square$, i.e., it's a subsemigroup of S. The case $n = 1$ gives $A \subseteq X$, so $A^+ \subseteq X$ by Theorem 2.

To show $X \subseteq A^+$, we use ordinary induction on n. Since $A \subseteq A^+$, $a_1 \in A^+$ whenever $a_1 \in A$. Assume inductively that $a_1\square\cdots\square a_n \in A^+$ for some n and some $a_1, \ldots, a_n$ in A, and let $a_{n+1} \in A$. Since $a_{n+1} \in A^+$, $a_1\square\cdots\square a_n\square a_{n+1}$ is in A^+ by (R) in the recursive definition of A^+. By induction, every member of X is in A^+. ■

EXAMPLE 8

(a) The subsemigroup $\{2\}^+$ of the semigroup $(\mathbb{Z}, +)$ consists of all "products" of members of $\{2\}$. Since the operation here is $+$, $\{2\}^+$ consists of all sums $2 + 2 + \cdots + 2$, i.e., of all positive even integers. Thus we have $\{2\}^+ = 2\mathbb{P} = \{2n : n \in \mathbb{P}\}$.

(b) More generally, the subsemigroup of $(S, \square)$ generated by a single element s is the set

$$\{s\}^+ = \{s^n : n \in \mathbb{P}\}$$

[or $\{s\}^+ = \{ns : n \in \mathbb{P}\}$ if $\square$ is $+$]. Such a single-generator semigroup is called a **cyclic semigroup**.

(c) The subsemigroup $\{2\}^+$ of $(\mathbb{Z}, \cdot)$ is $\{2^n : n \in \mathbb{P}\} = \{2, 4, 8, 16, \ldots\}$. Notice that our notation is deficient; we can't tell from the expression $\{2\}^+$ alone whether we mean this subsemigroup or the one in part (a).

(d) The subsemigroup $\{2, 7\}^+$ of $(\mathbb{Z}, \cdot)$ generated by $\{2, 7\}$ is the set

$$\{2^m 7^n : m, n \in \mathbb{N} \text{ and } m + n \geq 1\}.$$

(e) We show that the cyclic subsemigroup of $(\mathfrak{M}_{2,2}, \cdot)$ generated by the matrix

$$\mathbf{M} = \begin{bmatrix} 1 & 1 \\ 0 & 1 \end{bmatrix}$$

is

$$\{\mathbf{M}\}^+ = \left\{\begin{bmatrix} 1 & n \\ 0 & 1 \end{bmatrix} : n \in \mathbb{P}\right\}.$$

By part (b), it will be enough to show that $\mathbf{M}^n = \begin{bmatrix} 1 & n \\ 0 & 1 \end{bmatrix}$ for $n \in \mathbb{P}$. This is clear for $n = 1$. Assume inductively that $\mathbf{M}^n = \begin{bmatrix} 1 & n \\ 0 & 1 \end{bmatrix}$ for some $n \in \mathbb{P}$. Then matrix multiplication gives

$$\mathbf{M}^{n+1} = \mathbf{M}^n \cdot \mathbf{M} = \begin{bmatrix} 1 & n \\ 0 & 1 \end{bmatrix}\begin{bmatrix} 1 & 1 \\ 0 & 1 \end{bmatrix} = \begin{bmatrix} 1 & 1+n \\ 0 & 1 \end{bmatrix}.$$

The result now follows by mathematical induction. ■

The intersection of any collection of subsemigroups of a semigroup S is either empty or is itself a subsemigroup of S. Indeed, if s and t belong to the intersection, then they both belong to each subsemigroup in the collection, so their product $s \square t$ does too.

EXAMPLE 9

(a) Both $2\mathbb{Z}$ and $3\mathbb{Z}$ are subsemigroups of $(\mathbb{Z}, \cdot)$, so their intersection $6\mathbb{Z}$ is too. Likewise, the intersection $\mathbb{P} \cap 2\mathbb{Z} = \{2k : k \in \mathbb{P}\}$ is a subsemigroup of $(\mathbb{Z}, \cdot)$.

(b) Both $\mathbb{P}$ and $\{-k : k \in \mathbb{P}\}$ are subsemigroups of $(\mathbb{Z}, +)$. Their intersection is empty.

(c) Consider a nonempty subset A of a semigroup S. By Theorem 2, the intersection of all the subsemigroups that contain A [including S itself, naturally] is a subsemigroup that contains A. It must be the smallest subsemigroup containing A, i.e., A^+. This observation gives a way to define A^+ without describing its elements in terms of A, as we did in Theorem 3. ■

EXAMPLE 10

Lagrange's Theorem on page 472 can fail spectacularly for semigroups. Consider an arbitrary nonempty set S, and define $\square$ on S by $s \square t = t$ for all $s, t \in S$; i.e., the product of two elements is always just the second element. Then $(s \square t) \square u = u = s \square u = s \square (t \square u)$, so $\square$ is associative and $(S, \square)$ is a semigroup. *Every* nonempty subset of S is a subsemigroup, since it's closed under $\square$. ■

As with groups, the important functions between semigroups are the ones that preserve the semigroup structure. Let $(S, \bullet)$ and $(T, \square)$ be semigroups. A function $\varphi : S \to T$ is a **semigroup homomorphism** if

$$\varphi(s_1 \bullet s_2) = \varphi(s_1) \square \varphi(s_2) \qquad \text{for all} \quad s_1, s_2 \in S.$$

If φ is also a one-to-one correspondence, then φ is called a **semigroup isomorphism**. If there is a semigroup isomorphism of S onto T, we say that S and T are **isomorphic** and write $S \simeq T$ [or $(S, \bullet) \simeq (T, \square)$ if the operations need to be mentioned].

EXAMPLE 11

(a) Let $(S, \bullet)$ be $(\mathbb{P}, +)$, let $(T, \square)$ be $(\mathbb{P}, \cdot)$, and let $\varphi(m) = 2^m$ for m in $\mathbb{P}$. Since

$$\varphi(m + n) = 2^{m+n} = 2^m \cdot 2^n = \varphi(m) \cdot \varphi(n),$$

φ is a homomorphism of $(\mathbb{P}, +)$ into $(\mathbb{P}, \cdot)$.

(b) In Example 2 on page 488 we observed that MOD p is a group homomorphism of $(\mathbb{Z}, +)$ onto $(\mathbb{Z}(p), +_p)$, but that MOD p is not a group homomorphism of $(\mathbb{Z}, \cdot)$ onto $(\mathbb{Z}(p), *_p)$, for the very good reason that $(\mathbb{Z}, \cdot)$ and $(\mathbb{Z}(p), *_p)$ are not groups. But $(\mathbb{Z}, \cdot)$ and $(\mathbb{Z}(p), *_p)$ *are* semigroups and MOD p is a semigroup homomorphism. ■

EXAMPLE 12

Let U be any set. We define the complementation function φ from $\mathcal{P}(U)$ into $\mathcal{P}(U)$ by $\varphi(A) = A^c$ for $A \in \mathcal{P}(U)$. By a De Morgan law in Table 1 on page 25,

$$\varphi(A \cup B) = (A \cup B)^c = A^c \cap B^c = \varphi(A) \cap \varphi(B),$$

so φ is a homomorphism from the semigroup $(\mathcal{P}(U), \cup)$ into the semigroup $(\mathcal{P}(U), \cap)$. The function φ is a one-to-one correspondence of $\mathcal{P}(U)$ onto $\mathcal{P}(U)$, so it is an isomorphism of $(\mathcal{P}(U), \cup)$ onto $(\mathcal{P}(U), \cap)$. In fact, φ is its own inverse [why?], so φ also gives an isomorphism of $(\mathcal{P}(U), \cap)$ onto $(\mathcal{P}(U), \cup)$. Note that φ maps the identity $\varnothing$ of $(\mathcal{P}(U), \cup)$ onto the identity U of $(\mathcal{P}(U), \cap)$. ■

Isomorphisms get used in two different ways, as we saw when we looked at graph isomorphisms. Sometimes we want to call attention to the fact that two apparently different semigroups are actually identical in structure. At other times the identical structure is obvious, but we want to examine the various isomorphisms that are possible, for instance from S back onto itself, to see how much symmetry is present.

Exercises 12.5

1. $(\mathbb{N}, \cdot)$ is a semigroup.

(a) Is this semigroup commutative?

(b) Is there an identity for this semigroup?

(c) If yes, do inverses exist? If yes, specify them.

(d) Is this semigroup a monoid? A group?

2. $\mathcal{P}(U)$ is a semigroup with respect to symmetric difference $\oplus$. Repeat Exercise 1 for this semigroup.

3. Repeat Exercise 1 for the semigroup $\text{FUN}(\mathbb{R}, \mathbb{R})$ of real-valued functions on $\mathbb{R}$ under addition.

4. Repeat Exercise 1 for the semigroup $\text{FUN}(\mathbb{R}, \mathbb{R})$ under multiplication.

5. The set $\mathfrak{M}_{2,2}$ of 2×2 matrices is a semigroup with respect to matrix multiplication.

(a) Is this semigroup commutative?

(b) Is there an identity for this semigroup?

(c) If yes, do inverses exist?

(d) Is this semigroup a monoid? A group?

6. Let $\Sigma = \{a, b, c, d\}$ and consider the words $w_1 = bad$, $w_2 = cab$, and $w_3 = abcd$.

(a) Determine w_1w_2, w_2w_1, $w_2w_3w_2w_1$, and $w_3w_2w_3$.

(b) Determine w_1^2, w_2^3, and λ^4.

7. Let Σ be the usual English alphabet and consider the words $w_1 = break$, $w_2 = fast$, $w_3 = lunch$, and $w_4 = food$.

(a) Determine λw_1, $w_2\lambda$, w_2w_4, w_3w_1, and $w_4\lambda w_4$.

(b) Compare w_1w_2 and w_2w_1.

(c) Determine w_2^2, w_4^2, $w_2^2w_4w_1^2$, and λ^{73}.

8. Show that $\mathbb{R}^+ = \{x \in \mathbb{R} : x > 0\}$ is not a semigroup with the binary operation $(x, y) \to x/y$.

9. (a) Convince yourself that $\mathbb{N}$ is a semigroup under the binary operation $(m, n) \to \min\{m, n\}$ and also under $(m, n) \to \max\{m, n\}$.

(b) Are the semigroups in part (a) monoids?

10. (a) Show that $\mathbb{P}$ is a semigroup with respect to $(m, n) \to \gcd(m, n)$, where $\gcd(m, n)$ represents the greatest common divisor of m and n.

(b) Show that $\mathbb{P}$ is a semigroup with respect to $(m, n) \to \text{lcm}(m, n)$, where $\text{lcm}(m, n)$ represents the least common multiple of m and n.

(c) Are the semigroups in parts (a) and (b) monoids?

11. Describe each of the following subsemigroups of $(\mathbb{Z}, +)$.

(a) $\{1\}^+$ (b) $\{0\}^+$ (c) $\{-1, 2\}^+$

(d) $\mathbb{P}^+$ (e) $\mathbb{Z}^+$ (f) $\{2, 3\}^+$

(g) $\{6\}^+ \cap \{9\}^+$

12. Describe each of the following subsemigroups of $(\mathbb{Z}, \cdot)$.

(a) $\{1\}^+$ (b) $\{0\}^+$ (c) $\{-1, 2\}^+$

(d) $\mathbb{P}^+$ (e) $\mathbb{Z}^+$ (f) $\{2, 3\}^+$

13. Which of the semigroups in Exercise 11 are cyclic semigroups? Justify your answers.

14. Which of the semigroups in Exercise 12 are cyclic semigroups? Justify your answers.

15. If A is a subset of a monoid $(M, \Box)$ with identity e, the **submonoid generated by** A is defined to be $A^+ \cup \{e\}$. Find

(a) the submonoid of $(\mathbb{Z}, +)$ generated by $\{2\}$.

(b) the submonoid of $(\mathbb{Z}, +)$ generated by $\{1, -1\}$.

(c) the submonoid of $(\mathbb{Z}, +)$ generated by $\{0\}$.

(d) the submonoid of $(\mathbb{Z}, \cdot)$ generated by $\{1\}$.

(e) the submonoid of Σ^* generated by Σ, using concatenation on Σ^*.

16. (a) Give an example of a cyclic semigroup and a subsemigroup of it that is not cyclic.

(b) Give an example of a cyclic group that is not a cyclic semigroup.

(c) Give an example of a cyclic group that *is* a cyclic semigroup.

17. List the members of each of the following finite subsemigroups of $(\mathfrak{M}_{3,3}, \cdot)$.

(a) $\left\{\begin{bmatrix} 0 & 1 & 0 \\ 1 & 0 & 0 \\ 0 & 0 & 1 \end{bmatrix}\right\}^+$ (b) $\left\{\begin{bmatrix} 0 & 1 & 0 \\ 1 & 0 & 0 \\ 0 & 0 & 1 \end{bmatrix}, \begin{bmatrix} 0 & 0 & 1 \\ 1 & 0 & 0 \\ 0 & 1 & 0 \end{bmatrix}\right\}^+$

(c) $\left\{\begin{bmatrix} 0 & 2 & 3 \\ 0 & 0 & 4 \\ 0 & 0 & 0 \end{bmatrix}\right\}^+$

18. (a) Which of the subsemigroups in Exercise 17 are groups?

(b) Which are commutative?

(c) Which are cyclic?

19. (a) Find the intersection of the subsemigroups $2\mathbb{P}$ and $3\mathbb{P}$ of the semigroup $(\mathbb{P}, +)$.

(b) Is the intersection in part (a) cyclic? Explain.

(c) Repeat part (b) with the semigroup $(\mathbb{P}, \cdot)$.

20. (a) Find the intersection of the subsemigroups of $(\mathbb{P}, \cdot)$ generated by 2 and 3, respectively.

(b) Is the intersection in part (a) cyclic? Explain.

21. (a) Find the intersection of the three subsemigroups $4\mathbb{P}$, $6\mathbb{P}$, and $10\mathbb{P}$ of the semigroup $(\mathbb{P}, \cdot)$.

(b) Is the intersection in part (a) cyclic? Explain.

(c) Repeat part (b) with the semigroup $(\mathbb{P}, +)$.

22. Which of the following functions φ are homomorphisms from $(\mathbb{P}, +)$ to $(\mathbb{P}, \cdot)$?

(a) $\varphi(n) = 2^n$ (b) $\varphi(n) = n$

(c) $\varphi(n) = (-1)^n$ (d) $\varphi(n) = 2n$

(e) $\varphi(n) = 2^{n+1}$

23. Which of the homomorphisms in Exercise 22 are isomorphisms? Explain briefly.

24. Let Σ be the English alphabet. Define φ on Σ^* by $\varphi(w) =$ length of w. Explain why φ is a homomorphism from Σ^* with its usual operation to $(\mathbb{N}, +)$.

25. An element z of a semigroup $(S, \bullet)$ is called a **zero element** or **zero** of S in case

$$z \bullet s = s \bullet z = z \quad \text{for all } s \text{ in } S.$$

(a) Show that a semigroup cannot have more than one zero element.

(b) Give an example of an infinite semigroup that has a zero element.

(c) Give an example of a finite semigroup that has at least two members and has a zero element.

26. Let z be a zero element of a semigroup $(S, \bullet)$, and let φ be a homomorphism from $(S, \bullet)$ to a semigroup $(T, \Box)$.

(a) Show that $\varphi(z)$ is a zero element of $\varphi(S)$.

(b) Must $\varphi(z)$ be a zero element of $(T, \Box)$? Justify your answer.

27. Suppose that $(S, \bullet)$ is a monoid with identity e and that φ is a semigroup homomorphism from $(S, \bullet)$ to $(T, \Box)$.

(a) Show that $(\varphi(S), \Box)$ is a monoid with identity $\varphi(e)$.

(b) Must $\varphi(e)$ be an identity of $(T, \Box)$? Justify your answer.

12.6 Other Algebraic Systems

So far in this chapter we have been looking at sets with just one binary operation on them. A number of important and familiar algebraic structures, for example, $\mathbb{Z}$ and $\mathbb{R}$, have two binary operations, usually written $+$ and $\cdot$. The "additive" operation $+$ is typically very well behaved, while the "multiplicative" operation $\cdot$ is generally less so, and there are usually distributive laws relating the two operations to each other. In this section we briefly introduce rings and fields, which are the two main kinds of algebraic structures with two operations, and we give some examples and discuss the basic facts about homomorphisms of such systems.

There are many reasons for wanting to know about rings and fields, but our main motivation here comes from the study of fast computer arithmetic with large numbers.

EXAMPLE 1

Let's start with a very old problem and then work toward more recent developments. One basic tool in computer arithmetic is the Chinese Remainder Theorem, which was apparently first used to count an army in ancient China. Consider a small army, one known to have fewer than 1000 members. Observe that $7 \cdot 11 \cdot 13 = 1001$. The Chinese Remainder Theorem says that the numbers $0, 1, \ldots, 1000$ in $\mathbb{Z}(1001)$ correspond one-to-one with the triples in $\mathbb{Z}(7) \times \mathbb{Z}(11) \times \mathbb{Z}(13)$. The correspondence is

$$m \longleftrightarrow (m \text{ MOD } 7, m \text{ MOD } 11, m \text{ MOD } 13);$$

so if N is the army size and if we know $N \text{ MOD } 7$, $N \text{ MOD } 11$, and $N \text{ MOD } 13$, then in principle we can determine the value of N.

Ask the soldiers to group together in bunches of 7 and tell you how many soldiers are left over. Say it's 1. Then group them by 11's and find that there are, say, 2 left over. Finally, group them by 13's and find 8 left. This information is enough to determine N. It turns out to be 827.

One needs an algorithm to find N, or a staff of clerks. These days we would use a fast method based on the Euclidean Algorithm and the ideas in §4.7 [see Exercise 18 on page 514]. We could also use primes a lot larger than 7, 11, and 13. A 9-decimal-digit prime p fits into a 32-bit computer word, which means that computation mod p can be done quickly. The Chinese Remainder Theorem lets us use three such large primes p, q, and r to describe numbers up to $p \cdot q \cdot r - 1$, i.e., up to about 10^{30}. We can perform integer computations mod p, mod q, and mod r and then fit the results together at the end. In this case, snapshots of an object from three directions give us the object as a whole [at least mod $p \cdot q \cdot r$]. ■

EXAMPLE 2

Here's an example based on high school algebra. Suppose we want to multiply large numbers, for instance the 30-decimal digit numbers we get out of the Chinese Remainder Theorem. To illustrate the idea, let's multiply 32,011 by 278. [See Exercise 4 for another example.] Since

$$32{,}011 = 3 \times 10^4 + 2 \times 10^3 + 0 \times 10^2 + 1 \times 10 + 1$$

and

$$278 = 2 \times 10^2 + 7 \times 10 + 8,$$

if we let $p(x) = 3x^4 + 2x^3 + x + 1$ and $q(x) = 2x^2 + 7x + 8$, then $32{,}011 = p(10)$ and $278 = q(10)$. So $32{,}011 \cdot 278 = p(10) \cdot q(10)$. If we multiply the polynomials to get $p(x) \cdot q(x) = r(x)$, then we can evaluate $r(x)$ to get $32{,}011 \cdot 278 = r(10) = 8{,}899{,}058$.

This surely looks like the hard way to multiply. A key fact, though, is that the coefficients of $p(x)$ and $q(x)$ are less than 10, so it's easy to do arithmetic with them, and the coefficients of $r(x)$ aren't too big. If we were to do this all in binary, then the coefficients of $p(x)$ and $q(x)$ would just be 0's and 1's and, even if we were to use byte-sized coefficients, things would still be pretty fast.

The difficulty here is that it looks as if multiplying polynomials is harder and slower than multiplying numbers. There's a carry problem in either case, whether we multiply numbers or evaluate polynomials; but in this example, at any rate, it's not as bad in evaluating $r(10)$ as the carrying involved in the grade school procedure for finding $32{,}011 \cdot 278$. Obviously, it would take a careful analysis of implementation details to see which method was preferable for larger numbers.

In Example 11 we'll turn this idea around and see evaluation followed by the Chinese Remainder Theorem as a method for fast polynomial multiplication. ■

Before we can get to the applications, we need to look at some examples, set up notation, and establish some general facts.

EXAMPLE 3

(a) The sets $\mathbb{Z}$, $\mathbb{Q}$, and $\mathbb{R}$ are each closed under ordinary addition and multiplication. Both $+$ and $\cdot$ are commutative and associative for each of these sets. Moreover, these operations satisfy the distributive laws

$$a \cdot (b + c) = (a \cdot b) + (a \cdot c) \qquad \text{and} \qquad (a + b) \cdot c = (a \cdot c) + (b \cdot c).$$

(b) The set $\mathfrak{M}_{n,n}$ of $n \times n$ matrices with real entries is closed under matrix addition and also under matrix multiplication. Both operations are associative, and addition is commutative, but matrix multiplication is not commutative. For instance

$$\begin{bmatrix} 2 & 2 \\ 1 & 1 \end{bmatrix}\begin{bmatrix} 3 & -1 \\ -3 & 1 \end{bmatrix} = \begin{bmatrix} 0 & 0 \\ 0 & 0 \end{bmatrix} \neq \begin{bmatrix} 5 & 5 \\ -5 & -5 \end{bmatrix} = \begin{bmatrix} 3 & -1 \\ -3 & 1 \end{bmatrix}\begin{bmatrix} 2 & 2 \\ 1 & 1 \end{bmatrix}.$$

The distributive laws

$$\mathbf{A}(\mathbf{B} + \mathbf{C}) = (\mathbf{AB}) + (\mathbf{AC}) \quad \text{and} \quad (\mathbf{A} + \mathbf{B})\mathbf{C} = (\mathbf{AC}) + (\mathbf{BC})$$

are valid.

(c) For $p \geq 2$, the operations $+_p$ and $*_p$ on $\mathbb{Z}(p)$ are commutative and associative, and the distributive law holds by Theorem 4 on page 124. ■

Structures like the ones in Example 3 come up frequently enough to deserve a name. A **ring** $(R, +, \cdot)$ is a set R closed under two binary operations, generally denoted $+$ and $\cdot$, such that

(a) $(R, +)$ is a commutative group,
(b) $(R, \cdot)$ is a semigroup,
(c) $a \cdot (b + c) = (a \cdot b) + (a \cdot c)$ and $(a + b) \cdot c = (a \cdot c) + (b \cdot c)$ for all $a, b, c \in R$.

If $(R, \cdot)$ is commutative, we say the ring $(R, +, \cdot)$ is **commutative**. The ring $(\mathfrak{M}_{n,n}, +, \cdot)$ in Example 3(b) is not commutative if $n > 1$. The other rings in Example 3 are commutative. A ring always has an additive identity element, denoted 0. If it has a multiplicative identity that is different from 0, we usually call the multiplicative identity 1, and we say the ring is a **ring with identity**. Each ring in Example 3 is a ring with identity. The ring $(2\mathbb{Z}, +, \cdot)$ of even integers, with the usual sum and product, is an example of a commutative ring without identity.

EXAMPLE 4

The set POLY($\mathbb{R}$) of all polynomials in x with real coefficients is closed under $+$ and $\cdot$. Addition is easy. For example,

$$(2x^2 + 3x + 1) + (-5x^4 + 4x - 2) = -5x^4 + 2x^2 + 7x - 1.$$

Given polynomials $p(x)$ and $q(x)$, we just add the coefficients of x^k in $p(x)$ and $q(x)$ to get the coefficient of x^k in $p(x) + q(x)$. Addition is clearly associative and commutative; indeed, (POLY($\mathbb{R}$), $+$) is a commutative group.

Multiplication is a little more complicated. For instance,

$$\begin{aligned}(2x^2 - 3x + 1) \cdot (4x^3 - 2x^2) &= 2x^2 \cdot (4x^3 - 2x^2) - 3x \cdot (4x^3 - 2x^2) + 1 \cdot (4x^3 - 2x^2) \\ &= 8x^5 - 4x^4 - 12x^4 + 6x^3 + 4x^3 - 2x^2 \\ &= 8x^5 - 16x^4 + 10x^3 - 2x^2.\end{aligned}$$

We use the distributive laws several times and then collect terms with the same power of x.

The coefficient of x^k in the product

$$(a_m x^m + \cdots + a_1 x + a_0) \cdot (b_n x^n + \cdots + b_1 x + b_0)$$

turns out to be

$$a_0 \cdot b_k + a_1 \cdot b_{k-1} + \cdots + a_i \cdot b_{k-i} + \cdots + a_k \cdot b_0, \qquad (*)$$

using the notational convention that $a_i = 0$ for $i > m$ and $b_j = 0$ for $j > n$. One can check that this multiplication is both associative and commutative and that multiplication distributes over addition.

Distributivity here may seem to be a sure thing. Didn't we use it to *define* the product? No, we just used distributivity in motivating the definition of product. In fact, if we just start out defining $\cdot$ by $(*)$, a tedious proof shows that all the other good properties follow.

With these operations, (POLY($\mathbb{R}$), $+$, $\cdot$) is a commutative ring. So are the structures (POLY($\mathbb{Z}$), $+$, $\cdot$), (POLY($\mathbb{Q}$), $+$, $\cdot$), and (POLY($\mathbb{Z}(p)$), $+$, $\cdot$) that we get by taking polynomials with coefficients in $\mathbb{Z}$, $\mathbb{Q}$, or $\mathbb{Z}(p)$, respectively. [Of course, for $\mathbb{Z}(p)$ we use the operations $+_p$ and $*_p$ on the coefficients.] The identities in these rings are the constant polynomials $e(x) = 1$. By the way, the standard notation for these rings is $\mathbb{R}[x]$, $\mathbb{Z}[x]$, $\mathbb{Q}[x]$, $\mathbb{Z}(p)[x]$, etc., but we have given them names here that help one remember what they are. ■

The distributive laws make calculations in a ring behave very much like the arithmetic we are used to in $\mathbb{Z}$, allowing for the obvious fact that we need to watch out for noncommuting elements. For example, we have

$$\begin{aligned}(a + b)^2 &= (a + b) \cdot (a + b) = [a \cdot (a + b)] + [b \cdot (a + b)] \\ &= (a \cdot a) + (a \cdot b) + (b \cdot a) + (b \cdot b) = a^2 + a \cdot b + b \cdot a + b^2,\end{aligned}$$

but this is not $a^2 + 2(a \cdot b) + b^2$ unless $a \cdot b = b \cdot a$. See Exercise 12.

As another example, we get

$$(0 \cdot a) + (0 \cdot a) = (0 + 0) \cdot a = 0 \cdot a,$$

so

$$0 \cdot a = (0 \cdot a) + [(0 \cdot a) - (0 \cdot a)] = [(0 \cdot a) + (0 \cdot a)] - (0 \cdot a) = (0 \cdot a) - (0 \cdot a) = 0.$$

Similarly, $a \cdot 0 = 0$ for all a in a ring. One can also show that $(-a) \cdot b = -(a \cdot b) = a \cdot (-b)$ [Exercise 8].

We can never hope to divide by 0 and keep the ring properties. If we *could* divide by 0, say to get $a/0 = b$ for some a and b, then, from the algebraic rules for rings, we'd have $a = (a/0) \cdot 0 = b \cdot 0 = 0$, so $b = 0/0$. But then $b + b = \frac{0}{0} + \frac{0}{0} = \frac{0+0}{0} = \frac{0}{0} = b$, so $b = 0 = 1/b$ and things would go hopelessly wrong.

In the rings $(\mathbb{Q}, +, \cdot)$ and $(\mathbb{R}, +, \cdot)$ we can divide by every non-0 element. In $(\mathbb{Z}, +, \cdot)$, on the other hand, we can divide 6 by 3 successfully, but cannot divide 6 by 5 to get an answer that is still in $\mathbb{Z}$. A **field** is a commutative ring $(R, +, \cdot)$ in which the non-0 elements form a group under multiplication. In a field the inverse of a non-0 element a is usually written a^{-1} or $1/a$, and it has the property that $a^{-1} \cdot a = a \cdot a^{-1} = 1$. We also often write $\boldsymbol{b/a}$ for $b \cdot a^{-1}$, a notation that we justify by the fact that $(b/a) \cdot a = b \cdot a^{-1} \cdot a = b$.

Since the set of non-0 elements in a field is closed under multiplication, fields have the property

(ID) $\qquad$ if $a \cdot b = 0$, then $a = 0$ or $b = 0$.

This property might hold even if inverses do not exist, for example, in $(\mathbb{Z}, +, \cdot)$. A commutative ring with identity in which property (ID) holds is called an **integral domain**. These rings form an important class intermediate between fields and more general commutative rings with identity. They are the ones that satisfy the **cancellation law**

$$\text{if } a \cdot c = a \cdot d \text{ and } a \neq 0, \text{ then } c = d,$$

since if $a \cdot c = a \cdot d$, then $a \cdot (c - d) = 0$, so $a \neq 0$ implies $c - d = 0$.

One can show that every finite integral domain is a field [Exercise 16(c)].

EXAMPLE 5

(a) Groups are always nonempty, so the multiplicative identity in a field is always non-0. The smallest possible field is $(\mathbb{Z}(2), +_2, *_2)$ that has just two elements 0 and 1 and operations as shown in the tables.

$+_2$	0	1
0	0	1
1	1	0

$*_2$	0	1
0	0	0
1	0	1

(b) The set $\text{FUN}(\mathbb{R}, \mathbb{R})$ is a ring with $f + g$ and $f \cdot g$ defined by

$$(f + g)(x) = f(x) + g(x) \quad \text{and} \quad (f \cdot g)(x) = f(x) \cdot g(x)$$

for all $x \in \mathbb{R}$. The zero element is the constant function 0 defined by $0(x) = 0$ for all x. [See Exercise 5 on page 494 for the additive structure.] This ring is commutative, but is not an integral domain, even though it gets its multiplication from the field $\mathbb{R}$. For example, let $f(x) = 0$ for $x \leq 0$ and $f(x) = 1$ for $x > 0$, and let $g(x) = 1$ for $x \leq 0$ and $g(x) = 0$ for $x > 0$. Then $(f \cdot g)(x) = 0$ for every x, so $f \cdot g = 0$, but $f \neq 0$ and $g \neq 0$.

(c) The commutative ring $(\mathbb{Z}(4), +_4, *_4)$ is not an integral domain; $2 *_4 2 = 0$, but $2 \neq 0$.

(d) The polynomial rings $\text{POLY}(\mathbb{R})$, $\text{POLY}(\mathbb{Q})$, and $\text{POLY}(\mathbb{Z})$ are integral domains. It is not hard to see that

$$(a_m x^m + \cdots + a_1 x + a_0) \cdot (b_n x^n + \cdots + b_1 x + b_0)$$
$$= (a_m \cdot b_n) x^{m+n} + \text{terms with lower powers of } x.$$

If $a_m \neq 0$ and $b_n \neq 0$, then $a_m \cdot b_n \neq 0$; i.e., if $a(x) \neq 0$ and $b(x) \neq 0$, then $a(x) \cdot b(x) \neq 0$.

(e) The polynomial ring $\text{POLY}(\mathbb{Z}(4))$ is not an integral domain. Since $(\mathbb{Z}(4), +_4, *_4)$ itself is not an integral domain, there are non-0 constant polynomials whose product is the zero polynomial.

(f) It turns out that if p is a prime then $\mathbb{Z}(p)$ is a field [see Exercise 18]. Finite fields have become of great practical importance in recent years, with applications to encryption as well as to the kinds of error-correcting codes used in data storage and transmission. The fields $\mathbb{Z}(p)$ for p prime are the ones from which all other finite fields are built. See Exercise 20 for an example of a 4-element field based on $\mathbb{Z}(2)$.

The argument in part (d) shows that $\text{POLY}(\mathbb{Z}(p))$ is an integral domain in this case. ■

A **subring** of a ring R is simply a subset of R that is itself a ring under the two operations of R. A **subfield** of a field F is a subring of F that is itself a field; in particular, it contains the multiplicative identity 1 of F and is closed under taking inverses.

EXAMPLE 6

(a) The ring $(\mathbb{Z}, +, \cdot)$ has subrings $2\mathbb{Z}$, $73\mathbb{Z}$, and $\{0\}$ among others. In fact, the subrings of $\mathbb{Z}$ are precisely the rings $n\mathbb{Z}$ for n an integer, in view of Theorem 2 on page 473. The ring $(\mathbb{Z}(p), +_p, *_p)$ is *not* a subring of $\mathbb{Z}$; in fact, the two rings have quite different structures. For example, for each a in $\mathbb{Z}(p)$ we have $a +_p a +_p \cdots +_p a = 0$ if there are p terms in the sum, whereas $a + a + \cdots + a = pa$ in $\mathbb{Z}$.

(b) Given a field, every subring with identity is clearly an integral domain. In particular, the subring $(\mathbb{Z}, +, \cdot)$ of the field $(\mathbb{R}, +, \cdot)$ is an integral domain. It is not a field, since only 1 and -1 have multiplicative inverses in $\mathbb{Z}$. The field $(\mathbb{Q}, +, \cdot)$ is a subfield of $(\mathbb{R}, +, \cdot)$.

(c) The polynomial ring $(\text{POLY}(\mathbb{Z}), +, \cdot)$ is a subring of $(\text{POLY}(\mathbb{Q}), +, \cdot)$, which is itself a subring of $(\text{POLY}(\mathbb{R}), +, \cdot)$. ■

The appropriate mappings to use in studying rings are the ones that are compatible with both the additive and multiplicative structures. A **ring homomorphism** from a ring $(R, +, \cdot)$ to a ring $(S, +, \cdot)$ is a function $\varphi: R \to S$ such that

$$\varphi(a + b) = \varphi(a) + \varphi(b) \quad \text{and} \quad \varphi(a \cdot b) = \varphi(a) \cdot \varphi(b)$$

for all $a, b \in R$. The operations on the left sides of these equations are the operations in R; those on the right are in S. Thus a ring homomorphism is just a function that is both a group homomorphism from $(R, +)$ to $(S, +)$ and a semigroup homomorphism from $(R, \cdot)$ to $(S, \cdot)$.

EXAMPLE 7

(a) The function φ from $\mathbb{Z}$ to $\mathbb{Z}(p)$, defined by $\varphi(m) = m \text{ MOD } p$, is a ring homomorphism, as observed in the corollary to Theorem 2 on page 122.

(b) The function φ from $\mathbb{Z}$ to $\mathbb{Z}$ defined by $\varphi(m) = 3m$ is an additive group homomorphism, but is not a ring homomorphism, because $\varphi(m \cdot n) = 3mn$, whereas $\varphi(m) \cdot \varphi(n) = 3m \cdot 3n = 9mn$.

(c) The mapping φ of $\text{FUN}(\mathbb{R}, \mathbb{R})$ into $\mathbb{R}$ given by $\varphi(f) = f(10)$ is a ring homomorphism, since

$$\varphi(f + g) = (f + g)(10) = f(10) + g(10) = \varphi(f) + \varphi(g)$$

and

$$\varphi(f \cdot g) = (f \cdot g)(10) = f(10) \cdot g(10) = \varphi(f) \cdot \varphi(g).$$

More generally, for any set S and ring $(R, +, \cdot)$ we can make $\text{FUN}(S, R)$ into a ring just as we made $\text{FUN}(\mathbb{R}, \mathbb{R})$ into one, and each s in S gives rise to an **evaluation homomorphism** φ_s from $\text{FUN}(S, R)$ to R defined by $\varphi_s(f) = f(s)$ for all f in $\text{FUN}(S, R)$.

(d) We can think of $\text{POLY}(\mathbb{R})$ as a subring of $\text{FUN}(\mathbb{R}, \mathbb{R})$, since each polynomial defines a unique function and since one can show that two different polynomials must give different functions [i.e., have different graphs]. An evaluation homomorphism such as $\varphi_{10}: f \to f(10)$ from $\text{FUN}(\mathbb{R}, \mathbb{R})$ to $\mathbb{R}$ yields a homomorphism from $\text{POLY}(\mathbb{R})$ to $\mathbb{R}$, defined in this instance by $p \to p(10)$. The evaluation homomorphism $p \to p(0)$ assigns to each polynomial p its constant coefficient a_0.

We saw the evaluation homomorphism φ_{10} in Example 2. ■

Since ring homomorphisms are additive group homomorphisms, they have kernels. Suppose φ is a homomorphism from $(R, +, \cdot)$ to $(S, +, \cdot)$. The kernel of φ is the set $K = \{a \in R : \varphi(a) = \varphi(0)\}$ and, for $a, b \in R$, we have

$$\varphi(a) = \varphi(b) \iff a - b \in K \iff a + K = b + K.$$

Everything is in additive dress here, so $a + K$ is the coset $\{a + k : k \in K\}$ of the subgroup K of $(R, +)$. As before, φ is one-to-one if and only if $K = \{0\}$. The kernel of φ is a special kind of subring of R. It is an additive subgroup, of course, but it is also closed under multiplication not only by its own elements but even under multiplication by other elements in R:

$$a \in K, r \in R \quad \text{imply} \quad a \cdot r \in K \quad \text{and} \quad r \cdot a \in K.$$

The reason is that $\varphi(a) = \varphi(0)$ and, if $r \in R$, then

$$\varphi(a \cdot r) = \varphi(a) \cdot \varphi(r) = \varphi(0) \cdot \varphi(r) = \varphi(0 \cdot r) = \varphi(0)$$

and likewise $\varphi(r \cdot a) = \varphi(0)$.

An additive subgroup I of a ring $(R, +, \cdot)$ is called an **ideal** of R if $r \cdot a \in I$ and $a \cdot r \in I$ for all $a \in I$ and $r \in R$. [The term goes back to the 19th century, when ideals were associated with "ideal numbers."] Kernels of ring homomorphisms are ideals, as we noted in the last paragraph, and one can show [Exercise 10] that every ideal is the kernel of a homomorphism.

EXAMPLE 8

(a) If the ring $(R, +, \cdot)$ is commutative and $a \in R$, then the set $R \cdot a = \{r \cdot a : r \in R\}$ is an ideal of R. To check this, we observe that $r \cdot a + s \cdot a = (r + s) \cdot a \in R \cdot a$ and $-(r \cdot a) = (-r) \cdot a \in R \cdot a$ for every $r, s \in R$; so $R \cdot a$ is an additive subgroup of R. For s in R and $r \cdot a$ in $R \cdot a$, we have $s \cdot (r \cdot a) = (s \cdot r) \cdot a \in R \cdot a$; since R is commutative, we also have $(r \cdot a) \cdot s \in R \cdot a$. Thus $R \cdot a$ is an ideal in R. An ideal of the form $R \cdot a$ is called a **principal ideal**.

All the subgroups of $(\mathbb{Z}, +)$ are of the form $n\mathbb{Z}$ by Theorem 2 on page 473, so every ideal of $(\mathbb{Z}, +, \cdot)$ is principal. So are the ideals of $\text{POLY}(\mathbb{R})$, as it turns out, but such a situation is very special. For example, in the commutative ring $\text{POLY}(\mathbb{Z})$ consisting of polynomials with integer coefficients, the set of all the polynomials $a_n x^n + \cdots + a_1 x + a_0$ in which a_0 is even is an ideal that is not principal [Exercise 19].

(b) Ideals of fields are boring. Suppose I is a non-0 ideal of a field F. Let $0 \neq a \in I$. For every $b \in F$, we have $b = (b \cdot a^{-1}) \cdot a \in F \cdot a \subseteq I$, so $I = F$. That is, F has only the obvious ideals $\{0\}$ and F. ■

If R is a ring with ideal I, then the group $\boldsymbol{R/I}$ consisting of additive cosets $r + I = \{r + i : i \in I\}$ can be made into a ring in a natural way. We define

$$(\boldsymbol{r} + \boldsymbol{I}) + (\boldsymbol{s} + \boldsymbol{I}) = (r + s) + I$$

and

$$(\boldsymbol{r} + \boldsymbol{I}) \cdot (\boldsymbol{s} + \boldsymbol{I}) = r \cdot s + I.$$

We have already seen in Theorem 2(b) on page 490 that the addition on R/I is well-defined; we check multiplication. If $r + I = r' + I$ and $s + I = s' + I$, then $r - r' \in I$ and $s - s' \in I$ and hence

$$\begin{aligned} r \cdot s - r' \cdot s' &= r \cdot s - r \cdot s' + r \cdot s' - r' \cdot s' \\ &= r \cdot (s - s') + (r - r') \cdot s' \in r \cdot I + I \cdot s' \subseteq I. \end{aligned}$$

Thus $r \cdot s + I = r' \cdot s' + I$ and our definition of product is independent of the choice of representatives we take in the cosets $r + I$ and $s + I$. The rest of the properties of a ring are easy to check.

The Fundamental Homomorphism Theorem for groups on page 491 leads to a corresponding result for rings. Consider a ring homomorphism φ from R to S with kernel I. Then φ is an additive group homomorphism of $(R, +)$ into $(S, +)$, so we already know from the Fundamental Homomorphism Theorem that the mapping φ^* from R/I to $\varphi(R)$, defined by $\varphi^*(r + I) = \varphi(r)$ for $r \in R$, is a group isomorphism. Since

$$\varphi^*((r + I) \cdot (s + I)) = \varphi^*((r \cdot s) + I) = \varphi(r \cdot s)$$
$$= \varphi(r) \cdot \varphi(s) = \varphi^*(r + I) \cdot \varphi^*(s + I),$$

φ^* is in fact a ring homomorphism. Therefore, φ^* is a **ring isomorphism** between R/I and $\varphi(R)$, i.e., a ring homomorphism that is one-to-one and onto. We have shown the following.

Theorem 1 Let $\varphi: R \to S$ be a ring homomorphism with kernel I. Then the mapping $r + I \to \varphi(r)$ is an isomorphism of the ring R/I onto $\varphi(R)$.

EXAMPLE 9

(a) The ring homomorphism $n \to n \text{ MOD } 2$ of $\mathbb{Z}$ onto $\mathbb{Z}(2)$ has kernel $2\mathbb{Z} = $ EVEN. The ring $\mathbb{Z}/2\mathbb{Z}$ has just two members, EVEN and $2\mathbb{Z}+1 = $ ODD, with operations as shown in Figure 1. The ring isomorphism from $\mathbb{Z}/2\mathbb{Z}$ to $\mathbb{Z}(2)$ is EVEN $\to 0$, ODD $\to 1$.

Figure 1 ▶

+	EVEN	ODD
EVEN	EVEN	ODD
ODD	ODD	EVEN

·	EVEN	ODD
EVEN	EVEN	EVEN
ODD	EVEN	ODD

$+_2$	0	1
0	0	1
1	1	0

$*_2$	0	1
0	0	0
1	0	1

(b) The evaluation homomorphism $p(x) \to p(0)$ maps POLY($\mathbb{R}$) onto $\mathbb{R}$. Its kernel is the set of polynomials with constant coefficient 0, i.e., the principal ideal $x \cdot$ POLY($\mathbb{R}$) of all multiples of the polynomial x. The cosets are the sets of the form $r + x \cdot$ POLY($\mathbb{R}$) for constant polynomials r, $r \in \mathbb{R}$. The isomorphism from POLY($\mathbb{R}$)$/x \cdot$ POLY($\mathbb{R}$) to $\mathbb{R}$ is simply

$$r + x \cdot \text{POLY}(\mathbb{R}) \to r.$$

Here the r on the left side represents a constant polynomial, while r on the right side represents the value of that polynomial.

(c) More generally, the kernel of the evaluation map $p(x) \to p(a)$ is the principal ideal $(x - a) \cdot$ POLY($\mathbb{R}$) of multiples of $x - a$ [Exercise 5]. Its cosets are of the form $r + (x - a) \cdot$ POLY($\mathbb{R}$), and the isomorphism from POLY($\mathbb{R}$)$/(x - a) \cdot$ POLY($\mathbb{R}$) onto $\mathbb{R}$ is

$$r + (x - a) \cdot \text{POLY}(\mathbb{R}) \to r.$$ ■

EXAMPLE 10

Recall the additive group $\mathbb{Z}(2) \times \mathbb{Z}(3)$ in Example 9(b) on page 493, where

$$(m, n) + (j, k) = (m +_2 j, n +_3 k).$$

We can make $\mathbb{Z}(2) \times \mathbb{Z}(3)$ into a ring by defining

$$(m, n) \cdot (j, k) = (m *_2 j, n *_3 k).$$

The mapping $\varphi: m \to (m \text{ MOD } 2, m \text{ MOD } 3)$ is a ring homomorphism from $\mathbb{Z}$ onto $\mathbb{Z}(2) \times \mathbb{Z}(3)$. Its kernel is

$$\{m \in \mathbb{Z} : m \equiv 0 \pmod 2 \text{ and } m \equiv 0 \pmod 3\} = \{m \in \mathbb{Z} : m \equiv 0 \pmod 6\} = 6\mathbb{Z}.$$

Thus, by Theorem 1, the ring $\varphi(\mathbb{Z})$ is isomorphic to $\mathbb{Z}/6\mathbb{Z}$, i.e., is isomorphic to $\mathbb{Z}(6)$. That is,

$$\mathbb{Z}(2) \times \mathbb{Z}(3) \text{ is ring-isomorphic to } \mathbb{Z}(6).$$ ■

The ideas in Example 10 generalize. If $R_1, \ldots, R_n$ is a list of rings, not necessarily distinct, we can make the product $R_1 \times \cdots \times R_n$ into a ring by defining

$$(r_1, \ldots, r_n) + (s_1, \ldots, s_n) = (r_1 + s_1, \ldots, r_n + s_n)$$

and

$$(r_1, \ldots, r_n) \cdot (s_1, \ldots, s_n) = (r_1 \cdot s_1, \ldots, r_n \cdot s_n),$$

where the operations in the kth coordinate are the operations defined on the corresponding ring R_k. If $\varphi_1, \ldots, \varphi_n$ are homomorphisms from some ring R to $R_1, \ldots, R_n$, respectively, then one can check [Exercise 14] that the mapping φ from R to $R_1 \times \cdots \times R_n$ defined by $\varphi(r) = (\varphi_1(r), \ldots, \varphi_n(r))$ is a homomorphism. Its kernel is $\{r \in R : \varphi_k(r) = 0 \text{ for all } k = 1, \ldots, n\}$; i.e., it is the intersection of the kernels of $\varphi_1, \ldots, \varphi_n$.

Suppose now that $I_1, \ldots, I_n$ are ideals of R and that, for $k = 1, \ldots, n$, each φ_k is the natural homomorphism from R onto R/I_k given by $\varphi_k(r) = r + I_k$. Then the homomorphism φ described in the last paragraph is defined by $\varphi(r) = (r + I_1, \ldots, r + I_n)$ for $r \in R$. Since I_k is the kernel of φ_k, we obtain the following.

Theorem 2 Let R be a ring with ideals $I_1, \ldots, I_n$. Then $I_1 \cap \cdots \cap I_n$ is an ideal of R, and $R/(I_1 \cap \cdots \cap I_n)$ is isomorphic to a subring of $(R/I_1) \times \cdots \times (R/I_n)$.

In Example 10, with $I_1 = 2\mathbb{Z}$ and $I_2 = 3\mathbb{Z}$, the ring $\mathbb{Z}/(2\mathbb{Z} \cap 3\mathbb{Z}) = \mathbb{Z}/6\mathbb{Z}$ was isomorphic to the whole ring $(\mathbb{Z}/2\mathbb{Z}) \times (\mathbb{Z}/3\mathbb{Z})$, but in general $\varphi(R)$ is only a subring of $R_1 \times \cdots \times R_n$. For example, in $\mathbb{Z}$ we have $6\mathbb{Z} \cap 10\mathbb{Z} \cap 15\mathbb{Z} = 30\mathbb{Z}$ [check this], so $\mathbb{Z}/(6\mathbb{Z} \cap 10\mathbb{Z} \cap 15\mathbb{Z}) = \mathbb{Z}/30\mathbb{Z}$ has 30 members, while $(\mathbb{Z}/6\mathbb{Z}) \times (\mathbb{Z}/10\mathbb{Z}) \times (\mathbb{Z}/15\mathbb{Z})$ has $6 \cdot 10 \cdot 15 = 900$ elements. Exercise 9 gives another example.

Corollary If $p_1, \ldots, p_n$ are distinct primes, then the ring $\mathbb{Z}(p_1 \cdots p_n)$ is isomorphic to the ring $\mathbb{Z}(p_1) \times \cdots \times \mathbb{Z}(p_n)$.

Proof Consider the ideals $I_1 = p_1\mathbb{Z}, \ldots, I_n = p_n\mathbb{Z}$. The intersection $I_1 \cap \cdots \cap I_n$ is $p_1 \cdots p_n\mathbb{Z}$, because every number divisible by each p_k has to be divisible by their product. Now $\mathbb{Z}/I_k = \mathbb{Z}/p_k\mathbb{Z} \simeq \mathbb{Z}(p_k)$, so $\mathbb{Z}/I_1 \times \cdots \times \mathbb{Z}/I_n$ is isomorphic to $\mathbb{Z}(p_1) \times \cdots \times \mathbb{Z}(p_n)$, which has $p_1 \cdots p_n$ elements, just as $\mathbb{Z}/(p_1 \cdots p_n\mathbb{Z})$ does. By Theorem 2, $\mathbb{Z}/(p_1 \cdots p_n\mathbb{Z})$ is isomorphic to a subring of $\mathbb{Z}(p_1) \times \cdots \times \mathbb{Z}(p_n)$. Since the two rings have the same number of elements and since $\mathbb{Z}(p_1 \cdots p_n) \simeq \mathbb{Z}/(p_1 \cdots p_n\mathbb{Z})$, the Corollary follows. In fact, $m \to (m \text{ MOD } p_1, \ldots, m \text{ MOD } p_n)$ is an isomorphism of the ring $\mathbb{Z}(p_1 \cdots p_n)$ onto $\mathbb{Z}(p_1) \times \cdots \times \mathbb{Z}(p_n)$. ■

EXAMPLE 11

(a) The Corollary is a slightly special case of the Chinese Remainder Theorem, which was mentioned in Example 1 on page 501 and is itself a corollary of Theorem 2.

Chinese Remainder Theorem Suppose that the integers $m_1, \ldots, m_n$ are relatively prime in pairs, i.e., that $\gcd(m_i, m_j) = 1$ for $i \neq j$, and let $N = m_1 \cdots m_n$. Then the mapping

$$s \longrightarrow (s \text{ MOD } m_1, \ \ldots, s \text{ MOD } m_n)$$

is a one-to-one correspondence, indeed, a ring isomorphism, between $\mathbb{Z}(N)$ and $\mathbb{Z}(m_1) \times \cdots \times \mathbb{Z}(m_n)$.

(b) Polynomial interpolation, which has important applications in computer arithmetic, can be viewed in the context of Theorem 2. Here is a simple illustration. Suppose that we want to find $p(x)$ in $\text{POLY}(\mathbb{R})$ such that $p(1) = 5$, $p(4) = 8$, and $p(6) = 7$, and suppose we also ask that $p(x)$ have degree 2 or less so that $p(x) = ax^2 + bx + c$ for some unknown coefficients a, b, and c.

Evaluation at 1 is a homomorphism φ_1 from POLY($\mathbb{R}$) onto $\mathbb{R}$. We are given $\varphi_1(p(x)) = p(1) = 5$. Evaluation at 4 and at 6 give two more homomorphisms, φ_4 and φ_6, with $\varphi_4(p(x)) = 8$ and $\varphi_6(p(x)) = 7$. As noted in Example 9(c), the kernel of φ_1 is the ideal $I_1 = (x-1) \cdot \text{POLY}(\mathbb{R})$, while φ_4 and φ_6 have kernels $I_4 = (x-4) \cdot \text{POLY}(\mathbb{R})$ and $I_6 = (x-6) \cdot \text{POLY}(\mathbb{R})$.

By Theorem 2, the mapping

$$p(x) + I_1 \cap I_4 \cap I_6 \longrightarrow (p(x) + I_1,\ p(x) + I_4,\ p(x) + I_6)$$

is a ring isomorphism from the ring $\text{POLY}(\mathbb{R})/(I_1 \cap I_4 \cap I_6)$ into the ring $\text{POLY}(\mathbb{R})/I_1 \times \text{POLY}(\mathbb{R})/I_4 \times \text{POLY}(\mathbb{R})/I_6$, which is itself isomorphic to $\mathbb{R}\times\mathbb{R}\times\mathbb{R}$. The composite mapping from $\text{POLY}(\mathbb{R})/(I_1 \cap I_4 \cap I_6)$ to $\mathbb{R} \times \mathbb{R} \times \mathbb{R}$ is

$$p(x) + I_1 \cap I_4 \cap I_6 \longrightarrow (p(1), p(4), p(6)).$$

The interpolation problem is to find $p(x)$ to give the right image $(p(1), p(4), p(6))$. In our example, $(5, 8, 7)$ means $p(x) = -0.3x^2 + 2.5x + 2.8$.

Fast algorithms to solve the Chinese remainder problem can be adapted to solve the polynomial interpolation problem. Exercise 19 on page 514 shows an interpolation algorithm modeled on the Chinese Remainder algorithm in the previous exercise. Interpolation, in turn, gives the following method for multiplying polynomials. To find $a(x) \cdot b(x)$, we can evaluate $a(r)$ and $b(r)$ at lots of r's, compute $a(r) \cdot b(r)$ in each case, and then interpolate to find the $p(x)$ so that $p(r) = a(r) \cdot b(r)$ for each r. Techniques such as this, using clever choices of r's, are at the heart of the Fast Fourier Transform and other fast algorithms for computing with large integers. ■

A great deal more can be said about rings and fields. We have only introduced the most basic ideas and a few examples, but we hope to have given some feeling for the kinds of questions it might be reasonable to ask about these systems and the kinds of answers one might get. The study of groups, rings, and fields makes up a large part of the area of mathematics called abstract algebra. At this point you are in a good position to read an introductory book in this area.

Exercises 12.6

In these exercises, the words "homomorphism" and "isomorphism" mean "ring homomorphism" and "ring isomorphism."

1. Which of the following sets are subrings of $(\mathbb{R}, +, \cdot)$?

(a) $2\mathbb{Z}$ (b) $2\mathbb{R}$

(c) $\mathbb{N}$ (d) $\{m + n\sqrt{2} : m, n \in \mathbb{Z}\}$

(e) $\{m/2 : m \in \mathbb{Z}\}$ (f) $\{m/2^a : m \in \mathbb{Z}, a \in \mathbb{P}\}$

2. (a) For each subset in Exercise 1 that is a subring of $\mathbb{R}$, verify closure under addition and multiplication.

(b) For each subset in Exercise 1 that is not a subring of $\mathbb{R}$, give a property of subrings that the subset does not satisfy.

3. Which of the following functions are homomorphisms? Justify your answer in each case.

(a) $\varphi : \text{FUN}(\mathbb{R}, \mathbb{R}) \to \mathbb{R}$ defined by $\varphi(f) = f(0)$.

(b) $\varphi : \mathbb{R} \to \mathbb{R}$ defined by $\varphi(r) = r^2$.

(c) $\varphi : \mathbb{R} \to \text{FUN}(\mathbb{R}, \mathbb{R})$ defined by $(\varphi(r))(x) = r$. That is, $\varphi(r)$ is the constant function on $\mathbb{R}$ having value r at every x.

(d) $\varphi : \mathbb{Z} \to \mathbb{R}$ defined by $\varphi(n) = n$.

(e) $\varphi : \mathbb{Z}/3\mathbb{Z} \to \mathbb{Z}/6\mathbb{Z}$ defined by $\varphi(n + 3\mathbb{Z}) = 2n + 6\mathbb{Z}$.

4. Let $p(x) = 8x^4 + 3x^3 + x + 7$ and $q(x) = x^4 + x^2 + x + 1$.

(a) Evaluate $p(10)$ and $q(10)$.

(b) Compute the product $83{,}017 \times 10{,}111$ the hard way, by finding $p(x) \cdot q(x) = r(x)$ and evaluating $r(10)$. Notice the problem with carries.

5. (a) Show that if $p(x) = q(x) \cdot (x - a) + b$ is in POLY($\mathbb{R}$) with a and b in $\mathbb{R}$, then $b = p(a)$.

(b) Show that if

$$q_k(x) = \sum_{i=1}^{k} a^{i-1}x^{k-i} = x^{k-1} + ax^{k-2} + \cdots + a^{k-2}x + a^{k-1},$$

then $x^k - a^k = q_k(x) \cdot (x - a)$.

(c) Use part (b) to show that if

$$p(x) = \sum_{k=0}^{n} c_k x^k \in \text{POLY}(\mathbb{R})$$

and $a \in \mathbb{R}$, then $p(x) = q(x) \cdot (x - a) + p(a)$ for some $q(x) \in$ POLY($\mathbb{R}$). That is, prove the Remainder Theorem from algebra.

(d) Show that the kernel of the evaluation mapping $p(x) \to p(a)$ from POLY($\mathbb{R}$) to $\mathbb{R}$ is the ideal $(x - a) \cdot$ POLY($\mathbb{R}$) consisting of all multiples of $x - a$.

6. Find the kernel of the homomorphism φ from FUN($\mathbb{R}, \mathbb{R}$) to $\mathbb{R}$ in Example 7(c).

7. Consider the ring $\mathbb{Z}$. Write each of the following in the form $n\mathbb{Z}$ with $n \in \mathbb{N}$.

(a) $6\mathbb{Z} \cap 8\mathbb{Z}$ (b) $6\mathbb{Z} + 8\mathbb{Z}$

(c) $3\mathbb{Z} + 2\mathbb{Z}$ (d) $6\mathbb{Z} + 10\mathbb{Z} + 15\mathbb{Z}$

8. Show that in a ring $(-a) \cdot b = -(a \cdot b) = a \cdot (-b)$ for every a and b.

9. (a) Verify that the mapping φ from $\mathbb{Z}(12)$ to $\mathbb{Z}(4) \times \mathbb{Z}(6)$ given by $\varphi(m) = (m \text{ MOD } 4, m \text{ MOD } 6)$ is a well-defined homomorphism.

(b) Find the kernel of φ.

(c) Find an element of $\mathbb{Z}(4) \times \mathbb{Z}(6)$ that is not in the image of φ.

(d) Which elements in $\mathbb{Z}(12)$ are mapped to $(1, 3)$ by φ?

10. Let I be an ideal of a ring R.

(a) Show that the mapping $r \to r + I$ is a homomorphism of R onto R/I.

(b) Find the kernel of this homomorphism.

11. (a) Show that $(\mathbb{Z}(6), +_6, *_6)$ is not a field.

(b) Show that $(\mathbb{Z}(5), +_5, *_5)$ is a field.

(c) Show that if F and K are fields, then $F \times K$ is *not* a field.

12. Find 2×2 matrices that show that $(\mathbf{A} + \mathbf{B})^2 = \mathbf{A}^2 + 2\mathbf{AB} + \mathbf{B}^2$ does not generally hold.

13. Find 2×2 matrices that show that $(\mathbf{A} + \mathbf{B})(\mathbf{A} - \mathbf{B}) = \mathbf{A}^2 - \mathbf{B}^2$ does not generally hold.

14. If $R_1, \ldots, R_n$ are rings and if $\varphi_1, \ldots, \varphi_n$ are homomorphisms from a ring R into $R_1, \ldots, R_n$, respectively, then the mapping φ defined by $\varphi(r) = (\varphi_1(r), \ldots, \varphi_n(r))$ is a homomorphism of R into $R_1 \times \cdots \times R_n$. Verify this fact for $n = 2$.

15. (a) Show that if φ is a homomorphism from a field F to a ring R, then either $\varphi(a) = 0$ for all $a \in F$ or φ is one-to-one.

(b) Show that if I is an ideal of the ring R and if θ is a homomorphism from R onto a ring S, then $\theta(I)$ is an ideal of S.

(c) Show that if S is a field in part (b) then either $\theta(I) = S$ or $\theta(I) = \{0\}$.

16. Let R be a commutative ring with identity.

(a) Show that R is an integral domain if and only if the mapping $r \to a \cdot r$ from R to R is one-to-one for each non-0 a in R.

(b) Show that R is a field if and only if this mapping is a one-to-one correspondence of R onto itself for each non-0 a in R.

(c) Show that every finite integral domain is a field.

17. (a) Find an ideal I of $\mathbb{Z}$ for which $\mathbb{Z}/I$ is isomorphic to $\mathbb{Z}(3) \times \mathbb{Z}(5)$.

(b) Describe an isomorphism between $\mathbb{Z}(12)$ and $\mathbb{Z}(3) \times \mathbb{Z}(4)$.

18. Let p be a prime. This exercise shows that the commutative ring $\mathbb{Z}(p)$ is a field.

(a) Show that if $0 < k < p$ then there are integers s and t with $k \cdot s + p \cdot t = 1$.

(b) With notation as in part (a), show that $s \text{ MOD } p$ is the multiplicative inverse of k in $\mathbb{Z}(p)$. [Euclid gave us a fast algorithm to compute s and t, so we can compute effectively in $\mathbb{Z}(p)$, even if p is very large.]

19. Consider the ring $R =$ POLY($\mathbb{Z}$) of polynomials with integer coefficients. The constant polynomial 2 and the polynomial x both belong to R.

(a) Describe the members of the ideals $R \cdot 2$, $R \cdot x$ and $R \cdot 2 + R \cdot x$.

(b) Show that there is no polynomial $p(x)$ in R for which $R \cdot p = R \cdot 2 + R \cdot x$.

20. The set $\mathbb{B} \times \mathbb{B}$ can be made into a ring in another way besides the one described in the text. Define $+$ and $\cdot$ by the tables

$+$	(0, 0)	(1, 0)	(0, 1)	(1, 1)
(0, 0)	(0, 0)	(1, 0)	(0, 1)	(1, 1)
(1, 0)	(1, 0)	(0, 0)	(1, 1)	(0, 1)
(0, 1)	(0, 1)	(1, 1)	(0, 0)	(1, 0)
(1, 1)	(1, 1)	(0, 1)	(1, 0)	(0, 0)

$\cdot$	(0, 0)	(1, 0)	(0, 1)	(1, 1)
(0, 0)	(0, 0)	(0, 0)	(0, 0)	(0, 0)
(1, 0)	(0, 0)	(1, 0)	(0, 1)	(1, 1)
(0, 1)	(0, 0)	(0, 1)	(1, 1)	(1, 0)
(1, 1)	(0, 0)	(1, 1)	(1, 0)	(0, 1)

Verify that $\mathbb{B} \times \mathbb{B}$ is an additive group and that the non-0 elements form a group under multiplication. *Suggestion:* Save work by exhibiting known groups isomorphic to your alleged groups. [Finite fields such as this are important in algebraic coding to minimize the effects of noise on transmission channels—or scratches on a CD.]

Chapter Highlights

As usual: What does it mean? Why is it here? How can I use it? Think of examples. This chapter covers a lot of ideas. The main themes are subsystems, homomorphisms, and actions. As you review, try to see how individual topics connect with these themes.

CONCEPTS AND NOTATION

group acting on a set X, on $\text{FUN}(X, C)$
- orbit
- fix, $\text{FIX}_G(x)$, $\text{FIX}_X(g)$
- restriction

applications of groups
- digraph or graph automorphism, $\text{AUT}(D)$, $\text{AUT}(H)$
- coloring, equivalent colorings

algebraic system
- semigroup, monoid, group
  - identity, inverse, g^{-1}, $-g$
- ring, integral domain, field
  - zero, identity
- subgroup [subring, etc.], proper, trivial
  - subgroup $\langle A \rangle$ generated by A
  - subsemigroup A^+ generated by A
- cyclic group, order of an element

homomorphism, isomorphism, $\simeq$
- of groups, semigroups, rings
- kernel
  - normal subgroup, ideal
  - principal ideal

coset, gH, Hg, $g + H$

natural operation on G/K, natural operations on R/I

natural homomorphism $\nu : G \to G/K$ or $R \to R/I$

FACTS ABOUT ACTIONS

The G-orbits partition a set on which G acts.

$|G| = |Gx| \cdot |\text{FIX}_G(x)|$ for every x in the set X on which the group G acts.

The number of G-orbits in X is the average number of elements of X fixed by members of G.

The number of orbits of G on $\text{FUN}(X, C)$ is

$$\frac{1}{|G|} \sum_{g \in G} |C|^{m(g)},$$

where $m(g)$ is the number of orbits of $\langle g \rangle$ on X.

The number of G-classes of colorings of X with k colors is

$$C(k) = \frac{1}{|G|} \sum_{g \in G} k^{m(g)}.$$

GENERAL ALGEBRAIC FACTS

Cancellation is legal in a group. Cancellation of non-0 factors is legal in an integral domain.

A finite subset of a group that is nonempty and closed under the group operation is a subgroup.

Intersections of subgroups [subsemigroups, subrings, normal subgroups, etc.] are subgroups [subsemigroups, etc.].

The subsemigroup [subgroup] generated by A consists of all products of members of A [and their inverses].

The subgroups of $(\mathbb{Z}, +)$ are the cyclic groups $n\mathbb{Z}$. They are also the ideals of the ring $(\mathbb{Z}, +, \cdot)$.

The cosets of a subgroup partition a group into sets of equal size [only one of which is a subgroup].

Lagrange's Theorem: $|G| = |G/H| \cdot |H|$.

The order of an element of a finite group G divides $|G|$.

Normal subgroups are the kernels of group homomorphisms; ideals are the kernels of ring homomorphisms.

Homomorphisms take identities to identities and inverses to inverses.

Fundamental Theorem: If K is the kernel of a homomorphism φ on a group G, then $G/K \simeq \varphi(G)$. A similar statement is true for rings.

A group or ring homomorphism is one-to-one if and only if its kernel consists of just one element.

Every cyclic group is isomorphic to $(\mathbb{Z}, +)$ or to some $(\mathbb{Z}_p, +_p)$.

If $I_1, \ldots, I_n$ are ideals of R, then $R/(I_1 \cap \cdots \cap I_n)$ is isomorphic to a subring of $(R/I_1) \times \cdots \times (R/I_n)$.

The Chinese Remainder Theorem: If $\gcd(m_i, m_j) = 1$ for $i \neq j$, then

$$\mathbb{Z}(m_1 \cdots m_n) \simeq \mathbb{Z}(m_1) \times \cdots \times \mathbb{Z}(m_n).$$

The theory of rings applies to polynomial interpolation.

Supplementary Exercises

1. The graph H shown has automorphism group $\text{AUT}(H) = \{e, f, g, h\}$, whose members' actions are given in the table.

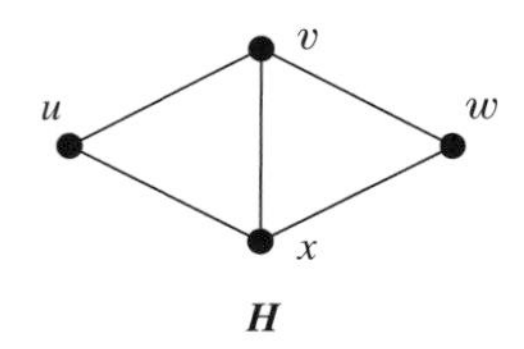

	u	v	w	x
e	u	v	w	x
f	u	x	w	v
g	w	v	u	x
h	w	x	u	v

(a) What are the orbits of $\{u, v, w, x\}$ under $\text{AUT}(H)$?

(b) Which of e, f, g, and h is gh?

(c) List the members of $\text{FIX}(v)$.

(d) Give the multiplication table for the group $\text{AUT}(H)$.

(e) How many essentially different ways are there to color the vertices of H with at most three colors?

(f) How many essentially different ways are there to color the edges of H with at most three colors?

2. Let G be the automorphism group of the graph H.

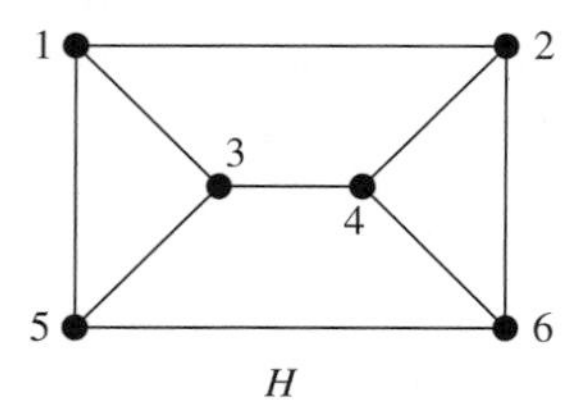

Thus G consists of: the permutation e; the flip h that interchanges 1 with 2, 3 with 4, and 5 with 6; the flip v that interchanges 1 with 5 and 2 with 6; and the half-turn rotation r that interchanges 1 with 6, 2 with 5, and 3 with 4.

(a) List the members of G that fix 3.

(b) List the members of $\text{FIX}_{V(H)}(v)$.

(c) List the members of the orbit of G that contains the vertex 2.

(d) How many orbits does $\langle v \rangle$ have in $V(H)$?

(e) How many essentially different ways are there to color the vertices of H with at most two colors?

3. A certain mystery group G with 24 members acts on the set $X = \{1, 2, \ldots, 59, 60\}$ in such a way that the orbit $G(5)$ has 2 elements.

(a) How many members g of G satisfy $g(5) = 5$?

(b) Suppose that G has six orbits on X. What is the average number of elements of X fixed by the members of G?

4. Let Σ be the alphabet $\{a, b, c\}$ and Σ^* the collection of all words using letters from Σ. Σ^* is a semigroup under concatenation.

(a) Give the smallest set of generators for Σ^* that you can.

(b) Is Σ^* commutative? Justify your answer.

(c) Does Σ^* have an identity? Justify your answer.

(d) List three members in the semigroup generated by cab.

5. Let L be a language over the alphabet $\Sigma = \{a, b\}$, with concatenation as multiplication.

(a) Under what conditions is L a subsemigroup of Σ^*? A submonoid?

(b) Suppose that L_1 is the subset of Σ^* defined recursively by (B) ab and ba are in L_1 and (R) $abwba$ is in L_1 for every w in L_1. Is L_1 a semigroup? Explain.

(c) Define L_2 recursively by (B) ab and ba are in L_2 and (R) $abwba$, $abwab$, $bawba$, and $bawab$ are in L_2 for every w in L_2. Is L_2 a semigroup? Explain.

(d) Define L_3 recursively by (B) $\lambda \in L_3$ and (R) awa, awb, bwa, and bwb are in L_3 for all w in L_3. Is L_3 a monoid? Explain.

6. (a) Give the addition and multiplication tables for $\mathbb{Z}(4)$.

(b) Is $(\mathbb{Z}(4), +_4)$ a semigroup? a monoid? a group? commutative?

(c) Is $(\mathbb{Z}(4), *_4)$ a semigroup? a monoid? a group? commutative?

(d) Is $(\mathbb{Z}(4) \setminus \{0\}, *_4)$ a group? Explain. Be specific; i.e., make it clear why this particular object is, or is not, a group.

7. The table shown is the multiplication table of a group G that has eight members.

	e	i	j	k	m	n	o	p
e	e	i	j	k	m	n	o	p
i	i	j	k	e	p	o	m	n
j	j	k	e	i	n	m	p	o
k	k	e	i	j	o	p	n	m
m	m	o	n	p	e	j	i	k
n	n	p	m	o	j	e	k	i
o	o	n	p	m	k	i	e	j
p	p	m	o	n	i	k	j	e

(a) What are the elements of order 2 in G?

(b) Describe a cyclic subgroup of G with 4 elements.

(c) Describe a subgroup with 4 elements that is *not* cyclic.

(d) List the subgroups of order 3 in this group.

8. The table describes the action of the group $G = \text{AUT}(H)$ on the set $E(H)$ of *edges* of the graph H.

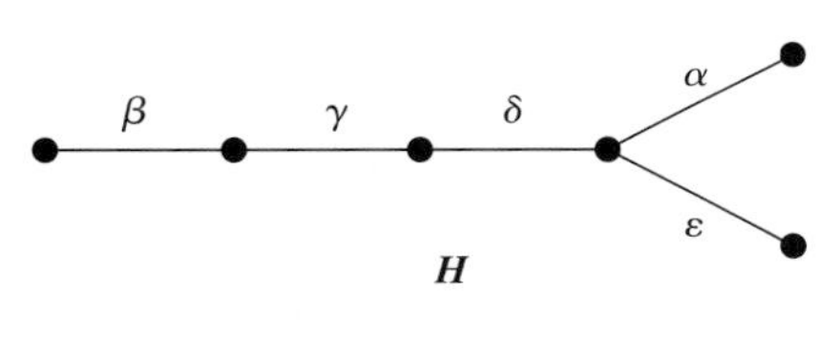

	α	β	γ	δ	ε
e	α	β	γ	δ	ε
g	ε	β	γ	δ	α

(a) List the orbits of g on $E(H)$.

(b) How many ways are there to color $E(H)$ with three colors if two colorings are considered the same in case one is obtained from the other by applying g or e?

9. (a) How many different ways are there to color the vertices of a square using 10 colors, if flips and rotations are allowed? [That is, two colorings are considered the same if one can be changed to the other by a succession of flips and rotations.]

(b) How many different ways are there if opposite vertices of the square must be the same color? [Look for an easy answer, and then explain it.]

(c) How many different ways are there if opposite vertices are not allowed to be the same color? *Hint:* Apply Theorem 1 on page 477, where X is the set of colorings of the vertices of a square that take different values on opposite vertices.

10. Consider colorings of six beads arranged on a circular string with colors black, red, and white. [Think of the beads as the vertices of a regular hexagon.]

(a) How many colorings are unchanged by the rotation R_2 that moves each bead two places along, all in the same direction?

(b) Give two examples of colorings unchanged by R_2 that are *not* equivalent, even if we allow flips as well as rotations.

(c) Describe the orbits of R_2 acting on the set of beads.

(d) How many essentially different colorings of the necklace with three colors are there if two colorings are regarded as equivalent in case one can be produced from the other by rotating the necklace or flipping it over?

11. (a) Give all subgroups of the additive group $(\mathbb{Z}(12), +_{12})$.

(b) Is the group $(\mathbb{Z}(12), +_{12})$ isomorphic to any subgroup of $\mathbb{Z}$? Explain.

(c) Is $\mathbb{Z}(12) \setminus \{0\}$ a semigroup under the operation $*_{12}$? Explain.

12. Let $G = \langle g \rangle$ be a cyclic group of order 6.

(a) List the subgroups of G, including the improper ones.

(b) Which of these subgroups are normal?

(c) Show that the mapping $\varphi\colon\ x \to (x^3, x^2)$ from G into the direct product group $G \times G$ is a homomorphism, and find its kernel.

(d) Show that G is isomorphic to the group $\langle g^3 \rangle \times \langle g^2 \rangle$.

13. Let φ be a semigroup homomorphism from the semigroup $(S, \bullet)$ to the semigroup $(T, *)$.

(a) Show that if U is a subsemigroup of S, then $\varphi(U)$ is a subsemigroup of T.

(b) If V is a subsemigroup of T, must $\varphi^{\leftarrow}(V)$ be a subsemigroup of S? Explain.

14. Show that if S is a semigroup, if A generates S, and if φ is a homomorphism defined on S, then $\varphi(A)$ generates $\varphi(S)$.

15. The collection $\mathcal{P}(S \times S)$ of all relations on a nonempty set S forms a semigroup with composition as operation, as we showed on page 448.

(a) Is this semigroup a monoid? Explain.

(b) Under what conditions is this semigroup commutative?

16. (a) Show that if φ is a homomorphism from a semigroup $(S, \bullet)$ to a semigroup $(T, \square)$ and if ψ is a homomorphism from $(T, \square)$ to a semigroup $(U, \triangle)$, then $\psi \circ \varphi$ is a homomorphism.

(b) Show that if φ is an isomorphism of $(S, \bullet)$ onto $(T, \square)$, then φ^{-1} is an isomorphism of $(T, \square)$ onto $(S, \bullet)$.

(c) Use the results of parts (a) and (b) to show that the isomorphism relation $\simeq$ is reflexive, symmetric, and transitive.

17. Ann tells Bob that, if M is a monoid and S is a subsemigroup of M that is also a monoid, then the identity element of S must be the same as the identity of M. Bob says that's not always true, but that, if e is the identity of S and φ is a semigroup homomorphism from S into some monoid T, then $\varphi(e)$ must be the identity of T. What's the story here?

18. Consider the algorithm shown.

Chinese Remainder Algorithm

{Input: Positive integers $m_1, \ldots, m_n$ with $\gcd(m_i, m_j) = 1$ for $i \neq j$, integers $b_1, \ldots, b_n$.}
{Output: An integer x with $x \equiv b_i \pmod{m_i}$ for $i = 1, \ldots, n$.}
Set $x_1 := b_1$.
for $i = 2, \ldots, n$ **do**
  Find s_i with $s_i \cdot (m_1 \cdots m_{i-1}) \equiv 1 \pmod{m_i}$.
  Set $x_i := x_{i-1} + s_i \cdot (m_1 \cdots m_{i-1})(b_i - x_{i-1})$.
return x_n. ■

(a) Prove by induction that $x_k \equiv b_i \pmod{m_i}$ for $i = 1, \ldots, k$ for each k in $\{1, \ldots, n\}$. Hence, in particular, x_n has the property claimed for x.

(b) Carry out the steps of the algorithm to find x with $x \equiv 4 \pmod{17}$ and $x \equiv 6 \pmod{11}$. To find s_2, you may use any method, such as the Euclidean Algorithm or simple inspection.

(c) Carry out the steps of the algorithm to find x with $x \equiv 6 \pmod{11}$ and $x \equiv 4 \pmod{17}$. Compare your answer with that for part (b).

(d) Suppose that we replace the line

$$\text{Set } x_i := x_{i-1} + s_i \cdot (m_1 \cdots m_{i-1})(b_i - x_{i-1})$$

by

$$\text{Set } x_i := (x_{i-1} + s_i \cdot (m_1 \cdots m_{i-1})(b_i - x_{i-1})) \text{ MOD } m_1 \cdots m_i$$

in the algorithm. Why might this be a desirable change? Would the modified algorithm still return a solution to the given list of congruences?

(e) Apply the algorithm as modified in part (d) to solve the system of congruences $x \equiv 6 \pmod{11}$, $x \equiv 4 \pmod{17}$, and $x \equiv -1 \pmod{73}$.

19. Consider the algorithm shown.

Polynomial Interpolation Algorithm

{Input: Real numbers $a_1, \ldots, a_n$ with $a_i \neq a_j$ for $i \neq j$, and real numbers $b_1, \ldots, b_n$.}
{Output: A polynomial $p(x)$ of degree at most $n - 1$ with $p(a_i) = b_i$ for $i = 1, \ldots, n$.}
Set $p_1(x) := b_1$.
for $i = 2, \ldots, n$ **do**
  Set $s_i = \dfrac{1}{(a_i - a_1) \cdots (a_i - a_{i-1})}$.
  Set $p_i(x) := p_{i-1}(x) + s_i \cdot (x - a_1) \cdots (x - a_{i-1}) \cdot (b_i - p_{i-1}(a_i))$.
return $p_n(x)$. ■

(a) Prove by induction that for each k in $\{1, \ldots, n\}$ the polynomial $p_k(x)$ has degree at most $k - 1$ and $p_k(a_i) = b_i$ for $i = 1, \ldots, k$. Hence, in particular, $p_n(x)$ has the property claimed.

[In fact, the notions of mod and gcd make sense for polynomials, and the Polynomial Interpolation Algorithm is simply a translation of the Chinese Remainder Algorithm into the polynomial setting, with the added advantage of an explicit formula for s_i. Moreover, this algorithm works not just for real numbers, but for elements of any field.]

(b) Carry out the steps of the algorithm to find $p(x)$ of degree at most 2 with $p(0) = 2$, $p(1) = 0$, and $p(3) = 8$.

(c) Is it possible that $p(x)$ might have degree less than $n - 1$?

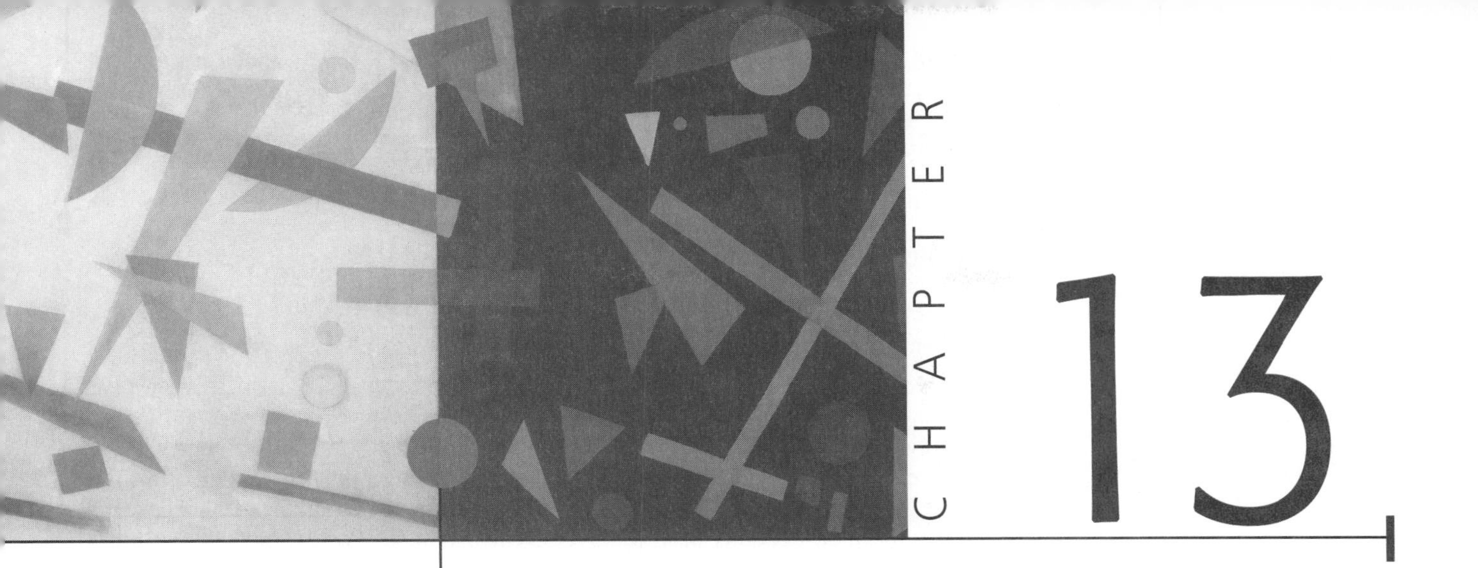

Predicate Calculus and Infinite Sets

This chapter contains topics that might well have been included in Chapters 2 and 5, but which we deferred until now to avoid breaking the flow of ideas in the core of the book. The informal introduction to $\forall$ and $\exists$ that we did give in Chapter 2 was adequate for everyday use, and even a moderately complete introduction to the predicate calculus would have been substantially more sophisticated than the level of the discussion at that stage. As it is, the account in the next two sections only begins to suggest the questions that arise in a formal development of the subject.

The reason for leaving infinite sets out of the chapter on counting is simply that the ideas and methods they require are completely different from those appropriate to finite sets. Moreover, finite intuition and infinite intuition are quite different and should be kept separate.

13.1 Quantifiers and Predicates

Chapter 2 emphasized logic as a working tool. We need to be able to untangle complicated logical statements, and the propositional calculus that we constructed from $\vee$, $\wedge$, $\neg$, $\rightarrow$, and $\leftrightarrow$ was aimed at this goal. That machinery is a nice, complete, self-contained theory of logic, but it is totally inadequate for most of mathematics. Part of the problem is that the propositional calculus does not allow the use of an infinite number of propositions, and the notation is even awkward for handling a large finite set of propositions. For example, we frequently encounter an infinite sequence of propositions $p(n)$ for n in $\mathbb{N}$. The informal statement "$p(n)$ is true for all n" means "$p(0)$ is true and $p(1)$ is true and $p(2)$ is true, etc." The only symbolism available from the propositional calculus would be something like $p(0) \wedge p(1) \wedge p(2) \wedge \cdots$, which is not acceptable. Similarly, the informal statement "$p(n)$ is true for some n" would correspond to the unacceptable $p(0) \vee p(1) \vee p(2) \vee \cdots$. To get around this sort of problem, we will use the **quantifiers** $\forall$ and $\exists$ introduced in §2.1. We need to develop the rules for using the new symbols and for combining them with the old ones. The augmented system of symbols and rules is called the **predicate calculus**.

Recall that the quantifiers are applied to families $\{p(x) : x \in U\}$ of propositions; the nonempty set U is called the **universe of discourse** or **domain of discourse**. The compound proposition $\forall x\, p(x)$ is assigned truth values as follows:

$\forall x\, p(x)$ is true if $p(x)$ is true for every x in U;
otherwise $\forall x\, p(x)$ is false.

The compound proposition $\exists x\, p(x)$ has these truth values:

$\exists x\, p(x)$ is true if $p(x)$ is true for at least one x in U;
$\exists x\, p(x)$ is false if $p(x)$ is false for every x in U.

Notice that $\forall x\, p(x)$ and $\exists x\, p(x)$ are propositions; each of them is either true or it's false.

EXAMPLE 1

(a) Compound propositions are not completely defined unless the universe of discourse is specified. Thus $\forall x\, [x^2 \geq x]$ is ambiguous until we specify its universe of discourse, which might be $\mathbb{R}$, $[0, \infty)$ or some other set. In general, the truth value depends on the universe of discourse. For example, this proposition is true for the universe of discourse $[1, \infty)$, but false for $[0, \infty)$ and for $\mathbb{R}$.

(b) General statements such as "$w + w = 2w$" and "if $x > 1$, then $x < x^2$" often come with implicit quantifiers. Thus we understand "$w + w = 2w$" to mean $\forall w\, [w + w = 2w]$ for some specific universe, unless w itself has been assigned a specific value, in which case the statement "$w + w = 2w$" is a statement about that value. Similarly, "if $x > 1$, then $x < x^2$" means $\forall x\, [x > 1 \rightarrow x < x^2]$ for some chosen universe. ■

We can also consider propositions that are functions of more than one variable, perhaps from more than one universe of discourse, and in such cases multiple use of quantifiers is natural.

EXAMPLE 2

Here is an example illustrating the multiple use of quantifiers and the importance of order when using both quantifiers $\forall$ and $\exists$. Let the universe of discourse consist of all people, living or dead, and consider

$$m(x, y) = \text{"}x \text{ is a mother of } y\text{."}$$

The truth value of $m(x, y)$ depends on both x and y, of course. The meaning of $\forall x\, m(x, y)$ is "every x is a mother of y," i.e., "everybody is a mother of y," which is false no matter who y is. The meaning of $\forall y\, m(x, y)$ is "x is a mother of everybody," which is also false, no matter who x is. The truth value of $\exists y\, m(x, y)$ depends on who x is, because its meaning is "x is a mother of somebody." On the other hand, $\exists x\, m(x, y)$ means "somebody is a mother of y," so it's true for every person y. [We won't worry about clones.] It would even be true for a person y with more than one mother, if that were possible.

Notice that in each of these cases only one variable matters in the final statement. For instance, $\exists y\, m(x, y)$ could be read as "x is a mother of some y," but it could just as well be "x is a mother of some z," because what it really means is that x is a mother of *somebody*, whether we use the label y or z or something else for that somebody. Thus $\exists y\, m(x, y)$ and $\exists z\, m(x, z)$ are both true or both false for a given person x.

The truth value of $\exists y\, m(x, y)$ depends on the value of the variable x. We can quantify it again, either with $\forall x$ or $\exists x$. The meaning of $\forall x\, [\exists y\, m(x, y)]$ is "every x is a mother of somebody," i.e., "everybody is a mother of somebody." The meaning of $\exists x\, [\exists y\, m(x, y)]$ is "some x is a mother of somebody," i.e., "somebody is a mother of somebody," which is equivalent to $\exists y\, [\exists x\, m(x, y)]$. It turns out that $\exists x\, [\exists y\, p(x, y)]$ and $\exists y\, [\exists x\, p(x, y)]$ always have the same truth value no matter what $p(x, y)$ is, since they both mean "$p(x, y)$ is true for some value of x and some value of y."

However, $\exists x\,[\forall y\, m(x, y)]$ and $\forall y\,[\exists x\, m(x, y)]$ have different meanings for our example. Since $\forall y\,[\exists x\, m(x, y)]$ means "everybody has somebody as a mother," it's true. Exercise 1 asks you to decide about $\exists x\,[\forall y\, m(x, y)]$. Hint: It's false.

Henceforth we will usually omit the brackets [ ] and write $\exists x\, \forall y\, p(x, y)$, for example, or even $\exists x\, \forall y\, \exists z\, q(x, y, z)$. Remember, though, that the order in which the quantifiers and their associated variables appear is important in general. ■

Let's analyze the proposition $\forall x\, p(x)$ more closely. The expression $p(x)$ is called a **predicate**. In ordinary grammatical usage, a predicate is the part of a sentence that says something about the subject of the sentence. For example, "_______ went to the moon" and "_______ is bigger than a bread box" are predicates. To make a sentence, we supply the subject. For instance, the predicate "_______ is bigger than a bread box" becomes the sentence "This book is bigger than a bread box" if we supply the subject "This book." If we call the predicate p, then the sentence could be denoted p(This book). Each subject x yields a sentence $p(x)$.

A **predicate** in our symbolic logic setting is a function that produces a proposition whenever we feed it a member of the universe; i.e., it is a proposition-valued function with domain U. We follow our usual practice and denote such a function by $p(x)$. The variable x in the expression $p(x)$ is called a **free variable** of the predicate. As x varies over U, the truth value of $p(x)$ may vary. In contrast, the proposition $\forall x\, p(x)$ has a well-defined truth value that does not depend on x. The variable x in $\forall x\, p(x)$ is called a **bound variable**; it is bound by the quantifier $\forall$. Since $\forall x\, p(x)$ has a fixed meaning and truth value, it would be pointless and unnatural to quantify it again. That is, it would be pointless to introduce $\forall x\, \forall x\, p(x)$ and $\exists x\, \forall x\, p(x)$, since their truth values are the same as that of $\forall x\, p(x)$. However, as we have seen in Example 2, multiple use of quantifiers makes sense for predicates that are functions of more than one variable.

EXAMPLE 3

Let $\mathbb{N}$ be the universe of discourse, and for each m and n in $\mathbb{N}$ let

$$p(m, n) = \text{“}m < n\text{.”}$$

We could think of the propositions $p(m, n)$ as being labeled by $\mathbb{N} \times \mathbb{N}$, in which case p would be a predicate defined on $\mathbb{N} \times \mathbb{N}$ that produces a proposition for each choice of (m, n) in $\mathbb{N} \times \mathbb{N}$. It seems more natural, though, to think of p as a proposition-valued function of two variables, m and n, both in $\mathbb{N}$, i.e., as what we will call a 2-place predicate. Another example of a 2-place predicate is the motherhood function in Example 2.

In that example and the present one, both variables m and n are free, and the meanings and truth values of $p(m, n)$ vary with both m and n. In the expression $\exists m\, p(m, n)$, the variable m is bound, but the variable n is free. The proposition $\exists m\, p(m, n)$ reads "there is an m in $\mathbb{N}$ with $m < n$," so $\exists m\, p(m, 0)$ is false, $\exists m\, p(m, 1)$ is true, $\exists m\, p(m, 2)$ is true, etc. For each choice of n, the proposition $\exists m\, p(m, n)$ is either true or false; its truth value does not depend on m, but depends on n alone. It is meaningful to quantify $\exists m\, p(m, n)$ with respect to the free variable n to obtain $\forall n\, \exists m\, p(m, n)$ and $\exists n\, \exists m\, p(m, n)$. The proposition $\forall n\, \exists m\, p(m, n)$ is false because $\exists m\, p(m, 0)$ is false, and the proposition $\exists n\, \exists m\, p(m, n)$ is true because, for example, $\exists m\, p(m, 1)$ is true.

There are eight ways to apply the two quantifiers to the two variables: $\forall m\, \forall n$, $\forall n\, \forall m$, $\exists m\, \exists n$, $\exists n\, \exists m$, $\forall m\, \exists n$, $\exists n\, \forall m$, $\forall n\, \exists m$, and $\exists m\, \forall n$. The first two turn out to be logically equivalent, in a precise sense that we will define in §13.2, since they have the same meaning as $\forall (m, n)\, p(m, n)$, where (m, n) varies over the new universe of discourse $\mathbb{N} \times \mathbb{N}$. Similarly, $\exists m\, \exists n\, p(m, n)$ and $\exists n\, \exists m\, p(m, n)$ are logically equivalent. The remaining four combinations must be approached carefully, as we saw in Example 2. In the present example, we have already observed that $\forall n\, \exists m\, p(m, n)$ is false. No matter what m is, $p(m, 0)$ is false, so $\forall n\, p(m, n)$ is false, and therefore $\exists m\, \forall n\, p(m,n)$ is also false. To analyze $\forall m\, \exists n\, p(m,n)$, note that, for each m, $\exists n\, p(m,n)$ is true because $p(m, m+1)$ is true. Therefore, $\forall m\, \exists n\, p(m,n)$ is also true. To analyze $\exists n\, \forall m\, p(m,n)$, note that $\forall m\, p(m,n)$ is false for each n because, for example, $p(n,n)$ is false. Therefore, $\exists n\, \forall m\, p(m,n)$ is false. We repeat:

for this example, $\forall m\, \exists n\, p(m,n)$ *is true, while* $\exists n\, \forall m\, p(m,n)$ *is false*.

The left proposition asserts, correctly, that for every m there is a bigger n. The right proposition asserts, incorrectly, that there is an n bigger than all m. ■

With these examples in mind, we turn now to a more formal account. Let U_1, $U_2, \ldots, U_n$ be nonempty sets. An **n-place predicate** over $U_1 \times U_2 \times \cdots \times U_n$ is a function $p(x_1, x_2, \ldots, x_n)$ with domain $U_1 \times U_2 \times \cdots \times U_n$ that has propositions as its function values. The variables $x_1, x_2, \ldots, x_n$ for $p(x_1, x_2, \ldots, x_n)$ are all **free variables** for the predicate, and each x_j varies over the corresponding universe of discourse U_j. The term "free" is short for "free for substitution," meaning that the variable x_j is available in case we wish to substitute a particular value from U_j for all occurrences of x_j.

If we substitute a value for x_j, say for definiteness that we substitute a for x_1 in $p(x_1, x_2, \ldots, x_n)$, we get the predicate $p(a, x_2, \ldots, x_n)$ that is free on the $n-1$ remaining variables $x_2, \ldots, x_n$, but no longer free on x_1. An application of a quantifier $\forall x_j$ or $\exists x_j$ to a predicate $p(x_1, x_2, \ldots, x_n)$ gives a predicate $\forall x_j\, p(x_1, x_2, \ldots, x_n)$ or $\exists x_j\, p(x_1, x_2, \ldots, x_n)$ whose value depends only on the values of the remaining $n-1$ variables other than x_j. We say the quantifier **binds** the variable x_j, making x_j a **bound variable** for the predicate. Application of n quantifiers, one for each variable, makes all variables bound and yields a proposition whose truth value can be determined by applying to the universes $U_1, U_2, \ldots, U_n$ the rules for $\forall x$ and $\exists x$ specified prior to Example 1.

EXAMPLE 4

(a) Consider the proposition

$$\forall m\, \exists n\, [n > 2^m]; \qquad \textbf{(1)}$$

here $p(m,n) =$ "$n > 2^m$" is a 2-place predicate over $\mathbb{N} \times \mathbb{N}$. That is, m and n are both allowed to vary over $\mathbb{N}$. Recall from our bracket-dropping convention that (1) represents

$$\forall m\, [\exists n\, [n > 2^m]].$$

Both variables m and n are bound. To decide the truth value of (1), we consider the inside expression $\exists n\, [n > 2^m]$ in which n is a bound variable and m is a free variable. We mentally fix the free variable m and note that the proposition "$n > 2^m$" is true for some choices of n in $\mathbb{N}$, for example, $n = 2^m + 1$. It follows that $\exists n\, [n > 2^m]$ is true. This thought process is valid for each m in $\mathbb{N}$, so we conclude that $\exists n\, [n > 2^m]$ is true for *all* m. That is, (1) is true.

If we reverse the quantifiers in (1), we obtain

$$\exists n\, \forall m\, [n > 2^m]. \qquad \textbf{(2)}$$

This is false because $\forall m\, [n > 2^m]$ is false for each n, since "$n > 2^m$" is false for $m = n$.

(b) Consider the propositions

$$\forall x\, \exists y\, [x + y = 0], \qquad \textbf{(3)}$$

$$\exists y\, \forall x\, [x + y = 0], \qquad \textbf{(4)}$$

$$\forall x\, \exists y\, [xy = 0], \qquad \textbf{(5)}$$

$$\exists y\, \forall x\, [xy = 0], \qquad \textbf{(6)}$$

where each universe of discourse is $\mathbb{R}$.

To analyze (3), we consider a fixed x. Then $\exists y\,[x + y = 0]$ is true, because the choice $y = -x$ makes "$x + y = 0$" true. That is, $\exists y\,[x + y = 0]$ is true for all x, so (3) is true.

To analyze (4), we consider a fixed y. Then $\forall x\,[x + y = 0]$ is not true, because many choices of x, such as $x = 1 - y$, make "$x + y = 0$" false. That is, for each y, $\forall x\,[x + y = 0]$ is false, so (4) is false.

Proposition (5) is true, because $\exists y\,[xy = 0]$ is true for all x. In fact, the choice $y = 0$ makes "$xy = 0$" true.

To deal with (6), we analyze $\forall x\,[xy = 0]$. If $y = 0$, this proposition is clearly true. Since $\forall x\,[xy = 0]$ is true for some y, namely $y = 0$, the proposition (6) is true.

In Example 3 of the next section, we will see that the truth of (6) implies the truth of (5) on purely logical grounds; that is,

$$\exists y\,\forall x\,p(x, y) \to \forall x\,\exists y\,p(x, y)$$

is always true. ■

We have already noted that an n-place predicate becomes an $(n - 1)$-place predicate when we bind one of the variables with a quantifier. Its truth value depends on the truth values of the remaining $n - 1$ free variables and, in particular, doesn't depend on what name we choose to call the bound variable. Thus, if $p(x)$ is a 1-place predicate with universe of discourse U, then $\forall x\,p(x)$, $\forall y\,p(y)$, and $\forall t\,p(t)$ all have the same truth value: true if $p(u)$ is true for every u in U and false otherwise. Similarly, if $q(x, y)$ is a 2-place predicate with universes U and V, then $\exists y\,q(x, y)$, $\exists t\,q(x, t)$, and $\exists s\,q(x, s)$ all describe the same 1-place predicate: the predicate that has truth value true for a given x in U if and only if $q(x, v)$ is true for some v in the universe V in which the second variable lies. On the other hand, the predicate $\exists x\,q(x, x)$ is *not* the same as these last three. The difference is that the quantifier in this instance binds both free variables.

EXAMPLE 5

Let U and V be $\mathbb{N}$, and let $q(m, n) =$ "$m > n$." Then $\exists m\,q(m, n)$ is the 1-place predicate "some member of $\mathbb{N}$ is greater than n," and so is $\exists k\,q(k, n)$. The predicate $\exists n\,q(m, n)$ is the 1-place predicate "there is a member of $\mathbb{N}$ less than m," which is the same predicate as $\exists j\,q(m, j)$ and has the value true for $m > 0$ and false for $m = 0$. But $\exists m\,q(m, m)$ is the proposition "$m > m$ for some m," which is quite different and has the value false. ■

Occasionally, we are confronted with a constant function in algebra, such as $f(x) = 2$ for $x \in \mathbb{R}$, in which the variable x does not appear in the right side of the definition of f. Although the value of $f(x)$ does not depend on the choice of x, we nevertheless regard f as a function of x. Similarly, in logic we occasionally encounter propositions $p(x)$ whose truth values do not depend on the choice of x in U. As artificial examples, consider $p(n) =$ "2 is prime" and $q(n) =$ "16 is prime," with universe of discourse $\mathbb{N}$. Since all the propositions $p(n)$ are true, $\exists n\,p(n)$ and $\forall n\,p(n)$ are both true. Since all the propositions $q(n)$ are false, $\exists n\,q(n)$ and $\forall n\,q(n)$ are both false. Those propositions $p(x)$ whose truth values do not depend on x are essentially the ones we studied earlier in the propositional calculus. In a sense, the propositional calculus plays the same role in the predicate calculus that the constant functions play in the study of all functions.

For the predicate calculus, we need to describe the logical expressions that we will be dealing with. In Chapter 2 we used the term "compound proposition" in an informal way to describe propositions built up out of simpler ones. In Exercise 15 on page 285 we gave a recursive definition of wff's for the propositional calculus; these are the compound propositions of Chapter 2. In the same way, we will use a recursive definition in order to precisely define "compound propositions" for the predicate calculus. Let us look more closely at what we did in the propositional calculus. Even if you have not read Chapter 7, you can follow the account below.

In Chapter 2 we used logical operators to build up propositions from the symbols p, q, r, etc., which we considered to be names for other propositions. The symbols could be considered to be variables, and a compound proposition such as $p \wedge (\neg q)$ could be considered a function of the two variables p and q; its value is true or false, depending on the truth values of p and q and the form of the compound proposition. The truth table for the compound proposition is simply a list of the function values as the variables range independently over the set {true, false}. For the propositional calculus, such variables are all that we need, but for the predicate calculus, we must also consider variables associated with infinite universes of discourse.

Suppose that we have available a collection of nonempty universes of discourse with which the free variables of all predicates we consider are associated. We define the class of **compound predicates** as follows:

(B_1) Logical variables are compound predicates.

(B_2) n-place predicates are compound predicates for $n \geq 1$.

(R_1) If P and Q are compound predicates, then so are

$$\neg P, \quad (P \vee Q), \quad (P \wedge Q), \quad (P \rightarrow Q) \quad \text{and} \quad (P \leftrightarrow Q).$$

(R_2) If $P(x)$ is a compound predicate with free variable x and possibly other free variables, then

$$(\forall x\, P(x)) \qquad \text{and} \qquad (\exists x\, P(x))$$

are compound predicates for which x is not a free variable.

If we delete (B_2) and (R_2), we obtain the recursive description of wff's for the propositional calculus in Exercise 15 on page 285. In our present setting there are compound predicates, besides those made from (B_1) and(R_1), that are propositions. If all the variables in a compound predicate are bound, then the predicate is a proposition. We extend our definition and call a compound predicate with no free variables a **compound proposition**. For example,

$$((\exists x(\exists z\, p(x, z))) \rightarrow (\forall y(\neg r(y))))$$

is a compound proposition with no free variables. In contrast,

$$(p(x) \vee (\neg \forall y\, q(x, y)))$$

and

$$((\exists z\, p(x, z)) \rightarrow (\forall y(\neg r(y))))$$

are compound predicates with free variable x.

The number and the placement of parentheses in a compound predicate are explicitly prescribed by our recursive definition. In practice, for clarity, we may add or suppress some parentheses. For example, we may write $((\forall x\, p(x)) \rightarrow (\exists x\, p(x)))$ as $\forall x\, p(x) \rightarrow \exists x\, p(x)$, and we may write $(\exists x \neg p(x))$ as $\exists x(\neg p(x))$. We sometimes also use brackets or braces instead of parentheses.

EXAMPLE 6

In Example 3 of the next section, we will encounter the compound proposition

$$\exists x\, \forall y\, p(x, y) \rightarrow \forall y\, \exists x\, p(x, y).$$

Let's verify that, with some additional parentheses, this is indeed a compound predicate. We won't ask you to make such verifications, but trust that you will recognize compound predicates when you see them.

The 2-place predicate $p(x, y)$ is a compound predicate by (B_2). Hence so are $(\forall y\, p(x, y))$ and $(\exists x\, p(x, y))$, by (R_2). By (R_2) again, $(\exists x(\forall y\, p(x, y)))$ and $(\forall y(\exists x\, p(x, y)))$ are compound predicates. Finally, by (R_1),

$$((\exists x(\forall y\, p(x, y))) \leftrightarrow (\forall y(\exists x\, p(x, y))))$$

is a compound predicate. ■

Exercises 13.1

As in Chapter 2, the truth values true and false may be written as 1 and 0, respectively.

1. Let $m(x, y)$ be as in Example 2. Write the meaning of $\exists x \forall y\, m(x, y)$ in words and determine its truth value.
2. As in Example 3, let $p(m, n)$ be the proposition "$m < n$," but with m and n varying over the universe of discourse $\mathbb{Z}$ instead of $\mathbb{N}$. Determine the truth values of the following.
   (a) $\exists n \exists m\, p(m, n)$ (b) $\forall n \forall m\, p(m, n)$
   (c) $\forall n \exists m\, p(m, n)$ (d) $\exists m \forall n\, p(m, n)$
3. Determine the truth values of the following, where the universe of discourse is $\mathbb{N}$.
   (a) $\forall m \exists n\, [2n = m]$ (b) $\exists n \forall m\, [2m = n]$
   (c) $\forall m \exists n\, [2m = n]$ (d) $\exists n \forall m\, [2n = m]$
   (e) $\forall m \forall n\, [\neg\{2n = m\}]$
4. Determine the truth values of the following, where the universe of discourse is $\mathbb{R}$.
   (a) $\forall x \exists y\, [xy = 1]$ (b) $\exists y \forall x\, [xy = 1]$
   (c) $\exists x \exists y\, [xy = 1]$
   (d) $\forall x \forall y\, [(x + y)^2 = x^2 + y^2]$
   (e) $\forall x \exists y\, [(x + y)^2 = x^2 + y^2]$
   (f) $\exists y \forall x\, [(x + y)^2 = x^2 + y^2]$
   (g) $\exists x \exists y\, [(x + 2y = 4) \wedge (2x - y = 2)]$
   (h) $\exists x \exists y\, [x^2 + y^2 + 1 = 2xy]$
5. Write the following sentences in logical notation. Be sure to bind all variables. When using quantifiers, specify universes; use $\mathbb{R}$ if no universe is indicated.
   (a) If $x < y$ and $y < z$, then $x < z$.
   (b) For every $x > 0$, there exists an n in $\mathbb{N}$ such that $n > x$ and $x > 1/n$.
   (c) For every $m, n \in \mathbb{N}$, there exists $p \in \mathbb{N}$ such that $m < p$ and $p < n$.
   (d) There exists $u \in \mathbb{N}$ so that $un = n$ for all $n \in \mathbb{N}$.
   (e) For each $n \in \mathbb{N}$, there exists $m \in \mathbb{N}$ such that $m < n$.
   (f) For every $n \in \mathbb{N}$, there exists $m \in \mathbb{N}$ such that $2^m \le n$ and $n < 2^{m+1}$.
6. Determine the truth values of the propositions in Exercise 5.
7. Write the following sentences in logical notation; the universe of discourse is the set Σ^* of words using letters from a finite alphabet Σ.
   (a) If $w_1 w_2 = w_1 w_3$, then $w_2 = w_3$.
   (b) If $\text{length}(w) = 1$, then $w \in \Sigma$.
   (c) $w_1 w_2 = w_2 w_1$ for all $w_1, w_2 \in \Sigma^*$.
8. Determine the truth values of the propositions in Exercise 7.
9. Specify the free and bound variables in the following expressions.
   (a) $\forall x \exists z\, [\sin(x + y) = \cos(z - y)]$
   (b) $\exists x\, [xy = xz \rightarrow y = z]$
   (c) $\exists x \exists z\, [x^2 + z^2 = y]$
10. Consider the expression $x + y = y + x$.
    (a) Specify the free and bound variables in the expression.
    (b) Apply universal quantifiers over the universe $\mathbb{R}$ to get a proposition. Is the proposition true?
    (c) Apply existential quantifiers over the universe $\mathbb{R}$ to get a proposition. Is the proposition true?
11. Repeat Exercise 10 for the expression $(x - y)^2 = x^2 - y^2$.
12. Consider the proposition $\forall m \exists n\, [m + n = 7]$.
    (a) Is the proposition true for the universes of discourse $\mathbb{N}$?
    (b) Is the proposition true for the universes of discourse $\mathbb{Z}$?
13. Repeat Exercise 12 for $\forall n \exists m\, [m + 1 = n]$.
14. Consider the proposition $\forall x \exists y\, [(x^2 + 1)y = 1]$.
    (a) Is the proposition true for the universes of discourse $\mathbb{N}$?
    (b) Is the proposition true for the universes of discourse $\mathbb{Q}$?
    (c) Is the proposition true for the universes of discourse $\mathbb{R}$?
15. Another useful quantifier is $\exists!$, where $\exists! x\, p(x)$ is read "there exists a unique x such that $p(x)$." This compound proposition is assigned truth value true if $p(x)$ is true for exactly one value of x in the universe of discourse; otherwise, it is false. Write the following sentences in logical notation.
    (a) There is a unique x in $\mathbb{R}$ such that $x + y = y$ for all $y \in \mathbb{R}$.
    (b) The equation $x^2 = x$ has a unique solution.
    (c) Exactly one set is a subset of all sets in $\mathcal{P}(\mathbb{N})$.
    (d) If $f: A \rightarrow B$, then for each $a \in A$ there is exactly one $b \in B$ such that $f(a) = b$.
    (e) If $f: A \rightarrow B$ is a one-to-one function, then for each $b \in B$ there is exactly one $a \in A$ such that $f(a) = b$.
16. Determine the truth values of the propositions in Exercise 15.
17. In this problem, $A = \{0, 2, 4, 6, 8, 10\}$ and the universe of discourse is $\mathbb{N}$. True or False.
    (a) A is the set of even integers in $\mathbb{N}$ less than 12.
    (b) $A = \{0, 2, 4, 6, \ldots\}$
    (c) $A = \{n \in \mathbb{N} : 2n < 24\}$
    (d) $A = \{n \in \mathbb{N} : \forall m\, [(2m = n) \rightarrow (m < 6)]\}$
    (e) $A = \{n \in \mathbb{N} : \forall m\, [(2m = n) \wedge (m < 6)]\}$
    (f) $A = \{n \in \mathbb{N} : \exists m\, [(2m = n) \rightarrow (m < 6)]\}$
    (g) $A = \{n \in \mathbb{N} : \exists m\, [(2m = n) \wedge (m < 6)]\}$
    (h) $A = \{n \in \mathbb{N} : \exists! m\, [(2m = n) \wedge (m < 6)]\}$
    (i) $A = \{n \in \mathbb{N} : n \text{ is even and } n^2 \le 100\}$
    (j) $\forall n\, [(n \in A) \rightarrow (n \le 10)]$
    (k) $(3 \in A) \rightarrow (3 < 10)$

(l) $(12 \in A) \to (12 < 10)$

(m) $(8 \in A) \to (8 < 10)$

18. With universe of discourse $\mathbb{N}$, let $p(n)$ = "n is prime" and $e(n)$ = "n is even." Write the following in ordinary English.

(a) $\exists m \, \forall n \, [e(n) \wedge p(m+n)]$

(b) $\forall n \, \exists m \, [\neg e(n) \to e(m+n)]$

Translate the following into logical notation using p and e.

(c) There are two prime integers whose sum is even.

(d) If the sum of two primes is even, then neither of them equals 2.

(e) The sum of two prime integers is odd.

19. Determine the truth values of the propositions in Exercise 18.

13.2 Elementary Predicate Calculus

One of our aims in Chapter 2 was to learn how to tell when two compound propositions were logically equivalent or when one proposition logically implied another. Questions like this become even more important, and also somewhat more difficult, when we allow ourselves to use quantifiers $\forall$ and $\exists$ to build up compound expressions. When we just had $\neg$, $\wedge$, $\vee$, $\to$, and $\leftrightarrow$ to deal with, we used truth tables to break our analysis into small steps. To get a similar breakdown for predicate logic, we need to determine the rules for how $\forall$ and $\exists$ interact with each other and with $\neg$, $\wedge$, $\vee$, $\to$, and $\leftrightarrow$.

The ideas of "proof" and "rule of inference" that we discussed in §2.5 for the propositional calculus can also be extended to the predicate calculus setting. Not surprisingly, more possible expressions means more complications. A moderately thorough account of the subject would form a substantial part of another book. In this section we will just discuss some of the most basic and useful connections between quantifiers and logical operators. Even such a brief account will be useful, though. The fact is that people commonly make serious mistakes where quantifiers, often implicit, are mixed with negation, so whatever we can do to avoid such blunders will be all to the good.

EXAMPLE 1

When we hear the television advertiser say, "All cars are not created equal," we know that what he *really* means is "Not all cars are created equal." The distinction would be even more obvious if we were comparing "All cars are not yellow" with "Not all cars are yellow." The first of these two statements is false—there *are* some yellow cars—but the second is true. The statements have different truth values. Given a universe of cars, if $y(c)$ is the predicate "c is yellow," then the first statement is $\forall c(\neg\, y(c))$, while the second is $\neg\,(\forall c \; y(c))$.

These are simple examples of predicates, in this case propositions, built using the quantifier $\forall$ and the logical operator $\neg$. More complicated predicates come up frequently in debugging computer programs that have branching instructions, to see how the programs will behave with different kinds of input data.

Our simple predicates do not have the same truth value for the universe of cars and the particular interpretation "c is yellow" for $y(c)$. Could they be logically equivalent if we had a different universe or a different meaning for $y(c)$? This is one kind of general question that we plan to answer. See Example 4(b) for a continuation of this discussion. ■

The truth value of a compound proposition ordinarily depends on the choices of the universes of discourse that the bound variables are quantified over, but there are important instances in which a compound proposition has the value true for all universes of discourse. Such a proposition is called a **tautology**. This definition extends the usage in Chapter 2, where there were no universes to worry about.

EXAMPLE 2

(a) An important class of tautologies consists of the generalized De Morgan laws; compare rules 8a–d in Table 1 on page 62. These are

$$\neg \forall x\, p(x) \leftrightarrow \exists x\, [\neg p(x)], \quad \textbf{(1)}$$

$$\neg \exists x\, p(x) \leftrightarrow \forall x\, [\neg p(x)], \quad \textbf{(2)}$$

$$\forall x\, p(x) \leftrightarrow \neg \exists x\, [\neg p(x)], \quad \textbf{(3)}$$

$$\exists x\, p(x) \leftrightarrow \neg \forall x\, [\neg p(x)]. \quad \textbf{(4)}$$

Let's think about these. To see that (1) is a tautology, note that $\neg \forall x\, p(x)$ has truth value true exactly when $\forall x\, p(x)$ has truth value false, and this occurs whenever there exists an x in the universe of discourse for which $p(x)$ is false, i.e., for which $\neg p(x)$ is true. Thus $\neg \forall x\, p(x)$ is true precisely when $\exists x\, [\neg p(x)]$ is true. This argument does not rely on the choice of universe, so (1) is a tautology.

De Morgan law (2) can be analyzed in a similar way. Alternatively, we can derive (2) from (1) by substituting the 1-place predicate $\neg p(x)$ in place of $p(x)$ to obtain

$$\neg \forall x\, [\neg p(x)] \leftrightarrow \exists x\, [\neg \neg p(x)].$$

The substitution rules in §2.5 are still valid, so we may substitute $p(x)$ for $\neg \neg p(x)$ and obtain the equivalent expression

$$\neg \forall x\, [\neg p(x)] \leftrightarrow \exists x\, [p(x)].$$

This is De Morgan's law (4) and, if we negate both sides, we obtain (2). An application of (2) to $\neg p(x)$ yields (3).

(b) Consider again the predicate $y(c)$ = "c is yellow," where c ranges over the universe of cars. De Morgan law (1) tells us that

$$\neg (\forall c\, y(c)) \leftrightarrow \exists c(\neg y(c))$$

is a tautology. We conclude that $\neg (\forall c\, y(c))$ and $\exists c(\neg y(c))$ must have the same truth value on purely logical grounds; we do not need to consider the context of cars. In Example 1 we decided that $\neg (\forall c\, y(c))$, i.e., "Not all cars are yellow," is true, so $\exists c(\neg y(c))$ must also be true. Sure enough, it's true that "There exists a car that is not yellow." ■

EXAMPLE 3

(a) The following compound predicate is true for every 2-place predicate $p(x, y)$:

$$\exists x\, \forall y\, p(x, y) \rightarrow \forall y\, \exists x\, p(x, y). \quad (*)$$

In other words, $(*)$ is a tautology. To see this, suppose that $\exists x\, \forall y\, p(x, y)$ has truth value true. Then there exists an x_0 in the universe of discourse such that $\forall y\, p(x_0, y)$ is true, and so $p(x_0, y)$ is true for all y. Thus, for each y, $\exists x\, p(x, y)$ is true; in fact, the same x_0 works for each y. Since $\exists x\, p(x, y)$ is true for all y, the right side of $(*)$ has truth value true. Since the right side is true whenever the left side is, $(*)$ is a tautology.

(b) The converse of $(*)$,

$$\forall y\, \exists x\, p(x, y) \rightarrow \exists x\, \forall y\, p(x, y),$$

is not a tautology, as we noted in Examples 2 and 3 on pages 516 and 517.

Here is another simple example. Let $p(x, y)$ be the 2-place predicate "$x = y$" on the two-element universe $U = \{a, b\}$. Observe that $\exists x\, p(x, a)$ is true, since $p(x, a)$ is true for $x = a$. Similarly, $\exists x\, p(x, b)$ is true, and so $\forall y\, \exists x\, p(x, y)$ is true. On the other hand, as noted in the proof of $(*)$, $\exists x\, \forall y\, p(x, y)$ is true only if $\forall y\, p(x_0, y)$ is true for some x_0. Since x_0 must be a or b, either $\forall y\, p(a, y)$ or

TABLE 1 Logical Relationships in the Predicate Calculus

35a.	$\forall x\, \forall y\, p(x, y) \Longleftrightarrow \forall y\, \forall x\, p(x, y)$	
b.	$\exists x\, \exists y\, p(x, y) \Longleftrightarrow \exists y\, \exists x\, p(x, y)$	
36.	$\exists x\, \forall y\, p(x, y) \Longrightarrow \forall y\, \exists x\, p(x, y)$	
37a.	$\neg \forall x\, p(x) \Longleftrightarrow \exists x\, [\neg p(x)]$	De Morgan laws
b.	$\neg \exists x\, p(x) \Longleftrightarrow \forall x\, [\neg p(x)]$	De Morgan laws
c.	$\forall x\, p(x) \Longleftrightarrow \neg \exists x\, [\neg p(x)]$	De Morgan laws
d.	$\exists x\, p(x) \Longleftrightarrow \neg \forall x\, [\neg p(x)]$	De Morgan laws
38.	$\forall x\, p(x) \Longrightarrow \exists x\, p(x)$	

$\forall y\, p(b, y)$ would be true. But $\forall y\, p(a, y)$ is false, since $p(a, y)$ is false for $y = b$ and, similarly, $\forall y\, p(b, y)$ is false. Thus, in this setting, the proposition $\forall y\, \exists x\, p(x, y) \to \exists x\, \forall y\, p(x, y)$ is false. Hence this compound proposition is not a tautology. ■

As in the propositional calculus, we say that two compound propositions P and Q are **logically equivalent**, and we write $\boldsymbol{P \Longleftrightarrow Q}$, in case $P \leftrightarrow Q$ is a tautology. Also, P **logically implies** Q provided $P \to Q$ is a tautology, in which case we write $\boldsymbol{P \Longrightarrow Q}$. Table 1 lists some useful logical equivalences and implications. We begin numbering the rules with 35, since Chapter 2 contains rules 1 through 34.

In Examples 2 and 3 we discussed the tautologies corresponding to rules 37 and 36. The remaining rules are easy to verify.

EXAMPLE 4

(a) To verify rule 35b, that is, to verify that

$$\exists x\, \exists y\, p(x, y) \leftrightarrow \exists y\, \exists x\, p(x, y)$$

is a tautology, we must check that this proposition has the value true for all possible universes of discourse. By the definition of $\leftrightarrow$, we need only check that $\exists x\, \exists y\, p(x, y)$ has the value true for a given universe if and only if $\exists y\, \exists x\, p(x, y)$ has the value true for that universe. Suppose $\exists x\, \exists y\, p(x, y)$ is true. Then $\exists y\, p(x_0, y)$ is true for some x_0 in the universe, so $p(x_0, y_0)$ is true for some y_0 in the universe. Hence $\exists x\, p(x, y_0)$ is true and thus $\exists y\, \exists x\, p(x, y)$ is true. The implication in the other direction follows similarly. Moreover, both $\exists x\, \exists y\, p(x, y)$ and $\exists y\, \exists x\, p(x, y)$ are logically equivalent to the proposition $\exists (x, y)\, p(x, y)$, where (x, y) varies over $U_1 \times U_2$, with U_1 and U_2 the universes of discourse for the variables x and y.

(b) Rule 38, applied to $\neg p(x)$ in place of $p(x)$, gives

$$\forall x \neg p(x) \Longrightarrow \exists x \neg p(x).$$

Then De Morgan law 37a applied to $\exists x \neg p(x)$ gives

$$\forall x \neg p(x) \Longrightarrow \neg \forall x\, p(x).$$

It is worth emphasizing that as we saw in Example 1, the reverse implication is false. ■

De Morgan laws 37a to 37d can be used repeatedly to negate any quantified proposition. For example,

$$\neg \exists w\, \forall x\, \exists y\, \exists z\, p(w, x, y, z)$$

is successively logically equivalent to

$$\forall w\, [\neg \forall x\, \exists y\, \exists z\, p(w, x, y, z)],$$

$$\forall w\, \exists x\, [\neg \exists y\, \exists z\, p(w, x, y, z)],$$

$$\forall w\, \exists x\, \forall y\, [\neg \exists z\, p(w, x, y, z)],$$

$$\forall w\, \exists x\, \forall y\, \forall z\, [\neg p(w, x, y, z)].$$

This example illustrates the general rule: The negation of a quantified predicate is logically equivalent to the proposition obtained by replacing each $\forall$ by $\exists$, replacing each $\exists$ by $\forall$, and replacing the predicate itself by its negation.

EXAMPLE 5

(a) The proposition

$$\forall x\, \forall y\, \exists z\, [x < z < y] \qquad \textbf{(1)}$$

states that for every x and y there is a z between them. Its truth value, which we really don't care about here, depends on the choices of the universes for x, y, and z [see Exercise 12]. The negation of (1) is

$$\exists x\, \exists y\, \forall z \{\neg [x < z < y]\}.$$

Since "$x < z < y$" means "$(x < z) \wedge (z < y)$," we can apply an elementary De Morgan law from Table 1 on page 62 to get

$$\neg [x < z < y] \Longleftrightarrow \neg (x < z) \vee \neg (z < y) \Longleftrightarrow (x \geq z) \vee (z \geq y).$$

Hence the negation of (1) is logically equivalent to

$$\exists x\, \exists y\, \forall z\, [(z \leq x) \vee (z \geq y)].$$

This states that for some choice of x_0 and y_0, every z is either less than or equal to x_0 or else bigger than or equal to y_0.

(b) Here is an example from calculus, but you don't need to know calculus to follow the example. By definition, a sequence $a_1, a_2, \ldots$ of real numbers has the real number L as a **limit**, and we write $\lim_{n\to\infty} a_n = L$, in case no matter how close we want to get to L we can get that close and stay that close by going far enough along in the sequence. In mathematical terms, $\lim_{n\to\infty} a_n = L$ means that for every $\epsilon > 0$ [ϵ measures how close we want to be to L, so ϵ could be very near 0] there is an N in $\mathbb{P}$ [telling how far out is far enough] such that $|a_n - L| < \epsilon$ [a_n is close enough to L, i.e., within ϵ of L] whenever $n \geq N$. Symbolically, this is

$$\forall \epsilon\, \exists N\, \forall n\, [n \geq N \to |a_n - L| < \epsilon],$$

with universes $(0, \infty)$ for ϵ and $\mathbb{P}$ for N and n. The sequence $a_1, a_2, \ldots$ has a limit in case $\lim_{n\to\infty} a_n = L$ for some L in $\mathbb{R}$, i.e., in case

$$\exists L\, \forall \epsilon\, \exists N\, \forall n\, [n \geq N \to |a_n - L| < \epsilon]$$

is true.

As you can imagine, sometimes one wants to say that a sequence does *not* have a limit. Using our De Morgan laws and the fact that $\neg(p \to q)$ is logically equivalent to $p \wedge \neg q$ by rule 10a on page 62, we can handle this negation. The compound proposition

$$\neg \exists L\, \forall \epsilon\, \exists N\, \forall n\, [n \geq N \to |a_n - L| < \epsilon]$$

is logically equivalent to

$$\forall L\, \exists \epsilon\, \forall N\, \exists n\, [\neg(n \geq N \to |a_n - L| < \epsilon)]$$

and hence to

$$\forall L\, \exists \epsilon\, \forall N\, \exists n\, [n \geq N \wedge |a_n - L| \geq \epsilon].$$

That is, no matter what L you try, there is a degree of closeness such that no matter how far out you go there will always be at least one term in the sequence that is not close enough to L. No wonder calculus seems hard sometimes. ■

EXAMPLE 6

Let the universe of discourse U consist of two members, a and b. De Morgan law 37a then becomes

$$\neg[p(a) \wedge p(b)] \Longleftrightarrow [\neg p(a)] \vee [\neg p(b)].$$

Except for the names $p(a)$ and $p(b)$, in place of p and q, this is De Morgan law 8b in Table 1 on page 62. ■

A general proposition often has the form $\forall x\, p(x)$, where x ranges over some universe of discourse. This proposition is false if and only if $\exists x\, [\neg p(x)]$ is true, by De Morgan's law 37a. Thus $\forall x\, p(x)$ is false if some x_0 can be exhibited for which $p(x_0)$ is false. As we pointed out after Example 12 on page 55, such an x_0 is called a **counterexample** to the proposition $\forall x\, p(x)$. Some illustrations are given in Example 13 on page 56. Here are a couple more.

EXAMPLE 7

(a) The matrices

$$\begin{bmatrix} 1 & 0 \\ 1 & 0 \end{bmatrix}, \quad \begin{bmatrix} 0 & 0 \\ 1 & 1 \end{bmatrix}$$

provide a counterexample to the [false] assertion "If 2×2 matrices $\mathbf{A}$ and $\mathbf{B}$ satisfy $\mathbf{AB} = \mathbf{0}$, then $\mathbf{A} = \mathbf{0}$ or $\mathbf{B} = \mathbf{0}$." This general assertion could have been written

$$\forall \mathbf{A}\, \forall \mathbf{B}[\mathbf{AB} = \mathbf{0} \rightarrow (\mathbf{A} = \mathbf{0}) \vee (\mathbf{B} = \mathbf{0})].$$

(b) The proposition "Every connected graph has an Euler circuit" is false. In view of Euler's theorem on page 234, any connected graph having a vertex of odd degree will serve as a counterexample to this assertion. The simplest counterexample has two vertices with one edge connecting them. ■

It is worth mentioning that implication 38 in Table 1 and its equivalent versions in Example 4(b) are valid because we have restricted our attention to nonempty universes of discourse. If we were to allow the empty universe of discourse, then implication 38 would be false. In that case, $\forall x\, p(x)$ would have the value true vacuously, whereas $\exists x\, p(x)$ would be false. It is true that everyone with three heads is rich. [You disagree? Give a counterexample.] But it is not true that there is a rich person with three heads. Here the universe consists of all three-headed people and $p(x)$ denotes "x is rich."

The second implication in Example 4(b) also fails for the empty universe, but it is a bit slippery to analyze. It is true that every three-headed person is not rich, but false that not every three-headed person is rich. Three-headed people have amazing properties.

Exercises 13.2

1. Consider a universe U_1 consisting of members of a club and a universe U_2 of airlines. Let $p(x, y)$ be the predicate "x has been a passenger on y" or equivalently "y has had x as a passenger." Write out the meanings of the following.

   (a) rule 35a (b) rule 35b (c) rule 36

2. Consider the universe U of all college students. Let $p(x)$ be the predicate "x likes broccoli."

   (a) Express the proposition "not all college students like broccoli" in predicate-calculus symbols.

   (b) Do the same for "every college student does not like broccoli."

   (c) Does either of the propositions in part (a) or (b) imply the other? Explain.

   (d) Write out the meaning of rule 37b for this U and $p(x)$.

   (e) Do the same for rule 37d.

3. Interpret De Morgan laws 37b, 37c, and 37d for the predicate $y(c)$ = "c is yellow," where c ranges over the universe of cars.

4. Let $p(m, n)$ be the predicate "$m \neq n$" on the domain of discourse $\mathbb{N}$. Give the truth value of each of the following.

   (a) $\forall n\, \exists m\, p(m, n) \rightarrow \exists m\, \forall n\, p(m, n)$

   (b) $\neg\, \forall m\, \exists n\, p(m, n) \rightarrow \neg\, \exists n\, \forall m\, p(m, n)$

(c) $\forall n\, p(m, n)$

(d) $\forall m\, [\neg \forall n\, p(m, n) \vee \exists n\, p(m, n)]$

5. Show that the following rules in Table 1 collapse to rules from Table 1 on page 62 when the universe of discourse U has two elements, a and b.

(a) rule 37d (b) rule 37b

6. (a) Show that the logical implication

$$[\exists x\, p(x)] \wedge [\exists x\, q(x)] \Longrightarrow \exists x\, [p(x) \wedge q(x)]$$

is false. You may do this by defining predicates $p(x)$ and $q(x)$ where this implication fails.

(b) Do the same for the logical implication

$$\exists x\, \forall y\, p(x, y) \Longrightarrow \forall x\, \exists y\, p(x, y).$$

[Compare this with the true implication of rule 36.]

7. For the universe of discourse $\mathbb{N}$, write the negation of $\forall n\, [p(n) \rightarrow p(n + 1)]$ without using the quantifier $\forall$.

8. Write the negation of $\exists x\, \forall y\, \exists z\, [(z > y) \rightarrow (z < x^2)]$ without using the connective $\neg$.

9. (a) Write the negation of

$$P = \forall x\, \forall y\, [(x < y) \rightarrow \exists z\{x < z < y\}]$$

without using the connective $\neg$.

(b) Determine the truth value of P when the universe of discourse for x, y, and z is $\mathbb{R}$ or $\mathbb{Q}$.

(c) Determine the truth value of P when the universe of discourse is $\mathbb{N}$ or $\mathbb{Z}$.

10. Give a counterexample for each of the following assertions.

(a) Every even integer is the product of two even integers.

(b) $|S \cup T| = |S| + |T|$ for every two finite sets S and T.

(c) Every positive integer of the form $6k - 1$ is a prime.

(d) Every graph has an even number of edges.

(e) All mathematics courses are fun.

11. The statement "There are arbitrarily large integers n such that $p(n)$ is true" translates into the proposition

$$\forall N\, \exists n\, [(n \geq N) \wedge p(n)]$$

with universe of discourse $\mathbb{P}$. Write the negation of this proposition using the connective $\rightarrow$, but without using the connective $\neg$. Your answer should translate into a statement that implies that $p(n)$ is true for only a finite set of n's.

12. (a) Choose universes of discourse for x, y, and z so that proposition (1) in Example 5 is true.

(b) Choose universes of discourse so that proposition (1) in Example 5 is false.

13. Our definition of compound predicate does not permit expressions such as $\exists x\, p(x, x)$, for 2-place predicates $p(x, y)$. Describe a predicate $q(x, y)$ such that

$$\exists x\, \exists y\, [p(x, y) \wedge q(x, y)]$$

is true if and only if $p(x, x)$ is true for some x.

14. In the case that the universe of discourse is empty, $\forall x\, p(x)$ vacuously has the value true regardless of $p(x)$, and $\exists x\, p(x)$ is false. Describe the situation for a universe with exactly one member.

13.3 Infinite Sets

Are there more rational numbers than there are integers? How about real numbers; are there more of them than there are of rationals? Are there fewer numbers in the interval $(0, 1)$ than in $(0, 2)$ or than in $\mathbb{R}$ itself? Mathematicians would say that the answers to these questions are no, yes, and no, respectively, but what do these answers—and the questions—mean? Read on, and leave your intuition behind; it won't help much here. The subject is fascinating, and we will just scratch the surface in this section.

Counting infinite sets may look like something only a theorist could love to do, but there is a practical side to the subject as well. If we are given some set S and we would like to examine its members with an algorithm, say with a "**`for s in S do`**" loop, then it's important to know whether the members of S can be listed in a sequence. If they can't, then no sequential algorithm can examine them all, and we should not look for one. The concept we'll need here, and which we will discuss further, is called "countability."

When we counted finite sets in Chapter 5, we knew what we meant by the size of a set—we could just count its elements—and we knew when two sets had, or didn't have, the same size. Counting infinite sets is another matter. The idea of enumerating the elements until we come to the end is not going to work, but we can still hope to compare sizes. The clue to the commonly accepted correct approach is the following elementary observation: Two finite sets are of the same size if and only if there exists a one-to-one correspondence between them. Following this guide, we define two sets S and T, finite or infinite, to be of the **same size** if there is a one-to-one correspondence between them. In this book we will not study the classification scheme for sets in detail, but we will distinguish between two kinds of infinite sets.

Sets that we can match up with $\{1, 2, \dots, n\}$ or with $\mathbb{P} = \{1, 2, \dots\}$ are especially important for both theoretical and practical reasons. The ones that match up with $\{1, 2, \dots, n\}$ are the finite sets of size n. Sets that are the same size as $\mathbb{P}$ are called **countably infinite**. Thus a set S is countably infinite if and only if there exists a one-to-one correspondence between $\mathbb{P}$ and S. A set is **countable** if it is finite or is countably infinite. One is able to count or list such a nonempty set by matching it with $\{1, 2, \dots, n\}$ for some $n \in \mathbb{P}$, or with the whole set $\mathbb{P}$. In the infinite case, the list will never end. As one would expect, a set is **uncountable** if it is not countable.

EXAMPLE 1

(a) The set $\mathbb{N}$ is countably infinite because $f(n) = n - 1$ defines a one-to-one function f mapping $\mathbb{P}$ onto $\mathbb{N}$. Its inverse f^{-1} is a one-to-one mapping of $\mathbb{N}$ onto $\mathbb{P}$; note that $f^{-1}(n) = n + 1$ for $n \in \mathbb{N}$. Even though $\mathbb{P}$ is a proper subset of $\mathbb{N}$, by our definition $\mathbb{P}$ is the same size as $\mathbb{N}$. This may be surprising, since a similar situation does not occur for finite sets. Oh well, $\mathbb{N}$ has only one element that is not in $\mathbb{P}$.

Figure 1 ▶

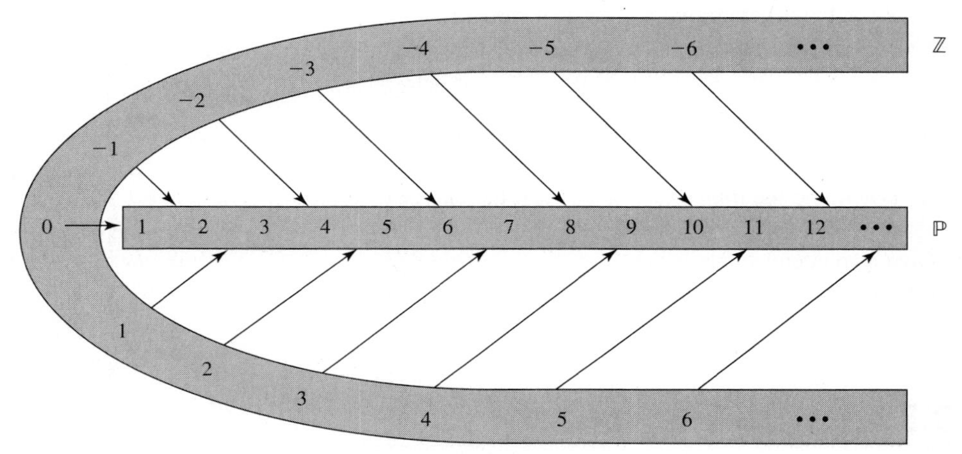

A one-to-one correspondence of $\mathbb{Z}$ onto $\mathbb{P}$

(b) The set $\mathbb{Z}$ of *all* integers is also countably infinite. Figure 1 shows a one-to-one function f from $\mathbb{Z}$ onto $\mathbb{P}$. We have found it convenient to bend the picture of $\mathbb{Z}$. This function can be given by a formula, if desired:

$$f(n) = \begin{cases} 2n + 1 & \text{for } n \geq 0 \\ -2n & \text{for } n < 0 \end{cases}.$$

Even though $\mathbb{Z}$ looks about twice as big as $\mathbb{P}$, these sets are of the same size. Beware! For infinite sets, your intuition may be unreliable. Or, to take a more positive approach, you may need to refine your intuition when dealing with infinite sets.

(c) Even the set $\mathbb{Q}$ of all rational numbers is countably infinite. This fact is striking, because the set of rational numbers is distributed evenly throughout $\mathbb{R}$. To give a one-to-one correspondence between $\mathbb{P}$ and $\mathbb{Q}$, a picture is worth a thousand formulas. See Figure 2. The function f is obtained by following the arrows and skipping over repetitions. We have $f(1) = 0$, $f(2) = 1$, $f(3) = \frac{1}{2}$, $f(4) = -\frac{1}{2}$, $f(5) = -1$, $f(6) = -2$, $f(7) = -\frac{2}{3}$, etc. Thus, at least in principle, it is possible to design a looping algorithm that goes through the whole set of rational numbers, one by one. No simple formula gives the nth term in our listing from Figure 2, but Exercise 16 describes a clever way to list $\mathbb{Q}$ that could be used as a basis for an algorithm. ■

Figure 2 ▶

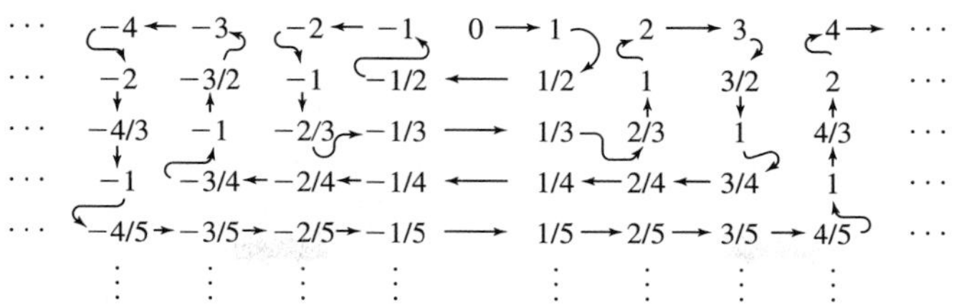

... −4 ← −3 −2 ← −1 0 → 1 2 → 3 4 → ...
... −2 −3/2 −1 −1/2 ← 1/2 1 3/2 2 ...
... −4/3 −1 −2/3 → −1/3 → 1/3 2/3 1 4/3 ...
... −1 −3/4 ← −2/4 ← −1/4 ← 1/4 ← 2/4 ← 3/4 1 ...
... −4/5 → −3/5 → −2/5 → −1/5 → 1/5 → 2/5 → 3/5 → 4/5 ...

EXAMPLE 2

Almost all of our examples of graphs and trees have had finitely many vertices and edges. However, there is no such restriction in the general definitions. Figure 3 contains partial pictures of some infinite graphs. The set of vertices in Figure 3(a) is $\mathbb{Z}$, and only consecutive integers are connected by an edge. Note that this infinite tree has *no* leaves, whereas every finite tree with more than one vertex has leaves. The set of vertices in Figure 3(b) is $\mathbb{Z} \times \{0, 1\}$; this tree has infinitely many leaves. The central vertex in the tree in Figure 3(c) has infinite degree; all the other vertices are leaves. There are only two vertices in the graph of Figure 3(d), but they are connected by infinitely many edges.

Figure 3 ▶

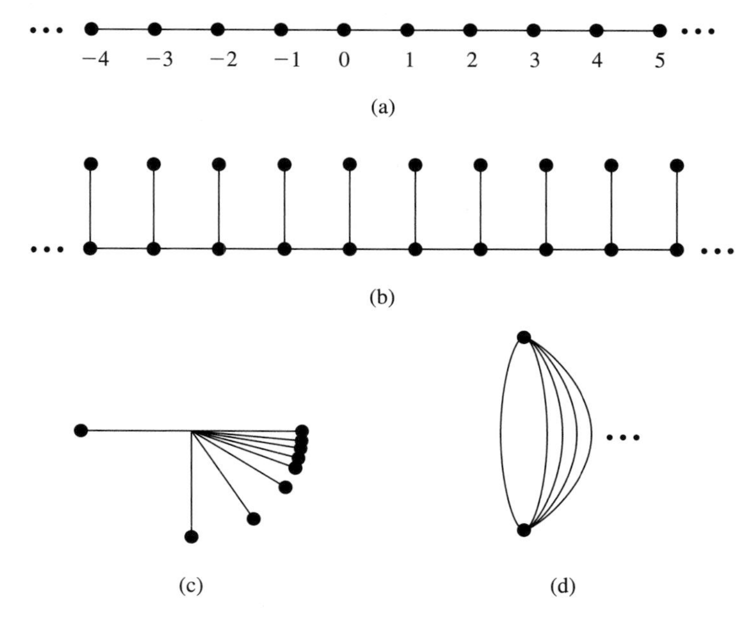

In all these examples, the sets of vertices and edges are countable. Graphs don't have to be countable, but it is hard to draw or visualize uncountable ones. ■

There *are* sets that are uncountable, i.e., not the same size as $\mathbb{P}$. Our next example gives two illustrations of a technique for showing that two sets are *not* the same size. Called "Cantor's diagonal procedure," the idea goes back to Georg Cantor, the father of set theory. You may find the result in part (b) more interesting, but the details in part (a) are easier to follow.

EXAMPLE 3

(a) We claim that the set $\text{FUN}(\mathbb{P}, \{0, 1\})$ of all functions from $\mathbb{P}$ into $\{0, 1\}$ is uncountable. Equivalently, the set of all infinite strings of 0's and 1's is uncountable. Obviously, $\text{FUN}(\mathbb{P}, \{0, 1\})$ is infinite, so if it were countable, there would exist an infinite listing $\{f_1, f_2, \ldots\}$ of *all* the functions in this set. We define a function f^* on $\mathbb{P}$ as follows:

$$f^*(n) = \begin{cases} 0 & \text{if } f_n(n) = 1 \\ 1 & \text{if } f_n(n) = 0 \end{cases} .$$

For each n in $\mathbb{P}$, $f^*(n) \neq f_n(n)$ by construction, so the function f^* must be different from every f_n. Thus $\{f_1, f_2, \ldots\}$ is not a listing of *all* functions in $\text{FUN}(\mathbb{P}, \{0, 1\})$. This contradiction shows that $\text{FUN}(\mathbb{P}, \{0, 1\})$ is uncountable.

(b) The interval $[0, 1)$ is uncountable. If it were countable, there would exist a one-to-one function f mapping $\mathbb{P}$ onto $[0, 1)$. We will show that this is impossible. Each number in $[0, 1)$ has a decimal expansion $.d_1d_2d_3\cdots$, where each d_j is a digit in $\{0, 1, 2, 3, 4, 5, 6, 7, 8, 9\}$. In particular, each number $f(k)$ has the form $.d_{1k}d_{2k}d_{3k}\cdots$; here d_{nk} represents the nth digit in $f(k)$. Consider Figure 4 and look at the indicated diagonal digits $d_{11}, d_{22}, d_{33}, \ldots$. We define the sequence d^*, whose nth term d_n^* is constructed as follows: if $d_{nn} \neq 1$, let $d_n^* = 1$, and if $d_{nn} = 1$, let $d_n^* = 2$. The point is that $d_n^* \neq d_{nn}$ for all $n \in \mathbb{P}$. Now $.d_1^*d_2^*d_3^*\cdots$ is a decimal expansion for a number a in $[0, 1)$ that is different from $f(n)$ in the nth digit for each $n \in \mathbb{P}$. Thus a cannot be one of the numbers $f(n)$; i.e., a is not in $\text{Im}(f)$, so f does not map $\mathbb{P}$ onto $[0, 1)$. Thus $[0, 1)$ is uncountable.

Note that we arranged for all the digits of a to be 1's and 2's. This choice was quite arbitrary, except that we deliberately avoided 0's and 9's, since some numbers, ones whose expansions involve strings of 0's and 9's, have two decimal expansions. For example, $.250000\cdots$ and $.249999\cdots$ represent the same number in $[0, 1)$. ■

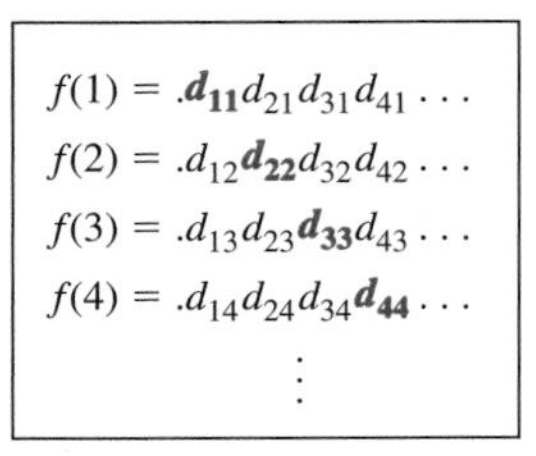

Figure 4 ▲

The proof in Example 3(b) can be modified to prove that $\mathbb{R}$ and $(0, 1)$ are uncountable; in fact, all intervals $[a, b]$, $[a, b)$, $(a, b]$, and (a, b) are uncountable for $a < b$. In view of Exercise 9, another way to show that these sets are uncountable is to show that they are in one-to-one correspondence with each other. In fact, they are also in one-to-one correspondence with unbounded intervals. Showing the existence of such one-to-one correspondences can be challenging. We provide a couple of the trickier arguments in the next example and ask for some easier ones in Exercise 3.

EXAMPLE 4

(a) It is easy to show that $(0, 1)$ and $(0, 2)$ are the same size; the function f defined by $f(x) = 2x$ gives a one-to-one correspondence from $(0, 1)$ onto $(0, 2)$. More generally, the linear function $f(x) = ax + b$ with $a > 0$ maps $(0, 1)$ one-to-one onto $(b, a + b)$.

(b) In fact, $\mathbb{R}$ is the same size as $(0, 1)$. The function $f(x) = 2^x/(1 + 2^x)$ whose graph is shown in Figure 5 provides a one-to-one correspondence from $\mathbb{R}$ onto $(0, 1)$, as would any other function whose graph had this same general shape.

Figure 5 ▶

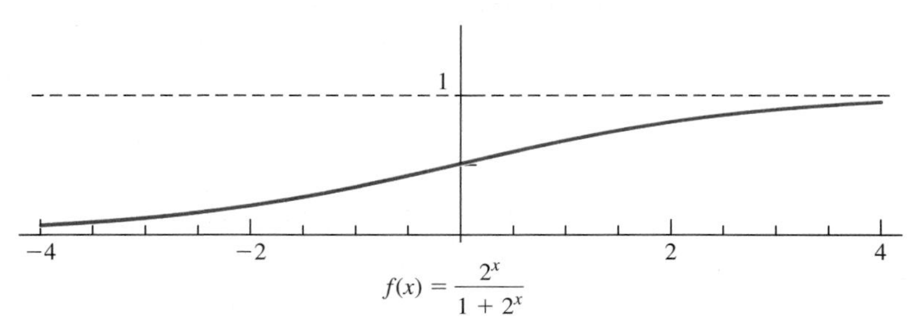

(c) We can show that $[0, 1)$ and $(0, 1)$ have the same size. No simple formula provides us with a one-to-one mapping between these sets. The trick is to isolate some infinite sequence in $(0, 1)$, say $\frac{1}{2}, \frac{1}{3}, \frac{1}{4}, \ldots$, and then map this sequence onto $0, \frac{1}{2}, \frac{1}{3}, \frac{1}{4}, \ldots$, while leaving the complement fixed. That is, let

$$C = (0, 1) \setminus \left\{ \frac{1}{n} : n = 2, 3, 4, \ldots \right\}$$

and define

$$f(x) = \begin{cases} 0 & \text{if } x = \dfrac{1}{2} \\ \dfrac{1}{n-1} & \text{if } x = 1/n \text{ for some integer } n \geq 3. \\ x & \text{if } x \in C \end{cases}$$

See Figure 6. ∎

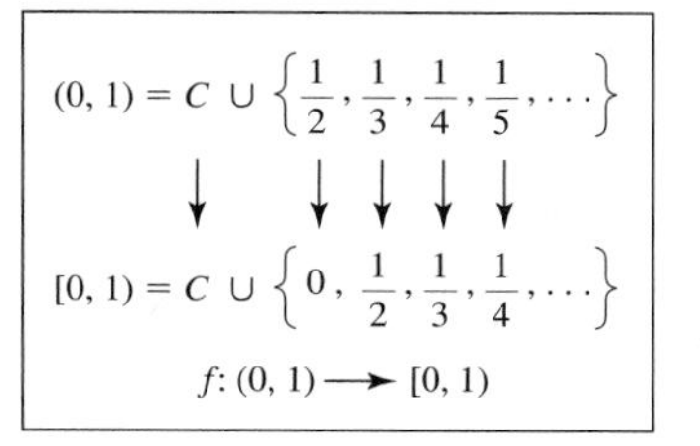

Figure 6 ▲

Knowing that a set is countable can be important. It would be truly annoying if we had to construct a new correspondence every time we had a new set that we wanted to show was countable. Fortunately, two basic facts often make the task much easier.

Theorem

(a) Subsets of countable sets are countable.

(b) The union of countably many countable sets is countable.

Proof

(a) It is enough to show that subsets of $\mathbb{P}$ are countable. Consider a subset A of $\mathbb{P}$. Clearly, A is countable if A is finite. Suppose that A is infinite. We will use the Well-Ordering Principle on page 131. Define $f(1)$ to be the least element in A. Then define $f(2)$ to be the least element in $A \setminus \{f(1)\}$, $f(3)$ to be the least element in $A \setminus \{f(1), f(2)\}$, etc. Continue this process so that $f(n+1)$ is the least element in the nonempty set $A \setminus \{f(k) : 1 \leq k \leq n\}$ for each $n \in \mathbb{P}$. It is easy to verify that this recursive definition provides a one-to-one function f mapping $\mathbb{P}$ onto A [Exercise 10], so A is countable.

(b) The statement in part (b) means that, if I is a countable set and if $\{A_i : i \in I\}$ is a family of countable sets, then the union $\bigcup_{i \in I} A_i$ is countable. We may assume that each A_i is nonempty and that $\bigcup_{i \in I} A_i$ is infinite, and we may assume that $I = \mathbb{P}$ or that I has the form $\{1, 2, \ldots, n\}$. If $I = \{1, 2, \ldots, n\}$, we can define $A_i = A_n$ for $i > n$ and obtain a family $\{A_i : i \in \mathbb{P}\}$ with the same union. Thus we may assume that $I = \mathbb{P}$. Each set A_i is finite or countably infinite. By repeating elements if A_i is finite, we can list each A_i as follows:

$$A_i = \{a_{1i}, a_{2i}, a_{3i}, a_{4i}, \ldots\}.$$

The elements in $\bigcup_{i \in I} A_i$ can be listed in an array as in Figure 7. The arrows in the figure suggest a single listing for $\bigcup_{i \in I} A_i$:

$$a_{11}, a_{12}, a_{21}, a_{31}, a_{22}, a_{13}, a_{14}, a_{23}, a_{32}, a_{41}, \ldots. \qquad (*)$$

Some elements may be repeated, but the list includes infinitely many distinct elements, since $\bigcup_{i \in I} A_i$ is infinite. Now a one-to-one mapping f of $\mathbb{P}$ onto $\bigcup_{i \in I} A_i$ is obtained as follows: $f(1) = a_{11}$, $f(2)$ is the next element listed in $(*)$ different from $f(1)$, $f(3)$ is the next element listed in $(*)$ different from $f(1)$ and $f(2)$, etc. ∎

Figure 7 ▶

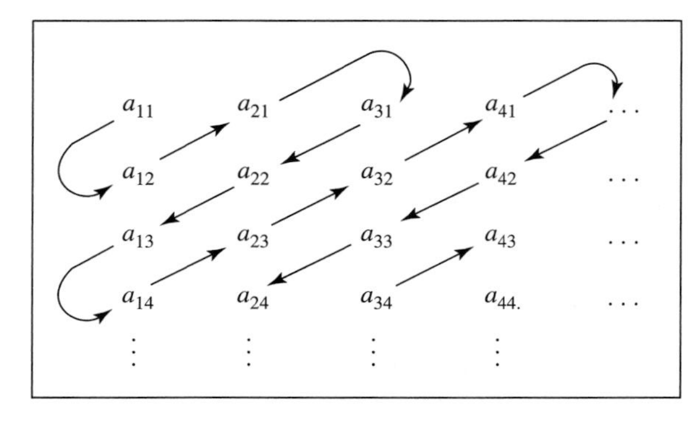

EXAMPLE 5

The argument in Example 1(c), which shows that $\mathbb{Q}$ is countable, is similar to the proof of part (b) of the theorem. In fact, we can use the theorem to give another proof that $\mathbb{Q}$ is countable. For each n in $\mathbb{P}$, let

$$A_n = \left\{\frac{m}{n} : m \in \mathbb{Z}\right\}.$$

Thus A_n consists of all integer multiples of $1/n$. Each A_n is clearly in one-to-one correspondence with $\mathbb{Z}$ [map m to m/n], so each A_n is countable. By part (b) of the theorem, the union

$$\bigcup_{n \in \mathbb{P}} A_n = \mathbb{Q}$$

is also countable. ■

EXAMPLE 6

(a) If Σ is a finite alphabet, the set Σ^* of all words using letters from Σ is countably infinite. Note that Σ is nonempty by definition. We already know that Σ^* is infinite. Recall that

$$\Sigma^* = \bigcup_{k=0}^{\infty} \Sigma^k,$$

where each Σ^k is finite. Thus Σ^* is a countable union of countable sets, and hence Σ^* itself is countable by part (b) of the theorem.

(b) It follows from part (a) that the set of all computer programs that can be typed on all the keyboards in the world [a finite set] is countable. Since each program can produce only a countable number of outputs, part (b) of the theorem says that the total number of outputs that can be produced by programs is countable. Hence, even if we allow irrational outputs such as $\sqrt{17}$ and $\pi/4$, there must be real numbers that cannot be computed by any program.

In fact, there must be uncountably many unobtainable outputs in part (b). The reason is that if S is a countable subset of an uncountable set U, then $U \setminus S$ must be uncountable, since otherwise U would be the union of two countable sets, and part (b) of the theorem would apply.

(c) Imagine, if you can, a countably infinite alphabet Σ, and let Σ^* consist of all words using letters of Σ, i.e., all finite strings of letters from Σ. For each $k \in \mathbb{P}$, the set Σ^k of all words of length k is in one-to-one correspondence with the product set $\Sigma^k = \Sigma \times \Sigma \times \cdots \times \Sigma$ (k times). In fact, the correspondence maps each word $a_1 a_2 \cdots a_k$ to the k-tuple $(a_1, a_2, \ldots, a_k)$. So each set Σ^k is countable by Exercise 15. The 1-element set $\Sigma^0 = \{\lambda\}$ is countable too. Hence $\Sigma^* = \bigcup_{k=0}^{\infty} \Sigma^k$ is countable by part (b) of the theorem. ■

EXAMPLE 7

Consider a graph with sets of vertices and edges V and E. Even if V or E is infinite, each path has finite length by definition. Let $\mathcal{P}$ be the set of all paths of the graph.

(a) If E is nonempty, then $\mathcal{P}$ is infinite. For if e is any edge, then e, $e\,e$, $e\,e\,e$, etc., all describe paths in the graph.

(b) If E is finite, then $\mathcal{P}$ is countable. For purposes of counting, let's view E as an alphabet. Since each path is a sequence of edges, it corresponds to a word in E^*. Of course, not all words in E^* correspond to paths, since endpoints of adjacent edges must match up. But the set $\mathcal{P}$ is in one-to-one correspondence with some *subset* of E^*. The set E^* is countably infinite by Example 6(a), so $\mathcal{P}$ is countable by part (a) of the theorem.

(c) If E is countably infinite, then $\mathcal{P}$ is still countable. Simply use Example 6(d) instead of Example 6(a) in the discussion of part (b). ■

Exercises 13.3

1. Let A and B be finite sets with $|A| < |B|$. True or False.

(a) There is a one-to-one map of A into B.

(b) There is a one-to-one map of A onto B.

(c) There is a one-to-one map of B into A.

(d) There is a function mapping A onto B.

(e) There is a function mapping B onto A.

2. True or False.

(a) The set of positive rational numbers is countably infinite.

(b) The set of all rational numbers is countably infinite.

(c) The set of positive real numbers is countably infinite.

(d) The intersection of two countably infinite sets is countably infinite.

(e) There is a one-to-one correspondence between the set of all even integers and the set $\mathbb{N}$ of natural numbers.

(f) The set of irrational real numbers is countable.

3. Give one-to-one correspondences between the following pairs of sets.

(a) $(0, 1)$ and $(-1, 1)$ (b) $[0, 1)$ and $(0, 1]$

(c) $[0, 1]$ and $[-5, 8]$ (d) $(0, 1)$ and $(1, \infty)$

(e) $(0, 1)$ and $(0, \infty)$ (f) $\mathbb{R}$ and $(0, \infty)$

4. Let $E = \{n \in \mathbb{N} : n \text{ is even}\}$. Show that E and $\mathbb{N} \setminus E$ are countable by exhibiting one-to-one correspondences $f: \mathbb{P} \to E$ and $g: \mathbb{P} \to \mathbb{N} \setminus E$.

5. Here is another one-to-one function f mapping (0,1) onto $\mathbb{R}$:

$$f(x) = \frac{2x - 1}{x(1 - x)}.$$

(a) Sketch the graph of f.

(b) If you know some calculus, prove that f is one-to-one by showing that its derivative is positive on $(0, 1)$.

6. Which of the following sets are countable? countably infinite?

(a) $\{0, 1, 2, 3, 4\}$ (b) $\{n \in \mathbb{N} : n \le 73\}$

(c) $\{n \in \mathbb{Z} : n \le 73\}$ (d) $\{n \in \mathbb{Z} : |n| \le 73\}$

(e) $\{5, 10, 15, 20, 25, \dots\}$ (f) $\mathbb{N} \times \mathbb{N}$

(g) $[\frac{1}{4}, \frac{1}{3}]$

7. Let Σ be the alphabet $\{a, b, c\}$. Which of the following sets are countably infinite?

(a) Σ^{73} (b) Σ^*

(c) $\bigcup_{k=0}^{\infty} \Sigma^{2k} = \{w \in \Sigma^* : \text{length}(w) \text{ is even}\}$

(d) $\bigcup_{k=0}^{3} \Sigma^k$ (e) $\bigcup_{k=1}^{3} \Sigma^{2k}$

8. A set A has m elements and a set B has n elements. How many functions from A into B are one-to-one? *Hint:* Consider the cases $m \le n$ and $m > n$ separately.

9. (a) Show that if there is a one-to-one correspondence of a set S onto some countable set, then S itself is countable.

(b) Show that if there is a one-to-one correspondence of a set S onto some uncountable set, then S is uncountable.

10. Complete the proof of part (a) of the theorem by showing that f is one-to-one and that f maps $\mathbb{P}$ onto A.

11. (a) Prove that if S and T are countable then $S \times T$ is countable. *Hint:* $S \times T = \bigcup_{t \in T} (S \times \{t\})$.

(b) Prove that if f maps S onto T and S is countable, then T is countable.

(c) Use parts (a) and (b) to give another proof that $\mathbb{Q}$ is countable. *Suggestion:* For (m, n) in $\mathbb{Z} \times \mathbb{P}$, define $f(m, n) = m/n$.

12. Show that if S and T have the same size, then so do $\mathcal{P}(S)$ and $\mathcal{P}(T)$.

13. (a) Show that $\text{FUN}(\mathbb{P}, \{0, 1\})$ is in one-to-one correspondence with the set $\mathcal{P}(\mathbb{P})$ of all subsets of $\mathbb{P}$.

(b) Show that $\mathcal{P}(\mathbb{P})$ is uncountable. *Hint:* Use Example 3(a).

14. Show that any disjoint family of nonempty subsets of a countable set is countable.

15. Show that if S is countable, then $S^n = S \times S \times \cdots \times S$ [n times] is countable for each n. *Hint:* Use Exercise 11(a) and induction.

16. Here's an elegant, explicit one-to-one correspondence of the set $\mathbb{Q}^+$ of positive rationals onto $\mathbb{P}$. Given m/n in $\mathbb{Q}^+$, where m and n are relatively prime, write $m = p_1^{m_1} \cdots p_k^{m_k}$ and $n = q_1^{n_1} \cdots q_l^{n_l}$ as products of primes. Define

$$f\left(\frac{m}{n}\right) = p_1^{2m_1} \cdots p_k^{2m_k} \cdot q_1^{2n_1 - 1} \cdots q_l^{2n_l - 1}.$$

In particular, $f(m) = p_1^{2m_1} \cdots p_k^{2m_k} = m^2$ for positive integers m.

(a) Calculate $f(\frac{1}{8})$, $f(\frac{1}{9})$, $f(\frac{1}{10})$, $f(\frac{1}{100})$, and $f(\frac{21}{20})$.

(b) What fraction is mapped to 23? to 24? 25? 26? 27? 28?

(c) Explain why f is one-to-one and why f maps $\mathbb{Q}^+$ onto $\mathbb{P}$.

This example was given by Yoram Sagher in the November 1989 issue of *The American Mathematical Monthly*, page 823.

17. The idea in Example 3(a) can be extended to show that the sets S and $\text{FUN}(S, \{0, 1\})$ are never the same size. Imagine that $\varphi\colon\ S \to \text{FUN}(S, \{0, 1\})$ is a one-to-one correspondence. Show that the function $f\colon S \to \{0, 1\}$ defined by $f(s) = 1 - \varphi(s)(s)$ for each s is not in the image of φ.

Chapter Highlights

As usual: What does it mean? Why is it here? How can I use it? Think of examples.

CONCEPTS AND NOTATION

quantifiers, $\forall$, $\exists$
universe [= domain] of discourse
predicate = proposition-valued function
n-place predicate
  compound predicate
  free, bound variable
  compound proposition
tautology
logical equivalence, implication, $\Longleftrightarrow$, $\Longrightarrow$
counterexample
infinite sets
  same size
  countable
  countably infinite
  uncountable

FACTS

$\forall$ and $\exists$ do not commute with each other:
  $\exists x\, \forall y\, p(x, y) \to \forall y\, \exists x\, p(x, y)$ is a tautology, but
  $\forall x\, \exists y\, p(x, y) \to \exists y\, \forall x\, p(x, y)$ is not.
$\mathbb{N}$, $\mathbb{Z}$, and $\mathbb{Q}$ are countable.
$[0, 1)$ and $\text{FUN}(\mathbb{P}, \{0, 1\})$ are uncountable.
$\mathbb{R}$, $[0, 1)$, and $(0, 1)$ are the same size.
Subsets of and countable unions of countable sets are countable.

METHODS

Use of generalized De Morgan laws to negate quantified predicates.
Cantor's diagonal procedure for showing certain sets are uncountable.

Supplementary Exercises

1. Here the universe of discourse is $\mathbb{R}$. That is, all the variables x and y are in $\mathbb{R}$. True or False.

(a) $\forall x \exists y [x^2 < y + 1]$ (b) $\exists x \forall y [x^2 < y + 1]$

(c) $\forall y \exists x [x^2 > y + 1]$ (d) $\exists y \exists x [x^2 < y + 1]$

(e) $\forall y \exists x [x^2 < y + 1]$

2. True or False. If false, provide an example [either by defining suitable predicates or by specifying a small universe of discourse, say $U = \{a, b\}$, and assigning truth values].

(a) $\forall y \exists x\, p(x, y) \Longrightarrow \exists x \forall y\, p(x, y)$

(b) $\exists x \forall y\, p(x, y) \Longrightarrow \forall y \exists x\, p(x, y)$

3. In this problem, the universe of discourse is $\mathbb{N}$.

(a) Write the negation of $\exists n \forall m\, [m > n$ or m is even] without using the connective $\neg$.

(b) Is your answer to part (a) true?

4. Determine the truth values of the following, where the universe of discourse is $\mathbb{N}$.

(a) $\forall m \forall n \forall p[mn = mp \rightarrow n = p]$

(b) $\forall m \exists n[m + n = 0]$

(c) $\exists n \forall m[m + n = m]$

(d) $\exists n \forall m[m + n = n]$

(e) $\forall m \exists n[n > m]$

(f) $\exists n \forall m[n > m]$

5. For each of the following sets, give the number of elements in the set if it is finite. Otherwise write "countably infinite" or "uncountable," whichever correctly describes the set.

(a) the set $\mathbb{N}$ of natural numbers

(b) the set $\mathbb{Z}$ of all integers

(c) the set $\mathbb{R}$ of all real numbers

(d) the interval $(0, 0.001)$

(e) the set $\mathbb{N} \times \mathbb{N}$

(f) the set $\mathbb{Q}$ of all rational numbers

(g) the set $\mathcal{P}(\mathbb{N})$ of all subsets of $\mathbb{N}$

(h) the set Σ^*, where Σ is a finite (nonempty) alphabet

(i) the set of all sequences of 0's and 1's of length 12

(j) the set of all infinite sequences of 0's and 1's

6. Determine the truth values of the following, where the universe of discourse is $\mathbb{R}$.

(a) $\forall x \exists y[x + y = 1]$

(b) $\exists y \forall x[x + y = 1]$

(c) $\forall x \exists y[x^2 + y^2 = 1]$

(d) $\forall x \exists y \forall z[z > y \rightarrow z^2 > x]$

(e) $\forall x \exists y \forall z[z^2 > x \rightarrow z > y]$

(f) $\forall x \exists y[x + y = 1] \leftrightarrow \forall y \exists x[x + y \neq 1]$

(g) $\forall x \exists y[x + y = 1] \leftrightarrow \exists x \forall y[x + y \neq 1]$

(h) $\forall x \exists y[x + y = 1] \rightarrow \exists y \forall x[x + y = 1]$

(i) $\exists y \forall x[x + y = 1] \rightarrow \forall x \exists y[x + y = 1]$

Dictionary

The words listed here are of three general sorts: English words with which the reader may not be completely familiar, common English words whose usage in mathematical writing is specialized, and technical mathematical terms that are assumed background for this book. For technical terms introduced in this book, see the index.

absurd Clearly impossible, being contrary to some evident truth.

all See **every.**

ambiguous Capable of more than one interpretation or meaning.

anomaly Something that is, or appears to be, inconsistent.

any see **every.**

assume Assume and suppose mean the same thing and ask that we imagine a situation for the moment.

axiom An assertion that is accepted and used without a proof.

bona fide Genuine or legitimate.

calls If algorithm A contains the instruction to use algorithm B, we say that A calls B.

cf. Compare.

class See **set.**

collapse To fall or come together.

collection See **set.**

commute Produce the same result in either order.

comparable Capable of being compared.

conjecture A guess or opinion, preferably based on some experience or other source of wisdom.

corollary See **theorem.**

define Often this looks like an instruction [as in "Define $f(x) = x^2$"] when it is merely a [bad] mathematical way of saying "We define" or "Let."

disprove The instruction "prove or disprove" means that the assertion should either be proved true or proved false. Which you do depends, of course, on whether the assertion is actually true or not.

distinct Different. Used to describe two or more objects. Also can mean "clear" or "not fuzzy," but we do not use the word that way.

distinguishable A collection of objects is regarded as distinguishable if there is some property or characteristic that makes it possible to tell different objects apart. In contrast, we would regard ten red marbles of the same size as indistinguishable.

e.g. For example.

entries The individual numbers or objects in ordered pairs, in matrices or in sequences.

every The expressions "for every," "for any" and "for all" mean the same thing. They all mean "for all choices of the variable in question," so they correspond to the quantifier $\forall$ in §2.1. The expression "for some" means "for at least one" and corresponds to the quantifier $\exists$ in §2.1. It is generally good practice to avoid the use of "any," which is sometimes misunderstood. For example, "If $p(n)$ is true for any n" usually means "If $p(n)$ is true for some n," not "If $p(n)$ is true for every n."

family See **set.**

fictitious Imagined, not actual.

inclusion We sometimes refer to the relation $A \subseteq B$ as an "inclusion," just as we may refer to $A = B$ as an "equality."

initialize Set the starting conditions.

inspection Something can be seen "by inspection" if it can be seen directly, without calculation or modification.

invertible Having an inverse.

irrational An irrational number is a real number that is not rational, i.e., that cannot be written as m/n for $m, n \in \mathbb{Z}$, $n \neq 0$. Examples include $\sqrt{2}$, $\sqrt{3}$, $\sqrt[3]{2}$, π, and e.

lemma See **theorem.**

loop A computer program or algorithm that is repeated under specified conditions.

matrices Plural of matrix.

max. For two real numbers a and b, we write $\max\{a, b\}$ for the larger of the two. If $a = b$, then $\max\{a, b\} = a = b$.

min. For two real numbers a and b, we write $\min\{a, b\}$ for the smaller of the two. If $a = b$, then $\min\{a, b\} = a = b$.

necessary We say a property p is necessary for a property q if p must hold whenever q holds, i.e., if $q \Longrightarrow p$. The property p is sufficient for q if p is enough to guarantee q, i.e., if $p \Longrightarrow q$. So p is necessary and sufficient for q provided that $p \Longleftrightarrow q$.

permute To change the order of a sequence of elements.

proposition See **theorem.**

redundant Unnecessary or excessive.

sequence A list of things following one another. A formal definition is given in §1.6.

set The terms "set," "collection," "class," and "family" are used inter- changeably. We tend to refer to families of sets, for example, to avoid the expression "sets of sets."

some See **every.**

sufficient See **necessary.**

suppose See **assume.**

terminate End.

theorem A theorem, proposition, lemma, or corollary is some assertion that has been or can be proved. The term "proposition" also has a special use in logic: see §2.1. Theorems are usually the most important facts. Lemmas are usually not of primary interest and are used to prove later theorems or propositions. Corollaries are usually easy consequences of theorems or propositions just presented.

unambiguous Not ambiguous. See **ambiguous.**

underlying set The basic set on which the objects [like functions or operations] are defined.

vertices Plural of vertex.

Answers and Hints

1.1 Answers

1. (a) 20. (c) 19. (e) 11.
(g) 10. (i) $10^{30} - 10^{29} + 1$.

3. (a) 0. (c) -5.

5. (a) 10. (c) 40. (e) 54.

7. (a) 17. (c) 7.

9. One could divide n by k to get n/k. If the answer is positive, then $\lfloor n/k \rfloor$ is the part of it to the left of the decimal point. If the answer is negative, then $\lfloor n/k \rfloor$ is the part of it to the left of the decimal point [a negative integer] minus 1. For example, $\lfloor 73/17 \rfloor = \lfloor 4.294\cdots \rfloor = 4$, and $\lfloor -73/17 \rfloor = \lfloor -4.294\cdots \rfloor = -4 - 1 = -5$.

11. (a) If t is an integer, then $\lceil x + t \rceil = \lceil x \rceil + t$ for every number x.

13. (a) 11. [They are 2, 3, 5, 7, 11, 13, 17, 19, 23, 29, 31.]
(c) Maybe. It turns out that the number of such primes is 470, and $3333/\ln 3333 \approx 410.89$. Not very close. Indeed, $470/3333$ and $1/\ln 3333$ differ by about 14%.

15. (a) $10^{30}/30 \ln 10 \approx 1.45 \times 10^{28}$. (c) (a) $-$ (b) $\approx 1.3 \times 10^{28}$.

17. Add $-m+1$ to everything. The number of integers between m and n is the number of them between $m + (-m+1)$ and $n + (-m+1)$, i.e., between 1 and $n - m + 1$. Apply Fact 1 or Fact 2.

19. (a) One example is $x = y = 1.5$.
(c) By definition, $\lfloor x \rfloor \le x$ and $\lfloor y \rfloor \le y$, so $\lfloor x \rfloor + \lfloor y \rfloor \le x + y$. Since $\lfloor x + y \rfloor$ is the largest integer less than or equal to $x + y$ and, since $\lfloor x \rfloor + \lfloor y \rfloor$ is an integer, $\lfloor x \rfloor + \lfloor y \rfloor \le \lfloor x + y \rfloor$.

1.2 Answers

1. (a) False. (b) False. (c) True. (d) True.
(e) False.
Give explanations.

3. (a) 1, 2, 15, 1, 13.

5. (a) $4 = 2^2$, $4 = 2^2$, $22 = 2 \cdot 11$, $14 = 2 \cdot 7$, 37.

7. (a) 10, 1, 10.
(c) The only multiple of 0 is 0, so $\text{lcm}(0, n)$ must be 0.

9. (a) They must be relatively prime, since $\gcd(m, n) = mn/\text{lcm}(m, n) = 1$.
(c) Since $\gcd(m, n)$ is a divisor of n, this is true if and only if $m|n$.

11. (a) Relatively prime, since $64 = 2^6$ and 729 is odd.
(c) Relatively prime, since $45 = 3^2 \cdot 5$ and $56 = 2^3 \cdot 7$.

13. (a) These are the odd ones, i.e., the ones that are not even.
(c) The same as (a).
(e) These are the ones that are not multiples of p.

15. (a) $\gcd(m, n)$ is a divisor of m and n, so it's a divisor of their difference.
(c) 2 and 1.

17. (a) Since $1 \le k^2 \le kl = n$, we have $k \le \sqrt{n}$.
(c) Since n is not prime, there are integers k and l with $1 \le k < n$, $1 \le l < n$, and $n = kl$. Say $k \le l$, and let p be a prime factor of k. By part (a), $p \le k \le \sqrt{n}$.

19. Since m, n, and $\gcd(m, n)$ are positive, s and t cannot both be negative. They can't both be positive, because if they were we'd have

$$\gcd(m, n) \ge 1 \cdot m + 1 \cdot n = m + n \ge \gcd(m, n) + \gcd(m, n) = 2 \cdot \gcd(m, n),$$

which clearly cannot hold.

21. If $x \le y$, then $\min\{x, y\} = x$ and $\max\{x, y\} = y$, so $\min\{x, y\} + \max\{x, y\} = x + y$. If $x > y$, then $\min\{x, y\} = y$ and $\max\{x, y\} = x$, so $\min\{x, y\} + \max\{x, y\} = y + x = x + y$. The equation holds in either case.

1.3 Answers

1. (a) 0, 5, 10, 15, 20, say. (c) $\emptyset$, {1}, {2, 3}, {3, 4}, {5}, say.
(e) 1, 1/2, 1/3, 1/4, 1/73, say. (g) 1, 2, 4, 16, 18, say.

3. (a) λ, a, ab, cab, ba, say. (c) $aaaa$, $aaab$, $aabb$, etc.
The sets in parts (a) and (b) contain the empty word λ.

5. (a) $\emptyset$. (c) $\emptyset$. (e) $\{2, 3, 5, 7, 11, 13\}$.

7. (a) $\{0, 4, 8, 12, 16, 20\}$. (c) $\{0, 1, 2, 3, 4\}$.

9. (a) ∞. (c) ∞. (e) ∞. (g) ∞.

11. $A \subseteq A$, $B \subseteq B$, C is a subset of A, and C, D are subsets of A, B, and D.

13. (a) aba is in all three and has length 3 in each.
(c) cba is in Σ_1^* and length$(cba) = 3$.
(e) $caab$ is in Σ_1^* with length 4 and is in Σ_2^* with length 3.

15. (a) Yes.
(c) Delete first letters from the string until no longer possible. If λ is reached, the original string is in Σ^*. Otherwise, it isn't.

1.4 Answers

1. (a) $\{1, 2, 3, 5, 7, 9, 11\}$. (c) $\{1, 5, 7, 9, 11\}$.
(e) $\{3, 6, 12\}$. (g) 16.

3. (a) $[2, 3]$. (c) $[0, 2)$. (e) $(-\infty, 0) \cup (3, \infty)$.
(g) $\mathbb{N}$. (i) $\emptyset$.

5. (a) $\emptyset$. (c) $\emptyset$. (e) $\{\lambda, ab, ba\}$.
(g) $B^c \cap C^c$ and $(B \cup C)^c$ are equal by a De Morgan law [or by calculation], as are $(B \cap C)^c$ and $B^c \cup C^c$.

7. $A \oplus A = \emptyset$ and $A \oplus \emptyset = A$.

9. (a) Make A very small, like $A = \emptyset$.
(c) Try $A = B \cup C$ with B and C disjoint.

11. (a) (a, a), (a, b), (a, c), (b, a), etc. There are nine altogether.
(c) (a, a), (b, b).

13. (a) $(0, 0)$, $(1, 1)$, $(2, 2)$, $\dots$, $(6, 6)$, say.
(c) $(6, 1)$, $(6, 2)$, $(6, 3)$, $\dots$, $(6, 7)$, say.
(e) $(1, 3)$, $(2, 3)$, $(3, 3)$, $(3, 2)$, $(3, 1)$.

1.5 Answers

1. (a) $4, -6$. (c) $12, 4$. (e) $28, 14$.

3. (a) 3. (c) 0.

5. (a) 1, 1, 1, 0, 0. See the answer to part (b).
(b) $f(n, n) = 1$ for even n, and $f(n, n) = 0$ for odd n. This can be checked by calculation or by applying the theorem on page 5, with $k = 2$.

7. (a) $f(3) = 27$, $f(1/3) = 1/3$, $f(-1/3) = 1/27$, $f(-3) = 27$.
(c) $\text{Im}(f) = [0, \infty)$.

9. $\{n \in \mathbb{Z} : n \text{ is even}\}$.

11. (a) The answers for $n = 0, 1, 2, 3, 4, 5$ are 0, 0, 1, 2, 3, 3. The remaining answers are 5, 5, 6, 7, 8, 60.

13. (a) $f \circ f(x) = (x^3 - 4x)^3 - 4(x^3 - 4x)$.
(c) $h \circ g(x) = (x^2 + 1)^{-4}$.
(e) $f \circ g \circ h(x) = (x^8 + 1)^{-3} - 4(x^8 + 1)^{-1}$.
(g) $h \circ g \circ f(x) = [(x^3 - 4x)^2 + 1]^{-4}$.

15. (a) 1, 0, -1, and 0.
(c) $g \circ f$ is the characteristic function of $\mathbb{Z} \setminus E$. $f \circ f(n) = n - 2$ for all $n \in \mathbb{Z}$.

1.6 Answers

1. (a) 42. (c) 1. (e) 154.

3. (a) 3, 12, 39, and 120. (c) 3, 9, and 45.

5. (a) $-2, 2, 0, 0$, and 0.

7. (a) 0, 1/3, 1/2, 3/5, 2/3, 5/7.
(c) Note that $a_{n+1} = \frac{(n+1)-1}{(n+1)+1} = \frac{n}{n+2}$ for $n \in \mathbb{P}$. Hence $a_{n+1} - a_n = \frac{n}{n+2} - \frac{n-1}{n+1} = \frac{2}{(n+1)(n+2)}$ for $n \in \mathbb{N}$.

9. (a) 0, 0, 2, 6, 12, 20, 30.
(c) Just substitute the values into both sides.

11. (a) 0, 0, 1, 1, 2, 2, 3, 3, $\dots$.
(c) $(0, 0)$, $(0, 1)$, $(1, 1)$, $(1, 2)$, $(2, 2)$, $(2, 3)$, $(3, 3)$, $\dots$.

13. (a)

n	n^4	4^n	n^{20}	20^n	$n!$
5	625	1024	$9.54 \cdot 10^{13}$	$3.2 \cdot 10^6$	120
10	10^4	$1.05 \cdot 10^6$	10^{20}	$1.02 \cdot 10^{13}$	$3.63 \cdot 10^6$
25	$3.91 \cdot 10^5$	$1.13 \cdot 10^{15}$	$9.09 \cdot 10^{27}$	$3.36 \cdot 10^{32}$	$1.55 \cdot 10^{25}$
50	$6.25 \cdot 10^6$	$1.27 \cdot 10^{30}$	$9.54 \cdot 10^{33}$	$1.13 \cdot 10^{65}$	$3.04 \cdot 10^{64}$

1.7 Answers

1. (a) No; S is bigger than T.
(c) Yes. For example, let $f(1) = a$, $f(2) = b$, $f(3) = c$, $f(4) = f(5) = d$.
(e) No. This follows from either part (a) or part (d).

3. (a) $f(2, 1) = 2^2 3^1 = 12$, $f(1, 2) = 2^1 3^2 = 18$, etc.
(c) Consider 5, for instance. 5 is not in $\text{Im}(f)$.

5. (a) $f(0) = 1$, $f(1) = 2$, $f(2) = 3$, $f(3) = 4$, $f(4) = 5$, $f(73) = 74$.
(c) For one-to-oneness, observe that if $f(n) = f(n')$, then $n = f(n) - 1 = f(n') - 1 = n'$. f does not map onto $\mathbb{N}$ because $0 \notin \text{Im}(f)$.
(e) $g(f(n)) = \max\{0, (n + 1) - 1\} = n$, but $f(g(0)) = f(0) = 1$.

7. (a) $f^{-1}(y) = (y - 3)/2$. (c) $h^{-1}(y) = 2 + \sqrt[3]{y}$.

9. (a) $(f \circ f)(x) = 1/(1/x) = x$.
(b) and (c) are similar verifications.

11. (a) All of them; verify this.
(c) $\text{SUM}^{\leftarrow}(4)$ has 5 elements, $\text{PROD}^{\leftarrow}(4)$ has 3 elements, $\text{MAX}^{\leftarrow}(4)$ has 9 elements, and $\text{MIN}^{\leftarrow}(4)$ is infinite.

13. If $s_1 \neq s_2$, then $f(s_1) \neq f(s_2)$ [why?]. Thus $g(f(s_1)) \neq g(f(s_2))$ [why?]. Hence $g \circ f$ is one-to-one.

15. Since f and g are invertible, the functions $f^{-1}: T \to S$, $g^{-1}: U \to T$, and $f^{-1} \circ g^{-1}: U \to S$ exist. So it suffices to show $(g \circ f) \circ (f^{-1} \circ g^{-1}) = 1_U$ and $(f^{-1} \circ g^{-1}) \circ (g \circ f) = 1_S$. One can show directly that $g \circ f$ is one-to-one [see Exercise 13] and onto, but it is actually easier—and more useful—to verify that $f^{-1} \circ g^{-1}$ has the properties of the inverse to $g \circ f$.

1.8 Answers

1. (a) The gcd divides 555,557 − 555,552 = 5, so is either 5 or 1. Since 555,552 is clearly not a multiple of 5, the gcd is 1.
(b) A similar argument applies to show the gcd is 1.

2. (a) Any sets with $A \cap C$ nonempty give a counterexample.
(b) Any sets with $A \cap C$ empty will work.

3. (a) The set consists of all multiples of 15.
(b) Presumably, $\mathbb{Z}$ is the universal set. Then any integers not divisible by 3 will work.
(c) Some small examples are 3, 5, and 6.

4. (a) 0, 0, 0, 0, 1, 1, 2, 2,14, respectively.
(b) $\lfloor \frac{k}{5} \rfloor = \lceil \frac{k}{5} \rceil$ if and only if $\frac{k}{5}$ is an integer, i.e., if and only if 5 divides k.

5. See the theorem on page 5, for example.

6. Domain is $[0, 2]$ and codomain is $[-4, 0]$. Solve $y = x^2 - 4x$ for x to get $x = (4 \pm \sqrt{16 + 4y})/2$. Since $x \leq 2$, we must have $x = (4 - \sqrt{16 + 4y})/2 = 2 - \sqrt{4 + y}$. Thus $f^{-1}(y) = 2 - \sqrt{4 + y}$ for $-4 \leq y \leq 0$.

7. $f^{\leftarrow}(\{0, 1\}) = \{n \in \mathbb{Z} : n^2 - 3 \in \{0, 1\}\} = \{-2, 2\}$.

8. (a) $[-1, 0]$. (b) $[-\sqrt{2}, -1] \cup [1, \sqrt{2}]$.
(c) $f^{\leftarrow}([-1, 0]) = [-1, 1]$.

9. (a) If n ends with digit 7, then $n = 10k + 7$ for some $k \in \mathbb{N}$. So $n^2 = 10(10k^2 + 14k + 4) + 9$ ends with the digit 9.
(b) There are 1,000 positive integers n such that $n^2 \leq 1{,}000{,}000$. Of these, 200 end with digits 3 or 7, so 200 of the squares end with digit 9.
(c) 100, 200, 0, 0, 200, 100, 200, 0, and 0, respectively. Note that no squares of integers end with digits 2, 3, 7, or 8.

10. (a) The first step, finding all divisors of m, is generally believed to be very difficult for large numbers m.
(b) There is no largest multiple of m.

11. (a) 1, 1, 1, 7.
(b) Theorem 3 fails pretty easily. Theorem 4(a) and (b) are both true. Give some examples. Can you find conditions when the equality

$$\gcd(\ell, m, n) \cdot \text{lcm}(\ell, m, n) = \ell mn$$

does not hold? The methods of the text don't extend from two to three integers, but a consideration of prime factorization does the job.

12. $40 \ln 10 \approx 92$, so roughly $\frac{1}{90}$ of the 40-digit integers will be primes.

14. Ann is right: $\gcd(2, 12) = 2 = \gcd(4, 6)$ and $\text{lcm}(2, 12) = 12 = \text{lcm}(4, 6)$, so both $m = 2, n = 12$ and $m = 4, n = 6$ are examples.

15. (a) 2, 30.
(b) If exactly one of m and n is negative, then $\gcd(m, n) \cdot \text{lcm}(m, n) = -m \cdot n$.

16. (a) No. The image of any function $f: \Sigma \to \Sigma^*$ will have at most two elements, whereas Σ^* is infinite.

(b) Yes. For example, define $g: \Sigma^* \to \Sigma$ by $g(\lambda) = \lambda$ and $g(w)$ = the one-letter word using the first letter of w, for $w \in \Sigma^*$ with $\text{length}(w) \geq 1$.

17. (a) $f(x_1) = f(x_2)$ implies $\sqrt{x_1 + 1} = \sqrt{x_2 + 1}$ implies (by squaring both sides) $x_1 + 1 = x_2 + 1$ implies $x_1 = x_2$.

(b) Consider $y \in T$. Then $y \geq 0$, so $x = y^2 - 1$ is in S and $f(x) = \sqrt{y^2 - 1 + 1} = y$.

(c) Yes. It is one-to-one and onto T by parts (a) and (b), hence invertible by the Theorem on page 42. The inverse is given by $f^{-1}(y) = y^2 - 1$ for $y \in T$.

(d) Since f maps S into T and $T \subseteq S$, $f: S \to S$. Thus $f \circ f: S \to S$ is defined, and $f \circ f(x) = \sqrt{\sqrt{x+1} + 1}$ for $x \in S$. Clearly, $\text{Dom}(f \circ f) = S$, and the codomain is any set containing $\{x \in \mathbb{R} : x \geq 1\}$.

(e) Yes. See Exercise 13 on page 45.

18. (a) We first show that $(A \oplus B) \oplus C \subseteq A \oplus (B \oplus C)$. Suppose that $x \in (A \oplus B) \oplus C$. Then either $x \in A \oplus B$ or $x \in C$, but not both. Suppose that $x \in A \oplus B$, so $x \notin C$. If $x \in A$, then $x \notin B$, so $x \notin B \oplus C$, and thus $x \in A \oplus (B \oplus C)$. If $x \notin A$, then $x \in B \setminus C$, so $x \in B \oplus C$, and thus $x \in A \oplus (B \oplus C)$. Suppose now that $x \notin A \oplus B$, so $x \in C$. If $x \in A$, then $x \in B$ [why?], so $x \notin B \oplus C$, and thus $x \in A \oplus (B \oplus C)$. This shows that $(A \oplus B) \oplus C \subseteq A \oplus (B \oplus C)$. To prove that $A \oplus (B \oplus C) \subseteq (A \oplus B) \oplus C$, we can simply interchange A and C in the argument.

(b) The set $A \oplus (B \oplus C)$ consists of the elements that are in an odd number of the sets A, B, and C. A member of A, but not of B or C, is not in $B \oplus C$, so it's in $A \oplus (B \oplus C)$; a member of A and B and C is also in $A \oplus (B \oplus C)$; but a member of A and B, but not C, is not in $(A \oplus B) \oplus C$.

(c) and (d) The set $A_1 \oplus \cdots \oplus A_n$ consists of the elements in an odd number of the sets $A_1, A_2, \ldots, A_n$. This fact can be proved most easily for general n by using the method of mathematical induction, discussed in Chapter 4.

19. Since $2 = 10^{\log_{10} 2}$, we have $2^n = 10^{n \log_{10} 2}$. For $n = 10^k$, this becomes $2^{10^k} = 10^{10^k \log_{10} 2}$. The exponent in the table entry for 2^{10^k} is the integer $\lfloor 10^k \log_{10} 2 \rfloor$, and the factor in front is 10^a, where $a = 10^k \log_{10} 2 - \lfloor 10^k \log_{10} 2 \rfloor$; so $1 \leq 10^a < 10$.

2.1 Answers

1. (a) $p \wedge q$. (c) $\neg p \to (\neg q \wedge r)$. (e) $\neg r \to q$.

3. (a) Parts (b) and (c) are true. The other three are false.

5. The proposition is true for all $x, y \in [0, \infty)$, but is false when applied to all $x, y \in \mathbb{R}$.

7. (a) $\neg r \to \neg q$.

(c) If it is false that $x = 0$ or $x = 1$, then $x^2 \neq x$.

9. (a) $3^3 < 3^3$ is false.

11. (a) $(-1 + 1)^2 = 0 < 1 = (-1)^2$.

(c) No. If $x \geq 0$, then $(x + 1)^2 = x^2 + 2x + 1 > x^2$. In fact, if $x \geq -\frac{1}{2}$, then $2x + 1 \geq 0$ and $(x + 1)^2 \geq x^2$.

13. (a) $(0, -1)$.

15. (a) $p \to q$. (c) $\neg r \to p$. (e) $r \to q$.

17. (a) The probable intent is "If you touch those cookies, then I will spank you." It is easier to imagine p of $p \to q$ being true for this meaning than it is for "If you want a spanking, then touch those cookies."

(c) If you do not leave, then I will set the dog on you.

(e) If you do not stop that, then I will go.

2.2 Answers

1. (a) Converse: $(q \wedge r) \to p$.

Contrapositive: $\neg(q \wedge r) \to \neg p$.

(c) Converse: If 3 + 3 = 8, then 2 + 2 = 4.

Contrapositive: If $3 + 3 \neq 8$, then $2 + 2 \neq 4$.

3. (a) $q \to p$. (c) $p \to q$, $\neg q \to \neg p$, $\neg p \vee q$.

5. (a) 0. (c) 1.

Note. For some truth tables below, only the final columns are given.

7.

p	q	part(a)	part(b)	part(c)	part(d)
0	0	1	1	1	1
0	1	1	0	0	1
1	0	1	0	0	1
1	1	0	0	0	0

9.

p	q	r	final column
0	0	0	1
0	0	1	1
0	1	0	1
0	1	1	0
1	0	0	1
1	0	1	1
1	1	0	1
1	1	1	0

11.

p	q	r	part(a)	part(b)
0	0	0	0	0
0	0	1	1	0
0	1	0	1	1
0	1	1	1	0
1	0	0	1	1
1	0	1	1	0
1	1	0	1	1
1	1	1	1	0

13. (b)

p	$p \oplus p$
0	0
1	0

(c)

p	q	r	$(p \oplus q) \oplus r$
0	0	0	0
0	0	1	1
0	1	0	1
0	1	1	0
1	0	0	1
1	0	1	0
1	1	0	0
1	1	1	1

15. (a)

p	q	$p \sim q$
0	0	0
0	1	1
1	0	1
1	1	1

(c) In view of part (b), it suffices to verify $p \vee q \Longleftrightarrow \neg p \rightarrow q$ and $p \vee q \Longleftrightarrow q \vee p$. These are rules 11a and 2a in Table 1 on page 62. Or use truth tables.

17. (a) No fishing is allowed and no hunting is allowed. The school will not be open in July, and the school will not be open in August.

21. (a) One need only consider rows in which $[(p \wedge r) \rightarrow (q \wedge r)]$ is false; i.e., $(p \wedge r)$ is true and $(q \wedge r)$ is false. This leaves one row to consider:

p	q	r
1	0	1

(c) One need only consider rows in which $[(p \wedge r) \rightarrow (q \wedge s)]$ is false; i.e., $(p \wedge r)$ is true and $(q \wedge s)$ is false. This leaves three rows to consider:

p	q	r	s
1	0	1	0
1	0	1	1
1	1	1	0

23. Let $p =$ "He finished dinner" and $q =$ "He was sent to bed." Then p is true and q is true, so the logician's statement $\neg p \rightarrow \neg q$ has truth value True. She was logically correct, but not very nice.

25. (a) Consider the truth tables. B has truth value 1 on every row that A does, and C has truth value 1 on every row that B does, so C has truth value 1 on every row that A does.

(c) We are given that $P \Longrightarrow Q$. Since $Q \Longrightarrow R$ and $R \Longrightarrow P$, by part (a), $Q \Longrightarrow P$. Thus $P \Longleftrightarrow Q$.

2.3 Answers

1. For each part, test about five examples [i.e., consecutive sequences of integers], unless you run into a counterexample, in which case you can stop. Why?

2. (a) See the discussion in Example 2.
(b) See the discussion after Exercise 7.

3. (a) Fine. One can use other variables, perhaps k and l, or i and j.
(b) This is fine if there's no need to name the original odd integers. It will depend on the situation, but it's always preferable to avoid unnecessary notation.

(c) This is bad. The integers m and n are probably different, so the same variable k cannot be used for both of them.
(d) This is just another way of writing part (a) and is fine.
(e) This is technically correct, but bad practice and confusing, because the notations x and y suggest that x and y can be more general real numbers. It is traditional to use letters in the middle of the alphabet for integers.
(f) Providing several examples is *not* sufficient to give a proof. See the discussion following the Silly Conjecture.

4. (a) There is nothing wrong here, but this sentence doesn't distinguish between rationals and irrationals. A second sentence is needed, as in part (b).
(b) This is fine.
(c) Though technically correct, this is a bad start because it doesn't distinguish between rationals and irrationals *and* because the choice of variables p and q is confusing. Rationals are often written $\frac{p}{q}$ [or $\frac{m}{n}$ or even $\frac{a}{b}$], but arbitrary real numbers are never written p or q. It wouldn't be illegal, but it would be confusing.
(d) This is pretty good. It would be better to specify what p and q are, as in part (b), though most readers would guess correctly that they are integers with $q \neq 0$. Still, it's better to be specific. This improves the communication between the writer and the reader.
(e) This is terrible and suggests confusion. The first half of the sentence specifies x, so the second half just tells us that $x \neq y$. This is what the sentence really says. Presumably, the author meant to tell us that the irrational number y satisfies $y \neq \frac{m}{n}$ for all choices of m and n in $\mathbb{Z}$ with $n \neq 0$. This is true and helpful, but this isn't what is stated in the clause, "let $y \neq \frac{p}{q}$ be an irrational number."

5. By the Division Algorithm, we can write $n = 3k + r$, where $r = 0, 1, 2$. Then $n^2 - 2 = 9k^2 + 6rk + r^2 - 2$. Since $9k^2 + 6rk$ is divisible by 3, it suffices to show that $r^2 - 2$ is *not* divisible by 3. Now we are reduced to three cases involving r. Since $r^2 - 2$ is either -2, -1, or 2 for $r = 0, 1, 2$, we are done.

6. n^2 is odd by Example 1, so $n^2 - 2$ is also odd.

7. (a) We can write $n = 7k + r$, where $r = 0, 1, 2, \ldots, 6$. Then $n^2 - 2 = 49k^2 + 14rk + r^2 - 2$. Since $49k^2 + 14rk$ is divisible by 7, it suffices to show that $r^2 - 2$ is *not* divisible by 7. For $r = 0, 1$, and 2, we get $r^2 - 2 = -2, -1$, and 2. None are divisible by 7; so far so good. Alas, for $r = 3$ we see that $r^2 - 2 = 7$, which is certainly divisible by 7.
(b) Any number that, when divided by 7, leaves remainder 3 will give a counterexample. Thus 3, 10, 17, etc., are counterexamples. If in part (a) we had continued, we would have discovered that $r = 4$ also does not work: $r^2 - 2 = 14$ is divisible by 7. So other counterexamples are 4, 11, 18, etc.
(c) The smallest three odd counterexamples are 3, 11, and 17.

8. Let x be a rational number, and let y be an irrational number. We can write $x = \frac{p}{q}$, where $p, q \in \mathbb{Z}$ and $q \neq 0$. We want to show that $x + y$ is irrational, i.e., that $x + y \neq \frac{m}{n}$ for all $m, n \in \mathbb{Z}$ and $n \neq 0$. This would be tough, perhaps impossible, to do directly. The way to tackle this is to give a proof by contradiction. Assume that $x + y$ is rational, say $x + y = \frac{m}{n}$ for some $m, n \in \mathbb{Z}$, where $n \neq 0$. Then we have $\frac{p}{q} + y = \frac{m}{n}$. Solving for y gives $y = \frac{m}{n} - \frac{p}{q} = \frac{mq-np}{nq}$, so y is a rational number, contradicting our assumption on y. Hence $x + y$ must be irrational.

9. If m and n are even integers, then there exist j and k in $\mathbb{Z}$ so that $m = 2j$ and $n = 2k$. Then $mn = 4jk$, which is a multiple of 4.

11. As in Example 4, an integer n has the form $5k + r$, where $k \in \mathbb{Z}$ and $r \in \{0, 1, 2, 3, 4\}$. Then $n^2 - 3 = 25k^2 + 10kr + r^2 - 3$. It suffices to show that $r^2 - 3$ is not divisible by 5. To check this, calculate $r^2 - 3$ for $r = 0, 1, 2, 3$, and 4.

13. (a) Factor $n^4 - n^2$ into $(n-1)n^2(n+1)$ and note that one of the three consecutive integers $n - 1$, n, $n + 1$ is divisible by 3. Or treat three cases: $n = 3k$, $n = 3k + 1$, $n = 3k + 2$.
(c) Apply parts (a) and (b).

15. (a) This statement is true and depends on the unique factorization of a number as a product of primes. If m and n are written as products of primes, then mn is a product of primes using only the primes used by m and n. Thus, if 3 is among the primes used by mn, it must have been used by m or n, or both. So $3|m$ or $3|n$. [Note that we've described the products of primes in words to avoid complicated notation: $m = p_1^{k_1} \cdots p_l^{k_l}$, $n = q_1^{j_1} \cdots q_m^{j_m}$, where $p_1, \ldots, p_l, q_1, \ldots, q_m$ are primes, $k_1, \ldots, k_l, j_1, \ldots, j_m$ are positive integers, etc., etc.]
(c) The statement holds if and only if d is a prime or $d = 1$.

2.4 Answers

1. (a) Give a direct proof, as in Exercise 9 on page 71.
(c) False. Try 2 + 3 or 2 + 5 or 2 + 11, for instance.

3. (a) True. Say the integers are n, $n + 1$, and $n + 2$. Their sum is $3n + 3 = 3(n + 1)$. This is a direct proof.
(c) True; $n + (n + 1) + (n + 2) + (n + 3) + (n + 4) = 5n + 10$. Alternatively, $(n - 2) + (n - 1) + n + (n + 1) + (n + 2) = 5n$. This is a direct proof.

5. This can be done using four cases; see Example 7.

7. Example 5 on page 70 shows that the set of primes is infinite. An argument like the one in Example 12 shows that some two primes give the same last six digits. This proof is nonconstructive.

9. (a) None of the numbers in the set

$$\{k \in \mathbb{N} : (n+1)! + 2 \le k \le (n+1)! + (n+1)\}$$

is prime, since if $2 \le m \le n+1$, then m divides $(n+1)!$, so m also divides $(n+1)! + m$. The proof is direct.

(c) Simply adjoin 5048 to the list obtained in part (b). Another sequence of seven nonprimes starts with 90.

11. (a) $14 = 2 \cdot 7$ and 7 is odd. So $14 = 2^1 \cdot 7$.

(c) $96 = 2 \cdot 48 = 2 \cdot 2 \cdot 24 = 2 \cdot 2 \cdot 2 \cdot 12 = 2 \cdot 2 \cdot 2 \cdot 2 \cdot 6 = 2 \cdot 2 \cdot 2 \cdot 2 \cdot 2 \cdot 3$, so $96 = 2^5 \cdot 3$.

13. (a) Prove this by cases: n has the form $3k$, $3k+1$, or $3k+2$ for some $k \in \mathbb{N}$. For example, if $n = 3k+1$, then

$$\left\lfloor \frac{n}{3} \right\rfloor + \left\lceil \frac{2n}{3} \right\rceil = \left\lfloor k + \frac{1}{3} \right\rfloor + \left\lceil 2k + \frac{2}{3} \right\rceil = k + (2k+1) = 3k+1 = n.$$

(b) Give a proof by cases.

2.5 Answers

1. (a) Rule 2a and Substitution Rule (b). (c) Rules 10a and 1 and rule (b).

3. (a) Rule 10a [with s for q, using rule (a)] and rule 11a [with s for p and t for q, using rule (a)] and rule (b).

(c) Rule 3a [with $\neg p$ for p, s for q, s for r, using rule (a)] and rule (b).

(e) Rule 3a again, using rule (a).

5. 1,2,3: Hypothesis

4: Rule 16, a tautology

5: 4, 1 and hypothetical syllogism (rule 33)

6: 5, 3 and rule 33

7: 6, 2 and modus tollens (rule 31)

7. (a) $\neg(p \to q) \to ((p \to q) \to p)$.

(c) $p \vee \neg p$.

9. (a) Rule 14 with $q \wedge r$ replacing q.

(c) Rule 8a with $\neg p \wedge r$ replacing p and $q \to r$ replacing q.

11. Interchange all $\vee$ and $\wedge$ in Example 8.

13. Consider the cases $\frac{p \quad q \quad r}{1 \quad 0 \quad 1}$ or $\frac{p \quad q \quad r}{1 \quad 1 \quad 0}$, and $\frac{p \quad q \quad r}{0 \quad 0 \quad 1}$ or $\frac{p \quad q \quad r}{0 \quad 1 \quad 0}$.

15. (a) Take the original proof and change the reason for A from "hypothesis" to "tautology." That is, the proof itself needs no change.

17. (a) See rules 11a and 11b.

(c) No. Any proposition involving only p, q, $\wedge$, and $\vee$ will have truth value 0 whenever p and q both have truth values 0. See Exercise 17 on page 286.

19. Use truth tables.

2.6 Answers

1. The argument is not valid, since the hypotheses are true if C is true and A is false. The error is in treating $A \to C$ and $\neg A \to \neg C$ as if they were equivalent.

3. (a) and (b). See part (c).

(c) The case is no stronger. If C and all A_i's are false, then every hypothesis $A_i \to C$ is true, whether or not C is true.

5. (a) With suggestive notation, the hypotheses are $\neg b \to \neg s$, $s \to p$, and $\neg p$. We can infer $\neg s$ using the contrapositive. We cannot infer either b or $\neg b$. Of course, we can infer more complex propositions, like $\neg p \vee s$ or $(s \wedge b) \to p$.

(c) The hypotheses are $(m \vee f) \to c$, $n \to c$, and $\neg n$. No interesting conclusions, such as m or $\neg c$, can be inferred.

7. (a) True. $A \to B$ is a hypothesis. We showed that $\neg A \to \neg B$ follows from the hypotheses.

(c) True. Since $(B \vee \neg Y) \to A$ is given, $\neg Y \to A$ follows, or equivalently $\neg\neg Y \vee A$.

9. (a) Let c := "my computations are correct," b := "I pay the electric bill," r := "I run out of money," and p := "the power stays on." Then the theorem is

$$\text{if } (c \wedge b) \to r \text{ and } \neg b \to \neg p, \text{ then } (\neg r \wedge p) \to \neg c.$$

1.	$(c \wedge b) \to r$	hypothesis
2.	$\neg b \to \neg p$	hypothesis
3.	$\neg r \to \neg(c \wedge b)$	1; contrapositive rule 9
4.	$\neg r \to (\neg c \vee \neg b)$	3; De Morgan law 8b
5.	$p \to b$	2; contrapositive rule 9
6.	$(\neg r \wedge p) \to [(\neg c \vee \neg b) \wedge b]$	4, 5; rule of inference based on rule 26b
7.	$(\neg r \wedge p) \to [b \wedge (\neg c \vee \neg b)]$	6; commutative law 2b
8.	$(\neg r \wedge p) \to [(b \wedge \neg c) \vee (b \wedge \neg b)]$	7; distributive law 4b
9.	$(\neg r \wedge p)$ $\to [(b \wedge \neg c) \vee \text{ contradiction}]$	8; rule 7b
10.	$(\neg r \wedge p) \to (b \wedge \neg c)$	9; identity law 6a
11.	$(\neg r \wedge p) \to (\neg c \wedge b)$	10; commutative law 2b
12.	$(\neg c \wedge b) \to \neg c$	simplification (rule 17)
13.	$(\neg r \wedge p) \to \neg c$	11, 12; hypothetical syllogism

(c) Let j := "I get the job," w := "I work hard," p := "I get promoted," and h := "I will be happy." Then the theorem is if $(j \wedge w) \to p$, $p \to h$, and $\neg h$, then $\neg j \vee \neg w$.

1.	$(j \wedge w) \to p$	hypothesis
2.	$p \to h$	hypothesis
3.	$\neg h$	hypothesis
4.	$\neg p$	2, 3; modus tollens (rule 31)
5.	$\neg(j \wedge w)$	1, 4; modus tollens (rule 31)
6.	$\neg j \vee \neg w$	5; De Morgan law 8b

11.

1.	$\neg\neg s$	negation of conclusion
2.	s	1; rule 1
3.	$s \to s \vee g$	addition
4.	$s \vee g$	2, 3;-modus ponens
5.	$s \vee g \to p$	hypothesis
6.	p	4, 5; modus ponens
7.	$p \to n$	hypothesis
8.	n	6, 7; modus ponens
9.	$\neg n$	hypothesis
10.	$n \wedge \neg n$	8, 9; conjunction
11.	contradiction	10

13. (a)

1.	$A \wedge \neg B$	hypothesis
2.	A	1; rule 29
3.	$A \to P$	hypothesis
4.	P	2, 3; modus ponens
5.	$\neg B$	1; rule 29
6.	$P \wedge \neg B$	4, 5; rule 34

2.7 Answers

1. $p \oplus q \Longleftrightarrow (p \vee q) \wedge \neg(p \wedge q) \Longleftrightarrow \neg(\neg p \wedge \neg q) \wedge \neg(p \wedge q)$.
Also, $p \oplus q \Longleftrightarrow (p \wedge \neg q) \vee (\neg p \wedge q) \Longleftrightarrow \neg(\neg(p \wedge \neg q) \wedge \neg(\neg p \wedge q))$.

2. If n is even, then $n = 2k$ for some $k \in \mathbb{Z}$, so $n^2 - 3 = 4k^2 - 3$, which is not a multiple of 4. If n is odd, say $n = 2k + 1$, then $n^2 - 3 = 4k^2 + 4k - 2$, which is also not a multiple of 4.

3. Show the contrapositive, using De Morgan law 8a. If $a \le 100$ and $b \le 100$ and $c \le 100$, then $a + b + c \le 300$; i.e., $a + b + c > 300$ is false.

4. "I heard something that could have been written by nobody but Stravinsky."

5. Prove the contrapositive. If $m < 8$ and $n < 8$, then $mn \le 7 \cdot 7 = 49 \le 56$.

6. (a) If $x \le y$, then $\max\{x, y\} = y$ and $\min\{x, y\} = x$, so $\max\{x, y\} \cdot \min\{x, y\} = yx = xy$. If $x > y$, then $\max\{x, y\} = x$ and $\min\{x, y\} = y$, so $\max\{x, y\} \cdot \min\{x, y\} = xy$.
(b) If $x \le y$, then $\max\{x, y\} = y$ and $\min\{x, y\} = x$, so $(\max\{x, y\} - \min\{x, y\})^2 = (y - x)^2 = (x - y)^2$. If $x > y$, then $\max\{x, y\} = x$ and $\min\{x, y\} = y$, so $(\max\{x, y\} - \min\{x, y\})^2 = (x - y)^2$.

7. (a) Rule 10a and Substitution Rule (a).
(b) Rule 3a and Substitution Rule (a).
(c) Rule 8c and Substitution Rule (b).
(d) Rule 1 and Substitution Rule (b).
(e) Rule 10a and Substitution Rule (a).

8. We want a proof of b from $r \to (d \vee b)$, $\neg d$, and r.

	Proof	*Reasons*
1.	$r \to (d \vee b)$	hypothesis
2.	$\neg d$	hypothesis
3.	r	hypothesis
4.	$d \vee b$	1, 3; rule 30 [modus ponens]
5.	b	2, 4; rule 32 [disjunctive syllogism]

9. It proves $P \to Q$ and then proves the contrapositive $\neg Q \to \neg P$, instead of proving $Q \to P$ [which is actually false in this example].

10. The "proof" shows the converse of what is to be proved. It starts by assuming what is to be shown.

11. Prove the contrapositive. If $n \leq 10$ and $m \leq n - 1$, then $n + 2m \leq 3n - 2 \leq 28 < 30$.

12. (a)

	p	q	r	$(\neg p$	$\wedge\ q)$	$\to$	$p \vee r$
	0	0	0	1	0	**1**	0
	0	0	1	1	0	**1**	1
	0	1	0	1	1	**0**	0
	0	1	1	1	1	**1**	1
	1	0	0	0	0	**1**	1
	1	0	1	0	0	**1**	1
	1	1	0	0	0	**1**	1
	1	1	1	0	0	**1**	1
step	1	1	1	2	3	4	2

(b) No. It is false in case p and r are false and q is true.

13. (a) False. One way to create an example is to arrange for A to contain B and C, and $B \neq C$.
(b) True. Consider $x \in B$. If also $x \in A$, then $x \notin A \oplus B$, so $x \notin A \oplus C$ and hence $x \in A \cap C \subseteq C$. Otherwise, $x \notin A$, so $x \in A \oplus B = A \oplus C$ and $x \in C \setminus A \subseteq C$. In each case, $B \subseteq C$. Similarly $C \subseteq B$.

14. (a) Tautology; see rule 19 in Table 2 on page 63.
(b) Not a tautology. Consider the $p = 0$, $q = 1$ line of the truth table.
(c) Same answer as for part (b).
(d) Tautology; see rule 20 in Table 2 on page 63.
(e) Not a tautology. Consider either line in the truth table with $p = 1$, $r = 0$.
(f) Tautology; see rule 17 in Table 2 on page 63. More accurately, combine rules 2b and 17 to get $(p \wedge q) \Longrightarrow (q \wedge p) \Longrightarrow q$.

15. We use the style in Example 3 on page 87.

1, 2, 3.	hypotheses
4.	negation of the conclusion
5.	3, 4; modus tollens rule 31
6.	2; commutative law 2a
7.	5, 6; disjunctive syllogism rule 32
8.	1, 4; disjunctive syllogism rule 32
9.	7, 8; modus ponens rule 30
10.	5, 9; conjunction rule 34
11.	10; rule 7b

16. Since n is not divisible by 2, $n = 2k + 1$ for some k in $\mathbb{Z}$. Thus $n^2 - 1 = 4k^2 + 4k$. Since n is not divisible by 3, $n = 3m \pm 1$ for some m in $\mathbb{Z}$, so $n^2 - 1 = 9m^2 \pm 6m$. Thus $n^2 - 1$ is a multiple of both 4 and 3, so it is a multiple of 12.

17. (a) Case 1: $n = 3m$. Then $n^2 - 1 = 9m^2 - 1$, which is *not* a multiple of 3. Case 2: $n = 3m \pm 1$. Then $n^2 - 1 = 9m^2 \pm 6m = 3m(3m \pm 2)$, which *is* a multiple of 3.
(b) This statement is logically equivalent to the one in part (a).

18.

	Proof	*Reasons*
1.	$p \to (q \to r)$	hypothesis
2.	$q \to p$	hypothesis
3.	$q \to (q \to r)$	1, 2; rule 33 [hypothetical syllogism]
4.	$(q \wedge q) \to r$	3; rule 14
5.	$q \to r$	4; rule 5b and Substitution Rule (b)

19. (a) If $m \cdot n \geq 8$ and $m \not\geq 3$, then $m \leq 2$, so $n \geq 8/2 = 4$.
(b) If $m^2 + n^2 \geq 25$ and $m \not\geq 3$, then $2^2 + n^2 \geq m^2 + n^2 \geq 25$, so $n^2 \geq 25 - 4$ and $n \geq 5$.
(c) If $m^2 + n^2 \geq 25$ and n is a multiple of 3 and $m \leq 3$, then $n^2 \geq 25 - 9 = 4^2$, so $n \geq 4$ and hence $n \geq 6$.

20. The "proof" shows the converse of what is to be proved. It starts by assuming what is to be shown. All the work is useless. In fact, $n^2 - n$ is always even.

21. (a) False. For an example, consider $f: [-1, 1] \to [0, 1]$ given by $f(x) = x^2$, and consider the sets $A_1 = [-1, 0]$ and $A_2 = [0, 1]$. Or consider a constant function.
(b) True. By definition of $f^{\leftarrow}$, $f(f^{\leftarrow}(B)) = \{f(x) : x \in f^{\leftarrow}(B)\} = \{f(x) : f(x) \in B\} \subseteq B$. One can also begin with y in $f(f^{\leftarrow}(B))$ and show that y must be in B.

3.1 Answers

1. (a) R_1 satisfies (AR) and (S).
(c) R_3 satisfies (R), (AS), and (T).
(e) R_5 satisfies only (S).

3. The relations in (a) and (c) are reflexive. The relations in (c), (d), (f), (g), and (h) are symmetric.

5. R_1 satisfies (AR) and (S). R_2 and R_3 satisfy only (S).

7. (a) The divides relation satisfies (R), (AS), and (T).
(c) The converse relation $R^{\leftarrow}$ also satisfies (R), (AS), and (T).

9. (a) The empty relation satisfies (AR), (S), (AS), and (T). The last three properties hold vacuously.

11. Yes. For (R) and (AR), observe that $(x, x) \in R \Longleftrightarrow (x, x) \in R^{\leftarrow}$. For (S) and (AS), just interchange x and y in the conditions for R to get the conditions for $R^{\leftarrow}$. There is no change in meaning.

13. (a) If $E \subseteq R_1$ and $E \subseteq R_2$, then $E \subseteq R_1 \cap R_2$. Alternatively, if R_1 and R_2 are reflexive and $x \in S$, then $(x, x) \in R_1$ and $(x, x) \in R_2$; so $(x, x) \in R_1 \cap R_2$.
(c) Suppose R_1 and R_2 are transitive. If $(x, y), (y, z) \in R_1 \cap R_2$, then $(x, y), (y, z) \in R_1$, so $(x, z) \in R_1$. Similarly, $(x, z) \in R_2$.

15. (a) Suppose R is symmetric. If $(x, y) \in R$, then $(y, x) \in R$ by symmetry, so $(x, y) \in R^{\leftarrow}$. Similarly, $(x, y) \in R^{\leftarrow}$ implies $(x, y) \in R$ [check] so that $R = R^{\leftarrow}$. For the converse, suppose that $R = R^{\leftarrow}$ and show R is symmetric.

17. (a) 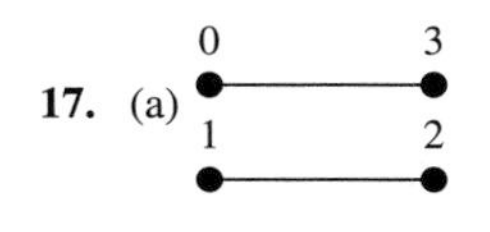

(c) See Figure 1(a).
(e) 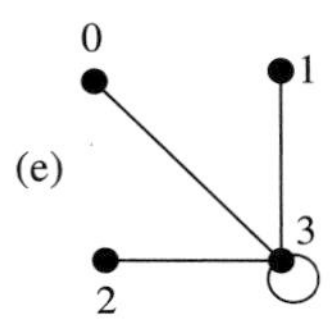

3.2 Answers

1. (a)

e	a	b	c	d	e	f
$\gamma(e)$	(x, v)	(v, x)	(v, w)	(w, y)	(w, y)	(y, x)

(c)

e	a	b	c	d
$\gamma(e)$	(x, w)	(w, x)	(y, z)	(z, y)

3. (a) and (e). Yes.
(c) No. There is no edge from t to x.

5. (a) $x\,w\,y$ or $x\,w\,v\,z\,y$.
(c) $v\,x\,w$ or $v\,z\,w$ or $v\,z\,x\,w$ or $v\,z\,y\,x\,w$.
(e) $z\,y\,x\,w\,v$ or $z\,w\,v$ or $z\,x\,w\,v$ or $z\,y\,v$.

7. Here is one: x • ⇄ • y ⇄ • z

9. (a) (v, w), (v, y), (v, z). Note that (v, z) is in the reachability relation, but not in the adjacency relation.
(c) All of them. The reachability relation is the universal relation.

11. (a) 2. (c) 3. (g) 3.
Parts (d) and (e) are not vertex sequences for paths.

13. (a) Edges e, f, and g are parallel.

15. (a) $A = \{(w, w), (w, x), (x, w), (y, y), (w, y), (y, w), (x, z), (z, x)\}$, whereas R consists of all sixteen ordered pairs of vertices.

17. (a) $c\,a\,d$ or $c\,b\,d$. Both have vertex sequence $y\,v\,w\,w$.
(c) There are four such paths, each with vertex sequence $v\,w\,w\,v\,y$.

3.3 Answers

1. (a) 1. (c) 2.

3. (a) $\begin{bmatrix} -1 & 1 & 4 \\ 0 & 3 & 2 \\ 2 & -2 & 3 \end{bmatrix}$. (c) $\begin{bmatrix} 5 & 8 & 7 \\ 5 & 1 & 5 \\ 7 & 3 & 5 \end{bmatrix}$.

(e) $\begin{bmatrix} 5 & 5 & 7 \\ 8 & 1 & 3 \\ 7 & 5 & 5 \end{bmatrix}$. (g) $\begin{bmatrix} 12 & 12 & 8 \\ 12 & -4 & 8 \\ 8 & 8 & 4 \end{bmatrix}$.

5. (a) $\begin{bmatrix} 1 & -1 & 1 & -1 \\ -1 & 1 & -1 & 1 \\ 1 & -1 & 1 & -1 \end{bmatrix}$. (c) Not defined.

(e) $\begin{bmatrix} 3 & 2 & 5 & 4 \\ 2 & 5 & 4 & 7 \\ 5 & 4 & 7 & 6 \end{bmatrix}$.

7. (a) $\begin{bmatrix} -15 & 45 \\ -5 & 15 \end{bmatrix}$. (c) $\begin{bmatrix} 18 & 54 \\ 6 & 18 \end{bmatrix}$.

9. (a) $\begin{bmatrix} 1 & 0 & 0 \\ 0 & 1 & 0 \\ 0 & 0 & 1 \end{bmatrix}, \begin{bmatrix} 1 & 0 & 0 \\ 0 & 0 & 1 \\ 0 & 1 & 0 \end{bmatrix}, \begin{bmatrix} 0 & 1 & 0 \\ 1 & 0 & 0 \\ 0 & 0 & 1 \end{bmatrix}, \begin{bmatrix} 0 & 1 & 0 \\ 0 & 0 & 1 \\ 1 & 0 & 0 \end{bmatrix}, \begin{bmatrix} 0 & 0 & 1 \\ 1 & 0 & 0 \\ 0 & 1 & 0 \end{bmatrix}, \begin{bmatrix} 0 & 0 & 1 \\ 0 & 1 & 0 \\ 1 & 0 & 0 \end{bmatrix}$.

11. (a) $\begin{bmatrix} 1 & 0 \\ n & 1 \end{bmatrix}$. (c) $\{n \in \mathbb{N} : n \text{ is odd}\}$.

13. (a) The (i, j) entry of $a\mathbf{A}$ is $a\mathbf{A}[i, j]$. Similarly for $b\mathbf{B}$, and so the (i, j) entry of $a\mathbf{A} + b\mathbf{B}$ is $a\mathbf{A}[i, j] + b\mathbf{B}[i, j]$. So the (i, j) entry of $c(a\mathbf{A} + b\mathbf{B})$ is $ca\mathbf{A}[i, j] + cb\mathbf{B}[i, j]$. A similar discussion shows that this is the (i, j) entry of $(ca)\mathbf{A} + (cb)\mathbf{B}$. Since their entries are equal, the matrices $c(a\mathbf{A} + b\mathbf{B})$ and $(ca)\mathbf{A} + (cb)\mathbf{B}$ are equal.

(c) The (j, i) entries of both $(a\mathbf{A})^T$ and $a\mathbf{A}^T$ equal $a\mathbf{A}[i, j]$. Here $1 \le i \le m$ and $1 \le j \le n$. So the matrices are equal.

15. (a) $\begin{bmatrix} 0 & 0 & 1 & 0 \\ 0 & 0 & 1 & 0 \\ 0 & 0 & 0 & 0 \\ 0 & 0 & 1 & 0 \end{bmatrix}$. (c) $\begin{bmatrix} 0 & 0 & 0 & 0 \\ 0 & 0 & 0 & 2 \\ 0 & 0 & 1 & 0 \\ 0 & 1 & 0 & 0 \end{bmatrix}$.

17. (a) (c)

19. (a) $\begin{bmatrix} 0 & 0 & 0 & 1 \\ 0 & 0 & 1 & 0 \\ 0 & 1 & 0 & 0 \\ 1 & 0 & 0 & 0 \end{bmatrix}$. (c) See Example 5(a) on page 110.

(e) $\begin{bmatrix} 0 & 0 & 0 & 1 \\ 0 & 0 & 0 & 1 \\ 0 & 0 & 0 & 1 \\ 1 & 1 & 1 & 1 \end{bmatrix}$.

21. (a) $\begin{bmatrix} 1 & 1 & 1 \\ 0 & 1 & 1 \\ 0 & 0 & 1 \end{bmatrix}$. (c) $\begin{bmatrix} 1 & 0 & 0 \\ 0 & 1 & 0 \\ 0 & 0 & 1 \end{bmatrix}$. (e) $\begin{bmatrix} 1 & 1 & 1 \\ 0 & 1 & 0 \\ 0 & 1 & 0 \end{bmatrix}$.

(g) $\begin{bmatrix} 0 & 0 & 0 \\ 0 & 1 & 0 \\ 0 & 0 & 0 \end{bmatrix}$. (i) $\begin{bmatrix} 0 & 0 & 0 \\ 1 & 1 & 0 \\ 0 & 0 & 1 \end{bmatrix}$.

3.4 Answers

1. (a) is an equivalence relation.

(c) There are lots of Americans who live in no state, e.g., the residents of Washington, D.C., so (R) fails for $\sim$.

(e) is not an equivalence relation because $\approx$ is not transitive.

3. Very much so. See Example 5 on page 97.

5. (a) The possibilities are

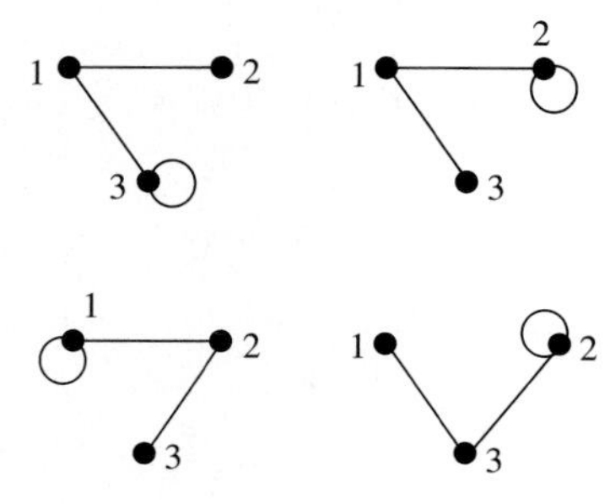

(c) The four equivalence classes have representatives

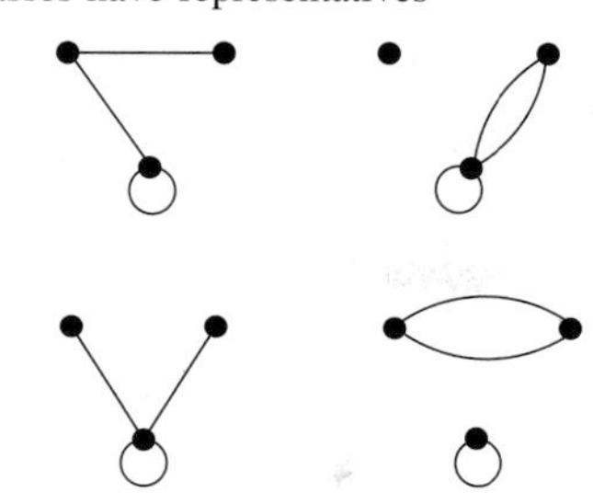

7. (a) Verify directly, or apply Theorem 2(a) on page 117 with $f(m) = m^2$ for $m \in \mathbb{Z}$.

9. (a) There are infinitely many classes: $\{0\}$ and the classes $\{n, -n\}$ for $n \in \mathbb{P}$.

11. Apply Theorem 2, using the length function. The equivalence classes are the sets Σ^k, $k \in \mathbb{N}$.

13. (a) Use brute force or Theorem 2(a) with part (b).

15. (a) Not well-defined: depends on the representative. For example, $[3] = [-3]$ and $-3 \le 2$. If the definition made sense, we would have $[3] = [-3] \le [2]$ and hence $3 \le 2$.
(c) Nothing wrong. If $[m] = [n]$, then $m^4 + m^2 + 1 = n^4 + n^2 + 1$.

17. (a) $\cong$ is reflexive by its definition, and it's symmetric since equality "=" and R are. For transitivity, consider $u \cong v$ and $v \cong w$. If $u = v$ or if $v = w$, then $u \cong w$ is clear. Otherwise, (u, v) and (v, w) are in R, so (u, w) is in R. Either way, $u \cong w$. Thus $\cong$ is transitive.

19. For one-to-one, observe that $\theta([s]) = \theta([t])$ implies $f(s) = f(t)$ implies $s \sim t$, and this implies $[s] = [t]$. Clearly, θ maps $[S]$ into $f(S)$. To see that θ maps onto $f(S)$, consider $y \in f(S)$. Then $y = f(s_0)$ for some $s_0 \in S$. Hence $[s_0]$ belongs to $[S]$ and $\theta([s_0]) = f(s_0) = y$. That is, y is in $\text{Im}(\theta)$. We've shown $f(S) \subseteq \text{Im}(\theta)$, so θ maps $[S]$ onto $f(S)$.

3.5 Answers

1. (a) $q = 6$, $r = 2$. (c) $q = -7$, $r = 1$. (e) $q = 5711$, $r = 31$.

3. (a) -4, 0, 4. (c) -2, 2, 6. (e) -4, 0, 4.

5. (a) 1. (c) 1. (e) 0.

7. (a) 3 and 2. (c) $m *_{10} k$ is the last [decimal] digit of $m * k$.

9.

$+_4$	0	1	2	3
0	0	1	2	3
1	1	2	3	0
2	2	3	0	1
3	3	0	1	2

$*_4$	0	1	2	3
0	0	0	0	0
1	0	1	2	3
2	0	2	0	2
3	0	3	2	1

11. Solutions are 1, 3, 2, and 4, respectively.

13. (a) $m \equiv n \pmod 1$ for all $m, n \in \mathbb{Z}$. There is only one equivalence class.
(c) $0 = 0 +_1 0$ and $0 = 0 *_1 0$.

15. (a) $n = 1000\mathbf{a} + 100\mathbf{b} + 10\mathbf{c} + \mathbf{d} = \mathbf{a} + \mathbf{b} + \mathbf{c} + \mathbf{d} + 9 \cdot (111\mathbf{a} + 11\mathbf{b} + \mathbf{c}) \equiv \mathbf{a} + \mathbf{b} + \mathbf{c} + \mathbf{d} \pmod 9$, or use Theorem 2 together with $1000 \equiv 100 \equiv 10 \equiv 1 \pmod 9$.

17. Like Exercise 15. Note that $1000 = 91 \cdot 11 - 1$, $100 = 9 \cdot 11 + 1$, $10 = 1 \cdot 11 - 1$, so $1000\mathbf{a} + 100\mathbf{b} + 10\mathbf{c} + \mathbf{d} \equiv -\mathbf{a} + \mathbf{b} - \mathbf{c} + \mathbf{d} \equiv 0 \pmod{11}$ if and only if $\mathbf{a} - \mathbf{b} + \mathbf{c} - \mathbf{d} \equiv 0 \pmod{11}$.

19. We have $q \cdot p - q' \cdot p = r' - r$, so $r' - r$ is a multiple of p. But $-p < -r \le r' - r \le r' < p$, and 0 is the only multiple of p between $-p$ and p. Thus $r' = r$, so $0 = (q - q') \cdot p$ and $q = q'$.

21. (a) By Theorem 3(a)

$$(m \text{ MOD } p) +_p (n \text{ MOD } p) = (m + n) \text{ MOD } p = (n + m) \text{ MOD } p$$
$$= (n \text{ MOD } p) +_p (m \text{ MOD } p).$$

Since $m, n \in \mathbb{Z}(p)$, $m \text{ MOD } p = m$ and $n \text{ MOD } p = n$.

3.6 Answers

1. (a) and (b). Eight edges and the sum is 16. Note that $\deg(v_1) = 6$ and $\deg(v_3) = 3$.

(c)

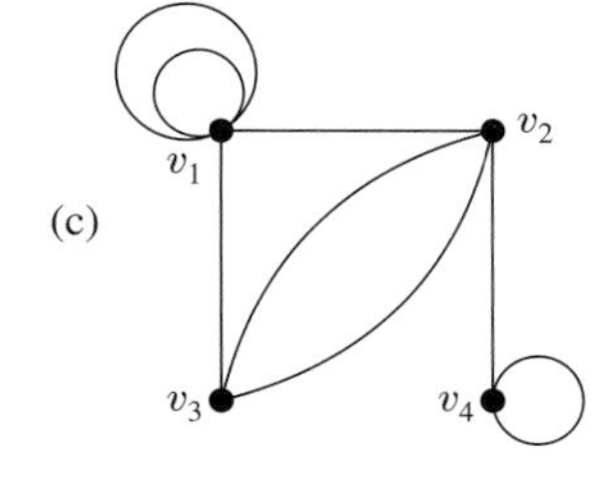

2. $\mathbf{A} = \begin{bmatrix} 3 & 5 & 7 \\ 4 & 6 & 8 \\ 5 & 7 & 9 \end{bmatrix}$ and $\mathbf{B} = \begin{bmatrix} 1 & 2 & 3 \\ 5 & 6 & 7 \\ 10 & 11 & 12 \end{bmatrix}$.

(a) $\begin{bmatrix} 3 & 4 & 5 \\ 5 & 6 & 7 \\ 7 & 8 & 9 \end{bmatrix}$. (b) $\begin{bmatrix} 6 & 9 & 12 \\ 9 & 12 & 15 \\ 12 & 15 & 18 \end{bmatrix}$.

(c) 19. (d) 54. (e) 18.

3. (a) This is an equivalence relation. For example, if $m + n$ and $n + p$ are even, so is $m + p = (m + n) + (n + p) - 2n$, so the relation is transitive. There are two equivalence classes, $\{n \in \mathbb{N} : n \text{ is even}\}$ and $\{n \in \mathbb{N} : n \text{ is odd}\}$.
(b) This is the equivalence relation $\cong$ in Example 5 on page 114. It is transitive because, if there's a path from u to v and one from v to w, then there's a path from u to w. The equivalence classes are sets $[u] = \{u\} \cup \{v \in V : \text{ there is a path from } u \text{ to } v\}$. These sets are called connected components in Chapter 6.
(c) This may not be an equivalence relation. As in part (b), R_3 is transitive, but it's not necessarily symmetric. There might be a path from u to v, yet no path from v to u. See, e.g., Figure 3 on page 102.

4. (a) Seven: $[0], [1], \ldots, [6]$.
(b) Some examples are: 1, 8, 15 in $[1]$; $-73, -66$, 4 in $[-73]$; 73, 80, 3 in $[73]$; -73, -66, 4 in $[4]$; 2^{250}, $2^{250} + 7$, $2^{250} + 14$ in $[2^{250}]$.

5. The (j, k) entry of $A + A^T$ is $a_{jk} + a_{kj}$, while the (k, j) entry is $a_{kj} + a_{jk}$.

6. (a) Check (R), (S), and (T) directly. Or apply Theorem 2 on page 117, using $f : S \to \mathbb{Z}(5)$, where $f(m) = m^2 \text{ MOD } 5$. Note that $f(m) = f(n)$ if and only if $m^2 \equiv n^2 \pmod 5$.
(b) The image values of f are 1, 4, 4, 1, 0, 1, 4, respectively, so the equivalence classes are $\{1, 4, 6\}$, $\{2, 3, 7\}$, and $\{5\}$.

7. The relation R is reflexive and transitive, but it is not symmetric. It is not an equivalence relation.

8. (a) $n \equiv 1 \pmod 2$ unless $2|n$; i.e., n is odd unless n is even.
(b) If $n \equiv \pm 1 \pmod 3$, then by Theorem 2 on page 122, $n^2 \equiv 1^2 \pmod 3 \equiv 1 \pmod 3$.

9. No. For example, we would need $f(1/2) = f(2/4)$, but $f(1/2) = -1$ and $f(2/4) = -2$.

10. (a) Yes. $m \equiv m \pmod 3$ or $n \equiv n \pmod 5$ is certainly true, so $((m, n), (m, n)) \in R$.
(b) Yes. $m \equiv p \pmod 3$ or $n \equiv q \pmod 5$ implies $p \equiv m \pmod 3$ or $q \equiv n \pmod 5$.
(c) No. Counterexamples are easy to find. For example, $((0, 0), (0, 1))$ and $((0, 1), (1, 1))$ are in R, but $((0, 0), (1, 1))$ is not. Give another example.
(d) No, because the relation is not transitive.

11. (a) Otherwise, there exist $x, y, z \in S$ with $x \approx y$, $y \not\approx z$, and $x \approx z$. Since $x \approx y$, $y \approx x$ by symmetry. Since $y \approx x$ and $x \approx z$, $y \approx z$ by transitivity, a contradiction.
(b) Apply part (a) to the equivalence relation $\equiv \pmod p$ on $\mathbb{Z}$.

12. (a) Yes. If $[n]_{10} = [m]_{10}$, then $n \equiv m \pmod{10}$; i.e., 10 divides $n - m$. Then 20 divides $2n - 2m$, so $2n - 3 \equiv 2m - 3 \pmod{20}$ and $[2n - 3]_{20} = [2m - 3]_{20}$.
(b) No. Imitating the proof for part (a) won't work. For example, $[1]_{10} = [11]_{10}$, but $[5 \cdot 1 - 3]_{20} = 2 \neq 12 = [5 \cdot 11 - 3]_{20}$.

4.1 Answers

1. (a)

$m = 13, n = 73$	q	r	$r \geq m$
Initially	0	73	True
After the first pass	1	60	True
After the second pass	2	47	True
After the third pass	3	34	True
After the fourth pass	4	21	True
After the fifth pass	5	8	False

(c)

$m = 73, n = 13$	q	r	$r \geq m$
Initially	0	13	False

3. (a) 0, 3, 9, 21, 45. (c) 1, 1, 1, 1, 1.

5. (a)

	m	n
initially	0	0
after first pass	1	1
after second pass	4	2
after third pass	9	3
after fourth pass	16	4

7. (a) 4, 16, 36, 64.

9. (a) If $m + n$ is even, so is $(m + 1) + (n + 1) = (m + n) + 2$. Of course, we didn't need the guard $1 \leq m$ to see this.

11. (a) Yes. If $i < j^2$ and $j \geq 1$, then $i + 2 < j^2 + 2 < j^2 + 2j + 1 = (j + 1)^2$.
(c) No. Consider the case $i = j = 0$.

13. (a) $b \geq 2$. (c) $b \in \mathbb{N}$.

15. No. The sequence "$k := k^2$, print k" changes the value of k. Algorithm A prints 1, 4, and stops. Algorithm B prints 1, 4, 9, 16, because "for" resets k each time.

17. (a) Yes. new $r < 73$ by definition of MOD.
(c) This is an invariant vacuously, because $r \leq 0$ and $r > 0$ cannot both hold at the start of the loop.

19. The sets in (a), (c), and (f) have smallest elements, (f) because $n! > 2^{1000}$ for all large enough n. The sets in (b) and (e) fail, because they aren't subsets of $\mathbb{N}$. The set in (d) fails because it is empty.

21. (a) This is an invariant: if $r > 0$ and a, b, and r are multiples of 5, then the new values b, r, and $b \text{ MOD } r$ are multiples of 5. Note that $b \text{ MOD } r = b - (b \text{ DIV } r) \cdot r$.
(c) This is an invariant: If $r < b$ and $r > 0$ on entry into the loop, then new $r <$ new b; i.e., $b \text{ MOD } r < r$, by definition of MOD.

23. (a) Yes. If $p \wedge q$ holds at the start of the loop, then p is true and q is true. If g is also true, then, since p and q are invariants of the loop, they are both true at the end; hence $p \wedge q$ is also true at the end.
(c) Not necessarily. For example, if p = "$n \geq 0$," q = "$3 < 2$," g = "$2 + 2 = 4$", and S = "$n := n + 2$," then p and q are invariants of the loop; but if $n = -1$ on entry to the loop, then $\neg p$ is true on entry, but false after one pass through.
(e) Yes.

25. (a)

$a = 2, n = 11$	p	q	i
initially	1	2	11
after first pass	2	4	5
after second pass	8	16	2
after third pass	8	256	1
after fourth pass	$256 \cdot 8$	256^2	0

(b) You need to show that if $q^i p = a^n$, then $(\text{new } q)^{(\text{new } i)}(\text{new } p) = a^n$. Consider the cases when i is odd and i is even. Since i runs through a decreasing sequence of nonnegative integers, eventually $i = 0$ and the algorithm exits the loop. At this point $p = a^n$.

27. The Well-Ordering Principle fails for $\mathbb{R}$ and $[0, \infty)$. You should provide subsets that do not have least elements.

4.2 Answers

Induction proofs should be written carefully and completely. These answers will serve only as guides, *not* as models.

1. This is clear, because both n^5 and n are even if n is even, and both are odd if n is odd.

3. (a) If $k^5 - k + 1$ is a multiple of 5 for some $k \in \mathbb{P}$, then, just as in Example 1,

$$(k+1)^5 - (k+1) + 1 = (k^5 - k + 1) + 5(k^4 + 2k^3 + 2k^2 + k),$$

so $(k+1)^5 - (k+1) + 1$ is also a multiple of 5.

5. Check the basis. For the inductive step, assume the equality holds for k. Then

$$\sum_{i=1}^{k+1} i^2 = \sum_{i=1}^{k} i^2 + (k+1)^2 = \frac{k(k+1)(2k+1)}{6} + (k+1)^2.$$

Some algebra shows that the right-hand side equals

$$\frac{(k+1)(k+2)(2k+3)}{6},$$

so the equality holds for $k + 1$ whenever it holds for k.

7. (a) Take $n = 37^{20}$ in Exercise 1.
(c) By (a), (b), and Exercise 1, $(37^{500} - 37^{100}) + (37^{100} - 37^{20}) + (37^{20} - 37^4)$ is a multiple of 10.
(e) By (c) and (d), as in (c).

9. The basis is "$s_0 = 2^0 a + (2^0 - 1)b$," which is true since $2^0 = 1$ and $s_0 = a$. Assume inductively that $s_k = 2^k a + (2^k - 1)b$ for some $k \in \mathbb{N}$. The algebra in the inductive step is

$$2 \cdot [2^k a + (2^k - 1)b] + b = 2^{k+1} a + 2^{k+1} b - 2b + b.$$

11. Show that $11^{k+1} - 4^{k+1} = 11 \cdot (11^k - 4^k) + 7 \cdot 4^k$. Imitate Example 2(d).

13. (a) Suppose that $\sum_{i=0}^{k} 2^i = 2^{k+1} - 1$ and $0 \leq k$. Then

$$\sum_{i=0}^{k+1} 2^i = \left(\sum_{i=0}^{k} 2^i\right) + 2^{k+1} = 2^{k+1} - 1 + 2^{k+1} = 2^{k+2} - 1,$$

so the equation still holds for the new value of k.

(c) Yes. $\sum_{i=0}^{0} 2^i = 1 = 2^1 - 1$ initially, so the loop never exits and the invariant is true for every value of k in $\mathbb{N}$.

15. (a) $1 + 3 + \cdots + (2n-1) = n^2$.

17. (a) Assume $p(k)$ is true. Then $(k+1)^2 + 5(k+1) + 1 = (k^2 + 5k + 1) + (2k + 6)$. Since $k^2 + 5k + 1$ is even by assumption and $2k + 6$ is clearly even, $p(k+1)$ is true.

19. *Hint:* $5^{k+1} - 4(k+1) - 1 = 5(5^k - 4k - 1) + 16k$.

21. *Hints:*

$$\frac{1}{n+2} + \cdots + \frac{1}{2n+2} = \left(\frac{1}{n+1} + \cdots + \frac{1}{2n}\right) + \left(\frac{1}{2n+1} + \frac{1}{2n+2} - \frac{1}{n+1}\right)$$

and

$$\frac{1}{2n+1} + \frac{1}{2n+2} - \frac{1}{n+1} = \frac{1}{2n+1} - \frac{1}{2n+2}.$$

Alternatively, to avoid induction, let $f(n) = \sum_{i=1}^{n} \frac{1}{i}$ and write both sides in terms of f. The left-hand side is $f(2n) - f(n)$, and the right-hand side is

$$1 + \left(\frac{1}{2}\right) + \left(\frac{1}{3}\right) + \cdots + \left(\frac{1}{2n}\right) - 2 \cdot \left[\left(\frac{1}{2}\right) + \left(\frac{1}{4}\right) + \cdots + \left(\frac{1}{2n}\right)\right]$$

$$= f(2n) - 2 \cdot \frac{1}{2} \cdot f(n).$$

23. *Hints:* $5^{k+2} + 2 \cdot 3^{k+1} + 1 = 5(5^{k+1} + 2 \cdot 3^k + 1) - 4(3^k + 1)$. Show that $3^n + 1$ is always even.

25. Here $p(n)$ is the proposition "$|\sin nx| \le n|\sin x|$ for all $x \in \mathbb{R}$." Clearly, $p(1)$ holds. By algebra and trigonometry,

$$|\sin(k+1)x| = |\sin(kx + x)| = |\sin kx \cos x + \cos kx \sin x|$$
$$\le |\sin kx| \cdot |\cos x| + |\cos kx| \cdot |\sin x| \le |\sin kx| + |\sin x|.$$

Now assume $p(k)$ is true and show $p(k+1)$ is true.

4.3 Answers

1. (a) $k = 2$. (c) $k = 12$.

3. (a) $n!$. Note that $3^n \ne O(2^n)$, but $3^n = O(n!)$; see Example 3(b) or Exercise 12.

5. (a) True. Take $C \ge 2$.
(c) False. $2^{2n} \le C \cdot 2^n$ only for $2^n \le C$; i.e., only for $n \le \log_2 C$.
(e) True, since $2^{n+1} = 2 \cdot 2^n = O(2^n)$ and $2^n = \frac{1}{2} \cdot 2^{n+1} = O(2^{n+1})$
(g) False. It is false that $2^{2n} = O(2^n)$; see the answer to (c).

7. (a) False. If $40^n \le C \cdot 2^n$ for all large n, then $20^n \le C$ for all large n.
(c) False. If $(2n)! \le C \cdot n!$ for all large n, then $(n+1)! \le C \cdot n!$ for large n, so $n + 1 \le C$ for large n, which is impossible.

9. (a) $k = 5$. (c) $k = 6$.
(e) $k = 2$. The sequence is the sequence t_n in Exercise 11(a).

11. (a) Clearly, $t_n \le n + n + \cdots + n$ [n terms] $= n^2$. In fact, $t_n = \frac{1}{2} \cdot n(n+1)$ as shown in Example 2(b) on page 139.

13. (a) We are given $s(n) = 3n^4 + a(n)$ and $t(n) = 2n^3 + b(n)$, where $a(n) = O(n)$ and $b(n) = O(n)$. Now $s(n) + t(n) = 3n^4 + [a(n) + 2n^3 + b(n)]$ and $a(n) + 2n^3 + b(n) = O(n^3)$, since $a(n)$, $2n^3$, and $b(n)$ are all $O(n^3)$ sequences. Theorem 2(b) with $a(n) = n^3$ is being used here twice.

15. (a) $s(n)/t(n) = n^4$ is not $O(n^3)$.

17. (a) Algorithm B is more efficient if $5n \log_2 n \ge 80n$; i.e., if $n \ge 2^{16} = 65{,}536$.
(c) $n \ge 2^{32} \approx 4.3 \times 10^9$.

19. (a) Let $\text{DIGIT}(n) = m$. Then $10^{\text{DIGIT}(n)}$ is written as a 1 followed by m 0's, so it's larger than any m-digit number, such as n. And 10^{m-1} is a 1 followed by $m-1$ 0's, so it's the smallest m-digit number.
(c) Use part (a). In detail, $\text{DIGIT}(n) - 1 = \log_{10} 10^{\text{DIGIT}(n)-1} \le \log_{10} n$, so $\text{DIGIT}(n) \le 1 + \log_{10} n = O(\log_{10} n)$.

21. (a) Since i starts as n and is replaced by $\lfloor \frac{i}{2} \rfloor$ in each pass through the loop, we have $i \le \frac{n}{2^k}$ at the end of the kth pass. The guard would fail for the next pass if $k > \log_2 n$, since then $i \le \frac{n}{2^k} < 1$.

23. There are positive constants A, B, C, and D so that

$$s(n) \le A \cdot n^5, \quad n^5 \le B \cdot s(n), \quad t(n) \le C \cdot n^2, \quad \text{and} \quad n^2 \le D \cdot t(n)$$

for sufficiently large n. Therefore,

$$\frac{s(n)}{t(n)} \leq \frac{A \cdot n^5}{n^2/D} = A \cdot D \cdot n^3$$

for sufficiently large n, so $s(n)/t(n) = O(n^3)$. Similarly, $n^3 \leq BC \cdot \frac{s(n)}{t(n)}$ for sufficiently large n, so $n^3 = O(s(n)/t(n))$. Hence $s(n)/t(n) = \Theta(n^3)$.

25. (a) Assume $c \neq 0$ since the result is obvious for $c = 0$. There is a $C > 0$ so that $|s(n)| \leq C \cdot |a(n)|$ for large n. Then $|c \cdot s(n)| \leq C \cdot |c| \cdot |a(n)|$ for large n.

4.4 Answers

1. (a) 1, 2, 1, 2, 1, 2, 1, 2, ...

3. (a) $\text{SEQ}(n) = 3^n$.

5. No. It's okay up to SEQ(100), but SEQ(101) is not defined, since we cannot divide by zero. If, in (R), we restricted n to be ≤ 100, we would obtain a recursively defined *finite* sequence.

7. (a) 1, 3, 8.
(c) $s_3 = 22$, $s_4 = 60$.

9. (a) 1, 1, 2, 4.
(c) Ours doesn't, since $t_0 \neq \frac{1}{2}$.

11. (a) $a_6 = a_5+2a_4 = a_4+2a_3+2a_4 = 3(a_3+2a_2)+2a_3 = 5(a_2+2a_1)+6a_2 = 11(a_1+2a_0)+10a_1 = 11 \cdot 3+10 = 43$. This calculation uses only two intermediate value addresses at any given time. Other recursive calculations are possible that use more.

13. Follow the hint. Suppose $S \neq \emptyset$. By the Well-Ordering Principle on page 131, S has a smallest member, say m. Since $s_1 = 2 = \text{FIB}(3)$ and $s_2 = 3 = \text{FIB}(4)$, we have $m \geq 3$. Then $s_m = s_{m-1} + s_{m-2} = \text{FIB}(m-1+2) + \text{FIB}(m-2+2)$ [since m was the smallest bad guy] $= \text{FIB}(m+2)$ [by recursive definition of FIB in Example 3(a)], contrary to $m \in S$. Thus $S = \emptyset$.

15. $\text{SEQ}(n) = 2^{n-1}$ for $n \geq 1$.

17. (a) $A(1) = 1$. $A(n) = n \cdot A(n-1)$.
(c) Yes.

19. (a) $\{1, 110, 1200\}$.

21. (a) (B) $\text{UNION}(1) = A_1$,
(R) $\text{UNION}(n) = A_n \cup \text{UNION}(n-1)$ for $n \geq 2$.
(c) (B) $\text{INTER}(1) = A_1$,
(R) $\text{INTER}(n+1) = A_{n+1} \cap \text{INTER}(n)$ for $n \in \mathbb{P}$
[or $\text{INTER}(n) = A_n \cap \text{INTER}(n-1)$ for $n \geq 2$].

4.5 Answers

1. $s_n = 3 \cdot (-2)^n$ for $n \in \mathbb{N}$.

3. We prove this by induction. $s_n = a^n \cdot s_0$ holds for $n = 0$, because $a^0 = 1$. If it holds for some n, then $s_{n+1} = as_n = a(a^n \cdot s_0) = a^{n+1} \cdot s_0$, so the result holds for $n+1$.

5. $s_0 = 3^0 - 2 \cdot 0 \cdot 3^0 = 1$. $s_1 = 3^1 - 2 \cdot 1 \cdot 3^1 = -3$. For $n \geq 2$,

$$\begin{aligned} 6s_{n-1} - 9s_{n-2} &= 6[3^{n-1} - 2(n-1) \cdot 3^{n-1}] - 9[3^{n-2} - 2(n-2) \cdot 3^{n-2}] \\ &= 2[3^n - 2(n-1) \cdot 3^n] - [3^n - 2(n-2) \cdot 3^n] \\ &= 3^n[2 - 4(n-1) - 1 + 2(n-2)] \\ &= 3^n[1 - 2n] = s_n. \end{aligned}$$

7. This time $c_1 = 3$ and $c_2 = 0$, so $s_n = 3 \cdot 2^n$ for $n \in \mathbb{N}$.

9. Solve $1 = c_1 + c_2$ and $2 = c_1 r_1 + c_2 r_2$ for c_1 and c_2 to obtain $c_1 = (1 + r_1)/\sqrt{5}$ and $c_2 = -(1 + r_2)/\sqrt{5}$. Hence

$$s_n = \frac{1}{\sqrt{5}}(r_1^n + r_1^{n+1} - r_2^n - r_2^{n+1}) \text{ for all } n,$$

where r_1, r_2 are as in Example 3 on page 161.

11. (a) $r_1 = -3$, $r_2 = 2$, $c_1 = c_2 = 1$, so $s_n = (-3)^n + 2^n$ for $n \in \mathbb{N}$.
(c) Here the characteristic equation has one solution, $r = 2$. Then $c_1 = 1$ and $c_2 = 3$, so $s_n = 2^n + 3n \cdot 2^n$ for $n \in \mathbb{N}$.
(e) $s_{2n} = 1$, $s_{2n+1} = 4$ for all $n \in \mathbb{N}$.
(g) $s_n = (-3)^n$ for $n \in \mathbb{N}$.

13. (a) By Theorem 1(a), there are constants C and D such that for $n \geq 2$ we have

$$s_n = Cr_1^n + Dr_2^n = \left(C + D\left(\frac{r_2}{r_1}\right)^n\right) r_1^n \leq (|C| + |D|) r_1^n.$$

(b) Any sequence $s_n = C2^n + Dn2^n$ with C and D different from 0 will work.
(c) By Theorem 1(b), we have $s_n = O(nr_1^n)$.

15. (a) $s_{2^m} = 2^m + 3 \cdot (2^m - 1) = 2^{m+2} - 3$.
(c) $s_{2^m} = \frac{5}{2} \cdot 2^m \cdot m$.
(e) $s_{2^m} = 7 - 6 \cdot 2^m$. (g) $s_{2^m} = (6 - m)2^{m-1}$.

17. $s_{2^m} = 2^m[s_1 + \frac{1}{2}(2^m - 1)]$. Verify that this formula satisfies $s_{2^0} = s_1$ and $s_{2^{m+1}} = 2s_{2^m} + (2^m)^2$.

19. (a) $t_{2^m} = b^m t_1 + b^{m-1} \cdot \sum_{i=0}^{m-1} \frac{f(2^i)}{b^i}$.

4.6 Answers

1. The First Principle is adequate for this. For the inductive step, use the identity $4n^2 - n + 8(n+1) - 5 = 4n^2 + 7n + 3 = 4(n+1)^2 - (n+1)$.

3. Show that $n^5 - n$ is always even. Then use the identity $(n+1)^5 = n^5 + 5n^4 + 10n^3 + 10n^2 + 5n + 1$ [from the binomial theorem]. Use the First Principle of induction to show that $n^5 - n$ is always divisible by 5. See Example 1 on page 137.

5. Yes. The oddness of a_n depends only on the oddness of a_{n-1}, since $2a_{n-2}$ is even whether a_{n-2} is odd or not.

7. (b) $a_n = 1$ for all $n \in \mathbb{N}$.
(c) The basis needs to be checked for $n = 0$ and $n = 1$. For the inductive step, consider $n \geq 2$ and assume $a_k = 1$ for $0 \leq k < n$. Then $a_n = \frac{a_{n-1}^2 + a_{n-2}}{a_{n-1} + a_{n-2}} = \frac{1^2+1}{1+1} = 1$. This completes the inductive step, so $a_n = 1$ for all $n \in \mathbb{N}$ by the Second Principle of Induction.

9. (b) $a_n = n^2$ for all $n \in \mathbb{N}$.
(c) The basis needs to be checked for $n = 0$ and $n = 1$. For the inductive step, consider $n \geq 2$ and assume that $a_k = k^2$ for $0 \leq k < n$. To complete the inductive step, note that

$$a_n = \frac{1}{4}(a_{n-1} - a_{n-2} + 3)^2 = \frac{1}{4}[(n-1)^2 - (n-2)^2 + 3]^2 = \frac{1}{4}[2n]^2 = n^2.$$

11. (b) The basis needs to be checked for $n = 0$, 1, and 2. For the inductive step, consider $n \geq 3$ and assume that a_k is odd for $0 \leq k < n$. Then a_{n-1}, a_{n-2}, and a_{n-3} are all odd. Since the sum of three odd integers is odd [if not obvious, prove it], a_n is also odd.
(c) Since the inequality is claimed for $n \geq 1$ and since you will want to use the identity $a_n = a_{n-1} + a_{n-2} + a_{n-3}$ in the inductive step, you will need $n - 3 \geq 1$ in the inductive step. So check the basis for $n = 1$, 2, and 3. For the inductive step, consider $n \geq 4$ and assume that $a_k \leq 2^{k-1}$ for $1 \leq k < n$. To complete the inductive step, note that

$$a_n = a_{n-1} + a_{n-2} + a_{n-3} < 2^{n-2} + 2^{n-3} + 2^{n-4} = \frac{7}{8} \cdot 2^{n-1} < 2^{n-1}.$$

13. (a) 2, 3, 4, 6.
(b) The inequality must be checked for $n = 3$, 4, and 5 before applying the Second Principle of Mathematical Induction on page 167 to $b_n = b_{n-1} + b_{n-3}$. For the inductive step, consider $n \geq 6$ and assume $b_k \geq 2b_{k-2}$ for $3 \leq k < n$. Then

$$b_n = b_{n-1} + b_{n-3} \geq 2b_{n-3} + 2b_{n-5} = 2b_{n-2}.$$

(c) The inequality must be checked for $n = 2$, 3, and 4. Then use the Second Principle of Mathematical Induction and part (b). For the inductive step, consider $n \geq 5$ and assume $b_k \geq (\sqrt{2})^{k-2}$ for $2 \leq k < n$. Then

$$b_n = b_{n-1} + b_{n-3} \geq 2b_{n-3} + b_{n-3} = 3b_{n-3} \geq 3(\sqrt{2})^{n-5}$$
$$> (\sqrt{2})^3(\sqrt{2})^{n-5} = (\sqrt{2})^{n-2}.$$

Note that $3 > (\sqrt{2})^3 \approx 2.828$. This can also be proved without using part (b):

$$b_n \geq (\sqrt{2})^{n-3} + (\sqrt{2})^{n-5} = (\sqrt{2})^{n-2} \cdot \left[\frac{1}{\sqrt{2}} + \frac{1}{2^{3/2}}\right] > (\sqrt{2})^{n-2}.$$

15. Check for $n = 0$ and 1 before applying induction. It may be simpler to prove "$\text{SEQ}(n) \leq 1$ for all n" separately from "$\text{SEQ}(n) \geq 0$ for all n." For example, assume that $n \geq 2$ and that $\text{SEQ}(k) \leq 1$ for $0 \leq k < n$. Then

$$\text{SEQ}(n) = (1/n) * \text{SEQ}(n-1) + ((n-l)/n) * \text{SEQ}(n-2) \leq (1/n) + ((n-l)/n) = 1.$$

The proof that $\text{SEQ}(n) \geq 0$ for $n \geq 0$ is almost the same.

17. The First Principle of Induction is enough. Use (R) to check for $n = 3$. For the inductive step from n to $n + 1$,

$$\text{FIB}(n+1) = \text{FIB}(n) + \text{FIB}(n-1) = 1 + \sum_{k=1}^{n-2} \text{FIB}(k) + \text{FIB}(n-1) = 1 + \sum_{k=1}^{n-1} \text{FIB}(k).$$

19. For $n > 0$, let $L(n)$ be the largest integer 2^k with $2^k \le n$. Show that $L(n) = T(n)$ for all n by showing first that $L(\lfloor n/2 \rfloor) = L(n/2)$ for $n \ge 2$ and then using the Second Principle of Induction.

21. Show that $S(n) \le n$ for every n by the Second Principle of Induction.

4.7 Answers

1. (a) 20. (c) 1. (e) 4. (g) 6.

3. (a) (20, 14), (14, 6), (6, 2), (2, 0); gcd = 2.
(c) (20, 30), (30, 20), (20, 10), (10, 0); gcd = 10.

5. (a) $\gcd(20, 14) = 2$, $s = -2$, $t = 3$.

a	q	s	t
20		1	0
14	1	0	1
6	2	1	−1
2	3	−2	3
0			

(c) $\gcd(20, 30) = 10$, $s = -1$, $t = 1$.

a	q	s	t
20		1	0
30	0	0	1
20	1	1	0
10	2	−1	1
0			

7. (a) $x = 21$ $[\equiv -5 \pmod{26}]$.
(c) No solution exists, because 4 and 26 are not relatively prime.
(e) $x = 23$ $[\equiv -3 \pmod{26}]$.

9. (a) $x \equiv 5 \pmod{13}$.
(c) Same as (a), since $99 \equiv 8 \pmod{13}$.

11. Assume that $a = s \cdot m + t \cdot n$ and $a' = s' \cdot m + t' \cdot n$ at the start of the loop. The equation $a' = s' \cdot m + t' \cdot n$ becomes $a_{\text{next}} = s_{\text{next}} \cdot m + t_{\text{next}} \cdot n$ at the end, and $a'_{\text{next}} = a' - q \cdot a = s' \cdot m + t' \cdot n - q \cdot s \cdot m - q \cdot t \cdot n = (s' - q \cdot s) \cdot m + (t' - q \cdot t) \cdot n = s'_{\text{next}} \cdot m + t'_{\text{next}} \cdot n$ at the end.

13. (a) $1 = s \cdot (m/d) + t \cdot (n/d)$ for some integers s and t. Apply Exercise 12 to m/d and n/d in place of m and n. A longer proof can be based on prime factorization.
(b) $k = s't - st'$ works. Note that $d \cdot s = ss' \cdot m + st' \cdot n$ and $d \cdot s' = s's \cdot m + s't \cdot n$; subtract.
(c) Let $x = t \cdot a/d$.

15. (a) Check for $l = 1$. For the inductive step, it suffices to show that if $a = m = \text{FIB}(l+3)$ and $b = n = \text{FIB}(l+2)$, $l \ge 1$, then, after the first pass of the while loop, $a = \text{FIB}(l+2)$ and $b = \text{FIB}(l+1)$. For then, by the inductive hypothesis, exactly l more passes would be needed before terminating the algorithm. By the definition of (a, b) in the while loop, it suffices to show that $\text{FIB}(l+3) \text{ MOD } \text{FIB}(l+2) = \text{FIB}(l+1)$. But $\text{FIB}(l+3) \text{ MOD } \text{FIB}(l+2) = [\text{FIB}(l+2) + \text{FIB}(l+1)] \text{ MOD } \text{FIB}(l+2) = \text{FIB}(l+1) \text{ MOD } \text{FIB}(l+2) = \text{FIB}(l+1)$, since $\text{FIB}(l+1) < \text{FIB}(l+2)$ for $l \ge 1$.
(b) Use induction on k. Check for $k = 2$. For the inductive step

$$\log_2 \text{FIB}(k+1) = \log_2(\text{FIB}(k) + \text{FIB}(k-1)) \le \log_2(2\text{FIB}(k)) = 1 + \log_2 \text{FIB}(k).$$

[This estimate is far from best possible. The Fibonacci numbers are the worst case for the Euclidean algorithm on page 175.]
(c) By part (a), GCD makes l passes through the loop. By part (b), $\log_2(m+n) = \log_2 \text{FIB}(l+3) \le l$ if $l \ge 2$.

4.8 Answers

1. (a) and (b). Both "n is odd" and "n is even" are invariants of the loop. If n is odd, then so is $3n$, and hence so is $10 - 3n$. Similarly, for n even.

2. (a) No, $3n < 2m$ is not an invariant. For example, try $n = 1, m = 2$.
(b) Yes. If $3n < 2m$ and $n > m$, then $3m < 3n < 2m$, so $m < 0$; thus $3 \cdot 3n = 9n < 6m < 4m = 2 \cdot 2m$.

3. (a) True.
(b) False. See Exercise 5 on page 143.
(c) and (d). False. See Theorem 1 on page 148.

4. (a) $O(n^3)$. (b) $O(n^5)$. (c) $O(n^4)$. (d) $O(n^6)$.

5. (a) For $n \in \mathbb{P}$, we have $n^2 \le (n+7)^2 < 3(n+7)^2$, so $n^2 = O(3(n+7)^2)$. By Example 5(e) on page 149, $3(n+7)^2 = O(n^2)$. Thus $n^2 = \Theta(3(n+7)^2)$.
(b) Suppose that $2^{2n} = O(3^n)$. Then there is a constant C such that $2^{2n} = 4^n \le C \cdot 3^n$ for all large enough n. Then $C \ge (\frac{4}{3})^n$, so $\log C \ge n \cdot \log \frac{4}{3}$; i.e., $n \le \log C / \log \frac{4}{3}$ for all large enough n, which is absurd.
(c) $n! = 1 \cdots n \le n \cdots n = n^n$, so $\log n! \le n \log n$. Take $C = 1$.

6. Suppose that there is a number C such that $n^n \le C \cdot n!$ for all large enough integers n. Then, observing that the first factor in $n!$ is 1 and the remaining $n-1$ factors are at most n, we get $n^n \le C \cdot n^{n-1}$, so $n \le C$ for all large n, which is absurd.

7. (a) $h_3 = 5$.
(b) The characteristic equation for this linear recurrence is $x^2 - 2x - 1 = 0$, with roots $1+\sqrt{2}$ and $1-\sqrt{2}$. Thus $h_n = C(1+\sqrt{2})^n + D(1-\sqrt{2})^n$ for suitable constants C and D. $n = 0$ gives $0 = C + D$, and $n = 1$ gives $1 = C(1+\sqrt{2}) + D(1-\sqrt{2})$. The solutions are $C = 1/2\sqrt{2}$ and $D = -1/2\sqrt{2}$. Thus

$$h_n = \frac{1}{2\sqrt{2}}\left[(1+\sqrt{2})^n - (1-\sqrt{2})^n\right].$$

8. (a) $a_n = 2 \cdot 3^n$. (b) $a_n = (-2)^n$. (c) $a_n = 3^n - (-2)^n$.

9. The defining conditions imply $b_1 = 7$, $b_2 = 11$, $b_3 = 19$, and $b_4 = 35 > 8 \cdot 4$. Assume inductively that $b_m > 8m$ for some $m \ge 4$. Then $b_{m+1} = 2b_m - 3 > 16m - 3 = 8(m+1) + 8m - 11 \ge 8(m+1)$, since $8m - 11 \ge 32 - 11 > 0$. The result follows by induction.

10. (a) 5. (b) $4^2 = 16$.

11. (a) For example, $s(n) = n^2$ and $t(n) = n^3$.
(b) For example, $s(n) = n^3$ and $t(n) = n^2$.

12. Prove the propositions $p(n) =$ "$3^n + 2n - 1$ is divisible by 4" by induction. Check that $p(1)$ is true. Observe that $3^{n+1} + 2(n+1) - 1 - [3^n + 2n - 1] = 2(3^n + 1)$. For all n, 3^n is odd, so $3^n + 1$ is even and $2(3^n+1)$ is divisible by 4. Now $3^{n+1} + 2(n+1) - 1 = [3^n + 2n - 1] + 2(3^n + 1)$, and if $p(n)$ is true, then each term on the right is divisible by 4. Thus the left-hand expression is also divisible by 4, and $p(n+1)$ is true whenever $p(n)$ is true. So all propositions $p(n)$ are true by induction.

13. The inequality holds for $n = 0$, and for $n \ge 1$, $(2n)! = \prod_{k=1}^{n}(n+k) \cdot n! \ge 2^n \cdot n!$, since $n + k \ge 2$ for $k \ge 1$. Or use induction.

14. (a) n^4. (b) n^3. (c) $n!$.
(d) $\sqrt{n}$. (e) n^4. (f) n^6.

15. (a) 1 and 5.
(b) Check the basis $2^5 > 5^2$. For the inductive step, one can use $2^{k+1} = 2 \cdot 2^k > 2 \cdot k^2$ [by the inductive assumption] $= (k+1)^2 + (k-1)^2 - 2 > (k+1)^2$, since $(k-1)^2 - 2 \ge 4^2 - 2 > 0$ because $k \ge 5$. Other algebraic approaches are possible.

16. One possible answer is the following.

GeneralDivisionAlgorithm(integer, integer)

```
{Input: integers m > 0 and n.}
{Output: integers q and r with m · q + r = n and 0 ≤ r < m.}
begin
q := 0
r := |n|
while r ≥ m do
   q := q + 1
   r := r − m
if n < 0 then
   if r = 0 then
      q := −q
   else
      q := −q − 1
      r := m − r
return q and r
end ■
```

17. The condition $n = m \cdot q + r$ is an invariant, but $0 \le r$ is not. The changed algorithm gives the correct answer if $\lfloor \frac{n}{m} \rfloor$ is even, but produces a negative r if $\lfloor \frac{n}{m} \rfloor$ is odd. It does increase q and keep $n = m \cdot q + r$ on each pass through the while loop, so those two conditions in the "somewhat more explicit" algorithm are not enough to guarantee a correct algorithm.

18. (a) Since $\gcd(s,t)$ divides s and t, it divides sm and tn, so it divides their sum, $sm + tn = 1$.
(b) By part (a), $\gcd(m,n)$ divides 1; hence $\gcd(m,n) = 1$. Thus m and n are relatively prime.
(c) $\gcd(s,n)$ also divides 1.
(d) Yes. In this case, $1 = tn$, so $n = \pm 1$ and $\gcd(0,n) = 1$.

19. By assumption, $s_n = t_n$ for all basis case values of n. Assume inductively that for some m we have $s_n = t_n$ whenever $0 \le n < m$. Then the recurrence condition defines s_m and t_m in the same way from the same values, so $s_m = t_m$. The claim follows by the Second Principle of Mathematical Induction. A proof can also be based on the Well-Ordering Principle.

20. Since the sequence is uniquely determined by the basis and recurrence relation, we need only verify that $0^2 + 1 = 1$, that $1^2 + 1 = 2$, and that

$$2((n-1)^2 + 1) - ((n-2)^2 + 1) + 2 = n^2 + 1 \quad \text{for} \quad n > 2.$$

These are the same calculations one would make in a proof by induction on n.

21. By definition, there is some constant C such that $a(n) \le C \cdot \log_2 n$ for all sufficiently large n. Hence, $n^2 \cdot a(n) \le C \cdot n^2 \log_2 n$ for the same constant C and all large enough n, so $n^2 \cdot a(n) = O(n^2 \log_2 n)$.

22. To see that S is nonempty, we need q' so that $n - m \cdot q' \ge 0$, i.e., so that $q' \le n/m$. Even if n is negative, there are such integers q', so S is nonempty. By the Well-Ordering Principle, S has a smallest element r, which has the form $n - m \cdot q$ for some $q \in \mathbb{Z}$. Clearly, $r \ge 0$, since $r \in S$. Since $r - m = n - m \cdot (q+1) < r$, $r - m$ cannot belong to S. So $r - m < 0$, whence $r < m$.

23. (a) Consider, for example, the set of primes larger than 17, or the set of powers of 19. Both sets are infinite, and none of their elements are divisible by primes less than 19.

(b) The set

$$\{n \in \mathbb{N} : n > 10^9 \text{ and } n \text{ is not divisible by primes less than } 19\}$$

is nonempty by part (a), and it has a smallest member by the Well-Ordering Principle on page 131.

(c) No. The argument in part (b) gives no clue as to the value obtained.

24. (a) $a_3 = 24$.

(b) Use induction. The claim is true for $n = 0, 1, 2, 3$ by inspection. If it is true for $n \ge 3$, then

$$a_{n+1} = 6a_n - 12a_{n-1} + 8a_{n-2} = 6 \cdot n \cdot 2^n - 12 \cdot (n-1) \cdot 2^{n-1} + 8 \cdot (n-2) \cdot 2^{n-2},$$

which simplifies to $2^{n-2}[8n + 8] = 2^{n+1} \cdot (n+1)$. That is, the claim is true for $n + 1$ if it's true for n, so the claim holds for all n by induction.

25. (a) The first seven are 1, 1, 3, 5, 9, 15, and 25, respectively.

(b) We use the Second Principle of Induction on page 168 to prove "a_n is odd" for all $n \ge 0$. This is clear for $n = 0, 1$; in fact, we've checked this for $n \le 6$. Assume the statement is true for $n = 0, 1, \ldots, k-1$. Then a_{k-1} and a_{k-2} are odd. Hence $a_k = a_{k-1} + a_{k-2} + 1$ is the sum of three odd integers, and so a_k is also odd. Now the assertion is true for all n by induction.

(c) As in part (b), we use the Second Principle of Induction to prove "$a_n < 2^n$" for $n \ge 1$. This is easily checked for $n = 1$ and $n = 2$. Assume that $a_n < 2^n$ is true for $n = 1, 2, \ldots, k-1$ (where $k \ge 3$). Then

$$a_k = a_{k-1} + a_{k-2} + 1 < 2^{k-1} + 2^{k-2} + 1 < 2^{k-1} + 2^{k-2} + 2^{k-2} = 2^k.$$

Hence the assertion for $n = 1, 2, \ldots, k-1$ implies the assertion for $n = k$. Thus the assertion holds for all n by induction.

26. Since $a_n = O(b_n)$, by definition there are a constant C' and a positive integer N such that $a_n \le C'b_n$ whenever $n \ge N$. Choose C to be at least as large as C' and all of the numbers a_k/b_k for $k = 1, \ldots, N-1$. Then $a_n \le Cb_n$ for every n in $\mathbb{P}$.

27. By Exercise 26, there exists C so that $|a_k| \le Ck$ for all k. So, for all n,

$$|b_n| \le |a_1| + |a_2| + \cdots + |a_n| \le C + 2C + \cdots nC = C \cdot \frac{n(n+1)}{2} \le C \cdot n^2.$$

So by the definition, b_n is $O(n^2)$.

28. By Exercise 26, there is a constant $C > 0$ such that $a_n \le C \cdot n$ for all n. We have $c_1 = a_0 + a_0$, $c_2 = c_1 + a_1 = a_0 + a_0 + a_1$, and in general $c_n = a_0 + \sum_{i=0}^{n-1} a_i$, as can be shown by induction. Thus

$$c_n \le a_0 + \sum_{i=0}^{n-1} C \cdot i \le a_0 + C \sum_{i=0}^{n-1} n = a_0 + Cn^2 \le (a_0 + C)n^2.$$

Or prove directly by induction that $c_n \le a_0 + Cn^2$, as follows. The case $n = 0$ is clear. Then $c_n = c_{n-1} + a_{n-1} \le a_0 + C(n-1)^2 + a_{n-1}$ [by the inductive assumption] $\le a_0 + C(n-1)^2 + C(n-1)$ [by the choice of C] $= a_0 + C(n^2 - n) < a_0 + Cn^2$.

5.1 Answers

1. (a) 56. (c) 56. (e) 1.

3. (a) 10. (c) $0 + 0 = 0$.

(e) $(1000!/650!)/(999!/649!) = 1000/650$.

5. (a) $\binom{20}{8} = 125{,}970$.

7. (a) 126.

9. (a) 0. (c) 2401.

11. (a) This is the same as the number of ways of drawing ten cards so that the *first* one is not a repetition. Hence $52 \cdot (51)^9$.

13. (a) $9 \cdot 10^6$.
(c) $5 \cdot 9 \cdot 10^5 = 9 \cdot 10^6/2 =$ answer to (a)−answer to (b). This is also $\lfloor (10^7-1)/2 \rfloor - \lfloor (10^6-1)/2 \rfloor$.
(e) $9 \cdot 9 \cdot 8 \cdot 7 \cdot 6 \cdot 5 \cdot 4 = 9 \cdot P(9,6)$. Choose the digits from left to right.
(g) $9 \cdot 9 \cdot 8 \cdot 7 \cdot 6 \cdot 5 \cdot 4 - 5 \cdot 8 \cdot 8 \cdot 7 \cdot 6 \cdot 5 \cdot 4 = 41 \cdot 8 \cdot 7 \cdot 6 \cdot 5 \cdot 4$.

15. (a) $13 \cdot \binom{4}{4} \cdot \binom{48}{1} = 624$. (b) 5108.
(c) $13 \cdot \binom{4}{3} \cdot \binom{12}{2} \cdot 4 \cdot 4 = 54{,}912$. (d) 1,098,240.

17. (a) It is the $n \times n$ matrix with 0's on the diagonal and 1's elsewhere.

19. (a) $n \cdot (n-1)^3$. If the vertex sequence is written $v_1\, v_2\, v_3\, v_4$, there are n choices for v_1 and $n-1$ choices for each of v_2, v_3, and v_4.
(c) $n(n-1)(n-2)(n-2)$. v_1, v_2, v_3 must be distinct, but v_4 can be v_1.

5.2 Answers

1. (a) $\frac{8}{25} = 0.32$. (c) $\frac{9}{25} = 0.36$.

3. (a) $\frac{5 \cdot 4 \cdot 3 \cdot 2}{5^4} = 0.192$. (c) $\frac{2}{5} = 0.40$.

5. Note that $\binom{7}{3} = 35$ is the number of ways to select three balls from the urn.
(a) $\frac{1}{35}$. (c) $\frac{3 \cdot \binom{4}{2}}{35} = \frac{18}{35}$. (e) 1.

7. (a) 0.2. (c) 0.6.

9. Here $N = 2{,}598{,}960$.
(a) $624/N \approx 0.000240$. (c) $10{,}200/N \approx 0.00392$. (e) $1{,}098{,}240/N \approx 0.423$.

11. (a) $\frac{1}{2}$. (c) $\frac{3}{36} = \frac{1}{12}$.

13. We have $P(E_1 \cup E_2 \cup E_3) = P(E_1 \cup E_2) + P(E_3) - P((E_1 \cup E_2) \cap E_3)$. Now $P(E_1 \cup E_2) = P(E_1) + P(E_2) - P(E_1 \cap E_2)$ and hence $P((E_1 \cup E_2) \cap E_3) = P((E_1 \cap E_3) \cup (E_2 \cap E_3)) = P(E_1 \cap E_3) + P(E_2 \cap E_3) - P(E_1 \cap E_3 \cap E_2 \cap E_3)$. Substitute. Alternatively, use the identity $E_1 \cup E_2 \cup E_3 = (E_1 \setminus E_2) \cup (E_2 \setminus E_3) \cup (E_3 \setminus E_1) \cup (E_1 \cap E_2 \cap E_3)$.

15. (a) $\frac{1}{64}$. (c) $\frac{15}{64}$. (e) $\frac{22}{64}$.

17. Let Ω_n be all n-tuples of H's and T's and E_n all n-tuples in Ω_n with an even number of H's. It suffices to show that $|E_n| = 2^{n-1}$, which can be done by induction. Assume true for n. Every n-tuple in E_{n+1} has the form (ω,H), where $\omega \in \Omega_n \setminus E_n$, or (ω,T), where $\omega \in E_n$. There are 2^{n-1} n-tuples of each type, so E_{n+1} has 2^n elements. This problem is easier using conditional probabilities [§9.1].

19. (a) If Ω is the set of all 4-element subsets of S, the outcomes are equally likely. If E_2 is the event "exactly 2 are even," then $|E_2| = \binom{4}{2} \cdot \binom{4}{2} = 36$. Since $|\Omega| = \binom{8}{4} = 70$, $P(E_2) = \frac{36}{70} \approx 0.514$.
(c) $\frac{16}{70}$. (e) $\frac{1}{70}$.

21. (a) The sample space Ω is all triples (k,l,m), where $k,l,m \in \{1,2,3\}$, so $|\Omega| = 3^3$. The triples where $k,l,m \in \{2,3\}$ correspond to no selection of 1. So $P(1\text{ not selected}) = \frac{2^3}{3^3}$ and answer $= 1 - \frac{8}{27} \approx 0.704$.
(c) $1 - (\frac{n-1}{n})^n$.
(d) $1 - (0.999999)^{1{,}000{,}000} \approx 0.632120$. This is essentially $1 - \frac{1}{e} \approx 0.632121$, since $\lim_{n\to\infty} (\frac{n}{n-1})^n$ is equal to e.

5.3 Answers

1. 125.

3. (a) 0.142.
(c) 0.78 since $1000 - |D_7 \cup D_{11}| = 1000 - (142 + 90 - 12) = 780$.

5. There are 466 such numbers in the set, so the probability is 0.466. Remember that $D_4 \cap D_6 = D_{12}$, not D_{24}. Hence $|D_4| + |D_5| + |D_6| - |D_4 \cap D_5| - |D_4 \cap D_6| - |D_5 \cap D_6| + |D_4 \cap D_5 \cap D_6| = 250 + 200 + 166 - 50 - 83 - 33 + 16 = 466$.

7. (a) $\binom{12+4-1}{4-1} = 455$.

9. (a) $x^4 + 8x^3y + 24x^2y^2 + 32xy^3 + 16y^4$.
(c) $81x^4 + 108x^3 + 54x^2 + 12x + 1$.

11. (b) There are $\binom{n}{r}$ subsets of size r for each r, so there are $\sum_{r=0}^{n} \binom{n}{r}$ subsets in all.
(c) If true for n, then

$$\sum_{r=0}^{n+1} \binom{n+1}{r} = 1 + \sum_{r=1}^{n} \binom{n+1}{r} + 1 = 1 + \sum_{r=1}^{n} \binom{n}{r-1} + \sum_{r=1}^{n} \binom{n}{r} + 1$$

$$= \sum_{r=1}^{n+1} \binom{n}{r-1} + \sum_{r=0}^{n} \binom{n}{r} = 2\sum_{r=0}^{n} \binom{n}{r} = 2 \cdot 2^n = 2^{n+1}.$$

13. (a) $\sum_{k=3}^{5}\binom{k}{3}=\binom{3}{3}+\binom{4}{3}+\binom{5}{3}=15=\binom{6}{4}$.

(b) Apply induction to $p(n)=$ "$\sum_{k=m}^{n}\binom{k}{m}=\binom{n+1}{m+1}$" for $n\geq m$, and apply Exercise 10(b).

(c) The set $\mathcal{A}$ of $(m+1)$-element subsets of $\{1,2,\ldots,n+1\}$ is the disjoint union $\bigcup_{k=m}^{n}\mathcal{A}_k$, where $\mathcal{A}_k$ is the collection of $(m+1)$-element subsets whose largest element is $k+1$. A set in $\mathcal{A}_k$ is an m-element subset of $\{1,2,\ldots,k\}$ with $k+1$ added to it, so $|\mathcal{A}|=\sum_{k=m}^{n}\binom{k}{m}$.

15. (a) Put 8 in one of the three boxes; then distribute the remaining 6 in three boxes. Answer is $3\cdot\binom{8}{2}=84$ ways.

(c) Since $1+9+9<20$, each digit is at least 2. Then $(d_1-2)+(d_2-2)+(d_3-2)=20-6=14$ with $2\leq d_i\leq 9$ for $i=1,2,3$. By part (b), there are 36 numbers. Another way to get this number is to hunt for a pattern: 299; 398, 389; 497, 488, 479; 596, 587, 578, 569; .... Looks like $1+2+3+\cdots+8=\binom{9}{2}=36$.

17. (a) Think of putting 9 objects in four boxes, starting with 1 object in the first box. $\binom{8+4-1}{4-1}=165$.

19. $\binom{p-l+1}{l}$. Note that there are no such sets unless $2l\leq p+1$.

5.4 Answers

1. (a) $\dfrac{15!}{3!4!5!3!}$.

(b) $\binom{15}{3}\binom{15}{4}\binom{15}{5}$.

3. (a) As in Example 8, count ordered partitions $\{A,B,C,D\}$, where $|A|=5$, $|B|=3$, $|C|=2$, and $|D|=3$. Answer $=\frac{13!}{5!\cdot 3!\cdot 2!\cdot 3!}=720{,}720$.

(b) 600,600.

(c) Count ordered partitions, where $|A|=|B|=|C|=3$ and $|D|=4$, but note that permutations of A, B, and C give equivalent sets of committees. Answer $=\frac{1}{6}\cdot\frac{13!}{3!\cdot 3!\cdot 3!\cdot 4!}=200{,}200$.

5. (a) $3^{10}=59{,}049$. (c) $\binom{10}{3}=120$. (e) $\frac{10!}{3!\cdot 4!\cdot 3!}=4200$.

7. (a) 625. (c) $5^3\cdot 2=250$.

9. There are $\frac{1}{2}\binom{2n}{n}$ unordered such partitions and $\binom{2n}{n}$ ordered partitions.

11. (a) $10\cdot\binom{8}{2}\cdot\binom{5}{2}\cdot\binom{2}{2}=2800$. Just choose a third member of their team and then choose three teams from the remaining 9 contestants.

13. Fifteen. Just count. We know no clever trick, other than breaking them up into types 4, 3—1, 2—2, 2—1—1, and 1—1—1—1. There are 1, 4, 3, 6, and 1 partitions of these respective types. This solves the problem, because there is a one-to-one correspondence between equivalence relations and partitions; see Theorem 1 on page 116.

15. (a) $\binom{9}{3\ 2\ 4}=\frac{9!}{3!\cdot 2!\cdot 4!}=1260$.

(c) $\binom{9}{2\ 2\ 3\ 2}=\frac{9!}{2!\cdot 2!\cdot 3!\cdot 2!}=7560$.

(e) $\binom{9}{0\ 2\ 3\ 4}\cdot 2^2\cdot 3^3\cdot 4^4=34{,}836{,}480$.

5.5 Answers

1. One way to do this is to apply Example 1(b) with $p=1$.

3. Here $|S|=73$ and $73/8>9$, so some box has more than 9 marbles.

5. For each 4-element subset B of A, let $f(B)$ be the sum of the numbers in B. Then $f(B)\geq 1+2+3+4=10$, and $f(B)\leq 50+49+48+47=194$. There are thus only 185 possible values of $f(B)$. Since A has $\binom{10}{4}=210$ subsets B, at least two of them must have the same sum $f(B)$ by the second version of the Pigeon-Hole Principle on page 213.

7. For each 2-element subset T of A, let $f(T)$ be the sum of the 2 elements. Then f maps the 300 2-element subsets of A into $\{3,4,5,\ldots,299\}$.

9. The repeating blocks are various permutations of 142857.

11. (a) Look at the six blocks

$$(n_1,n_2,n_3,n_4),\quad (n_5,n_6,n_7,n_8),\quad \ldots,\quad (n_{21},n_{22},n_{23},n_{24}).$$

(c) Look at the eight blocks (n_1,n_2,n_3), (n_4,n_5,n_6), ..., (n_{22},n_{23},n_{24}). By part (b), at least one block has sum at least $300/8=37.5$.

13. If $0\in\operatorname{Im}(f)$, then n_1, n_1+n_2, or $n_1+n_2+n_3$ is divisible by 3. Otherwise f is not one-to-one and there are three cases. If $n_1\equiv n_1+n_2 \pmod 3$, then $n_2\equiv 0 \pmod 3$ and n_2 is divisible by 3. Similarly, if $n_1\equiv n_1+n_2+n_3 \pmod 3$, then n_2+n_3 is divisible by 3. And if $n_1+n_2\equiv n_1+n_2+n_3 \pmod 3$, then n_3 is divisible by 3.

15. (a) $8^6=262{,}144$.

(b) Let A be the set of numbers with no 3's and B be the set with no 5's, and calculate $|A^c \cap B^c| = |(A \cup B)^c|$. Answer = 73,502.

(c) $8!/2! = 20{,}160$.

(d) 60.

17. (a) We show that S must contain both members of some pair $(2k-1, 2k)$. Partition $\{1, 2, \ldots, 2n\}$ into the n subsets $\{1, 2\}, \{3, 4\}, \ldots, \{2n-1, 2n\}$. Since $|S| = n+1$, some subset $\{2k-1, 2k\}$ contains 2 members of S by the Pigeon-Hole Principle. The numbers $2k-1$ and $2k$ are relatively prime.

(b) For each $m \in S$, write $m = 2^k \cdot l$, where l is odd, and let $f(m) = l$. See Exercise 24 on page 137. Apply the second version of the Pigeon-Hole Principle on page 213 to the function $f: S \to \{1, 3, 5, \ldots, 2n-1\}$.

(c) Let $S = \{2, 4, 6, \ldots, 2n\}$.

19. (a) By the remark at the end of the proof of the Generalized Pigeon-Hole Principle on page 217, the average size is $2 \cdot 21/7 = 6$.

5.6 Answers

1. 26^4. We can pick the first four letters any way we like, but then the last three are determined.

2. $|F \cap W| = |F| + |W| - |F \cup W| = 63 + 50 - |F \cup W| \geq 63 + 50 - 100 = 13$.

3. $26! - 2 \cdot 25! = 24 \cdot 25!$. We count the lists in which A and Z *are* together by imagining the letters A and Z as being tied together into a single big letter. Then there are 25! orderings for the resulting set of 25 letters, and each AZ letter can be ordered in two ways.

4. $5 \cdot 8 \cdot 8 \cdot 7 \cdot 6 \cdot 5$. There are 5 choices for the right-hand digit, then 8 choices for the left-hand one, avoiding 0 and the chosen 1's digit, after which there are 8, 7, 6, and 5 choices for the remaining four digits, say from left to right.

5. (a) $\binom{8+3-1}{3-1} = \binom{10}{2} = 45$.

(b) Same as (a). Just throw the fruits in two at a time.

(c) $1 + 2 + 3 + 4 + 4 + 3 + 2 + 1 = 20$.

6. (a) There are $\binom{15}{7} = \binom{15}{8}$ ways.

(b) There are $\binom{15}{5\ 5\ 5}/3!$ ways.

7. (a) There are 9^6 of them. Choose the first digit to be anything but 0; then choose each successive digit to be anything but the one just before it.

(b) Choose the rightmost digit first, from $\{1, 3, 5, 7, 9\}$, in one of 5 ways. Then choose the leftmost in one of 8 ways (avoid 0 and the digit just chosen), and the remaining digits in $8 \cdot 7 \cdot 6 \cdot 5$ ways. The answer is $5 \cdot 8 \cdot 8 \cdot 7 \cdot 6 \cdot 5$.

8. (a) There are $\binom{14}{5}$ ways.

(b) There are $\binom{14}{5} - \binom{11}{2}$ ways. Count the ways to get all three poisonous ones out, and subtract.

(c) There are $\binom{14}{5} - \binom{11}{5} - 3 \cdot \binom{11}{4}$ ways. Count the ways that don't get any poisonous ones, then the ways that get exactly one poisonous one, and subtract. Or $\binom{3}{2}\binom{11}{3} + \binom{3}{3}\binom{11}{2}$.

9. The easy way is to choose a flag for the pole that just gets 1, in one of 3 ways. Then order the 2 flags on the other pole in one of 2 ways. All told, there are $3 \cdot 2 = 6$ ways.

10. (a) $P(26, 8) = \frac{26!}{18!} = 26 \cdot 25 \cdots 19$.

(b) $\binom{5}{3} \cdot \binom{21}{5}$.

(c) $\binom{9}{3} \cdot 25^6$.

(d) $P(4, 4) = 4!$.

11. (a) $4 \cdot 20 = 80$.

(b) 41.

12. (a) $\frac{20!}{3! \cdot 6! \cdot 7! \cdot 1! \cdot 3!} = \binom{20}{3} \cdot \binom{17}{6} \cdot \binom{11}{7} \cdot \binom{4}{1} \cdot \binom{3}{3} = \binom{20}{3\ 6\ 7\ 1\ 3}$.

(b) $\binom{20}{10}/2$.

(c) $\binom{14}{4} = \binom{14}{10}$.

13. (a) $10^4 - 9^4$.

(b) $10^4 - 2 \cdot 9^4 + 8^4$.

14. 1; it's a certainty. Break the numbers into 10 groups, where the members of each group have the same last digit. By the Pigeon-Hole Principle, at least one of the sets has $\frac{21}{10} = 2.1$ or more members and hence has at least three members.

15. (a) $5!$ $[= P(5, 5)]$.

(b) $5^6 - 4^6$ [= the total number of words − the number without d].

(c) 5^4 [= the number of strings of length 4 to fill out the word].

16. $\binom{31}{2} \cdot \binom{18}{3}$ $[= 465 \cdot 816 = 379{,}440]$.

17. $800 + 800 + 800 - (250 + 250 + 250) + 50 = 1700$.

18. There are $\binom{20}{4}$ ways to draw 4 males and $\binom{20}{4}$ ways to draw 4 females. By the Product Rule, there are $\binom{20}{4}\binom{20}{4}$ ways to draw 4 of each. The probability is thus $\frac{\binom{20}{4}\binom{20}{4}}{\binom{40}{8}} \approx 0.305$.

19. (a) $1/52$.

(b) $1/52$. Reason 1: There are 52^2 possible draws xy, and 52 of them are xx with both the same, so the answer is $52/52^2$. Reason 2: Whatever card was chosen first, the probability is $1/52$ of getting it again.

(c) 0. This is *without* replacement.

(d) Approach 1: There are $P(52, 6) = 52!/46! = 52 \cdot 51 \cdot 50 \cdot 49 \cdot 48 \cdot 47$ strings with 6 different, but 52^6 possible strings, so the answer is $P(52, 6)/52^6$. Approach 2: We want the probability that the second card is not the first, and that the third card is neither the first nor the second, and so on. The answer is $\frac{51}{52}\cdot\frac{50}{52}\cdot\frac{49}{52}\cdot\frac{48}{52}\cdot\frac{47}{52}$. [Why does the answer in the first approach lead to 6 factors, but this answer leads to 5?]

(e) 1. That's what "without replacement" means. Alternatively, we could use the approaches in the last part to get $P(52,6)/P(52,6)$ or $\frac{51}{51}\cdot\frac{50}{50}\cdot\frac{49}{49}\cdot\frac{48}{48}\cdot\frac{47}{47}$.

(f) 1 minus the answer to part (d).

(g) Decide on which two should be the same, say the first and fifth or the second and fourth; there are $\binom{6}{2}=15$ ways to do that. Then find the probability that the other four are all different and different from these two. This is like part (d). Imagine picking the equal ones first and tying them together. The problem is then like picking 5 different cards from a 52-card deck, so there are $P(52,5)$ ways to do it. [Why not 4 from a 51-card deck? Sure, but then multiply by the 52 choices for the first (tied together) card.] Altogether, there are $\binom{6}{2}\cdot P(52,5)$ 6-card sequences with exactly two the same, so the answer is $\binom{6}{2}P(52,5)/52^6\approx 0.237$.

(h) There are $\binom{6}{4}$ ways to choose where the aces of spades go in the list, so $\binom{6}{4}\cdot 51^2$ strings meet this condition. Thus the probability is $\binom{6}{4}\cdot 51^2/52^6$.

(i) We count: 51^6 strings with no aces of spades, $6\cdot 51^5$ with one ace of spades, $\binom{6}{2}\cdot 51^4$ with 2 aces of spades, etc. The answer is $[51^6+6\cdot 51^5+\binom{6}{2}\cdot 51^4+\binom{6}{3}\cdot 51^3+\binom{6}{4}\cdot 51^2]/52^6\approx 1$. Alternatively, it's $1-(\binom{6}{5}\cdot 51+\binom{6}{6}\cdot 1)/52^6\approx 1-1.5\times 10^{-8}$, which we could also get from the Binomial Theorem, since $\sum_{i=0}^{6}\binom{6}{i}51^{6-i}=(1+51)^6$.

(j) We can use the approach in (i) to get $1-[51^6+6\cdot 51^5+\binom{6}{2}\cdot 51^4]/52^6\approx 1.36\times 10^{-4}$.

20. (a) 1/8.

(b) First approach: All words are equally likely, and there are $\binom{8}{2\,1\,4\,1}=\frac{8!}{2!\cdot 1!\cdot 4!\cdot 1!}=840$ words, so the answer is $2!\cdot 4!/8!=1/840$. Second approach: All 8! lineups of birds are equally likely, but different lineups give the same word. Indeed, $2!\cdot 1!\cdot 4!\cdot 1!$ give $W\,O\,L\,H\,O\,L\,L\,L$ (or any other word), so the answer is $2!\cdot 1!\cdot 4!\cdot 1!/8!$. Third approach: $P(\text{first letter is } W)=1/8$, $P(\text{next is } O \text{ given first is } W)=2/7$, $P(\text{next is } L \text{ given first two are } W\,O)=4/6$, etc., so the answer is $\frac{1}{8}\cdot\frac{2}{7}\cdot\frac{4}{6}\cdot\frac{1}{5}\cdot\frac{1}{4}\cdot\frac{3}{3}\cdot\frac{2}{2}\cdot\frac{1}{1}=\frac{2!\cdot 4!}{8!}$. [We haven't looked at such conditional probabilities yet, but this method turns out to be legal. We are in effect counting the ways we could assemble the birds in correct order.]

(c) We count the teams with no larks: $\binom{4}{4}=1$. Since there are $\binom{8}{4}$ teams, there are $\binom{8}{4}-1$ with at least one lark. How would the answer change if we were choosing a team of 3 birds instead of 4?

(d) We count the teams with no larks and with 1 lark: $\binom{4}{4}+4\cdot\binom{4}{3}=17$. [The factor 4 here comes from the 4 choices for the lark.] The answer is $\binom{8}{4}-17=53$. It would be a mistake to start by putting 2 larks on—in one of $\binom{4}{2}$ ways—and then to count the ways to fill out the team—$\binom{6}{2}$, and get $\binom{4}{2}\cdot\binom{6}{2}=90$, since then teams with more than 2 larks would get counted more than once.

(e) 5. (f) 4. (g) $4+1=5$.

(h) This is straight objects-in-boxes. There are $\binom{8+4-1}{4-1}=\binom{11}{3}=165$ possible distributions.

(i) It would make sense if we knew the probability of any given distribution. If all are equally likely, then the probability that the second one is the same as the first is $1/\binom{11}{4}$, but there's no reason to believe that all distributions are equally likely. They aren't, for example, in bridge. See Example 3 on page 206.

21. $4\cdot 5\cdot 2=40$, corresponding to the various numbers $2^a5^b7^c$ with $a\in\{0,1,2,3\}$, $b\in\{0,1,2,3,4\}$, $c\in\{0,1\}$.

22. (a) Color the outside edges one color and the "diagonal" edges the other.

(b) Pick a vertex, say v. At least 3 edges to v are the same color, by the Pigeon-Hole Principle. Say 3 are red, leading to vertices w,x,y. If any edge joining two of w,x,y is red, then we get a red triangle with v as one vertex. Otherwise, w,x,y are the vertices of a green triangle. [This exercise is in the general area of Ramsey theory.]

23. Consider the buckets $\{2i-1,2i\}$ for $i=1,2,\ldots,m$. We want to show that some bucket has at least 2 people in it. There are more people than buckets, so the Pigeon-Hole Principle wins again. [What if there were only m people?]

24. (a) $P(8,8)=8!$. Order matters.

(b) $8!-2\cdot 7!$. We count forbidden ways. Chain Bob and Ann together and treat them as a blob. There are 7! ways to order 6 students and a blob, then 2 ways to arrange the blob.

(c) $2\cdot 6!$. The blob is bigger, that's all.

(d) There are $\binom{6}{4}$ choices without either Pat or Chris and $\binom{6}{2}$ choices with both of them, so the answer is $\binom{6}{4}+\binom{6}{2}$.

(e) $\binom{8}{4}/2$. [Order doesn't matter here.]

(f) $\binom{6}{2}$ [ways to pick their companions].

25. (a) $\binom{40}{10}\cdot\binom{30}{10}\cdot\binom{20}{10}=\binom{40}{10\,10\,10\,10}$.

(b) The animals are independent. So it's like throwing 10 frogs into 40 boxes in any one of $\binom{10+40-1}{40-1}$ ways, then 10 gerbils, then 10 snakes. Altogether, $\binom{10+40-1}{40-1}^3$ ways to do it.

(c) If the animals have name tags, then there are 40 choices for each animal, so 40^{30} ways to do it.
(d) This is objects-and-boxes with a twist. The "objects" are the cages, and the "boxes" are the colors. There are $\binom{40+4-1}{4-1} = 12{,}341$ possible color distributions.
(e) $4^{40} \approx 1.2 \times 10^{24}$. There are 4 color choices for each cage.

26. (a) 70. Stick A and C inside B.
(b) 100.
(c) $100 \geq |B \cup C| = 70 + 65 - |B \cap C|$, so $|B \cap C| \geq 35$.
(d) 0.
(e) $|(A \cap B) \cup (A \cap C)| = |A \cap (B \cup C)| = |A| + |B \cup C| - |A \cup B \cup C| \geq 50 + 70 - 100 = 20$, and 20 is possible if $C \subseteq B$.

27. (a) There are 5^{12} words that do not use x and 6^{12} words altogether, so the probability is $\frac{5^{12}}{6^{12}}$. Or think of the word construction as a process consisting of 12 steps, for each of which the probability of choosing a letter other than x is $\frac{5}{6}$, and use the Product Rule to get $(\frac{5}{6})^{12}$.
(b) There are 5^{12} words that don't use x, of which 4^{12} don't use a either. The probability is $\frac{5^{12}-4^{12}}{6^{12}}$.
(c) Use Inclusion-Exclusion. Let X be the set of words that use x, and B the set of words that use b. We want $|X^c \cup B| = |X^c| + |B| - |X^c \cap B| = 5^{12} + (6^{12} - 5^{12}) - (5^{12} - 4^{12}) = 6^{12} - 5^{12} + 4^{12}$.
(d) First choose the 4 positions for the c's, in one of $\binom{12}{4}$ ways. For each such choice there are 5^8 ways to complete the word, so there are $\binom{12}{4} \cdot 5^8$ words containing exactly 4 c's.
(e) $\binom{12}{3\ 3\ 6} = \binom{12}{3}\binom{9}{3}\binom{6}{6}$.
(f) $\binom{6}{3}\binom{12}{4\ 4\ 4}$.
(g) $\binom{6}{3}[3^{12} - 3 \cdot 2^{12} + 3]$.

28. (a) There are three possible distributions: 4–1–1, 3–2–1, 2–2–2.
(b) The numbers of distributions of the three types listed in part (a) are $3\binom{6}{4\ 1\ 1}$, $3!\binom{6}{3\ 2\ 1}$, and $\binom{6}{2\ 2\ 2}$, respectively, so the total number of distributions with distinguishable people and cars is

$$3 \cdot \frac{6!}{4!} + 6 \cdot \frac{6!}{3!2!} + \frac{6!}{2!2!2!} = 540.$$

29. This is $NONSENSE$ again, and the answer is the same: $\frac{8!}{3! \cdot 1! \cdot 2! \cdot 2!}$. One way to see this is to count the words of length 7 with just two N's (that's $\frac{7!}{2! \cdot 1! \cdot 2! \cdot 2!}$), the number with no O's, the number with one S, the number with one E, and add these up. Alternatively, we can recognize that we get the words we want by making the words of length 8 and dropping off the last (forced) letter.

30. We follow the hint. There are $m!$ ways to arrange the flags on the long pole, and then there are $\binom{m+n-1}{n-1}$ ways to choose where to cut the pole into n separate poles, so the answer is $m! \cdot \binom{m+n-1}{n-1}$.

31. (a) 3^8.
(b) There are 2^8 strings with no a's in them and $8 \cdot 2^7$ strings with one a. Thus the answer is $2^8 + 8 \cdot 2^7 = 5 \cdot 2^8$.
(c) We use the Inclusion-Exclusion Principle. Let A be the set of strings with at most one a, let B be the set with at most one b, and let C be the set with at most one c. Then we want $|A^c \cap B^c \cap C^c|$. By part (a), $|A| = |B| = |C| = 5 \cdot 2^8$. Moreover,

$$\begin{aligned} |A \cap B| &= \text{the number with no } a\text{'s or } b\text{'s} + \text{ the number with no } a\text{'s and one } b \\ &\quad + \text{the number with one } a \text{ and no } b\text{'s} \\ &\quad + \text{the number with one } a \text{ and one } b \\ &= 1 + 8 + 8 + 8 \cdot 7 = 73 = |A \cap C| = |B \cap C|. \end{aligned}$$

Since $|A \cap B \cap C| = 0$, the answer is

$$3^8 - 3 \cdot 5 \cdot 2^8 + 3 \cdot (1 + 2 \cdot 8 + 8 \cdot 7).$$

32. $w_n = 3w_{n-1} + (4^{n-1} - w_{n-1})$. [Number of words that don't start with B plus number of words that do.] 0, 1, 6, 28, 120.

33. There are $\binom{16+3-1}{16} = \binom{18}{16} = \binom{18}{2}$ possible distributions. This is really an objects-in-boxes problem, since we can think of starting with 16 people and then assigning them to the categories M, W, C.

34. How many distributions are there of 8 distinct objects into four boxes so that the R box gets 2 objects, the E box gets 3 objects, the M box gets 2 objects, and the B box gets 1 object? There are $\binom{8}{2\ 3\ 2\ 1}$ distributions.

35. (a) $\binom{30}{5\ 5\ 5\ 5\ 5\ 5}$. (b) 3^{30}.

36. How many ways are there to distribute 9 distinguishable balls in three boxes so that the first box gets 3 balls, the second gets 4, and the third gets 2? The number is $\binom{9}{3\ 4\ 2}$.

37. (a) $\binom{20+3-1}{3-1} = \binom{22}{2}$.
(b) $\binom{13+3-1}{3-1} = \binom{15}{2}$. [Put 2 reds and 5 greens in first, then 13 more.]
(c) $\binom{22}{2} - \binom{16}{2}$. [Count ones with at least 6 blues, and subtract.]

38. (a) 15. (b) $\binom{40}{n}/\binom{100}{n}$. (c) $25+5=30$.

39. (a) 3^5.
(b) $3^5 - 3\cdot 2^5 + 3$. [This is Inclusion-Exclusion, working on the set of functions that include all three of a, b, and c in their images. Subtract the number that miss at least one of these values.]

40. (a) $\frac{70}{100} = 0.7$. (b) $\binom{45}{2}/\binom{100}{2} = \frac{45}{100}\cdot\frac{44}{99}$.
(c) $(70\cdot 30)/\binom{100}{2}$. (d) $\binom{100}{50} - \binom{55}{50}$.

41. (a) $\dfrac{5}{25} = 0.2$.
(b) $\dfrac{\binom{20}{3}}{\binom{25}{3}} = \dfrac{20}{25}\cdot\dfrac{19}{24}\cdot\dfrac{18}{23}$ [≈ 0.4957].
(c) 0.

42. (a) $\binom{10+3-1}{3-1} = \binom{12}{2}$.
(b) $\binom{10+2-1}{2-1} = \binom{11}{1} = 11$. [What are they?]
(c) Put 3 snakes on at first. Then select 7 more committee members, in one of $\binom{7+3-1}{3-1} = \binom{9}{2}$ ways.

43. (a) $3^8 = 6561$.
(b) $\binom{8}{5} = 56$.
(c) $\binom{8}{3}\cdot\binom{5}{3} = \binom{8}{3\,3\,2} = 560$.
(d) $\binom{8}{3}\cdot 2^5 = 1792$.
(e) $3\cdot 3^6 = 2187$.
(f) For $k = 1, 2, 3$, let A_k be the set of sequences in S with no k's. Then we want $|A_1^c \cap A_2^c \cap A_3^c|$. First we count the complement $A_1 \cup A_2 \cup A_3$, using Inclusion-Exclusion, to obtain $3\cdot|A_1| - 3\cdot|A_1\cap A_2| + |A_1\cap A_2\cap A_3| = 3\cdot 2^8 - 3\cdot 1 + 0 = 765$. Then the answer is $6561 - 765 = 5796$.

44. (a) This is another Inclusion-Exclusion problem. Not divisible by 2, 5, *or* 17 means not divisible by 2 *and* not divisible by 5 *and* not divisible by 17. As in Example 1 on page 197, for $k = 2$, 5, and 17, let $D_k = \{n \in S : n \text{ is divisible by } k\}$. Then $|D_2^c \cap D_5^c \cap D_{17}^c|$

$$= |(D_2\cup D_5\cup D_{17})^c| = 1000 - |D_2\cup D_5\cup D_{17}|$$
$$= 1000 - (|D_2| + |D_5| + |D_{17}|) + (|D_2\cap D_5| + |D_2\cap D_{17}| + |D_5\cap D_{17}|)$$
$$- |D_2\cap D_5\cap D_{17}|$$

[note the reversal of signs from our usual Inclusion-Exclusion formula]

$$= 1000 - (\lfloor 1000/2\rfloor + \lfloor 1000/5\rfloor + \lfloor 1000/17\rfloor)$$
$$+ (\lfloor 1000/10\rfloor + \lfloor 1000/34\rfloor + \lfloor 1000/85\rfloor) - \lfloor 1000/170\rfloor$$
$$= 1000 - (500 + 200 + 58) + (100 + 29 + 11) - 5$$
$$= 377.$$

(b) For the second part we want

$$|D_2 \cap D_5^c \cap D_{17}^c| = |D_2 \cap (D_5 \cup D_{17})^c|$$
$$= |D_2| - |D_2 \cap (D_5\cup D_{17})|$$
$$= |D_2| - |(D_2\cap D_5)\cup(D_2\cap D_{17})|$$
$$= |D_2| - (|D_2\cap D_5| + |D_2\cap D_{17}|) + |(D_2\cap D_5)\cap(D_2\cap D_{17})|$$
$$= |D_2| - |D_2\cap D_5| - |D_2\cap D_{17}| + |D_2\cap D_5\cap D_{17}|$$
$$= \lfloor 1000/2\rfloor - \lfloor 1000/10\rfloor - \lfloor 1000/34\rfloor + \lfloor 1000/170\rfloor = 376.$$

When you think of it, this is reasonable; there should be about as many even numbers as odd ones not divisible by 5 or 17. In fact, thinking in this way gives us another way to approach this problem. We can count the numbers not divisible by 5 or 17, subtract the number of those that are odd, i.e., the answer to part (a), and get the number that are even. Thus $|(D_5\cup D_{17})^c| = 1000 - |D_5\cup D_{17}| = 1000 - (|D_5| + |D_{17}| - |D_5\cap D_{17}|) = 1000 - (200 + 58 - 11) = 753$. Hence the number of even ones is $753 - 377 = 376$.

45. (a) $\binom{8+9}{6} = \binom{17}{6}$.
(b) $\binom{9}{6} = \binom{8}{0}\binom{9}{6}$, $\binom{8}{1}\binom{9}{5}$, and $\binom{8}{0}\binom{9}{6} + \binom{8}{1}\binom{9}{5} + \binom{8}{2}\binom{9}{4} + \binom{8}{3}\binom{9}{3}$.
(c) Same as (a). The sum counts the number of teams with i men (and hence $6-i$ women) for all i, so it's the total number of teams, $\binom{m+w}{t}$. Think of teams of size t from m men and w women.
(d) $\binom{8+9+5}{6}$; $\binom{17}{6} + \binom{5}{1}\binom{17}{5} + \binom{5}{2}\binom{17}{4}$.

46. (a) No. Each triple, such as (3, 1, 6), has probability $(\frac{1}{6})^3 = \frac{1}{6^3}$, but the number of triples that give each sum varies from sum to sum.
(b) 0, 0, 1, 3, $3+3$ [corresponding to one 1 and two 2's or two 1's and one 3], $3+6+1$ [corresponding to $1+1+4$, $1+2+3$, and $2+2+2$].

(c) First put an object into each box. There are then $\binom{3+3-1}{3-1} = \binom{5}{2}$ ways to distribute the other three. The constraint of at most 6 objects per box has no effect this time.
(d) Sure. Question (c) is really question (b) for sum 6. Think of each die as a box.
(e) Each die has at least 1 and at most 6. The question is like (c). Put 1 in each box; then distribute 5 more in any of $\binom{5+3-1}{3-1} = \binom{7}{2}$ ways. The problem gets more complicated for sum 9, since then each "box" contains at most 6 objects.

47. There are 10^{26} functions from Σ to D. Not all are *onto* D. If we let F_i be the set of functions from Σ to $D \setminus \{i\}$, then we want

$$|(F_0 \cup \cdots \cup F_9)^c| = 10^{26} - |F_0 \cup \cdots \cup F_9|.$$

Now $|F_i| = 9^{26}$, $|F_i \cap F_j| = 8^{26}$ if $i \neq j$, etc. Thus the answer is

$$10^{26} - 10 \cdot 9^{26} + \binom{10}{2}8^{26} - \binom{10}{3}7^{26} + \cdots + (-1)^9\binom{10}{9}1^{26} + (-1)^{10}\binom{10}{10}0^{26}.$$

48. (a) This is like putting 10 objects into three distinguishable boxes: first, second, and third. The answer is $\binom{10+3-1}{3-1} = 66$.
(b) Now put one object into each box first. The answer is $\binom{7+3-1}{3-1} = 36$.
(c) If we could tell the subsets apart (first, second, and third), this would be like part (b). But we can't. For instance, $3+4+3$ and $3+3+4$ are the same breakup, even though $(3,4,3)$ and $(3,3,4)$ are different ordered triples. To organize our thinking, let's count ordered triples (i, j, k) with $1 \le i \le j \le k$ and $i+j+k = 10$. There are 3 ways to choose i in $\{1, 2, 3\}$. Then the number of ways to choose j depends on how i was chosen. If $i = 1$, then there are 4 j's, corresponding to $(1,1,8)$ $(1,2,7)$, $(1,3,6)$, and $(1,4,5)$; if $i = 2$, there are 3, corresponding to $(2,2,6)$, $(2,3,5)$, and $(2,4,4)$; and if $i = 3$, there's only $j = 3$, corresponding to $(3,3,4)$. All told, there are 8 ways, but no good formula it seems.

49. (a) $\binom{k-1}{s}$. Just choose the other s numbers from $\{1, 2, \ldots, k-1\}$.
(b) By part (a) with $k = i+1$, this sum is the number of ways to choose $s+1$ numbers from $\{1, 2, \ldots, m+1\}$; i.e., it's $\binom{m+1}{s+1}$.

50. Let α be the set of words with two A's in a row and β the set with two B's in a row. We want $|\alpha \cup \beta|$. We have $|\alpha| = 4 \cdot 3 = |\beta|$ (choose where the A's start, then place the C, or view this as $4!/2!$), so $|\alpha| + |\beta| = 2 \cdot 4 \cdot 3 = 24$. Also, $|\alpha \cap \beta| = 3!$ (treat AA and BB as big letters). The number of winners is thus $24 - 6 = 18$, so the probability is $18/\binom{5}{2\ 2\ 1} = 18/30 = 0.6$.

51. (a) Count the ones in which ABC does occur. This string can appear in 3 places, and the other two letters can be anything, so there are $3 \cdot 5^2$ strings *with* ABC and $5^5 - 3 \cdot 5^2$ without it.
(b) Let α be the set of strings in which ABC does occur and β the set in which BCD occurs. We want $|\alpha^c \cap \beta^c|$. As in the answer to part (a), $|\alpha| = |\beta| = 3 \cdot 5^2$. Strings in $\alpha \cap \beta$ look like $ABCD*$ or $*ABCD$, so there are $2 \cdot 5$ of them. The answer is $5^5 - 6 \cdot 5^2 + 2 \cdot 5$. [This would be substantially more complicated if we were looking at strings of length 6. Why?]

52. The total number of ways to deal the cards is $\binom{52}{13\ 13\ 13\ 13}$, and Sam gets all four aces in $\binom{48}{9\ 13\ 13\ 13}$ of these deals, so the probability is $\binom{48}{9\ 13\ 13\ 13}/\binom{52}{13\ 13\ 13\ 13}$. Alternatively, the number of (equally likely) hands Sam can possibly get is $\binom{52}{13}$, of which $\binom{48}{9}$ give him all four aces, so the probability is $\binom{48}{9}/\binom{52}{13} = \frac{13}{52}\frac{12}{51}\frac{11}{50}\frac{10}{49}$. A natural interpretation for this last factored form is based on the concept of conditional probability, discussed in Chapter 9. Think of the four aces, S, H, D, C. Then $P(S \text{ in Sam's hand}) = 13/52$. Given that event, $P(H \text{ in Sam's hand}) = 12/51$. etc.

53. (a) 26^3. (b) $26^3 - 26 \cdot 25 \cdot 24$.
(c) Let a be the set of words with AA in them, and similarly for $b, c, \ldots, z$. We want $|a \cup b \cup c \cup \cdots \cup z|$. We have $|a| = |b| = \cdots = |z| = 2 \cdot 25 + 1$ (consider words $AA*$, $*AA$, or AAA). Since $|a \cap b| = 0$ and similarly for other pairs of letters, the answer is $26 \cdot (2 \cdot 25 + 1)$.

54. (a) 0, 1, 5, 21.
(b) $w_n = 2w_{n-1} + 2w_{n-2} + 3^{n-2}$. The terms here come from counting the allowable words of length n that start with B or C, with AB or AC, and with AA, respectively.

55. (a) $\binom{11}{4\ 1\ 2\ 4} = \frac{11!}{4!\cdot1!\cdot2!\cdot4!}$.
(b) $\binom{8}{1\ 1\ 2\ 4} = \frac{8!}{1!\cdot1!\cdot2!\cdot4!}$. [Treat the I's as a single letter.]
(c) $\frac{5!}{2!\cdot1!\cdot2!} = 30$. [The M goes in the middle, with two I's, one P, and two S's on each side of it.]
(d) $1/\binom{11}{4\ 1\ 2\ 4} = \frac{4!\cdot1!\cdot2!\cdot4!}{11!} \approx 0.00003$.

56. (a) $\sum_{i=2}^{n}(-1)^i\frac{1}{i!} \approx \frac{1}{e} \approx 0.368$. (b) Almost exactly $1 - \frac{1}{e} \approx 0.632$.

57. $|(A_i \cup \cdots \cup A_n)^c| = S_0 - S_1 + S_2 - \cdots + (-1)^n S_n$.

58. (a) To get the number with A at least 3 times, put 3 letters in the A box to start with and distribute the remaining 2 letters in $\binom{2+3-1}{3-1} = \binom{4}{2} = 6$ ways. We can't have 2 or more letters appear 3 times here, so the answer is $3 \cdot 6 = 18$.
(b) The total number of 5-letter multisets is $\binom{5+3-1}{3-1} = \binom{7}{2} = 21$, so the answer is $21 - 18 = 3$. [This is obvious as well from the fact that the distribution must be 2–2–1.]
(c) The number with exactly 3 A's is $\binom{5}{3} \cdot 2^2 = 40$, the number with exactly 4 A's is $\binom{5}{4} \cdot 2 = 10$, and the number with 5 A's is 1, so the answer is $3 \cdot [\binom{5}{3} \cdot 2^2 + \binom{5}{4} \cdot 2 + 1] = 153$.

(d) The total number of 5-letter words is $3^5 = 243$, so the answer is $243 - 153 = 90$. [We could also get this answer from (b). Choose the first pair, in one of 3 ways. Place it, in one of $\binom{5}{2}$ ways. Choose the second pair, in one of 2 ways. Place it, in one of $\binom{3}{2}$ ways. Divide by 2, because we can't tell which pair was chosen first. Alternatively, consider $3!\binom{5}{2\ 2\ 1}/2! = 3 \cdot \binom{5}{2\ 2\ 1} = 90$.]

(e) It would be possible to have two different letters show up three or more times. We'd need Inclusion-Exclusion to keep track.

59. We want partitions of 14 into four parts, each with at least 3 members. The only possibilities are $3+3+3+5$ and $3+3+4+4$. There are $\binom{14}{3\ 3\ 3\ 5}/3!$ partitions of the first type and $3!^3 \cdot 5 \cdot 4 \cdot 3$ ways to choose officers for each one. There are $\binom{14}{3\ 3\ 4\ 4}/2! \cdot 2!$ partitions of the second type and $3!^2 \cdot (4 \cdot 3 \cdot 2)^2$ ways to choose officers for each of them. Altogether, then, the total number of ways is

$$3!^3 \cdot 5 \cdot 4 \cdot 3 \cdot \binom{14}{3\ 3\ 3\ 5}\Big/3! + 3!^2 \cdot (4 \cdot 3 \cdot 2)^2 \cdot \binom{14}{3\ 3\ 4\ 4}\Big/2! \cdot 2!.$$

Alternatively, one could think of first choosing 4 sets of 3 officers, then adding the remaining 2 people to the sets so formed. The number of ways to choose officers if we were to distinguish the committees would be $14 \cdot 13 \cdot 12 \cdots 3 = 14!/2!$, so if we don't distinguish committees, it's $14!/2! \cdot 4!$. Then there are 4^2 choices for where to put the remaining 2 sociologists. The final answer expressed in this way is $4^2 \cdot 14!/2! \cdot 4!$. Believe it or not, these two answers are the same.

60. (a) As in Example 3 on page 199, the number of words with at least one letter in its original position is

$$5!\left(1 - \frac{1}{2!} + \frac{1}{3!} - \frac{1}{4!} + \frac{1}{5!}\right) = 120 - 60 + 20 - 5 + 1,$$

so the number of derangements is $60 - 20 + 5 - 1 = 44$.

(b) We must want a, l, and e to move, and also not want a p as the second or third letter. There are 4 words of form $p{*}{*}p{*}$; choose where e goes and then where a goes. Similarly, there are 4 of each of the forms $p{*}{*}{*}p$ and ${*}{*}{*}pp$, so we have 12 derangements altogether. There seems to be no obvious easier method here. A long computation with the Inclusion-Exclusion Principle gives this answer too, of course.

(c) The words must have the form ${*}a{*}aa{*}$; there are 3 derangements.

(d) 0. Where would the p's have to be?

61. (a) $\binom{p}{r}$ is an integer, because it is the number of r-element subsets of a set with p elements. Since $\binom{p}{r} \cdot r! \cdot (p-r)! = p!$ and since p divides the right side, p must divide the left side. But p does not divide $r!$ or $(p-r)!$, so p must divide $\binom{p}{r}$. *Note:* The fact that p is prime is crucial here. For example, $\binom{4}{2}$ is not a multiple of 4.

(b) The expansion of $(a+b)^p$ in the Binomial Theorem on page 200 holds. By part (a), p divides each of the terms $\binom{p}{r}a^r b^{p-r}$ for $0 < r < p$, so each of these terms is $\equiv 0 \pmod p$. Thus $(a+b)^p \equiv a^p + b^p \pmod p$.

62. (a) See Exercise 61(a). This uses the fact about primes.

(b) Of course, $1^p = 1$. If $k^p \equiv k \pmod p$, then $(k+1)^p \equiv k^p + 1^p \equiv k + 1 \pmod p$ by Exercise 61(b). The claim follows by induction.

(c) If $p = 2$, this just says that n^2 is even if and only if n is even; see Exercise 8(a) on page 127. If p is odd and $n \in \mathbb{P}$, then $(-n)^p = (-1)^p n^p \equiv (-1)n \pmod p$, by part (b).

(d) By (c) we have $p | n^p - n = n \cdot (n^{p-1} - 1)$, so if p does not divide n, then $p | n^{p-1} - 1$. Thus $n^{p-1} \equiv 1 \pmod p$.

6.1 Answers

1. (a) $s\ t\ v$ or $s\ u\ v$; length 2. (c) $u\ v\ w\ y$; length 3.

3. (a) is true and (b) is false.

5. (a) and (b) are now both true.

7. See Figure 1(c). u and w are vertices of a cycle, and so are w and x. But no cycle uses both u and x.

9. (a)

e	a	b	c	d	e	f	g	h	k
$\gamma(e)$	$\{w, x\}$	$\{x, u\}$	$\{t, u\}$	$\{t, v\}$	$\{u, v\}$	$\{u, y\}$	$\{v, z\}$	$\{x, y\}$	$\{y, z\}$

11. (a) $e\ b\ h\ k\ g$ and its reversal $g\ k\ h\ b\ e$. (c) $e\ c\ d$, $b\ h\ f$ and their reversals.

13. (a)

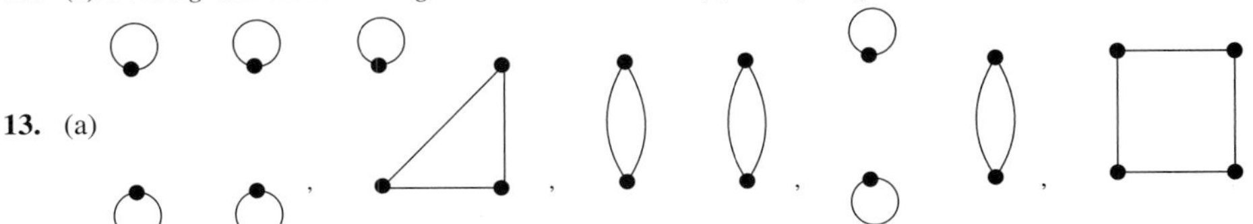

(c) There are none by Theorem 3, since $5 \cdot 3$ is not even.

15. (a), (c), and (d) are regular, but (b) is not. (a) and (c) have cycles of length 3, but (d) does not. Or count edges. (a) and (c) are isomorphic; the labels in the following figures show an isomorphism between (a) and (c).

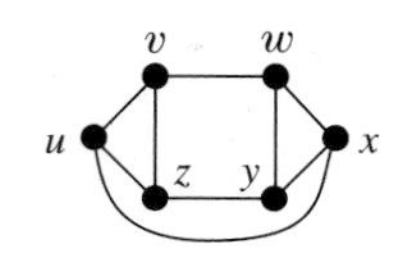

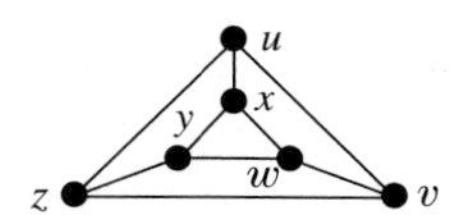

17. One isomorphism is indicated by the edge labels in the figure. Since the graphs have parallel edges, matching up vertices does not completely describe an isomorphism.

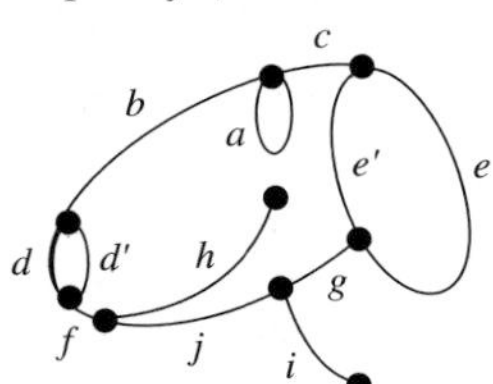

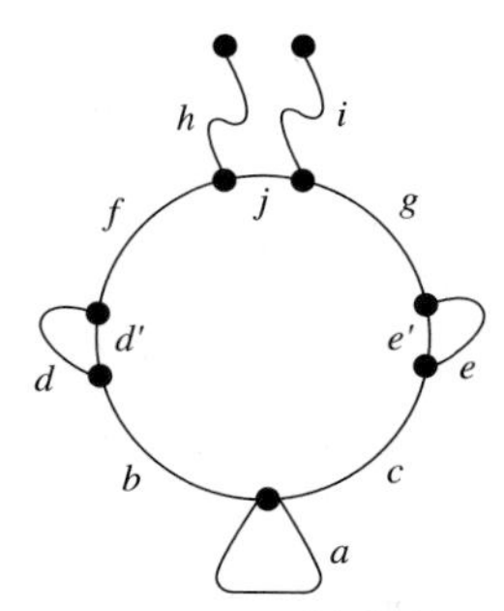

19. (a) $\binom{8}{5} = 56$. (c) $8 \cdot 7 + 8 \cdot 7 \cdot 6 + 8 \cdot 7 \cdot 6 \cdot 6 = 2408$.

21. (a) 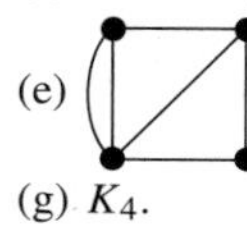

(c) No such graph. Use Theorem 3 [consider Exercise 6(b)].

(e)

(g) K_4.

23. Assume no loops or parallel edges. Consider a longest path $v_1 \cdots v_m$ with distinct vertices. There is another edge at v_m. Adjoin it to the path to get a closed path and use Proposition 1.

25. Use $|V(G)| = D_0(G) + D_1(G) + D_2(G) + \cdots$ and Theorem 3.

6.2 Answers

1. Only Figure 7(b) has an Euler circuit. To find one, do Exercise 3.

3. One solution starts with vertex r and simple closed path $C = r\,s\,v\,u\,r$. Then successive **while** loops might proceed as follows.

1. Choose v, construct $v\,y\,z\,w\,v$, obtain new $C = r\,s\,v\,y\,z\,w\,v\,u\,r$.
2. Choose s, construct $s\,t\,v\,x\,y\,u\,s$, obtain final $C = r\,s\,t\,v\,x\,y\,u\,s\,v\,y\,z\,w\,v\,u\,r$.

You should find your own solutions, starting perhaps at other vertices.

5. You will find that the closed paths all have length 4. After you remove any one of them, the remaining edge set will contain no closed paths.

7. (a) $v_3\,v_1\,v_2\,v_3\,v_6\,v_2\,v_4\,v_6\,v_5\,v_1\,v_4\,v_5\,v_3\,v_4$ is one.

9. (a) Add an edge e to G to get G^* with all vertices of even degree. Run EulerCircuit(G^*) to get C^*, then remove e from C^*. Alternatively, choose v initially to have odd degree, and use ClosedPath [with a modified input] to yield a starting path C from v that cannot be extended. The **while** loop remains the same.

11. $\{0, 1\}^3$ consists of 3-tuples of 0's and 1's, which we may view as binary strings of length 3. The graph is then

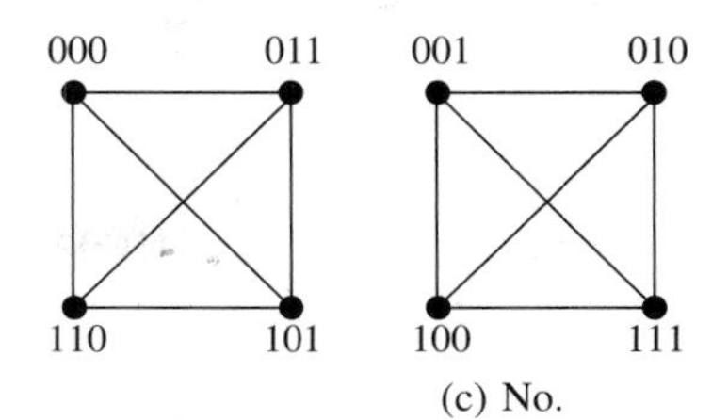

(a) 2. (c) No.

13. (a) Join the odd-degree vertices in pairs, with k new edges. The new graph has an Euler circuit, by Theorem 2. The new edges do not appear next to each other in the circuit, and they partition the circuit into k simple paths of G.

(c) Imitate the proof. That is, add edges $\{v_2, v_3\}$ and $\{v_5, v_6\}$, say, create an Euler circuit, and then remove the two new edges. Then you will obtain $v_3\, v_1\, v_2\, v_7\, v_3\, v_4\, v_5\, v_7\, v_6\, v_1\, v_5$ and $v_6\, v_4\, v_2$, say.

15. (a) No such walk is possible. Create a graph as follows. Put a vertex in each room and one vertex outside the house. For each door, draw an edge joining the vertices for the regions on its two sides. The resulting graph has two vertices of degree 4, three of degree 5, and one of degree 9. Apply the corollary to Theorem 1.

6.3 Answers

1.

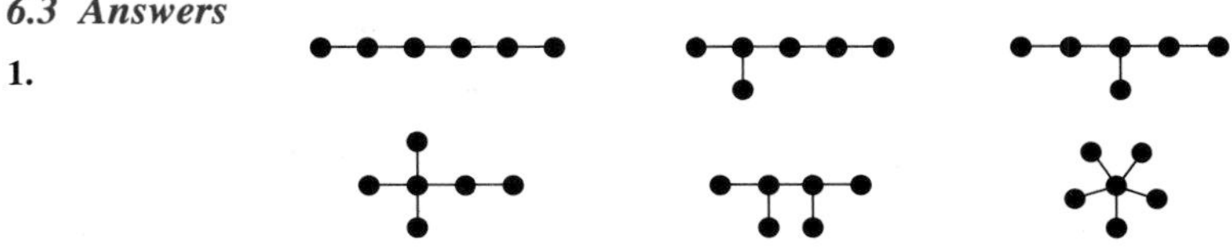

3. (a) 4.

(c) $4 + 2 \cdot 2 = 8$. Here are their pictures.

For (e), the answer is $8 \cdot 8 = 64$. Any spanning tree in (e) can be viewed as a pair of spanning trees, one for the upper half and one for the lower half of the graph. Each half of the spanning tree is a spanning tree for the graph in (c). Hence there are $8 \cdot 8 = 64$ such pairs.

5. All but edges e_1 and e_5.

7. (a) Since $2n - 2$ is the sum of the degrees of the vertices, we must have $2n - 2 = 4+4+3+2+1 \cdot (n - 4)$. Solve for n to get $n = 11$.

9. (a) (b) Prove by induction on n.

(c)$d_1 = 5$, $d_2 = 3$, $d_3 = 2$, $d_4 = \cdots = d_9 = 1$, and their sum is $16 = 2 \cdot 9 - 2$.

11. (a) Suppose the components have $n_1, n_2, \ldots, n_m$ vertices, so altogether $n_1 + n_2 + \cdots + n_m = n$. The ith component is a tree, so it has $n_i - 1$ edges, by Theorem 4. The total number of edges is $(n_1 - 1) + (n_2 - 1) + \cdots + (n_m - 1) = n - m$.

13. By Lemma 1 to Theorem 4, it must be infinite. Use $\mathbb{Z}$ for the set of vertices; see Figure 3(a) on page 529.

6.4 Answers

1. (a)

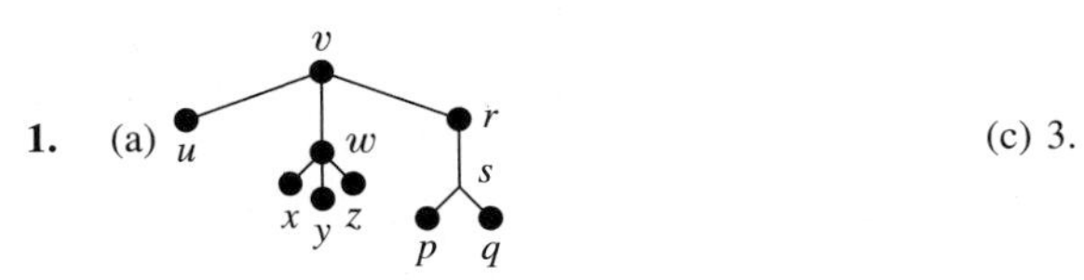

(c) 3.

3. (a) Rooted trees in Figures 5(b) and 5(c) have height 2; the one in 5(d) has height 3.

5. (a)

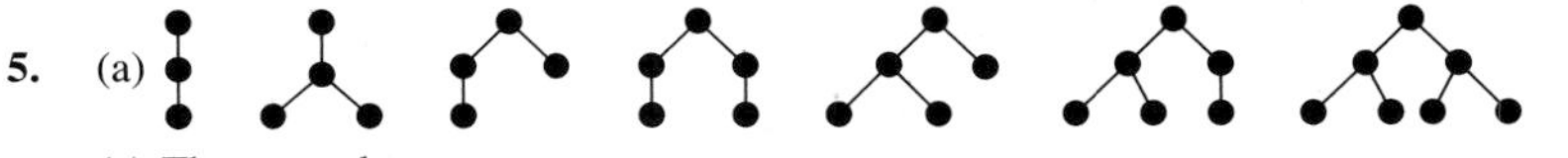

(c) The seventh.

7. (a)

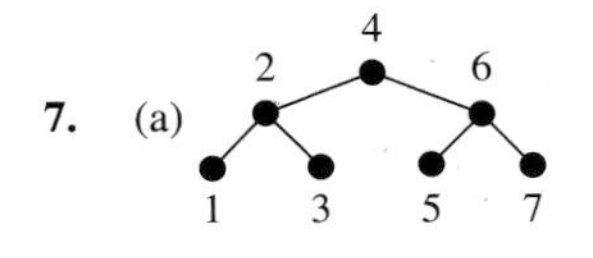

(c) The possibilities are

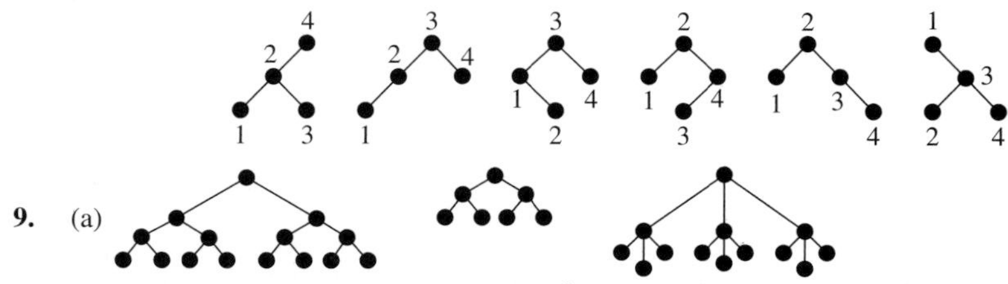

9. (a)

11. From Example 5, we have $(m-1)p = m^h - 1$, so $(m-1)p + 1 = m^h = t$.

13. There are 2^k words of length k.

15. (a) Move either Lyon's or Ross's records to the Rose node and delete the empty leaf created.

6.5 Answers

1. (a) In the vertex sequence for a Hamilton circuit [if one existed], the vertex w would have to both precede and follow the vertex v; i.e., the vertex w would have to be visited twice.

(b) In the vertex sequence for a Hamilton path [if one existed], the vertices of degree 1 would have to be at the beginning or end, since otherwise the adjacent vertex would be repeated. But there are three vertices of degree 1 in Figure 1(d).

3. (a) Yes. Try $v_1\ v_2\ v_6\ v_5\ v_4\ v_3\ v_1$, for example.

(c) No. If v_1 is in V_1, then $\{v_2, v_3, v_4, v_5\} \subseteq V_2$, but v_2 and v_3 are joined by an edge.

5. (a) Since there are $n!$ choices for the order in which the vertices in V_1 and in V_2 are visited and the initial vertex can be in either V_1 or V_2, there are $2(n!)^2$ possible Hamilton circuits.

(c) m and n even, or m odd and $n = 2$, or $m = n = 1$.

7. Here is the graph:

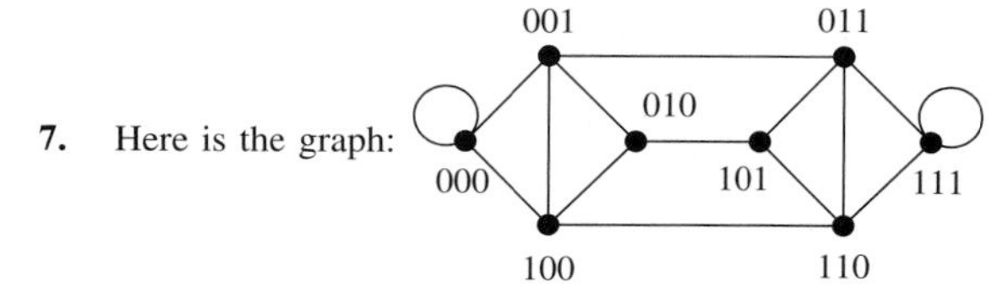

One possible Hamilton circuit has vertex sequence 000, 001, 011, 111, 110, 101, 010, 100, 000 corresponding to the circular arrangement 0 0 0 1 1 1 0 1. Although there are four essentially different Hamilton circuits in $\{0, 1\}^3$, there are only two different circular arrangements, 0 0 0 1 1 1 0 1 and 0 0 0 1 0 1 1 1, which are just the reverses of each other.

9. There is no Hamilton path because the graph is not connected. The graph is drawn in the answer on page 566 to Exercise 11 on page 238.

11. (a) K_n^+ has n vertices and just one more edge than K_{n-1} has, so it has exactly $\frac{1}{2}(n-1)(n-2)+1$ edges.

(b) Consider vertices v and w in K_n^+ that are not connected by an edge. Apply Theorem 3 on page 253.

13. (a)

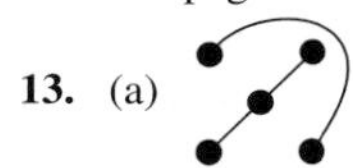

(c) Choose two vertices u and v in G. If they are *not* joined by an edge in G, then they are joined by an edge in the complement. If they *are* joined by an edge in G, then they are in the same component of G. Choose w in some other component. Then $u\ w\ v$ is a path in the complement. In either case, u and v are joined by a path in the complement.

(e) No. Consider

15. Given G_{n+1}, consider the subgraph H_0, where $V(H_0)$ consists of all binary $(n+1)$-tuples with 0 in the $(n+1)$st digit and $E(H_0)$ is the set of all edges of G_{n+1} connecting vertices in $V(H_0)$. Define H_1 similarly. Show H_0 and H_1 are isomorphic to G_n, so have Hamilton circuits. Use these to construct a Hamilton circuit for G_{n+1}. For $n = 2$, see how this works in Figure 5.

6.6 Answers

1. (a)

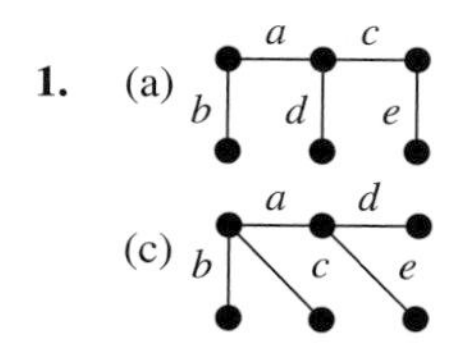

3. (a)

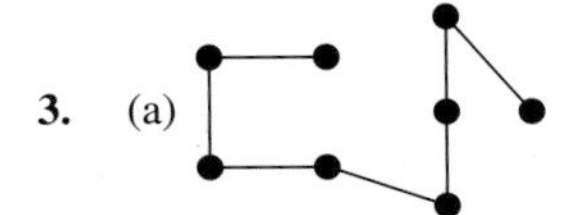

Both trees have weight 1330.

5. (a) There are four possible answers. The order choices are: e, d, a, b or b', c; or e, d', c, b or b', a.

7. (a) and (c)

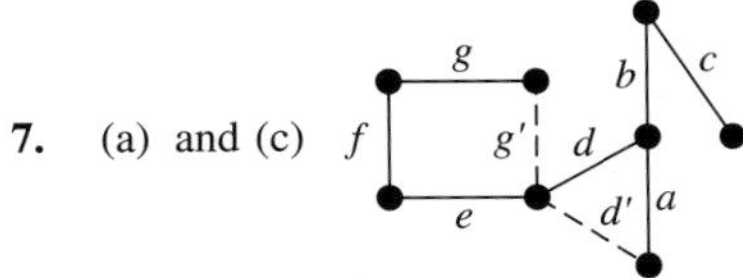

Either d or d' can be chosen and either g or g', so there are four possible answers to (a).

9. (a) $e_1, e_2, e_3, e_5, e_6, e_7, e_9$.

11. (a)

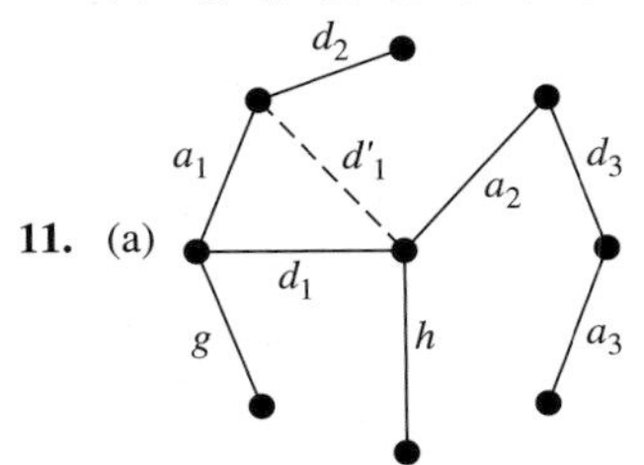

Edges a_1, a_2, a_3 can be chosen in any order. So can d_1, d_2, d_3. Edge d_1' can be chosen instead of d_1. The weight is 16.

13. 1687 miles.

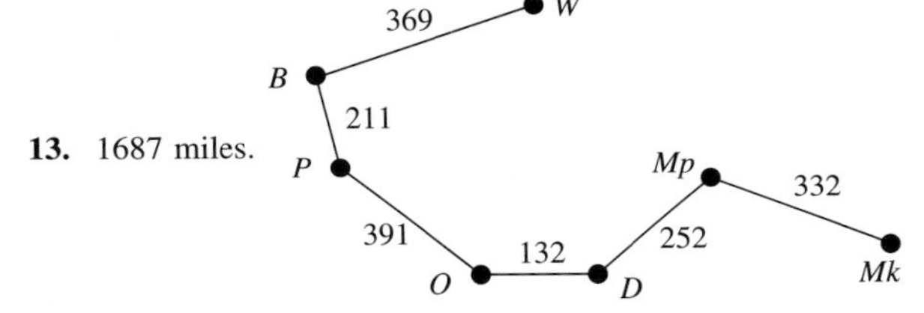

15. (a) How big will V get?
(b) Here is one possible algorithm.

PrimForest(weighted graph)

Set $E := \emptyset$.
Choose w in $V(G)$ and set $V := \{w\}$.
while $V \neq V(G)$ **do**
  if there is an edge $\{u, v\}$ in $E(G)$
    with $u \in V$ and $v \in V(G) \setminus V$ **then**
   Choose such an edge of smallest weight.
   Put $\{u, v\}$ in E and put v in V.
  else
   Choose $v \in V(G) \setminus V$ and put v in V.
return E ■

17. Follow the hint. Assume that G has more than one minimum spanning tree. Look at minimum spanning trees S, T with $e \in T \setminus S$. Then $S \cup \{e\}$ has a cycle that contains an edge f in $S \setminus T$. Then $(S \cup \{e\}) \setminus \{f\}$ is a spanning tree. $W(e) > W(f)$, contrary to the choice of e. Supply reasons.

6.7 Answers

1. (a) No. The graph has vertices of odd degree.
(b) One such path is $v\,v\,w\,u\,y\,x\,w\,z\,y$. [There are 11 others.]
(c) One cycle is $v\,v$. Another is $w\,u\,y\,x\,w$, and there are others.
(d) No. For example, the path $y\,u\,w\,y\,z\,v\,x\,y$ is simple, closed, but not a cycle, since it repeats a vertex.

2. (a) Five, since there are six vertices.
(b) No. Such a path can use at most two edges joined to y and so would miss w, x, or z.
(c) Same as (a).

(d) No. Such a path can use at most two edges joined to w and so would miss at least two among v, u, x, and z.

3. (a) False. See Lemma 2 on page 242.
(b) Could be either. See Figure 1(c) on page 239.
(c) True. See Lemma 1 on page 242.
(d) True. See Theorem 3(b) on page 241.

4. (a) No. Each vertex in the 2-element subset has odd degree 7.
(b) Yes. The only two vertices of odd degree are the ones in the 2-element subset. Each Euler path starts at one of these and ends at the other.
(c) No. Every path goes back and forth between the 4-element and 6-element sets of vertices, and since a Hamilton path could not contain more than 4 vertices from the 4-element set, it could not go through more than 5 members of the 6-element set.

5. (a) One of the four possible answers is

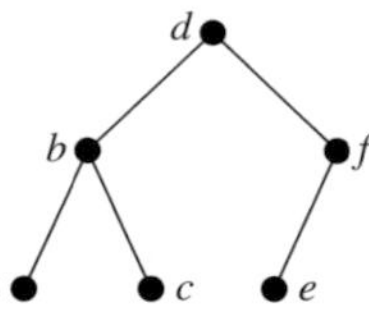

(b) c and d. There must be at least two nodes on each side of the root.

6. (a) Yes. The graph is connected, and all the vertices have even degrees, so Euler's Theorem guarantees an Euler circuit.
(b) Yes. The graph has no loops or parallel edges, and each vertex has degree at least half the number of vertices in the graph. Theorems 1 and 3 in §6.5 guarantee a Hamilton circuit.
[Note that in part (a) we have a couple of algorithms that would actually construct the circuit for us, whereas in part (b) we have no general method to do so.]
(c) 33. [The number of edges is $|E(G)| = \{\text{sum of degrees}\}/2 = 11 \cdot 6/2$.]
(d) 11.

7. (a) and (b). It consists of all four edges of weight 1, three of the edges of weight 2 [which ones must be chosen?], all three of the edges of weight 3, and one edge of weight 7. Its weight is 26.
(c) Since the draph has two vertices of odd degree, by Euler's theorem on page 234 it does not have an Euler circuit, but it does have an Euler path.
(d) No. There is no way to pass through the vertex of degree 1 or get back to it if one starts there.

8. (a) The diagonal edge of weight 3.
(b) The leftmost edge of weight 2.
(c) $2 + 4 + 2 + 1 + 2 + 3 + 4 = 18$.
(d) 28. [The only Hamilton circuit is the path around the outside edge.]
(e) $2 + 4 + 2 + 1 + 2 + 3 + 5 = 19$.

9. (a) $e_1, e_2, e_3, e_5, e_6, e_7$. (b) $e_6, e_5, e_7, e_2, e_1, e_3$.

10. (a) Go right of `foo`, left of `too` [and there is `goo`].
(b) Go right of `foo`, left of `too`, should be right of `goo` [but it's not there].
(c) There are four possibilities. Here is one.

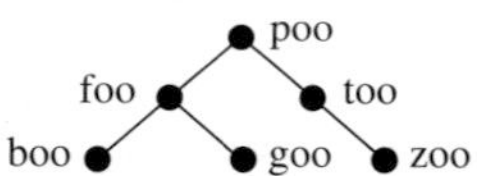

11. (a) There are four. Three of them correspond to subtrees of the tree in Figure 7 on page 249 having leaves as follows: T_1 with leaf aaa; T_2 with leaves aa and b; T_3 with leaves aa and ab. T_4 has three leaves, all children of the root.
(b) There are five, the four listed in part (a) and one with leaves ba and bb. This one is isomorphic to T_2 as a rooted tree, but not as an ordered rooted tree.

12. (a) Must hold.
(b) Must hold, because $t = 2^h$ by Example 5 on page 248.
(c) Must hold. By the same example, $n = 2^{h+1}-1$, so $n+1 = 2^{h+1}$ and $h+1 = \log_2(n+1) > \log_2 n$.
(d) Must hold.

13. Ann. Pick a vertex. Because the graph is a tree, there is a unique simple path from this vertex to any other. The sets of vertices for which this path has even length and for which this path has odd length form a bipartite partition of the vertices of the graph. [In the tree shown, the vertices in the two sets have been marked with the letters a and b, respectively.]

14. By Euler's Theorem, all vertices need to be of even degree, so m and n must be even. If, in addition, each vertex is visited exactly once by an Euler circuit, then each vertex must have degree 2. Thus $m = n = 2$. Thus $K_{2,2}$ is the only complete bipartite graph with an Euler–Hamilton circuit.

15. (a) There will be at least $\binom{n}{2} - (n-3) = \frac{(n-1)(n-2)}{2} + 2$ edges remaining. So by Theorem 2 on page 253, the subgraph is Hamiltonian.

(b) Select some vertex and remove all but one of the edges joined to it; i.e., remove $n-2$ edges. The resulting subgraph is the same as K_n^{++} in Exercise 12 on page 257, which is not Hamiltonian.

16. The possible degrees are $\{0, 1, \dots, n-1\}$. If some vertex has degree $n-1$, then no vertex has degree 0. Thus at most $n-1$ different degrees are possible. Apply the Pigeon-Hole Principle.

17. Let c be the number of components. A spanning forest for G consists of spanning trees $T_1, \dots, T_c$ for the components. Each T_i has as many vertices as its component does and has one less edge than that. Then $|V(G)| = \sum_{i=1}^{m} |V(T_i)| = \sum_{i=1}^{c}(|E(T_i)| + 1) \le |E(G)| + c$.

18. (a) A Hamilton path doesn't repeat vertices, so it must be acyclic. Since it's connected, it's a tree. Since it goes through or to each vertex, it must be a spanning tree.

(b) If we remove any edge e from H, we get a Hamilton path whose edges form a spanning tree for G, by part (a). Since T is a minimum spanning tree, $W(T) \le W(H \setminus \{e\}) = W(H) - W(e)$.

This fact is of practical importance in dealing with the Traveling Salesman Problem. We know at least two algorithms for finding the weight of a minimum spanning tree. If we can somehow find a Hamilton circuit whose weight is not too much larger than $W(T) + W(e)$ for every edge e in the graph, then that circuit has about as small a weight as is likely to exist for a Hamilton circuit, and it's time to stop looking for a better one.

19. (a) Theorems 1 and 3 in §6.5 say that, if G has no loops or parallel edges and if each vertex has degree at least $|V(G)|/2$, then G has a Hamilton circuit. Since $|V(G)| = 10$ and each vertex has degree at least half that, yet the graph has no Hamilton circuit, it must have a loop or parallel edges.

(b) Sure. Just string the vertices in a cycle, with each vertex of degree 2.

20. Recall that $T_{m,h}$ has $1 + m + m^2 + \cdots m^h = (m^{h+1} - 1)/(m-1)$ nodes, with m^k nodes at level k.

(a) The probability that a random node is at level k is $m^k/((m^{h+1} - 1)/(m-1))$, which for $m = 2$, $h = 3$ is $2^k/15$. For level 2, it's 4/15.

(b) For level 3, it's 8/15.

(c) The most likely level is level 3. In general, more than half the nodes are leaves.

(d) The probability that the node chosen is at level k is as above. So for $T_{2,3}$ the probabilities are 1/15, 2/15, 4/15, and 8/15, and thus

$$L(T_{2,3}) = 0 \cdot \frac{1}{15} + 1 \cdot \frac{2}{15} + 2 \cdot \frac{4}{15} + 3 \cdot \frac{8}{15} = \frac{34}{15}.$$

(e) For $T_{3,2}$, the probabilities are 1/13, 3/13, and 9/13, so

$$L(T_{3,2}) = 0 \cdot \frac{1}{13} + 1 \cdot \frac{3}{13} + 2 \cdot \frac{9}{13} = \frac{21}{13}.$$

21. (a) $999/1000 = 0.999$.

(b) $\frac{999}{1000} \cdot \frac{998}{1000} \cdot \frac{997}{1000} \cdot \cdots \cdot \frac{981}{1000} = \frac{P(999, 19)}{1000^{19}} \approx 0.826$. This is also $\frac{P(1000, 20)}{1000^{20}}$. Why?

(c) $20/1000 = 0.02$.

(d) 1000. The numbers could even all be the same, but the names all different. By the 1001st record, though, a collision is certain.

22. Both are correct. There are six vertices of odd degree. Ann's plan will make all of them have even degree, and her new vertex will have degree 6. Bob can hook the vertices together in three pairs with three new edges and make all vertices have even degree.

23. (a) Both have degree sequence 0, 0, 0, 3, 3, 2, 0, 1.

(b) No. For example, in the right graph, every pair of vertices is connected by a path of length ≤ 2. The left graph doesn't have this property. The left graph also has no adjacent vertices of degree 4, and there are various other differences between the two graphs.

(c) Yes. Examples of Hamilton circuits are easy to find.

(d) No. The hypothesis of Theorem 1 fails, because each graph has vertices of degree <4.5. The hypothesis of Theorem 2 fails, because $\frac{1}{2}(n-1)(n-2) + 2 = 30$, but each graph has only 19 edges. The hypothesis of Theorem 3 fails, because each graph has nonadjacent vertices of degree 3.

7.1 Answers

1. (a) Use induction on m. Clearly, 2^0 is in S by (B). If 2^m is in S, then $2 \cdot 2^m = 2^{m+1}$ is in S by (R).

(b) Use the Generalized Principle of Induction on page 273, where $p(n) =$ "n has the form 2^m for some $m \in \mathbb{N}$."

3. (a) $S = \{(m, n) : m \le n\}$.

(c) Yes. A pair $(0, n)$ is in S only by (B) or because $(n-1, n-1) \in S$, and a pair (m, n) with $0 < m \le n$ can only come from $(m-1, n)$.

5. (a) (B) λ is in S, (R) if $w \in S$, then $aw \in S$ and $wb \in S$.

(c) $\lambda \in S$ by (B), so a, aa, and aab are in S by (R).

7. (a) Two possible constructions are

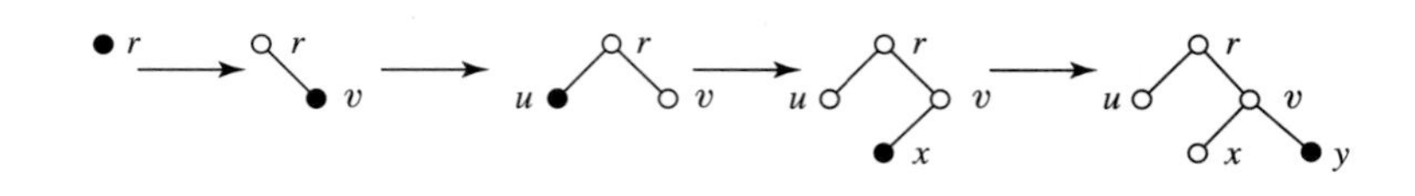

(b) First obtain the subtree with root v by hanging the trivial trees with vertices x and y. Then obtain the tree with root r by hanging the subtree with root v and the trivial tree with vertex u.

9. Use the Generalized Principle of Induction on $p(n)$ = "$n \equiv 0 \pmod 3$ or $n \equiv 1 \pmod 3$." Certainly, $p(1)$ is true. It suffices to show that $p(n) \Longrightarrow p(3n)$ and $p(2n+1) \Longrightarrow p(n)$. The first implication is trivial because $3n \equiv 0 \pmod 3$ for all n. For the second implication, show the contrapositive; i.e., $n \equiv 2 \pmod 3$ implies $(2n+1) \equiv 2 \pmod 3$. The contrapositive implication is easy to show directly, since $n = 3k+2$ implies $2n+1 = 3(2k+1)+2$.

11. (a) Use (B) and (R) to show that the sequence 1, 2, 4, 8, 16, 5, 10, 3, 6 lies in S.

13. We follow the hint. Let $p(w)$ = "$l'(w)$ is the number of letters in w." By (1) of Example 10(b), $p(w)$ is true if $w \in X = \{\lambda\} \cup \Sigma$. Suppose that $w = uv$ with $p(u)$ and $p(v)$ true. By (2) of Example 10(b), $l'(w) = l'(u) + l'(v)$ = (number of letters in u) + (number of letters in v) = number of letters in w, so $p(w)$ is true. [This argument is valid no matter how w is produced from members of Σ^* by (2).] Thus $p(w)$ is true for every w in Σ^* by the Generalized Principle of Induction.

15. (a) (2, 3), (4, 6), etc.

(c) No. For example, (3, 2) does not belong to S.

17. (a) Obviously, $A \subseteq \mathbb{N} \times \mathbb{N}$. To show $\mathbb{N} \times \mathbb{N} \subseteq A$, apply the ordinary First Principle of Mathematical Induction to the propositions

$$p(k) = \text{"if } (m, n) \in \mathbb{N} \times \mathbb{N} \text{ and } m + n = k, \text{ then } (m, n) \in A.\text{"}$$

(b) Let $p(m, n)$ be a proposition-valued function defined on $\mathbb{N} \times \mathbb{N}$. To show that $p(m, n)$ is true for all (m, n) in $\mathbb{N} \times \mathbb{N}$, it is enough to show:

(B) $p(0, 0)$ is true, and

(I) if $p(m, n)$ is true, then $p(m+1, n)$ and $p(m, n+1)$ are true.

19. (a) For w in Σ^*, let $p(w)$ = "length($\overleftarrow{w}$) = length(w)." Apply the Generalized Principle of Induction. Since $\overleftarrow{\lambda} = \lambda$, $p(\lambda)$ is clearly true. You need to show that if $p(w)$ is true, then so is $p(wx)$:

$$\text{length}(\overleftarrow{w}) = \text{length}(w) \quad \text{implies} \quad \text{length}(\overleftarrow{wx}) = \text{length}(wx).$$

In detail, suppose length($\overleftarrow{w}$) = length(w). Then

$$\begin{aligned} \text{length}(\overleftarrow{wx}) &= \text{length}(x\overleftarrow{w}) \\ &= \text{length}(x) + \text{length}(\overleftarrow{w}) \\ &= \text{length}(x) + \text{length}(w) \qquad \text{[by assumption]} \\ &= \text{length}(wx). \end{aligned}$$

(b) Fix w_1, say, and work with $p(w)$ = "$\overleftarrow{w_1 w} = \overleftarrow{w}\,\overleftarrow{w_1}$." To show $p(w)$ true for every $w \in \Sigma^*$, it is enough to show:

(B) $p(\lambda)$ is true;

(I) if $p(w)$ is true, then $p(wx)$ is true [for $w \in \Sigma^*$ and $x \in \Sigma$].

7.2 Answers

1. (a)

Test(20)
  Test(10)
    Test(5)= (false, $-\infty$)
  = (false, $-\infty$)
= (false, $-\infty$).

3. (a) $((x+y)+z)$ or $(x+(y+z))$. (c) $((xy)z)$ or $(x(yz))$.

5. (a) By (B), x, y, and 2 are wff's. By the (f^g) part of (R), we conclude that (x^2) and (y^2) are wff's. So, by the $(f+g)$ part of (R), $((x^2)+(y^2))$ is a wff.

(c) By (B), X and Y are wff's. By the $(f+g)$ part of (R), $(X+Y)$ is a wff. By the $(f-g)$ part of (R), $(X-Y)$ is a wff. Finally, by the $(f*g)$ part of (R), $((X+Y)*(X-Y))$ is a wff.

7. Blank entries below signify that the algorithm doesn't terminate with the indicated input n.

n	FOO	GOO	BOO	MOO	TOO	ZOO
8	8	40,320		4	4	8
9	9	362,880			4	8

9. Let $p(k)$ be the statement "ZOO$(n) = 2^k$ whenever $2^k \le n < 2^{k+1}$." Then $p(0)$ asserts that "ZOO$(n) = 2^0 = 1$ whenever $1 \le n < 2$," which is clear. Assume that $p(k)$ is true for some $k \in \mathbb{N}$ and consider n such that $2^{k+1} \le n < 2^{k+2}$. Then $2^k \le \lfloor n/2 \rfloor < 2^{k+1}$, so ZOO$(\lfloor n/2 \rfloor) = 2^k$. Since $n \ne 1$, the **else** branch of ZOO applies and gives ZOO$(n) = 2^k * 2 = 2^{k+1}$, so $p(k+1)$ is true. Hence all statements $p(k)$ are true by induction.

11.

Euclid$^+$(108,30)
  Euclid$^+$(30,18)
    Euclid$^+$(18,12)
      Euclid$^+$(12,6)
        Euclid$^+$(6,0) $= (6, 1, 0)$
      $= (6, 0, 1 - 0 \cdot (12 \text{ DIV } 6)) = (6, 0, 1)$
    $= (6, 1, 0 - 1 \cdot (18 \text{ DIV } 12)) = (6, 1, -1)$
  $= (6, -1, 1 - (-1) \cdot (30 \text{ DIV } 18)) = (6, -1, 2)$
$= (6, 2, -1 - 2 \cdot (108 \text{ DIV } 30)) = (6, 2, -7)$.

Sure enough, $108 \cdot 2 + 30 \cdot (-7) = 6$.

13. (a) Since $\gcd(m, n) = m$ when $n = 0$, we may assume that $n \ne 0$. We need to check that the algorithm terminates and that if Euclid$(n, m \text{ MOD } n) = \gcd(n, m \text{ MOD } n)$, then Euclid$(m, n) = \gcd(m, n)$. The latter observation follows from the equality $\gcd(n, m \text{ MOD } n) = \gcd(m, n)$, which was verified in the Proposition on page 172. To check that the algorithm terminates, we need to be sure that the second input variable n in Euclid($, n$) is eventually 0. This is clear, since at each step the new $n' = m \text{ MOD } n$ is less than n and greater than or equal to 0, so the values must decrease to 0.

(b) The algorithm terminates because it's the same algorithm as in part (a), but with extra outputs. You need to verify that $sm + tn = d$ at each stage of the algorithm.

15. (a) p and q are wff's by (B). $p \vee q$ is a wff by (R). $\neg(p \vee q)$ is a wff by (R).

(c) p, q, and r are wff's by (B). $(p \leftrightarrow q)$ and $(r \rightarrow p)$ are wff's by (R). $((r \rightarrow p) \vee q)$ is a wff by (R), so $((p \leftrightarrow q) \rightarrow ((r \rightarrow p) \vee q))$ is a wff by (R).

17. (a) $p, q \in \mathcal{F}$ by (B). $(p \vee q) \in \mathcal{F}$ by (R) with $P = p$ and $Q = q$. Hence $(p \wedge (p \vee q)) \in \mathcal{F}$ by (R) with $P = p$ and $Q = (p \vee q)$.

(c) If p and q are false, then $(p \rightarrow q)$ is true, so $r((p \rightarrow q))$ is false. Thus $(p \rightarrow q)$ cannot be logically equivalent to a proposition in $\mathcal{F}$, by part (b). This provides a negative answer to the question in Exercise 17(c) on page 85: $p \rightarrow q$ cannot be written in some way using just p, q, $\wedge$, and $\vee$.

19. (a)

ConvertToBinary(25); $25 \text{ MOD } 2 = 1$
  ConvertToBinary(12); $12 \text{ MOD } 2 = 0$
    ConvertToBinary(6); $6 \text{ MOD } 2 = 0$
      ConvertToBinary(3); $3 \text{ MOD } 2 = 1$
        ConvertToBinary(1) $= 1$
      $= 11$
    $= 110$
  $= 1100$
$= 11001$

(c) Replace $n \text{ DIV } 2$ and $n \text{ MOD } 2$ by $n \text{ DIV } 16$ and $n \text{ MOD } 16$, respectively. You would need new digits for 10, 11, 12, 13, 14, and 15; the standard ones are A, B, C, D, E, and F.

7.3 Answers

1. Preorder: $r\,x\,w\,v\,y\,z\,s\,u\,t\,p\,q$. Postorder: $v\,y\,w\,z\,x\,t\,p\,u\,q\,s\,r$.

3. Preorder: $r\,t\,x\,v\,y\,z\,w\,u\,p\,q\,s$. Postorder: $v\,y\,z\,x\,w\,t\,p\,q\,s\,u\,r$.

5. The order of traversing the tree is

$$r, w, v, w, x, y, x, z, x, w, r, u, t, u, s, p, s, q, s, u, r.$$

(a) Preorder: $r\,w\,v\,x\,y\,z\,u\,t\,s\,p\,q$. $L(w) = w\,v\,x\,y\,z$. $L(u) = u\,t\,s\,p\,q$.

7. The order of traversing the tree is

$$u, x, w, v, w, y, w, x, r, z, r, t, r, x, u, s, p, s, q, s, u.$$

(a) Postorder: $v\,y\,w\,z\,t\,r\,x\,p\,q\,s\,u$. (c) Inorder: $v\,w\,y\,x\,z\,r\,t\,u\,p\,s\,q$.

9. (a)

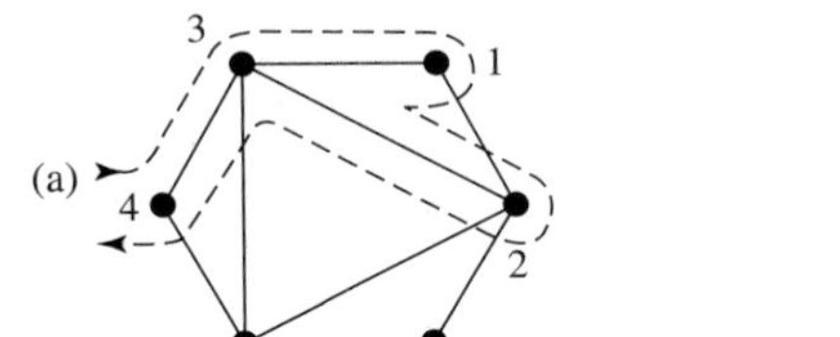

(c)

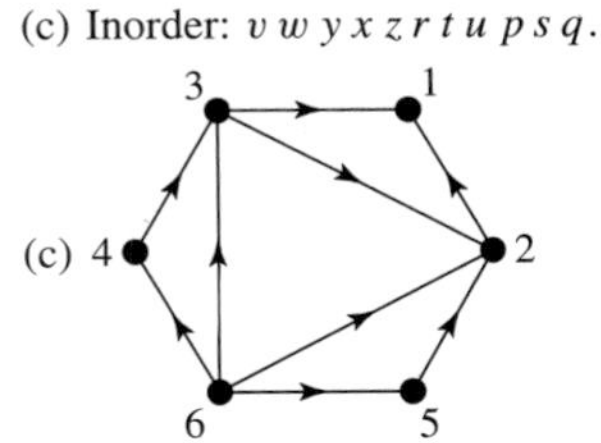

11. (a)

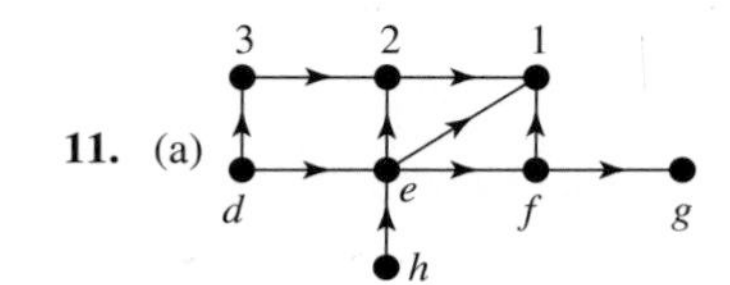

(c)

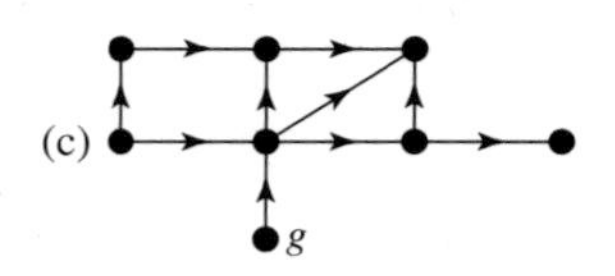

13. (a)

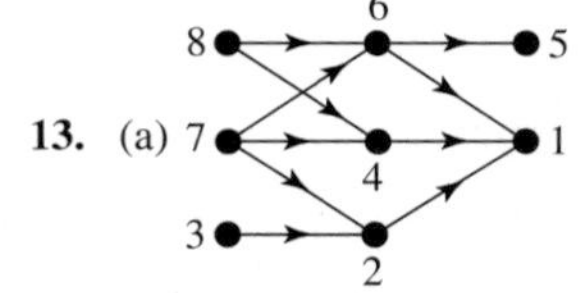

15. The number of descendants of v is a good measure. Each child w of v has fewer descendants than v has. The measure of a base case is 0.

17. Since the digraph is acyclic, there can be at most one edge joining each of the $n(n-1)/2$ pairs of distinct vertices.

7.4 Answers

1. Reverse Polish: $x\,4\,2\ \hat{}\ -\ y\ *\ 2\,3\ /\ +$. Polish: $+\ *\ -\ x\ \hat{}\ 4\,2\ y\ /\ 2\,3$.

3. (a) Polish: $-\ *\ +\ a\ b\ -\ a\ b\ -\ \hat{}\ a\ 2\ \hat{}\ b\ 2$. Infix: $(a+b)*(a-b)-((a\ \hat{}\ 2)-(b\ \hat{}\ 2))$.

5. (a) 20.

7. (a) $3\,x * 4 - 2\ \hat{}$.
(c) The answer depends on how the terms are associated. For the choice $(x-x^2)+(x^3-x^4)$, the answer is $x\,x\,2\ \hat{}\ -\ x\,3\ \hat{}\ x\,4\ \hat{}\ -\ +$.

9. (a) $a\,b\,c\ *\ *$ and $a\,b\ *\ c\,*$.
(c) The associative law is $a\,b\,c\ *\ * = a\,b\ *\ c\,*$. The distributive law is $a\,b\,c\ +\ * = a\,b\ *\ a\,c\ *\ +$.

11. (a) $p \to (q \vee (\neg\, p))$.

13. (a) Infix: $(p \wedge (p \to q)) \to q$.

15. (a) Both give $a\ /\ b\ +\ c$.

17. (a) (B) Numerical constants and variables are wff's.
(R) If f and g are wff's, so are $+\,f\,g$, $-\,f\,g$, $*\,f\,g$, $/\,f\,g$, and $\hat{}\ f\,g$.

19. (a) (B) Variables, such as p, q, r, are wff's.
(R) If P and Q are wff's, so are $P\,Q\,\vee$, $P\,Q\,\wedge$, $P\,Q\,\to$, $P\,Q\,\leftrightarrow$, and $P\,\neg$.
(b) Argue, in turn, that $q\,\neg$, $p\,q\,\neg\,\wedge$, and $p\,q\,\neg\,\wedge\,\neg$ are wff's. Likewise, $p\,q\,\neg\,\to$ is a wff. Thus $p\,q\,\neg\,\wedge\,\neg\,p\,q\,\neg\,\to\,\vee$ is a wff.
(c) (B) Variables, such as p, q, r, are wff's.
(R) If P and Q are wff's, so are $\vee\,P\,Q$, $\wedge\,P\,Q$, $\to\,P\,Q$, $\leftrightarrow\,P\,Q$, and $\neg\,P$.

21. The first operation is a unary operation and the second is a binary operation. In a reverse Polish calculator, the first key pops just one entry off the stack, whereas the second key pops off two entries.

7.5 Answers

1. (a) 35, 56, 70, 82.

3. We recommend the procedure in Examples 4 and 5.
(a) **1**, **3**, 4, 6, 9, 13 $\to$ **4**, **4**, 6, 9, 13 $\to$ **6**, **8**, 9, 13 $\to$ **9**, **13**, 14 $\to$ 14, 22. Weight = 84.

(c) Weight = 244.

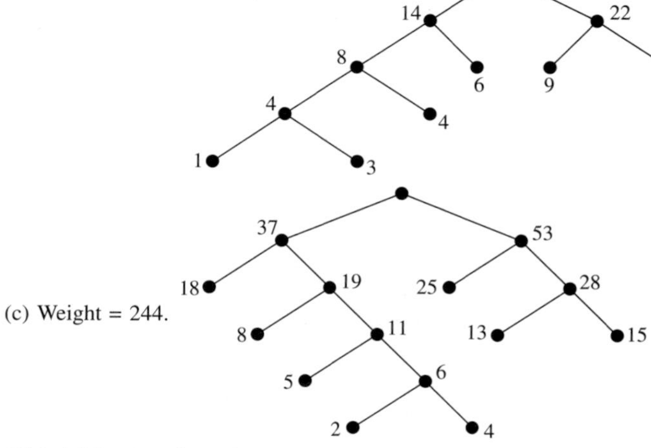

5. All but (b) are prefix codes.

7. (a)

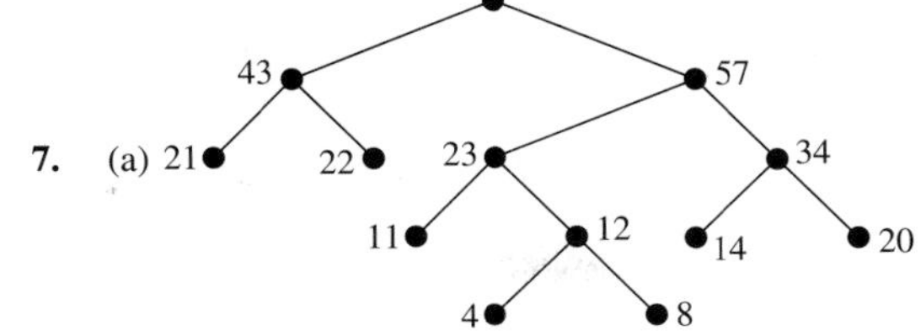

The corresponding optimal code is

letter	a	b	c	d	e	f	g
frequency	11	20	4	22	14	8	21
code	100	111	1010	01	110	1011	00

9. In both parts (a) and (c), no labeled vertex lies below another such vertex. See figures.

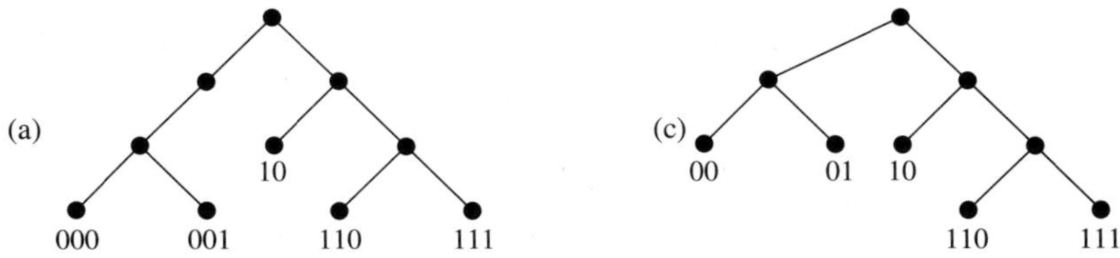

In (a), the vertex 0 has only one child, 00. The binary tree in (c) is regular.

11. (a) $(61+73-1)+(31+61+73-1)+(23+31+61+73-1) = 133+164+187 = 484$.
(c) $(23+31-1)+(61+73-1)+187 = 53+133+187 = 373$.

(e)

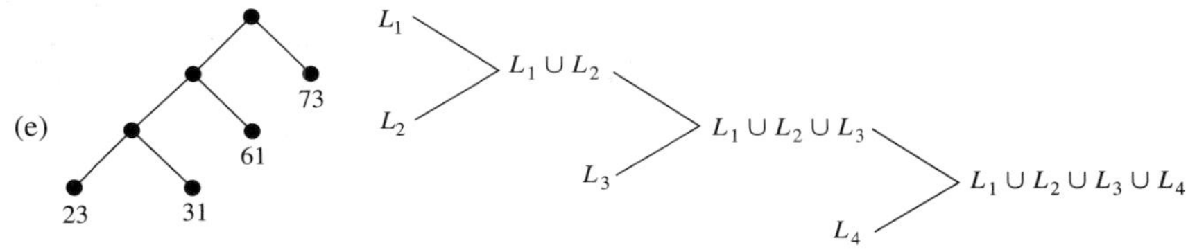

This merging involves at most $(23+31-1)+(23+31+61-1)+187 = 53+114+187 = 354$ comparisons.

13. (a)

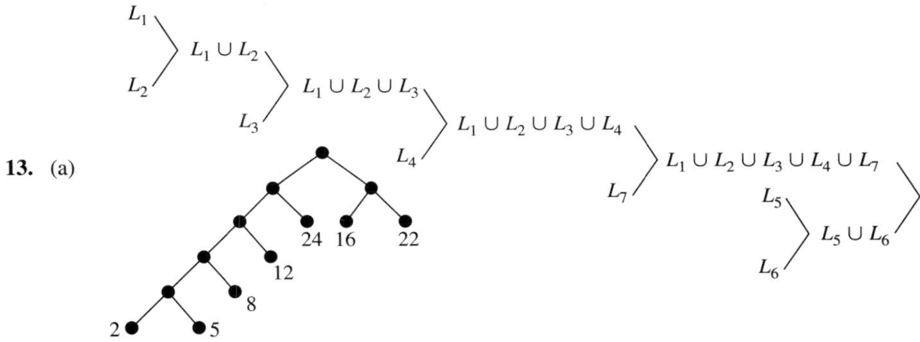

15. For example, in Exercise 1(a), the first tree has weight 35. The leaf of weight **21** $= 12 + 9$ was replaced by the subtree with weights 12 and 9, and the weight of the whole tree increased to 56, i.e., to $35 + \mathbf{21}$.

7.6 Answers

1. (a) $(1+2)*((3*4)-5) = 21$.

(b)

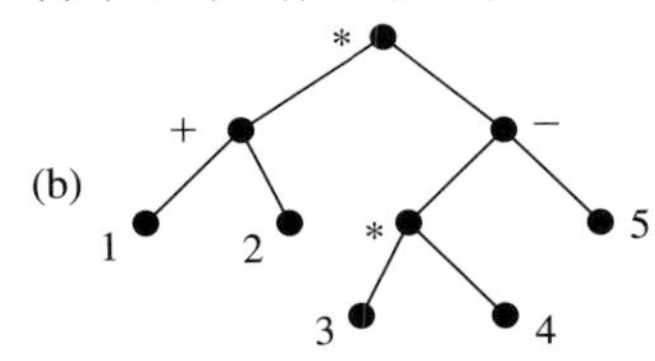

2. (a) λ, $bbaa$, $baba$.

(b) They all have even length, since λ has even length 0, and if w has even length, then so do awb and bwa.

(c) Yes, since $\lambda \to b\lambda a = ba \to bbaa \to abbaab$.

(d) They are the same, since λ has none of either, and if w has the same number of a's as b's, then so do awb and bwa. Note from part (a) that not all words with the same number of a's as b's are in S.

3. (a) (b) $r\,s\,u\,t\,v\,w\,y\,z\,x$.

4. (a)

or equivalent.

(b) $3 * (3 + 3 + 4 + 5) + 2 * (6 + 8) = 73$.

5. $/ * + 2\ 3\ + 4\ 7\ + 9 * 5\ 8 = \frac{55}{49}$.

6. (a) Yes. Yes. No. (b) 3. (c) 3.

(d) $y\,z\,u\,v\,s\,w\,x\,t\,r$. (e) $y\,u\,z\,s\,v\,r\,w\,t\,x$.

(f) There are many ways. For example, vertices $r, s, v, t, w, u, y, x, z$ could be labeled $1, 2, \ldots, 9$, respectively.

7. (a)

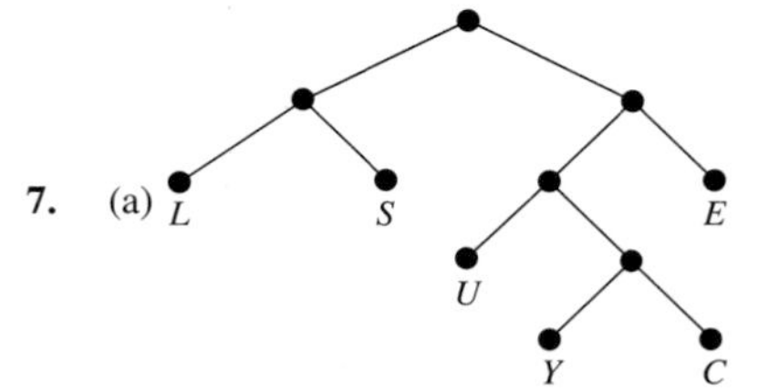

is one answer.

(b) Using the tree in part (a), the answer is 1 0 1 1 0 0 1 0 0 1 1.

8. (a)

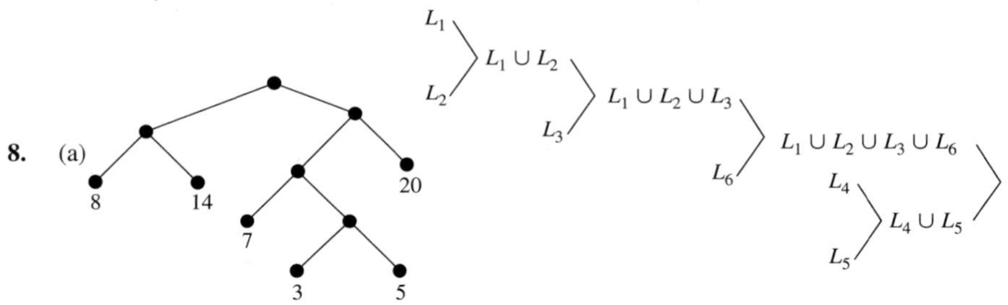

is one answer.

(b) The procedure requires at most $2(14 + 20) + 3(3 + 5 + 7 + 8) - 5 = 132$ comparisons, by part (a).

9. For w in Σ^*, let $p(w) =$ "w contains an even number of a's." We apply the Generalized Principle of Induction. For the base cases, $p(\lambda)$ and $p(b)$ are clearly true. If $p(w)$ and $p(u)$ are true, then w has $2m$ a's and u has $2n$ a's for suitable $m, n \in \mathbb{N}$. Then awa and wu have $2m+2$ and $2m+2n$ a's, respectively. Since $2m + 2$ and $2m + 2n$ are even, $p(awa)$ and $p(wu)$ are true. So the Generalized Principle of Induction shows that $p(w)$ holds for all $w \in S$.

10. (a) (B) Numerical constants and variables are wff's.

(R) If f and g are wff's, then so are $+ f\,g$, $- f\,g$, $* f\,g$, $/ f\,g$, $\wedge f\,g$, and $\ominus f$ [here $\ominus$ is the unary negation operation].

Compare Example 3 on page 280.

(b) The first two.

(c) $(3x + 1)(2x) \leftrightarrow 3\,x * 1 + 2\,x * *$ and $3x^2 - 5x + 4 \leftrightarrow 3\,x\,2 \wedge * 5\,x * - 4 +$.

11. One possible algorithm is the following.

```
SUM⁺(n)

if n = 0 then
   return (1, 1)
else
   return (SUM₁⁺(n − 1) + 1/[n·SUM₂⁺(n − 1)], n·SUM₂⁺(n − 1))
```

12. Suppose that L is a well-labeled subset of $V(G)$. Thus $L = \text{ACC}(L)$ and L has a sorted labeling. If $L = V(G)$, then $V(G)$ obviously has a sorted labeling that agrees with that on L. Suppose that $L \neq V(G)$, so Label chooses some v in $V(G) \setminus L$. We have already verified that TreeSort(G, L, v) well-labels $L \cup \text{ACC}(v)$ to agree with the L labels. Therefore, the recursive call to Label$(G, L \cup \text{ACC}(v))$ is legal and, moreover, the way it labels $V(G)$ will agree with the L-labeling if it agrees with the $L \cup \text{ACC}(v)$-labeling. The algorithm must terminate, because $L \cup \text{ACC}(v)$ is closer to the base case $V(G)$ than L is, since $|L \cup \text{ACC}(v)| > |L|$. Assuming that the recursive call gives a correct labeling, Label(G, L) does, too. Starting with $L = \emptyset$ is fair, since $\emptyset$ is vacuously well-labeled, and the result is a sorted labeling of $V(G)$, as desired.

13. (a) See part (c).
(b) For example, ba, bab, and bba. See part (c).
(c) B consists of all words in Σ^* in which all letters a precede all letters b.
(d) No. Let S_n be as in the discussion prior to Example 3 on page 270. Then $S_n = \{w \in B : \text{length}(w) = n\}$. For example, one can build ab via $\lambda \to a\lambda \to a\lambda b$ or via $\lambda \to \lambda b \to a\lambda b$. Find another example.

14. (a) The tree for one possible code is

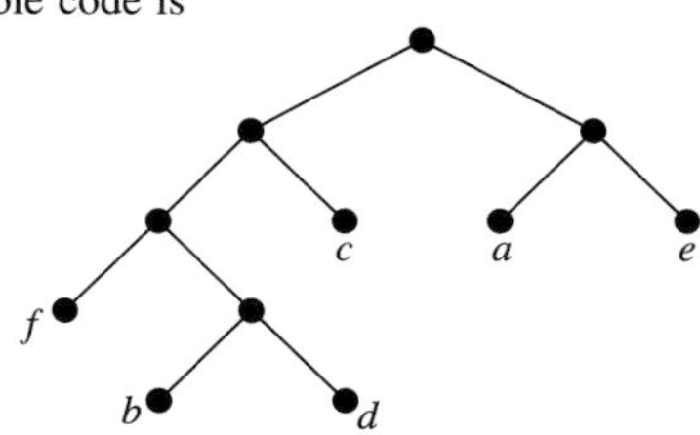

What matters is that a, c, and e have level 2, f has level 3, and b and d have level 4.
(b) $2 \cdot (20 + 30 + 33) + 3 \cdot 8 + 4 \cdot (4 + 5) = 226$.
(c) 0 0 1 1 1 1 1 0 0 0 1 1   0 0 1 0 1 1 1 1 0 0 0   0 1 1 0 0 0 0 1 1, using the answer to part (a). Whatever code is used, the encoded words will have lengths 12, 11, and 9 as bit strings.

15. (a) $3^3 * 17$. (b) $3^{n-1} * 17$.
(c) If $n = 1$, the base case, then ALGO returns 17, which is $3^{1-1} * 17$. If the recursive call is made, then $n \in \mathbb{P}$ and $n \neq 1$, so $n \geq 2$, thus $n - 1 \in \mathbb{P}$ and the recursive call is legal. If the recursive call produces the correct result, i.e., if ALGO$(n-1) = 3^{(n-1)-1} * 17$, then the algorithm returns $3 * 3^{n-2} * 17 = 3^{n-1} * 17$, as desired. The recursive calls are to a decreasing sequence $n, n-1, n-2, \ldots$ of positive integers. Since every such sequence is finite, the algorithm must terminate.
(d) For odd n, the new algorithm would return $17 * 3^{\lfloor n/2 \rfloor} = 17 * 3^{(n-1)/2}$. Consider $n = 5$, for example. For even n, it would not terminate if we allowed negative integers as inputs, or it would break down if we did not. Consider, for example, $n = 2$.

16. (a)

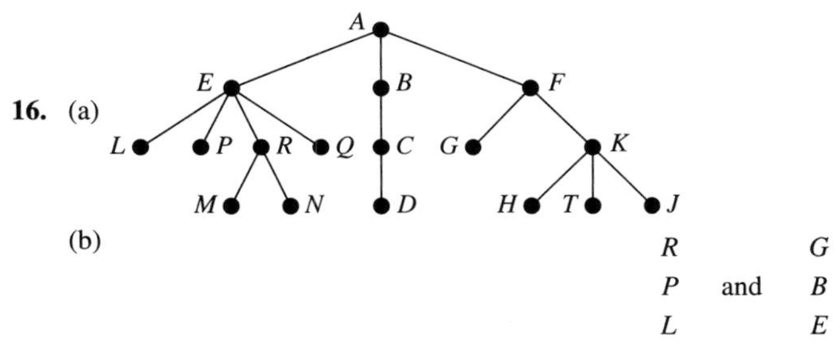

(b)

R		G
P	and	B
L		E

17. (a) $\lambda, ab, ca, aabb, ccaa, caba$, etc.
(b) Use the Generalized Principle of Induction on page 273. Clearly, length(λ) is even. If length(w) is even, then so are length(awb) = length(w) + 2 and length(cwa) = length(w)+2. By the Generalized Principle of Induction, length(w) is even for all $w \in B$.
(c) No. length(bb) = 2, but bb in not in B. In fact, no element in B can begin with b.
(d) Yes. We use the notation S_n, $n \in \mathbb{N}$, in the discussion prior to Example 3 on page 270. In our case, $S_n = \{w \in B : \text{length}(w) = 2n\}$. If a word in S_n ends in b, it is the word awb for a unique w in S_{n-1}. Similarly, if it ends in a, it is cwa for a unique w in S_{n-1}. No words in S_n end in c.

18. Use the Generalized Principle of Induction on n: If w is balanced and length$(w) = 2n$, then w is in S. Nothing is lost if we assumed w begins with the letter a; the other case is similar. The assertion for $n = 1$ is clear, since $ab \in S$. [The assertion for $n = 0$ is even clearer.] Now assume it is true for $n - 1$, $n \geq 2$, and consider a balanced word w of length $2n$. If w is of the form $a \cdots b$, then the middle string is balanced and we're done by induction. If w is of the form $a \cdots a$, then the initial string of length 1 has an excess of a's and the initial string of length $2n - 1$ has an excess of b's, so somewhere along the way there's an initial balanced string w_1 in S. [To make this more precise, for $j = 1, 2, \ldots, 2n$, let $f(j)$ be the number of a's minus the number of b's in the first j letters of w, and show that f must take the value 0 somewhere.] The remaining string w_2 obtained by removing w_1 from w is also balanced and in S, so by (R), $w = w_1 w_2$ is also in S.

19. One way would be to arrange the vertices of the tree in a list $L = (v_1, \dots, v_m)$ somehow and then execute the following.

```
Let M := L.
while there are a parent p with child c such that c comes
      after p in M do
   Interchange p and c in M.
   Label each vertex with its position in M.
      {E.g., give the fifth vertex in M the label 5.}
```

Checking the guard would require additional commands.

8.1 Answers

1. Sinks are t and z. The only source is u.

3. (a) $R(s) = \{s, t, u, w, x, y, z\} = R(t)$, $R(u) = \{w, x, y, z\}$, $R(w) = \{z\}$, $R(x) = \emptyset$, $R(y) = \{x, z\}$, $R(z) = \emptyset$.
(c) $s\,t\,s$ is a cycle, so G is not acyclic.

5. (a)

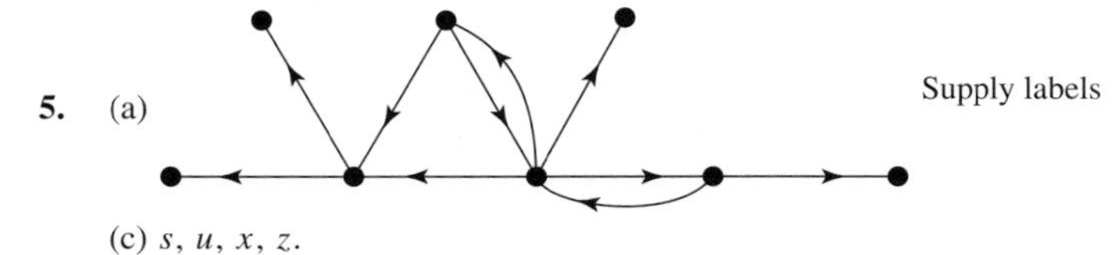

Supply labels

(c) s, u, x, z.

7. Use NumberingVertices. One example labels $t = 1$, $z = 2$, $y = 3$, $w = 4$, $v = 5$, $x = 6$, $u = 7$.

9. (a) One example is $r\,s\,w\,t\,v\,w\,v\,u\,x\,y\,z\,v\,r\,u\,y\,v\,s\,r$.

11. (a) One such digraph is drawn in Figure 3(b), where $w = 0\,0$, $x = 0\,1$, $z = 1\,1$, and $y = 1\,0$.

13. Show that $\overleftarrow{G}$ is also acyclic. Apply Theorem 2 to $\overleftarrow{G}$. A sink for $\overleftarrow{G}$ is a source for G.

15. (a) See the second proof of Theorem 2.

17. (a) In the proof of Theorem 1 on page 319 [given in the proof of Theorem 1 on page 227], choose a shortest path consisting of edges of the given path.
(b) If $u \neq v$, then Theorem 1 guarantees an acyclic path from u to v, and Corollary 2 says such a path has no repeated vertices, so it surely has no repeated edges. If $u = v$, then Corollary 1 says there is a cycle from u to u. Again, all vertices are different, so all edges are too.

19. Consider a cycle in a graph G, with vertex sequence $x_1\,x_2 \cdots x_n\,x_1$. Then $x_1, \dots, x_n$ are distinct and no edge appears twice. Assign directions to the edges in the cycle so they go from x_1 to x_2 to x_3 to ... to x_n to x_1. Since no edge appears twice, this assignment cannot contradict itself. Assign directions arbitrarily to the other edges of G. Then the edges in the cycle form a directed cycle in the resulting digraph.

Conversely, suppose it is possible to assign directions to edges in G so that a path with vertex sequence $x_1\,x_2 \cdots x_m\,x_1$ is a directed cycle. Then $m \geq 1$ and $x_1, \dots, x_m$ are all different. If $m \geq 3$, then the path is an undirected cycle by Proposition 1. If $m = 2$, the vertex sequence is $x_1\,x_2\,x_1$, but the edges from x_1 to x_2 and from x_2 to x_1 must be different because their directions are different. In this case the undirected path is simple, so it is a cycle. Finally, if $m = 1$, the path is a loop, so it's a cycle. In all cases the directed cycle is an undirected cycle.

8.2 Answers

1.

W	A	B	C	D
A	1.4	1.0	∞	∞
B	0.4	7	∞	0.2
C	0.7	0.3	7	∞
D	0.8	∞	0.2	7

W^*	A	B	C	D
A	∞	1.0	1.4	1.2
B	0.4	∞	0.4	0.2
C	0.7	0.3	∞	0.5
D	0.8	0.5	0.2	∞

3.

W	m	q	r	s	w	x	y	z
m	∞	6	∞	2	∞	4	∞	∞
q	∞	∞	4	∞	4	∞	∞	∞
r	∞	∞	∞	∞	∞	∞	∞	3
s	∞	3	∞	∞	5	1	∞	∞
w	∞	∞	2	∞	∞	∞	2	5
x	∞	∞	∞	∞	3	∞	6	∞
y	∞	∞	∞	∞	∞	∞	∞	1
z	∞	∞	∞	∞	∞	∞	∞	∞

W^*	m	q	r	s	w	x	y	z
m	∞	5	8	2	6	3	8	9
q	∞	∞	4	∞	4	∞	6	7
r	∞	∞	∞	∞	∞	∞	∞	3
s	∞	3	6	∞	4	1	6	7
w	∞	∞	2	∞	∞	∞	2	3
x	∞	∞	5	∞	3	∞	5	6
y	∞	∞	∞	∞	∞	∞	∞	1
z	∞	∞	∞	∞	∞	∞	∞	∞

5. (a) If the digraph is acyclic, the weights must be as shown below, with $\min\{a+1, b\} = 4$ and $\min\{c+1, d, e\} = 3$.

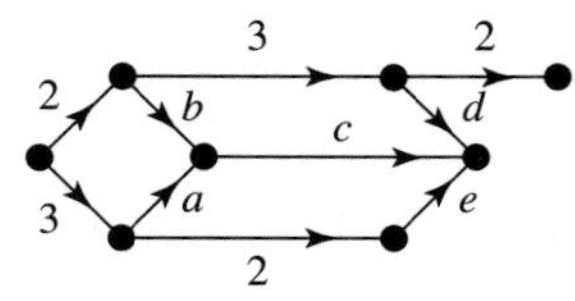

7. (a) The critical paths are $s\,v\,x\,f$ and $s\,v\,x\,z\,f$.

9. (a)

	s	u	v	w	x	y	f
A	0	2	7	5	5	11	15
L	0	2	9	7	5	11	15

(c) $s\,u\,x\,y\,f$ is the only critical path.

11. (a)

	m	s	q	x	w	r	y	z
A	0	2	6	4	10	12	12	15
L	0	3	6	7	10	12	14	15

(c) There are two critical paths: $m\,q\,w\,z$ and $m\,q\,w\,r\,z$.

13. (a) The two critical paths are $s\,u\,w\,x\,y\,f$ and $s\,t\,w\,x\,y\,f$. The critical edges are (s, u), (u, w), (s, t), (t, w), (w, x), (x, y), and (y, f).
(c) The edges are (u, v) and (x, z).

15. (a) Shrink the 0-edges to make their two endpoints the same.

17. (a)

W	u	v	w	x	y
u	∞	1	∞	∞	∞
v	∞	∞	3	-2	∞
w	∞	∞	∞	∞	∞
x	4	∞	∞	5	∞
y	∞	∞	∞	5	∞

W^*	u	v	w	x	y
u	3	1	4	-1	∞
v	2	3	3	-2	∞
w	∞	∞	∞	∞	∞
x	4	5	8	3	∞
y	9	10	13	5	∞

(c) There would be no min-weights at all for paths involving u, v, or x, because going around the cycle $u\,v\,x\,u$ repeatedly would keep reducing the weight by 1.

19. (a) $FF(u, v) = A(v) - A(u) - W(u, v)$.
(c) The slack time at v.

21. (a) $A(u) = M(s, u) =$ weight of a max-path from s to u. If there is an edge (w, u), a max-path from s to w followed by that edge has total weight at most $M(s, u)$. That is, $A(w) + W(w, u) \le A(u)$. If (w, u) is an edge in a max-path from s to u, then $A(w) + W(w, u) = A(u)$.
(b) Consider a vertex u. If (u, v) is an edge, then (u, v) followed by a max-path from v to f has total weight at most $M(u, f)$. That is, $W(u, v) + M(v, f) \le M(u, f)$. Hence

$$L(v) - W(u, v) = M(s, f) - M(v, f) - W(u, v) \ge M(s, f) - M(u, f) = L(u)$$

for every edge (u, v), and so $\min\{L(v) - W(u, v) : (u, v) \in E(G)\} \ge L(u)$. Choosing an edge (u, v) in a max-path from u to f gives $L(v) - W(u, v) = L(u)$, so $L(u)$ is the minimum value.

8.3 Answers

1. (a) $\mathbf{W}^* = \begin{bmatrix} \infty & 1 & 2 & 3 & 4 & 5 & 7 \\ \infty & \infty & \infty & 4 & 3 & 4 & 6 \\ \infty & \infty & \infty & 1 & 4 & 5 & 7 \\ \infty & \infty & \infty & \infty & 3 & 4 & 6 \\ \infty & \infty & \infty & \infty & \infty & 1 & 3 \\ \infty & \infty & \infty & \infty & \infty & \infty & 3 \\ \infty & \infty & \infty & \infty & \infty & \infty & \infty \end{bmatrix}$.

(c) $\mathbf{P}_0 = \begin{bmatrix} 0 & 2 & 3 & 0 & 0 & 0 & 0 \\ 0 & 0 & 0 & 4 & 5 & 0 & 0 \\ 0 & 0 & 0 & 4 & 5 & 6 & 0 \\ 0 & 0 & 0 & 0 & 5 & 6 & 0 \\ 0 & 0 & 0 & 0 & 0 & 6 & 7 \\ 0 & 0 & 0 & 0 & 0 & 0 & 7 \\ 0 & 0 & 0 & 0 & 0 & 0 & 0 \end{bmatrix}$, $\mathbf{P}_{\text{final}} = \begin{bmatrix} 0 & 2 & 3 & 3 & 2 & 2 & 2 \\ 0 & 0 & 0 & 4 & 5 & 5 & 5 \\ 0 & 0 & 0 & 4 & 4 & 4 & 4 \\ 0 & 0 & 0 & 0 & 5 & 5 & 5 \\ 0 & 0 & 0 & 0 & 0 & 6 & 7 \\ 0 & 0 & 0 & 0 & 0 & 0 & 7 \\ 0 & 0 & 0 & 0 & 0 & 0 & 0 \end{bmatrix}$.

3. (a)

M	$D(v_j)$						$P(v_j)$					
	v_2	v_3	v_4	v_5	v_6	v_7	v_2	v_3	v_4	v_5	v_6	v_7
$\{v_1\}$	**1**	2	∞	∞	∞	∞	v_1	v_1				
$\{v_1, v_2\}$	1	**2**	5	4	∞	∞	v_1	v_1	v_2	v_2		
$\{v_1, v_2, v_3\}$	1	2	**3**	4	9	∞	v_1	v_1	v_3	v_2	v_3	
$\{v_1, v_2, v_3, v_4\}$	1	2	3	**4**	9	∞	v_1	v_1	v_3	v_2	v_3	
$\{v_1, v_2, v_3, v_4, v_5\}$	1	2	3	4	**5**	7	v_1	v_1	v_3	v_2	v_5	v_5

no change now

(c)

M	$D(v_j)$				$P(v_j)$			
	v_2	v_3	v_4	v_5	v_2	v_3	v_4	v_5
$\{v_1\}$	2	**1**	7	∞	v_1	v_1	v_1	
$\{v_1, v_3\}$	**2**	1	7	4	v_1	v_1	v_1	v_3
$\{v_1, v_3, v_2\}$	2	1	6	**4**	v_1	v_1	v_2	v_3
$\{v_1, v_3, v_2, v_5\}$	2	1	**5**	4	v_1	v_1	v_5	v_3

no change now

5. (a) $\mathbf{W}_2 = \begin{bmatrix} \infty & \infty & \infty & \infty & 1 & \infty & \infty \\ \infty & \infty & \infty & \infty & \infty & \infty & 1 \\ \infty & \infty & \infty & 1 & \infty & 1 & \infty \\ \infty & \infty & 1 & \infty & 1 & \infty & \infty \\ 1 & \infty & \infty & 1 & 2 & \infty & \infty \\ \infty & \infty & 1 & \infty & \infty & \infty & 1 \\ \infty & 1 & \infty & \infty & \infty & 1 & 2 \end{bmatrix}$, $\mathbf{W}_4 = \begin{bmatrix} \infty & \infty & \infty & \infty & 1 & \infty & \infty \\ \infty & \infty & \infty & \infty & \infty & \infty & 1 \\ \infty & \infty & 2 & 1 & 2 & 1 & \infty \\ \infty & \infty & 1 & 2 & 1 & 2 & \infty \\ 1 & \infty & 2 & 1 & 2 & 3 & \infty \\ \infty & \infty & 1 & 2 & 3 & 2 & 1 \\ \infty & 1 & \infty & \infty & \infty & 1 & 2 \end{bmatrix}$,

$\mathbf{W}_7 = \begin{bmatrix} 2 & 6 & 3 & 2 & 1 & 4 & 5 \\ 6 & 2 & 3 & 4 & 5 & 2 & 1 \\ 3 & 3 & 2 & 1 & 2 & 1 & 2 \\ 2 & 4 & 1 & 2 & 1 & 2 & 3 \\ 1 & 5 & 2 & 1 & 2 & 3 & 4 \\ 4 & 2 & 1 & 2 & 3 & 2 & 1 \\ 5 & 1 & 2 & 3 & 4 & 1 & 2 \end{bmatrix}$.

7. (a) $\mathbf{W}_0 = \mathbf{W}_1 = \begin{bmatrix} \infty & 1 & \infty & 7 & \infty \\ \infty & \infty & 4 & 2 & \infty \\ \infty & \infty & \infty & \infty & 3 \\ \infty & \infty & 1 & \infty & 5 \\ \infty & \infty & \infty & \infty & \infty \end{bmatrix}$, $\mathbf{P}_0 = \mathbf{P}_1 = \begin{bmatrix} 0 & 2 & 0 & 4 & 0 \\ 0 & 0 & 3 & 4 & 0 \\ 0 & 0 & 0 & 0 & 5 \\ 0 & 0 & 3 & 0 & 5 \\ 0 & 0 & 0 & 0 & 0 \end{bmatrix}$,

$\mathbf{W}_2 = \begin{bmatrix} \infty & 1 & 5 & 3 & \infty \\ \infty & \infty & 4 & 2 & \infty \\ \infty & \infty & \infty & \infty & 3 \\ \infty & \infty & 1 & \infty & 5 \\ \infty & \infty & \infty & \infty & \infty \end{bmatrix}$, $\mathbf{P}_2 = \begin{bmatrix} 0 & 2 & 2 & 2 & 0 \\ 0 & 0 & 3 & 4 & 0 \\ 0 & 0 & 0 & 0 & 5 \\ 0 & 0 & 3 & 0 & 5 \\ 0 & 0 & 0 & 0 & 0 \end{bmatrix}$,

$\mathbf{W}_3 = \begin{bmatrix} \infty & 1 & 5 & 3 & 8 \\ \infty & \infty & 4 & 2 & 7 \\ \infty & \infty & \infty & \infty & 3 \\ \infty & \infty & 1 & \infty & 4 \\ \infty & \infty & \infty & \infty & \infty \end{bmatrix}$, $\mathbf{P}_3 = \begin{bmatrix} 0 & 2 & 2 & 2 & 2 \\ 0 & 0 & 3 & 4 & 3 \\ 0 & 0 & 0 & 0 & 5 \\ 0 & 0 & 3 & 0 & 3 \\ 0 & 0 & 0 & 0 & 0 \end{bmatrix}$,

$\mathbf{W}_4 = \begin{bmatrix} \infty & 1 & 4 & 3 & 7 \\ \infty & \infty & 3 & 2 & 6 \\ \infty & \infty & \infty & \infty & 3 \\ \infty & \infty & 1 & \infty & 4 \\ \infty & \infty & \infty & \infty & \infty \end{bmatrix}$, $\mathbf{P}_4 = \begin{bmatrix} 0 & 2 & 2 & 2 & 2 \\ 0 & 0 & 4 & 4 & 4 \\ 0 & 0 & 0 & 0 & 5 \\ 0 & 0 & 3 & 0 & 3 \\ 0 & 0 & 0 & 0 & 0 \end{bmatrix}$.

Also, $\mathbf{W}_5 = \mathbf{W}^* = \mathbf{W}_4$ and $\mathbf{P}_5 = \mathbf{P}^* = \mathbf{P}_4$.

9. (a)

$$\mathbf{D}_0 = \mathbf{D}_1 = [\ -\infty \quad 1 \quad 2 \quad -\infty \quad -\infty \quad -\infty \quad -\infty\ ]$$
$$\mathbf{D}_2 = [\ -\infty \quad 1 \quad 2 \quad 5 \quad 4 \quad -\infty \quad -\infty\ ]$$
$$\mathbf{D}_3 = [\ -\infty \quad 1 \quad 2 \quad 5 \quad 7 \quad 9 \quad -\infty\ ]$$
$$\mathbf{D}_4 = [\ -\infty \quad 1 \quad 2 \quad 5 \quad 8 \quad 13 \quad -\infty\ ]$$
$$\mathbf{D}_5 = [\ -\infty \quad 1 \quad 2 \quad 5 \quad 8 \quad 13 \quad 11\ ]$$
$$\mathbf{D}_6 = [\ -\infty \quad 1 \quad 2 \quad 5 \quad 8 \quad 13 \quad 16\ ]$$

11. (a) The algorithm would give

M	$D(v_j)$		
	v_2	v_3	v_4
$\{v_1\}$	5	**6**	$-\infty$
$\{v_1, v_3\}$	5	6	**9**

no change

whereas $M(1, 3) = 9$ and $M(1, 4) = 12$.

(c) Both algorithms would fail to give correct values of $M(1, 4)$.

13. (a)

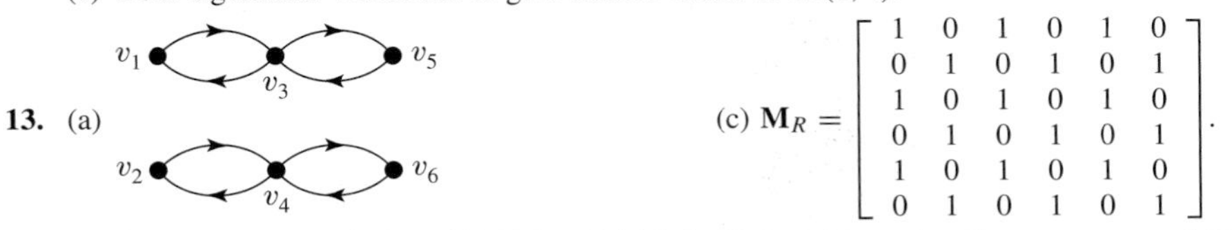

(c) $\mathbf{M}_R = \begin{bmatrix} 1 & 0 & 1 & 0 & 1 & 0 \\ 0 & 1 & 0 & 1 & 0 & 1 \\ 1 & 0 & 1 & 0 & 1 & 0 \\ 0 & 1 & 0 & 1 & 0 & 1 \\ 1 & 0 & 1 & 0 & 1 & 0 \\ 0 & 1 & 0 & 1 & 0 & 1 \end{bmatrix}$.

15. (a) Create a row matrix **P**, with $\mathbf{P}[j] = 1$ initially if there is an edge from v_1 to v_j and $\mathbf{P}[j] = 0$ otherwise. Add the line

Replace $\mathbf{P}[j]$ by k.

(b) Part of this exercise is solved in Example 4.

17. Initially, $\mathbf{W}[i, j] = 1$ if there is an edge from v_i to v_j, and $\mathbf{W}[i, j] = 0$ otherwise. Change the update step to

if $\mathbf{W}[i, j] < \min\{\mathbf{W}[i, k], \mathbf{W}[k, j]\}$, **then**
  Replace $\mathbf{W}[i, j]$ by $\min\{\mathbf{W}[i, k], \mathbf{W}[k, j]\}$.

or to

$\mathbf{W}[i, j] := \max\{\mathbf{W}[i, j], \min\{\mathbf{W}[i, k], \mathbf{W}[k, j]\}\}$.

The pointer portion could be omitted.

8.4 Answers

1. (a) Yes.
(b) No. There are four vertices of odd degree.
(c) c and h.
(d) a and g
(e) Use algorithm NumberVertices. One possible answer is $a = 9$, $b = 8$, $c = 3$, $d = 6$, $e = 4$, $f = 5$, $g = 7$, $h = 1$, $i = 2$.

2. (a) No. There is a loop at f.
(b) The sinks have empty successor lists. They are a and d. The sources are the vertices that appear on no successor lists. They are a, b, and g. Note that f is not a sink.

(c)

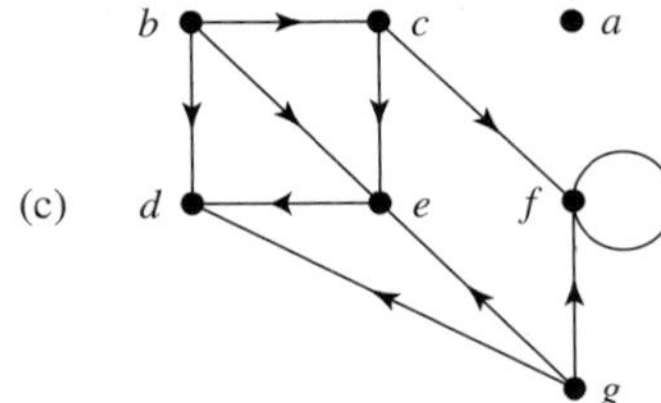

(d) No. The digraph is not acyclic. If the loop at f were removed, then it would be possible.

3. (a)

	start	a	b	c	d	e	f	finish
$A(v)$	0	1	6	7	2	4	12	17
$L(v)$	0	1	6	7	2	7	15	17
$S(v)$	0	0	0	0	0	3	3	0

(b)

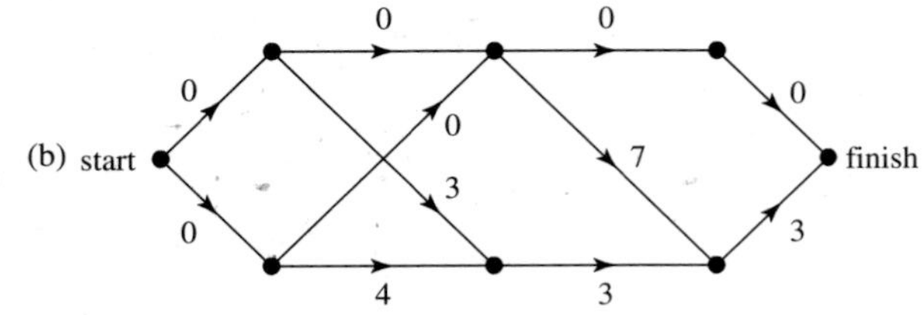

(c) start $\to a \to b \to c \to$ finish and start $\to d \to b \to c \to$ finish.

4. (a)

W^*	s	u	v	w	x	f
s	∞	2	7	4	5	8
u	∞	∞	∞	2	3	6
v	∞	∞	∞	∞	∞	2
w	∞	∞	∞	∞	∞	4
x	∞	∞	∞	∞	∞	6
f	∞	∞	∞	∞	∞	∞

(b) (s, u), (u, x), (x, f).

(c) $S(u) = 0$, $S(v) = 2$.

5. (a) One answer is the following.

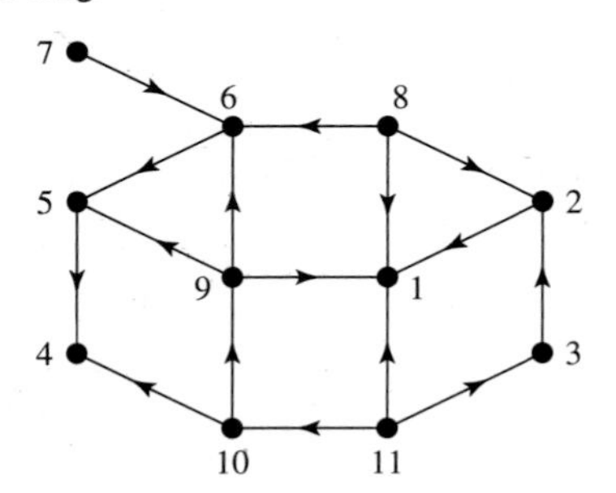

(b) The sinks are labeled 1 and 4. The sources are labeled 7, 8, and 11.

(c) No. There is no path from vertex 7 to vertex 1.

6. Float times are indicated next to the edges in the figure. Arrival and leave times are indicated by $A(v) - L(v)$ next to the vertex v. The only critical path is the one made up of edges with float time 0.

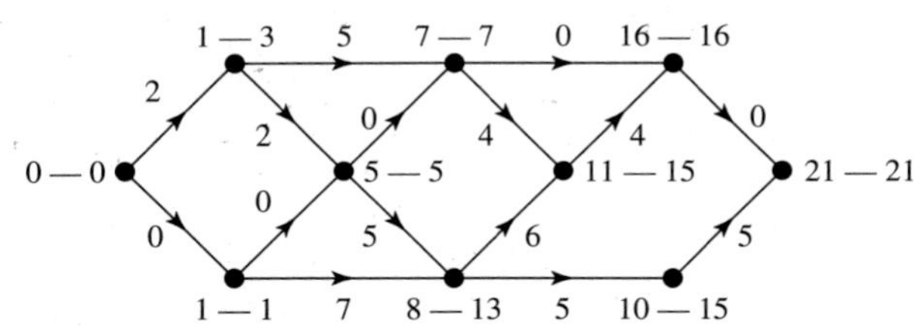

7. Bob is. If there is more than one critical path, then reducing the time for an edge that's in just one of the paths will not reduce the overall time.

8. Bob is correct again. Here's an example, where the vertices a and b both have 0 slack time, but the path $s \to a \to b \to f$ is not a critical path.

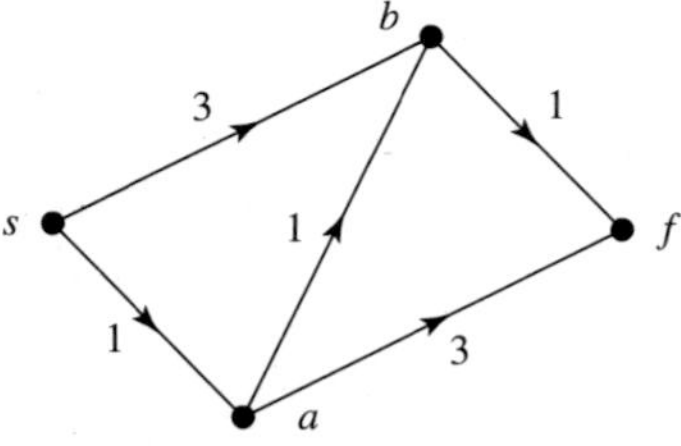

9. (a)

M	$D(v)$							$P(v)$						
	b	c	d	e	f	g	h	b	c	d	e	f	g	h
$\{a\}$	1	∞	5	∞	∞	∞	∞	a		a				
$\{a, b\}$	1	9	5	7	∞	∞	∞	a	b	a	b			
$\{a, b, d\}$	1	9	5	6	∞	∞	∞	a	b	a	d			
$\{a, b, d, e\}$	1	7	5	6	13	12	∞	a	e	a	d	e	e	
$\{a, b, d, e, c\}$	1	7	5	6	9	12	∞	a	e	a	d	c	e	
$\{a, b, d, e, c, f\}$	1	7	5	6	9	10	∞	a	e	a	d	c	f	
$\{a, b, d, e, c, f, g\}$	1	7	5	6	9	10	13	a	e	a	d	c	f	g

The minimum weight path from a to h has weight 13, and the path has vertex sequence $a\ d\ e\ c\ f\ g\ h$.

(b)

M	$D(v)$							$P(v)$						
	a	c	d	e	f	g	h	a	c	d	e	f	g	h
$\{b\}$	∞	8	∞	6	∞	∞	∞		b		b			
$\{b, e\}$	∞	7	∞	6	13	12	∞		e		b	e	e	
$\{b, e, c\}$	∞	7	∞	6	9	12	∞		e		b	c	e	
$\{b, e, c, f\}$	∞	7	∞	6	9	10	∞		e		b	c	f	
$\{b, e, c, f, g\}$	∞	7	∞	6	9	10	13		e		b	c	f	g

The minimum weight path from b to h has weight 13, and the path has vertex sequence $b\ e\ c\ f\ g\ h$.

10. (a) One pair is enough, since Warshall's algorithm solves for all starting points at once.

(b) $\mathbf{W}_0 = \begin{bmatrix} \infty & 1 & \infty & 5 & \infty & \infty & \infty & \infty \\ \infty & \infty & 8 & \infty & 6 & \infty & \infty & \infty \\ \infty & \infty & \infty & \infty & \infty & 2 & \infty & \infty \\ \infty & \infty & \infty & \infty & 1 & \infty & \infty & \infty \\ \infty & \infty & 1 & \infty & \infty & 7 & 6 & \infty \\ \infty & \infty & \infty & \infty & \infty & \infty & 1 & \infty \\ \infty & \infty & \infty & \infty & \infty & \infty & \infty & 3 \\ \infty & \infty & \infty & \infty & \infty & \infty & \infty & \infty \end{bmatrix}$ $\quad \mathbf{P}_0 = \begin{bmatrix} 0 & 2 & 0 & 4 & 0 & 0 & 0 & 0 \\ 0 & 0 & 3 & 0 & 5 & 0 & 0 & 0 \\ 0 & 0 & 0 & 0 & 0 & 6 & 0 & 0 \\ 0 & 0 & 0 & 0 & 5 & 0 & 0 & 0 \\ 0 & 0 & 3 & 0 & 0 & 6 & 7 & 0 \\ 0 & 0 & 0 & 0 & 0 & 0 & 7 & 0 \\ 0 & 0 & 0 & 0 & 0 & 0 & 0 & 8 \\ 0 & 0 & 0 & 0 & 0 & 0 & 0 & 0 \end{bmatrix}$

(d) $a\ b\ e\ f\ g\ h$.

11. (a) Its vertex sequence is $s\ u\ x\ y\ f$.

(b) s, u, x, y, f.

(c) 3 and 4, respectively.

(d)

M	$D(u)$	$D(v)$	$D(w)$	$D(x)$	$D(y)$	$D(f)$
$\emptyset$	2	7	5	∞	∞	∞
$\{u\}$	2	7	4	5	∞	∞
$\{u, w\}$	2	7	4	5	8	∞
$\{u, w, x\}$	2	7	4	5	8	9

12. Find an Euler circuit for the digraph shown.

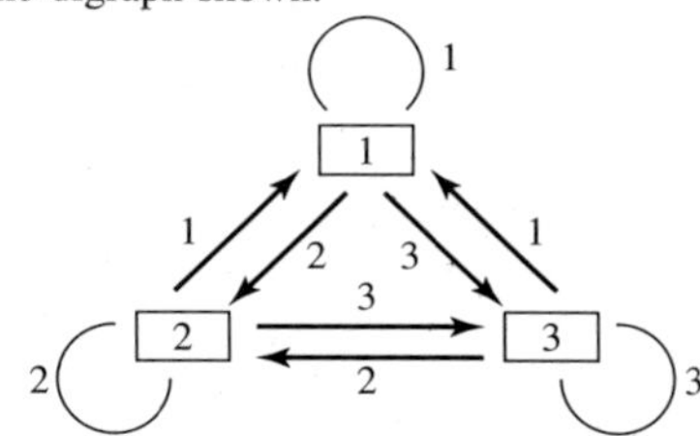

One Euler path starting at the box labeled 1 gives a de Bruijn sequence 1, 3, 3, 2, 2, 3, 1, 2, 1. Notice that all nine strings 1 1, 1 2, 1 3, ..., 3 3 occur along this [circular] sequence.

9.1 Answers

1. (a) $P(B_o) = P(R_o) = P(E) = \frac{1}{2}$; $P(B_o \cap R_o) = P(B_o \cap E) = P(R_o \cap E) = \frac{1}{4}$.

3. $\{S, L\}$ are dependent, since $P(S \cap L) = \frac{6}{36}$, whereas $P(S) \cdot P(L) = \frac{15}{36} \cdot \frac{15}{36}$.
$\{S, E\}$ are dependent, since $P(S \cap E) = \frac{3}{36}$, whereas $P(S) \cdot P(E) = \frac{15}{36} \cdot \frac{6}{36}$.
$\{L, E\}$ are dependent, since $P(L|E) = 0 \neq P(L)$. Similarly for $\{L, G\}$.

5. No. $P(B|A) = \frac{1}{2}$, whereas $P(B) = \frac{\binom{4}{2}}{2^4} = \frac{3}{8}$.

7. (a) 0.25. (c) No, $P(A|B) = 0.25 \neq P(A)$.

9. (a) $\frac{3}{11} \cdot \frac{2}{10} = \frac{3}{55}$. (c) $\frac{3}{11} \cdot \frac{8}{10} + \frac{8}{11} \cdot \frac{3}{10} = \frac{24}{55}$.
Equivalently, view this as drawing 2-element subsets. Then the answers are
(a) $\frac{\binom{3}{2}}{\binom{11}{2}} = \frac{3}{55}$. (c) $\frac{3 \cdot 8}{\binom{11}{2}} = \frac{24}{55}$.

11. (a) $\frac{4}{52} \cdot \frac{3}{51} \cdot \frac{2}{50} \approx 0.00018$. (c) $1 - \frac{48}{52} \cdot \frac{47}{51} \cdot \frac{46}{50} \approx 0.217$.

13. (a) $P(B) = \frac{1}{3} \cdot \frac{2}{3} + \frac{1}{3} \cdot \frac{2}{5} + \frac{1}{3} \cdot \frac{1}{2} = \frac{47}{90}$.
(c) $P(B \cap U_1) = \frac{1}{3} \cdot \frac{2}{3} = \frac{2}{9}$.

15. (a) $\frac{5}{9}$.

17. (a) $P(C) = P(E) \cdot P(C|E) + P(F) \cdot P(C|F) + P(G) \cdot P(C|G) = 0.043$.
(c) 1.

19. (a) $P(D) = P(N^c \cap D) + P(N \cap D) = 0.0041$, so $P(N^c|D) = \frac{P(N^c \cap D)}{P(D)} = \frac{0.004}{0.0041} \approx 0.9756$.
(b) Since $P(N^c) = 0.044$, $P(N) = 0.956$, so $P(N \cap D^c) = P(N) - P(N \cap D) = 0.9559$. Hence $P((N^c \cap D) \cup (N \cap D^c)) = 0.004 + 0.9559 = 0.9599$. This is the probability that the test confirms the subject's condition.

(c) $P(D|N^c) = \frac{P(D \cap N^c)}{P(N^c)} = \frac{0.004}{0.044} \approx 0.091$. Thus the probability of having the disease, given a positive test, is less than 0.10. The following table may help clarify the situation.

	D [diseased]	D^c [not diseased]
N^c [tests positive]	0.004	0.04
N [tests negative]	0.0001	0.9559

21. (a) $1-(1-q)^n$. We are assuming that the failures of the components are independent.
(b) $1-(0.99)^{100} \approx 0.634$. This is close to $1-\frac{1}{e} \approx 0.632$ because $\lim_n (1-\frac{1}{n})^n = \frac{1}{e}$.
(c) $1-(0.999)^{100} \approx 0.0952$. (d) $1-(0.9)^{100} \approx 0.99997$.

23. (a) $\frac{\binom{18}{5}}{\binom{20}{5}} = \frac{15 \cdot 14}{20 \cdot 19} = \frac{21}{38} \approx 0.55$.

25. No. For example, toss a fair coin three times and let A_k = "kth toss is a head." Then A_1, A_2, A_3 are independent, but $A_1 \cap A_2$ and $A_1 \cap A_3$ are not. Indeed,

$$P(A_1 \cap A_3 | A_1 \cap A_2) = \frac{1}{2} \neq \frac{1}{4} = P(A_1 \cap A_3).$$

27. (a) No. If true, then A and B independent and B and A independent would imply that A and A are independent, which generally fails by part (b).
(c) Absolutely not, unless $P(A) = 0$ or $P(B) = 0$.

29. (a) $$\frac{P^*(E)}{P^*(F)} = \frac{P(E|S)}{P(F|S)} = \frac{\frac{P(E \cap S)}{P(S)}}{\frac{P(F \cap S)}{P(S)}} = \frac{P(E \cap S)}{P(F \cap S)}.$$

9.2 Answers

1. (a) $\{3, 4, 5, 6, \ldots, 17, 18\}$.

3. (a) Using independence, we get $P(X=0) \cdot P(Y=2) = \frac{1}{8}$.
(c) $(\frac{1}{4}+\frac{1}{4}) \cdot (\frac{1}{4}+\frac{1}{2}) = \frac{3}{8}$.

5. (a) $\{0, 1, 2, 3, 4\}$.
(c) For $k = 0, 1, 2, 3, 4$, the answers are $\frac{1}{16}, \frac{1}{8}, \frac{5}{16}, \frac{1}{4}, \frac{1}{4}$, respectively.

7. (a) Value set for D is $\{0, 1, 2, 3, 4, 5\}$ and for M it is $\{1, 2, 3, 4, 5, 6\}$.
(c) $P(D \le 1) = \frac{16}{36} = \frac{4}{9}$ and $P(M \le 3) = \frac{9}{36} = \frac{1}{4}$. Since $\{D \le 1$ and $M \le 3\}$ has 7 elements, $P(D \le 1$ and $M \le 3) = \frac{7}{36}$.

9. The value set is $\{2, 3, 4, 5\}$. $P(X=2) = \binom{5}{2}\binom{5}{5}/\binom{10}{7} = \frac{10}{120} = \frac{1}{12}$ and $P(X=3) = \binom{5}{3}\binom{5}{4}/\binom{10}{7} = \frac{50}{120} = \frac{5}{12}$. Similarly, $P(X=4) = \frac{5}{12}$ and $P(X=5) = \frac{1}{12}$.

11. The value set for W is $\{1, 2, 3, \ldots\} = \mathbb{P}$. $P(W=1) = \frac{1}{6}$, $P(W=2) = \frac{5}{6} \cdot \frac{1}{6}$, $P(W=3) = \frac{5}{6} \cdot \frac{5}{6} \cdot \frac{1}{6}$, etc. In general, $P(W=k) = (\frac{5}{6})^{k-1} \cdot \frac{1}{6}$ for $k \in \mathbb{P}$.

13. (a) $(0.9)^{10} \approx 0.349$.

15. (a) $1-(0.98)^{50} \approx 0.636$.

17. Suppose χ_E and χ_F are independent. Since $E \cap F = \{\chi_E = 1$ and $\chi_F = 1\}$, we have

$$P(E \cap F) = P(\chi_E = 1 \text{ and } \chi_F = 1) = P(\chi_E = 1) \cdot P(\chi_F = 1) = P(E) \cdot P(F),$$

so E and F are independent.

Now suppose that E and F are independent. Since the value sets for χ_E and χ_F are both $\{0, 1\}$, we only need to check that

$$P(\chi_E = k \text{ and } \chi_F = l) = P(\chi_E = k) \cdot P(\chi_F = l) \quad \text{for} \quad k, l \in \{0, 1\}.$$

Since the pairs $\{E, F\}$, $\{E, F^c\}$, $\{E^c, F\}$, and $\{E^c, F^c\}$ are independent [see Exercise 20 on page 358], we have

$$P(\chi_E = 1 \text{ and } \chi_F = 1) = P(E \cap F) = P(E) \cdot P(F) = P(\chi_E = 1) \cdot P(\chi_F = 1);$$

$$P(\chi_E = 1 \text{ and } \chi_F = 0) = P(E) \cdot P(F^c) = P(\chi_E = 1) \cdot P(\chi_F = 0);$$

and similarly for $P(\chi_E = 0$ and $\chi_F = 1)$ and $P(\chi_E = 0$ and $\chi_F = 0)$. These four computations show that the random variables χ_E and χ_F are independent.

9.3 Answers

1. (a) $E(X) = (-1) \cdot \frac{1}{5} + 0 \cdot \frac{1}{5} + 1 \cdot \frac{1}{5} + 2 \cdot \frac{2}{5} = \frac{4}{5}$.
(c) $E(3X+2) = 3 \cdot E(X) + 2 = 3 \cdot \frac{4}{5} + 2 = 4.4$.

3. Using Theorem 4, $V(X^2) = E(X^4) - (E(X^2))^2 = 1 \cdot \frac{1}{5} + 0 \cdot \frac{1}{5} + 1 \cdot \frac{1}{5} + 16 \cdot \frac{2}{5} - (2)^2 = \frac{14}{5}$, so σ for X^2 is $\sqrt{14/5} \approx 1.67$.

5. $\mu_X = \mu_Y = 0 \cdot \frac{1}{4} + 1 \cdot \frac{1}{4} + 2 \cdot \frac{1}{2} = \frac{5}{4}$ and $\mu_{X+Y} = \mu_X + \mu_Y = \frac{5}{2}$. Also, $\sigma_X^2 = 0 \cdot \frac{1}{4} + 1 \cdot \frac{1}{4} + 4 \cdot \frac{1}{2} - \mu_X^2 = \frac{9}{4} - (\frac{5}{4})^2 = \frac{11}{16}$, so $\sigma_X = \sqrt{11}/4$. Similarly, $\sigma_Y = \sqrt{11}/4$. Since X and Y are independent, $V(X+Y) = V(X) + V(Y) = \frac{11}{8}$, so $\sigma_{X+Y} = \sqrt{11/8}$.

7. (a) The answer is rather obviously 3.5. To confirm this, use the answer to Exercise 9 on page 366, and calculate $\sum_{k=2}^{5} k \cdot P(X = k) = 7/2$.

9. (a) The mean deviation is $\sum_{k=1}^{6} |k - 3.5| \cdot \frac{1}{6} = \frac{1}{6}[2.5 + 1.5 + 0.5 + 0.5 + 1.5 + 2.5] = 1.5$, a bit smaller than $\sigma \approx 1.71$.

11. $E(X) = 5/13$. Note that $X = X_1 + X_2 + X_3 + X_4 + X_5$, where $X_i = 1$ if the ith card is an ace and $X_i = 0$ otherwise. What is $E(X_i)$?

13. (a) Let W be the waiting time random variable. If we imagine that all five marbles are drawn from the urn, then it is clear that the blue marble is as likely to be the first marble as it is the second marble, etc. That is, $P(W = k) = \frac{1}{5}$ for $k = 1, 2, 3, 4, 5$. [These equalities can also be easily verified directly. For example, $P(W = 3) = \frac{4}{5} \cdot \frac{3}{4} \cdot \frac{1}{3} = \frac{1}{5}$.] Thus $E(W) = \frac{1}{5}(1+2+3+4+5) = 3$.
(b) We expect $\frac{1}{5}$th of a blue marble on each draw, so we expect to wait 5 draws on average to get the blue marble. To verify this mathematically, let W be the waiting time random variable so that $P(W = k) = (\frac{4}{5})^{k-1} \cdot \frac{1}{5}$ for $k \geq 1$. Then $E(W) = \sum_{k=1}^{\infty} k(\frac{4}{5})^{k-1} \cdot \frac{1}{5}$, which turns out to be 5.
(c) It doesn't take as long to get to the blue marble if we keep removing red ones.

15. The mathematical expectation is $500,000 in the first case and $600,000 in the second case. From the strict point of view of maximizing expectation, one should prefer the 20 percent chance of winning $3,000,000. However, most of us would take the first choice because we regard the value of winning $1,000,000 as much more than one-third of the value of winning $3,000,000. In other words, we don't consider the dollars after the first million to be as valuable as the first million. A theory of "utility" takes this weighting of values into account.

17. (a) The expected number of hits at each official attempt is $\frac{1}{3}$, so the expected number of hits in 10 official attempts is 10/3. This is using the "nonsensical" approach used in Exercise 12.
(c) $1 - P(\text{at most 2 hits}) \approx 1 - (0.017 + 0.087 + 0.195) = 0.701$.

19. Let $Y = X + a$. Then $\mu_Y = \mu_X + a$, so $Y - \mu_Y = X + a - (\mu_X + a) = X - \mu_X$. Hence, by the definition,

$$V(X + a) = V(Y) = E((Y - \mu_Y)^2) = E((X - \mu_X)^2) = V(X).$$

By Theorem 1, $E(aX) = a \cdot E(X)$ and $E((aX)^2) = E(a^2X^2) = a^2 \cdot E(X^2)$. So, by Theorem 4,

$$V(aX) = E((aX)^2) - E(aX)^2 = a^2 \cdot E(X^2) - a^2 \cdot E(X)^2 = a^2 \cdot V(X).$$

21. (a) $\mu_S = E(S) = \sum_{i=1}^{n} E(X_i) = n\mu$ and, by Theorem 6, $V(S) = \sum_{i=1}^{n} V(X_i) = n\sigma^2$, so $\sigma_S = \sqrt{n} \cdot \sigma$.
(b) Use Exercise 19.

9.4 Answers

1. For $k = 2, 3, 4, \ldots, 12$, $F_s(k)$ is $\frac{1}{36}, \frac{3}{36}, \frac{6}{36}, \frac{10}{36}, \frac{15}{36}, \frac{21}{36}, \frac{26}{36}, \frac{30}{36}, \frac{33}{36}, \frac{35}{36}, \frac{36}{36} = 1$, respectively.

3. For $k = 1, 2, 3, 4, 5, 6$, $F(k)$ is $\frac{1}{36}, \frac{4}{36}, \frac{9}{36}, \frac{16}{36}, \frac{25}{36}, \frac{36}{36} = 1$, respectively. Note the interesting numerators. Can you explain this phenomenon?

5. The answer to Exercise 11 on page 366 gives $P(W = k) = (\frac{5}{6})^{k-1} \cdot \frac{1}{6}$ for $k \in \mathbb{P}$. For $k = 1, 2, 3, 4$, these probabilities are $\frac{1}{6}, \frac{5}{36}, \frac{25}{216}, \frac{125}{1296}$, respectively. So $F(1) = \frac{1}{6}$, $F(2) = \frac{11}{36}$, $F(3) = \frac{91}{216}$, and $F(4) = \frac{671}{1296}$. Using a geometric series, one can show $F(k) = 1 - (\frac{5}{6})^k$ for $k \in \mathbb{P}$.

7. Use the probability axioms on page 190 and the Theorem on page 191.

(a) $F(y) = P(X \leq y) \geq P(\emptyset) = 0$ and $F(y) = P(X \leq y) \leq P(\Omega) = 1$ for $y \in \mathbb{R}$.
(b) If $y_1 < y_2$, then $\{X \leq y_1\} \subseteq \{X \leq y_2\}$, so $F(y_1) = P(X \leq y_1) \leq P(X \leq y_2) = F(y_2)$. Why can't we conclude $F(y_1) < F(y_2)$?

9. (a) $F(69) - F(64) \approx \widetilde{F}(0) - \widetilde{F}(-1.7) \approx 0.455$.
(c) $1 - F(72) = 1 - \widetilde{F}(1) \approx 0.159$.

11. Use Exercise 10.

(a) $F(7700) = \widetilde{F}(\frac{7700-7500}{75}) \approx \widetilde{F}(2.7) \approx \Phi(2.7) \approx 0.996$.
(c) $F(7600) - F(7400) = \widetilde{F}(4/3) - \widetilde{F}(-4/3) \approx \Phi(4/3) - \Phi(-4/3)$, which is approximately 0.806 using the table, but is closer to 0.8176 using a better table.

13. (a) 600. (c) $[\mu - 2\sigma, \mu + 2\sigma] = [560, 640]$.

15. (a) Need $P(490 < X \leq 510)$. Observe $\mu = 500$ and $\sigma = \sqrt{1000 \cdot \frac{1}{2} \cdot \frac{1}{2}} \approx 15.81$, so $10 \approx 0.632\sigma$. Thus $P(490 < X \leq 510) \approx P(\mu - 0.63\sigma < X \leq \mu + 0.63\sigma) \approx \Phi(0.63) - \Phi(-0.63)$. This turns out to be approximately 0.47.
(c) $\mu = 500,000$ and $\sigma = 500$, so $P(490,000 < X \leq 510,000) = P(\mu - 20\sigma < X \leq \mu + 20\sigma) \approx \Phi(20) - \Phi(-20)$. This is very, very close to 1: $0.9999\cdots$, where the first 88 digits are 9. For all practical purposes, the event $\{490,000 < X \leq 510,000\}$ is a certainty.

Note. Each calculation above is a special of case of considering n tosses. Then $\mu = 0.5n$, $\sigma = \sqrt{npq} = \sqrt{n}/2$, and one needs $F(0.5n + 0.01n) - F(0.5n - 0.01n)$. This is equal to $\widetilde{F}(0.02\sqrt{n}) - \widetilde{F}(-0.02\sqrt{n}) \approx \Phi(0.02\sqrt{n}) - \Phi(-0.02\sqrt{n})$.

17. (a) $\frac{13}{12}$. Note that $X = X_1 + X_2 + X_3$, where $X_i = 1$ if the ith experiment is a success and $X_i = 0$ otherwise. So $E(X) = E(X_1) + E(X_2) + E(X_3) = \frac{1}{2} + \frac{1}{3} + \frac{1}{4}$.
(b) $\sigma_X \approx 0.81$.
(c) No; the probabilities of successive experiments are not the same fixed value p.

19. (a) Since y or $-y$ is nonnegative, we may assume that one of them, say y, is nonnegative. Now 1 = "area under the bell curve ϕ" = "area under ϕ to the left of y plus the area under ϕ to the right of y," which equals $\Phi(y)$ plus the area under ϕ to the right of y. By the symmetry of the graph of ϕ, the area under ϕ to the right of y is equal to the area under ϕ to the left of $-y$, i.e., to $\Phi(-y)$. So $1 = \Phi(y) + \Phi(-y)$.

21. (a) Since X is normal, the cdf $\widetilde{F}$ of $\widetilde{X}$ is equal to Φ. By Theorem 1(b) on page 380, $F(y) = \widetilde{F}(\frac{y-\mu}{\sigma}) = \Phi(\frac{y-\mu}{\sigma})$.

9.5 Answers

1. (a) 1, 0.5, and 0.25 [$= P(\{HHT, HTH\})$], respectively.
(b) Yes. $P(A \cap C) = \frac{1}{4} = \frac{1}{2} \cdot \frac{1}{2} = P(A)P(C)$.
(c) One possibility is that Y counts the number of heads in the first two tosses.

2. (a) 3/4.
(b) This is a standard Bayes' Theorem problem.

$$P(B|R) = \frac{P(B)P(R|B)}{P(B)P(R|B) + P(B^c)P(R|B^c)} = \frac{\frac{1}{4} \cdot \frac{2}{5}}{\frac{1}{4} \cdot \frac{2}{5} + \frac{3}{4} \cdot \frac{3}{5}} = \frac{2}{11}.$$

3. (a) 0.12 [$= P(B) \cdot P(A|B)$] and 0.58 [$= P(A) - P(A \cap B)$].
(b) No, since $P(A|B) \neq P(A)$.

4. (a) With self-evident notation, we use the Total Probability Formula on page 356 to obtain $P(W) = P(U_1) \cdot P(W|U_1) + P(U_2) \cdot P(W|U_2) = \frac{1}{2} \cdot \frac{1}{3} + \frac{1}{2} \cdot \frac{3}{5} = \frac{7}{15}$.
(b) By Bayes' Formula on page 356, $P(U_1|W) = \frac{P(U_1) \cdot P(W|U_1)}{P(W)} = \frac{\frac{1}{2} \cdot \frac{1}{3}}{7/15} = \frac{5}{14}$.

5. (a) Might. (b) Might.
(c) False, because $P(B)/P(A)$ is bigger than 1.
(d) Might, e.g., if independent.
(e) True, since $P(A^c) = 1 - P(A)$.
(f) True, by the Total Probability Formula on page 356.

6. $P(B|R) = 1/17$. The reason is that

$$P(B|R) = \frac{P(B \cap R)}{P(R)} = \frac{P(B) \cdot P(R|B)}{P(B) \cdot P(R|B) + P(B^c) \cdot P(R|B^c)}$$

$$= \frac{\frac{1}{13} \cdot \frac{3}{4}}{\frac{1}{13} \cdot \frac{3}{4} + \frac{12}{13} \cdot 1} = \frac{\frac{3}{4}}{\frac{3}{4} + 12} = \frac{3}{3 + 48} = \frac{1}{17}.$$

7. (a) $P(X = 0) = \frac{\binom{5}{0}\binom{3}{3}}{\binom{8}{3}} = 1/56$, $P(X = 1) = \frac{\binom{5}{1}\binom{3}{2}}{\binom{8}{3}} = 15/56$, $P(X = 2) = \frac{\binom{5}{2}\binom{3}{1}}{\binom{8}{3}} = 30/56$.
(b) $P(X = 3) = \frac{\binom{5}{3}\binom{3}{0}}{\binom{8}{3}} = 10/56$. Check by summing the answers to parts (a) and (b).
(c) $E(X) = 0 \cdot \frac{1}{56} + 1 \cdot \frac{15}{56} + 2 \cdot \frac{30}{56} + 3 \cdot \frac{10}{56} = \frac{15}{8} = 3 \cdot \frac{5}{8}$.
(d) $E(X^2) = 0 \cdot \frac{1}{56} + 1 \cdot \frac{15}{56} + 2^2 \cdot \frac{30}{56} + 3^2 \cdot \frac{10}{56} = \frac{225}{56}$, so $V(X) = E(X^2) - (E(X))^2 = \frac{225}{56} - \frac{225}{64} = \frac{225}{448} \approx 0.50$.
(e) No; $X + Y = 3$, so the bigger X is, the smaller Y is. For example, $P(Y = 3) = P(X = 0) = \frac{1}{56}$, but $P(Y = 3|X = 3) = 0$.
(f) $E(Y) = 3 - E(X) = \frac{63}{56} = 1.125$.

8. Let G be the event that the good shipment was selected, B the event that the bad shipment was selected, and N the event that no defective set was selected. We want $P(G|N)$. We have $P(N|G) = 1$ and $P(N|B) = \frac{\binom{8}{5}}{\binom{10}{5}} = \frac{2}{9}$. By Bayes' Formula on page 356, $P(G|N) = \frac{P(G) \cdot P(N|G)}{P(G) \cdot P(N|G) + P(B) \cdot P(N|B)} = \frac{\frac{1}{2} \cdot 1}{\frac{1}{2} \cdot 1 + \frac{1}{2} \cdot \frac{2}{9}} = \frac{9}{11}$.

9. (a) $L = 1$, $P(900 \leq X \leq 915) \approx 0.341$ [$= \Phi(1) - \Phi(0)$].

10. (a) 0.9 [$= 0.7 + 0.2$]. (b) 0.62 [$= 0.7 \times 0.8 + 0.3 \times 0.2$].
(c) 0.86 [$= 0.62 + 0.3 \times 0.8 = 1 - 0.7 \times 0.2$].
(d) It's a step function that's 0 to the left of 0, jumps up to 0.24 at $x = 0$, jumps to 0.86 at $x = 1$, and then jumps to 1 at $x = 2$.
(e) 0.37. First reason: $E(X^2) = 0 \times 0.24 + 1 \times 0.62 + 2^2 \times 0.14 = 1.18$, so $V(X) = 1.18 - 0.9^2 = 0.37$. Second reason: Say X_1 counts successes for the first experiment and X_2 counts them for

the second. By Theorem 7 on page 373 with $n = 1$, we have $V(X_1) = 0.7 \times 0.3 = 0.21$, $V(X_2) = 0.2 \times 0.8 = 0.16$. Since the experiments are independent, $V(X) = 0.21 + 0.16$.

(f) Independence doesn't matter in part (a), since $E(Y+Z) = E(Y) + E(Z)$ in general, but in (b) we need it in order to be sure that $P(X_1 = 1 \text{ and } X_2 = 0) = P(X_1 = 1) \cdot P(X_2 = 0)$, and similarly with X_1 and X_2 reversed. [We also used independence in the second argument for part (e).]

11. (a) $\binom{4}{3} \cdot (\frac{3}{4})^3 \cdot \frac{1}{4}$ [$= \frac{27}{64}$, a little less than a half].

(b) $pq = \frac{3}{4} \cdot \frac{1}{4} = \frac{3}{16}$.

(c) 300.

(d) $\displaystyle\sum_{k=290}^{310} \binom{400}{k} \left(\frac{3}{4}\right)^k \left(\frac{1}{4}\right)^{400-k}$.

(e) Since individual attempts are independent, $V(X) = 400V(Y)$, so by part (a) $V(X) = 400 \cdot \frac{3}{16} = 75$, and thus $\sigma_X = \sqrt{V(X)} = 5\sqrt{3}$. Or one can quote the theorem on binomial distributions that says that $\sigma_X = \sqrt{npq} = \sqrt{400 \times \frac{3}{4} \times \frac{1}{4}}$.

(f) $\widetilde{X} = \frac{X-\mu}{\sigma}$, $\mu = 300$, and $\sigma = 5\sqrt{3}$, so $290 \le X \le 310$ is equivalent to $\frac{290-300}{5\sqrt{3}} \le \widetilde{X} \le \frac{310-300}{5\sqrt{3}}$. Thus $L = \frac{10}{5\sqrt{3}} = \frac{2}{\sqrt{3}}$ [≈ 1.15].

(g) $P(290 \le X \le 310) \approx \Phi(1.15) - \Phi(-1.15) \approx 0.87 - 0.13 = 0.74$. [With a better table, we found that the value is about 0.75. The amazing thing is that we get such a large value—nearly three-fourths of the time she will shoot between 290 and 310 baskets, whereas the probability that she will get exactly 3 out of 4 baskets is only about a half. It's almost certain that she will shoot between 250 and 350 out of 400.]

12. (a) $P(X = -1) + P(X = 0) = \frac{1}{2}$.

(b) $P(X = -1) = \frac{1}{8}$.

(c) $P(Y = 1) + P(Y = 2) = \frac{1}{2}$.

(d) $\frac{1}{2}$ by part (a).

(e) $P(X \le 0) \cdot P(Y < 0) = \frac{1}{2} \cdot \frac{1}{8} = \frac{1}{16}$. Used independence.

(f) $1 - P(X > 0 \text{ and } Y \ge 0) = 1 - P(X > 0) \cdot P(Y \ge 0) = 1 - \frac{1}{2} \cdot \frac{7}{8} = \frac{9}{16}$. Used independence.

(g) $1 - P(X + Y = 4) = 1 - P(X = 2 \text{ and } Y = 2) = 1 - P(X = 2) \cdot P(Y = 2) = 1 - \frac{1}{8} \cdot \frac{1}{8} = \frac{63}{64}$. Used independence.

(h) $E(X) = -1 \cdot \frac{1}{8} + 0 \cdot \frac{3}{8} + 1 \cdot \frac{3}{8} + 2 \cdot \frac{1}{8} = \frac{1}{2}$.

(i) Since $E(Y) = E(X) = \frac{1}{2}$, $E(X - 3Y) = E(X) - 3 \cdot E(X) = -1$.

(j) By independence and Theorem 5 on page 372, $E(XY) = E(X) \cdot E(Y) = \frac{1}{4}$.

(k) $E(|X|) = 1 \cdot \frac{1}{8} + 0 \cdot \frac{3}{8} + 1 \cdot \frac{3}{8} + 2 \cdot \frac{1}{8} = \frac{3}{4}$.

(l) $E(X^2) = 1 \cdot \frac{1}{8} + 0 \cdot \frac{3}{8} + 1 \cdot \frac{3}{8} + 4 \cdot \frac{1}{8} = 1$, so $V(X) = E(X^2) - (E(X))^2 = 1 - \frac{1}{4} = \frac{3}{4}$.

(m) By independence, Theorem 6 on page 372, and Exercise 19 on page 374, $V(X-3Y) = V(X) + V(-3Y) = V(X) + 9V(Y) = 10V(X) = \frac{15}{2}$.

(n) $\sqrt{V(X+Y)} = \sqrt{V(X) + V(Y)} = \sqrt{2 \cdot V(X)} = \frac{\sqrt{6}}{2}$. Used independence as in part (n).

13. (a) $\left(\frac{5}{6}\right)^3 \cdot \frac{1}{6}$.

(b) 6. The "nonsensical" method in Example 4 on page 369 works here: You can expect 1/6 of a failure on each trial, so a whole failure on the sixth trial.

14. (a) $\{0, 1, 2, 3, 4\}$.

(b) The function is 0 to the left of $x = 0$; then it jumps to $\frac{1}{2}$ at $x = 0$ and is $\frac{1}{2}$ to the right of $x = 0$ and to the left of $x = 2$. Finally, it jumps to 1 at $x = 2$ and remains at that value to the right of $x = 2$.

(c) It means that

$$P(N \in I \text{ and } D \in J) = P(N \in I) \cdot P(D \in J)$$

for all intervals I and J. Equivalently, since the sample space in this case is finite, it means that

$$P(N = x \text{ and } D = y) = P(N = x) \cdot P(D = y)$$

for all x in $\{0, 1, 2, 3, 4\}$ and y in $\{0, 1\}$.

(d) 2.5 [$= E(N) + E(D) = 2 + 0.5$].

(e) 0.375 [$= \binom{4}{2}/16 = 3/8$, since $E(N) = 2$].

15. (a) $E(S) = np = 18000 \cdot \frac{1}{6} = 3000$.

(b) $\sigma = \sqrt{npq} = \sqrt{18000 \cdot \frac{1}{6} \cdot \frac{5}{6}} = 50$.

(c) We use $\widetilde{S}$ for the normalization of S. Then $P(17900 < S \le 18200) = P(\frac{17900-18000}{50} < \widetilde{S} \le \frac{18200-18000}{50}) = P(-2 < \widetilde{S} \le 4) \approx \Phi(4) - \Phi(-2) \approx 1 - 0.023 = 0.977$.

(d) We assumed that the tosses were independent, so S is a binomial random variable.

16. (a) This follows from independence and the fact that $(E \times \Omega_2) \cap (\Omega_1 \times F) = E \times F$.

(b) C is the event "the suit is a club," F is the event "the value is four," and $C \times F$ is the event "the card is the four of clubs." Also, $P(C \times F) = \frac{1}{52}$ and the product is $\frac{1}{4} \cdot \frac{1}{13}$.

(c) Ω_1 is the set $\{1, 2, 3, 4, 5, 6\}$ of possible values of the black die. Ω_2 is the same set of possible values of the red die. Also, $B_1 = \{1, 2, 3\}$, $R_2 = \{5, 6\}$, $P(B \cap R) = P(B_1 \times R_2) = \frac{1}{6}$, and the product is $\frac{1}{2} \cdot \frac{1}{3}$.

10.1 Answers

1. (a) Since the operations $\vee$ and $\wedge$ treat 0 and 1 just as if they represent truth values, checking the laws 1Ba through 5Bb for all cases amounts to checking corresponding truth tables. Do enough until the situation is clear to you.

3. We have

$$
\begin{aligned}
(x \vee y) \wedge (x' \wedge y') &= [x \wedge (x' \wedge y')] \vee [y \wedge (x' \wedge y')] && \text{distributivity} \\
&= [y' \wedge (x \wedge x')] \vee [x' \wedge (y \wedge y')] && \text{associativity and commutativity} \\
&= [y' \wedge 0] \vee [x' \wedge 0] = 0 \vee 0 = 0.
\end{aligned}
$$

5. (a) $\{a, c, d\} = \{a\} \cup \{c\} \cup \{d\}$.

(c) $f = f_a \vee f_c \vee f_d$, where f_a, f_c, f_d are defined as follows

	a	b	c	d	e
f_a	1	0	0	0	0
f_c	0	0	1	0	0
f_d	0	0	0	1	0

Note the similarity among parts (a), (b), and (c).

7. (a) The atoms are given by the four columns on the right in the table.

x	y	a	b	c	d
0	0	1	0	0	0
0	1	0	1	0	0
1	0	0	0	1	0
1	1	0	0	0	1

(c) In the notation of the answer to part (a), $h = a \vee b \vee d$.

9.

11. (a) If $a \leq x$ or $a \leq y$, then surely $a \leq x \vee y$ by Lemmas 4(a) and 3(a). Suppose $a \leq x \vee y$. Then $a = a \wedge (x \vee y) = (a \wedge x) \vee (a \wedge y)$. One of $a \wedge x$ and $a \wedge y$, say $a \wedge x$, must be different from 0. But $0 < a \wedge x \leq a$, so $a \wedge x = a$ and $a \leq x$.

(c) $a \leq 1 = x \vee x'$, so $a \leq x$ or $a \leq x'$ by part (a). Both $a \leq x$ and $a \leq x'$ would imply $a \leq x \wedge x' = 0$ by part (b), a contradiction.

10.2 Answers

1. $x'y'z' \vee x'y'z \vee xyz'$.

3.

x	y	z	(a) xy	(b) z'	(c) $xy \vee z'$	(d) 1
0	0	0	0	1	1	1
0	0	1	0	0	0	1
0	1	0	0	1	1	1
0	1	1	0	0	0	1
1	0	0	0	1	1	1
1	0	1	0	0	0	1
1	1	0	1	1	1	1
1	1	1	1	0	1	1

(a) $xyz' \vee xyz$.

(c) $x'y'z' \vee x'yz' \vee xy'z' \vee xyz' \vee xyz$.

5. (a) $x_1x_2x_3'x_4 \vee x_1x_2x_3'x_4' \vee x_1'x_2x_3x_4'$.

7. (a) $xz \vee y'$. Note $y' \vee y'z = y'z' \vee y'z \vee y'z = y'z' \vee y'z = y'$ and similarly $y' \vee xy'z' = y'$.

9. $x'y' \vee z \vee xyz \vee xy'z' = x'y'z \vee x'y'z' \vee z \vee xyz \vee xy'z'$. Now $x'y'z \vee z \vee xyz = z$ and $x'y'z' \vee xy'z' = y'z'$, so we get $z \vee y'z' = yz \vee y'z \vee y'z' = (yz \vee y'z) \vee (y'z \vee y'z') = z \vee y'$. This expression can also be obtained from a table of the corresponding Boolean function. The expression is not a single product of literals, by Exercise 12 [its function has the value 1 in more than four places], so any equivalent expression as a join of products of literals has at least two products, with at least one literal in each product. Thus the expression $z \vee y'$ is optimal.

11. (a) Find the minterm canonical form for E'. Then find $E = (E')'$ using De Morgan laws, first on joins and then on products.

13. (a) The Boolean function for $x'z \vee y'z$ takes the value 1 at three elements in $\mathbb{B}^3$. The Boolean functions for products of literals take the value 1 at 1, 2, or 4 elements in $\mathbb{B}^3$ by Exercise 12.

10.3 Answers

1. (a) $\{[((xy)z)' \vee x'] \vee [(z \vee y')']\}'$ simplifies to xyz with a little work.

3. (a) x y

5. (a)

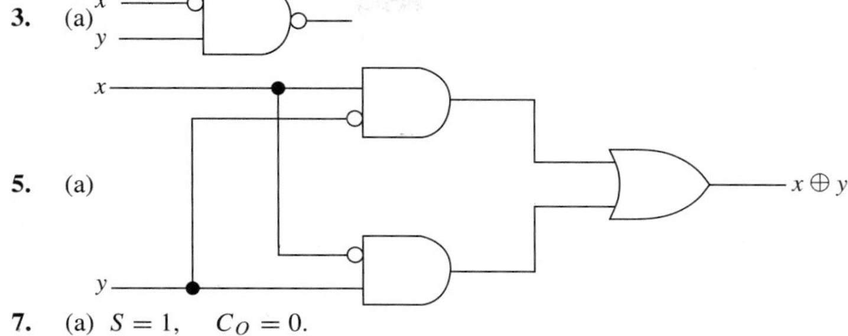

7. (a) $S = 1, \quad C_O = 0.$
(c) $S = 0, \quad C_O = 1.$

9. (a)

(c)

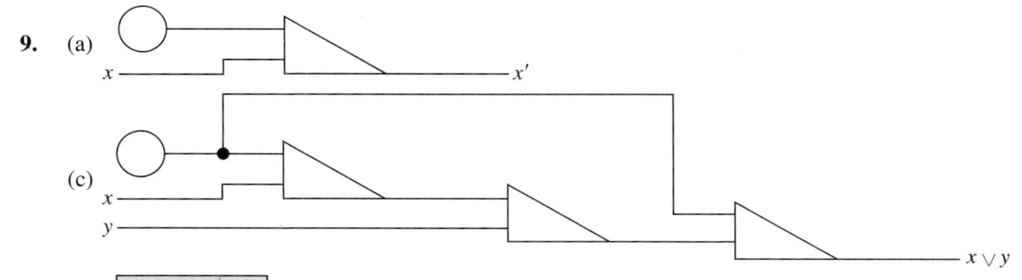

11.

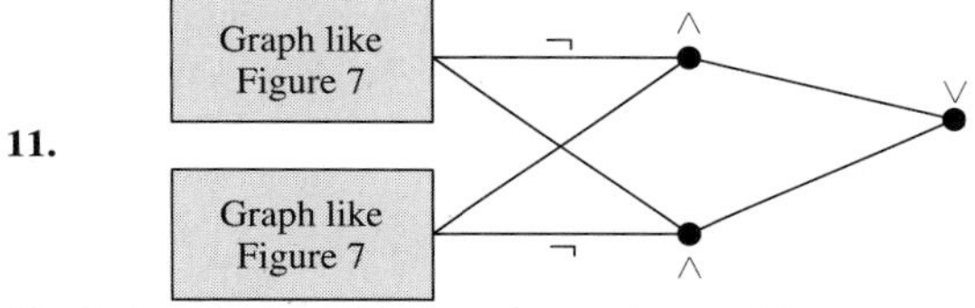

13. It is convenient to view the result as valid for $n = 1$. Apply the Second Principle of Mathematical Induction. A glance at a piece of Figure 8 shows that the result is valid for $n = 2$. Assume the result is true for all j with $1 \le j < n$. Consider k with $2^{k-1} < n \le 2^k$, and let $n' = 2^{k-1}$, $n'' = n - 2^{k-1}$. By the inductive assumption, there are digraphs D' and D'' for computing $x_1 \oplus x_2 \oplus \cdots \oplus x_{n'}$ and $x_{n'+1} \oplus \cdots \oplus x_n$. D' has $3(n' - 1)$ $\wedge$ and $\vee$ vertices, and D'' has $3(n'' - 1)$ $\wedge$ and $\vee$ vertices. Also, every path in D' and in D'' has length at most $2(k - 1)$. Now create D as shown.

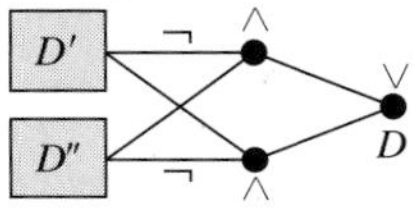

D has $3(n' - 1) + 3(n'' - 1) + 3$ $\wedge$ and $\vee$ vertices, i.e., $3(n - 1)$ such vertices. Moreover, every path in D has length at most $2(k - 1) + 2 = 2k$.

15. (a) The Boolean function for $E_1 \oplus E_2$ is $f = f_1 \oplus f_2$, where

$$f(a_1, a_2, \ldots, a_n) = f_1(a_1, a_2, \ldots, a_n) \oplus f_2(a_1, a_2, \ldots, a_n)$$

for each $(a_1, a_2, \ldots, a_n)$ in $\mathbb{B}^n$. This function has value 1 if exactly one of $f_1(a_1, a_2, \ldots, a_n)$ and $f_2(a_1, a_2, \ldots, a_n)$ does.

(b) The result is clear for the Boolean expression x_1, since the corresponding Boolean function is 1 at $(a_1, a_2, \ldots, a_n)$ if and only if $a_1 = 1$. Assume that the statement is true for $1 \le m < n$, and consider the Boolean function f for

$$(x_1 \oplus x_2 \oplus \cdots \oplus x_m) \oplus x_{m+1}.$$

By part (a), there are two cases where f takes the value 1. This happens if the Boolean function for $x_1 \oplus x_2 \oplus \cdots \oplus x_m$ takes the value 1 and the one for x_{m+1} takes the value 0, in which case $a_{m+1} = 0$ and an odd number of the values $a_1, a_2, \ldots, a_m$ are 1 [by the induction hypothesis]. This also happens if the Boolean function for $x_1 \oplus x_2 \oplus \cdots \oplus x_m$ takes the value 0 and the one for x_{m+1} takes the value 1, in which case $a_{m+1} = 1$ and an even number of the values $a_1, a_2, \ldots, a_m$ are 1 [again, by the induction hypothesis]. In both cases, an odd number of the values $a_1, a_2, \ldots, a_m, a_{m+1}$ are 1. A similar argument shows that f takes the value 0 when an even number of the values $a_1, a_2, \ldots, a_m, a_{m+1}$ are 1.

10.4 Answers

1. $xyz \vee xyz' \vee xy'z' \vee xy'z \vee x'yz \vee x'y'z = x \vee z$.

3. $xyz \vee xyz' \vee xy'z \vee x'y'z' \vee x'y'z = xz \vee xy \vee x'y' = xy \vee y'z \vee x'y'$.

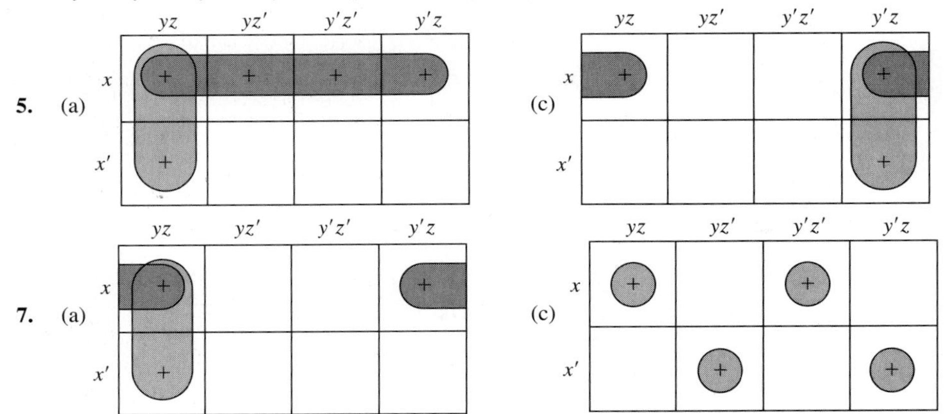

Each two-square block is essential. Each one-square block is essential.

9. (a) $z' \vee xy \vee x'y' \vee w'y$ or $z' \vee xy \vee x'y' \vee w'x'$.
(c) $w'x'z' \vee w'xy' \vee wxy \vee wx'z \vee y'z'$, not $w'x'z' \vee w'xy' \vee wx'y' \vee wyz \vee wxz'$, which also has five product terms, but one more literal.

10.5 Answers

1. One solution is to set $S = \{1, 2, 3, 4, 5\}$ and define $\phi(x_1, x_2, x_3, x_4, x_5) = \{i \in S : x_i = 1\}$.

3. (a) No. A finite Boolean algebra has 2^n elements for some n.
(b) No, since $|\text{BOOL}(n)| = 2^{2^n}$. For example, 8 does not have this form.

5. In each part, each of the maps is a one-to-one correspondence; see Example 4 on page 114. Thus we only need to verify properties (1)–(3) defining a Boolean algebra isomorphism.

(a) For $u, v \in B_2$, let $x = \phi^{-1}(u)$ and $y = \phi^{-1}(v)$. Then $\phi^{-1}(u \vee v) = \phi^{-1}(\phi(x) \vee \phi(y)) = \phi^{-1}(\phi(x \vee y)) = x \vee y = \phi^{-1}(u) \vee \phi^{-1}(v)$ and, similarly, $\phi^{-1}(u \wedge v) = \phi^{-1}(u) \wedge \phi^{-1}(v)$. Also $\phi^{-1}(u') = \phi^{-1}(\phi(x)') = \phi^{-1}(\phi(x')) = x' = \phi^{-1}(u)'$.
(b) Let $\phi = \phi_2 \circ \phi_1$. Show $\phi(x \vee y) = \phi(x) \vee \phi(y)$ and $\phi(x \wedge y) = \phi(x) \wedge \phi(y)$.

7. (a) $\phi(1) = 0$ and $\phi(7) = 1$ define a Boolean algebra isomorphism $\phi\colon D_7 \to \mathbb{B}$. One way to see that ϕ preserves the Boolean operations $\vee$ and $\wedge$ is to make tables for D_7 like those in Figure 1 on page 390 and compare. Also $1' = 7$ and $7' = 1$ in D_7, so

$$\phi(1') = \phi(7) = 1 = 0' = \phi(1)' \quad \text{and} \quad \phi(7') = \phi(1) = 0 = 1' = \phi(7)'.$$

Alternatively, if you show that D_7 is a Boolean algebra, then it must be isomorphic to $\mathbb{B}$, by the Remark after the Theorem. What are the atoms here?
(c) D_8 has 4 elements. However, 2 and 4 have no complements in D_8. For example, if $2' = z$, then $2 \wedge z = 1$ implies $z = 1$, whereas $2 \vee z = 8$ implies $z = 8$.

9. (a) $x \le y \Longleftrightarrow x \vee y = y \Longleftrightarrow \phi(x \vee y) = \phi(y) \Longleftrightarrow \phi(x) \vee \phi(y) = \phi(y) \Longleftrightarrow \phi(x) \le \phi(y)$.
(b) Suppose that a is an atom in B_1. Then $a \ne 0$ in B_1, and so $\phi(a) \ne 0$ in B_2, since ϕ is one-to-one. If $\phi(a)$ weren't an atom, then there would be z in B_2 with $0 < z < \phi(a)$. Since ϕ^{-1} is a Boolean algebra isomorphism [Exercise 5(a)], part (a) would yield

$$\phi^{-1}(0) < \phi^{-1}(z) < \phi^{-1}(\phi(a)), \quad \text{i.e.,} \quad 0 < \phi^{-1}(z) < a,$$

contradicting the assumption that a is an atom in B_1. For the converse, suppose that $\phi(a)$ is an atom in B_2. Then, from above and Exercise 5(a), $\phi^{-1}(\phi(a)) = a$ is an atom in B_1.

11. (a) It suffices to consider $\{[a_i, b_i)\}_{i \in \mathbb{P}}$ and show that each union $\bigcup_{i=1}^{n}[a_i, b_i)$ can be written as a finite *disjoint* union of such intervals. This is obvious for $n = 1$. Assume it has been done for $\bigcup_{i=1}^{n-1}[a_i, b_i)$; in fact, we can assume the intervals $[a_1, b_1), \ldots, [a_{n-1}, b_{n-1})$ are disjoint. If $[a_n, b_n)$ intersects none of these sets, we're done. Otherwise, let

$$I = \{i : 1 \le i \le n-1 \text{ and } [a_i, b_i) \cap [a_n, b_n) \ne \emptyset\},$$
$$a^* = \min(\{a_i : i \in I\} \cup \{a_n\}),$$
$$b^* = \max(\{b_i : i \in I\} \cup \{b_n\}),$$

and observe $[a_n, b_n) \cup \bigcup_{i \in I} [a_i, b_i) = [a^*, b^*)$ so that

$$[a^*, b^*) \cup \{[a_i, b_i) : i \notin I, 1 \le i \le n-1\}$$

expresses $\bigcup_{i=1}^{n} [a_i, b_i)$ as a disjoint union of intervals of the form $[a, b)$.

(b) Since $\mathcal{A}$ inherits the laws 1B, ..., 5B from the Boolean algebra $\mathcal{P}(S)$, you only need to show that if X and Y are in $\mathcal{A}$ then so are $X \cup Y$, $X \cap Y$, and X'.

(c) $\mathcal{A}$ has no atoms since nonempty members of $\mathcal{A}$ always contain smaller nonempty members of $\mathcal{A}$. For example, $[a, b)$ contains $[a, \frac{1}{2}(a+b))$.

13. (a) $\theta^{-1}(x \vee y) = \theta^{-1}(\theta(a) \vee \theta(b)) = \theta^{-1}(\theta(a \vee b)) = a \vee b = \theta^{-1}(x) \vee \theta^{-1}(y)$.

(b) Yes. Its inverse, of course, is θ.

(c) If $c \le d$, then $d = c \vee d$, so $\theta(d) = \theta(c \vee d) = \theta(c) \vee \theta(d)$, and thus $\theta(c) \le \theta(d)$.

(d) Repeat the argument for part (c) with θ^{-1} in place of θ, using part (a).

(e) Since θ maps onto B_2, there is an e in B_1 with $\theta(e) = 0_2$. Hence

$$\theta(0_1) = \theta(0_1) \vee 0_2 = \theta(0_1) \vee \theta(e) = \theta(0_1 \vee e) = \theta(e) = 0_2.$$

(f) Say $\theta(f) = 1_2$. Then

$$\theta(1_1) = \theta(1_1 \vee f) = \theta(1_1) \vee \theta(f) = \theta(1_1) \vee 1_2 = 1_2.$$

(g) We have $\theta(g) \wedge \theta(g') \le \theta(g)$, so, by part (d), $\theta^{-1}(\theta(g) \wedge \theta(g')) \le \theta^{-1}(\theta(g)) = g$. Similarly, $\theta^{-1}(\theta(g) \wedge \theta(g')) \le g'$. Thus $\theta^{-1}(\theta(g) \wedge \theta(g')) = 0$, so $\theta(g) \wedge \theta(g') = \theta(0) = 0$, by part (e).

(h) By part (f), $\theta(g) \vee \theta(g') = \theta(g \vee g') = \theta(1) = 1$.

(i) This follows from (g), (h), and Lemma 1 on page 392.

(j) Use part (i) and the De Morgan law twice, once in B_1 and once in B_2.

10.6 Answers

1. (a) $(0, 1, 0, 0)$. (b) $(0, 0, 1, 0)$. (c) $(1, 1, 1, 0)$.

2. $(x \vee z)y' = xy' \vee zy' = xy'z \vee xy'z' \vee xy'z \vee x'y'z$, so removing a duplicate gives $xy'z \vee xy'z' \vee x'y'z$.

3. (a) $(1, 0, 0, 0) \vee (0, 0, 0, 1)$. (b) $(1, 0, 0) \vee (0, 0, 1)$.

(c) $(0, 0, 1, 0, 0,) \vee (0, 0, 0, 1, 0)$. (d) $\{1\} \vee \{3\}$.

(e) $xy'z$ [it's already an atom].

4. (a)

$$\begin{aligned}(x \vee z)(y' \vee (xz)) &= xy' \vee zy'xxz \vee zxz \\ &= xy'z \vee xy'z' \vee xy'z \vee x'y'z \vee xyz \vee xy'z \\ &= xy'z \vee xy'z' \vee x'y'z \vee xyz,\end{aligned}$$

after eliminating duplicates.

(b)

	yz	yz'	$y'z'$	$y'z$
x	+		+	+
x'				+

5. $(0, 0, 0)$, $(0, 0, 1)$, $(0, 1, 0)$, $(0, 1, 1)$, $(1, 0, 0)$, $(1, 0, 1)$.

6. (a)

$$\begin{aligned}(xy \vee z')(y \vee z) &= xyy \vee xyz \vee yz' \vee zz' = xy \vee xyz \vee yz' \\ &= xyz \vee xyz' \vee xyz \vee xyz' \vee x'yz' \\ &= xyz \vee xyz' \vee x'yz'.\end{aligned}$$

(b)

	yz	yz'	$y'z'$	$y'z$
x	+	+		
x'		+		

(c) The optimal expression is $wy \vee x'y' \vee xz$ [not wx' or $y'z$].

7. (a)

	yz	yz'	$y'z'$	$y'z$
x			+	+
x'	+			+

(b) $xy' \vee x'z$.

8. (a) $xyzw'$ is one.

(b) $2^4 = 16$.

(c) Yes. *Every* finite Boolean algebra is isomorphic to some $\mathbb{B}^n$. [In this case $n = 16$.]

9. (a) $\varphi((0, 0)) = \emptyset$ and $\varphi((1, 1)) = \{a, b\}$.

(b) $\theta((0, 0)) = \emptyset$, $\theta((1, 0)) = \{b\}$, $\theta((0, 1)) = \{a\}$, and $\theta((1, 1)) = \{a, b\}$.

(c) 2.

10. A little algebra gives $f = ((x \vee y')' \vee x \vee z)' = (x'y \vee x \vee z)' = (y \vee x \vee z)' = x'y'z'$. One solution is shown, and you could also use two AND gates with negations on the inputs x, y, and z.

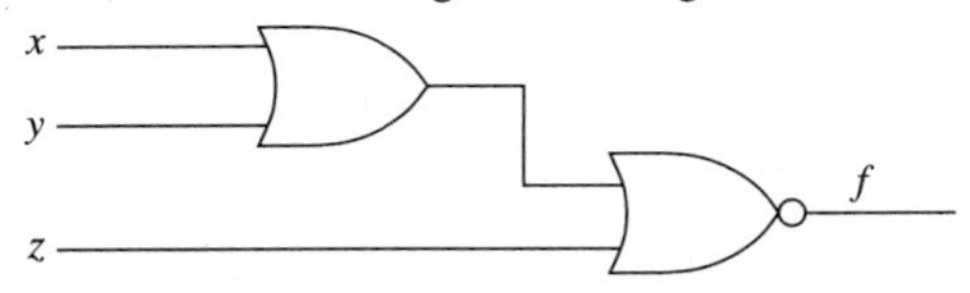

11. The optimal expression is $wy \vee w'z' \vee wx'z$. [Don't use the yz' block.]

12. (a) The best one can do is $wy' \vee wxz \vee wx'z' \vee x'y'z \vee xy'z'$, since each of the 2-square blocks must be used.
(b) $wxz \vee wx'z' \vee x'y'z \vee xy'z'$. The 4-square block isn't needed.
(c) This is another example where greed does not pay.

13. (a) $(y(x \vee z)) \vee xyz = xy \vee yz \vee xyz = xy \vee yz$, so one circuit is

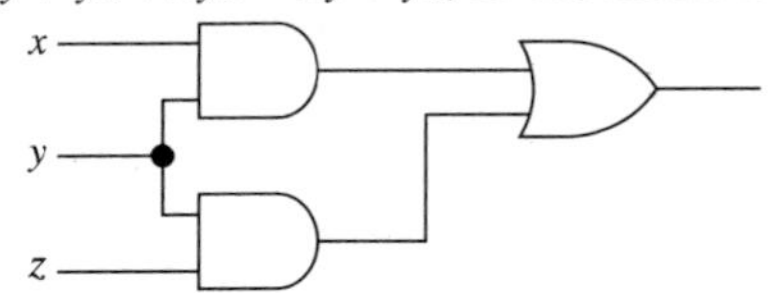

(b) Yes. Since $xy \vee yz = y(x \vee z)$, we can get this with an OR and an AND gate as shown.

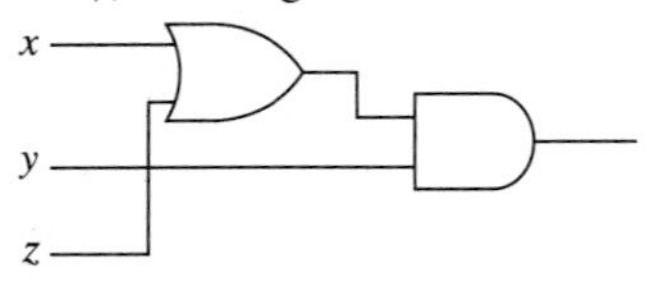

14. (a) Two examples are $\mathcal{P}(\{1, 2, 3, 4, 5\})$ and FUN$(\{1, 2, 3, 4, 5\}, \mathbb{B})$.
(b) $3! = 6$. [Watch the atoms.]

15. There are none; see the Theorem on page 419.

16. $\varphi(\emptyset) = (0, 0)$, $\varphi(\{a\}) = (1, 0)$, $\varphi(\{b\}) = (0, 1)$, $\varphi(\{a, b\}) = (1, 1)$, or interchange $(1, 0)$ and $(0, 1)$.

11.1 Answers

1. (a)

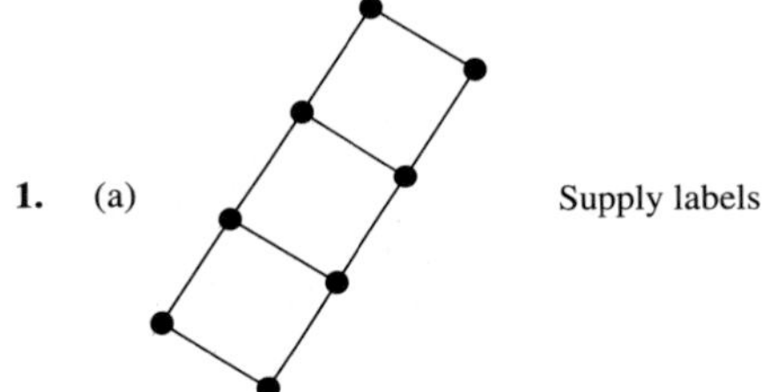

Supply labels

3. (a) h, o, p, q, r, z.
(c) B and C.
(e) f, z, p, does not exist.

5. (a) Yes. $a \vee b = \text{lub}(a, b) = \max\{a, b\}$, $a \wedge b = \text{glb}(a, b) = \min\{a, b\}$.
(c) 73.
(e) $\sqrt{73}$.

7. (a) Suppose that $\preceq$ is a partial order on S and that $\succeq$ is defined by $x \succeq y$ if and only if $y \preceq x$. Then $x \preceq x$, so $x \succeq x$. If $x \succeq y$ and $y \succeq x$, then $y \preceq x$ and $x \preceq y$, so $x = y$. If $x \succeq y$ and $y \succeq z$, then $y \preceq x$ and $z \preceq y$, so $z \preceq x$ and thus $x \succeq z$. Thus $\succeq$ satisfies (R), (AS), and (T).
(b) Clearly, $x \preceq x$, so (R) holds for $\preceq$. If $x \preceq y$ and $y \preceq z$, there are four possible cases:

$$x = y = z, \quad x = y \prec z, \quad x \prec y = z, \quad \text{and} \quad x \prec y \prec z.$$

Show that $x \prec z$ in each case, so (T) holds.

9. Not if Σ has more than one element. Show that antisymmetry fails.

11. (a) Use $x \vee y = y \vee x$ and $x \vee x = x$.

$\vee$	a	b	c	d	e	f
a	a	e	a	e	e	a
b	e	b	d	d	e	b
c	a	d	c	d	e	c
d	e	d	d	d	e	d
e	e	e	e	e	e	e
f	a	b	c	d	e	f

13. (a) No. Every finite subset of $\mathbb{N}$ is a subset of a larger finite subset of $\mathbb{N}$.
(c) $\text{lub}\{A, B\} = A \cup B$. Note that $A \cup B \in \mathcal{F}(\mathbb{N})$ for all $A, B \in \mathcal{F}(\mathbb{N})$.
(e) Yes; see parts (c) and (d).

15. (a) Only $\preceq$. $<$ is not reflexive and $\preceq$ is not antisymmetric.

(c) 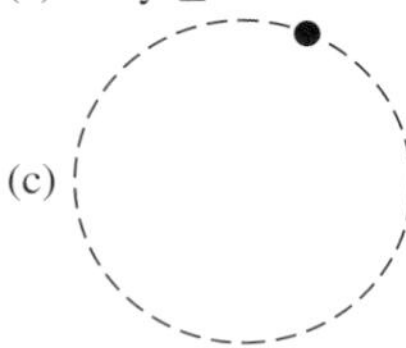

17. See Figures 2 and 7 or Exercise 16 for two different sorts of failure.

19. (a) Show that b satisfies the definition of lub$\{x, y, z\}$, i.e., $x \preceq b$, $y \preceq b$, $z \preceq b$, and if $x \preceq c$, $y \preceq c$, $z \preceq c$, then $b \preceq c$.

In detail: Since lub$\{x, y\} = a$, $x \preceq a$. Similarly $a \preceq b$, so $x \preceq b$ by (T). In the same way $y \preceq b$, and since lub$\{a, z\} = b$, we have $z \preceq b$.

Now suppose $x \preceq c$, $y \preceq c$, and $z \preceq c$. Then c is an upper bound for x and y, so $a = \text{lub}\{x, y\} \preceq c$. Then c is an upper bound for a and z, so $b = \text{lub}\{a, z\} \preceq c$. Thus $b \preceq c$ for every upper bound c of $\{x, y, z\}$, and we conclude that $b = \text{lub}\{x, y, z\}$.

(b) Show by induction on n that every n-element subset of a lattice has a least upper bound.

(c) Use part (a) and commutativity of $\vee$: $(x \vee y) \vee z = \text{lub}\{x, y, z\} = \text{lub}\{y, z, x\} = (y \vee z) \vee x$ [by part (a) again] $= x \vee (y \vee z)$.

11.2 Answers

1. (a) (1, 1), (1, 2), (2, 2), (2, 3), (2, 4), (3, 4), (4, 4) is one example.

3. (a) (0, 0), (0, 1), (0, 2), (1, 0), (1, 1), (1, 2), (2, 0), (2, 1), (2, 2).
(c) (3, 0), (3, 1), (3, 2), (4, 0), (4, 1), (4, 2).

5. (a) 000, 0010, 010, 10, 1000, 101, 11.

7. (a) No. If $w \in \Sigma^*$, then $w \preceq_{LL} wu$ for every $u \in \Sigma^*$.

9. (a) (c)

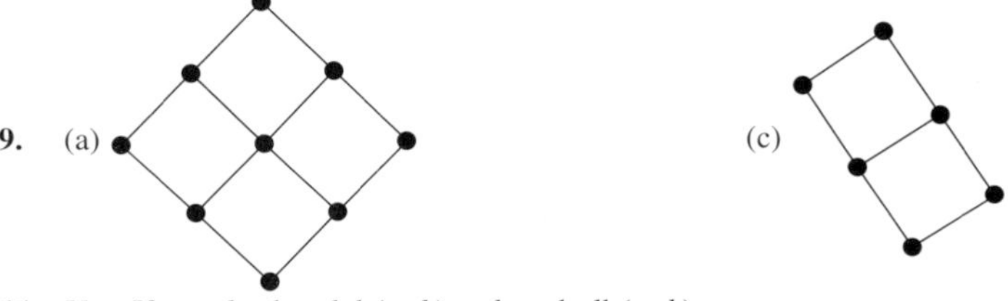

11. Yes. If $a \preceq b$, then lub$(a, b) = b$ and glb$(a, b) = a$.

13. (a) Transitivity, for example. If $f \preceq g$ and $g \preceq h$, then $f(t) \preceq g(t)$ and $g(t) \preceq h(t)$ in S, for all t in T. Since $\preceq$ is transitive on S, $f(t) \preceq h(t)$ for all t, so $f \preceq h$ in $\text{FUN}(T, S)$.
(c) $f(t) \preceq f(t) \vee g(t) = h(t)$ for all t, so $f \preceq h$. Similarly, $g \preceq h$, so h is an upper bound for $\{f, g\}$. Show that if $f \preceq k$ and $g \preceq k$, then $h \preceq k$, so h is the least upper bound for $\{f, g\}$.

15. (a) 501, 502, ..., 1000.
(c) Yes. Think of primes or see Exercise 17.

17. Consider a maximal chain $a_1 \prec a_2 \prec \cdots \prec a_n$ in S. There is no chain $b \prec a_1 \prec a_2 \prec \cdots \prec a_n$ in S, so there is no b with $b \prec a_1$. That is, a_1 is minimal. [Finiteness is essential. The chain $(\mathbb{Z}, \leq)$ is a maximal chain in itself.]

19. Antisymmetry is immediate. For transitivity consider cases. Suppose

$$(s_1, \ldots, s_n) \prec (t_1, \ldots, t_n) \text{ and } (t_1, \ldots, t_n) \prec (u_1, \ldots, u_n).$$

If $s_1 \prec_1 t_1$, then $s_1 \prec_1 t_1 \preceq_1 u_1$, so $(s_1, \ldots, s_n) \prec (u_1, \ldots, u_n)$. If $s_1 = t_1, \ldots, s_{r-1} = t_{r-1}$, $s_r \prec_r t_r$, if $t_1 = u_1, \ldots, t_{p-1} = u_{p-1}$, $t_p \prec_p u_p$, and if $r < p$, then $s_1 = u_1, \ldots, s_{r-1} = u_{r-1}$ and $s_r \prec_r t_r = u_r$; again $(s_1, \ldots, s_n) \prec (u_1, \ldots, u_n)$. The remaining cases are similar.

11.3 Answers

1. (a) $\begin{bmatrix} -8 & 13 \\ 2 & 9 \end{bmatrix}$. (c) $\begin{bmatrix} 31 & -16 & -6 \\ 29 & 4 & 26 \end{bmatrix}$. (e) $\begin{bmatrix} -1 & 7 \\ 8 & 0 \\ 16 & 0 \end{bmatrix}$.

3. The products written in parts (a) and (c) do not exist.

5. (a) $\begin{bmatrix} 1 & 10 \\ 11 & 19 \end{bmatrix}$.

7. (a) $\begin{bmatrix} 7 & 14 \\ 8 & 11 \\ 2 & -6 \end{bmatrix}$.

9. (a) $\begin{bmatrix} 0 & 0 & 1 \\ 1 & 0 & 0 \\ 0 & 1 & 0 \end{bmatrix}$. (c) $\begin{bmatrix} 1 & 0 & 0 \\ 0 & 1 & 0 \\ 0 & 0 & 1 \end{bmatrix}$.

11. (a) 2. (c) 2.

13. (a) $\mathbf{M}^3 = \begin{bmatrix} 3 & 20 & 2 & 3 \\ 0 & 8 & 0 & 0 \\ 2 & 9 & 1 & 2 \\ 0 & 4 & 0 & 0 \end{bmatrix}$.

(c) $f\,a\,b,\ f\,a\,c,\ f\,b\,d,\ f\,b\,e,\ f\,c\,d,\ f\,c\,e,\ f\,h\,j,\ k\,j\,d,\ k\,j\,e.$

15. (a) Simply remove the arrows from Figure 2 on page 440.

(c) $d\,d,\ e\,e,\ d\,e,\ e\,d,\ b\,b,\ c\,c,\ b\,c,\ c\,b,\ j\,j.$

17. (a) $\mathbf{I}^{-1} = \mathbf{I}$. (c) Not invertible.

19. For $1 \le k \le p$ and $1 \le i \le m$,

$$(\mathbf{B}^T\mathbf{A}^T)[k,i] = \sum_{j=1}^{n} \mathbf{B}^T[k,j]\mathbf{A}^T[j,i] = \sum_{j=1}^{n} \mathbf{B}[j,k]\mathbf{A}[i,j].$$

Compare with the (k,i)-entry of $(\mathbf{AB})^T$.

21. (a) In fact, $\mathbf{AB} = \mathbf{BA} = a\mathbf{B}$ for all $\mathbf{B}$ in $\mathfrak{M}_{2,2}$.

(b) $\mathbf{AB} = \mathbf{BA}$ with $\mathbf{B} = \begin{bmatrix} 1 & 0 \\ 0 & 0 \end{bmatrix}$ forces $\begin{bmatrix} a & 0 \\ c & 0 \end{bmatrix} = \begin{bmatrix} a & b \\ 0 & 0 \end{bmatrix}$, so $b = c = 0$. So $\mathbf{A} = \begin{bmatrix} a & 0 \\ 0 & d \end{bmatrix}$. Now try $\mathbf{B} = \begin{bmatrix} 0 & 1 \\ 0 & 0 \end{bmatrix}$.

23. (a) Consider $1 \le i \le m$ and $1 \le k \le p$, and compare the (i,k) entries of $(\mathbf{A}+\mathbf{B})\mathbf{C}$ and $\mathbf{AC}+\mathbf{BC}$.

(b) If $\mathbf{A}$ is $m \times n$, then $\mathbf{B}$ and $\mathbf{C}$ must both be $n \times r$ for the same r. For $1 \le i \le m$ and $1 \le k \le r$, show $(\mathbf{A}(\mathbf{B}+\mathbf{C}))[i,k] = (\mathbf{AB}+\mathbf{AC})[i,k]$.

11.4 Answers

1. (a) $\mathbf{A} * \mathbf{A} = \begin{bmatrix} 1 & 1 & 1 \\ 1 & 1 & 1 \\ 1 & 1 & 1 \end{bmatrix}$. Since $\mathbf{A} * \mathbf{A} \le \mathbf{A}$ is not true, R is not transitive.

(c) Not transitive. Note that $\mathbf{A} * \mathbf{A} = \begin{bmatrix} 1 & 0 & 0 \\ 0 & 1 & 0 \\ 0 & 0 & 1 \end{bmatrix}$.

3. (a) The matrix for R is $\mathbf{A} = \begin{bmatrix} 0 & 0 & 0 \\ 1 & 0 & 1 \\ 0 & 1 & 0 \end{bmatrix}$. The matrix for R^2 is $\mathbf{A} * \mathbf{A} = \begin{bmatrix} 0 & 0 & 0 \\ 0 & 1 & 0 \\ 1 & 0 & 1 \end{bmatrix}$.

(c) No; compare $\mathbf{A}$ and $\mathbf{A} * \mathbf{A}$ and note that $\mathbf{A} * \mathbf{A} \le \mathbf{A}$ fails.

5. (a) Matrix for R^0 is the identity matrix. Matrix for R^1 is $\mathbf{A}$, of course. Matrix for R^n is $\mathbf{A} * \mathbf{A}$ for $n \ge 2$, as should be checked by induction.

7. (a) $\mathbf{A}_f = \begin{bmatrix} 0 & 0 & 1 & 0 \\ 0 & 1 & 0 & 0 \\ 0 & 1 & 0 & 0 \\ 0 & 1 & 0 & 0 \end{bmatrix}$ and $\mathbf{A}_g = \begin{bmatrix} 0 & 0 & 0 & 1 \\ 0 & 0 & 1 & 0 \\ 0 & 1 & 0 & 0 \\ 1 & 0 & 0 & 0 \end{bmatrix}$.

(c) The Boolean matrix for $R_f^{\leftarrow}$ is $\begin{bmatrix} 0 & 0 & 0 & 0 \\ 0 & 1 & 1 & 1 \\ 1 & 0 & 0 & 0 \\ 0 & 0 & 0 & 0 \end{bmatrix}$. The Boolean matrix for $R_g^{\leftarrow}$ is $\begin{bmatrix} 0 & 0 & 0 & 1 \\ 0 & 0 & 1 & 0 \\ 0 & 1 & 0 & 0 \\ 1 & 0 & 0 & 0 \end{bmatrix}$. $R_g^{\leftarrow}$ is a function and $R_f^{\leftarrow}$ is not.

9. (a) R_1 satisfies (AR) and (S).

(c) R_3 satisfies (R), (AS), and (T).

(e) R_5 satisfies only (S).

11. (a) True. For each s, $(s,s) \in R_1 \cap R_2$, so $(s,s) \in R_1R_2$.

(c) False. Consider the equivalence relations R_1 and R_2 on $\{1,2,3\}$ with Boolean matrices

$$\mathbf{A}_1 = \begin{bmatrix} 1 & 1 & 0 \\ 1 & 1 & 0 \\ 0 & 0 & 1 \end{bmatrix} \quad \text{and} \quad \mathbf{A}_2 = \begin{bmatrix} 1 & 0 & 0 \\ 0 & 1 & 1 \\ 0 & 1 & 1 \end{bmatrix}.$$

Then $\mathbf{A}_1 * \mathbf{A}_2 = \begin{bmatrix} 1 & 1 & 1 \\ 1 & 1 & 1 \\ 0 & 1 & 1 \end{bmatrix}$. Since $(1,3) \in R_1R_2$ but $(3,1) \notin R_1R_2$, this relation is not symmetric.

13. Don't use Boolean matrices; the sets S, T, and U might be infinite.

(a) $R_1R_3 \cup R_1R_4 \subseteq R_1(R_3 \cup R_4)$ by Example 2(a). For the reverse inclusion, consider (s,u) in $R_1(R_3 \cup R_4)$ and show (s,u) is in R_1R_3 or R_1R_4.

(b) If $(s, u) \in (R_1 \cap R_2)R_3$, there is a t such that $(s, t) \in R_1 \cap R_2$ and $(t, u) \in R_3$. Since $(s, t) \in R_1$, we have $(s, u) \in R_1 R_3$, and since $(s, t) \in R_2$, we have $(s, u) \in R_2 R_3$. One example where equality fails is given by relations with Boolean matrices

$$\mathbf{A}_1 = \begin{bmatrix} 1 & 0 \\ 0 & 1 \end{bmatrix}, \quad \mathbf{A}_2 = \begin{bmatrix} 0 & 1 \\ 1 & 0 \end{bmatrix}, \quad \text{and} \quad \mathbf{A}_3 = \begin{bmatrix} 1 & 1 \\ 1 & 1 \end{bmatrix}.$$

(c) Show that $R_1(R_3 \cap R_4) \subseteq R_1 R_3 \cap R_1 R_4$. Equality need not hold. For example, consider R_1, R_3, R_4 with Boolean matrices

$$\mathbf{A}_1 = \begin{bmatrix} 1 & 1 \\ 0 & 0 \end{bmatrix}, \quad \mathbf{A}_3 = \begin{bmatrix} 0 & 0 \\ 0 & 1 \end{bmatrix}, \quad \mathbf{A}_4 = \begin{bmatrix} 0 & 1 \\ 0 & 0 \end{bmatrix}.$$

15. (R) R is reflexive if and only if $(s, s) \in R$ for every s, if and only if $\mathbf{A}[s, s] = 1$ for every s.

(AR) Similar to the argument for (R), with $(s, s) \notin R$ and $\mathbf{A}[s, s] = 0$.

(S) Follows from $\mathbf{A}^T[s, t] = \mathbf{A}[t, s]$ for every s, t.

(AS) R is antisymmetric if and only if $s = t$ whenever $(s, t) \in R$ and $(t, s) \in R$, i.e., whenever $\mathbf{A}[s, t] = \mathbf{A}^T[s, t] = 1$. Thus R is antisymmetric if and only if all the off-diagonal entries of $\mathbf{A} \wedge \mathbf{A}^T$ are 0.

(T) This follows from Theorem 3 and (a) of the summary.

17. Given $m \times n$, $n \times p$, and $p \times q$ Boolean matrices $\mathbf{A}_1$, $\mathbf{A}_2$, $\mathbf{A}_3$, they correspond to relations R_1, R_2, R_3, where R_1 is a relation from $\{1, 2, \ldots, m\}$ to $\{1, 2, \ldots, n\}$, etc. The matrices for $(R_1 R_2)R_3$ and $R_1(R_2 R_3)$ are $(\mathbf{A}_1 * \mathbf{A}_2) * \mathbf{A}_3$ and $\mathbf{A}_1 * (\mathbf{A}_2 * \mathbf{A}_3)$, by four applications of Theorem 1.

19. (a) To show that $R \cup E$ is a partial order, show

(R) $(s, s) \in R \cup E$ for all $s \in S$,

(AS) $(s, t) \in R \cup E$ and $(t, s) \in R \cup E$ imply $s = t$,

(T) $(s, t) \in R \cup E$ and $(t, u) \in R \cup E$ imply $(s, u) \in R \cup E$.

To verify (T), consider cases. The four cases for (T) are: $(s, t) \in R$ and $(t, u) \in R$; $(s, t) \in R$ and $(t, u) \in E$ [so $t = u$]; $(s, t) \in E$ and $(t, u) \in R$; and $(s, t) \in E$ and $(t, u) \in E$. The last two can be grouped together since, if $(s, t) \in E$ and $(t, u) \in R \cup E$, then $(s, u) = (t, u) \in R \cup E$.

11.5 Answers

1. (a) $\begin{bmatrix} 1 & 1 & 0 \\ 0 & 1 & 0 \\ 0 & 0 & 1 \end{bmatrix}$. (c) $\begin{bmatrix} 1 & 1 & 0 \\ 1 & 1 & 0 \\ 0 & 0 & 1 \end{bmatrix}$. (e) $\begin{bmatrix} 1 & 1 & 0 \\ 1 & 1 & 0 \\ 0 & 0 & 1 \end{bmatrix}$.

3. $\{1, 2\}$, $\{3\}$.

5. (a) $\begin{bmatrix} 1 & 1 & 0 & 0 & 0 \\ 0 & 1 & 0 & 1 & 0 \\ 0 & 0 & 1 & 0 & 1 \\ 0 & 1 & 0 & 1 & 0 \\ 0 & 0 & 0 & 0 & 1 \end{bmatrix}$. (c) $\begin{bmatrix} 1 & 1 & 0 & 0 & 0 \\ 1 & 1 & 0 & 1 & 0 \\ 0 & 0 & 1 & 0 & 1 \\ 0 & 1 & 0 & 1 & 0 \\ 0 & 0 & 1 & 0 & 1 \end{bmatrix}$. (e) $\begin{bmatrix} 1 & 1 & 0 & 1 & 0 \\ 1 & 1 & 0 & 1 & 0 \\ 0 & 0 & 1 & 0 & 1 \\ 1 & 1 & 0 & 1 & 0 \\ 0 & 0 & 1 & 0 & 1 \end{bmatrix}$.

7. (a) $r(R)$ is the usual order $\leq$.
(c) $rs(R)$ is the universal relation on $\mathbb{P}$.
(e) R is already transitive.

9. $(h_1, h_2) \in st(R)$ if $h_1 = h_2$ or if one of h_1, h_2 is the High Hermit. On the other hand, $ts(R)$ is the universal relation on F.O.H.H.

11. (a) Since $R \subseteq r(R)$, $t(R) \subseteq tr(R)$. Since $E \subseteq r(R)$, $E \subseteq tr(R)$. Thus $rt(R) = t(R) \cup E \subseteq tr(R)$. For the reverse containment $tr(R) \subseteq rt(R)$, it is enough to show that $r(R) \subseteq rt(R)$ and that $rt(R)$ is transitive, since then $rt(R)$ contains the transitive closure of $r(R)$. Now $r(R) = E \cup R \subseteq E \cup t(R) = rt(R)$, and $rt(R)$ is transitive by part (c) of the lemma to Theorem 3 on page 456. Thus $tr(R) \subseteq rt(R)$.

13. (a) By Exercise 12(a) and (b), $sr(R_1 \cup R_2) = sr(R_1) \cup sr(R_2) = R_1 \cup R_2$. Thus $tsr(R_1 \cup R_2) = t(R_1 \cup R_2)$. Apply Theorem 3.

15. Any relation that contains R will include the pair $(1, 1)$ and so will not be antireflexive.

17. (a) The intersection of all relations that contain R and have property p is the smallest such relation.
(c) (i) fails. $S \times S$ is not antireflexive.

11.6 Answers

1.

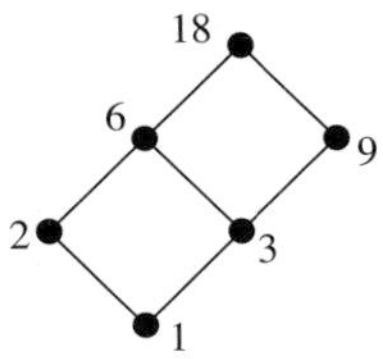

2. (a) Add vertices 8 and 12 to Figure 1 on page 427; then add edges from 4 to 8, from 4 to 12, and from 6 to 12.

(b) No. (c) Yes, 1. (d) 5, 8, and 12. (e) 1.

(f) {1, 5}, {1, 3, 6, 12}, {1, 2, 6, 12}, {1, 2, 4, 12}, and {1, 2, 4, 8}.

(g) No. For example, lub{8, 12} does not exist *in* the poset. Compare Example 9(b) on page 431.

3. (a) $0000 < 0001 < 0011 < 0111 < 1111$ or any other chain that adds one atom at a time.
(b) One example is $xy < x < x \vee y < 1$. [Since BOOL(2) is isomorphic to $\mathbb{B}^4$, its maximal chains have length 5. One could stick 0 on the front of this chain.]

4. (a) One possibility is a, aa, aab, $aaba$. There are infinitely many others.
(b) 00, 01, 10, 11 is the only choice.
(c) One example is (0, 0, 0), (0, 0, 1), (0, 1, 1), (1, 1, 1).

5. (a) $\mathbb{B} \times \mathbb{B}$ with product order is one example.
(b) How about $\mathbb{Z}$ with usual $\leq$?
(c) The matrix $\begin{bmatrix} 0 & 1 \\ 0 & 0 \end{bmatrix}$ for the $<$ relation on a 2-element set will do.
(d) One possible answer is this.

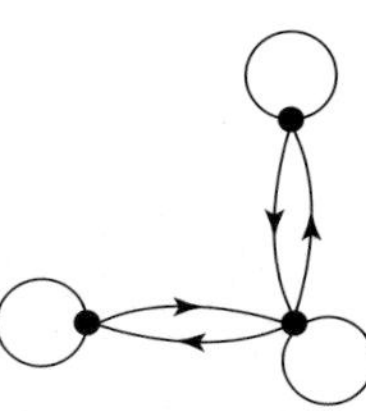

6. All but parts (e) and (g) are true. In connection with (e), any partially ordered set with more than one minimal element fails to have a smallest element. The set of negative integers, with the usual order, has a largest element, so (g) is false.

7. (a) 11, 001, 101, 00010 and 00010, 001, 101, 11.
(b) 001, 010, 011, 100, 101 [could leave off 001 and 101].
(c) 01, 010, 0100 [could just list 010].
(d) Infinitely many. [Every word that starts with 001 precedes 01.]

8. (a) $\begin{bmatrix} 1 & 0 & 0 & 1 \\ 0 & 0 & 0 & 1 \\ 1 & 0 & 0 & 1 \\ 0 & 0 & 0 & 1 \end{bmatrix}$.

(b) The relation is transitive, since $\mathbf{A} * \mathbf{A} \leq \mathbf{A}$.

9. (a) λ, *arc*, *bar*, *cab*, *car*, *bare*, *bear*, *beer*, *care*, *crab*, *brace*, *career*, *crabber*.
(b) λ, *arc*, *bar*, *bare*, *bear*, *beer*, *brace*, *cab*, *car*, *care*, *career*, *crab*, *crabber*.

10. (a)

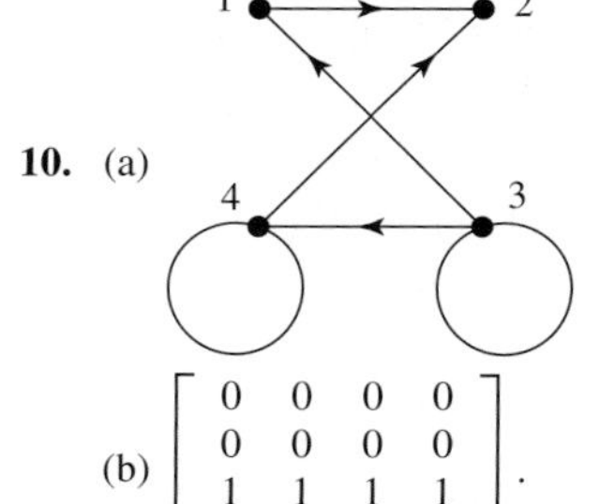

(b) $\begin{bmatrix} 0 & 0 & 0 & 0 \\ 0 & 0 & 0 & 0 \\ 1 & 1 & 1 & 1 \\ 0 & 1 & 0 & 1 \end{bmatrix}$. (c) $\begin{bmatrix} 0 & 1 & 1 & 0 \\ 1 & 0 & 0 & 1 \\ 1 & 0 & 1 & 1 \\ 0 & 1 & 1 & 1 \end{bmatrix}$.

(d) $\mathbf{A} \vee (\mathbf{A} * \mathbf{A}) = \begin{bmatrix} 0 & 1 & 0 & 0 \\ 0 & 0 & 0 & 0 \\ 1 & 1 & 1 & 1 \\ 0 & 1 & 0 & 1 \end{bmatrix}$.

(e) No. It's neither reflexive nor transitive.

11. (a) For instance, (1, 16) and (2, 15) are not related (i.e., comparable) to each other.
(b) The filing order does the job. Let $(s, t) \preceq (m, n)$ in case $s < m$ or $s = m$ and $t \leq n$.

12. [diagram: max, max, minimum] and [diagram: max, max, minimum] are two possibilities.

13. Yes. This follows from Theorem 1 on page 436, since $\{\mathbb{B}, \leq\}$ is a chain. Alternatively, we need to argue that if u and v are strings in $\mathbb{B}^{17}$, then either $u \leq v$ or $v \leq u$ in the filing order. If $u \neq v$,

then there's a first position in which u and v differ. If u has a 0 there, but v has a 1, then $u \preceq v$. If u has a 1, but v has a 0, then $v \preceq u$ in the filing order.

14. (a) a. (b) Yes, a. (c) g, h, f.
(d) No. (e) a, e, a, doesn't exist, c, g.

15. (a) Both equal A. The equality is an instance of a distributive law, which holds in some lattices.
(b) a and 1. Here the distributive law fails.
(c) A, A, x, and y.

16. Here are the answers in a table.

	R_1	R_2	R_3	R_4
R	Yes	Yes	Yes	No
S	No	Yes	Yes	Yes
AS	Yes	No	No	No
T	Yes	No	Yes	No
Partial order	Yes	No	No	No
Equivalence relation	No	No	Yes	No

17. (a) $\mathbf{A}_1 * \mathbf{A}_1 = \mathbf{A}_1$, $\mathbf{A}_1 * \mathbf{A}_2 = \begin{bmatrix} 1 & 0 & 1 \\ 1 & 0 & 0 \\ 1 & 0 & 1 \end{bmatrix}$ and $\mathbf{A}_1^T = \begin{bmatrix} 1 & 0 & 0 \\ 1 & 1 & 1 \\ 1 & 0 & 1 \end{bmatrix}$.
(b) It is reflexive and transitive, but not symmetric.
(c) It is not reflexive, symmetric, or transitive. For nontransitivity, look at the (2, 3) entry of $\mathbf{A}_2 * \mathbf{A}_2$.

18. (a) $251, 252, \ldots, 500$.
(b) Any four primes less than 500.
(c) $\{2, 4, 8, 16, 32, 64, 128, 256\}$, $\{257\}$.
(d) 216, 6, 15, not exist, 146.

19. (a) Yes; R is reflexive.
(b) No. $(6, 3) \in s(R)$, but $(6, 3) \notin R$.
(c) Yes; R is transitive.
(d) $\{2, 3, 4, 5, 6, 8, 9, 10, 12\}$. For example, $(12, 5)$ is in $tsr(R)$ because $(2, 12)$, $(2, 10)$, $(5, 10)$ are in R; hence $(12, 2)$, $(2, 10)$, $(10, 5)$ are in $sr(R)$, so $(12, 5)$ is in $tsr(R)$.

20. Yes. Restrictions of partial orders to subsets always make the subsets posets, because they're still reflexive, still antisymmetric, and still transitive. Every two members of Σ^* are comparable so, since Δ inherits the order from Σ^*, every two elements of Δ are also comparable.

21. (a) A one-to-one mapping φ of one lattice onto another lattice is a lattice isomorphism in case it satisfies the conditions $\varphi(x \vee y) = \varphi(x) \vee \varphi(y)$ and $\varphi(x \wedge y) = \varphi(x) \wedge \varphi(y)$ for all x, y. In fact, it is sufficient for φ to satisfy one of these conditions for all x, y. Alternatively, two lattices are isomorphic if they are isomorphic as posets. Two posets $(S_1, \preceq_1)$ and $(S_2, \preceq_2)$ are isomorphic if there is a one-to-one mapping φ of S_1 onto S_2 that preserves the order: $x, y \in S_1$ and $x \preceq_1 y$ imply $\varphi(x) \preceq_2 \varphi(y)$.
(b) D_{24} and D_{54} are isomorphic with L_1. The other two lattices are not isomorphic with any of the lattices in Exercise 15.

12.1 Answers

1. (a)

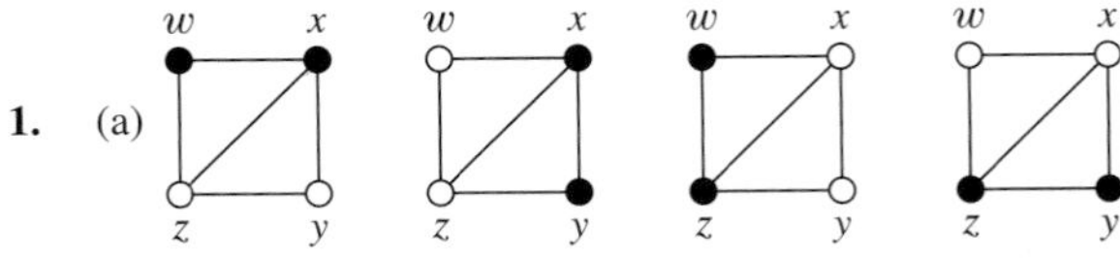

(c) Graph automorphisms can interchange w and y and they can interchange x and z.
(e) There are nine different colorings. Can you identify them?

3. (a) The group $\{e, r, r^2, r^3\}$ of rotations of the pyramid about the axis through its top vertex and the midpoint of its base.
(c) 4. (e) $4!/4 = 6$.

5. (a) 1. (c) 12.

7. (a) $\text{AUT}(F) = \{e, g\}$, where g interchanges z and w and fixes x and y.

9. (a) 2. What are they?

11. (c)

(e)

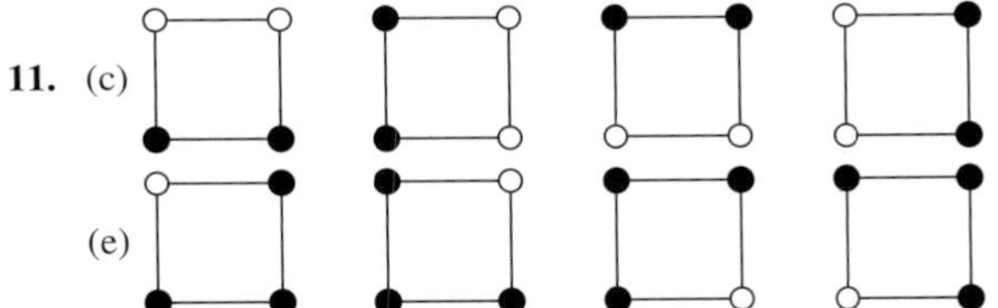

13. (a) Let g, f be in $\text{AUT}(D)$. Then (x, y) is an edge if and only if $(g(x), g(y))$ is an edge. Similarly, $(g(x), g(y))$ is an edge if and only if $(f(g(x)), f(g(y)))$ is an edge. Hence (x, y) is an edge if and only if $(f(g(x)), f(g(y)))$ is. So $f \circ g$ is in $\text{AUT}(D)$. Similar arguments show that $\text{AUT}(D)$ contains the identity and that g^{-1} is in $\text{AUT}(D)$ when g is.

15. (a) $G = \{e, g\}$, where e is the identity permutation of $V = \{u, v\}$ and $g(u) = v$, $g(v) = u$. Both e^* and g^* are the identity on the 1-element edge set E.

(b) Suppose not; i.e., suppose that $g^*(\{x, y\}) = h^*(\{x, y\})$ for every edge $\{x, y\}$ in E. Since $g \neq h$, there is a vertex z of H with $g(z) \neq h(z)$. Since H is connected and has at least two vertices, there is an edge $\{z, w\}$ in E with $z \neq w$. Since $\{g(z), g(w)\} = g^*(\{z, w\}) = h^*(\{z, w\}) = \{h(z), h(w)\}$ but $g(z) \neq h(z)$, it must be the case that $h(w) = g(z) \neq h(z) = g(w)$.

Now, if there were another edge $\{z, u\}$ attached to z with $u \neq w$, then the same argument would yield $g(z) = h(u)$, whence $h(u) = h(w)$, which is impossible. So z only has one edge attached to it, $\{z, w\}$, and, similarly, so does w. Since H is connected, this must be the only edge in H, contrary to the hypothesis.

12.2 Answers

1. (a) $\mathbb{Z}$. (c) $\mathbb{Z}$. (e) $\mathbb{Z}$.

3. (a) $3\mathbb{Z}$.

(c) Not a subgroup; 2 and 4 are in the subset, but $2 + 4$ is not.

5. $|\text{AUT}(D)| = 4$. $|\text{AUT}(D)(p)| = |\{p, r\}| = 2$, $|\text{FIX}(p)| = |\{e, f\}| = 2$, and $2 \cdot 2 = 4$; the cases for q, r, and s are similar.

7. (a) We know 1 is a generator. So is 2, since $2 +_5 2 = 4$, $2 +_5 2 +_5 2 = 1$, and $2 +_5 2 +_5 2 +_5 2 = 3$. Similarly, 3 and 4 are generators.

9. (a) $\varphi(n) = n + 3$ defines one.

11. (a) Observe that $(g_1 g_2)(g_2^{-1} g_1^{-1}) = g_1(g_2 g_2^{-1}) g_1^{-1} = g_1 e g_1^{-1} = g_1 g_1^{-1} = e$. Similarly, $(g_2^{-1} g_1^{-1})(g_1 g_2) = e$, or appeal to uniqueness of inverses.

(c) An easy induction shows that $(g_1 \cdot g_2 \cdots g_n)^{-1} = g_n^{-1} \cdots g_2^{-1} \cdot g_1^{-1}$. The inductive step uses $(g_1 \cdot g_2 \cdots g_n \cdot g_{n+1})^{-1} = g_{n+1}^{-1} \cdot (g_1 \cdot g_2 \cdots g_n)^{-1}$.

13. (a) Since $e \in G$ and $e(x) = x$, R contains all (x, x) and is reflexive. Since $g(x) = y \iff x = g^{-1}(y)$, $(x, y) \in R \iff (y, x) \in R$, so R is symmetric. For transitivity, note that the conditions $g(x) = y$ and $g'(y) = z$ imply that $g' \circ g(x) = z$.

15. (a) If g and h belong to the intersection, then they both belong to each subgroup in the collection, so their product gh does too. Since each subgroup contains the identity, so does the intersection. Every member of the intersection belongs to each subgroup, so its inverse does too. Thus its inverse also belongs to the intersection. Hence the intersection is closed under products and inverses and contains the identity.

(c) By part (a), $\bigcap\{H : H \text{ is a subgroup of } G \text{ and } A \subseteq H\}$ is a subgroup of G; call it R, say. Clearly, $A \subseteq R$. Since R is contained in every subgroup of G that contains A, it must be the unique smallest such subgroup.

Since R is a group that contains A, it also contains A^{-1} and all products that can be formed from elements of R, so $R \supseteq \langle A \rangle$. If we can show that $\langle A \rangle$ is a group, it will follow that $R \subseteq \langle A \rangle$ and hence that $R = \langle A \rangle$.

Clearly, products of members of $\langle A \rangle$ are still in $\langle A \rangle$. If $g = b_1 \cdots b_n \in \langle A \rangle$ with $b_1, \ldots, b_n$ in $A \cup A^{-1}$, then $g^{-1} = b_n^{-1} \cdots b_1^{-1}$ [see Exercise 11] with $b_n^{-1}, \ldots, b_1^{-1} \in A \cup A^{-1}$ too, so $g^{-1} \in \langle A \rangle$. Finally, $e = gg^{-1} \in \langle A \rangle$ for every g in $\langle A \rangle$, or we can view e as a "product" of 0 factors from $A \cup A^{-1}$.

17. Theorem 1 shows that each $|G(x_j)|$ must be a divisor of $|G| = 2^k$. So each $|G(x_j)|$ must be 1, 2, 4, ... or 2^k. Since $|X|$ is odd, for at least one j we must have $|G(x_j)| = 1$. By Theorem 1 again, $|G| = |\text{FIX}_G(x_j)|$, so $G = \text{FIX}_G(x_j)$. Hence $g(x_j) = x_j$ for all $g \in G$.

19. $\text{FIX}(\{w, y\}) = \{e, f\}$ since both $f(w)$ and $f(y)$ belong to $\{w, y\}$. However, $\text{FIX}(w) \cap \text{FIX}(y) = \{e\}$.

21. (a) For $h \in H$, $(g \cdot h)^{-1} = h^{-1} \cdot g^{-1} \in H \cdot g^{-1}$, so $\{f^{-1} : f \in g \cdot H\} \subseteq H \cdot g^{-1}$. Moreover, $h \cdot g^{-1} = (g \cdot h^{-1})^{-1}$ is in $\{f^{-1} : f \in g \cdot H\}$ for $h \in H$, so $H \cdot g^{-1} \subseteq \{f^{-1} : f \in g \cdot H\}$.

(b) $g \cdot H \to H \cdot g^{-1}$. Show this is one-to-one.

23. Given $g \in H$, we want to show that $g^{-1} \in H$. This will show $e \in H$, since $e = g \cdot g^{-1}$ for g in the nonempty set H. The elements $g, g^2, \ldots$ are all in the finite set H, so they cannot all be different. Thus there are integers $k, n \in \mathbb{P}$ with $k < n$ and $g^k = g^n$. Hence

$$e = g^k \cdot (g^{-1})^k = g^n \cdot (g^{-1})^k = g^{n-k}.$$

If $n - k = 1$, then $g = e = g^{-1} \in H$. Otherwise, $n - k \geq 2$, so $g^{-1} = g^{n-k-1} \in H$. [We have shown that $g^{-1} \in \langle g \rangle \subseteq H$.]

25. (a) $\langle a \rangle = \{e, a, b\}$.

(b) $\langle a \rangle \bullet c = \{e \bullet c, a \bullet c, b \bullet c\} = \{c, d, f\}$, and $c \bullet \langle a \rangle = \{c \bullet e, c \bullet a, c \bullet b\} = \{c, f, d\}$.

(c) $\langle c \rangle$, $\langle d \rangle$, $\langle f \rangle$.

(d) $|G|/|\langle d \rangle| = 6/2 = 3$. Also, see part (e).

(e) $\{e, d\}$, $\{a, c\}$, $\{b, f\}$.

27. $g \cdot H$ contains $g \cdot e = g$. So if $g \cdot H = H$, then $g \in H$. If $g \in H$, then H and $g \cdot H$ both contain g and so are not disjoint. By Proposition 2, $g \cdot H = H$ in this case.

29. (a) (R) Note that $g^{-1} \cdot g = e \in H$.
(S) Note that $g_1{}^{-1} \cdot g_2 = (g_2{}^{-1} \cdot g_1)^{-1}$.
(T) Note that $g_3{}^{-1} \cdot g_1 = (g_3{}^{-1} \cdot g_2) \cdot (g_2{}^{-1} \cdot g_1)$.

12.3 Answers

1. (a) From Example 1, the numbers are 6, 4, 4, 2, 0, 0, 0, 0, and their sum is 16.
(c) The sums agree; we have just calculated $|S|$ in the proof of Theorem 1 in two ways, using formulas (1) and (2).

3. $\text{FIX}(e) = \{w, x, y, z\}$, $\text{FIX}(g) = \{x, z\}$, $\text{FIX}(h) = \{w, y\}$, and $\text{FIX}(gh) = \emptyset$. So, by Theorem 1, there are $\frac{1}{4}(4+2+2+0)$ orbits under G. Indeed, they are $\{w, y\}$ and $\{x, z\}$.

5. For the orbit $\{w, y\}$, the automorphisms g and gh both move each element. For the orbit $\{x, z\}$, the automorphisms h and gh both move each element.

7. h takes each element of $\{w, x, u, v\}$ to itself, so h^* takes each 2-element subset of this set to itself. There are $\binom{4}{2} = 6$ such sets. The only other 2-element set mapped to itself is $\{y, z\}$, so $\text{FIX}_T(h^*)$ has 7 elements.

9. (a) 4.
(c) Apply Theorem 3, noting that $m(e) = 5$, $m(f) = 4$, and $|G| = 2$. The answer is $(k^5 + k^4)/2$.

11. (a) Here's the action.

	w	x	y	z
e	w	x	y	z
a	z	y	x	w

$m(e) = 4$, $m(a) = 2$, $|G| = 2$. Theorem 3 gives $C(k) = (k^4 + k^2)/2$.
(b) As in Example 10, the answer is $C(4) - 4 \cdot C(3) + 6 \cdot C(2) - 4 \cdot C(1) = 136 - 4 \cdot 45 + 6 \cdot 10 - 4 \cdot 1 = 12$. Alternatively, consider the 24 permutations of the labels, and note that two permutations give equivalent labels only if they are reverses of each other; e.g., the labels $wxyz$ and $zyxw$ are reverses of each other.

13. (a) Apply Theorem 3, noting that $m(e) = 4$, $m(r) = m(r^2) = 2$, $m(f) = m(g) = m(h) = 3$. The answer is $C(k) = (k^4 + 2k^2 + 3k^3)/6$.
(b) Apply Theorem 3 on page 482 to get $C(k) = (k^3 + 2k + 3k^2)/6$.

15. (a) $C(3) - 3 \cdot C(2) + 3 \cdot C(1) = 21 - 3 \cdot 6 + 3 \cdot 1 = 6$.
(b) 4.
(c) $C(5) - 5 \cdot C(4) + 10 \cdot C(3) - 10 \cdot C(2) + 5 \cdot C(1) = 120 - 5 \cdot 55 + 10 \cdot 21 - 10 \cdot 6 + 5 = 0$, of course.

17. (a) From Figure 12(b) we find: orbits under $\langle e \rangle$ are $\{w\}, \{x\}, \{y\}, \{z\}$; orbits under $\langle g \rangle$ are $\{w, y\}$, $\{x\}, \{z\}$; orbits under $\langle h \rangle$ are $\{w\}, \{x, z\}, \{y\}$; and orbits under $\langle gh \rangle$ are $\{w, y\}, \{x, z\}$. So $m(e) = 4$, $m(g) = m(h) = 3$, and $m(gh) = 2$.
(c) $C(2) = \frac{1}{4}(16 + 16 + 4) = 9$.

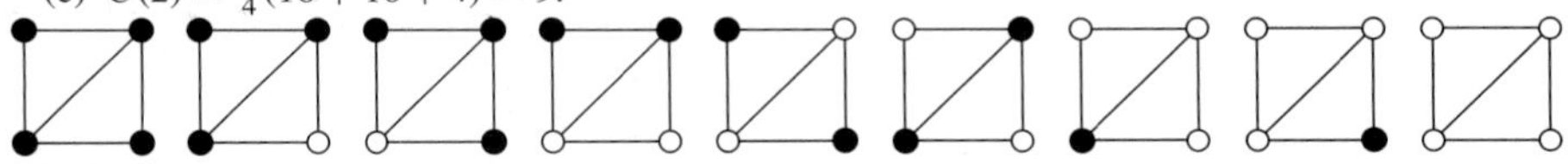

19. Following the suggestion:

Type	Number of that type	$m(g)$ when group acts on vertices	$m(g)$ when group acts on edges
a	6	4	7
b	6	2	3
c	3	4	6
d	8	4	4
e	1	8	12

(a) $C(k) = (k^8 + 17k^4 + 6k^2)/24$. For example, there are $6 + 3 + 8 = 17$ rotations with $m(g) = 4$.

21. (a) In this case, functions numbered n and $15 - n$ in Figure 10 are regarded as equivalent. So there are eight equivalence classes, which correspond to eight essentially different 2-input logical circuits.
(b) Now, as noted in the discussion of Figure 10, 2 and 4 are equivalent, as are 3 and 5. There are now six distinct classes. List them.

12.4 Answers

1. (a) and (c).

3. (a) Not an isomorphism, since it doesn't map $\mathbb{Z}$ onto $\mathbb{Z}$.
(c) Is an isomorphism, being a one-to-one and onto homomorphism.

5. (a) Associativity follows from associativity in $\mathbb{R}$: $[f + (g + h)](x) = f(x) + (g + h)(x) = f(x) + [g(x) + h(x)] = [f(x) + g(x)] + h(x) = (f + g)(x) + h(x) = [(f + g) + h](x)$ for all $x \in \mathbb{R}$. The zero function $\mathbf{0}$, where $\mathbf{0}(x) = 0$ for all $x \in \mathbb{R}$, is the additive identity for F. The additive inverses are just the negatives of the functions.

(b) Yes. Explain.
(c) $\varphi(f+g) = (f+g)(73) = f(73) + g(73) = \varphi(f) + \varphi(g)$.

7. (a) $\{0\}$.
(c) $5\mathbb{Z} = \{5n : n \in \mathbb{Z}\}$. (d) $\{0\}$.

9. (a) $(g,h) * (e,e_0) = (g \cdot e, h \bullet e_0) = (g,h) = (e \cdot g, e_0 \bullet h) = (e,e_0) * (g,h)$. $(g^{-1},h^{-1}) * (g,h) = (g^{-1} \cdot g, h^{-1} \bullet h) = (e,e_0)$, and $(e,e_0) = (g \cdot g^{-1}, h \bullet h^{-1}) = (g,h) * (g^{-1},h^{-1})$. [One of these is enough, by uniqueness of inverses.]
(c) The kernel is $\widetilde{G} = \{(g,e_0) : g \in G\}$.
(e) $\psi(h) = (e,h)$ for all $h \in H$.

11. (a) 4. (b) 4.

13. The pre-image of $\varphi(g)$ is gK, where K is the kernel of φ. By assumption, $|gK| = 1$. So $|K| = 1$, and φ is one-to-one as noted in the Corollary 2 to Theorem 1.

15. (a) Clearly, $eHe^{-1} = H$, and $(gh)H(gh)^{-1} = g(hHh^{-1})g^{-1}$. Also, if $H = gHg^{-1}$, then $g^{-1}Hg = g^{-1}(gHg^{-1})g = H$.
(c) $\{g \in G : gHg^{-1} = H\}$ is a subgroup [part (a)] of G containing A, and G is the smallest subgroup containing A. So $\{g \in G : gHg^{-1} = H\} = G$; i.e., $gHg^{-1} = H$ for all $g \in G$, so H is normal.

17. (b) $\begin{bmatrix} 0 & 1 \\ 1 & 0 \end{bmatrix} \cdot H = \{\begin{bmatrix} 0 & 1 \\ 1 & x \end{bmatrix} : x \in \mathbb{R}\}$, whereas $H \cdot \begin{bmatrix} 0 & 1 \\ 1 & 0 \end{bmatrix} = \{\begin{bmatrix} x & 1 \\ 1 & 0 \end{bmatrix} : x \in \mathbb{R}\}$.
(c) $\begin{bmatrix} y & z \\ 0 & 1/y \end{bmatrix} \cdot \begin{bmatrix} 1 & x \\ 0 & 1 \end{bmatrix} \cdot \begin{bmatrix} 1/y & -z \\ 0 & y \end{bmatrix} = \begin{bmatrix} 1 & xy^2 \\ 0 & 1 \end{bmatrix}$ is in H.
(e) Use the result of part (d) and the Fundamental Homomorphism Theorem.

12.5 Answers

1. (a) Yes. (c) No. Only 1 itself has an inverse.

3. (a) Yes. (c) Yes; compare Example 7(b).

5. (a) No.
(c) No; the zero matrix has no inverse, for example.

7. (a) *break*, *fast*, *fastfood*, *lunchbreak*, *foodfood*.
(c) *fastfast*, *foodfood*, *fastfastfoodbreakbreak*, λ.

9. (b) $(\mathbb{N}, \max)$ is a monoid because 0 is an identity. $(\mathbb{N}, \min)$ has no identity, so it is not a monoid. Check these claims.

11. (a) $\mathbb{P}$. (c) $\mathbb{Z}$. (e) $\mathbb{Z}$. (g) $18\mathbb{P}$.

13. $\mathbb{P} = \{1\}^+$, $\{0\} = \{0\}^+$, $18\mathbb{P} = \{18\}^+$.

15. (a) $2\mathbb{N}$. (c) $\{0\}$. (e) Σ^*.

17. (a) $\begin{bmatrix} 0 & 1 & 0 \\ 1 & 0 & 0 \\ 0 & 0 & 1 \end{bmatrix}$ and $\begin{bmatrix} 1 & 0 & 0 \\ 0 & 1 & 0 \\ 0 & 0 & 1 \end{bmatrix}$.
(b) $\begin{bmatrix} 1 & 0 & 0 \\ 0 & 1 & 0 \\ 0 & 0 & 1 \end{bmatrix}, \begin{bmatrix} 1 & 0 & 0 \\ 0 & 0 & 1 \\ 0 & 1 & 0 \end{bmatrix}, \begin{bmatrix} 0 & 1 & 0 \\ 1 & 0 & 0 \\ 0 & 0 & 1 \end{bmatrix}, \begin{bmatrix} 0 & 1 & 0 \\ 0 & 0 & 1 \\ 1 & 0 & 0 \end{bmatrix}, \begin{bmatrix} 0 & 0 & 1 \\ 1 & 0 & 0 \\ 0 & 1 & 0 \end{bmatrix}, \begin{bmatrix} 0 & 0 & 1 \\ 0 & 1 & 0 \\ 1 & 0 & 0 \end{bmatrix}$, i.e., the six "permutation matrices." This semigroup is isomorphic to the group S_3 of all permutations of a 3-element set.
(c) $\begin{bmatrix} 0 & 2 & 3 \\ 0 & 0 & 4 \\ 0 & 0 & 0 \end{bmatrix}, \begin{bmatrix} 0 & 0 & 8 \\ 0 & 0 & 0 \\ 0 & 0 & 0 \end{bmatrix}$, and $\begin{bmatrix} 0 & 0 & 0 \\ 0 & 0 & 0 \\ 0 & 0 & 0 \end{bmatrix}$.

19. (a) $6\mathbb{P} = \{6k : k \in \mathbb{P}\}$.
(c) No. For example, 6 and 12 are not both powers of the same member of $6\mathbb{P}$, so they cannot both lie in the same cyclic subgroup.

21. (a) $60\mathbb{P}$.
(c) 60 generates the additive semigroup $60\mathbb{P}$.

23. Only the function φ in (a) is a homomorphism. Though it is one-to-one, it does not map $\mathbb{P}$ onto $\mathbb{P}$, so it is not an isomorphism.

25. (a) If z' is also a zero, then $z' = z \bullet z' = z$.
(c) Each $(\{0, 1, \ldots, n\}, \cdot)$ is an example, where $k \cdot l = \min(k, l)$. Another example is $(\mathcal{P}(U), \cap)$, with U a finite nonempty set; the zero element is $\varnothing$.

27. (a) $\varphi(S)$ is closed under products, because $\varphi(s)\Box\varphi(s') = \varphi(s \bullet s') \in \varphi(S)$. And $\varphi(e)$ is an identity for $\varphi(S)$, because $\varphi(s)\Box\varphi(e) = \varphi(s \bullet e) = \varphi(s) = \varphi(e \bullet s) = \varphi(e)\Box\varphi(s)$ for all $\varphi(s)$ in $\varphi(S)$.

12.6 Answers

1. (a), (b), (d), (f).

3. (a) $\varphi(f+g) = (f+g)(0) = f(0)+g(0) = \varphi(f)+\varphi(g)$ and $\varphi(f \cdot g) = (f \cdot g)(0) = f(0) \cdot g(0) = \varphi(f) \cdot \varphi(g)$.

(c) For $r, s \in \mathbb{R}$, $(\varphi(r+s))(x) = r + s = (\varphi(r))(x) + (\varphi(s))(x)$ for all $x \in \mathbb{R}$. Hence $\varphi(r+s) = \varphi(r) + \varphi(s)$. Likewise, $\varphi(r \cdot s) = \varphi(r) \cdot \varphi(s)$.

(e) $\varphi((n+3\mathbb{Z}) \cdot (m+3\mathbb{Z})) = \varphi(nm+3\mathbb{Z})$ [definition of product in $\mathbb{Z}/3\mathbb{Z}$]
$= 2nm + 6\mathbb{Z}$ [definition of φ]

but

$$\varphi(n+3\mathbb{Z}) \cdot \varphi(m+3\mathbb{Z}) = (2n+6\mathbb{Z}) \cdot (2m+6\mathbb{Z}) \quad \text{[definition of } \varphi]$$

$$= 4nm + 6\mathbb{Z} \quad \text{[definition of product in } \mathbb{Z}/6\mathbb{Z}].$$

Consequently, φ is not a ring homomorphism.

5. (a) Evaluate both sides of the equation at a.
(c) By (b), $p(x) = \sum_{k=0}^{n} c_k x^k = \sum_{k=0}^{n} c_k \{q_k(x) \cdot (x-a) + a^k\} = (\sum_{k=0}^{n} c_k q_k(x)) \cdot (x-a) + \sum_{k=0}^{n} c_k a^k$. Let $q(x) = \sum_{k=0}^{n} c_k q_k(x)$.
(d) $p(x)$ is in the kernel if and only if $p(a) = 0$. Use part (c).

7. (a) $24\mathbb{Z}$.
(c) $3\mathbb{Z} + 2\mathbb{Z} = 1\mathbb{Z} = \mathbb{Z}$, since $1 = 3 \cdot 1 + 2 \cdot (-1) \in 3\mathbb{Z} + 2\mathbb{Z}$.

9. (a) Verify well-definedness directly, or apply Theorem 1 to the homomorphism $m \to (m \text{ MOD } 4, m \text{ MOD } 6)$ from $\mathbb{Z}$ to $\mathbb{Z}(4) \times \mathbb{Z}(6)$, as in Example 10. As noted in Exercise14, it suffices to check that $m \to m \text{ MOD } 4$ and $m \to m \text{ MOD } 6$ are homomorphisms on $\mathbb{Z}(12)$.
(c) (1,4) is one of the twelve; find another one.

11. (a) Since $2 *_6 3 = 0$, 2 has no inverse.
(b) Exhibit an inverse for each non-0 element. The inverses for non-0 elements in $\mathbb{Z}(5)$ can be read off of Figure 3 on page 123. [See Exercise 18 for the general argument.]
(c) $F \times K$ isn't even an integral domain since $(1,0) \cdot (0,1) = (0,0)$.

13. Use any $\mathbf{A}, \mathbf{B}$ such that $\mathbf{AB} \neq \mathbf{BA}$; e.g., $\mathbf{A} = \begin{bmatrix} 0 & 1 \\ 0 & 0 \end{bmatrix}$ and $\mathbf{B} = \begin{bmatrix} 0 & 0 \\ 1 & 0 \end{bmatrix}$.

15. (a) The kernel of φ is either F or $\{0\}$ by Example 8(b).
(b) Since I is a subgroup of $(R, +)$, $\theta(I)$ is a subgroup of $(\theta(R), +) = (S, +)$. If $\theta(r) \in \theta(R)$ and $\theta(a) \in \theta(I)$, then $\theta(a) \cdot \theta(r) = \theta(a \cdot r) \in \theta(I)$, since $a \cdot r$ is in the ideal I. Similarly, $\theta(r) \cdot \theta(a) \in \theta(I)$, so $\theta(I)$ is closed under multiplication by elements of $\theta(R) = S$.
(c) Use part (b) and Example 8(b).

17. (a) $I = 15\mathbb{Z}$.
(b) See Example 10. Let $\varphi(m) = (m \text{ MOD } 3, m \text{ MOD } 4)$ for $m \in \mathbb{Z}(12)$.

19. (a) $R \cdot 2 = \{a_0 + a_1 x + \cdots + a_n x^n \in R : \text{ every } a_i \text{ is even}\}$,
$R \cdot x = \{b_0 + b_1 x + \cdots + b_m x^m \in R : b_0 = 0\}$,
$R \cdot 2 + R \cdot x = \{c_0 + c_1 x + \cdots + c_r x^r \in R : c_0 \text{ is even}\}$.
(b) Suppose $R \cdot p = R \cdot 2 + R \cdot x$ for some $p \in R$. Since $2 \in R \cdot p$, p must be constant, and since $x \in R \cdot p$, p is 1 or -1. But then $R \cdot p = R$, a contradiction.

12.7 Answers

1. (a) $\{u, w\}$ and $\{v, x\}$. (b) f. (c) e and g.

(d)

	e	f	g	h
e	e	f	g	h
f	f	e	h	g
g	g	h	e	f
h	h	g	f	e

(e) Since $m(e) = 4$, $m(f) = m(g) = 3$, and $m(h) = 2$, the answer is $\frac{1}{4}[3^4 + 2 \cdot 3^3 + 3^2] = 36$.
(f) Since $m(e) = 5$ and $m(f) = m(g) = m(h) = 3$, the answer is $\frac{1}{4}[3^5 + 3 \cdot 3^3] = 81$. The values for m are different from those in part (e). Do you see why?

2. (a) e, v. (b) 3, 4. (c) 1, 2, 5, 6.
(d) 4. They are $\{1,5\}, \{2,6\}, \{3\}, \{4\}$.
(e) 24. The numbers of orbits of members of G are $m(e) = 6$, $m(r) = 3 = m(h)$, and $m(v) = 4$. Thus the number of colorings is

$$\frac{1}{4}[2^6 + 2^3 + 2^3 + 2^4] = \frac{1}{4}[64 + 8 + 8 + 16] = 24.$$

3. (a) $|\text{FIX}_G(5)| = |G|/|G(5)|) = 12$; see Theorem 1 on page 472.
(b) 6. This follows from Theorem 1 on page 477.

4. (a) $\{\lambda, a, b, c\}$.
(b) No. For example, $ab \neq ba$.
(c) Yes. λ is the identity.
(d) cab, $cabcab$, $cabcabcab$, for examples. λ is not an example.

5. (a) It's a subsemigroup if and only if the concatenation of every two words in L is a word in L. Associativity is inherited from Σ^*. It's a monoid if it contains λ as well.
(b) No. For example, L_1 contains ab but not $abab$.

(c) No. Again, $abab \notin L_2$, since $\lambda \notin L_2$.
(d) Yes. In fact, L_3 is the set of all words of even length, as one can show by induction on the length of a word in L_3. The identity is λ.

6. (a) See the answer to Exercise 9 on page 125.
(b) $(\mathbb{Z}(4), +_4)$ is a commutative group. The identity is 0 and inverses can be read from the table.
(c) $(\mathbb{Z}(4), *_4)$ is a commutative monoid, but not a group. The identity is 1, but 0 has no (multiplicative) inverse.
(d) $(\mathbb{Z}(4) \setminus \{0\}, *_4)$ is not even a semigroup, because it is not closed under multiplication: 2 is in $\mathbb{Z}(4) \setminus \{0\}$, but $2 *_4 2 = 0$ is not.

7. (a) j, m, n, o, p.
(b) The only one is $\langle i \rangle = \langle k \rangle = \{e, i, j, k\}$.
(c) The two such subgroups are $\{e, j, m, n\}$ and $\{e, j, o, p\}$.
(d) There are none, since 3 does not divide 8; see Lagrange's theorem on page 472.

8. (a) $\{\alpha, \varepsilon\}$, $\{\beta\}$, $\{\gamma\}$, $\{\delta\}$.
(b) 162. This number can be seen by inspection as $3 \cdot 3 \cdot 3 \cdot [3 + \binom{3}{2}]$ or by an application of Theorem 3 on page 482. The automorphism e has 5 orbits on $E(H)$ and g has 4, so the number of colorings is $\frac{1}{2}[3^5 + 3^4] = 162$.

9. (a) $\frac{1}{8}[10^4 + 2 \cdot 10^3 + 3 \cdot 10^2 + 2 \cdot 10] = 12{,}320$. See Example 8 on page 482.
(b) $\binom{10}{2} + 10 = 55$. There are 10 solid-color colorings and $\binom{10}{2}$ different 2-color colorings, since the two colorings with a given pair of colors are equivalent.
(c) The total number of such colorings, including ones that are equivalent, is $(10 \cdot 9)^2$; there are 10 ways to color each of the top two vertices, then 9 ways to color each of their opposites. Since e fixes all of them, other rotations fix none, horizontal and vertical flips fix $10 \cdot 9$ each, and diagonal flips fix none, the answer is

$$\frac{1}{8}[(10 \cdot 9)^2 + 2 \cdot 10 \cdot 9] = 1035.$$

10. (a) The unchanged ones are the ones for which the coloring is constant on the 2 orbits of R_2. There are 3^2 of them.
(b) Just use different colors on the orbits. Or easier yet, take two different solid-color colorings.
(c) The orbits of R_2 are the two sets consisting of alternate beads.
(d) The group acting on the necklace consists of the identity e, five rotations R_1, R_2, R_3, R_4, R_5, and six flips, three of which fix 2 beads and three of which fix no beads. The numbers of orbits for these operations are 6, 1, 2, 3, 2, 1, 4, 4, 4, 3, 3, 3, so the number of colorings is

$$\frac{1}{12}[3^6 + 3 \cdot 3^4 + 4 \cdot 3^3 + 2 \cdot 3^2 + 2 \cdot 3] = 92.$$

See Theorem 3 on page 482.

11. (a) $\langle 0 \rangle$, $\langle 1 \rangle = \langle 5 \rangle = \langle 7 \rangle = \langle 11 \rangle = \mathbb{Z}(12)$, $\langle 2 \rangle = \langle 10 \rangle$, $\langle 3 \rangle = \langle 9 \rangle$, $\langle 4 \rangle = \langle 8 \rangle$, and $\langle 6 \rangle$.
(b) No. All subgroups of $\mathbb{Z}$ have the form $n\mathbb{Z}$, and all of these are infinite except for $0\mathbb{Z} = \{0\}$.
(c) No. The set is not closed under the operation. For example, $3 *_{12} 4 = 0$.

12. (a) $\langle e \rangle = \{e\}$, $\langle g^2 \rangle = \{e, g^2, g^4\}$, $\langle g^3 \rangle = \{e, g^3\}$, and $G = \langle g \rangle = \{e, g, g^2, g^3, g^4, g^5\}$.
(b) They all are. The group is commutative.
(c) We need to check that $\varphi(xy) = \varphi(x) \cdot \varphi(y)$, i.e., that $((xy)^3, (xy)^2) = (x^3, x^2) \cdot (y^3, y^2)$. This is true, since $(xy)^3 = x^3y^3$ and $(xy)^2 = x^2y^2$ in G, and multiplication in $G \times G$ is coordinatewise. The kernel is just $\{e\}$, so the homomorphism is one-to-one.
(d) By the answer to part (c), φ is one-to-one, so $|\varphi(G)| = |G| = 6 = |\langle g^3 \rangle \times \langle g^2 \rangle|$. Since also $\varphi(G) \subseteq \langle g^3 \rangle \times \langle g^2 \rangle$, the sets $\varphi(G)$ and $\langle g^3 \rangle \times \langle g^2 \rangle$ are equal. Hence, φ is a one-to-one homomorphism of G *onto* $\langle g^3 \rangle \times \langle g^2 \rangle$, i.e., an isomorphism.

13. (a) If $x, y \in \varphi(U)$, then $x = \varphi(u)$ and $y = \varphi(v)$ for some $u, v \in U$. Since U is a subsemigroup of S, we have $u \bullet v \in S$, and thus $x * y = \varphi(u) * \varphi(v) = \varphi(u \bullet v)$ is in $\varphi(S)$. Associativity is inherited from T.
(b) Yes. If $x, y \in \varphi^{\leftarrow}(V)$, then $\varphi(x), \varphi(y) \in V$, so $\varphi(x \bullet y) = \varphi(x) * \varphi(y) \in V$ and $x \bullet y \in \varphi^{\leftarrow}(V)$. Associativity is inherited from S.

14. More generally, if $A \subseteq S$, then

$$\begin{aligned}
\varphi(A^+) &= \varphi(\{s : s \text{ is a product } a_1 \cdots a_n \text{ of members of } A\}) \\
&= \{\varphi(s) : s = a_1 \cdots a_n \text{ and } a_1, \ldots, a_n \in A\} \\
&= \{t : t = \varphi(a_1) \cdots \varphi(a_n) \text{ and } a_1, \ldots, a_n \in A\} \\
&= \varphi(A)^+.
\end{aligned}$$

15. (a) Yes. The identity element is the equality relation $\{(s, s) : s \in S\}$.
(b) $\mathcal{P}(S \times S)$ is commutative if and only if $|S| = 1$. If $x, y \in S$ with $x \neq y$, let $R_1 = \{(x, y)\}$ and $R_2 = \{(y, x)\}$. Then $R_1R_2 = \{(s, t) : (s, u) \in R_1 \text{ and } (u, t) \in R_2 \text{ for some } u \in S\} = \{(x, x)\} \neq \{(y, y)\} = R_2R_1$.

16. (a) More precisely, $\psi \circ \varphi$ is a homomorphism of $(S, \bullet)$ to $(U, \triangle)$, since $(\psi \circ \varphi)(s \bullet s') = \psi(\varphi(s \bullet s')) = \psi(\varphi(s) \Box \varphi(s')) = \psi(\varphi(s)) \triangle \psi(\varphi(s')) = (\psi \circ \varphi)(s) \triangle (\psi \circ \varphi)(s')$.
(b) For $t, t' \in T$, $t \Box t' = \varphi(\varphi^{-1}(t)) \Box \varphi(\varphi^{-1}(t')) = \varphi(\varphi^{-1}(t) \bullet \varphi^{-1}(t'))$, so $\varphi^{-1}(t \Box t') = \varphi^{-1}(t) \bullet \varphi^{-1}(t')$.
(c) $S \simeq S$, since the identity map 1_S is an isomorphism. $S \simeq T$ implies $T \simeq S$ by part (b). And $S \simeq T \simeq U$ implies $S \simeq U$ by part (a).

17. Ann is wrong, and Bob is wrong about the homomorphism. To see an example, let $M = \{0, 1\}$ with usual multiplication, and let $S = \{0\}$. The homomorphism $\varphi \colon S \to M$ given by $\varphi(0) = 0$ shows that Bob is wrong, too.

18. (a) By definition, $x_1 \equiv b_1 \pmod{m_1}$. Assume inductively that $x_{k-1} \equiv b_j \pmod{m_j}$ for $j = 1, \ldots, k-1$. Then

$$x_k = x_{k-1} + s_k \cdot (m_1 \cdots m_{k-1})(b_k - x_{k-1}) \equiv x_{k-1} \pmod{m_1 \cdots m_{k-1}},$$

so

$$x_k \equiv x_{k-1} \equiv b_j \pmod{m_j} \text{ for } j = 1, \ldots, k-1.$$

Moreover,

$$x_k \equiv x_{k-1} + 1 \cdot (b_k - x_{k-1}) \equiv b_k \pmod{m_k}.$$

The claim follows by induction.
(b) $s_2 = 2$ works, giving $x_2 = 72$.
(c) $s_2 = 14$ works, giving $x_2 = -302 \equiv 72 \pmod{11 \cdot 17}$.
(d) The change would keep x_i in the range $0 \le x_i < m_1 \cdots m_i$, in contrast to what happened in part (c), and would perhaps speed up the arithmetical calculations. The output would still be correct, since the proof in part (a) would still be valid.
(e) As in part (c), $x_1 = 6$ and $s_2 = 14$, but now $x_2 = 72$. The condition $s_3 \cdot 11 \cdot 17 \equiv 1 \pmod{73}$ is equivalent to $s_3 \cdot 41 \equiv 1 \pmod{73}$ and is satisfied by $s_3 = 57$. Thus

$$x_3 = 72 + s_3 \cdot 11 \cdot 17 \cdot (-1 - 72) \text{ MOD } (11 \cdot 17 \cdot 73) = 72.$$

[This time it didn't matter what s_3 was.]

19. (a) By definition, $p_1(a_1) = b_1$. Assume inductively that $p_{k-1}(a_j) = b_j$ for $j = 1, \ldots, k-1$. Then, for $j = 1, \ldots, k-1$,

$$\begin{aligned} p_k(a_j) &= p_{k-1}(a_j) + s_k \cdot (a_j - a_1) \cdots (a_j - a_{k-1})(b_k - p_{k-1}(a_j)) \\ &= p_{k-1}(a_j) + 0 = b_j \quad [\text{since } a_j - a_j = 0], \end{aligned}$$

and

$$\begin{aligned} p_k(a_k) &= p_{k-1}(a_k) + s_k(a_k - a_1) \cdots (a_k - a_{k-1})(b_k - p_{k-1}(a_k)) \\ &= p_{k-1}(a_k) + 1 \cdot (b_k - p_{k-1}(a_k)) = b_k. \end{aligned}$$

Since $p_1(x)$ is constant, it has degree at most 1, and if $p_k(x)$ has degree at most $k-1$, then adding to it a product of k linear factors to get $p_{k+1}(x)$ yields a polynomial of degree at most k. The claim follows by induction.
(b) We have $p_1(x) = 2$, $s_2 = 1$, $p_2(x) = 2 - 2x$, $s_3 = 1/6$, and finally $p_3(x) = 2 - 4x + 2x^2$.
(c) Yes. Consider interpolating $p(1) = p(2) = 3$ or $p(1) = 1$, $p(2) = 2$, $p(3) = 3$, say.

13.1 Answers

1. "Some person x is a mother of everybody," which is false.

3. (a) 0. Consider m odd. (c) 1. (e) 0. Consider $m = n = 0$.

5. (a) $\forall x \, \forall y \, \forall z[((x < y) \wedge (y < z)) \to (x < z)]$; universes $\mathbb{R}$.
(c) $\forall m \, \forall n \, \exists p[(m < p) \wedge (p < n)]$; universes $\mathbb{N}$.
(e) $\forall n \, \exists m[m < n]$; universes $\mathbb{N}$.

7. (a) $\forall w_1 \, \forall w_2 \, \forall w_3[(w_1 w_2 = w_1 w_3) \to (w_2 = w_3)]$.
(c) $\forall w_1 \, \forall w_2[w_1 w_2 = w_2 w_1]$.

9. (a) x, z are bound; y is free. (c) Same answers as part (a).

11. (a) x, y are free; there are no bound variables.
(c) $\exists x \, \exists y[(x - y)^2 = x^2 - y^2]$ is true.

13. (a) No. $\exists m[m + 1 = n]$ is false for $n = 0$.

15. (a) $\exists ! x \, \forall y[x + y = y]$.
(c) $\exists ! A \, \forall B[A \subseteq B]$. Here A and B vary over the universe of discourse $\mathcal{P}(\mathbb{N})$. Note that $\forall B[A \subseteq B]$ is true if and only if $A = \emptyset$.
(e) "$f \colon A \to B$ is a one-to-one function" $\to \forall b \, \exists ! a[f(a) = b]$. Here a ranges over A and b ranges over B. One way to make this clear is to write $\forall b \in B \; \exists ! a \in A[f(a) = b]$.

17. (a) True.
(c) False; e.g., 3 is in the right-hand set.
(e) False; the right-hand set is empty.
(g) True. (i) True. (k) True. (m) True.

19. (a) 0. One of $m, m+2$, and $m+4$ is always a multiple of 3.
(c) 1. (e) 0. Consider 3 + 5.

13.2 Answers

1. (a) Every club member has been a passenger on every airline if and only if every airline has had every club member as a passenger.
(c) If there is a club member who has been a passenger on every airline, then every airline has had a club member as a passenger.

3. Rule 37b says that "There does not exist a yellow car" is logically equivalent to "Every car is not yellow." In fact, both are false. Rule 37c says that "Every car is yellow" is logically equivalent to "There does not exist a nonyellow car." Both are false. Rule 37d says that "There exists a yellow car" is logically equivalent to "Not every car is nonyellow." Both are true.

5. (a) Rule 37d becomes $p(a) \vee p(b) \Longleftrightarrow \neg(\neg p(a) \wedge \neg p(b))$. This is rule 8c of Table 1 on page 62.

7. $\exists n[\neg\{p(n) \to p(n+1)\}]$ or $\exists n[p(n) \wedge \neg p(n+1)]$.

9. (a) $\exists x\, \exists y[(x < y) \wedge \forall z\{(z \leq x) \vee (y \leq z)\}]$.
(c) 0; for example, $[x < y \to \exists z\{x < z < y\}]$ is false for $x = 3$ and $y = 4$.

11. $\exists N\, \forall n[p(n) \to (n < N)]$.

13. One can let $q(x, y)$ be the predicate "$x = y$." Another way to handle $\exists x\, p(x, x)$ is to let $r(x)$ be the 1-place predicate $p(x, x)$. Then $\exists x\, r(x)$ is a compound predicate.

13.3 Answers

1. (a) True (c) False.
(e) True. Compare Exercise 1 on page 44.

3. (a) A function of the form $f(x) = ax + b$ will work if you choose a and b so that $f(0) = -1$ and $f(1) = 1$. Sketch your answer to see that it works. [For example, $f(x) = 2x - 1$ works.]
(b) Use g, where $g(x) = 1 - x$.
(c) Modify the suggestion for part (a). For example, $f(x) = 13x - 5$ works.
(d) Use $x \to 1/x$.
(e) Map $(1, \infty)$ onto $(0, \infty)$ using $h(x) = x - 1$ and compose with your answer from part (d) to obtain $h(1/x) = (1/x) - 1$.
(f) $f(x) = 2^x$, say. Sketch f to see that it works.

5. (a) Use a graphing calculator or the data

x	0.1	0.2	0.3	0.4	0.5	0.6	0.7	0.8	0.9
$f(x)$	−8.89	−3.75	−1.90	−0.83	0	0.83	1.90	3.75	8.89

7. Only the sets in (b) and (c) are countably infinite.

9. (a) We may assume that S is infinite. Let $f: S \to T$ be a one-to-one correspondence, where T is a countable set. There is a one-to-one correspondence $g: T \to \mathbb{P}$ since T is countable. Then $g \circ f$ is a one-to-one correspondence of S onto $\mathbb{P}$.

11. (a) Apply part (b) of the theorem to $S \times T = \bigcup_{t \in T}(S \times \{t\})$. Each $S \times \{t\}$ is countable, since it is in one-to-one correspondence with the countable set S.
(b) For each $t \in T$, let $g(t)$ be an element in S such that $f(g(t)) = t$. Show that g is one-to-one and apply part (a) of the theorem.
(c) By part (a), $\mathbb{Z} \times \mathbb{P}$ is countable. Since f maps $\mathbb{Z} \times \mathbb{P}$ onto $\mathbb{Q}$, $\mathbb{Q}$ is countable by part (b).

13. (a) For each f in $\text{FUN}(\mathbb{P}, \{0, 1\})$, let $\phi(f)$ be the set $\{n \in \mathbb{P} : f(n) = 1\}$. If $f, g \in \text{FUN}(\mathbb{P}, \{0, 1\})$ and $f \neq g$, then there exists $k \in \mathbb{P}$ so that $f(k) \neq g(k)$. Then k belongs to $\{n \in \mathbb{P} : f(n) = 1\}$ or $\{n \in \mathbb{P} : g(n) = 1\}$, but **not** both. Hence $\phi(f) \neq \phi(g)$; this shows that ϕ is one-to-one. ϕ maps onto $\mathcal{P}(\mathbb{P})$ because, given $A \in \mathcal{P}(\mathbb{P})$, its characteristic function χ_A belongs to $\text{FUN}(\mathbb{P}, \{0, 1\})$ and $\phi(\chi_A) = A$.
(b) Use Example 3(a) and Exercise 9.

15. For the inductive step, use the identity $S^n = S^{n-1} \times S$.

13.4 Answers

1. (a) True. (b) False. (c) True. (d) True.
(e) False. Consider any $y \leq -1$.

2. (a) False. With $U = \{a, b\}$, say, let $p(x, y)$ be the proposition "$x = y$."
(b) True.

3. (a) $\forall n \exists m[m \leq n$ and m is odd$]$. (b) No. Consider $n = 0$.

4. (a) 0. Consider $m = 0$. (b) 0. (c) 1. Consider $n = 0$.
(d) 0. (e) 1. (f) 0.

5. The sets in (a), (b), (e), (f), and (h) are countably infinite. The sets in (c), (d), (g), and (j) are uncountable. The set in (i) has 2^{12} elements. The sets in parts (j) and (g) are uncountable by Example 3(a) on page 529 and Exercise 13 on page 533, respectively.

6. (a) 1. (b) 0. (c) 0. (d) 1.

(e) 0. Given x, y, consider negative z with $|z|$ large.

(f) 1. (g) 0. (h) 0. (i) 1.

Index

N

O

P

V

W

X

Z